Encyclopedia of Aviculture

Glen Holland
and the World's Finest Aviculturists

hancock
house

ISBN-10: 0-88839-460-8
ISBN-13: 9780888394606

Cataloging in Publication Data

Holland, Glen, 1962–
Encyclopedia of aviculture / Glen Holland.

Includes bibliographical references and index.
ISBN 0-88839-460-8

1. Aviculture. I. Title.

SF461.H74 2006 636.6'8 C2006-903778-7

Printed in China: SINOBLE

Technical edting: Myles Lamont
Illustrations: Peter Karsten

We acknowledge the financial support of the Government of Canada through the Book Publishing Industry Development Program (BPIDP) for our publishing activities.

Published simultaneously in Canada and the United States by

hancock house

HANCOCK HOUSE PUBLISHERS LTD.
19313 Zero Avenue, Surrey, B.C. Canada V3S 9R9
(604) 538-1114 Fax (604) 538-2262

HANCOCK HOUSE PUBLISHERS
1431 Harrison Avenue, Blaine, WA U.S.A 98230-5005
(604) 538-1114 Fax (604) 538-2262

Website: **www.hancockhouse.com**
Email: **sales@hancockhouse.com**

CONTENTS

Publisher's Note

This is the Hancock House contribution to understanding and keeping birds in captivity, and to improving the role of both professional and amateur aviculturists in the conservation of wild birds.

This book is a work of love by many people and I want to thank them all, and set the scene as to how this project germinated and grew. When Glen Holland first sent me a manuscript in the mid-1990s on breeding birds in South Africa, I was impressed; but, since much of my own avicultural collection and experience was with Asian and South American birds, I felt left out. I asked Glen to expand the book, to go to the world's best and most experienced researchers in order to make this truly the finest aviculture encyclopedia. Time will tell if we have accomplished this.

But getting the lists of birds and the lists of the world's best aviculturists was daunting. A dozen key books were consulted for nomenclature. I was fortunate at the time to have on staff David Bender, an experienced aviculturist and former publisher, and an avid researcher and archivist of early records. Between us the project evolved, the writers came on line, the photographers emerged. But the task expanded from being just the world's common captive species, to the world's species. Overwhelming! The years passed, manuscripts were updated as new information became known, and the book grew. Then with David Bender's move to the U.S., I lost our chief aviculturist and encyclopedia editor.

A science student and keen aviculturalist, Myles Lamont, came on board and took charge of the birds; he also soon ended up working on the encyclopedia. Those who have come to know Myles recognize him as one of the upcoming aviculturists of the world. If he hasn't been working through the night on this book, he has been writing a paper, attending, or in some cases organizing, bird conferences around the world.

I am grateful to David and Myles for believing in this project and devoting so much to it. Thanks to both of you. Thanks also to our in-house editors and designers who have devoted many years to this project.

My own background was focused on raising pheasants, quail, and pigeons from ages ten to twelve when, to my great fortune, I met falconer Frank Beebe. The day was memorable. I was hiding in a blind, waiting to pull a drop trap onto wild flocks of band tail pigeons when Frank drove up to a nearby hillside and released a peregrine. I was mesmerized by the flight and pursuit. I went up to him blurting out questions. He suggested ways of modifying my quail trap to catch hawks. That night I called him to say I had already caught a small hawk. He was impressed and immediately came around to show me how to jess and keep the bird. Raptors and flying things have been my life ever since.

That was my introduction to falconry. Trapping accipiters and then peregrines and then eagles became my greatest sport — hunting with them was next. Being an ardent aviculturist from a young age, keeping and breeding hawks was a next step.

By the time I was out of high school I had decided to become a commercial pilot (I got my private pilot's license just as I turned 16) so I could have lots of time to chase birds. But, after three years of flying airplanes I decided to go to university to become a biologist. During the seven years of undergraduate and graduate school I studied bald eagles, continued to trap and hunt with raptors, and set up facilities to breed wild birds and mammals. This included the world's first major avicultural facilities devoted to breeding peregrines. By 1964 we had six breeding chambers and their maturing occupants.

Frank Beebe, who by this time lived next door to my breeding facilities, also set up his pair of Peales peregrines and they bred and produced young the year before my first three pairs raised offspring. For a number of years I had the only major peregrine breeding facilities in the world. During this period we were still able to access the Queen Charlotte Islands' Peales falcons for hunting, breeding, or airport-strike projects, and most years we went north and collected from two to five birds.

It was also during this time that I supplied a pair of Peales to Dr. Heinz Meng of New York, with the proviso that they not be hunted, as they represented too valuable a resource and had to be kept for breeding purposes. It was offspring from this pair that were the first peregrines released into the wild in North America; however, it wasn't until February 2001, when Heinz Meng and I were houseguests of the late George Kotsiopoulis in Florida, that I learned this. Furthermore, the incredible story of the release of those birds and their prolific breeding in the New York suburbs was also the subject of a book, *The City Peregrine* by Saul Frank, which documented the incredible return of the peregrine to North America. And I had published this book in 1994! I was flabbergasted to learn that the peregrines initially released into North America came from stock I had provided. Dr. Meng later gave that same breeding pair to Dr. Tom Cade at Cornell University as the founding stock for their incredible collection. The rest is history.

I tell this story to point out what many conservationists and preservationists do not understand. Namely, that many of the wild bird species of the world depend upon the practical efforts of aviculturists and dedicated groups like the falconers, gamebird breeders, and hunters who are prepared to go that extra mile to see conservation in action. Habitat preservation is essential, but the interim preservation of species is also a fundamental supportive requirement.

Over the years I have had an exciting, and at times quite satisfying, life field-studying and rearing thousands of birds, establishing the first major collections of captive puffins, murres, and other alcids, rearing hundreds of grouse in captivity and assisting the re-introduction of rare grouse into the wild, and more recently helping to develop very effective methods for rearing African turacos.

This encyclopedia is the work of thousands of dedicated aviculturists, those that have laboriously kept records and refined their techniques until they have succeeded. Now I am asking them to search their records for their best photographic shots for the encyclopedia. If I can get 5,000 or even 25,000 great shots, we will use them. These color images, planned to be Volume 2, will likely now be distributed free from a new website we are developing for this purpose.

We will be seeking participation in this site. If you have images or comments on this or other Hancock House publications, please contact me. Please also see www.hancockwildlfe.org.

— David Hancock
Publisher

ACKNOWLEDGEMENTS

With a book such as this, which is really a combination of contributions, there are many people to acknowledge. Let me start by saying that the response to my requests by aviculturists worldwide has been very generous and is greatly appreciated. I must go back in time however and thank my parents, Vic and Jean, for all the support they gave me in pursuit of aviculture as a child. Since I was nine, you have both spent much of your time helping me — driving all over the province to exchange birds, fetching dirty termite nests and plants in your clean car, building aviaries, understanding that my homework could wait until my birds were taken care of, and feeding my birds when I was away. Your support was a lesson in life for me with my children.

I have many aviculturists to thank but first I must again go back to my youth where I spent many hours working on a voluntary basis at the World of Birds in Cape Town, South Africa. Walter Mangold, the managing director, taught me much of the avicultural basics, which I have used for raising many of the more unusual birds that I have worked with. As a youngster, I was given the opportunity from Walter to raise species such as cranes, ibis, plovers, and raptors.

I must also thank Dr. Alan Abrey who also gave me the opportunity to work with many exotic and unusual birds, taught me much on the veterinary side, and contributed to this book.

Mike and Isobel O'Connor and Ernie Johnson played a crucial role in ensuring that the Pilanesberg indigenous aviaries were the success that they were. Your hours spent rearing chicks during your busy working day, much to the surprise of those you worked with, and helping to create the perfect landscape was all worthwhile when we discovered the impact which the "close encounters" concept had on our visitors.

To all the many contributing aviculturists who have received little more than acknowledgment for their work, both private and in the zoological world, your contributions are greatly appreciated and hopefully in this book, they will be of assistance to many more aviculturists.

I must here risk making particular reference to any one person, but the staff from the San Diego Zoo (in particular David Rimlinger, Michael Mace and Pat Witman), Neil Hamilton (ex Perth zoo), Liz Romer (ex Currumbin Sanctuary), Christine Shepard (Bronx Zoo), Charles Siegel (Dallas Zoo), Martin Vince (Riverbanks Zoo), and numerous private aviculturists such as Dick Schroeder, Josef Lindholm III, William Horsefield, Terry Irwin and others have contributed significantly toward this work. The great spirit of international assistance has been overwhelming.

To my "twitcher" friends in the Eshowe and Melmoth region of KwaZulu/Natal: the Lawries, Myburghs, Leitchs, Coleys and my Big Birding Day teammates, Hamish and Barry, thank you for the positive influence you all had on my life…even if you nearly pushed me into the "lunatic twitcher" category!

DEDICATION

They say behind every good man is a good woman. At the risk of putting myself in the "good man" category I dedicate this book to my wife Natalie who, together with my children Nikki and Kyle, has assisted, supported and somehow understood my hours of absence with the birds and behind the computer to complete this work.

With love, Glen

INTRODUCTION

Although initially this was to be a book on my avicultural experience with predominantly African bird species, the goal shifted to cover aviculture worldwide when I realized that many of the principles applied to aviculture globally. As a result, aviculturists internationally have contributed to the project. The avicultural principles described here are applicable to bird families rather than species and this gives all aviculturists a guide for success with the families and species requiring similar management. Realizing that there are three definite categories of aviculturists — hobbyists, commercial breeders and curators — responsible for public displays and/or species breeding programs, I have attempted to include not only species but also management principles that are applicable to all three categories of aviculturists.

In deciding on the introductory chapters and content of the book, I deliberately avoided chapters on topics such as aviary construction and avian nutrition. Not only are these topics vast enough to deserve entire books on their own, but they are also briefly covered in nearly every avicultural book on the shelf. I have always believed that factors often ignored, but which are of vital importance in aviculture such as live food, compatible species, stress management, and habitat selection (landscape), often do not feature enough in the management of our bird stock and I have, therefore, concentrated on these.

In more recent years the large numbers of wild-caught birds being exported has caused concern among conservationists, and many countries are now closed for exports. In addition, international laws affecting the import of wild-caught birds have been introduced. Methods used to capture, hold, and transport birds are often extremely primitive, and mortalities due to stress are high. Although most people, including many conservationists, would not vote against the capture of a small number of a species for breeding programs in recognized institutions, the controlling bodies lack the knowledge and manpower to regulate the activities. It is therefore easier to wrap it up in red tape and close all access to capture. The spread of disease is another issue facing both conservationists and aviculturists alike and is an issue that has not yet shown its full impact.

Having said this, however, and also after working on "both sides of the coin," we as aviculturists have much to answer for, and this is the main reason I started working on this book many years ago. If the true figures were available to show the number of birds, for example the tens (if not hundreds) of thousands of waxbills, exported off the African continent in recent years, two questions would come to mind: Where are all these birds now? Why is there still a demand for wild-caught birds? Unfortunately, I suspect the answers are that the majority of the birds are dead and that because the birds were cheap and readily available, we the aviculturists have not bothered to breed them and establish viable captive strains.

If aviculturists wish to maintain such birds in our collections, and to gain the respect of conservation bodies, we must begin in earnest to breed these birds. In recent years aviculture has taken leaps and bounds with many of the previously thought impossible to breed species of the past now being bred on a regular basis. This is the positive progress upon which we need to build. As imports become increasingly difficult, the pressure placed on aviculturists to breed the species has tremendous benefits for aviculture and the species.

There is nothing that gives us a better feeling of satisfaction than to see our birds successfully produce young. What we need now is to take it a few steps and generations further so that the captive strains that are now established in aviculture will fully supply the needs of the aviculture, and in some cases become a safeguard against the loss of the wild species through habitat destruction. A prime example here is the Congo peafowl, which has an uncertain future in its present-day natural habitat but is relatively secure in aviculture. In the case of the Campbell Island teal, which we breed in New Zealand, the wild population was down to approximately twenty birds with more than forty now breeding in captivity, and the first release onto a predator-free island in 1999. In 2006/07 150 captive birds were successfully released and are now breeding in the world. There is however no need to shroud aviculture under the veil of pure conservation, and aviculturists should stand proud of their hobby and their love of birds and bird breeding.

A very positive sign has been the formation of numerous specialist groups focused on achieving avicultural success, such as TAGs (Taxon Advisory Groups) and societies for families of birds such as Finches, *Poicephalus* parrots, turacos, hornbills, owls, and so on. Birds and bird watching have become very popular, and the important role they play in environmental education is an area in which aviculturists can play an important role. It is also important that we encourage and assist up-and-coming young aviculturists who will be responsible for the continuity so vital for the future for our passion and the birds.

Lastly, I would like to encourage all serious aviculturists to put pen to paper and share with all fellow aviculturists how success has been achieved. Aviculturists worldwide have been very generous in sharing the secrets of their avicultural success with me, and I hope that the information in this book will assist aviculturists in building on these successes.

— Glen Holland

ENVIRONMENTAL EDUCATION WITH BIRDS

Environmental education is about values, attitudes and behavior of people toward the environment in which we all live and share with countless other species. Aviculturists can play an important role in actively preparing young people for the future and for promoting an awareness of the beautiful species with which we work. Their knowledge and understanding of the fundamental need for the conservation of the environment is vital. They are the future guardians, and to equip them in order to take on this role is an important and essential part of their education.

But just as important, spiritually and morally we need the diversity and abundance of nature around us to give meaning and shape to our lives. The continued loss of our forests, wetlands, grasslands, alpine regions, and diversity of native wildlife have diminished our own existence. In New Zealand, for example, there are known to have been ninety-four species of terrestrial land birds. Today forty-six are now extinct. They include moa, piopio, giant eagle, laughing owl and huia, and of the forty-eight that remain, twenty-three are threatened, including kaka, kokako, Campbell Island teal, hihi and kakapo. The reality is that here and now we are in a position to actually do something to rectify the mistakes of the past. The future conservation of the environment is in our hands. It very much depends on the values we hold and the choices we make individually and as a community.

Not many students remember the math they did in school ten years ago, but they certainly do remember seeing, and perhaps touching, a live endangered species and learning the messages of their continued conservation. I have always aimed to provide powerful environmental education learning experiences, and in my understanding providing an experience where people can see and possibly touch live species certainly is the best option. When learning is based on active participation and providing people with the opportunity to see an endangered animal close up and interacting with nature first hand, it allows the optimum opportunity for learning and does change attitudes and values. The time is right for optimism. We are now in a position to take action, and to actually do something to rectify the mistakes of the past. It is also a time for reflection on where we are going, how well we are doing and what it is we value. What we can be certain of is that the future direction of the environment is in our hands.

"Close encounters of the avian kind" can prove to be an extremely powerful tool for environmental education. I witnessed this on a number of occasions but primarily in the indigenous aviaries that we constructed for the education department of the then Bophuthatswana National Parks Board, while working with farmers and communities in KwaZulu/Natal, and also when touring that same province with a hand-raised crowned eagle. Few people are as scared of birds as perhaps they are of reptiles, mammals, and other creatures, and few education tools make the message sink home like close contact with the real thing. The old proverb (modified slightly) "a bird on the hand is worth any number in the bush" most certainly applies.

In the previously mentioned indigenous aviaries, we had a large selection of hand-raised, tame, birds including hornbills, bee-eaters, barbets, and such. While in the aviary the visitor was almost ensured of a close encounter with one of the birds, for example, a barbet would land on a visitor and begin to inspect ears, mouth, pockets, and shoelaces. Such an encounter leaves a lasting impression on a person and very often sparks an interest that, in time, makes them passionate about birds. I remember watching a couple of children who regularly visited the indigenous bird aviary, walking around the aviary shouting out to their friends, "Come quickly; here is the amethyst starling." We need to install such appreciation in more of our youth.

A word of warning, however, is that in such a collection where people are in contact with tame birds, the aviary should only be open when there is somebody to keep an eye on the birds and the public. We had, apart from ourselves, a group of volunteers who assisted with this task. Without such supervision it is likely that birds will be harmed or even stolen. I have experienced both. The use of native birds is also preferable to foreign species and allows amateur ornithologists to sit quietly with their bird-watching guide while they get excellent views of the birds, which are very often otherwise just a flash in the distance. Comments from the visitors book in the indigenous aviaries included:

"A very genial way to make birds and mankind familiar with each other."

"Truly wonderful. The birds even tried to sit on my pen as I was writing."

"We need many more aviaries like this one to educate people re birds. Small paradise. Very educational."

In New Zealand a "Tuatara Roadshow" was run, the tuatara being an ancient endemic reptile about 40–50 cm (16–20 in.) long. This involved taking a single tuatara all over the country where, accompanied by a talk and slide show, people learned about — and some were able to touch — the live creature. Those involved with this program, including scientists and conservationists who have seen the impact on the public from a single tuatara, are absolutely supportive of the "hands-on" concept.

Another interesting possibility is to use feed stations for wild or released birds and involve visitors at the daily feeding time. One of the best-known examples of this is the daily lorikeet feed at Currumbin Sanctuary in Australia. Each day at feeding time, a large flock of rainbow lorikeets descends on the visitors to be fed. At the Mt. Bruce National Wildlife Centre we have a small flock of re-established kaka parrots, which also descend to feed stations in a public area to be fed. Many institutions now have large walk-through aviaries in which one is supplied with a small cup of nectar to feed the green-naped lorikeets, providing an excellent hands-on experience. Free-flight shows involving parrots, raptors, and other species have much the same effect. At Auckland Zoo a mixed collection of doves, finches, and other birds, provided little interest for visitors but replacing these with a pair of macaws that were flown free once per day had a large impact. It is important, however, to use these opportunities for environmental education as well. A show without a message is a wasted opportunity for conservation education.

Worldwide, many conservationists (some, perhaps, are preservationists) find it difficult to allow their own indigenous birds to be held in captivity. When it suits them, however, and I am one of them, to justify population and problem animal control they are very quick to state, "It is not the individual in the population that matters but the population as a whole," and conservation is defined as "the wise/sustainable use of natural resources." Surely an education facility taking a few chicks from the nests of common species, which will usually lay a second clutch soon after, and hand raising them for environmental education justifies the action.

Fortunately in Bophuthatswana in 1990 (now part of South Africa again) we were dealing with a few open-minded conservationists who could see the light and felt that this would be a great way to educate the Tswana children in the area. When the bird-watching fraternity (of which I am a member) got to hear of this, one of South Africa's leading ornithologists vowed to start a petition against our indigenous aviaries. The parks board said that if the "twitchers" did not like it, that was too bad and that their concern was for the less-privileged Tswana children. The amusing end to this story was that the man leading the objections and others like him were soon using the aviary as part of the education program for their bird-watching courses.

In New Zealand as well I have been extremely fortunate to be involved at the New Zealand Department of Conservation's rare and endangered bird breeding center at Mt Bruce. Although the prime objective here is captive breeding of rare and endangered species for research and release, interactive public displays and viewing of these otherwise difficult-to-see species, take place very successfully.

I would like to mention here, too, that aviculturists should not hide behind the "environmental education" and "breeding for conservation" banners, with a few run-down aviaries and a handful of poor signs with pictures of birds and writing that is difficult to read. I have seen this happen, and it only provides those who are against such projects with ammunition. If you are going to attempt such a project do it properly. In discussions with the public I was quite open about the fact that the Bophuthatswana project was for education and was not a breeding project. The birds we held were common species, and although we attempted to breed as many of them as possible, this was not the main aim of the project. I found people accepted this honesty. Small white-boards, which can be changed daily, moved around, etc., are an excellent way of informing the visitors of interesting events in the aviaries, for example, "The bald ibis nest in the middle of the cliff in this aviary has two chicks which hatched on 2/1/05. This brings to twelve the number of this rare species we have hatched so far this season."

The use of live "nest-cam" video where the visitor can, without causing any disturbance, watch a bird caring for its brood is a very powerful tool that keeps visitors enthralled. The video is set up on a tripod near the nest and the cables run some distance to a monitor where the public can view the action. This gives them the opportunity to view scenes that few people are lucky enough to experience live. Similarly, a "burrow-cam," where a small infrared camera sends a picture from a dark burrow or nest back to a monitor where people can observe the birds' nesting activities, is another very useful piece of equipment. We have found them particularly useful for elusive kiwi in a nocturnal house, but they can easily be positioned in parrot nests, or others.

Raptors are the ideal species for environmental education. Often described as majestic with their talons, "eagle eyes," and sharp beaks, they make the ideal education tools. Also, the fact that they occupy a position in the top of the food chain provides the ideal opportunity for describing how all the various species rely on each other in a natural system. I remember once taking a young, still very downy, wood owl to a group of preschool pupils. They all had a chance to touch the wide-eyed chick and to imitate his call and discuss what he ate. About five years later one of the young girls told me how she still remembered touching the owl. This young lady went on to study environmental law.

As a conservation officer with the Natal Parks Board, one of my prime targets was the farmers of KwaZulu/Natal who owned large tracts of forests and grasslands that were of great value for nature conservation. Having hand-raised two crowned eagles, I spent six months traveling through the province giving talks at such places as farmers' meetings and school groups. This was a very rewarding experience as the talks generated a lot of discussion and interest. My ultimate aim was the conservation of habitats, such as forests, that supported a wide range of biodiversity, but the crowned eagle provided the ideal focal point for this.

This reminds me of a case where a farmer/friend discovered a crowned eagle nest in a forest on his farm. Having built a small hide near the nest he would spend a few hours watching the birds whenever he could slip away. One day a fire caused tremendous damage to his crops and all the neighbors were called in to help control the blaze. Throughout this time he was called on the radio but could not be raised. Eventually, after all the drama, he appeared and apologized but his radio had been switched of for fear of disturbing the birds he was busy watching. Needless to say, on this farm and the farms of many like him, the invader problem plants are controlled, wetlands and marshes protected, and so on.

I have also been involved in environmental education with school children and communities and here again birds have proved a very valuable focal point. Examples here are the talks and slide shows I gave to various garden clubs where birds, butterflies, and planting native vegetation were the focus. An interest in birds soon leads to garden bird feeding and the planting of indigenous species. This interest often develops into an interest in all forms of nature, even to the point where people start to keep lists of all the birds they have seen in their gardens. While working as a conservation officer I regularly gave talks to groups of farmers' wives who in turn wielded a strong influence over the conservation policies and interest in the farms. Working with groups of school children in various habitats, they soon began imitating the calls of the various cuckoos and other birds. It is hoped this spark of interest grew into an enduring appreciation and respect for nature and, in particular, for birds.

Apart from the close encounters concept with live birds, interactive displays work well. Old incubators with a few dummy eggs, building a bird's nest with material that is supplied, hand rearing puppet gloves to use on dummy chicks that are fed pellet food, fitting rings (bands) to dummy birds' legs, artwork, work sheets on the species the group have learnt about, and so on, are all an excellent means of environmental education.

The behind the scenes work, such as food preparation, incubation, hand rearing, the bird hospital, etc., fascinates many visitors to bird parks. Providing visual access for visitors to these areas often proves to be the highlight of the visitors' experience. I will never forget hand feeding a group of young macaws, which were vigorously shaking the feeding spoon as they do. The visitors watching through the glass were absolutely fascinated and I regularly got stopped in the park to discuss what they had seen.

Finally, as a private aviculturist you may be wondering what all this has to do with you. Your aviary and your birds can bring pleasure to a lot of people if you give it some consideration. Don't always be in a hurry to chase the neighbors' kids away from your birds, but if they express an interest, show them a nest with eggs or chicks or any such point of interest. Also, if you have any chicks or hand-tame birds, a talk accompanied by these to a group of school pupils, retired people, etc., could bring great joy to their day. I recently had three bronze mannikin chicks, which I was hand rearing using a syringe. I took these to a group of young school children who thoroughly enjoyed watching them clamber over each other to be fed. This generated a lot of discussion and was also the focal point of their next art class.

At Pollsmoor Prison in South Africa some of the most hardened criminals are taught caring through the ownership of a hand-raised parakeet. The inmates spend a large portion of their day and pocket money caring for the needs of their pet bird.

Are we perhaps a little selfish, hiding away these wonderful creatures we are involved with every day?

LANDSCAPE

Most bird species are habitat specific and to ensure that the psychological and physiological requirements of the species are met, it is necessary to ensure that the birds are housed in the correct habitat. The more "at home" the bird is in captivity, the more likely it is to thrive and breed. This can best be achieved, for most species, in a naturally planted aviary, which also provides a pleasant feature where the aviculturist or park visitor can relax and enjoy the birds. A shallow, trickling water feature within the aviary surrounded by a few plants, scattered rocks, and stumps will greatly enhance the atmosphere within the aviary both for many species of birds and the aviculturist.

The disadvantages of a planted aviary are that they are more prone to rodent infestations and the birds are a little more difficult to view. The advantages of a natural habitat, however, far outweigh the disadvantages and, as already mentioned, the birds will thrive and breed more successfully where they feel at home. A planted aviary will not only provide natural nest sites but will also attract insects, which many species will feed on.

Vegetation in the aviary must be regularly pruned to avoid providing excessive cover where rodents and snakes may find refuge. The plants should not be allowed to grow through the wire, as they will soon cause damage, resulting in birds being able to escape or vermin to gain access. This also creates an untidy and badly managed image. By lightly pruning shrubs regularly, they will become quite dense, which will provide good cover and nest sites for the birds. Many species enjoy an open area of river sand where they can scratch and sunbathe.

Some species, such as the psittacines and ground hornbills, are extremely destructive to plants and here one is faced with the challenge of creating a habitat using indestructible materials, such as rocks and tree stumps. Even in suspended cages that are used to breed psittacines it is necessary to supply natural branches and sections of rotting logs, which the birds will systematically destroy. This assists in meeting the birds' instinctive desire to chew and avoids the boredom of captivity. Some species, such as chats, require rocky outcrops. Kingfishers and bee-eaters prefer a more open habitat with exposed branches from which they are able to "hawk prey." Flufftails and crakes choose dense reed and sedge thickets as their home. Natural planting plays an important role in providing a fresh supply of the birds' natural food, be it nectar, insects, or the plant itself, but equally important is the shelter that plants provide for birds to avoid aggression.

It becomes clear that it is essential for the successful aviculturist to study a bird's natural habits and then to attempt to replicate these in the aviary. If we look at one family in depth, some of the principles of creating the correct habitat can be demonstrated. Waxbills are habitat specific, and if the correct habitat is not provided the chances of successful breeding are slim. Those species from arid country, for example grenadiers, are not particularly fussy provided the aviary is not too lush. Others, such as St. Helena and orange-breast, need a few grass and reed thickets in which they will spend most of their day. Black-rumped (red-ears) nest in grass thickets on the ground. The natural vegetation allows the birds to carry out their instinctive behavior and displays, and also undoubtedly provides a mental stimulus for the birds to want to breed.

A variety of tall dense grasses are available, but I have found that those grasses that provide edible seed and seedheads, which can be used for nesting material when dry, are the ideal types. These can be grown in spring from seed or from a few clumps dug out of the wild, cut back and planted in the aviary. They are generally fast growing, particularly if well watered; they are easy to control. These grasses not only provide a good habitat but also are ideal plants for waxbills to nest in and the seed is a favorite food. In flights I plant the grasses down one side of the flight and hold it back with a tight loop of fishing line attached at each end of the flight. The grass is then all tucked behind the line.

To provide the perfect habitat for waxbills, the grass should be allowed to grow through a few dead/dry thorn branches attached to the side of the aviary. All the vegetation is thoroughly trimmed in spring and it will soon begin to take off with the warm weather. In my experience many of the waxbills are late breeders and only begin when the grasses begin to seed. Breeding then tapers off as winter approaches.

Suspending by the roots clumps of dry thicket under the shelter provides additional nest sites. A dense heather-type bush will do the trick. Having provided the correct nesting sites, ensure that the correct nesting material, teff grass and soft feathers, are made available. Feathers should always first be lightly dusted in a plastic bag containing a little bird-safe spray. In no time after providing material, the males will be seen courting the females with their dance, which is accompanied by a song. During courtship a feather or grass seed head is held like a flag in the beak. I do not provide any nesting receptacles but other breeders sometimes use baskets, boxes, and tins. I prefer natural nests as waxbills do not remove the feces of the young and the natural grass nests allow the feces to dry quickly, thus avoiding unsanitary conditions. I have found that in preference of manmade nests, my birds always prefer to build their own nests in their natural habitat. When purchasing plants for your aviary landscape it is important to check that no part of the plant is potentially toxic. As most nurseries use the "attracting garden birds" sales pitch, they should be able to warn of species of plants that are toxic to birds. Due to the wealth of information and lists of plants that are toxic to birds on the Internet, I will not list these here.

As it is not possible to go into great detail on the landscape required for every specie, and as many of these details would be repetitive, I have described six different habitats (A–F) and indicated which of these is suitable for each specie. Some species as indicated are less habitat-specific and will use a variety of terrain.

A: SUBTROPICAL

- At least 60% of the area planted to a mixture of evergreen trees and shrubs with scattered, rotting tree stumps.
- The aviary floor with a dense layer of leaf litter, particularly below the plants.
- Dense clumps of seeding grass on the edges of the "forest."
- A few unplanted, lawned areas between the thickets.
- A shallow stream of circulated water with occasional cascades.
- Mist sprays placed over a number of the densely planted areas.

Trees: Delicious monster, Bamboo Palm, Figs, Rubber Tree, Schefflera, Bamboo, Coffee Tree, Dracaena

Shrubs: Magnolia, Coprosma, Creepers, Tree Ferns, Ferns, Hydrangea, Rhododendron, Cissus creeper

Other: Bromeliads

B: WOODLAND

- Large sparse trees with good perching branches scattered through approximately 40% of the aviary.
- Clumps of dense shrubs and seeding grasses planted around the base of some of the trees.
- Most of the aviary floor a mixture of maintained short lawn and patches of dry sand.
- Occasional large boulder and dead tree stump/branch placed in sunny positions.
- Slow-flowing shallow stream with trickling circulated water.

Trees: Acacia, Mulberry, Boxwood, Beech, Broom, Juniper, Maple, Birch

Shrubs: Plumbago, Privet, Cape Apple, Hibiscus, Cotoneaster, Grevillea, Conifers, *Banksia spp.*

Grasses: Miniture Pampas, *Eragrostis spp., Panicum spp.,* Tussock grasses

C: GRASS THICKETS

- A mixture of tall, seeding grasses growing through clumps of dead dry branches, particularly thorn branches.
- Occasional clump of dense shrubs (up to 1.5 m in height) and one or two larger thorny trees.
- Areas of open, dry, sandy patches broken by an occasional large rock.
- Clumps of fine, dense, dry shrubs (heather-type plants) suspended by the roots in the shelter and planted in the aviary.
- A shallow pool of water, preferably with a trickle of circulated water.
- One or two mist sprays placed over a clump of shrubs and over the pool of water.

Trees: Buffalo Thorn, Boxwood, Lilac, Olive, Juniper, Prunus, Berberis

Shrubs: Spike Thorn, Hawthorn, European Hornbeam, Privet, Buddleia

Grasses: *Panicum spp., Setaria spp., Eragrostis spp.,* Tall seed producing, grasses.

D: ROCKY OUTCROP (KOPJIE)

- Large (at least 1.5 m high), uneven scattered mounds of fertile soil.
- Boulders, rocks, tree stumps, and dry branches scattered on the mounds.
- Plants such as small shrubs and succulents covering half of the soil.
- Half of the soil left exposed.
- The foot of the outcrop with a definite border surrounded by grass.

Shrubs: Plumbago, Broom, Hakea, Cotoneaster, Climbing Rose, Cabbage Tree

Succulents: Kalanchoe, Aloes, Geranium, Cactus, Euphorbias

E: SEMI-ARID

- The entire aviary floor consisting of well-drained river sand.
- A few scattered pebbles, rocks, and stumps.
- Occasional sparse shrub, succulents, and grass.
- A few large sparse, preferably thorn, trees planted in clumps on the edges of the aviary.
- Large, dead perching branches.
- Water supplied in a shallow standing pool surrounded by scattered pebbles.

Trees: Acacia thorn, Buffalo Thorn, *Grevillea spp.,* Boxwood, Aloe

Shrubs: Succulents, Broom, Hakea, Privet

F: AQUATIC

- Pools of water with shallow edges, preferably with flowing water linking the pools.
- Patches of sedges, reeds, and overhanging plants covering half of the fringes of the water.
- An occasional dead stump/branch entering the water from the bank.
- Open patches of lawn surrounding most of the water.
- An occasional shade tree on the lawn surrounding the water

Marginal Plants: Weeping Willow, Reeds, *Typha spp.,* Papyrus, *Scirpus spp.,* Phragmites, Ferns, Arum Lilies, Iris

Purely Aquatic: Water Lilies, Aponogeton, Lotus, Vallisneria, Amazon sword, Hornwort, Watercress

Floating: Water Lettuce, Salvinia, *Eichornia spp.,* Water chestnut, *Azolla spp.*

STRESS MANAGEMENT

Stress among the bird stock is a factor that few aviculturists take seriously enough. The most obvious example of how stress impacts negatively on birds is seen in the manner in which many captive-bred species settle down and breed far more successfully than their wild-caught counterparts. Most aviculturists keep birds with the ultimate aim of breeding them. One of the most important factors required for healthy birds and successful breeding are low stress levels. Newly imported birds often show illness and even die from the lengthy period of stress that they have endured.

PHYSIOLOGY OF STRESS

Animals respond to stress with a "flight or fight" reaction. When a stressor presents itself, the bird's body reacts with a range of physiological and behavioral responses. To ensure the mobilization of energy stores, corticosteroids are released into the blood. If the stress is acute, adrenaline is released, but in the case of longer term and less acute stress, the corticosteroids are released. In a situation where a bird really needs to defend itself or escape, the corticosteroids ensure the provision of glucose and mobilize energy for the bird's immediate use. Chronic stress results in prolonged elevated levels of circulating corticosteroids, and this can have a number of potentially adverse effects. Negative stimuli, such as capture, cause the release of adrenaline and other chemicals into the blood stream, which in turn affect the functioning of the bird's organs. Stress without any physical activity also results in an increased body temperature and respiratory rate. The heart rate is increased and at the same time the blood flow to the vital organs such as the kidneys and liver is decreased. This drop in the flow of blood to the vital organs is often termed "shock," and if untreated it is likely to lead to a circulatory collapse and death. In an extreme situation, sufficient stress could lead to a heart attack, or to some cardiac tissue damage, which makes subsequent heart attacks more likely to be fatal.

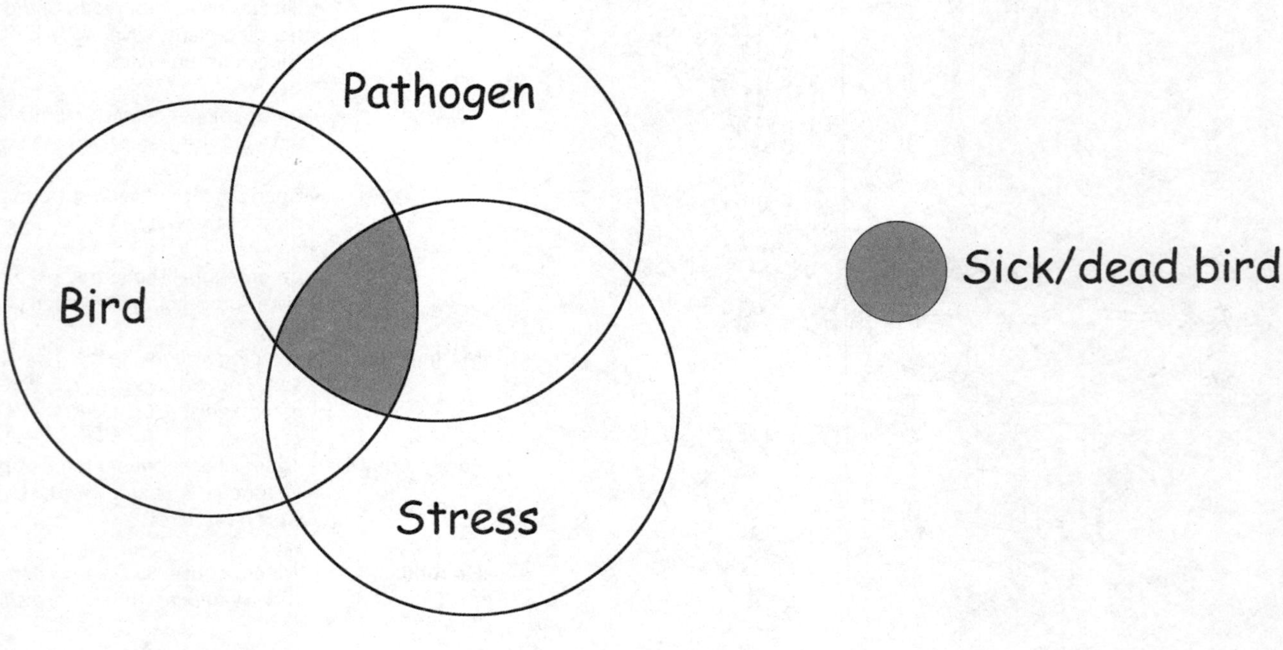

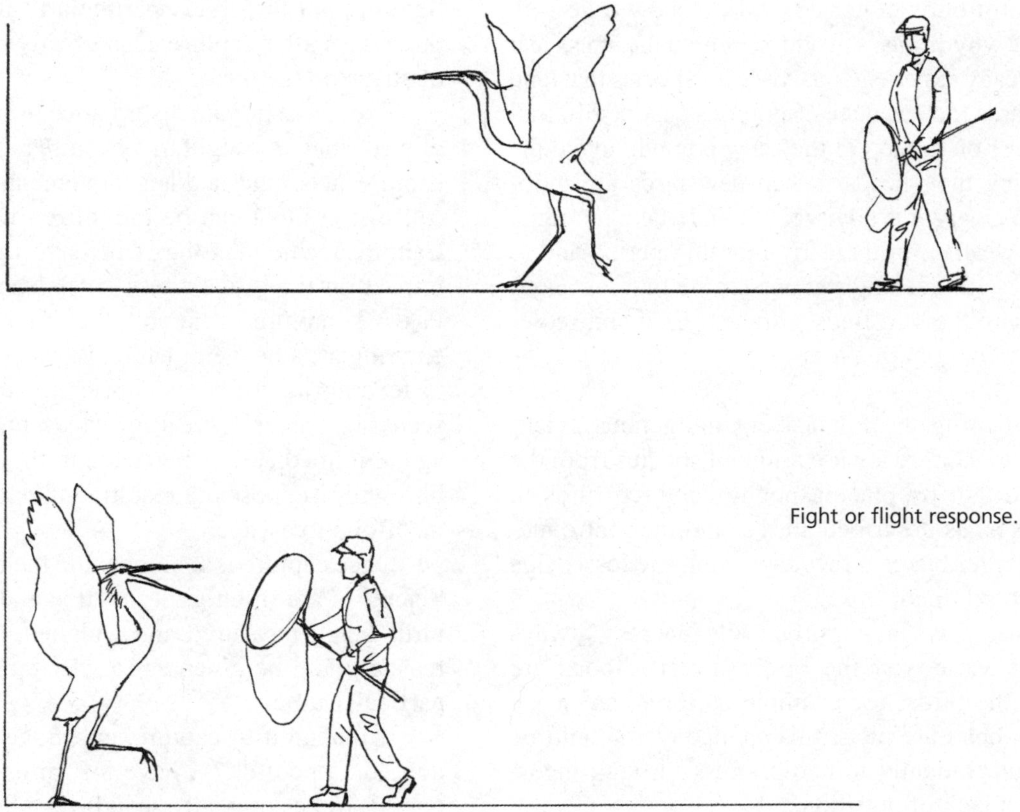

Fight or flight response.

Research has shown quite clearly that stress plays a role in decreasing egg production, shell thickness and in lowering the viability of the eggs produced. In many instances a usually aggressive, territorial male will cease his aggressive behavior when stressed or, alternatively, the male may become hyperagressive. Research has also proved that a pair of birds feeding a brood will feed less regularly when stressed and if stressed further or for prolonged periods the clutch will be deserted. I have always believed that it is important to place the bird in its correct habitat. Birds in the incorrect habitat or conditions are continually being subjected to stress.

Always remember that the bird is continually surrounded by a host of opportunistic pathogenic organisms just waiting for a drop in the bird's resistance, which will allow them to attack and infect the bird. High levels of adrenal activity associated with stress, even for short periods, can cause sufficient immunosuppression for a pathogen attack to be successful, whereas such an attack would otherwise be warded off by the immune defense systems. A combination of the bird + stress + pathogen will result in a sick, and possibly dead, bird. With effective management, stress factors can be minimized resulting in healthy, contented birds that will make ideal avicultural subjects.

Treatment: Stressed birds, other than those suffering from heat stress, that have become inactive and in shock should be placed in a secluded, warm environment, and with larger species 1% of the bird's body weight of Ringers Lactate can be administered to the crop. This is done with either a crop needle or a syringe with an attached piece of surgical tube. If the bird is still inactive after a few hours, up to 4% of the bird's body weight can be given as additional fluid in the first 24 hours. The supply of warmth is a vital factor in treating birds that are suffering from stress. A warm environment enables the bird to spend its energy on the required recovery process, rather than using its energy to have to support its body temperature. The warmth will also result in a more active bird, which will increase the chances of the bird beginning to feed and in turn will also boost the bird's strength and recovery process.

TYPES OF STRESS

Transportation: Birds that have to be transported or have been captured and are stressed but not in shock (they are still active), should be placed in transport boxes that offer a secluded, dark area in which the bird can hide and feel secure. In addition, the open areas on the transport box should be covered with fine gauze, which will avoid the bird panicking and injuring its beak and cere. The transportation of birds in, for example, a show cage with large exposed wire fronts will add greatly to the stress that the bird is already experiencing. Also, for species that tend to jump upwards (doves, quail, sandgrouse, etc.), a 10 mm layer of sponge on the top of the carry box will avoid the birds damaging their heads. When new birds are introduced into an aviary it is advisable to undertake introductions in fine weather, in the early morning period and to provide a few extra feeding stations for the first few days. This will ensure the bird finds sufficient food and settles down with a minimum of stress.

Capture: During the Pilanesberg indigenous aviary project I had to capture a wide range of species from the wild. I found that by placing newly captured birds in transport boxes, as described above, and then introducing them to larger boxes a few days later, the losses due to stress were minimal.

As diet also plays an important role in stress, always ensure that a variety of the bird's favorite foods are available. If the birds, for example softbills, are to be trained onto substitute diets, this changeover should be brought about gradually to avoid stress through incorrect diet and a lack of nutrition.

As well as a secure environment and the correct diet, I found that the provision of additional warmth and light around the clock, particularly for the first two to three days after capture, also greatly reduced the effects of stress on the birds.

Excessive human disturbance of nest sites, handling of bird stock, the sight of a capture net, the use of coarse capture nets, and sudden movements in the aviary all cause stress that can be minimized with some thought. Capture techniques that will have the lowest possible impact on the bird's stress levels should be employed. Passive capture, whereby the bird catches itself by entering a trap cage in which it has become accustomed to feeding for a few days prior to capture, is the ideal scenario. Never leave trap cages unattended, as once again, immediately when caught, the stress levels in the bird will rise, possibly resulting in damage to the bird in its efforts to escape.

Once captured, do not handle the bird unnecessarily. Scientists are often guilty of this and I have witnessed birds being measured and weighed, and afterwards no reason could be given as to why this was done at that particular time.

If planning to capture and move birds, this is best done on a cool day. I have lost birds that I have transported (to be surgically sexed) on a hot day.

Temperature: Birds that are either too hot or cold

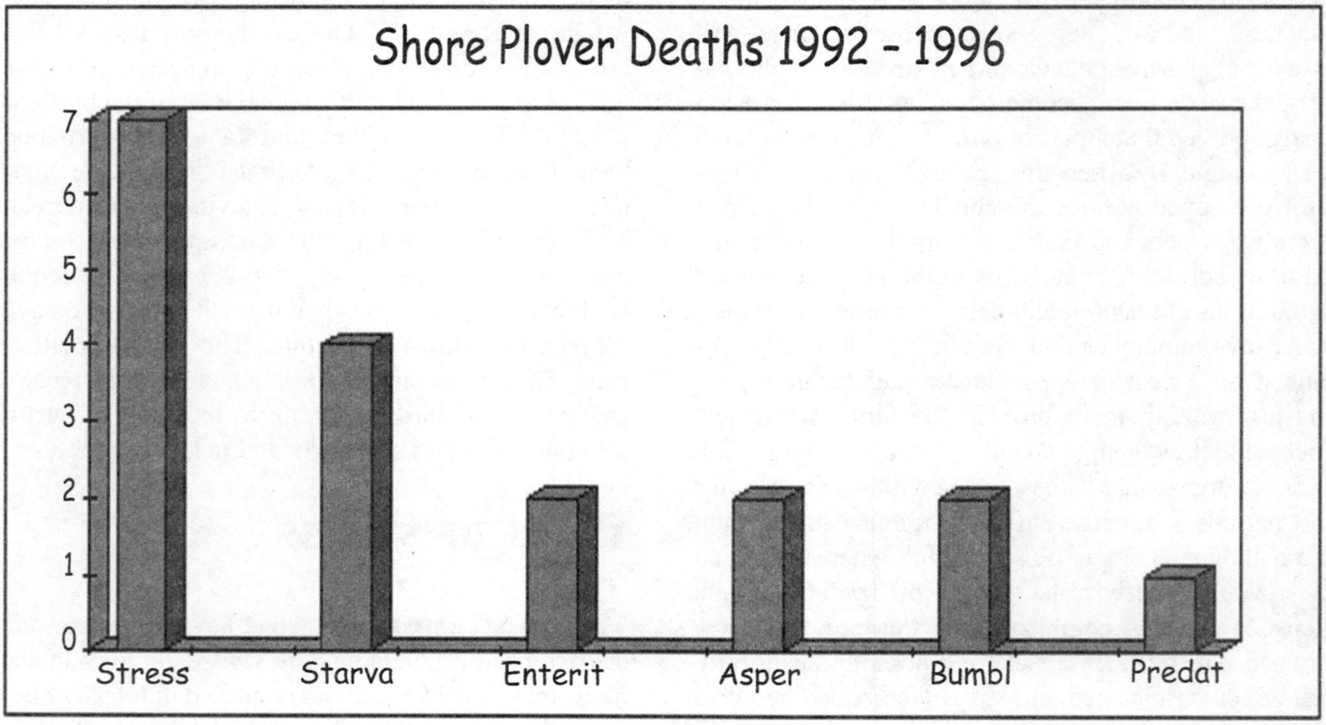

The following chart compiled by Dr. J. Pauli (consultant vet to the Mt. Bruce National Wildlife Centre) shows the important role that stress has played in the mortalities of captive, endangered shore plover over a six-year period:
The stress related mortalities in this species have been caused by territorial behavior, the presence of aerial predators, and excessive disturbance.

will suffer from stress. Heat stress decreases feed consumption, immune response and egg production. Cold stress also causes decreases in production by increasing feed demands, decreasing immunity and increasing disease susceptibility. The consequence of cold stress is visible in birds as they huddle to keep warm, which decreases feed consumption at a time when calorie demand is increased. It is also important to remember that when birds are confined, the air quality in confinement should be closely monitored.

Natural Elements: Nothing has a greater direct effect on bird's respiratory health than airborne pathogenic organisms. Lighting is another environmental consideration. Too little light suppresses immune function and decreases feed consumption of most species. Birds housed in unsanitary conditions will suffer when exposed to elevated ammonia levels, which irritate the eyes and respiratory tract. Even low levels of ammonia depress the growth rate, depress the immune system, and leave the bird more susceptible to disease. Placing a bird in the incorrect habitat, for example a grassland species placed in an aviary planted with shrubs, will add to the bird's stress and the chances of successful breeding will be greatly reduced.

Noise: Noise pollution can also have a stressful effect on birds. Sounds that irritate people, such as barking dogs, noises from heavy machinery and vehicles, have an even more profound effect on birds. Loud, unpleasant noise causes fright in birds that will result in the desire to flee. Of course, in captivity flight is not possible, and this results in high stress levels in birds. Absolute silence is also not natural and can result in stress and boredom. With birds kept indoors, such as ostrich chicks in a brooder or seedcrackers that were bred indoors at the Riverbanks Zoo, it is a good idea to have a radio or tape playing soft background music; this has a definite calming effect on the birds. If this is switched off (for example, in the ostrich brooder) signs of stress will soon begin to show.

Aggression/Compatibility: An incompatible species mix or pair of birds in the aviary also results in stress. Apart from direct persecution of smaller species by more dominant birds, factors such as nest robbing and the consumption of all the favorite foods by dominant birds all impact negatively on the submissive species or individuals. Breeding success will be directly affected under these circumstances. A prime example is a collection that I visited where the aviary included the small, rare pink-throated twinspot and large superb starlings that ate all the live food and would clearly not allow the delicate twinspots to breed. Similarly, social stress can result in a bird being unable to feed itself properly, interfere with incubation and chick rearing, and even in persecution to the point of death. In such cases, an "overenthusiastic" male is best moved to an adjacent aviary.

Overcrowding leads to increased aggression, direct competition for resources such as food, nesting material, and nest sites, and unsanitary conditions. All of this, particularly in combination, leads to increased stress among the stock. Apart from the number of birds in the aviary, the flight space is also important to consider and the requirements for this are directly related to the size of the birds. The presence of predators such as cats, dogs, rats, owls, hawks, and people all cause stress in the aviary.

Diet: Incorrect diet also results in stress, which in the long term will result in the bird most likely dying from a disease that was able to infect the bird in its weakened condition. Here again the stress resulting from incorrect diet will result in little interest in reproduction. Research on golden song sparrows has shown that birds subjected to continued stress and corticosteroids in the blood would store energy from proteins in a more accessible form — that of fat.

References:

Hicks-Alldredge, K.D. (DVM) November. 1996. Stress Management: *American Ostrich*.

Wingfield, J.C. & B. Silverin. "Effects of corticosterone on territorial behaviour of free-living male song sparrows." *Hormones & Behaviour* 20: 405-417.

CONTROL OF VERMIN

The control of pest species, be they plants, insects, or vermin requires a well-planned program with constant follow-up eradication programs, and seldom is the point reached where one can say that the problem species is eradicated. I find it appalling the number of times one hears of successful breeders who have had their entire collection wiped out by rats or other vermin. In my work in bird parks I have also witnessed rats destroying birds as large as fledged glossy ibis and wood duck, while dogs have been responsible for a killing frenzy where pheasants and waterfowl were killed. Other predators such as genets, otters, and stoats are more cunning and only take single birds every few days. In a large mixed collection this may take a while before the problem is noticed.

Snakes also offer their share of problems. My most recent, frightening experience was with a Mozambique spitting cobra that was curled up inside a conure nest box with the two dead parents, two live chicks, and two chicks in its belly. I opened the nest box with my young son standing next to me and in seconds we had a very upset 1.5 m cobra hanging half way out of the nest box. After some chaos, the snake was disposed of. Some snakes will feed on the bird's eggs and the smaller species are very often difficult to locate and remove.

The problems do not end with these few pests; otters, raptors, foxes, ferrets, mink, weasels, bobcats, cats, and a host of varying species all cause problems for aviculturists around the world. Fortunately, the larger predators are generally easy to trap and easily excluded with a good electric fence system.

Household cats also cause tremendous stress to the birds and will take the head off small finches and waxbills, which put their heads through the wire in their panicked state. Here again, electric fences offer a solution and a layer of larger mesh size placed a few centimeters above the aviary roof will make it difficult for the cat to move about. Cats are easily caught and taken to a pound; should the problem recur, more drastic measures will need to be employed. The golden rules with the control of vermin are:

- No one method alone will achieve control. Use a combination of exclusion, capture and in some cases destruction.
- Prevention is better than cure. Construct the aviary to exclude vermin and maintain a clear margin of at least 1m (3 ft.) around the perimeter of the aviary.
- Expect problems and be ready for them and have the correct eradication methods at hand.
- Conduct a biannual predator check of the aviary structure and mesh.
- Don't become complacent; vermin problems will usually recur.
- Maintain strict sanitation and remove any food spillage.
- Use the most time-effective method, for example., checking traps, etc., as part of the daily routine is time consuming. By making the top of trap boxes see-through, much time can be saved.

RATS AND MICE

Probably the most common vermin that cause problems in aviculture worldwide are rats and mice. Not only do they eat the birds' food and contaminate it with their urine and feces, which could lead to infections from pathogens such as Yersinnia, but they also take up residence in birds' nests. Rats, particularly when they have young, turn carnivorous and the results can be devastating. Once they have learned to kill, they will take ever-larger species and soon progress to birds the size of pigeons and ornamental waterfowl. Chicks in a nest make a favorite meal. I personally lost a small flock of quail finch overnight to a single pregnant rat. As they

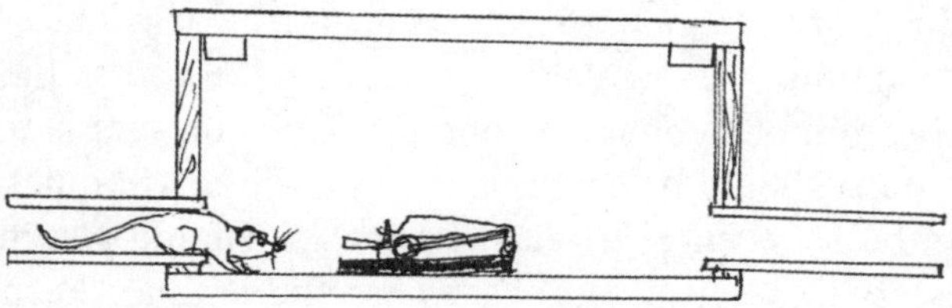

Bird-safe moustraps.

are largely nocturnal they also disturb the birds at night, which can result in injured and stressed birds. Rats are ingenious creatures and will climb relatively smooth surfaces, gnaw through solid objects, jump, and burrow to gain access to an easy food source. The Pacific rat is responsible for the extermination of more bird species than any other single factor in the last few hundred years. With planned and continuous eradication programs rats and mice can be beaten. A good example are the islands off New Zealand where rats and all other vermin have been totally eradicated to create a safe haven for endangered bird species. This cost many thousands of man-hours and phenomenal numbers of poison baits, but the desired results were achieved.

Control Methods: Poison, but always consider birds, children, and pets that may be able to gain access and, should you spill poison over yourself while working with it, scrub down immediately. Poison can also not be used in the presence of carnivorous birds, which are likely to eat the poisoned rats and/or mice.

Build good foundations, at least 60 cm (24 in.) deep of either solid material or a strong galvanized mesh, around the aviary perimeter. Ideally add a foot or so facing outwards in the bottom of the trench to create an L-shaped barrier that is highly effective.

Seal all gaps around doors, under roofing, etc.

Place electric wires (minimum of five) close together around the entire perimeter. Personally, I feel that electric fencing placed close to a smooth surface around the entire perimeter is the answer but, particularly in public displays, this is not always practical. Electric fencing will also exclude the larger predators that cause tremendous damage in a short space of time if they gain access to the bird collection.

Set walk-in traps under cover sheets/boxes to exclude birds. Walk-in, live traps are preferable to snap traps that kill the victim as, should birds gain access and get caught, they can be released unharmed. Wire cage walk-in traps that allow any number of rats to enter are well worth the investment. Ideally, these should be permanently set even when there is no sign of a problem, and bait should be regularly renewed. Use a number of traps simultaneously and place them along the routes that the rodents travel.

Trap boxes for mice — with two 32-mm (1-in.)sections of plastic pipe that protrude about 10 cm (4 in.) to exclude birds, and with a trap in the middle of the box — have proved extremely effective for me.

Poison placed in sections of pipe or double sectioned poison boxes is the best answer but clearly cannot be used in the presence of carnivorous birds such as raptors, hornbills, etc. A very effective method of eradication is to place numerous small containers of sherry in and around the aviary. The rats are attracted to this sweet drink and after a few days their first visit on their evening foraging will be to the "pub." At this stage a liquid poison is added to the sherry and the results are very effective.

Rats and mice can be flushed from burrows using a hosepipe and then killed as they emerge but this method will not eradicate the problem and is not a solution.

Rodent baffle.

It is possible to shoot rats at night using an air gun and a flashlight. A small portion of the population may be eliminated, but the disadvantages here outweigh the advantages. First, there is the risk of not only scaring but shooting a bird. Second, this method will not eradicate the problem, which is the required result. And I know of two cases in which birds died from lead poisoning after swallowing the lead pellets.

LARGER MAMMALIAN PREDATORS

Every country worldwide has its share of natural or introduced mammalian predators that attack birds. In South Africa it was genets, otters, monkeys, and mongoose; in New Zealand, the introduced ferret, stoat, and weasel have devastated the native bird populations with some species now only occurring on island safe havens. Domestic dogs and cats are possibly the two most common problem animals the aviculturist has to face worldwide. The two primary methods of control are exclusion by fencing and electric fencing, and destruction of the pests by trapping, shooting, and poisons. In New Zealand the eradication methods of small mammals is well studied and the results applied to many regions of the world. Research has found that some of the more common predators can jump heights as follows:

Mouse	300 mm (12 in.)
Ferret	360 mm (14 in.)
Weasel	420 mm (16.5 in.)
Ship rat	700 mm (27.5 in.)
Stoat	950 mm (37 in.)
Possum	1,200 mm (47 in.)
Cat	1,800 mm (70 in.)

For open enclosures with vulnerable species, such as pinioned waterfowl, a 2-m (6-ft.) high fence with 10 x 10 mm (4 x 4 in.) weld mesh is the first barrier for all predators. This is then combined with either electric wires on the side, near the base and the top, or with a metal sheet lip on the top of the fence. A good foundation reaching a depth of 60 cm (24 in.) is the next step in predator proofing an enclosure or aviary. Electric wires are a good deterrent but should not be trusted alone as many mammalian predators learn to get past these. The following fence design was tested against all of the species listed above and found to be successful. The fence is 2 m (6 ft.) high with a metal sheet overlap of 350 mm (13 in.) and two downward pointing sections of 50 mm (2 in.) each. A concrete foundation of 60 cm (24 in.) extends below ground level.

Traps, including live walk-in cage traps, Fenn (gin) traps, snap traps, etc., can all be used but be warned that the regulations regarding the use of these vary from one country to another. One constant is that all traps should be as humane as possible, kill quickly and be checked every day. Live traps should also not be left in sunny positions where the animal, even a pest, will cook to death. Baits vary widely from fresh meat and fish to eggs, day-old chickens, dog roll, and fruit. Ensure that traps such as Fenn's that can do considerable damage to fingers, are well out of the reach of people, pets, and birds. A combination of traps and good fencing is the ideal to ward off attack from these predators. In some countries, for example New Zealand, poisons such as 1080 are available for the legal control of species like possums and stoats. These are fed in extruded pellets (for possums) or injected into eggs (for stoats). It is illegal in most countries to use unregistered poisons for species on which the effects of the poison or secondary poisoning are unknown. Finally, to catch an animal, look at the situation and attempt to guess where it will run or climb and where the best place to catch it will be.

AVIAN PREDATORS

Hawks, owls, shrikes, and corvids all predate on birds and although not generally responsible for many deaths, their presence stresses the birds and can result in injuries as the birds rush about. They also manage to catch the odd bird against the wire, which they will kill. I remember as a young boy being able to obtain a pair of forest canaries that I managed to breed and, after much extra care, a single chick finally fledged only to have its head taken off by a fiscal shrike. The presence of these predators will suppress all breeding in the aviary. With waterfowl collections, which often do not have overhead cover, avian predators will soon learn to make use of these easy pickings. I would recommend that aviculturists explore every possibility before bringing out the shotgun, after all even these avian predators have their place in the environment. Although I have not used them, I believe that hot-wired perches near the aviary work well. Local falconers, using a bal-chatri trap, may be willing to trap and remove the offenders, which should then be moved a considerable distance and ideally held in captivity in their new location for a few weeks prior to release. A padded foothold trap can also be used for owls and other large raptors, particularly with a species that is willing to return to a kill or bait. A few of these can be slightly hidden around the bait. I recently had a little sparrowhawk that was visiting my aviary regularly and causing chaos with all the birds hiding away and none of the usual activity taking place. I tethered a strip of white plastic to a pole with a piece of fishing line; the pole extended above the roof of the aviary. In no time the aviary birds had got used to this plastic blowing in the wind, but despite being seen in the area, the sparrowhawk never returned to the aviaries. Where possible, large trees near the aviaries, which will provide ideal raptor perches, should be removed.

Other nonpredatory species such as doves and sparrows can also consume vast quantities of food, which can result in high feed bills. These birds also pose the threat of introducing disease and parasites.

At the Umgeni River Bird Park where there are a large numbers of parrot aviaries that are fed sunflower and other seed, the stage was reached where immediately when the birds were fed, a large flock of thick-billed (grosbeak) weavers descended and could be heard as they cracked and consumed all the feed. The aviaries had to be re-wired with a smaller mesh size to exclude these birds. At the Mt. Bruce National Wildlife Centre, mallards and sparrows were easily excluded from the pellet food of the endangered takahe by placing the food in a see-through feeder with a hinged lid. The takahe soon learned to lift the lid to reach the pellets inside. At the Patuxent crane center, a grid fitted to the crane feeder allows the cranes' bills to pass through but excludes wild birds and mammals.

To discourage house sparrows, remove their nests, cover the nest entrance hole with bird mesh, and trap the birds with a walk-in tunnel trap cage. Other larger species, such as sacred ibis and some waterfowl, can also reach these plague proportions and not only consume the food but also fight with the birds that belong in the enclosure. Once again, these birds are best trapped and moved. Seldom are we justified in destroying such species that have their place in nature and that we are in fact baiting by making food available.

In some countries, such as the United States, there are specific poisons that can be used, i.e., avicides, but of course one has to be extremely careful in the application of these. In New Zealand I have used a narcotic avicide to eliminate large numbers of sparrows, which posed a disease risk and ate all the food at a wild kaka parrot food station. This narcotic paste was mixed with breadcrumbs and took effect in 20–30 minutes of being consumed. There are also laws in most countries pertaining to the destruction of wildlife and often a permit is required to trap and move or to destroy a native species.

INVERTEBRATE PESTS

Invertebrates such as snails are often the host of diseases and internal parasites that infect the birds. Strict hygiene in the aviaries and ponds will assist in lessening these problems but they can seldom be eliminated. Avoid old standing water on the property, which will result in large numbers of mosquitoes that bite the birds and in some cases avian malaria can result. Some of the smaller bird species lose toes through excessive mosquito bites.

Bees and wasps will sting not only birds but also keepers. Bees are particularly a problem with species that use nest boxes, which are the ideal home for a swarm of bees out searching for a new home. When combined with a nectar diet, these are prime targets for bee invasions. I know of many lorikeet breeders who have their aviaries covered in a shade cloth, which excludes these pests. A good friend of mine received a call from home to say there were bees in his aviaries. When he arrived home he found four of his five hawk-headed parrots, a pair and two offspring, dead.

Wasps will sting larger species such as pheasants and cranes. Wasps can be sprayed at their nest site. Interestingly enough, many of the African waxbills and sunbirds deliberately place their nests close to the nest of wasps for protection.

Ants are another problem, with some aggressive species actually attacking young birds in the nest. Ants are also a problem with many of the food items placed out for birds. Fortunately, they are easy to control and numerous bait poisons can be supplied which the ants take back to their nest to destroy the entire colony. While visiting a fellow aviculturist who specialized in breeding twinspots, he took out a container of pyrethroid powder and liberally sprinkled this over the floor of the breeding cage to destroy ants. I was shocked but found out later that pyrethroid is a natural plant extract that is relatively harmless to vertebrates.

Flies can also become a problem at a food source but, fortunately, there are numerous effective control methods available. I would suggest contacting your local health department to find out the most effective control methods available in your area.

External parasites such as ticks, lice, and mites will all cause an irritation and a drain on birds' bodies. Particularly in times of stress will these parasites gain a stronghold on the birds. Here your local veterinarian should be able to provide an effective solution. When handling any of your birds or chicks from a nest, take the time to hold a wing open and view this against the light to show external parasites such as lice and mites.

PROPAGATION OF LIVE FOOD

The first consideration should always be where to breed the live food you wish to produce. There are numerous bird men who have nearly been evicted from their homes by wives who had enough of a house full of creepy crawlies. When breeding species requiring live food always consider the insects that are naturally and most easily accessible. This varies greatly depending largely on the year round temperatures in a particular country. Planting the correct flowering plants will entice insects into the aviary. In addition, by using a butterfly net to capture insects or placing an ultra-violet light in the aviary at night to attract insects, a good harvest can be available to the birds.

In Africa and Australia, termites are an excellent food source and are usually collected with their nest, which is broken open over a period of time to expose the white ants. Winged termites (alates) can be collected in large quantities when they emerge en masse on summer evenings and are easily attracted to a bright light. Insects can also be stored frozen in a double plastic bag which will help to prevent frost damage to the frozen insects. Turning sandy soil on the aviary floor or trays of fresh garden soil placed in the aviary will also offer the birds a variety of invertebrates.

MEALWORMS

by Glen Holland and Peter Karsten

Yellow mealworms are the most frequently used insect in aviculture. There are three stages in the life cycle: worms, pupae and beetles. The female lays about 450 eggs. Most colonies are maintained on a single container principle where all three stages are kept together for both breeding and collection of worms for feeding. Additional food such as sliced apple, cabbage, banana and carrot peels etc is added periodically. It is important to supply these fresh foods thinly sliced as large chunks will soon decompose and encourage mold. The container should be completely cleaned every three months. I have found the best method

Nutritional Values of Live Food*

Species	Ca (%)	P (%)	Mg (%)	Ca (mg/100g)	Cu (mg/kg)	Fe (mg/kg)	Mn (mg/kg)	Zn (mg/kg)	Moisture (%)	Ash (%)	Protein (%)	Fat (%)	Fibre (%)
Mealworms													
Tenebrio molitor	0.12	1.42	0.28	3.3	17.77	39.70	6.79	131.02	63	1.2	18.6	13.6	2
Giant Mealworm													
Tenebrio molitor	0.03	1.27	0.22		14.94	25.85	5.56	82.84					
Super Mealworm													
Zophobas morio	0.12	0.83	0.18		13.94	50.34	1.54	87.50					
Crickets													
Acheta domesticus													
Adult (6)	0.21	0.78	0.08	21.5	8.50	112.33	29.65	186.36	69	1.5	>20	5.7	3.3
Juvenile (8)	1.29	0.79	0.16		9.61	196.80	52.76	159.06					
Waxworm (5)													
Galleria mellonella	0.06	1.20	0.09	13.1	3.06	77.27	3.28	78.78	61	0.97	15.4	20	1.6
Fruit fly (4)													
Drosophila melanogaster	0.14	1.10	0.13		8.66	454.22	16.09	146.9	74	1.7	18	5.5	
Earthworms													
Lumbricus terrestris													
Wild-caught (6)	0.97	1.1	0.31	0.9	32.86	11,087.4	199.4	270.80		5.4	60	8.5	5-11
Commercial (6)	1.21	0.86	0.19		8.09	5,801.76	113.1	231.15					

* Whole animals as fed but dry weight for earthworms

of collection of the worms is a flour sieve with a mesh size that will allow the smaller worms to escape.

To start a colony place about 1.5 kg (3.3 lb.) of bran in a container, introduce about 200 mealworms and add 200 g (7 oz.) of rolled oats once a week (note the information on using wheat fines in this chapter). Add a slice of vegetable or fruit once a week. Cover the entire surface of the container with newspaper. Every second day about 20 ml (0.7 oz.) of water is sprinkled on the newspaper. Most colonies are maintained on the normal daylight hours. A more elaborate method involves separating the beetles into breeding containers and growing the worms in others.

The medium in which mealworms are grown has an effect on their food value. Many home grown colonies are kept in bran alone but to ensure success and worms of a high quality the following medium is suggested:

Bran	50%
Wholemeal flour	45%
Powdered multivitamin	3%
Calcium supplement	2%

The rate of growth is dependent on the ambient temperature. At 25°C (77°F) mealworms are mature after four months while increasing the ambient temperature to 30°C (86°F) will greatly increase growth rates. Temperatures above 35°C (95°F) are unsuitable and death occurs if temperatures rise above 40°C (104°F).

Relative humidity is important and development from a 3 mg hatchling is rapid at higher humidity and after twelve weeks weight gains are as follows:

Relative humidity at:	30% = 18 mg
	50% = 38 mg
	70% = 68 mg

A relative humidity above 70% results in the growth of dangerous molds. The colony can be maintained in glass, metal or plastic containers that have smooth, steep walls to prevent the escape of worms and beetles. Beetles' life spans are limited and they will not usually survive beyond the second egg laying and hatching stage. About 50% of mealworms usually survive pupation to become beetles.

When sieving the old medium it is advisable to do this outdoors, ensure that you are positioned up-wind and wear a facemask. The fine powder, which remains from exhausted cultures, can cause irritation to eyes, nose and even cause respiratory problems. Note below that using pollard instead of bran will negate these side effects. Using containers with an internal surface area of about 600 cm² (93 in.²) and a substrate depth of approximately 4 cm (1.6 in.), each container should yield 0.25–0.5 kg (0.55–1.1 lb.) of mealworms during the three week harvesting stage. Theoretically an output of 500 g (1.1 lb.) of mealworms per day could be achieved from a colony of 90 containers, provided that the supply of mature beetles is maintained.

Marcus Pollard from Tasmania describes a very successful method of managing a mealworm culture:

> "Considering the old methods of mealworm production a logical thing is to separate the three stages of the life cycle into different containers. Lo and behold the yields of mealies increase accordingly. The beetles are placed in a plastic box that is 20 cm wide x 30 cm long x 16 cm deep (8 x 12 x 6 in.). Into the bottom of this container a number of holes are drilled, just large enough to allow the bran to fall through but small enough to prevent the beetles from exiting the box. The beetles have about 5 cm (2 in.) of bran poured over them. On top of this bran a square of hessian (burlap) is placed which is used as a feeding table for the beetles. My beetles are only fed carrots.
>
> This "beetle box" is then placed inside another empty plastic box that is 30 cm wide x 44 cm long x 24 cm deep (12 x 17 x 9 in.). This then sits upon two coffee lids so that there is a clearance between the two boxes to allow the bran to fall through. The movement of the beetles causes the bran to fall into the bottom box where it is free from the ravages of the adult beetles. Every two days the beetle box is shaken like a sieve and the bran and eggs sieved through into the bottom box. New bran is placed on top to keep the depth around 5 cm (2 in.) and off you go again.
>
> This is repeated for about two weeks before another empty plastic box is inserted under your beetle box. Your box of sieved bran then goes onto the shelf, is labeled with the date, has a square of hessian placed on top of it and you then wait for about 40 days for the eggs to hatch. During this period you spray your culture everyday with water from some type of atomiser. When you see the tiny mealies you can then place some carrot out for them to nibble on. This allows you a range of mealworm sizes over an extended period of time.
>
> Pupae: When starting your colony you should place the desired number of mealworms into a container and allow them to naturally pupate. As the pupae appear it is important to remove them into a separate plastic box where they won't be eaten by your voracious larvae. The sedentary pupa is no

match for fast moving mealworms that like nothing better than a nice meaty feed. (After all the humble mealworm is often used to remove the flesh from specimens in forensic laboratories.) So as your mealworms turn into pupae, out they go on their own.

Beetles: The same is true for the beetles. As soon as they emerge from their metamorphosis they must be placed into your "beetle box" before they also start to eat the pupae. The advantage of this system is twofold: You will get far more mealworms for your beetle dollar, and that stockpile of dust won't be necessary because once you take your mealworms out you know there are no eggs left in the bran. The size of your system is also irrelevant as the guide that I have outlined can be applied on any scale. I started off by using a "mealworm fridge" (an old fridge with a light and thermostat replacing the electric motor) to supply my needs, then turned the back of my garage into a mealworm room.

In order to encourage adult mealworms to pupate, if necessary I cut back on moisture (i.e. carrots, greens, etc.) when they reach full size and, more importantly, place folded pieces of brown paper (grocery bags or animal feed bags) on top of the medium. Two or three folded pieces about the size of the surface, to create two pockets, work nicely. The larvae see this as a good site away from gnawing mealworms and pupate en mass. You can lift up one side of the pocket and look at the pupating larvae and pupae all gathered up cleanly in-between the sheets. Do not fold the paper accordion-style or you will dump most of the larvae when you inspect the inside. Raising the temperature to 29ºC (84ºF) should also be beneficial. Cutting back on moisture gives an alarm signal to better pupate than to die of dehydration.

Flies and waxworms will pupate when they run out of food and turn into smaller adults, but they do complete the cycle. This is why we see a given species of flies and other insects of different size in the imago stage. When I find more than I need for beetles I sometimes pick up a paper pocket and dump the clean larvae and pupae into a dish to harvest them as food for my birds. It is quick and efficient because the larvae are at maximum size and I do not need to disturb the culture."

Mealworms have a protein content of 21.5–24.5%. They have a 13% fatty acid content, which should not cause overweight problems in the birds consuming them. Although cooling can slow their development, they should not be frozen as this reduces their palatability and nutritional value.

A problem that exists is their reversed calcium/phosphorous value when compared to birds' requirements. The ratio in a mealworm is Ca:P 1:16–17 whereas birds require Ca:P 2:1. The Ca:P ratio is very much on the wrong side and the worst of the usual insects fed to birds. It is for this reason that a calcium supplement is added to the growth medium.

Mealworms fed on wheat bran contain relatively high levels of phytic acid which has been associated with calcium deficiencies. Phytic acid or cereal phytates are concentrated in the aleurone (third) layer of the seed coat of all cereal grains. The first three layers make up the bran that is commonly used for mealworm production. Cereal phytates have the property of being able to immobilize dietary calcium and magnesium and form insoluble complexes that are not readily absorbed and are prevented from entering birds' bodies.

Not all cereal grains have the same level of phytates in their aleurone. The highest levels are found in oats, followed by barley, rye, wheat, and lastly the millets. Oat bran is, thus, far less desirable that wheat. Naturally when a bird dehusks a seed the aleurone and other outer layers are peeled off and when a seed is sprouted 70% of the phytates are deactivated.

Pollard (wheat fines) is processed from the endosperm (starchy centre of wheat) and therefore has significantly reduced phytates levels. This product tends to be quite fine and easy to sieve and in addition, for those who suffer from hay fever when working with bran, pollard does not have this side effect.

References:

Martin, R.D., J.P.W. Rivers & U.M Cowgill. Culturing mealworms as food for animals in captivity. *International Zoo Yearbook.*

CRICKETS

Crickets are a valuable source of live food with a better nutrient balance than either mealworms or waxworms and are inexpensive to raise. "White" crickets have a lower roughage content as the chitin has not been formed, and as such are ideal for hand rearing young softbills. They are also maintained on a colony system but it is vital to ensure that sufficient food and water is available at all times otherwise cannibalism will occur.

They are kept in wooden, glass, metal or plastic containers with a coating of petroleum jelly or a similar product below the top of the container to ensure they do not escape. It is advisable to have a lid on the container, which should have an opening for air that is covered by gauze. As colonies thrive under warm conditions, in large containers a 40-watt light bulb can be used to increase the ambient temperature. A box measuring 60 x 80 x 40 cm (23 x 31.5 x 16 in.) is adequate to breed up to 500 crickets a week with the use of smaller back-up hatching containers.

Peter Karsten from Canada describes his success with a disused deep freezer, which is ideal for cricket breeding, with a one cubic meter (35 cu. ft.) unit able to produce 1,000 and more crickets per week. A pair of heat lamps can be mounted on the inside of the lid with thermostat controls for temperature. The lid is always left a little ajar for ventilation and the smooth walls prevent the escape of crickets. Egg cartons are provided for extra living space The sub-adult and adults are kept on the floor portion of the freezer. The containers for hatching eggs are placed on the elevated shelf of the compressor compartment.

Females have a 10–16 mm (0.4–0.6 in.) long ovipositor and will lay eggs in moist soil, peat moss/sand mixture and in other moist substrates that have air circulation. Floral foam ("Oasis") for wet flower arrangements is ideal. It has no organic substances that attract mold, absorbs plenty of water, and is porous.

Eggs are injected into the moist floral foam with females able to lay 200–300 eggs for 10–12 weeks. To prepare the foam, cut a shallow crater into the top to control the water when it is added, so that it does not run down the sides. The cubes of foam are best placed on a paper towel that will absorb excess water. Depending on humidity in the environment, water can be added daily or every two days. As any small crickets will drown in standing water, do not over water. The foam must be moist at all times and this is easily monitored by lifting the foam to check for moisture. A separate tray of food ensures the food remains dry. The egg-laden foam must be removed as soon as the crickets begin to dig out the eggs (within 48 hours) or they will dismantle the foam.

Eggs hatch in 10–14 days at 25–30°C (77–86°F). The warmer the temperature, the faster the development. The hatching larvae are about 2 mm long and white, until their exoskeletons harden and turn brown. Larvae shed their skin several times before they become adults and at 25°C (77°F) they mature in 5– 6 weeks and at 30°C (86°F) they mature at about four weeks. The foam is left in the separate rearing container and, together with fruit, acts as a source of moisture. The best fruit to add is a slice of orange but other types can be used. Always remember that crickets need good air circulation and must not be crowded. When crickets are 5–10 mm they are moved from the rearing container into the larger space in the bottom of the freezer.

Another method (although I would recommend the one above) is to use the bottom of the container covered with a layer of dry river sand. Take a number of inverted flower pots with a few entrance holes cut into the bases and place them in the container to house the crickets. Oviposition takes place in a 12-cm diameter shallow plastic flowerpot filled with moist peat, and on top of this is placed an inverted flowerpot for housing crickets. Water is supplied in a jar with a closed lid in which a hole is made and a wick (the type used in a paraffin lamp) inserted through the hole in the lid. It is important that the wick fits tightly in the lid or else the crickets will enter the jar and drown.

Although crickets have been successfully bred commercially using a bowl of chick starter mash for food, their nutritional value is dependent on their diet. A more complex diet as described below will produce a higher quality cricket:

20 g Dried milk	6 g Dried yeast
20 g Maize meal	40 g Wheat feed
10 g Grass meal	10 g Arachis oil
4 g Linseed cake meal	1 g Salt
10 g Casein	

Peter Karsten uses a mix of chick starter (non-medicated), wheat bran, and ground puppy or kitten chow (biscuits). Vitamin enriched, left-over softbill mixture is "recycled" through the cricket colony, given that the bird population is free of contagious diseases or parasites. (Fecal floatation is routinely done).

References:

Singh, P. & R.R. Moore. 1985. *Handbook of Insect Rearing* (Vol 1). Amsterdam: Elsevier.

COCKROACHES

Most cockroaches undergo a simple metamorphosis in that there are three stages in their life history: egg, nymph and adult. The eggs of cockroaches are assembled in a leathery bean-shaped capsule known as the egg sac. This egg sac is often seen partially protruding from the tip of the abdomen. In some species it is carried until just before hatching and in others it is dropped to undergo further incubation. Nymphs emerge from the sacs after varying periods of time depending on temperature and humidity. Molts occur at periods of several weeks until maturity is reached. With each molt the cockroach increases in size.

Glass or sheet metal tanks are the ideal containers. They must be fitted with a lid which has air holes covered by fine mesh gauze. To prevent escape a strip of petroleum jelly is applied slightly below the top of the container. River sand is used on the base and a strip of corrugated cardboard is placed in each container. The cardboard is rolled up loosely and stapled at each turn. Water is supplied by the same method as described for crickets. The rolled up cardboard is for the cockroaches to rest on and hide in, as they are accustomed to do during the daylight hours when they are rather sluggish.

Though cockroaches are omnivorous and will eat a great variety of foods, dry dog kibble has been found to be the most satisfactory. They are easily handled and do not decompose unless they become wet, and are sufficiently rough surfaced for the cockroaches to readily feed on them. Dog kibble is enriched with nutrients and the cockroaches thrive on them. The bowl of dog kibble is placed in the centre of the rolled up cardboard. The most important item in the breeding of cockroaches is the water jar. They can survive for months without food, but they must have a constant supply of water.

The cockroaches can be kept in their containers at room temperature or in a constant temperature cabinet maintained at 27°C (80.6°F). They will not survive temperatures above 33°C (91.4°F) and they develop too slowly at temperatures below 21°C (69.8°F). The containers should be cleaned once every few months and the rolls of corrugated cardboard replaced if they get too dirty or are chewed away.

References:

Tarshis, I.B. Breeding Cockroaches for Feeding Purposes. *International Zoo Yearbook.*

FRUIT FLIES

Fruit flies are an ideal food for small softbills, sunbirds and waxbills. The easiest way to breed fruit flies is to get a number of old ice-cream tubs and fill them up to the halfway mark with tightly folded newspaper. On top of the newspaper put layer upon layer of chopped banana until it almost falls over the sides of the tubs. Put the tubs of bananas in a warm place, preferably no less than 16°C (60.8°F), and place a container of fruit flies, available from one of the specialist suppliers, near the tubs and remove its lid.

In warmer climates the fruit flies will appear naturally on old fruit but in other countries it will be necessary to introduce them as described. Within two weeks the bananas will become mushy and be teeming with larvae. These will eventually pupate up the sides of the tubs and it is at this point that fresh tubs of bananas should be placed alongside the existing ones to allow the hatched-out flies somewhere to lay their eggs.

Eventually a rotation system can be devised making a tub available for the birds regularly, much to their delight. Winged fruit flies, rather than their wingless counterparts, are strongly recommended. The winged variety keeps the birds more active, and therefore healthier, as they have to chase after them. It also provides the opportunity to see the birds perform some captivating aerial maneuvers.

References:

Hinze, Ian. Fruit-Fly Culturing. 1992. *Estrilid Society Yearbook,* pg. 34.

WAXMOTHS

Waxworms offer an easy alternative to mealworms, which are generally larger, harder, and more difficult for smaller birds to consume. A variety of live food will also stimulate adult birds to want to feed their chicks. The moth measures 15–18 mm (depending on species) in length, the larvae are 24–28 mm, and the pupae 8–10 mm in length. The life cycle occurs in 6–8 weeks depending on the temperature which should be 28–30°C (82.4–86°F). The eggs hatch in 6–9 days and the larvae grow to full size in 18–24 days. The cocoon is formed in two days and the pupae rest for 12 days before the moth emerges. The moths live about three weeks and lay about 300 eggs.

Peter Karsten maintains his waxmoth in a warm and dark brooder cabinet with two heat lamps to back each other up. The cultures are set up in 2–6 liter plastic storage boxes; a metal fly screen is glued into the lid for larger larvae and fuel filter screen for hatching larvae. Formerly waxmoths were raised on discarded honey combs. Waxmoths are not fussy as long as the food mix is not dried or crystallized. A combination of chick starter, wheat bran, wheat germ and "liquid" honey (3:3:1:1 in volume parts) works well. Mixed baby cereal is a good addition to the mix. Honey can be increased if the mix goes dry in a dry climate. The mix is worked into a moist crumbly medium with a sturdy wooden spoon. Moths tend to lay their strips of eggs on the rim, under the lid, of the containers. Egg strips are pushed off with a knife or finger nail and placed on top of the food mix in a hatching container.

Hatching larvae are invisible to the naked eye. Lids must fit tightly otherwise the larvae will escape to find hiding places to pupate. Waxworms are cleanly harvested by removing them from the underside of the lid. Ensure that sufficient are left for the next generation of moths. Condensation causes mold and should be avoided. Waxworms will form cocoons under the lid or in the medium. The cocoons have a slot at one end to help the moth emerge. The cocoon can be opened here to remove the

An aviculturist in the making. *Photo: Richard Naisbitt*

Raptors are commonly used for educational exhibits. *Photo: M. Kettle*

Malaysian eagle owl leaving a lasting impression on its young handler.
Photo: David Hancock

Bird shows are popular at many zoos such as this one at Jurong Bird Park.
Photo: David Hancock

School students with a hand reared African grey hornbill *Tockus nasutus*. *Photo: Glen Holland*

Eric Kowalczyk at Woodland Park Zoo talking to the public about the plight of the Sulawesi wrinkled hornbill *Aceros cassidix*.
Photo: Myles Lamont

Live traps are much less stressful on birds as compared to being caught with a net. *Photo: Peter Karsten*

Allopreening is sometimes considered to be a stress relieving behavior, especially in Psittaciformes. *Photo: Myles Lamont*

A pekin robin *Leiothrix lutea* enjoying a bath, a usual sign of a healthy bird. *Photo: Peter Karsten*

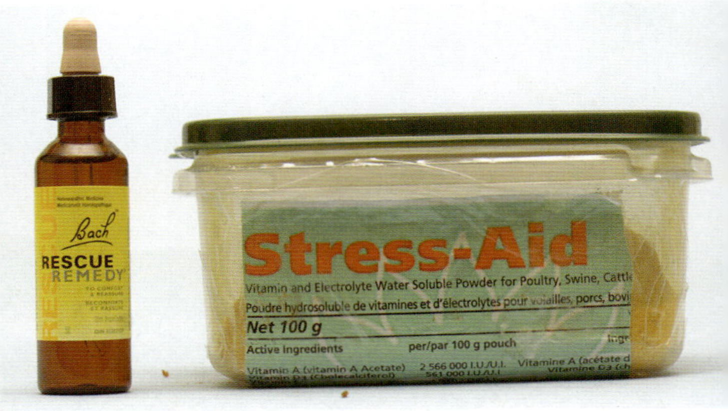

A few natural- and artificial-based medications exist for stress treatment in animals and birds. *Photo: Myles Lamont*

Sun or dust bathing is another sign of stress relief and enjoyment in an aviary, in this case an Erckel's francolin *Francolini erckelli*. *Photo: Myles Lamont*

An obviously stressed bird, perhaps due to illness, this pekin robin has been placed in a hospital cage until it has recovered. *Photo: Peter Karsten*

Various methods of rodent control exist, from poison baits to live and snap traps. *Photo: Myles Lamont*

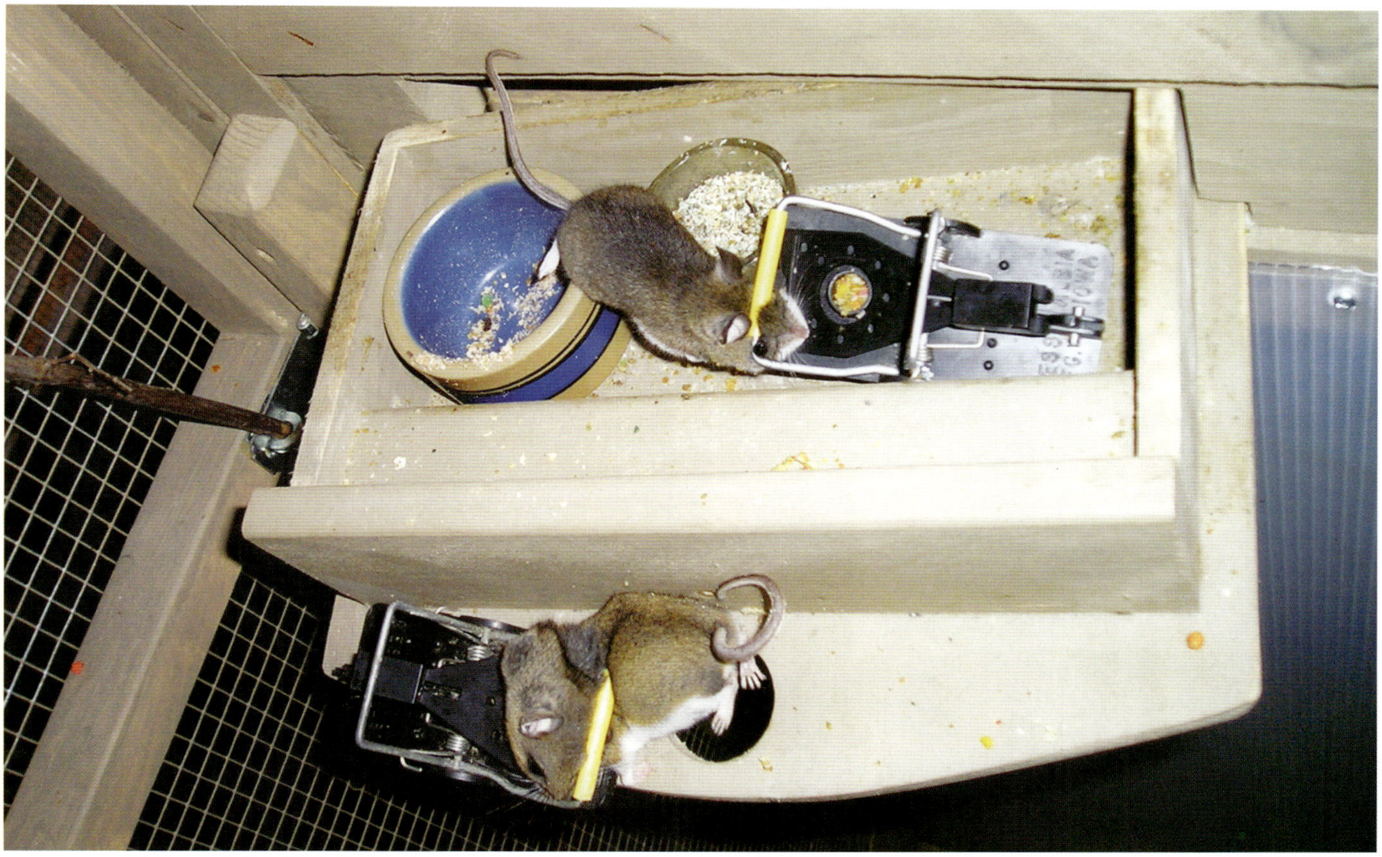

Rodents are animals of habit and will become accustomed to a particular feeding location, hence making their removal somewhat easier. *Photo: Peter Karsten*

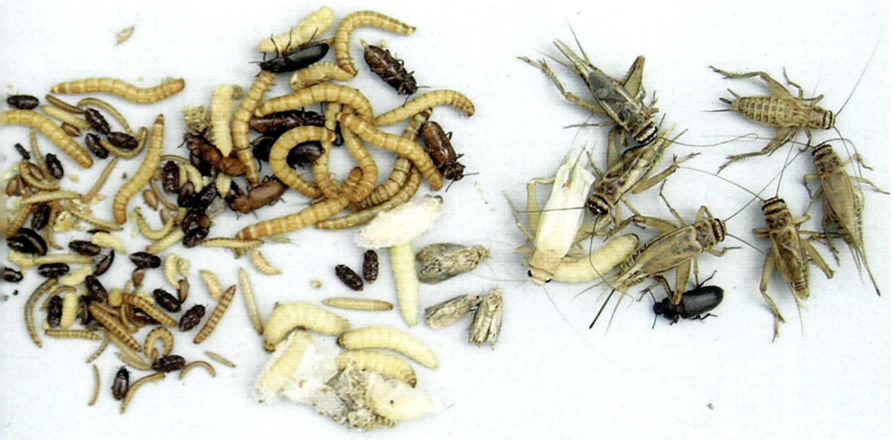

Various forms of live food, including lesser mealworms *Alphitobius* spp., greater mealworms *Tenebrio molitor*, lesser waxworms *Achroia grisella* and domestic crickets *Acheta domesticus*. *Photo: Peter Karsten*

Greater mealworm culture at Silver Pine Aviaries. *Photo: Myles Lamont*

Lesser mealworm culture at Silver Pine Aviaries. *Photo: Peter Karsten*

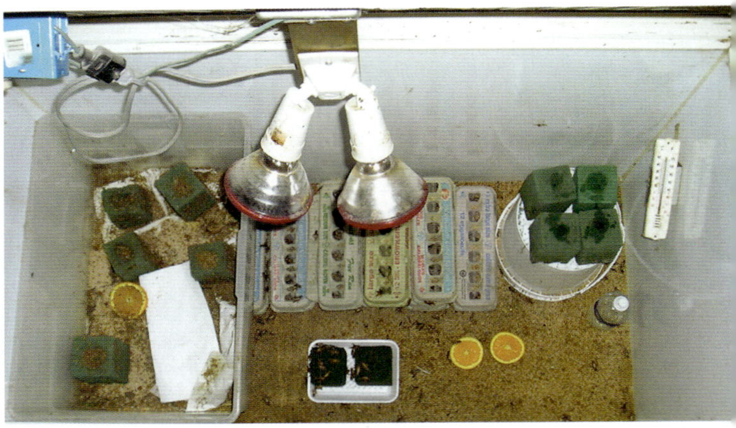

Cricket culture in a modified freezer. *Photo: Peter Karsten*

Free-choice live food in a smooth-sided container allows birds to have access to live food throughout the day. *Photo: Peter Karsten*

A modified UV insect trap captures live insects by blowing them into a collection basket that has had the electric screen removed. *Photo: Peter Karsten*

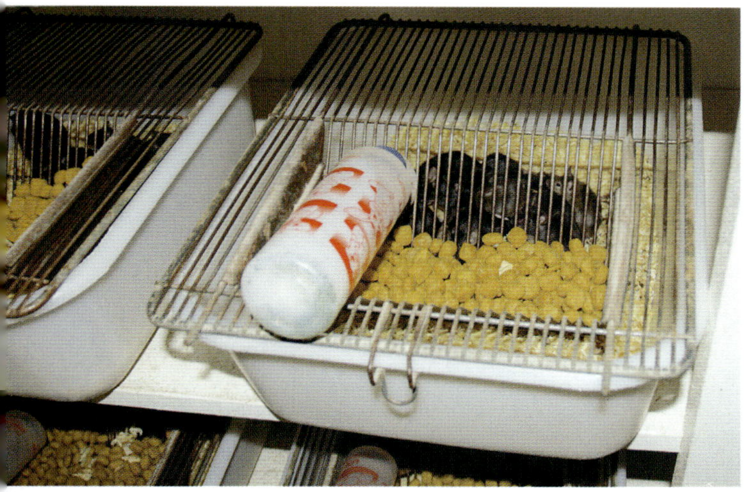

Mice are one of the easiest and cheapest live foods to produce.
Photo: Myles Lamont

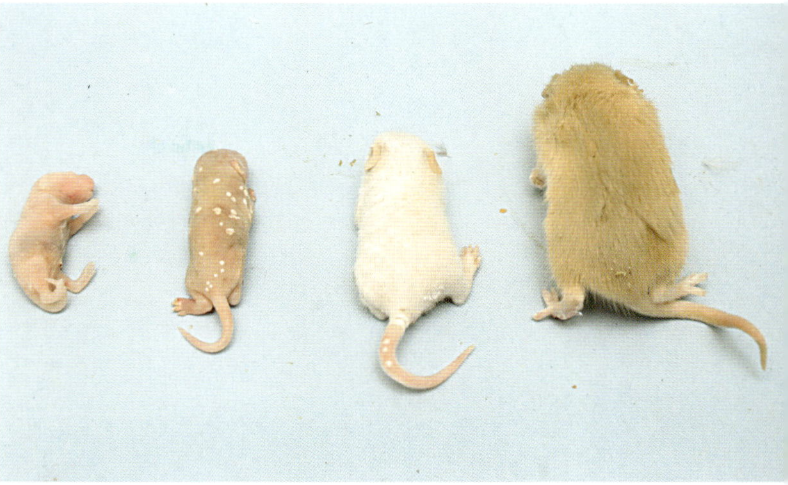

Various stages of growth in mice used for live food. *Photo: Myles Lamont*

Lesser mealworms congregate under sheets of paper facilitating collection.
Photo: Myles Lamont

Floral foam is an excellent medium for hatching cricket eggs. Here newly hatched crickets can be seen on the bottom of the container.
Photo: Peter Karsten

A modified fridge used for keeping insect cultures at a constant temperature and humidity at Silver Pine Aviaries. *Photo: Myles Lamont*

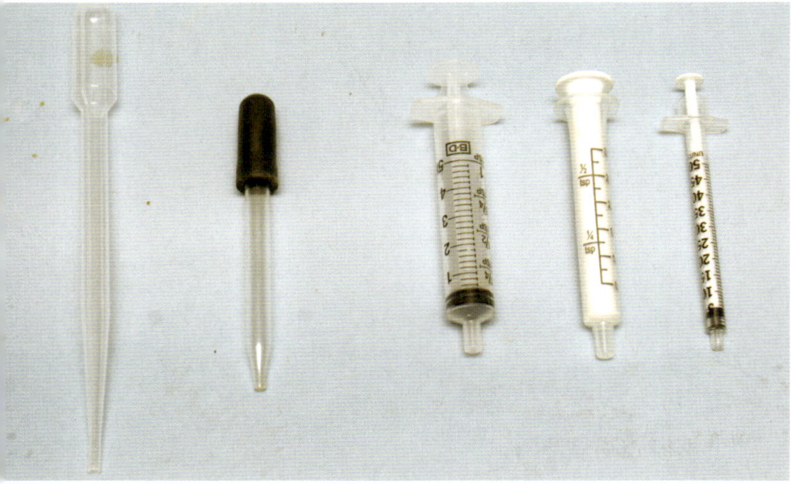

LEFT AND BELOW: Instruments for hand feeding chicks at various stages of development. *Photos: Myles Lamont*

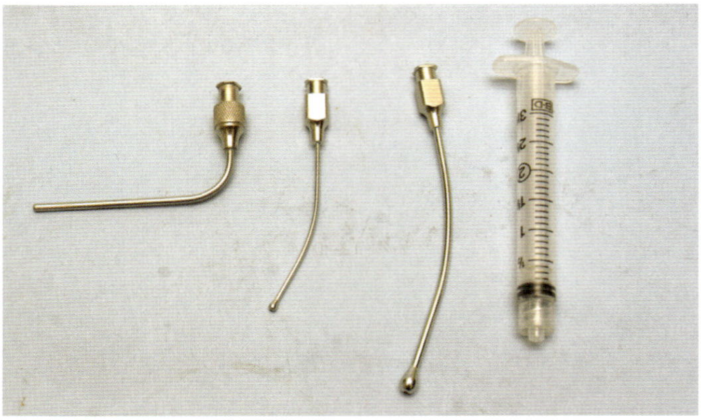

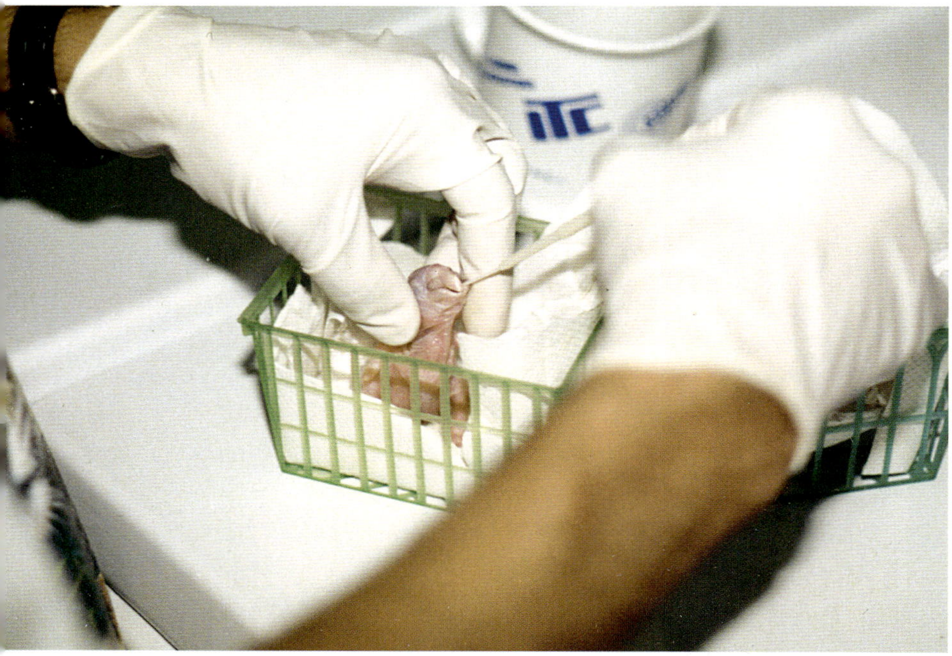

Hand rearing a young blue-and-gold macaw *Ara ararauna*. *Photo: David T. Longo*

Hand feeding an estrildid finch showing typical gape markings. *Photo: Glen Holland*

Hand feeding a bronze mannikin finch *Spermestes cucullatus*. *Photo: Glen Holland*

Hand rearing a red-faced parrot finch *Erythrura psittacea*. *Photo Glen Holland.*

Hand rearing a kingfisher. *Photo: Glen Holland*

Hand rearing a green wood hoopoe *Phoeniculus purpureus*. *Photo: Glen Holland*

Blue-winged Kookaburra *Dacelo leachii* chicks at Houston Zoo. *Photo: Myles Lamont*

Hand feeding a shrike. *Photo: Glen Holland*

Hand rearing a Florida sandhill crane *Grus canadensis pratenis* at Hancock Wildlife Research Center. *Photo: Myles Lamont*

Hand rearing a giant wood-rail *Aramides ypecaha* at Hancock Wildlife Research Center. *Photo: Myles Lamont*

A Dalmatian pelican chick *Pelecanus crispus* at San Diego Zoo.
Photo: Myles Lamont

Hand feeding a speckled mousebird *Colius striatus*. *Photo: Kateri Davis*

Hand rearing pekin robin *Leiothrix lutea* chicks in an artificial coconut fiber nest at Silver Pine Aviaries.
Photo: Peter Karsten

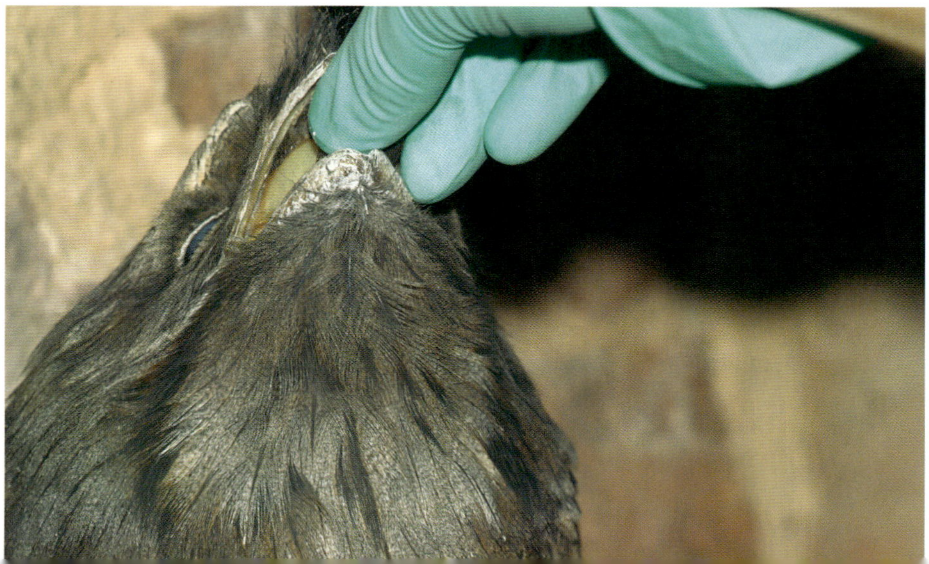

Hand feeding an adult tawny frogmouth *Podargus strigoides* at Oklahoma City Zoo. Most frogmouths require hand feeding even as adults. *Photo: Myles Lamont*

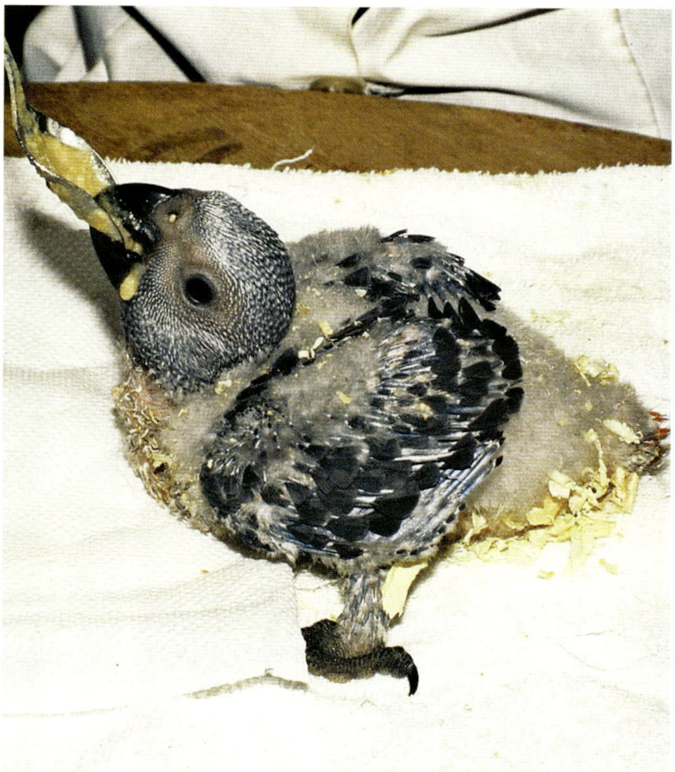

Hand rearing an African grey parrot *Psitticus erithacus*. *Photo: Glen Holland*

Hand rearing a celestial parrotlet *Forpus coelestis*. *Photo: Glen Holland*

Hand-reared rainbow lorikeet *Trichoglossus haematodus* chicks at Aquarium of the Pacific. *Photo: Myles Lamont*

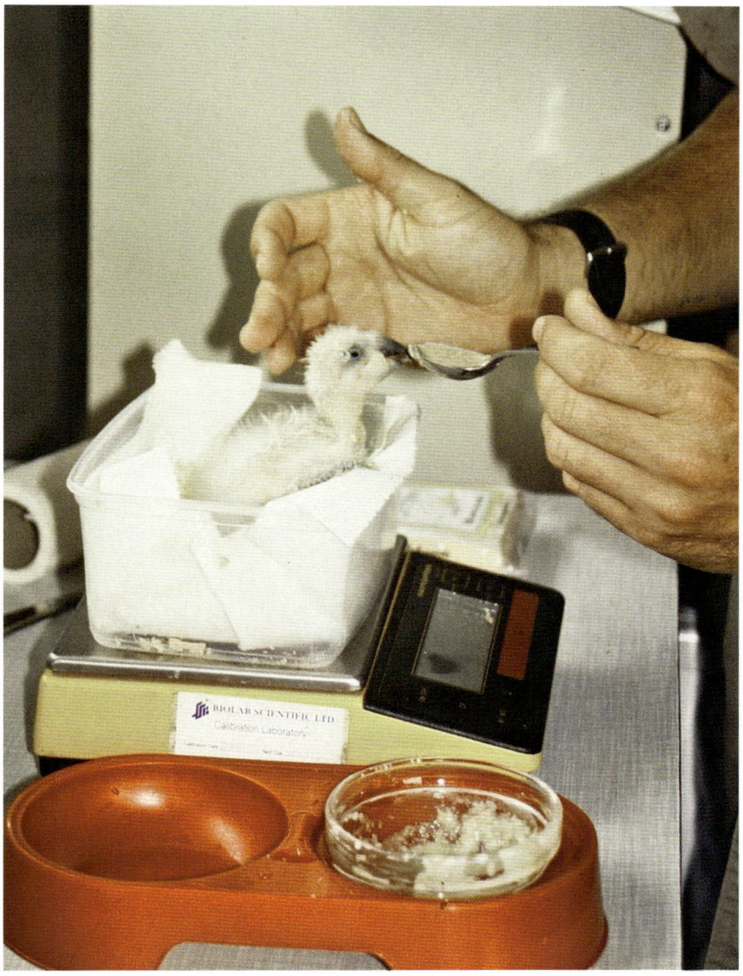

Hand feeding a cockatoo.

Chicks being pulled for hand rearing and banding. *Photos: David Hancock, Glen Holland*

A hand-reared lilac-breasted roller *Coracias caudata*.
Photo: Glen Holland

Glen Holland with a hand-reared African grey hornbill *Tockus nasutus*.
Photo: Glen Holland

A hand-reared southern yellow-billed hornbill *Tockus leucomelas*.
Photo: Glen Holland

Spotted eagle owl *Bubo africanus* chick eating a lizard. *Photo: Glen Holland*

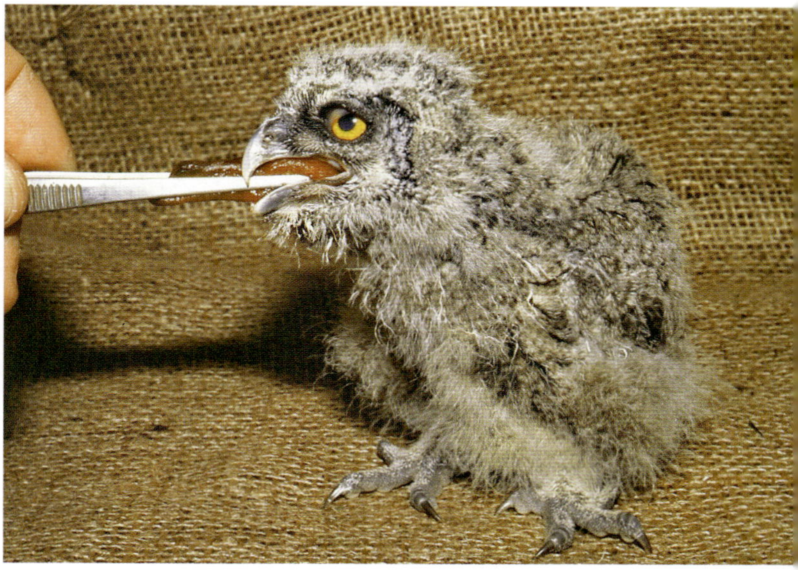

Hand feeding spotted eagle owl chick. *Photo: Glen Holland*

Barn owl *Tyto alba* chick being weighed on scale. *Photo: Glen Holland*

Rock Kestrel *Falco tinnunculus rupicolis* chick. *Photo: Glen Holland*

Barn owl *Tyto alba* at 2 days.

4 days

6 days

14 days

21 days

25 days

29 days

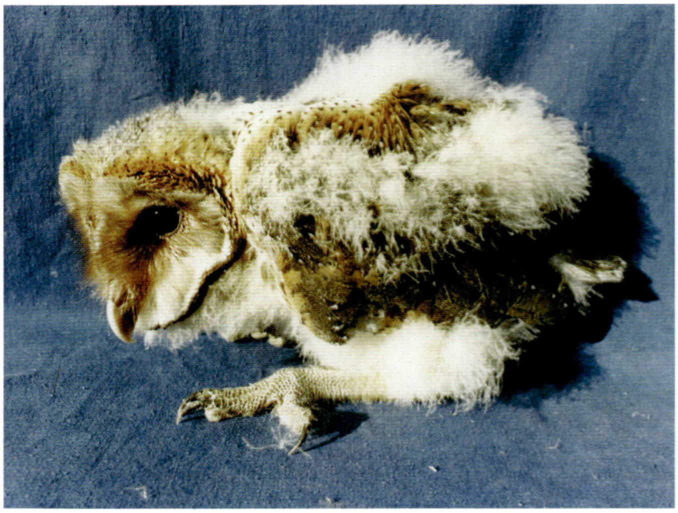

33 days

Photos: Glen Holland

Booted Eagle *Hieraaetus pennatus* chick.
Photo: Glen Holland

Jackal buzzard *Buteo rufofuscus* chick.
Photo: Glen Holland

Jackal buzzard *Buteo rufofuscus* chick.
Photo: Glen Holland

Vulture chick. *Photo: David Hancock*

Fostering ocellated turkeys *Agriocharis ocellata* under a bantam hen at Houston Zoo. *Photo: Myles Lamont*

Fostering blue-billed currasows *Crax alberti* under a bantam hen at Houston Zoo. *Photo: Myles Lamont*

Congo peafowl *Afropavo congensis* chick with a bantam chick as a feeding "teacher" at Houston Zoo. *Photo: Myles Lamont*

Species compatibility must also be considered with young birds. This grey peacock-pheasant *Polyplectron bicalcaratum* chick is showing typical toe pecking behavior towards a young Elliot's pheasant *Syrmaticus ellioti*. *Photo: Myles Lamont*

Birds can be mixed with mammals as well, as shown in this Australian exhibit with a Magpie goose *Anseranas semipalmata* and red kangaroo at Oklahoma City Zoo. *Photo: Myles Lamont*

Mixed sea duck aviary at Dry Creek Waterfowl. *Photo: Myles Lamont*

Sunda wrinkled hornbill *Aceros corrugatus*. *Photo: David Hancock*

Nest boxes. *Photo: Myles Lamont*

LEFT: A typical starling nest filled with various nesting materials. *Photo: Glen Holland*

BELOW: Removing a hornbill chick from a nest box at Houston Zoo. *Photo: David Hancock*

Various forms of nest boxes. *Photo: Matthew Vriends*

Various types of cavity nest boxes including some made by the birds themselves.
Photos: Myles Lamont

Nests

Trumpeter hornbill *Ceratogymna bucinator* with chicks in an artificial wood nest box with an access door. The mudded entrance hole can be seen in the background. Photo taken at HWRC. *Photo: David Hancock*

A common hoopoe *Upopa epops* nesting in a natural log hollow. *Photo: Nico Myburgh*

Red-billed hornbill *Tockus erythrorhynchus* pair in the late stages of nesting. *Photo: David Hancock*

An Oriental pied hornbill *Anthracoceros albirostris* inspecting a potential nest site at HWRC *Photo: David Hancock*

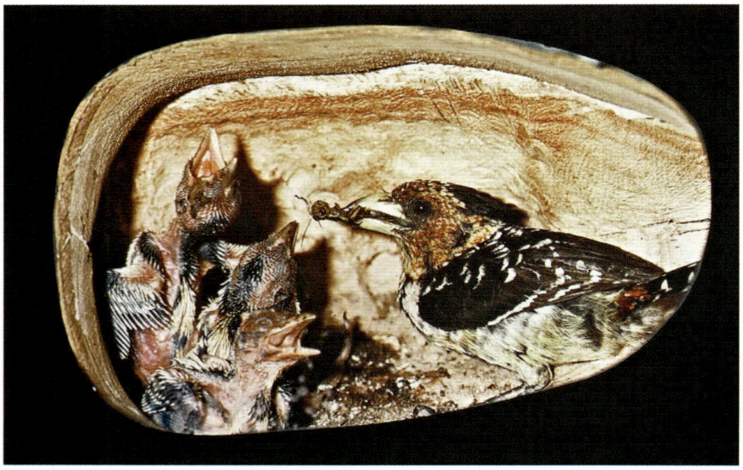

A crested barbet *Trachyphonus vaillantii* with young in a cavity box. *Photo: Nico Myburgh.*

A baby parrot in a nest box. *Photo: David Hancock*

A scarlet macaw *Ara macao* with a metal-rimmed entrance to prevent nest box destruction. *Photo: Myles Lamont*

An Acorn woodpecker *Melanerpes formicivorus* making use of an old palm log. *Photo: Myles Lamont*

Oak barrels serve as useful nesting sites, in this case for a pair of hornbills at Emerald Forest Bird Gardens. *Photo: Myles Lamont*

A palm log for aracaris placed under a covered partition at Emerald Forest Bird Gardens. *Photo: Myles Lamont*

Juvenile stork sp.
Photo: David Hancock

Black headed heron *Ardea melanocephala* chicks. *Photo: Glen Holland*

A blue heron *Herodios means* rookery. *Photo: David Hancock*

Waldrapp Ibis *Geronticus eremita* nesting platforms.
Photo: David Hancock

Various nesting site designs for smaller Passeriformes at Silver Pine Aviaries.
Photos: Myles Lamont

Cape white-eye *Zosterops pallidus* at nest. *Photo: Glen Holland*

Somber greenbul *Andropadus importunus* feeding chicks. *Photo: Glen Holland*

Nests can be reused and placed in the aviary once disinfected. *Photo: Myles Lamont*

Typical nest pan for *Serinus* species. *Photo: Glen Holland*

A green-bunting cissa *Cissa chinensis* on a nest. *Photo: Rosemary Low*

A yellow-legged honey creeper *Cyanerpes caeruleus* hen on a nest.
Photo: Kateri Davis

A male grosbeak weaver *Amblyospiza albifrons* displays at his nest. *Photo: Nico Myburgh.*

Olive sunbird *Cyanomitra olivacea*. *Photo: Glen Holland*

A typical woven bamboo hanging nest. *Photo: Myles Lamont*

A male malachite sunbird *Nectarinia famosa* at the nest. *Photo: Glen Holland*

Bowerbird nest.

ABOVE AND LEFT:
Bower bird nest.

A yellow-rumped cacique *Cacicus cela* at Dallas World Aquarium.
Photo: Myles Lamont

Troupial nests *Icterus icterus* at Woodland Park Zoo.
Photo: Myles Lamont

An underground nest entrance for shelducks. *Photo: Myles Lamont*

Elevated nest box for pinioned or wing-clipped birds.
Photo: Heidi Cline

A simple nest box for various species of geese.
Photo: Heidi Cline

A white-fronted goose *Anser a. albifrons* on a nest box. *Photo: Myles Lamont*

A female bufflehead Bucephala albeola inspecting a nest box.
Photo: Myles Lamont

A hollowed out log for larger species of waterfowl.
Photo: Heidi Cline

ABOVE, BELOW AND RIGHT: A Himalayan monal *Lophophorus impejanus* hen sitting on eggs.
Photo: Myles Lamont

An East African grey crowned-crane *Balearica regulorum* incubating.
Photo: David Hancock

A sandhill crane *Grus canadensis* chick. *Photo: David Hancock*

A pair of Atlantic brant *Branta bernicla bernicla* guarding their nest site. *Photo: Myles Lamont*

A Barrow's goldeneye *Bucephala islandica* leaving her nest. *Photo: Myles Lamont*

A typical waterfowl nest box for pinioned birds. *Photo: Myles Lamont*

Nests

Common bronze-winged pigeon *Phaps chalcoptera* in a basket nest.

Nicobar pigeon *Caloenas nicobarica* on a wire-framed nest platform.

White-cheek turaco *Tauraco leucotis* on nest.
Photo: Myles Lamont

Red-winged laughing thrush *Garrulax formosa* nest at Hancock Wildlife Research Center. *Photo: Myles Lamont*

A typical dove's nest with young.

Sunbittern *Eurypyga helias* chick.
Photo: David Hancock

Hazel Grouse chick.

Killdeer *Charadrius vociferus.*
Photo: Myles Lamont

Thicknee *Burhinus* sp.
Photo: David Hancock

Green-naped Pheasant Pigeon *Otidiphaps n. nobilis.*
Photo: Myles Lamont

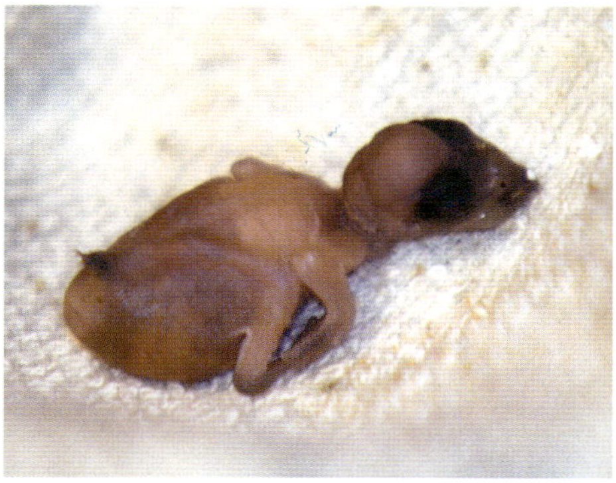

Yellow-knobbed Currasow *Crax daubentoni.*

A black swan *Cygnus atratus* cygnet.
Photo: Myles Lamont

Tragopan diet used at Woodland Park Zoo, Seattle.
Photo: Myles Lamont

Gallinaceous diet after being picked through by a pair of tragopans.
Photo: Myles Lamont

Frugiverous diet prior to chopping at Hancock Wildlife Research Center. *Photo: Myles Lamont*

A typical dry softbill diet. *Photo: Myles Lamont*

A typical moist softbill diet. *Photo: Myles Lamont*

Diet used for *Garrulax* spp. at Hancock Wildlife Research Center.
Photo: Myles Lamont

A diet of mealworms and smelt used for kingfisher
Photo: Myles Lamont

Various frugiverous diets used at Vogelpark Walsrode, Germany.
Photo: Simon Bruslund Jensen

Feed storage fridge at Miami Metro Zoo. *Photo: Myles Lamont*

Mazuri low iron proprietary pellet and papaya, a maintenance diet for Ramphastids at Emerald Forest Birds Gardens.
Photo: Myles Lamont

Dietary supplements used at Vogelpark Walsrode, Germany.
Photo: Simon Bruslund Jensen

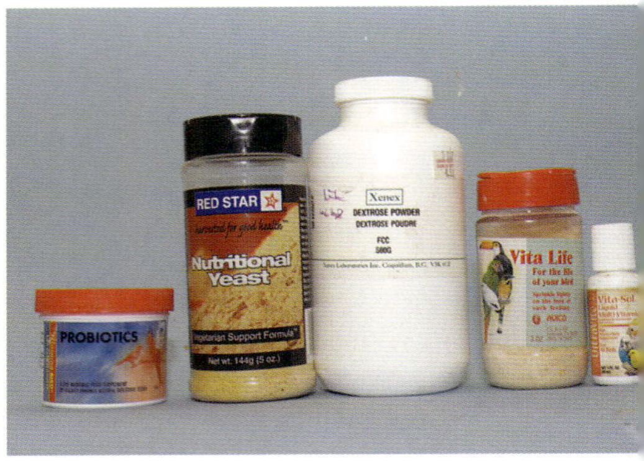

Various commercial proprietary diets exist for the aviculturist, making food preparation much easier. Food supplements are also available, however should be used with discretion. *Photo: Myles Lamont*

For the seminivorous species, a large variety of feed types are available. *Photo: Myles Lamont*

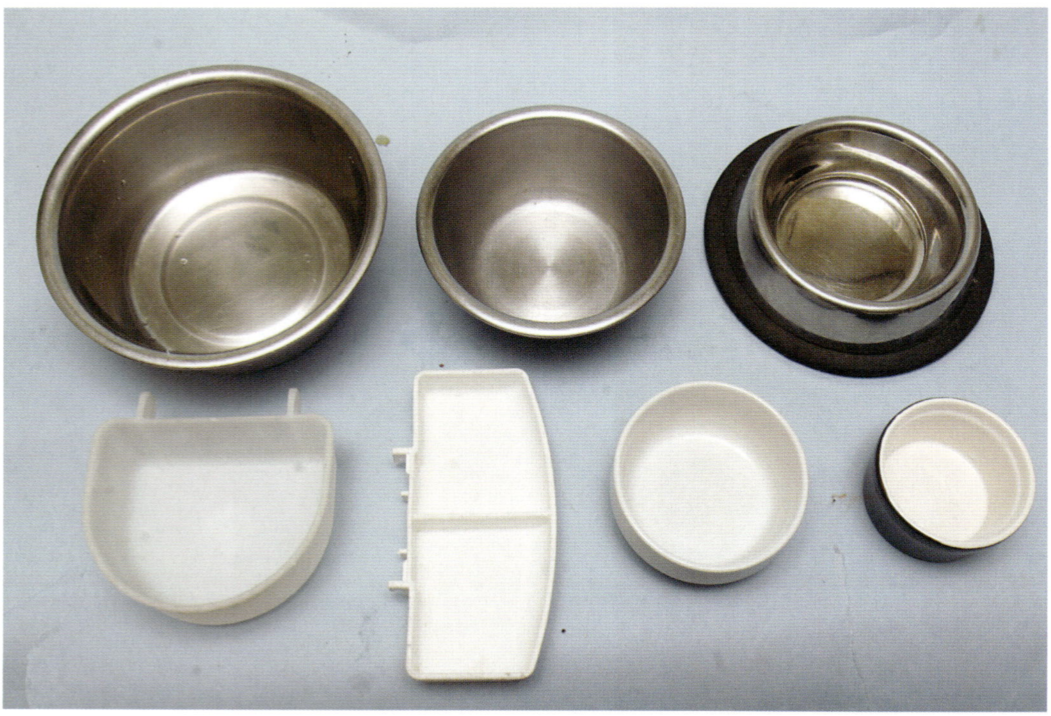

Various pet food dishes can be used for food presentation; however, stainless steel dishes are most commonly used and recommended because they are exceptionally durable and can be easily cleaned and disinfected. *Photo: Myles Lamont*

Food can be presented in a number of ways, usually dependent on the species and the diet it requires.
Photos: David Hancock, Myles Lamont

Ponds with a high salinity require areas where the birds can obtain fresh water with no salt content.
Photo: Myles Lamont

Mist systems have multiple functions, however are normally used as an artificial rain source. *Photo: Myles Lamont*

A drip line can be used to provide fresh water at all times.
Photo: Myles Lamont

Various forms of nesting material can be used, but selection is dependent upon the species in question.

FROM LEFT TO RIGHT: Commercial sphagnum moss, harvested wild moss, raffia, coconut fiber, harvested field grass and excelsior shredded "wood wool."
Photo: Myles Lamont

Nest material can be provided in a number of ways; however, offering the birds free choice is often the best method. Offering nesting materials from a hanging wire basket helps keep waste to a minimum. *Photo: Myles Lamont*

Incubation and Record Keeping

The aviculturist has a wide variety of incubation options, particularly in terms of artificial incubation. *Photo: Myles Lamont*

The early development of a turaco chick. Fertility is determined by the obvious eye spot and veins. *Photo: Myles Lamont*

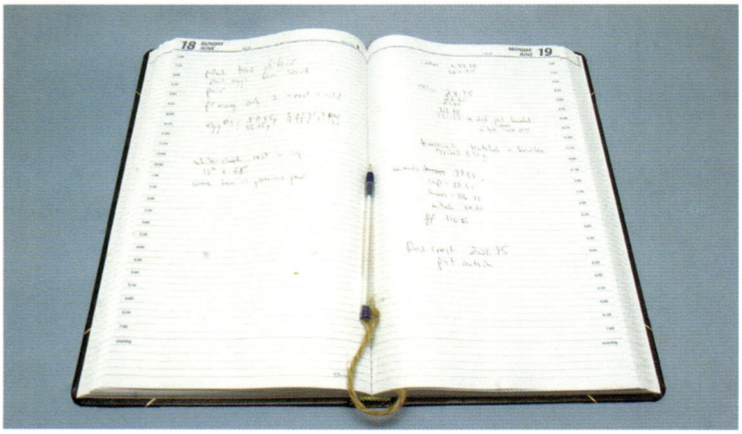

Good record keeping is essential for proper avian management, particularly when dealing with threatened species or those with few founder numbers. *Photo: Myles Lamont*

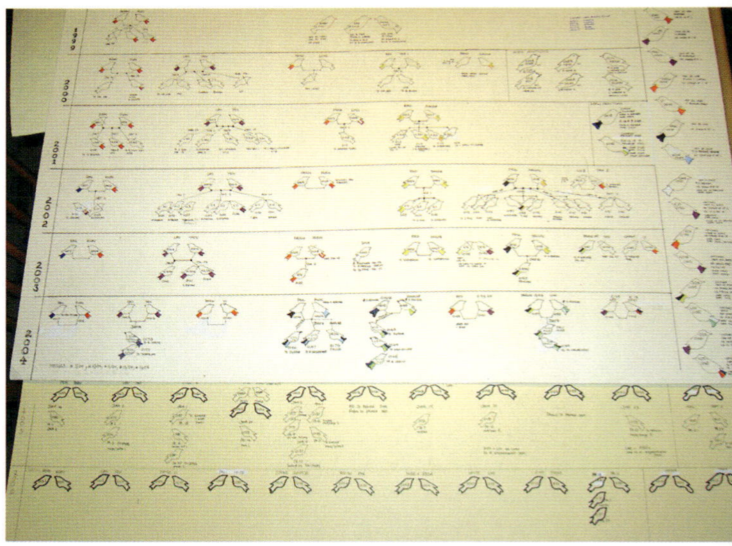

A studbook tree for pekin robins, developed by Peter Karsten, Silver Pine Aviaries. *Photo: Myles Lamont*

Typical incubation rooms where ambient temperature is kept consistent and is easily cleaned. *Photo: Myles Lamont*

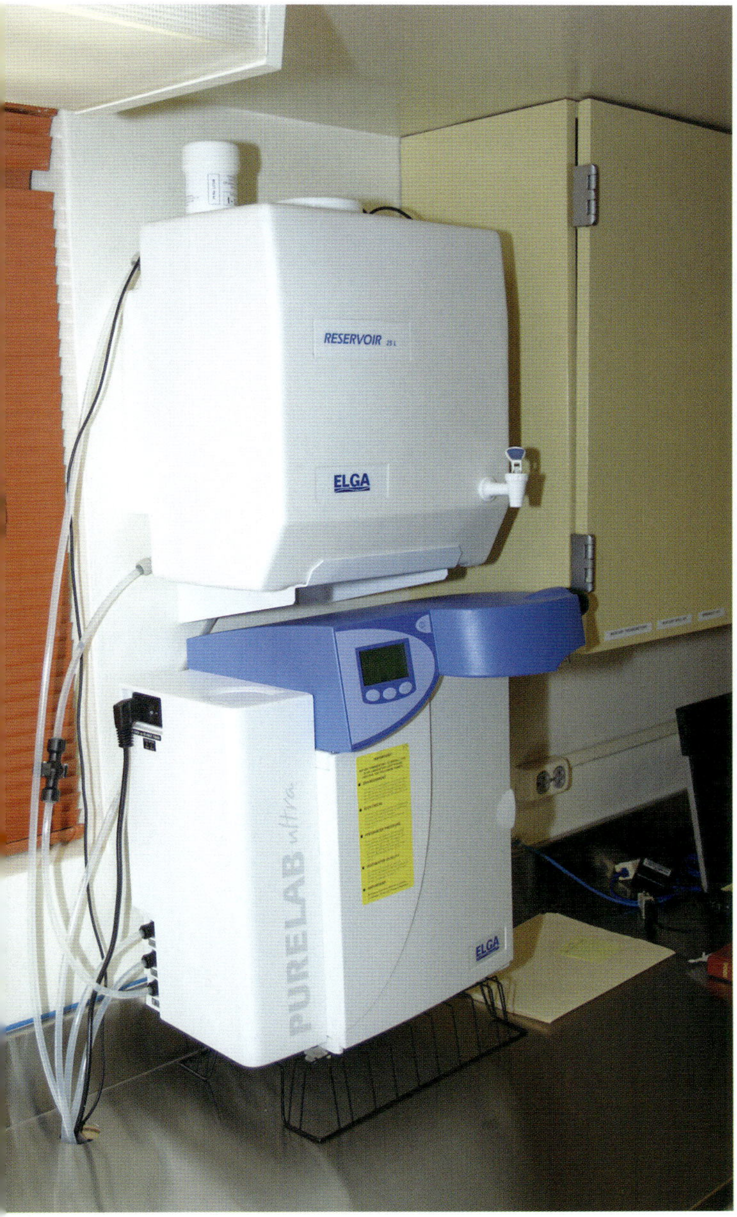

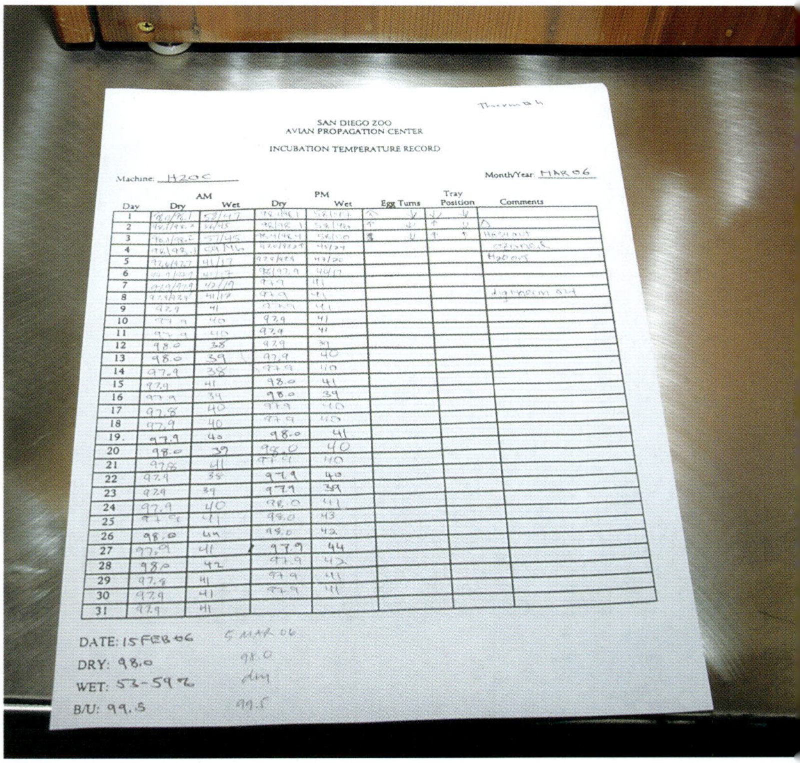

Incubator record sheet to monitor temperature and humidity fluctuations. *Photo: Myles Lamont*

A water purifier at San Diego Zoo to help ensure incubators are sanitary. *Photo: Myles Lamont*

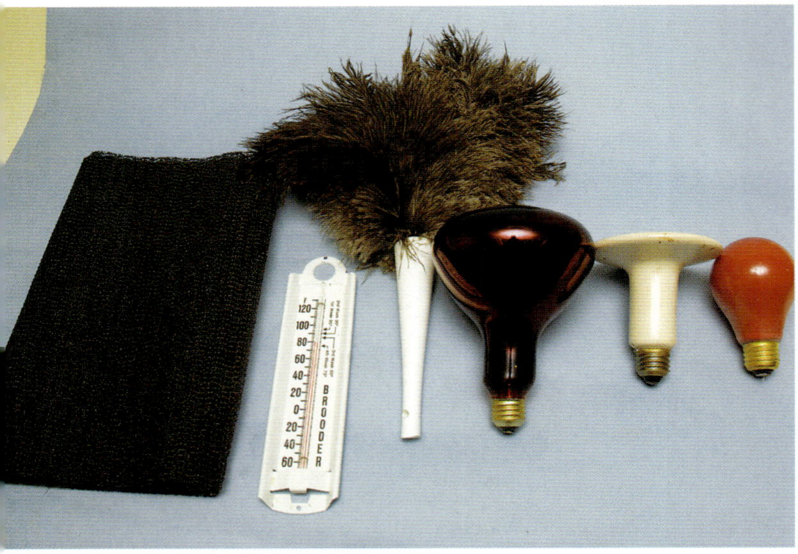

Brooder equipment: rubber drawer liner to prevent slipped tendons; thermometer; feather duster to give precocial chicks an artificial brood patch; infrared and ceramic heat lamp; and a red light bulb to prevent disruption of the birds' photoperiod. *Photo: Myles Lamont*

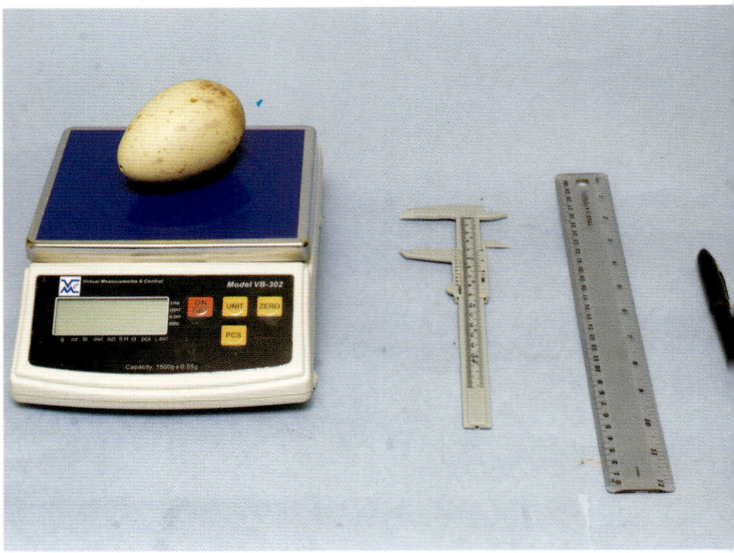

A scale and calipers to weigh and measure egg size and to monitor chick development. *Photo: Myles Lamont*

A typical brooder setup for Galliformes with radiant heating and water dish that prevents drowning. *Photo: David Hancock*

Toe pecking and other aggressive behaviors need to be monitored and aggressors removed.
Photo: Myles Lamont

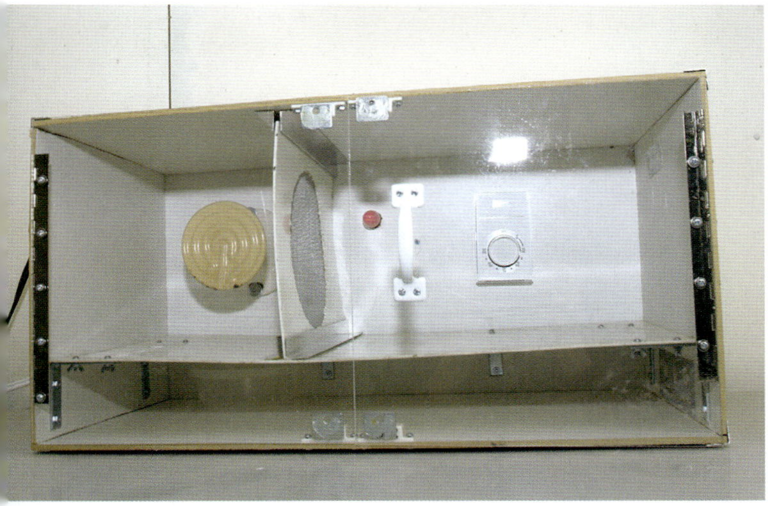

A homemade brooder for altricial species of birds, with a ceramic heater and thermostat. A water bowl is placed below the heat lamp for increased humidity. *Photo: Myles Lamont*

A stainless steel brooder with removable bottom allows for ease of cleaning. Two radiating heat bulbs are used in the event one burns out. *Photo: Myles Lamont*

As with incubators, numerous brooder models and designs are available to the aviculturist. The brooder on the left is treated with tinted plastic which prevents chicks from viewing their keeper to discourage imprinting. *Photo: Myles Lamont*

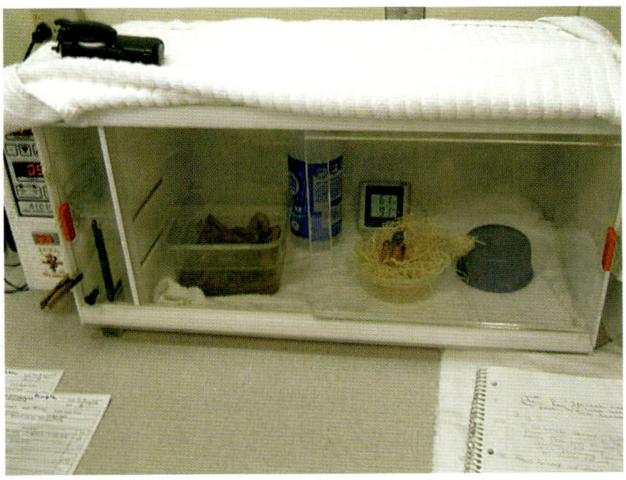

A typical brooder setup, with a royal starling chick. *Photo: Myles Lamont*

The Anseriformes present a different challenge to the aviculturist when it comes to incubation and brooding. Here, rubber matting is used to protect the delicate toes and webbing of these ducklings. *Photo: David Hancock*

An eider chick in the midst of hatching. Humidity plays a key role in a successful hatch.
Photo: David Hancock

Two types of waterfowl brooders: the wet brooder on the left and the dry brooder on the right. The wet method is more common as it is believed to aid in proper feather and motory development. *Photo: David Hancock*

Animal restraint and capture needs to be done with care. Nets of different sizes and mesh types should be chosen to help prevent injury.
Photo: Myles Lamont

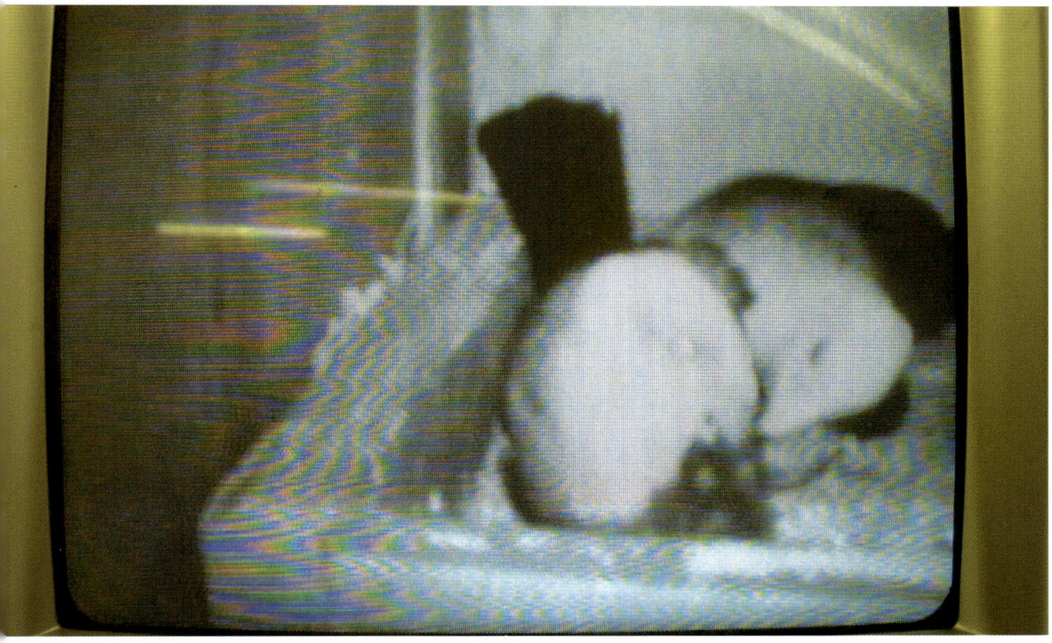

Pairs can be monitored via hardwired or wireless cameras. In this case, an infrared camera displaying a pair of red-crested turacos *Tauraco erythrolophus* feeding a chick. *Photo: Myles Lamont*

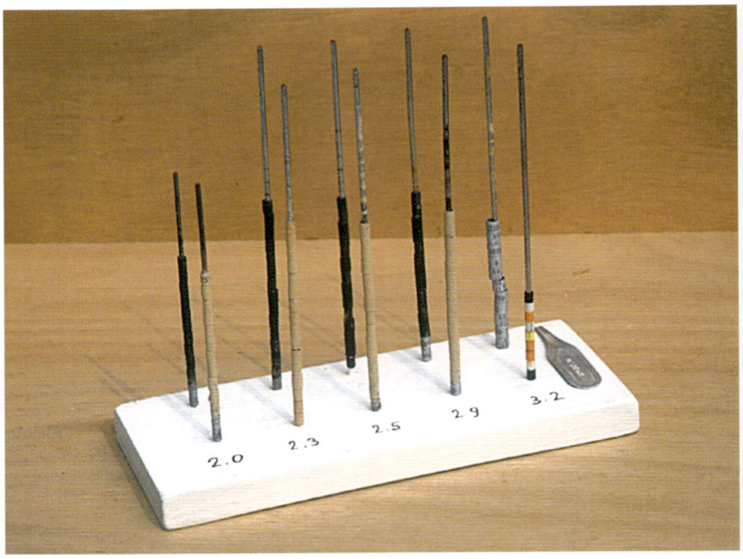

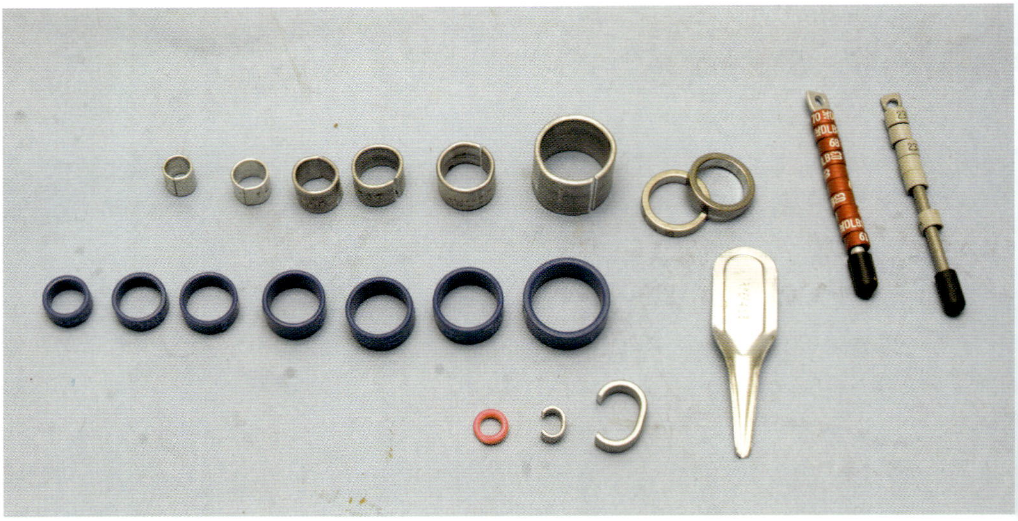

Banding of birds, particularly seamless bands, must be done at the correct age and by an experienced aviculturist to prevent damage to toes and legs. The correct size band must also be chosen and usually ordered well in advance to prevent missing any window of opportunity. *Photo: Myles Lamont*

Crooked toes due to improper incubation parameters can sometimes be corrected if caught early enough. In this case, masking tape is sufficient to correct a minor toe curvature. *Photo: Myles Lamont*

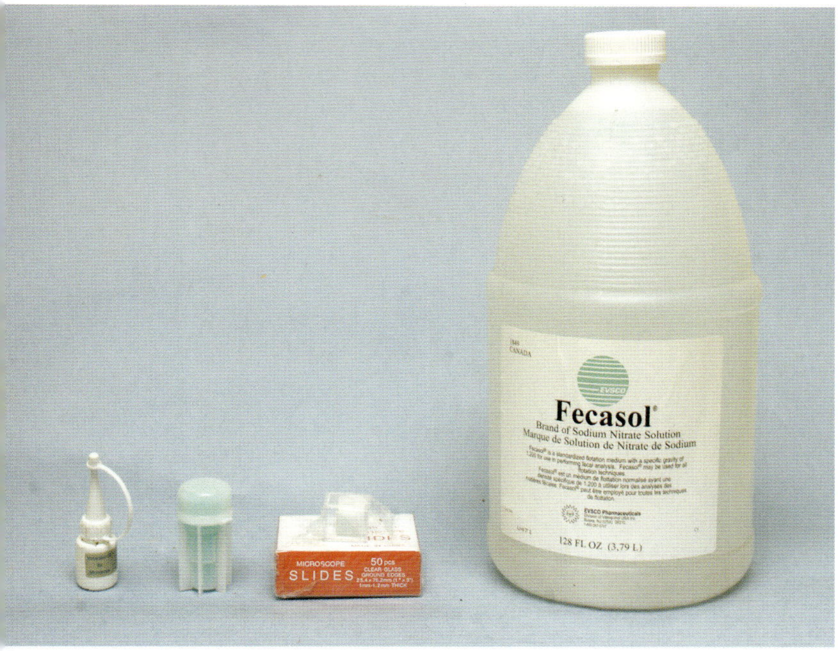

Fecal flotation equipment to monitor parasite loads. *Photo: Myles Lamont*

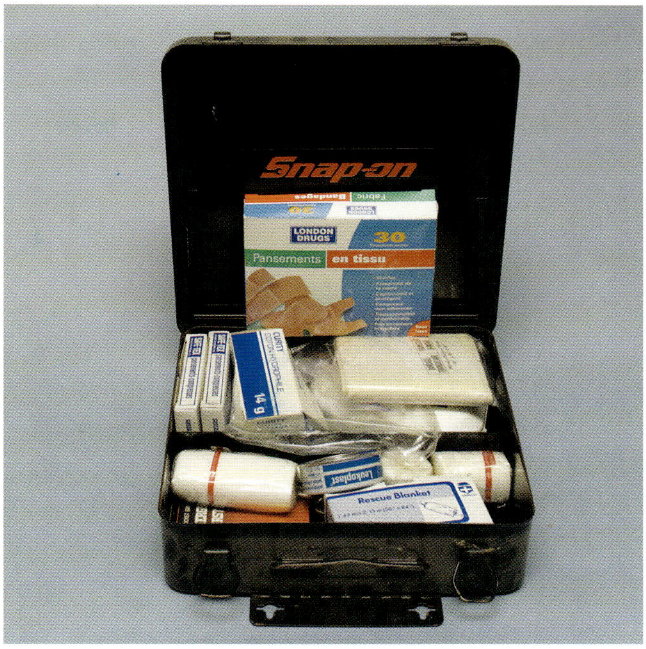

First aid for the aviculturist. *Photo: Myles Lamont*

Various disinfectants and topical antibiotics for minor flesh wounds.
Photo: Myles Lamont

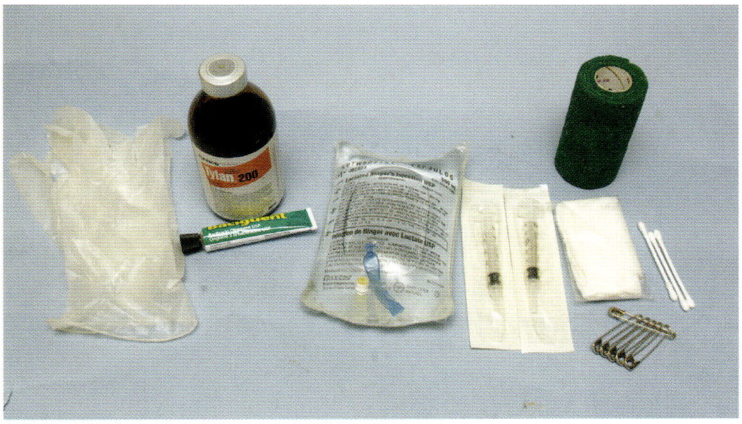

Ringers solution for rehydration along with various products to aid in medical treatment. *Photo: Myles Lamont*

In locations requiring some form of biosecurity, such as quarantines or nurseries, the appropriate footwear is needed to prevent the spread or introduction of disease.
Photo: Myles Lamont

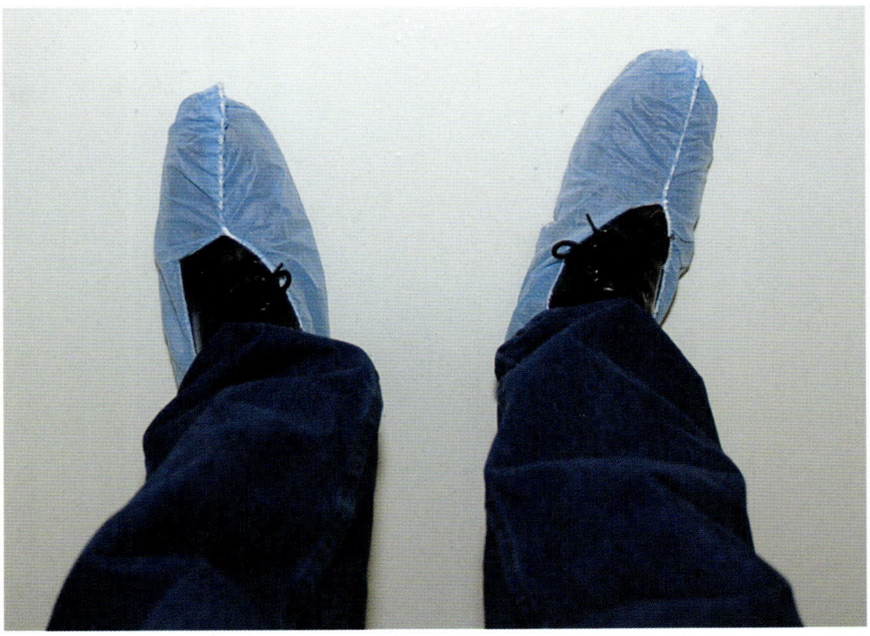

Photo: David Hancock

Waterfowl

Typical waterfowl aviaries have access to natural vegetation, nest sites and adequate space to prevent overcrowding and territorial issues. Photos taken at the collection of Paul and Lynn Dye.

Photos: Myles Lamont

A variety of aviaries ranging from small backyard aviaries, to larger park settings, as well as tropical indoor setups. Landscaping plays a key role in developing the aesthetics of an aviary which provides benefits for both the occupants and their keepers.
Photos: David Hancock

Various aviary designs, all of which are built for the needs of a particular group of birds. The construction of strong, predator-proof enclosures cannot be overstressed. *Photos: David Hancock, Myles Lamont*

Various waterfowl ponds, two of which were still being completed. Above right, Paul and Lynn Dye collection and Maynard Axelson collection, below. *Photos: Myles Lamont*

Impressive structures such as these are usually restricted to organizations with large budgets. The tent technique is quite common for large covered aviaries. Photos on the right were taken at Miami Metro Zoo; top left, Woodland Park Zoo in Seattle; and bottom left, Bloedel Conservatory in Vancouver, B.C. *Photos: Myles Lamont*

Water features at the Wings of Asia aviary at Miami Metro Zoo. The underwater viewing area allows the visitor to watch the waterfowl dive below the surface.
Photos: Myles Lamont

An open aviary for Ciconiiformes offering potential for mixing with hoofstock. *Photo: Myles Lamont*

A walkway through the conservation aviaries at Woodland Park Zoo, Seattle. *Photo: Myles Lamont*

A beautifully landscaped aviary for a pair of greater helmeted currasows showing the use of piano wire as a barrier, Woodland Park Zoo, Seattle. *Photo: Myles Lamont*

ABOVE AND BELOW: An outside and inside view of an observation deck showing the use of one-way glass to observe the birds' natural behavior, Paul and Lynn Dye collection. *Photos: Myles Lamont*

An indoor aviary at Dallas World Aquarium maximizes the use of its relatively small size by having multiple levels of observation creating the impression of a very large facility.
Photos: Myles Lamont

Some softbill flights at Houston Zoo. *Photos: Myles Lamont*

Few display techniques are as appealing to the visitor as the open-air exhibit but it presents several challenges to the aviculturist including the need to restrict flight and to prevent predators from entering the enclosure.

Wild birds also also must be prevented from spreading disease and consuming large amounts of feed.

ABOVE: Houston Zoo.
LEFT AND BELOW: San Diego Zoo.
Photos: Myles Lamont

Some typical pheasant-style aviaries in Europe making the most of a smaller location while still providing a good display.
Photos: David Hancock

This style of aviary causes visitors to continue walking past the exhibits without stopping to closely observe their inhabitants.

These row aviaries for waterfowl provide for easy maintenance; however, make disease control very difficult.
Photo: David Hancock

RIGHT AND BELOW: Two penguin exhibits. Note the holes in the bank walls for nesting sites.
Photo: David Hancock

A harlequin enclosure at Dry Creek Waterfowl.
Photo: Myles Lamont

Sea ducks make use of a large communal pond with smaller perimeter aviaries where the birds can establish nesting territories on their own.
Photo: Myles Lamont

Large covered aviaries at Dry Creek Waterfowl provide ideal habitat for nesting birds and allow for ease of flight.

A concrete pool is easy to clean and with a constant supply of fresh water helps to maintain feather condition.

Photos: Myles Lamont

Softbill flights at Emerald Forest Bird Gardens housing toucans and turacos (above) and a pair of hornbills (below).
Photo: Myles Lamont

Another large aviary for hornbills utilizing post and beam framing (right) as compared to steel tube framing used for this pair toco toucans (below).
Photo: Myles Lamont

Breeding aviaries for small ramphastids, in this case aracaris. *Photo: Myles Lamont*

The use of black wire removes much of the sense of division between the birds and the observer and makes photography much easier. *Photo: Myles Lamont*

A small backyard aviary such as this works well for those who have limited space. This type of design is ideal for smaller species of softbill and finch, in this case European insectivores and Asiatic babblers. Photo taken at the aviaries of Ed Storosinski, Mission, BC. *Photo: Myles Lamont*

One of many multi-unit aviaries at San Diego Zoo. *Photo: Myles Lamont*

Scripps Aviary, San Diego Zoo. *Photo: Myles Lamont*

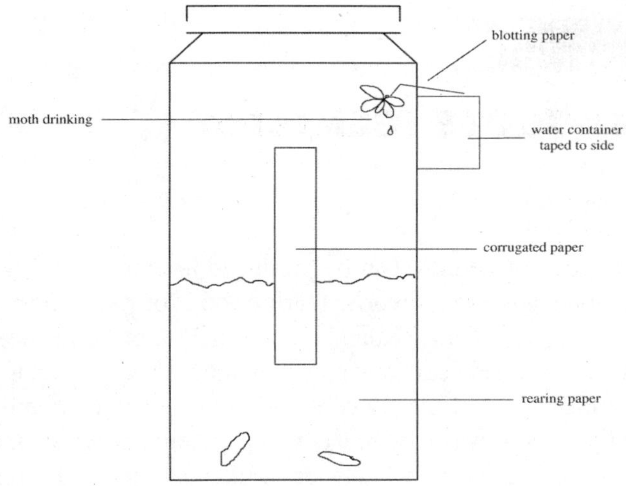

pupae. Some species such as pekin robins and starlings will eagerly feed the pupae and moths to their chicks.

Ian Hinze describes another method for producing smaller quantities of waxworm. The items required are a plastic sweets jar, blotting paper, pair of pointed scissors, closely woven metal gauze, strong sellotape, corrugated paper, small water container, wax-moth larvae, rearing food and non toxic glue.

1) Cut a square 3.8 cm (1½ in.) square out of the sweet jar lid and cover the hole from the underside with a piece of glued on, closely woven metal gauze.

2) In a bowl, mix together 170 g (6 oz.) Farley's Baby Rice Cereal and 453 g (1 lb.) wheatgerm. Add 340 g (¾ lb.) liquid honey and 100 ml (3.4 oz.) glycerin. Continue mixing until the whole becomes nice and crumbly moist, not wet. This is wax-moth rearing food.

3) Put rearing food into the jar up to the halfway mark. Insert some rolled up corrugated paper, 12.7 cm (5 in.) in height when stood upright, into the centre of the mixture leaving at least 8.9 cm (3½ in.) protruding. Place some wax-moth larvae (caterpillars) on top of the rearing food and then screw on the lid of the jar. The larvae will begin burrowing immediately and start to feed. NOTE: The excess rearing food can be put in a container and stored in a freezer for future use.

4) Put the jar into a warm place, such as an airing cupboard, and maintain a steady heat of around 27°C (80°F). Heat is essential to maintain a healthy culture and to encourage breeding. A dark place is also beneficial.

5) In due course the larvae will crawl into the corrugated paper and start to pupate. They spin a silky-like cocoon that eventually looks and feels like rice paper. When this happens, it is time to put fresh rearing food (such as the thawed-out food from the freezer) into a new jar. Remove the corrugated paper containing the pupae from the old jar and insert it into the food of the new one. IMPORTANT NOTE: If you do not supply corrugated paper or similar material to assist pupating, the larvae will amass at the top of the jar and possibly stop the flow of air.

6) Once the food in the old jar has been used up, dispose of the waste but make sure that there are no unpupated larvae in it. Clean the jar thoroughly in warm soapy water and then store for future use.

7) Emergence of Moths: On finding that you have moths in your new jar, you should cut a narrow horizontal slit just below the jar shoulder. The slit should measure approximately 20 mm wide x 1.5 mm deep. Cut a strip of blotting paper 12.7 cm (5 in.) long and wide enough to go through the slit in the jar. Using the sellotape, affix the water container to the side of the jar underneath the slit with the lower part outside and hanging into the water reservoir. The capillary action of the water rising up the strip will enable the moths to drink. IMPORTANT NOTE: It is only the moths that need to drink and not their larvae. When only larvae are in the jar the slit should be sealed up to prevent escape.

8) While the moths are alive make sure that the water is kept topped up. Because of the heat, water is lost through evaporation. Although unnoticed by the naked eye, the females will have laid hundreds of eggs and within a few days the microscopic larvae will emerge. They are probably first noticed when only 1 mm in length. They appear to move very quickly and are most noticeable when they crawl over the, now dead, moths. The growth from this stage is very rapid and within four weeks one only has to lift out the corrugated paper and unroll it to find masses inside. They can be anything from caterpillar-like grubs to creatures double the diameter of a mealworm and up to 3 cm in length. Unlike the mealworm, they have the added advantage of having been fed on a good

quality food which is passed on to the birds and have a skin as soft as any green caterpillar, making digestion of them easy.

9) If you are fortunate enough to have hatched out a large amount of larvae and they are big enough to handle, put some fresh rearing food into a warm place for at least half an hour and then break it in half. Take half the old rearing food containing larvae and add it to one-half of the fresh rearing food in a sweets jar. You will notice that the half containing larvae is warm. This is because, due to their activities, the larvae generate heat. Always check your thermometer, as the extra warmth may require one to turn the added heating down. Take the other half of the fresh rearing food and add that to the old rearing food that remains. You now have two cultures. Using this method, or by simply putting more than one roll of corrugated paper into each jar to put singly into new jars later, you can get at least six cultures going in only a few weeks.

   IMPORTANT NOTE: Never allow your jars to become overcrowded with larvae as there is the possibility of "sweating". If water can be seen running down the inside of the jars, immediate action is called for. Check that your heating is not too high, and also change the rearing food. The culture must never be allowed to get wet.

10) Once a culture is started there should never really be any need to purchase larvae again. Always allowing a certain number of larvae to pupate this way will provide continual rewards. Aim for as large a moth as possible as the larvae, in turn, will be that much better. By selective breeding this is easily accomplished. Very small moths should be discarded. At least six cultures are needed for the "average sized" collection of birds, but this should be achieved within a few weeks of obtaining the first culture. Try to get cultures growing at different stages, so you will not suddenly run out, and will be able to feed larvae to the birds all year round. Waxworms are ideal for helping to rear many species of bird, but especially those that require an all-insect diet while in the nest and during those nerve racking early days as fledglings.

References:

Hinze, Ian. 1992. *Estrilid Society Yearbook,* pg. 34–37.

MAGGOTS (GENTLES)

by David Holmes

Maggots, or gentles, can be produced in extremely large numbers and are a favorite rearing food for passerines in Australia and New Zealand. Historically a piece of meat such as a sheep heart was placed in a tin with an open top and coarse wire base. Once the meat was blown (covered in fly eggs) it was suspended over a container of bran and over the ensuing two weeks or so the maggots would fall into the bran. Bran is a vital part as the maggots in the meat still contain a gut full of rotting flesh but once they have spent a few days in the bran, the gut is cleared and once the maggot appears white it is safe to be harvested and fed to birds. They are usually fed in small glass bowls with straight sides that the maggots are unable to climb.

The risk of toxins to the birds, and probably the complaints about smelly meat from the families of bird breeders, has led to a far more hygienic method of maggot propagation. This method requires a wooden cabinet with a fly screen mesh front. A 40-watt bulb is fixed in the top of the box or two lamps on the back wall of the box about half way down.

A wooden sliding door is made on the side of the box. Four cups of bran are then mixed with 118 ml (½ c.) of full cream milk powder and the mix is then moistened with water to the point where it will retain a slightly crumbly form when squeezed in a fist. Flies are then introduced and after two days the mix should be blown. During this time keep turning the mixture otherwise the top dries out. The mixture can then be transferred to another warm place for the maggots to mature. At this stage another milk/bran mix is placed in the box to keep the cycle going.

When maggots have matured they can be transferred to a shallow dish containing a cup full of millrun or pollard bran. This is very fine, like coarse flour. Leave for two hours to clean the maggots and then sieve the maggots out. They are then placed in finely sieved "chick starter" and supplements, such as calcium and vitamins, are added. After an hour or so the maggots are fed out to the birds. If excess are produced they can be placed in a little medium in a fridge to slow down their development. It is necessary to let the odd container of fly blown mash to go the full cycle so you are always replacing your flies.

HAND REARING

Hand rearing birds is both a challenging and very rewarding practice. Typically it is undertaken for the following reasons: eggs and chicks have been abandoned or neglected by their parents; there is a desire for producing hand-tame birds as pets and for education; eggs and chicks have been removed to entice adults to produce further clutches and broods; and chicks that result from artificially incubated eggs have to be hand raised.

Hand rearing can be a very intense procedure involving hourly feeds around the clock or a relatively simple procedure where the precocial chicks feed themselves and need only be topped up four times per day. Most newly hatched chicks are naked and wet at hatching. The temperature is the most important initial consideration. For the first 12–24 hours, chicks are kept at a temperature a degree lower than that at which they were incubated. During this period the down will have dried, most of the yolk sac will have been absorbed into the body, and the chick should have passed its first green, very soft stool. During this initial period it is advisable to feed naked chicks a few drops of Ringers Lactate or Pedialyte solution obtainable from a pharmacy. Once the first stool has been passed the chick can receive its first feed.

Remember that the chicks of those species that feed by regurgitation, such as psittacines and seedeaters, will require the food to be warmed to 38°C (100.4°F). Food can be tested with a thermometer or against your cheek to ensure it is not too hot. Food that is too cold may be rejected by the birds and will not digest well. Food that is too hot will burn the chick's mouth and cause damage to the crop lining. Species that naturally feed a regurgitated, partially digested mass should initially receive food with a thin consistency. Thick food, which is difficult to digest, will result in sour crop or compaction, and food that is too thin will be digested too quickly. Insectivorous birds, such as bee-eaters, are able to digest small insects from the first feed and others, such as bulbuls, will take fruit immediately. For those precocial species, such as waterfowl and quail, that feed themselves, a little green food mixed into the food of the former and some live food in the diet of the latter will assist in stimulating feeding. Species that are not naturally fed by regurgitation do not require the food to be warmed. Commercial hand rearing formulas designed for psittacines can also be used as the basis of a hand rearing diet for a variety of passerines. For many species the hand rearing formulas lack the essential fats that growing birds require, and it is for this reason that substances such as a little peanut butter or vegetable oil are added to the diet.

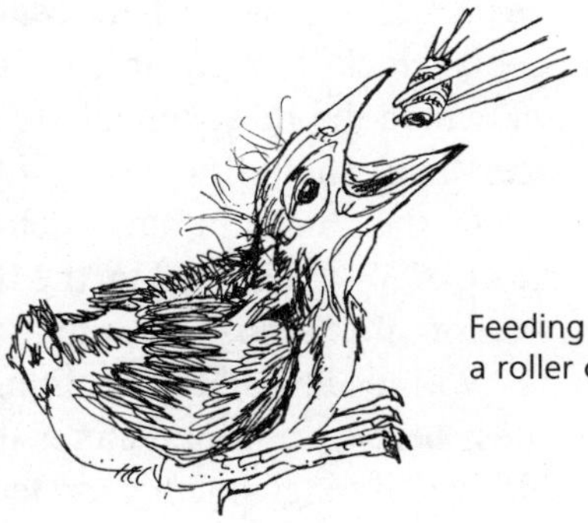

Feeding a cricket body to a roller chick.

Small altricial (helpless) chicks should be placed in a bowl lined with paper towel and some soft facial tissue for comfort. Most species are initially fed every hour. Many breeders advocate initial feeding through the night, but I have found that with one feed between 10 p.m. and 6 a.m., at about 2 a.m., I have not lost chicks and have the added advantage of maintaining my sanity after a reasonable night's sleep. The 2 a.m. feeds are dropped after the first week.

Home-made incubator.

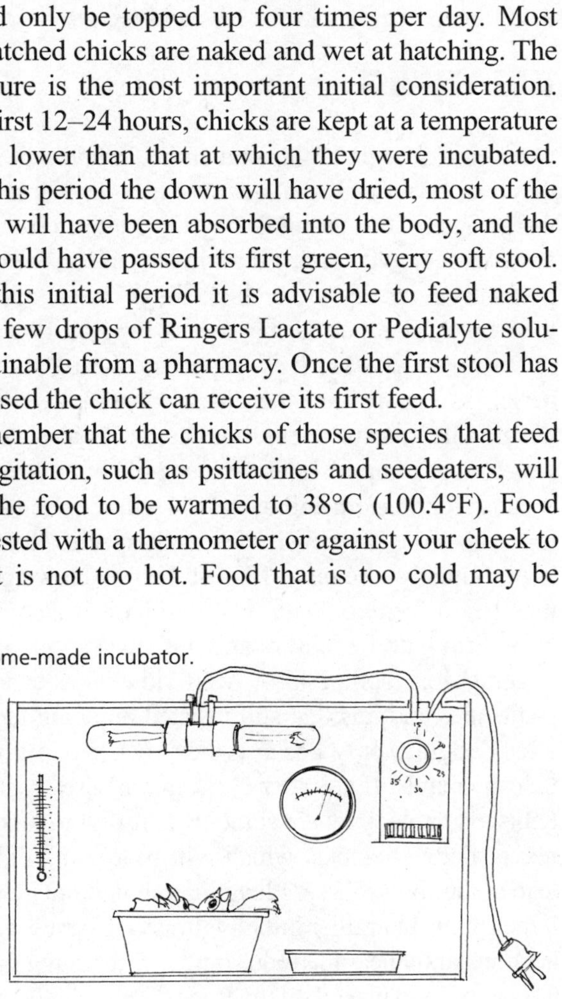

TEMPERATURE & HUMIDITY

As a rule of thumb for the first week the temperature, starting at 37°C (98.6°F), is dropped by 1°C (1.8°F) per day for most species. Once feathers are showing well, the temperature should be about 26°C (78.8°F), and by the time the entire body is well feathered the temperature should be about 24°C (75.2°F). Deaths within the first 10 days are often related to brooder temperatures that are either too low or too high. Sources of heat, such as heating elements, heating pads, oil heaters, and electric bulbs (preferably amber or red to soften the light), are all useful. Ideally, the source of heat should be thermostatically controlled. With precocial chicks that are able to position themselves in the area where the most suitable temperature exists, the brooder lamp should be suspended at a height that will not allow the chicks to burn their heads and to give a temperature of 37°C (98.6°F) on the floor of the brooder, directly below the lamp. Temperatures that are too high or too low will result in chicks panting, crying, and showing restless behavior. Temperatures that are too low result in chicks huddling together or lethargy, extended periods of sleep, and, once a little older, fluffed-out feathers. Chicks stressed by incorrect temperatures will not feed enthusiastically, their digestion will slow, and growth rates will be affected. Experience teaches the aviculturist to the point where, like any natural parent, the person rearing the chicks develops an instinct (gut feel) for the correct temperature the chicks require and also the ability to read the chick's body language, which will change in accordance with the temperature.

Humidity is also important — the ideal humidity being 50% or a little more. Small chicks in particular will dehydrate if their surrounding humidity is too low. Signs of low humidity are flaky, wrinkly, and/or reddish skin, crying, and decreased growth rate. Humidity is supplied by placing a bowl of water in the brooder. This should be measured to ensure the optimum humidity is maintained. High humidity can cause premature feather growth and the rapid growth of dangerous fungi.

NEST LINING

Chicks need to be kept on a suitable substrate that will ensure they remain warm and comfortable. The bedding needs to be absorbent to ensure the chicks remain clean and dry. It should also offer traction to avoid the chick slipping, which could result in splayed legs. The material should not be dusty as this could lead to respiratory problems and an irritation of the eyes as well. The chick should, particularly in the early stages, be comfortable. Very often paper towels and tissues are used in the early

stages with the chicks being moved onto pine shavings as they mature. One problem experienced with pine shavings is that some species will consume these, resulting in compaction. For example, a clutch of yellow-billed hornbills that I was rearing suddenly stopped feeding. After a few hours they regurgitated pellets of shavings. They were then placed on a bed of dry grass and feeding resumed as normal. Chicks of larger birds, such as raptors and ibis, should be placed in a container that is lined with twigs, a grass, and leaf bedding on which the chicks can rest. Without this coarse bedding, these chicks may develop deformities in the legs and feet.

WEANING

Once chicks begin to flap and climb, in favorable climates they can be moved into outside cages on warm days and inside again at night until they have adjusted to the new environment. After a few days in the cage, hand feeding is gradually reduced as the chicks begin to sample food on their own. This period may be as little as 10 days or as much as six weeks. It is important during the fledging period to ensure that the chicks do not lose too much weight due to insufficient food intake. As a rule of thumb, weight loss of up to 10% is acceptable but more than that is cause for concern and over 15% will require force-feeding. This does, however, vary with species, for example kiwi routinely lose up to 12% of their hatch weight before they begin feeding. With this in mind it is also necessary to be firm with the chicks and not always give in to their begging for food. I have seen ridiculous examples of psittacines and ibis that still insisted on being fed months after fledging, only because their owner had given in to their demands. If an older chick that is already independent is available, they are sometimes useful to mix with the newly fledged chicks, which will follow the older bird's lead at the food source. Fledglings should not be placed in large flights but rather initially limited to cages where they are constantly near the food source; with young psittacines a bowl of food placed on the floor of a small all-mesh cage that lacks perches will result in the chick walking through the food and it will soon start to sample the food. Favorite and easily consumed foods mixed in with the adult birds' diet will assist in the weaning of the chicks.

HAND REARING TOOLS

Tools used for hand rearing include bent spoons, primarily for psittacines, syringes for species naturally fed by regurgitation, and crop needles used both for these species and for dosing sick birds. Forceps tweezers are used for species requiring more chunky food items such as insects and fruit. Each of these techniques has its advantages and disadvantages:

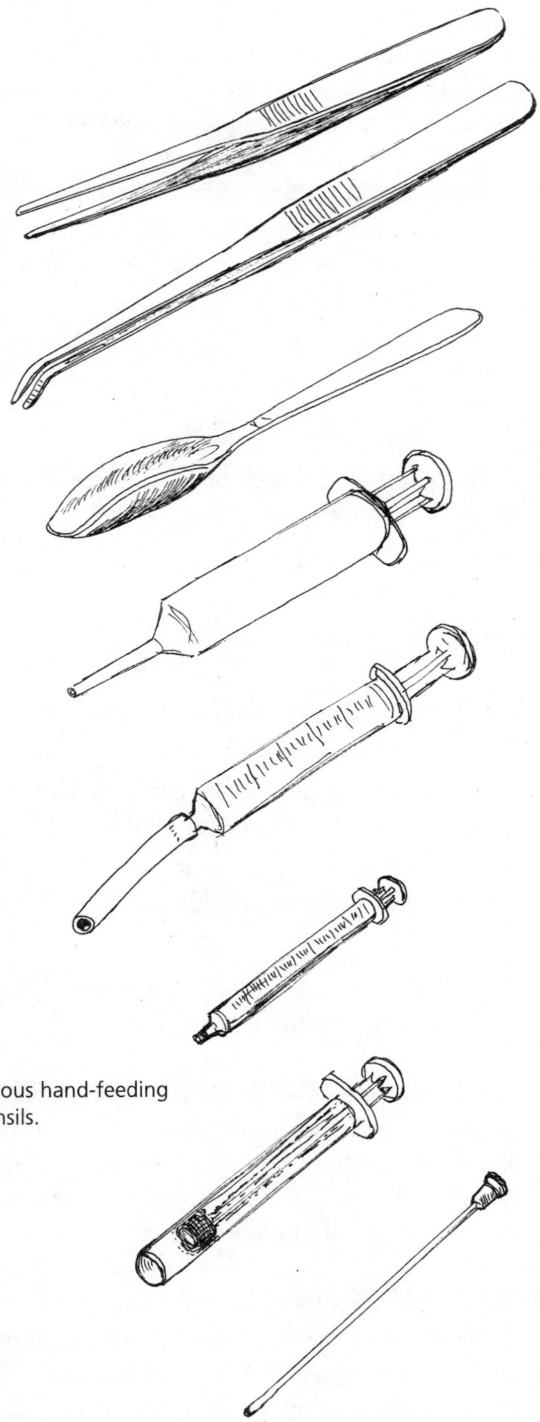

Various hand-feeding utensils.

BENT SPOON (primarily for Psittacines)

Advantages

- Spoon-feeding closely approximates the parents feeding and allows for a more personal experience.
- Birds taste and swallow the food that enters the crop.
- As the spoon touches the sensitive side of the mandible, it elicits a strong feeding response and the bird will swallow the food and stand less chance of choking.
- Spoons are easily washed and sterilized.
- The bird has the opportunity to refuse food.
- Many people find that spoon-fed babies wean easier
- Spoons allow bulkier foods, which are desirable for older birds.

Disadvantages

- Slow feeding technique.
- Food cools quickly in a spoon.
- This technique may be messy resulting in soiled birds.
- Difficult to feed tiny chicks with a spoon.

SYRINGES

Advantages

- Fast feeding technique.
- Not too messy resulting in clean chicks.
- Formula retains heat for a longer period than a spoon.
- Possible to measure the amount of food a chick consumes.

Disadvantages

- Beak deformities can occur where the food is fed from the side of the bill.
- Syringes are difficult to clean and sterilize.
- The rubber plunger must be kept well lubricated — a little vegetable oil works well.
- Only finely ground food will pass through the tip.

A syringe can also be used for feeding relatively coarse mixtures, for example when rearing doves, pigeons, and young ibis. In this case the entire narrow tip section is removed resulting in a tube with a plunger.

CROP NEEDLES: Personally I feel that the disadvantages here outweigh the advantages but crop needles do have their place for feeding sick or stubborn birds and a few specialized species.

Advantages

- The feeding process is quick.
- Difficult birds, e.g. sick or weak birds, can be fed quickly with minimal stress.
- Feeding is not messy.

Disadvantages

- Inexperience could result in food going into the lungs instead of the crop.
- The esophagus and crop could be scarred or punctured by the crop needle.
- The insertion of the crop needle can cause the chick to bring up the food, which could result in choking.
- The crop needle can disconnect and disappear into the crop of large birds.

A section of catheter or rubber tube attached to the tip of a syringe will perform a similar task to the crop needle. A section of tube with a large stock syringe is the most suitable apparatus for feeding newly hatched ibis, spoonbill, storks, and flamingos.

FORCEPS TWEEZER: In my work with species such as softbills and raptors, I discovered just how useful forceps are. Forceps are used to feed the more bulky foods such as insects, pieces of diced heart, and fruit. Not only can the food be firmly held, but it is also easy to dip the food into a little water or moistened rearing formula prior to feeding, and a quick pinch with the forceps tweezer not only kills insects but also pulverizes them as, for example, a parent kingfisher naturally would. As a rule those species that are fed with a forceps do not require the food warm. Feeding can be as quick as the feeder can operate. Some speed is sometimes required to avoid the chicks taking out their frustration on their siblings. It is, however, also necessary when feeding with a forceps to be careful of chicks rushing forward and puncturing the lining of the gape on the tip of a forceps.

Whatever instrument is used, it must be absolutely clean. After being washed in hot soapy water, the instruments must either be boiled for at least 15 minutes or else left to soak and be disinfected in Hibitane, Virkon, MicroDet, or another brand of antibacterial, fungal, and virucide solution. Remember also that we, the feeders of these chicks, also carry pathogens on our hands, which should also be well washed before feeding commences.

IMPRINTING

Imprinting and its effects is a factor that varies greatly between families and even species. There are times that imprinting is a desirable outcome to ensure the bird bonds with human companions in preference to other birds, for example when hand rearing psittacines as pet birds or other species to be dedicated to education. In such cases the birds are ideally raised alone, although rearing in groups does work as well, with the chick handled on a regular basis. With most species, however, if the birds are being raised for future breeding, imprinting needs to be considered. With most passerines, psittacines, coraciiformes, ciconiiformes, and galliformes for example, if the chicks are raised with others with a minimum of human contact, despite visual contact with humans from an early age, the birds will later recognize and breed with their own species. It is best to avoid excessive contact and wean such species as early as possible and leave them in a group with others of their species. With others, such as falconiformes and gruiformes, visual contact at an early age can be a problem and result in imprinting, and ultimately birds that will not breed. In such cases, methods such as puppet gloves, one-way shields to avoid visual contact between the chick and the person rearing, and crane suits that replicate an adult bird can be employed. The use of surrogate species to rear young can also cause the chick to imprint on the surrogate, resulting in incompatibility problems with its own species. For example, the use of bantams to raise pheasants or one species of parrot to raise another similar species often works well, but with others such as cranes and raptors, it is known that the young imprint on their surrogates and future breeding issues can result.

SIBLING COMPATABILITY

It is important to ensure that the siblings being reared are compatible. Here, one of two problems can result in

Sibling aggression in Turaco chicks.

the siblings having to be separated into individual containers. First, if there exists a large size difference, the smaller chick will be stood on and possibly even squashed to death. Second, some species such as bee-eaters, rollers, and hornbills are extremely aggressive toward their siblings and it may be necessary to separate them until all the birds are feathered, after which aggression subsides. With most large raptors and some other species, such as ground hornbills and wattled cranes, this sibling aggression lasts for up to two months. Unfortunately, it is not always possible to predict sibling rivalry.

References:

Ashington, F. 1994. Handfeeding Part 1 & 2. *Avizundum,* September pg 22-24 and October pg 4-5.

Dennison, D. & V. Dennison. 1993. Handrearing Part 1–5. *Avizundum,* April–July.

Kuehler, C. & J. Good. 1990. Artificial incubation of bird eggs at the Zoological Society of San Diego. *International Zoo Yearbook* 29: 118–136.

COMPATIBLE SPECIES

All too often this vital element in aviculture is ignored. The fact that two species or individual birds do not exhibit aggressive behavior does not mean that they are compatible. Ignorance of compatible species can be expected amongst aviculturists dealing with a species or perhaps even birds for the first time, but the problem often lies with the supplier of the birds who, despite his intimate knowledge of the species, promotes a sale to the inexperienced customer.

I recently received a phone call from a wealthy beginner who was extremely proud of his newly acquired waxbill collection and the pair of golden pheasants, which were such an attraction on the aviary floor. He was shocked to hear that the pheasants would most likely kill the waxbills newly fledged chicks and that the adult birds would feel threatened by such a large species within their territory.

Imagine the misery of having to live in a home with people who were larger than you, took all the choice food in the house, attacked you from time to time, killed and devoured your offspring, stole your bedding and even evicted you from your bed. As we experience in our own lives, conflict creates stress and in the same way stress affects us, it affects birds (see the chapter on stress management).

There are a few pointers to consider when you decide on suitable species to mix in an aviary. Firstly birds of the same family, for example the Pytilia family, are similar in behavior to the point where a female aurora finch would possibly accept the advances of a male melba finch. Her male will be aware of this, which will result in aggressive battles and ultimately even the death of a bird. Very often birds in the same family have similar courtship displays, calls, diet and nest site selection. All this results in competition and as in nature, only the fittest survive. The aviculturist who mixes the wrong species can be left feeling disappointed and guilty at having caused the death of a bird or wasted the breeding potential of a pair of birds.

Some species that require the same diet, particularly a specialized diet such as live food for rearing their chicks, will compete fiercely to obtain their preferred food. Good examples of similar species requiring the same food are sunbirds, which require a feeding tube for each pair in the aviary. Species which vary greatly however, for example, bee-eaters and drongos, will also compete for live food to the point where one of the species have to be removed.

Very often pairs are more aggressive than individuals and for public displays a single member of a family, for example, a single drongo, can be kept on display without serious compatibility problems. A soon as a mate is introduced, however, the bird's aggressive tendencies are increased ten-fold. Creating a number of feeding stations is one way of overcoming this particular problem to some extent. With basic maintenance foods such as seeds, which are plentiful, there is not usually a serious aggression problem. There may be a dominance ranking in the aviary but all the birds will get sufficient time to feed and meet their nutritional requirements.

As the parrot breeders will tell you, simply putting a pair of birds together with all their requirements catered to does not mean that the pair are compatible and will go on to produce offspring. With many of the birds, particularly the larger species such as cranes and swans, the pair has to be carefully observed for some time after they are first introduced to each other. It is generally, but not always, the male birds that persecute their mates. One of the families which always comes to mind when discussing this topic are the turacos and some cockatoo's which, with some pairs, even require special shelters where the females can hide from their mates. Even in established breeding pairs it is not unusual for the male to decide to "beat up" his female to the point where females are killed. Similarly with the psittacines I have had Meyers and Ruppells parrots suddenly turn on their mates. With some species, for example sunbirds and bowerbirds, where the male does not play an important role in breeding after mating has taken place, he can be removed to an adjacent aviary.

Coloration is also an important factor with many species. For example, the red-throated (Peter's) twinspot will attack most other species such as fire finches, which have a red coloration. Lilac-breasted rollers will also attack other species with a blue coloration and there are numerous other examples of this. Very often this problem only arises during the breeding season but it is an important factor to consider.

Many species are naturally colonial either breeding in tight groups, for example, sociable weavers, or in loose colonies, for example, lovebirds and black-rumped (red-eared) waxbills. With such species it is often a mistake to separate a single pair because the courtship and nesting behavior, including battles for tiny territories in the immediate vicinity of the nest, act to stimulate and synchronize the breeding. Some species, such as ground hornbills, moorhens, babblers and woodhoopoes, are co-operative breeders. A pair will have numerous helpers, often their own offspring from previous years, at the nest site, all of which will be submissive to the breeding pair. Even here however one has to monitor the number of birds in the aviary as limited space will result in fighting if the aviary is overcrowded.

Birds with similar breeding requirements are very often the biggest problem in a mixed aviary. Even if sufficient nesting sites and material are supplied, problems will occur. Prime examples are the cavity nesting birds such as rollers, hoopoes, hornbills and starlings. No matter how many nest logs and boxes are supplied, there will be fights over a few particular nest sites. If this becomes serious, it will be necessary to remove the less valuable species to allow at least some to breed. An example that comes to mind is of my flock of white-fronted bee-eaters that excavated nest holes in the sandbank I had constructed for them. A pair of brown-hooded kingfishers in the same aviary then began to excavate a nest in the same sandbank. While feeding the bee-eaters flying ants one day, a jealous kingfisher in flight hit one of the bee-eaters off my hand and killed it. Needless to say the kingfishers were removed.

Probably the easiest of the problems to detect when considering compatibility is the size of the birds, and even here I have seen aviculturists attempt the most ridiculous mixes of species. Birds will naturally associate with other birds of a similar size. Whether a party of softbills or a group of egrets and spoonbills hunting in water, they generally associate with others of their own size. The intimidation that birds experience when confronted by others much larger than themselves causes stress that, as described in the chapter on that topic, can have disastrous consequences. In large walk-through aviaries where the birds have plenty of space and habitat to escape each other, this problem is greatly reduced. For example, francolins can be housed with finches in a large aviary, but in a small flight the finches would be stressed by their presence.

Territorial aggression in adjacent males.

To maximize the breeding potential of the birds in the aviary, species that use different strata, habitats and nest sites can be mixed without competing. A basic example is a planted flight that may have a pair or two of quail of different size and species on the floor, quail finch, two pairs of doves, one small (Cape doves) and one large species (green-wing dove), a group of communal waxbills (orange-cheeks), a few other pairs of waxbills and finches, and a pair or two of small softbills such as white-eyes and sunbirds. Some species such as ibis, small cormorants, spoonbill, night herons and egrets can be bred as a colony and will even stimulate each other to nest.

CONVERSION TABLES

Millimeters to Inches

1.02	1/25	(0.04)
1.59	1/16	(0.0625)
3.2	1/8	(0.125)
6.4	1/4	(0.25)
9.5	3/8	(0.374)
10.0 (1 cm)	16/41	(0.39)
12.7	1/2	(0.5)
15.9	5/8	(0.625)
19.0	3/4	(0.75)
22.2	7/8	(0.875)
25.4 (2.54 cm)	1	
50.0 (5 cm)		(1.97)
100.0 (10 cm)		(3.94)

Meters to Feet

0.3048	1.0
1	3.3
4	13.1
7	23.0
10	32.8
20	65.5

Hectares to Acres

0.4	1.0
1	2.471
12.4	5

Liters to Fluid Ounce

0.0296 (29.6ml)	1
0.0888	3
0.118	4
0.148	5
1	33.814

Liters to Pints

	(US)	(UK)
1	2.11	1.76
2	4.234	3.52
5	10.57	8.8

Grams to Ounces

1	0.04
2	0.07
3	0.11
4	0.14
5	0.18
6	0.21
7	0.25
8	0.28
9	0.32
10	0.35

Parts Per Million to Milligrams

1 ppm = 0.1 mg/100g

Celsius to Fahrenheit

(n x 9 divided by 5 + 32)

0	32.0	**24**	75.2
1	33.8	**25**	77.0
2	35.6	**26**	78.8
3	37.4	**27**	80.6
4	39.2	**28**	82.4
5	41.0	**29**	84.2
6	42.8	**30**	86.0
7	44.6	**31**	87.8
8	46.4	**32**	89.6
9	48.2	**33**	91.4
10	50.0	**34**	93.2
11	51.8	**35**	95.0
12	53.6	**36**	96.8
13	55.4	**36.7**	98.0
14	57.2	**37**	98.6
15	59.0	**37.2**	99.0
16	60.8	**37.8**	100
17	62.6	**38**	100.4
18	64.4	**39**	102.0
19	66.2	**40**	104.0
20	68.0	**41**	105.8
21	69.8	**42**	107.6
22	71.6	**43**	109.4
23	73.4	**44**	111.2

Ostriches, Rheas, Cassowaries, Emus and Kiwis

01 Struthioniformes

01.01 Struthionidae Ostriches

Glen Holland

Introduction: Under the Sibley-Ahlquist-Monroe system of classification, ostriches, rheas, cassowaries, emus and kiwis are all included in the order Struthioniformes, which is divided into four families. The ostrich *Struthio camelus* is the sole member of the family Struthionidae. Five subspecies have been recognized, inhabiting widely separated areas of low rainfall, including semi-desert regions, in Africa. The nominate race *S. c. camelus* occurs in northern Africa and the Sudan; the race *syriacus* is found in the deserts of Syria and Arabia; the race *molybdophanes,* from northeastern Kenya and Somalia, has sometimes been regarded as a separate species; the race *massaicus* occurs naturally in eastern Kenya and eastern Tanzania; the race *australis,* in southern Africa. Ostriches favor short grass and avoid densely wooded areas where their view is obstructed. They are usually found in pairs in the breeding season, but are gregarious in the non-breeding period. There is a dominance hierarchy among individuals within a group. As ostriches have been bred in captivity for many generations, numerous varieties have developed, varying from the big red of east Africa to the blues and blacks of southern Africa and the 2/3 sized Australian birds which resulted from generations of inbreeding from birds released into the wild there many years ago.

Management: Being the largest birds in the world, they are well represented in public displays globally. In addition, their captive management has been thoroughly researched by those involved in the ostrich industry, which produces the birds for their skin, feathers, eggs and meat. For public display, a single male is best kept with between 1–3 females. Their pen should be open with a few large trees for shade and should be well drained, as the birds suffer if kept in damp conditions. They can either be maintained on small pastures and rotated regularly, or on larger dry lots. The larger the

Male ostrich in display.

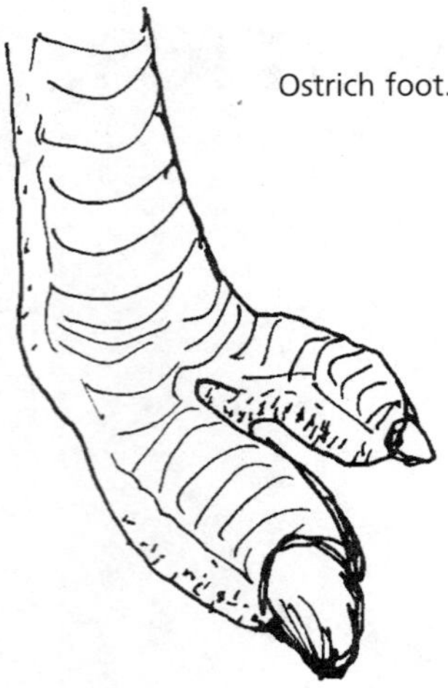

Ostrich foot.

pen, the easier it will be to maintain the birds in good condition; but in extremely large pens they become semi-wild and are difficult to manage. Although they may peck softly at visitor's hands, they will not cause any serious harm, provided that the low barrier around the pen is sturdy. If kept in small pens, the birds will destroy all grass cover and wear distinct paths by their continual pacing along fences. A trio of ostriches requires a pen of at least 500 m^2 (5,300 sq. ft.). Fences should be 2.2 m (7¼ ft.) tall. If they are made of deer fence, which measures 1.8 m (6 ft.), and are placed 40 cm (16 in.) from the ground, you will achieve the 2.2 m height. This gap prevents the birds' feet from hooking in the fence; it also allows a convenient escape route for keepers under attack. Always remember during construction that the birds will consume any pieces of wire, nails, and other building material dropped on the ground, usually resulting in death. Adjacent pens should have a corridor of 1 m (3¼ ft.) between them to avoid fighting. In many zoos, they are mixed with other large birds such as cranes and storks, and even antelope species, but compatibility will depend on the individual birds. At Auckland Zoo I kept a trio of ostrich with zebra, springbok and giraffe. The only real issue we had (apart from one of the birds swallowing the staff vehicle keys!) was that the male had to be conditioned to being locked away during cleaning. This conditioning was easily achieved with a food reward.

Breeding males can become extremely aggressive toward their keeper. The golden rule when dealing with mature birds is: never trust them. When the birds have to be handled they are best caught by the neck and head, either by hand or with a crook. The head is pulled downwards with the neck outstretched, thus avoiding the bird kicking forward. A cloth hood is then placed over the bird's head and upper neck. This procedure can pose a problem with adult males in breeding condition, which are physically strong and difficult to handle. Even chicks from two months old should be hooded when weighing.

A plentiful supply of small river pebbles should always be available for the birds to consume for digestive purposes; a lack of these can result in the death of a bird. The rule of thumb on the size of the stones is that they should measure approximately half the width of the widest, or upper, part of the front toe. This rule applies to all age groups. Ostriches roost squatting on the ground at night, and will also squat at times during the day. Clean water should be available in a shallow pool; on hot days both adults and chicks will bathe in this.

Diet: Feeding is done with a plucking action either while feeding on vegetation, or in captivity, on alfalfa or commercially prepared ostrich pellets. An adult being fed dry food alone requires 3.5 kg (7.7 lb) of food per day, and one kept on pasture 1 kg (2.2 lb) per day. Diets for newly hatched chicks to mature birds, kept on pastures, are as follows:

- Chicks 2 days–1 month: starter crumbles
- 1–3 months: starter pellets
- 3–4 months: equal portions starter and grower pellets
- 4–6 months: ¾ grower, ¼ starter pellets
- 6–12 months: ¾ grower, ¼ corn (maize)
- 12–18 months: corn and grower pellets (thin birds)
- Mature breeders: Breeder pellets (July–end of lay)
- Non-breeding season: ½ maintenance pellets, ½ corn

A new concept is to use cheap bulk foods such as corn, alfalfa and barley, to which a vitamin and trace element concentrate is added. This would be suitable for birds from eight months to sexual maturity.

Breeding: In courtship, the males make a deep booming sound; they also have a softer *booh* contact call. Oyster shell grit should be available to breeders. The breeding male is usually accompanied by a major female, that incubates the eggs, and minor females that will lay in any nest available, but will not assist with incubation. During the breeding season, due to the development of the testes, the facial skin and shins of the male flush bright red. Courtship displays involve wing flapping and erect feathers. The best known display is when the male squats with the wings spread and alternately raises and lowers them with the head and neck weaving laterally and the head thumping on the back. When the female is ready to copulate she solicits with the beak clucking, the wings partly spread with the primaries vibrating softly, and the head held low. For copulation, the female drops to the ground at the approach of the male. While snapping his bill and tapping his feet, the male places his right foot right of the center of the female's back, and his left foot on the ground beside her. He then drops gently onto her back.

The male chooses the nest site, making a few scrapes with his feet and beak; the female then completes the nest scrape. It is usually placed in an open position with a good view of the surrounding area. Females lay on alternate days, with a major hen laying 5–11 eggs and minor hens laying 2–6 eggs. The maximum number of eggs which can be covered is 21, and excess eggs lay scattered around the nest site. Incubation begins 16 days after the first egg is laid and lasts 45–46 days. Captive birds can be induced, by continually removing eggs, to lay more than 60 eggs. Females have a productive life of more than 30 years. Females mature at two years and males six months later. Young males should not be used on older females, as they may not be sufficiently aggressive to mate with them.

Eggs are collected daily and stored on their sides at 15–18°C (59–64.4°F), being turned once a day during storage. The incubator is set at 36.2°C (97.3°F) with a relative humidity of 18%. The eggs are turned at least three times a day, and aired once a day for 15 minutes. Many of the modern incubators have the eggs placed in an upright position, and in this case the air sac should be marked and placed on the upward side when the eggs are set. When nearing the end of the 42-day incubation period, the air sac should be checked; once the beak has entered the air sac, the eggs are placed in the hatcher, which is set at 35.5°C (96°F) with a slightly higher humidity of 20%. On average, the eggs should lose 15% of their initial weight during the incubation period. After 12 days of incubation, they can be candled to check for development. If the chicks have not broken the inner membrane by the 41st day, they should be gently assisted, care being taken to avoid bleeding and to break the membrane near the tip of the beak to allow the chick to breathe. This will also save mal-positioned chicks, which can be determined by feeling for the beak under the air sac membrane. If the tip of the beak can not be located, then a few small holes around the side of the shell will usually locate the beak. The membrane should be broken slightly to allow the chick to breathe and the chick can gradually be assisted over the next 12 hours. Such chicks may feel cold when the shell is held to the cheek; in this case they are best placed on a heater pad set at the same temperature that exists inside the brooder. Usually they hatch in 9–18 hours hours. Ostrich chicks lack an egg tooth, but break the shell with muscular movements. Should the chicks need to be fitted with a transponder, it is best placed in the large piping muscle within four days of hatching. Within a few hours of hatching, the navel area is best sprayed or dabbed with iodine or a similar anti-bacterial spray.

Tied legs to prevent 'splayed leg'.

Newly hatched ostrich chicks cannot walk well for the first 24 hours. They appear swollen, but usually absorb the excess fluids within a few days. Their down is spiky, resembling a hedgehog, and is buff colored, tipped black, with lines of spots down the sides of the neck. Chicks make a high pitched melodious trill, particularly when stressed. A radio playing soft music will help to calm chicks in the brooderand rearing pens.

When 24–48 hours old, they should be started on a good quality ostrich starter ration containing at least 26% protein. For the first 10 days, chicks are kept at 32°C (89.6°F) and thereafter at 30°C (86°F) until three weeks of age. Chicks must always be supplied with fine grit, which should be increased in size as they develop. Newly hatched chicks which show signs of spraddled legs should have the legs taped together between the tibia, separated by the width of the pelvis. Taping a small T-shaped piece of hard, flat plastic at the base of the foot and toe treats curled toes. The top of the T is placed under the hind part of the foot and the toe strapped down with the nail protruding from the thinner front section. This should be done for a week. From three weeks until eight weeks old, they are kept at a temperature of 25–28°C (77–82.4°F). The brooder pen should be equipped with a suspended heater lamp. For the first 10 days chicks are kept on carpet or burlap to avoid the consumption of litter and resultant development of intestinal obstruction problems. Thereafter, they are placed on rubber mats or river sand, with limited access to well-mowed grazing paddocks. Smooth substrates will result in the fatal spraddled legs.

Ostrich *Struthio camelus*

Week	Weight (kg)	Week	Weight (kg)	Week	Weight (kg)	Week	Weight (kg)
0	1.2	**9**	13	**18**	41.5	**27**	64
1	2	**10**	15.5	**19**	44.5	**28**	67
2	2.2	**11**	19.5	**20**	47	**29**	69.5
3	3	**12**	23.5	**21**	49	**30**	73
4	4.1	**13**	28.5	**22**	51	**31**	75.5
5	5	**14**	30	**23**	54	**32**	76.5
6	7.5	**15**	32.5	**24**	57	**33**	78
7	9.5	**16**	37	**25**	58.5	**34**	80
8	11.5	**17**	39.5	**26**	61	**35**	83

Initially, large quantities of green foods are avoided. During the first three weeks the chicks should have food and water ad lib. Some diced greens such as rye, alfalfa or clover which have been moistened and mixed in with the crumbles will stimulate the chicks to feed well. Refusal to eat or drink is a common problem with ostrich chicks. Placing a few slightly older chicks that are already eating with the newly hatched chicks can alleviate this. If these slightly older chicks are not available, then domestic poultry will also do the trick, but the disease transmission factor will be of concern here. Green feeding trays also attract the chicks to the food. At this stage, supplementing the pellets with fresh alfalfa or spinach will ensure that the chicks obtain adequate nutrients while minimizing excessive weight gain that may contribute to leg problems, particularly slipped tendons. Waterers should be cleaned daily, and a soluble poultry vitamin/trace-element powder added to the water for the first 2–3 weeks.

At two weeks, the chicks can be kept in outdoor pens but should be sheltered at night. They are best managed in groups of 25–50 birds. Juveniles can be maintained on this program until they reach sexual maturity. Breeder rations contain a high calcium level and should not be fed to juveniles. Remember that the chicks will consume any foreign objects, which may cause death through obstruction. Moldy, wet or spoiled food will cause diseases such as candida, which spreads fast and can be lethal to the chicks. At four months, they are maintained outdoors with shelter available for periods of inclement weather. They attain adult height at 12 months, but the first fertile eggs are only produced at two years. At about nine months of age, sex determination becomes easy when the penis emerges during urination.

Age in weeks	Food consumption
1–6 weeks:	300 g (10.6 oz.)
6–11 weeks:	700 g (24.7 oz.)
11–16 weeks:	1.3 kg (2.87 lb.)
16–21 weeks:	1.7 kg (3.75 lb.)
21–26 weeks:	2.1 kg (4.62 lb.)
26–31 weeks:	2.3 kg (5 lb.)

Contributors:

Walter Mangold, Managing Director, World of Birds, Cape Town, South Africa

Alison Schultz, Eggmonth Ostrich Farm, New Zealand

References:

Diamond, T. & J. Rozin. Artificial incubation and rearing of ostriches. *AFA Watchbird* 7 (4): 29–30.

Thornberry, Dr. F. D. Ostrich Production. *AFA Watchbird* 16 (4): 2.

01.02 Rheidae Rheas

David Bender

Introduction: The rheas of South America were formerly given their own order, but are now included with the ostriches and similar birds in the order Struthioniformes, while retaining their own family name, the Rheidae. The five recognized subspecies of the greater rhea *Rhea americana* inhabit the pampas of Brazil, Bolivia, and Argentina; the lesser rhea *R. pennata,* with three subspecies, has a disjunct distribution in the puna areas of Peru and the steppes of Argentina and Chile. Like most other members of the order, rheas are large, flightless birds. However, being considerably smaller than ostriches, cassowaries, and emus, they are much more popular in private avicultural collections. The greater rhea is the species most frequently found in captivity, the endangered lesser, or Darwin's, rhea being kept by only a few zoos.

Management: Mark Wright of Indiana bred rheas successfully for many years, and this section is based largely on his notes. Rheas are fairly calm birds, often becoming quite tame. Like many ratite birds, they are both polygamous and polyandrous. Males will mate with several females, and females will move from male to male. In addition to large amounts of green food, including broadleaved weeds, rheas will consume insects, rodents, and snakes. Adult rheas range in weight from 20–30 kg (45–70 lb.) and stand about 1.5 m (5 ft.) tall. A fence of 1.2 m (4 ft.) in height is sufficient to contain rheas. Being grazers, the pens, or pastures, for rheas should be fairly large, the actual size depending upon the number of birds kept. The sexes are similar, but males are generally slightly larger than females. They are basically grey in color, with black on the head and the back of the neck. A white mutation is fairly common in captivity. Rheas will swallow any pieces of wire, broken glass, nails, sticks, or other foreign objects found in their pens, resulting in stomach impaction and sometimes death. Special care should be taken to make sure their pens are free from such detritus. Rheas are compatible with other animals and birds, but will eat the eggs of ground-nesting birds. Breeding males can become aggressive when defending the nest. An aggressive bird can be fended off by grabbing his neck just below the head and pushing him away. Rheas should not be handled by their wings, which may result in injury to the bird. Proper respect should always be shown toward their powerful feet and legs.

Diet: Rheas will obtain much of their food through grazing. In addition to natural vegetation, they can be given chunky dog food, turkey pellets, and lettuce trimmings or other greens. During the breeding season, gamebird laying mash is mixed with the above feed. Chicks are offered finely ground dog food, turkey pellets, and gamebird starter, supplemented with leaf lettuce and grass. Their water should be medicated for five days per month with a coccidiostat, and periodically with Emtryl for blackhead.

Breeding: In the southern hemisphere, greater rheas start laying in September, however, captive birds in the northern hemisphere normally begin laying in April. The nest is a scrape in the ground, defended by the male, who does all of the incubation and chick-rearing. Several hens will lay in the nest of a male until he begins incubating, and then move on to another male. In captivity, this is not always possible. During the breeding season, males make a booming call which has been described as *uh hum;* they spread their wings in display to the females and attempt to herd them. Females indicate their readiness to mate by squatting. When mounting, a male will grasp the female by the back of her neck, often plucking feathers in the process. Females will lay eggs up to 3 m (10 ft.) from the actual nest scrape, and the male will roll the eggs into the nest with his beak. Eggs are laid every other day, and captive females will lay as many as 50–70 eggs in a breeding season. Rheas generally reach sexual maturity at two or three years of age, although younger females have been known to lay, and yearling males often show signs of displaying and nest defense.

Many breeders find that artificial incubation and rearing is more efficient under captive conditions, particularly when females continue to lay without having access to more than one male. Freshly laid eggs are butter yellow, fading to white as incubation progresses. Eggs should not be held for more than a week before being placed in the incubator. They are best incubated in groups, and hatching will be synchronized if eggs are in contact during incubation. This is particularly important in natural incubation, as the males will usually lead the chicks from the nest with 24–36 hours of the first chick hatching (Bruning 1971). The normal incubation period ranges from 34–40 days, the large variation being due to synchronized hatching. Mark Wright incubates rhea eggs at a temperature of 37°C (98.5°F) with a wet bulb reading of 84°F, or 65% humidity. The humidity level should be increased to at least 80% relative humidity during the hatching process, to prevent the membrane from drying out. After pipping, rheas hatch by first enlarging the initial pip, resting, and then shattering the egg shell by pushing with their feet and legs.

Newly hatched chicks are left in the incubator for 24 hours, and then placed in a brooder box with a heat lamp providing a temperature of 32.2°C (90°F). The bottom of the brooder should be lined with artificial turf as the chicks may ingest loose material such as straw, resulting in impaction. Grit and oyster shell are provided for chicks in the brooder. Chopped greens, such as lettuce, oats, barley or rape, is provided the first few weeks, with turkey pellets or small, dry dog or cat food gradually introduced (Wright 1988). In the wild, rhea chicks eat large amounts of greens, as well as numerous insects, while following the male. In captivity, chicks that initially refuse to eat may be stimulated by providing mealworms and crickets, but should not be allowed to rely solely on these food items (Bruning 1971). The rearing diet used by Bronx Zoo is similar to that used by Mark Wright. Chicks are started on turkey pellets and kale, adding small amounts of Gaines dog meal and, eventually, monkey chow as the chicks get older.

Greater and lesser rhea.
Rhea americana and Pterocnemia pennata

Exercise is important for young rheas, and will help to prevent leg problems. In the wild, chicks get plenty of exercise by following the adult male. In captivity, they should be provided with an outside area of mowed lawn on which to graze and exercise. This pasturage should include clover and other broad-leaved plants as well as grasses. After about a week in the brooder pen, chicks should be moved to this area, which should measure about 7.6 m^2 (25 sq. ft.), with a shelter from the rain, equipped with a heat lamp for use at night and during cool weather.

At this age, young rheas can be contained by fences made of 60-cm (24-in.) high, 1-inch mesh poultry netting, if predators are no problem. They can be confined to the shelter at night. Chicks of three months or older should have a larger run to provide adequate exercise; a minimum of 15 x 30 m (50 x 100 ft.) is recommended.

Rhea *Rhea americana*

Week	Weight (g)
0	439
1	595
2	1020
5	3600

Rhea Growth Chart. Data from Bowthorpe & Voss 1968. International Zoo Yearbook *8.*

References:

Bender, D.L. & M.B. Wright. 1976. Greater or common rhea. *A.P.&W.S. Magazine* 76–4: 5–8.

Bowthorpe, G. & G. Voss. 1968. Breeding the rhea and the emu at Winnipeg Zoo. *International Zoo Yearbook* 8: 146–150.

Bruning, D. 1971. Raising rheas in captivity. *Game Bird Breeders' Gazette* 20 (10–11): 7–12.

Wright, M.B. 1984. Raising rheas. *A.P.&W.S. Magazine* 84–6: 36–37.

Wright, M.B. 1988. Raising rheas the Wright way. *A.P.W.S. Magazine* 88–8: 6–8.

01.03 Casuariidae Cassowaries & Emus

David Bender & Liz Romer

01.03ai *Casuariini* Cassowaries

• Liz Romer

Introduction: Cassowaries belong to the order Struthioniformes, family Casuariidae, subfamily Casuariini. Double-wattled cassowaries *Casuarius casuarius* are distributed throughout Papua New Guinea and Northern Queensland, Australia. Because they need a year round supply of nutritious fruit, cassowaries depend on rainforest habitats with their large number of fruiting plant species. They can also be found in woodland and open forest areas when certain species of fruit are seasonally available. Their wild diet is predominantly fallen fruit, but they will eat anything they find on the ground, including dead rats, birds, live skinks, reptiles and even fungi. For most of the year cassowaries live alone. They appear to have a home range. Females tend to be the dominant sex. Cassowaries can be aggressive, particularly when guarding chicks. When fighting, they raise their feathers and bend their necks right under their bodies, roaring loudly. Raising their heads back up, they then charge each other, kicking with both feet at once. The bout is usually brief, with neither opponent sustaining much injury.

Management: Adult female cassowaries are normally slightly larger and more brightly colored than males. However, sexing by comparative size and color is not a viable method for determining the sex of immature birds. Because most cassowaries are sold or transferred as chicks, other methods must be employed. Many experienced ratite breeders are able to vent sex young birds, or even sedated older birds. This involves direct palpation of the penis or much smaller clitoris. DNA sexing of cassowaries is a good method in areas where this service is available.

Adult double-wattled cassowaries weigh up to 65 kg (143 lb.) and stand 1.4–1.7 m (4.5–5.5 ft.) high. They are very long lived, with a female of more than sixty-five years of age being recorded at the Healesville Sanctuary in Victoria, Australia. In captivity, cassowaries can be difficult to breed, as not all pairs are compatible. Some institutions have no problem keeping a pair together year round, even while the male is incubating; other institutions have never been able to get a pair together. Cassowaries have a reputation for being extremely aggressive, but aggressiveness varies between individuals. They can be aggressive toward keepers, and an unwary keeper could easily be gutted. Therefore it is important to take safety considerations into account when designing an enclosure. It is wise to have isolation facilities available in case a pair needs to be separated for any reason. However, such arrangements should still allow the pair to maintain visual and auditory contact. Additionally, accommodations need to be designed to enable servicing of the main enclosures while the birds are locked into holding areas to eliminate the risk of injury to the keeper. Ideally, feeding and watering should be performed without entering the enclosure. Providing suitable access to each enclosure for vehicle entry will facilitate the transfer and transportation of birds.

The minimum suggested size of an enclosure for a pair of cassowaries is 300 m² (3200 sq. ft.). Being a rainforest species, a large amount of shade is essential. Either a heavily planted enclosure or lots of shelter is acceptable. Anecdotal evidence suggests that cassowaries kept in enclosures with insufficient shade risk developing cataracts. In cooler climates, heated sheds may be necessary, especially in areas where the temperature regularly falls below 0°C (32° F). Fences should be 1.8 m (6 ft.) high, as numerous escapes have been recorded with lower fences. Chain link fences are commonly used. Solid concrete rendered walls and tin fences have also been used successfully. A poultry netting is not suitable for cassowaries, as toes can become easily entangled in the small mesh, and such light gauge wire is not strong enough to contain a frightened or aggressive bird. Fences should be free of obstacles or loose wires.

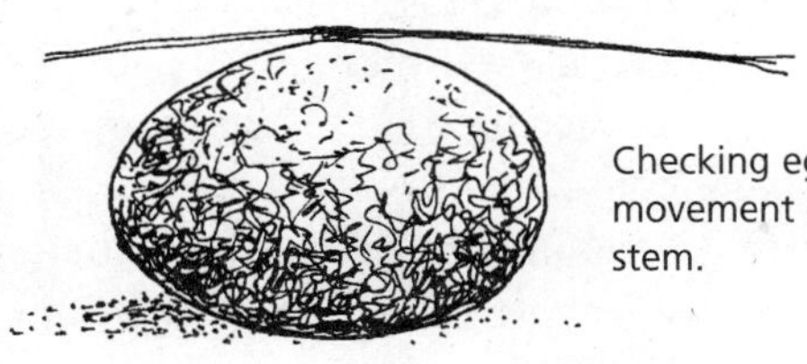

Checking egg fertility by movement of grass stem.

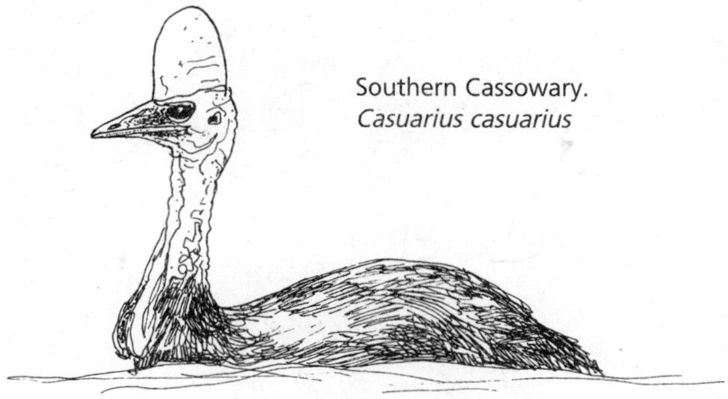

Southern Cassowary. *Casuarius casuarius*

Cassowaries love to bathe. They are good swimmers and will wade chest deep into water if given the opportunity. Provision of a pond of sufficient size for this purpose is recommended. The inclusion of ponds and muddy areas will greatly enrich the lives of these birds. Other behavioral enrichment activities for cassowaries might include scattering feeds and using as wide a variety of food items as possible.

Other species should not be housed with cassowaries. While some pairs may tolerate other species, many cassowaries have been known to attack and kill other animals that stray into their enclosures. Due to this aggressive nature, it is not advisable for keepers to enter occupied enclosures during times of high aggression, such as the breeding season. When it does become necessary to enter an enclosure with birds present, more than one experienced staff member should be in attendance, armed with brooms or other implements suitable for warding off an attack. Keepers should never deliberately incite aggressiveness in the birds.

Diet: Cassowaries are mainly frugivorous, although they also eat some animal matter. They have a very short digestive system and hence a fast passage of ingests. This results in a high dietary requirement, which can reach 10% of their body weight in daily intake. A large variety of fruits are fed, including tomato, banana, apple, pear, paw paw, watermelon, grapes, mango, plums, nectarines, cherries, kiwi fruit, figs, rockmelon, eggplant, and cooked sweet potato and carrot. Dead mice and rats, day old chicks, and some fish are readily taken.

As a general rule, most people feed 4–5 kg (8–11 lb.) daily, although this may vary seasonally. Additives such as multivitamin supplements, calcium carbonate, and wheatgerm have been used by some institutions which have bred cassowaries. Other common dietary items include brown bread, boiled rice, turkey pellets, trout chow, cotoneaster berries, canned raspberries,

Cassowary foot.

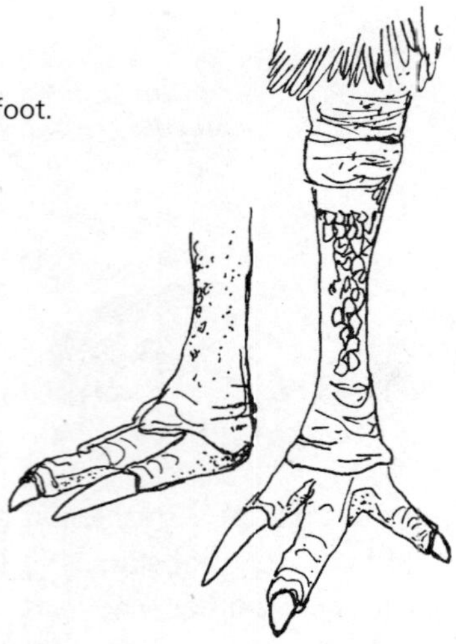

Eugenia berries, pyracantha berries, soaked raisins, horse cubes, kangaroo pellets, waterfowl pellets, and horse meat. Stones are also swallowed to aid digestion. It is noted that cooked vegetables are utilized metabolically more efficiently than raw, although if overcooked and mushy they will clog around the bill and be flicked away. In the USA a pelleted diet is used on birds from one month of age. When feeding, two food containers may be required for a pair. The food should be cut into pieces around the size of a billiard ball.

Breeding: Breeding usually begins at approximately four years of age, and may continue for at least 30 years. In the wild, cassowaries breed June to October, their reproduction coinciding with the maximum availability of fruit. As the breeding season approaches, the females become more tolerant of males. Each female eventually pairs with the male of her choice. The best combination appears to be an older hen with a younger cock. The pair remains together a few weeks until the female is ready to lay, and then selects a nest site. The nest consists of a shallow scrape in the ground augmented by only a few leaves. After egg laying, the female takes no more interest in the reproductive cycle, leaving her mate to incubate the eggs and rear the chicks. She may on occasion take on another mate. One female has been known to have three mates in one season. The chicks will remain with the male up to 16 months.

Bantam hen teaching chicks to eat.

Achieving breeding success in captivity revolves around pair compatibility. It is best to pair birds as juveniles, allowing the bond to develop as they mature. If this is not possible and the birds are already adults, it can take up to a year to get a pair together. Cassowaries generally resist change, and it takes time for them to adjust to new mates or new surroundings. When a prospective pair in adjacent enclosures begins to show interest in each other without overt signs of aggression, it is time to try putting them together. If they are not compatible, and aggression ensues, they should be separated after 15 minutes, with a subsequent introduction attempted later. As the breeding season nears, both sexes are likely to exhibit increased aggression toward the keepers.

The nests are often situated in strange and problematic places, such as near paths or on the side of a hill. The provision of a strategically placed hut or shelter will encourage the pair to nest in a more manageable location. Nests are always located in a shady spot with some leaf litter. High humidity is thought to be essential for this tropical bird. In drier climates, irrigation may be needed during the day, provided the temperature is not too cold.

Prior to breeding, calcium, vitamins B and E, and protein should be increased, although cassowaries have bred on a fruit only diet. Generally four, but up to nine, eggs have been recorded in captivity. Double clutching is possible by pulling the chicks at 1–2 weeks. Pulling eggs as laid increases the number of eggs laid, with up to 28 being recorded, although not all of these were fertile. As a general rule, the more eggs, the lower the fertility. The incubation period is 49–57 days, with smaller eggs tending to hatch sooner. A male incubating for more than 65 days requires a decision on whether to remove the eggs for examination. Eggs range in weight from 500–800 grams (1.1–1.76 lb). When the male starts incubating, it is often advisable to move the female from the enclosure if there is any aggression between the pair.

When the eggs hatch, the birds should be supplied with food cut into smaller pieces, although males have been known to break up the food items before dropping them in front of the chicks. This gesture is followed by a short burst of beak clapping to gain the chicks' attention and direct them to the food. The chicks soon learn to feed independently. Chicks can remain with the parents for up to 18 months, but are usually independent before then. If any aggression is shown by the adults toward the young, they should be immediately removed.

Artificial Incubation and Hand Rearing: Eggs should be set on their side and turned at least three times daily. Turning is not necessary during the final seven days of incubation. Cassowary eggs have successfully hatched at temperatures ranging from 35.5–36.7°C (95.9–98°F), with a preferred temperature of around 36.1°C (97°F). Humidity is a critical factor influencing hatching results. Since cassowaries occur in areas of naturally high humidity, their eggs may be particularly susceptible to embryo desiccation following excessive water loss. A wet bulb reading of 30.6°C (87°F) is recommended.

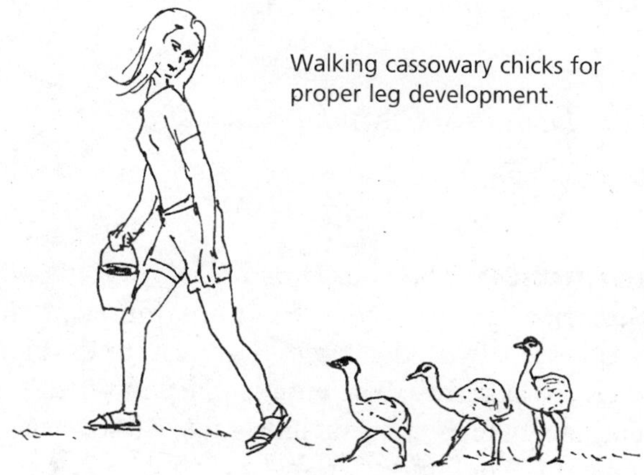

Walking cassowary chicks for proper leg development.

Double-wattled Cassowary
Casaurius casaurius

Day	Weight (g)	Day	Weight (g)	Day	Weight (g)
0	460	26	772	54	1454
1	440	28	837	56	1501
2	414	30	894	58	1520
4	416	32	987	60	1660
6	428	34	987	62	1682
8	445	36	1156	64	1742
10	458	38	1078	66	1742
12	503	40	1175	68	1918
14	570	42	1222	70	2043
16	565	44	1155	72	2180
18	613	46	1342	74	2231
20	673	48	1399	76	2260
22	724	50	1428	78	2270
24	760	52	1374	80	2412

A 12–15% weight loss in the egg during incubation is generally considered optimal, but weight losses of 11–20% have been recorded in eggs that have successfully hatched. Normal pip to hatch interval is from 1–24 hours. The eggs are too dark to candle, but fertility can often be determined by detecting egg movement. This can be accomplished by placing a piece of straw from a broom on top of the egg, attached with a piece of sticky material. Movement generally indicates the heart beat. When hatching, the eggs should be placed on the floor of the incubator on a towel or similar surface to provide adequate and secure footing for the chicks. Chicks have been housed in brooder boxes approximately 2 x 1 x 0.8 m (6.5 x 3.25 x 2.75 ft.) high with solid sides, and an overhead light usually set so there is a heat gradient. It is best to maintain the chicks initially at 30°C (86°F), as they are not at all cold-tolerant. A non-slip substrate such as astro turf should be used on the floor.

Chicks generally do not feed for the first 2–6 days, as the yolk is still being absorbed. To start the chick feeding, a day-old chicken may be used as a tutor. This also provides any single chicks with company, which appears to be important to their well being. Tapping the feeding dish with a pen or similar object will stimulate the interest of the chick. A shallow bowl of water should always be provided; most chicks will drink water before they begin to eat. It is important to weigh chicks daily at first to ensure they are not growing too quickly. As with other ratites, too rapid weight gain and too much protein, together with lack of exercise, can cause leg problems. Chicks should be exercised daily, preferably 2–3 times for 30 minutes. Chicks will follow staff as they jog around a suitable area. The diet for chicks is similar to the adult diet. Soaked sultanas and diced tomato and banana are good starting foods. These can be sprinkled with vitamin supplements if required. Mealworms will be readily eaten. Liquid calcium supplements have been used in the drinking water.

Individual Identification: In Australia, Trovan microchip implants are recommended, and should be inserted by a trained veterinarian. The preferred site is the left thigh muscle, although dorsally in the neck muscle is an acceptable alternate site. In determining the best placement for the implant, the enclosure design and the ability to scan the implant site should be considered. It is of vital importance that a proper record is kept of the actual site selected on each bird. Bands of any sort should not be used on cassowaries.

Species Requiring Similar Management:
Dwarf cassowary *Casuarius bennetti*
Single-wattled cassowary *C. unappendiculatus*

References:

Romer, L. (Editor). 1997. *Cassowary Husbandry Manual*. Currumbin Sanctuary, Queensland.

01.03aii *Dromaiini* Emus

Emus *Dromaius novaehollandiae*

- David Bender

Introduction: The emu *Dromaius novaehollandiae* is classified in the order Struthioniformes, which encompasses all of the large flightless birds known collectively as ratites. The emu is placed in the family Casuaridae, which also includes the cassowaries; it occupies the subfamily Dromaiini, of which it is the sole representative. The natural habitat of the emu includes plains and open woodlands ranging from semi-arid scrublands to alpine pastures and heathlands where suitable cover exists. They are the largest birds native to Australia, standing 1.5–1.85 m (5–6 ft.) tall and weighing up to 68 kg (150 lb.) when mature. They have long, powerful legs, and plumage similar in structure to ostriches and cassowaries. Females are generally slightly larger and heavier than males. In the wild, emus are fairly omnivorous, eating grasshoppers and other insects, caterpillars, wild fruits and berries, seeds and cultivated grain, grasses and weeds. Like ostriches, emus are now farmed commercially as alternative livestock.

Management: Originating from fairly dry, open areas where they can roam freely, emus require large lots in well-drained areas, and even then may, by pacing the fencelines, wear trenches which can fill with water during the rainy season. A pen of 24 m^2 (80 sq. ft.) is the minimum recommended size, but, as with all ratites, the larger the pen, the better. Pens should include some trees for natural shade, or artificial shelter, or both; in climates with cold winters, emus can be confined to their shelter during extreme weather. Fences should be constructed of heavy mesh or chain link, and should be around 1.8 m (6 ft.) high. A top rail is recommended, and posts should be relatively thin, as emus can injure themselves on heavy posts. Barbed wire should never be used for fencing emus.

While emus are not normally aggressive like cassowaries, they are powerful birds and should be handled accordingly. When moving emus, they should be approached slowly and quietly from the side or rear, and both wings grasped. After a brief initial struggle, the bird will settle down enough to allow the handler to propel it forward into a trailer or wherever it is to be moved. If some lift is required, the far wing can be released and the free hand used to boost the bird by the tailbone. To hold a bird still for treatment or sexing, the handler, after catching the bird as described, can lean over the bird's back, putting enough weight on the bird to prevent jumping. Emus should never be restrained in a chute or other holding device, as injuries may result when the bird fights against such restriction.

Diet: Emus are perhaps the most omnivorous of the ratites. Interestingly, the digestive tract of an emu more closely resembles that of a chicken than that of the more specialized ratites, such as ostriches and rheas. However, in selecting the proper diet for emus, particularly young, growing birds, it should be kept in mind that these birds develop much more slowly than poultry; therefore, high protein diets designed for rapid growth should be avoided. For adult birds, a diet consisting of game bird pellets, chicken scratch, dog food, and various greens ranging from alfalfa to fresh lettuce and cut up apples has been successfully used by some small breeders (Grey 1984). In 1964 and 1965 the Assiniboine Park Zoo in Winnipeg reared a number of emus. The adult diet used there consisted of chopped celery, including both leaves and stalk, lettuce, alfalfa leaves, chopped apples, and poultry breeder pellets, along with smaller quantities of monkey chow, dry dog food, rolled oats, and rabbit pellets. Dicalcium phosphate powder and a multivitamin mixture were added to the food, and oyster shell and grit offered *ad libitum* (Bowthorpe & Voss 1968). With the development of the commercial emu industry, specially formulated ratite pellets have become available. Larger operations can have feeds mixed to their own specifications.

Breeding: Emus cannot be accurately sexed visually, and should be vent-sexed for proper sex determination. In the northern hemisphere, emus breed from October to March. Most emus begin breeding when two years old, and many have been known to still produce chicks at more than 16 years of age. Male emus become friendlier, even to humans, as the breeding season approaches, however, they do not become aggressive. Both sexes will strut while displaying the neck feathers. The male grunts more frequently, and the female makes drumming sounds. There are no elaborate courtship displays. Males may peck at the back or neck of females to encourage them to mate. When ready to mate, a female will raise her tail feathers and sit down. The male will then squat on his haunches behind her and copulation, which takes less than a minute, is accomplished in a sitting position. Mating usually occurs in the morning or evening, and may occur as often as every day during the breeding season. Eggs are laid every three days, with up to two dozen or more being produced in a season if they are removed for artificial incubation.

Emu eggs are large, averaging about 550 g (1.2 lb) in weight, dark green in color, and with a rough, granular texture to the shell. The male emu incubates the eggs and broods the chicks. In the wild, a family party may stay together for up to 18 months, sometimes even accompanied by the female, if she has not moved on to another male.

If eggs are collected for artificial incubation, they should be set in groups, as chicks are much easier to rear together if they are of the same size and age. While emu eggs will hatch when incubated at temperatures ranging from 35.5–36.7°C (96–98°F), the incubation period will vary from 46–56 days, depending upon the temperature used. The optimum temperature for incubating emu eggs is 36.4°C (97.5°F), at which the eggs should hatch in 49–51 days. Initially, the humidity should be set at 24–35%, or a wet bulb reading of 70–75°F. The humidity can be adjusted to maintain the proper weight loss of eggs. Normal weight loss during incubation is around 15% of the fresh weight of the egg, but this may vary 13–17% and still result in a successful hatch. Eggs should be set with the air sac positioned toward the top, and should be turned at least three times daily; most incubators are set to turn eggs more often. Eggs are turned back and forth, not always in the same direction. Eggs are usually transferred to a hatcher several days before they are due to hatch. The hatcher temperature is normally set one degree lower than the incubator temperature.

A brooder area for emu chicks should be of sufficient size to allow the chicks to move away from the heat lamp, which should be adjusted to a height at which a temperature of 26.7–32.2°C (80–90°F) can be maintained at chick level. Up to 10 chicks can be brooded together, as long as they are of the same size and age, in order to prevent crushing or smothering while they sleep. In warm weather, emu chicks should be transferred to outside brooder pens during the day, as adequate exercise is essential to proper development. In common with many other avian species, emu chicks will chirp loudly and incessantly when they are uncomfortable. Crowding together may indicate insufficient heat, and the temperature in the brooder should be checked.

Water should be available to chicks on the second day, and food should first be introduced about 12 hours after they have learned to drink. Chicks are generally started by mixing chopped green vegetables with starter crumbles. If high protein poultry or game bird starter crumbles are used, a proportionately larger amount of greens should be added to avoid leg problems. Lower protein starter diets, of high quality, are best. Chicks

Emu
Dromaius novaehollandiae

should be supplied with grit, in the form of small pebbles or gravel. Fresh food should be offered twice a day. After the first week, pellets may be introduced into the feed. If good quality ratite grower pellets are not available, a mixture of equal amounts of 17% poultry laying pellets and rabbit pellets is recommended. Emus drink a lot, and should never be without a constant supply of clean, fresh water.

References:

Bowthorpe, G. & G. Voss. 1968. Breeding the rhea and the emu at Winnipeg Zoo. *International Zoo Yearbook* 8: 146–150.

Grey, P. 1984. Beginning luck with emus. *Game Bird Breeders' Gazette* 33 (4): 38–41.

Minnaar, P. & M. Minnaar. 1992. *The Emu Farmer's Handbook.* Groveton, Texas: Nyoni Publishing Co.

Minnaar, M. 1998. *The Emu Farmer's Handbook, Vol. 2.* Groveton, Texas: Nyoni Publishing Co.

Pizzey, G. 1980. *A Field Guide to the Birds of Australia.* NJ: Princeton University Press.

01.04 Apterygidae Kiwis

Glen Hollland

Natural Habits: The kiwi, along with ostriches, rheas, cassowaries and emus, are ratites– flightless birds of the order Struthioniformes. Kiwi retain only minute vestigial wings and have no tail. The three species are included in a single genus, which comprises the family Apterygidae. They are probably most closely related to the extinct moa. All are endemic to New Zealand.

The status of the various species was assessed by Rowe, who noted that in 1985 the little spotted kiwi *Apteryx owenii* was extinct on the mainland, but survived in an introduced population on an offshore island. In 2000, the first 20 birds were reintroduced to a predator-free site on the mainland and have since reproduced successfully there. The great spotted kiwi *A. haastii* is confined to the South Island, and is subject to habitat destruction and predation by introduced predators. The brown kiwi *A. australis* is the most widespread species, with several distinct subspecies. The nominate form, from the South Island, is endangered, but the Stewart Island race *A. a. lawryi* is still considered plentiful within its range. The North Island brown kiwi *A. a. mantelli,* abundant until recent decades, is rapidly decreasing in numbers due to extensive land development and predation within its range (Rowe 1985). This is the kiwi most often found in aviculture, although the other species and subspecies have occasionally been kept and bred in confinement.

Kiwi are largely nocturnal. They have excellent hearing and an exceptional sense of smell, their nostrils being uniquely placed near the tip of the bill. Their feathers are like branching hairs and there is no distinct molting season. Natural habitat varies from warm subtropical to cold alpine environments. They are generally found in rainforest. Most live as bonded pairs in fixed territories throughout the year. Both sexes contribute to the defense of their territory, which is maintained largely by their loud, whistling calls. Like mammals, some kiwi use scent posts to mark their territories.

Management: While there are slight morphological differences between the sexes, the most reliable method of sexing kiwi is the use of DNA sexing. If this service is unavailable, females can often be distinguished from males of the same age by their longer bills. In addition, the body mass of females is greater, although there can be a certain amount of overlap. The weight for males varies from 1.5–2.2 kg (3.3–5 lb.) whereas females weigh from 2.0–3.0 kg (4.4–6.6 lb.). Adults are long-lived; the average life span is about 20 years, with some birds reaching 30–35 years. An enclosure of about 300 m^2 (3250 sq. ft.) is ideal for a pair. Three or more kiwi in one enclosure causes stress, usually resulting in the death of one bird. Occasionally, two kiwi of the same sex will happily share an enclosure and burrows.

Kiwi require a solid, smooth perimeter fence from ground level to 1.2 m (4 ft.), with a concrete foundation to a depth of at least 60 cm (24 in.), to prevent escape. Wire fences may result in damage to the bill, which can result in death. Due to the threat of predators they are usually housed in enclosures with a 1.8 m (6 ft.) high fence. The enclosure should be well planted with shrubs and ferns to produce leaf litter and to supply berries as a natural food source. A small pond, no more than 20 cm (8 in.) deep and 60 cm (24 in.) wide, should be provided. The width is required to allow the birds to scoop up water with their long bills when drinking. This source of water will often be utilized by incubating males, or by females which have become uncomfortable prior to laying. The foraging area should include mounds and stumps, which will keep the birds busy and assist in exercising their muscles.

Although the birds will excavate their own burrows, artificial burrows should also be provided. These burrows should be cleaned as necessary, but not too often during the breeding season, with a little leaf litter added. Extra leaf litter placed in the enclosure will be taken into the burrow by the male. An infrared nest camera attached to a distant monitor will provide visitors an opportunity to view these elusive birds. Strict vermin control is essential around the perimeter of kiwi enclosures, as vermin will quickly invade burrows.

Nocturnal houses are currently the primary means of displaying kiwi to the public. Some institutions actually pull the birds out of their burrows to be held for public display, while others seal the burrow entrance, forcing the birds to remain in public view. Since kiwi are highly susceptible to stress, such practices should be strictly avoided. A nest camera, as previously described, combined with the reversed daylight regime of the nocturnal house, should provide visitors with a reasonable chance of observing the birds. Nocturnal houses should measure a minimum of 75 m^2 (800 sq. ft.) with a minimum depth of 2.5 m (8 ft.) between the viewing panel and the back wall of the enclosure. Ideally, the enclosure should be at least twice this size. The glass should extend for a maximum of 40% of the total perimeter.

Moving kiwi from indoor to outdoor enclosures will subject them to additional stress, and should be kept to a minimum, particularly if such changes are of short-term duration. Some keepers, however, recommend that captive kiwi be cycled between indoor and outdoor periods annually, or even more often. The current practice of housing growing juveniles indoors and then moving them to outdoor breeding pens at approximately three years of age appears to be working well. Nocturnal house burrows should measure 40 cm² (16 sq. in.) x 90–100 cm (36–40 in.). The roof should be hinged and the base left open. An arched entrance 30 cm (12 in.) high and 25 cm (10 in.) wide at the base is positioned on the side of the burrow toward the front third.

The floor of the nocturnal house should consist of a good mix of clean topsoil and leaf litter. It should be periodically forked, and new leaf litter added every 6–8 weeks. The planting of native ferns and shrubs will enhance the habitat. These plants can be maintained in good health with a grow light system and a sprinkler system which is switched on daily for 1–2 minutes, with a good soaking once a week. Some hand watering is required to avoid dry, dusty areas, which are detrimental to the plants as well as to the health of the birds. Extreme care should be taken during all maintenance procedures to ensure that no nails, pieces of rubber or wire, or any other foreign objects are left lying around, as these may be consumed by the birds, resulting in death. Sick kiwi should be x-rayed as a matter of course for ingested hardware. Proper ventilation and temperature control is obtained through an air conditioning system. The temperature should be 12–18°C (53.6–64.4°F). Kiwi become lethargic in heat and do not display well. Hot, damp conditions are also conducive to the development of diseases such as aspergillosis.

Diet: Due to their strong sense of smell, additives such as medications should not be included in the diet of kiwi. They are omnivorous, opportunistic feeders, with earthworms, beetles, invertebrate larvae and some fruits and berries forming the basis of their natural diet. Food should be restricted to 200 g (7 oz.) per bird per day, as captive kiwi, particularly in nocturnal houses, tend to suffer from obesity if over fed. The diet should be presented in a stimulating way and placed in different areas of the enclosure each day. At the Mt. Bruce Wildlife Centre, we divided the 200 g (7 oz.) of food into three large syringe covers, each of which is then pushed into the ground in the enclosure. These tubes are placed through a hole in the center of a round rubber mat measuring 50 cm (20 in.) in diameter. Each evening, which is of course morning for birds in a nocturnal house, the entire mat and food tube is removed for cleaning. The captive diet for two birds at Mt. Bruce consists of the following:

400 g	(14.1 oz.)	minced lean heart
120 g	(4.2 oz.)	mixed vegetables
500 g	(17.6 oz.)	water
50 g	(1.8 oz.)	insectivore diet
140 g	(4.9 oz.)	cat kibble

The mixture, excluding the meat, is allowed to soak overnight and the meat added in the morning. This is stored in a refrigerator and the food used for three days. Interestingly, the birds seem to prefer the food which has been refrigerated for three days over the fresh mixture. The captive diets currently used are as varied as the institutions holding these birds, and a two-year research project is currently underway to analyze the crop contents of dead birds, then formulate and test a captive diet.

Brown Kiwi
Apteryx australis

Little Spotted Kiwi
Apteryx owenii

Breeding: Unfortunately, the worldwide captive population of brown kiwi is not self-sustaining at the present time. The breeding programs are hindered by low fertility, poor hatchability, and the fact that many of the captive-laid eggs are smaller than those produced in the wild. Undoubtedly this boils down to inadequate nutrition, which will hopefully be corrected by the ongoing study mentioned above. Kiwi are monogamous, and males perform most of the parental care. This is a rare combination among birds, and somewhat limits the captive propagation potential of the species. Male kiwi are sexually mature at 18 months, while females do not usually lay until they are 3–5 years old. They have a long breeding season, starting in July and ending in March. The 2–4 eggs are laid in two clutches, with one captive pair historically producing seven eggs in a single season. Kiwi eggs are the largest in proportion to body size of any birds and have the highest percentage of yolk. Egg-laying females should be closely observed for loss of condition. Egg peritonitis is currently a common cause of mortality in female kiwi. Veterinarians who have worked with these birds recommend that breeding females be restricted to two clutches per season. If eggs are continually removed, females may become exhausted. Limiting their annual production will increase their total lifetime production.

As egg-laying approaches, a sharp increase in food intake by females over a period of several weeks will be noted, followed by a fast of about two days immediately prior to laying. Females carrying eggs walk with difficulty, and the shape of the egg is usually evident during this period. At this stage females have been observed soaking their bellies in standing water, a procedure which is thought to soothe the distended belly. The incubation in the North Island brown kiwi is carried out by the male alone for a period of 72–84 days. In other kiwi species the female participates in incubation.

Artificial Incubation and Rearing: Eggs are generally left for the male to incubate for the crucial first 30 days, then transferred to a still-air incubator. The following incubation protocol has successfully been used to hatch eggs set when fresh at Auckland Zoo.

Incubation Protocol: The incubation protocol was split among four age-related stages. Following is a breakdown of the parameters followed throughout each stage.

Turning is done in 45° stages, with the top center line of the egg beginning and returning to 12 o'clock each day, so that the maximum distance the center line travels is to 90° in alternating directions each day. For exemple, on Day 1 the egg is turned 90° clockwise in two stages of 45°, then a total of 90° counter-clockwise in the next two stages, bringing the center line back to its original position on the final turn of the day. On Day 2, the egg is turned counter-clockwise first.

- Stage 1: Day 0–Day10 (10 days duration)
  Temperature: 36.0–36.5°C (96.8–97.7°F)
  Relative Humidity: 60–65%
  Turning: None
  Cooling 1 hour per day (8:00–9:00 a.m.)
- Stage 2: Day 10–Day 55 (45 days duration)
  Temperature: 36.0–36.5°C (96.8–97.7°F)
  Relative Humidity: 60–65%
  Turning: 4 x 45° daily
  Cooling: 1 hour/day (8:00–9:00 a.m.)
- Stage 3: Day 56–Internal Pip (15 days duration)
  Temperature: 35.5–36°C (95.9–96.8°F)
  Relative Humidity: 55–60%
  Turning: None
  Cooling: None
- Stage 4: Internal Pip–Hatch (7 days duration)
  Temperature: 35.5–36°C (95.9–96.8°F)
  Relative Humidity: 55–60%
  Turning: None
  Cooling: None
- Total Incubation (77 days duration)
- Total Average Temperature = 35.93°C (96.7°F)
- Total Average Relative Humidity = 59.2%

As the developing embryo reacts strongly to light, the best policy is to incubate in a darkened incubator. After 36 days of incubation, it is possible to detect embryonic movement by the rocking motion of a straw or uncooked spaghetti which has been balanced across the top of the egg. By internal pip the egg should lose 12–16% of its original weight. Weight loss can be corrected by adjusting the incubator humidity accordingly, i.e., an egg losing weight too rapidly requires increased humidity, and vice versa.

The general pattern of hatching in kiwi is as follows. Four days before the actual hatching, the chick breaks into the air-cell and can be heard vocalizing. On the following day, a rupture appears in the air-cell end of the egg. The next day, further cracks appear, and on hatching day itself, the chick heaves violently and after a series of contractions, the end of the shell breaks open and the chick pushes out. The actual hatching takes thirty minutes or longer. The chick does not have an egg tooth. Occasionally, incubator-hatched chicks require slight assistance, particularly in instances where the membrane surrounding the chick contains live blood vessels, indicating that some of the yolk may not yet have been absorbed.

At hatching time, the egg should be placed onto a soft absorbent surface such as several layers of kitchen paper towel. After hatching, the brooder temperature should be reduced to 34°C (93.2°F), and then to 30°C (86°F) after 24 hours. The newly hatched chick has the abdomen grossly distended with yolk, and rests on the ground with its legs splayed out sideways. The chick must be handled with extreme care until it is walking well. At day two, the temperature should be maintained at 30°C (86°F), and slowly lowered to 26°C (78.8°F) over the next few days. When lowering temperatures, always monitor the chick for signs of shivering or overheating, and adjust accordingly. Some chicks stand within eight hours, others may take their first few ungainly steps when about 60 hours old, shuffling with their toes and metatarsi resting on the ground. On its fourth day the chick can be moved to the brooder and will walk, although somewhat unsteadily. Two days later it should walk quickly and confidently. At this stage coarse grit measuring 2–5 mm in diameter should be provided for gizzard stones, and a shallow bowl of water should be introduced. No additional heat is required after five days, as long as the room temperature does not drop below 18°C (64.4°F). The brooder should be kept semi-dark, as the chick will feed during the daytime within the first month. Kiwi chicks will lose up to 33% of their hatch weight within their first few weeks. The brooder floor is lined with a 40-mm layer of soil or peat, which will prevent the worms from escaping the tip of the bill. Ensure that worms and soil are obtained from a clean source. Leaf litter consisting of dry leaves, ferns and peat, should be changed in the burrow regularly. Droppings and mess from the food tray should be removed regularly. Soil must be cleaned every 10 days and should be kept moist, not wet.

When the chick is seven days old, worms can be introduced into the run and the burrow. After three days of feeding earthworms and mealworms, a sloppy artificial diet with chopped worms mixed in can be added. The adult diet can be used for chick rearing, but all components should be more finely diced. Once the chick is feeding well, the live food can be reduced, but not excluded. Initially, the chick will take about 50 grams (1.76 oz.) of food per day, but by a month of age this will have increased to 150–200 grams (5.3–7.0 oz.). The chick can be moved to an outdoor run with a shelter at 4–6 weeks. Ideally, each chick should have its own run measuring about 4 m x 5 m (3 x 16 ft.). Two birds of the same age or sex can, however, be raised together if careful observations are made to ensure that both birds are getting an equal share of the food. It is best to weigh the birds regularly and provide separate burrows, even though they will share accommodations up to a point.

Brown Kiwi *Apteryx australis*

Day	Weight (g)	Day	Weight (g)
0	350	**35**	505
5	300	**40**	540
10	270	**45**	580
15	300	**50**	640
20	320	**55**	710
25	360	**60**	730
30	465	**65**	800

References:

Rowe, B.E. 1985. The New Zealand Kiwis. *Avicultural Magazine* 91 (1–2): 59–63.

Nelson, A. 2002. Auckland Zoos Kiwi Incubation Protocol. Unpublished.

Tinamous

02 Tinamiformes

02.01 **Tinamidae: Tinamous**

02.01 Tinamidae Tinamous

Christine Sheppard

Introduction: Tinamous fall into the order Tinamiformes, and the family Tinamidae, which includes 47 species in nine genera. All tinamous are neotropical endemics of ancient lineage. In fact, tinamou share characters of palate and blood proteins with the reptiles. Both morphological and DNA evidence support the concept that tinamous and ratites diverged from a common ancestor, before the ratites themselves lost flight. The 29 species included in the genera *Tinamus, Nothocercus,* and *Crypturellus* inhabit forest floors in tropical and sub-tropical forests; their nostrils are located halfway down the bill. The 18 species encompassed by the genera *Rhynchotus*, *Nothoprocta, Nothura, Taoniscus, Eudromia,* and *Tinamotis* are steppe tinamous, inhabiting open, often arid habitats; their nostrils are at the base of the bill, presumably because these birds often dig when they forage.

Tinamous externally resemble the less colorful members of the Galliformes and have often been kept by aviculturists interested in that group, and managed like pheasants or partridges. However, tinamous are less popular and probably only two species can be considered reasonably well established in aviculture, the elegant crested tinamou *Eudromia elegans,* in which even a silver mutation has appeared in captive stock, and the Chilean tinamous *Nothoprocta perdicaria,* which has been reared in some volume experimentally as a table bird. However, the group as a whole has not been well studied. There is considerable diversity of diet and behavior among tinamou species, and long term management will require investigation of these differences and development of an appropriate range of management protocols. Species of steppe tinamou have appeared in collections more frequently than species of forest tinamou, and most of my own experience has been with the elegant crested tinamou, a steppe species from the pampas of Argentina and adjoining countries.

Management: Most tinamou species are monomorphic, although females are generally larger than males. Males have an intromittant organ, or penis; however, females may also have a small phallic-like organ, making vent sexing of young birds dependent on distinguishing between the two. During the breeding season, differences are very clear. Tinamous like to bathe in both water and dust and will also sun bathe. While not flightless, some typical avian adaptations for flight are reduced in tinamous. They have round wings, small in proportion to the rest of the body. Their flight is rapid, but labored, and usually of short duration. The tail, used by other species for braking and steering, is very small; as a result, tinamous seem to steer poorly, requiring relatively large landing spaces. Birds may run or turn at landing, to reduce speed. In comparison to flight-dependent birds, tinamous also have a small heart and circulatory system. If forced to fly or run repeatedly, they can easily become exhausted, although, as ground birds, they have stout legs and run well.

In captive situations, tinamous are clumsy fliers, likely to crash. This must obviously be taken into consideration when designing an enclosure and management routine. They generally spend all of their waking time on the ground, unless startled. Forest tinamou roost at night, preferring a perch on a slope, which allows

Red-winged Tinamou
Rhynchotus rufescens

Puna Tinamou
Tinamotis pentlandii

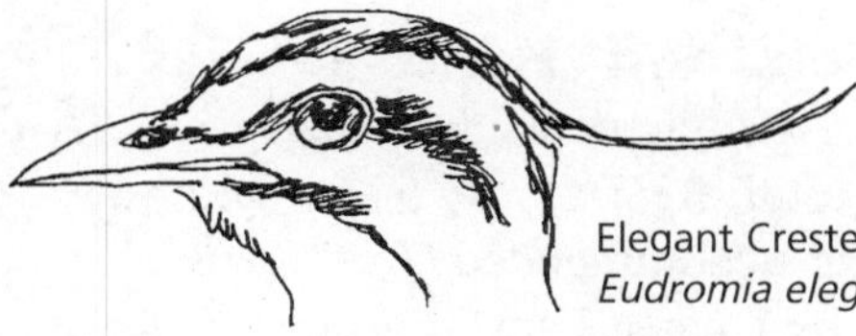

Elegant Crested Tinamou
Eudromia elegans

Hooded Tinamou
Nothocercus nigrocapillus

access without upward flight. Open habitat species typically roost under a bush. These preferences should be taken into account when arranging an enclosure. A-frames and other shelters can often be substituted for shrubs. Forest tinamou are generally solitary. In some species, pairs associate throughout the year; in others, birds come together only to mate. Grassland species, however, are often found in groups. Group size changes throughout the year, presumably as family groups first amalgamate, then split up for the next breeding season.

Breeding: Forest species may breed year-round. In other habitats, breeding tends to be restricted to the time of year when food is most abundant, which depends largely on temperature and rainfall. While there are exceptions, the general description of tinamou mating systems is simultaneous polygyny for males and serial polyandry for females. Either or both sexes may be territorial, depending upon the species. One sex is usually more aggressive than the other. Males advertise for females by calling and may be within earshot of one another. From one to four females may lay in a male's nest. Once the clutch is complete, and the male begins incubation, females usually move on to another male. In a captive situation, significant space would be required for more than one male to breed in an enclosure. In a space about 8 x 4 m (26 x 12 ft.), we found it difficult to maintain even 1.2 birds, because of aggression. If it is desired to allow a male to incubate and rear chicks, it is important to monitor laying activities. If a male begins to incubate before the hens have finished laying, asynchronous hatching may result. If this happens, the male may ignore hatchlings, and continue to incubate the remaining eggs. At the Hancock Wildlife Research Center in Canada, this problem was averted by keeping five males in separate pens and placing two or three females with one male until a full clutch of eggs had been laid, as indicated by the commencement of incubation. The females were then shifted to the next pen, and the process repeated. When the rotation was completed, and the fifth male was incubating a clutch of eggs, the females, if still laying, could be returned to the original male, whose brood was by that time old enough to be removed. Tinamous are not normally aggressive toward other species and do well in a community aviary.

Tinamous always nest on the ground. Steppe tinamou nest under shrubs, while forest tinamou nest among emergent tree roots. The nest is rudimentary, generally a layer of leaves or a scrape. Tinamou eggs are famous for their porcelain appearance and for their colors — green, blue, purple, and chocolate. It has been speculated that bright colored eggs are not a disadvantage, as they are usually covered, either by the male, or by grass or leaves. However, there has been no explanation for why these species should expend the energy required to pigment their eggs.

Diet: There is considerable variability in the diet of different wild species. As it does for pheasants, this may indicate a spectrum of requirements for different nutrients, but no actual requirements of any species have been studied. Wild birds consume combinations of insects, seeds, fruits and other plant parts, different species tending to take different proportions of these items. In collections, a good, balanced poultry ration will usually be accepted. We also offer a small amount of bird salad, a mixture of chopped fruits and vegetables, coated with powdered pellets to provide minerals and vitamins. Birds will forage within an enclosure, but these activities are not damaging.

Crested tinamou have large intestinal caecae, indicating probable digestion of some part of the ingesta by bacteria. The distribution of such caecae within the group and correlation of caecae with diet is not known. However, we have found it necessary to provide newly hatched birds with opportunities to inoculate their digestive tracts with appropriate bacteria. The ornithologist Alexander Wetmore (1926) observed that the crop and alimentary tract of an elegant crested tinamou chick collected in the wild were filled with the fecal matter of the parent. He speculated that the reason for this "curious circumstance" was either to supply partially digested vegetable matter or water in an arid region where succulent vegetation is scarce and moisture absent. Hindsight suggests that this habit may in fact be the natural method of passing vital digestive bacteria from one generation to the next.

Incubation and Hand rearing: Incubation ranges 16–20 days, depending on species. A dry bulb temperature of 37.5°C (99.5°F), with 65% relative humidity is recommended for artificial incubation. Chicks do best when kept in groups. If there are more than four chicks, they should be divided for better monitoring. A brooder box 40 x 60 cm (16 x 24 in.) with readily cleaned matting on the floor is good for the first stage. A temperature of 31°C (88°F) should be maintained for the first 4–5 days, then gradually lowered to about 27.5°C (82°F). In the wild, although chicks can pick up food on their own, males catch insects and present them to chicks; hand-reared chicks need similar attention and stimulation. A mixture of crumbled poultry pellets, tiny mealworms, ground hard-boiled egg and chopped greens works well, with tweezers used to point at or offer bits of food. Tapping on pans and sprinkling food on the brooder floor can also stimulate feeding. Fresh food should be offered several times a day. Ground egg is usually eliminated from the diet after about two weeks. By the third week, pellets can be offered.

A tiny drop, approximately 0.05 ml, of multivitamins containing vitamin D3 should be administered by mouth each morning when chicks are handled for weighing. Calcium carbonate can be dusted over food pans to supplement calcium. After about a week, chicks should be feeding themselves and should be moved to a larger space. Older chicks will usually go to a fresh food pan and younger chicks will follow, so placing newly hatched chicks in with chicks several days older can help the younger birds get started. Offering drops of water from the tips of the fingers is a useful technique initially. Floating greens in the water pan will stimulate interest. A stone or other object should be placed in the water pan to prevent chicks from climbing in. Feet should be monitored and cleaned if caked with food or feaces.

In the late 1970s and early 1980s, the Bronx Zoo reared a large number of crested tinamou. All chicks developed diarrhea in the first few days of life and were treated with antibiotics. Eventually, newly hatched chicks were given antibiotics prophylactically, before they developed symptoms. Reasoning that the large caecae in this species might indicate a need for inoculation with normal gut flora, we experimented by feeding newly hatched chicks with small amounts of adult stool. None of these chicks developed diarrhea and this practice has become part of our standard protocol. Of course, adults are checked for parasites at the start of the breeding cycle.

Chicks often lose weight for the first day or two. After that, during the first week, they gain 1–5 grams per day (see graph) for the first week. During week two, 4–8 grams per day is appropriate, and up to 10 grams a day by week three. Overly rapid weight gain can lead to airplane wing. This may indicate a need for more space and exercise; it can be corrected by taping the wing up for several days.

Crested Tinamou *Eudromia elegans*

Day	Weight (g)	Day	Weight (g)	Day	Weight (g)
1	44	**15**	98	**29**	210
2	43	**16**	100	**30**	230
3	42	**17**	110	**31**	215
4	43	**18**	120	**32**	222
5	44	**19**	125	**33**	240
6	46	**20**	130	**34**	242
7	50	**21**	147	**35**	270
8	56	**22**	152	**36**	290
9	60	**23**	145	**37**	270
10	68	**24**	150	**38**	270
11	70	**25**	135	**39**	280
12	80	**26**	132	**40**	285
13	87	**27**	200		
14	92	**28**	195		

The husbandry techniques described here can be applied to most tinamou species.

References:

Wetmore, A. 1926. Observations on the birds of Argentina, Paraguay, Uruguay, and Chile. U.S. *National Museum Bulletin* 133.

Currasows, Guans, Chachalacas and Megapodes

03 Craciformes

03.01 Cracidae Currasows, Guans and Chachalacas

Luiz Roberto Francisco

Significant reference, highlighted as (Houston Zoo) in this chapter is made to a curassow husbandry protocol by the Houston Zoo.

Introduction: Formerly included in the order Galliformes, the family Cracidae are now placed in their own order, Craciformes, along with the megapodes, which comprise a separate family.

Although the taxonomy of this family is still controversial, at the moment a total of 50 species and 11 genera are recognized. Distributed from the southwestern USA to the north of Uruguay and Argentina, the Cracidae can be divided in three natural groups: the curassows, the guans and the chachalacas. The curassows are the largest members of the family and comprise 14 species: 7 true curassows *Crax,* 4 razor-billed curassows *Mitu,* 2 helmeted curassows *Pauxi* and the monotypic nocturnal curassow *Nothocrax*. The guans, the most diverse of the cracids, comprising 24 species: 15 true guans *Penelope,* 4 piping guans *Pipile,* 2 sickle-winged guans *Chamaepetes,* and 3 monotypic taxa: wattled guans, *Aburria,* horned guan *Oreophasis* and highland guan *Penelopina*. And the chachalacas, the smallest members of the Cracidae, with 12 species, all in the genus *Ortalis* (Strahl and Schmitz 1997).

The members of this family range in size from that of a pheasant to a turkey. All have long necks and tails, strong legs with toes adapted to perch on branches, and a strong chicken-like bill. A distinctive feature of the cracids is the brightly colored bare skin, such us in the cere of the *Crax* curassows, dewlaps in the *Penelope* guans, and other kind of appendages such as wattles, and horny helmets, all which seems to have a function in sexual displays. Other significant morphological characteristic of the Cracidae include the lengthening of the trachea in many species, and the modification of the quills of the primary feathers of most guans, both related to sound production.

Most of the cracids are arboreal species. Chachalacas are generally found in open habitats of scrub and woodlands, guans in lowland rain forests and higher altitude cloud forests, and curassows in the rain forests.

Cracids play an important role in the regeneration of tropical forests as they disperse seeds and promote the biological cycle of many plants. Because of their great size, curassows are considered to be important components of the Neotropical forest's food chain (Brooks & Fuller 2006). Unfortunately, there has been a significant population decrease in many members of the family due to habitat destruction, hunting and wild-animal trafficking. With approximately half of the guans and curassows considered vulnerable or endangered, the Cracidae is considered the most endangered avian family in the Neotropics. (BirdLife International 2000). Of these listings, the most dramatic case is that of the Alagoas curassow *Mitu mitu,* probably extinct in nature but now well known from only a few individuals kept in a private collection (Bianchi 2006).

Because of the intense human influence on nature, the development of captive breeding programs is impor-

Northern helmeted curassow
Pauxi pauxi

Black curassow
Crax alector

Alagoas currasow
Mitu mitu

Horned guan
Oreophasis derbianus

tant to preserve the species. Various species of cracids have been kept in captivity with positive results, some species breeding particularly well. Efforts to promote the survival of the species in the wild must also be made, but can only be successful if the natural habitat can be preserved. No conservation program regarding the Cracidae can be carried out if their habitats cannot be protected, which means that a positive effort has to be made to prevent hunting, deforestation and illegal trade; additionally, environmental education programs should be devised.

Management: The captive management of Cracidae is not usually problematic, because they adapt well to captive conditions if their basic needs are satisfied. In spite of the many species coming from hot and sometimes humid areas, they are capable of surviving and breeding in colder climates if kept in appropriate aviaries, fed well and given proper health care. It is recommended to keep them in large and rectangular aviaries; the suggested minimum size for the larger species is 8 x 5 m (26 x 16 ft.), with a minimum height of 3 m (10 ft.). With a few exceptions, cracids are thought to be monogamous, and keeping them in pairs is recommended (see Breeding). A small pond for drinking and bathing should be inside each aviary. At Houston the substrate is "filter sand" - gravel approximately 0 .5-2 mm in size - which allows quick drainage and may be easily cleaned by raking or sifting. Sand may be used in combination with other substrates such as soil (for planting), sod, or bark chips. Substrates used in other collections include wood shavings, coarse straw, and volcanic gravel. Hard, unyielding surfaces, such as concrete, may cause or aggravate foot injuries, e.g. "bumblefoot," especially in growing birds. Cracids are avid dust bathers, and it is recommended to provide them with a large patch of fine sand under shelter for this purpose. In spite of the birds being more terrestrial than arboreal, it is important that perches are set up inside the aviary because the birds will make use of them. Perches could be made out of dead trees, which make an aviary both a more natural environment for the birds and more aesthetically pleasing. Damage to tail and wing feathers can be prevented by placing the perches a reasonable distance from the wire mesh. An aviary for cracids should be well planted, preferably including plants of the Bromeliaceae family and others as described for Habitat A. Curassows are forest birds; planting their cages offers advantages beyond aesthetic appearance. In tropical and subtropical climates plants provide additional shelter from sun and rain. Vegetation along cage sides or climbing vines serve as visual screens that provide rest from members of the pairs, to inhibit fighting between adjacently caged birds, and to provide privacy for nervous specimens. Flowering and fruiting plants may provide a seasonal dietary supplement, though care must be taken that no poisonous species are introduced (Houston Zoo). A small sheltered area should be incorporated into the back of the aviary, where they can feel more secure and build nests; this is also useful as an area in which the birds can be more easily caught, if necessary. In a cold climate, where temperatures may reach 0°C (32°F), it is advisable to enclose the sheltered area with wood or, preferably, brick walls. Its recommended that this indoor part make up a third of the total area of the aviary. Glass tiles could be installed in the roof to supply light and heat. It is not necessary to use an artificial heating system as long as temperatures remain at least 2-3°C (35.6-37.4°F). Cracids are somewhat nervous and may panic easily; nevertheless, they often become tame and confiding in captivity. If hand reared, guans and curassows of both sexes may become annoyingly aggressive to their keepers during the breeding season (their confidence and potential aggression may make them unsuitable inmates of public walk-through aviaries). A behavioral peculiarity of Cracidae, which often bewilders novices caring for them, should be explained: stereotyped head flicking. All members of the family make sudden, rapid head-shaking movements from an early age, which seem to increase in frequency with nervousness. Adaptation of this behavior is uncertain, but it is not aberrant or indicative of trauma or illness (Houston Zoo).

It is important to use the proper methods of capture and restraint for members of the Cracidae. Captive-bred birds are usually easier to handle than wild-caught individuals, but care must always be taken when handling these birds, especially the more strong curassows species, and their capture should be considered in advance. If possible, one should avoid catching birds during the breeding season, mainly because eggs still forming inside the female can cause severe internal damage or even death if broken during restraint. Curassows and other species with ornamental protrusions on their heads should be treated very carefully; when frightened, they usually fly up against the wire mesh of the aviary, which can cause serious wounds and sometimes even permanent mutilations. The best device for catching cracids in captivity is something like a large butterfly net, with the metal ring cushioned; and instead of netting, the bag can be made of cloth. Its diameter should be sufficient for the size of the bird's body, and the cloth sack should be deep enough to contain the bird when the handle is rotated.

It should always be taken into account that cracids are endowed with strong musculature and wide wings, and are therefore able to jump strongly. For these reasons, when trying to catch one of the birds, it should be driven off branches or perches and onto the ground. Once on the ground, the bird will be caught with the net described above and quickly and gently pressed against the ground, which will prevent the bird damaging itself by struggling while in the bag of the net.

Once caught, the bird should be held firmly with the wings closed against one's body, one arm surrounding the back and side of the bird, with the hand grabbing both legs in a bended position, and supporting the abdomen from underneath, and the other hand on its back. To avoid damage in the tarsus, one finger of the grabbing hand should be placed between them. Another method consists in wrapping the body of bird with a towel, with the legs stretched backwards. This method very conveniently immobilizes the bird's legs and wings, avoiding feather damage and allowing manipulations such us weighting. Covering the head with a hood makes the birds less stressed. Never press against the bird's chest, as this can hinder breathing and suffocate the bird. The legs should never be held together with the tail feathers, because if the bird abruptly moves its legs, there is danger of the tail feathers being pulled out, which can cause permanent damage such as the tail feathers never growing again. These birds should be manually restrained as briefly as possible, as excessive handling can cause severe stress, sometimes resulting in death. A common practice that must be avoided during manual capture of birds is grabbing hold of the bird's legs. This procedure is extremely dangerous because it can cause leg fracture, especially when the catcher grasps only one leg, which frequently rotates, causing a serious orthopedic injury.

Long transport or shipment requires a crate, either a commercial sky kennel or a custom-made wooden box. Whichever is used needs to be scaled to the size of a single bird (to reduce likelihood of injury, each bird will occupy a separate crate), providing just enough room to stand with raised head in a normal posture and enough room to turn around without extending the wings. The crate ceiling should be well cushioned to avoid injury should the bird jump in panic (foam rubber glued to the crate and covered with cloth, the edges of which are then stapled to the wood, is effective padding and prevents the bird from pecking at, and possibly ingesting, rubber). Ventilation slots should be numerous and, if large, covered with wire mesh in such a manner that no loose wires protrude to cause injury. To calm the bird, ventilation slots (and door in the case of a sky kennel) could be covered with porous, dark cloth (e.g. burlap) to darken the interior. A thick layer of hay or straw provides suitable substrate to prevent the bird from slipping. A perch should not be provided as it may cause injury if the bird panics (Houston Zoo). The transport time should be as short as possible, preferably during the cooler periods of the day, and always avoiding exposure to the sun.

Anesthetizing a bird is not generally complicated or difficult if basic procedures are followed. A combination of ketamine, at 10-30 mg/kg (4.5-13.6 mg/lb.), and xylazine, at 1-3 mg/kg (0.45-1.36 mg/lb.), produces excellent sedative and muscle-relaxing results, allowing safe orthopedic surgery. The use of a volatile anesthetic such as isoflurane is more convenient as it is safer and allows a fast induction with a reduced recovery time, and without the inconvenience of incoordination observed during recovery from anesthesia that resulted from injectible medicine.

Diet: A suitable diet is an important factor in the successful management and propagation of Cracidae, significantly influencing breeding results. In nature, the general trend in diet appears to be more leaves and less fruit in smaller species like chachalacas, to more fruit and less leaves in larger species like curassows. Similarly, animal matter seems to be more prevalent in the diets of smaller species (e.g. insects in the diet of Ortalis, snails in the diet of Aburria) than in curassows (Sick 1985). For this reason, there should be a balance of animal and vegetable foods. The base of the diet, however, should always consist of vegetables. The addition of animal protein promotes the healthy development of birds as well as decreasing the levels of cannibalism among chicks (Azeredo, R. 1998. Pers. comm.). Following are suggested diets for Cracidae:

Adult Diet: Estimated nutritional composition in dry matter (domestic chicken requirements in brackets):

Gross protein	15% (18%)
Metabolizable energy	3,000 kcal/kg
Gross fibre	4.8% (4.9%)
Fat	2.5% (3.4%)
Calcium	1.3% (1.5%)
Phosphorus	0.65% (0.7%)

Proportions of Foods used (in 100% dry matter):

Chick crumble (19% protein)	65%
Banana	14.5%
Corn	9%
Cabbage	5%
Papaya	5%
Calcium phosphate (dibasic)	1.5%

The expected consumption is only an estimate regarding the needed energy, which will vary in different climates, and according to the amount of exercise the bird takes. Diets of most species can be supplemented with small insects and earthworms. As well as providing an additional source of protein, pursuing insects can keep the birds busy, providing exercise and reducing their aggressive behavior. The food bowls for adults should be placed in rodent-proof place.

Rearing Diet: For a live bird weighing 2 kg (4.4 lb.), with an estimated requirement of 394 kcal of metabolizable energy/day, the expected daily consumption of a diet of the above composition would be 131 g (4.62 oz.) of dry matter. That would result in consumption close to:

Food Item	In 100% of dry matter	In 100% of natural matter
Chick crumble	110	126
Banana	65	20
Cabbage	4	39
Papaya	56	45
Boiled egg	33	15
Calcium phosphate	16	1.6
Total	284 g	246.6 g

It is important from the management point of view that the daily food amount is divided into two meals. This maintains the quality of food and also avoids excess feeding. It must be considered that the daily energy requirements may vary according to the species, climate, type and size of aviary or other factors. For feeding the chicks, it is advisable to use chicken or quail feeders. The chicks can reach the food with head and neck without being able to walk all over it, and better hygiene is maintained. Houston Zoo describes the diet as: "Captive diets are based on traditional game bird formulae with fresh fruit and additional protein sources. At Houston, poultry pellets have been found more appropriate to curassows' size than crumbles, with less wastage. In warm, humid climates it is important to avoid uneaten feed becoming a medium for the growth of mold and bacteria. Therefore rapidly spoiling items (e.g. hard-boiled egg) are not routinely used as a protein source. During breeding season, additional calcium supplement, such as crushed oyster-shell - "pullet-size" - available from poultry suppliers, is given freely. Catch-pans under feed bowls also minimize chances for decaying feed to contaminate cage substrate. Curassows swallow small, sharp stones ("grit") as a digestive aid. If cage substrate is not composed of such material, it should be provided from a clean source. In this connection, it should be mentioned that curassows are curious and may attempt to swallow inappropriate or dangerous items falling into their cages, such as nails, bits of wire, coins, or small hardware items. Special care should be taken to remove any and all foreign material following cage construction or repair. In addition to dietary staples, curassows may be offered "treat" foods (e.g. newborn mice, peanuts in the shell, live crickets, lizards) as available at times other than regular feeding. Such offerings allow keepers the opportunity for close inspection of the birds, encourage their confidence, and allow some behavior modification. Curassow "courtship-feed," i.e. the cock selects a choice morsel, holds it and calls his mate who plucks it from his beak. In the absence of delectable feed items, a symbolic gesture with a pebble or feather may be made. The opportunity to select attractive feed items enhances pair-bonding."

Breeding: The main objective for anyone working responsibly with birds in captivity is their reproduction. The members of the Cracidae are known to be monogamous birds, with the exception of the polygynous horned guan (Delacour & Amadon 2004). In captivity there are reports of curassows (bare-faced curassow) and guans (Highland guan) where a male mated with two females. However, cases like this are not the rule, and to avoid possible aggressions from the dominant pair, only one male and one female should be placed per aviary.

In chachalacas and guans the sexes are alike (with the exception of the Highland guan); curassows have varying degrees of dimorphism. The palpation of the trachea in the species of Crax, Ortalis and most of Penelopina, indicates the sex. In cases where the sexes cannot be told apart visually, a cloacal examination with the speculum is indicated. During the breeding season, the males show an easily visible intromittent organ that can also be everted by an experienced handler; in females it is possible to observe the *sphincter vaginae*. The development of genetic sexing techniques is also useful.

Once the pairs have been established in the aviaries, they generally show good reproductive results. However, there can be some exceptions, such as the case observed by the first author where several pairs of black-fronted piping guan *Pipile jacutinga* were kept together for years with no breeding results. He observed apathy in the sexual behavior of many males who simply showing no interest in the female. However, once a pair starts to produce fertile eggs, they usually will continue. For curassows, nests can be provided in the form of a basket with a diameter of 60 cm (24 in.) and a depth of 25-39 cm (10-15 in.), or a wooden box of similar size. Guans and chachalacas will use the same nest with dimensions scaled to their sizes. The nest can be indoors or outdoors, but always in a protected and high place where the birds feel secure. Houston provides the following advice that can also be applied to guans and chachalacas: "Once established, most curassow pairs remain compatible. Initial introduction may prove difficult, and there are individuals that never accept a mate. Aggression may occur when there is an age discrepancy between introduced birds, especially if one of the birds is less than two years old. Individuals should be placed in adjacent cages to become accustomed to the other's presence and to allow keepers to observe reactions. If no agonistic behavior is noted or if enticing behavior, e.g. courtship feeding, is attempted, the cock (or older bird) should be introduced to the hen's (or younger bird's) cage. Males are usually - but not always - attackers, and by introducing them to strange enclosures, their territorial defense response may be inhibited. Introduction should always be monitored, and if threatening interactions persist, birds should be separated, especially if an observer cannot be present. (Authors note: trimming the wing feathers of males helps reduce their aggressive behavior.) Unchecked aggression can lead to serious injury or death. The best sign of compatibility is courtship feeding, and the provision of "treat" items may induce the behavior, thus assisting pair establishment. Absence of courtship feeding indicates a weak pair bond, and, though breeding may occur, aggression is latent and threatens to erupt without warning. Birds that do not form pair bonds may be offered alternate partners. If alternates are not available or are unsuccessful, it is possible to allow cohabitation when the female is receptive, otherwise holding the birds in adjacent cages. In territorial species, the proximity of conspecifics or close relatives may prove distracting; curassows, especially cocks, may spend more time threatening a bird in the next cage than courting. Similar species should be separated by visual screens (e.g. reed fencing, vegetation) or by alternate cages housing unlike species. Visual isolation also provides nest privacy."

Egg Incubation: Chachalacas usually lay three eggs per clutch, guans between two and three, and curassows two; in all cases the eggs are laid 48 hours apart. If the eggs are collected soon after laying, a hen may re-lay in 3-4 weeks, up to six or seven clutches annually. Incubation is done only by females and the period varies between 23-24 days for Ortalis, 25-26 days for Chamaepetes and Penelopina, 26-27 days for Pipile and Penelope, 28-29 days for Alburria, 32-33 for Oreophasis, 27-28 for Nothocrax, 29-30 for Crax and Mitu, and 30-31 for Pauxi (Estudillo-Lopez 1997). In captivity some females may lay from the perches, or in the floor of the aviary. In these cases, and when maximizing production is a goal, artificial incubation or foster incubation is recommended. Domestic turkeys, bantam chickens and peacocks are some of the species used to incubate cracid eggs with good results, although, rearing by foster species is not recommended, as it is common for females to kill chicks of different species. Artificial rearing is the best option in these cases. Natural incubation and rearing should be attempted for any re-introduction project. Reserving the final clutch of the year for parent rearing might be an effective compromise with the desire to maximize production by removing eggs for artificial incubation.

At Curitiba Zoo, Brazil, eggs of the Cracidae have been successfully incubated at a temperature of 38.5-39°C (101.3- 102.2°F) and 60% humidity. At Africam Safari, Mexico, eggs of great curassow Crax rubra, crested guan Penelope purpurascens, Highland guan Penelopina nigra and horned guan Oreophasis derbianus have been incubated at 37.5ºC (99.5ºF), and different humidity according to the species weight loss.

Chick Rearing: A few hours after hatching, the chicks are already able to move around with plenty of agility. They should be placed on a non-slippery surface in a warm environment. Recommendations for their feeding have been given above. Cracidae chicks grow quickly with the appropriate diet. Eventually the growing chicks can be kept together in one aviary, but cannibalism may occur. Therefore it is important that the aviary is wide, and that the chicks receive a higher protein level in their diet. Houston Zoo advises the follow-

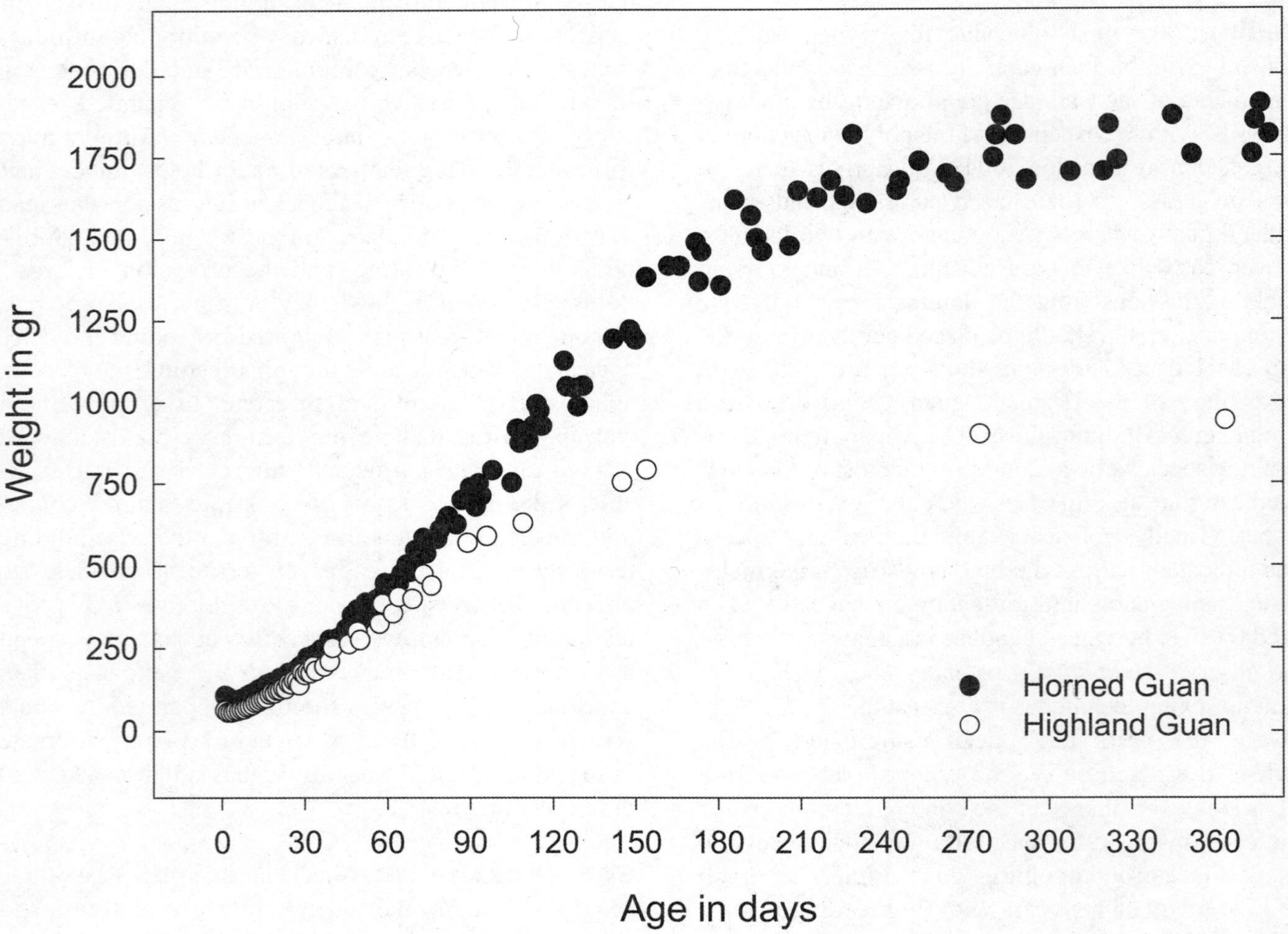

Figure 1: Average growth of horned guan *Oreophasis derbianus* and Highland guan *Penelopina nigra* for the first year in captivity.

ing: "The duration of time from the chick's entry into the airspace to pipping the egg is 24-72 hours, but once pip occurs, hatching may be completed in less than two hours (some chicks may take up to four hours). Normally a large amount of albuminous material is left behind in the shell. A betadine solution is applied to the chick's umbilicus to prevent infection, and it is left in the hatcher or placed under a broody hen for 24 hours before first feed to allow absorption of internalized yolk. Chicks may succumb to egg-yolk peritonitis if early feeding inhibits yolk absorption. If hatched by natural parents, no interference is attempted so long as chicks appear healthy."

Cracids feed their chicks from their beaks, so if a chick hatches singly it is important to offer food and water from the finger for the first few days until it feeds by itself from the floor. If possible, it is important to rear more than one chick together, even if they are from different species. As a last resource, a mirror can be used to give a single chick the needed company.

Artificially raised chicks usually start with a weight drop during the first 3-4 days for a total of around the 10%, after that the daily gain should be around 5-6% for the first two months, around 3-4% for months three and four, and around 1% or less until they achieve their adult weight in one or two years depending on the species.

After keeping the chicks in a brooder for the first weeks, when they are 25-30 days they are moved to a pen furnished with perches where they will be able to exercise better. At approximately six weeks, the diet is gradually changed to an adult ration.

Health management: Regular control of parasites is essential. Problems with ectoparasites can occur during natural incubation, as the presence of the female in the nest provides favorable conditions for the proliferation of mites. The nests should be inspected regularly. Care should be taken to avoid the presence of sparrows, pigeons, or other urban birds in the aviaries using a narrow wire mesh.

For the control of ectoparasites, the topical use of fipronil is recommended (Spelman 1999). Every six months, coprological examinations for endoparasites should be done, and positive cases submitted to clinical and environmental treatment. Parasite eggs can stay infectious in the soil for many years. Drugs commonly used are ivermectin, at the rate of 400 micrograms administered subcutaneously, and levamisol at 8 mg/kg (3.63 mg/lb.), subcutaneous. Every new animal should be submitted to a prophylactic protocol. It should be examined, dewormed, and kept under observation before coming into contact with the other birds or enclosures. It should be submitted to a coprological examination. Even with negative results, ecto-and endoparasitic drugs-could be administered.

Contributors: Carlos Eduardo do Prado Saad, Zootechnician, M.Sc. Animal Nutrition; Rogério Ribas Lange, D.V.M. M.Sc.; Soleane Maria Wichinievsky Francisco, Biologist; Zalmir Silvino Cubas, D.V.M.

Acknowledgements: I am very grateful for the co-operation of all contributors for their support in writing this chapter. Special thanks to Zalmir Silvino Cubas for his criticism and comments. I am also very grateful to Carmel Croukamp for the proper correction and translation into English.

References:

Bianchi, C.A. 2006. Alagoas Curassow (Mitu mitu). Pp: 26-28. In: Conserving cracids: the most threatened family of birds in the Americas. Brooks, D.M. (ed). Misc. Publ. Houston Mus. Nat. Sci., No. 6, Houston, TX.

BirdLife International. 2000. Threatened Birds of the World. Barcelona, Spaina: Lynx Edic. and Cambridge, UK: BirdLife Intl.

Brooks, D.M. & R.A. Fuller. 2006. Biology and conservation of cracids. Pp: 9-21. In: Conserving cracids: the most threatened family of birds in the Americas. Brooks, D.M. (ed). Misc. Publ. Houston Mus. Nat. Sci., No. 6, Houston, TX.

Delacour, J. & D. Amadon. 2004. Curassows and related birds. Second edition. Barcelona: Lynx Edicions and New York: The American Museum of Natural History.

Estudillo-Lopez, J. 1997. Los Crácidos, la familia de aves neotropicales más amenzada de extinción y su posible restablecimiento por la reproducción en cautiverio. Pp: 117-123. In: Strahl, S.D., S. Beaujon, D.M. Brooks, A.J. Begazo, G. Sedaghatkish & F. Olmos. The cracidae: their biology and conservation. Blaine, WA: Hancock House.

Spelman, L.H. 1999. Vermin control. Pp. 114-120 in M.E. Fowler & R.E. Miller (eds.), Zoo and Wild Animal Medicine-Current Therapy 4. Philadelphia: W.B. Saunders Company.

Strahl, S. & A. Schmitz. 1997. A taxonomic reference of the family Cracidae for the common use by ornithologists. Pp. 1-7. In: Strahl, S.D., S. Beaujon, D.M. Brooks, A.J. Begazo, G. Sedaghatkish & F. Olmos. The cracidae: their biology and conservation. Blaine, WA: Hancock House.

Todd, W.T, R. Plasse & C. Eckart. 1992. Curassow Husbandry: suggested protocol. Houston Zoo, TX.

03.02 Megapodiidae Megapodes

David Priddel, Robert Wheeler, Ann Gothe and Myles Lamont

Introduction: Megapodes, which comprise the family Megapodiidae, were traditionally included in the order Galliformes, but are now, on the basis of DNA evidence, placed in the order Craciformes. Megapodes differ markedly from other birds in that they rely on external sources of heat to incubate their buried eggs. Some species scrape up large amounts of leaf litter and soil to form mounds into which the eggs are laid and incubated by the heat produced by the decomposing organic matter. Other species bury their eggs in exposed sand to be warmed either by the tropical sun or geothermal sources. A few species are able to adopt either method.

The 22 extant species of megapodes are scattered from the Nicobar Islands in the Gulf of Bengal, through eastern Indonesia, the Philippines, New Guinea and Polynesia to mainland Australia. Most species have not been adequately studied in the wild and few have been bred in captivity. The most intensive captive breeding program has focused on the malleefowl *Leipoa ocellata.* In 1992, a colony of 16 breeding pairs was established at Western Plains Zoo, Australia, as a conservation initiative to produce progeny for reintroduction into rehabilitated habitat. The breeding program used artificial incubation techniques that were developed earlier for hatching eggs collected in the wild. Outside of Australia, the species most commonly kept and bred in captivity is the Australian brush-turkey *Alectura lathami*.

Australian Brush Turkey *Alectura lathami*

• Ann Gothe and Myles Lamont

Description: Like all members of the family Megapodiidae, Australian brush-turkeys *Alectura lathami* have developed a unique breeding system, in which eggs are incubated by external heat sources and chicks receive no parental care. Brush-turkeys are the largest of all megapodes, with weights of up to 3 kg (6.6 lb) in males and 2.5 kg (5.5 lb) in females. They are also one of the most common species of megapodes.

The Australian brush-turkey (henceforth brush-turkey) occurs along the east coast of Australia. It is the only species within the genus Alectura. There are two recognized subspecies: *A.l. purpureicollis,* inhabiting a smaller area along the Cape York Peninsula and *A.l.lathami* inhabiting the rest of the east coast, all the way to the Illawarra region, south of Sydney (Göth et al. In press). For many years, there also was a resident population of *A.l. lathami* on Kangaroo Island (off the coast near Adelaide in South Australia), apparently derived from one pair introduced in 1936. However, this population has now become almost extinct (R. Seymour, pers. comm.).

The two subspecies can be differentiated by the coloring of the wattles, with *A.l. lathami* having a yellow, and *A.l. purpureicollis* a white to mauve-colored wattle. *A.l. lathami* seems to be the only race that has been kept in captivity so far. Brush-turkeys, when seen in their natural habitat, cannot be mistaken for any other bird. Their large vertical tail, bright red head and yellow or pink wattle aid in identifying this species. With their black plumage, they usually blend in well with their surrounding, and the completely brown chicks are especially well camouflaged.

Brush-turkeys are not very vocal and only possess four different calls. Males utter deep, resonant so-called boom calls to advertise their territory, most often when they stand on top of their mound. While producing this call, they inflate their wattle. Other calls include some softer, very low and frequently repeated gulping sounds, which seem to indicate disturbance, a louder gulping sound, which seems to serve as contact call, and a high-pitched scream given by birds upon capture by a predator (Jones et al. 1995).

Breeding in the Wild: The species is well known for its mound-building capabilities, and is often considered a nuisance in Australia due to its habit of building mounds in parks and gardens. Mounds can become quite substantial over time, reaching widths of over 3.6 m (12 ft.) and are often over 1 m (3 ft.) high. Usually, they are composed of leaf litter and other organic material, and incubation heat is produced by the microbial decomposition of this material. Mounds are only used during the breeding season, which lasts from approximately July to February each year. It appears that the beginning of mound building is usually triggered by rainfall, as only wet organic material decomposes fast enough to produce sufficient incubation heat. Only the males build the mounds, whereas the females visit mounds for egg laying only. Some males own two mounds, but most occupy one only. Females produce one large egg at a time, in intervals of 4–5 days (Baltin 1969). Theoretically, they can choose on which mound to lay each time they produce a new egg, and observations so far indicate that they frequently change mounds during the breeding season. Each time they lay an egg, they have to copulate with the male mound owner. Mound choice in this species is thus equal to mate choice, and females can be described as polyandrous. Males are polygynous because they receive eggs from more than one female in their mound. In correct terms, the species thus has to be described as polygynandrous (Jones 1990).

Males tend the mound on a daily basis. They measure the temperature with a sensor located in their palate. If they find the mound temperature too low, they add more material onto the mound, if it is too hot, they open up the top to let it cool down. The mean incubation temperature in the mounds is 34ºC (93.2°F) (Booth and Jones. 2002). However, males differ considerably in their ability to maintain stable and suitable mound temperatures. For example, within some mounds, incubation temperatures differed up to 9°C between eggs, in others, this difference was only up to 1°C (Göth submitted).

Overall, brush-turkey eggs hatch at incubation temperatures ranging from 30–38ºC (86–100.4°F) (Booth and Jones. 2002). However, incubation temperature strongly affects the weight and thus survival chances of hatching. Those incubated at low temperature take much longer to hatch, and are lighter at hatching, most likely because they have used up most of the yolk during incubation (Göth and Booth. 2005). Most interestingly, incubation temperatures also affect the sex of the offspring, which is the first time this has been reported for any bird (Göth and Booth 2005). When eggs were artificially incubated at 31ºC (87.8°F) more males hatched, whereas at 36ºC (96.8°F) more chicks were females. At the average temperature of 34ºC (93.2°F) the proportion of males to females was 1:1. The mechanism behind this skewed sex bias still has to be discovered, but most likely, it is caused by a biased embryo mortality at low and high temperatures

Artificial Incubation: Most of the suggestions given here arose from A. Goeth's long-term research project on the behavior of brush-turkey chicks. The following points need to be considered:

Incubation lasts for 45 days at a temperature of 34°C (93.2°F), which is the best temperature for artificial incubation.

Eggshells are very thin, and eggs are thus very fragile. They should be touched with gloves only, to prevent potential infection with human introduced bacteria. Candling the eggs with a strong light gives a good idea about how far developed the embryo is (Wong 1998). Heartbeats of embryos can be monitored in a digital heart rate monitor (Buddy egg monitor®, Avian Biotech, Tallahassee, Florida) from when eggs are about 15 days into incubation.

Eggs are not turned during incubation. They should be left in the same position as they are found in the incubation mound, which is with their blunt end upwards (unless the birds have accidentally moved them after laying).

Eggs need to be incubated in a moist medium. Most eggs so far have been incubated in soil collected from those areas of the incubation mounds where the eggs had been found. Some attempts to incubate them in vermiculite have been successful, but for unknown reasons, this method seemed less reliable than incubating eggs in soil.

Highest hatching rates of 80–90 % were achieved in a Brinsea Octagon 200® incubator (Brinsea, UK), in which the heat is rotated evenly, and in which humidity can be kept very high by filling an internal water reservoir to the top and creating a large surface area for evaporation by inserting pieces of plotting paper into this water reservoir. The same incubator can also be purchased with an additional (expensive) humidity control unit, which may be even more effective.

High soil humidity seems important, and boxes containing the eggs and soil should also be sprayed with water once a day.

The soil (or vermiculite) should be exchanged every month, and soon after a chick has hatched or a rotten egg has exploded. Otherwise, harmful bacteria will accumulate in the soil and kill eggs or chicks. If this advice is followed, several eggs can be incubated together in the same box.

If eggs need to be transported, this is best done in a thermo-insulated box filled with a mixture of vermiculite (wetted and warmed up) and polystyrol bean bag filling.

Some zoological gardens have also had chicks hatching successfully from incubation mounds built by captive males in their aviaries (Fleay 1937, Sutter 1965, Baltin 1969, Flieg 1970, Raethel 1988, Ruempler 1988, Hume 2002). In colder climates, it may be useful to install a heat plate underneath the mound, to increase mound temperatures. This resulted in chicks hatching from mounds in the Zoo of Amsterdam (Jacobi 1970), and has also resulted in successful captive breeding in the closely related *Talegalla jobiensis* (Robiller et al.

1985). Males need to be provided with sufficient mulch as building material. In the wild, several tons of organic material are incorporated into the mounds. Occasional wetting of the material will help decomposition and thus heat production. In the St. Louis Zoo, heat production was improved when the male was provided with bushels of bison dung that he incorporated into his mound (Flieg 1970). In the wild, mounds are usually built in locations that are partly shaded by trees (Jones 1987). Aviaries should thus have partly shaded areas as potential mound locations. Some males start building their mounds at a very young age. In Goeth's study, four of six males built their first mound in captivity when they were between 130 and 270 days old; the other two showed no mound building behavior before they were released at the age of 11 months (Göth and Astheimer, submitted).

Housing: After hatching, chicks usually sit motionless in the incubator, just above where their egg had been buried. They are wet and their feathers are still enclosed in waxy sheaths. Such young chicks should either be left in the incubator until they are dry or be moved to a resting box that is equipped with a heat lamp. They are still attached to the chorioallantoic (inner egg) membrane, which should be moved together with the chick without damaging the umbilical cord. After some hours, the chord will constrict and the membrane will detach from the bird.

Even though chicks are equipped with fully developed wing feathers, they do not use these until they are two or more days old. After hatching, they rest motionless for many hours, most likely to fill their lungs with air and dry their plumage. At such young age, it is not unusual for chicks to fall over and lie on their back. On average, it takes chicks 40 hours to dig themselves out of the soil (Göth 2002).

Each chick should be kept in its own resting box until its second day of life. Transport boxes for small pets (cats etc) are very suitable, as they can be cleaned easily and a socket for a light bulb can be installed at the back. The box should be covered with a dark cloth as chicks respond strongly to any disturbance. Water and a few small seeds should be provided, even though chicks do not usually feed during their first days while they live on their internal yolk reserves. A branch to roost on is sometimes used by the chicks during the night. Boxes should have a firm lid, as the chicks try to climb out at dusk, when they would usually go up a tree to roost for the night.

Chicks should be transferred to their aviary when at least 40 hours old. They can be transported in their resting box or a black cotton bag. Young chicks can be kept in groups. They usually feed in close proximity to each other and quickly establish a rank order with little or no fighting. The aviary should be equipped with the following:

1) Cover, as young chicks are very shy. In the wild, they spend their first weeks hiding in thickets. Newly born chicks are excellent escape artists and hence a well-protected aviary is essential to keep them from escaping if they are to be left with their parental counterparts. Half-inch wire was then used on the entire aviary to prevent further attempts at escape. Chicks often are nervous at a young age and providing them with areas in which to hide and escape aids in preventing stress.

2) Mulch to scratch in. Chicks spend most of their time scratching for food, even if provided with additional food.

3) Some shade, best provided by plants. A sunny spot with sand or dry soil. Brush-turkeys love to take sun and sand baths. While sunbathing, they lie down, stretch their wings out and sometimes even roll over on their back.

4) Small trees or shrubs as roosting sites during the night. Small chicks roost on the outer end of densely foliaged shrubs or small trees. Older birds will attempt to roost as high up as possible, and it is thus advisable to provide horizontal roosting branches approximately 80 cm below the ceiling of the aviaries. These branches should be strong enough to handle a 2.5 kg (5.5 lb) bird, as juveniles will have almost reached their adult weight by the age of six months (see below).

In Goeth's study, up to seven young chicks were kept in an aviary 3 x 6 x 3 m (9.8 x 19.6 x 9.8 ft.) high. From the age of approximately two months, only four birds were kept in aviaries of the same size, two males and two females. At the age of 22 weeks, the males started to fight with all other birds and had to kept in an aviary of their own.

Observation of captive brush-turkeys in zoos also showed that males are highly aggressive towards other females and especially other males. It is thus unethical to keep two males in the same aviary when they are older than approximately 22 weeks. Females will also be chased by males, and if they are not provided with adequate space and cover, males can run them to death. A false sense of division within the aviary can be created by the placement of brush piles, hedges and shrubs, which seems to work well for this species. By creating visual barriers, the male often seems to be much less abusive if the female is out of sight. These brush piles will also serve to provide an escape route in the event that the male becomes belligerent. The addition of elevated platforms and perches will also help the hen to escape if need be.

Due to the male's extreme territorialism of his mound, young males should be removed in advance of the breeding season. The addition of smaller Galliformes such as quail and partridge can be beneficial in that they can distract an aggressive male and in a way serve as a form of environmental enrichment.

The recommended aviary size is ~46.5 m^2 (~500sq ft), however in the case of brush-turkeys the larger the better, as small aviaries can prove to be deadly for females and immature males.

Overall, brush-turkeys incubated and raised in captivity can become quite tame, but only if not being handled. In Goeth's study, the birds had to be handled on a monthly basis for weighing and other measurements, and this made them rather shy. It appears that the birds do not become accustomed to such disturbance. For weight measurements, it is thus advisable to use a digital scale that the birds can be lured on to and that can be read from the a distance.

Leg bands for individual identification should have a diameter of 9 mm for hatchlings, 11 mm for 3–4 week old birds, and 16 mm for subadults and adults. Celluloid/Darvic PVC wrap-a-round (spiral) bands need to be glued tight, as the birds otherwise open the spiral with their strong beaks and remove the bands.

Diet: In the wild, both young and adult brush-turkeys are omnivorous and include a wide range of seeds, fruit, buds, fungi, invertebrates, tubers and carrion in their diet. Adults were also observed eating frogs and rats, as well as many types of man-made food (noodles, canned pet food, minced meat, raw meat, poultry feed, breakfast cereal etc). In captivity, brush-turkeys grow quickly, and a high-protein diet is thus essential. In Goeth's study, chicks developed well with a basic diet of chicken starter crumble food that was high in protein. All chicks fed unaided soon after being transferred to the aviary. In addition, they received the following vitamin treatments:

1) Vitamin B powder and calcium powder added to the starter mix twice a week.

2) Liquid mix of fresh sugar-free A, D and E vitamins and trace elements added to the drinking water once a week (Dufoplus: vitamin A & E product, Ioford: vitamin D, iodine and iron mixture). Such sugar-free water preparations offer the best option for administering fat soluble vitamins (A, D and E) because their potency is assured.

All vitamins were provided throughout the 11-month period during which brush-turkeys were raised in Goeth's study. The fine-grained chicken starter crumble food was gradually replaced by larger chicken pellets, until by the age of three months, the birds received chicken pellets only. Corn kernels were occasionally offered as a treat, but the birds also developed well without them. In addition, the chicks found some invertebrates in the mulch provided to them.

In the St. Lous zoo, brush-turkey chicks were raised on the Purina Game Bird F & M in the fall, and in winter, on 50% of the above, 25% Trout Chow No. 4 and 25% small seed. As a breeder ration in the spring and summer they received 50% Trout Chow and 50% Game Bird F & M (Flieg, 1970). Other publications mention poultry, pheasant, chicken, and pigeon food, rice, boiled eggs, all kinds of chopped fruit, soya-bean sprouts, carrots, mealworms, termites, and other insects (summary in Dekker 1990).

Health: To prevent disease, the birds in Goeth's study were also treated with the following substances, as recommended by an expert bird veterinarian:

1) A water cleanser (KD cleanser) added to the drinking water once a week. This contains organic acids and protects health by acidifying the contents of the crop. Its action is confined to the crop, and it is completely inactivated as the crop contents pass into the highly acid environment of the stomach. The water cleanser also keeps the drinking water clean of sludge and germs.

2) During long wet-weather spells, an alternative water cleanser was added to the drinking water (Megamix). This is a natural acid (citric acid) that acidifies the contents of the food being stored in the crop and helps prevent diseases caused by high humidity (E. coli, thrush, molds, coccidiosis and worms).

3) A special treatment against coccidiosis was given for two days each month (Carlox). It was assumed that coccidiosis is the most dangerous disease for young brush-turkeys, as this is also the case for young chicken.

4) To prevent illness and fatality due to worms and lice, a wormer (S76) was added to the drinking water every three weeks. This substance kills roundworms, blood parasites (microfilaria), blood sucking lice, mites and flies.

Where brush-turkeys are kept in the vicinity of other galliforms or ground-living birds, it is also advisable to vaccinate them against fowl pox (at two weeks and 12 weeks), infectious laryngo-tracheitis (at two weeks and 8 weeks), and infectious bronchitis (at five weeks and 12 weeks).

Growth: It is normal for chicks to lose weight during the first days after hatching, while they use up their internal yolk reserves. After this initial weight loss, however, they grow quickly. During the first two months or so males and females grow at a similar rate, but thereafter males gain more weight. From the age of three weeks, the brown plumage present at hatching will be gradually replaced by the black feathers typical for adults. By the age of three months, the birds look like miniature subadults, except that the red head and yellow neck are less colorful.

Injuries, Handling and Transport: Young chicks usually try to stick their head or beak through the wall of the enclosure. If the wall is made of wire, they can easily hurt themselves by getting stuck or rupturing the skin around their beak. It is advisable to cover the wire at the base of the enclosure with shadecloth or a similar material, to a height of approximately 40 cm (15.7 in.).

Injuries can also result when the birds crash into the side or roof of their enclosure when panicked or alarmed. The birds become accustomed to people around them, but physcial handling causes them to panic. The skin on top of their head is rather thin and ruptures easily. In Goeth's study, one such head wound was stitched together by a vet and healed successfully. Such incidences can potentially be limited by clipping one wing to stop the birds flying against the roof of the enclosure.

Brush-turkeys have very strong legs and so far, no incidence of broken legs in captivity is known. If they have to be caught, it is best to hold them firmly by their legs, as the large claws can also cause injuries to the handler. They should never be caught by their tail, as they will shed all tail feathers as an anti-predator response.

Brush-turkeys can be transported in well-ventilated cardboard boxes lined with newspaper. It's best to provide them with a stick that has been attached to the floor so they can hold onto this during transportation. Only one bird should be kept per box. Adults especially can harm each other with their large, powerful feet. Chicks tend to flee from everything that moves and may get frightened by the movements of a second chick in the box.

References:

Baltin, S. 1969. Zur Biologie und Ethologie des Talegalla-Huhnes (Alectura lathami Gray) unter besonderer Berücksichtigung des Verhaltens während der Brutperiode. *Zeitschrift für Tierpsychologie* 26:524 - 572.

Booth, D. T., and D. N. Jones. 2002. *Underground nesting in the megapodes. in Avian Incubation: behaviour, environment, and evolution.* (D. C. Deeming, Ed.). Oxford: Oxford University Press.

Dekker, R. W. R. J. 1990. Conservation and management of megapodes (Galliformes; Megapodiidae). in Conservation and Biology of Megapodes (Megapodiidae, Galliformes, Aves), Thesis, University of Amsterdam (R. W. R. J. Dekker, Ed.). Rodopi, Amsterdam.

Fleay, D. H. 1937. Nesting habits of the Brush Turkey. *Emu* 36:153-163.

Flieg, M. 1970. Breeding the yellow-wattled brush turkey in North America. *Avicultural Magazine* 76.

Göth, A. 2002. Behaviour of Australian Brush-turkey (Alectura lathami, Galliformes: Megapodiidae) chicks following underground hatching. *J. Ornithol*. 143:477-488.

Göth, A. submitted. Incubation temperatures and sex ratios in Australian brush-turkey Alectura lathami mounds. *Journal of Animal Ecology*.

Göth, A., and L. Astheimer. submitted. Ontogeny of mound building in Australian Brush-turkeys (Alectura lathami): the role of learning, testosterone and body weight. *Journal of Avian Biology*.

Göth, A., and D. T. Booth. 2005. Temperature-dependent sex-ratio in a bird. *Biology Letters* 1:31-33.

Göth, A., K. P. Nicol, G. Ross, and J. Shields. in press. Present and past distribution of Australian brush-turkeys Alectura lathami in New South Wales - implications for management. *Pacific Conservation Biology* 11.

Hume, D. 2002. Breeding the Australian Brush-turkey at the Big Swamp Wildlife Park. *The Pheasant* 36:14-17.

Jacobi, E. F. 1970. Die Zucht von Talegallahühnern (Alectura lathami Gray) mit elektrischer Bruthitze. *Zoologischer Garten* 39:129-132.

Jones, D. N. 1987. Selection of incubation mound sites by the Australian Brush-turkey Alectura lathami. Ibis 130:251-260.

Jones, D. N. 1990. Social organization and sexual interactions in Australian Brush-turkeys Alectura lathami: Implications of promiscuity in a mound-building megapode. *Ethology* 84:89-104.

Jones, D. N., R. W. R. J. Dekker, and C. S. Roselaar. 1995. *The Megapodes*. Oxford: Oxford University Press.

Raethel, H. S. 1988. *Hühnervögel der Welt*, 4th edition. Neumann-Neudamm, Melsungen.

Robiller, F., R. Gerstner, and K. Trogisch. 1985. Naturbrut von Halsbandtalegalla oder Jobi-Maleo Talegalla jobiensis Meyer 1874. *Gefiederte Welt* 109:214-216.

Ruempler, G. 1988. Nachwuchs bei den Buschhühnern Alectura lathami im Allwetterzoo Münster. *Gefiederte Welt* 112:254-255.

Sutter, E. 1965. Zum Wachstum der Grossfusshühner (Alectura und Megapodius). *Ornithologischer Beobachter* 62:43-60.

Wong, S. 1998. Estimating the hatch dates of Australian brush-turkey embryos by candling. *Wildlife Research* 25:669-676.

Malleefowl *Leipoa ocellata*

• David Pridell and Robert Wheeler

The malleefowl is peerless among the megapodes in that it is the only species adapted to arid habitats, all others being confined to moist, tropical or temperate environments. The malleefowl, a ground bird weighing about 2 kg (4.4 lb.), was once abundant in the drier parts of the Australian continent, occurring from the Indian Ocean in the west to the Great Dividing Range, just short of the Pacific seaboard, in the east. Excessive land clearance for agriculture, the introduction of the European red fox *Vulpes vulpes,* and the spread of exotic herbivores such as rabbits *Oryctolagus cuniculus* and goats *Capra hircus* have markedly reduced the malleefowl's abundance and distribution.

Management: Malleefowl are prone to stress, trauma and injury, particularly fractured legs, if not handled or housed appropriately. These shy, cryptic birds rely principally on camouflage and concealment to protect themselves from predators. If stress is to be avoided, it is essential that any pens housing malleefowl contain substantial amounts of cover. The aviaries should be landscaped as described for Habitat B in the section on LANDSCAPE, but should be a rather dry version of this habitat with plenty of sandy areas. Shrubs or small trees growing within their pens offer megapodes the best shelter and provide important natural sources of food such as flowers and seeds. Freshly cut branches and foliage placed within the pens will suffice, but these will have to be replenished regularly, thereby imposing unnecessary stress on the birds each time this is done. Malleefowl of all ages roost on the outer foliage of tall shrubs and trees in preference to larger limbs or branches. Small trees growing within each pen make ideal roost sites for these birds.

Most injuries to captive malleefowl result from birds crashing into the side or roof of their enclosure when panicked or alarmed. To reduce the potential for such injury, it is important that disturbance be kept to an absolute minimum. Routine inspections and feeding should be carried out during the middle of the day when the birds are generally lethargic and resting under dense cover. Whenever a person enters a pen containing malleefowl, live invertebrates, such as mealworms, should be thrown to the birds to distract and pacify them. Feeding malleefowl in this manner provides an opportunity to check the birds without inducing stress. The most potentially damaging time to enter any malleefowl pen is at dusk when the birds are normally very vigilant and active.

Malleefowl should be handled only when absolutely necessary and this should always be undertaken with extreme caution as the risk of injuring the bird is high. Some birds may also be reluctant to eat for one or two days after capture. If it is essential that they be handled, malleefowl should be captured at night by quietly removing them by hand from their roost as they sleep. Once disturbed from their roost, a long-handled net may be necessary to capture the birds, but the risk of injury is increased. Captured birds should be kept in the dark and returned to the pen the same night. When being returned, each bird should be placed into its own temporary roost box — a small, open cardboard box fastened to a tree or other solid structure. Malleefowl will usually remain in these boxes until dawn. Birds returned directly to their roost may fly away in panic and injure themselves. Unencumbered birds left on the ground may do likewise.

Malleefowl can be transported in well-ventilated, cardboard boxes with straw or some other non-slip substrate covering the floor. Several young birds can be placed together in the same box without undue risk. Older birds, however, need to be placed singly in each box, as their large and powerful feet can cause severe injury to travel companions. Birds that have been treated for injuries, including those that have undergone surgery, should be allowed to recuperate in the familiar surroundings of their pen. Confinement in small holding cages, although providing the opportunity for intensive monitoring and care, is best avoided if further trauma is to be averted.

Diet: Malleefowl are omnivorous, readily eating a diverse range of foods, including seeds, buds, flowers, leaf shoots, herbs, fungi, invertebrates and lerps (the sugary secretion of sap-sucking insects). In captivity, they have been purported to eat several types of unnatural food, including poultry feed, breakfast cereal, noodles, young mice, canned dog food, minced meat and dehydrated dog biscuits that have been soaked in water.

Captive malleefowl require a varied diet. Grain, such as cracked maize, sorghum and wheat, should be provided, together with green vegetables, fruit, and invertebrates, supplemented with vitamins and calcium. Cereal grain should not be placed in feeders but broadcast throughout the pen. This will not only encourage malleefowl to forage, which is an important consideration if the birds are to be released as part of a reintroduction program, but germinating seed will provide the birds with fresh greens, as well as additional cover if a reasonable number of seedlings survive to maturity. Malleefowl relish most invertebrates; live food such as mealworms, termites, beetles, grasshoppers or cockroaches should be provided throughout the year.

Upon hatching, malleefowl chicks have about three days in which to begin feeding before their stored yolk supply is exhausted. Chicks are naturally inquisitive, and most feed unaided soon after leaving the nest or incubator. Those chicks that show little initial interest in feeding are invariably keen to peck at live invertebrates. In captivity, very young malleefowl chicks should be fed small larvae or termites until they have enough coordination to deal with larger, more mobile prey. Malleefowl are able to obtain all their moisture requirements from the food they eat and from droplets of rain and condensation that form on leaf tips. Consequently, they can survive without potted water. However, malleefowl will drink water if it is available, and the availability of clean water may have an important role in maximizing egg production.

Breeding: In winter, malleefowl prepare their incubation mound by digging a small depression in the ground which, over the next few weeks, is gradually enlarged until it measures approximately 1 m (3 ft.) deep and 1 m (3 ft.) wide. Into this crater, malleefowl deposit substantial quantities of leaf litter which has been laboriously raked from the area surrounding the nest. Once this pile of litter has been moistened by winter rains, it is covered with soil to form a mound typically more than 3 m (10 ft.) across, and up to 1 m (3 ft.) high. A nest chamber is subsequently prepared within the top of the litter layer, and by spring the first eggs are laid. The female usually produces eggs throughout the months of October to December, during the austral spring and summer, but if food is plentiful she may continue to lay for several more weeks. On the other hand, if food is scarce the period of laying will be greatly curtailed.

The hard-working male tends the nest throughout spring and summer, regulating the temperature of the egg chamber by adding or removing soil. Malleefowl are able to gauge the temperature of the nest by probing the loose soil with their opened beak. The pair may continue working the nest well into autumn, abandoning it only when the weather cools to such an extent that they can no longer keep the egg chamber warm. On abandoning the nest, the birds often dig out the entire mound, destroying in the process any remaining eggs that have not yet hatched.

Nests tend to be located on well-drained, friable soils. Malleefowl seldom construct an entirely new nest, preferring instead to renovate an existing one. Nests are often used over many successive seasons, even decades, although not necessarily by the same pair. If insufficient rain falls during autumn and winter, the birds may not begin to build a nest or they may abandon a partially completed one. Water has been artificially added to several unfinished nests during unseasonably dry winters, but to no avail, for they remained unused throughout that breeding season.

The minimal size of an aviary suitable to house a breeding pair of malleefowl is about 20 x 10 m (66 x 33 ft.). A pen of this size allows the birds to calmly withdraw to the far side when a keeper enters. In smaller pens, the birds have insufficient space into which they can retreat; consequently, they tend to pace along the fence in great agitation when danger threatens. During the brief non-breeding season, adult malleefowl may tolerate the presence of conspecifics, but when nesting, malleefowl are highly territorial, so only a single pair can be housed in each breeding pen. Adults have continued to work their mound with complete indifference to the emergence of their offspring from the nest chamber, neither helping nor attacking the young as they struggle to free themselves from the soil. However, there has been one recorded incidence of an adult killing a juvenile malleefowl that had not been removed from the pen.

Nests in adjoining or nearby pens should be separated by at least 100 m (328 ft.) so that the birds cannot see other malleefowl from the immediate vicinity of their nest, nor can they be intimidated by the closeness of the booming territorial call of their neighbors.

To encourage first-time breeding, a crater should be dug and ample leaf litter placed in the pen before the potential breeders are released into the enclosure. Preparing a crater for the young adults may not only stimulate them to nest earlier than could otherwise be expected, but it also encourages the birds to construct their nest at a location determined by the keeper rather than somewhere inconvenient, such as against a fence or near a gate. The nest can be situated in an open or partly shaded location. Supplies of litter should be provided early each winter to aid nest construction.

Two-year-old malleefowl have been known to construct a nest, but females rarely begin egg laying until they are at least three years old. Some females may take up to six years to begin laying. Often a delayed start to breeding is an indication that the birds have not formed a strong pair bond with their assigned partner. In such circumstances, it is advisable to change the pairings in the hope that a more productive pair will be formed. In the wild, malleefowl have been known to breed for at least 12 years. Sexes are identical, and sex can only be

Malleefowl
Leipoa ocellata

determined by surgical examination of internal reproductive organs.

The number of eggs laid and the rate of production can vary greatly from pair to pair and from season to season. In average years most females lay 16–20 eggs, although in particularly good seasons clutches of up to 35 eggs have been recorded. In drought years, malleefowl lay few or no eggs, with some pairs failing to complete construction of their nest. The interval between successive eggs ranges from 4–17 days, the period increasing as the season proceeds. When the female is ready to lay, both birds will usually open the mound to expose the egg chamber. If the weather is inclement, however, the male may chase the female from the mound.

Malleefowl *Leipoa ocellata*

Day	Weight (g)	Day	Weight (g)
0	117	**113**	400
1	115	**159**	460
2	116	**258**	730
3	114	**284**	765
4	118	**332**	876
7	128	**390**	918
12	155	**434**	1012
19	178	**480**	1018
26	201	**622**	1076
48	280	**711**	1219
56	307	**799**	1272
68	330	**901**	1208
90	390	**1099**	1450

Young birds less than six weeks old often exhibit antisocial behavior including threat displays similar to those of the adult, but seldom is there any actual physical aggression. The incidence of losses as a result of cranial damage caused by young birds hitting the roof of the aviary when flying up in fright can be greatly reduced by clipping one wing. Birds of this age can be held in relatively high densities; up to nine individuals have been held in pens measuring 4 x 2 m (13 x 6.5 ft.). Beyond this age, however, densities may need to be reduced to avoid the possibility of feather picking. At about seven weeks of age, young malleefowl begin to develop the plumage pattern and coloration of the adult. This transition involves the simultaneous eruption of large numbers of feathers. In small or overcrowded pens, the emergence of these pinfeathers can stimulate the development of feather-picking habits, resulting in some individuals being harassed by the more dominant members of the group. Although reported incidences of feather picking are rare, birds of this age should be inspected regularly for early signs of feather picking, especially if the holding pens are small.

Artificial Incubation and Rearing: Not surprisingly, megapode eggs are unlike all other avian eggs. They are relatively large with a high yolk content, do not need to be rolled or turned during incubation, are thin-shelled, and can be extremely fragile. Incubation periods are generally protracted and the chicks are highly precocious, being fully feathered at hatching and able to fly and thermoregulate efficiently within hours of emerging from the nest. There is no parental care or rearing of the young. Despite their fragility, malleefowl eggs have been transported successfully in insulated boxes containing layers of foam rubber. Each layer was 150 mm (6 in.) thick and had a series of cylindrical compartments cut into it, each measuring 85 mm (3½ in.) in

Malleefowl checking tempurature of soil.

diameter by 120 mm (4¾ in) deep. These compartments, packed with goose down, each accommodated a single egg.

Average hatch rates for wild eggs brought into captivity have been about 60%. However, hatch rates have been lowest for eggs 10–22 days old when collected. This suggests a vulnerable period in the development of the embryo when being transported, at least over long distances, can be fatal. As a precaution, it is recommended that only eggs more than one month old be moved. This is best achieved by excavating nests monthly. The first time that eggs are found they should be marked and replaced. When the nest is next visited one month later, only the marked eggs should be collected. Malleefowl eggs weigh about 180 g (6.3 oz.) and, although they are large, the thin shell allows them to be candled successfully, at least in the early stages of chick development.

An incubator suitable for rearing malleefowl eggs is the Rx1, manufactured by the Lyons Electric Company in California. The incubator should be maintained at a temperature of 33–34°C (91–93°F) and a relative humidity of at least 80%, as the relative humidity within the egg chamber of malleefowl nests can exceed 90%. High humidity is critical for the successful incubation of malleefowl eggs. Low humidity not only causes dehydration of the embryo and developmental abnormalities, but it can lead to difficulties in hatching when the egg membrane dries out and becomes difficult to tear. If a chick has broken the shell but appears to be constrained by a dry membrane, the membrane should be softened by wetting it with a small amount of warm sterile water.

Each egg placed within an incubator should be held within a separate container. An open-topped, cylindrical polystyrene container, measuring 113 mm (4½ in) high x 72 mm (3 in.) internal diameter, of the type commonly used to hold a small bottle of beer, has proved ideal for this purpose. The top of the container should be fitted with a transparent but permeable covering strong enough to prevent the hatchling from escaping. The container should incorporate a drain hole or two. Fluid that spills from the ruptured egg on hatching should be collected in a Petri dish, or something similar, placed underneath the container. Eggs should always be incubated with the blunt end uppermost.

Malleefowl eggs begin incubation as soon as they are laid, consequently hatching is not synchronous. Incubation usually takes 62–64 days, but periods as long as 78 days have been recorded. The rate of weight loss in the first five weeks after laying is small, typically less than 0.3 g per day. In the week before hatching weight loss can increase to as much as 1.5–2 g per day. About this time, the embryo is generating so much heat that the egg is warm to touch. The shell of megapode eggs is relatively thin compared to those of other birds of comparable size. As calcium from the shell is absorbed by the developing embryo the shell becomes even thinner and extremely fragile, so extra care is needed when handling late-stage eggs.

There is no requirement to turn or roll megapode eggs. No air sac forms within the egg and no internal pipping takes place, so megapode chicks must switch from chorioallantoic to pulmonary respiration immediately upon breaking the eggshell. Megapode chicks have no egg-tooth, the embryo rupturing the shell by exerting pressure with its feet and back. Upon hatching, malleefowl chicks begin to breathe through their lungs, but as these are initially full of fluid, the young birds are incapable of strenuous activity. At this stage, the umbilical blood vessels have not yet constricted and, if these are damaged, injury can result in severe blood loss and death. The chicks should not be moved from the incubator or released from their containers until several hours after hatching. As a rule of thumb, each chick should remain in its container until the feathers on its head and upper body are dry and fluffy. By this time, the chorioallantoic membranes and associated blood vessels will have dried, the umbilicus should have sealed, and the lungs emptied of fluid. Wild-hatched malleefowl chicks can take up to 15 hours to dig their way out of the nest, so an extended stay in the incubator after hatching is no more restrictive than the natural process of emergence from underground. Once fully dry, the chick should be removed from the incubator and allowed to roam in an outside pen containing plenty of cover, scattered seeds and small invertebrates. In captivity, malleefowl can be susceptible to erysipelas caused by the organism *Erysipelothrix rhusiopathiae*. This disease can be prevented by vaccination with Eryvac at the rate of 1 ml per bird, delivered subcutaneously.

Ring Size: Adult birds: size 12 metal bands (internal diameter 14 mm) for most, but large adults require size 13 (internal diameter 16 mm).

Similar Species: All megapodes harness external sources of heat to incubate their buried eggs and all species within the family share many similarities. The construction of the nest and the method of incubation, however, vary considerably between species, and most megapodes exhibit highly specialized nesting behavior. The incubation mound of the malleefowl is the most complex and aberrant of any megapode. This species has developed a highly sophisticated method of incubation that, in the early part of the breeding season, relies on heat produced by decaying vegetation to warm the eggs, but becomes increasingly dependent on incident solar radiation as the season progresses.

The husbandry techniques developed for malleefowl have general applicability to other megapodes, but each species is likely to require specific incubation and captive-rearing protocols. The unique reproductive adaptations of the megapodes, particularly the absence of egg brooding during incubation, highly precocious hatchlings, and the lack of parental care of the young suggest that these species are well suited to artificial incubation and captive rearing. In practice, however, malleefowl have proved to be no less difficult to maintain or breed in captivity than other highly specialized avian species.

Francolin, Pheasants, Grouse, Turkeys, Guineafowl & Quail

04 Galliformes

04.01 Phasianidae Francolin

Glen Holland and Gary Robbins

Natural Habits: Francolins belong to the order Galliformes and the family Phasianidae. The 41 species comprise the single genus *Francolinus*. Although most species are called francolins, a few are known as spurfowl, and others are alternatively referred to as partridges. Thirty-six of the species, including several with subspecies, are found in Africa; the remaining five species are native to southern Asia. Francolins often occur in trios or in family groups which include the young of the year. They are seldom found far from thickets and long grass, where they take refuge if disturbed. They prefer to hide rather than flush, and maintain contact with raucous calls. As they are territorial, groups can be located in the same area repeatedly. Their raucous calls are repeated in quick succession in the early morning and at sunset, the call from one covey stimulating a response from all other coveys in the area.

Management: In Africa, the francolins are the equivalent of the partridges found in many other regions of the world. My experience has largely been with the crested francolin *Francolinus sephaena* and the Natal francolin *F. natalensis*. The crested francolin is an attractive species which adapts well to captivity. Sexes are alike in plumage, but only the male has spurs, and he is larger than the female. Wild caught birds are extremely nervous. For this reason, the following steps should be followed to prevent losses of newly captured birds: First, the birds must be transported in a wooden box which has a soft foam/sponge layer under the lid, and a soft nylon gauze on the inside of the wire front. This will prevent the birds from damaging their heads and beaks. They should then be introduced into a small flight with a dense thicket in the shelter. The food and water must be placed adjacent to the thicket, and some food scattered on the ground. Before being released, the flight feathers on one wing of these birds should be clipped to keep them from flying up and damaging their heads on the roof of the aviary. For the first month after introduction, the birds will hide in the thicket continually, but their confidence will soon grow and they can then be moved into their permanent aviaries.

The permanent aviary for francolins should include thickets of tall grass and clumps of dense shrubs inter-

Black Francolin
Francolinus francolinus

spersed with open sandy areas where they will sun and dust bathe as a group. Either Habitat B or C will suit most francolins. Although relatively hardy birds, in severe climates where temperatures drop to freezing for extended periods, they should be housed indoors during winter. As the birds are territorial, it is best to keep one pair or trio of a species per aviary. I have mixed crested francolins with both Natal and Coqui francolins *F. coqui* in a large planted aviary. While no interspecific aggression was noted under these circumstances, potential aggression would pose a problem if mixing was attempted in a smaller flight. Mixing similar or closely related species, such as Natal and Cape francolin *F. capensis,* would undoubtedly result in aggression. Harrap (1964) kept a pair of Natal's in a 6 m (20-ft.) square aviary with pairs of Swainsons *F. swainsonii* and crested, and found that the crested would attack and kill the Natal chicks as they emerged from the nest. It is always preferable to house only one pair of each species in pens and have suitable solid partitions between pairs, as the males can run up and down the wire constantly trying to fight with their next door neighbors. When the breeding season arrives, such males may be more interested in fending off intruders than mating with a female.

Internal parasites such as gape and threadworms are a common cause of poor production, and while the birds appear to be in good health, they could well have some underlying disease which makes them unproductive or even infertile. Worms will upset a bird's nutrition, so all birds should be wormed regularly. Fecal checks quarterly are a good method of determining internal parasite loads. Avian tuberculosis, aspergillosis, and coccidiosis are common infections where birds are kept on the same ground for many years. Vermin such as sparrows and mice should be excluded by the aviary mesh to help prevent parasites and infections such as listeria, yersinia, and salmonella. Any sign of external parasites such as fleas, mites, lice, etc., should be met with immediate treatment, as these can become a constant irritation, affecting the health of the birds. There are many suitable brands of insecticides available on the market today. All birds should be treated for internal parasites prior to the breeding season.

Stress will also have a negative impact on the birds. Birds which are confined to pens that are too small, lack sufficient light, or in which the occupants spend their day running back and forth along the partitions cannot be expected to produce fertile eggs. The presence of potential predators such as cats and dogs can prevent birds from settling down to breed, and even noisy children or constant traffic can take a toll on breeding results.

Diet: In the natural state, the bulk of the winter diet of francolins consists of bulbs, tubers, roots, seeds, berries, and grass leaves. In the summer, this diet is supplemented with insects, including beetles, locusts, and grasshoppers, as well as ticks, spiders, and small molluscs. Some species have been recorded catching prey such as frogs. Such prey is usually beaten to death before being devoured. In captivity, francolins should be fed on a 50/50 mix of gamebird pellets and mixed poultry grain with some millet. Gary Robbins recommends the following formulated gamebird diets for maintaining, breeding, and rearing francolins:

Breeding	Breeder pellets	17% protein
Non-breeding	Maintenance pellets	13% protein
First 14 days	Starter crumbles	28% protein
14 days–6 weeks	Grower crumbles	24% protein
Over 6 weeks	Grower pellets	20% protein

Francolins dig consistently and will consume any sprouting seeds and fresh grass shoots they can find. Once settled in the aviary, they soon learn to feed on a variety of foods such as softbill mixtures and fruit. They are also fond of insects, and the other method of taming these birds is, in association with a call or whistle, tossing them the odd mealworm, which they will come out of cover to fetch. They will soon learn to respond to this call for the reward.

Breeding: As a general rule, partridge and francolin have good fertility from 2–7 years of age. However, indications of sexual maturity do not automatically lead to the desired breeding results; good husbandry is all-important. The age of the parents has a marked influence on hatchability. Most francolins are sexually mature at a one year, but good results are not normally obtained until the second year in many species. The birds will be stimulated to lay by warmer weather and longer days, usually from October to January in the Southern Hemisphere and April to July in the Northern. Although francolins, as with commercially bred partridges, can be induced to lay by increasing the length of daylight in indoor enclosures, this is not recommended, as the birds produced in this manner may be of lower quality.

The crested francolin lives in family coveys of 2–5 birds. The basis of their social organization is a male/female monogamous family system, which appears to be the general pattern for a number of francolin species. The presence of a moderate degree of sexual dimorphism in plumage probably acts as a visual stimulus during courtship, with the male being attracted to the characteristic plumage of the female.

Crested are willing nesters and will usually lay the first breeding season after being introduced to the aviary. The nest is a shallow depression placed in a thicket of grass. The usual clutch is 5–8 eggs, which are cream colored. Incubation is by the female alone and lasts for 19–21 days, beginning when the last egg is laid. Incubating hens sit extremely tight, particularly during the final stages of incubation. The chicks are usually raised successfully by the adults, with low mortality in captivity, but they experience a high rate of mortality from predators in the natural state. Commercial poultry rearing mash or crumbles, or preferably, a rearing feed formulated especially for game birds, and insects such as termites should be supplied for the chicks. The chicks also enjoy diced greens and a light scattering of millet seed. The adults hold food in the tip of the beak and call the chicks to be fed. Chicks of most species begin to fly at 10 days of age. Young males show the first signs of spurs at a month of age, and by two months the chicks are independent and should be removed. Crested francolin chicks develop powers of flight at an early age and roost in vegetation off the ground. Most other species roost on the ground. Shelley's francolins *F. shelleyi,* for example, roost on the ground in tight groups in thick grass. They fly rather than walk to the roosting site to avoid being tracked to the roost by predators.

Mallet (1964) reported on breeding red-billed francolins at the Jersey Zoo, now known as the Durrell Wildlife Conservation Trust:

> "The original pair arrived in May, but no courtship behavior was observed until early the next year, when the cock would puff out his feathers and make short runs around the hen. Actual copulation was not witnessed. The first egg, about the size of a bantam's but more sharply pointed, was laid on the 26th of February. It was a dull biscuit brown in color. Eventually six eggs were laid which were placed under a silkie when it became clear the francolin hen was not going to sit. Five eggs proved fertile, hatching after an incubation period of 22 days. The young were described as closely resembling golden pheasant chicks in size and color. Initially, they were fed chick starter crumbles and mealworms, but after a few days hard-boiled egg and chick mash was added to the diet. (It should be noted that formulated gamebird diets are now available and should form the basis of the rearing diet.) The chicks were also given bread and milk, along with maw seeds and chopped greens. At 9 days of age they started to eat millet and canary seed. By 8–9 weeks of age they were fed on seeds and grower pellets. During rearing, they were kept in a moveable wire run on fresh lawn. When they began molting into adult plumage, in May, their beaks and feet turned from brown to the red coloration of the adult birds. By August the young males began to show spurs, and by September they were displaying to and attempting to tread the young females."

Erckel's Francolin *Francolinus erckelii*

Age in days	Chick #1	Chick #2
Day 1	19.8	14.2
Day 3	22.7	17.1
Day 4	20.2	17.6
Day 12	39.7	28.4
Day 20	65.2	42.5
Day 27	76.5	54.3
Day 34	113.4	88

Born on Oct. 3rd, 2005; Measured in grams

Artificial Incubation and Rearing: Most francolin species will incubate their own eggs and rear the chicks, but if double clutching is desired, or if the birds fail to incubate, the eggs can be removed for artificial incubation and rearing. When the eggs are removed, the hen will begin a second clutch almost immediately. Eggs which have been collected should be marked with a non-toxic pen such as those used to write on plastic bags, or a very soft pencil which should be held at a sideways angle while writing gently on the eggs. All eggs are best stored at about 12.71°C (55.1°F). Fluctuations in temperature can be very damaging to eggs, resulting in poor hatching percentages. The humidity should be 65–75%, and the temperature should never reach 21.1°C (70.1°F) during storage, as this is the temperature at which embryos will begin to develop. When growth begins before proper incubation has commenced, the embryo will either die or be too weak to succeed when the normal incubation period begins. Eggs in storage should be turned at least twice a day. They should be stored either on their sides or large end up. For best results, they should be set as soon as possible after they are laid. It is not always wise to wait until the full clutch has been completed. Ideally, eggs should not be stored longer than a week, but should be set in small batches so that groups of chicks of the same age hatch periodically.

The eggs should be incubated at a temperature of 37.5°C (99.7°F) with the thermometer bulb positioned 12.5 mm (½ in.) above the eggs in a forced-air incubator. In a still-air incubator, the temperature should be 38.8°C (101–102°F) just above the upper surface of the egg. The relative humidity should be 55% for either type of incubator. The total weight loss during incubation should be approximately 15%; see the section on KIWI for details on calculating weight loss. The eggs should be turned either manually or automatically through 180 degrees at least five times per day. On hatching, the chicks are placed in a brooder, as described for QUAIL. It is important to ensure that the chicks do not become too hot under the heat lamp, as we have noted that this is a cause of cannibalism. Gary Robbins recommends corrugated paper as a substrate, for two reasons. First, the chicks can gain a secure foothold on the corrugations and, second, food can be sprinkled over the floor and will not be scattered into corners as it would be on a flat surface. Other people use rolled oats very successfully as a substrate; this not only provides a good foothold but gives the chicks something to peck at. I have used paper covered with a fine layer of river sand for single clutches, but this would be impractical if a large number of birds were being reared. Chicks should not be moved often as this causes stress, which can lead to health problems.

They should be fed formulated gamebird starter crumbles and mash. One way to stimulate chicks to begin eating is to sprinkle a small amount of starter on their backs. This catches their interest and induces them to start pecking and feeding. Parent francolins usually feed their chicks from the tips of their bills, and chicks can prove stubborn, refusing to eat. There are two ways to overcome this problem. The first is to feed a few mealworms from the tip of a forceps every hour through the day. The second is to place a coturnix quail chick hatched with the francolins in the brooder to teach them to eat. For these chicks, a 50:50 mix of gamebird starter and insectile mix can be scattered on the brooder floor. Once chicks are eating well, the feed should be confined to a container. I have read that crumbles that are too finely ground will lead to cannibalism in young francolins, as this fine food collects on the chicks' nostrils and other chicks in turn begin to peck at it. I have also used breadcrumbs and boiled egg that can be prepared in a blender and frozen to avoid having to make the mixture daily. I combine one large cup of breadcrumbs and two hard-boiled eggs with the shells intact. The diet for chicks should have a 28% protein content. After about 10–14 days I reduce, and eventually phase out, the egg food, and begin to introduce some larger food; at six weeks of age, they can be transferred to the adult diet. Francolin chicks are rather slow growing, and unlike, for example, Coturnix chicks, cannot be safely moved outdoors until they are six weeks old.

A quantity of insects such as termites or small mealworms can be added to the chick diet, but are not essential if the chicks are on a formulated diet. Termites are beneficial because the acid they contain reduces bacterial activity in the other food which has been consumed by the chicks, and thus reduces the chances of illness. The sight of moving insects usually stimulates the chicks to begin feeding. The hobbyist with just a few chicks in a brooder should try putting them outside in the evening with a light bulb on for half an hour and watch their antics as they chase the small insects that soon appear. They should of course not be left unattended where the local cat may gain access. Once they are feeding well, greens such as diced spinach, chickweed, or duckweed *(Lemna)* is added to the diet. The water bowl initially should be shallow with small pebbles in the water to prevent slipping. Care must be taken to ensure that damp areas do not occur around the water bowls, as this may lead to infections such as coccidiosis. *Aspergillus* spores, coccidiosis, and *E. coli* are the most common problems encountered when rearing francolin chicks, and good hygiene is essential to success. Over-crowded brooder cages is another cause of cannibalism and stress, which will result in mortalities. Francolin chicks grow slowly and it takes patience and dedication to raise them successfully.

An alternative method of hatching and rearing francolins, described by Gary Robbins, is to place the eggs under small bantams, which will successfully raise the chicks. As with parent rearing, there are several important considerations which must be resolved before using this method. If the eggs or chicks are from a particularly rare or valuable species, one should have a good indication that the hen being used is a reliable mother that has historically performed well. Before entrusting eggs to any broody to incubate, it is wise to test the bird to make sure she is going to sit tight. This can be accomplished by placing some dummy eggs under her for a few days before committing the more valuable eggs. A nest box measuring 30

Silkie hen rearing francolin chicks.

cm (12 in.) square inside and approximately the same height should be adequate for a small broody. If the box is too large she will probably move around let the eggs roll away from her and become chilled. If the box is too small, she will be cramped and uncomfortable, which will make her restless. The front of the box can have a piece of sacking which is secured at the top.

For additional information or rearing, consult the artificial rearing sections in the chapters on GUINEAFOWL and QUAIL.

Ring Size: 8 mm for most species, but larger species such as the Cape francolin *F. capensis* take a ring size up to 11 mm.

Species requiring similar management:

Guineafowl *Numida, Guttera, Acryllium* spp.
Spurfowl *Francolinus* and *Galloperdix* spp.
Partridge *Alectoris* and *Perdix* spp.

References:

Harrap, K.S. 1964. Breeding the Natal francolin. *Avicultural Magazine* 70 (4): 146–147.

Mallet, J. 1964. Breeding the red-billed francolin. *Avicultural Magazine* 70 (2): 72–73.

Robbins, G.E.S. 1998. *Partridges and Francolins: their conservation, breeding and management*. Berkshire: World Pheasant Association, Reading.

04.01 Phasianidae Hill-Partridge *Arborophila* spp.

Myles Lamont

Introduction: The genus *Arborophila* is composed of 21 species and a number of subspecies, most of which are still under some taxonomic debate. Of these 21 species, 10 are considered to be endangered or threatened with extinction. Indeed one of the world's most endangered birds is the Sichuan hill-partridge.

The majority of the *Arborophila* inhabit sloped, montane areas, mostly avoiding areas that have been disturbed by human encroachment; in fact *A. javanica* appears to be limited strictly to old growth forest and is thereby particularly vulnerable to human encroachment. Habitat varies depending on the species in question, as this genus is widespread, from the slopes of the Himalayas to the tropical forests of Java. The genus as a whole usually is restricted to hill country, inhabiting dense forest, mainly sticking to areas with dense underbrush and bamboo thickets.

Of the genus, only about six species can be found in captivity, with the most frequently available being the common hill-partridge *A. torquela* and the chestnut-bellied *A. javanica*. Others species that can be found include:

Brown-breasted hill-partridge *Arborophila brunneopectus*
Collared hill-partridge *Arborophila gingica*
Orientalis white-faced hill-partridge *Arborophila orientalis*
Sumatrana grey-breasted hill-partridge *Arborophila orientalis*

Species of this genus were imported sporadically into North America during the mid 1900s and a few imports have occurred over the last decade, but specimens soon disappeared. At the time of writing, there are only four zoological facilities in the New World working with this genus, all keeping *A. javanica*. Fortunately, this genus has been more successful in Europe, with a fair number of birds in private and zoological collections. This particular family of Galliformes, along with many francolin and Old World quail, has never really been well established in captivity, especially in North America.

Management: Hill-partridge are fairly tolerant of other species in their aviary, so they can be used in a mix exhibit. They do well with species such as doves, finches and small Passeriformes that do not pose a threat to their eggs. Other ground birds such as pheasants, and even smaller species such as the peacock pheasants have been found to be compatible; however, closely related

Hill-partridge
Arborophila torqueola

species such as the quail and other partridge species should be avoided (Tjoen, W. 2006. Pers. comm.).

Unlike most partridge species that inhabit the dry savannah or desert plains of Africa and western Asia, most species of hill-partridge dwell in densely forested slopes from elevations of +10,000 feet, to the sea level lowlands of tropical Indonesia. Due to this, some species will fare better than others in the temperate climates of the northern US, Canada and Europe. However, these tropical dwelling species require some sort of protection during the winter months, preferably heated quarters adjacent to the breeding pens where the birds can be shifted into winter aviaries without having to be caught. Non winter-hardy species should be treated similar to the roul-roul partridge (see chapter on this species). For breeders in warmer, tropical climates, specimens can be left outside all year long, providing temperatures do not go below ~5°C (40°F).

Hill-partridge can be fairly flighty and should therefore be provided with numerous areas in which to hide. Providing sufficient cover is essential to ensure hens feel secure enough to begin egg production and remain on the nest for the entire incubation period. Aviaries should be well planted with additions of small boulders and logs, as this not only serves as a more stimulating enclosure, but also is more aesthetically pleasing to the eye. Substrate should be comparable to the forest floors of their natural environment. A 20–30 cm (8–12 in.) bottom layer of sand, with a 5 cm (2 in.) layer of clean peat covered with leaves of a hardwood species would be ideal. Ensure that the aviary is well drained and there is not any mold production on leaves during the wet seasons.

Nest construction in Galliformes is usually quite primitive; an exception to this rule is the roul-roul and hill-partridge. Due to their unique nesting behavior, the addition of appropriate nest materials needs to be provided. Small sticks and twigs, bamboo leaves, along with flexible grass species should suffice. Hay and even straw should be avoided, as once it becomes wet it is prone to mold and a potential source of aspergillosis.

Diet: The wild diet is not well known, but being recognized as an omnivorous group, most captive diets include a mixture of game bird pellets, mixed grains, insects and chopped fruit. The diet of the roul-roul is adequate for the hill-partridge species.

The recommended diet for roul-roul by Lint (1981) is:

- 5 g canary or millet seed
- 5 g choice of tomatoes, figs, dates, melons, grapes, eugenia, pyracantha, contoneaster, mulberries and species of the Rubus family
- 9 g chickweed or dandelion greens
- 3–4 mealworms
- Chick-size oyster shell

A successful diet used in Asia includes: (Tjoen, W. 2006. Pers. comm.)

- 20% protein chick starter
- Grated carrot and shredded lettuce
- Unpolished rice (de-husked)
- Crushed maize
- Mealworms and crickets

A European feed by Teurlings called Sierhoenders korrel II is recommended by some European breeders.

Breeding: The normal clutch size is usually 4–6 eggs with an incubation period of approximately 20–21 days. The genus as a whole is generally accepted to be monogamous, so pairs rather than trios are recommended. Pairs do not usually produce fertile eggs until they are two years of age. As with the bamboo partridge, they are quite noisy for their size, and if multiple pairs are kept in adjacent aviaries they will often call in unison when excited (Witlox, Dann. 2006. Pers. comm.).

During breeding season the female will begin to assemble nesting material, picking it up and throwing it behind her in a similar fashion to cranes. Once she has assembled sufficient material, the female will begin the actual nest construction. She will intertwine the nest material into a dome shape, then line the nest with softer material such as bamboo leaves (Witlox, Daan. Pers. comm.). Chicks can be left for the female to raise, as they make very good parents. Males can also be left with chicks, however, it is necessary to keep careful watch against aggression towards the chicks. The hen will lay a second clutch if the first one is pulled. If chicks are artificially reared, they require hand feeding similar to peacock pheasants (see chapter on peacock pheasants). Young quail chicks can be added into the brooder as "teachers" and this method has proven to be quite successful (Tjoen, W. 2006. Pers. comm.).

New bloodlines of the entire genus are greatly needed, especially for those of the *A. o. orientalis*. If more become available, every effort should be taken to ensure that the genetic integrity of each species is maintained. The use of studbooks, seamless bands and microchips are greatly encouraged.

Leg Band Size: 8–9 mm

References

BirdLife International (2001) *Threatened birds of Asia: the BirdLife International Red Data Book*. Cambridge, UK: BirdLife International.

Feed: www.teurlings.nl.

International Species Information System: www.isis.org.

Johnsgard, P.A. 1988. *The Quails, Partridges, and Francolins of the World*. Oxford, UK: Oxford University Press.

Lint, KC, A. Lint. 1981. *Diets for birds in captivity*. Blandford Press.

Madge, S., McGowan, P. 2002. *Pheasants, Partridges, and Grouse*. Princeton, NJ: Princeton University Press.

Nijman, V. 2003. Distribution, habitat use and conservation of the endemic Chestnut-bellied Hill Partridge (Arborophila javanica) in fragmented forests of Java, Indoneseia. *Emu* 103, 133-140.

Robbins, G.E.S. 1998. *Partridges & Francolins, their Conservation, Breeding & Management*. World Pheasant Association.

Robbins, G.E.S. 1984. *Partridges: Their breeding and management*. The Boydell Press for the World Pheasant Association.

04.01 Phasianidae Roul Roul Partridge *Rollulus roulroul*

Glen Holland

Natural Habits: The roulroul is included in the family Phasianidae of the order Galliformes. The very attractive roulroul partridge *Rollulus roulroul* hails from the tropical forest areas of Thailand, Malaysia, Sumatra and Borneo. In aviculture they are often referred to as crested wood partridge. Seldom found far from the dense thickets in which they take refuge when disturbed, roulrouls are strictly territorial. They may be found as pairs, trios, or family groups, spending their day searching for food among fallen leaves. The mean group size of 60 sightings in the wild was 2.67 ranging from one to thirteen birds. They are sexually dimorphic. The prominent color of the male bird is blue-black merging into brilliant dark green on the rump. The eye color is reddish brown. An area of naked crimson skin, shaped like a harlequin mask, surrounds the eye. The female is predominantly dark green, of a slightly lighter hue than the male, with brown wings. The harlequin mask in her case is brown, with a red ring around the eye. The eye itself is similar in color to that of the male. Females lack the crest which gives the species its alternative common name, crested wood partridge.

Management: Roulrouls should be housed as individual pairs in aviaries 1 m (3 ft.) wide by 3 m (10 ft.) long by 2 m (6½ ft.) high with a shelter at the back end. The floor in the shelter should be concrete, while the rest of the aviary bottom should be natural soil in which miniature bamboo, grass thickets, Napier grass or similar reeds, and other hardy dense plants have been planted. The miniature bamboo does not last very long in the aviary as the birds break the shoots off to form their nests. I have found that placing more than one pair of mature roulrouls in the same aviary eventually results in one pair becoming dominant and bullying the other birds until they lose condition and have to be removed from the aviary. Roulroul partridges are prone to helminthic and coccidial infections and following quarterly fecal checks for parasite eggs they should be dosed as required for worms. By occasionally offering broiler starter crumbles containing an added coccidiostat, outbreaks of coccidiosis can be controlled or prevented. Should a bird become infected, a coccidiostat should be administered in water as well. Roulroul partridges are particularly subject to stress; when stressed, they succumb very easily. For this reason I handle the chicks a great deal, in the hope of reducing the stress factor when adult. Half-grown chicks have been known to develop stress symptoms and die merely from being separated from the parent birds.

Because of their tendency to suffer from stress and their susceptibility to disease, roulroul partridges are considered difficult to rear. Nevertheless, these delight-

Male Crested Wood Partridge
Rollulus rouloul

ful little partridges become very tame in captivity and are an attractive addition to any aviary.

When roulroul chicks reach the perching stage, they often fly up and damage their heads on the wire mesh lid of the rearing cage. To prevent such injuries, I use shade cloth underneath the lid of the container and provide a branch cut from a tree, with twigs of different diameters for perches. The chicks enjoy clambering over the twigs in the rearing cage. In the aviary itself, roulrouls roost high up, just below the roof of the shelter. Originating from tropical climates, they are not comfortable at low temperatures for extended periods. A combination of cold and wet weather can result in dead birds. In severe climates, roulrouls are best housed indoors during the coldest months.

Diet: The staple diet of roulrouls in captivity consists of broiler starter crumbles, fruit, live food, and sprouted seeds. A light bulb, connected to a time switch, can be suspended in one corner of the aviary and set to come on the first four hours every night to attract insects. In areas of the world where they occur, white ants or termites can be supplied to these birds, and are relished. The folic acid found in termites and ants is thought to be very valuable to this species. Roulrouls are fond of greens, particularly cut up chickweed and clover. They should be offered mulberries, guavas and papayas in season, and apple, cut into small pieces, all year round. Laying pellets are a good food for these birds, and their diet should be supplemented with oyster shell grit during the breeding season. Live food should be considered a necessity with this species. They do find odd insects on the aviary floor, but mealworms and waxmoth larvae will work wonders in providing tame and confident birds, which are easily visible.

Breeding: During courtship, the male becomes extremely vocal and begins feeding the female from the tip of his bill. He performs a lateral display in which the wing closest to the female is stretched low to the ground as he circles her. He will repeatedly approach the female and cut her off while walking. Throughout pair formation and the subsequent bond, much play is made of the rapidly vibrated, short tail. Once the pair is formed, the male generally follows the female wherever she leads, and it is she that performs the majority of tail shaking. More typical of partridge displays, the male, standing alongside the female and parallel to her, straightens his legs so that his body is raised up. There is little accompanying change in feather posture, but a major effect of the display is to emphasize the red legs. Courtship feeding is very common and the male will pick up any tiny piece of food, especially animal food, and twitter until the female takes it from his bill. Two other displays occur which are unique to roulrouls amongst the other partridges. Males will run alongside a female and then tuck their head under the female's body and lift slightly. Males will also crouch with the head lowered and beak open while twittering. Females do not respond by offering food.

Both sexes of roulroul participate in building a dome-shaped nest of coarse grass blades and a few leaves, with a side entrance. Material is moved using a sideways throwing and building technique. One bird will enter the nest space and sit there while pulling in leaves from within reach, tucking them down the sides of its body. It will, after a time, emerge and walk away from the nest throwing leaves and twigs over its back. As this bird emerges, its partner enters the nest and begins sideways building. Males spend approximately twice as much time building as the females. The eggs are relatively large for the size of the bird, white in color, pyramidal in shape and usually laid every third day. Fertility is very good. In the Southern Hemisphere, roulrouls start laying in October and may continue to the end of December if the eggs are regularly collected. The incubation period is approximately 19 days, and managed by the female alone. If the female is allowed to incubate her own eggs, she will pull some vegetation across the front of the nest, often concealing herself from view. The male stands guard close to the nest throughout the incubation period.

Both sexes are model parents and share in rearing the young, feeding them by calling softly and then offering food from the tip of the bill. After hatching, the brood splits between the parents. If only one chick hatches from a clutch it is tended by both parents, but in larger broods about half go to each parent. Each parent restricts their care specifically to its own set of young. Incorrect diet and inbreeding can result in chicks hatching with crooked toes and lack of pigmentation in the plumage. Feeding a commercial extruded avian diet along with diced fruit, live food, and sprouted seed to the breeding pairs will result in improved hatching success, fewer problems with crooked toes, and reduced loss of pigment in the chicks. When normal-colored chicks occur in the same brood as those showing pigment losses, a marginal deficiency is indicated. The normal color for chicks is reddish brown. A medium shade of gray is symptomatic of the loss of pigment. When such chicks molt into adult plumage, the gray feathers are replaced with normal-colored feathers.

Parents are very courageous in defense of their young and will chase off any birds that approach their young. Feeding from the parent's bill is gradually reduced with the young moving to feeding themselves and scratching for food by a week old. Odd pieces of food continue to be taken from the parent's bill for up to six months. Chicks have been noted to gape for food exposing the yellow skin within the mouth but adults do not respond to this by feeding.

Brooding is solicited by a chick walking under the parent and pressing upwards with its head, back and lifted wings, exerting pressure by straightening the legs as well as wing lifting. This pressure seems to induce the brooding response (crouching) in the parent. At five days the primaries are beginning to show; by seven days they are well grown and the secondaries and wing coverts are showing. After 14 days, the youngsters are able to roost beside their parents on a low branch. At this age they can be sexed, as the females' wings are a light chestnut brown; in the male, the brown is a darker shade. By three months of age they develop full adult plumage. At this time they can safely be removed from the parents. If the father becomes aggressive toward them, they should be removed earlier, always keeping in mind that removing them too early can result in potentially fatal stress.

Incubation and Rearing: A forced-air incubator is recommended for incubating roulroul eggs. Eggs should be set only after a number have accumulated, as roulroul chicks should always be raised in groups. Single chicks seldom thrive. Newly hatched roulroul chicks have down on the outside of their legs, which is soon lost. Chicks can be reared in a fish tank or brooder with solid, not mesh, sides, and an overhead light for warmth, as described for artificial incubation and rearing of pheasants. As roulroul chicks can fly at an early age, the brooder must be covered at all times. Moving chicks from one brooder to another can cause enough stress to result in dead birds. Chicks up to the age of 14 days should be left in one type of brooder. Shallow containers containing moistened rearing food or broiler starter crumbles should be placed in the brooder. A shallow dish with water, to which a vitamin supplement has been added, must also be supplied. As the chicks like to walk through the water much as rail chicks do, some coarse grit or marbles in the water bowl will prevent drowning.

The chicks can be encouraged to eat by offering them commercial rearing food, slightly moistened, on the end of a spoon. Alternatively, small mealworms held in tweezers can be tapped against the floor to stimulate their interest. John Vermaak has described his method of rearing roulroul chicks using small mealworms coated with a multivitamin powder, termites, grated hard-boiled egg yolk mixed with whole wheat bread crumbs, and broiler starter crumbles offered ad-lib. Small live food sprinkled on the rearing food will also encourage feeding. Once the chicks have learned to eat, their growth is rapid. Young birds can also be imprinted to feed from the hand for the first five days, after which they will feed themselves. It does not work to place a young roulroul with other gamebird chicks as the roulroul will spend all of its time trying to feed from the bill of other gamebirds and will eventually fade away. Incubator-hatched young can also successfully be introduced to a family group with newly hatched young of approximately the same age.

A word of warning to anyone wishing to rear these birds: the brooder cage should never be left in direct sunlight. These dark chicks, adapted to life in the shade on the forest floor, suffer from exposure to excessive sunlight and will quickly succumb to heat stress.

Species requiring similar management:

Bamboo partridge *Bambusicola* species

Madagascar partridge *Margaroperdix madagascarensis*

Francolins *Francolinus* species

Quail species

References:

Davidson, G.W.H. 1986. Breeding Behaviour of Crested Wood Partridges in Captivity. *Avicultural Magazine* 92 (1): 18–22.

Rutgers, A. & K.A. Norris. 1970. *Encyclopaedia of Aviculture*. Vol. 1. Poole, Dorset: Blandford Press: 205-206.

Wright, A. 1989. Roul Roul Partridge. *Avizandum* Magazine. November 1989.

04.01 Phasianidae Pheasants in General

Don Bruning, Richard Olsen and Myles Lamont

Introduction: The pheasants and partridges belong to the order Galliformes, and the family Phasianidae, which comprises, under the S-A-M system of classification, 44 genera and 176 species, and is the largest and most familiar family in the order. This total includes the 17 species, in six genera, of grouse, which are treated separately. While the family is distributed from Africa across Asia down to Australia and New Zealand, the greatest diversity occurs in Southeast Asia, where the majority of species occurs. The family also includes the two species of turkeys, which, along with many of the grouse species, are the only New World representatives of the family. The family Phasianidae was formerly divided into two subfamilies, the Perdicinae and the Phasianinae. While the new classification does not recognize the validity of these divisions based on DNA evidence, the family can still be divided along traditional lines, with 105 species in 20 genera relegated to the partridge group and 51 species in 16 genera regarded as pheasants.

Pheasants are medium to large sized terrestrial birds with plump bodies, short bills, and short rounded wings. Tails can be long or short, depending upon genus. These birds are not generally good fliers, but they can burst into flight when disturbed and fly up with rapid wingbeats and then glide off to land some distance away. Throughout most of their range, man has hunted pheasants as food for thousands of years. The pheasants are probably the best known and most commonly recognized members of the order Galliformes. Man has kept and bred many species in captivity over the years. The most familiar to everyone is the domestic chicken, which is a staple in human diets around the world. The chicken was developed by early man to provide food and eggs by domestication from wild junglefowl, probably the red junglefowl *Gallus gallus*.

Pheasants have been, and still are because of their appearance, among the most popular birds kept in aviaries. The majority of species are dimorphic with males being larger, more colorful, and more aggressive. The male in most species is brightly colored and highly adorned, while the female is more camouflaged and drabber in color so that she can not be detected by predators as she incubates her eggs. The bright colors of the males are generally used to attract females and to threaten away other males. In many species, males develop elaborate plumage for display to potential mates. Perhaps the best known example of this is the long tail, or train, of the peacocks *Pavo* spp. The displays and calls of most pheasants are spectacular. The males of many species have colorful and extendible wattles or other adornments in addition to colorful plumage, also used in courtship displays or threats toward rival males. Some species have crests, horns, or ruffs that are presumably used for the same purposes as the wattles and plumage.

In spite of the facts that pheasants have been so popular and that some of the species have been kept in captivity for hundreds of years, relatively little is known about the biology of most species in the wild. Until recently, only a few studies of pheasants in their native habitat have occurred. Even pheasant biology, for the most part, is based on research done on chickens, with a small part based on introduced game birds like ring-necked pheasants *Phasianus colchicus*. The pheasants are less morphologically specialized than are the other families of Galliformes. These are terrestrial birds that vary greatly in size from the large peacocks down to the small partridges. Many species have one or more spurs on their legs that are used for defense and can be formidable weapons against each other. Pheasants have bare legs, unlike grouse whose legs are feathered to protect them from the cold and snow. Their feet are not flexible for perching, like those of Cracids, nor extremely large, for digging, like those of the Megapodes. Pheasant feet are adapted to scratching around on the litter of a forest floor looking for seeds or insects.

The short, stout wings give ample lifting power for a fast takeoff, but the birds are not adapted for sustained flight and consequently tire quickly when required to fly for any significant distance. Pheasants are relatively clumsy fliers, but are good runners; when threatened, most would rather hide or escape through the underbrush than fly. It is important to consider this when designing either an exhibit or a breeding facility for pheasants. These birds need good cover, so either plantings or other barriers should be included in a pheasant facility. Most pheasants also require night perching which gets them well off the ground and away from potential predators.

Chicks are precocial and most are able to fend for themselves within a very short time, however the female plays an important role in leading chicks to food, teaching them about predators and brooding them during the first few days. Parental care is better developed in some species, such as peacock pheasants. While many temperate species produce large clutches of eggs, more tropical species produce smaller clutches. At the same time, clutch size is reduced as parental care increases. This trend reaches its peak with the tropical Malayan peacock pheasant that provides significant parental care for several weeks and only produces one egg per clutch.

The social structure of pheasants varies greatly; some are generally solitary while others are quite gregarious seasonally. Probably the greatest success of introduction within the bird world has been the ring-necked pheasant. Originally from Asia, they now inhabit much of North America, where they were introduced for hunting. This bird has adapted to habitats ranging from eastern hardwood forests to the open prairies of central North America.

Phasianids generally roost at night high up in trees, if any are available. In open grassland habitats, roosting is usually by hiding in low brush or dense shrubbery. Pheasants in groups will generally quickly develop a peck order or dominance hierarchy, while solitary pheasants tend to defend a territory that may consist of only a display ground or a larger resource territory. Groups of pheasants usually only occur seasonally or with young birds because adult birds become very territorial during the breeding season.

In most species in the wild, the calls are confined to specific periods of time during the day and only in specific seasons of the year. While the display and courtship calls are the best known, some pheasants have a great variety of calls to communicate with each other. Man is only now beginning to document and record some of the calls of a few species. Since many of these calls are for very local communication, they are quite low and soft and thus have gone largely unnoticed. There are louder calls, however, such as that of the male argus pheasant *Argusianus argus,* which under the right conditions can be heard for several kilometers. Many species have several different calls by which parents and chicks communicate. There may be calls to alert chicks to danger, to call them to scatter and hide, to give the "all clear" signal, to gather them under the hen, and to alert them to food. There are also calls used for identification and location purposes. There are probably many other calls whose functions will some day be identified and classified. Adult calls also can cover many of these same purposes, alert/alarm, identification, threats, and many other communications, in addition to the most obvious calls of courtship and territory advertisement, nest and courtship feeding, and post-laying calls. Wing whirring, the rapid beating of wings, is another means of communication used by some species.

Management: Pheasants should be kept in enclosures with sufficient ground surface areas and artificial or naturally planted areas of cover so that they can walk, run, and hide as they prefer. Food and fresh water should be provided daily. Most pheasants can readily be transferred to commercial pellets as their food staple, however, all pheasants do not eat the same things in the wild. Efforts should, where possible, be made to provide some of the type of diet that the species naturally eats.

Pheasant hen using elevated nest site.

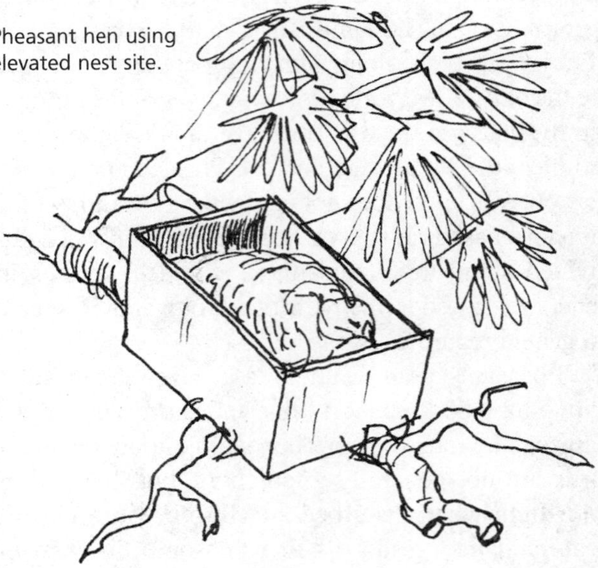

Fresh water should always be available, as well as shelter from inclement weather. While most temperate species can take substantial cold, most tropical species cannot. Therefore, the temperature requirements of each species must be understood. Many aviculturists who keep pheasants provide a sheltered area and bedding for the temperate species during rainy or winter weather. This is good, but care must be taken to avoid any bedding that carries with it high volumes of aspergillus spores. The aspergillus fungus can be lethal for any animal under stress, and in sufficient quantities can affect a seemingly healthy animal. Coccidia and avian pox also

Growth Chart for Temminck's Tragopan *Tragopan temmincki*

Date	Age (days)	Hand-reared Weight (g)	Parent-reared Weight (g)	Notes
June 20th	1	34.1		
June 22nd	3	31.4		
June 26th	7	35.7		
June 28th	9	43.98		
June 29th	10	50		
July 1st	12	58.35		
July 4th	15	74.21		
July 8th	19	99.05		
July 12th	23	118.34		
July 17th	28	148.45		
July 21st	32	185.63		
July 27th	38	205.39		
July 30th	41	230		

cause mortalities among chicks, but most modern rearing diets contain a coccidiostat, and a vaccine is available for pox.

Delacour divided the pheasants into sixteen groups by general characteristics. Even with the recent taxonomic considerations, this division of the pheasants is still a useful tool in discussing each group.

The blood pheasants *Ithaginis cruentus* are probably closer to partridges than they are to most other pheasant groups. There are thirteen or fourteen subspecies of a single species recognized, and the forms grade into each other clinically. Blood pheasants are not as dimorphic as most pheasants, and this is one of the ways in which they are more like partridges. These are high altitude birds that range from India and Nepal across the Himalayas eastward into China. As a group they are not currently threatened, although four of the subspecies are considered vulnerable, and are listed on Appendix II of CITES (Convention on International Trade in Endangered Species). For further information on these birds, see the section on BLOOD PHEASANTS written by Piet and Trijn Kreeft.

Genus *Tragopan,* the tragopans, or horned pheasants, are another Himalayan mountain group that consists of five species, with two subspecies for each of two species. The five species range from the western tragopan *T. melanocephalus,* found in a small portion of the northwestern Himalayas, Northern Pakistan, northwestern India and southwestern Tibet, to the Cabot's tragopan *T. caboti,* found in central and southeastern China. The most widespread is the Temminck's tragopan *T. temminckii,* found from northeastern Myanmar (Burma) and northern Vietnam to northern and central China, while the least widespread, the Blyth's tragopan *T. blythii,* is now found only in northeastern India and northern Myanmar. Two species, the Cabot's and the Western, are considered endangered, and the Blyth's is considered rare. The Satyr tragopan *T. satyra,* from northern India and Nepal, and Temminck's, are not immediately threatened in the wild, but are declining in numbers as habitat is slowly lost.

Tragopans are largely vegetarian, eating a great variety of plant material. They also, however, eat a large quantity of insects. They breed in the spring and produce 3–5 eggs that are frequently laid in nests made of old twigs and lined with grass, from 2–6 m (6½–20 ft.) off the ground in evergreen vegetation; however, occasionally nests are built on or under tussocks of grass. Some birds migrate to higher altitudes for breeding each spring.

Tragopans have thrived and bred well in captivity, especially in northern, cold-weather aviaries and zoos. They generally do not do well in tropical climates. There are studbooks for the rare Blyth's and the endangered Cabot's. Both of these species need to be managed carefully in captivity. At the same time, renewed

Impeyan Pheasant *Lophophorus impejanus*

Date	Age (days)	Hand-reared Weight (g)	Parent-reared Weight (g)	Notes
May 23rd	1	45.55		4.90g shell
May 24th	2	42.7		
May 25th	3	40.45		
May 28th	6	44.7		
May 31st	9	58.32		
June 3rd	12	69.2		
June 6th	15	85.12		
June 8th	17	96.05		
June 12th	21	141.65		
June 15th	24	158.2		
June 19th	28	171.55		
June 22nd	31	193.5		
June 26th	35	211.04		
June 29th	38	232.3		
July 1st	41	256.2		
July 4th	44	281.24		
July 8th	48	335.3		
July 13th	53	389.32		
July 17th	57	422.78		
July 23rd	63	490		
July 26th	66	545.2		
July 30th	70	607		

efforts are needed to secure more native habitats for all tragopans, and to learn more about their behavior and biology in the wild.

The koklas pheasant *Pucrasia macrolopha* comprises, like the blood pheasant, a monotypic genus, with the single species having nine or ten subspecies scattered in populations from Afghanistan and Pakistan across India and Nepal to China. Subspecies are found in scattered areas all over northeastern, southeastern, central, and southwestern China. These birds breed in the spring to early summer and have an incubation period of 26–27 days. Koklas are Himalayan birds, but from lower altitudes than the tragopans and blood pheasants. Koklas eat largely plant matter ranging from seeds, including acorns, to berries and buds. A great variety of insects and worms are also taken. While these birds do well in captivity, no organized studbook or management plan has been established. Very little is known about the birds in the wild, or about the effects of habitat loss and hunting on these birds.

Genus *Lophophorus,* the monals, are also Himalayan birds. There are three recognizable species. The Himalayan monal *L. impejanus,* or Impeyan pheasant, is found in eastern Afghanistan through Bhutan, northeastern India, southern Tibet, and extreme northern Burma. This species migrates to different altitudes in winter and summer. Breeding and nesting occur in the coniferous and mixed forests of the Himalayas between April and June, with the nest a scrape on the ground with 3–5 eggs. Incubation is around 27 days. Monals are great diggers. It is not suprising that their diets consist of seeds, tubers, roots, shoots, and berries. They also consume a number of insects and their larvae found while digging. All three species are listed on appendix I of CITES. The Sclater's monal *L. sclateri* is listed as rare, and very little is known about it, although a related form, probably a subspecies, has recently been discovered, while the Chinese monal *L. lhuysii* is listed as endangered and is the subject of ongoing research by Dave Rimlinger of the San Diego Zoo and his Chinese colleagues. It is hoped that enough will be learned to initiate conservation plans for this species in the wild, as well as through a captive-breeding program. Currently only a few specimens are held in captivity, but no organized management plan exists at present. Habitat loss is a major threat to many populations of these birds in the wild, and efforts are needed to provide more protected areas for them.

The genus *Gallus,* the junglefowl, is divided into four species, one of which as five subspecies. Three species are limited in range: the green *G. varius,* to Java; the Ceylon *G. lafayettei,* to Sri Lanka; and the gray *G. sonnerati,* to India. The red junglefowl *G. gallus,* whose wild range extends from India across southeastern Asia all the way down through Sumatra and Java, is also the ancestor to all of our domestic chickens. These domestic junglefowls have been the target of volumes of research on all aspects of their biology. In fact, most scientific information that we have on birds in general is based on information from domestic chickens. Junglefowl are generally forest birds, but they also come out into scrub and marginal habitats. They have a very diverse diet ranging from seeds and other plant parts to insects, invertebrates, and even small reptiles. Junglefowl lay 3–12 eggs, although for most species in the wild, the number of eggs is 3–6. None of the junglefowl are endangered, however there is a decline in the numbers and ranges of gray, green, and Ceylon junglefowls due to habitat loss. While these species do breed well in captiv-

Lady Amherst Pheasant *Chrysolophus amherstiae*

Date	Age (days)	Hand-reared Weight (g)	Parent-reared Weight (g)	Notes
May 6th	0	hatched		shell weight- 0.1g
May 7th	1	22.67		
May 8th	2	25.5		
May 9th	3	25.6		
May 10th	4	28.34		
May 17th	11	59.53		
May 20th	14	73.7		
May 28th	26	129.05		
June 8th	37	200.05		banded with 11mm
June 12th	41	173.01		
June 15th	44	261.1		
June 19th	48	287.83		
July 1st	60	351.55		
July 17th	76			
July 30th				

Himalayan Snowcock
Tetraogallus himalayensis

ity, there is no organized captive breeding program for any of the junglefowl at the present time. There has been some concern in recent years that a carefully managed group of "wild type" red junglefowl is needed. There is clearly a need for managed groups of the three non-domestic species, and a need to assure that each species will not loose its wild habitat.

The blue pheasants, genus *Lophura* are a large group of twelve species and at least thirty-two subspecies. The total range of the genus is from the western Himalayas across southeastern Asia to Sumatra and Borneo. Each species, however, is more restricted in range, with the kalij *L. leucomelanos* having nine subspecies occupying the broadest range, from the western Himalayas to the top of the Malayan peninsula. The silver pheasant *L. nycthemera,* with fifteen subspecies, ranges from southeastern China to Myanmar, and down to Vietnam, Cambodia, and Thailand. Both of these species are very common, and neither is threatened at the present time, even though there are some subspecies that may be vulnerable.

The next group in this genus includes species which are quite rare and vulnerable in the wild, with very restricted ranges. This includes the nearly extinct imperial pheasant *L. imperialis,* the Edwards's pheasant *L. edwardsi,* and the recently discovered Vietnamese pheasant *L. hatinhensis,* which may eventually prove to be a subspecies of either the imperial or Edwards's pheasant. Recent genetic work proves a long-suspected close relationship between the Edwards's pheasant and Swinhoe's pheasant *L. swinhoei,* the blue pheasant occurring on Taiwan. All of these species are vulnerable, and only the Swinhoe exists in any real numbers. The Salvadori's pheasant *L. inornata* is also vulnerable; this species is restricted to a small range in southern Sumatra. The Atjeh pheasant *L. hoogerwerfi,* from the mountains of northern Sumatra, is even less well known, and presumably, much rarer. Many authorities regard this as a subspecies of Salvadori's pheasant.

The three species of the blue pheasant group known as firebacks have a much broader range, with two of the three having a number of subspecies. The crestless fireback *L. erythropthalma* has one subspecies from the lowland portions of the Malayan peninsula and Sumatra and another distinct subspecies in Borneo. The Bornean subspecies *L. e. pyronota* is thought to be quite endangered, as is the Sumatran population of the nominate subspecies. The crested fireback *L. ignitis* is found in the same areas as the crestless, although it is much more widespread in Borneo, where two subspecies, the nominate *L. i. ignita,* known as the lesser Bornean crested fireback, and the greater Bornean crested fireback *L. i. nobilis,* are found. The mainland subspecies, Veillot's crested fireback *L. i. rufa,* ranges from the Malay Peninsula to the Isthmus of Kra and on to Sumatra, where it eventually integrates with Delacour's crested fireback *L. i. macartneyi,* a subspecies restricted to the southeastern corner of the island. The other species of fireback, the Siamese fireback *L. diardi,* ranges from Myanmar through Thailand and across Laos and Vietnam, with no recognized subspecies.

Firebacks are birds of the lowland forest, a habitat which has been greatly reduced throughout southeastern Asia. Several subspecies and populations are rare and endangered, and all others are vulnerable and declining. There is a great need for the development of protected areas within the range of each species and subspecies before they have completely disappeared from the wild.

The final species in the blue pheasant group is the Bulwer's pheasant *L. bulweri,* or wattled pheasant, from Borneo. This beautiful pheasant has seldom been col-

Red Junglefowl Cock during breeding season.

Red Junglefowl Cock.
Gallus gallus

Great Argus (*Argusianus argus*) in display.

lected from the wild, and is not well established in captivity. This is unfortunate, as the species is vulnerable due to habitat loss and hunting.

While most of these species breed in the spring, breeding is not so clearly seasonal in the more tropical birds like the Bulwer's and the firebacks. The kalij and silvers, being more temperate species, have larger clutches, normally 6–9 eggs, with reports of up to 12–15 eggs. Incubation for kalij is recorded as 20–22 days, but for silvers it can be up to 26 days. Swinhoe's and Edwards's have smaller clutches of 4–7 eggs and a more uniform incubation period of 21–22 days. The firebacks and Bulwers also have clutches of 4–7 eggs, but with an incubation period of 23–25 days. The Salvadori's pheasant lays only two eggs to a clutch, and incubation is 22 days.

Most of the species in this group have bred well in captivity when given the opportunity and several, including the rare and endangered Swinhoe's and Edwards's, are well established in zoos and private pheasantries. The crestless firebacks seem to be a bit more difficult, especially when using artificial incubation. The Bulwers, however, are currently the most puzzling because the females will lay eggs for two years and then stop egg production. Re-pairing and moving birds has stimulated a couple of females to breed again, but only for another two years. Studies of this species in the wild may be essential to solving this puzzle, and such studies are now underway.

While the diets of these 12 species vary, they are all considered omnivorous, depending on a lot of insects and invertebrates as well as roots, tubers, seeds, berries, fruit, leaves, shoots, and a variety of other plant parts. In the wild, each species probably feeds on different proportions of live animal and plant food to make up its natural diet.

The three, or four, species of the genus *Crossoptilon,* the eared pheasants, are found in Tibet and China. The white-eared pheasant *C. crossoptilon* has four or five subspecies, depending upon which authority is followed, ranging from Tibet to central China. The disputed subspecies is often recognized as a full species, Harman's eared pheasant *C. harmani.* Several subspecies of the white-eared pheasant are currently being kept in captivity, and every effort should be made to avoid hybridization. The Tibetan white-eared pheasant *C. c. drouyni,* is pure white, except for the dark tail and crown characteristic of the species. The nominate race, known as the Sichuan white-eared pheasant *C. c. crossoptilon,* is similar, but has gray wings. The Yunnan white-eared pheas-

Growth Chart for Mikado Pheasant *Syrmaticus mikado*

Date	Age (days)	Hand-reared Weight (g)	Parent-reared Weight (g)	Notes
May 6th	0	pipped		
May 7th	1	22.67		
May 8th	2	25.35		
May 9th	3	28.34		
May 10th	4	29.01		
May 17th	11	39.68		
May 20th	14	36.85		
May 28th	26	52.32		
June 8th	37	64.3		
June 12th	41	94.15		
June 15th	44	96.52		
June 19th	48	99.55		
July 1st	60	158.21		
July 17th	76	246.65		Banded with 13mm
July 30th	89	338.86		

ant *C. c. lichiangense,* which occurs south of the nominate form, is unknown in aviculture, as is the dark slaty gray Harman's eared pheasant, which is found west of the range of the Tibetan. The northernmost subspecies, Dolan's white-eared pheasant *C. c. dolani,* is pale ashy gray. The white-eared pheasant is listed as vulnerable, with one subspecies listed as endangered.

The brown-eared pheasant *C. mantchuricum* has a much more restricted and fragmented range in northeastern China, and is also listed as endangered. The blue-eared pheasant *C. auritum* has a substantial range in west central China, and is not currently endangered or threatened. Both the brown- and blue-eared pheasants are readily available in captivity. All of the eared pheasants are great diggers, feeding on bulbs, tubers, and roots, as well as on leaves, berries, and other plant parts. Insects and invertebrates make up a small part of their diet. Eared pheasants breed in the spring, from April through June. They lay large clutches of eggs, ranging 4–14 in numbers and have an incubation period of 24–28 days. All forms that have been kept in captivity have bred well, but lack an organized long-term management plan.

The monotypic cheer pheasant *Catreus wallichi* is found in the Himalayas from eastern Afghanistan and northeastern Pakistan to west central Nepal. This species occurs on steep, grassy hillsides covered with shrubs and stunted trees, often inhabiting disturbed areas. This species is considered vulnerable and endangered and has had a major reintroduction effort. This is a largely vegetarian species that feeds on roots, tubers, bulbs, seeds, leaves, and probably some earthworms and insect larvae. The cheer pheasant breeds from April through June. Nests are usually located in clumps of grass and contain 9–12 eggs. Incubation is 25–26 days. There is a great need for areas where habitat can be protected and hunting restricted if this high altitude species is to have a chance of recovery in the wild. Cheer have bred well in captivity and, as a result, thousands of eggs and chicks have been returned to the wild. However, the success of this reintroduction program has been very limited.

The genus *Syrmaticus* consists of a group of five species of long-tailed pheasants with seven additional subspecies. These birds occur in Myanmar, China, Taiwan, and Japan. The Hume's pheasant *S. humiae* has two recognized subspecies, the nominate western form inhabiting the hills of extreme eastern India and northwestern

Grey Peacock Pheasant *Polyplectron bicalcaratum*

Date	Age (days)	Hand-reared Weight (g)	Parent-reared Weight (g)	Notes
April 28th	1			
April 29th	2	25.5		0.1g shell
April 30th	3	19.84		
May 1st	4	22.67		
May 2nd	5	20.55		
May 3rd	6	19.9		
May 4th	7	19.85		
May 5th	8	19.95		Starting to eat on own
May 6th	9	22.65		Showing right wing threat behavior
May 7th	10	22.85		
May 8th	11	24.5		
May 9th	12	25.5		
May 10th	13	28.34		
May 11th	14	31.18		
May 12th	15	33.25		
May 13th	16	34.1		
May 14th	17	36.85		
May 17th	20	45.35		
May 20th	23	53.86		
May 23rd	26	63.25		
May 25th	28	64		
May 28th	31	82.25		
June 1st	35	100.3		
June 8th	42	114.15		
June 15th	49	151.35		
June 19th	53	170.05		
June 28th	62	215.85		
July 22nd	86	333.15		Moved outside

Myanmar, and the eastern race *S. h. burmanicus* inhabiting the mountains of northeastern Myanmar, southwestern Yunnan province, China, and northern Thailand. This species, along with the Elliot's pheasant *S. ellioti* of China, is considered rare and endangered, as is the mikado pheasant *S. mikado* from Taiwan. The Reeves' pheasant *S. reevesi,* from northern and central China, although more widespread, is also rare and endangered. The only species of this group that is not currently threatened is the copper pheasant *S. soemmerringi* from Japan. Three of the five recognized subspecies are fairly common in aviculture, the scintillating copper *S. s. scintillans,* the nominate Soemmerring's copper *S. s. soemmerringi,* and Ijima's copper *S. s. ijimae.*

These long-tailed pheasants are generally birds of the forest and thick undergrowth, nesting from mid-March through July. They lay clutches of 5–10 eggs with an incubation period of 24–26 days. All five species are mainly vegetarian, feeding on seeds, leaves, shoots, berries, and occasionally insect larvae. Most of these birds have done well in captivity, however more protected areas of natural habitat are needed to assure their survival.

The true pheasants comprise the genus *Phasianus,* which includes the well known ring-necked pheasant *P. colchicus,* which has been introduced as a game bird in many areas around the world, and the green pheasant *P. versicolor,* which is sometimes considered conspecific with the ring-necked pheasant. The ring-necked pheasant is represented by 30 subspecies with ranges from portions of Europe across Asia to China and Taiwan, and Korea. These subspecies form several natural groups which share certain characteristics of plumage coloration. The black-necked, or Caucasian pheasant, includes the four westernmost subspecies. The next six subspecies are referred to as the white-winged pheasants. Two subspecies are known as Kirghiz pheasants, and a single subspecies is called the Tarim pheasant. The group known as the gray-rumped pheasants includes 17 subspecies found in Mongolia, Manchuria, Korea, and throughout China, with a single representative on the island of Taiwan. The green pheasant replaces the ring-necked pheasant in Japan; three subspecies are recognized. The introduced populations throughout the world consist of hybrids of the various races of both species.

While these pheasants feed mainly on plant materials such as leaves, buds, seeds, and fruit, they also feed on invertebrates and insects. The breeding season varies throughout the range of these birds due to the wide variation of climates and habitats included in their range. Most subspecies are birds of open grassland and shrubby woodlands that are not currently threatened in the wild. Clutches of 9–14 eggs are known, with an incubation time of 23–25 days. These birds have done extremely well in captivity and huge numbers have been introduced throughout North America. They have also become established in many other areas around the world.

Elliot's Pheasant *Syrmaticus ellioti*

Date	Age (days)	Hand-reared Weight (g)	Parent-reared Weight (g)	Notes
May 5th	1	22.67	22	20
May 6th	2	23.36	25	22
May 7th	3	26.45	28	25
May 8th	4	28.54	28	28
May 9th	5	32.36	34	34
May 10th	6	36.87	37	36
May 11th	7	39.65	43	39
May 12th	8	44.56	48	45
May 13th	9	48.62	51	48
May 14th	10	52.14	53	51
May 15th	11	54.51	59	56
May 16th	12	63.26	65	62
May 17th	13	66.32	70	65
May 18th	14	72.64	76	71
May 19th	15	78.22	81	78
May 20th	16	83.55	86	84
May 21st	17	85.74	91	90
May 22nd	18	89.23	96	96
May 25th	23	93.25	99	98
June 6th	30	185.23	198	195
June 15th	41	217.05	227	223
June 23rd	49	275.63	284	282
July 1st	57	311.85	322	317

The ruffed pheasants of the genus *Chrysolophus* are extremely popular in captivity, being among the most beautiful of an exceptionally attractive family of birds. Next to the ring-necked pheasant, the golden pheasant *C. pictus* and the Lady Amherst pheasant *C. amherstiae* are probably the best known members of the family. Both are found in the mountains of western China, where they are still considered relatively common and not threatened. Both species breed well in captivity. The diet for these birds is largely plant materials such as leaves, shoots, buds, and flowers, with some insects and spiders also taken. Both species breed in early spring, from March/April until June, with clutches of 5–12 eggs and an incubation time of 22–24 days. Both species have been bred for many generations in captivity, and currently, significant emphasis is placed on color varieties, especially of the golden pheasant. There are questions as to what are natural color variations and what variations have been selectively bred in captivity. The females of the two species look similar and hybrids have been commonly produced. Female lady Amhersts have more green legs than the legs of the female golden. Pure strains need to be captively managed and protected areas established for both species in the wild.

The genus *Polyplectron,* the peacock pheasants, consists of seven species, some with subspecies, ranging from northern India through Myanmar and Thailand to Malaysia and Vietnam and on to the Philippines, Borneo, and Sumatra. All species except the gray peacock pheasant *P. bicalcaratum* are restricted in their range. Five subspecies of the gray are recognized. The northernmost is the Himalayan gray peacock pheasant *P. b. bakeri* from Sikkim, Bhutan, and northeastern India. The distribution of Lowe's gray peacock pheasant *P. b. bailyi* is unknown, but presumed to be western Assam or the eastern Himalayas. The nominate subspecies *P. b. bicalcaratum* is known as the Burmese gray peacock pheasant, but also occurs in Thailand, western Tonkin, and Laos. Ghigi's gray peacock pheasant *P. b. ghigii* inhabits central and northern Vietnam, eastern Tonkin, and parts of Laos. The final subspecies, the Hainan gray peacock pheasant *P. b. katsumatae* is restricted to the island of Hainan.

The bronze-tailed peacock pheasant *P. chalcurum* from Sumatra, with two subspecies, is listed as vulnerable. The Germain's peacock pheasant *P. germaini* from southern Vietnam, once considered a subspecies of the gray, is listed as rare. The Bornean peacock pheasant *P. schleiermacheri* and the Palawan peacock pheasant *P. emphanum* from the Philippine island of Palawan are both endangered. The Bornean was thought to be nearly extinct, but recent surveys have shown that they are rare but probably occur over a large area of Borneo. Both the Malayan peacock pheasant *P. malacense* and the Rothschild's, or mountain, peacock pheasant *P. inopinatum* are vulnerable. The latter species, while numerous

Hume's Bartailed Pheasant *Syrmaticus humiae*

Date	Age (days)	Hand-reared Weight (g)	Parent-reared Weight (g)	Notes
May 14th	1	19.84		
May 15th	2	19.12		
May 16th	3	20.47		
May 17th	4	22.67		
May 20th	7	23.88		
May 23rd	10	25.06		
May 24th	11	26.37		
May 25th	12	27.65		
May 28th	15	36.75		
June 1st	19	40.17		
June 4th	22	46.12		
June 8th	26	53.15		
June 12th	31	75.47		
June 15th	34	93.35		
June 19th	38	116.75		
June 20th	39	120.13		
June 22nd	41	130.35		
June 26th	45	148.32		
June 29th	48	164.63		
July 1st	50	182.31		
July 8th	57	197.36		
July 17th	65	224.98		
July 30th	78	264.51		

locally in the mountains of the Malayan peninsula, is clearly threatened by development in its limited range. The Malayan is often overlooked and not considered rare, but over 95% of its lowland forest habitat in Thailand and Malaysia is gone. Clearly more protected areas are needed for all these species. While these tropical birds can breed throughout the year, they probably only breed seasonally due to rains or other natural conditions. Most species produce two eggs per clutch, but the Malayan, and possibly the Bornean, produce only one. The incubation period is 21–23 days. Peacock pheasant chicks receive and require more parental care than most other pheasants. As a result, they need more intensive care when being hand reared in captivity.

International studbooks have been established for both the Malayan and mountain peacock pheasants in conjunction with the Malaysian Wildlife Department. Similar management programs are needed for several of the other species. The Palawan, which is endangered, has done extremely well in captivity, but is not being managed at the present time. Peacock pheasants feed on insects, invertebrates, seeds, leaves, buds, shoots, and a variety of other plant parts. For more detailed information on rearing techniques for these rare birds, see the section entitled PEACOCK PHEASANTS, by Richard Olsen.

The genus *Rheinardia* has but a single species, the crested argus R. ocellata, a very rare bird with two subspecies. The northern form, Rheinard's crested argus *R. o. ocellata,* found in Vietnam, is quite rare, but the zoo in Vietnam has started a captive-breeding program with a few individuals. The southern subspecies, the Malay crested argus *R. o. nigrecens,* is now found in a couple of scattered reserves in the high mountains of peninsular Malaysia. The Malaysian Wildlife Department has collected a trio of birds from which they hope to develop a captive-breeding program. These birds nest from March to June producing only one or two eggs per clutch, with an incubation time of 24–25 days. Both subspecies, being endangered, need protected areas and careful management of any captive populations.

The great argus *Argusianus argus* is found in Malaysia, Sumatra and Borneo, with three recognized subspecies. The Malayan great argus *A. a. argus* occurs in peninsular Malaysia and Thailand, Tenasserim, and on the island of Sumatra. The distinct Bornean subspecies *A. a. grayi* is found only on the island of Borneo. A third form, the double-banded great argus *A. a. bipunctatus*, is known only from a single feather of unknown origin. It has been regarded by some authorities as a separate species, or as an extinct form from Tioman Island. In addition, there are indications of more than one color phase, geographic variation, or subspecies in Borneo. These birds eat a variety of plant materials and invertebrates. They nest from March to July, producing 1–3 eggs per clutch, with an incubation time of 24–25 days. Male argus pheasants have a loud booming courtship call that attracts females to their display ground. Although not common, the great argus is bred regularly in captivity, but there is no long-term management plan for the species. Very few of the Bornean subspecies are currently in captivity.

Elliots Pheasant *Syrmaticus ellioti*

Day	Chick #1	Chick #2
1	22	20
2	25	22
3	28	25
4	28	28
5	34	34
6	37	36
7	43	39
8	48	45
9	51	48
10	53	51
11	59	56
12	65	62
13	70	65
14	76	71
15	81	78
16	86	84
17	91	90
18	96	96

By Myles Lamont

The peafowl, genus Pavo, range from India, where the Indian blue peafowl P. cristata is quite common, to Java, where the nominate, or Javanese, subspecies of the green peafowl P. muticus is vulnerable and endangered due to habitat loss. Two other subspecies of green peafowl are recognized. The Burmese green peafowl P. m. spicifer is restricted to Bangladesh and northeastern Myanmar, while the Indo-Chinese green peafowl P. m. imperator ranges over southern China and Southeast Asia from the Salween-Irrawaddy divide on the border of Myanmar and Yunnan through southeastern Myanmar and Thailand to Vietnam. These subspecies are also vulnerable and endangered.

Peafowl are omnivorous, feeding on berries, fruits, grains, and other plant materials, invertebrates, insects, reptiles including snakes, and small mammals. Breeding is dependent upon the wet season, but can occur during any month if conditions are right. Usually 3–6 eggs are produced, and incubation lasts for 28–30 days. Peafowl breed well in captivity, and many color variations of the

Indian blue peafowl have been produced by aviculturists. While the blue peafowl is common in captivity and still abundant in the wild, the green peafowl, due to its endangered status, needs to have a long term captive management plan inaugurated to keep the subspecies pure. This should be done in conjunction with the protection of some suitable habitat within its natural range.

The genus *Afropavo,* with its single species, the Congo peafowl *A. congensis,* is the only pheasant species not native to Asia or the surrounding islands. This rare pheasant is represented in captivity by a small, carefully managed population, and is covered more extensively under the separate heading CONGO PEAFOWL.

Diet: There is great variation in feeding styles and types of food preferred by most pheasants. Generally most pheasants are opportunistic and will take a variety of food items such as seeds, leaves, flowers, stems, buds, shoots, tubers, roots, and other plant parts, as well as animals ranging from insects and their larvae through other invertebrates and even small reptiles and amphibians. This also can include eggs of other birds and reptiles. Each species has its own preferences and needs. Some are entirely vegetarian, while others feed almost exclusively on animal matter. In the wild, these birds spend the majority of their time foraging. While most species forage on the ground, either scratching, digging, or picking on plants, some actually forage up in trees and bushes. Bill shape and structure can give good clues to which species are diggers and which eat more seeds or insects. The heavy and strong bills of the monals and eared pheasants, for example, clearly point to their digging habits. The captive dietary requirements are discussed more thoroughly in the chapter entitled DIET, ARTIFICIAL INCUBATION, AND REARING OF PHEASANTS, by Richard Olsen.

Breeding: The diversity of breeding styles is very extensive in phasianids. While most species are monogamous, at least within a single breeding season, nearly one-third of all species are polygynous or promiscuous. This is a much higher percentage than in birds generally. While most species of phasianids are very dimorphic a few species exhibit almost no dimorphism. As a rule, the most dimorphic birds exhibit the most elaborate courtship displays. These displays can be largely visual, like that of the Indian peafowl, or both visual and vocal, as in the argus pheasant, or largely vocal, as in the chukar *Alectoris chukar,* and most other partridges.

Clutch size also varies greatly, from the Malayan peacock pheasant with its single egg clutch to the temperate species which can lay 10–15 eggs, and on to the gray partridge *Perdix perdix,* which can produce 20 or more in each clutch. Generally, small clutch size means relatively larger eggs and more parental care. Most nest sites are on the ground, with some consisting of no more than a scrape in the ground, while others can build a more elaborate nest, with the maximum being a fully covered dome nest built by some partridge species, such as the roulroul, or crested woodpartridge, *Rollulus roulroul* of southeastern Asia. Then there are the tragopans, which build their nests well off the ground in trees and shrubs. The seasonality of breeding also varies greatly, with the more temperate species responding to day length and breeding only during spring and summer. Tropical species, while capable of breeding throughout the year, generally breed when triggered by a rainy or dry season, depending upon the species in question.

References:

Del Hoye, J., A. Elliot and J. Sargatel (eds.). 1997. *Handbook of the Birds of the World.* Vol. 4. New World Vultures to Guinea Fowl. Barcelona: Lynx Edicions.

Delacour, J. 1951. *The Pheasants of the World.* London: Country Life Limited, and New York: Charles Scribner's Sons.

McGowan, P. and P. Garson. 1995. *Pheasant Status Survey and Conservation Action Plan 1995–1999.* Gland, Switzerland: IUCN.

Diet, Artificial Incubation and Rearing

Richard Olsen

Diet: Pheasant habitats range from steamy, equatorial, sea-level jungles to cold, sparse, forbidding mountain fastnesses, even in excess of 5000 m (15,000 ft.) in the Himalayas; various species of pheasants successfully occupy virtually every type of habitat between these extremes. Surely such phenomenal diversity in habitat (and possibly resulting morphology) must pose an intractable problem for a would-be captive pheasant nutritionist. Some believe we doing our captive charges a great disservice in not even attempting to tailor-make the captive diets of individual species (or at least genera) to reflect their respective wild food resources.

Prima facie, this argument seems unassailable and, indeed, in a perfect world, wherein no scientific stone went unturned and conservation funds grew on trees, it would have to be accorded some weight when trying to properly manage the different species of pheasant in captivity. It is unfeasible because we woefully lack: the scientific knowledge of actual wild food items and their respective nutritional analyses, something absolutely crucial to such a scheme, and; the management resources requisite to the implementation of such a multifaceted plan. We don't know enough about the actual wild food choices of any pheasant, save the ringneck, to begin to try to offer individual species idiosyncratic, habitat-specific food items, or their exact nutritional equivalents, in captivity.

The primary purpose of breeding pheasants in the ecological holocaust in which we find ourselves, aside from the mere enjoyment of aviculture, is to prevent the extinction of as many species of endangered pheasant as we possibly can. And there are precious few resources with which to accomplish this task. So, however interesting nutritional speculation may be, I contend it is time to get on with the job at hand, i.e., breeding pheasants, especially endangered species, in large numbers (while maintaining studbooks so that our efforts may break with precedent and not necessarily be in vain). One thing is certain in this dispute: it is (luckily) quite unnecessary to recreate actual wild diets in captivity, as the countless pheasants, representing every genus and nearly every species that have been successfully bred in captivity have literally proven. For no one would disagree that the vast majority of these specimens was fed essentially the same diet: some form of proprietary poultry feed. Apparently, actual wild diets need not be mimicked in captivity as long as adequate sources, however "unnatural," of the amino acids, vitamins, and minerals actually essential to pheasants' health, as well as sufficient carbohydrates and fats, are somehow provided.

Wouldn't isolation in new habitats, which contributed to interspecies phenotypic differences, also, over time, alter their basic dietary needs? Not necessarily. Apparently, it is truer to say that they have "learned", or evolved, the ability to extract the dietary requirements of a pheasant from the wild resources at hand, than to say that pheasants changed their basic requirements to conform to the nutritional exigencies they came to encounter. To take a familiar example, this is much as humans did when extending their range into wildly varying habitats. Few would dispute that Eskimos and Aborigines have the same universal, human nutritional needs, however disparate the available food sources are in their respective habitats. Similarly, the changes brought about in pheasants challenged by new environmental resources tended to be more behavioral than metabolic or morphological.

The most thoroughly studied and understood animal on the planet, with the possible exception of *Homo sapiens,* is the domestic chicken. And fortunately for the phasianist, the wild progenitor of the chicken is a pheasant: the red junglefowl *Gallus gallus*. Given this connection, the truly enormous body of knowledge on chicken nutrition available to us appears to be all we have to know when formulating a diet for pheasants. The vast majority of pheasant species achieve productivity (though possibly not optimum longevity: see below) in captivity when fed nothing but proprietary feed made for chickens. The few exceptions that come to mind serve to prove the rule, so to speak: koklas *Pucrasia macrolopha* and blood pheasants *Ithaginis cruentus* appear to need more roughage for proper digestion than concentrated commercial feeds alone

Koklass Pheasant *Pucrasia macrolopha*

Date	Age (days)	Hand-reared Weight (g)	Parent-reared Weight (g)	Notes
May 5th	0	pipped		shell weight- 0.1g
May 6th	1	25.5		
May 7th	2	22.67		
May 8th	3	23		
May 9th	4	22.8		
May 10th	5	19.84		
May 17th	12	34.01		
May 20th	19	42.52		
May 23rd	22	54.85		
May 28th	27	73.01		
June 4th	37	deceased		

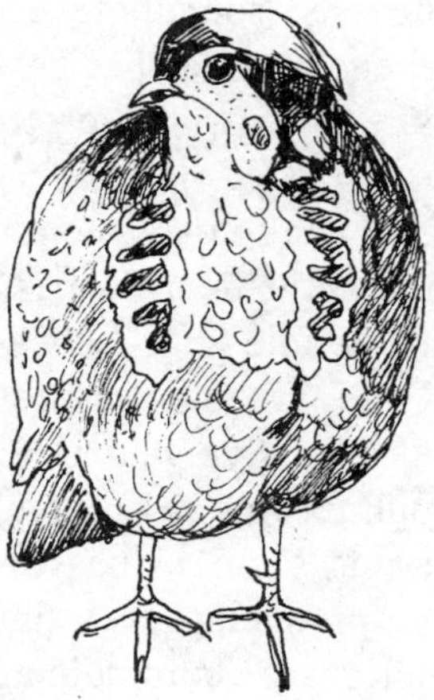

Temminck's Tragopan
Tragopan temminckii

can provide, and Bulwer's wattled pheasants *Lophura bulweri* may possibly be difficult to breed in captivity, although not to keep, due to some strange deficiency in their captive diet.

Most wild birds seek seasonally available sources of higher protein prior to and during the breeding season to aid them in spermatogenesis, ovulation and egg production. Higher protein intake may also serve as a specific stimulus or "trigger" for the onset of such functions, especially in equatorial species not governed by photoperiod. Hence, most successful phasianists feed a poultry maintenance diet outside of the breeding season, and a poultry breeder formulation from a month or two before the onset of expected laying and on through the rest of the reproductive cycle. The main difference between these two proprietary formulations (aside from greater calcium fortification in the latter) is in the percentage of crude protein. Maintenance diets tend to be around 15–18% protein, whereas breeder diets run 20–28%.

Chicken feed is based on a truly colossal corpus of scientific publications on what is, after all, basically the red junglefowl, whereas so-called pheasant formulations are derived from a vastly more modest body of data on ringnecks. Exhaustive research has shown that there exists very little documentable difference between two randomly selected pheasants from wildly disparate habitats regarding their (allegedly) respective nutrient requirements. Nutritionists, of course, would not find this conclusion surprising for, as stated above, due to common ancestry, nearly all monogastric animals, however seemingly diverse, are forged from the same mold when it comes to basic nutritive requirements. These bagged, proprietary concentrates come in both pelleted and so-called "crumble" form, and while which type one uses is a matter of personal preference, one must always ascertain which form a new bird is used to eating. This is because pheasants are very neophobic when it comes to food and, as they rely primarily on sight, not taste or smell, when selecting food items, do not accept new-looking foods quickly.

As a commitment of "the good life" provided by the unique nutritional completeness of proprietary poultry feeds, and especially as this renders foraging activity redundant, obesity can become a problem with many species of pheasant if these concentrates are fed *ad libitum*. Obesity is not usually obvious enough to even be suspected by the breeder, but there is a growing body of anecdotal evidence that excess fat reduces sexual activity in males and egg production in females, and promotes premature mortality. This phenomenon is best documented in tragopans, but it would appear that many species, although probably not all, can easily, given the confines and overabundance of captivity, become too fat to reproduce normally or realize their genetic potential for longevity.

It has been my experience, as well as that of some taxidermists (who know what the interiors of many species of captive pheasants look like), that obesity rarely, if ever, becomes a serious problem with captive tropical pheasants. Only temperate species seem to put on excessive fat and, although this has not been scientifically proven by experiment, evolutionary logic also lends credence to the observation that there seems to be a northern/southern dichotomy in this regard. Northern birds make hay while the sun shines, as it were. When food is abundant, they have evolved the ability to store it in the form of body fat to fortify them against the privation and cold of the coming winter. This is true of most northern mammals and non-migratory birds, so why not pheasants? And by spring they are lean and

Brooder with rounded corners to prevent group suffocation.

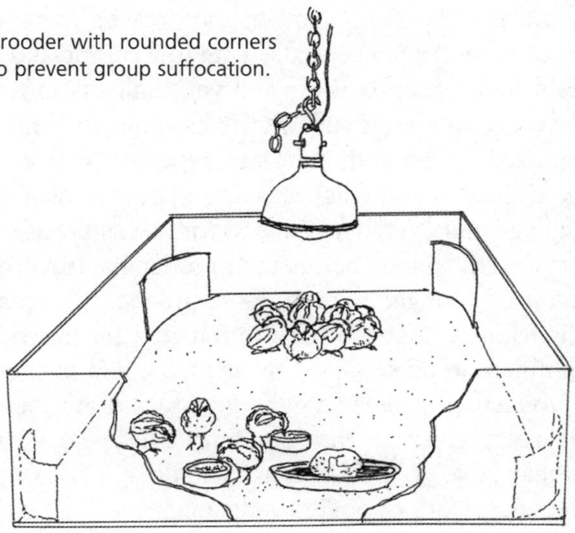

eager to breed again, quite unlike the sedentary "perch potatoes" we often inadvertently produce in captivity.

Some temperate zone species should not be given unlimited access to poultry feed. In my opinion, roughly one-half to two-thirds as much they would normally eat, if left to their own devices, should be offered. This may seem cruel, but will actually result in healthier, more productive birds. You can kill them more easily with uninformed kindness than with a more benignly spartan feeding regime.

This is not to say that temperate zone pheasants must live a hungry (not to mention humdrum) existence in captivity, subsisting entirely on what they perceive to be an inadequate supply of boring proprietary concentrate. Although poultry feeds, due to their nutritional completeness, must constitute the essential core of their diet, the daily intake of all captive pheasants can and should (if you can afford the management time and costs) be supplemented with "treats" such as chopped greens (kale is best), diced fruits and vegetables, mealworms, etc. These relished items seem to add atavistic vigor to our captives' often otherwise rather pedestrian step, presumably by rendering them less bored with their daily routine. Furthermore, there is scientific evidence that naturally occurring, "organic" vitamins (especially vitamin E) are slightly more efficacious than their largely synthesized counterparts found in most proprietary feeds and vitamin supplements. And perhaps most importantly, the feeding of "treats" serves to tame the individuals with which you are working. In my opinion, pheasants that are afraid of their keepers are often not relaxed enough to breed normally.

As I alluded earlier, although pheasants may achieve maximum production on poultry formulations, these same feeds may not be conducive to longevity. The one problem with nearly all poultry feeds is that they are too "hot." So another good reason to supplement commercial concentrates with low caloric and low protein foods such as fruits and vegetables is to lower the average intake of such nutrients. Poultry feeds are formulated for birds that are not expected to live very long. Chickens and ringnecks are marketed at a very young age and even those spared for use as breeders or layers are culled long before their productive life is over in favor of younger, slightly more productive, replacements. Hence, their diets are formulated for maximum weight gain in those raised for the table and phenomenal production in breeders and layers, with no thought whatsoever given to the long-term effect such "hot" food may have on their health. "Middle-age" is not part of the vocabulary of poultry nutritionists.

If one wants one's charges to realize their genetic potential in terms of longevity, something especially important with rare and/or endangered species, one has to tone down or dilute the energy and protein with which commercial diets are fortified. The main thing to suffer from having to break down so many complex carbohydrates, fats and proteins is the liver. Just as a force-fed goose will produce pate de foie gras, a pheasant too may develop an enlarged and ultimately dysfunctional liver on feeds designed exclusively for birds not expected to live long enough for this to become a problem. Though many pheasants seem to possess "cast-iron" livers, thriving and producing well into their twenties, others die "mysteriously" in middle age, with premortem blood work simply showing elevated liver enzymes and postmortem histology reporting nonspecific hepatic granulation, or something of the sort. These are probably cases of gradual burnout on overly generous helpings of poultry concentrates.

If we are ever to put the breeding husbandry of endangered pheasants on a scientific footing, we must try to strike a balance between optimal annual production and maximum longevity. The "Golden Mean" may achieve neither of these extremes, but might ultimately compensate for the possibility of slightly less production per year by production over a longer period of time. The only proprietary feed known to me which attempts to address this issue is a line of feeds called Mazuri that is primarily formulated for zoos by the American company Purina Mills. Mazuri brand formulations are largely concerned to prolong the life span of zoo exhibits and may go slightly overboard in this regard, in that yearly productivity could somewhat suffer. But Mazuri feeds, which are high in roughage and are relatively lower than most commercial feeds in energy and protein, can easily be supplemented in the breeding season with high protein items such as trout chow, insect larvae, etc., which, due to the relative brevity of this annual assault, probably won't create acute obesity or permanently compromise liver function. But much long-term research would be required (albeit probably never forthcoming) to determine how to most scientifically strike the exact balance sought.

Artificial Incubation: The most common question about this topic is: Why do it? Can't the female pheasant hatch her own eggs and rear her young? Indeed, the best incubator and brooder is a female pheasant. Though some females do bungle the job in captivity, most do splendidly. The problem is, most will only do it once a year with but one clutch of eggs. If you remove the eggs for artificial incubation as the female lays them, pheasants will usually lay multiple clutches and the breeder will realize many more chicks in a year than would be possible by natural means. Nevertheless, with common species, from which maximum production may not be the goal, letting the female do what comes naturally can be the best choice. To maximize the number of offspring, however, one must assume the role of mother oneself. Fortunately, this is not particularly difficult with pheasants as compared with, say, altricial birds. Most serious problems with incubation and/or rearing arise from inbreeding depression, which manifests itself in weak embryos and/or chicks which fail to thrive. This is a major problem with several species in captivity and the reason studbooks have become so important.

Provided you have unrelated stock, pheasant eggs can easily be hatched, with a hatch rate approaching 90% to be expected. You don't need fancy or expensive incubators (albeit such are more reliable and easier to control); you just need to follow two simple guidelines. If you carelessly stray outside of the parameters outlined below, you will have very little success, whereas if you don't, failure will be a rather rare disappointment.

First, pheasant eggs should be incubated within 1/3°C (1°F) of 37.5°C (99.5°F). Any decent machine can hold the temperature within this range, but they must be carefully monitored with accurate thermometers. Each species of pheasant has an exact incubation period (e.g. 28 days for tragopans), and, although some books list a range (say 27–29 days for this species), if conditions are right, 90% will hatch on the precise day that species is supposed to hatch. If your chicks hatch later (temperature too low) or sooner (too high) than this, you can bet the temperature is off; and although some will still hatch, your hatch rate or percentage will plummet.

The second, and more frequently bungled, factor that must be controlled more or less exactly to realize a good hatch rate is relative ambient humidity. Measuring the exact humidity in your incubator is notoriously difficult because hygrometers (whether of the old-fashioned "wet bulb" or the more modern hair or electronic variety) tend to be erratic and undependable. I gave up on trying to accurately measure actual humidity levels when I realized that all one really needs to know is the rate at which your fluid-filled eggs are being desiccated or losing fluids. If they are drying out too fast, raise the humidity, and if they are retaining too much moisture, lower the humidity. But to accurately determine this, knowing the actual, numerical humidity level is irrelevant. If the temperature is correct, nearly all pheasant eggs from non-inbred parents will hatch if they lose, via transpiration through the pores of the shell, 16% of their original laid-weight over the entire incubation period. For example if an egg weighs 40 grams (1.4 oz.) when put into the incubator, and should hatch in, say, 22 days, then it should lose a total of 6.4 grams (16%) in those 22 days or, since the progression is linear, 0.29 grams per day. If it loses less than 0.29 grams per day (GPD), your incubator is too damp, whereas if it loses more than 0.29 GPD, the air in the incubator is too dry. It's as simple as that.

Fluid loss, per se, cannot be directly measured (outside of a laboratory setting), but the egg's daily weight loss is a direct function of fluid loss. One needs a relatively accurate gram scale (+/–0.1 gram, at the very least) to measure this. Employing a technique called candling, many breeders still indirectly measure fluid loss not via weight loss but rather through air gain. In candling, one peers directly into the egg by means of a strong light and judges whether the air cell, which grows as a direct function of fluid loss, is growing too fast or too slowly. Some veteran breeders can probably rival a gram scale in determining fluid loss by this method. But if you don't have twenty years of candling experience under your belt, weighing eggs takes all of the guesswork out of determining just how much humidity a given egg requires. Once you determine this, however, you are on the horns of an insoluble dilemma if you have more than one species in your incubator. Do you raise the humidity to suit one species or lower it to cater to another, or compromise on a humidity ideal for none? And suppose you have several species to worry about? In this case, worry is about all you can do. You'll simply have to resign yourself to a suboptimal hatch. If this is unacceptable to you because you are working with rare and/or endangered species, you'll have to do as I do and set up three incubators: one with no added humidity, one with high humidity, and one with a kind of average humidity suitable for most eggs.

With this array of choices, one can tailor make the environment not only to the requirements of each species, but of each egg, as well (for eggs idiosyncratically vary as to shell thickness and porosity within the same species, and even within the same clutch, for that matter). And although none of the three incubators may have perfect humidity settings for a given egg, nature's neat, linear progression need not be reproduced artificially — it is only the end result, the 16% weight loss overall that counts — so one can switch eggs from one environment to another, losing too much here, not enough there, until the proper overall weight loss is achieved in the end. If one must err when it comes to humidity, always err on the dry side. I believe more pheasant embryos are lost because of too damp an environment than from any other single cause. When the humidity is too high, most chicks will drown in the egg's own fluids when respiration commences during the last 72 hours of incubation. It is notoriously unrewarding to attempt to revive dead embryos! Whereas eggs that end up too dry often merely imprison the chick in its own dried out membranes, preventing normal hatching, but at least allowing the aviculturist an opportunity to carefully extricate the chick manually — and alive. Such rescue operations are a very tricky business, however, and far beyond the scope of this piece.

Ideally, although this is unnecessary if incubation conditions have been perfect, eggs should be in very high (near saturation point) ambient humidity at the time it is seen by candling that the chick has entered the air cell and is ready to pip (i.e., peck a hole in the shell). This is to ensure that the outside air that now enters the egg doesn't dry out the membranes surrounding the chick and imprison it. For this you will need yet another incubator used exclusively as a hatcher, unless all of your eggs are due hatch on the same day. I keep the newly hatched chicks in the hatcher for 24 hours before transferring them to a brooder. Chicks are still absorbing nutrients and fluids from the yolk during this period and require no external source of food or water for some 48 hours, although I try to induce them to eat and drink starting at about 36 hours, which gives me time to detect problem consumers, and try to intervene before it is too late.

Brooding and Rearing: The artificial brooding and rearing of pheasants is quite straightforward and not at all the daunting task it is with many genera. With the exception of peacock pheasants (see the chapter on these), pheasant chicks are so precocious as to take most of the work and guesswork out of these procedures. There are but a few simple rules to follow when brooding pheasant chicks, but follow them you must. Temperature is the most crucial variable in the health and survival of the chick. For the first week after hatching, all pheasant species must have access to a source of radiant heat which produces 36.5°C (98°F) in some part of the brooder space. It is neither necessary nor desirable to have the entire brooder at this temperature, though the room in which the brooders are kept should not be allowed to fall below 20°C (70°F) or so. Although they must be watched carefully for the first week to see that they utilize the heat source, chicks quickly learn to seek heat when they need it and make forays away from the source to eat, drink, and explore when they don't. This closely mimics their behavior in the wild, wherein the female only broods them intermittently, as needed. I use a simple 60 watt, yellow light bulb as the source of heat but, of course, a wide variety of options is available. The bulb hangs above the chicks' heads such that the temperature on the brooder floor immediately below it is 36.5°C (98°F). Light bulbs, needless to say, produce rather intense light 24 hours per day. Although this is indeed quite unnatural, I have detected no ill effects from this continuous photoperiod. Some breeders, rightly or wrongly, disagree with me on this point and accordingly employ heat sources that do not emit light.

It is a rule of thumb that the temperature near the heat source can be lowered 3°C (5°F) per week. To accomplish this, I have the light bulbs on rheostats and place a thermometer on the brooder floor beneath the light. By this rule of thumb, one can gradually lower the brooder temperature and judge when the chicks are feathered and thermoregulated enough to thrive without added heat; first of all in the ambient temperature of the brooder room itself and then, eventually, in the night and day fluctuations they must endure outdoors. Common sense must, however, always prevail. There is no substitute for observation: the behavior of the chicks will tell you if they are too hot (holding wings out, gaping, shunning heat source) or too cold (cheeping, seldom leaving heat source). And if you must err, always err on the side of caution — the hot side. As long as the temperature in the entire brooder does not rise above 38°C (100°F), being hyperthermic is merely uncomfortable for the chick. Chilling causes deadly immunosuppression and the accumulation of fluid in the lungs, leading to pneumonia and other fatal illnesses.

There are nearly always as many schools of thought on the proper feeding of pheasant chicks, as there are phasianists. One thing is clear, however: the very simplest regimen is consonant with success. Most pheasants chicks can be fed nothing but proprietary chick crumbles and thrive. Exceptions to this rule are blood pheasants and koklass which, for proper digestive function, seem to require greater bulk or roughage in their diet than commercial concentrates can provide. Also, peacock pheasants need hand-fed live food during their first week of life (see the chapter on this unique genus). However, within limits, the more slowly a pheasant grows, the better (except in commercial production of ringnecks).

In the wild, chicks do not generally have access to unlimited protein and furthermore, get plenty of exercise. I never feed chick crumbles of greater than 20% protein *ad libitum* to any species.

Even having said that commercial concentrate is all you really need to feed pheasant chicks, I believe there are rather compelling, if not scientifically conclusive, arguments for offering greater dietary variety than crumbles alone can provide.

1) Overly rapid growth is always a risk with pheasants and rather low protein and carbohydrate foods such as diced vegetables and greens serve to lower the average intake of nutrients.

2) Although a threat to all species, feeding excess protein is particularly risky with tragopans and large, long-legged genera such as argus and peafowl. Such birds tend to develop severe, often fatal, leg deformities if they are allowed to grow too fast. Until they are nearly fully developed, I offer the above taxa no more than 16% protein commercial feed, further "watered down" with greens and other plant food.

3) Most pheasants, as most birds, are rather neophobic as adults when it comes to newly introduced food items. As feeding habits are established in infancy, any "treats" you wish to offer your birds later should be offered young.

4) There are said to be elusive, subtle, ineffable virtues to naturally occurring vitamins and minerals, which cannot be duplicated artificially, and there is even some scientific evidence for this. So, even though proprietary chick crumbles are indeed complete, nutritionally balanced diets, their processed, often synthesized, constituent ingredients might possibly be bettered by their organic counterparts.

5) I believe that tame birds are synonymous with relaxed birds; birds, therefore, more likely to breed in captivity than continuously stressed, anxious specimens. Stress induced secretions of glucocorticoids have been shown to suppress reproduction in many animals. Pheasants seem to have a natural fear of humans, yet the feeding of "treats" can utterly inure them to their keepers if started young enough.

6) The wild, though no picnic, is an endlessly complex and fascinating environment for something so naturally precocious as a pheasant chick. Hence, chicks consigned to the sterile, simplistic, numbingly boring environs of a typical brooder box literally need to "get a life." In my experience, any real-life-like, stimulating features you can introduce into the otherwise entirely stultifying brooder environment, such as greater variety in food items, will help somewhat to reduce infant mortality.

How much time, energy and resources can be expended in the service of any of the above, probably pseudo-scientific, suggestions will vary with individual circumstances (and I don't always practice what I preach here either). If all you can manage is feeding chick crumbles, while minding the protein level, you will be successful, too (albeit possibly not quite as). Brooder units are box-like structures easily built at home by the breeder. For the first two weeks or so of life, suitable dimensions are, for example, 0.6 x 0.3 x 0.9 m (24 x 12 x 36 in.). The height is to discourage chicks from jumping out, often to their death. The bottom should be a sliding tray for ease of waste disposal. Above this tray should be wire mesh (so called "hardware cloth") with openings up to 1.2 cm (½ in.) on a side. Protruding down from the top, near one corner, is a light bulb controlled by a rheostat. Although there are many variations possible on this theme, the above are all the salient features of a brooder.

Some aver that chicks raised on wire, up and away from their feces and other disease vectors, become akin to hothouse plants. Such an unnatural substrate, it is said, prevents them from developing antibodies by way of ubiquitous, minor environmental challenges and that hence, they easily fall pray to common pathogens and parasites when finally exposed to the rigors of the real world. My experience supports a different logic here. Young chicks must be protected from environmental, pathogenic challenges until their immune systems are developed enough to adequately respond to them. Wire mesh, just like hygiene, helps serve this purpose, and I have personally never had a disease problem when exposing young birds raised on wire to the greater challenges the outdoors presents to them. In extremely wet and/or humid climates, however, where pathogens and vectors of all types especially thrive, moving chicks from wire to the real world could pose a problem.

For the first week or so, when feces are in any case relatively sterile, I spread paper towels over the whole of the wire floor and sprinkle crumble everywhere. This tends to enhance the speed at which chicks start eating, for virtually everywhere they instinctively peck, they hit food. Offering small mealworms at this time also gets them eating heartily straightway. Once they come to recognize the crumbles as food, I put the food in a small container and remove the towels. You must observe and be sure that your chicks are eating within 72 hours of hatching or they will soon be in real trouble.

I once read the words of wisdom of an aging phasianist, summing up the total of what he had learned as "Chicks can't swim!" Pheasant chicks have an uncanny ability to drown themselves in all but the very shallowest of water. This is truly a case where Murphy's Law rules; if anything can go wrong, it will. Every precaution must be taken not to offer chicks an expanse of water sufficient to drown themselves in. The water used must be bacteria free (and this doesn't mean water straight from your well).

So-called cannibalism is a big problem with some species, and can be a problem with most, although "cannibalism" is a bit of a misnomer, in that chicks never try to actually eat each other. The problem takes two forms:

1) Actual aggressive attacks with intent to drive away or harm. Introducing new chicks into the territory of others usually causes this, especially if there are age and/or size discrepancies.

2) Idle pecking which "unintentionally" results in injury due to repetition. Once a bloody wound is apparent, the victim must be isolated, as the sight of blood will stimulate serious attacks.

As none of this ever occurs among parent-raised chicks, I believe that the close confinement and unchallenging nature of the brooder environment are the causes of this odious practice. Since the brooder environment is a necessary evil, there are no easy solutions to this problem. Nothing works better than acute observation and intervention based on the diligent application of common sense. Suffice to say you will always need more brooders than you think you will.

Rearing: As chicks lose their down and feather out into juveniles, their need for supplemental heat and tiny chick crumbles is gradually reduced over a period of several weeks. Exactly when chicks are ready to leave the closely controlled environment of the brooder room varies between species. The flexible application of the previously explained "rule of thumb" regarding heat reduction will serve as a guide. When moving juveniles out of doors, it is best to construct a kind of "halfway house" with a heat source, e.g. a heat lamp, in the shelter area of an outdoor run to ensure a smooth transition to the vicissitudes of outdoor weather. Juveniles simply thrust outside to fend for themselves experience a higher mortality rate than those shown a bit more coddling. The heat source is gradually turned down until it is no longer needed, a period of perhaps two weeks, depending on weather. Adult maintenance pellets (15–18% protein) should be gradually mixed with their familiar chick crumbles until they are accepted. Also if well water is used for your adult birds, it too should be gradually added to the bacteria-free water you have been using in the brooder room.

There is a very low mortality rate among pheasants once they have survived for 10 weeks or so. Prior to this, however, pathogenic bacteria, even in the hopefully high hygienic brooder environment, can and do sometimes overwhelm a chick's fledgling immune system. Even so, a water soluble, broad-spectrum antibiotic usually works wonders in short order. Of course there is no substitute for prevention through hygiene. Chicks have very underdeveloped immune systems and just as with human children, often succumb to the slightest pathogenic challenge. The importance of hygiene is underscored when chicks die without presenting symptoms. Sudden infant death syndrome (SIDS) occurs among baby birds, too, and can only be prevented, not cured! In general, however, pheasants are excellent candidates for artificial brooding and rearing, and an 80% or more survival rate is to be sought. If you can't achieve this with some species or the offspring of particular pairs, but you can with others, begin to suspect vertically transmitted disease, i.e., directly from the usually asymptomatic female into the egg.

Blood Pheasant *Ithaginis cruentus*

• Piet & Trijn Kreeft

Introduction: The blood pheasant *Ithaginus cruentus* belongs to the order Galliformes and the family Phasianidae, which includes the partridge, francolin, quail, tragopans, pheasants and guineafowl. Fourteen subspecies of the blood pheasant are widely distributed through an area extending from Nepal to northwestern China. They are found in high country along the snow line. Outside of the breeding season, they live in large groups of 10–20 birds, which separate in spring to form pairs. Occasionally they form trios of one male and two females. Even less commonly, two males are found with a single female.

Management: Blood pheasants are friendly birds that become very tame in captivity. They tolerate small songbirds in their aviary, but are aggressive to other fowl-like birds. They can withstand cold weather, but at extreme temperatures, –20°C (–4°F) or lower, they require shelter and protection from the wind. Their aviary should consist of two sections. The first section should be about 10 m² (108 sq. ft.) with a corrugated roof that is closed on three sides. In this section a perch should be installed at a height of about 1.5 m (5 ft.) below which is placed a wooden box covered by mesh to collect all feces. The box should be constructed so that it can be easily cleaned. The food tray should also be placed in the sheltered section over a container and grid to prevent the birds from feeding on spilled or contaminated food. The aviary floor should be covered in river sand. In spring, a nestbox measuring 30 x 30 x 10 cm (12 x 12 x 4 in.) should be suspended in the aviary approximately 1.5 m (5 ft.) from the ground. The second section of the aviary, about 3 m (10 ft.) wide and 10 m (33 ft.) long, is exposed to the elements. The substrate should consist of a layer of river sand covering the natural earth floor. This outside portion of the aviary should be planted with scattered grass tussocks and shrubs of gorse *Ulex europaeus,* elder *Sambucus* spp., and pine *Pinus* spp. The outdoor and indoor sections should be connected by a sliding door, enabling the keeper to lock the birds in the shelter at night or during inclement weather.

Blood pheasants are reasonably hardy after they become well established, and are compatible with many other species. We have kept and bred these birds for seven years, during which we have experienced several disease problems. Internal parasites, including coccidia, must be treated. Young blood pheasants are prone to gastric problems, particularly in warm weather when they drink a lot, or if their green food is too wet. Such problems can be difficult to combat. Antibiotics have not helped, but introducing the proper intestinal flora by mixing fresh feces from the adult birds into their drinking water often alleviates the problem. Females laying in cold weather are subject to eggbinding. We strongly believe that preventive medications should not be given to these birds. In the past, breeders have resorted too often to such measures in an effort to prevent diseases or conditions to which the birds otherwise might have built up a natural resistance or immunity. We recommend that birds only be treated when actually sick. The presence of internal parasites can be detected by regular fecal checks for the parsite eggs.

Diet: Blood pheasants are fed a formulated pellet *ad lib.* They are also given barley and chopped alfalfa *ad lib.* Each evening they are fed a small handful of millet. Once a week a fresh pine branch is placed in the aviary. No vitamins or other additives are mixed into the water as these birds have an exceptionally sensitive gastric system that is easily upset by such additives. No fruit is allowed, but plenty of roughage is required. Blood pheasants are easily upset by sudden changes in diet. Young birds should not be fed large quantities of egg white.

Blood Pheasant
Ithaginis cruentus

Breeding: In courtship, the male approaches the female with his tail spread out and his crest puffed up. He produces a hissing sound and then whirls around the female for a few minutes. Both males and females emit the loud territorial calls, which may be heard all year round. They also have a short, loud alarm call. Blood pheasants lay 3–6 buff eggs with scattered brown blotches. The removal of the first clutch will sometimes result in a second clutch, increasing production to10–12 eggs per pair, per season. One-year-old females generally only lay 2–3 eggs. Females usually lay in the nest box, but sometimes a hen will dig a small scrape in the corner of the aviary where she will lay. The birds are sensitive to disturbance, and will desert a nest site if eggs are removed. Once a nest has been deserted, it is not unusual for a female to drop her eggs from a perch. It is therefore advisable to allow the female to complete the clutch before removing the eggs for artificial incubation and rearing.

A still-air incubator is used for artificial incubation. The incubation period is 28–29 days. The temperature is set at 37.8°C (100°F), measured at the top surface of the egg. A low relative humidity of 40–45% is required. The eggs are turned three times per day, by hand, over the point of the egg.

Two days prior to hatching, when the chick's beak has penetrated the air cell, the egg is placed in a hatcher and no longer turned. Once the chick has pipped externally, the egg is positioned with the pip facing upward. Newly hatched chicks are left in the hatcher for 24 hours. They are then placed in a brooder box measuring 50 x 90 x 40 cm (20 x 36 x 16 in.), with a thermostatically controlled heating element placed overhead. The

Malaysian Peacock Pheasant (*Polyplectron malacense*) with one chick.

temperature at the chicks' head height is maintained at 37°C (98.6°F) for the first week and then gradually reduced by 3°C (5.4°F) per week.

The bottom of the brooder is initially lined with soft paper towel on which some turkey starter is sprinkled. A section of clear plastic, measuring 4 x 10 cm (6 x 4 in.) is glued onto the front of the brooder box. This is sprayed three times per day with a syringe to wet it. The rearing food is then sprinkled onto this and the chicks will peck at it. As soon as the chicks are feeding well, the paper towel is removed, leaving them standing on a base of 1.0 cm (0.4 in.) square weldmesh. The food is then offered in a shallow container, and fine oyster shell grit is added to the diet. The water is supplied in a drinking fountain, and must be changed twice per day. In the second week, finely cut green food consisting of grass, alfalfa, and nettle is also given. The chicks are not difficult to rear, but they are inclined to pick at each other. To avoid this problem, and to keep the chicks busy, fronds of fresh pine are placed in the brooder from the second week. The chicks will spend much of their time picking at these. Fresh fronds are offered daily. In addition to the rearing mash, turkey starter crumbles are also offered in the second week.

After about five weeks, when they are fully covered in feathers, the chicks are moved outdoors into a completely covered brooder pen measuring 1 x 3 m (3 x 10 ft.). This is placed about 60 cm (2 ft.) off the ground and fitted with a floor of 1.5 cm (0.6 in.) square weldmesh that is covered with branches. A box containing river sand is then placed in the pen, as the chicks love to sand bathe. At about two weeks of age the males begin to show gray wing covert and chest feathers and the females begin to show brown plumage. At about 14–16 weeks the young are gradually weaned onto the adult diet.

Ring Size: 9 mm

Species Requiring Similar Management:

Tragopans
Monals

Peacock Pheasants *Polyplectron* spp.

- Richard Olsen

Introduction: Peacock pheasants belong to the order Galliformes and the family Phasianidae. The genus *Polyplectron* comprises a spectacular group of pheasants that most modern taxonomists divide into seven species: the Malay peacock pheasant *Polyplectron malacense,* the Bornean *P. schleiermacheri,* the Palawan *P. emphanum,* the Sumatran bronze-tailed *P. chalcurum,* the Rothschild's mountain *P. inopinatum,* the Germain's *P. germaini,* and the gray *P. bicalcaratum,* this last species being by far the most common in captivity. All are relatively small, secretive denizens of the dense, dipterocarpaceous forest floor, with species ranging from the foothills of the Himalayas to Borneo. They are thought to be primarily insectivorous in the wild, although fruit, seeds and other plant products are opportunistically taken. Although their proper behavior and breeding biology in the wild remains little known, it can safely be inferred from their pronounced sexual dimorphism that no species is likely to be monogamous.

Management: While all species of peacock pheasant except the gray are of tropical origin, they can withstand temperatures down to freezing or slightly below with no apparent ill effects. Nonetheless, I believe that a fully enclosed shelter area is required for their well being in all but the warmest climates, and supplemental heat in most. This shelter should contain the only perches and be no smaller than 4 m^2 (40 sq. ft.) or so.

Peacock pheasants can, in general, be successfully and humanely housed and bred in smaller aviaries than are required by larger pheasants, the minimum being perhaps 2 x 3 m (6 x 10 ft.). However, this does not always prove wise in specific cases, as the males of some species have a tendency in captivity to be aggressive toward females. In species prone to mayhem, I prefer to give the hen as much cover and space as possible to avoid the cock: never less than 14 m^2 (150 sq. ft.). This precaution is especially well advised in the Sumatran bronze-tailed, the Malay and, presumably, in the closely related Bornean, which is currently unavailable to aviculture. Should actual aggression be observed, or the hen seen to be perching unduly, the safest strategy is to separate the pair in contiguous pens until such time as the hen lays her first clutch. Warring pairs can normally be safely reunited 10 days or so after the first clutch is sacrificed. However, overtly aggressive males should be separated again immediately after the next clutch is laid, and allowed subsequent conjugal visits commencing a week or so before each expected clutch. Although cautious management is the hallmark of the successful peacock pheasant breeder, this should not be taken to mean that this is a particularly difficult genus with which to work. I have experienced potentially lethal aggression in only 20–30% of potential pairs, even in the three species mentioned above. In any case, proper management, as outlined, should prevent actual mortality, for when the hen is about to ovulate, she is receptive to the cock's advances.

Despite a largely undeserved reputation for being delicate in captivity, peacock pheasants are actually no more prone to disease than are other pheasants and, therefore, mortality among adults is low. Indeed, individuals have been known to live and even reproduce for more than 20 years! However, juvenile mortality is somewhat elevated in these species; they are slow to develop and should be brooded more carefully, and some 25% longer, than other pheasants. What premature mortality does occur in mature individuals is generally due either to aspergillosis or mechanical injury, especially when working with wild-caught specimens.

Diet: A wild diet need not be reproduced in captivity, except nutritionally. Peacock pheasants appear to thrive on a proprietary poultry pellet or crumble of about 16% crude protein, with greater protein only being offered in the breeding season. At this time, I offer a pheasant breeder pellet of about 28% protein as the staple, with supplementary trout chow (Purina brand: 45% protein), mealworms (+/– 60% protein) and a vitamin /mineral mixture. Diced fruits and vegetables are always taken if offered, and raw peanuts are a relished treat that help to keep the birds tame. Obesity in captivity seems to be an ever-present problem with this genus.

Breeding: Peacock pheasants should generally be kept exclusively in pairs, for even though they almost certainly have a non-monogamous mating system in the wild, running more than one female with a male proves very difficult in captivity due to female intra-gender aggression. Occasionally, females that have been reared together will tolerate each other when mature, even in the presence of a male. But normally, if there is a shortage of males, they must be ushered from pen to pen to service each hen serially. Though time consuming, this can be easily accomplished and rarely poses problems.

The Malay and Bornean are unique among Galliformes in laying only one egg to a clutch, while the other peacock pheasants lay two. Hens can be expected to lay in their first or, more normally, second year and produce 3–6 clutches, or more, per season. Fertility is rarely a problem with this genus, unless the male is less than two years of age.

Congo Peafowl
Afropavo congensis

Artificial Incubation & Rearing: Peacock pheasant eggs are not difficult to artificially incubate and, if the parents are not inbred, a 90% or more hatch rate should be achieved. All species require a temperature of 37.5°C (99.5°F) and most, an ambient relative humidity of roughly 50%. The exceptions to this are the Malay and, presumably, the closely related Bornean, which generally need higher ambient humidity to hatch. Regardless of an individual species' humidity requirements, however, all eggs will hatch at a high rate if their weight loss is monitored so that they lose 16% of their initial-laid weight over the entire incubation period. The period of incubation is 22 days for all but the Palawan, which is unique among pheasants in hatching in a mere 18 days.

In the wild, the peacock pheasants' breeding strategy consists of the intensive rearing of a very small clutch, rather than, as with most pheasants, the more haphazard care of a larger brood, and this poses problems for the aviculturist. Although the chicks are fully precocious, peacock pheasants are singular in having no instinct to pick-up food from the ground for the first several days to a week following hatching. Left to their own devices, then, most individuals will starve to death, even in the presence of suitable food, during this period. They do, however, have a strong instinct to snap at food held in their mother's beak at this time. In nature, this presumably guarantees good initial nutrition, as well as teaches them which of the myriad potential wild food items to actually eat.

In captivity, offering the chicks mealworms or the like from a beak must simulate this parental rearing strategy. I crush the head of the worm with the tweezers and offer the hatchlings small numbers at two-hour intervals, twelve hours per day, for the first week. Most individuals will be picking up chick crumble and mealworms from the floor of the brooder before the end of this period. To avoid some of the issues related to the concerns over nutritional value of mealworms, young crickets can be alternated with mealworms and offer a far more nutritious meal.

Congo Peafowl *Afropavo congensis*

• Glen Holland & Myles Lamont

Natural Habits: The Congo peafowl *Afropavo congensis,* which is placed in the family Phasianidae in the order Galliformes, is the only pheasant native to Africa. Known only from the Ituri Forest in Zaire, the species has a limited distribution. Like many pheasants, this species is sexually dimorphic. The male is dark, with areas of glossy black, violet blue, and green. It has a crest of velvety black feathers and a tuft of white bristles, but lacks the tail train of the Asian peafowls. The attractively patterned female is largely rufous brown, with much green over the back and wings. Pairs perform a series of loud duets, which are initiated by the male and carry on for some time. When the moon is full they call late into the night. They are naturally found in pairs or small groups, and inhabit humid lowland forests with a dense undergrowth.

Management: In their natural habitat, Congo peafowl dwell on the forest floor and avoid bright sunshine. Their aviary should therefore be well planted with clumps of dense shrubs and a few taller trees. The aviary floor should be covered with a layer of leaf litter and a few scattered rotting logs. Habitat A, as described under the LANDSCAPE chapter, will provide an ideal aviary environment for Congo peafowl. Sufficient roosting branches must be available at various levels. The aviary should measure at least 10 x 5 x 2.5 m (33 x 16 x 8 ft.) high. They are rather inactive during the day, being more active at dawn and dusk, particularly during the breeding season. Post-mortem examinations have shown that capillariosis, coccidiosis, aspergillosis, and malaria have been the most common causes of death in this species in captivity.

Diet: Congo peafowl take a similar diet to other pheasants. A basic mixture of laying pellets, crushed maize, wheat, barley, and oats should be supplemented with a separate mix of fruit, vegetables, and soft food. In addition, live food such as mealworms and crickets should be offered regularly.

Breeding: Congo peafowl are a difficult species to breed. Of the initial 32 eggs laid at Antwerp and Rotterdam Zoos, only four hatched, and only two survived longer than a month. More recently, however, successes have been more frequent, and the studbook for this species indicates over 100 birds in captivity in 2005.

In courtship, both birds sit face to face on a branch while bowing head, neck, and back, with the tail spread at a 45° angle. Males also display on the ground. The first display involves the tail erect as with a peacock; in the second display, he flattens his body and wings laterally. No nest is made and the eggs are laid on the flat surface or slight hollow of a branch, generally higher than 1.5 m (5 ft.) from the ground. In captivity, a flat basket has been accepted as a nest site. The usual clutch consists of 2–3 eggs, rufous brown to cream in color, laid at intervals of 1–3 days. Incubation is carried out by the female alone for 27 days. She sits continuously on the nest, pressed flat, motionless, with her eyes closed, leaving only for a short period at dusk to eat and drink. While the female incubates, the male stays in the vicinity of the nest, sometimes even roosting on a perch next to the female. When the chicks hatch, they remain in the nest for a day or two before tumbling to the ground where, initially, the male takes care of them. Chicks are kept hidden in the most secluded parts of the aviary, where both parents care for them. They either offer the chicks food from their bills, or the food is dropped in front of the chicks. A rearing food and small insects should be supplied for this purpose. By the second evening, the young scramble onto a low branch where they are brooded by the female during the night.

After a week the chicks are able to fly, and at a month of age they roost next to the female, who no longer broods them. The male is strictly territorial and will attack anybody who threatens the brood. The female does most of the brooding of the chicks and the male supplies most of the food. By two months of age the parents loose interest in the chicks. At this stage the sexes can already be determined. Since a second clutch is often laid, it is recommended that the first brood be removed as soon as they are independent. They are able to breed at a year, but are fully mature by 18 months. Considering the conservation status of this species, it should not be mixed with other species which may disturb breeding attempts.

Species Requiring Similar Management:

Currasow *Cracidae*
Wood partridge *Odontophoridae*
Peafowl
Monal

References:

Verheyen, W.N. 1964. The Congo peacock Afropavo congensis at Antwerp Zoo. *International Zoo Yearbook* 4: 87–9.

Verheyen,W.N. 1965. Breeding the Congo peacock *Afropavo congensis* at Antwerp Zoo. *International Zoo Yearbook* 5: 127–128.

Grouse

• David Hancock

Introduction: Grouse are not well represented in either institutional or private collections, although generally speaking some of the species breed relatively easily in captivity; the difficulty is keeping them alive to breeding age.

The largest number of captive grouse have been either wild-caught or captive-bred for restoration efforts on the part of government agencies. Generally these people have not had experience dealing with live specimens, determining habitat suitability, or minimizing the dispersal activities of the release stock (Toepfer and Anderson, 1990). Whether captive propagation is being undertaken as a hobby, part of an interpretive zoological program directly to increase the reserve of an endangered species genetic reservoir, or ultimately to enhance the species status in the wild, the following summary should serve as a useful introduction to the topic.

Review of Species in Captivity: In spite of the IUCN CBSG policy that encourages government agencies in all countries to advocate the capture of wild birds so that they be made available to experienced aviculturists, many countries have discouraged acquisition of breeding stock. This is the prime reason that little progress has been achieved with the tetraonids. Many species are annually hunted by the thousands but no birds are available for avicultural research, an unfortunate dichotomy.

Of the 16 species of grouse listed by Johnsgard (1983) nearly half are not regularly represented in avicultural collections. Yet four species or subspecies are suffering from very reduced habitat, threatening their existence. Many other local or regional populations of other species have been, or are threatened with, severe habitat alteration. There is a great need to improve grouse avicultural techniques and begin to establish genetically meaningful captive reserves. It is ironic that in North America we find the world's rarest grouse species. The further irony is that the world's last extinct grouse, the heath hen, is also a close relative of the now endangered Attwater's prairie chicken.

Of the Eurasian species, the black-breasted hazel hen *Bonasa sewerzowi,* Caucasian black grouse *Tetrao mlokosiewiczi,* sharpwinged grouse *Dendragapus falcipennis,* and the black-billed capercaillie *Tetrao parvirostris* have only occasionally been represented in captivity and little progress has been achieved on their aviculture. In North America little avicultural progress has been achieved with the sage grouse *Centrocercus urophasianus* or white-tailed ptarmigan *Lagopus leucurus,* and beyond some birds being held for scientific experimentation, both are probably only represented in the author's collection at this time. The circumpolar rock ptarmigan *Lagopus mutus* is more commonly represented in European than North American collections (Aschenbrenner 1944 in press).

The capercaillie *Tetrao urogallus* is represented in over 200 aviaries in Norway and Sweden. This is due to recent government sponsorship of breeders to rear captive birds for reintroduction into the wild. Also on the positive side, the next most widely represented captive-bred species in Europe and America is the North American ruffed grouse *Sonasa umbelius* with perhaps 60–75 breeders keeping them. The Eurasian counterpart, the hazel hen *B. bonasia,* is becoming popular although probably less than a third of breeders presently house a pair.

Following the ruffed grouse in popularity is the North American dancing grouse, the greater prairie chicken *Tympanuchus cupido* and the sharptailed grouse *T. phasianellus*. These favorites are also probably represented in approximately 30–40 collections at this time. The blue grouse *D. obscurus* spruce grouse *D. canadensis* and black grouse *T. tetrix,* are probably present in 20–30 collections at any one time. The willow ptarmigan *L. lagopus,* including its UK race the red grouse *L. 1. scoticus,* is probably in fewer than 15 collections. Excluding the recent and probably temporary Scandinavian experience with capercaillie, probably no more than 150–200 breeders worldwide presently have a pair of grouse.

Following from the Norwegian capercaillie project mentioned earlier there is a pilot commercial operation to sell processed capercaillie meat to the restaurant market that has had some limited success. However, the major market for most grouse breeding projects is producing the breeder stock for aviculturists, scientific research, or for reintroduction efforts. Some black grouse have been bred for shooting preserves in Czechoslovakia; red grouse and capercaillie are presently undergoing captive breeding evaluation for restocking possibilities in the UK; and in North America the southern race of the pinnated grouse, the Attwater prairie chicken *T.c. attwateri* is presently undergoing exhaustive field research by the USF&WS and some initial captive studies in a belated effort to save this highly endangered species.

Some current captive and field research has been initiated by both Chinese and Russian biologists to preserve the geographically restricted black-breasted hazel hen and the black-billed capercaillie, but these projects are greatly hampered by the lack of funds and a firm commitment on the part of key government agencies.

Large scale, specialized grouse farming is approaching a reality but cannot take that final leap forward until vaccines are developed against common poultry diseases or immunity is developed in the breeding stock. The author's experience in rearing over 550 grouse two seasons ago, only to have over 300 die in a few days due to a clostridium outbreak from contaminated commercial feed is indicative of the "surprise" problems (Grouse News #5 June 1993).

At the time of writing, there are currently no European species of grouse or ptarmigan in the Western hemisphere save one.

Source and Purpose of Birds: The source and purpose of the birds should play an important role in the choice of breeding stock. For example, if the objective is to release birds into the wild in a stocking or a reintroduction program, then after the problems of determining the desirability of the potential habitat, one must consider the appropriate subspecies for this particular area. It is the obligation of aviculturists to keep the sub specific lines pure. One must also consider whether the birds' availability is compatible with the desired time for a release. For example, is the stock's breeding cycle at a time of minimal dispersion or migration.

For a lekking species you might start with introducing males at a suitable or historic lek so that they can immediately "identify" with the area. Wild trapping a subpopulation or releasing a group of pen-reared and already dancing males could help establish the new lek by reducing "searching" dispersal. One or more females might be housed in the immediate vicinity to encourage the males to stay and display. Alternatively, if the season only permits that females and their broods are available, then they could be reared on site and then released into the adjacent habitat directly or through pens that initially let the chicks come and go, but retain the adults. Or several wild broods might be translocated.

The point is that the birds must not only be released into suitable year-round habitat, but that the birds must be conditioned to finding food and shelter in these habitats and be released at a time to minimize dispersal tendencies. You want them to settle down to breeding and rearing young as quickly as possible.

Areas rich in predators should be avoided, or a heavy trap and relocation program should be initiated prior to release to avoid a high mortality rate amongst released birds.

Obviously cleanliness is a major key to grouse culture. The problem is that grouse generally live in remote areas devoid of humans and their farm animals. These isolated grouse populations have simply not had the evolutionary opportunity to develop resistance to the many organisms and parasites of domesticated poultry. It is interesting to note that the species most frequently kept successfully are those with the historically closest association with the farm, the ruffed, sharptailed, and black grouse. It is likely that they have developed, through repeated exposure, the greatest resistance to disease and parasites.

Pens: Aviculturists achieve the necessary cleanliness by brooding, rearing and wintering the birds on wire-bottomed pens. Indeed this is by far the easiest method, with the least risk of disease transmission, although to me, distasteful, as it seems to inhibit some of the natural behaviors that I so enjoy seeing. A common compromise is to winter the birds on wire and then, and only then, to put the birds briefly into vegetated and ground breeding pens just prior to egg laying. Since many species will lay eggs in the wire-bottomed pens they obviously lay in these "short season" breeding ground pens. Cleaning under these wire bottomed pens can be done weekly, monthly, yearly or never according to one very successful breeder. I have found, in our wetter coastal climate, that spilled food even one meter below the pen is a dangerous aspergillus culture bed.

A slight compromise, but one that still requires daily cleaning, is to put down a .25–.35 m (9.8–13.8 in.) layer of sand on which the birds walk. Shrubs or trees grow up through the sand but it is very easily cleaned of the daily droppings. The sand permits effective drainage to reduce dampness, and it inhibits earthworms and various insects that carry pathogens and parasites. Combining wire-floored brooding and winter pens with sand-floored breeding pens is double insurance. Allowing natural sunlight into the pens is another method of sterilization due to ultraviolet light.

Some grouse breeders very effectively over-winter birds in individual wire pens 0.5 x 1.0 x 0.5 m (1.6 x 3.3 x 1.6 ft.) high. Pairs are often housed in adjacent pens to maintain pair bonding or familiarity. The bottoms of these elevated pens can be either heavy plastic or wire. I prefer the 2.5 cm (1 in.) square plastic coated weld wire for strength and reduced aggravation on the birds' feet. After weld wire, galvanized wire is satisfactory for most species, however ptarmigan seem to be subject to foot problems more so than others. Prairie chickens temporarily held on concrete have shown to have problems with bumble foot.

The pen size seems less important to the birds than to the breeder in most cases. The classic wire-bottomed breeding pens pioneered by New York State in the 1930s and 1940s and used in rearing thousands of ruffed grouse were only 1.8 x 2.4 m (6 x 8 ft) and about 0.8 m high, and were elevated about 0.5 m (1.6 ft.) off the ground (Bump et al. 1947). My basic sand bottomed breeding pens are 2.4 x 8.5 x 3.6 m high (8 x 28 x 12 ft. high). The sides have a 1.0 m (3.3 ft.) high wooden barrier between pens (about 0.2 m (7.8 m) is buried in sand) with all internal side and roof meshing above this of 1.0–2.5 cm (0.4–1.0 in.) nylon. The top is roofed over for a distance of 3.0 m (9.8 ft.) to provide the only protection from the elements. The feed and water containers go under this roof.

All my brooder, breeding, and wintering pens contain many tree limbs for perches. Since the basic pens are nearly 4 m (13 ft.) high, the top level of perches is well above the keeper's head and gives the birds a great deal of security. The perches are about 0.3–0.6 m (11.8–23.6 in.) below the top netting so that owls, etc. cannot reach perching birds. The sharp-tails and white-tailed ptarmigan nightly seek the highest perches. The sage grouse is the only grouse species to not use the perches extensively. The many perches allow exercise and physical separation of birds as well as removing the birds from direct contact with the ground. The frequent climbing and flying to perches also strengthens the chicks' and adults' legs and wings, and improves general fitness.

Grouse, along with many galliformes will eat and kill most or some pen vegetation. The various species of cedar and juniper seem most resistant to being eaten but sharptails will ring even these species. If elderberries are given a chance to get started they can survive and provide great beauty, excellent cover, and fine edible berries. It is essential to ring plant stems and trunks with wire or nylon mesh to ensure their survival. When branches become strong enough to support the birds' weight they will perch on the branches and strip all leaves.

Because I wish to enjoy and study the birds behavior I keep all my birds full-winged. In the heavy direct flying species (capercaillie, sage, and black grouse), to prevent damage and loss from unexpected "spooking" and the birds flying into posts and roofs, I have stretched polynetting below the inside roofs or solid ceilings. l have also placed drop nettings 2.0 m (6.5 ft.) in front of side walls when the pen is more than 10 m (32.8 ft.) long to absorb the shock of fast-flying birds. The drop netting hangs from below the upper layer of perches near the ceiling to 1.0 m (3.3 ft.) off the ground and cushions the throwback effect of a bird fast flying into a wall mesh. A simpler technique is to clip the primaries on one wing of the strong flyers or aggressive males.

Outer perimeter pen security is maintained by burying heavy fox wire 0.5 m (1.6 ft.) deep, providing a 1.2 m (3.9 ft.) lower solid wooden barrier with a double upper mesh with steel wire on the outside and nylon mesh on the inside of the vertical supports. Just above ground level at the top of the wooden barrier we run an electrified wire. We also run llamas in a peripheral enclosure to intimidate or scare off coyotes, foxes, and dogs. The internal pen divisions are all nylon mesh that does little damage to birds. All pen doors open into alleyways to give a double door security.

Lek Pens: For the capercaillie, which have potentially very aggressive males that are substantially larger than females it is easy to incorporate selective "bolt" holes that permit the smaller hens access to the males' quarters when they choose, but enable them to seek privacy in the larger laying pen, which may house 5–50 hens. Each laying pen may be serviced by one or more adjoining single male pens of 6–10 m^2 (19.7–32.8 sq. ft.) each. Pens for all lekking species can be 6 x 6 m (19.7 x 19.7 ft.) to several times this dimension. Thirty square meters will adequately house 1:3 (male to female ratio designation) capercaillie. Other lekking species, like the three prairie grouse or the black grouse, can be housed from 1:1 to 1:10 sex ratios in similarly sized and larger enclosures.

With the prairie grouse I also prefer to start the breeding season with several males in each pen. The dominant- male quickly takes command and lesser males either move out of the way (sage) or assume the non-aggressive female position to avoid conflict (prairie chickens and sharp-tails). If a severe beating occurs I then remove either the alpha male if it is very aggressive, or the beta male(s) and give the alpha male first options for the season. If there is only one male you have a more limited genetic distribution. However, if you do not need to know the exact parentage then to increase the gene mixing after a week you may exchange males. This can be repeated again another week later. This weekly shift of males also insures that less fertility will be lost if one male is ineffective, or disliked by the hens.

In other species — blue grouse, ruffed grouse, and black — two or more females can effectively run with each cock. One exception is the spruce grouse, and particularly the Franklin race *D. C. Franklinii* where some females will kill each other at the onset of breeding. In ruffed grouse, running 1:2 or even 1:4 it is preferable to having several pens, even quite small, but accessible through bolt holes. This just offers more privacy and escape. The additional pens also seem to reduce or maybe spread out the males' attention and helps to eliminate hens getting killed by aggressive males. The problem of over-aggression in the males is partly overcome with extra space, but particularly by offering escape routes, cover, and many perches where the females can seek solitude.

While our high pens give security and escape to females, they can cause a problem if insufficient ground cover is not provided for the nesting hen. She may be too exposed and vulnerable to attack when on the ground, and may seek high perches from which to lay eggs. Placement of soft styrofoam pads below the perch sites can be effective, but provision of more ground cover is best. The trick is to apply this brush or wood cover late in the season to prevent sheltered, dark, damp aspergillus culture sites.

Most of the ptarmigan and hazel grouse seem to prefer imposed monogamous pairing but, as with the spruce grouse or other species, by housing females in adjacent pens you can (daily or every two days) easily move a single male to the different pens to service two females. Increasing the female ratio has the downside of reducing the genetic diversity but allows greater breeding density within limited pen space, or the utilization of otherwise unpaired hens.

At nesting season I always dig out two or three 8–10 cm (3.1–3.9 in.) scrapes per pen to soften the soil under bushes or lean-to supports. Again it is important that the females can nest with minimal disturbance from the males, so we introduce "salvaged" Christmas trees (we get 125 –150 free trees the day after Christmas) at the beginning of April for additional nesting cover. Placement of a 0.7 m-long (2.3 ft.) wooden lean-to against a wall forms both an escape "hide" and a nest site.

In the event that a breeding male is lost and his hens show "squatting" behavior, it is possible to bring the hens to another displaying male and hold them while the male mounts and copulation occurs. This has been done for a few species including ruffed, capercaillie and blue grouse.

Incubation and Rearing: Most species of commonly reared grouse will incubate and rear their own young. However, if maximizing the production is important then you would pull all eggs of the first clutch(es) and leave the hen to sit the final set. The problem I find with letting the hen sit is that because we keep the pens so clean there is little natural food, bugs, or greens available for the chicks, and in the larger enclosures little of the supplemented food is found per unit of ground covered by the chicks. This can be overcome by confining the hen and brood to smaller pens or parts of pens where the feed trays are more apparent.

The incubation of grouse eggs is generally straight forward following standard poultry criteria: The eggs should lose 15% weight over the incubation period (this is generally attained by 50–53% relative humidity); the temperature is maintained at 37°C (99.5°F) in forced-air incubators; frequent egg turning is preferred (every 15 minutes is better than every 60 minutes), and turning is stopped three days prior to hatching. Broody bantams are excellent for incubation, but due to the possibility of poultry acting as disease vectors, it is recommended that eggs are pulled prior to hatching and placed in a hatcher.

Capercaillie eggs have traditionally been difficult to hatch in either tilting or rolling tray incubators, and placement under banty hens for a minimum of 10 days is recommended. My own intensive plotting of egg weight loss, with the required prolonged daily open incubator, has resulted in both delayed and reduced hatching success. So be careful about too much data collection! Generally speaking, grouse eggs have high fertility and hatchability with minimal attention if the temperature and humidity are properly maintained and if the male is in good breeding condition.

Hatching is facilitated by moving the eggs to a separate maximum humidity hatcher 2–3 days before pipping and hatching respectively. As soon as several chicks hatch — but not too frequently as you do not want to open the hatcher more than necessary and reduce the humidity — the wet chicks are moved to another dry hatcher to facilitate drying for 16–24 hours. From here the chicks are moved to the first-stage brooder. Since these brooders have concentrated heat at one end, the chicks quickly learn to select the temperature of choice by distance from the lamps. This same principle of "self regulated" temperature control is applied in the second and third stage brooders.

If you experience a delayed hatch from lower incubation temperature, and if the drawdown of fluid has been too great or the relative humidity of the hatcher is too high and there is a delay from pipping to full hatch, it is sometimes possible to assist the chick out with forceps. It is important not to damage the peripheral blood vessels inside the shell so as to prevent bleeding. If there is any danger to the chick of contamination from the surrounding hatcher and brooder, and particularly the newly exposed umbilical cord, I suggest dabbing the area with iodine on a swab. This area is very vulnerable to infection entry. We also swab the toe punched wound if this has been necessary for identification purposed to plot blood lines.

I use a three-stage brooder program. The first stage brooder 0.3 x 1.0 x 0.3 m [11.8 x 39.3 x 11.8 in.] high with two ordinary lamps at one end is used from hatching to up to 8–10 days of age. The floor of the brooder is covered with paper towelling that is changed at each feeding. The second stage brooder is also portable but kept indoors, has a wire bottom and is 0.5 x 1.25 x 0.3 m (1.6 x 4 x .98 ft.) high and is used up to 4–6 weeks of age. The final, third stage brooders also double as wintering pens, and are 1.0 x 5.0 x 3.0 m (3.3 x 1.6 x 9.8 ft.) high. Half the roof is enclosed. These are wire-bottomed, outdoors, and have the option of heat lamps if the weather is cold. In all brooders the heat source is ordinary light bulbs on rheostats to control heat. The 40–60 watt bulbs are dipped into blue paint to over half height so the brightness is toned down. This means that the birds have light 24 hours per day, as with many northern grouse species.

Feeding: Considerable research has been done an grouse nutrition in the wild and most recently in captivity (Aschenbrenner in Press 1994). This latter work suggests a great breakthrough in grouse culture. Except for newly hatched chicks, which eat a high protein diet based on insects, the non-migratory grouse have evolved an efficient year-round use of abundant local vegetation that is generally low in protein (often only 4–8 percent) and high in fibre. Aviculturists, often following the lead of the poultry industry, have generally been led into providing too nutritious a diet in too great abundance.

The challenge facing grouse breeders is getting back to either a natural diet or one offering suitable levels of the protein/fibre rations, the subsequent correct digestive tract pH, and foods that are nutritious, while keeping the birds physiologically and psychologically fit. Aschenbrenner (pers. comm.) recently wrote that using a totally natural diet the past couple of years has enabled him to rear capercaillie chicks literally without loss. On a natural diet focusing on conifers and dandelions he also has capercaillie hens producing up to 50–57 eggs per season. He attributes this incredible survival improvement largely to the lower protein levels, with their subsequent lower gut pH levels, which inhibits pathogenic organisms. Can we derive commercial feeds that accomplish this? This would certainly seem a valid avenue of research.

Most grouse breeders use as much natural green vegetation (conifers, willow, dandelions, to name but a few) as possible and supplement this with available fruits. The bulk diet, usually available in unlimited amounts, is generally commercially available chicken or game bird cereal pellets: maintenance ration (12–16% protein) in winter, breeder ration (18–20%) in spring, and starter crumbles (28–30%) for 1–8 day old chicks, and grower pellets (20–24%) for growing chicks. The lower protein levels are probably still too high for all but the tiny chick grouse. The addition of an abundance of greens and shrub buds in early spring is a sure stimulus to breeding and likely increases fertility and hatchability. We start the breeders onto dandelions as soon as they are available in March or April.

Many breeders supplement the chick starter with chopped egg and various greens and fruits. Some use mealworms if the chicks are slow to start eating on their own. Hanging greens from the sides of the brooders invites jumping and feeding. By first dipping leaves into water and then into crumbles, these hanging feeders stimulate the eating of the crumbles. Little water will be consumed if the chicks are eating much fruit and greens. Appropriately sized grit should be available from hatching. Early in the season we try to keep hatches to every three days so chicks have companions as a stimulus to feed and to reduce imprinting to humans.

Along this line of reasoning, after purchasing several imprinted rare pheasants that don't mate naturally, we started to place chick brooders in proximity to adult birds of that species to assist in species identification by the chicks. At the third-stage brooder level we ran adults on the ground below the chicks.

We give five feeds daily to the chicks through to four weeks of age. After that it is gradually reduced to two feeds per day by 12 weeks. Adults are fed once daily.

Food presentation is very important. The objective is avoid feeds getting moldy or contaminated by feces or water or air disseminated pathogens. Water should be changed daily or an automatic watering system used that inhibits contamination and facilitates cleaning of the cups. Feeds should be off the ground, fruits on removable plates, greens in wire bins that hold leaves off the floor, pellet feeders set over wire bases so birds don't pick up dirty or moldy feed.

Transport and Permitting: Adult grouse are strong, durable and easy to transport and can readily be moved via express mail or air express. In the USA the commercial cardboard shipping crates were designed in cooperation with the post office; this is a very safe, efficient, and rapid delivery system. Grouse provided with half an apple, which is largely water, or half a head of lettuce, will not drink or need water. The IATA regulations for containers are not nearly as satisfactory for the birds but must be followed for international shipments.

As with many endangered bird species, one of the greatest threats to the species survival is the abundance of paper work required to comply with thoughtless anti-conservation endangered species, CITES or other regional or national government regulations.

Summary: Grouse represent one of nature's most prolific and important elements in the northern ecosystem. The ultimate survival of several wild tetraonids now depends upon the level of wisdom that man brings to bear on the conservation or "wise use" of this resource. Paramount in this struggle will be both the speed and willingness of governments to not only maintain suitable habitats, but in the most urgent manner to facilitate the live capture of breeding stock, to ease for established breeding facilities the deleterious international laws restricting the movement of both endangered and presently common species across borders to facilitate breeding programs and interchange of genetic diversity. And finally their survival in many emergency situations will depend upon the dedicated investment of more time and money by aviculturists to further expand avicultural expertise and establish reservoirs of captive breeding stock suitable for release into the wild.

I particularly wish to thank Chris Merker, Department of Biology, University of Eastern Washington, Cheney WA and David Bender, Hancock Wildlife Research Center, for their contributions to, and review of, this manuscript.

Contributors:

Fred Laub. Private Aviculturist. Langley, British Columbia, Canada.

References and Suggested Readings

Aschenbrenner, Hans. 1994 in press. Grouse of the World: their biology and captive management. Blaine, USA: Hancock House Publishers.

Bump, Gardiner, Robert W Darrow, Frank C. Edminster, Walter F. Crissey. 1947. The Ruffed Grouse: Life History, Propagation, Management. New York:

Hancock, David. 1993. Game Bird Feed Possibly Contaminated. Grouse News June 5: 3–5. World Pheasant Association.

Johnsoard, Paul A. 1983. The Grouse of the World. University of Nebraska Press.

Toepfer, John E. and Raymond K. Anderson. 1990. Translocating Prairie Grouse: What have we learned? Transcript 55th North American Wildife & Nat. Res. Conference.

Ocellated Turkey *Agriocharis ocellata*

• Myles Lamont

The ocellated turkey is the second of only two turkey species in the family Phasianidae. Historically it has been placed within the same genus as its cousin, the American wild turkey *Meleagris,* however, has more recently been placed in its own genus *Agriocharis*.

The ocellated turkey is endemic to the Yucatan Peninsula, mainly occurring in Mexico, but also entering into adjacent areas of Guatemala and Belize. It can be found commonly in certain reserves and parks, however, for the most part is considered to be generally rare. It is considered to be near threatened by the IUCN Red List.

Although physically resembling the common wild turkey, the ocellated is slightly smaller. Its plumage is much more brightly colored than the wild turkey, being mainly metallic green and copper with bright blue ocelli in the tail and upper-tail coverts feathers, from which it derives its name. Facial skin is mainly blue, with orange-yellow caruncles that congregate over the lore. During display, a fleshy blue horn is extended on the crown. Interestingly, the beard and snood seen in *Meleagris* is absent in this species.

Sexes are similar, but males are larger and females can only be confused with juvenile males. Males will also develop large spurs and a red orbital eye ring, as opposed to an orange one in females. Males are approximately 100 cm (39 in) in length and weigh 4,300–5,000 g (9.5–11 lb), while females approximately 80 cm (31.5 in) and 2,600–3,100 g (5.7–6.8 lb).

Ocellated turkeys naturally inhabit tropical oak and pine woodlands, broadleaf forest and will use cultivated cornfields and cleared areas to some extent. Feeding usually occurs in these open grasslands and savannahs, while forested areas are used for daily preening, roosting and nesting. They are usually observed in these open clearings and roadsides, presumably for the availability of grit and agricultural grains. They soon lose their fear of humans if left unpersecuted, and wild birds have been known to attack humans intruding into their territories.

This species is quite gregarious and is usually seen in small flocks consisting of a hen and her previous year's offspring, but can be found in large gregarious groups during the winter months. They roost communally in trees during the night, and disperse during the day. Males do not reach sexual maturity until at least two years of age, three being the most common. Although considered polygamous, pairs can be seen together even outside of the breeding season.

Avicultural Management: This species has had a true rise and fall in its avicultural history. The earliest notes of it being held date back to England in 1848, however London Zoo received three birds as a gift from Honduras in 1857, although no breeding success was achieved.

Further importations (less than two dozen birds) were sent to Dr. Delacour in Cleres and Mr. Spedan Lewis of England in 1938 by Charles Cordier; the species bred at Leckford in 1940 but never became established.

After the Second World War, Cordier and Dr. D.S Neville attempted to capture wild birds a second time for a few American zoological facilities. In 1946, five American zoos each received a single pair of young birds between the ages of 6–8 months. As it turned out, some of these pairs were actually multiples of the same sex.

The first successful breeding in North America occurred in 1948 at San Diego Zoo using artificial insemination (due to the breeding male having an injured leg.) Over the next 28 years, 148 poults were hatched and raised at San Diego Zoo. The progeny was distributed to both Europe and other private and public institutions in North America. Few importations occurred again, and the majority of captive stock is suffering from poor genetic diversity. Due to sexes coming into breeding condition at different intervals, obtaining fertile eggs is a challenge in itself and has been the main reason for the use of artificial insemination.

Diet wise, turkeys can be managed similarly to others in their family. A diet consisting of game bird pellet, mixed grains, insects and fruits can be fed. The diet used by San Diego Zoo during their first successful breedings consisted of green oat pellets, canary seed, millet seed, milo, maize, bread, egg yolk, various greens and ground meat. Oyster shell was also offered.

The ocellated appears to have a higher intake of foliage and fiber content as compared to its American cousin. Captive diets should, therefore, take this need into consideration.

Average egg weight at San Diego Zoo was 47.19 g (1.6 oz) and approximately 66.76 x 44.2 mm (2.6 x 1.7 in) in size; in France it is 65 g (2.3 oz) and 66 x 45 mm (2.6 x 1.7 in); and in Germany 60 x 46 mm (2.3 x 1.8 in). Egg weights vary 62–68 g (2.2–2.4 oz). Eggs are similar in color to the wild turkey, being cream with brown speckling. Average clutch size is 7–12 eggs, with an incubation period of 28 days. Laying interval is normally one day.

Artificially incubated eggs are kept at 37.7–38.3°C (100–101°F) with a relative humidity of 50%, gradually being increased a few days prior to hatching to approximately 75% RH. Brooder temperature should be 34.4–35°C (94–95°F) for the first week and gradually lowered by 2.7°C (5°F) each week until the poults are approximately four weeks of age. Newborn chicks are approximately 42–46 g (1.5–1.6 oz) in weight.

Food need not be offered for the first 24 hours from hatching, after which a high quality game bird crumble should be offered, as well as finely chopped leaf material. The addition of small insects such as fruit flies and buffalo worms can be used to stimulate feeding, otherwise quail or other non-aggressive precocial chicks can be used as feeding "teachers." Domestic poultry have been used successfully at some zoological institutions for the hatching and rearing process of the turkey poults (Hannah Bailey. 2006. Pers. comm.). It is important to ensure that food is being eaten, as a high amount of protein is needed during this early stage. As the chicks age, larger live food items can be added such as mealworms, locusts and crickets. According to Lint, at approximately five weeks of age, the turkey poults change their feeding habits from live food, to favoring the starting crumbles, boiled eggs and small greens. Poults can be difficult to rear to maturity even in the most favorable conditions, as they are prone to disease, cold, drafts and dampness.

If artificial insemination techniques are to be attempted, it is recommended that hens be inseminated on two consecutive days using at least 0.03 ml of good quality semen to obtain a good fertility rate. Multiple inseminations are recommended to ensure fertilization has occurred.

Bantam hens can be used as fosters and can be placed in heated brooder rooms with a recommended temperature of 30°C (86°F) for the first week and can then be gradually decreased and eventually introduced outdoors. Caution should be taken not to give too many chicks to the hen, as smaller breeds are not able to sufficiently brood young poults and they may chill.

If birds are to be left to their natural tendencies, hens should be monitored to insure they are sitting tightly, as there are conflicting experiences on the reliability of this species as a dedicated parent.

According to Perron, other large Galliformes can be housed with *A. ocellata* with no interspecies aggression; in this case they were housed with green peafowl *Pavo muticus imperator,* however the aviary was quite substantial in size.

Chicks can be banded near seven weeks of age using an 18 mm band.

References:

Benchley, B. 1954. Breeding the Ocellated Turkey at the San Diego Zoo. 122–125. Avicultural Magazine. 60: 122–125.

Delacour, J. 1952. Breeding Ocellated Turkeys at the San Diego Zoo. *Avicultural Magazine.* 58:148–150.

Lastere, R. Raising the Ocellated Turkey. *AFA Watchbird* 8–9.

Leopold, A.S. 1959. *Wildlife of Mexico: the gamebirds and mammals*. Berkeley: University of California Press.

Lint, K.C. Ocellated Turkeys in Display. Zoonooz. 27(No.8):5.

Motta Gill, J. Incubation of Ocellated Turkey eggs using the common turkey as a foster mother. *AFA Watchbird* 20(1): 11–14.

Monroe, B.L. and Sibley, C.G. 1993. *A world checklist of birds*. New Haven: Yale University Press.

Perron, R. 1989. Breeding the Ocellated Turkey in North Germany. *Avicultural Magazine*. 95(4):158–161.

04.02 Numididae Guineafowl

Glen Holland

Introduction: The six species of guineafowl, which belong to the order Galliformes and the family Numididae, are terrestrial birds endemic to the continent of Africa. Some species, such as the black guineafowl *Agelastes niger* and the white-breasted guineafowl *A. meleagrides,* occur in dense forests. Others, like the plumed guineafowl *Guttera plumifera* and the closely related crested guineafowl *G. pucherani,* are found in humid forest and dense scrub. The vulturine guineafowl *Acryllium vulturinum* prefers arid acacia scrub, and the helmeted guineafowl *Numida meleagris* inhabits a wide variety of open-country vegetation, including savannas, woodland margins, and dry steppe country. Domestic guineafowl, of which numerous strains and color varieties have been developed for the table and as ornamental garden birds, are descended from the helmeted guineafowl. In Africa, wild guineafowl are used as a food source. They are caught in nooses baited with grain, or sometimes caught with poisoned bait.

Natural Habits: Guineafowl are usually found in flocks which disperse into pairs during courtship and incubation, only to reassemble into flocks when the chicks are a few weeks old. Guineafowl have rather raucous rattling calls, particularly when disturbed. Much of their day is spent scratching for food and establishing their dominance and hierarchy within the flock.

Management: The helmeted guineafowl is the easiest species to maintain and breed in captivity. There are a number of domestic mutations such as silver, royal purple and Rhodesian runners, all of which are as easily bred as domestic chickens. Unfortunately, the domestic mutations are generally unable to fly as well as their wild progenitors, and there is some concern among conservationists that these weaker strains will adversely affect wild populations if they are allowed to interbreed. This is a legitimate concern, as feral populations of some domestic strains have already become established in areas with suitable habitat. As they originate from warm dry climates, they are not able to withstand severe weather conditions for extended periods.

The forest-dwelling guineafowl, of which the crested is the only species regularly available to aviculture, require Habitat A, while the vulturine and helmeted guineafowls are more at home in Habitat B, as described in the chapter on LANDSCAPE. The aviary should be large and sparsely planted with shrubs and a few dense patches of tall grass. At least half the aviary floor should be open sandy soil. Here the birds will spend hours dustbathing and basking in the sun, often lying on their sides with the legs stretched out behind. These activities are social affairs, and the entire group joins in. Even the aviary for forest species, which should be well planted with shrubs, must also have patches of open sand where the birds can scratch, sand bathe, and lie in the sun. A few thickets of grass at the base of the shrubs will provide nesting sites.

Vulturine Guinefowl
Acryllium vulturinum

Day	Weight (g)	Day	Weight (g)
0	35.93	**32**	190
2	34.11	**34**	210
4	35.07	**36**	235
6	41.7	**38**	250
8	46.67	**40**	275
10	53.6	**42**	295
12	59.61	**44**	310
14	63.9	**46**	320
16	74	**48**	335
18	84	**50**	345
20	94	**52**	354
22	110	**54**	365
24	122	**56**	372
26	130.5	**58**	380
28	140.75	**60**	382
30	170		

Male helmeted guineafowl can become extremely aggressive during the breeding season. Some specimens will not hesitate to attack domestic poultry, dogs, and even people. This is more likely to occur with hand-reared birds. At the Hancock Wildlife Research Center in Canada the male of a pair of vulturine showed signs of aggression toward the keeper during the breeding season. He would stalk anyone who entered the enclosure but never actually attack, preferring to retreat when his posturing was challenged. As the hatching date approached, he would flop down on the ground when the keeper was in the vicinity of the nesting hen, in a manner reminiscent of the agonistic behavior of a crane (D. Bender. Pers. comm.). During aggressive displays, as well as during courtship, males arch the wings above the back and compress them on their sides. This makes them look larger from the side. At the beginning of the breeding season, fighting between males is common. Males also feed females as part of their courtship rituals. Vulturine guineafowl are the least aggressive, and, although occasional excited chases do take place within the group, males seldom have to be removed due to serious aggression.

I have not noticed any serious interspecific aggression between crested and vulturine guineafowl, and in a large planted aviary, groups of both species can be kept together. Vulturines have also shared aviaries with various species of pheasants and quail without incident. Sexes are alike, and surgical or DNA sexing is advised to ensure an equal sex ratio within the flock. In common with other terrestrial species, if positive results are received from fecal checks, guineafowl should be dewormed at least twice a year, preferably before and after the breeding season.

Diet: This section should be read in conjunction with the dietary section for FRANCOLINS. Guineafowl can be maintained on a mixed poultry diet consisting of either grain or pellets, or preferably both, supplemented with a small amount of sunflower seed and a daily ration of greens. Domestic guineafowl can be kept free range, allowing them to dig for roots, bulbs, and insects; they will even take small vertebrates such as mice. Occasionally, high protein treats such as grated boiled eggs or pet foods can be offered, and are relished. In the natural state, the forest species such as crested are highly insectivorous, but also feed on fallen fruits and some seeds.

Breeding: For maximum production, guineafowl should be housed as pairs or trios. The females tend to hide their nests in grass thickets. They may disappear during incubation only to reappear after three weeks with a brood of chicks trailing behind them. If kept in close confinement where vegetation is lacking, they may lay in a box of straw, but often the eggs are simply scattered on the ground with no serious attempt at incubation. An exceptional vulturine female at Tracy Aviary in Salt Lake City produced over 100 eggs within 12 months, which resulted in 54 chicks. Guineafowl will sometimes eat their eggs, particularly when housed in small pens without natural cover for nest sites. If eggs are being eaten or scattered about, or if artificial incubation is to be used, the eggs should be collected up to three times a day, and stored in a cool place until 4–6 eggs have accumulated. Each female will lay 6–14 eggs. The eggs are buff colored and heavily pitted. Incubation takes 24–26 days. Since captive guineafowl are often not very successful parents, artificial incubation and rearing is recommended. Ideally, the first clutch should be removed for artificial incubation and the second clutch left for the female to incubate. Hens which prove to be model parents can be trusted to hatch and rear first clutches in future years.

Incubation and Rearing: While eggs are being stored, they must be turned on a daily basis. After 4–6 eggs have accumulated, they should be placed in an incubator at a temperature of 37.2°C (99°F) and a relative humidity of 55–59%, which should be increased to 66% at hatching. Egg weight loss should be approximately 14–15% over the incubation period. For information on calculating weight loss, as well as on incubation hygiene, see the section on KIWI. For information on egg storage and brooding of chicks, see the section on PHEASANTS. The latter section also describes methods of using broody bantams, which are also suitable for incubating guineafowl eggs. The reason for incubating eggs in batches is that individually hatched chicks do not thrive as well as those reared in groups. In addition, the management of a group of chicks of similar size is easier than a group of mixed sized birds. The chicks should be left in the incubator for the first 12–18 hours after hatching. They can then be removed to a brooder box with a heat lamp placed at one end. The temperature directly under the lamp should be 36.5°C (98°F). This allows the chicks, by moving closer to or further away from the lamp, to choose the temperature at which they are most comfortable.

The bottom of the brooder box should be covered in corrugated paper or wood shavings. Finer material, such as river sand or sawdust, can be ingested by the chicks, resulting in crop impaction. Such materials should only be used after the chicks are 10 days old. A shallow dish of broiler starter mash should be supplied, and a shallow water dish containing small pebbles to prevent the chicks from slipping should be placed in the center of the brooder box. Poultry water founts with small open-

Crested Guineafowl
Guttera pucherani
with eggs.

Helmeted (left) and Vulturine Guineafowl.
Numida meleagris and Acryllium vulturinum

ings for drinking can also be used for the first 10 days or so. When the chicks are first placed in the brooder, they can be taught where the water source is by dipping their beaks in the water. Dishes should never be placed near the sides of the brooder, as the chicks will run through them continually, contaminating the food and water. Brown bread crumbs and boiled egg yolk can be blended together and sprinkled over the mash. This can be prepared in bulk and frozen to avoid a time consuming daily task. Live food such as termites or small mealworms can be sprinkled on top of the mash to stimulate the chicks to begin feeding. Fresh food dishes should be supplied three times a day. The section on FRANCOLINS contains further material applicable to the rearing of guineafowl. The diseases described for francolins will also affect guineafowl. By rearing the chicks in pens with wire floors, protozoan and endoparasite infestations can be avoided.

At about 10 days of age the chicks can be transferred to cage a made of welded mesh, which can be placed outside in dappled sun, on a patch of grass, in good weather. It is important to introduce green food at an early age; chicks that have been starved of greens will gorge themselves when grass suddenly becomes available. This can cause compaction of the crop, which may result in death. Chicks must have a source of shade while outdoors. At night the cage can be brought indoors and a brooder lamp placed over one end. When moving the cage, care must be taken to avoid catching or damaging the chicks' toes in the bottom mesh as it is put down. It has been suggested that a slightly moist substrate will help prevent incorrect development of the toes.

Chicks begin to fly within two weeks and are strong fliers when a month old. I have found that pinioning does not work well with guineafowl; pinioned birds are able to flutter enough to scramble out of open-topped pens. At about a month old, when they are well feathered, the chicks can be placed in a flight with a shelter from bad weather and a perch for them to roost above the ground. The secret to raising guineafowl chicks is good hygiene combined with plenty of exercise, and green food added to the prescribed diet. The chicks can be de-wormed at 6–8 weeks if fecal checks show positive parasite egg counts. Adults should have fecal checks for internal parasites at least twice per annum, and preferably, quarterly. Chicks will resemble the adults by six months of age and are sexually mature at two years.

Ring Size: 11 mm

Species Requiring Similar Management:

Peafowl Pavo and *Afropavo* spp.
Pheasants, all genera
Partridges, all genera
Francolins, all genera

Contributor: Dr. A. Abrey, Managing Director of Umgeni River Bird Park.

References:

Dekker, D. 1970. Breeding vulturine guineafowl *Acryllium vulturinum* at Amsterdam Zoo. *International Zoo Yearbook* 10: 98.

Roles, D.G. 1993. Vulturine guineafowl at Tracy Aviary. *AFA Watchbird* 20 (3): 32.

04.03 Odontophoridae Quail

Glen Holland

Introduction: Quail belong to the order Galliformes. The Old World quail are included with the pheasants and partridges in the family Phasianidae, while the New World species are separated into the family Odontophoridae. Numerous species from both families are kept and bred in captivity. The Old World species fall mainly in the genus *Coturnix,* which includes the common quail *C. coturnix*, with Eurasian and African subspecies; the Japanese quail *C. japonica;* the harlequin quail *C. delegorguei* from Africa and Madagascar; the widespread blue-breasted quail *C. chinensis,* commonly referred to in aviculture as the Chinese painted quail and sometimes separated into its own genus, *Excalfactoria*. This little quail, often erroneously called the button or king quail, should not be confused with the buttonquail of the genus *Turnix,* which belong to their own family and order (see BUTTONQUAIL). Other species of *Coturnix* quail occur in Africa, Asia, Australia, and New Zealand.

Most of the New World quail regularly kept in aviculture are North American species such as the California quail *Callipepla californica;* the scaled quail *C. squamata* from the southwestern US and northern Mexico; the elegant quail *C. douglasii,* a western Mexican species; Gambel's quail *C. gambelii* from the southwestern US and Mexico; the mountain quail *Oreortyx pictus,* a native of western US; the Montezuma, or Mearns, quail *Cyrtonyx montezuma,* which has subspecies in the southwestern US and Mexico; and the bobwhite quail *Colinus virginianus,* with subspecies scattered throughout North America. Other species from the genus *Colinus* are sometimes kept, and rarely, some of the Central and South American wood-quail and tree-quail from such genera as *Odontophorus* and *Dendrortyx,* respectively.

In most species, the sexes can be easily determined by color differences. In addition, males are generally more vocal than females. They are territorial, seasonal breeders, producing clutches of 5–20 eggs. Quail make an interesting addition to aviaries, where few other birds utilize the terrestrial strata. They are compatible with most species other than their own kind.

California Quail *Callipepla californica*

Natural Habits: Originating from the west coast of North America, the California quail is well represented in avicultural collections. They inhabit primarily areas of short, dry scrubland, but also thrive in well-vegetated areas. They are adaptable, often living in close proximity to urban areas. A common game bird species, they have been introduced to New Zealand, where they continue to thrive despite all the introduced predators.

Management: California quail are easily sexed, the male having much brighter colors, with a facial mask of black contrasting sharply with a white forehead, eye stripe, and collar. In the female, the head is blackish brown, lacking white. The crest of the male is larger than that of the female, forward bending, and consisting of broadly spatulate black feathers. The crest of the female is short and relatively straight. One of the best known quail in aviculture, their needs are not difficult to provide. The aviary should be landscaped as described for Habitat C. These quail are rather aggressive toward other quail, particularly during the breeding season, but can be kept as a breeding flock if the following conditions are met. The flock must include a minimum of 12 birds, females must outnumber males by a ratio of 2:1, and all birds in the flock must be introduced into the aviary simultaneously. Unfortunately, this species has the habit of roosting high in the aviary. For this reason, they are not suitable to keep with small finches, which panic when these comparatively large birds fly around. They are best housed with doves, pigeons, large softbills, and small waterfowl. Extended periods of combined cold and wet weather will have a negative impact on the birds. They are able to withstand periods of freezing temperatures as long as they have sufficient shelter to keep them dry. They are fond of sun bathing and sand bathing, and a pair will often engage in these activities simultaneously. A patch of dry sand should be available for all quail species, as dust bathing is often a communal affair, usually started by one bird, with others soon joining it in close proximity.

Diet: The diet should consist of a mixed finch seed diet, a commercial soft food which is moistened with a little water, and some insects such as white ants and mealworms. Green food is also taken, but in a planted aviary additional greens need not be supplied. Dry mash, as fed to poultry, and pelleted feeds specifically manufactured for game birds, are also eaten by quail; such products are used exclusively by commercial quail breeders.

Breeding: Males become increasingly vocal at the onset of the breeding season, which peaks over the summer months. The males will also courtship feed the females at this time. California quail generally are good parents, quite capable of raising their own chicks. The practice of removing all eggs for artificial incubation has unfortunately led to some strains which no longer show an interest in incubating. If birds fail to incubate, the alternative is to artificially incubate the eggs and hand-rear the chicks. Eggs can also be placed under a small bantam for the initial incubation period and then transferred to a hatcher at 22 days. For most aviculturists, artificial incubation is considered an option only when females fail to incubate their own eggs, or if large numbers of young are desired. Females that lay without incubating may produce over 60 eggs in a season.

Despite the fact that natural incubation results in relatively few offspring, it is such a pleasure to watch a mother quail and her brood of chicks that this should always be the preferred method of propagation. The nest scrape is made under a grass tussock or in dense vegetation on the ground, and a clutch of 10–20 eggs is laid. Incubation begins with the final egg. Some birds, possibly as a result of artificial rearing, fail to build proper nests, scattering their eggs instead throughout the aviary. The incubation period is 23 days, with incubation carried out primarily by the female, the male only occasionally assisting for short periods. Once hatched, the brood should have access to a shallow dish of water, lined with coarse grit, and a second dish of chick-rearing mash. A hard-boiled egg with the shell included can be blended with two slices of wholemeal bread. A tablespoon of this mixture can be sprinkled over the rearing mash twice per day. Insects such as termites or small mealworms should also be offered twice per day. In the natural state, the chicks feed almost entirely on live food for the first week. When parent rearing, the female will initially call the chicks and offer them food from the tip of her bill. Within 4–5 days of hatching, however, the chicks begin to rely less on the female and will pick up food for themselves. By two weeks of age the chicks begin to feather out, and shortly thereafter the young males can be recognized by their plumage. When they become independent, the chicks can be removed from the aviary and the female allowed to produce a second clutch, which she may already have begun laying.

Artificial Incubation & Rearing: In the US these birds, like many other quail species, are mass-produced for the table. Numerous books have been written on commercial quail and game bird production, and should be consulted by anyone wishing to rear these birds in volume. The information contained here is to provide the aviculturist with some of the basics of artificial incubation and rearing. Those wishing to breed game birds on a large scale are referred to the *Commercial and Ornamental Game Bird Breeders Handbook* by Woodard, Vohra, and Denton, published by Hancock House Publishers.

When collected, eggs should be placed, large end up, in a tray of sterilized river sand and stored in a dark cupboard at a temperature of 10–16°C (50–60.8°F), ideally 13°C (55.4°F), with 70% humidity. They can be kept for up to 10 days in this manner, and need not be turned during storage. They should be set only after batches of six or more eggs have accumulated. With a reasonable level of fertility, this will result in manageable groups of chicks of the same size and age. Prior to storage, the eggs should be fumigated by mixing 80 ml of formalin with 40 g of potassium permanganate per 2.8 cubic meters (100 cubic feet). A toxic gas is produced from this mixture, and extreme caution should be used when undertaking this procedure.

Incubation Periods of Quail Species

Species	Incubation
Chinese, button, Cape	16 Days
Japanese, brown, harlequin, rain, stubble	18 Days
Jungle bush (singing), red, painted bush	21 Days
Elegant, bobwhite, Douglas	22 Days
California, crested, crested bobwhite, barred, scaled	23 Days
Black-throated bobwhite, mountain	24 Days

Either still-air or forced-air incubators are satisfactory for quail eggs. A steady temperature of 37.5°C (99.5°F) on the top surface of the egg and a relative humidity of 55% are optimal. The eggs should be turned at least 4–5 times per day during incubation. At 21–22 days, the eggs are removed from the incubator and placed in a hatcher set at 37°C (98.6°F), with a relative humidity of 65–70%. At this stage it is no longer necessary to turn the eggs. After pipping, eggs usually take 10–18 hours to hatch. The chicks are left in the hatcher for 8–12 hours.

When the chicks have dried off and become active, they can be transferred from the hatcher to an indoor brooder box measuring about 80 x 60 cm (32 x 24 in.). A brooder lamp is placed over one corner of the brooder box and should be adjusted to supply a temperature not exceeding 38°C (100.4°F) on the brooder floor directly below the lamp. This will allow the chicks to select the temperature at which they are most comfortable. Food and water receptacles should never be placed beneath the brooder lamp. The floor should be covered in an absorbent, soft material that the chicks are not able to eat. Chilling is probably the most dangerous, and often fatal, problem in starting quail chicks. Chicks chilled in transit may appear normal on arrival, but start dying two or three days later. If resting chicks huddle closely together under the heat lamp, the temperature is probably too low, and should be adjusted accordingly. If resting chicks remain a short distance apart, they are comfortable. If chicks stay at the opposite end of the brooder box, as far from the heat lamp as possible, they

Pink-eared Duck
Malacorhynchus membranaceus

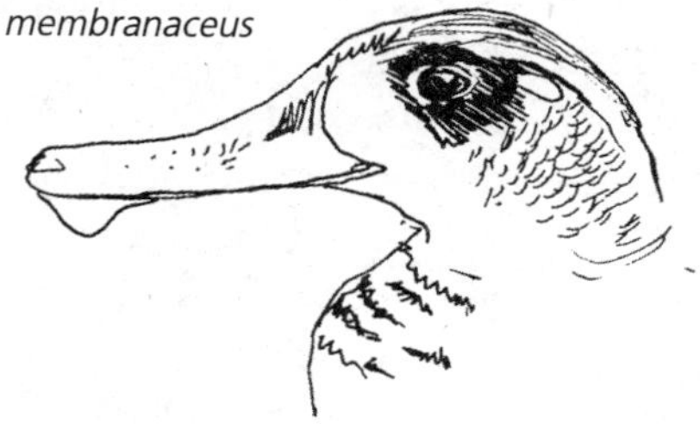

Tundra (Whistling) Swan
Cygnus columbianus

Musk Duck
Biziura lobata

Red-headed Duck
Aythya americana

King Eider
Somateria spectabilis

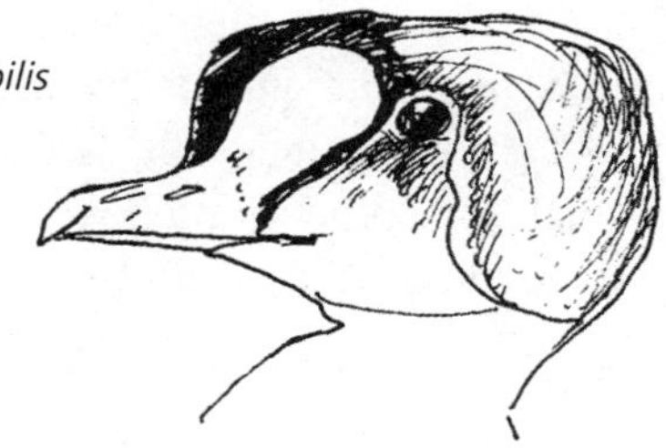

Common merganser
Mergus merganser

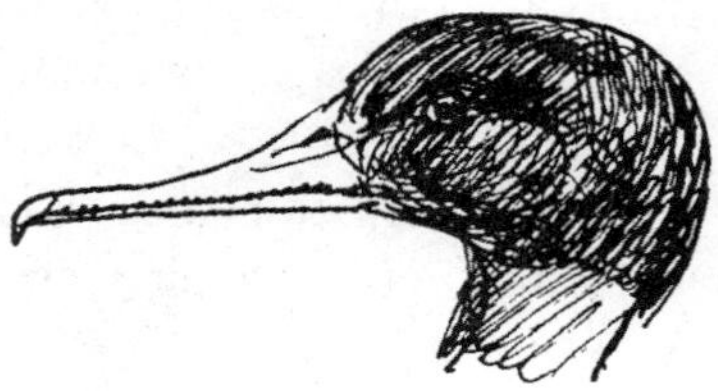

05.01 Anseranatidae **Magpie Goose** *Anseranas semipalmata*

Heidi Weingartz

Management: The magpie goose is an aberrant species from Australia and southern New Guinea that does very well in captivity in warm climates. Magpie geese enjoy large, shallow water expanses where they can dabble in the mud and uproot water plants. They do not swim very well, as they have much reduced webs on their feet. They will seldom be seen swimming in deep water, except when they feel threatened. They should not be kept in a small area, as they will destroy all vegetation in a short time.

Diet: Magpie geese consume large amounts of natural vegetation, as well as small organisms, which they find in the mud. Their diet can be supplemented by poultry laying pellets, wheat, and floating pellets.

Breeding: Magpie geese generally start breeding in the latter half of the summer in areas where the summer rainfall is regular. They need high humidity and plenty of food sources. For breeding purposes, the males prefer to take two or more females each, and they bond for life. Two or more females lay eggs in one nest, which the whole family group constructs over a period of weeks. They prefer to nest in bamboo or a thick reed bed, either on an island or at the water's edge. The nest is usually well concealed, as they bend the reeds down on top of each other, creating a very sturdy floating mass of plant material, often as high as 50–80 cm (20–32 in.). They use their sharp claws to climb up the nest. Two females together can lay 12–18 large white eggs. The eggs are rough-shelled and carry the strong odor typical of magpie goose feathers, an odor which is often detected when these birds flap away in fright from an observer. The male incubates the eggs and protects the nest area, while the females stay nearby. If the male needs to leave the nest to eat, one or more of the females relieves him of

Magpie Goose
Anseranas semipalmata

his incubation duties. The male usually has a favorite female, with whom he mates more often and spends more time. She is normally the one to relieve him at the nest. Breeding males are very aggressive and will quickly kill any other males that are housed in the same pen, even if the area is extremely large. Most males even show aggression toward intruding females. They have a very well armed beak, which they use to peck at the back of the neck of their victim. Even young birds can be aggressive toward each other, often picking on one bird in particular.

Magpie geese are good parents, with the male and the females all helping to rear the goslings. If the young are left with the parents for rearing, enough natural food, both plant and animal matter, must be available. Natural food can be supplemented with starter crumbles. It is recommended that the goslings be left with the adults, because hand rearing can, and usually does, result in permanent imprinting on humans. However, if the weather becomes too cold, the goslings will require some protection, as they are not very cold hardy until fully feathered. The goslings can easily be dry-reared and need starter crumbles, *Daphnia,* bloodworms, and green food in the form of duckweed for the first ten days. Thereafter, they can be weaned onto pellets and green food. They should be rung at about 10 days maximum, as their awkward feet grow very fast. Their unusual skin breaks easily and they bleed profusely, so care must be taken with their feet. They can be pinioned at one week of age. Catching the goslings in a big pond for pinioning is very difficult as they dive and hide very well, and this should be borne in mind when deciding whether or not to leave them with the adults.

05.02 Dendrocygnidae Whistling Ducks

Heidi Weingartz

Management: The eight main species of whistling-ducks all show great character and are a pleasure to keep. However, originating mainly from the tropics in the wild, they are quite sensitive to cold weather, especially unexpected cold spells. Therefore, if kept in areas where below-freezing conditions occur, they need to have either a lot of natural shelter or, preferably, a shade-cloth covering, at least at night. Because whistling-ducks tend to spend long hours resting on the banks of ponds, they can be kept in an area where water is at a premium and ponds are small. They enjoy perching on logs or on branches that overhang the water. A thick branch or log placed in a sunny spot will give them a perch to warm up on in the morning.

Diet: Whistling-ducks enjoy dabbling in shallow water and can easily be kept on a diet of poultry laying pellets, wheat, and floating pellets. They will find their own green food in the garden around the pond.

Breeding: Whistling-ducks form very strong pair bonds, mating for life. After the breeding season, the families come together, forming huge flocks which roost, forage, and rest together. Just prior to the breeding season, the pairs split off from the flock, each going its separate way to breed. Aggression at this time is variable, depending on the species. The most severe aggression is shown by black-billed, or West Indian, whistling-ducks *Dendrocygna arborea,* which are known to kill other birds that enter their territory. The spotted whistling-ducks *D. guttata* and plumed whistling-ducks *D. eytoni* show little aggression; and even milder aggression, or none at all, is shown by the other five species. Whistling-ducks use a number of different nest types. Raised nest boxes are very popular with most of the species, as are hollowed out tree stumps. The lesser whistling-ducks *D. javanica,* white-faced whistling-ducks *D. viduata,* and plumed whistling-ducks normally use natural grass nests. Tufts of long grass placed close together near the water's edge are most suitable. The male and female take turns at incubation and defend the nest very well. Natural incubation is normally successful but, if the ambient relative humidity is less than 60%, the eggs need to be sprayed with clean water, in the box, daily from the commencement of incubation. For the spotted whistling duck, 70–75% relative humidity is recommended. The average clutch size for whistling-ducks is 12–16 eggs. All whistling-ducks are good parents, and can be relied upon to rear their own ducklings. Both parents partake in rearing their young, and will protect them defiantly. However, parents with young should be confined to a relatively small space, with only the family present; otherwise, other species of ducks might victimize the ducklings. Whistling-ducks also have a tendency to walk great distances with their ducklings, often resulting in the ducklings getting lost, stranded on the wrong side of a fence, or caught up in long grass. Restricting them prevents this hazard and facilitates feeding, in the form of starter crumbles. When whistling-ducks are young, they are very sensitive to cold, so if cold or wet weather is imminent, it is preferable to take the ducklings indoors for hand rearing. They are very easy to dry-rear, using starter crumbles for the first three weeks and then weaning them onto pellets. They are prone to getting their hocks caught in the wire of cages from 2–3 weeks of age, if the mesh size is bigger than 13 x 13 mm (½ x ½ in.). From three weeks of age, they should be kept outdoors unless cold weather is expected, when they should be taken indoors at night. Having well-developed legs and feet, they need to be rung at about 2–3 weeks of age. Even if pinioned correctly, due to their broad wings, pinioned whistling ducks can often clear a fence of up to 1.5 m (5 ft.) high; therefore, regular feather clipping may be necessary.

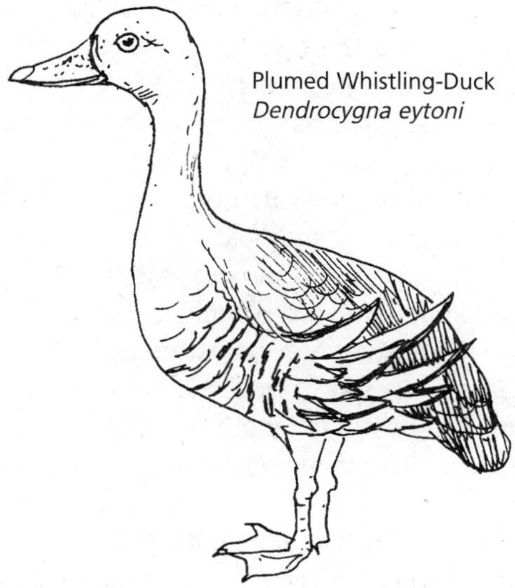
Plumed Whistling-Duck
Dendrocygna eytoni

Ring size: 11–13 mm, depending upon species.

05.03 Anatidae Ducks, Geese, Swans

Heidi Weingartz & Paul Dye

05.03a Stifftails

Oxyurinae

• Heidi Weingartz

The stifftailed ducks comprise the subfamily Oxyurinae in the family Anatidae. The seven species in the genus *Oxyura* include the well-known ruddy duck *O. jamaicenis* of North and Central America and the West Indies; the masked duck *O. dominica,* with a range extending from the southwestern US to South America and the Greater Antilles; the Andean duck *O. ferruginea* from South America; the white-headed duck *O. leucocephala* of the southern Palaearctic; the maccoa duck *O. maccoa* from the mountains of eastern and southern Africa; the Argentine blue-bill lake duck *O. vittata* of southern South America; and the blue-billed duck *O. australis* of southern Australia. This subfamily also includes the musk duck *Biziura lobata* of Australia and the black-headed duck *Heteronetta atricapilla* from South America.

Management: Stifftails are among the more difficult waterfowl to keep, both as adults and as ducklings.

Adults have great difficulty in manoeuvring on land with their short legs, and the pens must be adapted accordingly. The water level must reach the edge of the pond so that they can easily get out, even if they are carrying a huge load of eggs. The food tray must be within 5 cm (2 in.) of the ground/water so that they can reach it. There must be natural grasses and boxes or logs placed on the ground so that the females can get to them. They should not be required to climb steep banks to get to the nesting areas. Also, the nesting areas must be near the water so the females can escape from predators quickly. If they have to travel slowly over land in an open-topped enclosure, an owl can take them very easily.

Diet: Although they can be maintained on poultry laying pellets and wheat, it is advisable to supplement their diet heavily with floating pellets, because it is much easier for them to eat while on the water. They will also dive to the pond bottom to filter the mud for live food.

Breeding: Male stifftails can very easily handle 1–4 females each, and it is definitely advisable to keep more females than males because the males are very aggressive toward each other. Once the female has laid her eggs, the male will start trying to mate with his next favorite female, relentlessly. All stifftail eggs are bluish-white in color, and very large for the size of the bird. The females lay 8–14 eggs. In the case of the North American ruddy duck, the eggs are so big that once a female has laid seven eggs, she has laid her own body weight in eggs. If there is enough natural food in the pond, the ducklings may be left with the female. She will not, however, bring them to a bowl of starter crumbles, so it is essential that there is natural food available. The ducklings may take some crushed floating pellets. Alternatively, the ducklings should be hand-reared, and this process is also very difficult. If they are to be hand reared, the ducklings should not be allowed to hatch under the female, as they will be too wild. They should be hatched in an incubator and then transferred to a brooder pen where they can be dry-reared for one week, after which they should be wet-reared. While in the brooder for the first week, they must not be able to see other birds. They need one drinker with frozen Daphnia and bloodworms in the water and one with finely ground floating pellets in the water. They also need starter crumbles on a piece of disposable paper towel. They need to be encouraged to start filtering the food in the water. Once in the wet tank, they must still have the two drinkers for another two weeks, as well as floating pellets and bread in the water. They must be introduced to the outdoors slowly, as they are sensitive to cold, wet weather. They should be pinioned at two weeks and ringed at 3–4 weeks of age.

Ring Size: 10–11 mm.

05.03b Swans
Cygninae

• Heidi Weingartz

Management: The subfamily Cygninae contains seven species in two genera, six in the genus *Cygnus,* of which one species, the tundra swan *C. columbianus,* has two subspecies, and one in the genus *Coscoroba.* Housing the various species of swans depends largely on the amount of space available and on the individual characters of the swans. Their aggressive tendencies make housing them rather difficult; therefore some general guidelines will be given.

Black swans *Cygnus atratus* tend to be the most aggressive of all the swans, not only toward other birds of their own size, but to any bird sharing the same pen, even smaller species of ducks. It is therefore advised that black swans be housed on their own. They will breed in a pen 5 x 10 m (16 x 33 ft.), with a pond 3 m (10 ft.) in diameter and 75 cm (30 in.) deep. Whooper swans *C. cygnus* and trumpeter swans *C. baccinator* are also very aggressive, especially during the breeding season and, because of their size, can do a lot of harm if they attack another bird. However, their aggression is normally directed toward other birds of their own size. Therefore, although not suitable for mixed collections (unless they are on a huge lake), if placed in a separate pen with a few ducks or geese, they generally behave themselves. During the breeding season it would be best to keep a close watch to detect any increase in aggression before it becomes serious. Mute swans *C. olor,* whistling swans *C. columbianus columbianus,* and Bewick's swans *C. c. bewickii* are all suitable for mixed collections. Once again, during the breeding season, aggression may increase, but in these swans it is generally confined to the area around the nest site, and only toward other large birds, which should get out of their way. The Coscoroba swan *Coscoroba coscoroba* and black-necked swan *Cygnus melanocorypha* are the most placid of all the swans, and are most suitable for mixed collections, but they too will defend a nest site and surrounding territory defiantly. They are very sensitive species and may lose a fight even if they started it. They are also the smallest of the swans.

Swans form extremely strong pair bonds and are devoted parents. If a brood of cygnets has been parent-raised, they should never be taken away from the adults and placed in another pen where they can see or hear the adults, as both adults and cygnets have been known to pine to death. If one member of a pair of swans dies, the surviving swan should be moved to another pen immediately, otherwise it could pine for the lost mate in the old pen, and this can be fatal.

Diet: Swans enjoy eating green food and filtering mud more than anything else, so an earth-bottomed pond with surrounding reeds and grass is ideal. They also eat laying pellets, wheat, and floating pellets.

Breeding: Swans can often be encouraged to nest in a particular spot by providing nesting material in that place, in the form of cut dry grass or hay. They prefer a shady spot near the water. An average clutch is 4–7 eggs. Whenever possible, the cygnets should be left with their parents for as long as six months. The adult swans, which both tend the brood, are normally excellent parents. Provide green food, starter crumbles, and floating pellets up to three weeks of age, and then wean the cygnets onto laying pellets, wheat, and floating pellets. If, for some reason, hand rearing the cygnets becomes necessary, they should be dry-reared for one week before starting regular swimming lessons of one minute at a time. Only at about 3–4 weeks of age should they be left outside. They need starter crumbles and green food up to three weeks of age and laying pellets thereafter. Ringing of swans is usually best done at three weeks of age.

Ring Size: 25 mm for mute, trumpeter, and whooper swans; 23 mm for black, whistling, and Bewick's swans; 19 mm for black-necked and Coscoroba swans.

05.03c Geese & Ducks
Anatinae

• Heidi Weingartz

05.03ci Aserini *Geese, Sheldducks & Pygmy Geese*

Management: The true geese of the genus *Anser* and the brant geese of the genus *Branta* share a large number of attributes and will thus be treated together here. The true geese include such species as the greylag goose *Anser anser,* the snow goose *A. caerulescens,* and the bar-headed goose *A. indicus;* the brant geese include the Canada goose *Branta canadensis,* with several subspecies, the brent goose *B. bernicla,* also with several subspecies, the barnacle goose *B. leucopsis,* the red-breasted goose *B. ruficollis,* and the aberrant Hawaiian goose, or nene, *B. sandvicensis.*

All of these geese enjoy wandering through large fields of grass, grazing and calling to each other. They also enjoy dabbling in shallow earth-bottomed ponds, and, except for the Hawaiian goose, they all require water at least 50 cm (20 in.) deep in which to mate. They prefer to be in large gaggles but keeping just a pair of each does not preclude breeding. Most of these species come from very cold areas with large variations in day length in the different seasons. This makes keeping them in warm climates, such as those found within the tropics, where the day length differs little between winter and summer, rather difficult. These day length discrepancies tend to cause irregularities in molting, resulting in problems maintaining the birds in breeding condition. However, they may adapt to a change in climate over a few years, and once molting regularly and quickly, may begin to

Red-breasted Goose
Branta ruficollis

breed. An exception again is the Hawaiian goose, which comes from a warm climate and has difficulty in coping with temperatures below 0°C (32°F). Hawaiian geese tend to breed in the mid-winter, while the others tend to breed from late spring to mid-summer, when the day length is longest.

Diet: These geese can be supported on a diet of laying pellets and wheat, and enjoy floating pellets as a treat. Plentiful green food is of utmost importance from day one.

Breeding: Most of these geese will make a natural grass nest between tufts of grass or under a shelter. Some, such as the greylag goose, snow goose, Ross's goose *Anser rossii,* bar-headed goose, and barnacle goose will use a large goose box, or log, if available. The true geese tend to have larger broods of goslings of 6–8, and the goslings are generally light in color (yellow), and are very large. The brant geese tend to have smaller broods of 3–5, and goslings are generally smaller in size and darker in color (gray and brown). Both types of goslings have very long, thick down for withstanding low temperatures. Care must therefore be taken not to overheat them if they are being hand reared. As a rule, hand rearing these birds should be avoided whenever possible, as imprinting on people is inevitable, and this makes it very difficult for the birds to accept a mate of their own species. Goslings reared individually are more likely to become imprinted on humans than those reared with others of their species. When hand rearing, care should be taken to avoid overfeeding the goslings at the crumbles or pellets stage, otherwise drop-wing or angel-wing will occur. Also, they tend to get their hocks caught in the wire cage if the mesh is larger than 13 x13 mm (1 x 1 in.). If left to be parent-raised, starter crumbles must be provided for two weeks, and then laying pellets. Green food must always be available. Family groups must be penned separately, especially in the case of the brant geese, as the goslings are so small. De-worming should be started at three weeks of age. They should be rung at two weeks of age, as the legs grow very fast.

Ring Size: 13 mm (red-breasted goose) to 19 mm (greylag goose).

Cape Berren Goose *Cereopsis novaehollandiae*

• Heidi Weingartz

Management: The Cape Barren goose *Cereopsis novaehollandiae* is an Australian species that is very aggressive; pairs need to be housed alone, with no other birds for company. They are often used as watchdogs, as they defend their territory viciously and sound alarm with their deep pig-like grunting when an intruder is near.

The minimum size of a pen for a pair of Cape Barren geese is 6 x 8 m (20 x 26 ft.). They have little need for water, preferring to step over a stream rather than through it. However, the goslings can be seen bathing occasionally, so a small pond or stream can be included in the pen for their benefit. They require green grass or lucerne to eat all year round, so this must be provided in the pen. It is best to screen off their pen, otherwise they will be pacing up and down the fence trying to fight with whoever is on the other side. The danger they pose to other birds cannot be overstressed; it is extremely important to keep them away from other birds, especially during the breeding season. It is common to have difficulty in pairing a male and female, and even bonded breeding pairs occasionally fight, and may need to be separated for a period of time. When this becomes necessary, the dominant bird, whether male or female, should be removed from the pen and kept separate, but in view of the victimized bird, for a month at a time. After a month, the dominant bird can be re-introduced into the pen. This technique is usually successful, although it may have to be repeated a few times.

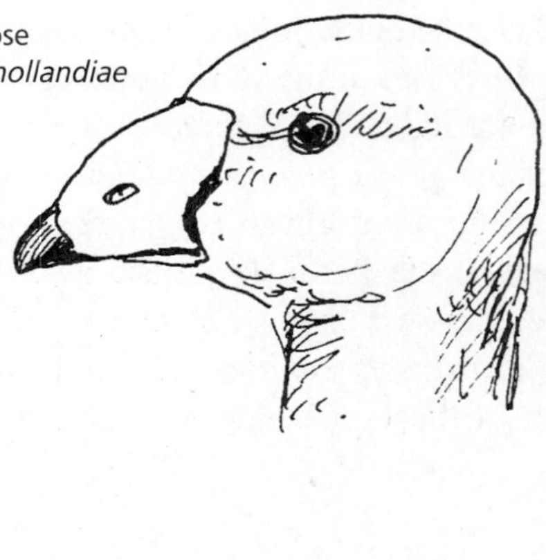

Cape Barren Goose
Cereopsis novaehollandiae

Diet: These geese consume large amounts of green food, usually in the form of grass, and a basic diet of laying pellets and wheat is very suitable.

Breeding: Cape Barren geese tend to breed in autumn, although they have been recorded breeding successfully at almost any time of the year. Increased aggression and frequent mating on dry ground is a sure sign that eggs are on the way. A simple lean-to shelter of about 1 m^2 (10 sq. ft.) in the pen, with some dry, cut grass underneath, is quite sufficient for a nesting site. A grass screen on one or more sides will encourage the use of this shelter. The female will form a shallow depression in the grass and lay a clutch of 4–7 large white eggs. The female does all the incubation. These birds can withstand freezing temperatures, even at breeding times. The goslings are large and strong when they hatch and, therefore, it is recommended that they be left with the adults. Both male and female protect the goslings with great force. Starter crumbles provided in bowls, and green food are their only requirements. The crumbles should be discontinued at two weeks of age, as the goslings grow very fast and are prone to drop-wing and angel-wing. They should be rung at two weeks as the large feet grow very quickly. This is also a good age at which to de-worm them for the first time, as they are very prone to worm infestations. All birds in surrounding pens should be de-wormed at the same time. As the goslings approach 3–4 months of age, the adults may become aggressive toward them. If this occurs, they should be removed to a pen where they cannot see or hear their parents. Hand rearing of Cape Barren geese should be avoided whenever possible, as imprinting is inevitable. If the goslings need to be hand reared, the starter crumbles should be limited to avoid excessive growth too quickly. They can be dry-reared, but need access to grass from the first day onward.

Ring Size: 19 mm

Sheldgeese

• Heidi Weingartz

Management: Sheldgeese tend to be very aggressive and care must be taken when they are in a mixed collection. The most aggressive species are the Magellan, or upland, goose *Chloephaga picta,* the Andean goose *C. melanoptera,* the Egyptian goose *Alopochen aegyptiacus,* and the blue-winged goose *Cyanochen cyanopterus.* To avoid injuries to other birds, these four species are best housed separately. A pen measuring about 5 x 8 m (16 x 26 ft.) is sufficient for a pair. The other species, the ruddy-headed goose *Chloephaga*

Egyptian Goose
Alopochen aegyptiacus

rubidiceps, the ashy-headed goose *C. poliocephala,* and the Orinoco goose *Neochen jubata,* can be kept in a mixed collection; however, strong territorial behavior should be expected before and during the breeding season. A healthy bird will be able to escape their aggression, but a weak or injured bird may be further harmed if these geese decide to attack it. Also, when kept in a mixed collection, the enclosure should include as much area as possible. They need plenty of land to run and graze, and they enjoy wading in shallow water which has water plants growing on the edges.

Diet: Their green food diet can be supplemented with laying pellets, wheat, and floating pellets.

Breeding: When it comes to nesting, although all species are capable of making a natural nest in the grasses, they generally prefer to use large nest boxes in captivity, when available, and even large logs. Orinoco and Egyptian geese often choose a raised nest site, but the other species generally choose a site at ground level. Pairs are very protective of their nest sites and a large territory surrounding them. This can be very unfortunate for any passing bird, which they regard as a threat. The females are usually very good sitters, but if the relative humidity is less than 50–60%, it may be necessary to spray the eggs under her with clean water once a day. After the goslings hatch, it is advisable to separate the family from a mixed collection for two reasons: (a) to prevent the adults from hurting other birds, and (b) to prevent the goslings from getting lost or hurt during a territorial fight.

Sheld-goslings tend to grow too fast on unlimited starter crumbles, so a small bowl of crumbles each, once a day, is enough. They must be able to graze large amounts of green grass. Both male and female partake in rearing the goslings, and pair bonds are so strong it is not uncommon for a surviving bird to pine for its lost mate. If hand reared, goslings should be started on a dry-rearing system but, during the day, from their second day, they must be placed outside on grass. Their starter crumbles must be limited, and grazing encouraged. This is especially critical in the Andean goose, where each gosling must get only about one tablespoon of crumbles per day for 2–3 weeks. After this time, pellets may be offered together with the grass. Care must be taken to avoid imprinting. Like all other geese, their legs grow very fast and they should be ringed by two weeks of age. They should also be pinioned at that age.

Ring Size: 13–19 mm, depending upon species.

Shelducks *Tadorna* spp.

• Heidi Weingartz

Introduction: The shelducks of the genus *Tadorna* have a reputation for being aggressive, but this is not a fair generalization for all of the species. The paradise shelduck *T. variegata* is most certainly a bully, and should always be kept separate from any other waterfowl. The other five species of shelducks, however, are suitable for mixed collections, as long as the area they are housed in is not too small. In decreasing order of aggressiveness they are the ruddy shelduck *T. ferruginea,* the Cape, or South African, shelduck *T. cana,* the European, or common, shelduck *T. tadorna,* the Australian shelduck *T. tadornoides,* and the radjah shelduck *T. radjah.*

Diet: Shelducks enjoy dabbling in shallow water for organic matter, and green food is very important in their diet. They do well on laying pellets, wheat, and floating pellets.

Breeding: It is a fallacy that shelducks will not breed in captivity without a tunnel in the ground. This is not at all the case. They are quite happy to use boxes, hollowed out logs and, as observed in the Cape shelduck, even make natural grass nests occasionally. The Southern Hemisphere shelducks tend to breed in late winter and early spring, whereas the Northern Hemisphere shelducks tend to breed in mid-summer. Shelducks are usually very reliable sitters and are very good parents when it comes to protecting their ducklings, so it is good to leave the ducklings with them, if the increased aggression levels at this time can be tolerated. If left to be parent raised, the ducklings should be fed starter crumbles in flat bowls placed near the pond, floating pellets on the water, and plenty of green food, such as duck-weed, kikuyu grass, water lettuce, or finely chopped spinach. The parents will be very protective over the ducklings and may try to clear the whole pond of other ducks. If the enclosure does not provide plenty of space for the other ducks to escape this aggression, then hand rearing the ducklings should be considered. They are very easy to hand rear using the dry-rearing system, as long as they receive sufficient green food along with the crumbles. They are very prone to getting their hocks caught in wire mesh larger than 13 x 13 mm (1 x 1 in.) from a very young age. Shelducks and their young are also prone to worm infestations because of their love for green food, so regular de-worming from a young age is essential. Ringing and pinioning are recommended at 10 days to two weeks of age.

South African Shelduck
Tadorna cana

Day	Weight (g)	Day	Weight (g)
0	62	16	442
1	63	17	495
2	64	18	535
3	74	19	610
4	88	20	714
5	110	21	789
6	120	22	838
7	145	23	893
8	163	24	948
9	190	25	976
10	220	26	1028
11	252	27	1037
12	286	28	1055
13	309	29	1077
14	356	30	1098
15	396		

Ring size: 11–15 mm, depending upon species.

Pygmy Geese *Nettapus* spp.

• Heidi Weingartz

Management: The housing, feeding, and breeding of pygmy geese, genus *Nettapus,* is totally different to that of any other waterfowl. Pygmy geese are best housed in a flighted condition in an aviary. Keeping them full-winged will increase the chances of breeding success, as they are said to prefer utilizing high nest sites, 2 m (6½ ft.) or more above ground level. There are, however, proven records of pinioned pairs breeding. Each pair should be housed separately, as the males are very aggressive during the breeding season. They are compatible

African Pygmy-goose
Nettapus auritus

with other small species, but keeping them in mixed aviaries may complicate feeding systems. Another advantage of housing each pair separately is that one can leave their ducklings with them to be parent raised very easily. Each pair should have a pond of at least 4 m^2 (43 sq. ft.) surface area, and a similar area of land. Perches over the water are very popular with these birds.

An aviary for pygmy geese must be covered with shade cloth for temperatures below 10–12°C (50–53.6°F), as all three species originate from tropical regions and are very sensitive to cold temperatures and frost. The cotton pygmy goose *Nettapus coromandelianus* is found in southern Asia and Australia; the green pygmy goose *N. pulchellus,* in eastern Indonesia, New Guinea, and Australia; and the African pygmy goose *N. auritus,* in Africa and Madagascar. When these birds are kept in areas where temperatures fall below –5°C (23°F), it is recommended that they be kept indoors during the winter.

Diet: In the wild, pygmy geese consume vast amounts of green food, including various parts of water plants, such as leaves, stems, flowers, roots, and seeds. In captivity, if natural water plants are not available, green food must be supplemented with spinach, lucerne (alfalfa), or duckweed. The latter is very popular, and is the best option, if it can be obtained. They should also be fed on a mixture of small seeds such as millets fed dry or in water. If fed wet, this food must be changed every day. Some pygmy geese in captivity eat laying pellets or laying mash, but very few. Mealworms can be used as a treat for the adults.

Breeding: It is very difficult to get wild-caught pygmy geese to breed in captivity. Captive-bred specimens usually start breeding at about two years of age, but will occasionally breed successfully as yearlings. They tend to favor a hollowed out log, but there are also records of natural nests having been made in grass and reeds. It is best to give them a choice of logs, raised to about 2 m (6½ ft.) in height, as well as one down on the ground. The logs should be about 50 cm (20 in.) tall, 25 cm (10 in.) in diameter, with an entrance hole up to 10 cm (4 in.) in diameter. Females usually will lay up to 10 eggs, and they are normally excellent sitters. If hatched in a high nest, the ducklings, when dried off, will jump out of the box to the ground below, where the female is calling them. The same food given to the adults is offered to the ducklings, as well as another shallow bowl, with gently sloped edges, filled with water, blocks of frozen *Daphnia,* and bloodworms, which must also be changed daily. Stress is a big factor with pygmy goose ducklings, and it is recommended that medication against coccidiosis be added to the food and water mixture for the first week. Ringing is possible at about three weeks of age.

Ring Size: 4.5 mm

05.03cii Anatini

Perching Ducks *Aix* spp.

• Heidi Weingartz

Management: Most beginner collections start with a pair of Carolina, or wood, ducks *Aix sponsa* and/or a pair of mandarin ducks *A. galericulata.* This is no surprise, as both species are stunningly colored, and are usually freely available. Mandarins and wood ducks are most often kept together, and it is often a concern to keepers that they will hybridize. However, fertile eggs

Mandarin Duck
Aix galericulata

cannot be produced from a mandarin-cross. Of course, friendships between the two species can form, due to the similarity in plumage of the females, and if this occurs they should be separated from each other. Mandarins tend to be shyer than other perching ducks, and their pen should be such that they cannot hide away too much. Being perching ducks, they enjoy a well-wooded pen; a branch over the water will prove a popular perching site.

Diet: All perching ducks (except the pygmy geese) can be maintained on poultry laying pellets, wheat, and floating pellets. They will help themselves to a small amount of green food. Wood ducks are said to be adept at catching small fish, so one should avoid putting valuable fish in any pond accessible to this species.

Breeding: Wood duck males are capable of mating with more than one female in a season, and this species should be kept in trios if maximum production is required. Mandarin males tend to be more loyal, however, and this species should be kept in pairs. Once the eggs have been laid, the male sits outside the nest box, protecting his female and her nest. Although quite capable of making nests in vegetation on the ground, both wood ducks and mandarins prefer a nest box which is raised from the ground approximately 40–50 cm (16–20 in.), with a ladder leading up to it. If the ducklings are parent raised, the family will need to be kept separately to prevent the ducklings from getting lost or hurt by other birds, and to make sure they get some starter crumbles. Once the ducklings are feathered up, they can be released. If hand rearing is contemplated, they should be dry-reared. The brooder must be surrounded with disposable paper towel so that they cannot see out. It must also have a cover, as these ducklings can climb well, and will otherwise be escaping from the brooder within hours of being placed in it. They can be reared on starter crumbles for 2–3 weeks, and then weaned onto pellets. They tend to over-eat slightly, so care should be taken to prevent drop-wing. They should be ringed at two weeks of age.

Ring Size: 9 mm for both mandarins and wood ducks. The other perching ducks use ring sizes ranging from 8–25 mm.

Anas Species *Anas* spp.

• Heidi Weingartz

The genus *Anas* contains more than 40 species of ducks, including the common mallard *A. platyrhynchos* and related species, most of the species of teals, the gadwall *A. strepera* and its subspecies, and the wigeons, shovelers, pintails, black ducks, and other species. The genus has a worldwide distribution, with representatives occurring in North and South America, Europe, Asia, Africa, Indonesia, Australia, New Zealand, and on various other islands throughout the world. They can be found in climates ranging from tropical to temperate.

Management: Ducks from the mallard complex, and the teals, are all dabbling ducks, making them easy to keep in most garden pond set-ups. They require only shallow water to dabble in, and most of them are very compatible with other species. Most species of teals are, petite ducks with bright colors. The mallard complex is comprised of ducks that are slightly larger, all of which originally derived from the mallard duck. They have certain traits of the mallards, namely a tendency to hybridize freely with closely related species, and also a tendency to bully other birds. The teals and mallard complex have in common the brightly colored green, blue, or purple speculum on each wing, often utilized in courtship displays.

Cape Teal *Anas capensis*

Day	Weight (g)	Day	Weight (g)
0	24	**16**	124
1	23	**17**	132
2	25	**18**	148
3	28	**19**	154
4	33	**20**	165
5	39	**21**	172
6	46	**22**	186
7	50	**23**	189
8	59	**24**	205
9	64	**25**	207
10	72	**26**	215
11	79	**27**	230
12	87	**28**	245
13	97	**29**	253
14	101	**30**	258
15	107		

Diet: The *Anas* spp. can all be maintained on poultry-laying pellets, wheat and floating pellets. They will help themselves to whatever green food is available in the pen.

Breeding: Various types of nest boxes and logs should be provided as a choice for these birds. Also, many of them enjoy making natural nests among tufts of grass, which should be included in their aviary. Examples of grass nesters would be the garganey teal *A. querquedula,* the African black duck *A. sparsa,* the European green-winged teal *A. crecca crecca,* the Cape teal *A. capensis,* and the Cape shoveller *A. smithii.* These nests are well camouflaged, and care must be taken when walking through the grass in their pens. Most *Anas* species are summer breeders, exceptions being the Cape teal, the Hottentot teal *A. hottentota,* and the chestnut teal *A. castanea,* which can breed at any time of the year, as well as the African black duck, which breeds in winter. The ducklings of all *Anas* species are very small at hatching, therefore it is easier and more successful to hand rear them, using the dry-rearing system. A screen around the brooder will prevent the ducklings from running up and down the wire. To start them eating, their beaks should be dipped in the drinker and food many times the first day. By the second day, they should be eating unassisted. Starter crumbles is all they need for the first three weeks, and then they can be weaned on to laying pellets.

If they are to be parent raised, it is preferable to separate the female and ducklings from the rest of the collection. They should be moved out of sight and hearing of other birds, or else they may pace up and down the wire, trying to get back. A small pond can be provided, but the edges must be rough and sloped to allow ducklings to get out easily. They should be fed only starter crumbles for three weeks, then poultry laying pellets. In certain species, such as the chestnut teal and the New Zealand brown teal *A. aucklandica chlorotis,* the male can be left with the female and ducklings. Pinioning should be performed when they are two weeks old, and most species should be ringed at 2–3 weeks of age.

Ring Size: 7–13 mm, according to species.

Pochards

• Heidi Weingartz

Management: The pochards are among the easiest waterfowl to keep. This group, which includes the genera *Netta* and *Aythya,* contains some of the most enjoyable species of ducks. Their bodies are designed for spending most of their time in the water, diving to the pond bottom to sift the mud for organic morsels of food. They are streamlined, and their short, stout legs are placed widely apart to aid in keeping their buoyant bodies under the water for great lengths of time. They prefer deeper water than the dabbling ducks; their pools should be at least 50 cm (20 in.) deep, but deeper is better. Due to their inability to run from danger with their widely placed legs, they often prefer to rest during the day on an island, where they feel safe. All species of pochard are very compatible with other species, and are therefore suitable for mixed collections. However, there are a number of species that are capable of hybridizing in captivity. Generally, similar looking species should be kept separately, such as the European pochard *Aythya ferina,* the redhead *A. americana,* and the canvasback *A. valisineria.*

Diet: In the wild their diet is high in protein, as they eat a lot of aquatic organisms, but in captivity they can be easily maintained on a diet of poultry-laying pellets, wheat, floating pellets, and green food.

Breeding: Pochards very often choose to make a natural nest in soft vegetation near the water, such as grass, irises, reeds, or ferns. These natural nests are well camouflaged and are protected by the incubating female. Other popular nest sights are caves made of rocks, wooden boxes, either raised or on the ground, and hollowed out logs. Once the female has laid her eggs and starts to incubate them, the male has no further interest in the female or the youngsters. Average clutch sizes for the smaller species are 10–12 eggs; for the larger species, 8–10 eggs. Female pochards are very good parents, protecting their broods well. If the ducklings are left with the mother, their diet must be supplemented with starter crumbles for 2–3 weeks. Pochards of all species are very easy to hand rear using a dry-rearing system. They need only starter crumbles for 2–3 weeks. To produce very strong birds, the ducklings can be boosted with frozen *Daphnia,* bloodworms, and shrimps for their first week. Also, they will relish a handful of mealworms once a day. Pochards have a tendency to get their hocks caught in the wire if the mesh is bigger than 13 x 13 mm (1 x 1 in.). They also tend to overeat, causing them to grow very quickly. It is important to get them outdoors and swimming by three weeks of age so that they can take the weight off their legs as soon as possible. They need to be pinioned and ringed between two and three weeks of age.

Ring size: 10–13 mm, according to species.

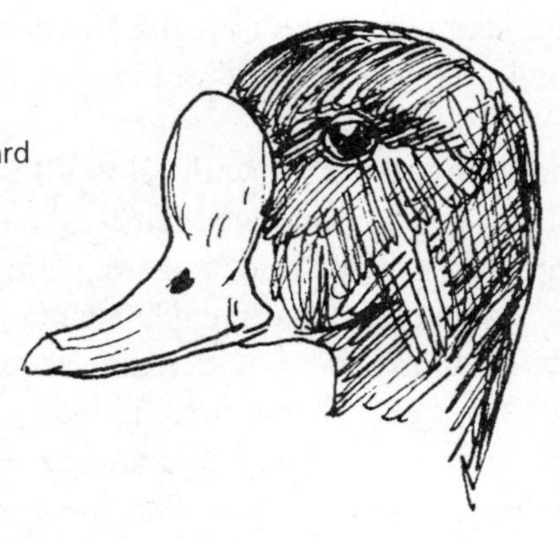

Rosy-billed Pochard
Netta peposaca

Seaducks

• Paul Dye

Introduction: The term "seaducks" refers to a number of species belonging to the tribes Anserini and Anatini, subfamily Anatinae, family Anatidae, order Anseriformes, which are usually found in coastal areas, as opposed to inland waterways and freshwater lakes and ponds. Seaducks are diving ducks, feeding and often swimming underwater. Various species are adapted to widely differing habitats. Extremes would include the eiders *Somateria* and *Polysticta* spp., the long-tailed duck, *Clangula hyemalis,* the scoters, *Melanitta* spp., of the American and Eurasian far north and the steamer ducks, *Tachyeres* spp., of the American far south and the Falkland Islands. Between these extremes are found the mergansers, *Mergus, Mergellus,* and *Lophodytes* spp., the bufflehead, *Bucephala albeola,* common goldeneye, *B. clangula* and Barrow's goldeneye, *B. islandica*, and the harlequin duck, *Histrionicus histrionicus* of the mostly temperate regions. Some, such as the harlequin and steamer ducks, have only recently been studied extensively in the wild, and one, the Brazilian merganser, *Mergus octosetaceus,* is among the most rare of birds. Most travel to fresh water areas to reproduce and a few rarely, if ever, travel far out to sea.

Seaduck species that have become more common in aviculture during the last 30 years include the bufflehead, the goldeneyes, the common eider *Somateria mollissima* and several of the mergansers. Although still not considered particularly easy to breed, improved feeds, along with avicultural information sharing, is enabling more skilled aviculturists to succeed with them. Other species, such as the arctic eiders, longtails, scoters, and harlequins, remain a real challenge and are only occasionally bred in captivity. To my knowledge, the steamer duck most frequently represented in captivity is the Magellanic flightless steamer duck *Tachyeres pteneres*. The only breeder in the USA to successfully breed steamer ducks is Mike Lubbock, who was kind enough to provide input on the species for this chapter.

Management: As with all wildfowl, seaducks have evolved to accommodate their wild habitat — an almost completely aquatic one, and frequently turbulent due to wind exposure. This aquatic adaptation necessitates a captive environment with easy access to clean, cold water for their swimming, diving, copulating, and to some extent, feeding and resting needs. Water depth need not be extreme; in fact there are advantages to keeping ponds wadeable in order to better manage individuals requiring treatment for parasites or illness. Loafing areas at the edge of the water must be easy to keep clean, otherwise soiled and broken breast feathering will result. Areas of rock or smooth concrete will meet this requirement and can be made even more attractive to the birds by being shielded from intense midday sun. A floating log with a central shaft anchored to the pond bottom, if balanced properly, will rotate occasionally due to waterfowl activity, and can provide a self-cleaning loafing site.

Many of the seaducks are among the more difficult species to propagate, being easily affected by captivity-related stress such as exposure to native predators, domestic pets, loud children, etc. Such stress will undoubtedly affect the health of these waterfowl, and will most certainly curtail any breeding activity. It is imperative to surround ponds with predator-proof fencing, and to cover the area with good quality aviary netting if avian predators are a local problem.

One should carefully observe changes in behavior as spring approaches, because excessive aggression on the part of one bird can negatively affect the breeding potential of many others. A good example is the king eider *Sometaria spectabilis*. Several propagators have observed that having more than one male in the same enclosure frequently reduces both nesting attempts and fertility.

Diet: Most seaducks are by nature carnivorous, subsisting on a wide variety of invertebrates, crustaceans, shellfish, shrimp, krill, and small fish species. Not too many years ago this diet was difficult to duplicate in captivity, and in more remote areas of the world may still present quite a challenge. In the USA and parts of Europe, the Purina Company now supplies, under their Mazuri brand name, a seaduck diet with a protein content of approximately 21%, formulated from fish meal, meat meal, and blood meal, in addition to the usual grain and vegetable matter, and supplemented with vitamins and minerals. This diet consists of extruded pellets, approximately 7.5 mm (0.25 in.) in diameter, that float when presented on the water, allowing them to be fed either directly on the water or dry in a feed container. Clarke and Butcher is another European supplier of specialty feeds for seaducks. This diet has proven to be a real time saver; however, it must be kept fresh and dry, as it degrades with time and molds rapidly in a wet environment.

Many successful propagators provide supplemental feedings (treats) of freeze-dried krill or live mealworms or crickets to keep their seaducks tame, thus providing an opportunity to evaluate their condition. On occasion a zoo with access to a steady supply of small bait-sized fish will provide it the same way.

Several feed companies also provide grain, soybean, and fishmeal-based waterfowl starter feeds that have greatly facilitated the feeding of young seaducks. The Mazuri waterfowl starter is 20% protein, and is comprised of extruded pellets, approximately 3 mm in diameter, that float when presented on the water. More information on this feed will be provided in the rearing section.

Breeding: Because of the diverse nature of the seaducks, their courtship and nesting habits vary, species by species, and must be kept in mind when planning accommodations for these birds. For example, king eiders breed more reliably if no more than one male is kept in an enclosure with 1–3 females. On the other hand, spectacled eiders *S. fischeri,* American eiders, *S. mollissima dresseri,* and northern eiders *S. m. borealis,* as well as long-tailed ducks, or oldsquaws, seem to reproduce better if maintained in flock situations, with three or more, males and females in the same enclosure, provided they're not crowded and can attain some degree of privacy for copulation and nest site selection. Harlequins, buffleheads and goldeneyes are generally more productive if kept no more than one pair to an enclosure, although sporadic success has been experienced with 2–3 pairs in a fairly large enclosure. Steamer ducks, being very aggressive, can only be kept as single pairs. Their enclosure needs to be at least two-thirds water, and the pond should be at least 1 m (3 ft.) deep at the deepest end.

While the eiders prefer a relatively open nest site, scoters, long-tailed ducks, and harlequins prefer to hide their nest in relatively thick vegetation, concealed nest boxes, and driftwood or brush piles. The buffleheads, mergansers, and goldeneyes prefer a vertical nest box, simulating a hollow tree, for their nest. Hollowed out logs and wooden nest boxes seem to work equally well for this purpose. Steamer ducks prefer a large ground nest box with a porched entrance on the side.

Incubation and Rearing: The goal of hatching the greatest percentage of eggs possible is usually accomplished in one of, or a combination of, three methods. These include natural incubation by the female seaduck, fostering under broody bantams, and temperature and humidity controlled incubators. Most female seaducks tend to be reliable at incubating their own eggs, which seems to have the added benefit of encouraging repeat nesting, often at the same location, in subsequent years. Due to the relatively high value of these species, it is advisable to remove all, or all but one, of the pipping eggs to an incubator for hatching. This only risks the one remaining duckling if the enclosure or mother duck proves to be inadequate for parent rearing. In the case of arctic species, second clutches are not common, and could compromise the health of the female. This seems to be especially true for spectacled eiders, which rarely, if ever, leave the nest to feed during incubation. Some species, such as bufflehead, the mergansers and goldeneyes will frequently re-nest if their first clutch is collected within a day or two of the start of incubation.

Many breeders find good success by leaving the complete clutch of eggs under the mother duck for 1–2 weeks, and then transferring them either to an incubator or to a broody bantam. This has the advantage of getting the eggs off to a good start during the most sensitive early incubation period. Modern incubators should then do a good job of completing the incubation process. However, if the electric power source or the incubator itself is unreliable, the broody bantam hen provides an excellent alternative. Bantams seem to have the additional advantage of compensating for unusual extreme perturbations in local ambient temperatures and humidity.

The success of early attempts to breed seaducks in captivity was limited by the dietary requirements of these species; breeders experienced difficulty in getting young ducks off to a strong start. Tadpoles, salaman-

Common Eider
Somateria mollissima

Spectacled Eider
Somateria fischeri

King Eider
Somateria spectabilis

ders, crickets, fly larvae, and mealworms have all been used to supplement standard waterfowl starter feeds, with varying levels of success. The difficult aspect of this approach was trying to provide the animal-based protein required while still offering a balanced diet. Another problem sometimes resulted from the reluctance of some seaducks to adapt from live food to a permanent prepared diet.

The development of the new waterfowl starter feeds described earlier greatly simplified this task. For species such as longtails, which are sometimes reluctant to begin eating, the use of soft paper toweling as a dry brooder substrate for the first 2–3 days will often stimulate eating, since the spilled starter pellets bounce and appear to be alive as the ducklings move about. There are still advantages to providing treats such as millet or mealworms after the ducklings begin eating, as the young birds will eagerly await these treats, providing an ideal opportunity to observe their overall condition.

Species Requiring Similar Management: Some of the fresh water diving species, such as canvasbacks *Aythya valisineria,* and the other pochards, *Aythya* and *Netta* spp., will benefit from the same management techniques described for seaducks. Canvasbacks are somewhat unique in that the female initiates pair bonding by making her choice from a number of displaying suitors. A mistake in choosing which male to isolate her with may well result in her killing that male. As with most of the pochards, canvasbacks prefer to nest in reed or iris beds over or adjacent to the pond.

Contributor: Mike Lubbock, Sylvan Heights Waterfowl II, Scotland Neck, NC.

Paul Dye passed away in 2006. May this chapter serve as a reminder of his passion for seaducks.

Mergansers & Goldeneyes

• Heidi Weingartz

Management: The mergansers and goldeneyes, including the bufflehead *Bucephala albeola,* are some of the most striking species regularly kept in waterfowl collections, with their black, white and iridescent plumage. They are best appreciated in deep ponds with crystal clear water, allowing one to observe their diving prowess as they play and search for food under water. Many of them occur in wooded areas in the wild, so this must be borne in mind when designing their pen. The Barrow's goldeneye *B. islandica* and the goosander, or common merganser *Mergus merganser* tend to be aggressive toward smaller birds when confined to a small area, but all of the other species are compatible. The mergansers will all catch fish if they are present, so one must remember not to put prize fish in the pond with these ducks.

Diet: Green food does not play a big role in the diet of mergansers and goldeneyes. They can be maintained on poultry-laying pellets, wheat, and floating pellets. They will also catch any edible organisms in the pond, such as small frogs and young crabs, not to mention fish. Mealworms can be used as a special treat for the adults.

Breeding: Being inhabitants of forests and wooded areas, these waterfowl seek out nest sites in hollowed out logs and wooden boxes raised off the ground. They require a dark hollow of any sort, and therefore will not normally make a natural grass nest. The merganser's average clutch size is 8–14 eggs, whereas the goldeneyes lay 6–8. The females can generally be relied upon for incubation, and only the females tend to the ducklings. It has been observed that some of the mergansers carry the ducklings on their backs when they are tired of swimming. If the ducklings are left with the female, they will need starter crumbles as a supplement, and some mealworms if there is no live, natural food, such as fish or insects, in the pen. They are easy to hand-rear but hygiene is of utmost importance in these diving ducks. They very easily pick up enteritis and then an oral general antibiotic has to be administered immediately. The first sign of enteritis is a fouled cloaca. They can be reared on starter crumbles alone, but it is much easier if mealworms are available to get them started. At first the worms should be cut in half and the ducklings encouraged to pick at them. After 24–36 hours they should be pecking at whole worms. After 48 hours the worms are moistened and put into the crumbles to aid in weaning the ducklings onto the crumbles. Worms must be given for 10–14 days. At three weeks of age the ducklings can be weaned onto laying pellets. Pinioning should be done at two weeks and ringing at 3–4 weeks. Paul Dye discusses mergansers and goldeneyes further in his section on SEADUCKS.

Ring size: 9–13 mm, depending upon species.

Artificial Incubation of Waterfowl Eggs

• Helen Gummer

Why should we artificially incubate eggs? There are a number of reasons to incubate eggs, the most obvious being to increase the production of the birds. Most waterfowl species will lay at least a second clutch, if not more, if their eggs are removed. This can double or treble the production of youngsters in a season. Another reason to artificially incubate is if the laying female has a poor track record of incubating eggs and/or rearing the youngsters. Yet another reason would be if there was a predator problem in the area which would put an incubating duck under great threat. In this case it would be better to remove the eggs, thereby saving both the eggs and the female.

1) Collection of the eggs: It is advisable, if at all possible, to allow the female to incubate her eggs for 7–10 days before collecting them because the eggs are in their most sensitive and delicate stage during the first few days of incubation. During this time, excessive movement or irregularities in temperature and humidity can easily damage the embryo. When the eggs are due to be collected, it is better to collect them while the female is off the nest, to prevent her from defecating over the eggs and possibly cracking some of them as she jumps off the nest. They should be placed in a large flat bowl which is padded with a soft towel, and then covered with a towel to preserve their warmth as long as possible while being transported to the incubator room. One should avoid sharp movements while carrying the eggs. If the eggs are badly soiled, they will need to be cleaned before being placed in the incubator. The eggs can be rinsed in water which is slightly warmer than the eggs themselves, and then dipped in a solution of egg disinfectant. If the eggs are not dirty, place them directly in the incubator.

2) Setting the eggs: If the eggs are freshly laid and the female has not yet commenced incubation, the eggs can be stored outside the incubator for a period of up to three weeks, under ideal circumstances. During this time, the eggs must be turned 180 degrees at least twice daily. The relative humidity should be in the region of 50% and the temperature below 20°C (68°F). If incubation has already commenced, the eggs will have to be placed in the incubator immediately after collection. Each egg should be numbered consecutively with a soft pencil, and the following details recorded: egg number, species, sire and dam, nest details, date of collection, and date on which incubation commenced. Before the eggs are placed in an incubator, their viability should be assessed. They can be candled by shining a bright light through the egg in a dark room to check fertility and ascertain if the embryo is alive. Any egg with an embryo which has died should not be placed in the incubator, as it will be a source of bacteria for the other eggs.

3) Incubation requirements: Waterfowl eggs should be incubated at a temperature ranging from 37.3–37.5°C (99–99.5°F). The relative humidity for most species should be 60%. There are some species that occur in very humid climates, such as New Guinea, which should be incubated at about 70% relative humidity. Certain species from drier climates will still hatch at a relative humidity of 50%. More importantly In most species, the eggs should lose 15% of their fresh weight during incubation. Plotting the weight loss on a graph can monitor this. If it is found that the egg is not losing the correct amount of moisture, the humidity can be adjusted accordingly. If the egg is losing too much moisture, increase the relative humidity in the incubator and vice versa. Waterfowl eggs need to be turned regularly so as to prevent the embryo from developing incorrectly. Set the incubator to turn the eggs 180 degrees very slowly every two hours. When lying in the incubator, the eggs should not be level on their side, the pointed end of the egg must be pointed slightly downward. It is important to simulate the cooling of waterfowl eggs, which occurs when the female leaves her nest to swim and eat. She is often off the nest for up to half an hour, and although she covers the eggs with down, the temperature in the nest does drop. When the female returns to the nest, she usually has cold water on her breast feathers and this further reduces the temperature. It is sufficient to

simulate this twice daily. If the incubator does not allow for this, then leave the incubator door open for a few minutes twice daily. Just before closing the door, spray clean cool water on the eggs. An alternative is to remove the egg tray from the incubator for 15–20 minutes.

4) Hatching: During the hatching process, it is important for the eggs not to be turned. The hatchling itself must rotate inside the eggshell, and if the egg is turning, this will be more difficult. It is best to remove the egg from the incubator just after the hatchling has pipped, i.e., made the first crack in the shell. If possible, the egg should be moved to a hatcher where the temperature is about 37°C (98.6°F) and the relative humidity is between 80–90%. This high humidity prevents the hatchling from sticking to the membranes during hatching. Different species have different times which they take to hatch. For example, stifftails normally hatch within two hours of pipping, but swans can take up to two days. If it appears as though a hatchling is struggling to hatch, it may be helped by peeling away some of the egg shell with pair of sterilized tweezers. It is also important not to rush the process as, during hatching, all the blood in the blood vessels surrounding the embryo must be reabsorbed into the hatchling's body; if by rushing it the embryo is made to bleed, the chances of survival are drastically reduced. It is important to remove the egg from the incubator during the hatching process because a lot of bacteria and other dirt are released from inside the egg which can affect the other eggs adversely. If a hatcher is not available, then the same incubator will have to be used, but will have to be cleaned and disinfected after each clutch of eggs has hatched. Using the same incubator for hatching also makes it difficult to increase the humidity for the hatching embryos, but spraying the hatching eggs with water every two hours can partly solve this problem.

5) Drying: Once a hatchling has come out of an egg, it is in a wet, tired, and disorientated state for a few hours during which it needs a dry, clean, and warm place to rest and find its feet. A drying box needs a wire platform on which the hatchlings can lie and needs to be kept at around 34°C (93.2°F). The platform area should be no more than 30 x 30 cm (12 x 12 in.). The hatchlings should remain there until they are dry and walking properly, which can take up to 12 hours. During this drying period, it is not necessary to give the hatchlings food or water. Once dried, the hatchlings can be moved to a brooder cage with food and water.

6) Hygiene: During every part of the incubation process, hygiene is of utmost importance. All equipment needs to be cleaned and sterilized at least once a month during the breeding season. If an egg should accidentally hatch or explode in an incubator, it is necessary to clean and fumigate it immediately, before infection of other eggs takes place. Hatchers and dryers should be cleaned after each clutch of eggs has been through it. After cleaning with a water and disinfectant solution, the equipment can be fumigated with a Formalin and Potassium Permanganate solution (5 ml Formalin and 3 g $KMnO4$ for an incubator which holds 100 duck eggs). This solution is very volatile and should be used with great care. One must never expose the skin and eyes, and always ensure good ventilation during this process. The solution should be placed in the incubator and the door closed quickly. Five minutes later the door of the incubator can be opened and left ajar for 10 minutes to air out. Embryos can be damaged by fumigation, so ensure that there are no viable eggs near to the equipment being fumigated. Once aired out it is ready for use. Formaldehyde tablets can also be used. This alternative is safer to use if there are eggs in the incubator at the time of fumigation.

Buttonquails

06 Turniciformes

06.01 Turnicidae: Buttonquails

06.01 Turnicidae Buttonquails

Glen Holland

Introduction: Once included with the cranes in the order Gruiformes, the buttonquails, or hemipodes, have now been assigned their own order, Turniciformes, on the basis of DNA studies. The order includes a single family, the Turnicidae, containing 17 species. Sixteen of these species belong to the genus *Turnix,* and the single remaining species to the genus *Ortyxelos*. This latter species is called the lark-quail, or quail-plover *Ortyxelos meiffrenii,* the term buttonquail generally being confined to the genus *Turnix*. These buttonquail are not true quail, and should not be confused with the blue or painted quail of the genus *Coturnix,* which are often referred to aviculturally as button quail. Buttonquail are widespread across Africa, Asia, northern and eastern Australia, and many of the eastern islands. They occupy open, short grassland but can also be found in savanna where sufficient ground cover is available. They occur in pairs or small parties, and are particularly numerous in some years. When disturbed, they explode into the air, flying a short distance with tremendous speed. The female makes a deep *ooom* call and the pair makes a series of peeping sounds between them.

Management: My experience is primarily with the kurrichane, or little, buttonquail *Turnix sylvatica* which is a fascinating species. The females are larger and more colorful, particularly on the chest, than males. Buttonquails provide an interesting challenge in aviculture. In contrast to the true quails, a single female is kept to every two or three males. The males are either housed separately in flights awaiting the introduction of a female, or in a large planted aviary where three males and one female can be kept together. These birds should all be placed in the aviary simultaneously. When accustomed to each other, the males will live in relative harmony; however, they will attack and kill any new male that is introduced. I have also heard of Australian black-breasted buttonquail *T. melanogaster* which have been housed with finches without any problems until a new finch was introduced. The latter was set upon and killed by the quail. While I have not experienced this with the African species, it would be worth watching any introductions into an aviary containing buttonquail. Aviary landscape is as described for the harlequin quail. The black-breasted buttonquail digs so vigorously that food and water bowls have to be placed above ground level to prevent them from being buried. Pairs often dust and sun bathe together. Cold and wet climates with temperatures remaining below 12°C (53.6°F) for extended periods are not suitable for these birds; in such climates, dry sheltered areas are essential if these birds are kept. In areas where daytime temperatures regularly drop below 8°C (46.4°F) for extended periods, some additional warmth must be supplied.

Diet: As described for the harlequin quail, but buttonquail are more insectivorous than most true quail.

Breeding: In this species the female calls and courts the male. Males are solely responsible for incubation and the subsequent rearing of the chicks. When two birds are initially placed together, the female fights with the male; once she has established her dominance, both birds settle down and live amicably. When the female is ready to breed she solicits males by making a deep *ooom* call. She accomplishes this by inflating her crop with air and then standing in a horizontal position, her body vibrating with the effort as she calls. This sound carries far and is uttered both day and night when the female is ready to mate. When a male responds to her call, she will initially attack him. Provided he stands his ground, the pair will be feeding together within a few minutes. That night they will roost side by side in typical quail fashion. The other males tend to wait on the sideline while a pair are together. Mated pairs have an unusual habit of walking around in slow motion, placing one foot in front of the other with slow jerky movements while simultaneously rocking their bodies back and forth.

The female regularly feeds the male, a gesture he seldom reciprocates. During copulation, the male continually pecks at the female's crown while dropping his wings and flitting them over her flanks. The nest site is chosen by the female. The nest consists of a scrape in the ground, placed at the base of a tussock of grass, and lined with grass. As soon as a full clutch of eggs has been laid, the male begins incubation.

Within a day of the first male starting to incubate, the female will commence her *ooom* call and choose another of the males which have been waiting their turn. Her new mate now becomes the dominant male, but within seven days he too will be sitting on a clutch of eggs. This procedure is then repeated with the third male. The usual clutch consists of 3–4 cream colored eggs speckled with rust, brown, and slate colored markings. The incubation period is 13–14 days. Once the chicks have hatched, the male is usually a model parent. I have had experiences with females held in small flights with a single male that have killed all the young at hatching. Ideally, the female should be placed in a new aviary with a new mate as soon as the previous male begins incubation.

The chicks are tiny creatures that can pass through the usual wire aviary netting. For the first week, the chicks take all their food from the beak of the male. They are fed largely on live food and it is quite amusing to watch the speed at which the male feeds termites to the chicks as they stand in a semi-circle around him. Generally, broods of two are raised. After feeding, the chicks nuzzle against the male's flanks to stimulate brooding. Chicks over a week old soon begin feeding on seed and rearing mash. They are able to fly at 10 days, are fully grown at 35 days, have molted into adult plumage at 56 days, and begin to breed at 103 days. Chicks should be removed as soon as they are independent, as females are temperamental and will occasionally attack and kill chicks. It is possible to remove a male with his brood to a separate brooder box. Leaving older broods in the aviary will make life impossible for younger broods.

Incubation and Rearing: Artificial incubation and rearing is as described for the CALIFORNIA QUAIL, but buttonquail chicks require far more live food and attention. Initially, insects are fed every two hours with a pair of tweezers. Chicks should not be placed into outdoor pens until at least two-thirds grown. Stress is often the cause of death in young buttonquail.

Ring Size: 3 mm

Species Requiring Similar Management:
All buttonquail *Turnix* species.

Contributor: Colin Wintle, Bromley, Zimbabwe.

Woodpeckers, Barbets, Toucans

07 Piciformes

07.01 Picidae Woodpeckers

Martin Vince

Introduction: There are more than 200 species of woodpeckers, including those known as flickers, sapsuckers, piculets and wrynecks, all belonging to the family Picidae in the order Piciformes. They are all highly insectivorous but, to varying degrees, they also eat fruits, nuts, seeds, sap and some nectar. Woodpeckers range in size from 7.5–55 cm (3–22 in.) in total length, and can be found worldwide, except in the polar regions, Australia, Madagascar and oceanic islands. They live in all types of forest, as well as scrub and grassland, and search for insects at every level, including the ground itself. Most woodpeckers use their zygodactyl feet (two toes pointing forward, two pointing back) to perch on vertical tree trunks, searching for food or excavating a cavity in which to sleep or nest. They are generally highly vocal and strictly territorial during the breeding season.

Management: Woodpeckers can be bred in aviaries of a minimum of 2.5 m x 2 x 2.5 m (8 x 6 x 8 ft.) high. All species are very active, however, and fully occupy larger spaces if they are available. To appreciate the long, undulating flight of the larger species, aviaries as long as 6 m (20 ft.) are best, although they are not essential for breeding. Woodpeckers are destructive by nature and can be expected to ruin a wooden aviary as effectively as a parrot. Indeed, it is best to build the aviary as if for a medium-sized parrot, such as an Amazon species. Metal should be used as much as possible; exposed wooden surfaces can be protected by nailing on aluminum sheeting. Thin metal sheeting is adequate for woodpeckers, since they lack the dexterity and strength of parrots, and can only hammer surfaces uniformly.

Habitat "A", as described in the section on LANDSCAPE is ideal for forest species, and habitat "B" is suitable for species from open wood land. Most woodpeckers spend a lot of time running up, or across, tree trunks, their stiffened tails being used for support. It is

Rufous woodpecker in bark lined test box.

Black-headed Woodpecker
Picus erythropygius

important to furnish the aviary with plenty of vertical, bark-covered surfaces on which the birds can climb and perch comfortably. Horizontal branches should also be provided, including very thick ones onto which the birds can hop. Horizontal branches are particularly important for the wrynecks and piculets; these species perch more like a passerine than a typical woodpecker. For the majority of woodpeckers, vertical trunks should be provided at both ends of the aviary so that the members of a pair can roost separately if they wish. Small roofs installed over roosts will protect sleeping birds from the rain. Many woodpeckers will roost in nest boxes, or cavities that have been excavated specifically for this purpose. Two nest boxes or nest cavities can be provided for roosting. Boxes should have bark fixed to the inner vertical surfaces for the birds to perch on.

Winter protection is recommended for all woodpeckers, even those species that come from temperate climates. Once acclimated, all woodpeckers should be able to safely tolerate temperatures as low as 10°C (50°F), but none of them should be placed in freezing or drafty conditions. Woodpeckers that roost in nest cavities or boxes are far more protected from the cold, much like their relatives, the aracaris and toucanets. Species or individuals that do not roost in a cavity should be protected from severe weather with a shelter and heat source. A night-light to allow feeding after dark is also beneficial.

Diet: The major part of the diet should be a medium or coarse grade insectivorous mixture that includes live foods.

85% insectivorous diet (described below)
15% diced fruit

Woodpeckers are said to eat nectar in the wild, but it is probably not a major part of their natural diet. My personal preference is to offer only a small amount of nectar as a conditioner every second or third day during the breeding season.

Insectivorous diet:

60%	pellets
15%	hard-boiled egg
10%	live food
10%	pureed apple
5%	tofu

The bulk of the insectivorous mixture is made up of small pellets. In the US, companies such as Marion Zoological and Kaytee market tiny pellets. They are the size of canary seed, and are actually intended for finches and canaries but, when properly prepared, they make a satisfactory food for insectivores, as well as other softbills. If very small pellets cannot be obtained, any good quality softbill pellet can be used, provided that it is first ground into small pieces the size of poultry starter or canary seed. A coffee grinder, set for a coarse grind, will quickly reduce the pellets to the correct size without turning them to powder.

In the UK, a proprietary insectile mixture such as Prosecto™ may replace the pellets. The balanced, researched nutrition of a good quality pellet, however, is difficult to ignore, and if possible, proprietary pellets should be included in the avian diet. Hard-boiled egg

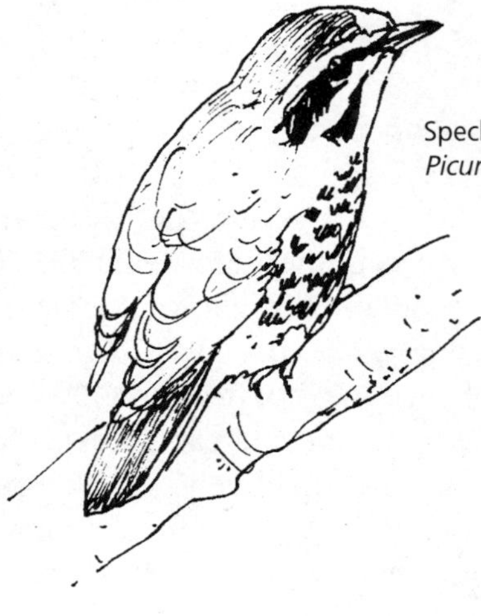

Speckled Piculet
Picumnus innominatus

should be chopped or mashed very finely. All insectivores currently kept in aviculture, even newly imported ones, will eat the white, which presumably resembles certain nutritious white foods utilized in the wild, such as wax moth larvae (waxworms) and ant pupae. Live foods such as crickets about 2.5 cm (1 in.) long, medium mealworms, and waxworms should be included in the insectivorous mixture. If the birds are in the process of being acclimated to an artificial diet, the movement of the live food will stimulate the birds to eat. Initially, the live food will be eaten to the exclusion of most other foods. To provide the birds with more balanced nutrition, the insects can be doctored by dipping them in a sticky honey or nectar solution and then rolling them in the insectivorous mixture so that a light coating adheres to them. Once the birds start to accept the artificial diet, live insects become less important, and inanimate foods such as frozen crickets and locusts can be used. Supplements formulated for insectivores, such as Nekton Tonic I, can be added. This is probably not essential in a balanced diet, but may be beneficial, especially when newly imported birds are being acclimated. To help the acclimation process, sprinkle the insectivore supplement on the insects after they have been dipped in the nectar or honey solution. [GH: Watching a pair of wood peckers at Walsrode, I noticed the food bowl was placed adjacent to a natural log. This allowed the birds to cling to the log while reaching into the food bowl.]

Breeding: Woodpecker eggs are roundish and glossy white. Clutch sizes are 2–14 eggs. The incubation period for the smaller species is 11–14 days, with fledging occurring around three weeks of age. The larger species have an incubation period of 21 days, and take up to five weeks to fledge. The woodpeckers currently in aviculture are remnants of previous importations, or rehabilitated birds that are unfit for release. There has not been much captive breeding. This is arguably because woodpeckers, apart from the white-backed in Sweden, have not been the focus of coordinated breeding attempts, rather than because they are unusually hard to keep or breed.

Woodpeckers are cavity nesters, like the related toucans and barbets. The cavity is excavated by both sexes in the trunk or thick branch of a tree, or sometimes in the wood of man-made structures such has fence posts. Cavities tend to be made in rotten wood, but soft, live wood is also used. An entrance hole of approximately 2.5–7.5 cm (1–3 in.) diameter runs into a tunnel 15–76 cm (6–30 in.) long, depending on the species. The entrance tunnel terminates at the nest chamber, which is lined, inadvertently, with wood chips and debris from

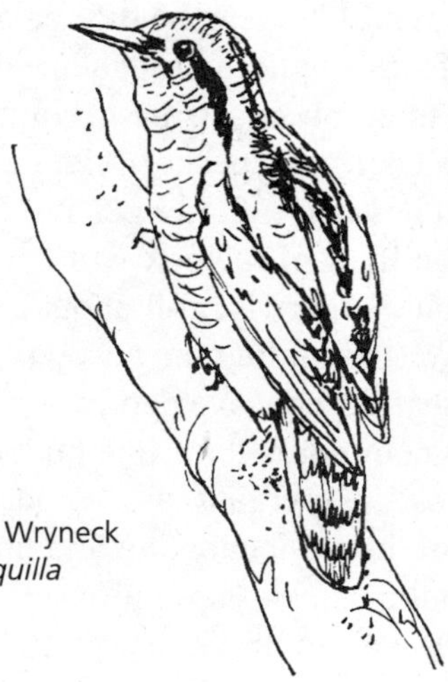

Eurasian Wryneck
Jynx torquilla

the excavation. The two wryneck species, genus *Jynx,* are the most primitive woodpeckers. They do not excavate their own cavities but use the abandoned holes of barbets and other woodpeckers. Some of the flickers excavate holes in the ground or in termite mounds.

In aviaries, woodpeckers will usually accept palm logs for nesting; they are soft enough for the birds to excavate, and can often be obtained from breeders of toucans. They can, however, be expensive, and alternative woods are equally good. A rotten or semi-rotten trunk 20–30 cm (8–12 in.) in diameter and about 1.5 m (5 ft.) long is suitable for medium to large species. If softwood is not available, a nest log can be made as follows: Cut the log into two pieces by slicing down its length. Hollow out the two halves with a chainsaw, then glue and screw them back together, pressing wood shavings into any cracks that admit light. Cut a suitable entrance hole in the side of the log, near the top, and hang the log in a sheltered position high in the aviary.

A pre-excavated cavity such as this may be all that is required for some species of woodpeckers; however, it is often a good idea to pack the log with natural material to encourage the pair to excavate their own cavity. New partners also seem to bond more easily as they excavate their nest together. The log can be packed with a mixture of wood shavings, sphagnum moss, and mud. The woodpeckers will be able to excavate this material fairly easily, giving them access to the chainsaw-prepared cavity. Many species excavate a new cavity each year. Re-packing the log at the beginning of each breeding season can therefore help stimulate the reproductive cycle.

Nest boxes have been used successfully by some aviculturists in breeding woodpeckers, although I have only tried breeding them in logs, getting as far as the egg stage with the black-headed woodpecker *Picus erythropygius* and the chick stage with the rufous woodpecker *Celeus brachyurus*. A tall, grandfather-clock-style parrot nest box can be offered as well as a log. The inner surfaces should be roughened or bark-covered for the birds to climb on easily. More than one box can be provided, since woodpeckers often use them for roosting as well as nesting. A 2.5 cm (1 in.) layer of wood shavings on the floor of the box will help stabilize the eggs, which would otherwise roll about on a smooth, unlined surface. When building the nest box, it is also helpful to gouge a shallow concave depression into the floor of the box, as is commonly done in a budgerigar box.

In most, or perhaps all, woodpecker species both sexes share in the incubation and chick rearing. Insects are fed to the chicks, especially soft, white ones such as ant pupae, wax moth larvae, and freshly molted mealworms. Crickets are worth trying, although I have not found them to be popular with woodpeckers. In some woodpecker species, ants are a major part of the adult and chick diet, and are thought to be regurgitated to the young. The provision of ants, at least during breeding, may therefore be important in determining whether a particular species needs them or their pupae to successfully rear chicks.

Artificial Incubation: Based on research conducted at the San Diego Zoo, the recommended incubation temperature for woodpecker eggs is 37.5°C (100.6°F) (Kuehler & Good 1990). The relative humidity should be approximately 55% at the beginning of incubation. The humidity should be adjusted so that when the chick hatches, or more precisely, when it enters the air space, the egg will have lost 12–15% of its weight when freshly laid.

Pair of white-backed woodpeckers.

Artificial diets as described for shrikes are recommended for hand rearing woodpeckers.

Common Flameback
Dinopium javanense

Day	Weight (g)	Day	Weight (g)
1	4.65	**20**	68.03
2	4.53	**22**	67.75
4	5.83	**24**	69.75
6	9.45	**26**	70.75
8	13.18	**28**	67.75
10	18.93	**30**	65.75
12	29	**32**	60.5
14	41.2	**34**	62.75
16	48.1	**36**	61.5
18	56.3	**38**	62

Ring Size: 3 mm for the smaller species, up to about 16 cm (6½ in.), and 4.5 mm for the larger species. Extremely small or extremely large species are seldom available to aviculture, but would require smaller or larger ring sizes, accordingly.

Species Requiring Similar Management:

Woodpeckers
Piculets
Wrynecks

Reference:

Kuehler, C. & J. Good. 1990. Artificial incubation of bird eggs at the Zoological Society of San Diego. *International Zoo Yearbook* 29: 118–136.

White-backed Woodpecker

Dendrocopus leucotus

• Christer Larsson

Introduction: The Swedish Society for Nature Conservation decided in 2002, in cooperation with Nordens Ark in Sweden, to initiate a captive breeding programme for the white-backed woodpecker *Dendrocopus leucotus*. The aim is to produce young woodpeckers for release to restore a wild population. The white-backed woodpecker is one of Sweden's most threatened bird species; and it has been on the Red Data list for decades and is listed as CR — critically endangered. The woodpecker is found in deciduous forests or mixed forests rich in deciduous trees. The changes in forest management throughout the 1900s have led to the percentage of deciduous trees decreasing and at the same time decaying wood has become scarce. The species' natural distribution has decreased to 2–4% of earlier recorded distribution and the bird is now three isolated populations that have fallen below the sustainable population threshold of 25 pairs. In Finland the species has seen similar declines to the Swedish population but in Norway the woodpeckers are a little more numerous, with population estimated at about 1000 pairs in 1995.

As the loss of the woodpeckers would mean a loss of focus on forest restoration, the Swedish Society for Nature Conservation decided to focus on a long-term captive breeding program, and at the same time pay attention to protecting valuable habitats and increasing the percentage of deciduous trees in the forests. The society harvested 81 chicks between 1995 and 2001. Fifty-one of these were released into the wild in Sweden, others placed in the nests of great-spotted woodpeckers *Dendrocopos major* and some transferred to Nordens Ark to initiate a captive breeding program. It is envisaged that this will be a 25–30 year program and that the project will be run and assessed in five-year periods. The aim of the programme at Nordens Ark is to produce sufficient numbers of young to restore populations to more than 25 pairs.

At the ouset there was substantial doubt as to how successful captive breeding would be. Initial problems occurred with infestations of internal parasites, which resulted in high mortality amongst the flock. With parasite control now in hand and an excellent husbandry and hygiene routine in place, this mortality has been greatly reduced. The first pair were placed in a 80 m^2. (860 sq. ft.) aviary with a mean height of 5.5 m (18 ft.) in the public section at Nordens Ark. The aviary houses several natural trees and, in addition, trunks of birch and aspen have been placed in the aviary. The trees were quickly perforated by the woodpeckers and need to be replaced on a regular basis. Visitors have close access to the aviary but the woodpeckers tend to ignore the people and go about their activities of drumming and searching for food. They fly around the circular aviary like projectiles immediately below the roof netting and then suddenly land on a tree trunk. Both the male and female came as nestlings from Vest-Agder in Norway in 1999. They showed some copulation activity in spring 2001, but no egg laying occurred. They were nearly four years old when on March 26, 2002 they began to excavate a hole (approximately 25 cm (10 in.) in diameter) in a section of aspen. It was primarily the male that excavated the hole while the female cleaned the nest of chips. The nest was complete on April 20. The first copulation was witnessed on April 1 but at least 40 were observed up until the female started egg laying on the April 23.

Incubation was shared by the sexes for 14 days, with the male usually incubating at night. On May 10 it was evident that the adults were feeding chicks. There was always at least one parent in the nest hole for the first eight days. When the chicks were 14 days old they climbed to the nest opening where they were fed. Two female chicks fledged on June 3 and a male chick the following day. The nestlings were fed 2–3 times every hour. They began searching for their own food within a week of fledging. The young woodpeckers soon learned to climb the tree trunks and fly between trees and around the aviary at high speed. Unfortunately one youngster later hit the netting and broke its neck. A number of more conventional aviaries measuring approximately 18 m^2 (195 sq. ft.) are now used for off-display breeding. The breeding is the second on record with the first having taken place at Moscow in 2000.

Christer Larsson from Nordens Ark provided the following information on the hand rearing of the chicks:

> "All the birds we keep as breeding pairs and all the birds that have been released have been taken from wild nests in Norway. This, of course, means that we do not know the exact age of the birds. Another problem is that we sometimes get the young woodpeckers at very different ages. The weight can vary from 60–80 g when they arrive here. The weight of adult birds is 95–115 g. During the hand rearing the chicks are fed every second hour from 07.00 in the morning to 19.00 in the evening; sometimes more frequently and sometimes with a 'siesta' in the middle of the day. When our hand-reared chicks start to leave their 'nest boxes' the weight must be 85–90 g. Immediately they are very active and search for food. The following days (up to one week) they lose weight before they learn how to find and handle live food. The diet consists of meal worms, zophobas worms, crickets and mixture for insectivorous birds. Sometimes tinned cat-food and small slices of different meat are added. During every feeding time some of the food is dipped in mineral powder and vitamins."

07.02 Lybiidae African Barbets

Glen Holland and Dick Schroeder

Introduction: Barbets belong to the order Piciformes and form the family Capitonidae that includes 81 species and includes 10 species of tinkerbirds. Barbets are found in most tropical regions. The largest number of species is found in Africa and Asia; they are also found in South America. Barbets occur in a wide range of habitats from dry woodland and scrub to dense tropical forests. My experience has been primarily with African species. Usually seen in pairs or family groups, which are strictly territorial, their calls are generally loud and used for the maintenance of territories. With some species such as the black-collared, a few birds congregate and create an awful din as they duet. They are dependant on water. Some species such as the crested (Le Vaillants) are common visitors to garden bird food tables. Barbets are popular hosts for honeyguides, and the black-collared barbet, for example, is parasitised by the lesser honeyguide and will immediately attack a honeyguide that enters into the vicinity of the nest. They have been recorded to have killed a lesser honeyguide, that was caught while in the nest.

Management: Barbets adapt well to captivity and will nest readily. When newly acquired they are delicate and need additional warmth and some treatment for stress for the first week. During this stage they are best confined to a flight cage with a water-soluble anti-stress solution in the drinking water. They are largely frugivorous and are able to swallow large pieces of fruit whole. Barbets are also fond of soft food and insects, which are essential when they are raising chicks. In flights, all barbets must be kept as sexed pairs of individual species, but in large planted aviaries it is possible to keep pairs of different species together.

Their aviaries should be landscaped in accordance with habitat "B" for most species. A few species, however, originate from forest habitats and their aviary should be landscaped in accordance with habitat "A". In his article on breeding the red-headed barbet M.D. England wrote about the bad reputation for aviary destruction that barbets have:

> "As a family, barbets have in captivity a reputation for being very destructive to woodwork of aviaries, to such an extent that they have never been very popular and my friends warnings, when I said that I was hoping to study the family, led to my housing my first two species in metal framed aviaries. However, these forebodings proved unjustified and in my opinion the bad reputation is quite undeserved. Aviary damage can nearly always be avoided by providing plenty of rotting logs to which the birds can devote their almost inexhaustible boring propensity. It may not be inappropriate here to mention that one should not allow oneself to be over optimistic when the birds appear to be boring at a nest hole. Holes are bored for roosting, to find food and, apparently, just for something to do, while some species bore, in the autumn, a hole which will be used for nesting the following summer."

Barbets will not remain in good condition if kept in cages for long periods. They are not generally openly aggressive to other species but will defend their nest ferociously against intruders. In two instances I saw a crested barbet *Trachyphonus vaillantii* fall to the ground fighting with a glossy starling which attempted to inspect its nest, and in the second instance, one fell to the ground fighting with a sparrow-weaver whose nest the barbet had investigated. They should, therefore, be kept with

Black-collared Barbet
Lybius torquatus

species of their own or a larger size. Double-toothed *Lybius bidentatus* and others of the larger species are rather combative and should not be kept with small softbills or seedeaters. More nests should be provided than pairs of barbets. Larger species will also steal eggs from the nests of smaller species. I have kept crested, black-collared *Lybius torquatas,* and pied *Tricholaema leucomelan* in a mixed collection in an aviary with a floor space of 850 m^2 (1,017 sq. yd.) . They should not be kept in flight-type aviaries with other cavity nesting species such as starlings as they will compete for nests, resulting in aggression. Provided enough nesting facilities are supplied, this aggression will not be a problem in a large aviary. Barbets will use nest boxes to roost in, both during the day and night, but for breeding a more natural log should be supplied. They always sleep in a nest log at night and without this, they will not thrive. I have seen a nest log in the wild in which a pair of black-collared and their three fledged chicks all slept together.

Adult pairs of black-collared are often accompanied by their previous season's chicks. In an aviary the chicks can be left with the parents for up to eight months, but the adults should be rung to avoid confusion when the chicks are removed. The previous brood of chicks in some species also assists in feeding and the removal of feces from a new brood. With pied and crested, for example, at a month old the chicks should be removed to avoid aggression from the adults. Chicks are best color ringed at fledging, which will assist in avoiding confusion with their parents at a later stage. They are fond of bathing in shallow water or in a mist spray that is placed over a dense shrub in the aviary.

Barbets are generally from warm areas and although, due to their habit of roosting in nests, they are able to withstand overnight temperatures down to freezing, they require warm daytime temperatures of 14°C (57.2°F) minimum. In wet climates they should be provided with plenty of shelter. M.D. England writes on keeping the red-headed barbet in England: "In winter my aviaries are thermostatically controlled at a minimum temperature of 10°C (50°F) although one or two of last winters "Norfolk nights" beat the heating and brought it down to little more than 4.4°C (40°F)."

Countries with severe climates, particularly where cold temperatures are combined with wet conditions, will require that they be housed indoors with some additional warmth.

Diet: Barbets feed on fruit and insects, the percentage of each varying among species, the season, and the availability of food. Most species become largely insectivorous when they have chicks in the nest. In captivity they feed on paw-paw, banana, grapes, soaked raisins and sultanas, oranges, apples, figs, and commercial softbill mixtures. Grated boiled egg can be added to the softbill food. Some insects such as mealworms, locusts, moths, and winged termites must be added to the diet. I have not found them keen to take minced meat of any kind. If the diet is not kept varied, and insects excluded from the diet, the colors on the birds tend to fade. A multivitamin and trace element powder should be added to the diet twice per week. Nectar such as described for sunbirds is also enjoyed by barbets.

Breeding: In addition to duet calls, courtship includes the male feeding berries and insects to the female. He leans towards her with his tail cocked in her direction, and mutual bill touching occurs. The female solicits copulation by crouching, stretching her body, and gaping. Often after copulation the male lures the female to the nest by tapping round the entrance of the cavity with his bill. In the same way, males and females guide newly hatched chicks to the sleeping cavity. Their nest is excavated in a dead stump by both sexes. I gave one of my pairs of crested a section of a palm stump, which they immediately accepted. Once they had excavated their nest, a wedge was cut into the nest and a hinge put on this. As soon as I heard the trilling calls from the chicks, the nest was opened and the chicks removed for hand rearing. The wedge was then replaced and the pair raised a second clutch. It is advisable that an access hole with a diameter of 10 cm (4 in.) be cut into the back or side of the log and hinged to allow for nest inspection and removal of chicks for hand rearing.

Nest logs should be slightly soft internally yet firm externally. They should measure about 1 m (39 in.) in length and about 20 cm (8 in.) in diameter. A section of a dead palm trunk provides an ideal nest log. The nest log can either be placed in an upright position or at an angle with the top leaning outwards. In this instance the birds will enter on the underside of the log. The nest log should be positioned at 1.5–2 m (59–79 in.) off the ground. A small pilot hole can be started about 20 cm (8 in.) from the top of the log and this will assist in starting the birds off. Often the birds start a number of test holes before a nest is completed and it is not unusual for even a complete nest to be abandoned. The nests should not be removed at the end of the season as the birds roost in the nests throughout the year. No nesting material is added to the nest. A little dust from the bird's excavations lines the base of the nest. I bred pied barbets from a pair of hand-raised birds that nested in their first season. Numerous attempts were made at building a nest, which was finally completed in a dead log in the center of the aviary.

Pied barbets are capable of excavating into hard wood including acacia thorns, hence the old name of acacia barbet. Once the female had shown her acceptance of the nest the male commenced feeding her at the nest entrance. Incubation was shared and lasted 15 days. During warm weather the birds spend as much time off the eggs as the time spent incubating. The usual clutch is 2–4 white eggs. When the chicks hatched, the female spent most of her time brooding and it was up to the male to provide food for the family. The male would fly to my hand to fetch insects and then straight back to the nest to feed his family. Some fruit was also taken to the nest. As the chicks grew, the female spent increasing periods out of the nest with one of the parents visiting the nest approximately every five minutes. After a month the chicks' heads appeared at the entrance of the nest to beg for food. The two chicks fledged at 32 days and although they were feeding themselves after two days, they continued to be fed by the parents for a week. While feeding chicks, the quantity of live food offered should be greatly increased. Once chicks appear at the entrance the adults try to lure them out of the nest by refusing to feed them, but perching close to the nest with food in their beaks.

M.D. England wrote an interesting breeding record with this species from England:

> "When mine arrived they had an obvious dislike for the cold of a British April — or rather at 15.5°C (60°F) at which my birdroom was kept, and they never looked happy until I put an infra-red lamp on them, although they did well in the start. After a week they were given access to a flight which measured 7 x 1.35 m (24 x 4.5 ft.) and despite the miserable weather they both immediately improved. They are not happy in close confinement, and even in this case the 'confinement' consisted of 2.4 x 1.5 x 1.35 m (8 x 5 x 4.5 ft.). Needless to say I gave them no companions. It was not long before I was fairly sure that I had two birds of the same sex, probably males. One was aggressively dominant to the extent that the other began to skulk in the shelter, seeming unwilling to go out into the flight when the other bird was there and only dared to snatch a bit of food while fearfully 'looking over its shoulder.' My pessimism grew when one of them took over a nest box at the far end of the flight while the other started boring a hole in a rotten tree stump in the shelter. However, gloom eased a little when relations improved sufficiently for both of them to take turns at boring.
>
> The method used was interesting: each would attack the wood, hammering away until a piece was loosened at one end; this was then seized and torn off and, when a large beakful was free, it was carried to the far end of the flight and dropped on to an ever-increasing heap. Amazingly small pieces of wood were swallowed and regurgitated on to the heap; practically nothing remained beneath the hole to give the site away to enemies. Hopes really did soar when mating and courtship feeding were seen (the dominant bird was the male), and one bird began spending long periods in the hole. Some very aggressive chasing still occurred but mating took place in a most casual manner, with no preliminary display, after the two birds had one of their rare periods of sitting peacefully side by side preening. Far more often than not it was unsuccessful. Any hope of breeding disappeared when one morning there was a gaping hole in the side of the log; they had opened up the nest cavity, whether from inside or outside we could not tell. So that was that and summer was fast passing. A new log was put up in place of the old but they ignored it and seemed as though even their very half-hearted desire to breed had left them. In a final attempt to stimulate them I cut a hole in the bark and exposed the soft wood within — immediately they

Foot of a barbet.

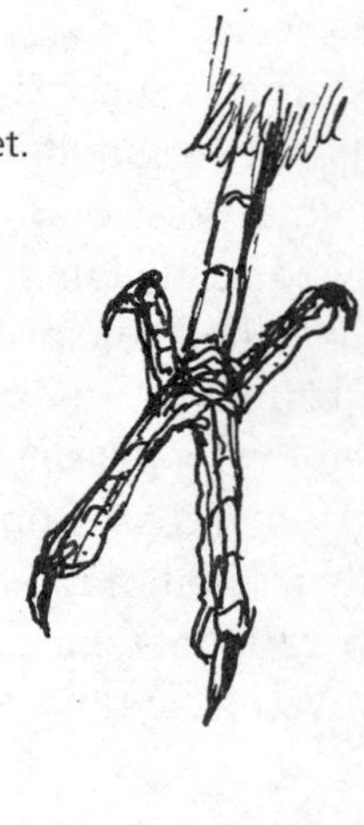

started to bore. Again they took it in turns, and again they carried their winnings to the far end of the flight, where they filled their bath with them.

Although it was impossible to be quite sure when it started, there was little doubt that incubation was taking place, since frequent nest-relief was seen. The relieving bird did not always wait for its mate to leave the nest, but clung to the tree outside the hole (presumably making a noise which I could not hear) or went inside and down to the cavity. Sometimes they were in there together for quite long periods. Hatching was presumed to have happened when on the August 26 the birds food consumption increased. For the first few days chopped lettuce and chickweed was the favorite food, but the youngsters' menu was gradually increased until it was being fed on the widest variety of chopped food that I have ever experienced. All of the food was regurgitated by the parents and included grated cheese (little), chopped hard boiled or scrambled egg yolk, soft food, chopped raisins and currants, apples, mashed carrot, maggots (few), mealworms and small locusts. Until September 9 (that is, for a fortnight) one parent was almost continually in the nest, and from September 1, feces were seen to be carried away; presumably it had been swallowed prior to this. On September 9, and thereafter, both parents were off most of the time, though both continued to feed at frequent intervals — surprisingly frequent considering that regurgitation was employed. During this time persecution of the female by the male intensified, to be interrupted occasionally by brief periods of truce when courtship feeding and mating took place. On September 22 the youngster was looking out of the hole and was being fed by the male with egg yolk in beakfuls, not swallowed and regurgitated, using a rapid 'pumping' action with the bill inserted far into the youngsters' gape. On the September 23 the youngster fledged."

A pair of double-toothed at Frankfurt Zoo was busy with a sixth brood 14 months after starting with the first. Nine young fledged from the first five broods, of which eight survived to maturity. San Diego Zoo have also kept and bred this species and have raised about 30 over the past eight years. Here, they are housed in a large walk-through aviary with a variety of African species such as starlings and robins.

Hand Rearing: Barbets are relatively easy to raise on a diet of equal portions of commercial rearing food and beef-flavored cat kibble which has been moistened with water. Small pieces of pinkie mice should be used in the first week. Insects such as bee larvae, waxmoth larvae, and small crickets can be given with the remainder of the diet consisting of grated apple and a few raisins that have been soaked overnight. Initially the chicks are fed hourly from 6 a.m. to 10 p.m. and maintained at a temperature of 35°C (95°F). The temperature is reduced by 1°C (1.8°F) per day until a temperature of 25°C (77°F) is reached at which stage the quills and feathers will begin to appear. Once this stage is reached the chicks should be fed every 1½–2 hours from 6 a.m. to 8 p.m. Once the feathers begin to develop and the birds are able to consume more at a feed, insects and diced fruits are fed with a forceps tweezer. The rearing food mixture and beef-flavored cat kibble are moistened to form a firm porridge texture and fed with a thin syringe of which the entire narrow tip is cut off. A calcium supplement and a few drops of multivitamin are added to the diet on a daily basis.

The chicks are not aggressive and can be raised together. They are best kept in a parakeet-type nest box, which can easily be opened for feeding. To ensure hygienic conditions, the base of the box should be covered in a thick layer of wood shavings, which can easily be replaced twice per day. The chicks are dull versions of the adults. Once half grown the chicks of many species become quite vocal, making calls similar to those of the adults. Crested barbet chicks, for example, *chirrr* in the brooder continually.

Crested Barbet
Trachyphonus vaillantii

Day	Weight (g)	Day	Weight (g)
0	4.5	20	39
1	4.4	22	40.5
2	4	24	39.5
4	6.5	26	42.5
6	9	28	46
8	10.5	30	52
10	15.5	32	50
12	19.5	34	50
14	23.5	36	53.5
16	28.5	38	50.5
18	32.5	40	55

* Information provided by the San Diego Zoo.

Hand-raised chicks remain hand tame, and some that I had in a walk-through aviary gave many hours of enjoyment to visitors whose hair, ears, shoe laces, and anything else of interest were inspected by the barbets.

Ring Size: 4.5 mm (0.2 in.) for the species mentioned.

D'arnaud's barbet *Trachyphonus darnaudii*

• Dick Schroeder

Introduction: There are five species in the family Trachyphonus. They have some interesting habits. One is the ability to "sing" duets and the other is the fact that several of the species nest underground. The D'Arnaud's is one of these subterranean nesters. They live in the open wooded grasslands of Sudan, Somalia, Kenya, and Tanzania. They require flat terrain with soil suitable for nesting. Not much larger than a lovebird and a mottled black and yellow, they have the start of a black crest, giving them the appearance of wearing a witch's hat or a dunce cap. Their tails are long, as are their legs. Sexes are alike. Their interesting vocalizations are a good reason for keeping these birds. One bird will start and the next will fall in. Some of the notes are long and almost musical, one bird harmonizing with the other.

Maintenance: D' Arnaud's barbet *Trachyphonus darnaudii* is a great bird for a mixed species aviary. Never aggressive to others, it is in fact far too passive to be housed with even smaller species of barbets, like the pied *Tricholaema leucomelaina* which in itself is not aggressive either. They require an aviary with a flat dirt floor and dirt suitable for excavating a nest: neither too sandy nor too heavy, although the latter is preferable to a sandy soil that could collapse. Obviously the soil should drain well, especially in areas of summer rain, as this is the time when these birds nest. An aviary that floods and has standing water will not work for breeding. Also, it seems that they are easily disturbed by mice, who also like underground burrows, especially if something else was kind enough to dig it for them.

Diet: They are more insectivorous than the small South American barbets, but they do take some fruit. Mine are usually housed with turacos, so they receive the fruit mix for the turacos, which is made up of diced papaya, apple, banana, grapes, melon, and other seasonal fruit. Also in the mix is frozen/thawed mixed vegetables and soaked Kaytee Softbill Pellets. To this mix I add mealworms and crickets that have been dusted with Avico Vita Life, vitamin and mineral powder. This works well as the turacos ignore the live food and the barbets make very little dent in the fruit; however, they do consume the soaked pellets.

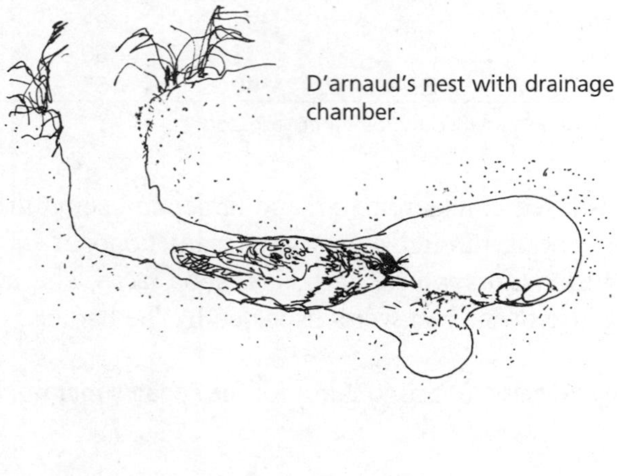
D'arnaud's nest with drainage chamber.

Breeding: As stated above these barbets nest underground. In fact they have always had a nest box or log available wherever I have housed them and still they use a hole in the earth that they excavate themselves. They fly to the far end of the aviary to drop the beakful of dirt. Both birds build the nest, which apparently has a nest chamber above the bottom and off to the side so that water does not accumulate inside. They lay 2–4 white eggs, but I have never had more than two chicks fledge from a nest. It is of course impossible to monitor what is going on underground. The nest is always near a clump of grass or ground cover. In fact, it is often difficult to find. The first chicks I raised I knew existed because the parents were carrying crickets off somewhere in the aviary, but they used a different route and a different direction each time. I never found the nest until I saw a newly fledged chick standing outside of it. It was in a clump of fountain grass. I have never had more than one successful nest per pair in a breeding season, so I do not know whether the young would help feed a second clutch or not. They do retire to the nest at night with the parents, long after fledging, so I suspect that this would be the case. I have never seen any parental aggression with these birds either. I do remove the chicks prior to the next breeding season.

Species Requiring Similar Management:

Barbets
Tinker Barbets
Starlings
Woodpeckers
Toucanet
Aracari

References:

England, M.D. 1973. Breeding the red-fronted Barbet. *Avicultural Magazine*. Vol. 79: 9-13.

England, M.D. 1975. Breeding the red-headed Barbet. *Avicultural Magazine*. Vol. 81: 121-129.

Hilty, Steven L. and W. Brown. 1986. *Birds of Columbia*. NJ: Princeton University Press.

Fry, C. Hilary, Stuart Keith and Emil K. Urbane. 1988. *Birds of Africa*. Vol. III. NJ: Princeton University Press.

07.03 Ramphastidae New World Barbets and Toucans

Glen Holland and Dick Schroeder

07.03a New World Barbets Capitoninae

• Glen Holland

Black-spotted Barbet *Capito niger*

• Dick Schroeder

Introduction: The little South American members of the family Capitonidae are found in the Guianas, Columbia, Venezuela, northern Bolivia, and Amazonian Brazil. Being just over 20 cm (8 in.) long they are more often heard than seen in their native jungle. Their call is a frog-like, braying kind of sound. It is very distinct from the sounds of the other aviary birds. They are dimorphic with the yellow breast of the hen being covered with black spots and streaks, while the male's is just yellow. They have red heads and throats; the back is black with yellowish streaking and spots. Heavy bills are black.

Maintenance: The black-spotted barbet *Capitio niger* is an excellent addition to a mixed species aviary, they are only aggressive to other barbets. The fact that they are cavity nesters makes them excellent candidates for successful breeding in these conditions as well. In fact, all that I have raised over the years have been in this type of environment. These are very active birds, spending much of the day hammering at logs or perches. They are really too small to be very destructive, but I always keep several partially rotted logs in their aviaries for this purpose. I see only minimal damage to any of the aviary woodwork.

Diet: They eat fruit as well as insects and even small lizards or pinkie mice. I feed them my standard softbill mix, consisting of a variety of diced fruit: mainly papaya, apple, grapes, melon, banana, frozen/thawed mixed vegetables, and dry as well as soaked Kaytee Exact Softbill Pellets. Mealworms and crickets are added on top of this mix. The live food is sprinkled daily with Avico Vita Life, vitamin and mineral powder. The crickets are fed frozen/thawed.

Breeding: I have never had these barbets nest in an artificial nest box. They have always excavated their own nest in logs provided. On one occasion I had a nest in a hollow oak limb that I had provided for that purpose, but most often they will make their own hole in the dead trunk of a century plant *Agave americana* flower stalk. These can be 6 m (20 ft.) or more in length and up to 25 cm (10 in.) in diameter at the base. Many times there will be a hole from a native woodpecker using it for this very purpose. I cut them into 1.8-m lengths (6 ft.) and either hang them in the aviary or stand them upright and wire them to the side to prevent them from falling over. Almost immediately the barbets will be over to inspect this and by the next day will have hammered a hole in the hard outer shell into the soft, balsa-like interior. Sometimes they just start at the top and make a tunnel down from there. These logs are only good for one breeding season as they soon are riddled from top to bottom and side to side with tunnels and passageways. The entrance hole is the same diameter as the bird's body, so there is no way a larger intruder can enter.

The eggs are white and usually 2–4 are laid. I can only approximate the incubation time, based on when the hen was suspected to be sitting until a chick is heard or shells are found. It appears to be 14–16 days, but nesting as they do it is impossible to check the log. They always have a curve to the tunnel so even with a flashlight nothing can be seen. While this species feeds primarily on fruit, the youngsters are fed lots of crickets and mealworms for the first week to 10 days, then more and more fruit and soaked pellets are offered. Crickets are rubbed on the perch to remove the legs, then beaten until soft enough to feed. This species has not been a problem eating other birds' eggs, as have some of the large Asian species. I have had turacos and plovers nesting in the same aviary without a problem. It is interesting to note that the first clutch of the season will help feed a second, and on one occasion, a third clutch, with all of the weaned youngsters flying food into the nest log. These same young need to be removed before the next breeding season.

M.D England provides a very informative account

of the breeding of the red-headed barbet *(Eubucco bourcierii),* a species naturally occurring in Columbia, Ecuador, Costa Rica, and Panama:

"As breeding time approaches, the diet of the red-headed barbet makes a radical change. Fruit is almost ignored and live food is sought. What little courtship there is consists of the male offering meal-worms to his mate, and even maggots, which they normally do not like and tend to bring up undigested, are eaten in limited quantities. Locusts are ignored except for feeding young. I have three species with three different methods for preparing locusts for food: the red-headed holds the prey in its feet while the legs and wings are torn off. The red and yellow holding the insect by a leg or wing and shaking it until he body flies off the grasped member and the golden-whiskered *(Megalaima chrysopogon),* which seizes and manipulates the locust in its massive bill, occasionally throwing it up to catch it another way around, and finally swallowing the lot.

While the nest contains youngsters a good deal of hammering is heard from within. When the young are small this can only be by the parent; indeed then it only occurs when one of the adults is in the nest hole. Later, however, I suspected that the growing chicks did it as well and I thought that — as in some other species, sawdust was being made to absorb feces (which were never carried away by the parents). The entrance hole is a very tight fit being usually only a little over 3 cm (1 ¼") in diameter. Male and female take equal turns at boring the hole, incubation and feeding the young, which is done with food carried in the bill. The first sign of hatching was the distressing unsatisfied search for something not included in the diet, which I was supplying. The pair seemed unwilling to take mealworms to the nest and obviously thought nothing of the assortment of crawling things, which I laboriously found in the garden. However I did see them take two small green grasshoppers to the nest so, in desperation, I gave them some very small locusts, which they had steadfastly refused to touch before. Immediately both parents became very excited and flew down to the food dish before I had even closed the door; the answer had been found.

After three weeks the locust diet was supplemented by the adults with mealworm pupae, white worms, chopped sultanas and small pieces of apple rolled in a protein supplement. The first out was a male and the parents then tried very hard to make the other to follow suit by holding food just in front of reach, even allowing them to nibble at it but never to grasp it. Instead of coming out they retied to the nest cavity and stopped appearing at the hole, giving me grave fears that they would starve. I need not have worried because failing to entice them out, the male from within and the female from without worked desperately to cut a hole at nest level. When I later arrived I could see the two nestlings through a hole which was already large enough for them to emerge through. They later fledged and I erected ladders from the floor in all directions to perches above because at first the young could not fly. Like the young of other barbets I have bred, they appeared disinterested in food, and accepted it with reluctance after one of their tireless parents had prodded all around their face with encouraging "clucks". The first brood were removed as soon as it was certain they were feeding themselves because another nest hole was being bored by the parents, chiefly by the male. Regular incubation of the second clutch commenced five weeks after the first brood left the nest."

07.03b Toucans

Ramphastinae

• Martin Vince

Introduction: Toucans are found only in the tropical forests and wooded areas of the New World. They range from southern Mexico to the northern parts of South America, with their center of abundance being the Amazon region. Toucans are largely frugivorous, showing a preference for dry, fibrous fruits such as *Ficus* berries; the are also semi-predatory and use their powerful, serrated bills to capture and eat small birds, reptiles, and amphibians. Forty-one species in six genera make up the subfamily Ramphastinae of the family Ramphastidae, order Piciformes. There are 11 species in the genus *Ramphastos,* which includes most of the large species made famous by television and print media. However, the majority of toucan species are actually much smaller birds about the size of a jay. They include six species of toucanets in the genus *Aulacorhynchus,* 13 species of aracaris of the genus *Pteroglossus,* six in *Selenidera,* and one in *Baillonius*. In addition, there are four medium-sized species, known as mountain-toucans, in the genus *Andigena,* living in the Andes.

The toco toucan *Ramphastos toco* is the most familiar, and the largest species at 60 cm (24 in), including the bill of up to 21.5 cm (8½ in.) long. The small toucanets and aracaris range from 25–38 cm (10–15 in.). Most species are monomorphic, i.e., cannot be definitively sexed visually and must, therefore, be sexed surgically or genetically. However, interesting differences between the sexes can often be detected by comparing body weights and the length and shape of their beaks. Males tend to be bigger-bodied and slightly heavier, with longer, straighter bills. The bills of females are shorter and fatter, with more of an arched appearance. Toucanets of the genus *Selenidera* are dimorphic, with the females having mostly chestnut underparts. The green aracari *Pteroglossus viridis* is also dimorphic.

Management: Toucanets and aracaris can be kept with robust birds of a similar, or preferably larger, size. However, all toucans must be matched very carefully with their companions, since they will always look for an opportunity to kill or injure unsuspecting birds; chicks of any species will always be preyed upon. The larger species, such as toco and keel-billed toucans *Ramphastos sulphuratus,* should not be kept with any other birds except very large, tough species such as seriemas and curassows. Even these combinations should only be attempted in the knowledge that the toucans will dominate the aviary, reducing the breeding prospects of any other species almost to zero. Exceptionally large enclosures, of the kind found in only a few zoos, could hold multiple pairs of toucans and other avian species. But again, the strongest toucans will prey uncontrollably on all other birds, just as they do in the wild. They should be given sole occupancy of an aviary, or accommodated with other specimens very carefully.

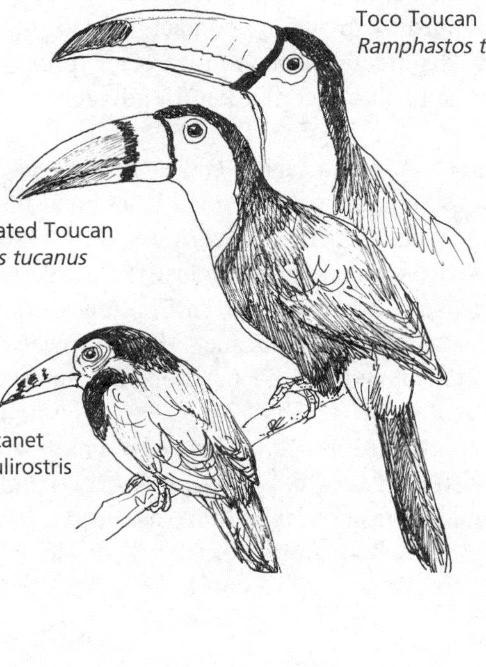

Privacy and a large aviary are essential for breeding success. The back and both ends of each of the successful toucan breeding aviaries at Riverbanks are solid, assuring the birds that they cannot be approached from any direction except the front. This helps the birds to feel less threatened than if they were in an open-sided aviary. Mature bushes and trees as described for Habitat "A" are also important, enabling nervous birds to hide and feel secure. A solid-sided aviary also protects the birds from cold winter winds. A partly covered roof helps to eliminate driving, freezing rain, as well as pathogens that may be introduced by the feces of wild birds or mammals. To help protect the aviary from predators and nuisances such as feral cats, opossums, raccoons, and foxes, electric wire designed for animal fencing can be strung around the aviary roof and exterior floor. A small light fixed near the aviary is also a simple and valuable fixture, allowing the birds to find a perch if startled after dark.

Healthy toucans can withstand temperatures as low as –20°C (–6.7°F) for very short periods of time. Here in South Carolina, winter nights often fall well below freezing, but daytime temperatures that rise to at least 5°C (10°F) prevent the birds from becoming ill. In our climate we are able to keep toucans outside year-round, which we firmly believe is one of the secrets of our success. See AIR QUALITY. In the northern states, Canada, and western Europe, where winter days are almost as cold as the nights, toucans are far less likely to be able to stay outside year-round, and therefore less likely to breed. In spite of the favorable climate in South Carolina, toucans are vulnerable to the lowest winter temperatures and should be offered protection. Frostbitten toes are the greatest risk, as well as frost damage to the beak. Many of the toucanets and aracaris sleep in their nest log and do not necessarily need a separate shelter. For the larger *Ramphastos* species that do not sleep in their log, we provide a heat box constructed of 10 mm (½ in.) plywood. This rectangular box, measuring 75 x 45 x 45 cm (30 x 18 x 18 in.), is mounted horizontally to the back of the aviary. Part of the box is open to give the birds access to a 250 watt heat lamp located inside. The heat lamp should be shrouded in wire mesh to prevent the birds from burning themselves. We have not seen the birds enter these boxes, but they do stand on top of them and receive sufficient heat to prevent frost-damaged toes.

Sunbathing seems to be an important part of toucan biology, apparently being necessary to bring birds into peak condition, particularly for breeding. Indeed, considerable thought must be given to the location of the aviary in order for the birds to have access to several hours of direct sunlight. This is also very important during the winter, when early morning sunlight helps to rapidly warm the birds. The aviary should face south. We try to find a balance between foliage, with the seclusion it provides, and access to the sun. Depending on the temperament of the birds, one tall, bushy tree inside the aviary will help them feel less exposed; at the same time, they can fly to the extremities of the tree to sunbathe. A substantial, foliated tree inside the aviary will also help to obstruct an aggressive bird, usually the male, in the event that the pair does not always live together amicably. Trees outside the aviary will need to be inspected and perhaps trimmed thoughtfully to allow more sun to penetrate. Trimming evergreen varieties is especially important since they will obviously occlude the winter sun.

Good aviary hygiene and design are the simplest and best defenses against the vast majority of avian diseases. The aviary must be designed carefully to facilitate quick and thorough cleaning and disinfection. Good drainage is one of the keys to a disease-free toucan aviary. The ideal floor covering is smooth concrete that slopes to a drain, enabling proper cleaning and disinfection. This type of floor covering is not very attractive and, for display purposes, materials such as river rock or rock chips can be placed on top of the floor. Planters in the concrete floor should also be made for trees and bushes. Yersiniosis, or pseudotuberculosis, is a major cause of toucan deaths in aviaries worldwide. The disease is normally passed on to the birds as they eat food that has been defecated upon by mice and rats. All Ramphastids are highly susceptible to this disease, and many aviculturists have seen their birds transformed from apparent good health to death within only a few hours. Appropriate antibiotics can help affected birds, but they must be administered immediately, and in practice the rapid onset of the disease makes early detection virtually impossible. Survivors from an afflicted aviary could benefit from a prophylactic antibiotic treatment. But of course prevention is far better than a cure. Careful steps must be taken to prevent this devastating disease from occurring in the first place:

Metal pen prevents rodents from reaching food tray.

1) Raise food pans at least 1.2 m (4 ft.) above the ground to help keep vermin away from them. Do not place branches or objects near the food that vermin might use as ladders.

2) Ensure that the aviary floor is kept free of food and feces. Fallen food on a damp floor will attract vermin and support disease. Ensure that the aviary floor is well drained.

3) Disinfect the floor at least twice a week, rotating between two or three different types of disinfectant. Before disinfection, it is important that the floor be visibly free of organic material, soaps or detergents, since these substances greatly or completely reduce the effectiveness of most disinfectants. Carefully read the manufacturers' instructions. Synthetic phenols, for example, are toxic for several days after use, and must be completely rinsed away before the birds can safely contact treated surfaces. All disinfectants have important pros and cons; consult a veterinarian for advice.

For aracaris and toucanets, enclosures of 1.8 x 2.4 m (6 x 8 ft.) are large enough for good health and breeding. As with most birds, the higher the aviary, the safer the toucans feel. At Riverbanks, green aracaris, curl-crested aracaris *Pteroglossus beauharnaesii,* saffron toucanets *Baillonius bailloni,* and emerald toucanets *Aulacorhynchus prasinus* have bred in indoor and outdoor enclosures ranging in heights of 2.7–3.6 m (9–12 ft.). The larger toucans need more space. Keel-billed toucans have bred in aviaries 6.4 x 3 x 3.6 m (21 x 10 x 12 ft.) high. Toco and chestnut-mandibled toucans *Ramphastos swainsonii* have bred in aviaries 8.5 x 3 x 3.6 m (28 x 10 x 12 ft.) high.

Established birds, housed in large breeding aviaries, seldom damage their bills. Those most at risk are young birds. Their bills are not yet fully developed, and the protective outer skin is very soft. Juveniles are prone to flying into mesh in an uncoordinated way, and can easily damage their bills, leaving notches and creases. It is only from approximately 12 weeks of age that toucans seem to develop enough coordination to land on wire mesh instead of flying, beak-first, into it. A skilled avian veterinarian, using sympathetically colored acrylics, can repair deep notches in the bill. Over time, notches and creases should grow out. However, beaks grow from the head of the bird, and the time needed for serious damage at the end of the beak to be completely repaired may be considerable. Although such damage is only cosmetic and does not normally have any medical implications, it is an important blemish that should be minimized or prevented completely.

While young birds remain in a large aviary, either with their parents or a prospective mate, they are less likely to injure their bills. But when they are moved to a smaller holding aviary, they are more panicky and can easily fly into the wire. Young birds will feel more secure if they can hide in thick, tall plants at least 3 m (10 ft.) above the ground. Special care must be taken when servicing small aviaries. One must pay attention to the birds without looking directly at them, and immediately withdraw if they are unsettled by your presence. Gradually, they will learn to tolerate the keeper. However, in holding aviaries lower than approximately 3 m (10 ft.), no amount of care will reduce the inclination of young birds to fly into the wire mesh, particularly if the aviary is open and devoid of good foliage. Whether or not bill damage occurs will depend on the type of mesh used and the temperament of the young birds. Since 1977 Riverbanks has been breeding toco toucans behind 2.5 x 5 cm (1 x 2 in.) mesh. The same mesh houses the breeding keel-billed toucans and chestnut-mandibled toucans. To date, no bill injuries have occurred. Riverbanks breeds, or has bred, green aracaris and emerald toucanets behind 2.5 x 1.27 cm (1 x ½ in.) mesh with no bill injuries. We have kept eight-week-old toco toucans in holding aviaries constructed of 1.5 x 1.2 cm (1 x ½ in.) mesh. But the toco is considerably larger than an aracari or toucanet, and the nervous young tocos damaged their bills so seriously that they had to be repaired, under anesthesia, with acrylic. The young toucans were able to push their bills through the mesh such that a significant notch was cut into their upper mandibles. As well as the mesh being of an inappropriate size, the aviaries were also small, low, and open. To prevent such bill damage, solid material should be fixed to the upper half of holding aviaries, or the use of ZooMesh™ might be considered. ZooMesh™ is a relatively new type of mesh that is available in the US. This is an exceptionally fine mesh made of thin stainless steel wire. The wire is woven rather than welded, and produces a fabric-like material that is so fine it is almost invisible from 3 m (10 ft.) away. The mesh is best painted matt black, and substantially resembles mosquito/insect netting of the type used to screen porches or windows. It can prevent bill damage in toucans; it is also aesthetically appealing in a display aviary. The drawbacks of the material are that rolls are only available in 2.1 m (7 ft.) widths, joins are very conspicuous, and it is weaker than welded mesh. Overall, 2.5 x 5 cm (1 x 2 in.) mesh in a large aviary appears to be safe. Holding aviaries for young toucans can be made of ZooMesh™. Alternatively, this material can be stretched over larger mesh, or solid sides can be installed in the aviary.

Dietary Iron: Certain birds, such as toucans, seem to have evolved particularly efficient mechanisms for iron absorption and utilization. The reason for this is not clear, although it could simply be that their natural foods are low in iron and other minerals because they tend to grow in nutrient-poor soils. It is also possible that wild toucans consume the tannins found in pools of water, as well as the mineral inhibitors found naturally in fruits such as *Ficus* berries. See TEA, below. The source of the iron may also be another factor governing iron absorption. It is thought that so-called "heme" iron, from meats and animal products, may be in a more bioavailable form than iron from fruits and plants, and therefore more harmful to the bird. An excess of meat in the diet may therefore be damaging, since the natural diet is substantially fruit, and therefore low in iron, or at least low in iron bioavailability. At Riverbanks, the toucan diet has no added meat, except some pinkies for chick rearing. The remarks above represent unproved theories on a subject about which people rarely fully agree. But one thing that everyone is agreed upon is that the artificial diet should be low in iron.

Most artificial diets probably provide far more iron than is available to the birds in the wild. The excess iron is stored in body tissues, particularly the liver, which becomes enlarged and damaged in the process. It is likely that the toucan liver starts absorbing iron from the first days of life, eventually causing hemosiderosis. In itself hemosiderosis is probably innocuous, common and even normal; but in time, sufficient iron accumulates to cause the disease hemochromatosis (iron storage disease). Sooner or later all toucans seem to succumb to this disease, but with a low iron diet, hemochromatosis can be relegated to old age instead of being a major killer of young birds, as it once was. The current treatment for hemochromatosis is drug therapy combined with phlebotomies (blood letting) over approximately a three-week period. This is normally a stressful treatment for a bird and tends to be a last resort. The more one learns about iron, the more difficult it is to say definitively what "low iron" actually is. But generally, pellets at or lower that 150 ppm can be considered "low." Having said that, the toucans at Riverbanks, and other major toucan collections, have bred for many years on pellets that contain approximately 200 ppm of iron. Simultaneously, iron storage disease has become a thing of the past. If you have any doubts about your current brand of food, it is well worth calling the manufacturer or having the food tested independently to determine its iron content. The cost for a laboratory analysis is usually very low indeed.

In humans, it is thought that foods with a high acidic content, such as citrus, can facilitate iron absorption by lowering the Ph of the stomach. No studies have yet been completed to determine whether this also applies to birds, but in the absence of contrary evidence, it is reasonable to assume that it does. Conventional wisdom therefore currently recommends against feeding toucans acidic fruits such as citrus, tomatoes, and strawberries.

Tea: Also in humans, it is thought that tannins (such as ordinary tea) bind iron and other nutrients, making them unavailable to the body. It has been suggested that toucans in the wild drink from pools, or rain-filled hollows, where tannins from bark and fallen leaves leach into the water. Riverbanks is one of several zoos that routinely uses tannin (tea) in the diet, in the hope of reducing iron absorption. The efficacy of tea has not yet been studied in birds. But at worst, tea seems to be harmless, and at best, it may be beneficial. Pellets can be soaked overnight in liquid tea, or weak tea can be provided as the sole source of drinking water. Tea leaves can also be lightly sprinkled directly onto the fruit. It is likely that tea also binds other nutrients; to prevent widespread nutrient depletion, we offer tea for one month, followed by a month without tea. This one-month-on/one-month-off cycle is continued permanently, and finds a compromise between minimizing iron absorption and depleting the body of other nutrients.

Diet: Maintenance diet for all members of the subfamily Ramphastinae:

55% diced fruits
43% low iron pellets
2% Science Diet Canine Maintenance Kibble (soaked overnight in water or tea)

The fruits and pellets are mixed together and offered in the same dish. The exact composition of the fruit mixture is not very important, although, as mentioned above under Dietary Iron, the avoidance of citrus fruits is recommended. A few grapes can also be included, since they are, in fact, not especially high in iron, and are certainly a favorite toucan food.

Riverbanks' fruit mixture is based on apples for economy:

62% apples
12% grapes
14% banana
12% seasonal fruit, or additional apple

A fairly large, low-iron pellet should be chosen for the toucan diets and offered in the same dish as the fruits. My personal preference is to also include a tiny pellet (manufactured for finches and canaries) with the fruits and large pellets. Although this is not a common practice for birds as large as toucans, it is beneficial because the tiny pellets stick to all of the fruit pieces, ensuring good nutrition with every mouthful. In places like South Carolina, where the temperature on winter nights can fall to well below freezing, the guarantee of good overall nutrition is reassuring. It is important to remember that commercially produced fruits are mostly water and, in themselves, are not whole foods. Consumption of a substantial amount of nutritious low iron pellets is therefore essential for good health, breeding, and longevity.

Toco Toucans *Ramphastos toco*

Day	Toco Toucan Weight (g)	Day	Toco Toucan Weight (g)	Day	Toco Toucan Weight (g)
1		17		33	
2	33.8	18	166.9	34	420.1
3		19		35	
4	41.9	20	207.6	36	432.4
5		21		37	
6	52.5	22	240	38	464.2
7		23		39	
8	64.4	24	276.6	40	477.5
9		25		42	507
10	71.7	26	290	44	533.6
11		27		46	555.5
12	97.3	28	336.1	48	560.2
13		29		50	500.3
14	123.3	30	399.3	52	510
15		31		54	527.4
16	143.2	32	410.4		

Dog kibble is also a worthwhile addition to the diet. Many kibbles, however, are high in iron, and even the

broad, flat, short horizontal plane but it rapidly changes shape, becoming longer and narrower and flatter in a vertical plane. The nostrils opened one at a time at 8–10 days of age at the base of the beak, and were obvious as holes that were not there the previous day. The chick presumably breathes through its mouth until the nostrils open! In the adult bird, the nostrils are situated behind the base of the beak and are not easily seen unless one is looking at the bird from directly above and behind.

From day 8 onward, a transition from the liquid diet to the adult diet can be made. Very small pieces of sliced pinky mice, soaked Hills Science Diet Canine pellets, toucan pellets, and diced fruit can all be fed. All pieces are dipped in warm Kaytee formula before being fed, and offered from forceps. In my excitement I read the handraising section on TANAGERS in the chapter by Martin Vince. The hand rearing method for toucans is the same up to day 8, but toucans do not require waxworms. Tanagers benefit from waxworms after Day 8, but not before. Having misunderstood that they were not necessary for toucans, I supplied them to the chick, which loved them. Nekton MSM calcium/mineral supplement was also added once per day.

The chicks have most peculiar pads on the legs which have small hooks, and the feet are initially very small and out of proportion. Young chicks do not use their feet but rather perch on their elbows with their feet sticking up in the air, using the broad tail base as a support. These pads presumably offer a grip to the chick in the nesting chamber where big feet would simply get in the way. The pads eventually fall off once the chicks have left the nest.

While feeding the little toucan one day, her feathers suddenly puffed up, her wings stretched out, and she keeled over to one side. Her beak opened wide and I thought she was about to die. I pulled the overhead infrared lamp away from her to get a better look at what was happening, and she immediately regained her composure! I pulled the lamp back toward her and she immediately repeated her bizarre posture. Voila! She was loving the heat and how obvious anyway; toucans love to sunbathe and she was instinctively reacting to the sudden warmth."

Species Requiring Similar Management:

Toucans *Ramphastos* spp.
Toucanets *Aulacorhynchus, Baillonius,* and *Selenidera* spp.
Aracaris *Pteroglossus* spp.
Mountain-toucans *Andigena* spp.
Hornbills, order Bucerotiformes

Contributor:

William Horsfield, Amazona Bird Farm, South Africa.

References:

Jennings, J. 1999. Toucans in aviculture at the millenium. *AFA Watchbird* 26 (5): 30–31.

Hornbills

08 Bucerotiformes

08.01 Bucerotiformes Hornbills

Glen Holland

Introduction: Hornbills are now given their own order, the Bucerotiformes, which is divided into two families. The typical hornbills are placed in the family Bucerotidae; the ground-hornbills, in the family Bucorvidae. The fifty-four species of typical hornbills are distributed through Africa, India, Indochina and Malaysia, Burma, New Guinea, Borneo, Sumatra, Sulawesi, and the eastern islands, while the two species of ground-hornbills are endemic to Africa. They range from 100 g to 4 kg (3.5 oz. to 8.8 lb) in mass, and males are up to 17% heavier and 20% longer than females. The bill length of males is up to 30% longer and the casque always larger than females. Most species are relatively common in a range of habitats varying from dry woodland to dense evergreen forests.

Hornbills have been kept in zoos for more than 250 years. These birds are featured increasingly in avicultural collections, and in recent years good breeding results have been achieved with many species. In 1993 an IUCN/ICBP Hornbill Specialist Group was formed. Membership includes many of the hornbill scientists and many others in the field and in zoos. This led to numerous other groups being formed, including the CBSG Hornbill Global Captive Action Plan Group and the AZA North American Hornbill Taxon Advisory Group in the United States.

I would like to acknowledge the AZA Hornbill Husbandry and Management Guideline (1997) that was used as a considerable reference for this chapter.

Asiatic Hornbills

The species described here are not confined to mainland Asia, as many are endemic to the Philippines, Indonesia, Sulawesi, Sri Lanka, Sumatra, Borneo, New Guinea, and various smaller islands. Four genera are recognized. The genus *Anthracoceros* includes the Malay black *A. malayanus,* the Indian pied *A. coronatus,* and the Asian pied *A. albirostris albirostris*, along with its subspecies, the Sunda pied *A. a. convexus*. The genus *Buceros* includes the great hornbill *B. bicornis,* the rhinoceros *B. rhinoceros,* and the rufous hornbill *B. hydrocorax*. The group known as tarictic hornbills are placed in the genus *Penelopides*. Formerly regarded as a single species, this group is now split into four different species, the Visayan *P. panini,* the Luzon *P. manillae,* the Mindanao *P. affinis,* and the Mindoro *P. mindorensis*. The fourth genus is *Aceros*, which includes the Sulawesi wrinkled *A. cassidix,* the white-crowned *A. comatus,* the Sunda wrinkled *A. corrugatus,* the Papuan wreathed or Blyth's *A. plicatus,* the wreathed *A. undulatus,* and the writhe-billed *A. leucocephalus*. These hornbills are primarily birds of the tropical forest canopy. During most of the year they are found in pairs or small groups; however, outside of the

Rhinoceros Hornbill
Buceros rhinoceros

Black Hornbill
Anthracoceros malayanus

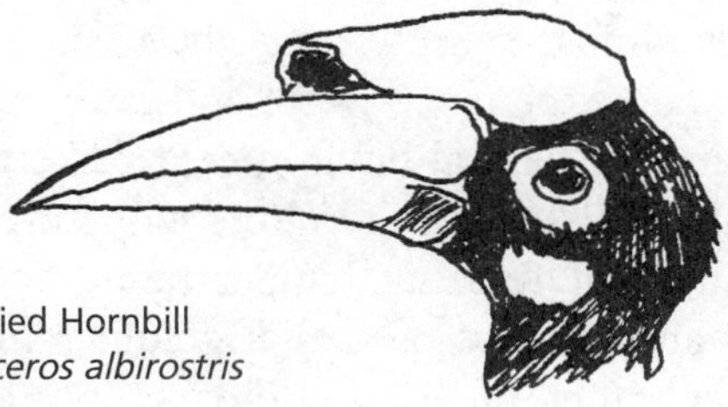

Oriental Pied Hornbill
Anthracoceros albirostris

Great Indian Hornbill
Buceros bicornis

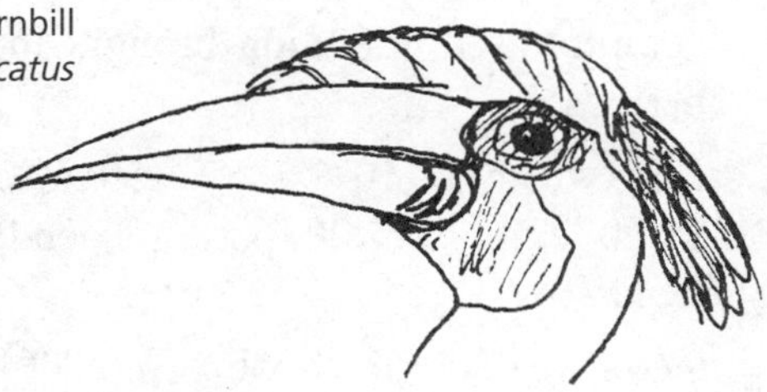

Blyth's Hornbill
Aceros plicatus

White-crowned Hornbill
Berenicornis comatus

breeding season they often congregate in large flocks for short periods of time. The colors of some species, such as the southern Sulawesi hornbill *Penelopides exarhatus sanfordi,* and the immense size and color of others, such as the great Indian hornbill and the rhinoceros hornbill, make them popular species for public displays.

Management: Although the management of the Asiatic species differs relatively little from that of the African species, considering their popularity and the conservation efforts behind breeding many of these species, I feel it is worth covering the captive husbandry of these birds separately. The larger species, such as the great Indian and rhinoceros hornbills, require aviaries measuring approximately 12–20 m (40–66 ft.) in length, 5 m (16 ft.) in width, and 3.5 m (12 ft.) in height. Although they have been bred in smaller aviaries, they need space to be exhibited at their best and to exercise sufficiently. The smaller species, such as the southern Sulawesi hornbill, the wrinkled hornbill and the tarictic hornbill require aviaries measuring 10 x 4 m x 3 m (33 x 13 x 10 ft.). Again, they have been bred in smaller aviaries, but generally thrive better with increased space. The aviaries should be planted with a variety of small trees that will offer perches to the birds. Shrubs and climbers should also be planted, as well as some grass tussocks on the floor, as in Habitat A. As a rule, hornbills do little damage to the vegetation. Large stumps and branches should be strategically placed around the aviary to ensure the birds do fly and not just bounce around the aviary. Care should be taken to position perches so that the birds' flight path is not obstructed. Placing perches at different heights provides more perching options, and hornbills often jump back and forth between different levels. A stable perch should be placed where the male can extend his wings and has sufficient room above his head to mount the female during copulation. Some of the natural perches should have small, forked side-branches that hornbills can use to clean the inside of their bills. A sandy area around the feed station will facilitate aviary cleaning and hygiene. The public should be able to view the hornbills on only one or two sides of the enclosure to prevent excessive disturbance. Aviaries are generally covered with chainlink, or weldmesh with a 25 mm^2 (1 sq. in.) aperture. The size of the wire-mesh should be small enough that the hornbill's bill cannot pass through it. Some hornbills can alter the shape of the casque and beak, apparently through rubbing, and mesh size should preclude this possibility. It should also be suitable to exclude undesirable animals such as predators and vermin. Care must be taken to choose a suitably strong mesh. Trumpeter hornbills *Ceratogymna bucinator* housed in

14-gauge 1.27 cm (0.5 in.) wire-mesh tore it to shreds. Sunda wrinkled hornbills *Aceros corrugatus* systematically opened a 1.27 cm (0.5 in.) galvanized poultry mesh. A vinyl-coated chicken wire proved unsuitable for ground hornbills. It was found in a cross-institutional study of 36 pairs of great hornbills *Buceros bicornis* that "solidness" of the roof was the only environmental feature that correlated with breeding success: the more solid the roof, the more reproductive behaviors were performed by females, resulting in higher reproductive success. Entirely open-topped (thus outside) enclosures were least successful; partially transparent enclosures were more successful than enclosures with entirely transparent roofs. (AZA Guidelines 1997.)

Harro Strehlow has reported that, unfortunately, with the tarictic hornbills, zoos have often housed pairs representing different species or subspecies, resulting in hybrid offspring. To complicate matters, these hybrids usually resemble one or the other parent species. Conspecifics, other hornbills or even other avian taxa showing territorial behavior, especially during the breeding season, should not be housed in aviaries adjoining the hornbills and visual contact should be minimized. For example, fighting between a male toco toucan *Ramphastos toco* and a Papuan wreathed hornbill *Aceros plicatus* in adjacent enclosures at Rotterdam Zoo resulted in loss of the tip of the toucan's bill. Enclosures housing hornbills with other species have had variable success. Several factors can play a role, including the species and individuals involved, amount of vegetation and space, and perching possibilities. Reproductive condition was reported by several hornbill managers to be an important factor. For example, a pair of Sunda wrinkled hornbills *Aceros corrugatus* housed in the 150 x 90 x 20 m (492 x 295 x 164 ft.) rainforest exhibit at Arnhem Zoo (Netherlands) was compatible with the approximately 100 other species with which it was housed until it began breeding. The parents developed a marked preference for feeding the offspring live food, taking a huge toll on smaller birds as well as reptiles and amphibians in the exhibit, resulting in the pair being moved elsewhere. Many species will catch sparrows and other small birds that enter the aviary; this is not a pleasant experience for visitors, and such birds may pose a health risk to the hornbills. Blue jays, European starlings, mockingbirds, and blackbirds have all been recorded as prey of captive hornbills. A word of warning: do not place rat bait in an area where poisoned rodents may gain access to hornbill aviaries, as the hornbills are likely to feed on them. Several hornbill managers mentioned problems with secondary poisoning from rodenticides, although in some cases assurances had been made that these compounds were safe for birds.

Damaged casques or bills can be a problem, especially if the broken appendage becomes infected. Fowler (1986) recommended using fiberglass and epoxy to repair bills. Two people reported successful repair of bird bills using material developed for hoofstock hoof repair. Hornbills are highly susceptible to pseudotuberculosis, a disease caused by the bacterium *Yersinia pseudotuberculosis*. This disease results in very rapid and usually fatal inflammation of the liver (peracute hepatitis). The formol killed vaccine "Pseudovac" to protect pseudotuberculosis-sensitive animals is available. Vaccine protectiveness declines after nine months, however some Dutch zoos have chosen to vaccinate only once per year, before the winter, as pseudotuberculosis is not considered a problem in the Netherlands in the summer and fall months. (AZA Guidelines 1997.)

Although exact tolerable minimum and maximum temperatures are not known for any hornbill species, hornbills almost entirely range between 30°N to 30°S latitude, and experience warm and fairly stable temperatures in their natural environment. Like most tropical to sub-tropical species, hornbills will suffer from frostbite if held in freezing temperatures, and their large casques and bills are particularly vulnerable. The hoofstock department at the Rotterdam Zoo developed the following temperature management scheme for tropical species. A similar approach to hornbill management may be useful:

- Frost (under 5°C [41°F]): animals remain inside or come outside for a short period (approximately an hour) if it is not windy or rainy.
- Cold weather (5°C to 10°C [40–50°F]): animals allowed outside for a couple of hours, or less if windy or rainy. Young animals or animals in compromised condition brought in after shorter periods.
- Fair weather (10°C–16°C 50–60.8°F): outside 11:00 a.m. to 3:00 p.m., dependent on other factors.
- Good weather (above 16°C [60.8°F]): outside the entire day. Hornbills may be sprinkled with water as refreshment for short periods during the day in warm temperatures. (AZA Guidelines 1997.)

Diet: Regular hand feeding of hornbills may have a negative effect on pair-bonding between hornbills, as a hand-fed bird could eventually become more interested in the keeper than in his/her partner. Ideally non-breeding hornbills should be provided with fresh food two times a day (early morning and mid-afternoon), and

Hand Rearing Data for the Southern Sulawesi Hornbill (Mace and Azua 1997)

Day	% intake per day	No. feeds	Diet	Housing	Observations
1	20	2 hours	Chopped pinkies & cricket abdomens dipped in Pedialyte with CaCO3.	Brooder at 33.3–34.4°C (92–94°F)	Chicks in plastic container with paper towel and tissue.
2	25	-	-	Decrease temp. by 1°C per day	Naked & pink eyes closed.
3	30	-	-	-	-
4	35	3 hours	Chopped fuzzies, papaya & crickets dipped in Nekton-S with bonemeal.	-	-
5	35	-	-	-	Body color darkening: skin drying and peeling.
6–7	40	-	-	-	-
8–9	45	-	-	31°C (87.8°F)	-
10–14	50	-	-	-	-
15–19	-	4 hours	-	Wooden box brooder 29.4°C (85°F)	Eyes opening; pin feathers developing on back.
20–24	-	3/day	Waxworms added.	-	Black color observed on the predominantly bone-colored maxillae on some chicks.
25–29	ad lib	2/day	Banana & skinned mice added.	-	Body feathers developing.
30–39	-	-	Soaked dog food and diced grapes and mixed fruit added.	Indoor pen 23.9–25.6°C (75–78°F)	Chicks kept in plastic tub with a towel in indoor pen; grabbing food items.
40–49	-	-	Nebraska Bird of Prey meat added to the diet mix.	Moved to an outdoor pen with heat lamps.	Chicks observed to cast pellets, perch on branches and eat on their own; body plumage developed.
50–59	-	1/day	-	Supplemental heat removed.	Approximate fledgling age
60	-	-	-	-	Body plumage and flight feathers predominantly black; all chicks with some yellow facial coloration resembling adult female.

to overcome; lack of contact with conspecifics in an early stage of life may result in no interest for the opposite sex later when the physiological full breeding potential is reached. Hornbills have been seen displaying towards their keepers instead of their mates. Hand-reared great hornbills *Buceros bicornis* at the Rostock Zoo have been difficult to pair with conspecifics; one female even killed a male she was housed with.

Hand-raised Sulawesi Hornbill
Penelopides exarhatus sanfordi

Day	Weight (g)	Day	Weight (g)
1	15	**17**	170
3	20	**19**	195
5	22	**21**	210
7	30	**23**	260
9	42	**25**	300
11	65	**27**	330
13	100	**29**	370
15	120	**31**	380

If hand rearing is undertaken, some steps that can be taken to reduce the likelihood of abnormal imprinting are:

- puppet rear (minimizing direct contact with humans)
- play conspecific calls
- keep at least two (preferably conspecific) chicks together
- give visual and auditory access to adults as soon as possible
- hold juveniles in conspecific flocks, as hornbills join youngster groups or remain in family groups in the wild after fledging. (AZA Guidelines 1997.)

A longevity record for great hornbills is 47 years, and it is suspected that rhinoceros hornbills should have a similar life span. Smaller species such as those in the genera *Aceros* and *Penelopides* have survived for up to 34 years. A male Indian pied hornbill imported in 1969 by a New York wildlife dealer died at the Hancock Wildlife Research Center in British Columbia in 2000.

African Hornbills

Some species, such as the trumpeter hornbill *Ceratogymna bucinator,* the silvery-cheeked hornbill *C. brevis,* and the large hornbills in the genus *Buceros* are canopy dwellers; others, particularly members of the genus *Tockus,* regularly descend with dipping flights and high pitched calls to feed on the ground. Some, including the red-billed hornbill *T. erythrorhynchus* and similar species, have characteristic "tok, tok, tok" calls, often made as a duet. The ground hornbills are large birds uniquely adapted to a life of walking, with occasional flights between nests, roost sites, and the feeding grounds.

African Gray Hornbill *Tockus nasutus*

Management: The African gray hornbill *Tockus nasutus* is the second most common species in collections. The male gray hornbill has a cream colored patch, which is absent in the female, on the casque. Gray hornbills have soft whistling calls. These are the most aggressive of the three species of small hornbills I have worked with, and are difficult to mix in any collection where breeding is desired. They will rob nests, devouring eggs and chicks alike. In an aviary they will corner, kill and eat small birds such as waxbills and finches. Their long beaks allow them to reach into hollows to rob nests. I have lost clutches of both amethyst starlings and hoopoes to this species. Their aviary should be landscaped as described for Habitat B under the chapter on LANDSCAPE.

Hornbills are intelligent birds and, particularly if hand-raised, will give the aviculturist hours of entertainment. They prefer a planted aviary with plenty of rocks, stumps and crannies to keep them busy throughout the day. They will spend much of the day strutting about on the aviary floor. As they enjoy sand bathing, a patch of dry sand should be provided. A mist spray placed above a bush in which the hornbills can sit and enjoy the water is appreciated. They also like to leaf bathe in wet vegetation.

Diet: Gray hornbills are to a large extent carnivorous

and will prey on mice, lizards, insects, small birds, etc. Occasionally they will take some fruit and soft foods. In captivity they are best fed on ox heart and chicken strips, pinkie mice, locusts, crickets, bonemeal, insects, and a small quantity of insectivorous soft food and fruit. Dead chicks of smaller species, such as finches, will not go to waste if given to the gray hornbills. The food is swallowed whole, but only after some degree of mutilation.

Breeding: This species has bred in at least 14 zoos, bird parks, and private collections. Harro Strehlow reported in March 2001 that of 176 chicks hatched, only 19 died. Vogelpark Walsrode was the first to breed the species in 1981. The most successful zoos are Cincinnati, with 35 chicks, and Chester, with 29 chicks. The clutch size is large, and often five or six young are hatched. A natural nest log should be supplied for nesting. During courtship, both sexes call with their beaks pointing upward while rocking backward and forward on the perch, opening and closing their wings. To commence nesting, a male will fly to the entrance hole with food in his beak, trying to entice the female into the nest. Throughout the breeding period, the male feeds the female. The process of sealing the female in the nest takes about two weeks.

Gray hornbills are particularly fond of strips of bark, which they will add to the nest throughout the breeding cycle. Females begin to lay about a week after being sealed in the nest. The normal clutch consists of 3–6 white eggs. The incubation period is 18 days. Thus, about 25 days after the nest is sealed, a sudden demand for live food should become obvious, indicating that the chicks have hatched. The male flies to the nest with individual food items which the female grabs. She squeals while a tug of war takes place before she is given the food. These squeals can be heard from some distance. The female breaks out of the nest 28 days after the chicks hatch, and the nest is resealed. The female then assists the male in feeding the chicks in the nest for another three weeks. When the chicks are ready to fledge, the seal is again broken. To lure the chicks out of the nest, the female will sit close to the nest with food in her bill.

The chicks fledge at 45–50 days and begin feeding themselves a week later. Youngsters attain adult plumage at 5–6 months. In one instance at Chester Zoo, the female died while there were chicks in the nest. The entire nest was removed, cleaned, given new bedding, and then replaced in the aviary with the male. Fortunately the chicks were large enough to reach the entrance and receive food, and survived to adulthood after being fed by the male alone. On two occasions a second clutch was laid a short time after the young fledged. In Zurich, the female sealed again 18 days after leaving the cavity; two young eventually fledged from this second nest. In August 2000, six young fledged at Niendorf. The female sealed in again, but left the cavity after five weeks without any signs of eggs or chicks. Chester Zoo also records an instance in which a female whose first clutch failed nested a second time in the same season. On both occasions she molted her feathers. I have experienced this same phenomenon with a red-billed hornbill. These small hornbills require a cockatiel sized nest box, or a natural log 30 cm (12 in.) deep below the entrance hole, and with a diameter of 20 cm (8 in.). The entrance should be oval-shaped, measuring about 5 cm (2 in.) wide and 8 cm (3 in.) high. The base of the nest should be covered with a layer of wood chips. Kemp gives the longevity for this species as 20 years.

Hand Rearing: As described for the yellow-billed hornbill.

Ring Size: 8 mm

Yellow hornbill chick at 12 days old.

Red & Yellow-billed hornbills *Tockus* spp.

Management: The information included in this section is equally applicable to both the yellow-billed hornbill *Tockus flavirostris* and the red-billed hornbill *T. erythrorhynchus,* two species which require similar management. Red-billed are the most common hornbills in captivity. These birds are real characters and, particularly in the evening, the members of a pair will chase each other around and lock their beaks together in play. With the yellow-billed, males bills are usually about

9cm (3 ¾ in.) and the bill of the female about 7.5 cm (3 in.). Both sexes have a pinkish red fleshy area around the eyes and on the throat; it is redder in males, and in the breeding season it becomes redder still. Another characteristic of the yellow-bill is the long eyelashes and the way the eyes seem to swivel around in a chameleon like-way. Despite their large bill, tails and wings, *Tockus* hornbills have rather small bodies. Although red-billed are less aggressive than yellow-billed or gray hornbills, they are opportunistic feeders. If a clutch of eggs or brood of young is easily accessible, they too will rob nests. On occasion they will catch and kill recently fledged chicks, although this seems more out of curiosity than aggression. Yellow-billed hornbills should not be mixed with finches as they will catch and eat these small birds. Red-billed hornbills can, however, be mixed with the larger finch species, such as the red-headed finch *Amadina erythrocephala,* as well as with other seed-eaters such as weavers. In one mixed collection, tambourine doves were breeding freely on their exposed nests with no interference from red-billed hornbills. In contradiction to this observation, however, I have read an article in which the author noted that this species had been observed killing small seedeaters to feed their nestlings, so one must be extremely careful when mixing them with other birds. Their aviary should be landscaped as described for gray hornbills.

Diet: Both red-billed and yellow-billed hornbills are omnivorous and, while favoring insects, also eat quantities of fruit. In captivity they feed on fresh bird of prey diet, soaked cat cubes, insectivorous softbill food, and fruits such as apples and grapes. They are particularly fond of soaked raisins. A powdered multivitamin is added to the food twice weekly. Mealworms, flying ants, crickets, and locusts are cherished. I have tried feeding them strips of meat, but have found that they play with these rather than eat them, except during the breeding season, when they will feed small strips of bird of prey diet to their young. Red-billed hornbills also eat dry seeds, and I have found them to be fond of crushed corn.

Breeding: The first attempted captive breeding of yellow-billed hornbills took place at Berlin Zoo in 1926. Red-billed hornbills were, however, the first species to be successfully bred in captivity, with a pair breeding at Frankfurt Zoo in the same year. The peak of the breeding season in the Southern Hemisphere is from November to February, although captive birds in the Northern Hemisphere readily adapt their breeding seasons to nest between April and September. At the onset of the breeding season, the males will begin feeding the females.

African Grey Hornbill at a nest site.

Courtship display involves the birds bouncing up and down with raised wings while dueting with their "wot, wot, wot" calls. Either sex may initiate the display. Once the birds have mated, the female enters the nest and the pair seals the entrance hole with mud from the outside and with the female's droppings from the inside. This takes about two and a half weeks. To facilitate the sealing process, clay should be provided in a bowl. The birds will mix this with sand and excrement. The nest is lined with leaves, which are added throughout the nesting cycle.

When the entrance is sealed, the female lays the clutch and molts her flight and tail feathers. The male feeds the hen through a narrow slot left in the entrance seal, and she in turn feeds the chicks. This routine continues until she emerges from the nest, at which stage the chicks are half grown. The chicks then re-seal the nest with their droppings, and both parents feed them until they are ready to emerge. Yellow-billed hornbills lay 3–5 white eggs, which hatch after 24 days. The female leaves the nest 21–26 days after the chicks hatch. The chicks fledge at 45 days. The entire nesting cycle from the time the female enters the nest to the fledging of the chicks takes about 80 days. At Bush Gardens, Tampa, a pair bred five times in ten years. Fourteen chicks hatched, of which twelve eventually fledged. From 1979 through 1987, San Diego Wild Animal Park reared 31 young to maturity; 10 others did not survive.

While in the nest, the chicks are constantly cleaning house, discarding bits of waste materials through the feeding slot. They also squirt their feces through this opening. Due to the strong competition among hornbill chicks, it is usual for only half the clutch to survive.

Since incubation commences with the first egg, the interval at which eggs are laid results in a large size difference among the chicks. It is therefore advisable to remove the smaller chicks for hand rearing before they are smothered by their larger siblings. In the wild, pairs often use the same nest for many seasons and, where possible, captive birds should be left with the same nest in the aviary. Aggression toward the chicks, on the part of either parent, can be a problem; chicks should therefore be separated at ten days to two weeks after fledging, and even during this period the female should be carefully observed for signs of aggression. Red-billed females remain sealed in the nest for 58 to 87 days. In one case, the female left the nest after 45 days and the chicks had to be hand raised. Nestling time is given as 41–57 days.

In another instance, a pair of red-billed nested in a donation box and I removed the female and took her three chicks for hand rearing. At this time the female had not completed her molt, and could not fly well. Within a month, however, she was back in the donation box on a second clutch. Steve Horne in the UK also recorded regular double clutching from this species, with one pair raising 38 chicks over a six-year period. A pair at the Hancock Wildlife Research Center in Canada has also produced two broods annually for several years. Third clutches are occasionally recorded. Frankfurt Zoo holds the record, with 15 young in 1980, and a total of 89 chicks reared from 1970 through 1986.

In captivity they will breed in a natural log or a wooden box with an internal diameter of approximately 25 cm (10 in.). The depth of the nest should allow an incubating female to reach the entrance slot with the tip of her beak to receive food from the male, approximately 25 cm (10 in.). The nest is lined with a few green leaves, or not at all, in some cases. It is important to supply mud, which the male will use to seal the nest. The male seems to get into a plastering frenzy and will use any material available, including foods with a porridge consistency. He will also plaster any crevice he finds. I even had one plaster up the side of a book kept in the aviary for visitors' comments.

When feeding, the males take all the choice foods, such as insects, straight to the nest. Males call to the females and then offer them food. Should the male take too long to bring food to the nest the female will tap on the inside of the nest; if this does not produce a response, she will begin calling to him. The male will also indicate any threats by calling to the female.

Red-billed Hornbill
Tockus erythrorhynchus

Hand Rearing: As a number of chicks die in the nest as a result of sibling rivalry, and because hand-raised birds are far more pleasing to the aviculturist, I believe in removing some, if not all of the clutch, for hand rearing. Hornbill chicks, even when incubator hatched and hand-fed from day one, are easy to rear. If hatched from the egg, they should be kept in a brooder at an initial temperature of 34°C (93.2°F) and a relative humidity of 60%. As they begin to develop, the temperature can be dropped by ½°C (1°F) per day until it reaches 28°C (82.4°F). This should remain constant until the chicks are about four weeks of age, with feathers covering most of the body. In warm climates, where room temperatures exceed 28°C (82.4°F), the chicks can be removed from the brooder when well feathered.

(This should be read in conjunction with the table describing the hand rearing of the Sulawesi hornbill) Feeding is done with a pair of forceps, and foods used include insects, pinkie mice, ox heart strips, small reptiles such as skinks, and commercial rearing food. They are voracious feeders and will latch on to anything offered, while at the same time giving their squealing calls. To avoid fighting and injury to the smaller chicks, they should be separated as soon as the feathers start to appear. They are kept in containers lined with paper towel, which is replaced on a daily basis. Chicks should not be kept on wood shavings, as they will ingest this material. This results in compaction and they will refuse any food until this blockage is regurgitated. If the chicks refuse to feed it is probably because they are about to regurgitate a pellet. Even at this young age they use their droppings to seal any openings around the container. Kemp gives longevity for red-billed as 18 years.

Ring Size: 8 mm

Species requiring similar management:

Bradfield's hornbill *Tockus bradfieldi*
Monteiro's hornbill *T. monteiri*
Toucans *Ramphastos* spp.
Aracaris *Pteroglossus* spp.

Trumpeter Hornbill *Ceratogymna bucinator*

Natural Habits: Trumpeter hornbills *Ceratogymna* (often listed alternatively as *Bycanistes*) *bucinator* favor tall forests and riverine vegetation, particularly the large wild figs on which they feed. They are found in pairs and family groups while breeding, and in large flocks the rest of the year, with up to two hundred birds being recorded at a roost during the non-breeding season. Their loud braying cries are characteristic, and much chattering occurs between birds in a group. After feeding, pairs often separate from the flock to spend some time sitting in the sun while mutually preening.

Management: These large hornbills are best given a tall, well-planted flight to themselves. At Gettorf Zoo a pair bred successfully in an aviary measuring 4 m (13 ft.) x 2.8 m (9 ft.) x 2.7 m (9 ft.) high, although somewhat larger dimensions, such as 8 m (26 ft.) x 3 m (10 ft.) x 3 m (10 ft.), are preferable. The aviary should be planted with clumps of dense vegetation as described for Habitat B, and a number of thick branches on which they can perch must be included. A mist spray is appreciated on hot days. As they seldom descend to the ground, food should be offered on a platform 1.5 m (5 ft.) from the ground. A captive bird was recorded by K. Brouwer as living at least 21.7 years.

At San Diego Zoo, a silvery-cheeked hornbill *Ceratogymna brevis,* a species which is only slightly larger than the trumpeter hornbill, was observed flying down and catching a newborn Suni antelope. Thanks to the quick intervention of horrified staff members, the prey was left unharmed. Clearly these larger hornbills cannot safely be mixed with smaller species of birds or mammals!

Diet: They are omnivorous, taking large quantities of fruit and some vertebrates such as mice and fledgling birds. The captive diet should consist of a variety of diced fruits, soaked dog cubes, softbill food, diced hard-boiled egg with the shell, live food, and a small amount of diced ox heart. Due to their high iron content, only small quantities of cat cubes should be used, or commercial low-iron softbill mixtures can be substituted for the cat food. The live food is increased during the breeding season. A male at a nest in a forest which I managed was recorded as supplying 10% of the nestling diet in the form of live food, including wasp's nests and wood lice. I have watched trumpeter hornbills in the wild robbing the nest of a drongo. On another occasion, a male brought a large lizard to the nest.

Breeding: Twenty-one collections are known to have bred this species. Despite nesting attempts at London Zoo in 1936, the first successful breeding took place at Natal Zoo in South Africa, where two young fledged in 1970. Since that time more than 150 trumpeters have hatched in captivity, making this species much more common in collections. Courtship includes the male feeding and preening the female, which strengthens and maintains the pair bond. Copulation takes place on a branch with the female leaning forward with the tail slightly raised. The male entices the female into the nest by repeatedly probing his beak, filled with food, into the nest. Breeding pairs will use a large hollow tree stump, or barrel, 70 cm (28 in.) high and with a 50 cm (20 in.) internal diameter. An entrance hole 15 cm (6 in.) in diameter should be positioned about 40 cm (16 in.) from the base of the nest. At Chester Zoo, the nest was filled with composted bark, which the birds then removed, the female completing the work inside the nest. This served as a stimulus for breeding, but in the natural state, where the same nest is used each year, this does not occur. A bowl of wet clay is supplied for the birds to use in sealing the nest entrance.

Throughout the breeding cycle, bark and small sticks are brought to line the nest. During nest inspections, the male flies about excitedly and calls loudly. The entrance is sometimes partially broken and then resealed by the female. The female is completely sealed into the nest within 10 days of the commencement of sealing, and does not emerge until three months later. She calls and vibrates her bill on the side of the entrance hole when waiting to be fed. The 2–4 white eggs hatch after 29 days, which is about 36 days after the female is sealed into the nest. Once nesting begins, live food such as crickets, locusts, mealworms, chopped day old chicks, and baby mice should be fed in larger quantities three times per day. Chester Zoo recorded the birds consuming 100 locusts, 300 crickets, and 50 g of mealworms per day during the peak of the breeding season. Fruits are also fed in large quantities, smaller fruits being fed whole and with the seeds. The male is wary of approaching the nest in the presence of people. In the natural state, chicks from a previous season will sometimes assist the male in feeding both the female and the chicks in the nest. The female and chicks squirt their feces from the nest, and with each visit the male cleans the area around the nest entrance before departing.

Males visit the nest with food every half-hour. The birds emerge almost exactly three months after the female is first sealed into the nest. Unlike the smaller hornbills described above, the female does not emerge prior to the chicks, but breaks open the nest when the

Trumpeter Hornbill
Bycanistes bucinator

chicks are ready to fledge. The male becomes very excited when the family begins to break out of the nest, and will fly about calling continually. The chicks fly well immediately upon fledging, and will accept food from both parents. Fledglings are easily distinguished by the blaze of chestnut feathers on the forehead and over the eyes. They begin to feed themselves two weeks after emerging from the nest.

As with other hornbills, the larger chicks dominate and smaller chicks often wither away. In captivity, these younger chicks should be removed for hand rearing. At Gettorf Zoo, one pair produced seven chicks in four seasons, with two chicks in three of the seasons. Each time two chicks fledged, they ultimately proved to be a pair.

Hand Rearing: As described for the yellow-billed and Salawesi hornbills, but trumpeter chicks are able to take most live food whole.

Ring Size: 12.5 mm

Species requiring similar management:
Silvery-cheeked hornbill *Ceratogymna brevis*
Pied hornbill *Anthrococeros albirostris*

08.02 Bucorvidae Ground Hornbills

Glen Holland

Introduction: Two species of ground hornbill, the Southern *Bucorvus leadbeateri* and the Abyssinian *B. abyssinicus,* occur in southern and eastern Africa, respectively. They are normally found in groups of 2–5 birds, but up to 8 have been recorded in a group. Groups are strictly territorial. Each group will usually have only one adult female. They traditionally inhabit open savanna, but have adapted well to agriculture and are frequently found on farms. They feed while walking on the ground, hence the name of the BBC video on this species, *The Walking Birds*. Many native African people believe that these birds will bring floods if harmed, and this widespread belief has enabled these hornbills to survive where few other creatures have been successful. The Zulu name "Ingududu" is descriptive of the call, "du, du, dududu." There are a number of other calls, such as a deep grunting and squealing sound made when males feed the females and during aggressive displays.

Management: The southern ground hornbill, which is found from northern Angola to Kenya, and southward to the Cape Province, is glossy black with distinctive white primaries and red wattles, throat and face. In the adult female, a blue patch occurs on the throat. Immatures are not as dark as adults, with the facial skin a light beige color. The Abyssinian ground hornbill, which is found from Gambia to Kenya, differs by having the skin on the throat entirely blue in the female and red in the male. They have a deep booming call, usually given in duet. The voice of the male is deeper in tone than that of the female.

At present, the Abyssinian ground hornbill is better represented in zoological collections than the southern species. Recently, however, conservation bodies in South Africa have allowed the collection of second eggs, facilitating the establishment of a number of groups of this species in captivity.

A suitable aviary for ground hornbills should be about 20 m (65 ft.) x 10 m (33 ft.) x 5 m (16 ft.) high, and should be landscaped with boulders, stumps, clumps of grass, trees, and a few shrubs. As these birds are rather destructive, it may be necessary to provide protection around the base of the plants. A few rotting logs to destroy will keep them busy and, hopefully, divert their attention from the plants. They can either be maintained as a group or, alternatively, in pairs. In captivity, pairs are likely to prove more successful, unless the group has a large range such as that provided at the San Diego Wild Animal Park, where ground hornbills roam a 51 ha (126 acre) East African exhibit which also accommodates other species such as secretary birds and marabou storks. With continual struggles for dominance within both sexes of a group, birds which are ousted will be persecuted, with no escape in confinement. Due to their inquisitive and rather aggressive nature, ground hornbills are best given their own enclosure under normal captive conditions.

Diet: Ground hornbills are carnivorous, feeding largely on insects, lizards, snakes, baby birds, rodents, and small mammals. The food is caught on the ground and swallowed whole. In captivity they are fed on day-old chicks, mice, soaked cat pellets, diced hard-boiled eggs with the shell, and insects. Larger food items such as chicks may be broken into smaller pieces, and are sometimes carried in the beak for some time before being swallowed.

Breeding: The first captive breeding of the northern ground hornbill took place during 1953, when two chicks hatched at Basel Zoo, Switzerland. In the wild, there is an extremely high mortality rate among chicks, with groups in the natural state usually succeeding in rearing only one chick every 6–7 years. In captivity, however, with manipulation and hand rearing of second eggs, these birds can be far more prolific. For example, in seven years at the Dallas Zoo, a pair laid 14 eggs of which 9 were fertile. All nine hatched and were successfully raised, with the second egg of each clutch being removed for hand rearing. This pair nested in a metal barrel measuring 47 cm (18 in.) in diameter and 74 cm (29 in.) high, with an entrance hole of 27 cm (11 in.). The nest was placed on the ground. A pair at the San Diego Wild Animal Park nested in a drainage culvert which was 45 cm (18 in.) in diameter and 76 cm (30 in.) deep. Many pairs prefer to excavate their own nest in a bank of compacted soil, a process which takes about two weeks. Prior to nesting, the birds can be clearly seen searching the entire enclosure for a suitable nesting site. The male collects quantities of leaf litter, which the female then arranges in the nest. This collection and supplying of leaf litter continues throughout the nesting cycle.

Southern Ground Hornbill
Bucorvus leadbeateri

In courtship, the male feeds the female. Sometimes, while facing each other, the birds clap their beaks together. Prior to copulation, the male, vocalizing softly, approaches the female, walks around her striking the ground with his beak, and then, after completing one circuit, mounts. The clutch usually consists of 2 eggs, which are chalky white.

Incubation is by the female alone and begins with the first egg. The incubation period is 37–40 days, and as the eggs are laid 4–5 days apart, the older chick has a definite advantage over its younger sibling. Second chicks are usually ignored, and soon die. These second eggs or chicks should be removed for hand rearing. The pair at Dallas Zoo was of an unusually placid temperament, allowing the two chicks to be alternated between the parents and the zoo staff on a weekly basis. Initially, the female seldom leaves the nest, but as the chick develops, her foraging time increases. Chicks fledge at 80–90 days, but remain dependent upon the adults for food for up to a year. Immatures can be left with the adults for a few years, but are best removed when adult plumage is attained.

Incubation and Hand Rearing: The hand rearing of ground hornbills has not been without problems. Initially, many of the chicks fed the diet described above died at 5–7 days. These deaths were attributed to obstruction caused by the exoskeletons of insects. Other chicks were lost due to visceral gout, indicating kidney damage, probably from a diet that was too high in protein. Eggs are incubated at a temperature of 37.2–37.5°C (99–99.5°F) with a relative humidity of 56%, which is increased to 60% at hatching. Newly hatched chicks vary in weight from 52–70 grams. By two weeks of age they weigh around 400 grams. They should be weighed daily, and the rate of gain should remain constant, approximately 15% per day. Initially, the chicks are placed in a brooder at 32°C (89.6°F). For the first 12 hours, they are given ¾ cc of warm Ringers Lactate every 2–3 hours.

After their first droppings are seen, the following schedule should be followed:

- Days 1–4: Chopped pinkie mice plus minced ox heart with mineral and vitamin additives, fed every 3 hours round the clock. Ringer's Lactate is still given between feeds if the volume of droppings is small.

- Days 5–8: Quantities are gradually increased, and the chicks fed every four to five hours, including during the night.

- Days 9–12: Feedings every five hours during the day and once at night.

- Days 13–30: Three feedings per day. After day twelve, soaked cat cubes are added to the diet. After a further week, roughage in the form of chopped rodents, mice with fur, and day old chicks with the skin removed, as chick down causes impacting, are added. Both the droppings and the abdominal distention must be closely monitored, as these will provide the first indications of impacting. If there are any signs of slow bowel movements, Duphalac in warm Bovril water is given instead of food until the droppings return to normal.

References:

I would like to acknowledge the AZA Hornbill Husbandry and Management Guideline (1997) that was used as a considerable reference for this chapter.

Abrey, Dr. A. 1993. Handraising Southern Ground Hornbills. *Avizandum Magazine*.

Buay, J. 1991. Breeding and hand rearing the Tarictic hornbill. *International Zoo Yearbook* 30: 180–187.

Falzone, C.K. 1988. Maintaining and breeding the Abyssinian Ground Hornbill at the Dallas Zoo. *International Zoo Yearbook* 28: 246–249.

Fowler, M.E. (1986). Editor. Zoo and wild animal medicine. Second edition. Saunders Company, Philadelphia. ISBN: 0-7216-1013-7. 1127 Pp.

Golding, R.R. & M.G. Williams. 1986. Breeding the great Indian hornbill. *International Zoo Yearbook* 24/25: 248–252.

Kemp, A.C. (1995). The Hornbills: *Bucerotiformes*. In the Bird families series. Oxford University Press, Oxford. ISBN: 0-19-857729 X. 302 Pp.

Mace, M. & J. Azua. 1997. Breeding and hand rearing the southern Sulawesi hornbill. *International Zoo Yearbook* 35: 247–253.

Martin, P.K. 1982. Redbilled Dwarf Hornbills. *AFA Watchbird* 9 (3): 33–36.

Penny, C.G.1975. Breeding the Abyssinian Ground Hornbill at the San Diego Wild Animal Park. *International Zoo Yearbook* 15: 111–115.

Poonswad, P. (1994). Nest site characteristics of four sympatric species of hornbills in Khao Yai National Park, Thailand. Ibis 137: 183–191.

Poonswad, P., Tsuji, A. and Ngarmpongsai, C. (1983). A study of the breeding biology of hornbills (Bucerotidae) in Thailand. *In*: Proceedings of the Jean Delacour/ International Foundation for the Conservation of Birds symposium on breeding birds in captivity: 239–265.

Reilly, S.E. 1988. Breeding the rhinoceros hornbill. *International Zoo Yearbook* 27: 263–269.

Sheppard, C. and Worth, W. (1997). Report on AZA Hornbill Workshop- Phoenix 1997. Wildlife.

Sigler, E.U. & M.S. Myers. 1992. Breeding the wrinkled hornbill. *International Zoo Yearbook* 31: 147–153.

Thompson, D.R. 1996. Great pied hornbill. *AFA Watchbird* 23 (1): 31–32.

Wilkinson, R. & W. McLeod. 1990. Breeding the African grey hornbill at Chester Zoo. *Avicultural Magazine* 96 (4): 167–170.

Wilkinson, R. & R. Merry.1991. Breeding the trumpeter hornbill at Chester Zoo. *Avicultural Magazine* 97 (2): 67–73.

Hoopoes and Woodhoopoes

9 Upupiformes

09.01 Upupidae Hoopoes *Upupa* spp.

Glen Holland

Introduction: Along with the woodhoopoes and scimitarbills, hoopoes belong to the order Upupiformes. The typical hoopoes, of which there are only two species, the Eurasian hoopoe *Upupa epops* and the African hoopoe *Upupa africana,* are the only members of the family Upupidae. The Eurasian hoopoe has eight subspecies, including several that occur in Africa, while the African hoopoe has only two subspecies, the nominate form on the continent, and the race *marginata* on the island of Madagascar. Hoopoes are widespread, occurring in much of Africa, Madagascar, Europe, western Asia, India, eastern Siberia, northern and eastern China, Assam to Indochina, Malaysia and Sumatra.

Hoopoe
Upupa epops

Hoopoes prefer an open, dry habitat where they spend much of their time strutting about on the ground, probing into the soil with their long beaks. They are regular visitors to gardens. The sexes appear alike outwardly, but in flight the solid white upper half of their secondary wing feathers can distinguish males. In females, white stripes alternate with black in the upper half of the secondaries. Their varied pattern of rust, black and white gives them a rather striking appearance, particularly when the crest is raised. The long bill is used to probe into soil and crevices in search of insects and worms. Their legs are short, and their feet have three toes forward and one back, making them remarkably agile. Their name is derived from the characteristic *hoop-hoop* call made mainly during the breeding season.

Management: Wild-caught hoopoes are delicate, and difficult to convert to artificial diets. I know of one instance in which it took two months before a hoopoe would begin to accept substitute food items. I have only kept birds that have been handraised, and no problems were experienced feeding these. The ideal aviary for a pair of hoopoes should have a large natural earth substrate where they can dig for insects. They enjoy lawns, but the grass must not be allowed to grow too long, as this will hinder their movement. They also enjoy feeding on patches of bare soil, where they will also indulge in dust bathing or sand bathing. Habitat B, as described under LANDSCAPE, is ideal. I would recommend an aviary of at least 10 m x 6 m (33 x 20 ft.) in area. As they are not suited to extended periods of severe temperatures, they must be housed indoors during extremely cold weather. If this becomes necessary, it is important that they still be maintained on a natural floor such as river sand, peat and leaf litter. Problems with split skin on the feet have been experienced when the birds have been housed on a cement floor. These cuts can lead to infections from pathogens, resulting in such conditions as bumblefoot. As they have been recorded killing finches, it is advisable to keep hoopoes with species of similar size. They are highly territorial, and should only be kept in sexed pairs or same sex groups.

Diet: Weaned birds are fed on ox heart strips, mealworms, small crickets, wax moth larvae and winged termites, where these are available. They spend most of their day digging in the aviary lawns and find some natural food in this way. I experimented with feeding earthworms, but, although they played with these, they soon lost interest and left them. Food is held in the tip of the beak and tossed into the back of the throat with a flick of the head. Large insects are killed by beating them against the ground.

Breeding: Courtship includes the male feeding the female; sometimes the food is passed back and forth between the members of a pair before the female finally swallows it. Copulation usually takes place on the ground. In a large, mixed species walk-through aviary, I had a yearling pair that laid in a hollow log about 1 m (3 ft.) from the ground. Unfortunately, the clutch was lost to predation by a yellow-billed hornbill *Tockus leucomelas*. At Harewood Bird Gardens, hoopoes have been bred in both natural logs and nesting boxes measuring roughly 25 x 25 x 35 cm (10 x 10 x 14 in.), with an entrance hole of 6–7 cm (2½ in.). The base of the nest should be lined with a mixture of peat and river sand. Two, and sometimes three, broods are raised in a season; in 1993, Harewood raised 22 young from two pairs.

The clutch consists of 4–7 eggs, that are pale blue initially, but change to an olive color before hatching. Only the female incubates the eggs, which hatch after 17–18 days. The male feeds the female and chicks in the nest for the first week, after which the female will leave the nest to assist in gathering food. Observations of one of Harewood's breeding pairs revealed that the two birds took food to their nestlings 40 times during a 15-minute period. It has also been recorded that chicks from one brood assist with feeding the next (co-operative breeding). The chicks fledge at 27–30 days of age.

Kenton Lint provides an interesting account of breeding the Thailand hoopoe *Upupa epops longirostris* at the San Diego Zoo:

"In early March two birds were noted to be particularly interested in each other and subsequent mating was observed. A daily ration from 6–8 mealworms was fed to each. The male would take a mealworm in his bill, advance to the female and present the grub. He would then wait until she had accepted and devoured his offering. The male was a particularly tame bird and would fly to the aviary wire begging for a mealworm. If this was given he would return to the female repeatedly until her appetite was satisfied. After the feeding she would fly to a perch and rest. Only then did the male feed himself. It appears that the female will only accept food from the male once she is preparing to breed in the breeding season.

We observed that the nesting site was chosen by the male, a box placed in a recess in the aviary wall. The male ran around, raising and lowering his crest, fluttering about, at the same time producing soft calls to attract the females attention. This interesting courtship was observed for approximately a fortnight, during which time the female was indifferent to the advances of the male. March 27 the male was seen paying particular interest in the nest site He would fly suddenly into the hole, enter and exit, after which he would walk round and round, displaying and calling to the female to consider his choice. Several days passed with the female giving little notice to the males behavior. However on the April 6, the female was seen to peer into the nest. The following morning 10 cm (4 in.) of pine shavings was placed in the nest — which appeared to spur the female's interest. Both birds were then observed to enter the nest where they spent the greater part of the day. Further observations showed that the pair would come to the ground, feed, then return to the nest. We could now expect the laying of eggs within a short period.

On April 11 the female had one egg. We believe that the following two eggs were laid on successive days. Only the female incubates for the 18-day incubation period. She leaves the nest only to drink and for short intervals during the morning and evening. As is that case with hornbills, she is fed in the nest by the male. On May 1, I passed the hoopoe aviary and heard a faint squeaking coming from the nest. This sound was reminiscent of that made by baby mice. I immediately placed a container of small mealworms in the aviary. The male carried eight mealworms in separate trips to the female. For a fortnight following hatching the male continued to feed the female who would not allow him to enter the nest. She would raise her crest, wing and body feathers in order to block the nesting hole. On the sixteenth day the male was first seen in the nest feeding the young. Shortly thereafter the chicks could be seen moving around the nest begging for food. On fledging the young were perfect replicas of their parents. One fledged on day 28 and the other the following day. We believe the young were reared primarily due to a two-hour feeding of live food that consisted of mealworms and some small earthworms in soil. Our observations indicated that the majority of the earthworms were not consumed by the adults but were carried to the nest to provide food for the young."

It should be noted that modern practice would be to offer a wider variety of live food such as small crickets, waxmoth larvae, etc. These can be gut loaded and dusted in supplements such as calcium and vitamins. Of the three wild hoopoe nests I have found, the first was in a pile of stones, the second down a termite hole in the ground, and the third in a hollow tree stump. Hoopoe nests are notoriously foul smelling, and it has been suggested that this is a defense against predators. Nest sanitation is not always the best, and addled eggs often remain behind after the chicks have fledged. Nestlings produce a hissing sound when threatened, and this can be a further deterrent to predators. At least I have been a bit hesitant to insert my hand into a dark hole in the African Savannah where something was hissing at me!

Hand Rearing: When threatened, hoopoe chicks excrete a foul smelling dark liquid. Upon being removed for hand feeding, they soon accept the keeper as foster parent, and this behavior will cease. A chick's beak is much shorter than that of an adult, and it has a wide soft gape, which makes feeding easy. They are rather sluggish feeders in comparison with other softbill chicks.

If chicks are fairly well developed when taken for hand rearing, they will be extremely stubborn and it will be necessary to force-feed them for some time. During this process they should be wrapped in a cloth to prevent struggling, as well as to protect the handler from being squirted with their foul liquid repellent. When force-feeding, the food should be placed at the back of the throat and the chick transferred to a dark container, where it will later swallow the food. This is done hourly between 6:00 a.m. and 6:00 p.m.

After about four days the chicks begin to take food without force-feeding. As soon as one begins to utter their "sweet" call, all the chicks begin to respond. Chicks have a large gape and will take 3–4 pieces of food in quick succession. After feeding, their container should be covered with a cloth. They are best kept on wood shavings, which should be changed daily to avoid smelly, unsanitary conditions. When the chicks approach fledging age, the gape begins to shrink and they will only take 1–2 pieces of food per feeding. They are fed with a forceps and diet is as described for BEE-EATERS and SHRIKES.

Upon being placed in an aviary, young hoopoes remain very tame for a few months and will follow people crying *sweet, sweet* and begging for food. The one problem with hand-reared hoopoe chicks is that they are easily trodden upon, particularly in public displays. As they mature, however, they soon lose their tame nature to some extent, and are no longer in danger of being stepped on.

Ring Size: 4.5 mm

Reference:

Lint, K.C. 1964. Breeding the Thailand Hoopoe. *Avicultural Magazine* No. 70: page 119–121.

Woolham, F. 1994. Hoopoe breeding program — a remarkable success. *Cage & Aviary Birds* 22/10/94.

09.02 Phoeniculidae Woodhoopoes *Phoeniculus* spp.

Glen Holland

Introduction: Woodhoopoes belong, along with hoopoes and scimitarbills, to the order Upupiformes, but are separated into the family Phoeniculidae. The five species of woodhoopoes comprise the genus *Phoeniculus;* the three species of scimitarbills in the genus *Rhinopomastus,* formerly included in this family, are now placed in the family Rhinopomastidae. Unlike the common hoopoe, with a range extending into

Green Woodhoopoe
Phoeniculus purpureus

Europe and Asia, the woodhoopoes and scimitarbills are confined to the African continent. The red-billed, or green woodhoopoe *Phoeniculus purpureus,* is the species most often found in aviculture, and it is upon my experience with this species that this section is based. These birds occur in savanna, open woodland, riverine vegetation within arid thorn bush areas, and in wooded gardens. Woodhoopoes are adapted to clambering about in trees and on tree trunks. Their long tail is often used to steady the birds while searching along tree trunks for insects. They feed on various insects, larvae dug from under the bark with their long beaks, and spiders. They move as a group of 4–12 birds while feeding. The males are slightly larger than the females, with a larger and longer beak. Immatures are dull versions of the adult, but with the throat buff and the bill black. They are noisy birds and the Zulu name *Uhlekabafazi,* meaning "laughing women," describes them well. One bird will begin with a cackling call and soon the entire group joins in to make quite a commotion. While calling, the birds sway backward and forward.

Management: Hand-raised birds become very tame and are real characters in the aviary. At the slightest provocation, they will begin to cackle loudly. Like mynas, they also have the strange habit of probing their beaks into cracks, between lips, fingers, or the pages of a book, and trying to pry these open. One I had was very efficient at opening pockets and stealing cash notes. This same bird often played a game of cat and mouse with the yellow-billed hornbills and would cackle loudly at them when excited. Woodhoopoes are unable to maintain their body temperature at night, and for this reason a nest log should always be available to them. Their aviary should cover at least 60 m² (646 sq. ft.), and the landscape must include a number of dead stumps, with the bark attached, mixed in with the aviary plants as described in the woodland Habitat B in the LANDSCAPE chapter. They can also be housed in Habitat A, another habitat in which they occur naturally.

Hand-reared Woodhoopoe Chicks
***Phoeniculus* spp.**

Day	Chick 1 (g)	Chick 2 (g)	Day	Chick 1 (g)	Chick 2 (g)
1	3.2	3.1	**16**	40.8	41.5
2	3.1	3.4	**17**	44	44.5
3	4.1	4.0	**18**	47.5	45.4
4	5.1	5.4	**19**	48.8	51
5	6.4	7.8	**20**	48.6	55
6	9.4	10.2	**21**	52	57
7	12.8	12.9	**22**	55	58.5
8	15.2	17.4	**23**	57.5	60
9	19.5	27.7	**24**	59.5	63.2
10	21.7	25.8	**25**	61.4	60.2
11	25.5		**26**		59.5
12	28	32	**27**	59.8	60.1
13	31.2	34.8	**28**	59.5	61.5
14	35	36	**29**	60.1	63
15	36.5	38.5			

As woodhoopoes are strictly territorial, new birds should never be introduced into an existing family group. New birds introduced into an established group will most likely be found dead in a matter of hours. Young birds removed for hand rearing will not be accepted by their parents if introduced back into the aviary. Should it become necessary to form a new breeding group, all of the birds to be included in the group should be introduced into their new aviary simultaneously and then carefully observed.

Woodhoopoes are the first birds to have a proven chemical defense against predators. Their preen glands are able to produce a foul smelling and tasting secretion. By smelling their nest cavities, reseachers are able to predict quite reliably how recently the cavity was last occupied. When disturbed at their nest cavity, woodhoopoes turn their backs to the threat and secret a drop of the foul smelling liquid to deter the predator. In tests with cats that had been left to go a little hungry, the cats were offered two bowls of food. One was laced with the preen gland secretion of a woodhoopoe and proved to be the food the cats most obviously disliked to taste and

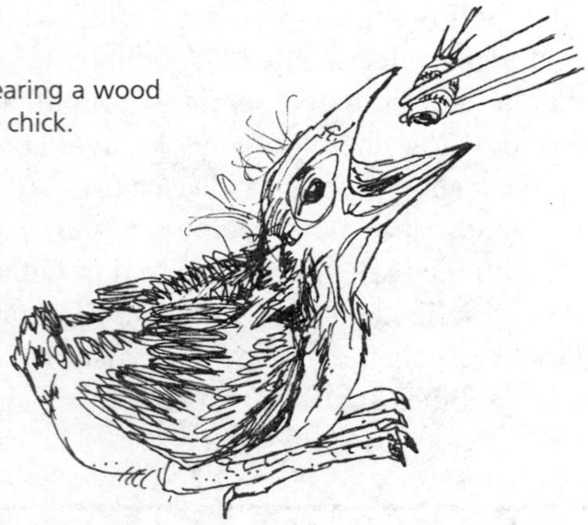
Hand rearing a wood hoopoe chick.

smell. It is interesting to note that the hand-raised birds I have had did not excrete this smell, presumably because they did not feel threatened.

Diet: The captive diet should consist of thin slices of ox heart, insects such as mealworms, fly larvae, crickets, and moths. Woodhoopoes will also pick at commercial softbill mixtures, boiled egg, fresh bonemeal, and minced meat. A multivitamin and trace elements are added to the diet twice a week. In a research project on this species, the birds were fed by placing a softbill mixture which included boiled egg in a 50 mm (2 in.) diameter clear plastic tube which had a number of perches and holes down the sides. The birds would perch on the tube and then probe their beaks into the plastic tube to retrieve the food, which included some live food.

Breeding: Woodhoopoes are co-operative breeders. Each group includes only one breeding pair. Only the female incubates, and the rest of the family group feeds her. The entire group also assists in raising the chicks. In captivity, the birds will be more productive when kept in pairs. The nest is usually a vacated barbet or woodpecker nest, and is often in a very hard wood. I have also seen a nest in an old metal pipe. In captivity, they should be given a natural nest log to use for roosting and nesting. I had one bird that used a donation box as his roosting site. Copulation lasts up to two and a half minutes. The eggs number 3–4 and are olive-green in color. The incubation period is 16 days. While only two chicks are reared in the average clutch, woodhoopoes are usually double brooded. The beaks of chicks are much shorter than those of the parents, and their gapes are large and yellow. They attain sexual maturity at 18 months to two years. This species has been successfully bred by a number of zoos, particularly Berlin, Frankfurt, and the Bronx Zoo in New York.

In a report on breeding this species at Winged World (England) Clive Roots provides the following information: "The green woodhoopoes were housed with a white-headed woodhoopoe and many other softbills including red-fronted barbets, blue-winged pittas, grey-hooded kingfishers, snowy and white-browed robin-chats. They never showed any signs of aggression towards these birds, and in fact seemed to enjoy the company of the other woodhoopoe.

Several large rotten tree trunks were provided for them, and within these nesting cavities had been chiseled out from the back and then covered, leaving only a small access hole visible in the front of the trunk. The birds however chose a small plywood box approximately 25 cm (10 in.) square, which had been suspended from the roof of he compartment. Although the tree trunks were not used for nesting , the woodhopoes certainly made good use of them, clambering around the crevices searching for insects, using their tails in the manner of the woodpeckers. They used the nest box regularly for roosting in, and until they had settled down, which took several weeks, they sought refuge in it whenever their compartment was entered.

As far as we could tell only the hen incubated the eggs, and she commenced sitting on April 8. Due to the position of the nest box and the small size of the entrance hole it was impossible to see inside without disturbing the birds. After 17 days it was apparent that some eggs had hatched, as the hen appeared more frequently and both birds began regular trips to the box with live food. Maggots, crickets, mealworms and a few small locusts were fed for the first week and some raw meat mixture included thereafter. Sixteen days later fully fledged chicks began appearing at the entrance hole and ventured out two days later. There were four chicks and they were about half the size of the adults. They were fed by both parents for just over three weeks, by which time they seemed to be completely independent."

Hand Rearing: Woodhoopoes are not difficult to hand rear. I have used a mixture of equal portions of insects such as grasshoppers, wax moths, moths, winged termites, mealworm larvae and beetles, and moistened ox heart strips with a multivitamin and calcium supplement added once per day. The insects can be rubbed in a little soaked and ground cat kibble. The food is fed every 1½ hours from 6:00 a.m. to 8:00 p.m., using a forceps. The chicks are rather sluggish in comparison to most softbill chicks. In common with the typical hoopoes, they excrete a foul smelling liquid when stressed. See the section on hand rearing TANAGERS for further advice.

Green Woodhoopoe *Phoeniculus purpureus* Feeding Schedule

Days of Age	Times Fed Between (7 a.m.–7 p.m.)	Items Fed
1–5	5–6	2–4 tiny items of pinky each.
6–12	6–7	2–5 pieces of pinkie, and Gaines dog meal.
13–21	4–5	5–8 pieces of pinkie, dogfood, crickets and mealworms.
22–45	2–3	Eating out of dish

Donald Bruning and Joseph Bell gave an excellent account of hand rearing this species:

> "Following a successful fledging of a single chick in May, it drowned in a pool in the aviary. Notwithstanding this tragedy, the pair was determined to nest and by September they were back on eggs. The site chosen this time was a hollow in a dead tree branch that extended across the display just above the water surface, easily accessible for both observation and management. Two chicks hatched on October 1 and by midmonth they could be seen regularly extending their heads from the hollow to be fed by their parents. Rather than run the risk of losing this clutch, they were removed from the nest hollow at the point when they were fully feathered and ready to fledge. Placed in a small cage equipped with a heat lamp and a nest hollow similar in size and conformation to their natal home, it was necessary to force-feed them for about 10 days before they began to pick up food on their own. Daily weights of the chicks and measurement of the food offered to them were valuable tools in monitoring their progress.
>
> Following this, three clutches were unfortunately found broken. When five eggs were discovered in February it was decided to remove them for artificial incubation. Two of these proved fertile and hatched on subsequent days. The chicks were kept in the incubator at 35°C (95°F) with a fairly high humidity of above 80% to avert any dehydration problems at this early stage of development. Very small pieces of pinkie mice and Gaines dog meal were offered from the tips of forceps at intervals of two hours (see the chart for further details). To date 15 green woodhoopoes have been reared at the Zoological Park; four of these were hatched in the incubator. A number of other hatchlings from the very prolific pair did not survive the delicate period as nestlings or subsequent "breaking-off" to independence."

Ring Size: 4.5 mm (.2 in.)

Species Requiring Similar Management:

Scimitarbills *Rhinopomastus* spp.
Babblers
Liocichlas *Liocichla* spp.
Laughingthrushes *Garrulax* spp.

Contributor: Dr. S.G. Stadler, Curator of Birds, Frankfurt Zoological Gardens

References:

Bruning, D and J. Bell. 1976. hand rearing the Green Woodhoopoes at the New York Zoological Park. *Avicultural Magazine*. 82: 4–7.

Roots, C. 1968. Breeding the Green Woodhoopoe at the Winged World. *Avicultural Magazine*. 74: 184–185.

Trogons

10 Trogoniformes

10.01 **Trogonidae: Trogons**

10.01 Trogonidae Trogons

Glen Holland

Introduction: Jeff Claffy introduces this fascinating group well:

> "Peaceful, calm, sluggish, quiet, patient, solitary, confiding, sedate, dignified, and gentlemanly; these are the adjectives various authors have utilized to describe trogons. Chiefly classified by the peculiar arrangement of their toes, trogons are remarkable birds that also make excellent aviary occupants. The distribution of family Trogonidae is widespread in both the New and Old World. The range of the 40 species of trogons includes Southeast Asia, Sub-Saharan Africa and the tropics of the Americas, with the majority of species occurring in South America. Australia has no trogons and fossil evidence indicates that at one time Europe did. Trogons have no clear relatives and appear to be a very ancient, static group, some recent genetic studies suggest a distant relationship to the coraciiformes, perhaps even places them within this assemblage. (Gilliard 1958). The heterodactyl foot is the defining family characteristic. The first and second digits point backward while the third and fourth face opposite. At a glance, this arrangement of phalanges appears similar to the zygodactylus psittacines. However in parrots the first and fourth toes are backwards with the second and third facing forward. Other morphological traits of trogons are: a short neck, large head, prominent eyes, small feet, and a long, wide, square tipped tail. The bill is small and slightly curved with basal bristles. The perching posture is very upright with the tail pointing directly beneath, sometimes pointing slightly forward. Abdomen feathers hide small feet when the bird is perched. The sexes are dimorphic with males more colorful than females. Usually the female possesses the same colors as the male although not as brightly or completely colored. The plumage is soft, delicate and loose. Males share all nesting duties. They nest in holes, which both sexes carve, in decaying trunks, occupied termitaries, large arboreal wasps nests, rarely arboreal ants nests, or compact root masses of epiphytes. (Stiles & Skutch 1989). Trogons are found alone or in pairs and have a ventriloqual call. The diet includes fruit, berries, blossoms, insects and even small lizards. A widely accepted generalization is that New World trogons are primarily frugivorous whereas Old Word species are mainly insectivorous. Insects are taken while the bird is in flight."

Management: Dietar Reinke provides information from Walsrode where the pairs are housed in various sized flights, which have attached heated shelters, and in large tropical houses. Temperatures range from just above 20–30°C (68–86°F) depending on the time of year and sunshine. Humidity is maintained reasonably high through the use of a daily mist spray. They have also been housed in pheasant aviaries at ambient temperatures but have access to a heated room that is has a minimum ambient of 15°C (59°F). These pheasant aviaries are less densely planted than the tropical house aviaries but still have sufficient plants to create the correct environment for these forest birds. High altitude, mountain trogons such as red-headed *Harpactes erythrocephalus,* and Javan *Apalharpactes reinwardtii* trogons, as well as golden-headed quetzals *Pharomachrus auriceps* suffer somewhat if temperatures get too high. They then start open-bill breathing.

These species are perhaps better placed in a indoor-outdoor facility.

Jeff Claffey reports that the white-tailed trogons *Trogon viridis* at the National Aquarium in Baltimore are displayed in the South American tropical rainforest exhibit, which is a mixed species, walk-through aviary. The aviary is a three-sided glass pyramid 18.3 m (60 ft.) tall enclosing approximately 35,450 m^3 (116,600 $ft.^3$) of space. Sharing the aviary with the trogons are approximately 45 individual birds of 18 species. The avian collection is heavily skewed towards passerines with six species of tanagers plus a small mix of cotingas, pigeons, grosbeaks and finches. Additionally there are three species of hookbills, blue-crowned motmots, sun bitterns and scarlet ibis. Also free roaming are a pair of Hofmann's two-toed sloths and a pair of golden lion tamarins, several green iguanas and a variety of turtles, toads and frogs.

Trogons have disproportionately small feet and a small tarsometatarsus bone. Therefore perching needs to be deceivingly small. Trogons fare better suited with perching appropriate for tanager size birds instead of pigeons sized ones. Banding the small tarsi can be problematic as well. Before banding, be prepared with two people and a variety of band options. At Walsrode they do not have banding problems and also use closed rings for chicks.

The trogons are fond of bathing and do this in two ways. Sometimes they will sit in water and bathe as do most birds, but they have been observed to almost completely submerge while bathing. The second method is in similar fashion to bee-eaters whereby the bids fly across a pool of water and dip into the water while on the wing. This is done repeatedly. Jeff Claffey reports that at Baltimore, "Our birds enjoy a very light misting directly from a hose. Also the birds on exhibit will sit adjacent to the waterfall and bathe in the mist."

The plumage of trogons is very fine, soft, delicate, and extremely loose. The way feathers seem to fall out with a mere brushing is reminiscent of leafbirds and handling should be kept to a minimum. It is unusual to see any social interaction between adult pairs or young birds even when held in a group. They have not been seen to allopreen or even sit together. Male and female will call to each other. In the wild in Africa I have heard this between the male and female Narina's trogon *Apaloderma narina* where the male makes a clear repetitive *hoo-hoo, hoo-hoo...* call and the female answers with a softer and more guttural version of the male's call. I have also heard similar calls between a pair of red-headed trogons in a forest in Thailand. Other than that, interactions are limited to the shared preparation of the nest cavity

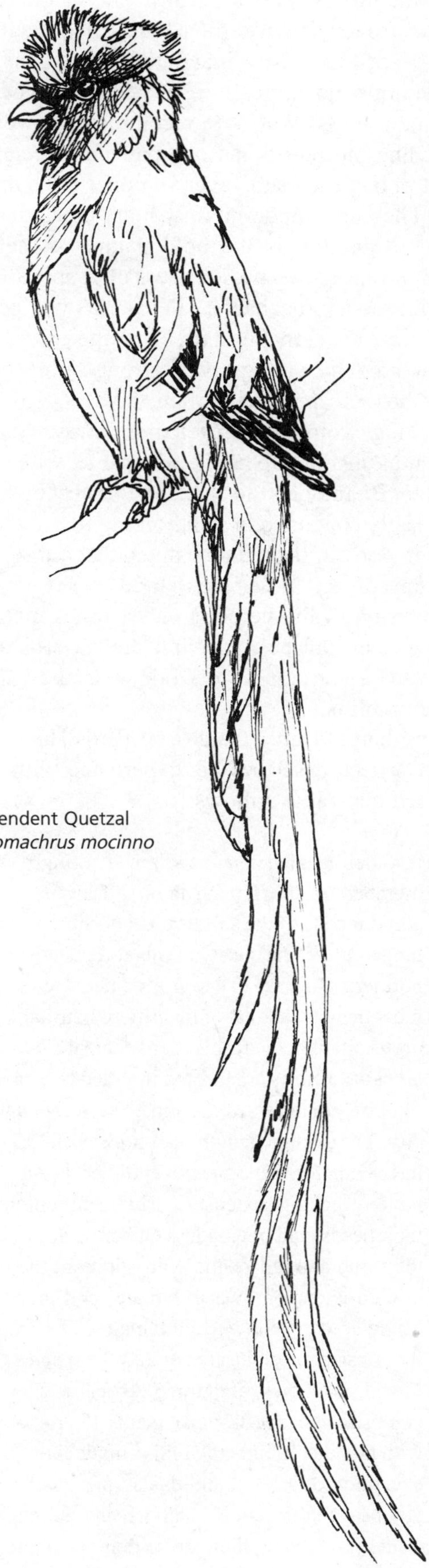

Resplendent Quetzal
Pharomachrus mocinno

and mating. At Walsrode different species such as the Javan trogon (Reinwardt's blue-tailed trogon) and the red-headed have been integrated without any issues and, in addition, no aggression has been observed when pairing new birds. Walsrode recommend however that for breeding, no more than one pair per species should be kept in one enclosure, even in rather large ones.

The most important husbandry concern at the Baltimore Aquarium is the bold nature of the adults and young. The adult pair is fearless of people and the young are far too approachable. The parents will perch on railings near the active nest amidst large crowds. The drive to attend to the young is very strong. While checking the nest box, keepers will often have one of the parents checking from an arms length away. Staff use this parental instinct to their advantage when the chicks fledge. By moving the chick and/or nest box, the parents are easily contained in an adjoining flight when they follow in pursuit. It should be noted that during any manipulation of the nest, the parent never exhibited any aggression to the perpetrator. At times management of the species can be tricky but the rewards are twofold. Trogons are a delight to work with; they are keen and tolerant birds.

Robert J. Berry, Curator of Birds Houston Zoological Gardens describes his experience with the golden-headed quetzal as follows:

> "Golden-headed quetzals are strikingly beautiful members of the Trogon family. They inhabit humid cloud forests of the sub-tropical portions of the Andes range in Venezuela, Colombia, Ecuador, Peru and northern Bolivia. (This is a colder, wetter environment than the habitat of their more familiar and glamorous relative, the resplendent quetzal *Pharomachrus mocinno*.) Nests are located in rotten tree cavities and the birds themselves are rather secretive and sedentary. The average length of *P. auriceps* is 35 cm (13.8 in.). Females are somewhat duller versions of the males, lacking extensive iridescent coloration on their heads, and having less elaborate development of the rump and tail feathers. In addition, the beak coloration is dark brown in females and juveniles, and bright golden yellow in adult males. The Trogon family consists of eight genera and 34 species (20 New World, 11 Asian and three African). The quetzals comprise the neo-tropical genus Pharomachrus. As with all other members of this unique family, with the exception of a few individuals, quetzals have generally proven difficult to establish and maintain in captivity. As a result, they are seldom available and few specimens have been exhibited in zoos or maintained

Red-headed Trogon
Harpactes erythrocephalus

in private collections. We were, therefore, not surprised when a review of the literature in 1981 indicated that of the 34 trogon species, only one successful breeding has been reported in captivity: the resplendent quetzal in a private collection in Costa Rica in 1941 (Delacour 1943).

Upon her arrival, our female quickly proved that quetzals are, indeed, delicate and tedious birds to maintain. The zoo quarantine period was difficult for both this female and her keepers. The diet instructions received from Hendee consisted of cubed whole wheat bread, puffed rice cereal, chopped banana, avocado, soaked dog chow and a vitamin/mineral supplement. These were the only items they had been able to persuade her to eat. We soon learned, to our dismay, that these were eaten none too well.

This particular specimen was unbelievably sedentary. In fact, I often wondered when I observed her through the window of the quarantine ward, if she was indeed alive or whether her keepers, in fear of my not uncommon fits of rage, had taken her carcass to a taxidermist and were simply moving her realistically stuffed body from perch to perch on a daily basis. As the Houston Zoo had previously received a degree of notoriety in the press for exhibiting fake coral snakes, my suspicions in this direction were not totally unfounded. In March of 1982, a juvenile male was received from Hendee. This bird was placed in our quarantine building in a flight adjacent to the female. By then, with our now somewhat vast experience associated with the stresses and traumas of trogon husbandry, the male's conditioning and quarantine were relatively uneventful. This was perhaps due to the fact that he was a younger bird and seemed to adapt more easily to his captive environment. He accepted a broader range of diet items, occasionally flew around his cage and, even more amazingly, often vocalized. His animation did little, however, to stimulate the female, and I continued to think she might actually be a realistically mounted taxidermy specimen. Both birds completed their quarantine in mid-April and were subsequently transferred to the Tropical Bird House.

During their quarantine we noticed that the female's left eye was becoming cloudy and obviously she could not see well. Examination revealed that a cataract had developed. Due to her visual impairment, the decision was made not to release the pair into our large, mixed species, walk-through aviary as was originally intended. (This proved to be a prudent decision, as by spring of the following year the female was totally blind in her left eye). The pair was, ultimately, placed in a 3 x 3 x 2.7 m (10 x 10 x 9 ft.) high glass-fronted, planted exhibit. The back wall of this exhibit was wire and looked out into the walk-through aviary. The exhibit contained a small pool with running water, a skylight and supplemental fluorescent lighting. After trying several wooden nest boxes and fiberglass simulated logs, in which the pair showed little interest, a natural palm log with a hollow core was installed. The 1.8 m (5.9 ft.) tall log was mounted vertically with the base resting on the floor of the exhibit. The hollow core was filled with leaves, pine needles and bark mulch to within 1 cm (½ in.) of the entrance hole and a perch was attached just below the exterior opening. The top was covered with a removable lid.

Through a slow and tedious process of coaxing, the diet now accepted by the pair consisted of ZuPreem Bird of Prey Diet, avocado, grapes, raisins, soaked dog chow/cat chow mix, bananas, papaya/cantelope mix and tomatoes. Although mealworms and crickets were offered regularly, few were eaten. Chopped pinkies were also fed but seldom touched. Moths and other flying insects were totally ignored. We became concerned that if the quetzals nested, providing the right nestling food would be a difficult problem. Most field observations indicate that chicks are reared on only live foods up to 10 days of age. We, therefore, persisted in trying to find other foods that would be accepted, but to no avail.

The pair appeared to adapt well to their exhibit. Their plumage became brighter after each molt and the male's tail feathers lengthened. The two birds, however, were never observed sitting together, allopreening or otherwise sharing an interest in each other or their nesting site. I still wondered whether the female's body cavity was stuffed with cotton balls or excelsior."

Diet: At Walsrode the birds receive the following diet:

- Dried insectivore mix (ORLUX Insect Patee Premium at Walsrode).
- Frugivore pellets (NutriBird Beo komplett).
- Mixed fruit diced into small cubes, berries, in particular blueberries.
- Live insects including mealworms, wax moth larvae and crickets — these are always taken first.
- The trogons have also been seen to take small fish from the food trays. A Javan trogon has also been observed eating very young leaves of a weeping willow. As the birds are in mixed species exhibits, no quantitative data on the diet is available.

At the Baltimore Aquarium, due to the wide gape trogons have and their ability to swallow berries whole, a variety of berries are added to the standard frugivore diet. The base diet consists of a variety of fruits and vegetables mixed with Tropical Bits manufactured by Marion Zoological. To this diet, blueberries, raspberries, and other in-season berries are regularly added. Extra corn, peas and small grapes are also provided because of their berry-like size and shape. Amazingly, when the pair of white-tailed were acclimatizing to the exhibit they were seen to swallow whole parrot fruit pellets. The white-tailed also relish waxworms, superworms and mealworms, and preferred in that order. The Trogons did not eat silkworms or walking stick insects. Locusts are eaten on occasion with small, soft locusts preferred over larger, harder ones. Surprisingly, the trogons will even eat pinkie mice. While feeding the young, the parents are given bowls containing free choice amounts of fruit such as papaya, mango and melon, plus berries and insects.

The *Roberts Book of South African Birds* reports that the African Narina Trogon is a predator of small vertebrates and invertebrates:

"They are a perch-and-wait predator, flying to snatch prey from foliage or a branch, and returning to perch. They rarely take prey from the ground; in one observation a bird sat still on a branch moving only its head from side to side, and up and down, and then dropped to the ground to snatch an item, barely alighting before returning to the branch. In another observation a bird dropped to the ground to take a lizard. Food consists mainly of invertebrates, including smooth-skinned caterpillars and they occasionally take hairy caterpillars. Other invertebrate prey includes spiders, mantids, cicadas, tree grasshoppers, beetles, flying termite alates and moths, and occasionally small arboreal lizards, and chameleons. Vertebrate prey is beaten on a perch before swallowing. Caterpillars are softened before swallowing; in one observation a bird landed on a branch carrying a hairy caterpillar that it worked backwards and forwards 3–4 times through its beak until holding it at one end and attempting to swallow it. One stomach contained grasshoppers, ladybirds, small beetles and a hairy caterpillar."

Breeding: Until recent times they have been infrequently bred, with many reporting them difficult to maintain in captivity, let alone breed. As with many species, the secrets of breeding these magnificent birds are now being discovered and documented. Walsrode Bird Park had the distinction of breeding four species: Diard's *Harpactes diardii,* red-headed and Javan trogon in 2003, and white-tailed Trogons since 1995 with various success; second generation breeding was achieved in this species. Houston Zoo bred golden-headed quetzals (Berry 1987). Jeff Claffy also reports on breeding the white-tailed trogon at the National Aquarium, Baltimore, USA, and in Mexico, Dr. Jesus Estudillo-Lopez has bred the resplendent quetzal *Pharomachrus mocinno*. A government breeding station in Cuba has managed to rear a few Cuban Trogons *Priotelus temnurus* very recently. (J. Rodriguez in litt. to D. Rinke.)

Prior to 2002, Dieter Rinke, zoological director of Walsrode Bird Park, reports that in 2002 the first Javan was bred. In 2001 a female Javan was lost through egg binding that resulted in an operation to remove the egg, but ultimately the bird died. A new female was sourced and moved with the male into a large flight at the end of 2001. In February the pair were seen flying to an old tree trunk and knocking at the bark as would a woodpecker. The tree proved to be too hard but they did not give up until the tree was extensively damaged. Clearly the pair were happy with their diet and environment, so staff brought in another nest log with a nest box attached. The nest cavity had been filled with wood shavings up to the entrance hole. The nest was immediately accepted and the pair took turns at throwing out the shavings to create a depression that was lower than the nest entrance. This behavior had not been observed with either the Diard's or red-headed but, soon after this, pairs of both of these species were supplied with similar nests and both immediately began to nest. Walsrode only use natural nest logs that are placed vertically.

Trogon foot.

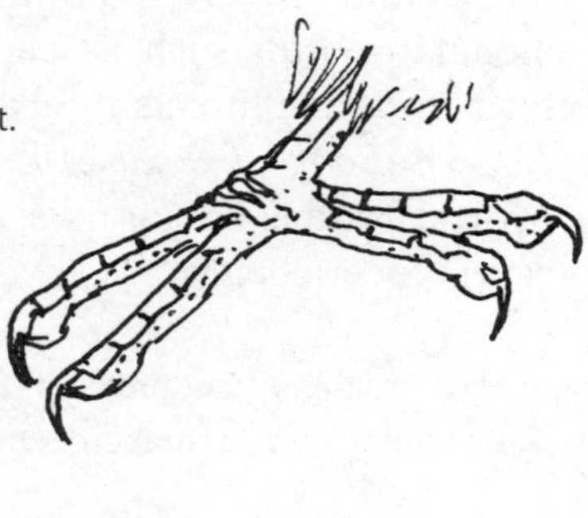

In all cases the nest depression was situated so that the brooding bird could be seen in the nest and it in turn could see out to its environment. Trogons generally lay 1–2 eggs (usually the latter) and occasionally three. There is talk of clutches of four but these have not been documented. Eggs are a yellowish to cream color. The incubation period is about 20 days. It is difficult to confirm the start of incubation and the male will spend periods of time sitting in the nest both pre and post laying of the first egg. Incubation tends to start after the second egg is laid. At Walsrode, due to the lack of success with parent rearing, the eggs are removed for artificial incubation and rearing.

The pair takes turns in incubating, with the male incubating during the day and the female at night. The birds swap places in the nest at approximately 7 a.m. and 6–8 p.m. The daytime incubation of the male is disrupted briefly when the female relives him for up to an hour to allow him to feed. The only aggression ever seen was when a rose-crowned fruit dove *Ptilinopus pulchellus* landed near the nest and the male was seen to chase it away. The parents mainly feed the chicks insects that they chew up thoroughly first. Chicks have been lost after swallowing hard pieces of plant material, and also nesting material and for this reason they are now artificially raised. Second clutches are not uncommon and at Walsrode the Javan changed their nest hole after the second clutch while the Diard's changed to a new nest after their first clutch.

Jeff Claffey reports that the pair of white-tailed at the Baltimore Aquarium nested four times. Six chicks survived to fledging; the first chick was hand raised from day five and the five remaining chicks were all parent raised. In spring 2000 the pair began investigating an artificial tree trunk for a nest site. The tree was a concrete trunk with several large branches dotted with planter pockets. The whole structure was veneered in cork bark to create the illusion of a living tree. Live epiphytic plants covered the entire fabrication. Both the male and female were flying to the underside of a "branch" and biting of chunks of cork bark. This was recognized as nesting behavior and a nest box was promptly installed. The box was mounted at approximately a 15° angle facing the ground on the underside of the branch that the birds were investigating. Cork bark initially covered the entrance to the box. Almost immediately the birds excavated the entrance and entered the man-made cavity.

The nest box was constructed from standard plywood and measured 22.9 wide x 20.3 deep x 40.6 cm high (9 x 8 x 16 in.) lined on the interior and exterior with cork bark. With the lining the interior dimensions shrunk to a 10.2 cm (4 in.) diameter cylinder, 35.6 cm (14 in.) in length. The 6.4 cm (2.5 in.) entrance hole was centered 26.7 cm (10.5 in.) from the cavity bottom. The nest was placed 5.7 m (18 ft.8 in.) above the ground. Since installation the pair used the nest box continually and of four known nesting attempts, three resulted in chicks fledging, although one attempt required keeper intervention. Successful clutches were laid in late March or early April of 2000, 2001 and 2002. The number of chicks surviving per clutch was one, two and three respectively. A nesting cycle in September 2001 resulted in a dead chick estimated at three days old. In each successful nesting period there have been instances where at least one of the chicks had fallen or climbed out the nest box. The first time this occurred a chick of an estimated five days old was successfully hand raised. Staff have two theories as to why the chicks may leave the nest too early. The first theory was that the 15° angle of the nest box was incorrect and this has been altered to 45°. The second theory was that the nest cavity might be too small for the adult and chicks. Placing chicks back in the nest box did not irrevocably disrupt the nesting cycle. Post fledging, the adults and offspring are placed in a small, heavily planted walk-in holding cage measuring 2.44 x 2.44 x 2.44 m (8 x 8 x 8 ft.). They are kept in the holding cage until the fledglings are strong fliers and completely independent of the parents.

Observations of wild nesting narina trogons are well documented in the Roberts Book of South African Birds:

> "Courtship display follows the finding of a suitable nest cavity by the male. The male pursues the female with slow, gentle, undulating butterfly flight, raising wings almost to vertical with each stroke, exposing the white undersides When perched, the male calls repeatedly, fluffs out breast and belly feathers, and flicks tail up and down or fans it wide open to show white outer feathers. Courtship may continue for several days. The pair copulates frequently until clutch is laid. In one observation the male flew 10 m (32.8 ft.) with slow, fluttering flight showing erect , red breast feathers to the female that was perched on bare, sunny branch, landed directly on her back and copulated. The nest site is inspected frequently by the mated pair before final acceptance by the female. Antagonistic encounters between trogons and other hole-nesting species, including green woodhoopoe and black-bellied starling in the Eastern Cape, suggest competition for limited nest sites. Four of six nests located in the wild were chimney-shaped and slanting upwards at an angle of 20–90°. Nest cavity entrance variable in size, but capacious interior; may be used year after year, in one instance for more than 13 years. The entrance diameter of six nests varied from 75–120 mm (3–5 in.); depth of cavity 200–600 mm (8–24 in.); height above ground of 2–16 m (39–52 ft.). Most nests are 3–5 m (9.8–16 ft.) above ground level.
>
> Soon after hatching the male feeds twice as often as does the female, delivering food about once every 20 minutes. The adults approach the nest swiftly, occasionally perching within 1–2 m (3–6.5 ft.) of the nest before entering with food. In general large food items fed to nestlings include caterpillars (carried crosswise in bill), butterflies and moths, cicadas, bladder grasshoppers, stick insects up to 150 mm (5.9 m) and mantids. Adults are very unobtrusive and mostly silent around the nest. The nestling period (n=2) was 25 and 28 days. Fed by both adults until flight feathers fully grown at about 60 days; thereafter stay near parents for several months, occasionally fed by either ad. Nestlings hiss and gape with lateral movements of the tongue, violently biting any object near the bill."

Robert Berry gives the following account of breeding the golden-headed quetzal:

> "In early April 1985, the male was noticed pulling at fibers around the entrance to the nest log and he gradually began entering and removing materials from the cavity. The female showed no interest in this cavity until July of 1985 when she was also observed pulling at the loose fibers around the nest hole. I was finally convinced after three and a half years that she was actually a real living bird. On August 15, 1985 the female began entering the log

for brief periods while the male perched at the entrance. By September 12, the female was remaining in the log for longer periods of time. When the nest was opened for inspection, we discovered that all of the material had been removed from the cavity right down to the ground level.

We did not wish to disturb the pair's efforts, but knowing that if eggs were laid and hatched, we would never be able to reach the eggs or chicks due to the depth of the cavity, we debated whether to repack the hollow. On September 16, the log was refilled. This time the lower half was packed with dried leaves and the top half with clean pine shavings. The male almost immediately began removing the shavings and by the following day he had excavated to a depth of 20.3 cm (8 in.). Thereafter digging slowed. On September 26, the male was seen dropping a leaf from the hole to the outside, which meant he was down to the leaf level. Three days later, both birds were seen removing leaves. Excavation continued until the cavity eventually reached a depth of 45.7 cm (18 in.) below the nest. A keeper's arm could barely reach the nest from the removable top of the log.

On the morning of October 5, the male was noticed to be staying inside the nest. That afternoon, when he left the nest to feed, a quick inspection was made before the female could enter the log. A single pale blue egg was observed. Another inspection four days later revealed a second egg had been laid. At this point, the birds were left strictly alone other than to be fed and watered, and a bamboo screen was erected across the outside of their exhibit glass to provide the pair with a greater sense of security. Both parents shared incubation duties. The female was generally in the nest early in the morning when keepers arrived and then again later in the day just before keepers left. The male sat during the day with occasional periods during mid-day when both birds were out together. On October 25, the first egg hatched. The second egg hatched the following day. The incubation period was, therefore, estimated at 17–20 days. The day the first chick hatched, both parents were inside the nest together from early morning to mid-afternoon when the male came out to feed. The female was not seen at all that day. Foods taken by the parents changed slightly at this time and in order of preference were: avocado, dog chow, grapes, pinkies (only sporadically), and ZuPreem Bird of Prey Diet (occasionally). More mealworms were consumed than normal but not in the number one would expect for birds feeding newly hatched chicks. Other live food in the form of crickets, spiders, and waxworms were totally ignored.

On October 27, both chicks appeared to be healthy; their eyes dark and closed; and their skin a bright, rosy pink. Their length was estimated at 8 to 10 cm (3–4 in.) from beak to rump. By October 30, banana, apple, and a little more bird of prey diet were being consumed. The chicks appeared to be doing well; both were plump and feather tracts were appearing. Satisfied that they were progressing normally, the nest was checked less frequently

On November 5, six days later, we were shocked to discover that one of the chicks was missing from the nest. The carcass was never found. The following day the female was seen entering the nest after having bathed. She did not exit the log to feed from early morning, nor did the male enter; obviously the chick was not being fed. Routine attempts to get the female off the nest failed. Tapping the nest log gently at the base finally brought her to the entrance hole. After four or five minutes she flew out. The remaining chick was found to be cold, dehydrated and very thin. No apparent growth had occurred in seven days. The chick was immediately removed and transferred to a heated brooder. Weight was recorded at only 26 grams. Due to the chick's weakened and debilitated condition the nestling was taken to my home for around-the-clock intensive care and hand rearing. Although the chick constantly vocalized, there was little or no feeding response. Initial feedings were of soaked cat/dog chow hydrated with 5% dextrose at hourly intervals. Smaller additions of papaya, avocado and peeled grapes were made. Often the chick had to be force fed as a normal response still could not be elicited. The chick frequently stayed in a head-down, arched-back position, often pushing backwards and flipping over. I had little hope that the chick would survive. If so, I knew it would be an uphill struggle. I was not disappointed. Hand rearing is often a rewarding experience but not with a chick that refused to gape or beg for food, and not when most feedings were forced, spit out or regurgitated.

On day 15, mucous was noticed accumulating in the chick's oral cavity. Culture results revealed *Enterobacter cloacae, Proteus mirabilis* and *E. coli*. Treatment was begun and consisted of 0.15cc chloromycetin and 0.25cc 5% dextrose administered daily by gavage. The mucous accumulation continued for about five days making feeding difficult. This condition cleared up just in time for a new problem to appear. The chick, whose eyes had opened on day 17, suddenly kept its left eye closed for two days, and upon reopening of the eye, the pupil was cloudy. This cloudiness persisted almost until the bird was independent. In my fatigued state, and as the chick was presumed to be a female, I began to wonder whether "cataract development in the left eye" might be some subtle form of sexual dimorphism as no medical reason for this condition could be determined. For some time afterward we feared that the bird would be blind in that eye. Today, however, there seems to be no visual impairment, and a short term dietary deficiency due to the bird's initial weakened condition is believed to have been the primary cause. Feather development appeared to proceed normally although there was depigmentation in areas of the developing primaries. Iridescent coloration was apparent by day 29. By day 33, the bird was approaching the size of the adults. On day 34, the youngster began perching on its food bowl. The chick was returned to the zoo and a gradual weaning process was begun. It was not until age three and a half months (mid-February) that we were confident that the chick was consuming enough food by itself to be independent.

In April 1986, a second clutch of two eggs was laid and two chicks hatched in early May. Efforts were again made to allow the parents to rear their young. Both chicks appeared to be developing normally when on day 12 one of the chicks appeared to have heavy bruising on the cranium and back. Both chicks were immediately pulled for hand rearing. The bruised chick deteriorated rapidly and expired the following day. Post-mortern revealed an enlarged heart and liver believed associated with protein deficiency. With a fortified diet, the remaining chick developed normally and reached full independence at seven weeks of age and is currently housed with its older sibling. In August 1986, the pair produced their third clutch, which consisted of only a single egg. This egg hatched on September 7 1986 and efforts were again made to have the parents rear their offspring to fledging. Unfortunately, it became necessary to pull this chick for hand rearing on day 14. The youngster has now reached independence and is believed to be a female. Hopefully, through a continuing process of trial and error, we will ultimately be able to locate some suitable form of live food that the adults will readily accept. Our initial efforts with this species have been very challenging and rewarding. The first two offspring are a male and a female and we are currently negotiating to acquire an adult female from a European collection and an adult male from Canada in order to expand our work with these intriguing birds. At least we now know they will breed in captivity.

What led to our success? A "little" avicultural experience, a lot of luck, a significant degree of effort, a modicum of intuition, and an infinite amount of patience. Luck — because the female's visual impairment caused us to place the pair in an individual exhibit rather than introducing them into our mixed species walk-through aviary as was originally intended. Unbelievably shy and nervous birds, I now sincerely doubt that this pair would have adapted to a mixed species environment to the point that they would have attempted to nest. They appear totally intimidated by any other bird larger than a finch that approaches them. Effort — because these birds did not adapt easily to captivity and great care had to be devoted to their conditioning and acclimation. Diets had to be closely monitored and any changes that were necessary in their environment had to be done gradually over a period of time. Intuition — because development of an insight into their behavioral needs was critical to their maintenance and breeding. For example, providing a nesting site that they would accept was mainly a matter of trial and error. Realizing that their beaks were not strong and adapted for digging, providing a type of fill for the nesting cavity that they could easily excavate was more a matter of intuition. The excavation of the nest site appears to be a critical phase of their courtship and pairing. Patience — because we worked with these birds for almost three and a half years before we were able to bring them into breeding condition.

The real success of this project, however, is due to efforts of the conscientious and talented keeper and support staff working with me. It is through their diligence, dedication, and commitment that our avian breeding programs in Houston have met with a significant degree of success and achievement."

Artificial Incubation and Hand Rearing: Jeff Claffey: Newly hatched chicks are completely bald with eyes closed as in all altricial species. Feathering occurs within 12–13 days and the earliest estimated date of fledging is 15 days. The young male white-tailed will start to develop some of the green and violet iridescent plumage at 30–35 days old. The single hand-reared trogon was fed a diet of: Lafeber's Nutri-Start Hand feeding Formula, a variety of fruits and veggies (mango, banana, blueberry, grape, strawberry, peas, corn, blackberry, cantaloupe melon, and pear), and insects (newly shed mealworms, waxworms, chopped pinky mice, crickets, and small locusts). Slowly softbill pellets from Marion Zoological Inc. were introduced into the diet.

White-tailed Trogon
Trogon viridis

Day	Weight (g)	Day	Weight (g)	Day	Weight (g)
5	23	**13**	39	**21**	45.5
6	24.5	**14**	37.5	**22**	46
7	27	**15**	40.5	**23**	47
8	29	**16**	42	**24**	46.5
9	28.5	**17**	45	**25**	46
10	29.5	**18**	44	**26**	47
11	32.5	**19**	45	**27**	48
12	35.5	**20**	44	**28**	49

Weight gain of a white-tailed trogon hand raised from approximately 5 days.

Dieter Rinke reports that at Walsrode incubation and hatching was successful using similar to conditions used for parrots: 37.3–37.5°C (99.1–99.5°F), humidity 50–55 % and during hatching 75%. New hatchlings are given the first feed 18 hours after hatch. This consists of small segments of pinkie mouse or rat. As the skin of the pinkies becomes rather tough after two days, only new-born are used for this purpose. Prior to each feed the food is initially dipped in lactobacillus. The young are fed on this diet every 90–120 minutes for the first two days between 6:30 a.m. and 8:30 p.m. After the first two days the feeding interval is increased to every 2–3 hours and stopped only when chicks cease begging. The food is lightly dusted with a mineral and vitamin mix (Korvimin). From day seven onwards the diet is supplemented with ox heart, small insects and fruit. This diet gradually replaces the pinkie diet. From day 28 when the young start to feed independently they are also given mealworms and insectivore diet with the finely chopped fruit. Survival rates on this diet have been good.

References:

Berry, R.J. 1987. Breeding the Golden-headed Quetzal. *AFA Watchbird.* 1987: 24–31.

Claffey, J. 2002. Breeding the White-tailed Trogon at the National Aquarium in Baltimore. *AFA Watchbird.* Volume XXIX, Number 4, pages 12–15.

Delacour, J. 1943. Quetzals. *Avicultural Magazine*. 5th Series. Volume 8, Number 6: 165–168.

Gillard, T.E. 1958. *Living Birds of the World.*

Rinke, D., W. Müller. M. Magnus. 1996. Erfahrungen mit Weißschwanz-Trogonen (Trogon viridis). *Gefiederte Welt* 120: 345–347. [Management and breeding of White-tailed Trogons]

Rinke, D., B. Marcordes, B. 2002. Nachwuchs bei drei Arten asiatischer Trogone. *Gefiederte* Welt 126: 402–406 [Management and first breeding of three species of Asian trogons]

Roberts Birds of Southern Africa, 6th edition. 1993. John Voelcker Bird Book Fund.

Skutch, A.F. and Stiles,G. 1989. *A Guide to the Birds of Costa Rica.*

Rollers, Motmots, Kingfishers, Bee-eaters

11 Coraciiformes

11.01 Coraciidae Rollers

Glen Holland

Introduction: Rollers belong to the order Coraciiformes and the family Coraciidae. The eleven species are distributed through Africa, Madagascar, Southern Europe, Asia, India, Australia, New Guinea and the eastern islands, and Sulawesi. The single Australian species is known as the dollar bird. Characterized by bright colors and a heavy flight, they nest in a natural hole in a tree. The sexes are alike. Incubation of the 3–6 white eggs is by the female alone. Most rollers are found either singly or in pairs in open woodland. They are territorial, and the members of a pair communicate with each other continually, particularly when breeding. They have favorite perches, usually exposed dead branches, to which they repeatedly return to survey the surrounding area. Their calls include a gruff bark like that of a small dog, as well as a variety of chattering calls used during display, aggression, or when threatened. They may call from a perch or in flight.

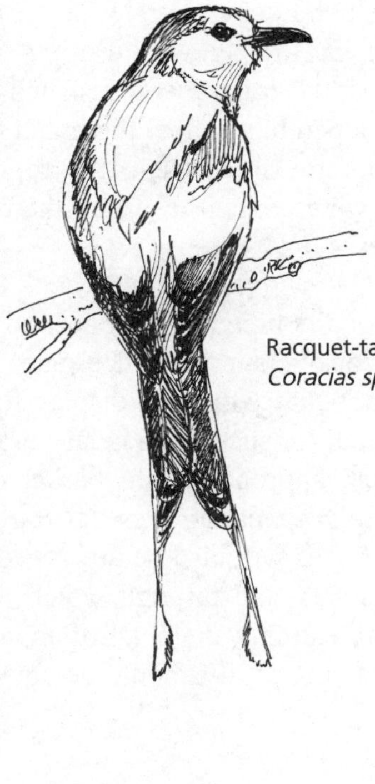

Racquet-tailed Roller
Coracias spatulatus

Management: Rollers should be included, whenever possible, in any African bird exhibit. They are beautifully colored birds and are active in an aviary. The lilac-breasted roller *Coracias caudata* is the species most commonly featured in aviculture, but several others are sometimes available and have occasionally been bred in zoos or private collections. These include the racquet-tailed roller *C. spatulata,* the Abyssinian roller *C. abyssinica,* the rufous-crowned roller *C. naevia*, the European roller *C. garrulus,* the Indian roller *C. benghalensis,* the broad-billed roller *Eurystomus glaucurus,* and the dollar bird *E. orientalis*. While this chapter refers basically to the lilac-breasted, other rollers should receive similar treatment in captivity.

As a general rule, rollers should only be kept together in sexed pairs. Aggression between birds of the same sex, particularly males, can result in mortalities. In contradiction to this, I have seen racquet-tailed rollers breeding as a small group in a very large planted aviary, but groups should not normally be housed together under typical aviary conditions. Odd birds, especially hand-reared males, often become aggressive toward their keepers; I know of two instances in which such birds have drawn blood. When introducing a pair, both birds should be in comparable condition, otherwise the stronger bird will immediately dominate the weaker one. Different species of rollers should not be mixed. The members of a pair are very vocal, and their raucous calls made when greeting each other are accompanied by bowing movements. In the lilac-breasted roller, at least, this display, accompanied by calls, may also be made by hand-raised birds when greeting their owner.

Aggression reaches a peak in the nesting season. When their nest is approached, a pair will become increasingly vocal in response to the perceived threat. If the nest is interfered with, the pair will swoop low over the head of the intruder, uttering loud calls. In an aviary, the nest will be defended against intruding birds. They are particularly aggressive toward other cavity nesting species such as hornbills and starlings. If kept as a sexed pair in a mixed collection of similar size birds, they generally take very little notice of the other species around them. They cannot be mixed with small birds such as sunbirds and waxbills, as they will corner, kill and devour them.

Rollers are particularly fond of bathing in a summer shower or under a sprinkler. They will perch under the spray with both wings outstretched until absolutely drenched. They are also fond of sand baths; a dry, dusty sand patch should be supplied for this purpose. Habitat B, with a number of dead protruding tree stumps, is ideal for rollers. They are hardy birds, but prone to foot problems in cold climates; shelter must be provided accordingly. Although they can withstand cold temperatures for short periods, a combination of cold and wet weather for an extended period will prove fatal to them. In such climates they must be housed indoors during the cold months. Rollers are not suited to cage life; when caged, they soon lose condition.

Roller chick at a few days of age.

Roller chick at two weeks.

Diet: In captivity, the diet should consist of diced strips of ox heart and bird of prey diet, winged termites, crickets, grasshoppers/locusts, moths, pinkies and young mice, and the occasional small lizard. Some aviculturists have described a diet of equal portions of fruit and meat mix. Although fruit was available for other occupants of my aviary, I never observed my rollers eating any. Live prey is hawked from an exposed perch or caught on the ground. The roller will then fly back to the perch, where the prey is bashed against the branch until dead. Some prey may also be run backward and forward through the beak, then tossed into the air and caught again. After being pulverized, the prey is swallowed whole. Even dead food is usually given a few good bashes against the perch before being consumed. Like other carnivorous birds, rollers regurgitate pellets comprised of such material as the hard exoskeletons of insects, bones, scales, and the fur or feathers of their prey. The ox heart and bird of prey diet strips can form the basis of the diet, with enough live food added to allow for the natural process of pellet formation. In one of the aviaries in which I kept rollers, we once had some mouse holes which we flooded with water. As the mice evacuated the holes, the rollers would snatch them up, bash them against a perch, swallow them and return for more. After about three mice each, the rollers could hardly move, and were content to watch the remaining mice scamper away.

Breeding: Most pairs breed in their second year, but I have had a one-year-old pair reproduce successfully. In the wild, courtship includes dipping display flights, but these are not essential for successful aviary breeding. In nature, rollers often appropriate old barbet or woodpecker nests; in captivity, natural nest logs are the best bet, although nest boxes will also be utilized. They prefer a nest with a small hole through which they must squeeze to enter. An entrance diameter of about 7 cm (3 in.) is ideal. The nesting cavity should be about 20 cm

(8 in.) in diameter with a depth of about 30–40 cm (12–16 in.) below the entrance hole. The top of the log should preferably have a hinged lid to allow access.

The nest is lined with a mixture of fine wood shavings and river sand. Two or three white eggs are incubated by the female alone, and hatch after 17 days. While the female incubates, the male sits in a tree close by, giving warning calls at the first sign of danger. He also feeds the female while she is in the nest. Newly hatched roller chicks are voracious feeders, and there is strong competition between the siblings in the nest. The chicks have a strong musky smell. They are extremely vocal, and their cries can be heard some distance from the nest.

Competition between the chicks in the nest is so strong that usually only one chick survives. As with the Micronesian King Fisher, here the aviculturist can intervene, either by removing the clutch of eggs for artificial incubation, or by pulling the newly hatched chicks for hand rearing. A single chick can be left for the parents to rear but, if the entire clutch is removed, most pairs will soon produce a second clutch. I once took two successive clutches from a pair for hand rearing, and the hen laid a third clutch which I left for her to incubate and raise. Other than the obvious advantage of increasing production through hand rearing, there is nothing as beautiful as having a hand raised roller fly to the hand to fetch an insect. Hand reared rollers are always a great attraction in a walk-through display open to the public. When parent-raised chicks fledge, the male soon becomes aggressive and will chase the female in an attempt to start a second clutch. Should he direct his aggression toward the fledglings, they should be removed and fed by hand until independent.

J. Simmons of Cotswold Wildlife Park reported the first successful breeding in England as follows:

> "The male kept flitting to a coffin-type nest box measuring 20 cm^2 x 60 cm deep (8 x 8 x 24 in.), calling to the female in a series of harsh chattering notes. On April 7 the female roller spent most of the day in the box, as she did for the next two days. Between April 9–21 there was always one of the pair in the box and it was assumed that they were incubating. On April 29 the birds were acting as if they had chicks; extra live food was supplied to the diet. All the food was dusted with calcium, Abidec drops and a multi-vitamin powder. Over the next few days activity seemed to die down and on May 8 an inspection of the box revealed nothing, no sign of eggs or chicks.
>
> The rollers carried on going in and out of the box during the rest of May and on June 28 quite distinct

Dollarbird
Eurystomus orientalis

European Roller
Coracias garrulus

chick-like noises could be heard from the box. The food was as before, with the addition of cockroaches. As the days went by the noise from the box increased. Both parents carried food, mostly live food, into the box. On July 21 the box was inspected and one really healthy looking chick was observed. It was nearly as large as the parents, fully feathered with the same markings but much duller colors, with a very short tail and no tail streamers. On July 23 the chick left the box. It spent most of the next day on the aviary floor beneath the box but by the next day was perching alongside its parents, who both continually fed it. On July 1 the male roller suddenly became more protective towards the youngster and spent the next two weeks attacking any of the keepers who ventured into the aviary.

This pair of rollers was housed with hoopoes, green woodhoopoes, bamboo partridge and bronze-winged pigeon. It is interesting to note that during the nesting and rearing periods the rollers showed no aggression towards the green woodhoopoes, even allowing them to perch on top of the nest box which contained a young bird."

Hand Rearing: [This section should be read in conjunction with the section on KINGFISHERS.] Rollers, even when hatched from the egg, are not difficult to hand rear. Newly hatched chicks weigh about 6.5 g (0.23 oz.) and are kept in a brooder at 34°C (93.2°F). They will snap up any morsel that is touched on the tip or side of the beak. Food should initially consist of equal portions of pinkie mice, with a multivitamin and calcium powder added once a day, and live food such as mealworm larvae and beetles, crickets, moths, winged termites, diced mice and lizards. Although I used ox heart, the current accepted practice is to use bird of prey diet and pinkie mice in place of this. Initially, the chicks are placed on paper towel. At first, they take one or two pieces of food per hour, but this soon increases to 3–5 pieces. Feeding starts at 6:00 a.m. and continues through to 8:00 p.m. As the chicks begin to mature, they are fed every two hours. They will take increasingly large pieces until, when half grown, they will easily swallow pinkie mice.

By this stage the chicks weigh about 50 g (1.8 oz.) and the temperature can be reduced to 28°C (82.4°F). The chicks are now placed on top of the brooder in a plastic container lined with wood shavings. Pellets comprised of chitin, bones, and fur are regurgitated. As the quills begin to appear, the chicks become very aggressive and it is necessary to place them in separate containers to prevent them from snatching each other by the leg or wing and hanging on like terriers. When well feathered, this aggressive behavior diminishes and they can be put together again. The immature birds are dull versions of the adults, but lack the two elongated outer tail feathers. They obtain adult plumage by eight months of age.

Ring Size: 6 mm (0.23 in.)

Species Requiring Similar Management:

Roller species
Kingfishers
Kookaburras
Bee-eaters

References:

Simmons, J. 1979. Breeding the Lilac-breasted Roller at the Cotswold Wildlife Park. *Avicultural Magazine*. 85: 61–62.

Sweeney, R. 1993. Keeping and breeding the lilac-breasted roller. *AFA Watchbird*. 20(3): 38–41

11.02 Momotidae Motmots

Elizabeth Prouse & Angela Hardy

There are 10 species of motmot in the family Motmotidae. Only one species, the keel-billed motmot *Electron carinatum,* is threatened. As part of the Order Coraciiformes, motmots are related to bee-eaters (Meropidae), kingfishers (Alcedinidae), and most closely to todies (Todidae). Motmots get their name from the Mexican pronunciation of their call which sounds like *moat moat*. All motmot species are indigenous to Central and South America. However, fossil evidence shows that they once ranged as far as Europe. The blue-crowned motmot *Momotus momota* is the most geographically widespread. It is also the species most common in aviculture. Motmots are medium sized birds, 16–48 cm (6.5–18 in.) long and weigh 50–150 g (1.75–5.35 oz). They are large-headed, brightly feathered birds with rounded wings, a robust body, and short legs. All species have long tails. Many, such as the blue-crowned motmot, have a racquet-shaped tail. All 10 species are monomorphic, however males may be slightly larger than females.

Motmots are quick and agile in short flights, but tire quickly over long distances. This prevents them from crossing large bodies of water, thus limiting their range. In fact, populations of blue-crowned motmots on Trinidad and Tobago islands a short distance off the coast of South America are effectively isolated and are markedly different in appearance from the mainland birds. Motmots are omnivores with a diet consisting of fruit, insects, invertebrates, and very small vertebrates. Their basic hunting technique is similar to that of a kingfisher and other Coraciiformes. They perch motionless, and then make a quick dash to grab insects in flight. Afterward, the prey is taken back to a perch where it is beaten and crushed before being consumed. They also use their long, serrated bill to sift through soil and leaf litter in search of invertebrates. Even prey that

Rufous Motmot
Baryphthengus martii

is found on the ground is typically taken back to a perch before being crushed and eaten. Blue-crowned motmots eat a higher percentage of fruit than most other smaller motmot species. Fruits such as nutmeg, incense, palm and figs are eaten in the wild. Since these are usually swallowed whole, motmots fulfill an important seed-dispersal role in tropical and sub-tropical forests.

Management: Motmots in captivity are relatively hardy and can tolerate lows of 10°C (50°F). Motmots are territorial and should not be kept in large groups. At Riverbanks Zoo, we maintain a breeding pair of motmots in an indoor aviary 2.4 m high x 5.0 m long x 0.9 m wide (8 x 15 x 3 ft.). Adding a few plants to the aviary will naturalize their environment and provide privacy. However, soil in potted plants will need to be covered with wire mesh to prevent the birds from up-rooting plants. Motmots make nice additions to large community aviaries. They are usually not aggressive to their mates or to other species. However, motmots are opportunistic feeders and will prey upon unguarded eggs, chicks, or small sleeping birds. Motmots have also been known to kill birds as large as a Pekin robin *Leiothrix lutea* and should not be kept with birds smaller than themselves.

Blue-crowned motmots at Riverbanks Zoo have been successfully kept in a large exhibit with birds such as: troupials *Icterus icterus;* sunbitterns *Eurypyga helias;* boat-billed herons *Cochlearius cochlearius;* green aracaris *Pteroglossus viridis;* and curl-crested aracaris *Pteroglossus beaharnaesii*. Motmots should have regular access to a mister or rain system to keep them clean and hydrated. A water bowl should also be offered. However, it is not actually clear whether motmots use standing water for either drinking or bathing.

Diet: This is the Riverbanks diet for one bird per day. Half of this diet is fed in the morning; half in the afternoon. In practice, motmots eat very little and can even give the impression that they have eaten virtually nothing in a given day.

- 24 1-inch crickets (10 g).
- Hills Science Diet dog kibble, soaked overnight in water (12g).
- Nebraska Brand Birds of Prey Diet™ (small pieces) (16g).
- Diced fruits (90g).
- Marion Zoological Tropical Bits™ (10g).
- 1 pinkie mouse per bird twice a week.

Crickets are fed Purina Game Bird Chow at least two days before being offered to the birds. This enhances the nutrition of the live food; and since poultry meal is high in calcium, it also helps to correct the inverted calcium:phosphorus ratio common to many commercially produced live foods, such as crickets. That is to say, commercially produced live foods are often dangerously high in phosphorus and low in calcium. In addition to feeding the crickets with a nutritious poultry meal (a process sometimes called "gut loading") we also sprinkle lab grade calcium carbonate powder on the crickets before feeding.

Rain and fruit appear to be their primary sources of water. It is important to mist motmots, both to keep them clean and so they can drink the water preened from their feathers. This equally applies to bee-eaters which are also in the order Coraciiformes. Interestingly, we have noticed that motmots at Riverbanks eat all of their fruit when we do not mist them; and they eat almost none of their fruit when they are misted every one or two days.

Breeding: The blue-crowned motmot was first bred in 1966 at the Kilverstone Wildlife Park in England. It is still bred from time to time, especially in zoos, although it is not easy to breed. No records were found for any other species of motmot having been bred in captivity.

Blue-crowned Motmot
Momotus momota

Blue-crowned motmots nest in burrows, dug either in the ground or in an embankment. Males do most, but not all, of the excavating. Females help more with the finishing process of the nest cavity. The tunnel leading to the nest chamber can be between 0.6–4 m (2–12 ft.) in length, and may be straight or twisting. There may also be a side "escape" tunnel. Motmots normally excavate their tunnels at least one month before nesting. In the wild, this may be due to easier digging conditions caused by the rainy season. At Riverbanks Zoo, motmots usually excavate their tunnels in January or February, and lay in March.

Surprisingly, motmots will accept a manmade nest tunnel, which can be made using plastic, corrugated drainage pipes and a wooden nest box. At Riverbanks Zoo a successful artificial nest was made by connecting three pipes, each 10 cm (4 in.) in diameter, to a closed wooden box that is a 0.3 m (1 ft). Two of the pipes are 0.9 m (3 ft.) long and are joined by a "Y" piece that fits into a hole in the side of the box. A third tunnel is 1.5 m^2 (5 ft.) long and bends at a 90° angle before entering the nest box. The tunnel entrances fit into three separate 10 cm (4 in.) holes in a piece of plywood that is fixed vertically to create an embankment. The whole set-up is suspended two feet above the floor in an aviary that is 2.4 m high x 4.5 long x 0.9 wide (8 x 15 x 3 ft.). The tunnels and nest box can be opened and cleaned between uses, and are packed with fresh soil prior to nesting.

Blue-crowned motmots usually lay 2–4 round, white eggs, 1–2 days apart. Incubation begins just before the last egg is laid and continues for 18 days. Both sexes incubate, but switch every 12 hours. Each successful hatching at Riverbanks occurred promptly on day 18 of incubation. Chicks are fed live food almost exclusively for the first 12 days. Fruit is given to the chicks on approximately day 13, in addition to ever increasing amounts of live food. The chicks grow quickly and consume amazing quantities of food. During the first two weeks, a clutch of two chicks will eat 15 pinky mice and a dozen pink worms (earthworms) every day, in addition to the regular diet. Soon after, the number of pinky mice will increase to 25 per day. Chicks can be noisy, and are often heard close to feeding time.

There is a notable decrease in food consumption a few days prior to fledging, which occurs at 28–32 days. Fledglings weigh about 85 g (3 oz), and already have adult plumage. Though they lack the long racquet tail, it will take only a few weeks to develop. Motmot fledglings can fly quite well, although they are nervous and uncoordinated, and should be approached with great care. Motmot chicks wean very quickly, and begin feeding themselves as soon as three days after fledging. They can be removed from the aviary 2–3 weeks after fledging, by which time they are completely independent. Siblings can be housed together for some months, although males will become increasingly aggressive towards each other.

For information on artificial incubation and hand rearing it is suggested that the section on BEE-EATERS be used as a reference.

Contributors: Special thanks to Martin Vince, Clinton Carroll and Bob Seibels for their help in writing this text.

11.03 – 11.05 Alcedinidae, Halcyonidae, and Cerylidae Kingfishers

Glen Holland and Liz Romer

Introduction: The kingfishers belong to the order Coraciiformes and include three families: the Alcedinidae, the Halcyonidae, and the Cerylidae. More than 90 species are distributed widely throughout the world. For this section I have concentrated on two African kingfisher species, one representing the fish eaters, which are associated with, and dependent upon, aquatic habitats, and one representing the woodland kingfishers, which feed largely on invertebrates and small vertebrates, and often live away from aquatic habitats. Significant reference has been made to the A2A husbandry manual for the Guam (Micronesian) kingfisher.

Malachite Kingfisher *Alcedo cristata*

• Glen Holland

Introduction: The malachite kingfisher *Alcedo cristata* is widespread in Africa. Few species can match or beat the beauty of these little jewels. They are true

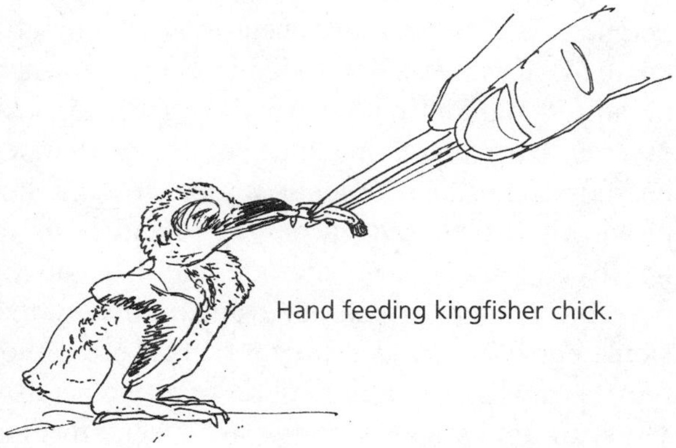

Hand feeding kingfisher chick.

fishers and are found in pairs or perched singly on vegetation along riverbanks and lakes. Their call is a soft, high-pitched *tseep* made in flight.

Management: Sexes are alike and surgical or DNA sexing is advised to ensure true pairs. These beautiful little kingfishers deserve an aviary to themselves featuring a small flowing stream that can be regularly stocked with fish, allowing visitors the opportunity to watch them fish. When a separate aviary cannot be provided, they are compatible with small seedeaters, sunbirds, and softbills such as white-eyes. Cyperus and other reeds should be planted along the edges of the water. In addition, a few exposed dead branches over the stream will provide ideal hunting perches. Habitat F will suit these tiny kingfishers. Aviaries do not have to be large and a 6 x 2 x 2 m (20 x 6½ x 6½ ft.) flight will suffice. Do not use glass on displays housing kingfishers, as these birds are prone to killing themselves by flying into it. Due to territorial aggression, only one pair should be kept per aviary. Although they are able to withstand overnight temperatures down to freezing, they require daytime temperatures in excess of 12°C (53.5°F). In countries where severe climates persist for extended periods, they are best housed indoors with some additional warmth during the winter months.

Malachite Kingfisher
Alcedo cristata

Diet: Ideally, about 50 minnows should be stocked into the aviary stream once a week. In addition to live fish, they are fed on small dead minnows that can be stored frozen, strips of fish and heart. A multivitamin is added to the food twice weekly.

Although I have not had the opportunity to trial this, small lizards and skinks may also be taken.

Breeding: To encourage breeding, a sandbank should be provided, preferably with a pool of water in front of it. Starting a tunnel will encourage the birds to excavate a nest. Alternatively, a pipe of 5 cm (2 in.) diameter and about 30 cm (12 in.) long can be built into the mound to serve as an access tunnel to the nest chamber. The pair shares incubation for 14 days. Adults indicate clearly when chicks have hatched; instead of turning the fish headfirst into the beak, the fish are carried by the tail section with the head pointing forward.

Fledglings are generally duller than the adults and have a slightly shorter beak, which is black, as are the legs. One breeder reported 13 chicks raised in captivity from a single pair that reared four clutches in quick succession. The aviary became crowded and the entrance pipe to the nest was blocked to stop further breeding! As the young birds started to mature they were attacked by the adult male and had to be removed.

Hand Rearing: Hand rearing is much the same as described below for the MICRONESIAN the chicks are still very small they are best fed on small minnows that are broken in half and slightly crushed. Initially, only one food item is taken, but when half grown they begin to take two, and later three food items per feeding. As quills start to appear, ox heart strips or pieces of pinkey mice are introduced. I lost one chick from a brood because initially I fed them on flying ants, which were too coarse, causing compaction. As soon as the ants were replaced with minnows, the chicks progressed well.

The food is fed with a forceps. It is necessary to hold the food at the fore-end and push it into the throat. This prevents the excited chicks from knocking the food out of the tweezer. Care must be taken to avoid injury to the chicks from the tip of the tweezer as they lunge forward to swallow the food. As the chicks reach maturity they will not take food unless it is placed well back in the throat, triggering this forward lunge. By the time the chicks are well covered in quills, minnows of about 30 mm (1.18 in.) can be fed whole, alternated with diced pinkie mice or ox heart strips. When minnows were not available, thin strips of fresh fish were substituted.

A moistened rearing food and a multivitamin and calcium supplement are added to the moistened meat once per day. The chicks make a rapid *chirr-chirr-chirr* call while being fed. The larger chicks seem to be slightly aggressive and dominant over the smaller ones. It is best to place each chick in an adjacent separate compartment. After fledging, the chicks make a short *chik-chik-chik* call when hungry, and while being fed they flap their wings rapidly. This wing flapping appears to be a form of exercise and often continues beyond feeding. Within five days of fledging, the birds instinctively begin to bash food against the perch. They do not take food well at this stage, but knock it about with the beak open. Small fish or strips of ox heart/pinkie mice should be offered with the forceps, then dropped onto a plate containing shallow water, prompting the chicks to fly down to fetch it. Some live minnows in shallow water will also stimulate the natural feeding behavior. A strong perch should be situated beside and above the source of food.

Ring Size: 2.5 mm

Species Requiring Similar Management:

Giant kingfisher
Belted kingfisher
Common kingfisher
Pied kingfisher
British kingfisher
The larger species should not be mixed with birds smaller than themselves. As predators they may learn to feed on small birds.

Brown-hooded Kingfisher *Halcyon albiventris*

• Glen Holland

Introduction: Brown-hooded kingfishers *Halycon albiventris* are usually found in pairs inhabiting riverine vegetation, but may also be found some distance away from water in open woodland and in gardens. Their call consists of a melodious choo, *choo, choo, choo, choo,* often in the form of a duet and accompanied by the birds flitting their wings. The call is repeated rapidly when disturbed. Members of a mated pair also make soft twittering contact calls when not in sight of each other. Males are brighter in color than females.

Management: Their aviary should be landscaped as described for Habitat B, and should include a pool of water into which they can dive. A few dead branches and upright tree stumps will provide ideal hunting perches. As they naturally occur in warm climates, adequate shelter must be provided in countries where temperatures remain below 10°C (50°F) for extended periods, particularly if cold and wet. In temperatures near freezing, an additional source of heat should be provided. Brown-hooded are woodland birds and are not fishers. They will sit motionless on an exposed branch for long periods, interrupted only by the occasional bob of the head as they watch for prey below. The prey is caught in a straight glide from the perch.

They are extremely aggressive when breeding and I have had a bee-eater knocked off my hand and killed instantly by this species. I have, however, kept them with pied kingfishers *Ceryle rudis* with no problems. This was due in part to the different food sources used by the two species, and because the pied were too young to compete for nest sites. Brown-hooded and other woodland kingfishers are best housed with birds of the same or a larger size than themselves, such as *Tockus* hornbills, shrikes, starlings and, weavers.

Diet: They should be fed on 40% pinkie mice and ox heart strips, 30% bird of prey diet with a multivitamin powder added once per day, and 30% live food such as mealworms, winged termites, grasshoppers, locusts,

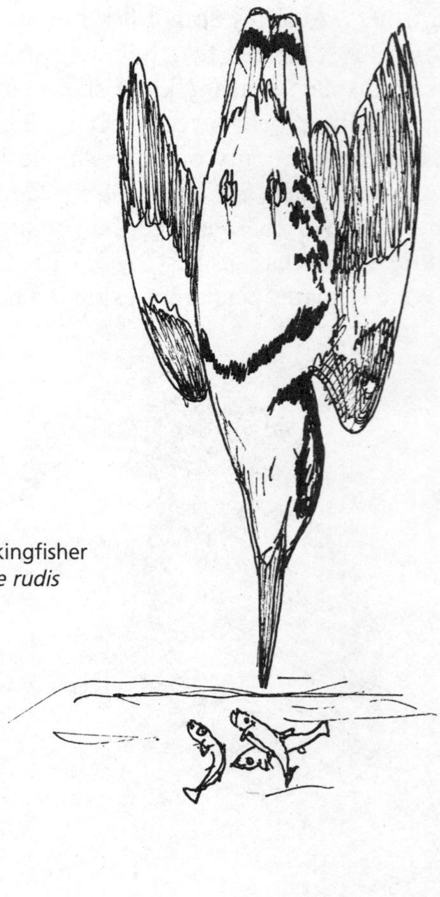

Pied kingfisher
Ceryle rudis

Common Kingfisher
Alcedo atthis

praying mantises, millipedes, and crickets. If a small fish is placed in very shallow water they will take it. They will also take small dead fish fed by hand, but will not dive into deep water after fish.

Breeding: Unfortunately, these kingfishers become rather aggressive during the breeding season and should be carefully monitored to prevent them from picking on other birds. A brown-hooded pair will excavate their own nest into a sandbank. For additional information on building sandbanks, read the section on BEE-EATERS. The tunnel is approximately 70–90 cm (2–3 ft.) long, built 1–3 m (3–10 ft.) above the ground, with a nest chamber at the end. Both sexes incubate the 4–6 eggs for 14 days. Removing a clutch of young will stimulate the pair to produce a second clutch. Hand-raised chicks must never be reintroduced into the parent's aviary, as they will be attacked immediately. Should successful breeding occur, the chicks must be removed when independent.

Hand Rearing: All species of woodland kingfishers can be hand raised on 50% diced pinkie mice and ox heart strips dipped in commercial rearing food, with a multivitamin and calcium supplement added once a day. The remaining 50% of the diet should consist of insects, and young rats that have been cut in pieces. Small lizards are a real treat for the chicks and are devoured whole. Initially, chicks should be fed hourly and provided with an artificial source of heat. Newly hatched chicks are kept at 36°C (96.8°F) and after three days this is reduced by 1°C (1.8°F) per day until an ambient temperature of 26°C (78.8°F) is reached. By this time the chicks will be half-feathered, and feeding every two hours will suffice. All food is fed with forceps, and chicks will usually take three food items per feeding.

The chicks are able to swallow quite large pieces of food. While their eyes are still closed, the food is touched on the side of the gape, and will immediately be seized. Chicks become agitated and aggressive when food is presented, but I have not found it necessary to separate them. They are extremely vocal when hungry. The chicks are kept on sawdust. Due to their pungent droppings and regurgitated pellets, this bedding must be changed daily. (This should be read in conjunction with the section on hand rearing the MICRONESIAN KINGFISHER.)

Ring Size: 4.5 mm

Species Requiring Similar Management:

- Sacred kingfisher
- Woodland kingfisher
- Striped kingfisher
- White-breasted kingfisher
- Rollers
- Bee-eaters

Micronesian Kingfisher
Todirhamphus cinnamominus

• Micronesian Kingfisher Manual

This section is quoted directly from the extensive Micronesian Kingfisher (Halcyon Cinnamomina Cinnamomina) Species Survival Plan Husbandry Manual. 1998. Edited by Beth Bahner, Aliza Baltz and Ed Diebold

The adult female resembles the male except that the upper breast is paler, as are the chin and the throat, with the rest of the underparts and underwing coverts white. The recommended minimum enclosure size for breeding pairs of kingfishers is 3 x 2.4 m (10 x 8 ft) with a height of 3 m (10 ft.). Although pairs have bred in enclosures as small as 1.8 x 2.4 x 2.4 m (6 x 8 x 8 ft.) problems with pair compatibility may be exacerbated by these small enclosure sizes. Pairs *Halcyon cinnamomina cinnamomina* should be provided with the opportunity to put some distance between themselves when not breeding. Birds have escaped from enclosures that use piano wire barriers; therefore, this type of barrier should not be used with this species. There have been several cases in which kingfishers have attacked their images reflected in glass cage fronts. Birds housed in glass-fronted enclosures should be monitored carefully to assure that this does not occur and cause injury to the bird. As a forest bird, the Micronesian kingfisher has a natural shyness and needs adequate cover. Marshall (1989) found that kingfisher nest cavities were found in areas with approximately 80% canopy cover. Their aviary should be landscaped as for Habitat A. Kingfishers must be provided with water dishes that are large enough to permit bathing, a minimum dish size of 12.7 cm (5 in.) deep and with a diameter of 45.7 cm (18 in) is recommended with a water depth of 5.1–7.6 cm (2–3 in). The water dish should be elevated approximately one foot off of the floor for easier access by the birds.

Aggression can be problematic and at the Philadelphia Zoo a Micronesian kingfisher killed a superb fruit dove *Ptilinopus superbus* and, over a two-year period the Pittsburgh Aviary lost one silver-eared mesia *Leothrix argentauris,* three shama thrush *Copsychus malabaricus,* one dhyal thrush *Copsychus s. saularis,* and two fairy bluebirds *Irena puella* to kingfisher aggression. Given the current state of the captive population, it would be inappropriate to risk housing this species with any other potentially aggressive species, leaving single-species maintenance the best option for the Micronesian kingfisher. Birds should also be monitored to ensure that neighboring species are not causing a disruption as well. Another concern about mixed species exhibits is the difficulty of monitoring and controlling food intake. Given the weight control problems, incidence of hemosiderosis and other nutritional concerns, kingfishers are best housed alone. Because of their propensity for intra-specific aggression, adult Micronesian kingfishers should be held either in breeding pairs or as singletons. Young birds can often be housed together if monitored carefully for aggression. While adults have occasionally been held in same-sex groups without incident, this practice is not recommended. In addition, because several pairs have failed to breed until neighboring pairs or singletons were removed from sight, it is recommended that all breeding pairs be visually isolated from conspecifics.

Feeding: Food dishes should be shallow and open to allow birds to feed on the wing. It is preferable to place feed dishes above ground level in a central location to facilitate feeding. Pairs should be provided with more than one food dish in different locations, to minimize the potential for aggression over food. It is also recommended that birds be fed twice daily. Observations have shown that kingfishers prefer to feed early in the morning and in the evening. Food cups that attach to the side of the cage (as used for psittacines) should not be used.

Breeding: The following protocol for introducing birds to each other is recommended in an attempt to decrease the potential for aggression. Prior to placing the birds in the same enclosure, the prospective pair should be placed in adjacent cages for two weeks of acclimation and observation prior to introduction. If the pair appear compatible, to minimize problems that could arise from territorial defense, potential mates should be introduced into a neutral enclosure. All pairs should have nest logs available at the time of introduction. Several successful pairs have begun striking at the nest within one hour of introduction. This may be: 1) important to initiate the pairing process and 2) a good early indicator of the potential for success. All pairs should be observed for at least one hour after introduction

In captivity pairs produce multiple clutches in a year, and there have been females that lay eggs almost continuously throughout the year. Because of potential problems with eggshell thinning, more than four clutches per year are not recommended. For pairs that do not stop breeding on their own, removal of nest log(s) may be necessary. Removal should occur in the "non-breeding" season, approximately August through

Micronesian Kingfisher Nutrition Survey: 1994 and 1996 Significant Intake Results

Items Offered	1994 Intake Ratings	Offered by	1996 Intake Ratings	Offered by
Mice (pink or fuzzy)	Good (>80% offered)	13	Good (76.7–100% of offered)	5
Insects (mealworms, crickets waxworms)	Good (>80% of offered)	13	Good (>80% of offered)	6
Green anoles	Excellent (100%)	2	Excellent (100% of offered)	4
Complete foods (dog food, bird of prey)	Poor (<40% of offered)	5	Poor (<40% of offered)	4

November. Several observers have suggested that the process of excavating a nest cavity plays a substantial role in pair formation or the maintenance of pair bonds. Micronesian kingfishers excavate a nest cavity by "alternating thrusts of the beak at the chosen site in flights initiated from a nearby perch. Not having the foot structure necessary for perching on the vertical trunk, a bird flaps its wings vigorously (and awkwardly) as it attempts to deliver more than one blow per flight. Usually a bird is restricted to only one or two blows and returns to the perch after each attempt. A call is given upon leaving the perch with each new attempt." Although the Micronesian kingfisher does not appear to have an elaborate courtship display the male does courtship feed his mate. Courtship feedings occur coincident with successful pair copulations and have not been observed during the early stages of pair formation, incubation, or care of the young. The male may offer the female a variety of food items but the female does not always accept what he offers.

Age of Egg Production/Breeding (years)

	Minimum	Minimum fertile	Average	Maximum (as of 6/97)
Females	0.67	0.75	2.12± 1.03	10.83
Males	0.83	0.92	2.33±1.08	>13

Most copulations and attempts occur within three weeks of egg laying, and mounting frequency peaks during the week preceding oviposition. Mountings have been observed during brooding for the pair observed by Marshall on Guam and have been noted to occur in captivity as well. Clutch sizes are 1–3 eggs. Incubation begins with the laying of the first egg, and both sexes participate in incubation duties. Eggs should be candled at least one week after egg laying. During the first week of incubation, eggs can be quite fragile and easily punctured or broken. Although it is possible to determine egg fertility earlier than seven days, candling the eggs before this time is not recommended because of the potential risk to the developing chick.

Clutch sizes are 1–3 eggs; 64% of clutches consist of two eggs, 32% are one egg clutches and only 4% are three egg clutches. Out of 631 eggs produced, 254 (40%) were fertile and 194 (76% of fertile eggs) hatched with 116 (60.56) viable offspring produced. Of the 116 viable offspring 96 (47:49) lived to reproductive age. In captivity, pairs have not been very successful at fledging offspring from the nest. As of

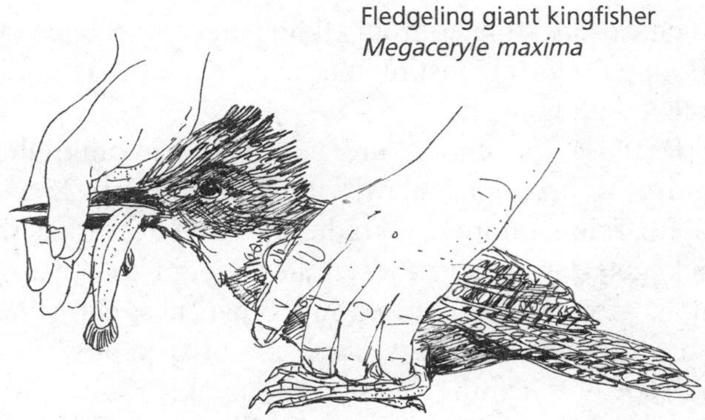

Fledgeling giant kingfisher
Megaceryle maxima

December 31, 1995, 69% of parent-reared chicks have been lost in the nest. Of those chicks that are lost prior to fledging, 75% disappear from the nest and the cause of death is unknown, although cannibalism has been suspected; 90% are lost within the first 10 days. Hand rearing has been more successful with only 19% lost before 30 days of age; 76% of those are lost within the first 10 days. As a result, hand rearing is recommended. In light of aggression by parent birds, it is recommended that all parent-reared chicks be removed from the enclosure in which they are reared within two weeks of the time they fledge. In the interim period the chick(s) should be monitored closely for signs of parental aggression.

Artificial Incubation and Rearing: The recommended protocol for pulling eggs for hatching and hand rearing is to remove the eggs approximately 20 days after laying. Eggs left with non-experienced pairs should be candled to determine fertility at least one week after laying of the last egg and removed from the nest approximately two days prior to hatching, preferably after the chick has penetrated the air sac. The egg(s) should be placed in the hatcher set at a wet bulb temperature of 31.1–32.2°C (88–90°F) and dry bulb of 37.8°C (100°F) with a relative humidity of 64%. After hatching, follow the protocol outlined in Table 1. High humidity appears to be very important to successful hatching. In cases where the egg is removed prior to penetration of the air sac, set the wet bulb at 30–31.1°C (86–88°F) for a humidity of 58% and adjust after air sac penetration. The egg should be turned approximately three times/day in the hatcher until the air sac is penetrated. Care should be taken to avoid removing the egg more than four days prior to hatching.

The San Diego Zoo rearing protocol describes changes in diet for each developmental stage and provides a description of a diverse diet recommended for Micronesian kingfisher chicks. The San Diego Zoo uses a Micronesian kingfisher hand puppet for hand rearing and chicks are isolated from all human contact once eye slits open. Not all institutions take steps to minimize contact with humans.

Because the chicks are naked at hatching, they require a brooder/isolette temperature of 35.5°C (96°F). Temperature can gradually be decreased as the chick gets larger and the feathers begin to develop. Chicks are typically housed in a bowl (nest cup) lined with a non-slip substrate, such as toilet paper, paper towel or washcloth. These are quickly soiled and should be replaced regularly. Clutch-mates can be housed in the same brooder/isolette but should not be placed in the same nest cup since aggression has been observed between siblings. As the chicks get older and more mobile, birds that are housed within the same brooder should be monitored to assure that no aggressive interactions occur. When chicks begin perching at approximately 21 days of age, they can be moved to larger quarters. At approximately 30 days of age, chicks will begin to attempt flying but are not yet completely independent and self-feeding. Fledging occurs for parent-reared chicks at 33 (±2.74) days after hatching (N=18 parent-reared birds).

Table 1: Artificial Incubation Parameters Used at Selected Institutions

Institution	Temp (°F)	Weight loss (%)	# Egg Turnings (per day)
Houston	100/84–86	14–16% loss	12 (auto)
San Diego	99.5–86.0	13% loss (av.)	5 or 12 (auto)
St. Louis	99.5/85		5 or 12 (auto)
Cincinnati	99.5/86		4 (auto)
Audubon	99.5/87	9–10% loss	4 (hand)

Begin chicks on chopped pinkies (whole pinkies or eviscerated) and chopped anoles. To minimize problems with dehydration, pinkies are dipped in dextrose or Pedialyte. The typical signs of dehydration are dry, baggy skin and dry fecals. A well-hydrated chick is pink with the skin fitting well on the body. Calcium carbonate and vitamin supplements can also be added to the pinkie diet. Additional diet items are added as the chicks grow, beginning as early as one week of age at some facilities. Additional food items can include cricket abdomens, chopped whole crickets, mealworms, and waxworms. As the chick grows, diet items can also increase in size and whole insects can be fed. Analysis of pellets cast by young chicks revealed that they do not digest many of the components of soft food.

Chicks should be weighed in the morning before the first feeding. Normal hatch weight is in the range of 4.4–7.1 (0.16–0.25 oz.). Chicks should be fed 6–8 times per day initially, beginning first thing in the morning and continuing approximately every 1½–2 hours throughout the day. After 7–10 days, additional feedings are dropped until the chick reaches independence. The frequency of feedings should be adjusted to assure that the chick follows the expected weight curve. For young chicks, small food items are offered by touching them to the side of the chick's beak. They quickly learn to respond to the touch and gape or lunge for the food item. Blunt-ended toothpicks work well to feed young chicks until tweezers or forceps become practical. Chicks begin to refuse food after reaching a weight of 60–70 g (2.1–2.5 oz.) at 20–25 days. Force-feeding is not necessary as long as the chicks remain on the typical weight curve. Hand feedings can be reduced to 3–4 per day and food items can be left with the chick between feedings. Chicks typically decrease in weight at this time down to as little as 50–55 g (1.76–1.9 oz.). By day 26, the chick can often be encouraged to pick up food on its own by moving it with forceps. Birds are capable of self-feeding at 32–35 days and, generally, they can be weaned by 40–55 days of age.

Hand-raised White-Breasted Kingfisher *Halcyon Smyrnensis*

Day	Weight (g)	Day	Weight (g)
1	6.88	22	91
2	7.1	24	80
4	9.45	26	74.5
6	14	28	71
8	20.05	30	67
10	41	32	68.5
12	41	34	74.5
14	51.9	36	79
16	67.5	38	84
18	81	40	83
20	88.5		

Hand-raised Micronesian Kingfisher *Halcyon cinnamomina cinnamomina*

Day	Weight (g)	Day	Weight (g)
1	5.19	22	57.38
2	5.41	24	65.13
4	6.69	26	66.85
6	8.87	28	68.24
8	12.19	30	63.57
10	20.71	32	61.1
12	20.71	34	56.34
14	26.35	36	56.34
16	33	38	53.7
18	41.35	40	54.61
20	49.08		

Kookaburra *Genus Dacelo*

• Glen Holland

Introduction: In Australia there are two species of kookaburra. The laughing kookaburra *Dacelo novaeguineae,* the world's largest kingfisher, is native to eastern Australia and has been introduced to Tasmania and southwestern Australia and New Zealand (Forshaw 1992). It is a clever and persistent predator, usually preying, from a strategic perch, upon creatures on the forest floor. The laughing kookaburra is found in a variety of habitats including open eucalyptus forest, open rainforest, parks, suburban gardens, fenced paddocks, and sugar cane fields with telegraph poles, all providing adequate perches for foraging opportunities (Shields 1994). It is particularly able to exploit habitats altered by European development. Anyone conducting an outside barbecue soon discovers that the laughing kookaburra is prepared to add steak, chops, and sausages to its diet (Shields 1994). The normal diet includes insects and their larvae, earthworms, snails, snakes, lizards, frogs, freshwater crayfish, small birds, and fish.

Laughing kookaburras are extremely sedentary, residing in delineated territories, that may contain a mated pair or a family group comprising the adult pair with one or more offspring from previous years acting as helpers at the nest. They have been known to live for up to 20 years (Shields 1994). The sexes are similar, but females are slightly larger than males. The laughing kookaburra is well known for its famous laughing call.

Across northern Australia the smaller, more colorful blue-winged kookaburra *D. leachii* replaces the laughing kookaburra. It is more distinctive than its more common relative with brighter colors, more exotic habitat requirements, and a more spectacular song. It generally prefers more moist habitats than the laughing kookaburra, being found mostly in gallery forests, mangroves and along inland streams. It is more wary than its close relative, frequently flying away when approached, in contrast to the laughing kookaburra, which in some instances will even seek out human contact (Shields 1994). Female blue-winged kookaburras are similar to males with the exception of the tail. The tail of the male is dark blue, while that of the female is rufous evenly barred with blue-black.

Laughing Kookaburra
Dacelo novaeguineae

Management: Kookaburras are not commonly kept in aviculture in their native land. Some Australian zoos display and breed kookaburras, with the blue-winged kookaburra probably being the more commonly bred species. Laughing kookaburras often come to zoos as juveniles to be hand reared and rehabilitated back to the wild, but are seldom kept for display. Outside of Australia this situation is reversed. The blue-winged kookaburra is relatively rare in European and North American zoosx and private collections, while the laughing kookaburra is well-established, having been captive bred for many generations. Kookaburras are best kept in pairs in captivity. They have been maintained in aviaries 2 x 4 x 2 m (6½ x 6½ x 6½ ft.) high, but generally prefer a bit more space. At Currumbin Sanctuary we have bred blue-winged kookaburras in aviaries 4 x 6 x 3 m (13 x 20 x 10ft.) and 7 x 10 x 6 m (23 x 33 x 20 ft.). In both cases they have been housed with other species, including Australian kestrels *Falco cenchroides,* Pacific bazas *Aviceda subcristata,* and bush stone-curlews *Burhinus grallarius*. Due to their predatory nature, they should not be housed with smaller species. The aviaries are best landscaped with natural perches of various thickness. Kookaburras also like to bathe, and a reasonably sized water dish is recommended.

Diet: In captivity they are primarily fed a meat based diet. The Currumbin Sanctuary meat mix for kookaburras is listed below.

Coarse meat mix ingredients:

- 2 kg lean mince/pieces
- 1 cup of Wombaroo Insectivore mix
- 2 cups wholemeal bread
- 1 cup soaked dog chow
- 6 teaspoons calcium
- 3 teaspoons vitamin E
- 2 grated carrots
- Preparation: Blend the wholemeal loaf and dog chow, and then thoroughly mix in the other ingredients.

This mixture is supplemented with mice and mouse pinkies, chopped day-old chicks, mealworms and whitebait.

Breeding: In the wild, laughing kookaburras normally nest in hollow logs or termite nests, while blue-winged kookaburras nest in hollowed-out termite nests or bottle trees. In captivity, we have been successful in breeding blue-winged kookaburras in both hollowed-out logs and nest boxes measuring approximately 30 cm (1 ft.) square. The opening is in the front, with a lower lip preventing the eggs from accidentally rolling out. Laughing kookaburras have been consistently bred using natural logs or nest boxes of a similar size. Some breeders provide a nest box that is more rectangular, perhaps 30 cm (1 ft.) x 40 cm (1 ft. 4 in.) x 30 cm (1 ft.) high, such as a wood duck *Aix sponsa* nest box turned on its side. Birdworld Bird Park has successfully used a box with a sloped shape measuring 90 cm (3 ft.) in length and 35 cm (14 in.) in both height and width, with a 15 cm (6 in.) diameter entrance hole, and a perch below, located near the top of the front panel (Sweeney 1995).While we have found the blue-winged kookaburra tolerant of nest inspection, Shepard (1989) reports that laughing kookaburras are intolerant of nest inspection during incubation.

Spring is the usual breeding season for kookaburras. Blue-winged kookaburras commonly lay a clutch of 2–4 eggs, while Sweeney (1995) reports that the most common clutch size for laughing kookaburras in captivity is two or three eggs. In both species, both parents share the incubation duties. Blue-winged kookaburra eggs weigh approximately 22–23 g (0.8 oz.). The incubation period for both species is 24–26 days. Blue-winged kookaburras have fledged at approximately 38 days of age, while Widman and Vorous (1961) reported that laughing kookaburras at the National Zoological Park in Washington, DC first left the nest at 30 days of age.

During breeding, fresh food is offered up to three times a day. The primary diet when the chicks are very young is pinkies and coarse meat mix, well combined. As the chicks grow older, larger food items are offered, including cut up day-old chicks and gutted and portioned mice. Mealworms are also offered. Young are only removed when required as surplus stock for exchange, as no aggression has been seen between the parents and offspring. In 1998, following the successful fledging of a chick, our blue-winged kookaburras nested a second time. This is the first instance of double clutching in one season recorded for this species.

Artificial Incubation and Hand Rearing: We have artificially incubated wild laughing kookaburra eggs brought to Currumbin Sanctuary. Two chicks hatched in October 1998 and were successfully reared. They were incubated at 37.2°C (99°F) with a wet bulb reading of 29°C (84°F). The chicks were kept in a brooder with an initial temperature of 35°C (95°F). The temperature was lowered 0.6°C (1°F) every second day after pin feathers were observed emerging. By the time the feathers were breaking out, the chicks were kept at room temperature. Eyes were fully open at 18 days. Fledging occurred at 35 days.

Growth Observations on *D. novaeguineae* & *D. leachii*

Day	No. 1	No. 2	No. 3
1	21.3	20	
2	27.1		
3			33.42
4		47.5	
5	61.8	56.5	
6	69.7	77.5	
7	88.8	89.9	
8	105.2	107.4	
9	121.7		
10			103.88
11		158.5	
12	170.7	172.5	
13	185.5	188.3	
14	196.6	196.8	
15	207.7	eyes half open	
16	219.5	232.4	
17	245.4	252.3	
18	257.4	eyes fully open	
20		295.3	212.37
		first laugh heard	
21	292.2	302.8	
22	305.7		
25		315.7	
26	327.2	321.5	
27		310.6	259.47
28		313.9	
29		313.4	
32		312	
33	317.6	306.04	
34	303.3	292.6	267.37
35	290.5	fledged	fledge day 38

Nos. 1 and 2, laughing kookaburra, incubator-hatched wild eggs, hand raised 1998.
No. 3, blue-winged kookaburra naturally incubated and raised, 1994

The chicks were fed eight times per day for the first week, seven for the second week, five for the third week, and then four times per day until they were self-feeding. The diet consisted of small pieces of lean meat, ox heart, day-old chicken, and mouse, dipped in Wombaroo Insectivore mix and water mixed to a slurry. At one feeding each day one pinch of Enzyplex enzyme complex was added to the feed. Balanced Cal was also added to the feed once daily.

Ring Size: 9 mm

Acknowledgments: Jim Stokman, Jan Leicester and Cathy Mofflin for data collection.

References:

Bahner, B., A. Baltz E. Diebold. 1998. AZA Micronesian Kingfisher *(Halcyon cinnamomina cinnamomina)* Species Survival Plan Husbandry Manual.

Forshaw, J. 1992. Kingfishers and their relatives in Australia. *AFA Watchbird* 19(5):42–46.

Shepard, M. 1989. *Aviculture in Australia.* Black Cockatoo Press.

Shields, J. 1994. *Cuckoos, Nightbirds and Kingfishers of Australia.* Angus & Robertson Publishers.

Widman, W.F. H.M. Vorous. 1961. Hatching and rearing of kookaburras in the National Zoological Park, Smithsonian Institution, Washington, DC *Avicultural Magazine* 67(5):166–168.

11.06 Meropidae Bee-eaters

Glen Holland

Introduction: Bee-eaters belong to the family Meropidae in the order Coraciiformes, and are closely related to kingfishers, rollers, and motmots. The 24 species of bee-eaters are found primarily in Africa, with some found in southern Europe, South-East Asia, Australia and the eastern islands. They have certain preferred perches, usually a dead branch from where they hawk their prey. Although colonial, the birds disperse into individual hunting areas when feeding. They nest in pairs or colonies in sandbanks along river courses. Within the nesting colony are a number of clans that consist of a nesting pair, old parents, brothers and sisters, and even cousins. Each clan is made up of about five birds. Not all species, however, are colonial; some, such as the little bee-eater *Merops pusillus,* occur in pairs.

Management: My experience with these birds is based on a group of eight white-fronted bee-eaters *M. bullockoides* and two little bee-eaters, all of which were hand reared. Bee-eaters are one of the most challenging families that I have worked with, and I would only recommend them for public displays or to the individual aviculturist who can afford the extensive time and attention they require. Newly acquired birds are delicate and only accept live food. Teaching these birds to accept an artificial diet is a difficult and time-consuming process; I know of one aviculturist who force-fed his little bee-eaters ox heart once each day for a month before they would take this food on their own. Such pressure is likely to lead to stress-related health problems and ultimately death. No serious aggression occurred within my group of white-fronted, and I have never seen bee-eaters initiate aggression against any other species. To avoid competition from aggressive softbills, they are best housed with smaller species such as waxbills and finches. Due to the competition for food, nesting sites, and hunting perches, it is advisable to keep only one species of bee-eater per aviary. Kingfishers also compete with bee-eaters for food and nest sites.

As bee-eaters spend much of their time on the wing, they must be kept in large, sparsely planted aviaries with plenty of large dead, exposed branches to provide a selection of perching sites. These should be positioned in sunny locations in the aviary. Habitat B, as described in the chapter on LANDSCAPE, will suit bee-eaters. These are birds of the sun; in their natural habitats, temperatures of 40°C (104°F) are common in the summer months. Although they can withstand overnight temperatures down to 0°C (32°F), their winter ranges are usually dry, warm, and sunny during the day. Extended periods of combined cold and wet conditions will result in the death of these birds. In adverse climates they should be housed indoors through the winter. A few infrared lights placed above strategic perches will benefit the birds.

The bee-eater aviary should be planted with nectar-producing plants to attract insects. By selecting plants that bloom at different times of the year, the availability of this natural food source can be extended. Honeysuckle *Lonicera* spp., red-hot poker *Kniphofia,* canary creeper *Tropaeolum peregrinum,* and any other flowers that provide nectar to attract bees and other insect pollinators are useful. An open pool of water will furnish a bathing site for them. They do not settle in the water to bathe, but will drop from a perch into the water and then fly out, returning to the original perch. This process may be repeated 4–6 times during a typical bathing session. A suitable perching branch near or overhanging the water is essential.

Bee-eaters are also fond of sand bathing. In colonial species this is a communal activity; most of the colony will join in and have a good "chat" while so engaged. A dry patch of light, dusty soil must be provided for this. While sand bathing, they lie on their bellies with their wings outstretched, baking in the sun. Bee-eaters also enjoy sunning themselves from a perch. On particularly cold mornings, they will sit motionless with the head cocked to the side and all the feathers of the head, neck, and back puffed out to expose the skin to the warming rays of the sun.

White-throated bee-eater
Merops albicollis

Diet: Their natural diet consists of flying insects. They eat many bees and wasps, but also take butterflies, grasshoppers, flies, etc. Most of the prey is caught on the wing, but I have occasionally seen them land on the ground to pick up insects. Sometimes the prey is too small to be seen by the human eye. The bird will seemingly be pursuing a phantom insect through the air. They are extremely fast and agile. If prey is dropped from the perch, it is usually caught again before it reaches the ground. In common with rollers and kingfishers, they kill their prey by bashing it against a perch. A bee-eater that has captured a venomous bee or wasp takes it to a perch and prepares it by beating its head against a perch and then gripping it by the tip of the abdomen and rubbing it against the perch to express the venom and the sting. The insect's head is pulverized and sometimes broken off in the process. The prey is tossed into the throat and swallowed whole. As noted earlier, newly acquired birds are reluctant to accept artificial diets. Captive-bred or hand-raised birds that have grown up with substitute diets are much easier avicultural subjects. Even hand-raised birds will, however, always take live insects in preference to artificial diets.

At Frankfurt Zoo, the following feeding regime is followed: morning, live moths; before noon, live moths, mealworms, and house fly larvae; and in the afternoon, live moths and live house flies.

I fed my birds by hand twice per day in the warm months when insects were plentiful in the aviary and three times per day during the colder months. Artificial food was always offered first, followed by insects. I fed 50% ox heart strips and 50% insects such as mealworm larvae and beetles, grasshoppers, winged termites, and moths. Even dead food is beaten on the perch. Hand-raised bee-eaters will fly to the shoulder or hand to receive food, but prefer to feed on the wing. The food can be thrown into the air while calling the birds with a repetitive whistle to which they are conditioned, or it can be held up with a forceps and released as the birds snatch it while flying past. In a mixed species aviary, one advantage of tossing food into the air is that the bee-eaters will nearly always beat the other aviary inhabitants to the insects. If the food is held in the hand, larger and more aggressive species will have an advantage over the bee-eaters.

At San Diego Zoo, the bee-eaters receive a plentiful supply of mealworms in an open-topped perspex tray, to which they swoop down to take one worm at a time. In addition, beehives are placed in the aviary. These are replaced as soon as the birds have depleted the stock. In a planted aviary, the natural influx of insects greatly assists in maintaining these birds in peak condition.

Breeding: To encourage my birds to breed, I constructed a sandbank by piling sand behind some scaffold planks set at an angle to create a backwards sloping face. The resulting sandbank measured 1.2 m (4 ft.) high and 1.6 m (5 ft.) deep. This was well-watered and planted on top with shrubs selected because they had a shallow root system. After a month, the scaffold was removed, exposing the face of the sandbank. At the time of writing, I had not yet bred bee-eaters, but the white-fronted had excavated in the sandbank. A problem arose when a pair of brown-hooded kingfishers *Halcyon albiventris* also moved into the sandbank, evicting the bee-eaters. The kingfishers were subsequently removed from the aviary. White-fronted bee-eaters lay a clutch of

2–5 white eggs. The incubation period is 20 days and the chicks fledge 27–29 days after hatching.

At San Diego Zoo, a colony of carmine bee-eaters *M. nubicus* successfully nested in an artificial sandbank that consisted of a large slatted wooden box filled with sand, with every second slat on the front face removed. The birds excavated their tunnels in the exposed areas. A photo of such a nestbank, used to breed carmine and European *M. apiaster* bee-eaters in Italy, can be found on page 186 of the *Avicultural Magazine,* Volume 76 (1970). There are several other records of bee-eaters breeding in captivity. At Frankfurt Zoo, 15 little bee-eaters were bred over a seven-year period. A sandbank, as described, and a plentiful supply of live food were essential for success.

A report by Rhonda Barbut on breeding rainbow bee-eaters at Perth Zoo reads as follows:

> "During November 1999, the bird-keeping staff at Perth Zoo created mounds at the back of the exhibit. One mound contained an artificial wooden nestbox. The rainbow bee-eaters investigated these burrows but made no attempt to nest in them; however they started to burrow into the flat ground at the front of the exhibit. But this was also not used as a nesting site. Another burrow was dug behind the grass, under a small paperbark tree *Melaleuca* spp in January 2000. Both sexes contributed to building this burrow. The roots of the paperbark provided enough support for the bee-eaters' burrow so that it was able to withstand the collapse during summer storms. In early January 2000 the keepers noticed reproductive behavior between the two birds. The pair was observed tidying up outside the entrance to the burrow by removing any ants from the area. Shortly afterwards both birds were observed taking live food into the burrow. Chick vocalisations were heard and on February 3, 2000, the chicks were observed at the entrance to the burrow. On February 9, the first chick emerged from the burrow and stayed on the ground for approximately 24 hours. Three days later the second chick fledged. It appeared healthy and remained on the ground for seven hours. By February 13, the third and last chick emerged. All three chicks resembled the adults but were a little less colorful. The parents were very vigilant and were careful when entering the burrow. They gave the alarm call in response to the presence of both the keepers and the public. When they had food they always flew to a nearby post and waited for a short time before entering the burrow with the food. The parents continued to feed the chicks until March 2 when the young were seen feeding on live food by themselves."

Clive Roots from England gives the following account of breeding the little bee-eater:

> "A wall for tunnel nesting was constructed with gaps approximately 30 cm² (1 sq. ft.) A retaining wall was then built about six feet behind this rock face and the gap between filled with soil and planted. Four birds were placed into the exhibit but as the first pair came into breeding condition, they would not tolerate the presence of the others. Their first nesting attempt was unsuccessful when the first part of the nest tunnel collapsed in shallow soil. Shortly afterwards a group of five carmine bee-eaters were added to the aviary but in the spring the little bee-eaters became so aggressive that it was necessary to remove the several times larger carmine bee-eaters for their own safety.
>
> In the middle of March the pair had excavated a new nest and towards the end of the month, because of the long period of absence of one birds, it became apparent that incubation had commenced. As far as we could tell incubation commenced on April 4, and on the 25th of the month the first of many visits to the tunnel with food commenced. Both parents fed the nestlings with mealworms, mealworm pupae and blowflies. Crickets and locusts were also offered but ignored. From entering the nest hole with food and reappearing again, no more than four seconds elapsed, and often only three. We naturally assumed that the tunnel was short, possibly 45 cm (18 in.) long, but we never took the risk of inspecting the

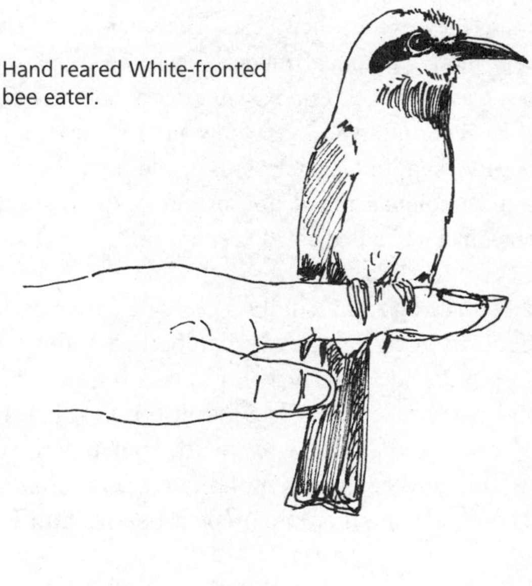

Hand reared White-fronted bee eater.

> entrance hole or tunnel during incubation and rearing for fear of upsetting the birds.
>
> During the early days of rearing, two dozen trips with food, on several occasions daily, were observed, yet towards the latter part of the rearing period the food carrying visits became less frequent and we were concerned that some of the nestlings may have been lost. Even a few days before the young eventually emerged we witnessed no more than a handful of visits with food daily. After a nestling period of exactly 28 days three nestlings appeared. Two of the nestlings were poor fliers with one being replaced in the nest only to emerge again two days later. The parents appeared to ignore the chicks, which could not fly well, and both were removed. One died but the other was successfully hand reared." [Author: note an earlier comment I made on the competition between nestlings and boosting production by hand rearing smaller chicks.] Hybrids have occurred in a mixed colony of European and carmine bee-eaters.

Regarding the breeding success of the White-fronted Bee-eaters at the San Diego Wild Animal Park, Michael Mace, Debbie Marlow and Debbie Gungle provide the following comment:

> The San Diego Wild Animal Park had been successful in breeding this species since 2000. In that period of time we have produced 62 offspring. The exhibit measures 30 x 15 x 14 ft. We offer two different types of nesting areas. The first is a structure that measures 8 x 4 x 4 ft. Three of the sides are constructed of plywood. The fourth side, which the birds have access to, is made of wooden slates. The entire box is filled with condensed sand. One third of the wooden slates are removed to allow the bee-eaters to tunnel in the sand as they excave their nests. The second nest site is the more traditional model consisting of plastic tubes that lead to a nestbox. We offer a diet of crickets, mealworm larvae, waxmoths and live bees from a beehive kept in the enclosure. In addition, climate, adequate colonial population of birds (20-40 specimens), diet and a dedicated keeper staff.

Hand Rearing: The chicks I collected for hand rearing ranged in age from newly hatched to about a week old. They were fed hourly from 6:00 a.m. to 8:00 p.m. The food was presented using a thin forceps. hand rearing diet consisted of 50% ox heart, moistened, with a multivitamin and calcium supplement added once a day, and 50% insects such as mealworm beetles and larvae, winged termites, moths, butterflies, and grasshoppers. Insects were killed and slightly crushed with a tweezer before feeding. While the chicks were still tiny, the larger insects were broken into small pieces. The chicks were voracious feeders and would attempt to swallow half the tweezer as well. They were extremely aggressive toward each other. In nature, a maximum of two chicks may survive, but usually only a single chick fledges. Their beaks are very sharp and they become frenzied when hungry, often latching onto the beak, wings, or legs of their nestmates. Smaller chicks are literally thrown around the nest. To circumvent this problem, I placed the chicks into separate boxes as soon as they began to develop feathers. One of the chicks was an obvious runt, but eventually caught up with, and was indistinguishable from, its siblings. After fledging, the young bee-eaters lose this aggressive tendency.

Bee-eater chicks should be kept at a temperature of 28°C (82.4°F) until the feathers start to appear on the chest. If the chicks are not warm enough, they will become sluggish and will not feed properly. They also refuse to eat when they are about to regurgitate a pellet. Once the pellet has been expelled, feeding resumes as usual.

On one occasion, we observed an adult pair of wild little bee-eaters carrying food to their nest. Upon opening the nest chamber, we discovered a chick which was clearly not their own. This subsequently proved to be a young greater honeyguide *Indicator indicator,* a species that often parasitizes little bee-eaters. This particular nest had been excavated in a pile of building sand. Fortunately, we managed to convince the builder to leave the pile intact until we could remove the chick for hand rearing.

Ring size: 3 mm

Species Requiring Similar Management:

Bee-eaters
Kingfishers
Rollers
Motmots

Contributors:

Dr. S.G. Stadler, Curator of Birds, Frankfurt Zoological Gardens.

Patricia Witman, Animal Care Manager, Avian Propagation Center, San Diego Zoo.

References:

Barbut, R. 2002. Successfully breeding one of the most challenging species of birds in captivity at Perth Zoo. *Australian Birdkeeper* 15(1).

Roots, C. 1969. Breeding the Little Bee-eater at the Winged World. *Avicultural Magazine* 76:186–188.

San Diego Zoo Avian Propagation Center Bee-Eater Hand Rearing Protocol

Day	Brooder temp.	Frequency	Diet	Intake	Misc.
1	AB Newlife at 96.0°F 80.0°F wetbulb.	Every hour (13x).	Alternate chopped: pinkies 2-week-old crickets waxworms and half mealworms. Begin and end with bee larvae moistened with pedialyte.	ad lib	Do not house chicks together until fledged due to aggression.
2	AB Newlife at 96.0°F 80.0°F wetbulb.	Every hour (13x).	See day 1.	ad lib	CaCO at 1% Apetate at 1 ml/50g fed previous day.
4	Add nomad matting.	Every 1½ hours (9x).	See day 1.	ad lib	Ghost feed as eyes open.
7	AICU with water.**	Every 1½ hours (9x).	See day 1.	ad lib	no sunning-cavity/tunnel nester
8	AICU with water.**	Every 2 hours (7x).	Change to distilled water feed mix: pinkies, 2-week-old crickets waxworms half mealworms.	ad lib	
12	AICU with water**	Every 2 hours (7x)	Change to whole crickets without heads.	ad lib	"move down tunnel"***
16	Move to box brooder with water for humidity	Every 3 hours (5x)	See day 12.	ad lib	
21	Move to Howdy****	Every 3 hours (5x)	See day 12.	ad lib	
23	Outside with heat during the day.	As above	Adult diet: crickets, mealworms (regular and giant) waxworms, alternating with Protavit and Nekton I.	ad lib	
	Inside with heat at night	As above	As above.	ad lib	
30	Outside without heat at night	As above	As above.	ad lib	Wean 6 weeks ater fledging.

This hand rearing protocol is what San Diego used for both white-throated *Merops albicolli* and white-fronted bee-eaters.
First 15 days: average weight gains 32–55%, range 17–92%

Apetate (liquid vitamin B complex for children): Use 1cc/50g food fed.

Calcium carbonate: Use 1% of amount fed previous day.

* Ghost Feed: When we don't have a puppet, sometimes we still "hide" from the chick by using one-way glass or mylar. Keepers will also use a shroud-like covering over themselves to try to fool the chick.

**AICU w/ water: Animal Intensive Care Unit, an acrylic brooder, made by Lyon Electric in San Diego, with water in the reservoir for humidity.

***"Move down tunnel:" Something I read about wild bee-eaters; at this stage chicks may meet the parents half way down the nest cavity.

****Howdy: A small wire cage, 30 x 30 x 60cm (1 x 1 x 2 ft.), used to transition a chick from a brooder to a flight cage. Typically used for ~5 days. Chicks placed in "howdy" when they first jump out of their nestcup; when chicks appear to be strong and steady, the door would be opened with perches stuck into it, leading them out into a small flight cage 1.2 x 3.6 x 2.4 m high (4 x 12 x 8 ft. high).

Mousebirds

12 Coliiformes

12.01 **Coliidae: Mousebirds**

12.01 Coliidae Mousebirds

Glen Holland

Natural Habits: Mousebirds are the sole members of the family Coliidae and the order Coliiformes. They are also known as colies after their Latin genus *Colius*. The six species of mousebirds are all found only on the African continent, the only bird order that is strictly endemic to that continent. Some species are highly gregarious, occurring in large groups at all times, while other species occur in small groups or as pairs in the breeding season. They are birds of the open woodlands and tangled thickets, but are seldom found in true forests. They occur in groups of 4–12 birds, which follow each other from tree to tree. Highly gregarious, they live in flocks year round and form tight bonds with other members of their flock. They are swift, direct flyers. Mousebirds have a peculiar build, having the outer toes on each side reversible, either forwards or backwards. Their claws are strong and curved to assist in climbing in vegetation. The bill is short and soft around the nostrils, and the feathers are hair-like on the body with a long, stiff tail averaging about 22 cm (8.7 in.) and a short crest. The blue-naped *Colius macrourus* is regarded as the prettiest of the group, as it is adorned with a patch of blue feathers on the neck and soft parts at the base of the bill and the naked skin around the eyes is colored red. Blue-naped also have the longest feather crest of all colies. The birds whistle and chatter amongst themselves almost incessantly. Allopreening and allofeeding are common behaviors.

Management: There are many reasons why mousebirds make interesting avicultural subjects. Their personalities, movements, and behavior cause them to stand out in the avian world despite their lack of gaudy colors or their inability to mimic. Mousebirds, especially the speckled, have an inquisitive, bold nature and always need to be aware of what is going on around them. For the purposes of describing captive management, I will be concentrating on the speckled mousebird *C. striatus,* blue-naped and red-faced *C. indicus*. I have also kept, but not bred, the white-backed *C. colius,* which are much like the speckled in behavior. They are strange birds, which spend most of their time clambering like mice in foliage in search of fruit, berries, flowers, leaves, and the occasional insect. The gape is large and most berries are swallowed whole. While clambering about, resembling rodents in a bush, they utter soft contact calls between members of the flock.

Mousebirds are miserable birds in cold weather, particularly when the cold is accompanied by wet. On cold nights and mornings, the group will cluster together vertically on a branch, with their bellies touching. Their metabolic rate and body temperature drops overnight inducing a natural state of torpor and returns to normal as they warm with the early morning sun. Their metabolic rate can be reduced by an amazing 90%. When the morning sun reaches them, they puff out their feathers to absorb the warm rays, and only when the group has "defrosted" will they move off and begin their daily activities. Clustering is done by mousebirds in the daytime also, simply because they enjoy the company of others. In captivity they will hang from the wire with their legs stretched up above their heads. In severe climates where daytime temperatures regularly remain below 16°C (60.8°F), I would recommend that they be housed indoors with an additional source of warmth during the winter months. Dr. Schifter bred mousebirds for many years in Vienna and held his birds indoors permanently:

> "I had the best breeding results when mousebirds were kept in large indoor aviaries with high temper-

atures usually ranging from 20–30°C (68–86°F). Colies can, however, withstand colder temperatures, provided it is dry."

They are best kept in a planted aviary as a group. Habitat B will suit mousebirds. It is important to introduce all the birds of a flock at the same time. Once a group is established new birds should not be introduced, as they will immediately be attacked. It is possible to have a few pairs breeding in a large planted aviary with the young of these pairs living harmoniously with their parents for many seasons. They can form groups of up to 20 birds with no signs of aggression. It is interesting, however, that these young birds will not breed while in the company of their parents. Dr. Schifter records having two pairs and a spare male red-faced mousebird that lived together and bred harmoniously.

Aggression is usually violent, bloody, and often fatal within a short time if not stopped. They will attack the toes and tails first and will often gang up on the victim. I once had a visitor in South Africa release a young red-faced into a walk-through aviary containing a group of speckled. The red-faced was found dead and badly plucked a few hours later. Any time birds have been separated from the flock, they need to be introduced again very carefully. It is possible to keep more than one species in a large aviary but, again, they should all be introduced simultaneously. When housed together it is very interesting to observe that specimens of the red-faced and speckled never keep together, although there is some allopreening between blue-naped and red-faced colies. Because red-faced and speckled mousebirds sometimes even attack each other, they are best separated in winter or whenever placed in smaller or indoor aviaries. I have never noticed any aggression between mousebirds and other species. They have been suspected of stealing the young of small seedeaters but I have not found any proof of this. They do, however, definitely steal nesting material, which can disturb other breeding birds.

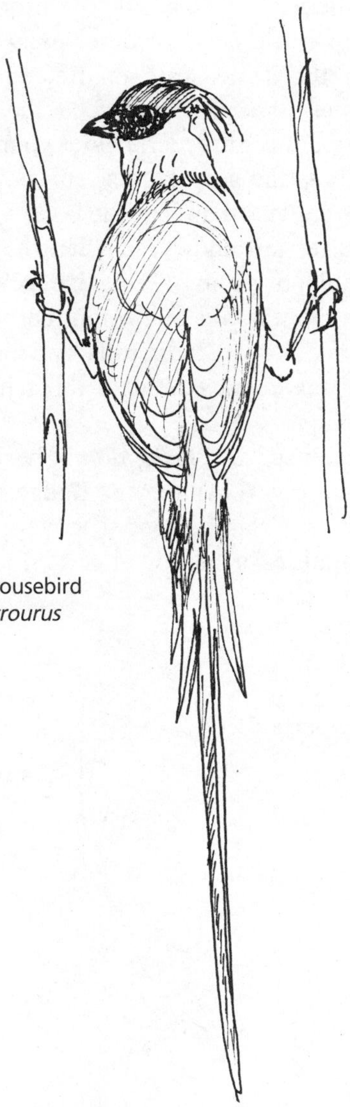

Blue-naped Mousebird
Urocolius macrourus

When newly acquired, the birds are fragile, nervous, and require additional warmth particularly at night. During this period they readily feed on soft fruits such as paw-paw. Deaths are mainly due to a combination of the birds being cold and stressed, not to a reluctance to feed. They are fond of bathing, particularly if a mist spray is placed near a shrub; they will dart in and out of the wet leaves until they look like drowned rats. A sand bath is a social event and the entire group will readily join in. A patch of dusty dry soil should be available for this activity. Sun bathing is also a regular activity. Allopreening is common within the group and the bird that is receiving the attention appears to go into a trance-like state with the eyes closed and the feathers puffed out.

Diet: In captivity they feed on a variety of soft fruits such as diced apple, figs, grapes, paw-paw, banana, soaked raisins, diced apple, and half oranges on a spike. Blue-naped and red-faced prefer bananas whereas the speckled prefer the other fruit. Green leaves form an important part of the diet and the aviary should be planted accordingly. They also soon learn to take commercial softbill mixtures and will readily take nectar that has been put out for nectivores. The food passes through their system quickly, which results in the birds eating large quantities for their size. A plentiful supply of food must therefore be ensured. Approximately half a cup of food should be supplied per bird per day. Food is often held in a foot while feeding. No doubt the odd small insect found in the aviary is also fed to the chicks, but my birds did not show much interest in mealworms, gentles, or flying ants.

Breeding: The first species bred in captivity appears to be the speckled in England in 1912. In a planted aviary with some dense thickets of vegetation, they will nest readily through most of the year in warm climates, and seldom have any problem raising their chicks on the diet prescribed for adult captive birds. There are some differences in nest building, the speckled using more grass and rootlets than the blue-naped and red-faced, which prefer twigs and that will breed in nesting sites fixed in the aviaries, sometimes almost without nesting material. The nest building duties are shared between the male and female, although the male is usually much more eager to bring nest material to the nest where the female pushes it into place. Later on, material, especially green leaves, will be brought to the nest, even during incubation and until the young hatch. Dr. Schifter describes the nest building behavior of his red-faced which were held in indoor aviaries:

> "Before supplying baskets I had observed where the birds had tried to deposit dry and green leaves of lettuce, small twigs, and other nesting materials. Then I fixed the baskets at these places and the birds seemed very pleased to build their nests there. In the following years, the birds often bred two or even three times in the same nest, but sometimes they also changed from one basket to another. Baskets had to be removed to be cleaned as the birds did not keep them clean."

As I have always bred mousebirds in well-planted aviaries I have never found the need to supply baskets, but clearly they assist in indoor situations.

Courtship includes mutual feeding, a bobbing dance by the male, and allopreening. In display the male bobs up and down adjacent to the female, sometimes with a leaf in his beak. Mousebirds are both communal and co-operative breeders and are multiple brooded. A pair of red-faced was recorded to have nested six times producing eight young in a season. My speckled nested three times a season, raising 2–3 chicks per brood. Two females have been recorded laying in the same nest.

The 2–3 and, exceptionally, four eggs are laid on consecutive days. The eggs of all the species are similar in size but red-faced and blue-naped eggs are more pointed and reddish-brown speckled. The eggs of speckled are creamy white without blotches. The incubation period is a very brief 11 days and the chicks fledge at 12–16 days. Red-faced generally have a slightly longer nesting period than the other two described. Both parents share incubation with several changes during the day. The female sits on the nest during the night. There are no behavioral ceremonies when exchanging incubation duties but the male sometimes brings green leaves or other nesting material to the nest. As the parents commence incubation with the first egg, young hatch at intervals, which sometimes results in smaller chicks being squashed to death. These doomed chicks can be saved by removal for hand rearing. The newly hatched chicks are curious looking creatures weighing 1.5–2 g (0.05–0.07 oz.) and measuring about 35 mm (1.4 in.) in length. The young of the speckled are covered in sparse down, even on the head and neck, while those of the blue-naped are nearly naked. Red-faced chicks have a little down on the body only. Young are fed by regurgitation, but young speckled will also take food from the bills of the parents. They are independent at 5–6 weeks. Sometimes the young are still fed when the parents have started incubating their next clutch. They occasionally return to the nest to sit beside the brooding parent and will also visit the nests of other pairs where they will behave in the same way, unaffected by their hosts, although the hosts will not feed the visiting young. Being co-operative breeders, older young from previous nests will help the parents raise subsequent broods.

Young are less than half the body size of the adults, with a shorter tail and crest when they fledge but are soon capable of flying. Full size is only reached at about two months old but young can be identified by the lack of color on their bills and facial skin. Juveniles' bills take approximately three months to match the adult color.

If threatened, the chicks do not hesitate to leave the nest prematurely. When they do fledge, they are not able to fly well and clamber about in vegetation. Independence is attained a month after fledging. The chick's

Mousebird foot.

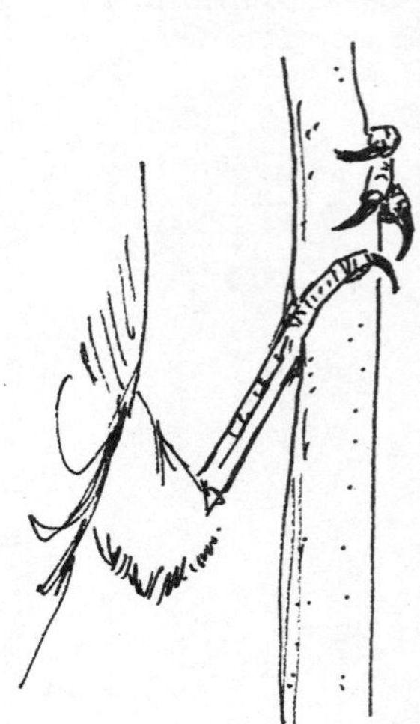

habit of leaving the nest prematurely results in the public often picking up the "deserted" chicks. These chicks, when hand reared, make very interesting avicultural subjects. Young, although capable of breeding at six months, will usually breed at one year of age. To maximize the production of these birds, some species of which are now scarce in aviculture, DNA-sexed pairs should be set up in suitable aviaries for the breeding season. All birds, including juveniles, should be banded to ensure that genetics are considered when pairing birds. Juveniles can be flocked to allow for maximum social interaction but, again, a careful watch should be maintained until the integration is known to be successful.

Hand Rearing: They are best hand raised using a 1-ml syringe, which has the entire tip removed. Small chicks are fed on mashed banana, finely diced apple (with the skin), moistened commercial rearing food, and a few insects, such as diced mealworms, which are dipped in the moistened commercial rearing food. The fruit given should include the seeds of fruits such as figs and guavas or the skin of apple for roughage. Initially the chicks are rather delicate and make soft *chirring* sounds when hungry. Food is taken with a sudden jerk of the body, the stimulation for which occurs when the back of the throat is touched.

During feeding the chicks attempt to hide their nest mates by covering them with a wing, and at fledging age they even hold each other back with their long sharp toes. They are rather unique when small, as defecation has to be stimulated by touching the cloaca with a solid object such as a tweezer. This is a messy process and the droppings are loose, not in a fecal sac, and it is best done on a piece of paper towel to avoid soiling the nest. By the time they are two-thirds feathered this stimulation is no longer necessary. Older chicks have to be force fed at first, but within two days they begin to beg for food. The food passes through the birds quickly and appears to be only partially digested before it is excreted. Because of this, the chicks must be fed hourly from 6 a.m. to 8 p.m. The chicks must also receive additional warmth at night, and in the day during cold weather. If cold, they become sluggish, refuse to feed, and soon die.

Hand-raised birds become very tame and enjoy company. They like to be held in the hand and then tickled around the head, particularly under the chin. When receiving this attention, they role the head back to expose their chin and close their eyes with ecstasy as they receive a tickle. They remain tame indefinitely and will fly onto their owner as he enters the aviary. I once had a red-faced which we nicknamed "The Vulture". If he did not receive enough attention, he took

Hand rearing a mousebird.

great delight in biting the skin on one's neck or ear and twisting this as hard as he could. Hand-raised birds will spend long periods playing with "toys" such as paper towel, paper clips, feathers, and pencils. In some countries mousebirds are becoming popular as pets. If the owner is able to spend most of the day with the bird I feel the birds' psychological needs will be met. I would, however, hate to see such a sociable creature cooped up and left alone in a cage all day while its owner is at work.

Longevity records for captive-bred birds are 12–13 years. An albino speckled mousebird has been recorded in the wild.

Ring Size: 4.3 mm

References:

Davis, K.J. 1999. Mousebirds for the New Millennium. *AFA Watchbird* 26 (5).

Schifter, H. 1965. Breeding the red-faced mousebird. *Avicultural Magazine* 72: 79–83.

Schifter, H. 1969. Breeding experiences with blue-naped and speckled mousebirds. *Avicultural Magazine* 75: 54–62.

Cuckoos

13 Cuculiformes

13.01 Cuculidae Old World Cuckoos

Glen Holland

Introduction: Cuckoos belong to the order Cuculiformes. The Old World cuckoos, koels, malkohas, ground-cuckoos, and couas comprise the family Cuculidae, with around 80 species in numerous genera. The American cuckoos of the family Coccyzidae number 18 species in four genera. Other families within this order include Opisthocomidae, the hoatzin; Crotophagidae, the anis and the guira cuckoo; Centropodidae, the coucals; and Neomorphidae, the roadrunners and New World ground-cuckoos. They are generally rather secretive birds, inhabiting dense forests and woodland, and are more likely to be heard than seen. Many species of cuckoos are parasitic, each with its own particular host species. Among the non-parasitic cuckoos are the couas, the coucals, the anis, the guira, and the hoatzin. Cuckoos are highly vocal, with characteristic calls which are often easy to imitate.

Management: Cuckoos are a relatively specialized family with only a few species adapting well to captivity. Even here it is generally hand-raised individuals that are most suited to aviculture. I have hand-raised a number of cuckoo species, but have only had one species, the great spotted, which thrived in captivity. The main hosts for the great spotted cuckoo *Clamator glandarius* are crows, but cavity-nesting species such as starlings and hoopoes have also been recorded. This large cuckoo is more like a crow in its feeding habits, and my bird voraciously ate even an artificial diet. The smaller, more delicate species, such as the Diederic cuckoo *Chrysococcyx caprius* and Klaas's cuckoo *C. klaas,* are far more difficult to raise and maintain in captivity. The great spotted that I had was brought in as an immature, in a weakened state, and was force-fed for three days. It then gained strength and began to take food voluntarily. Large quantities of ox heart strips, moistened with water with some calcium and a multivitamin added, insects, pinkie mice, and beef-flavored, tinned dog food were all consumed. The food was fed with forceps. The bird was released in a mixed aviary where it thrived, behaving much like a young crow. It showed no signs of aggression toward other species, but performed terribly when hungry, begging for food and consuming 4–6 large mouthfuls at each feeding. It remained tame even when fully colored at the age of six months.

The aviary should be planned according to the habitat from which the species originates. Some species, such as the emerald *C. cupreus,* occur in forests and require a densely planted aviary as described for Habitat A, while others, such as the great spotted, inhabit open woodland; for these species, a more open habitat with a few trees will suffice. Their aviary should be landscaped as described for Habitat B in the chapter on LANDSCAPE. Great spotted cuckoos are not territorial, so several specimens can be kept together. Due to their size, possible nest robbing tendencies and diet, they are best mixed with larger species such as hornbills and rollers. They have extremely sharp claws, which scratch terribly, especially when they clamber about on the keeper trying to beat other birds to a tasty morsel. The more tropical cuckoo species are not particularly hardy and begin to suffer at temperatures below 14°C (57.2°F). In countries with severe climates, they are best housed indoors with an additional source of warmth during winter.

Diet: Although the cuckoo I reared joined the other birds in eating soaked dog cubes mixed with bonemeal, it was always ready to snap up any insects and ox heart strips provided for birds such as bee-eaters. I once observed it in the process of swallowing a small fledgling, which may have been pulled out of a nest by the hornbills, although it would not surprise me if the cuckoo was responsible. Most cuckoos are particularly fond of caterpillars. Branches from a bush infested with caterpillars can be placed in the aviary. The cuckoos will make quick work of this favored food source.

Breeding: Each cuckoo species has its own particular host, or hosts. Two hand-raised female great spotted cuckoos have been recorded as laying 15 and 16 eggs, respectively. Eggs are usually laid every two days. The eggs are rather rounded, and of a pale greenish-blue ground color with rust and lilac markings. Females will lay more than one egg in a nest. The incubation period is 14 days. The chicks do not evict the host's chicks but, as they hatch first, by the time the host's eggs hatch, the cuckoo chicks have a head start and receive most of the food. They fledge at 22–26 days.

Hand Rearing: As described for SHRIKES.

Asian Emerald Cuckoo
Chrysococcyx maculatus

Ring Size: 6 mm

Species Requiring Similar Management:

Cuckoos, all genera
Coucals *Centropus* spp.
Drongos *Dicrurus* spp.
Shrikes, all genera

Reference:

Rowan, M.K. 1983. *The Doves, Parrots, Louries and Cuckoos of Southern Africa.* John Voelcker Bird Book Fund and David Philip, Capetown and Johannesburg.

13.02 Centropodidae Coucals

Glen Holland and Neil Hamilton

Natural Habits: Coucals share the order Cuculiformes with cuckoos, hoatzins, and roadrunners, but are given their own family, the Centropodidae. This family consists of the single genus *Centropus,* which includes 29 species found in Africa, southeastern Asia, the eastern islands and one species in Australia. Their natural habitat is dense thickets, wetlands, tall grasslands, and riverine vegetation. Coucals also inhabit gardens with plenty of dense shrubs and hedges. Sexes are alike, and they are found either singly or in pairs that aggressively defend their territories against others of their own kind. Their territories are maintained largely through their calls, which have been likened to water running from a bottle — a loud descending *Doo-doo-doo-doo.*

Management: The focus of this section is on Burchell's coucal *Centropus superciliosus burchellii,* an African species, and the pheasant coucal *C. phasianinus,* from Australia and New Guinea, but the captive management for other species would be similar. The pheasant coucal, also known as the swamp pheasant, a common species in the wild, is becoming established in captivity.

The weight of mature female pheasant coucals ranges is 400–450 grams (14.1–15.9 oz.). Adult males are slightly smaller, with a weight range of 350–380 grams (12.3–13.4 oz.). These weights may vary seasonally. In the Burchell's coucal, the sexes do not exhibit such obvious weight differences. Since the sexes are otherwise alike in appearance, coucals should be surgically or DNA sexed to guarantee true pairs if breeding is desired.

In many ways, these birds can be likened to giant shrikes. They feed on grasshoppers and other insects, reptiles, frogs, small birds, nestlings, and mice. While small prey is normally swallowed whole, larger prey is held in the foot and torn apart with the beak. Being large, aggressive, and carnivorous, coucals cannot be safely mixed with smaller species of softbills or seedeaters. They are best kept with other large softbills of a similar nature, such as gray hornbills and long-tailed shrikes. Habitat F, as described in LANDSCAPE, would be ideal for coucals. Their aviary should include clumps of tall grass and a few reeds such as Napier grass, *Phragmites,* or bulrush, preferably planted around a shallow pan of water. They will also use scattered clumps of dense shrubs.

Neil Hamilton found the pheasant coucal to be very hardy, friendly and, at times, strange in its habits. This species has great character, adapts well to captivity, and makes a fine display bird for the general public. The first pheasant coucals to arrive at Taronga Zoo were two young birds that had been hand reared at Currumbin Sanctuary in Queensland after being collected in the wild. Upon completing quarantine, they were moved to a large display aviary that already housed a number of birds of the following species: Torresian imperial pigeon *Ducula spilorrhoa,* bar-shouldered dove *Geopelia humeralis*, bleeding heart pigeon *Gallicolumba* sp., Victoria crowned pigeon *Goura victoria,* Radjah shelduck *Tadorna radjah,* and nankeen kestrel *Falco cenchroides*. The bar-shouldered doves were especially prolific, having rapidly increased in numbers to around 20 birds. When the dove population stabilized, and breeding had seemingly stopped, it was observed that the coucals were feeding on their eggs and nestlings, keeping the population in check. This aviary measured about 12 m (40 ft.) in length, 4.5 m (15 ft.) in width, and 2.7 m (9 ft.) in height. It was framed with 50 ml pipe and covered with 1.6 mm heavy grade wire, and had a metal sheet shelter area.

Coucals select favorite perches in the aviary where they will sit and sun themselves in the early morning. While they are soaking up the sun, their head and neck feathers are puffed out and their wings are drooped. Their flight consists of a few quick wing beats and then a glide. Coucals do bathe but, unlike other birds, they simply walk through the water until their plumage is soaked. They are relatively hardy birds and can withstand cold, damp weather for short periods, but in severe climates, which may have temperatures near freezing for extended periods, coucals require additional shelter and warmth.

Diet: In captivity, coucals are fed on minced chicken, pets mince, ox heart and some live food such as mealworms, crickets, various other insects, mice, and frogs. This natural food is particularly necessary if breeding is desired. The daily diet at Perth Zoo consists of a coarse mix made up of minced beef heart, crushed hard boiled eggs, ground dog kibble, and white bait (*Galaxias* minnows), with S.A.37 vitamin powder added. Other foods include mice, day old chickens, and mealworms. During the breeding season, the chicks and mice are diced, and pinkies, moths, and earthworms are also provided. While young are being reared, fresh food is offered 4–6 times a day.

Breeding: Coucals have no elaborate courtship display, but do practice courtship feeding. The male approaches the female with an item of food that he gives her, and copulation ensues shortly thereafter. The nest of Burchell's coucal is a ball-shaped structure constructed of thick blades of grass and twigs, lined with leaves, placed in a dense shrub or creeper. At the Johannesburg Zoo I saw a nest that had been built about 50 cm (20 in.) above the ground in a thick clump of grass.

The pheasant coucal has an eclipse plumage for about six months of the year, during which the feathers are of chestnut and rufous brown coloring. From about July to January the birds are in their breeding plumage. During this period, the body is black, with brown wings and tail. The coucals at the Perth Zoo molt into their breeding plumage any time from February to July. In August, they began calling loudly. Neil Hamilton describes the call as very similar to that of the gibbons at the zoo. It is also around this time that the birds start to explore for nesting sites among the small bushes and grass tussocks provided for this purpose. They prefer to use fresh plants and fresh plant material to build their nests, but will use dead material if nothing else is available. When the cup-shaped nest is completed, the birds will continue to line the inside of the nest with fresh green material every few days. Both male and female assist in the nest building. Nests are usually placed from about 50 cm to 1 m (20 in. to 3 ft.) above the ground, sometimes slightly higher. A good nest will be completely covered, with a side entrance.

A normal clutch consists of 4–5 white eggs. Incubation is undertaken by the male alone. An incubating male seldom leaves the nest unattended, even to eat. The eggs hatch after an incubation period of 14–16 days.

At around 14 months of age, the pheasant coucals at Perth Zoo started to build a nest under the lowest grass tussocks, about 40 cm (16 in.) from the ground. Material used for the nest consisted mainly of long strands of grass from the surrounding grass tussocks. These were woven together to form a cup-shaped nest. The occasional piece of fresh green tree fern or leaf litter was placed in the middle of this nest. Even after a clutch of eggs was being

Pheasant Coucal
Centropus phasianinus

incubated by the male, the female continued to lay eggs all over the aviary floor. The chicks hatched, but died early or disappeared for unknown reasons. It was then decided to hand raise all further offspring.

To successfully and consistently breed coucals, removing the female when incubation commences is the best option, as females may interfere with an incubating male. Between 1987 and 1990, Neil Hamilton reared 17 young birds from a single pair at Perth Zoo by this method. The following year, eggs were removed, as the stock numbers had built up and no orders were pending. Coucals have been bred from pairs kept together throughout the breeding season, but not as successfully as when the female is separated during incubation. From 1992 through 1998, two pairs were maintained at the zoo for display purposes only, but still managed to raise one or two young per year. When the female was left with the male during the entire reproductive process, she offered little assistance in rearing the chicks.

Newly hatched coucal chicks are unusual in appearance; their skin is black and leathery with a few long white down feathers. They fledge at 14–16 days of age, when they are about half the size of adults. They almost seem to fledge prematurely, as they are unable to fly well for some time after leaving the nest. Immature coucals are pale versions of their parents, with most of the plumage covered by light barring. Unlike the adults, the chicks have a short stumpy tail. They utter a *kurrrch* sound repeatedly. Young coucals reach independence when around 44 days old.

It is extremely important to supply fresh food regularly after the chicks have hatched to prevent the parents from feeding the weaker chicks to the stronger ones. Sometimes the larger chicks will devour their smaller siblings on their own.

Hand Rearing: Coucal chicks are not easy to hand raise, and natural foods are essential. All food is offered in a forceps. Diced bird of prey diet, pinkie mice, and ox heart strips are fed after being dipped in a commercial hand rearing formula with a multivitamin and calcium supplement added. The bulk of the hand rearing diet must consist of crickets, grasshoppers, baby mice, chopped reptiles (if a common skink or lizard is available) etc. The bones, skin, and feathers should not be removed; the chicks will form pellets from this material. Young coucals have a large gape and are easy to feed once they learn the technique. Initially they require force-feeding, but soon begin to beg for food. When begging, they make a *chirring* sound with the mouth open and the tongue, which has a black marking on it, vibrating. The first food item is usually swallowed immediately, but the second and third are held in the beak for a while before being tossed into the back of the mouth. Their droppings are rather messy and smelly, and the bedding material needs to be changed regularly.

Fledged chicks enjoy the sun and when placed in the sun after a night spent indoors, they immediately raise the feathers on the back. At the same time the wings are dropped downward and forward, exposing the skin of the back to the warm sun and giving the birds obvious pleasure. Fledglings have sudden bursts of energy, and will dash about for a few minutes as if being chased by a predator. Hand-reared birds remain very tame; one that I raised would put its head down and boom out its call if a visitor showed it any attention. I have not experienced any serious problems with aggression toward people, but did once have a tame coucal bite my lip after I had imitated its call while it was sitting on my shoulder.

Ring Size: 8 mm

Species Requiring Similar Management:

All *Centropus* species
Shrikes

Reference:

Rowan, M. K. 1983. *The Doves, Parrots, Louries and Cuckoos of Southern Africa.*

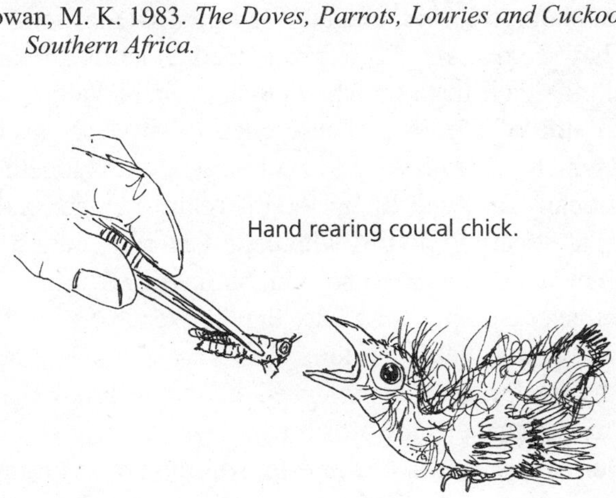

Hand rearing coucal chick.

13.03 Crotophagidae New World Cuckoos

Tiffany Thompson

Guira Cuckoo *Guira guira*

The guira cuckoo, Guira guira, belongs to the order Cuculiformes and is a member of the Cuculidae family. The species is native to South America and it seems that this is one species that deforestation has not had a negative impact on so far. The Firefly Encyclopedia of Birds states that the birds will nest anywhere and that houses, or any type of building, are good support structures for nests. The sexes are alike in appearance. This species of cuckoo is not a parasitic nester, and within the species they are communal nesters, with many females using the same nest.

Management: Guira cuckoos thrive in an aviary setting and are not difficult to breed provided they are offered adequate food. To appropriately manage this species, Discovery Cove divides the male and females into two groups and houses them separately due to the fact that they will constantly lay eggs. When the birds are in nesting mode a female may lay so many eggs that her body becomes depleted of calcium, causing her to become egg-bound.

They are very curious birds; therefore maintaining a safe environment for them is crucial. They will explore any type of opening, so is very important to make sure that irrigation boxes, or areas that you do not want them to explore, are covered. They do not tend to be aggressive so it is easy to keep them in mixed species aviaries. In Discovery Cove's aviaries they live with toucans, rollers, guinea fowl, hornbills, and turacos, just to name a few. Hand-raised birds are very relaxed and interactive with people and make wonderful educational animals. They also make great interactive aviary birds and will often sit on people's head or shoulders, even without the incentive for food.

They seem to enjoy being scratched under the bill and will stretch their head up while being scratched along the ventral side of the neck. These cuckoos are very social birds, which makes it easy to keep large groups together. Occasionally they will fly by and hit a staff member in the head. They seem to do this with certain staff members, in certain areas (no nests are around, no feeding stations are present, so we are not sure why they are territorial of this area) but they have never done this to a guest. Discovery Cove has so far never experienced any aggressive behavior from the cuckoos towards the guests.

Guiras spend a lot of their time on the ground foraging, so we have conditioned ours to stay off pathways used by the staff members and the guests since the cuckoos often will not move or look out for oncoming traffic. They do prefer access to sunny areas and will often fluff up their back feathers and drop their wings to sun themselves. Average adult weight is 100–120 g (3.5–4.2 oz.) and the birds have a naturally slender build.

Diet: In their natural environment, this species primarily eats insects. In a zoological setting they will eat insects such as large and small mealworms, waxworms and crickets, but they will also eat pinkie mice, bird of prey diet, soaked dog food, as well as a variety of other soaked pelleted diets (such as pelleted parrot diets). Occasionally they will eat grapes, but it is not a preferred food item. When the females are laying eggs they require increased calcium in their diet due the fact that they will lay many eggs and they will become egg-bound if calcium is lacking in their diet.

Breeding/Rearing: We have observed the birds laying eggs year-round and many females will lay eggs in one nest. Up to 16 eggs have been counted in a single nest, with five females living in the aviary at that time. The eggs are about the size of a ping-pong ball and range in color from powder blue to a blue-greenish, with white spots/splashes all over them. Birds may randomly toss eggs out of the nest, and we have yet to document any patterns for this behavior.

The nest is usually built out of small sticks, and is often high in a tree. At Discovery Cove the nests have been found in the crooks of palm fronds that have broken off, towards the top of the tree against the trunk, as well as in the very top of a tree. When the nests are built in a tree, they are usually far out on a limb towards the spindly edges, where it is very difficult for the keepers to access. Both males and females assist in building the nest and once they start building, the nest is completed quickly, with the egg-laying females and the males working at it diligently until it is completed.

Once the eggs are laid the birds will defend the nest vigorously. Anyone trying to access the nest may get hit in the head by the defensive bird. Many times there is more than one bird participating in this defense and it continues while the keeper is checking the nest. Both sexes incubate the eggs.

When the chicks hatch, both sexes feed them and there is a large increase in the amount of food eaten. We feed pinkie mice and bugs and, depending on the number of chicks, we have increased the diet from 12 mice (0.5-1.0 g each) per bird to 20 mice per bird in a day's time, the majority of these being fed to the nestlings. The adults may lose some weight during this period. Our adults tend to offer food to the chicks even when the chicks are not begging, so the chicks grow rapidly. We have pulled parent-raised chicks at five days old that were 65 g (2.3 oz.), and the chicks usually hatch at 10–20 g (0.3–0.7 oz.).

When the parents are tending the nest they become very dirty and odiferous. Our birds are interactive with guests and the guests have often commented on how bad the birds smell, even though the bird is a meter (3 ft.) or so away.

The cuckoos have proven to be very easy to hand raise and they grow quickly. The eggs are not difficult to artificially incubate and we have had successful hatchings using a 37ºC (99ºF) dry bulb with 60% relative humidity (in Central Florida). The first feeding we give the chick is 0.5cc pedialyte, and then we begin adding solids to the diet at the second feeding. For the second and third feedings we offer the chicks 0.5–1.0 gram of pinkie mouse viscera and after that they are fed as much as they want (see chart below) approximately every 2–3 hours. They will beg between feedings, sometimes as soon as half an hour after a feeding.

We start seeing pinfeather growth around 4–6 days old. They have a tendency to get splayed legs so it is best to keep them in a tight nest cup and we often keep two or three chicks together. Since they grow so quickly, they also have a tendency to get droopy wings, but we have never had to tape their wings and they have grown out of it once the blood feathers start to dry up.

We start a newly hatched chick out at 35–37ºC (95–98ºF) in a brooder (w use “The Brooder” from Dean's Animal Supply, Inc.) and the temperature is dropped quickly, depending on the chick’s behavior. They are often out of a brooder and into a cage with perches by 10–15 days old, with a heating pad and/or a heat lamp as the heating source. They do not learn to feed out of a bowl quickly and therefore are usually hand fed for several months.

Ring Size: 8mm

References:

Perrins, C. 2003. *The Firefly Encyclopedia of Birds*. New York: Firefly Books Ltd.

Dean’s Animal Supply, Inc., Orlando, Florida. TheBrooder.com

Average Amount of Food Fed per Feeding

	Chick 1	Chick 2	Chick 3	Chick 4
Day 1	0.5g PV	0.5g PV	1 g V	0.25 PV
Day 3	3g PV, T	2.9 PV	2.3 g T	3 g PV
Day 5	5 g T	5.3 g T	4g T	4g T
Day 8	7g P	* 4 P	*2 T, 2 DF	7g P
Day 11	*4 P	*1–4 P	*2 P	7g P
Day 15	*4 P	*2P, 1 DF. 1 MW	*2–3 P	* 2 P
Day 20	ND	*2P, 2 DF	* 3 P	ND
Day 25	ND	*4 P, 2 DF	* 3 P	ND

(PV = pinkie mice viscera, T= pinkie mice midsection, P= whole pinkie mice, DF = soaked Dog Food, MW = Mealworm, WW = Waxworm). One to two pieces of food are dipped in a 50/50 calcium/vionate powdered vitamin mixture.
The amounts are in grams (g).
* Food was not weighed prior to feeding

Random Weights/Brooder Temperatures

	Chick 1	Chick 2	Chick 3	Chick 4
Day 1	20.54g/98F	11.7g/98F	19g/95F	17g/98F
Day 3	35g/97F	19.1g/97F	30.7g/95F	21g/95F
Day 5	44g/95F	34.9g/96F	47g/94F	50g/94F
Day 8	60g/92F	55.5g/91F	72g/89F	74g/89F
Day 11	76g/89F	68.5g**	79g/85F	88g/85F
Day 15	76g/**	69.0g	75g/**	84g/**
Day 20	87g	74g	ND	103g
Day 25	ND	85g	76g	

ND = no data
** = moved out of brooder into small cage with perches, heat lamp and a heating pad. Once the chicks have learned to perch and feed from a bowl, they are moved into a larger cage. When the chicks have learned to eat on their own they are moved into a small aviary, and once they are maintaining a good weight they are moved in the large interactive aviaries

Parrots

14 Psittaciformes

14.01 Psittacidae: Parrots

Psittacine Enrichment

Glen Holland

This section is based on a combination of work produced by Field & Thomas, 2000, on enrichment for psittacines at Edinburg Zoo, an animal watch report of kea activity budgets (Butler 2003 unpubl.), various forms of kea enrichment at Auckland Zoo (Friedman and Jenkinson unpubl.) and a similar research project undertaken at Melbourne Zoo on parrots and cockatoos by Harriet Elson (unpubl.) in 2001.

Field & Thomas advise that "environmental enrichment is, or should be, an integral part of any husbandry programme. The concepts and philosophies of enrichment are well documented, yet the application of such techniques to birds has been limited. Even now when more enrichment is being provided for birds, much of it is centred on parrots, possibly as a means of alleviating undesirable behaviors rather than providing enriched environments. The need to provide enrichment for all primate species is widely accepted. The application of such techniques to birds has, however, been limited. Parrots have, on a number of occasions, been likened to primates in terms of their ecological niches and, more pertinently, for their psychological and social needs in captivity. Parrots have been long recognised as having lively minds, and manipulative and problem solving abilities. Psittacines express abnormal behaviors in captivity; feather plucking, self-mutilation and stereotypic behaviors are typical examples. These conditions arise from inadequate environments and poor husbandry regimes. Psittacine enclosures are commonly designed using concrete and metal for the superstructures and furnishings, perhaps because of the destructive nature of these species, and the added expense of such heavy-duty materials may have led to the often restricted size of aviaries.

This allows them to demonstrate their extensive range of behaviors, thereby increasing the expression of those that are socially desirable and preventing those that are abnormal or socially disruptive, thus facilitating captive breeding. In addition, the various techniques employed should improve the physiological well-being of the birds by providing exercise and increasing the amount of aviary space used by them. These in turn would also provide an exciting display for the public. Successful enrichment/stimulation requires forethought and a strategy of implementation, and will benefit both bird and keeper if driven by at least some degree of imagination. Many of the best enrichment devices/programs are the result of some clear lateral thinking. The occasional introduction of new perching or new foods, although they may be novel items, is not effective and parrot keepers have been doing this for decades. The essential aspect of a successful program is that all these techniques are incorporated into an overall strategy that includes feeding, presentation of food, aviary furnishings and the siting of enclosures, all in full consideration of the daily husbandry routines.

Enrichment techniques are often used to replicate the time budget of species in the wild, but the lack of data for parrots means that this may not be possible. Birchall (1990) suggested that wild parrots may use 90% of their time foraging for food and preening their partners. Before the enrichment programme was initiated at Edinburgh Zoo, all food was chopped into small pieces and offered once daily in a bowl; but now a range of food items is presented in a wide range of forms in an attempt to extend the time spent feeding and to stimulate foraging activity. An ideal situation would be to offer all food in some form of puzzle feeder or enrichment device. The strategy is now to provide the birds with their daily nutritional requirements as early as possible and, at some point later in the day, to offer additional food in a puzzle feeder or in dishes placed around the enclosure. Only fruit and vegetables are provided as enrichment foods. These are surplus to daily requirements but only amount to around 100 g (3.5 oz.) per bird. No extra seed or pulses are given.

Flower ropes for lories: To encourage natural foraging behavior in lories, flowers are strung between the strands of hemp rope and the rope is hung from the aviary roof or from a perch. At Edinburgh Zoo this has been used for chattering lories *Lorius garrulus flavopalliatus* and blue-streaked lories *Eos reticulata* and always results in intense activity, with the flowers being drained of any pollen and subsequently picked to pieces. This technique has also proved successful with vasa parrots *Coracopsis vasa*. Care must be taken to ensure that the flowers have not been previously sprayed with any chemicals.

Foray trays: At Edinburgh Zoo galvanised trays are attached to the mesh of the enclosure and filled with a few sunflower seeds or peanuts mixed with bark, requiring the parrot to forage for seeds. The trays can occasionally be filled with bark alone which lends some degree of unpredictability to the techniques (Mason 1995).

Hanging treatball: This enrichment is a treatball manufactured for dogs that can be purchased cheaply from a pet shop for the large ones. The ball is made of thick plastic, with a hole at the top where food can be pushed through. The hole has a slide that can be adjusted to make the opening smaller. At Auckland Zoo a few changes have been made to the ball so it can hang off the ground. The treatball has had two holes drilled in the sides where bolts have been inserted and we have attached a chain to these bolts. The purpose of the ball is for it to hang from a branch or perch with the opening facing upwards. The birds have to tip the ball 180 degrees before any food will fall out. The treatball has been balanced in such a way that when the bird lets go of the ball to retrieve their food, the ball returns to its original position. This way only small amount of food may be retrieved at each attempt, therefore taking longer for the birds to feed before they became full.

Roll treatball: This is basically the same as the hanging treatball, but is a smaller size (which can also be purchased from a pet shop for minimal cost for the small ones) and is placed on the ground. To retrieve the food inside the ball has to be rolled around the enclosure.

Peanut scatter: This is peanuts still in their shells hidden around the enclosure. This can occupy the birds for some time, firstly by finding the peanuts, and then by breaking into them before eating them.

Roll cages filled with fruit: These are cages made of small gauge weld mesh. The wire is shaped into a square measuring about 10cm^3 (3.9 cu. in.). Ensure all cut edges are ground down to avoid sharp edges. One end can be opened and closed. This end is used to load food and is kept closed with wire or "D" bolt. The roll cages are filled with large pieces of fruit, and the birds use their long, strong beaks to manipulate and move the fruit to get any food to eat.

Hanging cage or bag filled with fruit: This is the same as the fruit cages but the cage is hung up so that the birds have to figure out how to hang onto the cage and extract the whole fruit at the same time. The cage swings, making it slightly more challenging than roll cages that are just placed on the ground.

Food in water: Most enclosures have a small pool where the birds can bathe. Seeds can be thrown into the pool to encourage foraging. Kea in particular enjoy dipping and playing with their food, or simply a twig, in water.

Puzzle feeders: These are a means by which the handling of food items can be increased. Such techniques are particularly suitable for the manipulative skills of parrots and stimulate extractive foraging.

Spray systems: For many parrot species a fine shower of water on a hot day creates a flurry of activity as they perch or hang, often dashing back and forth, under the spray to bathe. This is often accompanied by shrieks of sheer delight, particularly where tropical rainforest species are held in dry climates that lack regular rain during the warmer months.

Savoury ice block: This is a small container half filled with water and topped up with other food items such as corn cornels, sunflower seeds, cashew nuts, etc. A wire rod with a bend at each end is pushed into the middle of the mixture and placed into freezer until solid. The ice block can be hung from a branch or perch. As a change, rather than using food, low joule cordial with no sugar or preservatives is used to make the ice block.

Hanging corn on the cob: Fresh (with husk removed) or frozen corncobs are skewered with a metal rod and hung with chain from a branch. The corn cobs are hung high enough so that the birds can reach them, but at a stretch. The birds can also access the corn by perching on the branches they are hanging from and pulling the food up to them using their beak and feet.

Fruit kebabs: Fruit pieces are speared onto pieces of thick wire that are hung around perches or pieces of rope for the birds to manipulate and remove the fruit. Cuttings from conifers are occasionally wound around the kebab to make the removal of fruit more difficult.

Nuts on wire: Shelled peanuts were initially provided, but offering one or two unshelled peanuts in the daily diet is enough to increase the handling time. Other whole nuts were introduced as part of the diet and these stimulated varying degrees of interest. Drilling a hole through nuts and stringing them on wire produced considerable foraging activity by a pair of keas.

Pine cone kebabs: Pine cones are strung together to create a kebab. Pieces of fruit can be pushed into the pine cones, which are then strung up for the birds to play with.

Log feeder: The log feeder is a length of wood with holes drilled in one side. Food is placed in the holes, which are then covered with starch paper, and the log is suspended in the aviary.

Bamboo toys: The bamboo is cut at about 30 cm (11.8 in.) lengths or on the outer side of each section throughout the bamboo length, making it into a chamber. Holes are drilled in two or three places in the side of the bamboo chamber and small amounts of food such as honey are placed into the bamboo. Other treats can also be forced in with the honey, such as corn, popcorn, seeds, sultanas, etc. The birds have to strip the bamboo to get to treats and this can occupy them for hours.

Container within a container filled with fruit: This is simply any empty container fitted into another larger empty container. The smaller of the two containers is then filled with fruit. The challenge for the birds is to then get the lid off the first container, pull the smaller container out of the larger one, and then remove the lid from this container to reveal the fruit inside.

Kong toy: This is a toy made for dogs and can be purchased at pet stores. Food items are stuffed into the toy, which the birds will then need to try to extract. To make it more challenging for the birds the toy can be adjusted so it hangs, or it can be put in the freezer the night before.

Cored cabbages: Several holes are made in a whole cabbage by removing cores. Fruits, such as sultanas, are pushed into the holes and the cabbage is suspended in the aviary. This system worked well with green-winged macaws *Ara chloroptera* which tear up the cabbage to obtain the sultanas. This technique was copied directly from the primate section at Edinburgh Zoo.

Stainless steel mirrors: These are similar to plastic mirrors sold commercially in pet stores. The stainless steel mirrors serve the same purpose by entertaining the birds when they see their reflection in them. Stainless steel is more robust than plastic, better able to handle the kea's strong beak and is still reflective.

AVIARY DESIGN AND FURNISHINGS

Size, space and position: Aviary position should be given careful consideration. As well as reviewing the environmental parameters for an enclosure, attention should be paid to the view that the birds will have. Perches located high in the enclosure will provide good views and, consequently, an extra form of stimulation for the birds. Some of the high perches should be under cover so that the birds can have a feeling of security.

Plantings: Plants in parrot enclosures rarely survive, but this is really dependent upon the individual birds rather than a species as a whole. However, the provision of regular browse, and leafy browse in particular, has two notable benefits. First, it satisfies the desire of the parrots to chew and strip the bark, providing both psychological and physical benefits as well as diverting their attention away from the destruction of fixtures and fittings and any live plants within the aviary. Second, the provision of plenty of broad-leafed browse or conifer provides essential cover and security for the birds; this has been particularly beneficial with thick-billed parrots *Rhynchopsitta pachyrhyncha*.

Substrates: Too many aviaries have a concrete or sand floor to allow easy cleaning, in the mistaken belief that this is the most hygienic floor covering. Bark chippings make an ideal substrate for many aviaries and provide excellent foraging material. Fecal build-up can be controlled by the daily removal of feces and contaminated substrate. The bark can also be easily replaced if it becomes too dusty or full of seed husks. Although some species rarely go to the floor others, such as Patagonian conures and parakeets, will spend considerable time on the floor foraging through bark, an activity that is unavailable with a concrete floor.

When renovating the kea enclosure at Edinburgh Zoo, mounds of soil were positioned at various locations around the aviary. The mounds provide cover for the birds when they are at ground level and break up the monotony of the enclosure floor. This design has also greatly increased the amount of time the birds spend foraging, as they dig through the mounds for hidden food items.

Furnishings: Perhaps one of the most crucial elements in an aviary is the perching. In addition to providing roosting sites, effective perching can not only enhance the look of the aviary, but also provide essential exercise and stimulation for the birds. Perches should be natural; that is, they should be complex and vary in size and diameter. A main branch with all its off-shoots is an ideal perch. In addition, the perches should not be rigid but should be a mixture of some sturdy branches and others which are free to move. Perches may be suspended from the roof of the aviary. The benefit of this approach is that as the perch moves, and in order for the bird to retain its balance, it has to grip with its feet and flap its wings, providing exercise opportunities. In terms of aviary design, bigger access doors or panels facilitate the ability to move large perches into and out of the enclosure.

Perches should be positioned a reasonable distance apart to encourage the birds to fly; some birds, particularly parrots, will walk rather than fly from perch to perch. The height of the perch is also critical. A variety of perches at different heights is the ideal, and some must be positioned above head height of the viewing public. Parrots will benefit if vertical perches are provided and others are set at "beyond the horizontal." This is particularly appropriate at the ends of the flight space to provide a more challenging place to alight, and in conjunction with hanging feeders.

Food hoppers: These are specially designed food holders. The birds must step onto a platform allowing the lid to be opened to retrieve the food inside. These are also to stop pests getting into the food, such as sparrows and mice that are not heavy enough to open the lid. These feeders also allow the public to experience the intelligence of psittacines. This feeder was specifically designed and made for the kea at Auckland Zoo.

Ropes and chains can be used as alternatives to wooden perches and branches. A chain suspended in an aviary can provide some stimulation but keepers should be aware that there are reports of parrots trapping their toes between the links, although this has not been recorded at Edinburgh Zoo. Hemp rope is a natural material that is an excellent alternative to perches; the rope moves freely and, like a branch, provides exercise for the parrots. Rope is easy to attach in aviaries and tends to be more durable than branches. Chewing the rope also provides stimulation and some birds will use the strands for nesting. The various foraging devices described in this article are easily hooked around or woven into rope. As the birds attempt to manipulate food items, for example, fruit kebabs, the whole exercise is made considerably more difficult ensuring that the bird has to grip with its feet and flap its wings to maintain balance. This simple procedure provides excellent opportunities for the parrots. Placing rope in the aviaries during the rehabilitation of parrots following an illness provides exercise and helps to rebuild muscle tone.

Novel objects: Many simple and everyday objects can provide stimulation to birds, particularly parrots, and at Edinburgh Zoo unusual items are occasionally placed in the aviaries. For example, the green tops of pineapples, sometimes with a little pineapple left attached or with food items hidden between the leaves, have been tried. The pineapple tops can be suspended from or speared onto branches. Parrots are exceptionally destructive and access to a wide variety of material, from cardboard boxes and phone books to rawhide chews, provides ample opportunity to tear and destroy. Care should be taken that none of the items offered are toxic.

Using the described techniques at various locations within the aviary or simply using more feed bowls greatly increases the use of available space by the birds. A study carried out at Edinburgh Zoo indicated that the availability of feeding stations in the centre of the enclosure had two effects, depending upon the nature of the bird. Birds that spent time at the back of the enclosure would come forward to feed and birds that spent time interacting with visitors at the front of the aviary showed a reduction in this behavior. Increasing the number of feeding stations and the use of fruit kebabs and other devices have been particularly valuable in the husbandry of colonies of vasa parrots and Patagonian conures *Cyanoliseus patagonus patagonus* at Edinburgh Zoo. Providing numerous feed stations prevents any dominant birds from restricting access of conspecifics to food dishes, which appears to lessen aggression within the group.

Experiments in cooperation of kea suggest that dominant individuals will allow subordinates to cooperate in tasks that benefit only the dominant birds. In captivity, the life expectancy of kea is 40+ years and because of their intelligent, playful and social nature, kea require considerable mental stimulation in the form of play/food objects and companions. In the past, captive kea have been prone to stereotypic behavior (pacing, head flicking, etc.) due to lack of stimulation and boredom. Research at Auckland Zoo included pre and post enrichment activity budgets. Table 1 presents comparisons between total minutes of 17 individual behavioral enrichments (BE) showing the longest time of use from the top, to the shortest time use for the group at the bottom.

Table 1

Enrichment	Minutes /activity
Forage box	48
Roll cages	41
Bamboo treat tubes	41
Hanging roll cages	31
Hanging roll treat balls	27
Large ice cubes	26
1 container inside another	23
Scattered peanuts	25
Pine cones	20
Kebab corncobs	25
Roll treat balls	16
Peanut kebabs	15
Roll treat balls	14
Frozen Kong's	14
Bamboo logs	7
Fruit kebabs	5
Kebab frozen bananas	0.6

Total time spent per BE for the average for the group of 3.

The group spent the longest time (48 minutes) with the forage box. One bird was still feeding/foraging from the box at the end of that observation day (3:10 p.m.). Also popular were hanging enrichments, metal cages and the bamboo treat tubes (toys). One enrichment, frozen bananas, has been dropped from this enrichment list. As can be seen from the *Animal Watch Report,* the kea showed next-to-no interest in them.

Since a comprehensive BE program has been introduced at Auckland, activity has increased (from Baseline 27% to BE 49%) with a combination of enrichments, staff feeding, foraging, manipulation and locomotion, hence inactivity has decreased. Out of sight has greatly decreased from 27% (previous study) to 11% (BE study). When the kea are visible the remarks of the public are very positive. The placing of the second staff feeding tray near the front has also contributed to the visibility of the kea.

Another research project on the benefits of psittacine enrichment at Melbourne Zoo, Australia by Harriet Elson looked at whether changes occur in the activity budgets of captive parrots and cockatoos during short-term enrichment programs. When I listened to Harriet present her findings at a conference I was struck by the enormous difference that enrichments had made to some of the smaller psittacine species. Data from studies of wild psittacines was examined in order to identify the behaviors that the captive birds performed at lower levels than their wild conspecifics. Enrichments were then designed to allow the expression of some of those behaviors in captivity. The study group consisted of two golden-shouldered parrots *Psephotus chrysopterygius,* three scarlet-chested parrots *Neophema splendida,* two princess parrots *Polytelis alexandre,* two Australian king parrots *Alisterus scapularis* and seven bourkes parrots *Neopsephotus bourkii.*

The group was scan-sampled eight times an hour from dawn until dusk (7:30–6:00) for five days. Activity budgets were compiled, both for the entire parrot group and for each species, in their unenriched baseline enclosures. The activity budgets indicated what percentage of daylight hours was spent in different activities by each bird. These baseline results were compared with a) activity budgets calculated in the same way from data taken after the parrots' enclosures were enriched; b) activity budgets of wild parrots. Various enrichments were added to the enclosures and renewed daily or when necessary. All of the enrichments only contained food materials that were already present in the daily diet of the birds, with the exception of flowers and fresh browse. The birds were fed their usual diet in feeding dishes throughout the study.

When the baseline results were compared with the available wild data there were three main areas of difference between the two groups. The wild birds spent more time in social interaction, locomotion and, in particular, feeding and foraging than the captive parrots. While opportunities for locomotion and social interaction are largely limited by enclosure size and management strategies in the captive environment, foraging enrichments can be used in all types of husbandry set-ups and so, were the focuses for this study. While the captive group spent a mean of 18% of daylight hours in these feeding behaviors, the wild parrots spent a significantly greater mean of 40% in such behaviors.

When enrichments were added to the enclosures, levels of inactive and maintenance behaviors decreased significantly, as did the amount of time spent feeding from food dishes. Levels of foraging and locomotive behavior increased significantly. Foraging (feeding behaviors not including taking food from dishes) increased significantly for all the species studied, although species differed both in their levels of activity and in their preferences for particular enrichments. These preferences are detailed in table 2 below:

Table 2: Species-specific use of particular enrichments indicated preferences not only in foodstuffs, but also in forms of food presentation. For example, dry seed mix was presented in forage trays, in seed blocks, in the millet sprays (in part) and stuck to hanging branches; different parrot species exhibited strong, apparently wild-type, preferences for different forms of presentation. This indicates that it is the performance of the foraging activity, rather than the ingestion of the food, that is the main purpose of the behavior. When enrichments were provided, levels of inactive and maintenance behaviors decrease significantly, as well as the time spent feeding from food dishes. Levels of foraging behavior increased significantly for all species and rose from 14% to 30% of daylight hours for the whole group.

SOCIAL ENRICHMENT

Although every attempt is made to ensure that pairs are compatible, policy that will come into force in the future should facilitate the more natural pairing of parrots through free mate choice. The choice of aviary at Edinburgh Zoo also involves consideration of species housed in the adjacent enclosures. This is particularly important for the salmon-crested cockatoo *Cacatua moluccensis,* which appears to benefit from the presence of other species of white cockatoos. The display and interaction between pairs that are housed not directly adjacent but in visual and auditory contact, serves as useful stimulation; however, it is also essential that individuals have the opportunity to avoid such contact.

NESTING ENRICHMENT

The choice, construction and maintenance of a nest is a major part of bird activity. If they are provided with a limited choice of materials and sites, or nestboxes made of metal or wire trays, the birds are denied the opportunity to display important elements of their behavioral repertoire. Whatever the species, a choice of sites and nesting materials should be provided. Many parrots destroy their nestboxes and, therefore, are provided with metal nestboxes. The destruction of the box, however, is a display of adaptation and modification of the environment. If the nestbox is destroyed completely, it is obviously unacceptable. A variety of both nestbox design and location should be available to allow the birds to make the appropriate selection.

The use of these various techniques, in isolation and as a full program, has been reviewed in a number of under-graduate research projects and the enrichment techniques were shown to increase the behavioral diversity of the parrots. Introduction of the foraging techniques has indeed increased the amount of time spent feeding and foraging in captive parrots without adversely affecting the daily nutritional requirements. Murray (1993) and Smith (unpubl.) reported a reduction in undesirable behaviors, particularly excessive aggression between pair members. Visual and auditory contact between cockatoos has been established, together with improvements in diets and husbandry; both pairs of salmon-crested cockatoos are now parent-rearing, although it is difficult to identify which enrichment technique has been of the greatest benefit.

Successful enrichment can be achieved easily and simply. However, although enrichment is an integral part of any husbandry regime, it is the first task to be dismissed when shortages of staff or time occur. The program at Edinburg Zoo ensures that enrichment has a high priority by determining that enrichment considerations underlie the planning of aviaries and that the time is allocated in the daily routine by incorporating enrichment techniques into the basic husbandry procedures. In order to achieve this the techniques used must be simple; for example, although the log feeder was shown to be effective enrichment, it has not been used since the study because the application is too time consuming.

Table 2

	Golden-shouldered	Princess	Scarlet-chested	King	Bourkes
1	Seeded branch 18% (119 mins.)	Forage Tray 9% (59 mins.)	Millet 9% (59 mins.)	Swinging Fruit Branch 13% (86 mins.)	Seeded Branch 11% (73 mins.)
2	Millet 2% (13 mins.)	Flower Rope 6% (40 mins.)	Forage Tray 6% (40 mins.)	Flower Rope 6% (40 mins.)	Forage Tray 7% (46 mins.)
3	Forage Tray 2% (13 mins.)	Swinging Fruit Branch 5 % (33 mins)	Flower Rope 4% (26 mins.)	Seeded Branch 5% (33 mins.)	Millet 1% (7 mins.)
4	Flower Rope 2% (13 mins.)	Seed Block 4% (26 mins.)	Sprouted Seeds 3% (20 mins.)	Seed Block 3% (20 mins.)	Fruit Branch <1% (<7 mins.)

Enrichment is best when it is simple in application and design. Effective enrichment is not only about providing the odd puzzle feeder or the introduction of a novel object but also should be part of an overall strategy that is incorporated into the design of aviaries and the daily husbandry regime. Only then does environmental enrichment achieve the aim of providing individual birds with more control over their environment and the opportunity to display a diverse behavioral repertoire.

References:

Butler, Beverley. 2003 unpublished. Animal watch report on comparisons between 17 behavioural enrichment items for kea at Auckland Zoo.

Elson, Harriet. 2001 unpublished. Effects of Environmental Enrichment on Parrots and Cockatoos in captivity at Melbourne Zoo.

Field, D.A. & R Thomas. R. 2000. Environmental Enrichment for Psittacines at Edinburgh Zoo. *International Zoo Yearbook* 37: 232–237

Friedman, Melanie & Todd Jenkinson. 2003 (unpublished report). Kea Enrichment at Auckland Zoo.

14.01 Psittacidae Lories

Jos Hubers

Introduction: The 55 species of lories belong to the order Psittaciformes. Under the traditional systems of classification they were assigned to their own family, the Loriidae, but under the S-A-M system, based on DNA relationships, they are included with all other parrots and cockatoos in the family Psittacidae.

In general, lories are striking and noisy birds. Certain species, like the well-known rainbow lorikeet *Trichoglossus haematodus mollucanus* from Australia, exist in such large numbers that you can hear them coming from far away. In roosting places where they gather in large numbers, like in Townsville, the birds are active and noisy the whole night. As they depend on flowering trees and shrubs, you won't see them that often in open plains unless they have to fly from one feeding area to another. Therefore, lories can be found in forests and also in gardens. During feeding between the blossoms, the generally colorful birds seem to have enough protective coloring to make it hard to observe them. In Australia where some lories have become accustomed to people, they are regularly seen on feeding trays in gardens. A well-known example of this can be seen at the Currumbin Bird Sanctuary where large numbers of lories are fed daily by the hands of tourists, the most common being rainbow lorikeets, and in a lesser quantity, scaly-breasted lorikeets *T. chlorolepidotus*.

Management: Lories are known to nearly every aviculturist. However, they have never become very popular. This is unfortunate, as they are, with some adjustments, very grateful and pleasant birds to keep. Although most lory species are bred in sufficient numbers in aviculture, wild-caught birds are still imported. As is the case with virtually all birds imported from the tropics to cooler parts of the world, lories must be acclimatized very carefully to lower temperatures. This is especially true for the smaller species like, for instance, the fairy lorikeet *Charmosyna pulchella* and the whiskered lorikeet *Oreopsittacus arfaki*. When properly acclimatized, they will thrive under these lower temperatures. It appears that the birds do well with a lower temperature at night and a little bit higher temperature during the day. An aviculturist in Switzerland kept these birds in indoor aviaries heated to 20°C (68°F); however, they had the opportunity to go outside where it was only a few degrees above zero. The birds were in excellent condition. Of course the bigger species, too, should be acclimatized very carefully, after which they are very strong and can certainly withstand some degrees below zero. However, you should not force birds to stay outside in the cold. You will see that if the indoor shelter is heated, they will still go outside even if it is –10°C (14°F). The advantages of a heated indoor accommodation are that the nectar won't freeze, and as lories in aviculture do not have a set breeding season, there won't be a problem with young birds suffering from the cold. The section on PSITTACINE ENRICHMENT provides ideas that can easily be adapted for the environmental enrichment of lorikeets.

Problems with lories caused by changing over to a different diet are rare. If it does happens, it usually is because they were initially used to a rather sweet food, but this has nothing to do with the quality. A well-known fact is that with acclimatization of lories, as well as during changes of diet, the fungal disease candidiasis, caused by the fungus *Candida albicans,* often shows up. This is because the birds become less resistant through the stress resulting from the change. Another important point is that infections caused by the parasite *Trichomonas* regularly occur. This usually shows up when the resistance of the bird is low. Ronidazol is an effective medicine against these infections. Lories, like all birds, are also susceptible to *E. coli* and *Salmonella* infections. During stressful situations, and particularly with newly imported birds, they should be given probiotics to help prevent infections. When infections occur, often a vet will give Baytril (Bayern product = enrofloxacin), a widely used antibiotic that is highly effective against many bacteria.

Because of the liquid droppings of these nectar-eating birds, aviaries and cages must be designed or adapted for fast and easy cleaning. For this reason, it is logical to avoid timber construction, and to think more of metal aviaries, probably in combination with plastic. Depending on where you live, you can keep them in suspended aviaries. In cooler climates, indoor aviaries with tiles on walls and floors are easy to clean with a hose. Smaller species can also be kept in cages made out of plastic, which can be cleaned with a sponge.

The best results are achieved by keeping lories in separate pairs. In principle, colony breeding is possible with every species, but the aviary for a breeding group must be of the right size, which is usually too big to be practical for most aviculturists. Colonies of mixed species can be kept together if there are no nest boxes supplied. However, you must supply feeding stations in several places to prevent aggression. On occasion, colonies of *Trichoglossus* species have been successful for long periods where breeding has been encouraged. No new birds can be added; the balance can suddenly change resulting in aggression and birds having to be removed from the aviary. New birds should be introduced into an established group only with great care, as the "old" birds will test them out. Occasionally it might happen that a bonded pair, and this does not necessarily have to be of the same species, wants to start a family somewhere on the floor. It is better to catch and remove these birds, as there is a fair chance that sooner or later there will be a fight.

Lories like to bathe, and should be given the opportunity to do so. One way to accomplish this is by using a sprinkler. From time to time, lories like to chew from fresh branches, which should be supplied regularly. It is useless to plant shrubs in an aviary for these birds, as, like many parrots, they completely destroy them in no time. Lories will appreciate toys, such as plastic balls hanging on a piece of cord, as they are very active and playful birds. It has been suggested that this helps prevent them from plucking their own feathers, which is not a rare occurrence in captive lories.

Diet: Many books can be written about lory diets. It has been said that there are as many different diets as there are lory breeders. This, of course, is a little bit exaggerated, but it tells something about this subject. The main distinction between lory diets is that between commercial and home made ones. Prior experience is a factor in choosing a diet, and many people rely on a certain diet that has worked well for them in the past, or for the breeder from whom their birds are originally obtained. Of course, cost is an important consideration, too. There are many commercial products on the market, which differ greatly in price. Among the more popular brands are Nekton, CeDe, Aves Product, Avico, Shepherd's, and Avi Plus. Unfortunately, the price does not always tell you much about the quality. Of course you can always make your own lory mixture. A satisfactory diet does not have to be formulated by scientists. In principle, lory feeding does not need to be very expensive, as the ingredients are based on cereals, which are cheap, and other ingredients, such as vitamins and minerals, which are not dear if bought in bulk. It appears that you can breed the bigger species, especially, on nearly every diet, but the delicate smaller species might give some problems. If you have a diet that works well for the smaller species like the fairy lorikeet, it will certainly be suitable for the bigger species as well. Some lory species, particularly those in the genera *Trichoglossus* and *Eos,* like to consume some seeds. Seeds, however, should never form the main part of any lory diet. A few species are known to give better breeding results if seeds are added to the diet. Among these species are the emerald lorikeet *Neopsittacus pullicauda,* the Musschenbroek lorikeet *N. musschenbroekii,* and the iris lorikeet *Trichoglossus iris*. There is an ongoing discussion regarding the relative merits of dry and wet mixes for lories.

Low (1998) provides some excellent information on the dietary value of pollen:

> "Pollen is a very important food source for lories. Analysis of pollen grains shows great variation in composition, presumably dependent on plant species and locality from which it is collected. Protein levels (dry weight) vary 7–40% and carbohydrates (sugars and starch, dry weight) 13–39%. Pollen contains up to 16 different vitamins, up to 18 enzymes and coenzymes, up to 16 minerals and up to 18 amino acids, plus many other substances. The best pollen is unprocessed and relatively fresh from honey farms. As an alternative or even as enrichment, although pollen may be present in the diet, fresh flowers are an excellent source of pollen and some nectar. The flowers are best picked the early morning and fed

> directly or late in the afternoon and left in water overnight. The later will ensure a fresh supply of pollen that has not been reduced by wild birds."

It is certain that in nature lories consume much less moisture than in aviculture. Still, we have good breeding results in both cases. Mr. Neff from Germany is one of the most successful lory breeders in years, using either a wet mix or a combination of wet and dry mix. He has bred large numbers, especially of the smaller species like the whiskered lorikeet. However, he has also been very successful with the larger species, such as the purple-naped lory Lorius domicellus. For this reason, we recommend his diet.

Wet mix:

- Add to one liter of lukewarm water:
- 30 grams of pollen
- 65 grams of oats flour
- 50 grams of honey
- 5 grams of yeast
- 0.3–0.5 grams of calcium
- 3 drops of vitamins

Stir this mixture just before giving it to the birds. Give small quantities, preferably in flat dishes as it settles rather quickly.

Dry mix:

- 1000 grams of dry mix consists of:
- 430 grams of pollen
- 360 grams of oats flour
- 140 grams of glucose
- 50 grams of yeast
- 20 grams of calcium (ground egg shells from boiled eggs can be used)

Breeding: Low (1998) describes the display of lorikeets well:

> "Most of the ritualized components of the display of lorikeets are not unique to them. Behavior such as head bobbing, lateral swaying, wing fluttering and of course, courtship feeding, occur in other unrelated parrots. An exaggerated swaying walk is a pronounced feature of some of the lories, such as *Trichoglossus* and *Eos* species. Lorius species and some *Trichoglossus* (e.g, Iris), when on the ground jump high, up and down, or in a small circle, in a very excited manner. Probably the only element of display that is unique is that sometimes described as "hiss-ups." The bird will suddenly extend itself vertically upward while simultaneously sticking out the tongue and uttering a hissing sound. The courtship and precopulatory display of lories is energetic and exaggerated. It is often designed to show off some colorful part of the plumage, especially the underwing coverts."

As mentioned earlier, lories are not confined to a fixed breeding season. Still, there are breeding pairs that only produce one clutch in spring while others of the same species produce three clutches a year. So it does not just depend on the species, but the individuals as well. Lories accept both horizontal and vertical nest boxes. With vertical nest boxes, it is better to fix wire or small pieces of wood against the inside so the birds do not have to jump down, with a risk of damaging the eggs. Also, the boot or L-shaped form is suitable. If using vertical nest boxes, a height of 30–45 cm (12–18 in.) is recommended, and a square of 18 x 18 cm (7 x 7 in.) to 25 x 25 cm (10 x 10 in.), depending on the size of the lory. Horizontal nest boxes seem to be preferred by the smaller species. The size can be between 15 x 20 cm (6 x 8 in.) and 15 x 30 cm (6 x 12 in.), with a height of about 18 cm (7 in.). However, when the birds are ready for breeding they are not too fussy and they will take almost any size nest box. It is not recommended with smaller birds to offer a nest box that is too big; it appears that the birds leave their droppings behind in their nest box. Wood shavings are most commonly used as nest filling.

Lories lay an average of two eggs per clutch. Clutches of one or three eggs are seldom found. A few species produce larger clutches as a standard. Examples include the perfect lory *Trichoglossus euteles,* the purple-crowned lorikeet *Glossopsitta porphyrocephala,* and the little lorikeet *G. pusilla*, which all lay clutches of 3–4 eggs.

Regardless of the species, incubation periods for lorikeets range from 23–26 days. However, the incubation time of the Stella lorikeet *Charmosyna papou,* for example, is hard to determine, as these birds stay in the nest box for a couple of days before incubation starts. Because of this habit, there are cases on record of more than 30 days elapsing before the eggs hatched. Another exception is the purple-crowned lorikeet with an incubation period of less than three weeks.

Four Species of Lorikeets

Day	Scaly-breast Weight (g)	Emerald Weight (g)	Swainson Weight (g)	Musk Weight (g)	Day	Scaly-breast Weight (g)	Emerald Weight (g)	Swainson Weight (g)	Musk Weight (g)
0	4		6	3	**26**			74	27
1	4.6	3.7			**27**	67	33		
2	5	4.1	11		**28**			82	
3		4.8			**30**			93	
4	6.6	5.7	14.5	5.5	**31**		35		41
5		6.7			**32**	80		100	
6	9.1		18		**34**			105	49
7			8.7		**35**	88	37		
8	11		22.5	6.5	**36**			112	52
9		11.2			**38**			123	
10	13.6	14.1	27		**39**	93	35		54
12	17.7	19.6	30	8	**40**			131	56
14	22		36		**42**	96		136	57
15		24		10	**43**		35		
16	28		43	12	**44**			142	
18	34	26	51		**46**		33	146	
20	41	27	55		**48**			150	62
22	47		58		**52**			157	61
23	52	30		22	**54**			160	61
24			63.5						

Scaly-breasted lorikeet - *Trichoglossus chlorolepidotus*
Emerald - *Neopsittacus pullicauda*
Swainson's Lorikeet - *Trichoglossus haematodus moluccanus*
Musk Lorikeet - *Glossopsitta concinna*

Recommended Dietary Items

Species	Nectar	Fruits	Vegetables & Sprouted Seed	Green food	Seed	Flowers
Chalcopsitta	**	**	*			*
Eos	**	**	*	*	*	*
Pseudeos	**	**			*	*
Trichoglossus	**	**	*	*	*	*
Iris, Mount Apo	*	**	**	*	**	
Glossopsitta	**	*				*
Neopsittacus	*	*	*	*	**	*
Lorius	**	**	*			*
Vini	**	*				*
Charmosyna	**	**		*		*
Oreopsittacus	**	*		**		*

** Indicates main dietary items.
* Indicates also eaten.
From the Hancock House *Encyclopedia of the Lories* by Rosemary Low (1998)

Young lories hatch with just a little bit of primary down. Within two weeks the eyes will slowly open and they will get their secondary down, which is usually gray or gray-white. Within a few weeks they look like nice thick fluff balls. Then the primaries and tail feathers come through. We have found that it is necessary to change the wood shavings in the nest box when the chicks are about 20 days old, as by that time the shavings are becoming thoroughly wet, which can result in super cooling (GH recommends that the nest boxes are checked for excess moisture every week). After about 7–9 weeks, depending on the species, the young fledge. At this time they already resemble the parents, however, they are a bit duller. Chicks of those species in which adults have yellow to orange-red beaks fledge with black beaks. Shortly after fledging, they learn to eat by themselves. Nevertheless, they are getting additional food from the parents. Sometimes the hen lays another clutch of eggs while the young are still in the nest.

It is also important to know that lories can easily be

foster reared by other lory species. This allows the breeder to transfer chicks from exceptionally large broods to hens that only have one or two chicks. The maximum number of chicks a pair can accommodate adequately is four. Smaller species can be raised by bigger species, but the size difference should not be absurdly great. The larger-billed birds may have difficulty feeding exceptionally tiny chicks. So, no eggs from a fairy lorikeet, for instance, under a black lory *Chalcopsitta atra*. It is equally important that the age difference of chicks reared together not be too great. When fostering small species with large, or vice versa, actual chick size is more critical than age, as larger chicks might damage smaller ones.

For some reason, young lories are often plucked by their parents. Several methods have been tried in an effort to solve this problem. I have found that the most successful method is to place wire — 2 x 5 cm (¾ x 1 in.) mesh, for example, depending upon the size of the species — above the youngsters, through which they can still be fed by the parents. I have also been successful with a few pairs by simply placing the youngsters on the floor of a small box with no top. Apart from these options, foster parenting and hand rearing are also good solutions to the problem.

Hand Rearing: Although I try to work as much as possible with foster parents, there are probably no other bird species that can be hand reared as easily as lories. In the first few weeks, they are similar to other species; however, as they progress it becomes easier because they quickly learn to eat. Once they are five weeks old, they can be convinced to eat by themselves by keeping their beaks in lukewarm nectar. Soon they know where to find the food dish.

The easiest ways to hand feed lory nectar are with a folded little teaspoon or by using a syringe. It is better, especially in the first days, to add lactobacillus on a regular basis, for example, in the form of a probiotic. It is better to give this to the young every day during rearing time. During the first 3–4 days it is necessary to feed the young every 2–3 hours. If the regular feedings are scheduled between 6:00 a.m. and 11.00 p.m., a single night feeding is enough. The same nectar formula fed to the adult birds can be used for the young. If necessary, the protein as well as the calcium content can be increased to stimulate better growth.

If birds are to be deliberately removed from the nest for hand rearing it is best to leave them with the parents for at least two weeks. This will result in less work for the aviculturist, greater enrichment for the parent birds and chicks, with a far better weight for the age when compared to chicks hand raised from the egg. For example, a musk lorikeet that was hand raised weighed 12 g (0.4 oz.) at 16 days old, and the parent-raised bird weighed 17 g (0.6 oz.) when it was removed from the nest at the same age. The bird hand raised from the egg weighed 61 g (2.1 oz.) at 47 days old and the partially parent-raised bird weighed 85 g (3 oz.) at the same age.

Species Requiring Similar Management: Only the hanging parrots, *Loriculus* spp., and to a lesser extent, the fig parrots *Cyclopsitta* and *Psittaculirostris* spp., can be compared with lories, and then only because their diets are similar. For this reason, many lory breeders also diversify by keeping these birds.

Ring sizes for lories as given by Low (1998):

Chalcopsitta	7.5 mm
Eos and *Pseudeos*	7 mm
Small *Trichoglossus*	5.5 mm
T. chlorolepidotus	6 mm
Other *Trichoglossus*	6.5 mm
Lorius chlorocercus	7.5 mm
Other *Lorius*	8 – 8.5 mm
Glossopsitta concinna	5.5 mm
Other *Glossopsitta*	5 mm
Small *Charmosyna*	4.5 mm
Other *Charmosyna*	5.5 mm
Neopsittacus	5.5 and 4.5 mm
Vini	5.5 mm
Oreopsittacus	3.5 mm

References:

Low, R. 1998. *Hancock House Encyclopedia of the Lories*. Surrey, Canada & Blaine, USA: Hancock House Publishers.

14.01 Psittacidae Cockatoos

William Horsfield

Introduction: Cockatoos belong to the order Psittaciformes and the family Psittacidae. Under the traditional system of classification, they were placed in a separate family, the Cacatuidae, but under the DNA-based S-A-M system, this separation is not recognized. They are distinguished from other parrots by the presence of a gall bladder; by the position of their carotid arteries; by the absence of blue or green colors in their plumage, due to the lack of Dyck texture, a structural composition of the feather barbs which, through Tyndall-scattering of light, produces color in plumage; by an ossified orbital ring and bridged temporal fossa; by an erectile crest; and, in most species, by downy hatchlings.

The 21 species of cockatoos may be subdivided on the basis of color. The black cockatoos of the genera *Probosciger* and *Calyptorhynchus* are, as the designation implies, basically black in plumage coloration. The species in the genus *Cacatua* are basically white, with a few exceptions, and the monotypic genera *Callocephalon* and *Eolophus* are basically gray or gray and pink. The cockatiel *Nymphicus hollandicus* is multicolored.

Five of the six cockatoo genera have different chromosome numbers, as shown below:

palm cockatoos *Probosciger*: chromosome no. = 74
black cockatoos *Calyptorhynchus*: chromosome no. = 78
gang-gang cockatoo *Callocephalon*
white cockatoos *Cacatua*: chromosome no. = 82
galah *Eolophus*: chromosome no. = 76
cockatiel *Nymphicus*: chromosome no. = 72

Cockatoos range in size from the small cockatiel, measuring 33 cm (13 in.) in total length and weighing 80–100 grams (2.8–3.5 oz.), to the massive male Goliath palm cockatoo *Probosciger aterrimus goliath*, measuring 55–60 cm (22–24 in.) in length and weighing 1,057 grams (2.3 lb.) Sexual dimorphism is pronounced, slight, or absent, depending upon species. When disturbed, cockatoos make a hissing, puffing noise. If truly alarmed, they screech loudly or, in the case of palm cockatoos, even growl.

Cockatoos are found throughout Australia and Tasmania, as well as in New Guinea, the islands of Indonesia, the Solomon Islands, and the Philippines. Some, such as the galah, or rose-breasted cockatoo *Eolophus roseicapillus* and the bare-eyed cockatoo *Cacatua sanguinea*, are regarded as agricultural pests in Australia, and permits are regularly issued to destroy them. These species may be seen outside the breeding season in flocks of many thousands, literally blocking out the sun as they swarm overhead. Red-tailed black cockatoos *Calyptorhynchus banksii*, although rare and valuable in international aviculture, are regularly seen in magnificent flocks in excess of 5,000 birds in Australia's north. They are so abundant that they are around the sixth to tenth most common psittacine species in Australia, and therefore among the most common parrots in the world. For some of the cockatoo species and subspecies, however, especially those confined to the Pacific islands, some of which only occur on single islands, such as the citron-crested cockatoo *Cacatua sulphurea citronocristata* on Sumba, the picture is far from rosy. Their numbers have been decimated by trapping for the avicultural trade as well as by habitat destruction, and some face extinction in their natural habitat in the new millennium.

Management: Cockatoos are among the most interesting of aviary subjects. With their spectacular crests and raucous calls, they never fail to entertain. They are extremely destructive and will destroy all vegetation in their flights. They need to be kept behind heavy-gauge welded mesh, with individuals of certain species, especially Moluccans *C. moluccensis* and tritons *C. galerita triton*, capable of chewing through anything less than 2.5 mm diameter wire. I use 25 x 25 mm (1 x 1 in.) x 2.5 mm Twilweld welded mesh. Although they certainly do breed in suspended aviaries, I feel that the cockatoos need more space, including access to the ground. Many of them spend hours digging in the soil in the wild. A thick base of coarse river sand is recommended as a floor covering for cockatoos, allowing them to dig in a relatively hygienic substrate. Biannual, random, microscopic analysis of feces is essential for internal parasite detection and control in these birds. Avian malaria is another parasite that is commonly found in cockatoos (see AMAZONS). An alarming aberrant form, in which the *Haemoproteus* parasite encysts in the cardiac and skeletal muscle, has recently been observed in various birds, including a cockatoo. The bird develops a sudden and dramatic fatal allergic reaction to the cyst.

A non-stimulating environment may lead to excessive screeching, plucking, and self-mutilation in these active birds.

Cockatoos fly vast distances in the wild, and in depriving them of this we are perhaps instigating hostile behavior. They are also generally highly sociable and gregarious. Most are found in family groups, if not flocks of thousands, where continuous interaction between pairs and individuals takes place with much bickering, screeching, and chasing.

The frequency of mate aggression is most likely also related to the lack of foraging opportunities, coupled with frustration. Jurgen Glissmann, a successful SA breeder of cockatoos, including Leadbeater's *C. leadbeateri,* has never experienced mate aggression in his birds. He attributes this to his very large, 4 x 8 x 3 m (13 x 26 x 10 ft.)high, conventional aviaries with river sand on concrete floors.

Mate mutilation in captivity is usually because the female is not responding to advances from the male. Hand-raised birds that have been kept as pets in isolation from other cockatoos and have become vicious and "ready to breed" are not ideal candidates for breeding at all. There is nothing worse than finding a mutilated female dead or barely alive early in the morning. This behavior occurs far too often, even in proven breeding pairs that seem very well bonded and compatible. It needs serious consideration, and preventative measures must be taken. These may include:

1) Providing large conventional aviaries, at least 4 x 1.8 x 2 m high (13 x 6 x 6½ ft.), for all of the larger species, with a stimulating view, if possible, and adjacent vegetation that can grow into the aviary for chewing.

2) Providing two nestboxes, each with two entrance holes. This enables the hen to exit through one hole if the cock is guarding the other and refusing to allow her to leave. Hollowed out natural logs are nests of preference; if made of hardwood and capped top and bottom with flat metal sheeting, they will last well. If metal or plastic nests are used, they should be lined with wood so the birds can chew.

3) Clipping one wing of the cock bird to reduce his speed, or both wings if he persists in chasing the hen. Such wing-clipping needs to be maintained between each molt.

4) Providing separate feeding stations in case the cock dominates a single area, causing the hen to become malnourished.

5) Providing fresh branches and destructible objects, like pine cones and coconut shells, regularly.

6) Using non-cockatoo neighbors, which are less likely to induce territorial behavior and trigger tantrums in the male. (Some breeders have advocated the total opposite — housing the same species alongside each other. I have seen many such setups, and unless the aviaries are exceptionally large, this system often results in disaster. Housing the same or related species in adjacent aviaries may work in public parks, as someone usually notices birds fighting. However many attacks occur at night, often around a full moon.

7) Situating bright spotlights above aviaries, which can be turned on at night if birds are heard fighting.

8) Avoiding the pairing of mature males with young females.

9) Avoiding the use of aggressive ex-pet males for breeding purposes.

10) Never giving a killer another chance, or benefit of the doubt. Such birds will often kill again, and should have one wing permanently clipped, or even pinioned. Breeders should always be suspicious of adult males for sale at reasonable prices.

11) Removing fledglings when weaned, as parents often attack same sex young if wishing to breed again.

12) Providing a smaller adjacent flight to house the male temporarily until aggressive behavior subsides.

13) Where occasionally possible, allowing the male to choose his own female in an aviary containing possible mates.

14) Installation of monitoring camera. This allows behavior exhibited in the absence of humans to be observed, and can often provide early warning signs of mounting agitation, e.g., the female sitting on the aviary floor is a typical sign of attempting to avoid the male.

15) Providing vertical perches (I use dead palm tree trunks planted in the aviary floor) and allow only one bird to perch. Aggressive birds don't often jump onto another bird, and this is therefore a safe perching spot.

Cockatoos are very fond of bathing and should have a mist sprayer system or baths for this purpose. They produce large quantities of powder down and this is believed to irritate the skin, encouraging self-plucking if they cannot bathe for extended periods.

Diet: Captive-bred cockatoos present few problems when being encouraged to eat a varied diet or a total extruded diet. The originally wild-caught adults are often more difficult to convert onto a healthy diet, and are inclined to stick to a staple sunflower mix if allowed to do so. These birds need to be persuaded to eat a healthy diet, otherwise the typical dietary deficiencies will manifest themselves.

At Amazona Farm I have used a number of feeding options over the years. At present, I feed a mixture of seeds, sprouts, fruits, vegetables, legumes, nuts, and commercial soft food, which I vary according to age, breeding status, and climatic seasons. I have also used a system of flush feeding with great success. I have used a complete extruded diet (Kaytee, Nutribird, and Pretty Bird) in the off-season. When the birds are due to start breeding, the diet is changed; I then use a mixture of soaked seeds, boiled and soaked pulses, fresh vegetables and fruit, and a commercial soft food such as Aviplus, with a vitamin and mineral supplement. This change is effected in conjunction with warm weather and good rains to stimulate the pairs to go down. It should be noted that the black cockatoos do require some fat in their diet and are usually fed a ration of dry sunflower in addition to the sprouts. Many breeders in Australia also offer the black cockatoos a few mealworms and in the case of gang-gangs, some breeders even offer a dead young mouse for each pair once a week. These mice are usually devoured by the pair!

Manufactured diets are used by many breeders and certainly reduce labor costs while claiming nutritional balance, thus eliminating the guesswork associated with many varied diets. There is also no risk of the pesticide and herbicide residual contamination often found on fresh produce. One of the main drawbacks, however, is that the different colors and shapes of the pellets do not provide the same degree of visual and sensory stimulation provided by a more natural, varied diet. Also, there is a shortage of long-term studies on additions like chemical preservatives and artificial flavorings and coloring agents. More recent organically formulated complete diets (e.g., Harrison's UK) have addressed this issue. It is safe to say that while complete diets are generally accepted for domestic canine and feline animals, opinion is still divided as far as they pertain to captive parrots.

Once breeding has commenced, if eggs are removed from problematic pairs for artificial incubation, some pairs of cockatoos will lay many clutches in a single season. As they will also breed at any time of the year, the actual breeding seasons may also overlap. It is essential for these continuous laying hens to have the correct amounts of calcium, phosphorous, and vitamin D3 in the diet. Low levels of calcium and vitamin D3 and high levels of phosphorous are typical causes of dangerous seizures. Seeds are high in phosphorous. Complete diets are balanced in this regard, but specialist veterinary advice is advised when varied diets are used, to ensure correct balance of Ca:P and other essential elements.

Obesity is often encountered in cockatoos, particularly galahs. This is as a direct result of poor exercise and overfeeding, especially too much sunflower seed in the diet. These birds develop fatty tumors or lipomas, which may grow to the size of tennis balls. Surgical removal is required, but unless the diet is corrected they will reappear. The best breeding results are obtained from galahs described as lean. Healthy galahs should certainly breed by three years of age at the latest.

I provide all the cockatoos with a variety of wild grasses, fresh branches, flowers, and other vegetable matter, much as I do for the MACAWS and AMAZONS (see those chapters). They spend many hours relishing these daily treats. Fresh spring blossoms like peach and Brazilian pepper are favorites.

Palm cockatoos need a much higher percentage of fat in their diet, and to provide this they can be given a variety of nuts, fresh coconut, fresh palm nuts (espe-

Young cockatoo chick.

cially *Cocus plumosa* and *Pandanus* spp.), and dates, daily. I have had good breeding success with this species, and put this down largely to the huge effort we make to provide these dietary items.

Most cockatoos will chew on the charcoal logs that are placed in the bottom of all my conventional aviaries. Charcoal is renowned for absorbing impurities. High dosages of sanitizing chlorine commonly used in South African drinking water are now being positively linked to various cancers and other ailments in humans. Wild parrots would not have to contend with chlorine in their water, and this information must surely pertain, in my opinion, to our captive birds as well. For these reasons, I have recently installed a water-purification system on the main municipal water line that supplies my home and aviaries.

Breeding: White cockatoos lay 2–3 pure white eggs per clutch. Galahs often lay four, and even up to seven, eggs, and are sexually mature at a mere 12 months of age. Incubation periods vary within the same species. For Goffin's *Cacatua goffini* and bare-eyed cockatoos, the incubation period is 24–25 days, and the pip-to-hatch interval is 36–48 hours. For the galah, incubation is 22–24 days and pip-to-hatch interval 24–36 hours. Greater sulphur-crested *C. galerita galerita* and tritons have a 27–28 day incubation period; the pip-to-hatch interval is 36–48 hours for the greater, and 24–48 hours for the triton. The pip-to-hatch interval for the medium sulphur-crested *C. g. eleonora* is the same as for the triton, but the incubation period is about a day shorter, at 26–27 days. The incubation periods for umbrella *C. alba* and lesser sulphur-crested *C. sulphurea* are 28 days and 24–25 days, respectively; both species have a pip-to-hatch interval of 24–48 hours. With citron-crested, the incubation is 25–26 days and pip-to-hatch 36–48 hours; with Leadbeater's, 26 days and 24–48 hours, respectively. The Moluccan, or salmon-crested, cockatoo has in incubation period of 28–29 days, with a pip-to-hatch interval of 24–48 hours. The incubation period for the black palm cockatoo is 29–31 days, and pip-to-hatch may take from 48–72 hours.

Adult white male cockatoos have dark brown to black eyes; females have light brown to bright red eyes. Unfortunately, there are exceptions, and birds should either be surgically sexed or DNA-typed to determine the correct sex. Females are usually smaller and narrower across the head. Immature birds often have gray feathers in the crest, which are not retained after the first molt. Typical mating of the citron-crested consists of the female crouched down on perch and the male mounting her back, while twisting his head from side to side and raising and lowering his crest. The male may dismount, mount in reverse, climb right over a few times, turn 360 degrees, occasionally grasping the female's beak, and then finally position himself for copulation, all with fairly slow and deliberate movements. During this performance the female crouches almost trance-like with her head arched upward and backward, often pushing her tongue in and out and making quiet noises. After mating, the female shakes herself vigorously and both birds fly and screech a few times, then settle down to preen.

Incubation is by both sexes in alternating shifts, although the female will usually incubate on her own if the male is removed. Sadly, many captive-bred endangered cockatoos have been sold as pets. Many of these will develop so-called "bad habits" and will at some stage probably re-enter the breeder market. These birds will generally be highly unsuitable for breeding programs, due to their years of isolation from other cockatoos and their imprinting on humans, and will remain problematic. Breeders must make concerted efforts to retain young birds for future breeding. It is imperative that these birds are well socialized in groups, preferably in large walk-through aviaries where they can interact in typical flock situations and begin the process of mate selection and bonding activities. A disadvantage of this free-choice mate selection is that siblings or related birds may well bond together, in which case forced pairing may be necessary. Trapping for the pet trade has decimated so many of the once prolific species. Lessers, citron-cresteds, and umbrellas are prime examples. Birds that are hand-raised for breeding should not be allowed to become fixated on the hand-feeder. To prevent imprinting, they should only be handled when being fed, weighed, etc., and then left alone. They should always be raised with at least one of their own species, or a closely-related species, or placed near an adult of their own kind. Ideally, they should be nicely steady but not silly tame. After being placed in outdoor socializing aviaries, such birds may come and land on the caretaker, perhaps nibbling on a finger, but will not allow anyone to handle them.

Neville and Lorette Steyn of Blanco have a pair of free-flying tritons that breed every year on their front porch. The male was originally wild caught and the female hand raised. Both are tame, but attack without hesitation if their nest log is approached. The young are essentially wild, and have to be caught when they fledge, as the male drives them away when the female is due to lay again. At one time all the young cockatoos the Steyns bred, from Moluccans to Goffin's, were allowed to fly free in one huge flock, and it was quite a sight. Unfortunately, they demolished neighbors' TV aerials and washing lines, and regularly flew 10 km (6 mi.) to the radar tower at George airport, causing chaos; as a consequence, they have to be mostly contained these days.

While some pairs make excellent parents, many of the originally wild-caught cockatoos have poor breeding success. This is partly due the fact that they have remained very shy and nervous, often crashing onto eggs and chicks when the nest is approached by humans. It is also due in part to their dietary fussiness and tendency to feed their chicks poorly, or exclusively on certain items, such as sunflower seeds, resulting in stunted and malnourished chicks. Such pairs are candidates for complete extruded diets while raising young. As aviculturists, we have a responsibility to ensure that when the wild-caught generations finally die out in our aviaries, we have managed to secure the species within our collections with our own captive-bred progeny. The challenge is to raise these birds in such a manner that they in turn will be the forefathers for generations to come.

Artificial Incubation and Hand Rearing: Cockatoo eggs are relatively easy to artificially incubate. They are treated the same as any other eggs at Amazona Farm (see AMAZONS and MACAWS). However, there is a higher incidence of damaged eggs with these birds. Tiny nail punctures and even small cracks can be sealed using quick-drying nail varnish. Provided this is done promptly, these eggs usually develop normally and hatch successfully. Careful candling of eggs is important, as some of these small fractures will not be seen otherwise. If there is any albumen leaking out of the damaged egg, then the inner shell membrane has been damaged and there is little chance of hatching, as air enters the egg and a bacterial infection normally kills the embryo approximately two-thirds of the way through development. However, especially with rare eggs, it is always worth a try.

Chicks about to hatch announce their imminent arrival with loud vocalizing and are fed almost immediately, or as soon as they are dry. They exhibit the characteristic pumping cockatoo feeding response, accompanied by loud *chick chick chick chick chick* as they feed. They have long, wispy down, which is bright yellow in the white species, and are hatched with ears open, unlike other parrots. Palm cockatoos are naked at hatch, except for two rows of tiny down along the back.

Hand raising is also straightforward, with few problems. Temperature of brooders is critical with these newly hatched chicks. Cockatoo chicks can tolerate high temperatures but pant vigorously and are far more content at lower temperatures than newly hatched amazons and macaws. They all start off at 37.2°C (99°F), but are quickly reduced over 2–3 weeks to 28°C (82.4°F). It is important to use very soft tissue paper as a substrate for the first few weeks, as they actively jump up and down in their tubs, often clinging onto the rim of the container with their beaks, both before and after being fed. This can cause raw friction lesions on wingtips and toes, which can then bleed excessively. Bleeding is immediately stopped using Ferric Chloride drops. Tiny strips of Elastoplast on soft gauze with Cicatrin powder are used to dress the lesions. Occasionally, older birds may accidentally break a nail or even beak-tip and Ferric Chloride usually stops the bleeding. However, the bird grinding the broken beak tip often hinders clotting, and it may have to be immediately cauterized by a veterinarian. In extreme emergencies where there is profuse bleeding from the beak-tip, the bird should be wrapped in a towel, taken to the car, and while holding the beak shut to prevent the tongue getting burnt, the beak-tip should be very carefully cauterized using the red hot cigarette lighter. This is momentarily painful but highly effective, and while definitely not for the faint hearted, the procedure will stop your bird from losing blood and may well save its life.

Once they are moved out of the brooder units, the chicks are placed in plastic tubs on clean pine wood shavings. Occasionally, certain individual birds grasp onto and swallow these shavings while soliciting for food. These chicks need to be kept on absorbent towels until they are slightly older, otherwise they may end up with impacted crops. A few pieces of shavings normally pass through without problems. Partially parent-raised chicks should not be introduced to the incubator-hatched nursery. Psittacine Beak and Feather Disease Virus (PBFDV) is easily transmitted in this way, as are many other serious viral conditions, such as Avian Polyomavirus (APV) and herpes viruses. Birds that have

White Cockatoo
Cacatua alba

PBFDV are also more prone to APV. According to latest research, adult cockatoos will not die from APV even if they are latent carriers. However, they can pass it onto their young; should there be an outbreak in a nursery, there can be high mortalities. There has been rapid progress in avian medicine in the last five years, and vaccination options and programs should be discussed with your veterinarian. For additional information on this subject, see MACAWS.

Cockatoos take longer to wean than Amazons do and may cry and beg to be fed at the weaning stage. They lean forward and whine while slowly lifting their heads up and down when they see the hand feeder. They may also grasp onto the side of the weaning cage and pump furiously while begging vocally. It is better to stay out of sight at this stage, as this behavior is usually a sign that the baby is becoming spoilt. Chicks will also learn to eat faster if they have an older bird to learn from. Excessive begging can also indicate malnutrition, inadequate feeding volume, low percentage solids, continuous light without darkness, or infection. These possibilities should be investigated, and ruled out or corrected as necessary. Once weaned, cockatoo chicks can be moved into outdoor aviaries. Due to their sensitive natures, it must be carefully ascertained that they continue to eat at this stage.

Palm Cockatoos *Probosciger aterrimus*

Day	Palm Weight (g)
0	20
5	40
10	70
15	100
20	155
25	230
30	300
35	375
40	430
45	410
50	525

David Rogers provides the following account of hand rearing a red-tailed black cockatoo at Auckland Zoo, New Zealand (recorded in 2001):

Feeding: Kaytee Exact hand rearing formula was used with a small amount of smooth peanut butter added twice daily to provide some fat content. With the appropriate amount of boiled water added to a small drinking glass (140ml–150ml), it was swirled around to cool in conjunction with adding the feeding formula. (The formula temperature is assessed by touching the glass onto your cheek; it should feel quite warm. Note: a bird's body temperature is higher than the human body temperature.)

From hatch to day 20 the formula was progressively been thickened from a watery mix to one that is creamy thick.

Red-vented *Cacatua haematuropygia*

Day	Red-vented Weight (g)
0	15
5	35
10	80
15	145
20	220
25	270
30	310
35	325
40	
45	
50	

Feeding Method: Syringes were used, as it has been documented that some cockatoo chicks haven't advanced at weaning time as quickly as they should have with the use of other feeding tools coinciding with an extended period of dependence. This has been documented in the book *A Guide to Incubation and Handraising Parrots* by Phil Digney. Published by ABK publications, 1998.

A good feeding response was prompted before actual feeding. This was initiated by touching thumb and forefinger on either side of mandible, imitating a beak-like touch. With the syringe ready in the opposite hand, and while the chick was giving a good "pumping" response, the chick was allowed to feed while a reasonable pace was maintained expressing food from the syringe. The direction aimed for was towards the bird's back right of the tongue. (Caution: the trachea is at the base of the tongue; it is important not to flood the mouth.

The mandibles were allowed to push against the forefinger and syringe tube rather than the syringe tip. (The syringe tip can sometimes be sharp causing potential damage to mucous membranes. Filing the tip smooth may be required.)

Syringes: Initially 1 syringe was used with progression onto 10 ml then a 20 ml.

Push-lock syringes were found to be much better than the luer-lock (screw type). The luer-lock was quite messy; it didn't seem to be able to cleanly deliver the food, resulting in dribbling down the lower mandible.

The disposable syringes lasted two to three days. At night the syringe was filled with a surgical disinfectant/water solution and left overnight till the next feed the following morning.

During the day, feeding equipment was cleaned using boiling water, although when I fed peanut butter, some micro-shield was used to help wash off the nut oil.

Syringe rubber plungers becme sticky. A drop of vegetable oil ensures smooth syringe action again.

Brooder: Hatch to day 14 a Brinsea Hatchmaker incubator/hatcher was used. The chick was kept inside a small plastic container padded with soft tissue. From day 14 a 'Brinsea' octagon TLC-4 brooder was used.

An ice cream container with wood shavings contained the chick. Daily, or as required, the container/shavings were cleaned or changed. Micro-shield who used to clean the container.

A small bowl of water was placed into the brooder to provide humidity. The bowl was used, rather than the brooder humidity container provided, because of the travelling involved.

By day 29 the ice cream container was getting too small. The container was changed to a larger round one about the size of a standard dinner plate. The edge of the container was padded with a towel. Paper towels were placed on the bottom and then shavings added.

Transportation: Brooder matting was placed in the brooder to stop container movement. A strip of wood under the brooder on the car seat greatly reduced travel movement.

Imprinting: Handling was kept to minimum with feeding and weighing done as quickly as possible. Talking was avoided around the chick, and the brooder was kept in a room with little people activity.

Brooding Parameters for Cockatoos

Day	°C	°F	Day	°C	°F
0–6	36	98.6	24	24.2	75.5
7	33	91.4	26	24	75.2
8	33	91.4	28	23.8	74.8
9	31	87.8	29	23.9	75
10–13	30	86	33	22.7	72.5
14	28.2	82.5	34	22.2	71.8
14	28	82.4	35	22	71.6
15	27.2	80.8	38	21.4	70.5
16	26.5	79.7	40	21	69.8
17	25.4	77.8	45	19.5	67.1
20	24.6	76.2	46	18*	64.4
22	24.4	75.9			

- Day 34: Down to three feeds. With the chick progressing well other members of staff on the exotic bird section were encouraged to have a go at feeding.
- Day 46: Brooder used only at night with temperature between 17.5°C–18.0°C. During the day the chick was in a cardboard box with shavings.
- Day 47: At feeding time the chick was exercising the wings whenever taken out of the brooder/box.
- Day 48: A reduction in the volume of formula was taken, with sometimes no feeding response.
- Day 49: Fecal sample and oral swab taken, both results returned clear.
- Day 51: When in the brooder, chick was starting to perch on the side of the container.
- Day 53: Sunflower oil was added at a ratio of 1 ml per 20 ml of formula. This was a result in a drop in body weight over the previous two days; a loss of 10 g each day.
- Day 63: Down to two feeds per day.
- Day 64: First night not in brooder. Majority of feathers developed. Only a little natal down remained around back of neck. Chick at peak.
- Day 66: Moved to parrot cage, low perches, corn, peas, sprouts – sunflower, mung bean, oats, etc. Acacia tree with seed offered.
- Day 67: Contact with Ryan Watson (Adelaide Zoo) over concern of weight loss. Poor feeding response for the previous three days.

    - Weight loss can be up to 25%.
    - Be concerned if weight gets down to 450grams, then force feed.
    - 10 days of poor feeding normal. Then may go a bit sulky and resume feeding again.
    - Recommended sprouted sunflower instead of dry, as problems can sometimes arise where hard seeds haven't been learned how to be cracked properly.
    - Peas, corn, corn on the cob.
    - Day 63 for peak weight and drop in feeding response about right. 665 g–670 g (23.4–23.6 oz.) good weight.

- Day 69: First sign today of chewing on solid food. Some minimal although definite signs of interest in a piece of corn on the cob.

- Day 73: First flight across from the feeding perch.
- Day 74: Three secondary coverts from L wing pulled for DNA sexing.
- Day 75: Moved into weaning cage inside aviary with other RTB cockatoos. This was during the day between feeds with chick returning to parrot cage inside brooder room for the night.
- Day 78: A noticeable regular pick-up in feeding response, although not a lot in the volume of formula taken.
- Day 79: Vocalising at p.m. feed to birds in aviary and vice versa.
- Day 82: DNA results showed the chick to be a hen.
- Day 83: Put a small piece of catheter tube onto syringe approx 4 cm (1.5 in.) long to assist with reducing a feeding mess created by a strong jogging/pumping feeding response.
- Day 84: Big reduction in feeding mess/spillage due to extension.
- Day 85: Tube slipped off syringe and now in crop!
- Day 89: Paraffin oil added to formula to help chick in passing the small piece of tube. (Feces were checked every morning for a couple weeks but the tube was never found. The chick may have passed the tube during the day whilst in the weaning cage in the aviary.)
- Day 103: Only one wisp of natal down on chick at the back of the neck.
- Day 121: Evidence of chick having had a go at eating peanuts (out of shell). These were not consumed but approx 10–12 broken up.
- Day 122: Chewed peanuts found early a.m. First true day at self-feeding.
- Day 123: Peanuts found well chewed. Plenty food chewed during the day.
- Day 129: First day into a small aviary approx. 5 x 1.5 x 2.2 m (17.6 x 5 x 7.3 ft.)
- Day 131: One feed per day, has been chewing on solid food now for a good week. Favors peanuts out of shell, sprouted – linseed, hulled oats plus other foods such as peas, corn kernels, etc. Whole corn on the cob chopped into segments. At regular intervals food was offered by hand to the chick, as well as two dishes full of food always being hung at perch level.
- Day 135: Remaining in outside aviary from now on.
- Day 137: Reluctant to feed, not fed. No more feeds from here on.
- Day 138: 16 g (0.5 oz.) weight increase without being fed.
- Day 139: Temporarily back into brooder room to dry after heavy rain. (Just sat out in the heavy rain getting absolutely soaked.)

Weight Gain Chart

Date	Day	Weight	Date	Day	Weight
11/6/01	1	21.72g	**7/7**	27	410g
12/6	2	25g	**9/7**	29	460g
13/6	3	30g	**12/7**	32	510g
15/6	5	39g	**15/7**	35	540g
16/6	6	48g	**19/7**	39	590g
17/6	7	55g	**22/7**	42	659g
18/6	8	65g	**28/7**	48	655g
20/6	10	80g	**30/7**	50	651g
21/6	11	105g	**4/8**	55	658g
23/6	13	120g	**5/8**	56	656g
24/6	14	140g	**10/8**	61	667g
25/6	15	150g	**11/8**	62	647g
26/6	16	170g	**13/8 2 feeds/day**		
27/6	17	200g	**13/8**	64	665g- peak today
28/6	18	225g	**21/8**	72	0830h–523g 1610h–518g
29/6	19	245g	**21/8**	72	0830h–523g 1610h–518g
30/6	20	265g	**14/9**	96	0840h–482g 1625h–495g
1/7/01	21	310g	**14/10**	126	0830h–486g 1615h–489g
3/7	23	310g	**14/10**	126	0827h–530g 1627h–538g
4/7	24	335g	**28/10**	140	1530h–531g
6/7	26	390g	**7/11**	150	1608h–553g

Ring sizes:

Galah, Goffin's, lesser, bare-eyed, and red-vented *C. haematuropygia*: 10 mm

Major Mitchell (Leadbeater's) and citron-crested: 11 mm

Umbrella and slender-billed *C. tenuirostris*: 12 mm

Palm *P. a. aterrimus*: 12 mm

Greater sulfur-crested (galerita, fitzroyi and triton) and Moluccan: 14 mm

Black, including Goliath palm *P. a. goliath*: 14 mm

Contributors:

C.C. Kingsley, Savannah Game Services, Box 14275, Sinoville, Gauteng, RSA.

J. & E. Daw, Eve Daw Aviaries, Everton, Kwazulu-Natal, RSA.

References and Suggested Reading:

Arndt, T. 1996. *Lexicon of Parrots*. Vol. 1. Arndt-Verlag, Bretten del Hoyo, J., A. Elliott & J. Sargatal (eds.). 1997. *Handbook of the Birds of the World*. Vol. 4. Sandgrouse to Cuckoos. Barcelona: Lynx Edicions.

Connors, N. 1998. Black cockatoos–Calyptorhynchus Species. Proceedings IV International Parrot Convention, Loro Parque, Tenerife, Spain.

Dingle, S. 1998. Breeding the blue-eyed cockatoo. *AFA Watchbird* 25 (5): 4–5

Frey, W.N. 1998. Male aggression in cockatoos. *AFA Watchbird* 25 (5): 45–46

Low, R. 1993. *Cockatoos in Aviculture*.

McGrath, J. 1998. My lifetime of experiences with galahs in both field and aviary. *AFA Watchbird* 25 (5): 18–26

Rodgers, D. 2001. Hand rearing a Red-tailed Black Cockatoo at Auckland Zoo. (unpublished)

Schubot, R.M., K.J. Clubb & S.L. Clubb. 1992. Psittacine aviculture: perspectives, techniques and research. ABRC, Loxahatchee, Florida.

Taylor, M. 1998. American Zoo and Aquarium Association's Palm Cockatoo SSP Husbandry Manual.

McWatters, A. 1988. *Australian Aviculture* 52 (11):

Wilson, L. 1999. Your questions answered. *Parrots*. Issue 28, June/July.

Vriends, M.M. 1997. Five Indonesian cockatoos: their biology and captive management. *AFA Watchbird* 24 (2): 28–33.

14.01 Psittacidae Kea and Kaka

Glen Holland

Introduction: The kea *Nestor notabilis* and the kaka *Nestor meridionalis* belong to the order Psittaciformes and the family Psittacidae. They were formerly assigned to their own subfamily, the Nestorinae, which is not recognized under the S-A-M system of classification. Both species are endemic to New Zealand. The kea is the world's only alpine parrot. Kea are found only in the South Island of New Zealand where they occur in mountainous areas. Although associated with the bushline, they are frequently found much higher, and occasionally venture into lowland forests. Kaka are smaller than kea, and are found on both the North and South Islands, inhabiting forested areas, as opposed to the alpine habitat of the kea. The South Island birds are bigger and brighter and at the time of writing there are only six birds in captivity. The captive program for North Island kaka includes about 60 birds and provides birds for three mainland release sites, with reintroduction programs looking very promising to date. The kaka was once among the most common species in New Zealand, but the population has diminished substantially due to introduced predators. Now generally found in mainly male groups of 3–6 birds the females having been killed on the nest by introduced predators larger flocks were once common. Today their main strongholds are on offshore, predator-free islands.

Management: The kea and kaka are large, active, inquisitive and, particularly in the case of the kea, destructive birds in an aviary. An aviary for a pair should measure a minimum of 6 x 3 x 2.5 m (20 x 10 x 8 ft.). It is essential that the aviary be constructed of heavy-duty materials capable of withstanding intensive and prolonged biting, digging, and pulling; steel pipe framing covered with chain link (18 mm sq. mesh x 2.5 mm gauge) is recommended. Smaller gauges are not sufficiently robust to contain kea, whereas larger mesh sizes permit the entry of rodents and wild birds, substantially increasing the risk of predation and/or disease. Because of their tendency to chew almost any object within reach, kea are highly susceptible to heavy metal poisoning from galvanized wire products, old lead-based paints, and various objects within the aviary environment. To prevent birds from digging out and rodents and other mammals from digging in, the aviary should have a concrete foundation extending well below ground level, preferably to 60 cm (24 in.). Their aviary should include sufficient shelter for the birds to be able to escape severe weather, but indoor accommodation is only required in climates where temperatures remain below freezing for extended periods, but even under these conditions additional heating is not required.

To enhance the bird's captive environment and minimize the risk of developing foot problems, a variety of substrates should be provided. A combination of large rocks arranged in stable piles and a range of natural logs offering perches of varying diameters and textures is ideal. Rotted logs should also be supplied for the birds to tear apart. These and the log perches will need to be replaced as soon as they are destroyed. The inclusion of "toys" in a kea aviary is essential for the bird's mental well being. See the section on PSITTACINE ENRICHMENT. Plastic ball cocks, lengths of chain, old leather boots, and similar items will keep kea amused for hours. Avoid objects containing substances that are potentially toxic, such as some paints, metals, glues, etc., as well as those that could cause a gastro-intestinal obstruction if swallowed; for example, the laces should be removed from the old boots. Because of their ability to destroy things, chewable play objects will need to be replaced at frequent intervals. If given the opportunity, kea will bathe frequently and eat most of their food in water. In addition to a basic source of drinking water, the aviary should ideally contain a pond that is suitable for bathing but easy to keep clean.

At Paradise Park, over the years the kea have been maintained in different enclosures, with varying results. The experiences at Paradise Park suggest that, within reason, the size of the aviary is less critical than insuring that it contains a variety of items to stimulate the kea, which are playful and inquisitive birds, and facilitate breeding. A typical aviary for kea at Paradise Park measures 4.8 x 1.8 x 3.4 m (16 x 6 x 11.3 ft) high. Outside the breeding season all kea are maintained together in order to give them the opportunity to socialise, and to exercise in and explore a larger enclosure, which measures 45 x 7.2 x 4.5 m (150 x 24 x 15 ft.) high, with up to 60 other psittacines at any one time.

All enclosures are constructed using 2.5 cm^2 (1 sq. in.) weldmesh. The breeding aviaries have a metal frame while the larger enclosure has a timber frame, which is covered with a secondary strip of 2.5 cm^2 (1 sq. in.) weldmesh to protect it against damage. The breeding aviaries have a sand substrate and are simply furnished with natural branches and natural fibre rope. Leafy branches, usually ilex oak *Quercus ilex,* are provided regularly and the birds tend to destroy these either during play or for constructing and lining their nest. The communal aviary has an undulating floor comprising 90% grass, 5% sand and 5% rocks, and this enclosure is more elaborately furnished with a profusion of logs, branches and ropes. Most enclosures are entered daily to present food and to clean up spilled food, and once per week for a thorough cleaning of the floor area.

Kea are intelligent, playful, and social animals requiring considerable mental stimulation in the form of play objects and companions. For this reason, kea must be held in groups of no less than two individuals. Kaka can be housed and bred in colonies and at Auckland Zoo up to eight kaka have been held in a single aviary with a number of the females laying eggs and producing chicks. In the long term, however, if production is a priority, individual pairs will be more productive. Kaka males have a longer upper mandible than females, but immature birds are best DNA sexed. Young birds up to a year old show a yellow ring around the eye and pointy wisps on the tips of their tail feathers. Juvenile kea can be distinguished from adults by the yellow-orange color around the cere, which they retain until approximately two years of age. Although males generally have larger bodies and bills than females, size is not a reliable method of determining sex and the birds should be either surgically or DNA sexed.

Diet: Kea are opportunists that are generally willing to eat a wide variety of foods. Consequently, the nutrient requirements of captive kea can potentially be supplied through a range of different food combinations. It is therefore not surprising that experienced kea breeders hold varying opinions on the optimum diet for this species. Some breeders base their diet on specially formulated parrot feeds (pellets or crumbles) designed to supply all of the bird's nutrient requirements. Others claim better results from feeding a more traditional diet based on a wide range of natural ingredients. While this approach is regarded as less scientific, it is likely that most animals offered a variety of foods will use their "nutritional wisdom" to select foods that provide the nutrients they need at the time. Cuttlefish bone should be available at all times.

The following foods are commonly fed to captive kea:

- Parrot pellets or crumbles.
- Sunflower, peanuts, and walnuts, fed in limited amounts.
- Sprouted seeds, cooked beans and corn
- Cheese, 25 g (0.9 oz.) per bird/day, increased to 75 g (2.6 oz.) when rearing chicks.
- Fresh greens such as silverbeet (spinach), thistle, chickweed, etc.
- Fruit and vegetables, including oranges, apples, grapes, carrots, sweet potatoes, etc.

Historically, kea have often been given bones to chew on, but although there is no proven need to add these to the diet, it is interesting how many psittacine breeders are now feeding such items. Feeding excessive quantities of high protein and fatty foods such as sunflower, nuts, etc. can lead to obesity and gout.

Birds feeding chicks can be offered a nectar mix made as follows:

- 2 cups rolled oats
- 1 tablespoon Complan (human protein/dietary supplement)
- 2 cups cold water

Simmer slowly until thickened and add brown sugar to sweeten to taste. When cool, add one slice of wholemeal bread and a multivitamin at the recommended dosage to each litter of mix. Offer 400 g (14 oz.) of this mixture per pair per day. Surplus mixture may be stored in a refrigerator for up to 48 hours.

At Paradise Park every day the kea are offered a small quantity of seed mixture, 100–150 g (3.5–5.3 oz.) per bird, together with a larger quantity of whole or chopped vegetables and fruit.

One of the following seven items is also added to this standard diet on a rotational basis:

- Chopped cheese or grated boiled egg.
- Millet sprays and chunks of wholemeal bread soaked in nectar.
- Mixed nuts (the larger ones are partially cracked).
- Corn on the cob and sultanas soaked in a vitamin solution.
- Cooked chicken wings.
- Boiled rice.
- Peanut butter on wholemeal bread, and peas and sweetcorn.

In addition, each pair is offered a small quantity of nectar three times per week just before and during the breeding season. Calcium supplementation is also provided during the egg-laying period. Vitamin supplements are added to the food on a daily basis: SA-37 for three days, Psittamix for two days and Avimix for two days. Abidec vitamin supplement is added to the drinking water once per week. The food is offered in stainless-steel bowls once per day and each bird has its own feeding station. Although the birds usually feed together, there are periods when the female will prevent the males from reaching the food, and if this occurs individual water stations are also provided. Kea bathe even in the coldest weather and, therefore, a suitably large water container should be provided.

Kaka receive a diet similar to that described for kea. Unlike kea, kaka have not been recorded eating meat and are therefore not offered bones. Their nectar can consist of commercial lorikeet nectar or jam water (one cup per liter of water) with some pollen added. In nature, kaka are highly insectivorous and spend much of their day searching for grubs in rotting wood. Some birds may take insects such as mealworms and waxmoth larvae in captivity. Both species spend a considerable time digging and are therefore susceptible to internal parasites such as gapeworm. Regular fecal checks are advised.

Kea
Nestor notabilis

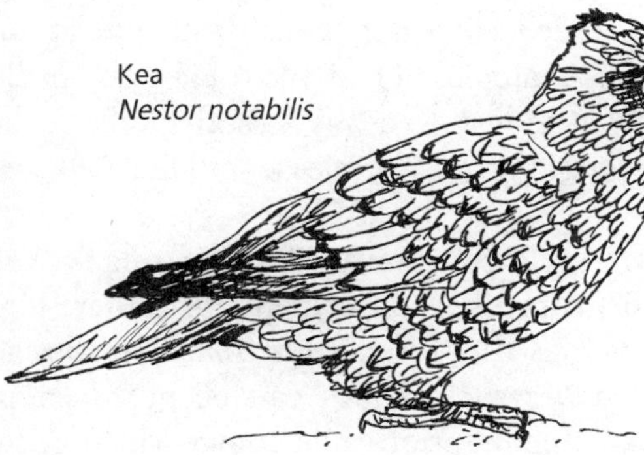

Breeding: In the wild, kea do not usually breed until five or six years of age. In captivity, however, some three-year-old birds have bred and raised young successfully. The breeding cycle, both in the wild and in captivity, commences in mid-winter when the female prepares the nest site with twigs, leaves, etc. Females in the southern hemisphere typically begin laying in early July. Up to six eggs may be laid, but the typical clutch size is 3–4 eggs, which the female alone incubates for 25 days. The male feeds his mate on the nest and later on in the rearing process he also feeds the chicks. The chicks spend up to 12 weeks or more in the nest. Once the chicks have left the nest, both parents feed them for a further three months. Because of the long period associated with rearing chicks (approximately four months from start of incubation to chicks fledging) it is uncommon for kea to rear more than one brood in a season. However, if the eggs fail to hatch or are damaged, or if the chicks die or are removed, pairs will generally re-nest almost immediately. For example, when three seven-day-old chicks were removed as part of a hand rearing experiment, the female laid a second clutch within two weeks.

Wild kea usually nest in rock cavities. In captivity, most pairs breed on the ground rather than using the traditional parrot nest log or box. A semi-natural nest site can be constructed by placing a plywood box of approximately 1 m^2 (3 sq. ft.) at ground level and lining it with rocks, inside and out. A door is cut in one wall of the plywood box to allow for nest cleaning and observations. The birds gain access to the nest area through a 20 cm (8 in.) diameter concrete pipe approximately 1.5 m (5 ft.) long.

At the time of writing, Paradise Park maintains three kea pairs aged 8–21+ years old. The internal measurements of the nestbox, which is at ground level and built from concrete blocks, are about 30 x 30 x 52.5 cm (1 x 1 x 20.7 in.) high, although successful breeding has also occurred in nest boxes measuring 40 x 75 x 55 cm (15.7 x 29.5 x 21.7 in.) high. The nest boxes have a wood-shaving substrate to a depth of about 3.75 cm (1.5 in.). The most important element of nest box design is the access tunnel, which should allow the female to restrict access by the male. Access tunnels that are 17.5 cm wide, 17.5 cm high and up to 90 cm long (7 in. x 7 in. x 3 ft.) fulfil this requirement, although in practice a tunnel, which is 37.5 cm (14.8 in.) long, is usually adequate. To facilitate nest inspection the concrete nest boxes are built outside, and to the back of, the enclosure, with the tunnel entrance in line with the rear of the enclosure. One female did not lay eggs in a concrete nestbox but began egg laying within days of being provided with a wooden nestbox measuring 55 x 47.5 x 40 cm high (21.7 x 18.7 x 15.7 ft.) with a wooden access tunnel.

At Paradise Park pairs are separated from the group during the breeding season, although when a single male was maintained with two females, one of the two females laid eggs. In hindsight, this could be attributed to the young age of second female (four years). Polygamy has been observed in wild kea and the future development of a large enclosure, containing perhaps three or four females of breeding age, may facilitate reproduction. Since 1983, 106 chicks have hatched and 62 young have been raised to independence, although 78 of the chicks were the offspring of one pair. The large number of chicks is a direct result of egg removal for artificial incubation which was initiated in May 1983 when one female lost her entire upper mandible while rearing her first two chicks. It has never been fully established how this damage occurred to her beak. One chick was removed immediately for hand rearing and the other was left with the parents to determine how they would cope under these circumstances. For about four weeks the male fed both the female and the chick, which was successfully parent-reared to independence. The female was observed feeding the chick after it had left the nest. The second chick was successfully hand-reared to independence.

Although the damaged beak of the female has not fully regrown, there has been a partial hardening of the area and she has no difficulty eating. The same pair successfully raised three chicks in 1984 but failed to rear chicks in 1985, when four eggs were abandoned in the nest after a disturbance. In 1986, four chicks died at 10 days of age after ants invaded the nest, and in 1987, one chick was hand reared but died from an *Escherichia coli* infection and three chicks died in the nest at 6–7 days of age. Since 1988 eggs have been removed from this pair for artificial incubation and, as a result of re-clutching, a large number of eggs have been laid. Through a process of trial and error with both successes and failures, appropriate techniques for artificially incubating eggs and hand rearing kea chicks have been developed. With only a few exception, all kea that have been reared to independence have been tagged for identification purposes using 11 mm stainless-steel leg rings (size V) and a microchip. In addition, Swiss Institute-style sexing rings have been fitted after sexing by laparoscopy.

Kea usually establish pairs in the breeding aviaries around mid-November and the first eggs of the season are invariably laid on or around January 1. A clutch commonly contains four eggs. At Paradise Park, kea lay their eggs at 2–3 day internals and incubation begins once the third egg is laid. The next cycle of egg-laying usually occurs within 24–38 days. If the eggs are to be removed for artificial incubation it is preferable to leave them under the hen for the first 10 days of incubation. However, if the female does not restrict access of the male to the nest he may destroy the eggs and it is often necessary to remove the eggs as soon as they are laid: by doing this, one female laid 22 eggs in the space of 12 weeks and only three eggs were damaged by the male.

Progeny should be removed from the breeding aviary well before the next breeding season. Conflict and injuries are likely to occur when juveniles interfere with the nest site and attempt to interact with the breeding pair. The adult male can become particularly aggressive toward his male progeny. Young birds can be transferred to a colony situation where they learn to socialize with a larger group of birds. Birds introduced into male-female groups will eventually select their own mates. However, natural pair selection will not always be appropriate in a controlled breeding situation because the birds do not necessarily choose mates that will result in maximum retention of genetic diversity within the captive flock.

Kaka are perhaps more standard in their breeding requirements, nesting like most large psittacines in wooden or metal boxes with wood shavings or wood chips in the bottom. Many pairs produce two clutches per year. Up to six eggs have been recorded in a clutch, with one female, which had her first clutches removed, recorded as laying three clutches totalling 16 eggs in one season, of which 12 proved fertile and 10 hatched. Her second and third clutches each consisted of six eggs, of which four and all six hatched, respectively. The incubation period is 30 days, and fledging takes 60–65 days. Previously recorded as being sexually mature at four years, numerous examples of two-year-old birds breeding have now been recorded. Recently a captive-bred female bird of one year old bred after release into the wild.

Hand Rearing: Incubation and hand rearing for kaka differs little from that described for other large psittacines, such as Amazon parrots *Amazona* spp. and cockatoos *Cacatua* spp. Chicks can be heard calling from the egg 72 hours prior to hatching, which occurs within 24 hours of the first external pip. Chicks are left in the incubator for the first four hours to dry, and then moved to a brooder. I have hand-raised kaka from the egg and found them to differ little from other large psittacines but for the fact that the chicks are rather fussy and often reluctant to take a full crop. See the growth chart for the amount of food consumed per feed.

The eggs are maintained in forced-air incubators at 37.2°C (99°F) with an initial humidity of 50%, which is altered, as necessary, in accordance with the loss of mass of the egg. The incubation period is usually 23–26 days and the mean mass at hatch is 18.0 g (0.6 oz.) range 15.0–20.53 g (0.5–0.7 oz.). Kea chicks grow quickly and only require a high temperature of 30–35°C (86–95°F) in a heated brooder for a short period of time. The chicks hatch with a white down but quickly develop a dark dense down. This insulating layer is undoubtedly an asset in the high-altitude wild habitat but, in the warmth of a typical parrot rearing unit, if the temperature is too high at this stage of development, about 10–12 days onwards the chick may die. When the chicks reach about 25–30 days of age they are transferred to a box measuring 120 x 60 x 60 cm high (4 x 2 x 2 ft.) which has a dull emitter infra-red lamp suspended over one end, allowing the highly mobile chicks to select the part of the box that best meets their temperature requirements.

Hand Rearing Diet

The diet fed to kea chicks at Paradise Park has been developed over a number of years by John Heath, a private aviculturist with a special interest in hand rearing. The high levels of animal protein or fat that are present in a typical hand rearing diet, which probably result in high levels of *Clostridium* bacteria, are not suitable for kea chicks. The diet developed by John Heath takes this into account and is comprised of 250 g (8.8 oz.) Nectarblend, 250 g (4.2 oz.) Budgie Protein food, 150 g (5.3 oz.) fruit salad (infant dessert), 120g Vegetable hotpot (infant dinner), one level teaspoon Ace-High, one level teaspoon Nutrobal and three teaspoons Avipro Paediatric. It is also possible to use EMP or CEDE eggfood, or a similar parakeet rearing food with, or instead of, the Nectarblend and/or budgie protein food because they all contain about 16% protein. The vegetable hotpot (infant dinner) has a suitable protein content for this hand rearing diet, but if other products are used in its place the protein should comprise at least 18% of the dry mix (4.7% wet mix) for most parrot chicks, although this may be less important for the kea. Whatever mix is provided, it is important that vegetable rather than animal protein is used.

To prepare the hand rearing diet the Nectarblend, budgie protein food, fruit salad and vegetable hotpot are sieved and the residual coarse material is ground to a powder in a coffee grinder. The powder is sieved again and any coarse material is reground twice. Any coarse material that remains is discarded. The powder is and the Ace-high, Nutrobal and Avipro Paediatric are mixed thoroughly and stored in a sealed polythene container at 4°C (39.2°C). The feed is prepared by mixing about two parts powder with four parts warm water at 40°C (104°F). The mixture was fed until the crop was 2/3–3/4 full. For the first few feeds after hatching the mixture must be extremely watery, but after the first two days it can be slightly thicker. Any unused food should be discarded. When the chicks are about 15–20 days old the initial mixture can be supplemented with 250 g (8.8 oz.) sesame seed and 62.5 g (2.2 oz.) hulled pumpkin kernels and this mix may be offered until the chicks are weaned.

Since 1996 Prettybird hand rearing formula has been fed to the chicks at seven days of age with no detrimental effects, However, attempts to feed this diet from hatching have resulted in problems with crop clearing and the subsequent death of chicks. Adult kea have been observed taking food into the nestbox well before the chicks are ready to fledge and, therefore, hand rearing chicks are offered seeds and fruit at 6–7 weeks of age. Weaning is comparatively straightforward and can usually be accomplished in 14–21 days if chicks have access to seeds and fruit. If they do not, weaning can take up to six weeks. Maintaining the chick with an independent kea that can act as a "teacher" has also encouraged weaning.

Kea Development

When fully feathered, at around nine weeks, young birds can be distinguished from the adults by their yellow, rather than black cere. The birds can usually be sexed at this stage of development by the size and shape of the beak: the upper mandible of the male is 12–14% longer than that of the female. The sex of a bird is normally confirmed by laparoscopy or DNA. As soon as the young are weaned they are placed in a large outdoor enclosure containing plenty of furnishings. Rope, however, is not provided because the kea tend to chew the fibers, which then become entangled around their toes and legs. For up to six months after weaning, the young exhibit a begging response when they see a familiar person. There are reports of kea in the wild spending a considerable amount of time with their parents after fledging and the begging response observed at Paradise Park may be related to this natural behavior pattern. At the end of the season the young birds join the adults in the communal aviary.

Kea *Nestor notabilis* & Kaka *Nestor meridionalis* (includes food consumption for kaka)

Day	Weight (g)			Day	Weight (g)		
	Kea	Kaka	food		Kea	Kaka	food
1	17.2	15	10.5	**19**	234	253	172
2	19.3	16	11	**20**	274	260	160
3	22.5	20	21	**21**	314	269	173
4	27.2	25	25	**22**	394	287	125
5	31.4	29	29	**23**	440	313	194
6	36.2	38	31	**24**	452	335	178
7	45.8	44	56	**25**	496	347	185
8	48.2	56	25	**26**	554	350	175
9	56	71	60	**27**	576	372	190
10	66.7	80	74	**28**	606	394	215
11	77.9	101	157	**29**	-	410	129
12	93.4	120	178	**30**	-	407	129
13	106.7	148	157	**31**	-	411	124
14	123.2	152	241	**32**	-	436	173
15	145.4	174	135	**33**	-	430	106
16	182	199	130	**34**	-	440	131
17	182	217	150	**35**	-	451	139
18	206	227	105	**36**	-	459	138

In most cases it has not been possible to establish the age at which kea first lay eggs because the birth date is known for only two of the females, the female which laid at 46 months proved infertile while the other laid four fertile eggs at 98 months. Without further data, however, the reasons for these differences can only be speculative. Recently two female kaka of a year old, paired to older males, have successfully bred.

Kaka *Nestor meridionalis*

Age in Days	Feeds/day
1	10
2–4	9
5–7	8
8–12	6
13–26	5
27–40	4
41–50	3
51–85	2
85–100	1

Ring Size: Kea: 12 mm. Kaka: 10 mm.

Contributor: Tony Pullar, National species co-ordinator for kea. Dunedin Botanic Gardens, New Zealand.

References:

Pullar, T. A captive management plan and husbandry manual for kea.

Folwell, D. 1989. Guidelines for breeding Kaka in captivity. Unpublished paper.

Woolcock, D. 2000. Husbandry and management of Kea (Nestor nobilis) at Paradise Park, Hayle. *International Zoo Yearbook* 37: 146–152.

14.01 Psittacidae Fig Parrots

Jos Hubers, Alison Beutel and David Carr

Introduction: The five species of fig parrots belong to the order Psittaciformes and the family Psittacidae. The two species of the genus *Cyclopsitta* (Opopsitta) and the three in the genus *Psittaculirostris* are found primarily in New Guinea and Australia. Very little is known about the behavior of fig parrots in the wild. Most of what is known relates to the subspecies of the double-eyed fig parrot *Cyclopsitta diopthalma* from Australia. In general, fig parrots are observed in pairs or small flights, the members constantly in contact with each other by a kind of twittering. However, during feeding, which they do most of the day, they are very quiet and can hardly be heard. On rare occasions, very large flights can be seen. Flocks of the Edwards's fig parrot *Psittaculirostris edwardsii* of northeastern New Guinea, probably the most common species, can number up to 400 birds (Beehler 1986). The mating displays of fig parrots are beautiful to watch. The birds make splendid slow motion movements while rubbing their necks against each other's bodies.

Management: A number of significant factors should be taken into account before a decision is reached to keep fig parrots. One of the most important is feeding, but I will discuss this aspect later. Potential buyers usually have the idea, from seeing fig parrots, that these birds can easily be kept in small cages. Keeping them in this manner should never be considered. They may appear to be quiet birds, but in fact they are very lively, once they have adequate space. They can easily become obese through lack of sufficient exercise, particularly if this continues over a period of time, or in combination with an improper diet. Another disadvantage of keeping them in facilities that are too small is that the outer sides of the primary feathers, as well as the tail feathers, wear rapidly from rubbing against the sides of the cage. To avert this problem, an aviary, rather than a typical parrot breeding cage, is the best solution. The length of the aviary is not as important as the width. If a number of climbing perches are placed at different heights, fig parrots will fanatically jump from one perch to another.

Red-browed fig parrots *Cyclopsitta diophthalma macleayana* have been successfully housed and bred in Australia in several different types of enclosures. These include outside flight aviaries measuring approximately 2 m long x 0.9 m wide x 2 m high (6.5 x 3 x 6.5 ft.), as well as heavily planted outside aviaries measuring 5 m long x 2 m wide x 2 m high (16 x 6.5 x 6.5 ft.), approximately one-third of which is sheltered. They have also been bred in inside-suspended cages measuring 193.75 cm long x 87.5 cm wide x 118.5 cm high (76 x 34 x 46 in.) with solid back walls and a double barrier between adjoining cages.

Fig parrots should not be housed in spacious aviaries that are not heavily planted with trees, as these birds can sustain serious injuries from collisions with the wire. These robust little parrots build up considerable momentum in flight and appear to be unable to turn quickly enough to avoid crashing into the aviary wire. Heavy planting will break up this momentum.

Fig parrots really enjoy chewing. In a planted aviary, Habitat A, as described in LANDSCAPE, will suit these birds best. An appropriate captive environment is important, because with all species of fig parrots, the upper mandible often grows too long. If this happens, it is recommended that the upper mandible be allowed to grow to at least 8 mm (0.3 in.) in length, because it is then very easy to correct the problem by hand. I have done this with many fig parrots for several breeders. Just take the part of the mandible that has grown too long, and bend this down 90 degrees, using thumb and forefinger. This does look strange, but the tip that is too long breaks exactly at the right place, and usually in the natural V-shape.

Like any other birds, newly purchased fig parrots require extra attention. This especially applies to imported birds. Fig parrots have a reputation for being difficult birds to acclimatize, therefore, several basic rules should be followed. When one acquires these birds, one should find out what food they were being fed, as well as the temperature at which they were being kept. Any necessary changes in diet or temperature should be made slowly. At least part of the diet used by the previous owner should be continued for some time as the new diet is being introduced. Gradual temperature changes are equally important. Upon observation, fig parrots usually look perfectly feathered. However, when unaware that they are being observed, birds that are not healthy will soon fluff up their feathers. Increasing the temperature will do wonders for such birds. Fig parrots only have a thin layer of feathers and can not withstand low temperatures for a very long period of time, especially when they have become accustomed to constantly high temperatures. We always prefer to lower these temperatures step by step, and never to less than 15°C (59°F).

Overhead mist sprinklers are attached to all outside aviaries for watering plants and cooling the birds in hot weather. Fig parrots love to bathe, so it is recommended that they be supplied with fresh bath water daily. The bath dish should not be placed too close to the feeding bowls, as there is a chance that some food will end up in the bath water, since they usually make a mess while feeding. However, we do not have to make the kinds of adjustments to the aviary that we find necessary for lories, as fig parrot droppings are more solid. I have not noticed that fig parrots are particularly susceptible to certain diseases. However, the area around the feeding tray must be cleaned regularly to prevent mold and bacterial growth.

The best way to breed fig parrots in captivity is to keep them in pairs, although there are reports that they are reasonably tolerant when maintained in a colony. However, if the colony system is used, the aviary must be sufficiently large to accommodate the occupants without bickering over food or nest boxes, as with any other birds that are colony bred. Adjoining aviaries should be separated by double wire as these birds are prone to chewing neighbors' toes, especially during breeding periods. At Currumbin Sanctuary, fig parrots have been successfully housed with various ground-dwelling birds such as noisy pittas *Pitta versicolor* and species of quail. However, Perth Zoo has had pairs cause injuries to black-breasted buttonquail *Turnix melanogaster,* parrotfinches *Erythrura* sp., and pekin robins *Leiothrix lutea.* Single birds seem to be less aggressive than pairs.

Except for Desmarest's *Psittaculirostris desmarestii,* all species of fig parrots are sexually dimorphic. In addition to differences in plumage coloration, females quite often have flatter and rougher heads than males, a consideration that is particularly important when trying to determine the sex of the otherwise similar Demarest's.

Diet: Fig parrots did not get their name for no reason.

In the wild, figs constitute the main portion of their diet. They are methodical feeders, returning to the same fruit until its seed supply is exhausted. They have also been observed feeding on exposed insect larvae. However, other fruit, pollen, and nectar, as well as insects, are also consumed in lesser quantities, so it is important to give them a good replacement diet. A 100% commercial diet ready for use for fig parrots does not exist at this time. However, it is possible to choose for a suitable combination. The basic diet consists of all kinds of fruit. Often it is not the fruit that makes a diet successful, but the additives. Most of the fruit used for feeding, for example, has a relatively low to extremely low protein content, and the diet must be adjusted to compensate for this fact. Apple can be used as a dietary basis, as fig parrots always like to take it. When cut into small pieces, apple can easily be mixed with other ingredients. Figs can be given, in dried form, but are not necessary for successful breeding. Besides fruit, soaked and sprouted seeds can also be given; normal dry seeds should only be offered in a limited quantity, and not when there are young in the nest.

Another important fact is that fig parrots consume a significant amount of vitamin K1 through fresh figs. Vitamin K1 is an important factor in the congealing process. In the case of fig parrots, internal bleeding, even on a relatively small scale, is difficult to stop if there is a deficiency of vitamin K. It is always a good idea to add a vitamin K supplement (Konakion) through the food. A veterinarian can be consulted to determine the required dosage.

I would now like to outline the diet I use with very good breeding results with both the *Psittaculirostris* and the *Cyclopsitta* species. When there are no young being fed, the adults get a mixture of small seed, such as that fed to grass parrots, along with a small quantity of sunflower seeds, no more than a handful. Sprouted sunflower seeds are better, but so far I have not had any problems with dry sunflower seeds. I remove all seeds as soon as the young hatch. Further, I give them Aves lory nectar, but this could just as well have been any other brand, or a home-made mix. I provide as much of this as they can eat. The *Psittaculirostris* species take more in comparison to the *Cyclopsitta* species.

I also cut apples into small pieces, with which I mix some fruit syrup. In this way the apple becomes sticky, after which I add Aves dry powder fruit mix into it, in a ratio of 1:10, powder to fruit. This powder has been developed specifically to mix with fruit for fruit-eating birds so they will still consume the ingredients that may be missing from fruit alone. As mentioned earlier, fruit in the shop often has a low protein content. Therefore, I use the apple as bait for the birds to eat from the fruit mix. Some other fruit can be mixed in, or fruit salad can be used. It is not necessary to add this to be a successful breeder, but it is very pleasant for the birds to have some variation in the diet. As not everybody has a fruit mix at his disposal, one can also try substituting egg food. The addition of vitamin K1 is best made in the lory nectar.

The daily diet per bird used at the Currumbin Sanctuary is as follows:

- 1 Turkish fig, covered in water and microwaved on high until boiled for 1 minute
- 1 tablespoon Avione dry lorikeet mix
- ½–1 tablespoon finch seed
- A variety of fruit pieces, such as apple, pawpaw, passion fruit, cherry tomatoes, banana, pear, grapes, rock melon, corn on the cob, and guava
- 1 tablespoon greens mix, consisting of finely chopped silver beet, lettuce, broccoli, peas, corn, and grated carrot
- 10 mealworms

During the breeding season, the proportions are increased. Native figs and mealworms are added to the diet after chicks have hatched. These items are fed liberally, the amount increasing as the chicks grow. It has also been thought important to ensure an adequate supply of vitamin K, calcium, and zinc; this can be accomplished by using Nekton Q, a supplement specifically designed to meet the requirements of fig parrots. When chicks are in the nest, the feeding frequency is also increased to compensate for the demands of the chicks on the parents. At this time, fresh food is offered up to three times daily. Calcium is added to the diet at the onset of the breeding season by fortifying the figs with Calcivet at a rate of 0.1 ml per bird per day. The fig species include Port Jackson fig *Ficus rubiginosa,* weeping fig *F. benjamina,* Moreton Bay fig *F. macrophylla,* small-leaved fig *F. obliqua,* and sandpaper fig *F. coronata.*

An equally successful alternative diet used by Vogelpark Walsrode in Germany for Psittaculirostris fig parrots is given below:

- 3 tablespoons cooked rice
- 1 medium-sized potato, cooked and cut into pieces
- 1 large apple, peeled, cored, and steamed
- 1 steamed carrot
- 1 banana

- 2 tablespoons Humana Spezial, a special powdered diet for babies, which is free of lactose, fructose, and refined sugar
- 2 tablespoons soft oat flakes
- ¼ teaspoon Korvimin, a powdered vitamin/mineral supplement for birds
- 10 ml honey
- ½ teaspoon olive oil or wheatgerm oil
- 1/8 lettuce or Chinese cabbage
- Approximately 100 ml water

All of the above ingredients are mixed in a blender until they reach a paste-like consistency. This amount is sufficient for approximately 10 birds. It is fed twice daily in a bowl. In addition, one tablespoon of sprouted seeds and pulses and half a dried fig, which has been soaked, are given once a day to each pair. The sprouted seed and pulse mix includes sunflower seed, mung beans, and a seed mix intended for pigeons. These are soaked in five liters of water, to which two Chinosol tablets have been added to prevent the growth of mold. After soaking approximately eight hours, the mixture is thoroughly rinsed with clean running water, then kept in a sieve and allowed to sprout for 24 hours. In addition, maize (corn) and rice that have been boiled for 10 minutes are offered.

Every other day, 5 ml of one of the following supplements is added to the food: powdered yeast, Muschelkalk (ground shells which provide calcium), powdered soya malt, or Nekton Tonic K (vitamin/mineral supplement for seed-eating birds). The seeds and pulses are removed from the diet two days before chicks are due to hatch. Only the paste is provided until the chicks are 14 days old. From Day 15 onward, 10 blanched mealworms per day are added. When the chicks fledge, the basic fig parrot diet of paste, seeds, pulses, and figs is offered and the mealworms are reduced.

David Carr reports his breeding diet consists of the following:

- One or two figs per bird (including young)
- 10 g mixed fruit
- 5g Golden Delicious apple or Nashi pear
- 25 mealworms
- 4 blueberries
- 4 g madeira cake
- 4 g Finch Soft Food (Passwell product)

"In addition, all seed became soaked seed and all food was removed at 8:00 p.m. to avoid spoiled food being available the following morning. A small dish of mealworms and Finch Soft Food was left so it was available at first light. Complete diet of fresh food was placed in aviary at 7:30 a.m. each morning. Only supplement given was two drops of Calcivet (Vetafarm) on each fig, in the drinking water and in the soaked seed at the prescribed dosage, two weeks prior to expected egg laying and until young fledged. Mixed grit, cuttle bone and a clay-based mineral block were all fed and eagerly taken, particularly when egg laying was in progress."

Breeding: The nest of wild red-browed fig parrots is in a hole excavated by the birds in a rotten tree trunk or a dead limb in a living tree. Active nests have been found from August to January. One nest was found 12 m (40 ft.) above ground in the decayed branch of a living tree. The hollow had been excavated by the birds. The entrance hole was 38 mm (1½ in.) in diameter. An access tunnel turned downward and extended to a depth of 23 cm (9 in.). The bottom of the hollow was lined with small wood chips, obviously from the excavation process (Forshaw 1992). No nest containing more than two chicks or eggs has been found in the wild, although three eggs per clutch is not uncommon in captivity. The female does most of the excavating, and after she is able to fit inside the hole she will spend each night and most of the day in the hollow. Courtship feeding is frequent, and copulation is accompanied by a shrill chattering from the male. Allofeeding is common. Incubation commences with the laying of the first egg and lasts 21–23 days.

Only the female broods, sitting very tightly, and is fed by the male during the early morning and again in the late afternoon. Males will feed the chicks, which fledge between 36–42 days after hatching and appear to become independent at 60–65 days of age. After 10 months, the first red feathers start to appear on the males and the adult plumage is acquired by 14 months after leaving the nest. It is assumed that once adult plumage has been acquired the birds are sexually mature.

If the birds are in good condition and the temperature is not too low, fig parrots can breed year around. The minimum recommended temperature for keeping fig parrots, and certainly the minimum at which they should be allowed to breed, is 15°C (59°F). The normal clutch consists of two eggs that hatch after about 23 days of incubation. On hatching, the young have only a bit of down. With a total of 12 clutches laid by David Carr's Macleays, nine consisted of three eggs and three of two eggs.

Red-browed chicks hatch with a sparse white down covering their bodies. The eyes begin to open around day six and dark feathers begin to show on the back. By day 12, pin feathers are present on the wings. Ringing is performed when the chicks are about 12 days old, using the band sizes recommended below. From day 21 onward, colors are very obvious and tail feathers begin to emerge. Most body feathers have lost their sheaths by this stage, and the chicks appear well-feathered. Fledging weight is approximately 35–40 grams. The nestling period for the *Psittaculisostris* species is about 60 days, but the *Cyclopsitta* species fledge after a shorter period. Red-browed chicks usually fledge between 40–45 days. They are still fed by the parents for a few weeks after fledging. I have sometimes left the young with the parents until the next brood of chicks has hatched. Some fig parrots have been known to lay up to three clutches in one breeding season, usually going back to nest soon after chicks have fledged. When the young fledge, they look exactly like the adult female; after just over a year, the young males are fully colored. The young should be monitored closely after being removed from the parents to ensure that they do not stop eating.

Either horizontal or vertical nest boxes can be used for fig parrots. The advantage of horizontal nest boxes is that the young put their droppings in a corner, ensuring that they themselves remain on a dry nest filling for much longer. I would advise a nest box with a reasonably big surface, for instance 40 cm long, 18 cm deep and 18 cm high (16 x 7 x 7 in.) for the larger species, and 30 cm, 15 cm, and 15 cm (12 x 6 x 6 in.), respectively, for the smaller species.

Currumbin Sanctuary has used a variety of nesting boxes with success. Pairs have also bred in *Pandanus* logs into which they have excavated their own hollows. The drawbacks with this are the inaccessibility of the chicks for checking and weighing. The nesting boxes most commonly used are basic parrot breeding boxes similar to those used for budgerigars. Although it has often been suggested that fig parrots require the stimulation of digging out a nest cavity for themselves in order to breed, we have not found this to be the case.

At Perth Zoo the fig parrots used a natural palm log on a stand. When ants became a problem, the staff applied grease under the stand to deter them. Lory Lore Avian Breeding Centre provides fig parrots with three nest boxes, a lovebird nest box, a Gouldian finch nest box, or a cockatiel nest box, with about 2 cm of sawdust in the bottom. Most of their fig parrots have preferred the L-shaped boxes. The females remove most of the sawdust, leaving only a thin layer of material for the chicks to sit on.

Nest boxes should be disinfected prior to breeding and fresh nesting material supplied at the beginning of the breeding season. After the chicks have hatched, the nesting substrate should be changed on a regular basis to minimize bacterial growth and the build-up of excrement. Nest boxes should be thoroughly cleaned and disinfected, and the nesting material replaced, between clutches. Wood shavings or dry sphagnum moss are suitable nesting materials. Currumbin Sanctuary has in past years used heat sterilization (autoclave) on nesting material prior to use.

David Carr gives the following account of his successful breeding of the Macleay's fig parrot:

> "Although being housed as individual pairs, the birds seemed to be stimulated by close contact with other birds, my cages being only centimeters apart. The female without doubt was the dominant of the pair, however males were extremely persistent in approaches to their mates, often commencing with offering to preen the female and then enticing her to reciprocate. However if she didn't want to play, he knew it immediately.
>
> Birds greeted each other with much excitement, constant chattering and wing shuffling and, in particular, what I would describe as wing shivering. This occurred at any significant event, such as, introductions, courtship, egg laying, chicks hatching or fledging. My assumption is, in fact, that the event was not always the cause of celebration but rather the reunification of the pair after a period of absence. Copulation was only observed on one occasion. Courtship consisted of lots of chattering and perch hopping. The male then mounted, left leg on the perch and right leg on his mate. Actual cloacal contact lasted 130 seconds.
>
> The female carried out incubation in all occasions. I was initially unable to ascertain if the male assisted in rearing of the young, however the male did enter the nest regularly. The male was observed on numerous occasions with head in the nest entrance apparently feeding young, as was the female. I believe this to be the case, although once the young had fledged, they begged to the male regularly with limited success. A nest camera would answer some of these questions and more. For the 2000–2001 season, an infrared camera was purchased and fitted to a nest box. This delivered answers to many questions and all findings were confirmed in 2001–2002, when all four pairs were filmed.
>
> As suspected, the female, carried out all incubation and brooding. Males entering the nest at night, sitting immediately inside the box at the end of the 20cm (8 in.) entrance tunnel.
>
> Hatchlings were fed on their backs and tightly brooded for the first week. After this time brooding became irregular, confirming suspicions as to why young die in cold climates.

The relationship between male and female altered immediately upon the young fledging, to the point that some males had to be separated, aggression being observed from the female, directed to the male, usually with the purpose of keeping him away from the young. There was also some minor aggression from individual males directed at the young, although this generally appeared approximately three weeks after fledging, at which time you could separate the young. Young became independent approximately two weeks after fledging at which time they were separated from the parents.

Neither parent was seen brooding young during the day after approximately ten days, which probably explains deaths of young in outdoor enclosures. It is also interesting to note that one female who had a nest with a spout measuring 30 cm (11.8 in.) and an inside diameter of approximately 40 mm, appeared to be brooding young; however, night time inspections with a torch revealed that she in fact slept in the spout immediately outside the nesting chamber. This was confirmed when cameras were installed, confirming some females spend large amounts of time in the nest chamber but only brood occasionally."

David reported that out of a total of 33 eggs, 23 (69%) were fertile, with individual clutch variations from 100% (three out of three) down to 50% (one out of two). Out of a total of 23 fertile eggs, 18 chicks emerged, representing variations from 100%, down to 0%. Chick survival rate (hatch to independence) proved to be the most satisfying result with seven out of seven in the first year and nine out of 11 in the second year.

The average percentages of fertility, hatchability, and fledging rate were summarized for the period of 1988–2001 at Currumbin Sanctuary: 77.5% fertile eggs, 85% hatch rate, and 42% fledge rate. Chicks have been successfully reared by alternate fig parrot pairs. The captive red-browed are intended to be used as an analogue species for the eventual raising of another subspecies, Coxen's fig parrot *C. d. coxeni,* in conjunction with recovery plan guidelines.

In the early years of the breeding program, many chicks at Currumbin died with full crops. Post mortems often failed to reveal the cause of death as the chicks were too autolysed. Many chicks died as the result of bacterial infection. Some have also died of trauma inflicted either by the adults or possibly by rodents. Cases of rickets and parasites were also experienced, and there have been indications of pneumonia and genetic malformation. It is therefore imperative to ensure a high quality diet with adequate protein and calcium for the parent birds. Regular checks for parasites are also necessary to maintain a healthy population. Hygiene standards must be kept at a high level to reduce the possibility of bacterial infections. In recent years, swabbing each chick's umbilicus with a betadine solution via a cotton tip has proved effective as a preventative to bacterial infections. Chicks are weighed daily for the first 10 days and monitored closely for any signs of possible illness such as weight loss or reduced vigor. Thereafter, chicks are weighed every other day until they fledge.

Artificial Incubation and Hand Rearing: Fig parrot eggs are set in an incubator at a dry bulb temperature of 37.2–37.5°C (99–99.5°F) and a wet bulb temperature of 29.9–31.1°C (86–88°F) for a relative humidity of ~62%. The hatcher temperature is maintained at 36.6–36.9°C (98–98.5°F) with a wet bulb reading of 32.2–33.3°C (90–92°F) for a relative humidity of ~70–76%. Embryos begin to develop at a temperature of about 36°C (97°F) and continue to develop at temperatures up to about 38.8°C (102°F). The higher the temperature within this range, the faster the development. The optimum temperature will be the one at which the chick develops at exactly the same rate it would if incubated by the natural parents. Chicks that develop too quickly or too slowly tend to be weak at hatching time, if the embryos survive long enough to hatch at all. The humidity in the incubator is the controlling factor in determining how much weight or fluid an egg will lose during incubation. As the egg develops, fluids evaporate through the pores in the shell. The amount of fluid lost through evaporation is directly related to the humidity in the incubator (Jordan 1989). As indicated above, the relative humidity is increased at hatch time. Chicks will normally hatch within 24–36 hours of pipping. If they have not hatched by 48 hours there is an obvious problem and assistance may be required.

Newly hatched chicks are kept at hatching temperatures for at least 2–6 hours after hatching. They are then placed in a brooder set at 36.35°C (97.5°F) for the first 4–5 days, after which the temperature should be reduced to 35°C (95°F) for the next 5–9 days. The temperature should then be further reduced to 33.85°C (93°F) and maintained at this level until the chicks have developed sufficient down feathers to no longer be considered naked babies.

For artificial rearing of fig parrot chicks either syringes or bent spoons may be used, but due to the tiny size of the chicks it is easier to use syringes. This method also allows the keeper to accurately monitor the amount of food each chick is consuming. The use of separate feeding utensils for each chick is also recommended. For the first six hours of life chicks should only be fed an electrolyte solution such as lectade. This gives the chick time to absorb the yolk sac while at the same time preventing dehydration. The electrolyte solution should be warmed to between 37.7–38.8°C (100–102°F). Newly hatched chicks should only be given about 0.1–0.2 ml per feed. They should be fed every two hours. When feeding solely liquid, special care must be taken not to aspirate the chick (Voren & Jordan 1992).

The first feed after the yolk sac has been absorbed should still be of a watery consistency, but should now consist of the substitute diet outlined below. It is extremely important to ensure the correct temperature of the feed and to monitor closely the emptying of the crop. The food should be warmed to a temperature of 37.7–38.8°C (100–102°F). If the food is too hot when fed it will cause painful crop burns; if the food is too cool, the chicks will refuse to eat. Chicks should never be fed when there is still food present in the crop from the previous feeding, as the old food will ferment and cause crop stasis and bacterial/fungal infections. During the first week, chicks should be fed every 2–3 hours, depending upon the rate at which the crop empties between feedings. As the chicks progress, the consistency of the food is thickened and the interval between feedings adjusted accordingly. Before changing the consistency of the food or the feeding interval, it is necessary to evaluate crop movement and chick development (Voren & Jordan 1992).

Fig parrot hand rearing diet (from Pauline Courtney):

- 4 tablespoons Lakes Hand rearing Formula
- 1 tablespoon pre-cooked tinned baby carrot
- 1 tablespoon banana custard baby food
- 1 tablespoon pureed apple baby food
- 2 tablespoons Nekton K powder
- 1 pinch of Enzyplex enzyme powder

All ingredients are blended to a fine mixture with mineral water. As chicks grow and the formula is fed at a thicker consistency, it may be easier to revert to the use of a bent teaspoon to administer the food. This also encourages the chick to start to feed for itself by picking at food, and facilitates the feeding of a much thicker formula during the last few weeks prior to weaning.

Fig parrot hand rearing diet (from Currumbin Sanctuary):

- Prettybird Hand rearing Formula (22% protein, 10% fat)
- Nekton Q, at a rate of 0.33 g per 10 g formula pureed fig

The formula is made using boiled, filtered water. The concentration of formula is increased, and the amount of pureed fig increased, as the chicks develop.

Red-browed
Cyclopsitta diophthalma macleayana

Day	Weight (g)	Day	Weight (g)
1	4.62	**28**	38.09
3	7.1	**30**	39.14
5	9.42	**32**	43.62
12	20.11	**34**	41.83
15	24.1	**36**	38.83
17	27.3	**38**	39.02
19	34.15	**41**	37.15
21	34.14	**45**	
25	39.96	**50**	

Chicks in brooders are placed in a small container padded with tissue paper with a small square of "gripper" pad placed on top of the tissue. The gripper provides a firm substrate that the chicks can grip, helping to keep the toes open and prevent clenched feet. Chicks are usually kept as clutches (3) and therefore use each other to prop themselves upright. If there is only one chick, sufficient padding must be provided for the chick to lean against to stay upright. Chicks are normally kept in the brooder box until fully feathered. They can then be transferred to small cages with perching while hand-feeding continues until they are taking sufficient food on their own.

Salvadori's
Psittaculirostris salvadorii

Day	Weight (g)	Day	Weight (g)
1	4	**32**	48
3	5.5	**34**	63
12	16.6	**36**	68.7
15	20	**38**	73
17	21.8	**41**	77
19	23	**45**	86
25	42	**50**	94
30	39.9		

When chicks approach fledging age, they become more curious and begin to chew and nibble. At this stage, food should be placed in the cage with the fledglings to encourage them to eat, but they will generally not eat sufficient quantities to sustain themselves. Hand feeding should continue until they can maintain their weight on their own. As with most parrot species, the weaning process is very long and drawn out. Fresh food should be provided and hand feeding reduced as the chicks learn to eat. The last formula feed of the day should not be eliminated unless the chick has a full crop of weaning food at the normal feeding time. Patience is vital, as these chicks are dependent on hand feeding and need to remain strong if they are to wean and thrive.

Ring Size: Psittaculirostris spp.: 6 mm; Cyclopsitta spp.: 5 mm; After about 12 days fig parrots must always be banded with metal bands as these birds are vigorous chewers and will destroy bands made of softer material.

Contribution: Personal notes on the Macleay's fig Parrot, David Carr, Yarra Valley, Victoria, Australia.

References:

Harmon, A. 2002. Avian husbandry notes for red-browed fig-parrot *Cyclopsitta diophthalma macleayana*. Currumbin Sanctuary unpublished report.

Jordan, R. 1989. Parrot Incubation Procedures. Black Cockatoo Press.

Muller, M. & N. Neumann. 1998. Keeping and breeding fig-parrots of the genus Psittaculirostris at Vogelpark Walsrode. Avicultural Magazine 104(2).

Voren, H. & R. Jordan. 1992. Parrots: hand feeding and nursery management. Osborne Editions Pty. Limited.

14.01 Psittacidae Rosellas *Platycercus* spp.

Terry Irwin

Introduction: Rosellas belong to the order Psittaciformes and the family Psittacidae. Also known as broad-tailed parakeets, there are eight species and a number of subspecies of rosellas included in the single genus *Platycerus*. They are all endemic to Australia. The members of this genus are among the best known parrots in captivity. Their bright colors and readiness to breed have endeared them to aviculturists the world over. Their plumage pattern is distinctive in that in all species the feathers of the mantle have their centers black and their edges colored. There is also a prominent cheek patch extending to the throat. Rosella species can be divided into two categories according to the cheek color. These are the white- or pale-cheeked and the blue-cheeked, as follows:

WHITE or PALE CHEEK	BLUE CHEEK
Eastern rosella *P. eximius*	Green rosella *P. caledonicus*
Mealy rosella *P. adscitus*	Pennant's parakeet *P. elegans*
Brown's rosella *P. venustus*	Adelaide rosella *P. adelaidae*
Stanley rosella *P. icterotis*	Yellow rosella *P. flaveolus*

Species: Most rosella species have one or more alternative common names, and the green rosella is no exception, also being known as the Tasmanian rosella. As this name implies, the species is endemic to the island of Tasmania and the islands of the Bass Strait. It is the largest of the genus, measuring 37 cm (14½ in.) in total length. This bird has never been well known in aviculture, possibly due to the fact that it is not brilliantly colored. A cinnamon mutation has been recorded for this species.

The Pennant's parakeet or crimson rosella inhabits the Eastern side of Australia. It measures 36 cm (14 in.) and is the most strikingly colored of the genus with its crimson plumage and blue cheek patches. The subspecies include *P. e. nigrescens,* which is slightly smaller, and *P. e. melanoptera,* which is darker red. Some taxonomists include the yellow and Adelaide rosellas as subspecies of the Pennant's. There are various mutations including the popular blue, lutino, cinnamon, and a dilute yellow.

The yellow rosella, regarded by some as a subspecies of the Pennant's, is accorded full specific status under the S-A-M system of classification, although the Adelaide is not. It is smaller than the Pennant's, measuring 33 cm (13 in.). Endemic to southeastern Australia, it occurs along timbered watercourses where it is closely associated with the River Red Gum. This beautiful rosella has never been a common aviary bird even in Australia. Pied and yellow mutations of this species have been recorded.

The Adelaide rosella is generally regarded as a subspecies of the Pennant's. However, Forshaw considers it a hybrid population and Juniper and Parr classify it as a cross between the Pennant's and yellow rosellas. It is a popular aviary bird and pairs produce a rosella identical to the Adelaide, and the young when mated breed true. Lutino, pied, and cinnamon mutations exist.

The Brown's, or northern rosella, as the alternative name implies, inhabits northwestern and northern Australia. This very attractive black-headed rosella is the most sought after, and expensive, of the rosellas. It is also one of the smallest at 28 cm (11 in.). This species is notoriously aggressive, and often difficult to re-mate. Although there are no recognized subspecies, the amount of blue and white on the cheeks varies between western and northern populations. No mutations have been recorded.

The mealy, or pale-headed, rosella is confined to eastern Australia. This species is well-established in aviculture, and measures 30 cm (12 in.) in length. Two subspecies are recognized, the nominate race from the north, also known as the blue-cheeked rosella, has more blue suffusion on the cheeks. The intensity of the blue suffusion reduces further south, where the race *palliceps* occurs. This subspecies has a very pale head, lacks the blue throat, and the upper parts are a richer yellow.

The eastern rosella is commonly referred to as the golden-mantled rosella, however, this name should properly only apply to the subspecies *cecilae,* which has broader brighter yellow edges to the mantle feathers. The subspecies *diemenensis,* also known as the Tasmanian eastern rosella, has larger white cheek patches and more extensive and darker red on the head and breast. The three races vary from 29–34 cm (11½–13½ in.) in length. Unfortunately, there has been much mixing of the races in captivity and pure subspecies are now rare outside Australia. It is the most popular and well known of the rosella family. Many mutations have been bred, including dilute yellow, cinnamon, lutino, black, red, and pied.

The Stanley, or western rosella is the smallest rosella, measuring 25–30 cm (10–12 in.). It inhabits the southwestern tip of Australia from Dongara in the north to Israelite Bay on the south coast. It is a very placid-natured bird that is very popular in captivity. It is also far less aggressive toward other males of its species or other birds. The subspecies *xanthogenys* is generally paler than the nominate race, with more extensive red borders to the feathers of the mantle. The tail is blue without traces of green. Yellow, or lutino, and blue mutations exist.

Management: Rosellas should be kept in flights of at least 3 m (10 ft.) in length, although the smallest species, the Stanley, does well in shorter flights. They are sometimes kept in suspended cages, but as they like to spend a lot of time on the ground, conventional aviaries are preferable. Because of their ground feeding habits, rosellas should be wormed regularly.

Rosellas are extremely aggressive birds, so no attempt should be made to keep more than one pair together. It is also preferable not to house similar species, or even other rosellas, next to each other. Bickering between males in adjoining aviaries can become so intense that they ignore the females, resulting in infertile eggs.

Diet: A basic seed mixture of canary, striped sunflower, various millets (Japanese, pannicum, etc.), and oats should be offered ad lib. The sunflower should not exceed more than about 25% of the mixture, to avoid obesity. Rape and niger in very small proportions may also be added. In addition to their dry seed mixture, they should get a supply of fruits and vegetables. These can include corn on the cob, peas, pumpkin, grated beetroot, grated par-boiled carrot, papaya, apples, oranges, guavas, etc. These can be experimented with to establish preferences. Often certain items are refused by some but eaten by others. The importance of green food for rosellas cannot be over-emphasized. Shredded cabbage, Swiss chard, chickweed, seeding grasses, carrot tops, fresh branches with leaves intact, etc. should be fed in abundance, especially when the birds are feeding young.

Breeding: In order to breed rosellas, the pairs must be compatible. This may sound obvious, but many pairs that tolerate each other do not breed because they are not compatible. There is no mutual preening in rosellas to indicate compatibility, but the male feeding the female is a sure sign that breeding will take place. Adult, fully colored birds can often be visually sexed. Generally, the males are bigger, brighter, bolder birds, with larger flatter heads. However, if there is any doubt they should be either surgically sexed or DNA sexed.

Males can sometimes turn nasty when paired with a much younger hen that does not come into breeding condition at the same time he does. Also, for no apparent reason, a male might turn on its mate after having bred successfully together for two or three years. With older birds, very often if a lost female is replaced, it will not be accepted by the male, which may even kill it. If any unusual aggression is noticed, the birds must be separated immediately.

Nesting boxes should measure 20–25 cm^2 (8–10 sq. in.) and 60 cm (24 in.) high, and have a 7 cm (2¾ in.) entrance hole. They should be hung as high as possible in the darkest part of the shelter. Similarly sized hollow logs can also be used. Some breeders hang their boxes at an angle to prevent the birds from dropping down directly onto the eggs and possibly breaking them. If egg breaking is a problem with certain pairs, it is better to give them an L-shaped box.

Rosellas can lay 8–9 eggs, but most clutches contain 4–6. The incubation period varies 18–21 days. With very light sitters, this period can be extended to over 30 days. The chicks fledge after about 35 days. After leaving the nest, they should be left with the parents, who continue feeding them, for about two weeks, or until they are seen to be completely independent.

Although rosellas have been hand reared in an attempt to produce affectionate pet birds, this has not been successful. Rosellas are by nature flighty birds and even hand-reared babies soon revert back to this condition when left on their own. Also, only the females seem to tame slightly, the males never losing their aggressiveness.

Ring Size: The Stanley should be rung with a 5 mm ring; all the others, 6 mm.

Species Requiring Similar Management:

Ringnecked parakeet *Psittacula krameri*
Other *Psittacula* species
Crimson-wing
King Parrot

References:

Forshaw, J.M. 1990. *Parrots of the World.* Third (revised) edition. Melbourne: Lansdowne Editions.

Low, R. 1992. *Parrots: their breeding and care.*

Silva, T. 1991. *Psittaculture.* chapter 21.

14.01 Psittacidae Eclectus *Eclectus roratus* spp.

Kirsty Chalmers

Natural habits: The eclectus parrot *Eclectus roratus* spp. is found in tropical rainforest habitat throughout New Guinea, the Indonesian archipelago, the Solomon Islands, and the northern Cape York Peninsula of Australia. Some 8–10 sub-species of eclectus parrot are generally recognised. The most popular of these in aviculture are the nominate sub-species, grand eclectus *Eclectus roratus roratus* and the sub-species vosmaer eclectus *E. r. vosmaeri,* red-sided eclectus *E. r. polychloros,* and Solomon Islands eclectus *E. r. solomonensis.* Less commonly encountered are the Aru eclectus *E.r. aruensis,* Macgillivray's or Australian eclectus *E.r. macgillivrayi,* Biaki eclectus *E.r. biaki,* and Cornelia eclectus *E.r. cornelia.* The Riedel's or Tanimbar eclectus *E. r. riedeli* is not known to have been held by aviculturists. It is believed that the species originally emerged in New Guinea and later spread to the surrounding islands and the northern tip of Australia, where it differentiated into separate sub-species. Each sub-species inhabits a different part of the overall range, and they vary in characteristics such as size, conformation, color, vocalizations and age of sexual maturity.

Eclectus parrots are fairly common in their natural habitat, but have declined in some areas due to habitat destruction. Only small numbers of *E.r. cornelia* and *E.r. riedeli* remain in the wild (Taylor, 50). Eclectus generally prefer to inhabit lowland secondary forests and open woodland below 914 m (3000 ft.), and are less common in primary forest, possibly due to the higher humidity in these areas (Nogare, 58). They feed from the forest canopy; their diet consists primarily of leaves, stems, fruits, nuts and blossoms. In some parts of their range, they are considered to be pests of human food crops, particularly grains.

More than any other parrot species, the eclectus shows a marked sexual dimorphism. The male is bright emerald green with a yellow-tipped orange upper bill. Males of the different sub-species are hard to distinguish. Although there tend to be subtle hue variations between sub-species, size is really the only reliable visual clue to male sub-species, with the grand eclectus being the largest and the Solomon Islands and Biaki eclectus the smallest sub-species. The female eclectus is scarlet and maroon, with a large blue area on her breast, and a black bill. There are distinct color differences between females of different sub-species. In the female grand eclectus, the breast feathers are purple, while they are lavender in the Vosmaer eclectus. Female Vosmaer eclectus also have a bright yellow band at the tail tip and on the under tail coverts. Females of the Aru, red-sided, Macgillivray's, Solomon Island and Biaki sub-species have cobalt blue feathers on the breast and blue periorbital rings around the eyes. The Cornelia and Riedel's eclectus are completely red-fronted with no blue feathers on the breast at all. The extreme color difference between male and female is thought probably to be due to the need for camouflage; the female's coloring may disguise her in the dark nest cavity and in the areas of deep foliage, which she frequents. The male eclectus prefers higher locations in the canopy, and his coloring may help him blend in to the treetop foliage. The plumage of both sexes is unusual, with its tight interlocking of feathers creating a fur-like appearance.

Eclectus nest high in the canopy of large rainforest

trees, 20–30 m (65.6–98.4 ft.) above the ground. They make their nests in cavities deep in the tree's trunk, excavating into the centre of the trunk where the wood is the softest, then downward for 1 m (3 ft.) or more to modify the cavity into a nest chamber. The pair defend the nest site from competitors and predators, and the female tends to stay near the nest, even when she is not nesting. During nesting, the female alone incubates the eggs and broods the chicks, while the male feeds her. In the wild, eclectus parrots will breed all year round, and dominant breeding pairs usually have helpers (offspring of previous years or unpaired adult birds) to assist them in the rearing of their young.

Management: Eclectus parrots have been kept in captivity as pets and in aviculture for centuries. They have been widely bred by aviculturists since the 1970s. As with all bird species, it is important that the aviculturist set management goals to provide for the eclectus' essential needs. Suitable housing, perches and nest-boxes, along with a healthy, balanced diet, are all essential. Keeping records is also vital, as these provide the tools for the control and improvement of management. Healthy, well cared for birds can reach ages of 30–50 years in captivity.

Although they may be noisy under certain circumstances, eclectus parrots are not known for the prolonged periods of screaming exhibited by some parrot species. Where a number of birds are kept, they will call to one another. Their behavior is quite different from other medium-sized parrots, such as macaws and Amazons, in that eclectus are circumspect and will usually study a situation before taking action. This behavior has given rise to a reputation for being somewhat slow. When extremely frightened, the eclectus tends to remain calm and unmoving, rather than screaming and attempting to flee. The fact that the bird is stressed may not be obvious to a casual observer. This behavior has led to the term "eclectus freeze".

As with all animals in a captive situation, the basic survival needs that occupy the time of wild eclectus have largely been removed, as food, water and nesting sites are all provided for them. Mental stimulation is the key to avoiding boredom and the behavioral problems which may result. As well as natural branches to chew and eat, eclectus will enjoy a range of toys, such as ropes and swings (which should be free of loose strings that may entangle toes or feet) and wooden toys for chewing. Plastic toys should be able to withstand normal bill pressure without shattering into sharp pieces. The section on PSITTACINE ENRICHMENT provides more information on physical and psychological stimulation for these birds.

Eclectus are communal in the wild and may be kept this way in a large aviary. The advantage of this type of management is that birds can choose their mates. Disadvantages are the risk that parents will lose contact with their young when they leave the nest, and this type of housing also makes it difficult to maintain accurate breeding records. One way of overcoming the latter problem is to band young birds before they leave the nest. Colony-style housing is only workable if all pairs are compatible.

Most aviculturists keep eclectus as single pairs, either in walk-in flights or suspended aviaries. Desborough suggests a minimum size for a suspended aviary of 90 cm wide x 90 cm high x 1.8 metres long (35 x 35 in. x 5.9 ft.), but recommends a size of 1.2 metres wide x 1.2 metres high x 2.4–3 metres feet long (3.9 x 3.9 x 4.9–9.8 ft.) Hunt suggests 3.5 metres long x 1 metre wide x 2 metres high (11.5 x 3.3 x 6.5 ft.) as the minimum size for a walk-in flight, and keeps his own red-sided eclectus in steel aviaries 4.5 metres long x 1.2 metres wide x 2 metres high (14.7 x 3.9 x 6.5 ft.), with good results. The aviary should be partially roofed for shelter from bad weather, but have some open wire to allow access to sunshine and rain for general health and feather condition. Eclectus can be kept successfully in hot areas, as long as there is sufficient humidity. In dry areas, sprinklers can be used daily for short periods to increase humidity.

Perches placed at either end of the aviary will provide the opportunity for flying and maintaining muscle tone. Use soft woods for perches (e.g. pine, eucalyptus) so that they are not too hard and slippery for roosting or copulating birds. Natural branches are ideal perches and will provide stimulation in the form of something to chew on. Toys should be located to the sides of the aviary, where they don't interfere with flight. Take care not to use anything that could entangle or injure your birds. Aviaries should have access to water for bathing.

Diet: Eclectus are an arboreal species and their natural diet comes from the forest canopy. Their bill is not designed to open hard nuts but it is perfect for cutting into hard skinned fruits, cracking open fruit seeds and small seeds, snipping off buds and stripping soft bark. In the wild, eclectus eat soft foods such as fruits, nuts, seeds, leaf buds, blossoms and nectar. It is essential that they eat a diet high in vitamin A/beta-carotene and fibre. It is preferable that vitamin A should be available in the diet, although vitamin supplements should be used with care if necessary. The problem with vitamin A is its toxicity in large amounts, and it is much safer to give foods or supplements containing its precursor, beta-carotene (pro-vitamin A), which the body can convert to vitamin A as needed.

A good diet for eclectus consists primarily of soft foods. These should comprise around 80% of the diet. The main food item is fresh vegetables, such as carrots, corn on the cob, zucchini (courgettes), squash, broccoli, celery, peas in the pod, spinach, lettuce, tomato, peppers, silverbeet, cooked kumara (sweet potato), white potatoes, beans and rice. Thawed frozen vegetables may also be used. Generally speaking, highly colored red and orange fruits and vegetables tend to provide the best source of natural beta-carotene. Parrot pellets, sprouted seeds (e.g. sunflower, mung beans, wheat, and barley) will add a nutritious food component to the soft food mix. Wild greens such as chickweed, seeding grasses, sow thistle and dandelion should also be offered regularly. Suitable fruits for eclectus include grapes, pawpaw (papaya), orange, apple, banana, mango, passionfruit, pomegranate, kiwifruit, pear and watermelon. About 20% of the diet should consist of dry food in the form of dry seed mix (e.g. white millet or canary seed) and soft, raw nuts (e.g. shelled almonds, pine nuts, peanuts, or walnuts). Eclectus will also readily eat spray millet, and can be offered treats such as dog biscuits, chicken bones, and boiled egg and biscuit mix from time to time. Hard-boiled eggs crushed with the shells will provide extra protein and calcium, especially for breeding birds. Cooked corn is one of the eclectus' favourite foods and is particularly sought after by parents feeding chicks.

Each bird should be given ¾–1 cup of soft food mix plus ¼ cup of dry food daily. Offer food early in the morning, and feed parents with chicks first. In hot climates, remove the soft food mix after four hours and replace it with a fresh serving. Soft food bowls should be removed daily and not left in the cage overnight. Provide fresh water daily. It is essential to store and prepare all foods safely. Dry foods should be kept in airtight containers to prevent contamination by vermin or airborne pathogens or moisture. Wet foods should be covered and refrigerated until use. Choose food and water bowls made from materials that are easy to disinfect (e.g. stainless steel) and clean these daily.

To sprout seeds, wash them thoroughly, then soak them overnight in a large clean container (e.g. a stainless steel pot). An antibacterial solution (e.g. Aviclens ®) can be added to the water to slow fermentation and reduce bacterial content. On day two, rinse the seeds and place them in a sieve to drain the remaining water. Leave them to sprout for 24 hours. On day three, they are ready to use. Rinse them once more before use. Sprouts can be stored for a short time in the fridge, but should be used before the root is over 12 mm long, at which point they are past their prime and become drier.

Like many parrot species, eclectus parrots are prone to behavioral problems such as feather plucking, which may be brought on by nutritional imbalances or stress. Causes may include insufficient aviary space, a cold environment, a dominant partner, strangers in the vicinity, the breeding season, a change of diet or boredom. To prevent boredom, place tree branches and toys in aviaries to entice birds to chew on these instead of on their feathers. Also, provide natural vitamins and minerals. If plucking occurs, the cause needs to be identified and rectified immediately to prevent a permanent habit forming.

Nogare, who keeps Solomon Islands eclectus, has observed that the eclectus hen starts to chew on wood and excavate a nest hole as she matures. If not provided with soft wood to chew, she will chew her toenails or feathers instead. She will also defend her space and may be quite aggressive in doing so. By giving the female a dedicated space and leaving this undisturbed, you will avoid aggression problems. She has also noticed a periodic "hyper-vigilant" behavior in the hen, which is also due to maturing and may be related to defending the nest from predators. The female's eyes become glassy, large and dark, and she lunges and bites when suddenly disturbed. Problems can be avoided if you move more slowly, and speak more quietly around her when she is exhibiting this condition, which may last for 10–14 days. Male eclectus may overpreen when the female is on the nest and this appears to be because, in the wild, the male would need to spend large amounts of time foraging to feed the female and chicks. When given greater opportunities for exercise, this behavior will be reduced.

She has also observed that both male and female tend to be afraid of sinuous dark brown objects and items with zig-zag patterns or reptile-skin textures, which she has interpreted as resulting from the eclectus' instinctive fear of predators.

Breeding: It is generally recognised that birds should be paired only with mates of the same sub-species. Apart from maintaining the integrity of the various sub-species, different sub-species also have slightly different patterns of breeding behavior and tend to mature and breed at slightly different ages. Females of each sub-species vary in plumage, while males differ only in size. The only reliable way to select mates visually is to match male and female for size as closely as possible. The next step is to obtain a compatible pair. This may be easier said than done, as the female tends to be dominant, except during courtship, and may be quite aggressive towards the male. To help avoid pairing problems, some breeders prefer to raise young unrelated pairs together. Others prefer to pair adult birds. When pairing older birds, you will need to be patient and vigilant. Desborough recommends setting the birds to be paired up in adjoining cages. Set the cages up so that when the birds feed, they stand next to one another. When they have become accustomed to one another, put them into a new cage together, set it up for breeding and observe them for compatibility. Food should be provided at both ends of the aviary to ensure equal access for both birds. Provide wood for them to chew as this appears to form part of the pre-mating ritual. Sparring is common between newly introduced males and females, and while serious injury is rare, newly paired birds should nevertheless be closely observed. Some pairs may be incompatible or take a long time to develop a pair bond. Switching mates is often effective, and random mate swapping has been observed in colony situations. Keeping only eclectus parrots in the immediate vicinity can help to strengthen pair bonding, but a visual barrier is necessary to prevent stress and squabbling between neighbors if the aviaries are immediately adjacent.

Common mistakes include giving the new pair an aviary that is too small. This will affect the birds' natural behavior and cause stress, which in turn leads to illness and behavioral problems, such as feather plucking and mate aggression. Placing a male with a female without prior socialization side by side is an invitation to disaster, as is placing a bonded pair ready to nest into an aviary without a nestbox. Once a pair has bonded, they can be given a nestbox. The best indication of bonding is mutual feeding (mutual preening is less common). Mutual feeding is critical to successful chick rearing; the male must feed the female so she can incubate the eggs and feed the chicks while he gathers and delivers food to them. If the nestbox is provided too early, the hen may spend most of her time in it and jeopardize this bonding process. Again, good observation is essential and you need to ensure that the pair has a nestbox in place by the time egg-laying is due. Look for loose droppings and a bulge around the hen's vent.

Age of sexual maturity depends on the sub-species. Solomon Islands eclectus mature as early as two years, while red-sided eclectus *E.r. polychloros* mature at about three years old. Vosmaer eclectus *E. r. vosmaeri* are only mature at 4–5 years old. The eclectus male in breeding condition makes a deep *konk konk konk* call, while rapping his bill against the female's. Courting females make an almost metallic sound.

Eclectus need a fairly large nest box – about 30 cm^2 (11.8 sq. in.) and 1.2 m (3.9 ft.) high. Provide large wood shavings in the nestbox, as the hen needs large pieces of wood to shred to make her nest. As eclectus will sometimes mate in the nestbox, it should be large enough to allow this. Hens may also enter the nestbox and drop down onto the eggs and break them. Some breeders recommend an L-shaped nestbox to help overcome this problem. Another option is a nestbox with a ledge beneath the entrance, as this allows the hen to look out of the box without leaving her eggs (as she would in the wild). The hen will spend time in the nestbox chewing and removing nesting material until her first egg is laid. In the wild, the pair works together to excavate their nest and chewing wood is thought to be an important part of the stimulation required by the female for ovulation. Working together on the nest is also part of the bonding process leading up to mating.

Eclectus normally lay two white, semi-glossy eggs per clutch, usually 2–3 days apart. Incubation may commence as soon as the first egg is laid or when the last egg has been laid, and lasts 28–30 days. Only the hen incubates and is thought to turn the eggs about once an hour. She sits tight and leaves the nest only a few times daily to eliminate and to drink water. She is fed on the nest by the male. If you have a close bond with your pair, they will probably allow you to candle their eggs and handle the chicks. If not, remove the birds from the nesting area before checking eggs or chicks, as the female can be very aggressive in defending them. Egg fertility can be checked with a candling torch by day five. The safest approach is to candle the eggs without handling the eggs, as this may disrupt egg development or introduce bacteria. Even if the eggs are infertile, it is advisable to leave them with the hen for the full incubation period to avoid conditioning her to abandon the nest. Leaving the eggs with her will also help to avoid over-laying, which is a common cause of calcium depletion.

Chicks usually hatch one day apart and weigh about 16 g (0.56 oz.) at hatch. The female may feed the chicks soon after hatching, or may wait 12–24 hours for the residual yolk sac to be absorbed. The average hen sits very tightly once the chicks hatch and feeds the chicks on regurgitated food brought to the nest by the male. Corn on the cob and greens are particularly popular with parents feeding newly hatched or one-week old chicks, so it is advisable to provide a few pieces of these 2–3 times daily, in addition to the regular soft food. Parents feed the germ of the corn and the juice from dandelions to the young chicks. As the chicks grow, all types of food are regurgitated, from pellets to soft foods and small seeds. The female eclectus will not spend any length of time away from the nest until the chicks are about two months old. Chicks usually fledge at 70–80 days, at which stage the female may become very aggressive towards them. It may be necessary to remove either the female or the young birds at this point. Predominance of chicks of one sex is not unusual, and it is more common for them to be all males. Once chicks have attained their independence and been removed from the aviary (or been removed from the nest for hand rearing), it is advisable to empty the nestbox and disinfect it, before refilling it with large wood shavings. Female eclectus will usually wait 4–6 weeks before breeding again.

Once bonded, eclectus are prolific year-round breeders, although it is common for the female to lay infertile eggs for the first year or so, while the pair masters mating techniques. Another common problem is for the female not to incubate her eggs properly, or for the pair not to feed their first few clutches of chicks. These problems are usually overcome in time, as inexperienced birds learn how to rear correctly in subsequent clutches. Artificially incubated and reared birds especially may have difficulty with parenting skills, e.g. they may brood but neglect to feed the chicks. Swicegood reports experience with one such female. In her attempts to encourage the female to care for her chicks, she tried various strategies, such as swapping chicks for older chicks that begged louder and longer, and hand feeding the chicks, but returning them to the nest in the hope that their hungry cries at feeding time would stimulate the hen to feed them. Finally, after a year, the hen suddenly started feeding her new chicks and successfully reared them.

You will know that there is a parenting problem if the chicks cry continuously during the first few days. Check the nestbox daily to see whether they have food in their crops. If the parents are not feeding the chicks, you may need to pull the chicks for hand rearing. Be circumspect, however, as the eclectus hatchling has a very tiny crop, and can accommodate only a small amount of food per feed. A crop the size of a half pea is normal in an eclectus chick during the first few days. It is also important to let first-time parents raise their chicks for as long as they will feed them, to encourage them to complete the full rearing process. Pulling a pair's chicks early on may inadvertently teach them to leave their chicks before they are ready to fledge. Although not a

Eclectus Parrot
Eclectus roratus

common problem, some males will destroy eggs or newly hatched chicks. Swicegood suggests caging the problem male within the aviary when the eggs are due to hatch, or using indestructible fake eggs to help break the egg destruction habit.

A balanced, nutritious diet is vital for the pair to produce healthy eggs. After the chicks hatch, their parents need a good diet in order to raise healthy youngsters. Parents that are not adequately fed may stop feeding their chicks, while those on a poor diet will usually produce stunted chicks or chicks with nutritional problems, e.g., deformed bones, stress bars on feathers, etc.

The tendency of eclectus to nest continuously can also cause problems. If left to her own devices, the hen will continue breeding, regardless of season, and will not even stop to molt. This may result in problems such as winter chicks requiring hand rearing, or nutritional (especially calcium) deficiencies in the female, which may have an adverse effect on her long-term breeding prospects. To prevent over-breeding, you may remove the male for several weeks. The only disadvantage of this is that the female will probably call noisily for her mate.

Problems may also arise when holding breeding birds indoors, and you may find that increasing the birds' access to natural light will stimulate production of reproductive hormones and promote breeding behavior. If a pair shows no interest in breeding, you may find that a move to a new aviary with a new nestbox may be all that is needed to stimulate a pair to breed.

New hatchlings are blind and almost naked, apart from a few wisps of white down, which turns to dense dark grey down at about two weeks. The first feathers start to appear at about four weeks. Juveniles can be sexed as soon as the feathers on the mantle appear, usually between 3–4 weeks, Juvenile plumage is duller than that of the adult, and young birds can also be distinguished by their brown bill and dark grey iris. It may be several years before adult eye color is attained. The bill also changes color gradually. Young birds can be removed from the aviary when they are independent and eating without help from their parents

Artificial Incubation and Rearing: Hand rearing may occasionally be a necessity, e.g., for winter-hatched chicks, while many breeders use hand rearing to raise tame birds as pets. The advantage of hand rearing by choice is that you may leave chicks with the parent birds for the first 10–14 days, before removing them for hand rearing. At this stage, their eyes have not yet opened and they have not had a chance to imprint on their parents or siblings. This will make them easier to hand feed and introduce to humans, ideal for birds destined to be companion animals. It is best not to sell unweaned chicks, as they tend to be imprinted on their first hand feeder and do not adapt well to the new one. Weaned birds appear more interested and ready for human companionship.

For artificial egg incubation, the temperature should be 37.3°C (99.14°F) , with a relative humidity of 60%. As the brooding female is thought to turn the eggs about once every hour, eggs should be hand turned to simulate natural incubation. Jody Burgess recommends turning eggs 45 degrees every hour between 6 a.m. and 11 p.m. and every two hours between 11 p.m. and 6 a.m. Once the chick has internally pipped, place the egg in a hatch incubator, lower the temperature to 37°C (98.6°F) and raise relative humidity to 80–90%. After hatching, chicks are left for 3–4 hours to dry off, and relative humidity is lowered to 70%. Maintain this temperature and humidity for the first five days, then gradually lower the temperature to 32°C (89.6°F) and the relative humidity to 55% by day 10. If the chick is cold, it will appear lethargic and the crop will be slow to empty. If the temperature is too warm, the wings may droop and the bill may be open. Respiration rate will also be increased. Burgess recommends using red light bulbs in the brooder room; he has found that the brighter the environment, the more the chicks fidget, which expends energy needed for growth. A snug tub also helps to restrict the chick's movement and leads to better weight gain. Chicks should be carefully checked each time you remove them from the brooder. Look for any restrictions around the toes and feet, infected feather follicles, hardened food stuck in feathers, soiled vent, etc.

As usual, a variety of hand-feeding techniques may be used. For chicks hand reared from day one, Swicegood recommends using a curved-tip, glass eye dropper for the first week, as the new hatchling has a tiny crop that can hold only a few drops per feed. Using an eye dropper ensures that a small, exact measure of formula is dispersed gradually and evenly into the chick's mouth. This also helps guard against aspiration of food. For older chicks, spoon feeding is the closest to parent-feeding, and has the advantage of fostering the bond between chick and rearer. However, if you have a number of chicks to feed, syringe feeding is also effective and saves time. Some breeders advise against crop feeding eclectus chicks, as they have found that chicks tend to become frustrated with this technique. Crop injuries are also more of a risk with this feeding technique. Burgess uses crop feeding exclusively, with excellent results. One of the advantages of crop feeding is that it is less messy than other methods (with less chance of bacterial growth in the chick's environment), and, because feeding is controlled by the hand feeder, it also does not require a strong feeding response from the chick. Eclectus chicks are slower feeders than most other parrot chicks, and have an unusual feeding response, in that they do not "pump" (bob their heads) when fed. It is possible to stimulate the chick's feeding response by rubbing the side of its bill several times, pulling out to the tip of the bill with your thumb on one side and your index finger on the other. This is similar to the motion the hen uses with her bill to stimulate the chick to feed. Because they don't bob their heads, chicks also tend to build up excessive air in the crop. They can be "burped" like a human baby to remove excess air.

An all-purpose bird-rearing formula, such as Kaytees Exact, is suitable for hand rearing eclectus. However, Burgess has found that eclectus parrot chicks tend to need a higher fat content in the diet than other bird species, so recommends adding a very small amount of peanut butter to each feed. He has also noted that eclectus chicks tend to have a slower crop movement than other species, so recommends adding a probiotic (such as acidophilus) to the feed; he suggests using a children's formula. This helps to establish the chick's gut bacteria (particularly important in cases where the chick has been completely hand reared and not exposed to the parents' gut bacteria through regurgitation of food), and should eliminate the slow crop problem. Formula should be offered at 40°C (104°F) and never any hotter than 45°C (113°F). Initially, the formula should be just thick enough to form a strong droplet (the consistency of fruit nectar), but thicken the formula daily. By three weeks, it should be the consistency of pancake batter.

The hand rearing chart outlines chick weights and feeding information for the first 30 days of chick rearing *E.r. polychloros*. (Information supplied by Jody Burgess.)

Weaning usually begins at 5–6 weeks, and finishes by 11–12 weeks. Fledging weight is around 360–400 g (12.7–14 oz.), depending on the size and sub-species of eclectus. Introduce a variety of foods during the weaning stage to encourage healthy eating habits. As well as chopped fruit and vegetables, suitable foods for birds just starting to wean include cooked brown rice or pasta, sections of loosened corn on the cob and boiled eggs chopped with the shell (for extra protein and calcium). Continue to formula feed chicks 2–4 times daily, feeding smaller amounts of formula at each feed until the chick is eating most of its solid food and loses interest in the formula. If the chick cries constantly for food during weaning, you may need to feed a formula higher in fat content and grains to satiate the appetite, or add high-fat soft food such as peanut butter or ground sunflower seeds to the formula. Fresh food and water should be offered at least five times per day to avoid health problems in the young bird, whose immune and digestive systems are still developing. Do not leave food in the brooder for longer than 1–2 hours, due to the risk of bacterial and fungal growth. This applies to dry food also, as this can easily become contaminated if moisture is carried to the dish by the chick, or simply due to natural moisture in the air. To encourage the weaning chick to eat from the dish, place small bites of food in its bill by hand. Warm soaked pellets, bread or cereal may also encourage feeding.

Socializing is imperative for all young birds. If the bird is intended to be a pet, positive handling experiences will condition it to human companionship. Juveniles should also be socialized into their species group.

Eclectus *Eclectus roratus roratus*

Day	Chick Weight (g)	Number of Feeds	Amount per Feed	Feed Intervals
1	17	10	0.1cc	2 hourly 6 AM–12 AM
2	18	10	0.2cc	2 hourly 6 AM–12 AM
3	18	10	0.3cc	2 hourly 6 AM–12 AM
4	24	10	0.4cc	2 hourly 6 AM–12 AM
5	32	10	0.4cc	2 hourly 6 AM–12 AM
6	41	7	0.6cc	3 hourly 6 AM–12 AM
7	54	7	0.7cc	3 hourly 6 AM–12 AM
8	62	7	1ml	3 hourly 6 AM–12 AM
9	65	7	1.5ml	3 hourly 6 AM–12 AM
10	69	7	2ml	3 hourly 6 AM–12 AM
11	78	4.5	5ml	8 AM, 12 PM, 4 PM, 8 PM top up at 11 PM
12	89	4.5	7ml	8 AM, 12 PM, 4 PM, 8 PM top up at 11 AM
13	108	4.5	8ml	8 AM, 12 PM, 4 PM, 8 PM top up at 11 PM
14	118	4.5	9ml	8 AM, 12 PM, 4 PM, 8 PM top up at 11 PM
15	132	4.5	10ml	8 AM, 12 PM, 4 PM, 8 PM top up at 11 AM
16	148	4	12ml	8 AM, 12 PM, 4 PM, 8 PM
17	162	4	12ml	8 AM, 12 PM, 4 PM, 8 PM
18	180	4	15ml	8 AM, 12 PM, 4 PM, 8 PM
19	193	4	17ml	8 AM, 12 PM, 4 PM, 8 PM
20	201	4	20ml	8 AM, 12 PM, 4 PM, 8 PM
21	213	4	25ml	8 AM, 12 PM, 4 PM, 8 PM
22	225	4	25ml	8 AM, 12 PM, 4 PM, 8 PM
23	240	4	25ml	8 AM, 12 PM, 4 PM, 8 PM
24	255	4	30ml	8 AM, 12 PM, 4 PM, 8 PM
25	270	4	30ml	8 AM, 12 PM, 4 PM, 8 PM
26	278	4	30–35ml	8 AM, 12 PM, 4 PM, 8 PM
27	283			8 AM, 12 PM, 4 PM, 8 PM
28	283	4		8 AM, 12 PM, 4 PM, 8 PM
29	295	4		8 AM, 12 PM, 4 PM, 8 PM
30	305	4		8 AM, 12 PM, 4 PM, 8 PM

This is particularly important if you wish to pair hand-reared birds for breeding in the future. The best way to accomplish this is to raise and cage young males and females together so that they can develop normal interactions with each other.

References:

Bacon, Constance. *Handfeeding eclectus babies.* www.geocities.com/eclectusgroup

Burgess, Jody. 2003. Personal communication on hand rearing eclectus parrots.

Desborough, Laurella. 1995. Eclectus parrots. *Avizandum*, September 1995: 3–4.

Desborough, Laurella. 1995. 6 components of success in breeding eclectus parrots. *Bird breeder*, 67(9): 12–21.

Hunt, Chris. 2002. The eclectus parrot. *Australian BirdKeeper*, 15(4): 209–213.

Lee, Garry. 2001. The enchanting eclectus parrots in Australian aviaries. *AFA Watchbird*, 28(1): 42–45.

Low, Rosemary. 1986. *Parrots: their care and breeding.* Rev. ed. London: Blandford Press, 140–143.

Nogare, Mary. 2001. Treasure of the Solomons. *AFA Watchbird*, 28(1): 56–63.

Perdomo, Jose. *The eclectus: winged flowers*. www.eclectus.com

Swicegood, Carolyn. 2001. Breeding the eclectus parrot. *AFA Watchbird*, 28(1): 8–13.

Swicegood, Carolyn. *Land of Vos eclectus*. www.landofvos.com

Taylor, Graham. 2001. The eclectus parrot Eclectus roratus. *AFA Watchbird*, 28(2): 49–50.

14.01 Psittacidae New Zealand Parakeets *Cyanoramphus* spp.

Kirsty Chalmerss

Natural Habits: New Zealand is home to at least three species of *Cyanoramphus* parakeet, popularly known by their Maori name, *kakariki* (green). Of these, the Antipodes Island parakeet *C. unicolor* and the yellow-crowned kakariki *C. auriceps* are endemic to New Zealand, while sub-species of the red-crowned kakariki *C. novaezelandiae* are found in New Caledonia and on Norfolk Island, as well as on offshore islands of New Zealand. Two sub-species, formerly found on Lord Howe and Macquarie Islands, are extinct. There is a fourth kakariki species, the orange-fronted kakariki *C. malherbi,* which overlaps in habitat with yellow-crowned kakariki, but is found only in two forested valleys in the Canterbury high country of New Zealand's South Island.

The Antipodes Island parakeet is the largest *Cyanoramphus* species at 29–32 cm (11–13 in.) and 130 g (4.6 oz.)and is found only on the inhospitable Antipodes Island and its associated islets, 770 km (478.5 mi.) south-east of New Zealand. The island is small, only 5 x 8 km (3.1 x 5 mi.), and covered in sub-antarctic vegetation, including tussock grasses, which are the parakeets' main food source. Although numerous on the island, the parakeets depend on the predator-free status of Antipodes Island for their survival, and are classed by the IUCN as vulnerable. As with all kakariki species, male and female plumage is identical, but the sexes are easily distinguished by size — males are bigger, heavier and have much heavier, broader bills than females. These parakeets rarely flock, and are mainly solitary or seen in small family groups. Despite being strong fliers, they prefer to walk and climb through vegetation, where their green plumage provides them with excellent camouflage. They spend much of their time on the ground, where they feed on tussock and sedge leaves, which they chew to extract the juices. They also eat seeds, berries and flowers, and occasionally visit the neighboring penguin colonies to forage for scraps of carrion. They have even been seen to kill and eat nesting gray-backed storm petrels. Nests are also made on the ground, in burrows up to 1 m (39.4 in.) deep in the peat under tussock grasses. In New Zealand, a limited number of aviculturists currently hold Antipodes Island parakeets as part of the New Zealand Department of Conservation's captive management program.

Yellow- and red-crowned kakariki have a wider natural distribution. Yellow-crowned kakariki are the smaller of the two species at 23–25 cm (9–10 in.) and 40–50 g (1.4–1.8 oz.) and are found throughout New Zealand, mainly on offshore islands, but also in areas of thick forest on the main islands. Red-crowned kakariki 25–28 cm (9.8–11 in.), 70–80 g, (2.5–2.8 oz.) survive only on offshore islands, a few areas of heavy forest on the North Island, and on Stewart Island. Both species were regarded as serious pests of crops, orchards and gardens in the 1870s and 1880s, and their numbers were subsequently severely reduced by shooting and trapping, as well as by habitat loss from widespread deforestation. Introduced mammalian predators (especially rats, cats and stoats) have also had a huge impact on kakariki populations and remain a serious threat to both these species, especially the red-crowned kakariki, which spends more time foraging on the ground than the yellow-crowned, which prefers to feed in the forest canopy. Kakariki diet varies seasonally and includes a wide variety of seeds, fruits, flowers and buds, nectar, leaves, shoots, and occasionally invertebrates. Both species are usually solitary or seen in pairs, but form small flocks in autumn and winter. Kakariki are cavity nesters, which makes brooding hens and chicks particularly vulnerable to predators. Nests are usually made in holes in the branches and trunks of trees (particularly decaying trees), but red-crowned kakariki will also nest in rocky crevices and burrows in the ground. Nests are lined with small pieces of wood, feathers, moss and grasses. Both species breed well in captivity and are commonly held by aviculturists. In New Zealand, holders of kakariki must have a permit from the Department of Conservation. Red-crowned kakariki are a CITES Schedule I-listed species, so international trade in this species is strictly controlled. Outside of New Zealand, they are commonly held by aviculturists, and there are a number of mutations, including lutino, pied and cinnamon.

Ever since they were first described in 1857, debate has raged over whether the orange-fronted kakariki is a distinct species, or simply a color morph of the yellow-crowned kakariki. Based on recent DNA research, it now seems likely that the orange-fronted kakariki is indeed a distinct species, and a critically endangered one at that. In 2000, it was estimated that only 150–500 individuals remained. Studies of wild orange-fronted kakariki indicate that they are similar in behavior and ecology to yellow and red crowned kakariki, however this species is not currently kept in aviculture.

Management: Amongst the many characteristics that make them appealing to aviculturists, kakariki are attractive, intelligent and fairly friendly little birds that do not make a lot of noise, apart from a soft chattering call, and require relatively little space. Provided that they are well cared for, they will thrive in captivity. An outdoor aviary, measuring around 4 x 2 x 2 m (13 x 7 x 7 ft.) (high), and incorporating an all-weather shelter, is appropriate for housing kakariki. For the slightly larger Antipodes Island parakeet, 5 x 2 x 2 m (16 x 7 x 7 ft.) is recommended. Wooden frames are fine for kakariki aviaries, and the birds will not chew them if they have plenty of branches in the aviary to occupy them. The aviary should incorporate plenty of flying space; one perch high up at each end will leave plenty of space free for flying. Natural branches make the best perches and kakariki will also enjoy chewing on the bark. These little birds are incredibly active and inquisitive and will use the whole aviary. They enjoy climbing along perches and mesh, which they manage to do without using their bill as a hook, unlike other parrots. Because they are also avid ground foragers (especially Antipodes Island and red-crowned kakariki), they will enjoy having access to some soil or sand substrate, which should be cleaned daily and replaced regularly to limit their exposure to internal parasites and other disease-causing pathogens. Planted turf or tussocks will also be enjoyed, and should especially be provided for Antipodes Island parakeets, which are almost completely ground dwelling. However, be aware that kakariki are quite hard on vegetation and there is a good chance that any plants you place in the aviary will be demolished. The aviary shelter should preferably be about 1 x 1 x 2 m (3.3 x 3.3 x 6.6 ft.) and enclosed on three sides to protect the birds from bad weather and to provide a dry shelter for the feeding and nesting facilities. Kakariki are adapted to temperate conditions — mild, humid summers and cool, wet winters — so will not be troubled by cold weather, but will need protection from extreme weather conditions, such as frost, gales, torrential rain or severe heat. Except in very hot climates, permanent nestboxes should be provided, as kakariki will use them for roosting all year round. One pair should be kept per aviary, and double wire between adjoining kakariki aviaries is a good idea to prevent any problems with cock birds biting one another's toes through the mesh. Some breeders have reported problems with aggression between kakariki and other species, while others have had no problems. Auckland breeders, Kurt and Rita den Hertog, keep a mixed species aviary, where red-crowned kakariki, king parrots and finches lead a peaceful coexistence.

As long as your kakariki have access to a good quality diet and shelter from the elements, you should find that they are relatively hardy. As with other species, good hygiene in and around the aviary is essential to their health. Water dishes should be cleaned and replenished daily and clean food dishes used at each feed. Food and fecal waste needs to be removed daily to prevent the build-up of disease-causing pathogens in the aviary. One of the most common health problems with kakariki is worm infestation, especially if they have access to soil or sand on the ground. In these cases, you should either worm your kakariki regularly, or do regular fecal checks for internal parasites. I use a suspended aviary for my kakariki and regular fecal checks have so far found no problems with worm infestation. Sow thistle (puha) and dandelion both have antibiotic properties, so regular "doses" of these will also help to prevent worm infestation. A problem encountered with Antipodes Island parakeets in particular is that, because of their naturally tame and trusting nature, they tend to run around underfoot, which can easily lead to accidents. Their tail feathers also tend to become frayed and broken from climbing up and down the aviary mesh. All kakariki are very prone to heat stress, so should not be caught up or moved in hot conditions. In very hot summers, 28–30° C (82–86° F) and up, it may be advisable to remove nestboxes and allow breeding at cooler times of the year, as chicks have been known to die from overheating in nestboxes. It should also be mentioned that although red-crowned kakariki generally weigh 70–80 g (2.5–2.8 oz.) in the wild, captive birds tend to weigh a lot less, with average captive weights at about 70 g (2.5 oz.) for cocks and 55 g (2 oz.) for hens.

Diet: Kakariki are predominantly herbivorous and will eat a wide range of plant matter, but particularly favor seeds. They like to hold their food in one foot to eat and are amazingly dexterous. They are also quite messy eaters and will frequently kick less interesting food items out of their dish in order to get at their favorite foods more easily. Soaked or sprouted seed, mixed with chopped fruit and vegetables, makes a suitable base diet. I use a sprouted mix of millet, canary seed, oats, maize and sunflower, which I combine with thawed frozen vegetables (diced carrot, beans, peas and corn) and chopped fresh fruit or vegetables. I feed about two tablespoons of mix per bird per day, plus one or two small pieces of fruit or vegetables spiked on branches in the aviary (e.g., apple, pear, sweetcorn, banana, tomato, kiwi fruit, cooked kumara, etc). A dry mix of seed and grit should also be available at all times, as these active little birds have quite large appetites for their size. They also enjoy millet spray, which should be offered occasionally for variety. Limit the amount of sunflower fed as, unless it is sprouted, it is very fattening, and long-term heavy intake can also cause metabolic problems. Most fruit and vegetables will be acceptable, but I have found that fruits containing a lot of seeds are particularly favored, e.g., most berries, tamarillos, kiwifruit, feijoas, tomatoes, persimmons, etc. Greens are also an important component of the diet and will usually be relished. I have found silverbeet and beetroot leaves to be particular favourites (lettuce to a lesser extent), while spray-free garden weeds will be enjoyed whole, e.g. chickweed, dock, dandelion, sow thistle (puha) or seeding grasses. Dandelion flowers also make a popular treat. Fresh, leafy branches should be offered whenever possible, and you will find that kakariki enjoy devouring the whole plant. Fruiting *Coprosma* and *Pittosporum* species and flowering *Metrosideros* (pohutukawa) species are all particularly enjoyed, while I have also found Australian species such as flowering bottlebrush and gum trees with nuts to be acceptable. All kakariki species will benefit from occasional treats of invertebrates, e.g., mealworms, rotting logs containing insects, etc. Cheese is also suitable as an occasional treat, and some breeders recommend offering small pieces of meat, but be sure to remove uneaten pieces promptly. Chicken bones can be offered but must be very well-cooked due to the risk of transmitting diseases such as salmonellosis. Cuttlefish should be provided to supplement dietary calcium, although kakariki do not particularly like it, so try breaking it into small pieces and mixing it into their food. Avian multivitamins will help to keep your birds in peak condition and should particularly be given to breeding birds. The den Hertogs recommend adding a few drops of wheatgerm oil to food during the breeding season, as a natural source of vitamins and minerals. For pairs rearing chicks, a softened egg and bread mix also makes a suitable high-protein supplement, while I have found that apple, silverbeet and millet spray all make nutritious and easily manageable foods for fledglings. In the case of Antipodes Island parakeets, green food forms a major component of the rearing diet, especially during the first two weeks, so provide plenty of greens such as silverbeet to pairs with chicks.

Fresh water should be supplied at least once daily, and if you have multiple aviaries, it is worth installing a water system that can flush the water several times per day. Kakariki will drink water and are also extremely fond of bathing, so provide a water container big enough to accommodate this (I find a pot plant saucer about 20 cm (8 in.) in diameter and about 2 cm (¾ in.) deep works very well). They are extremely vigorous bathers and will splash water over quite a wide area, so place the bathing dish well away from food containers, sand areas, etc.

Breeding: Pairing kakariki is a relatively straightforward business. Aggression is not a problem, even when birds do not take to one another. A bonded pair will usually breed within 5–6 months, so if your pair fails to do so, re-pairing is advisable. It is usually fairly obvious when a pair have not bonded well, as they will interact very little with one another. As with many other bird species, it is usually best to breed from birds that are at least one year old, although younger birds may be capable of breeding. Most kakariki breed from about October to January. However, they are generally willing breeders in captivity and will breed all year round if left unchecked. When ready to breed, the hen will beg food from the cock by crouching low on a perch, vibrating her wings and making a puppy-like soliciting call. The cock will feed her by regurgitation, and will continue to do so while she incubates their eggs. It is best to provide a choice of nestboxes at varying heights in different parts of the aviary initially. Extra nestboxes should be removed once the hen has chosen one, to prevent her from laying eggs in more than one nestbox. A 20 x 20 x 35 cm deep (8 x 8 x 14 in.) nestbox, with a hole 6.5 cm (2.6 in.) in diameter is adequate for kakariki. Larger boxes will be used but eggs may get scattered or trampled. A short perch below the entrance hole will allow the cock to feed the hen on the nest. For nesting material, use a mix of damp peat moss and untreated wood shavings or bark, which should be well compacted to a depth of 50–75 mm (2–3 in.). Antipodes Island parakeets need a slightly larger nestbox—15 x 20 x 53 cm (6 x 8 x 21.3 in.) deep, with a 6.8 cm (2.7 in.) diameter entry hole near the top. The hen will strip the inside of the box to reline the nest as it becomes fouled, so soft, untreated pine makes the best nestbox material. A good alternative, successfully used at Wellington Zoo, is to screw soft wood battens onto the inside of the nestbox. These can then be replaced after breeding.

The number of eggs laid varies. Red- and yellow-crowns usually lay 5–9 white eggs, which the hen alone incubates for about 20–23 days. However, up to 11 eggs is not uncommon, and it is a good idea to remove a few of these, as kakariki are not normally able to fledge more than 6–7 young. Eggs are usually laid 48 hours apart, and the hen starts incubation after laying her second egg. Antipodes Island parakeets usually lay 2–6 white eggs, which the hen incubates for about 28 days. The cock feeds the hen and chicks by regurgitation, but normally transfers food to the hen for the chicks.

The tiny hatchlings are covered in light gray down and are initially blind. Their eyes open at about 10 days, and pin feathers should become visible a few days later. Chicks are almost fully feathered by about day 40, and will fledge at 40–50 days. In the wild, red-crowned kakariki chicks will often leave the nest a week before they are fully fledged, and are fed on the ground by their parents until they can fly. The hen may also start a new clutch before her previous young are fully independent, in which case the cock usually takes over feeding them. Aggression from the cock towards male offspring has been observed and fledglings should be removed from the aviary at about eight weeks old to prevent problems.

Although several clutches may be laid during a successful breeding season, limit your hens to 2–3 clutches per year to prevent exhaustion and to maintain the quality of eggs and offspring. Kurt den Hertog has observed that kakariki that try to breed while molting tend to lay infertile eggs, probably due to the physical demands of molting, so this period should be avoided. Kakariki are generally steady birds, so nestbox inspection does not usually cause problems, although it is advisable to keep such disturbances to a minimum. Also, be aware that red- and yellow-crowned kakariki will interbreed, so if you do wish to keep both species together, keep a single-sex group to prevent hybridization. Hybridization is not a good avicultural practice, and should be avoided to ensure pure lines of these species for the future.

Hand Rearing: Kakariki are normally excellent parents, so hand rearing is not normally required, but it is an option if, for any reason, the parents fail to rear the chicks. Hand rearing is also preferred where the offspring are intended to be tame or pet birds, and in this case, it is best to leave the chicks with the parents for the first 8–10 days. The following guidelines are based on my own experience of hand rearing a red-crowned kakariki chick from the age of 8–10 days.

The TLC brooder (made by Brinsea) was used for the duration of the hand rearing process and hygiene was relatively easy to maintain, through regular use of Avisafe® disinfectant and hot water. The base of this brooder can be removed to facilitate cleaning. Use of a towel for padding and a container to hold the nestling also helped to keep the brooder clean. Equipment used for formula preparation and feeding was rinsed in hot water and soaked in a mild bleach (Sterinova) immediately after feeding. After not less than one hour, sterilized equipment was removed from the solution and thoroughly rinsed before being left to air dry.

The hand rearing formula used was KayTee Exact bird-rearing formula. This is a general-purpose hand rearing formula, suitable for a range of species, including most parrots. Spoon feeding was used throughout the rearing process (syringe feeding was tried at first but the chick resisted this quite actively). Feeding was done on a flat surface with a clean tea towel to provide a non-skid surface for the chick. Tissue paper and cotton buds dipped in warm water were used to gently clean spilled food from the chick, and these were discarded after use.

The accompanying hand rearing chart gives full details of daily feed intervals, volumes and number of feeds. Initially, the formula consistency used was a sloppy mix, similar to thin yogurt. From day 15, the formula was thickened slightly and by day 19, formula had been thickened to the consistency of tomato puree.

No major feeding problems were encountered during the hand rearing process and the chick took readily to spoon feeding right from the start. The first feed was initially offered at 6:00–6:30 aa.m. and a final feed was given at 10:30–11:00 p.m. However, she consistently fed reluctantly first thing in the morning, and took larger feeds in the afternoon and evening, so the first feed was adjusted to between 6:45–7:30 a.m. Daily formula intake started to fall from day 33, as the chick prepared to wean. Formula feeds were steadily reduced over the next few days as she started to eat solid foods, and were discontinued altogether from day 45, as the chick was readily self feeding and had lost interest in the formula.

The chick was initially placed in a small tub, which, when lined with toilet paper, provided plenty of support and warmth. A small square of flannel was provided as substrate to allow a gripping surface for the chick's claws and to prevent foot splaying (this was subsequently replaced with rubber non-skid matting, which provided a better gripping surface). A towel was used to line the brooder and prevent the tub from sliding around when moved. By day 30, the chick needed more space, so the tub was removed and she was given access to the whole (lined) brooder. The brooder temperature was initially set to approximately 29°C (84.2°F). From day 14,

Hand Rearing Chart for *Cyanoramphus* spp.

Age (Days)	Average Brooder Temp. (°C)	Weight	Av. per Feed Volume (ml)	Estimated Daily Feed volume (ml)	No. of Feeds	Feed Intervals (hours)
10	29.2		3.7–5	19.7	5	2–2.5
11	29.2	34.3	3.7–5	33.5	8	2–2.5
12	29.2	36.3	3.7–5	29.4	7	2–2.5
13	29.2	39.5	3.7–5	29.1	7	2–2.5
14	28.4	44.1	3.7–5	37.5	7	2–2.5
15	28.4	50.0	5–7.5	34.4	6	2.5–3
16	27.8	53.3	5–7.5	32.4	6	2.5–3
17	27.0	55.9	5–7.5	37.9	6	2.5–3
18	26.8	57	5–7.5	44.4	6	2.5–3
19	26.2	60	10	42.4	5	3–4
20	25.6	62.9	10	48.2	5	3–4
21	25.4	65.6	10	51.3	5	3–4
22	25.4	66.5	10	43.2	5	3–4
23	25.4	68.2	10–13	44.1	4	5
24	25.4	69.8	10–13	42.9	4	5
25	25.4	70.8	10–13	37	4	5
26	24.8	69.5	10–13	43	5	5
27	24.6	69.3	10–13	45.3	4	5
28	24.4	70.7	10–13	38.3	4	5
29	23.8	69.6	10–13	43.7	4	5
30	23.8	71.5	10–13	40.4	4	5
31	23.8	71.6[1]	10–13	41.5	4	5
32	23.8	70.7	10–13	41.8	4	5
33	23.8	70.0	5–10	37.4	4	5
34	23.8	69.4	5–10	34.7	4	5
35	23.2	70.7	5–10	32.7[2]	4	5
36	22.4	67.7	5–10	31.7	4	5
37	ambient[3]	68.3	5–10	24	3	6
38		66.1	5–10	20	3	6
39		63.8	5–10	20	3	6
40		62.5	3–5	6.3	2	12
41		59.6	3–5	6.7	2	12
42		59.4	3–5	4.1	2	12
43		57.3	5.4	5.4	1	
44		56.3	2.6	2.6[4]		
45		55.6				
46		54.7				
51		54.2				
52		53.3				
64		53.2				

[1]peak weight
[2]solid foods introduced
[3]chick moved to weaning cage at ambient temperature (~23°C) during day
[4]formula feeds discontinued

temperature was lowered approximately 0.5°C (0.9°F) daily, down to 20.6°C (69°F) by day 38 (average daily temperatures are shown in the accompanying hand rearing chart). One problem encountered was the effect of ambient temperature on the brooder temperature. As hand rearing occurred during summer, ambient daytime temperature was generally 20–25°C (68–77°F), but was occasionally as high as 30°C (86°F). At times, it was necessary to partially open the brooder door, and use a fan to lower the temperature, especially after a hot car trip home in the afternoon. However, the chick was never observed to experience any discomfort as a result of temporarily raised temperatures.

By day 35, the chick was starting to nibble at different materials and appeared ready to try solid foods. The first foods used were apple and silverbeet, cut into small strips. The chick was seen nibbling at both foods very soon after they were offered. More of these foods were offered as necessary (3–4 times daily). On day 38, mixed soaked seed was offered, along with a spray of millet. The chick immediately started cracking open the millet seed, after which she readily ate the soaked seed. She was now offered a mix of soaked seed, chopped apple and silverbeet twice daily, and fed herself with no hesitation. Other foods were gradually introduced for variety and nutrient balance, including lettuce, corn, peas, carrot, kiwifruit, pear, orange, grapes, potato, kumara, cooked beans, rice, lettuce, spinach, etc. Small branches of fruiting Coprosma species were regularly offered, and she eagerly ate the berries, leaves and gnawed on the bark. By day 45, she was completely self feeding. It appears that this species will almost wean itself if offered suitable solids at the appropriate time.

On day 38, the chick was moved into a small weaning cage, with two low perches. The cage was cleaned daily with Avisafe® and hot water, and the floor was lined with absorbent paper and rubber matting. She was initially returned to the brooder (switched off) overnight, but use of the brooder was discontinued from day 41. On day 57, she was moved to a large portable aviary. Initially, the weaning cage was placed inside the aviary, with the doors open to allow her out to investigate. Her food was placed in the familiar cage, but plenty of Coprosma browse was placed in the aviary to encourage her to venture out, which she did almost immediately.

This section summarizes plumage and behavioral development. Weights are shown in the accompanying hand rearing chart.

- Day 10: Good down covering, pin feathers showing on the wings. Chick vocal and active, and quite feisty when handled. Immediately exhibited very strong feeding response, when introduced to spoon feeding.
- Day 11: Starting to open eyes fully.
- Day 13: Complete gray down covering, tail pin feathers visible.
- Day 14: Able to hold feet straight under body, but still unstable on legs.
- Day 16/17: Wing and tail pin feathers starting to open. Able to grip with claws and to stand up on the non-skid matting in tub, but still unsteady on feet when out of tub. Began to preen for first time and to sleep with head on wing, rather than flopped forward on chest. Increasingly active and curious about surroundings over this period.
- Day 20: Tail about 2.5 cm (1 in.) long. Standing well and using legs to stretch. Tried climbing for the first time.
- Day 22: Feathers coming through on chest, back, wing coverts, and abdomen. Wing feathers opening and wing shape visible. Red tinge visible on forehead. Very mobile, able to grip strongly onto a finger and climb up clothing, using bill to grip and wings for balance. Started to climb out of tub when placed back in brooder after feeds.
- Day 24: Tail about 3 cm (1.2 in.) long.
- Day 25: chick covered with a mix of green feathers and gray down over most parts of body, although flanks remained fluffy.
- Day 26: Wings and back almost completely feathered, abdomen 50% feathered. Able to perch steadily on finger. Starting to flap wings to prepare for flight.
- Day 27: Head feathers opening.
- Day 28: About 66% of body feathered. Tail about 4 cm (1.8 in.) long. Patches of fluffy down still on flanks and parts of the abdomen.
- Day 30: Red feathers next to left ear opening. Walking, climbing, wing flapping. Vocalizations had deepened slightly.
- Day 31: Red feathers next to right ear visible. Abdomen fully feathered. Crop still bare.
- Day 33: Tail about 5 cm (2 in.) long. Attempted first flight, travelling about 90 cm (3 ft.). Continued to practise through vigorous wing flapping.
- Day 34: Red eye stripes developing.
- Day 36: Flew about 2 meters (6.5 ft.) across the

room at feed time. Flying attempts now a regular occurrence whenever young bird taken out of the cage to feed.

- Day 37: Red crown feathers open.
- Day 39: Call starting to resemble "chattering" of adult kakariki. Flying well over short distances.
- Day 40: Almost all head feathers open. Tail 6–7 cm (2.4–2.8 in.) long.
- Day 42: Right wing trimmed.
- Day 43: Attempting to bath in small water dish, and eagerly bathed when offered saucer filled with water. Bathed and preened regularly, and continued to explore surroundings avidly.
- Day 45: Plumage almost fully developed and in excellent, glossy condition.
- Day 50: Tail about 10 cm (4 in.) long. (Based on weight and bill size, young bird sexed as female).
- Day 52: Observed for the first time using claws to hold food while eating.

No major behavioral problems were encountered during the rearing process. Imprinting was not regarded as a problem, since the aim was to achieve a tame bird that would be comfortable with people. Thus, every opportunity was taken to socialize her and accustom her to handling. She remained tame and easy to handle, although she tended to be flighty around strangers.

References:

Batty, Michael. 2002. Personal communication.

Chalmers, Kirsty. 2003. Hand rearing a red-fronted kakariki at Auckland Zoo in *Australian BirdKeeper,* 16(7): 385–389

Den Hertog, Kurt & Rita. 2003. Personal communication.

Digney, P. 1998. *A guide to incubation and handraising parrots*. South Tweed Heads, NSW: ABK Publications.

Dyer, IS. 1979. *Kakariki*. Bristol: Isles d'Avon.

Gerstenfeld, S.L. 1989. *The bird care book*. Rev ed. Reading, Mass.: Perseus.

Goudswaard, R. 1988. Breeding the Antipodes Island unicolor or green parrakeet *Cyanoramphus unicolor* at the Wellington Zoo, New Zealand. *Avicultural Magazine,* 94(3): 120–128.

Greene, TC. 1999. Aspects of the ecology of Antipodes Island parakeet *(Cyanoramphus unicolor)* and Reischek's parakeet *(C. novaezelandiae hochstetteri)* on Antipodes Island, October-November 1995. *Notornis*, 46(2): 301–310.

Heather, B & Robertson, H. 1996. *The field guide to the birds of New Zealand.* Auckland: Viking.

Holland, Glen. 2002. Personal communication.

Jakob-Hoff, Richard. 2002. Personal communication.

Kearvell, JC & Grant, AD. 2003. The orange-fronted parakeet *(Cyanoramphus malherbi)* is a distinct species: a review of recent research into its taxonomy and systematic relationship within the genus *Cyanoramphus. Notornis*, 50(1): 27–35.

Mt Bruce National Wildlife Centre. 1998. Feeding records for yellow-fronted parakeets *(Cyanoramphus auriceps)*. Unpublished notes.

Potter, John. 2002. Personal communication.

Rogers, David. 2002. Personal communication.

Warne, J. The Antipodes parrot *(Cyanoramphus unicolor)*. www.parrot.co.nz (website of the Parrot Society of New Zealand).

14.01 Psitticidae Grass Parakeets *Neophema* spp.

Kirsty Chalmers

The Neophema grass parakeets are an Australian genus of small, ground-feeding parrots. Six species are recognised: the turquoisine parrot *N. pulchella,* splendid or scarlet-chested parrot *N. splendida,* elegant parrot *N. elegans,* blue-winged parrot *N. chrysostoma,* rock parrot *N. petrophila,* and the endangered orange-bellied parrot *N. chrysogaster*. Although previously classified as a member of the Neophema genus, the Bourke's parrot *Neopsephotus bourkii* has recently been placed in its own genus.

Neophema parakeets are a very popular aviary genus worldwide. Amongst their attractions for aviculturists are their small size and quiet behavior, which make them perfect for the suburban environment, where larger parrot species may be too big and noisy. Because of their small size, Neophemas do not demand large aviary space and are relatively inexpensive to maintain. They also breed freely in captivity. They are visually attractive, especially the colorful turquoisine and splendid parrots. In addition to the normal color phases described below, many color mutations are available for all Neophema species, and new ones are continually being developed.

Elegant Parrot *Neophema elegans*

The elegant parrot is about 22 cm (8.8 in.) long, with an

adult weight of around 42–51 g (1.5–1.8 oz.). The male has an olive-colored crown and upper parts, with a frontal band of dark blue. His under parts are yellow, with a small orange patch on the abdomen, and a greenish tinge on the breast. The wing coverts are blue, and dark blue on the under wing. The tail is blue above and yellow underneath. The bill is black and the iris brown. The female is distinguished from the male by her duller plumage, and the absence of the orange patch on the abdomen. Immature birds resemble females but lack the blue frontal band. Adult plumage is assumed at about four months.

Elegant parrots have a wider distribution than most other *Neophema* species. They are found in the extreme west of Western Australia, and over much of South Australia and western Victoria. During the breeding season, they are found in pairs or small groups, but form large flocks at other times. Their preferred habitat includes grassy plains, open woodland, wooded water courses, and high heathland. They are also abundant over the agricultural lands in their range. Elegant parrots are nomadic in habit and follow the availability of food supplies.

Orange-bellied Parrot *Neophema chrysogaster*

The adult male orange-bellied parrot has bright green upper parts, a blue frontal forehead band, edged with pale blue, a yellowish-green face, throat and breast, and yellow abdomen with a central orange patch extending to the vent. The orange patch is often not visible at rest. The adult female is noticeably duller than the male, and has a much duller frontal band that lacks the pale blue upper edging. The female's orange belly patch is also duller and often smaller than that of the male. The bill is grey-brown, and the iris brown. The average weight of an adult bird is 45–50 g (1.6–1.8 oz.) and the length is 20 cm (8 in.). (Jolly 2003.)

It is a migratory species, which breeds only in coastal southwest Tasmania and spends winter in coastal Victoria and South Australia. The breeding habitat is eucalypt forest and extensive moorland plains dominated by button grass and intersected by wooded teatree and paper-bark, creeks and river estuaries within the Tasmanian Wilderness World Heritage Area. Its preferred habitat on the mainland is salt marshes, coastal heath, grasslands and pastures, estuaries and coastal sand dunes, all within 10 km (6.2 mi.) of the coast.

The orange-bellied parrot is one of Australia's rarest bird species, currently classified as critically endangered by the IUCN. It is strictly protected by CITES and a variety of Australian commonwealth and state laws. The wild population of orange-bellied parrots is believed to number around 200 individuals. The main reason for their decline is habitat destruction, coupled with predation by foxes and cats, competition for food from introduced finch species and competition for nest sites with starlings. Since 1983, a recovery team has conducted a wide range of activities aimed at recovering the population and conserving its habitat. These measures have prevented further decline in wild breeding numbers and have permitted a small amount of population growth in recent years.

Captive orange-bellied parrots are restricted to authorized holders — currently only three Australian institutions: Healesville Sanctuary (Victoria), Taroona (Tasmania) and Adelaide Zoo (South Australia). All three groups are involved in captive breeding as part of the Orange-bellied Parrot Recovery Program. The captive breeding program aims to act as a safeguard against catastrophic decline in the wild population, increase wild population numbers through reintroductions, and to provide opportunities for research and public education. (Jolly 2003.)

Turquoisine Parrot *Neophema pulchella*

The turquoisine parrot is 20–21 cm (8–8.3 in.) long and an adult bird weighs about 40 g (1.4 oz.) The male is easily distinguished by the dark red bar down his wings, which is absent in the female. The upper parts are green in both sexes, while the underparts are uniformly yellow. The face and upper wing coverts are turquoise blue, while the under wing coverts and outer edges of the flight feathers are dark blue. The underside of the tail is yellow. The bill is black and the iris brown. Apart from lacking the red wing bar, the female's plumage is generally duller than that of the male. The female may be confused with the female splendid parrot, but has deeper blue on the face and wings. Immature birds resemble the female, but males have more blue on the face and usually a hint of a red wing bar. Juveniles undergo their first molt at about five months old.

Turquoisine parrots are found in south-eastern Australia, and occur in scattered locations from south-eastern Queensland to northern Victoria. They are usually seen in pairs or small groups, in timbered grasslands and open forests, where they forage for food in the early morning and late afternoon, spending the rest of the day in the shade. Turquoisine parrots declined in the wild last century, due to the introduction of sheep, which resulted in the destruction of the grasses on which they feed. However, their numbers have recovered since the 1970s and the species now seems to be doing well in the wild.

Splendid or Scarlet Chested Parrot

Neophema splendida

The male splendid parrot is brightly colored, with a blue head, scarlet breast and yellow underparts. The upper parts and tail are green, while the under wing coverts and outer edges of the flight feathers are dark blue. The bill is black and the iris brown. Apart from her pale green breast, the female's plumage is similar to that of the male, but duller. Female splendids closely resemble turquoisine females, but have less intense blue wings and cheeks than the turquoisines. Immature birds resemble the female but are even duller. However, young males usually show more blue on the face than females and start to develop their red breast plumage about 2–3 months after leaving the nest. Adult plumage is assumed at 5–6 months. Young female splendids can be identified by a white band on the inside wings.

Splendid parrots inhabit the interior of southern Australia and are a fairly nomadic species. They are usually found in small groups but sometimes occur in quite large flocks. They are irregular visitors to some areas and locally common in others.

Management: The Neophema species commonly held in aviculture are all very similar in behavior and have the same basic management requirements. They generally do better in small aviaries than in large ones. In particular, it is advisable to have shorter rather than longer flights, as both adults and young birds are able to pick up speed quickly, and newly fledged young may get up enough speed to injure or kill themselves against the end mesh. The same problem may occur with adult birds if they are disturbed at night. A screen of soft branches at the end wires will show birds the end of the aviary and help to prevent this problem. An enclosure of 2.1–3 m long by 800 mm–1.8 m wide by 1.8–2 m (7–10 x 2.6 x 6 x 6–7 ft.) is ample for a breeding pair. The aviary should include an enclosed shelter (about one-third of the total aviary length) in which the birds can be shut during bad weather. If possible, situate the shelter at the southern end of the aviary, leaving the warmer northern end open to natural weather conditions. Martin suggests a shelter area 1.2–080 mm, 1.8 metres high (4 x 2.6 x 6

Turquoisine parrot-courtship display.

ft.) and divided from the main flight at ground level with 450 mm (1.5 ft.) high tin or steel, to keep vermin out of the shelter. He also recommends using lead-free paint to brighten up dark shelter walls, as dark areas will discourage birds from entering (the shelters are where the birds should breed, feed and retreat if disturbed or during bad weather, so they must be comfortable using them).

Coarse river sand makes a suitable aviary substrate, but needs to be kept very clean. Neophemas are ground feeders, and are consequently prone to worm infestation problems. The use of suspended aviaries is an ideal method of overcoming this problem, and most Neophemas will breed successfully in suspended aviaries.

Use natural branches of varying thickness for perches. This provides exercise for the birds' feet and also allows them to choose the most comfortable width for roosting. Swinging perches are particularly good for exercise, especially in keeping balance on landing. Martin has also found swinging perches useful in managing male aggression; they make it more awkward for cocks in neighboring flights to squabble, as they have to hang on the side wire.

A shallow water dish should be provided on the aviary floor to allow birds to bathe. Neophemas will also use wet foliage to bathe, but will not normally bathe in rain.

Neophemas mix well with other species such as finches, barrabands (superb parrots), king parrots and princess parrots, but only one pair of Neophemas should be kept per aviary, due to the probability of aggression and the possibility of cross-breeding. Young Neophemas may also be aggressive to birds of other species. Orange-bellied parrots are not compatible with yellow robins, which are physically aggressive towards orange-bellied parrots when breeding. (Jolly 2003.)

Diet: All Neophema parrots are ground-feeding seed eaters. A suitable diet comprises a good quality seed mix, supplemented with fresh foods for balance and variety. (The Healesville diet offers an example of a suitable diet for Neophema parrots). In addition to dry seeds, sprouted seeds are rich in nutrients and should be provided as a regular part of the diet. A mix of canary seed, white millet, hemp, niger, small sunflower seed, linseed and hulled oats provides a good dry seed mix, and millet sprays are also relished. Green food should be included in the diet all year round. Silverbeet, lettuce, spinach, chickweed, dandelions and milk thistles are all recommended, but make sure that plants have not been sprayed with herbicides. Grasses may also be planted in turfed aviaries, and Neophemas will enjoy nibbling at these. Cuttlefish and grit should be available ad lib to maintain bill health and meet the birds' general calcium needs; Neophemas will enjoy a dish of grit and washed river sand. During the breeding season, a mix of bread and hardboiled egg will help to provide for the birds' increased protein and calcium requirements.

Because Neophemas are primarily seed eaters, it is vital to ensure that the seed you feed is of good quality. You can test this by soaking some seeds and allowing them to sprout. If 70–80% of the seed sprouts, then the quality is satisfactory. Dry seed should be clean and free of dust or weevils, and should be stored in sealed bins in a cool, dry location. Offer dry seed ad lib, as this will keep for several days if not wet or contaminated. Fresh foods are best fed late in the day, as any leftovers will be eaten first thing in the morning and will not dry out in the heat of the day. Offer a small portion of soaked seed, mixed with diced fruit and vegetables (corn, peas, carrots, apple, etc.), one tablespoon of wet mix per pair daily. Increase this amount for birds with young.

To sprout seeds, wash them thoroughly, then soak them overnight in a large clean container (e.g., a stainless steel pot). An antibacterial solution (e.g., Aviclens®) can be added to the water to slow fermentation and reduce bacterial content. On day two, rinse the seeds and place them in a sieve to drain the remaining water. Leave them to sprout for 24 hours. On day three, they are ready to use. Rinse them once more before use. Sprouts can be stored for a short time in the fridge, but should be used before they are over 12 mm (½ in.) long, at which point they are past their prime and become drier.

A shallow dish should be provided for drinking water; as these birds are ground feeders, it should be placed on the ground, well away from perches (where it may become contaminated by feces), and changed daily. A calcium block may be added to the drinking water if your birds do not eat cuttlefish and do not get sufficient calcium, especially during the breeding season. Although not all species will drink much water, most birds will use it to bathe and Dennison has noted that her splendid parrots like to use the water dish to dunk food.

Healesville Sanctuary Diet: Orange-bellied Parrot *Neophema chrysogaster,* from Jolly, 2003, Orange-bellied parrot husbandry manual.)

Ad lib daily

- water
- cuttlefish
- orange-bellied parrot seed mix
- fine shell grit
- greens

Daily per bird/non-breeding season (post molt March/April to August/September)

- 5–10 sunflower seeds
- 4 g fly pupae
- 4 g apple
- 10 g Sprouted seed
- 30 g fresh corn

Daily per bird, breeding season (August/September to post independence of young).

- 2 g apple
- 5 g sprouted seed
- Orange-bellied parrot seed mix: 40% canary seed, 15% Japanese millet, 10% white millet, 20% Red panicum, 15% niger.
- Greens: milk thistle, millet plants, winter grass, chickweed, silverbeet, spinach, endive, and dandelions.

Handling and Transport: It may be necessary on occasion to catch and restrain your Neophemas, e.g., for health checks or when transferring birds between aviaries. Hand nets should be used to catch birds. Lightweight nets with a 60 cm (23.6 in.) handle and 30 cm (11.8 in.) diameter padded hoop with woven mesh are preferred over cloth nets, as they move through the air faster and more easily, and they are not as visible to the bird. Care should be taken to avoid causing undue stress or trauma to the bird being caught.

To handle Neophemas, grip the bird around the

shoulders and neck between the thumb and forefingers, and use your free hand to examine and manipulate the bird's wings and feet. Weigh birds by placing them in a cloth bag and putting them in an open container on small digital scales (the empty bag weight is deducted to give the bird's weight).

When transporting Neophemas, use well-ventilated, individually compartmented boxes, which should not be large enough to allow excessive fluttering (which could lead to injury). The box should be dark on the inside, as birds will settle down quickly in a darkened environment. Only lightweight construction is necessary, due to the weight of the species and their limited chewing capabilities. Three-ply sheet is an ideal material for transport boxes. Transport boxes used by Healesville Sanctuary measure 65 x 23 x 18 cm (26 x 9 x 7 in.), each consisting of three compartments with dimensions 20 x 20 x 16 cm (8 x 8 x 6.5 in.).

Birds being transferred to new aviaries can be released into aviaries at any time, provided there is ample light for them to see their new surroundings. The earlier in the day, the better, as this gives the bird time to find food, water and shelter prior to roosting for the night. (Jolly 2003.)

Health: In general, Neophemas are very hardy and can adapt to varied climatic conditions. Splendid parrots are more susceptible to sudden death when apparently in good health than other Neophema species. The usual cause is damp, cold conditions, so it is advisable during very cold periods to confine splendids to their shelter at night and provide warmth. As mentioned previously, Neophemas are ground feeders and susceptible to infestation by ingested parasites. Suspended aviaries can avoid this problem altogether, but in walk-in flights, the substrate needs to be kept clean to minimise parasite problems. Routine fecal testing and deworming are advisable for birds held in such aviaries. Psittacine Beak and Feather Disease was found to be endemic in the first captive breeding group of orange-bellied parrots in Tasmania in 1986, however, this disease is no longer considered to be a high priority health problem in the captive colony, and no birds have died from it since 1990. (Jolly 2003.)

A common problem among all captive parrots is obesity, which has adverse effects on the birds' fertility and egg-laying ability, and can promote heart disease. Obesity can be fully controlled with the implementation of the correct diet and adequate exercise. Longevity of Neophemas in captivity is likely to be the same as in the wild. Orange-bellied parrots live 8–10 years in the wild, with the oldest recorded bird in captivity at 13 years. Reproductive productivity, however, declines after six years of age. (Jolly 2003.) The only common problem with these quiet little birds occurs when more than one species of Neophema parrot is kept in an aviary. Martin reports that when he housed elegant parrots in an aviary with young Bourke's parrots, turquoisine parrots and splendid parrots, the elegant parrots dive-bombed the other birds, hitting their heads with their bills, and causing cerebral haemorrhage and death in a number of specimens. For this reason, it is inadvisable to house more than one species of Neophema parrot in an aviary, or alongside an aviary housing other Neophema parrots.

Breeding: Pairing is generally straightforward, however it is vital to monitor the new pair's behavior closely and watch for any developing problems. Pairing may be more difficult with the feisty turquoisine parrots, as males have a tendency to chase females, so females should be observed closely for signs of stress.

As previously mentioned, breeding birds should be housed as single pairs. The only Neophema parrots that breed best in colony situations are rock parrots and blue-winged parrots. As well as the aggression problems, which will almost certainly be encountered when more than one species of Neophema parrot is housed in an aviary, species hybridisation may also occur, as most species of Neophema will interbreed with one another if given the opportunity.

All Neophemas perform the same basic courtship display. This involves the cock strutting and hopping around the hen, bobbing up and down, stretching up to his full height and drooping or raising his wings and fanning his tail, while giving a twittering call. His head bobs up and down as he regurgitates food to her. Mating usually occurs on a perch or on the ground.

Neophemas are cavity nesters. In the wild they use holes in trees, stumps or logs on the ground to make their

Comparison of *Neophema* Average Breeding Data

Species	Eggs per Clutch	Incubation Period	Fledging Age	Clutches per Season
Splendid parrot	3–7	18–20 days	28–31 days	2
Elegant parrot	4–6	18–24	28–30 days	2
Turquoisine parrot	4–6	18–25	28–30 days	2
Orange-bellied parrot	3–6	20–21	28–35 days	1–2

nests. In captivity, they need to be provided with nest boxes once it becomes clear that a pair has bonded and are ready to breed. The most commonly used nest box is a closed plywood box, 30 cm deep and 15–20 cm^2 (12 in. and 6–8 sq. in.) with an entrance hole just large enough to admit the birds, 5.5 cm diameter (2.25 in.). Healesville and Taroona sanctuaries use plywood nest boxes 70 x 20 x 15 cm deep (28 x 8 x 6 in.), with side or end openings, and a 4cm (1.75 in.) diameter entrance hole. (Jolly 2003.) The entrance hole should have a perch just below it, protruding about 6 cm inside and 7.5 cm outside (2.4 in. and 3 in.); this allows both birds to perch at the entrance, so that the cock can easily feed the hen on the nest. The nest box should be hung just below roof level, about 4 cm (1.5 in.) beneath the roof, allowing a little space for the lid to be propped open to allow hot air to escape in warmer climates. Martin also recommends an open top nest box 20 cm^2 (8 sq. in.) and 30 cm (12 in.) deep, with the front and back tapering to a wedge shape at the top. Instead of an entrance hole, the box is open at the top. A ladder made from wire-clad timber is provided for the birds to climb up inside. This style of nest box needs to be hung about 15 cm (6 in.) from the roof. The advantage of the open top nest box is that it holds warmth at the bottom in colder weather, and allows hot air to escape in hot weather. The nest box should have an inspection door, either hinged or a removable slide above the nesting chamber. There are a number of advantages in being able to inspect the nest: the condition of the incubating hens can be checked, the number of eggs and date laid can be recorded, development of nestlings can be monitored, and dead chicks can be removed. (Jolly 2003.) The bottom of the nest box should be filled to a depth of 60–75 mm (2.5–3 in.) with damp peat or a mix of peat with untreated pine shavings. The nest box substrate should be dampened with a few drops of water, as Neophemas like humidity.

Neophemas are usually fertile 8–12 months old, but breeding results will be better with hens over one year old, and cocks over 18 months old. There are slight differences between Neophema species in terms of number of eggs laid and length of incubation (see Table 2 for detail) however, the average incubation period is 18–25 days. This is affected by climate conditions and may even last up to 30 days in colder weather. Average clutch size is 4–6 eggs, but up to 10 is not uncommon. Seven or more eggs or chicks are usually more than the pair can adequately care for, so it is advisable either to remove or foster out excess eggs in these cases. Incubation is done solely by the hen, with the cock visiting the nest to feed her. It is common for females to lay a single clutch during their first season, however double clutches are the norm in subsequent seasons. In good weather, the hen may even produce three or more clutches. Although it may be tempting to allow her to continue producing chicks, it is best to stop her breeding after two clutches, as non-stop reproduction may jeopardize her future breeding prospects. Martin reports once allowing a Bourke's parrot to rear 16 young in one season, with the result that she only reared two the following. One way to maximize breeding is to keep spare hens and cocks in holding aviaries, then, when the original hen is stopped from breeding, you can give the cock another hen.

Infertile eggs are fairly common, but candling after day eight will usually show whether the eggs are fertile or not. Only the hen broods the eggs and chicks. Once the chicks have hatched, she feeds them on her own for the first four days or so, then the cock also helps feed them. She broods them closely for about 10 days. They fledge after roughly four weeks but are fed by their parents for a further few days before they become fully independent.

Neophemas breed readily in captivity and are relatively problem-free. However, it is always advisable to keep written records of breeding. These will help to pinpoint problems as they develop, and will also provide data for use in planning future breeding management. Detailed and accurate observations may also provide new avicultural information. Records should include: specimen identification and ancestry; nest box information; ages of birds; diets; general behavior; number of eggs laid, hatched and chicks reared. Always note the date and time of observations, and environmental conditions, such as temperature and weather conditions, as these will all have an influence on breeding outcomes. Martin has found that it is not uncommon for a majority of hens to be produced for two seasons in a row, then a majority of cocks in the third season. This also occurs in the wild. There is also a theory that hens can control the fertility of their eggs; e.g., heavy rainfall is associated with plenty of food, and may prompt a hen to lay six eggs of which 4–6 are fertile. However, in a dry season, she may not have a single fertile egg.

Young Neophemas are covered with long white down on hatching. They are ready to fledge at about 28–31 days. At this stage, you should ideally leave them with their parents for a further four weeks, until they become fully independent. However at, and sometimes before, independence, the male adult will usually start to attack male offspring. As soon as the male is noticed to persecute any of the young, they should be removed to a separate aviary.

When the young of the first nest fledge, the female may be already incubating a second clutch. For this reason, it is advisable to keep two nest boxes, as the second clutch may be laid in a different box, and the male will finish feeding the young of the first clutch.

Artificial Incubation and Hand Rearing: The information in this section is taken from the recommendations in the Orange-bellied parrot husbandry manual. (Jolly 2003.)

If possible, cross fostering should be attempted ahead of hand rearing, as this genus cross fosters very well, and has a much higher success rate than artificial incubation and hand rearing. Hand rearing requires experience, dexterity and a great deal of equipment and time. Avoid taking newly hatched chicks for hand rearing, if at all possible, as chicks fed by their parents or foster parents for even just a few days have a much better chance of survival. Chicks at pin feather stage adapt more quickly to their new environment and diet, and require fewer feeds per day and little artificial heat.

Always thoroughly clean the incubator and all equipment prior to commencement of incubation. Run the incubator for at least 24 hours prior to use, and test and make any necessary adjustments prior to placing the eggs in the incubator. It is not advisable to put eggs straight into an incubator at full temperature — they should be started 5–6 degrees below the recommended incubation temperature and be gradually warmed over a few hours to 36.92°C (98.2°F). The correct incubation temperature for Neophema parrots is 36.9°C (98.2°F) ± 0.5°C (0.9°F) with 50% humidity. Healesville Sanctuary uses a Marsh Rolex fan-forced incubator for incubation of small bird eggs.

Eggs should be weighed prior to being placed in the incubator. Care should be taken when handling them not to cause undue shock to the embryo. Eggs should be labeled and dated, and placed with the blunt end uppermost and inclined on a 45° angle to the horizontal. If the incubator does not have an automatic turning device, the eggs need to be turned manually from 45° inclination in one direction to 45° in the other, commencing 24 hours after the start of incubation. Eggs should be marked with an arrow on each side, so that correct rotation can be achieved — if eggs are continually turned in the same direction the suspensory ligaments (chalazae) that hold the yolk in the centre lose their centering effect on the yolk, resulting in embryonic death. Egg turning is essential to bring the germinal disc into contact with fresh nutrients.

The generally accepted weight loss incurred during the incubation period is 16%. This is calculated by dividing the initial weight of the egg by 6.25 to obtain the preferred 16% overall weight loss, which in turn should be divided by the number of days in the normal incubation period (e.g., 21 days for orange-bellied parrots). This information should be calculated for each egg prior to commencing incubation. Eggs should be candled at five days and again at eight days to determine fertility. Regular weighing will help determine whether eggs are progressing at the correct pace. Egg turning should cease four days prior to estimated hatch date, and the egg should be removed from the incubator and placed in a brooder box lined with paper towel. About 2–3 days before hatching, the chick pierces the air cell membrane (internal pipping), At this stage, the brooder temperature should be dropped by 1–1.5°C (1.8–2.7°F) and humidity increased. During this period the chick begins to use its lungs to breathe in the air cell, and prepares to hatch.

Hand Rearing: Parrot chicks are not able to regulate their own body temperature for the first few weeks of their life, so should be reared in a brooder box. A good brooder box should have the following features: easy to clean and disinfect, perspex front for viewing without opening, thermostatic temperature control, heating pad (chicks should not be able to come into direct contact with the heat source) and adequate ventilation.

The brooder temperature should be maintained at 36.1°C (97°F) with 30% humidity for the first two to three days, then gradually dropped to 32.2°C (90.1°F) over the next three weeks. Once they are feathered and moving about, chicks can be moved to a larger brooder box and kept at about 28°C (82.4°F). If the temperature is too high, the chick may exhibit: panting, restlessness, hyperactivity, poor growth gate, and the skin may appear red. If the temperature is too cold it may cause: poor gut mobility, crop stasis, digestive disorders, failure to feed or beg, inactivity, shivering, increased risk of respiratory disease, or death.

Hand Rearing Diets: The basic rules in parrot hand rearing formulas are:

- It should be simple to prepare.
- It should not contain any ingredients that will deteriorate or promote bacterial or fungal growth when stored.
- It should provide all the nutritional requirements to rear healthy chicks.
- It should have about 18–20% protein content, which is the optimum range for rearing Neophema chicks.

*A number of handrearing diets are listed in appendix 1.

Formula should be fed at 37–38°C (98.6–100.4°F). If too hot, the formula may burn the chick's crop, if too cold, it may be refused or chill the chick. The formula should be the consistency of thick soup. If it is too diluted, the chick will receive inadequate nutrition and its growth may be retarded. If the formula is too thick, it can result in slow gut transit time and dehydration. A more watery consistency is recommended for the first few days. Chicks 0–2 days old should be fed six times daily. As the crop stretches, allowing more volume per feed, feeds can be reduced to four per day within 3–4 days. Feeds can be further reduced to three daily once the crop emptying time is prolonged. As feathers emerge and digestion becomes further prolonged, feeds can be reduced to two daily until weaning.

Orange-bellied Parrot
Neophema chrysogaster

Age (days)	Weight (g) (bird 1)	Weight(g) (bird 2)	Age (days)	Weight (g) (bird 1)	Weight (g) (bird 2)
6		9.2	**20**	37.7	38.9
7	12.5		**21**	40.5	
8			**23**		43.2
9		12.8	**24**	43.8	46.3
10	15.9	16.6	**25**	47	46.6
11	18.9		**26**	46.7	46.6
12		20	**27**	47	46.9
13	24	24.1	**28**	46.8	
14	27.7		**32**		53.9
16		29.7	**33**	51.5	
17	33.7	30.5	**35**		47
18	34	33.2	**36**	44.5	46.3
19	35.9	36.1	**37**	43.5	

Courtesy Jocelyn Hockley, Healesville Sanctuary.

Parent-raised chicks grow more rapidly in the first few weeks, but healthy hand-reared chicks will catch up as weaning approaches. Weaning is a normal physiological response: when birds are old enough and have adequate weight gain, they will wean rapidly. Weaning should not be attempted too early as chicks may lose too much weight and become malnourished and more susceptible to disease. Learned behavior skills may be required for weaning, as some chicks learn that persistent begging results in continued hand feeding rather than becoming self-reliant. Younger chicks that are housed with older ones will usually self-feed earlier. Weaning should begin after the chick has reached an age and level of behavior at which it is physically ready to wean. At this time the chick should have reached its peak weight, and have most of its body feathers, except for the crop and back. At this stage, the chick is usually ready to sample solid foods, and should be offered a variety of soft foods, as well as water to ensure adequate hydration. Weight loss of 10–20% is considered normal during weaning, but hand feeding should resume if chicks become thin. Weaning off is not usually difficult in Neophemas; hand raised orange-bellied parrot chicks tend to start picking at the floor of the brooder box at about 3½ weeks old, at which time they are offered soaked seed. Once they are eating soaked seek well, they can be introduced to hard seed mix.

Imprinting problems can arise when hand rearing, particularly when rearing single chicks. It is better to hand rear chicks in groups of two or more to ensure adequate socialization.

Most hand rearing failures are not the result of incorrect formula, but of bacterial infection due to poor hygiene, or poor brooder facilities. Cleanliness is very important for hatchlings, as their immune system is immature and not able to withstand common bacteria. Hands should be washed between feeding each chick and care should be taken to thoroughly clean all feeding equipment after use.

Appendix 1: Neophema hand rearing diets:

Diet #1: Healesville Sanctuary

- 1 part Heinz high protein baby cereal
- 1 part canary starter
- 1 part ground Eukanuba dry dog food
- Mix ingredients with boiled water, and allow formula to cool to body temperature before feeding.

Diet # 2: Martin (A) – for chicks up to pin feather stage

- 110 g high protein baby cereal
- 110 g egg & biscuit mix
- 60 g full cream milk powder
- 1 dessertspoon of equal quantities: maize meal, ground rice, arrowroot biscuit, millet meal
- Mix ingredients with boiled water.

Diet #3: Martin (B) – for older chicks

- 225 g high protein baby cereal
- 60 g sunflower meal
- 110 g crushed arrowroot biscuits
- 110 g unprocessed wheatgerm
- 1 tsp Glucodin glucose supplement
- 1 dessertspoon of equal quantities: maize meal, ground rice, arrowroot biscuit, millet meal

- 1 drop per chick of Pentavite vitamin supplement
- Mix ingredients with boiled water, feed four times daily.

Diet #4: Stan Sindel

- 1 part ground chicken starter crumble
- 1 part ground egg & biscuit
- 1 part sunflower meal
- 1 part Farex baby cereal
- 1 tsp per 4 cups of formula multivitamin & mineral powder
- 1 tsp per 4cups of formula calcium supplement
- Mix with hot water.

* from Jolly, 2003. Orange-bellied parrot husbandry manual.

References

Anderson, Troy. Neophemas: care and management. www.parrotsociety.org.au

[Dennison, Vera]. 1999.The splendid grassparakeet (Neophema splendida). *Avizandum,* 11(8): 16–17.

Dennison, Vera. 2000. Turquoisine. *Avizandum,* 12(2): 24, 35.

Dennison, Vera. 2003. Elegant parakeets. *Avizandum,* 15(1): 16.

Dick, Paul. 2002. Neophemas: small but choice. *New Zealand Birdz.* Issue 2: 6–8.

Dingle, Sheldon. 2000. Turquoisine parakeet Neophema pulchella. *AFA watchbird.* 27(12): 42.

Jolly, Lorraine. 2003. Orange-bellied parrot husbandry manual. Unpublished report.

Low, Rosemary. 1986. *Parrots: their care and breeding.* Rev. ed. London: Blandford Press, 140–143.

Martin, Toby. 1997. *A guide to Neophema and Psephotus grass parrots: their mutations, care & breeding.* South Tweed Heads: Australian Birdkeeper Publications.

Vriends, Matthew. 2000. The five popular Australian grass parakeets (an overview). *AFA watchbird.* 27(12): 45–48.

14.01 Psittacidae African Parrots

Glen Holland

Africa is home to 10 endemic parrot species and eight LOVEBIRD species, as well as a subspecies of the ringnecked parakeet of the Asian genus *Psittacula.* Like all parrots, they belong to the family Psittacidae and the order Psittaciformes. Other than the subspecies of *Psittacula krameri* mentioned above and the well-known gray parrot *Psittacus erithacus,* all African parrots except the lovebirds are included in the genus *Poicephalus*. This genus includes such popular avicultural subjects as the Senegal parrot *Poicephalus senegalus* of western Africa; Meyer's parrot *P. meyeri* of central, eastern, and southern Africa; the red-bellied parrot *P. rufiventris* of eastern and southeastern Africa, the brown-headed parrot *P. cryptoxanthus* of southeastern Africa. The red-fronted, or Jardine's parrot, *P. gulielmi,* from western and central Africa; the brown-necked parrot *P. robustus*, of which the nominate race is known as the Cape parrot; and the Ruppell's parrot *P. rueppellii* from Namibia are increasing in numbers in aviculture. The two remaining species, the yellow-fronted parrot *P. flavifrons* from the mountains of Ethiopia; and the Niam-niam parrot *P. crassus* from cental Africa, are available less frequently. Few habitats in Africa are safe, and the future of many of these species is uncertain. As the management of most of these parrots is similar, and many fine publications are available that are devoted entirely to them, I will refer largely to three species to describe the basic principles of captive management. The three representative species are the gray parrot, or African gray as it is known outside Africa, the Cape parrot *P. r. rubustus,* and Meyer's parrot, with a few supplementary notes on Jardine's parrot.

Large scale unsustainable trapping of African parrots has continued for far too long. A good example is the southern population of brown-headed parrots.

The reasons for the decline of the brown-headed parrots in Southern Africa are almost a cliché: habitat fragmentation, habitat destruction, and trapping. The game reserves are islands separated by seas of urban areas and agricultural lands. BHPs are specifically an arboreal species; they require trees for food and trees, especially old or dead trees, for nesting. In poor rural communities such trees are cut up for firewood. Usually this is the only source of heat and excess wood is made into charcoal and sold. In such a poverty-driven society, a few dollars can mean the difference between having a meal and going hungry for a week.

The Maputo central market deals in everything, including brown-headed parrots. You can find five birds crammed into tiny cages measuring 50 x 30 cm (20 x 12 in.) sitting on top of one another. Many have open wounds from being pecked and presumably from being roughly handled by the trappers. The vendors take great delight in vigorously shaking the cage to show that the birds are still alive. This of course, causes huge stress to the parrots. There is no food in these cages and water is available from a cut beer can. The price for one bird is around seven US dollars, which is very expensive for a Mozambiquan.

The main trade would seem to be from South African tourists who probably buy them out of sympathy and smuggle them back into South Africa, feeling that they have done the right thing. Having made some money out of this deal the trader will order more parrots from the trappers and the terrible cycle continues. They are trapped using bird lime, a plant extract that when chewed becomes sticky and is smeared onto branches where the parrots roost. As the feet stick the bird flaps its wings, which also get stuck. The birds can then be plucked from the branches. This also explains why many of the birds on sale are missing toes. These are probably hacked off to speed up harvesting. The trappers are usually kids and the birds are held in cages until someone arrives to buy the whole consignment. These consignments are then transported to the capital for sale. All of the birds caught apparently come from north of the Limpopo as all those on the south have already been caught.

Taylor estimates that at any one time there are about 200 brown-headed parrots on sale in Maputo and at roadsides along the coast. The turnover in the markets is 2–4 weeks, either in sales or deaths. This amounts to 2,600–5,200 birds per year. Stuart believes a strong pair bond exists in the species and that a parent will not enter a nest if its partner is missing. Therefore, the capture of one parent not only results in one individual being lost to the population, but the loss of the chicks as well. This trade goes on despite the fact that Mozambique is a signatory to CITES, which prohibits the sale of brown-headed parrots except under license. The legal annual export quota of approximately 400 birds is caught during the brown-headed parrot breeding season and requests (by the trappers themselves) to shift the trapping period to outside of the breeding season has been refused as it then does not coincide with the hunting season! The brown-headed parrot is listed on CITES Appendix II. With an estimated 20,000 birds in Mozambique at the last Atlas publication in 1999, and with virtually no individuals left outside protected areas in South Africa, that will satisfy the market trade for 4–8 years at the most. Unless local, national, and international communities take notice of the plight of this species, the silent decline will continue.

The African Gray *Psittacus erithacus*

Introduction: Of the three subspecies of the gray parrot *Psittacus erithacus,* the nominate form is most popular in aviculture, although the Timneh gray *P. e. timneh* is also well represented. To many people the African gray parrot is the symbol of aviculture. They are often touted as the finest talking parrots, ranking with the greater hill mynah *Gracula religiosa intermedia* as the most accomplished avian imitators of the human voice and other sounds. In the wild, gray parrots inhabit tall evergreen forests, where they are found in flocks. They are fast fliers, often calling loudly while in flight. Like the CAPE PARROT, they have communal roosts where they occur in large numbers, chattering among themselves until late after sunset. They favor the flesh on the fruit of the oil palm, and are often persecuted due to their habit of raiding agricultural crops growing near their forest habitat. These parrots are usually caught by inhumane methods by uneducated trappers who have little regard for the birds themselves. During export from their countries of origin, injuries and mortalities due to unhygienic conditions, post capture stress, and bad handling are common. Many of these wild-caught birds are passed on to ignorant customers as ideal pets but subsequently succumb to stress or incorrect diets. As aviculturists, we have much to answer to in regard to the trade in this species, a practice that is, fortunately, on the decline.

Management: In common with most parrots, grays like to spend much of their day chewing. They are destructive in an aviary, but their psychological needs must be catered to, and green branches of fruit trees, willows, gum trees, etc. should be supplied on a regular basis. Their destructive tendencies also necessitate the frequent replacement of perches. In recent years, the gray parrot has increasingly been housed and bred in suspended cages. A flight for a pair of grays should be at least 3 m in length and 1.2 m wide and high (10 x 4 x 4 ft.). Although many African grays are being bred for the pet trade very successfully in cages measuring approximately a cubic meter (3.3 cu. ft.) in a dark indoor environment, I do not believe that such commercial production measures conform to the standards of most aviculturists, who strive to meet the needs of, and enjoy, the birds in their care. The gray parrot has certainly proved, over the last few decades, to be far more easily bred in captivity than originally believed possible. One of the primary requirements is to let the birds select their own mates, thus ensuring compatibility. Bonded pairs become regular and reliable breeders. Grays can be kept either as individual pairs or in a colony system in large aviaries. Pairs kept in a colony situation are not as productive as pairs kept separately, and nest boxes should not be placed too close together. It is often best to buy young birds for breeding. This results in a longer wait before breeding takes place, but much pleasure will be derived from the birds during this time. Their calls consist of a pleasant series of whistles and clicks. When threatened, wild-caught birds in particular will emit an unpleasant growling sound. Grays are excellent mimics, and will imitate sounds, voices, and pets to perfection.

Diet: As described for MEYER'S PARROT. Wild-caught grays will often eat nothing but sunflower seed and peanuts, but by mixing a little sprouted sunflower with a finely chopped mix of fruits and vegetables they will gradually learn to eat these other foods. Boiled whole yellow corn and corn on the cob is a favorite food. Peanuts, which have a high fat content (and the shells may contain toxins), should be limited but not necessarily excluded. In the USA where there was a big move away from all foods that contained any fat, many breeders found that their breeding tapered off. Fat is particularly important during the breeding season. A calcium/phosphorous and vitamin D supplement (I.C.D. granules) is recommended for the African parrot species, which often suffer from a low calcium level in their system. Additionally, grays are prone to vitamin A deficiencies. The symptoms of this condition include thickened skin on the corners of the mouth, nasal, sinus and throat abscesses, and feather plucking. A veterinarian should be consulted if this condition arises. Gray parrots also enjoy large chicken bones, such as a drumstick, as well as table scraps such as rice, grated cheese, whole wheat bread, etc. Avocados should be strictly avoided, as they are toxic to psittacines. My birds are very fond of whole meal bread with a thin spread of molasses. A commercial parrot soft food should be added to the fruit and vegetable mix. More recently many breeders have moved their birds on to a formulated pellet diet with some excellent results.

Breeding: Some breeders house all of their birds together for a non-breeding period and then separate the pairs for the breeding season. This provides the birds with an opportunity to select a compatible mate, and some will pair off and even lay eggs in the communal aviary. While some pairs nest only once per year, at the same time each year, other pairs will have 2–3 clutches per breeding year. Fertility and hatching rates are usually good once pairs have been established. A nest box 35 cm^2 and 70 cm deep (14 sq. in x 28 in.), constructed of strong wood, is suitable. A large natural stump will also be enjoyed. The nest should be fitted with an internal ladder of weld mesh that will assist the bird to ease onto, rather than dropping onto, the eggs below. The base of the nest should be covered with a 10 cm (4 in.) layer of wood shavings. Some breeders line the entire interior of the nest with weldmesh to prevent the birds chewing the wood. This is a particularly safe way of having a nest box that is positioned on the outside of a suspended cage. Adding to safety, at least two latches should be fitted to the inspection door. In such instances, however, it is important to regularly check the nest for exposed sharp pieces of wire and nails which may become exposed. Nests should be positioned in such a way that direct sunlight is prevented from entering the nest cavity. Grays often take some time, even a few years, to settle down and begin to show an interest in breeding. Experimentation with different pairs will help to determine their needs under given conditions. All the African parrots prefer a secluded position in which to breed. Jordan and Pattison describe that in an experiment where some pairs were placed in rather dark accommodations while others were placed in direct sunlight, the pairs in the darkened area produced far better results. While materials such as shade cloth can be used, it is important that the birds obtain sufficient sunlight to synthesize vitamins and minerals in the diet. Grays are sexually mature at 3–4 years of age. Courtship includes a display during which the wings droop, as well as courtship feeding of the female by the male. The normal clutch consists of 3–5 white eggs, which are incubated for 28–30 days. The newly hatched chicks are pink and covered with white down. At 12 days of age the skin color begins to darken. Chicks fledge at 12 weeks and start sampling food shortly thereafter. They should be left with the parents for three months, and will then be fully independent. There is a great demand for African gray parrots in the pet trade, and young birds intended for that market should be hand-raised.

Hand Rearing: It is possible to artificially incubate gray parrot eggs and hand rear the chicks from day 1, but this is not usually necessary, and should be left to experienced aviculturists. It is of course an essential procedure when adults desert the eggs or chicks. It is also an option that may stimulate the parents to produce another clutch. Hand-reared greys become very tame and are a pleasure to manage. Although they do imprint on people to a certain extent, hand-reared grays will accept a mate and breed successfully, particularly when raised in the company of other birds. Ideally, chicks should be left in the nest for the parents to start, and removed for hand rearing when the first pin feathers begin to emerge. By 10 days of age, baby birds have developed sufficient immunity and resistance to infections, and few problems will be encountered.

When removing chicks from the nest for hand rearing, or eggs for artificial incubation, it may be necessary to use a thick towel to push the parent off the eggs or chicks and avoid being bitten. Alternatively, a section of 110 mm (4.5 in.) drain pipe, cut at a 45° angle at the bottom end, can be inserted over the eggs or chicks, allowing safe passage for a hand to remove the nest contents. As newly hatched chicks are unable to regulate their body temperature, incubator hatched chicks, or chicks not fed by their parents, should be placed in a thermostatically controlled brooder set at a temperature of 34°C (93.2°F), with a relative humidity of 50%. Brooders should be constructed of material that is easily cleaned and sterilized. Young chicks are best kept on paper towels in flat-bottomed plastic bowls; older chicks can be kept on towels or coarse wood shavings in a cardboard box or a polystyrene container in which ventilation has been provided. Once pinfeathers have appeared, the temperature should be gradually decreased to 28°C (82.4°F). A good room temperature for feathered chicks is 25°C (77°F).

Chicks that are too hot will pant, spread out in the brooder, and become listless with no appetite. As dehydration can result, the temperature should be adjusted accordingly. On an extremely hot day, chicks may not digest their food properly. Under these circumstances, extra water should be added to the food. This will allow for the loss of water and prevent dehydration. When hand rearing a number of chicks in a room, a fan, extractor fan, or ionizer will assist in removing feather dust from the environment. It is advisable to keep a record of the chicks' weight gain. Weight should increase daily; if the weight remains constant or decreases, a potential problem is indicated. In addition to monitoring the weight gain, it is important to take the overall appearance of the chicks into consideration.

The feeding schedule for African gray chicks from hatch to weaning should be as follows:

Day 1–3: 9 feeds over 18 hours
Day 4–12: 6 feeds over 18 hours
Day 13–21: 4 feeds per day
Day 22–45: 3 feeds per day
Day 46–70: 2 feeds per day
Day 71–85: 1 feed per day

Although a variety of home made rearing formulas have proved successful, commercial hand rearing formulas, such as Kaytee and Avi-Plus, are fortunately now readily available. These are mixed with warm water. Many breeders now use a commercial hand rearing formula as a base to which they add blended fruit and vegetables, usually in the form of baby food. Other frequent additions are a small amount of peanut butter, and supplementary calcium and phosphorous at a ratio of 2:1. Formulated hand rearing diets such as those mentioned will not require the addition of vitamins, minerals, or trace elements.

When hand rearing from day 1, which is necessary when eggs have been artificially incubated, Carine Lavril of the Parrot Breeding Centre recommended that the chicks first be placed in a brooder set at 36.5°C (97.7°F) with a separate bowl of water to maintain humidity and prevent dehydration. Some breeders suggest that chicks not be fed until they have produced their first droppings from the digested egg yolk. Many other breeders begin feeding immediately with equal success. Some breeders feed only Ringers Lactate or pedialyte until the first droppings are passed, and then introduce the hand rearing formula.

It is important to follow strict hygiene at all stages. The food should be allowed to stand for a while to ensure that it is thoroughly saturated. Either a bent steel spoon or a syringe can be used as a feeding utensil, and should be washed and thoroughly disinfected between feeds, using Vikron, Hibitane, or other products from the XD 10 range. Before feeding, the formula should be warmed to a temperature of about 38°C (100.4°F), but never more than 40°C (104°F). To prevent burning, the temperature should always be tested before feeding. Food warmed in a microwave should be thoroughly stirred to eliminate hot spots.

A syringe with a stainless steel crop needle or a section of intravenous drip tube can be used to feed chicks

Brown-headed Parrot
Poicephalus cryptoxanthus

African Grey Parrot
Psittacus erithacus

that are sick or those that were removed from the nest too late to accept their keeper as a surrogate mother. When feeding chicks by this method, it is important that proper care is taken to avoid forcing food down the trachea or damaging the lining of the mouth or throat. If possible, one should have an experienced aviculturist demonstrate this procedure before undertaking it for the first time. By placing a finger on each side of the gape, the chick will be stimulated to respond by instinctively bobbing the head. During this response, the glottis is simultaneously closed, ensuring that no food enters the trachea. The syringe is placed into the beak on the chick's left side and should point inward toward the chick's right side of the throat. A full crop will feel a little tight when lightly prodded with a finger.

African Gray *Psittacus erithacus*, Jardines *P. gulielmi*, Cape *P. robustus*

Day	Gray	Jardine	Cape
0	13.5	10	9
1	15.5	11.7	12
3	18	13.5	19
5	23.5	18.5	26.8
7	32.5	34	38.2
9	44	43	52.5
11	60	57	68
13	78	70	82.1
15	99	85	100
17	122	92	119.6
19	148.5	108	138
21	173	118	146.5
23	200	132	173.6
25	230	149	189.4
27	259	166	206.8
29	287	178	222
31	314.4	189	235.4
33	342.5	205	249.1
35	362.5	228	251
37	381.5	240	258.8
39	396	252	262
41	410	260	271
43	426.5	268	277.5
45	441	276	281.5
47	455.5	291	287
49	468.5	295	290
51	480	302	292

The crop should be nearly, or completely, empty at each feeding. If the crop is not empty at the scheduled feeding time, the chick might be either too hot or too cold. This condition can also result from the food being too hot when fed, or not fluid enough for proper digestion. An infection might also be indicated, in which case a veterinarian should be consulted. If an air bubble can be seen on the side of the neck, the chick is over full. When chicks begin exercising their wings, it is time to move them into a cage with a weldmesh bottom. From this point on they are fed twice daily, and soft food, fruit and vegetables, and a small amount of seed can be offered. Chicks will learn to eat faster and easier if placed with older chicks that are already weaned. However, groups of chicks must be monitored for fighting, even at this young age. As the weaning process progresses, hand feedings can be gradually reduced to one feed in the evening to ensure that the chicks do not spend the night hungry. When the chicks are eating well on their own, hand feeding can be discontinued.

Hand-feeding formula should not be offered to the chicks in a dish, as they will quickly become soiled. Instead, a rather dry mixture of the hand feeding formula can be added to some soaked seed and diced fruit. Weaning is a stressful period, during which weight losses of 10–15% are not unusual. If greater weight losses occur, however, it is advisable to resume hand-feeding twice a day until the bird returns to good condition.

African grays are highly intelligent, and pet birds require plenty of attention to prevent boredom. Some bird owners take their pet birds for automobile rides to add some excitement to their routine. These birds spend much of their time chewing vegetation, and both pet and aviary birds should be supplied with green branches, which they will demolish with great enjoyment. See the section on PSITTACINE ENRICHMENT for further details and ideas to keep captive birds stimulated.

The other subspecies of the African gray parrot require similar management. The Timneh, or maroon-tailed gray, is smaller than the nominate red-tailed gray, and is easily distinguished by the maroon tail with dark edges and the overall body coloration, which is much darker than the nominate form. Another distinguishing feature is the horn-colored upper mandible. The other subspecies, *P. e. princeps,* is found on the islands of Principe and Fernando Po, and closely resembles the nominate race. This subspecies is seldom available in aviculture.

Ring Size: A 10 mm ring should be placed on the chick when it weighs about 150 grams (5.3 oz.).

Cape Parrot *Poicephalus robustus*

Introduction: Like the gray parrot, the Cape or brown-necked parrot *Poicephalus robustus* has three geographical races. The nominate form *P. r. robustus,* with a restricted range in extreme southeastern Africa, is known as the Cape parrot. The subspecies *suahelicus* has a larger range, occurring in Mozambique, Zimbabwe, Malawi, Zambia, southern Tanzania, southeastern Zaire, Angola, northern Namibia, and the Transvaal. This subspecies is often referred to as the gray-hooded parrot, as the head is more silvery in color, with a slight pink tinge. The West African race *fusciscollis* is similar to *suahelicus,* but with more blue in the green body plumage. The head feathers are more grayish, tinged with brown. This form is found from Gambia and southern Senegal to northern Ghana and Togo.

The head, neck, and throat of the Cape parrot are dusky olive brown. The chest, belly, and rump are grass green. The flight and tail feathers are a dark slate color, and the scapulars and wing coverts are an oily dark green. The edge of the bend to halfway down the wing is bright orange. Males tend to have darker wings than females, which can be further distinguished by a more golden olive color on the head and neck. Females also have a coral pink band across the forehead and crown. The iris is dark reddish brown. The huge beak, which gives these birds a top-heavy appearance, is ivory colored. Males have larger beaks than females, even as juveniles.

These birds occur in flocks of up to 20 birds outside the breeding season, but separate into pairs for breeding. The flocks leave the roosting sites at sunrise to feed, returning again at sunset. Their flight is fast, and they call continually during flight. Captive birds will learn to whistle and talk.

The Cape parrot has declined in numbers dramatically in the wild state. This subspecies is confined to evergreen forests, while the northern subspecies also venture into open woodland and riverine vegetation. The Cape parrot suffers from the same impacts as those described for the brown-headed parrot in the introduction to this section, in addition to the historical shooting of this species due to their habit of raiding orchards.

Management: Cape parrots are highly prized as avicultural subjects. Newly imported birds suffer from stress, and mortalities can be high if precautions are not taken. They should be provided with additional warmth, seclusion, and an electrolyte anti-stress formula in their drinking water. Even birds that have been in captivity for many years may be extremely nervous, and any stress makes them more prone to illness. For wild-caught birds, a planted aviary of at least 3 x 1.8 (10 x 6 ft.), located in a secluded spot, is recommended. Captive-bred birds can, however, be housed in suspended weldmesh cages. If this arrangement is used, vegetation should be planted around the exterior of the suspended cages to give the birds a sense of privacy and security. The increasing interest in this species has resulted in more successful breeding in recent years.

Diet: As described for MEYER'S PARROT. Cape parrots are also fond of nuts, which should be fed in cold weather and the breeding season. They will also enjoy a mature sunflower seed head, which they quickly demolish to extract the green seeds. When small fruits are fed, this species will often discard the pulp and eat only the kernel. As the breeding season approaches, wheat germ granules and high protein cereals should be added to the diet.

Breeding: Cape parrots are undoubtedly one of the most difficult *Poicephalus* species to breed in captivity. The first successful breedings were recorded by Basle Zoo in Switzerland in 1964. To encourage breeding, pairs should be housed in close proximity to other pairs and have a private nest box area. They should be provided with a choice of nesting sites, including a nest box and a natural log. The dimensions should be 30 x 30 x 60 cm (12 x 12 x 24 in.), and the entrance hole, which the birds will enlarge as required, should be approximately 8 cm (3 in.) in diameter. Males often spend long periods alone in the nest, and while both sexes spend much of their time in the nest, only the female incubates. In display, males will drop their wings and call in a manner similar to the gray parrot.

During nest inspections the pair will move into the distant corner and emit a soft growling sound. The 2–4 eggs are laid at three-day intervals. They are rounded in shape. The incubation period is 27–28 days. Once the chicks have hatched, the parents become very vocal and aggressive during nest inspections. As they tend to stand over, and on, their chicks when under threat, nest inspections should be kept to a minimum. As the chicks develop, the parents will become more tolerant of nest inspections.

Newly hatched chicks are naked, but by two weeks of age they are covered in fuzzy white down and the large beak is already characteristic. Their eyes open at 25 days. During nest inspections, the chicks make loud screeching noises. Due to the strength of the beak of this species, it is not advisable to use rings other than stainless steel. At six weeks the chicks are docile and can be handled. The head and wings are covered in feathers, and the pinkish patch on the forehead is obvious. Juvenile males have more color on the head than juvenile females.

In common with all *Poicephalus* parrots, the immatures resemble the more brightly colored adult — in this case, as in the Rueppell's parrot, the female. All immatures have the pink on the forehead, but lack the coloring on the bend of the wing. At six months of age, all youngsters lose the coral pink on the forehead, but at nine months this color returns on the females.

The chicks fledge at 65–74 days, but return to the nest at night. They begin feeding themselves almost immediately, but should be left with the parents for two months. Cape parrots are usually double clutched. One exceptional pair in the United States, belonging to the subspecies *suahelicus,* produced 18 chicks from six clutches of three eggs each. In this instance, the chicks were removed for hand rearing.

Parents often pluck the chicks in the nest. When this occurs, the chicks should be removed for hand rearing. Hand-raised Cape parrots become very tame and affectionate, and will learn to talk and imitate sounds. Although they make ideal pets, I feel that until this species becomes well established in aviculture, even hand-fed birds should not be wasted as pets but paired up for breeding.

Hand Rearing: As described for the AFRICAN GRAY. Growth chart also covered under the hand rearing section on grays.

Ring Size: A 9.5 mm stainless steel ring should be fitted at 18–20 days.

Jardine's Parrot *Poicephalus gulielmi*

The Jardine's parrot *P. gulielmi* requires similar management to the Cape parrot. There are three subspecies of Jardines parrots that are commonly held in aviculture. The greater Jardine *P. gulielmmi massaicus* from Kenya is a large, rather upright, bird with a more rounded forehead and a beak that appears smaller than those of the other subspecies. The black-winged *P. gulielmmi. gulielmi* is from the Cameroon area in West Africa, and the lesser *P. gulielmmi fantiensis* occurs in the Nigerian area of West Africa. These subspecies should always be held and bred seperately. In recent years, Jardines have been more freely available, and are now established in aviculture, although still not common. Newly imported specimens are initially rather delicate, and require extra care. Wild-caught Jardines are rather nervous, and should be placed, communally if possible, in a secluded aviary with plenty of shelter where they can hide. Food and water should be provided in the shelter where the birds prefer to perch, and an anti-stress formula added to the water for the first 10 days. They will learn to accept a varied diet by using the same method described for the AFRICAN GRAY. Like other *Poicephalus* species, they enjoy nectar, which is taken from flowers. Within a few months they tame down and will no longer panic when approached. For nesting, they should be given the option of a natural log or a nest box measuring 30 x 30 x 60 cm (12 x 12 x 24 in.). The typical clutch consists of three, occasionally four, white eggs, which hatch after an incubation period of 27–29 days. Hatching take about 48 hours from the first pip.

Wild-caught Jardine's parrots will breed in their fourth year, but captive-bred birds have been recorded laying at two and one-half years. They can be expected to breed two or three times per year but, despite incubating well, they are often not reliable parents. Initially, two clutches per year are normal, but established pairs will produce up to four clutches. One pair at the Parrot Breeding Centre in South Africa produced over 30 chicks in less than three years. The male will often join the female in the nest box while she is incubating. Chicks removed for hand rearing make extremely interesting and intelligent pets. The diet for Jardines is as described for the CAPE PARROT, and the hand rearing is as for the AFRICAN GRAY. Growth chart also covered under the hand rearing section on grays.

Ring Size: 8 mm placed on the chick at 19–21 days.

Meyer's Parrot *Poicephalus meyeri*

Introduction: Since all small *Poicephalus* species require similar management, I have chosen the Meyer's parrot *P. meyeri* and brown-headed parrots *P. cryptoxanthus* as representatives of the smaller members of the genus. The species name of the brown-headed *cryptoxanthus,* means "hidden yellow" and refers to the bright yellow of the under-wing coverts. In some individuals this yellow extends to the carpal edge of the wing and is visible in a resting individual. In many cases yellow patches appear on the median and lesser wing coverts and even on the hind neck, making these birds individually distinct. The techniques described for these species can be applied with equal success to Ruppell's *P. rueppelli,* Senegal *P. senegalus,* Niam-niam *P. crassus,* and red-bellied *P. rufiventris* parrots.

Meyer's and brown-headed parrots prefer open woodland, but also use forest patches. They may occur in quite dry areas, where they are found mainly in the vegetation along watercourses. They usually form small flocks that appear at water holes soon after sunrise to drink. They appropriate vacant woodpecker or barbet nest holes as nesting sites, often occupying the same site for many years. A high-pitched call is emitted during flight or when perched, and pairs often utter a two-note duet. When under stress, they make a growling sound. Captive birds imitate human voices and whistles, as well as the calls of other birds. Six subspecies are distributed through northeastern, eastern, south-central and southwestern Africa.

Management: These small parrots are well established in aviculture. Newly acquired birds are rather nervous, but their captive-bred progeny are much steadier. They can be housed in flight aviaries or suspended cages with a minimum recommended size of 1.8 m (6 ft.) in length and 0.9 m (3 ft.) in width and height. To avoid aggression, they should be kept one pair per aviary. Like other *Poicephalus* species, they are not noisy birds. Outside of the breeding season the members of a pair show little evidence of bonding, and some birds sometimes even become aggressive toward their mates. They are great chewers, and will soon demolish any wood that they can reach. Aviaries should therefore be constructed of metal tube frames covered with weldmesh wire. Suspended cages, which allow better hygiene, should also be constructed of weldmesh. I have found that all of the smaller *Poicephalus* parrots are able to exercise sufficiently and reproduce well in suspended cages of the recommended size. The rear end of the cage should be covered on the roof, back, and the side from which inclement weather normally arrives.

Perches should be of hard wood. Natural branches are preferable, as these will cater to their exercise requirements. Perches of varying thickness will ensure that the feet are properly exercised as well. Green branches from a fruit, gum, or willow tree, or a piece of sugar cane or Napier grass, should be offered regularly for the birds to strip. Although this is a messy procedure, the nutrients and enjoyment that the birds derive from this activity makes the extra work worthwhile. The birds will also enjoy chewing on a partially rotten branch. Toys such as pine cones, a piece of rope hanging from the top of the cage, shells, and pebbles will also provide mental stimulation for the birds. A bare enclosure with dowel rods for perches and nothing to occupy the birds' attention will result in bored and overweight birds, unsuitable for breeding. Although these parrots are relatively hardy and can withstand overnight temperatures down to freezing, they do require warm days. In climates where the daytime temperature remains below 10°C (50°F) for extended periods, particularily if accompanied by wet conditions, they should be housed at least partially indoors during the most severe months.

Diet: *Poicephalus* parrots feed on soaked and sprouted sunflower seed, oats, millet, apples, grapes, carrots, guavas, boiled corn and beans, spinach, cabbage, and sugar cane. In addition, they require a mixture of fruits such as papaya, including the seeds, oranges, and apples. Hard beans should be soaked overnight and then brought to a boil for a few minutes before being cooled for feeding. A commercial parrot soft food or extruded parrot pellets should be mixed with the diced vegetables and fruit. Once a week I add hard-boiled egg, with the crushed shell, to the soft food. I have found that all of the small *Poicephalus* species require the same quantity of food, which is 170 ml (2/3 cup) of dry sunflower seed mix covered by 115 ml (½ cup) of soft food, fruit, and vegetable mix. They are fond of nuts, but these should be limited because of the high fat content. High fat foods should, however, not be totally excluded from the diet. During the breeding season in particular some high fat foods such as peanuts and sunflower should be added to the diet in limited amounts. Some pairs enjoy squares of whole wheat bread with a very thin spread of molasses or peanut butter. Calcium, vitamin, and mineral supplements are must for all the African parrots. Calcium supplements should contain some vitamin D3 as this vitamin is vital to the absorption of the available calcium. A teaspoon of diced cheese is relished by the birds but limited to once per week.

William Horsfield describes the following diet as given to his birds at the Amazona Bird Farm:

> "The brown-headed parrots at Amazona bred exceptionally well on a conventional diet of mixed seeds that were mostly soaked overnight, but not sprouted. These include sunflower, oats, wheat, barley, red and white sorghum. Sprouts are offered separately and include mung beans. A finer seed combination of mixed wild bird and mixed canary was offered but was not eaten in appreciable quantities and was totally removed from the diet at a later stage due to wastage.
>
> All manner of fresh vegetables and fruits are offered every day and these are diced finely in a food processing machine and fed in the early morning. The seed mixture is fed at midday. Birds are rationed so that they eat almost everything in their bowls by the evening. Breeding pairs have unlimited food. No supplements are added to fresh water that is changed if it is soiled during the day. Gardena sprinkler systems supply misted water for bathing. A vitamin and

mineral supplement like Avi-Plus Twinpack or Hagen Prime is sprinkled on the fruit/veg mix on alternate days. Probiotics are also added to this mixture.

Fresh flowering branches are relished, e.g., peach blossoms, Cape honeysuckle, Tecomaria, Coral Tree Erythrina spp and fruits like loquats and mulberry on the branches are eagerly appreciated. Hawthorn Pyracantha and Cotoneaster are also favorites. Although small birds, they enjoy cracking open large pods and seeds and items that are normally not thought to be given to a species of this size. Diet is one of the main ways in which aviary birds can be stimulated and we always make an effort in this regard. Feeding times are the highlight of their day, and surprise beats routine hands down. Plain old pellets day in and day out must be terribly boring and I personally only like to see them used as a supplement to a diet and not fed exclusively."

See the section on PSITTACINE ENRICHMENT for further ideas on creating a stimulating environment for the birds.

Chunks of food are held in one foot while the birds bite off pieces. In common with all *Poicephalus* species, water is sucked up when drinking without throwing the head back. A fine spray of water will also stimulate the birds to begin bathing and, in general, adds some enrichment to the birds day. There are records of these parrots consuming live food when they have chicks in the nest. Most aviculturists, however, breed this species without the addition of live food.

Breeding: The sexes are similar, but females have smaller heads and beaks. It is not unusual to have a pair of these birds set up for three or four years before they breed. Successful breeding of this species becomes more common every season. In the wild, brown-headed begin to mate in April, the beginning of winter. Strong evidence indicates that they use the same nest site each year, which consists of a hole in an old tree, dead or alive, and that they maintain a strong pair bond throughout the year and throughout their lives. These cavities are highly prized commodities. One observer watched hornbills and squirrels investigate cavities containing brown-headed parrot chicks without visible alarm from their parents, but within hours of other chicks leaving the nest some squirrels had moved in! Tony Cavalho has observed brown-headed feeding and breeding in coconut palms in Mozambique. They apparently eat the small, green coconuts on the budding inflorescences. In Zavora he saw them breeding in dead coconut palms. These trees had been struck by lightning and lost all their leaves and up to three pairs were nesting in one tree 6–8 m (19.6–26.2 ft.) above the ground. Although each nest faced a different direction, there appeared to be no aggression among the pairs nesting in the same tree. The ever growing and poverty-stricken population of rural southern Africa, view all trees as either a potential fuel supply or an impediment to growing food. Notwithstanding the capture and sale of the birds, their habitat, including all the nest sites, has quite simply disappeared. It has changed into buildings, fences, carvings, firewood, and unsustainable agricultural and grazing practices.

In captivity they will breed through most of the year in favorable climates, breaking only for the molt. Courtship in the Meyer's parrot includes allopreening and courtship feeding. In addition, the male runs backward and forward along the perch, completing a 360° turn in the process. The display also includes lifting the wings over the back and tilting the body toward the other bird. In the Senegal parrot both the male and female move back and forth along the perch, opening and closing their beaks and emitting soft crying sounds. The male then lowers his wings and the birds lift one foot and touch each other. During this display the male interrupts to feed the female. If the female is ready, copulation follows.

These birds should be given the choice of a nest box of 25 x 25 x 50 cm (10 x 10 x 20 in.) with an entrance hole 5 cm (2 in.) in diameter, or a natural log of similar dimensions. They prefer a rather tight entrance hole, which they will shape to their satisfaction. It is possible that the removal of the nest box at the end of the breeding season and its replacement at the beginning of the next season may provide the stimulus needed by *Poicephalus* parrots, which are reluctant to breed. It is not uncommon for them to choose a nest low down, and I know of one pair of Senegals that nest every year in a burrow that they excavated under the concrete floor of their aviary shelter. The clutch consists of 2–4 rather rounded white eggs. Some pairs are multiple brooded. Although nest inspections and ringing will be tolerated, if the adults were wild-caught these activities should be kept to a minimum. The stress of nest inspections and ringing chicks has caused some pairs to maim their young. If this problem is experienced with a particular pair, their nest should not be checked for about 40 days after the first egg is laid, at which stage the chicks will be ready to be removed for hand rearing.

The incubation period is 25 days, and the newly hatched chicks are covered in a light gray down. Both sexes take turn feeding the chicks. Well-developed chicks are fed at the nest entrance. They fledge at 12 weeks. Parent-raised birds are rather nervous initially, and care must be taken to not panic them, as they may injure themselves against the sides of the aviary. They should be removed when they are completely independent. Hand-reared chicks often make very good pets, but some become vicious. Such birds should be given a mate for breeding purposes. The young will begin breeding at 3–4 years of age.

Meyer's *P. meyeri*

Day	Meyer's
0	5.2
1	6.4
3	8.7
5	12.9
7	18.8
9	26
11	32.8
13	40.2
15	48.7
17	56.2
19	61.7
21	70
23	79.8
25	86
27	90.6
29	97

William Horsfield provides the following account of breeding the brown-headed parrot:

> "I found them to prefer natural logs as opposed to wooden boxes and these also lasted longer as they are more difficult to destroy. They are great chewers and the top and bottom of the *Syringa* logs (which are usually closed with soft pine) should be capped with metal sheeting to prevent them chewing through. Nest-log dimensions used with great success at Amazona Farm are 40–45 cm (16–18 in.) high x 25 cm (10 in.) diameter with 5 cm (2 in.) entrance hole. This is just large enough to allow a bird to squeeze through and also reduces light into the nest. The round inspection hatch is approximately 15 cm (6 in.) in diameter and allows the keeper to easily remove chicks carried in hand for ringing, etc., without having to squeeze too tightly. A welded-mesh ladder is stapled securely onto the inside of the log to allow easy access, as flat as possible so as to reduce the chance of the birds hooking a toe and getting trapped inside. In ambient temperatures over 30°C (86°F), the temperature in the nest will rocket with no air flow and excess heat being given off by the chicks, and we always open the inspection hatch to allow the chicks more ventilation. A section of welded mesh is secured to prevent escape. Even small chicks can be seen standing on their tip-toes trying to reach up to the fresh air of the entrance hole while panting and gasping for breath in such heat. Added ventilation can certainly save lives from heat exhaustion or trampling of weaker chicks by stronger ones trying to get fresh air.

As nesting material I offer untreated eucalyptus chips that are rinsed in VIRKON-S or the recently marketed VIRUKILL as nesting substrate, which the birds chew up into fine pieces before the female lays eggs. Using moss, bark, rotten wood and other natural substrates increases the risks of introducing fungal spores into the nest, which flourish in the hot, humid conditions of the nesting chamber and may lead to Aspergillus infections. Up until she is about to lay, both birds renovate their home to such an extent that they often kick and scratch most of the chips up and out of the nest entrance. I have observed them emerge from the nest and give themselves a good shake, and, have seen quite large amounts of nesting material being shaken from their feathers. For this reason the nest needs to be topped up with substrate to prevent the eggs rolling around on the floor of the nest and being damaged. Once the hen has laid, the pair cease their redecorating and the hen incubates alone with the male standing guard near the nest. Prior to laying, the lower abdomen of the hen becomes noticeably swollen. The hen sits very tightly and only vacates the nest to eat, drink and defecate. Voluminous droppings more than 10 times the normal size indicate that she does this only when she is really desperate to relieve herself. The male supplement-feeds the incubating hen, and her periods off the nest are brief. Clutch size in younger hens is usually 2–3 and in older hens 3–4. Eggs are pure white and not as rounded as the smaller *Poicephalus*, but typically elliptical and laid on alternate days with an occasional three-day gap. Incubation is 28 days and the first chicks often hatch on the same or consecutive days indicating that the hen only starts to sit tight a day or two after the first egg is laid. Chicks are covered with longish wispy, slightly off-white down and weigh approximately 4 g (0.14 oz.) at hatch.

They are initially fed a smooth, cream-colored consistency by the hen and which gradually becomes coarser and filled with more solid bits of seeds, fruits and vegetables as they get older and are fed by both parents. The chicks are quiet in the nest and only vocalize when they are being fed or when soliciting for food. They lie very still if inspected by the keeper or if older and fully feathered, try to crawl under each other to hide their heads. They leave the nest at approximately nine weeks and the upper beaks of the chicks are a light horn color and the iris is very dark and hard to distinguish from the pupil. The beak changes to the adult gray coloration within six months and the eyes lighten till about 12 months. Health problems are generally few with this group but at least bi-annual, if not quarterly, fecal checks should be undertaken to check for internal parasites."

Dr. D.M. Allwright, a particularly successful breeder of the red-bellied parrot, averages 10 chicks per pair per year, most of the young being hand raised. Wild-caught birds have been known to desert their young from the first clutch, so chicks should be carefully monitored and removed for hand rearing if necessary. A nest box measuring 20 x 20 x 40 cm (8 x 8 x 16 in.) will be accepted by red-bellied, but it is best to give the pair an options of a box, a natural log, or a sisal stump. Pairs become aggressive when breeding, and adjacent aviaries should be double-wired to prevent injuries to neighboring birds.

In display, the male approaches the female with the head and wings lowered while making a peculiar grating sound. Once copulation has taken place, the male feeds the female at frequent intervals. He continues to make the grating sound and regularly changes the side from which he serves his mate. Mating is a long-winded procedure, continuing for at least 20–30 minutes. Prior to laying, females spend about a month preparing the nest box. The 2–4 white eggs hatch after 26 days, and the fledgling period has been recorded as 80 days. Incubation is performed by the female alone, beginning with the first egg. The male often joins her in the nest box at night. Dr. Allwright had two wild-caught pairs that produced 10 clutches in two years; nine clutches consisted of three eggs and the other contained only two. Of these, 24 were fertile and 21 eventually hatched.

The chicks take about 36 hours to hatch after first pipping the shell. Parent-raised red-bellied, in common with other *Poicephalus* parrots, are extremely nervous when they fledge. The chicks' eyes open at two weeks of age, and they are independent at 8–10 weeks of age. Immatures vary considerably in color, but resemble the brightly colored male. Sometimes young females have paler plumage than young males, but often they are identical. They may be duller, have an orange-colored forehead and orange-brown irises. Young males have a broader head than young females. After the first molt, at 6–8 months, the males are brighter than the females. By the second molt, at just over a year, the females lose their orange feathers. The iris becomes bright red by two years of age. Dr. Allwright has had a 3½-year-old pair produce fertile eggs, but they did not prove to be good parents. William Horsfield had a hand-raised male paired with a mature female that laid a clutch of infertile eggs when the male was 12 months old, followed by a fertile clutch at 13 months.

These species are sociable outside of the breeding season and can be kept in a colony situation in a large aviary without problems. Juveniles can also be kept in groups and this is highly advisable in terms of them learning all the behavioral tricks through this type of interaction. They are curious, playful, and intelligent and get up to all manner of delightful mischief. They approach new objects in their aviary cautiously and suspiciously but soon inspect, initially with extended tongue and then by clamping the object in one foot and chewing on it. Fresh branches with buds are relished and objects like pine cones and sugar cane stems are chewed to bits. Some individuals may be more dominant and signs of plucking are usually an indication that the plucked birds are being harassed. The keeper must intervene at this point and separate the culprits from the group for a while.

Hand Raising and Artificial Incubation: William Horsfield describes his techniques as follows:

"Eggs are incubated at 37.5°C (99.5°F) and at a RH of approximately 55%. They are automatically turned in the Grumbach incubators and turning is stopped when the chick pips into the airspace or chips externally, whichever is first. Some chicks chip the shell before they enter the airspace. Humidity is increased at this point to 70% until the chick hatches and it is then placed in a brooder at 37.2°C (99°F) and fed Kaytee Exact every two hours. As with other Poicephalus, chicks often lie on their backs while being fed. I initially feed with a 1.0 cc tuberculin syringe with a tip attachment so as to deliver watery formula directly into the side of the beak. I add Medpet's probiotic formulation, Enteroplus, to my hand-raising formula and believe this to be beneficial to incubator-hatched chicks. Feeding intervals are decreased as is the temperature until the chicks are starting to get pin feathers and are being kept at approximately 25–28°C. They are then placed into plastic tubs on untreated pine shavings (changed daily) and kept in a heated room where they are fed 3–4 times per day. They are rung using a 8 mm stainless-steel ring at 18–21 days.

Once they are fully feathered, they start to try to climb out of the plastic tubs and we then place them into small wire bottomed weaning cages measuring 50 x 50 x 60 cm (20 x 20 x 24 in.) until they are independent. They are offered a wide selection of soaked seeds and a mixture of finely diced fruits and vegetables, on which they soon start to nibble. Boiled maize or corn-on-the-cob are favorite weaning foods as are softened (boiled) beans and peas. They wean at approximately 12 weeks at which point they can be moved to larger cages, weather permitting."

Ring Size: 8 mm for all small *Poicephalus* species.

Contributor: William Horsfield, Amazona Bird Farm, Durban, South Africa.

References:

Allwright, D.M. 1993. The redbellied parrot. *Avizundum Magazine,* July 1993: 8–10.

Garvin, C. 1993. Breeding the Rueppell's parrot. *AFA Watchbird* 20(3): 8–9.

Harris, F. & R. Harris. 1986. Breeding the brown-headed parrot. *AFA Watchbird* 13(1): 22–24.

Lavril, C. 1993. Handrearing at the Parrot Breeding Centre, Parts 1 & 2. *Avizundum Magazine*, October 3–7, and November 1993: 4–7.

Lavril, C. 1994. Jardine's parrots. *Avizundum Magazine,* May 1994: 20–23.

Low, R. 1986. *Parrots: Their Care and Breeding*. 2nd edition.

Low, R. 1995. Gentle Cape parrots are excellent breeders. *Cage & Aviary Birds,* 27/05/95.

Sharples, A. 1988. Breeding the Cape parrot. *AFA Watchbird* 15 (6).

Taylor, S. 1993. Poicephalus parrots: Senegals and their relatives. *AFA Watchbird* 20(3): 27–31.

14.01 Psittacidae **Lovebirds** *Agapornis* spp.

Glen Holland

Introduction: The nine species of lovebirds belong to the family Psittacidae and the order Psittaciformes. All are included in a single genus, and all except one, the Madagascar, or gray-headed lovebird *Agapornis cana,* originate from Africa. In their natural state, most species are found in dry woodland areas. They are gregarious and occur naturally in flocks of 10–300 or more, the larger flocks either being found at a good food source or near a water hole. Flocks are extremely noisy and easily located by their chattering calls given in flight. In many areas of abundance they are caught as a food source. The four eye-ring lovebirds ,which include black-cheeked *Agapornis nigrigenis,* masked *A. personata,* Fischer's *A. fischeri,* Nyasa or Lilian's *A. lilianae,* and peach-faced lovebirds *A. roseicollis,* are not sexually dimorphic. Madagascar, or gray-headed lovebird *A. cana*, Abyssinian *A. taranta* and red-faced *A. pullaria* are sexually dimorphic. Black-collared lovebirds *A. swindernia* are all but unkown in aviculture and are not covered in this section.

Black-cheeked *A. nigrigensis* is Africa's most threatened lovebird. Its entire range is localized to one area of southwest Zambia, occupying about 2,500 sq. km (965 sq. mi.). It is recorded that in 1926 trade in wild-caught birds resulted in the export of about 16,000 wild-caught birds. It is thought that the current wild population is about 8,000–10,000 maximum. One of the current threats to the wild birds is the change from farming a popular food source of millet and sorghum to maize. The World Conservation Union (IUCN) has recognized the threat to the viability of the wild population and possible extinction and has recommended a need, "to ascertain the present distribution of the species; to identify and evaluate the population's recovery; and to prepare a conservation strategy." A conservation project for the species was initiated in 1998. This includes working very closely with the rural population in the area and education on the threats to the species.

Management: Entire publications have been written on the management of lovebirds and, in particular, on the peach-faced lovebird, its various mutations, and the genetics involved. For the description of captive management, I will be concentrating on the black-cheeked lovebird, with additional comments and advice where management varies for the different species. Lovebirds are active birds, shrieking to their mates as they fly. Their calls consist of a variety of harsh shrieks and chattering. Peach-faced and black-masked lovebirds, in particular, have loud calls and can be a problem in urban areas. Pairs often duet with rapid calls.

Unfortunately, many of the *Agapornis* species have been hybridized. The continued practice of indiscriminate hybridization will result in a very limited "pure" gene pool in several of the rarer species, and will act to the detriment of aviculture. It is important that captive breeders husbandry practices are sound and do not pollute captive bloodlines with these undesirable hybrids or other weak stock. Unfortunately the production of hybrids was an acceptable practice only a few short years ago, and there are already many captive inter-specific crosses between black-cheeked and black-masked lovebirds. The pure black-cheeks do not have blue feathers on the rump. Only recently have pure black-cheeks become reasonably well established in aviculture. Aviculturists could play an important role in the survival of this species if pure, viable captive-bred populations are maintained. Studbooks and other forms of captive management should be initiated to assist in the management of our captive gene pools. Many of the captive strains of the Nyasa lovebird have become "polluted" through hybridization with the Fisher's lovebird. In New Zealand I am horrified by the number of Fischer's and peach-faced, even on the show bench, which have dark cheeks from hybridization with black-masked.

While lovebirds can be considered hardy, they originate from areas where frosts are probably the coldest weather they experience, and even then the peach-faced, for example, are comfortably tucked up in the middle of a sociable weaver's nest where temperature fluctuates little. They are fond of bathing and a wide-diameter shallow water bowl will cater to their bathing. A fine mist spray placed in the aviary on hot days will also encourage a flurry of bathing.

Lovebirds are rather destructive to vegetation, and the landscape of an aviary occupied by these birds is usually sparse, consisting of a coarse sand floor, plenty of dry branches for perches, with green branches and bunches of seeding grasses offered for the birds to strip. Trying to maintain lovebirds in a small, planted, garden aviary would be a frustrating experience. From an aesthetic standpoint, the rather sparsely furnished aviary is preferable.

Lovebirds can be aggressive, usually with the females being the more dominant and belligerent birds. Since some pairs or individual birds may be combative, a constant supply of vegetation to strip may provide entertainment and keep these birds busy. The introduction of new birds to a cage or colony can pose a problem. Established dominant birds have been known to attack or even kill newly introduced birds. Even a temporary situation, such as placing lovebirds in a carrying box, can result in this aggressive behavior. Rule number one: never crowd lovebirds! In the community cage, fledged youngsters should be removed after they are weaned to avoid any confusion with parent birds, or other breeding pairs. While the majority of species cannot be housed with birds other than those much larger than they are, I have housed single pairs of both gray-headed (Madagascar) and red-faced with estrildids, doves, and quail without any problems.

It seems important to mention that single-pair cage breeding is much preferred, and often more successful than colony breeding. I have noted this with several species, particularly the gray-headed (Madagascar) lovebird. A colony situation can be used to establish pairs by placing up to six pairs in a colony cage with six or seven nest boxes. When pairs are established and are sleeping in the boxes with their chosen mates, the aviculturist can block the entrance hole and transfer the box, still containing the pair of birds, to their new breeding cage. Once paired, most lovebirds will not accept another mate unless the original mate is gone. Remember also that lovebirds can and will bond with members of the same sex. It is therefore suggested that all stock is surgically or DNA sexed to determine if pairs are true. With species that are sexually dimorphic, like the gray-headed, Abyssinian, and black-winged, a pair of birds can be placed together and will usually bond and breed successfully. Should the pair fail, however, the mates can be swapped around again. When kept in communal situations, most species tend to get along quite well, and will sit side by side preening each other or feeding each other. Although the practice is not recommended, and would be illegal in many areas, lovebirds have been successfully kept at liberty, breeding outside the aviary.

Diet: The basis of the diet is an equal portion of millet, whole oats, and sunflower. This dry seed diet is, however, insufficient by itself, particularly if breeding is desired. Green food such as chickweed, dandelion, clover, or diced spinach should be mixed with a moistened soft food, the latter of which will supply the bird's protein requirements. Mixed finch seed, wheat, barley, and/or oats can be propagated in containers of clean river sand. Once they have produced a flourish of green leaves, a handful of these should be placed in the aviary after the roots have been thoroughly washed. Sprouted Japanese millet, oats, mung beans, beans, and sunflower should be added to the diet throughout the year. In the USA, many of the breeders are now using formulated pellets as a basis for their diets. Diced fruits, grated carrot, broccoli and beetroot, and seeding grasses are also devoured. Whatever greens remain after feeding will be taken into the nest. Once or twice a week each pair can be given a quarter slice of wholemeal bread. Many of these foods provide not only nutrition but also a form of psychological and physical enrichment to the birds, particularly in a cage. Providing more than one feeding site will help dissipate aggressive behavior in large colonies. Sterilized eggshells, cuttlebone, and grit should be made available at all times. Breeding birds should receive their soft food, sprouted seed and green foods, and water twice per day and in far greater quantities than the non-breeding birds.

Breeding: Courtship is evident by certain behavioral observations. A bird interested in mating will often extend the wings above the body, dilate the pupils of the eyes, and fan the tail while rubbing its beak on the perch. Head nodding and bowing to each other is also quite common as the male approaches the female for mating. The male often hops over the female and continues his "dance" on the other side. These behavioral moves continue until the female accepts the male and allows him to mate. Copulation then follows, lasting a few minutes. Most lovebird species are not really fussy about the boxes in which they nest. Many aviculturists prefer L-shaped boxes, as they help to prevent females from breaking eggs when entering the box. When breeding in colonies is contemplated, the aviary should always be furnished with at least one more nest box than the number of pairs of birds. In this way, even the last pair to nest will have a choice of nesting sites. Each nest box is placed on the same level, but vertical pieces of board can be placed between the boxes to provide privacy and avoid aggression. The nest boxes should be supplied with a base of teff grass, to which the birds will add a few leaves, pieces of bark, and additional coarse grass.

It is important that plenty of nest material is available. The black-masked and peach-faced in particular add large quantities of material to their cup nest, and spend days on end adding and subtracting the materials they choose. Peach-faced are rather unique in that, unlike the other species, which carry the material in the beak, they tuck strips of material into the feathers on the rump and then fly up into the nest with their "cargo". The female does most of the nest building, using grass, fresh twigs from fruit or willow trees, and fresh green grasses such as kikuyu. Palm fronds are a favorite nesting material and should be supplied as fresh and green as possible. It is particularly impressive to watch a peach-faced strip a large piece of palm frond and return to the nest with these trailing from her rump. This green nesting material also assists in maintaining a relatively high humidity in the nest. One of the reasons for low hatch rates can be low humidity, which encourages too great a weight loss in the eggs. A lack of fresh nesting material may result in no breeding taking place.

Lovebirds can prove extremely prolific and as the chicks become independent, they should be removed to start another colony. They breed most of the year but will take a break after a few clutches, or during periods of extreme temperatures. It is not uncommon for a successful colony to suddenly stop breeding for a period, particularly during periods of excessive heat and dryness. They will often produce up to four clutches of chicks before taking a break to molt.

The clutch numbers 3–6 white eggs, which are incubated by both sexes for 21–23 days. Usually the first two will hatch on the same day, as females often wait until the second egg is laid before they begin incubating. Initially the female remains in the nest after the chicks have hatched, receiving food from the male, from which she in turn feeds the chicks. After about 10 days, the male begins to feed the chicks directly, and although the male continues to feed her, the female starts to fetch food herself. Once independent, chicks should be removed to avoid aggression from the adults and interference with the next clutch. I have twice heard of breeders experiencing problems with adults attacking newly fledged chicks. The answer to this problem is to place the newly fledged young in weldmesh cages within the aviary, which permits the parents to continue to feed them, yet protects them from possible aggression from the adults. Chicks fledge at six weeks and are independent 2–3 weeks after fledging. The eggs and chicks of similar species can be fostered to other pairs even of different species. Pairs that repeatedly produce infertile clutches can be used to raise the chicks of other pairs. Chicks that are fostered should be added to a nest of chicks of a similar size.

One breeder who has had excellent success with black-cheeks kept the breeding areas rather dark, and it is possible that this provided the security that the birds required. This applies to many of the African parrot species as well. A fine-mist spray placed in the aviary can be switched on for short periods on hot, dry days; the humidity created will help to ensure good hatching success.

Two species that require slightly different management are the red-faced lovebird and the Abyssinian lovebird. Both of these species, as well as the gray-headed lovebird, have soft twittering calls and are sexually dimorphic. The red-faced lovebird is not well established in aviculture. Although they adapt well to captivity, they have very specific breeding requirements and are seldom bred. They are rather nervous birds and

Fischer's Lovebird
Agapornis fischeri

best housed in aviaries where disturbance will be minimal. These birds are not suited to combined cold and wet environments and, as their breeding tends to coincide with the rains, success in outdoor aviaries is unlikely in severe climates.

To breed these birds, a nesting system that will allow the birds to excavate their own nest is recommended. In the wild, nests are excavated into a termite nests a few meters above the ground. In captivity, a number of attempts at substitute nests have been made, most of which involve a soft material such as peat or cork into which the females have burrowed. It is important that the sides of the nest are covered in a more solid material that the birds will not be able to burrow through. A lovebird nesting box filled with compacted peat or cork would seem to be the answer, but it would be interesting to experiment with an artificial termite mound made of a soil mixture with a low cement content to add some binding properties. The tunnel should be started and left for the birds to complete. One breeder recommends that horizontal rather than vertical boxes are preferred.

Red-faced lovebirds require a diet with a high proportion of seeding grasses and weeds, and they will also take live food such as mealworms, which, together with soft food, are in great demand when the birds are nesting. One successful breeding record talks of the male taking in excess of 300 mealworms per day. In flights they should be kept as single pairs, but in large aviaries, provided sufficient nesting sites are available, a few pairs can be housed together.

The Abyssinian lovebird originates from montane forest where freezing temperatures at night are common. They descend to low savanna to feed. Unlike most other lovebirds, they are solitary nesters, so it is advisable to allow the pair a flight to themselves. A flight cage of 0.9 x 0.9 x 1.8 m (3 x 3 x 6 ft.) is recommended, but they have also bred successfully in smaller cages. As they are fond of bathing, fresh water should be supplied daily. This species has been found to have a higher vitamin B requirement than other lovebirds. As they invariably roost in the nest, a nest box should always be available to them. Long, slanted boxes with an internal diameter of 10 x 12 cm (4 x 5 in.) have produced the best results with this species, but they will also accept conventional L-shaped boxes. The floor of the box should have a central concave area to ensure that the eggs remain in the center. Unlike the other lovebird species, they do not use much material to shape the nest. A small quantity of shavings and moss should be provided, and the birds will then proceed to throw most, if not all, of this out. They usually produce two, and occasionally three, broods per year. Chicks resemble the adults after 4–5 months. Unlike other lovebirds, they are born naked with little or no down.

The Gray-headed Madagascar is one of my favorite species and, while stationed in Zululand, South Africa, I had a pair of these in most of my finch aviaries and found them quite prolific. It was interesting that while in the tropical/humid areas, others and I bred this species in large numbers. In the higher altitude areas, which lacked the high humidity and were extremely cold in winter, few of this species were bred, with many reporting chicks dead in shell (presumably because of low relative humidity). Fred Barnicoat provides a good account of his experience with this species:

> "I housed my pair in an aviary 2.3 m long and 2 m wide (7 ft. 8 in. x 8in.), and 1.3 m (4 ft. 4 in.) of the length was covered by the shelter, which had a glass fiber front to ensure it is relatively light. By using the large door of the shelter to close the birds in at night, these lovebirds were soon trained to sleep under cover; and like many other birds they do not revert to sleeping in the flight if given the chance. They made some of their nesting material from the leaves of a mulberry tree. The lovebirds nibbled around the outline of each of the leaves to form a piece of greenery shaped like a large letter C. Most of these pieces are taken into the nest boxes, but some are dropped on the ground. The skill they show in cutting out these regular shapes is amazing.
>
> Before I got around to building them some boxes they went down in the L-shaped Gouldian boxes I normally use for Gouldians, and they have continued to find these attractive. In these boxes I have tried a few centimeters of peat and wood shavings, which so far they have for the most part removed. My most

successful nest was produced when a pair took over a box that I had filled with grass for a pair of Gouldian finches. They enlarged the hollow I had shaped, and gradually the grass was reduced to very tiny pieces and the remains were eventually scratched into a tiny pile in the corner by the chicks. The pair nested twice, in May and August.

The female disappears into the box and must surely be fed by the male. She will remain in the box when it is taken down for inspection and will have to be lifted out or moved aside if the eggs are to be seen. The females also appear to pluck some of their feathers when they nest.

The chicks are covered in short white down and hatch after an incubation period of 22–23 days. Incubation commences with the first egg and the chicks show a considerable size difference between the largest and smallest, but the latter usually survive without any problems. Several fanciers to whom I have spoken maintain that this lovebird breeds best when several pairs are kept separately but in nearby aviaries where they can hear each other, and that a single pair seldom breeds."

Hand Rearing: Individually hand-raised birds can make excellent pets. The basic principles of hand rearing lovebirds are as described for other psittacine species. I personally prefer to use a feeding spoon, but have used a syringe, which is equally successful with lovebirds. Newly hatched lovebird chicks are extremely small, demand very regular feeds and are rather difficult to handle. If rearing lovebirds as pets, it is best to let the parent rear for 10–12 days and then pull the chicks. It is best not to leave chicks for hand rearing with the parents too long as they can be stubborn and challenging to feed and tame. When small, lovebird chicks, as with some of the lories, will lie on their backs. While this is quite normal it is easier to feed them if held upright in one hand while feeding with the other.

Claire Peche of World of Birds in Cape Town, South Africa, gave an interesting account of some dietary problems encountered while hand rearing the Lilian's or Nyasa lovebird:

> "The reason we had to take the chicks out of their boxes before fledging was that there seemed to be too many adults in the small aviary that housed them. Fledgling chicks had to face aggression from adults other than parents, and were treated as unwelcome new intruders. [G.H.: I have seen the fledglings of this species being raised in separate cages placed inside the breeding aviary, which prevented them from being bitten by other birds while still allowing the parents to feed them. This may be a suitable solution in warm climates.] It was a difficult learning period when the normal commercial, parrot hand-rearing diet failed on the Lilian's. Although they ate willingly, they didn't seem to absorb the nutrition of the food, and they kept on losing weight more visibly with every day. Experimentally, I started to add boiled egg for extra protein but it made no difference. Then I tried cottage cheese, which didn't work either. Also peanut butter didn't help fatten them up again. While young were continually being produced, I only managed to rear a minimal number with a continuous struggle. I then experimented with adding digestive enzymes to the parrot hand-rearing formula. It worked! Immediately 10 chicks were reared without difficulties or losses."

Avian nutritionist, David Dennison, comments as follows:

> "We really must commend Claire for her efforts and dedication. Claire was using a parrot hand rearing formula, which is intended for parrots and is not meant for small parakeets. Lovebirds have a higher energy requirement than parrots and parakeet hand-rearing would be more suitable. When using parakeet hand-rearing formula, digestive enzymes would not be necessary. On a more general note: When baby birds being hand reared on a commercial diet are thin, it is hardly ever due to a lack of protein. Baby lovebirds need no more than 20% protein for good growth, and even less once they are feathered. If baby birds are inclined to be thin and always hungry on a certain diet, the addition of a small amount of sunflower oil will increase the total energy value of food, which may be what a particular species requires. Another tip: in most cases one can use commercial hand-rearing diets from beginning to end, but when baby birds are nearing the weaning stage,

Hand rearing lovebird chick.

they often refuse to eat and lose condition. At this stage it may help to add grated fresh apple and carrot to the formula. This would decrease the total amount of protein and the extra fiber would prepare their gizzards for more solid adult food."

Jeffie Delaere provides the following guideline on weight gain for peach-faced lovebirds:

"General consensus is two grams per day of age. The averages I go by are:

1–6 days, 2–15 g (0.07–0.5 oz.)
7–13 days, 15–25 g (0.5–0.9 oz.)
14–20 days, 25–35 g (0.9–1.2 oz.)
21–27 days, 35–45 g (1.2–1.6 oz.)
28–34 days, 45 g & up (1.6 oz.)

Most of my babies will reach 60 g (2.1 oz.) by the time they are five weeks and weaning. They should never lose more than 15% of their body weight while weaning. We have lots of birds that weigh over 60 g (2.1 oz.), so they tend to keep growing a little longer than average birds. After four weeks most babies actually drop a few grams in preparation for flying at about six weeks.

The birds were fed as follows:

02/04/2003: 7-10-1-4-7-10 o'clock,
from 16/04/2003: 7-11-3-7-10 o'clock
from 25/04/2003: 7-12-7-10 o'clock.

When four weeks old they get a small container with some seed, a piece of millet spray, etc. available. They initially play with these items, but soon learn they are edible (after approximately one week). Till about three weeks they are kept in a brooder (thermostatically controlled, ceramic heating element), and then in a large with a nestbox allowing them to exercise feet and wings.

Lovebird chicks should be weaned by six weeks old. From about 32 days on, a variety of foods should be offered to the chicks and spread out on the substrate that the chicks are on to encourage testing, and ultimately feeding. If food is only placed in a bowl it can take the chicks quite some time to learn where to find and test the food".

As a child I owned a pet peach-faced lovebird, "Mushi," that made an excellent companion.

Lovebird Websites:
http://www.africanlovebirdsociety.com/
http://www.parrotparrot.com/lovebirds/breeding.htm
http://www.aviancompanions.com/links/lovebirdsociety.html

Ring Size: 4.5 mm for most species; 4 mm for red-faced, Madagascar, and black-cheeked.

References:
Barnicoat, F. 1980. Breeding the Grey-headed lovebird. *Avicultural Magazine* 86: 74–80.
Peche, C. 2002. Hand-rearing Lilian's (Nyasa) Lovebirds. *Avizandum Magazine* 14(2). 21–22.
Warburton, L. 1997. The Black-cheeked Lovebird of Zambia. *Avizandum Magazine*. October 1997: 5–8.

14.01 Psittacidae Asian Long-tailed Parakeets

Glen Holland

It is interesting to note that while many species, such as the African and Indian ringnecks, are common in the natural state, there are some species still considered common that are under severe pressure in parts of their range. These pressures arise from the loss of habitat and tall nesting trees, and collection for the pet trade. While bird watching in Thailand, for instance, I was taken to a an area where moustached *P. alexandri* were common until recent times, but local people had found a patch of tall trees where the birds nested and, due to the harvesting of chicks for the pet trade, the colony was no longer self sustaining.

The genus *Psittacula* has 13 species and numerous sub-species, which are described in the Table 1 on the following page.

Management: It is now common practice to breed many of the psittacula in suspended "Noegel" cages but many breeders still use aviaries that include the ground.

Although many such species may now be considered domesticated, they naturally spend much of their day in the canopy, and aviaries should be built as tall as possible. Minimum recommendations on aviary size are supplied in Table 3, but these are only the minimum to avoid obesity and ensure some level of breeding success. Note the success that taller aviaries made to the breeding of the echo parakeet. As with all psittacines a fair amount of enrichment, as described in the section on PSITTACINE ENRICHMENT, should be provided for this family. They are all naturally destructive to some degree and enrichment needs to include both psychological and physical stimulation. The Alexandrine is the most destructive of the family and should never be without a half rotten stump or two to work on.

It is essential that in cold climates the birds have access to and are encouraged to use shelter. Frostbite is a common problem and will result in toes lost if the birds do not use their shelter. I have housed a number of species of psittacula in large walk-through aviaries where they display beautifully. One pair of Derbyan's in particular did very impressive flight displays as they lapped around a large walk-through Asian aviary where they were housed with a variety of fruit pigeons *Ducula* spp, crowned pigeons, softbills such as pekin robins, fairy blue-birds and pheasants. It should be noted that although some species, such as ringnecks, can be kept as a colony in a larger aviary, when breeding psittacula other than ringnecks, they become aggressive and serious wounds often result following belligerence between the birds. They are best and most productively held as pairs in separate breeding aviaries. It is also not advisable to mix different species of psittacula in one aviary as, even though the birds may appear to settle, at some stage aggression and resulting wounds will occur. It has been recorded, however, that the plum-headed are extremely gentle birds and pairs have even been housed with finches without any issues.

Lyrae Perry writes on off-season fighting:

> "maintaining strength and breeding condition of the birds is of utmost importance. Therefore the breeding pairs should be housed in smaller aviaries in the breeding season and should be well flighted in the off-season. In other words, they should be removed from the breeding aviary and put into the largest cage or aviary possible so they can exercise and fly around. We have found that flighting the birds without nest boxes, and resting them in the off-season, has improved production in all that are held in small aviaries or cages. Moving them is actually stimulating for the birds and not negatively stressful as once thought. When resting your birds you want to minimize squabbles, so its best to put the females in one flight together and the males of the same species together in an adjacent flight. In large aviaries up to 10–15 birds can be placed together in the off-season. The birds should all be observed frequently to be sure that each is getting enough to eat and drink. For that reason it is best to put more than one food and water bowl in each aviary. Nest boxes should never be left in resting flights."

Sexing of young psittacula is best done by DNA feather samples, but most adults show sexual dimorphism (see Table 2) by 18–24 months of age. The nominate race of moustached parakeet *P. alexandri alexandri* and two sub-species *P.a. kangeanensis* and *P.a dammermani* are the only psittacula where the sexes are not clearly dimorphic as adults. Both males and females have coral red bills and look so similar in appearance that most are taken for adult males of other five moustached subspecies.

Table 2

Common name	Sexual Dimorphism
Alexandrine *Psittacula eupatria*	The female lacks the black and pink neck ring, being mainly green with the wing coverts red; the tail is slightly shorter than that of the male.
Derbyan *Psittacula derbiana*	The female is distinguished by the bill, which is entirely black as oposed to the males which has a red upper mandible with a yellowish tip and a black lower mandible.
Long-tailed *Psittacula longicauda*	Male has a wide black moustache stripe, the top of the head is emerald green, and has a rose-pink color between the crown and moustache. Female has a dark green moustache stripe, darker green crown and the color of the cheeks is more orange than pink.
Echo *Psittacula eques*	In the male the bill is red; black in female. Mature males have a ring around the neck, similar to that of a male rose-ring parrot.
Rose-ring (ringneck) *Psittacula krameri*	Male has a black ring encircling the neck, edged with pink on the back of the neck. Female is entirely green, with shorter central tail feathers than the male.

Slaty-headed *Psittacula himalayana*	Adult males have a maroon alar or wing patch.
Malabar *Psittacula columboides*	Male has a bluish collar below the black nuchal collar, a facial mask of bluish green, and red upper mandible. Female lacks the bluish collar below the black nuchal collar, has a gray head and both mandibles xare black.
Moustached *Psittacula alexandri*	Adult males have a red upper and black lower mandible and pink chest. Female has a black upper mandible, pink chest but the head is bluer.
Plum-headed *Psittacula cyanocephala*	Male has a plum-colored head with a bluish-purple tinge. Female has a yellowish gray-green head. Young males begin displaying at five months.
Blossom-headed *Psittacula roseata*	Male differs from male plum-head with forecrown, cheeks and ear coverts pink. Female differs from female plum-head by having a small maroon wing patch.
Emerald-collared *Psittacula calthorpae*	Lower mandible brownish in male, upper is red; in the female the bill is black. The female has duller gray areas than the male and the mantle is a greener shade of gray.

Diet: Lyrae Perry writes on the diet of psittacula:

> "Asian parakeets are not primarily seedeaters in the wild. They eat a variety of plant materials, flowers, fruits, vegetables, grain, and possibly insects in the wild. In captivity they can be fed a seed diet, but this will be insufficient to raise chicks. There are the occasional pairs that will feed babies on a seed diet, but these are the exception, not the rule. It's not because they are bad parents, it is because the volume of food necessary to adequately feed the average clutch of 5–6 chicks is just incredible. The parents would be exhausted cracking two or three cups of tiny seed each day just to feed the little ones. So we make it easy for the parents with chicks and give them soft foods first thing in the morning. The soft foods can be any of the following: frozen mixed vegetables (thawed before he birds get them), cooked brown rice or whole grains of any kind, seasonal vegetables like squash, snow peas, snap beans, and spinach. We feed grated carrot and chopped celery every day to our Asian parakeets. The parent birds will appreciate extra celery for the hatchlings after the third day. It has been my experience that the parents rarely choose to feed fruit to the chicks. They prefer to feed vegetables instead.
>
> Since vegetables alone will not provide enough protein for the growing chicks, the birds should be offered the proper foods for complete protein groups. This is easily accomplished by feeding cooked legumes along with corn or preferably brown rice. It should be noted that legumes are not digestible unless completely and thoroughly cooked, and we have had evidence that undercooking beans may be linked to visceral gout in birds. Some of the seed coatings on legumes are toxic when undercooked, primarily the red beans. Legumes are cooked well enough if they are the texture that is pleasing to humans, which is very easy to do. (GH: I have read that by soaking the beans overnight, washing them with clean water the following morning and then bringing them to the boil before cooling again, avoids lengthy periods of boiling and provides the ideal bean texture. Soy beans can be dangerous as they contain growth inhibitors that remain active if not sufficiently treated with heat. They need to be cooked twice as long as other beans.)
>
> Some of the most nutritious legumes to use are lentils, black-eyed peas, and black beans. Black beans were not favored by Lyrae Perry's birds but lentils and black-eyed peas were. When cooking legumes it's best to cook them separately, since the different sizes require different cooking times. If you cook lentils and black-eyed peas together, you end up with a pot of black-eyed peas in thin lentil "gravy." This is because the lentils take one-third the time to cook as the black-eyed peas and they just cook down to a fine mush. You might want to sprout the lentils instead. It is easy and the birds love them."

The echo parrots in Mauritius are fed the the following diet: apple, pear, orange, chilli, capsicum, silverbeet, corn, beans, melon, tomato, eggplant, watermelon, squash, cucumber, kiwi fruit, sprouted mung beans, sprouted sunflower and sprouted chick peas. Kaytee parrot pellets and small amounts of hemp, plain canary and millet. I have spoken to a number of breeders who give their birds a sprouted buckwheat, millet and sun or safflower mix and, in addition, add a little commercial or homemade soft food to the rearing diet of their birds. The homemade soft foods tend to consist of a boiled egg and wholemeal bread based crumble. One boiled egg, with the shell, to two thick slices of wholemeal bread is paced in a blender. A number of successful breeders also supply their birds with dry millet and limited dry sunflower, particularly in the non-breeding season. In addition the birds should have access to cuttlefish bone at all times.

Breeding: The female is the dominant of the psittacula pair and it is for this reason that males are always extremely cautious when approaching a female or during courtship. Display is pronounced in the males; it is a sort of ritual necessary to overcome their fear of the female. Table 3 provides most of the details required for setting up pairs for breeding.

The base of the nest should be filled with about 7.5–9 cm (3–4 in.) of untreated wood, such as pine shavings. The young psittacula are generally independent by three weeks of age (post fledging) but if the adults are not showing any signs of wanting to re-nest, the young should be left for at least five weeks. Although some species have been known to breed before fully colored at two years of age, it is usual for them to commence breeding at three years. Ringnecks are early nesters usually beginning to lay between mid-February and mid-March in the Northern Hemisphere and mid-August to mid-September in the Southern Hemisphere.

Rosemary Low and George Smith describe the courtship well:

> "As they are slow to form a pair bond, a male and a female ringneck should be put together several months before the start of the breeding season. Courtship commences in the late autumn or early winter. Prior to this, the male ringneck has cautiously preened the feathers of the female's wing butts. In preening he stands some distance away, and only by stretching his neck is he able to reach the female. After preening the female, he pulls himself upright with the wing butts held far from the body, but with only the wing tips in place. His eyes blaze. So far back, and high, does his head move that his near-side foot lets go of the perch to give the head more movement backward. He raised-foot salutes with a quick waver in the air and then his head is shot rapidly forward for another pecking-kiss of a preen. Copulation starts several weeks before egg laying."

George Smith also describes the display of the male long-tailed:

> "The whole body, but more particularly the head, is gracefully weaved in front of the female so that first one cheek and then his other is brought into her immediate gaze. The eye is blazed until it becomes a circle of white, then regurgitatory feeding occurs. During this, as with many of the psittacula, the head appears to be swirled around two or three times. When the movement is closely observed the head is seen to get its swirling movement because of combining the movements of normal vertical up and down head-regurgitation with the head swaying. Dignified walking and jumping are included in the display. Young females often lay small clutches. Most psittacula are single-clutched with occasional double clutching, but second broods are most likely after an unsuccessful attempt at the first clutch."

Plum-headed Parakeet
Psittacula cyanocephala

Rosemary Low reports that ringnecks often prove to be long-lived with one record of a wild-caught African ringneck that survived to a day short of 50 years in captivity. Although psittacula generally do not make good pets, I have seen the odd ringneck and Alexandrine that had been hand raised and maintained an affectionate bond with their owners. Many hand-raised ringnecks, however, loose their tameness after about six months.

Psittacula are not difficult to hand rear using commercial hand-rearing formulas. In the following section Ryan Watson provides detail on the hand rearing of the echo parrot.

Echo Parakeet *Psittacula eques*

• Ryan Watson

Before human settlement, the Mascarenes and the Seychelles had up to 12 parrot species, including possibly seven endemic species belonging to the Genus Psittacula (Jones, pers. comm.). Today only two parrot species remain, the Seyschelles black parrot *Coracopsis barklyi*

Table 3: Recommended Aviary Sizes

Species	Minimum Aviary Size	Nest Box Internal Dimensions	Clutch Size	Incubation (days)	Fledging	Adult Plumage (months)
Alexandrine *Psittacula eupatria*	1.2x3.6x2.5 m high (4x12x8.3 ft.)	30x30x90 cm deep (12x12x26 in.)	2–4	28	50–55	27
Derbyan *Psittacula derbiana*	See Alexandrine	See Alexandrine	2–4	26–27	54–60	15–20
Long-tailed *Psittacula longicauda*	1.2x3x2 m high (4x10x6.6 ft.) See Echo notes on height.	20x20x60 cm (9x9x18 in.)	2–5	23–24	60–70	18+ Males may start displaying at 5 months old.
Echo *Psittacula eques*	See Long-tailed	See Long-tailed	1–4	23–24	60–65	See Long-tailed
Ringneck *Psittacula krameri*	See Long-tailed	See Long-tailed	3–6	23–24	50–60	See Long-tailed
Slaty-headed *Psittacula himalayana*	See Long-tailed	See Long-tailed	3–5	23–24	48–52	See Long-tailed
Malabar *Psittacula columboides*	See Long-tailed	See Long-tailed	3–4	23–24	52–58	See Long-tailed
Emerald-collared *Psittacula calthorpae*	0.9x2.7x1.8 m high (3x9x6 ft.)	20x20x45 cm (9x9x18 in.)	3–5	23–24	+-50	See Long-tailed
Moustached *Psittacula alexandri*	See Emerald	See Emerald	3–4	23–24	51–58	See Long-tailed
Plum-headed *Psittacula cyanocephala*	See Emerald	See Emerald	3–4	23–24	49–55	See Long-tailed
Blossom-headed *Psittacula roseata*	See Emerald	See Emerald	3–4	23–24	49–55	See Long-tailed

Distribution and Size of *Psittacula* spp.

Common name	Latin name	Distribution	Length
Alexandrine	*P. eupatria* (5 subspecies)	Sri Lanka, India, Afghanistan, Assam Burma, Thailand and Indochina	60 cm (23")
Blyth's	*P. caniseps*	Nicobar is.	56 cm (22")
Derbyan	*P. derbiana*	NE Assam, SE Tibet	50 cm (20")
Long-tailed	*P. longicauda* (5 subspecies)	Malaysia, Borneo, Sumatra, Andaman Is., Nicobar Is. Riau, Archipelago	42 cm (16")
Echo	*P. echo*	Mauritius	36 cm (14")
Rose-ring (ringneck)	*P. krameri* (5 subspecies)	Senegal to Sudan and Somalia (African) Pakistan, India, Buram and Sri Lanka (Indian)	40 cm (15") African is smaller
Slaty-headed	*P. himalayana*	Afghanistan, Assam, China and Indochina	40 cm (16")
Malabar	*P. columboides*	SW India	38 cm (15")
Intermediate	*P. intermedia*	N. India	36 cm (14")
Moustached	*P. alexandri*	Java, Bali, India, China, Islands, incl. Andaman Simular, Babi, Nias	33 cm (13")
Plum-headed	*P. cyanocephala*	India, Sri Lanka	33 cm (12")
Blossom-headed	*P. roseata* (2 subspecies)	Assam, Buma and Indochina	30 cm (12")
Emerald-collared	*P. calthorpae*	Sri Lanka	29 cm (11")

and the echo parakeet *Psittacula eques*. In 1986 the wild population of the echo fell to an estimated 8–12 individual birds. Thanks to the intensive management of these few remaining birds, the population increased to an estimated 16–22 individuals by 1993. At the end of the 2002–2003 breeding season the wild population of echo parrots was estimated to be between 170–190 birds.

Management of the echo began in 1974 but did not become intensive until the early 1990s. Successful captive breeding was first achieved in 1994 (Jones, pers. comm.). Between 1974 and 1992, 20 birds were collected from the wild to form the captive breeding program. Eight of these birds were derived from harvested eggs, nine from harvested chicks, one was a sick chick rescued from a failing nest, one an abandoned egg hatched and raised in captivity, and one a non-breeding male (Jones, pers. comm.). Mortality was very high with only three of the 20 harvested birds surviving to 1993.

At the beginning of the breeding program, Indian ringnecks were used as a model species but proved to be inadequate. For the first twenty years of the breeding program the two species were kept in the same conditions and managed the same way. The Indian ringnecked thrived and successful breeding was a regular occurrence. However, under the same conditions the echo did not thrive and did not breed. These conflicting results stimulated the whole new approach to the methodology of keeping and breeding this species.

The three remaining birds consisted of one unsexed juvenile and a pair that were harvested during the 1991–1992 season (Jones, pers. comm.). Leading up to the 1993–1994 breeding season, this pair was observed courting and went on to successfully breed (Jones, pers. comm.). A world first! The successful breeding from this pair was no fluke, it came after many important changes were implemented to the echo husbandry techniques. The key changes were in the areas of diet and housing. The diet changed from a primarily seed-based offering of sunflower seed, canary seed, millet seeds, wheat and nuts, to a much more appropriate diet of fruits, vegetables, greens and nutritionally balanced parrot pellets. This change in diet drastically reduced obesity and other dietary related issues with echos. A large flight aviary with generously spaced out and partitioned breeding aviaries attached, was constructed for the echos. Previously the echos were housed side by side in small breeding aviaries that measured 3 x 1 x 1.8 m (10 x 3.3 x 6 ft.). The new facility is much more in tune with this species' natural ecology. In the wild, echos are strictly arboreal, so the height of the new facility was raised to 4 m (13.3 ft.). Anecdotal evidence indicates that highly arboreal bird species feel more secure in captivity when the height of their enclosure is considerably higher than that of an average person. The attached breeding aviaries were spread out with at least 2 m (6.6 ft.) between each and were solidly partitioned to avoid visual contact between pairs. In the wild, there is considerable distance between active nesting sites. Since these husbandry changes were made, successful captive breeding of this species has resulted every year since the 1994 breakthrough.

The intensive management of the echo is essential, and although certain aspects of the recovery process will cease, some will remain indefinitely. Recovery of the echo from near extinction has been very labor intensive and has required a multi-faceted approach in order to be successful. When numbers decline as low as the echo parakeets' did in the late 1980s and early 1990s, every individual bird is important and all measures should be taken to ensure that every bird has the best chance of survival. The practical aspect of the echo recovery program involves three teams: the field team, the captive breeding team, and the hand-rearing and release team. At this stage, all three of these teams play an important role in the recovery process, but by the end of the 2005–2006 breeding season, the captive breeding and artificial propagation of echos should be finished. However, the management of the wild population will continue indefinitely.

The primary goal over the last ten years, and until the 2005–2006 breeding season, is to produce as many birds as possible. The second priority is to have as many of these birds fledge from wild cavities as possible. Many measures are taken to achieve these goals, measures such as predator control, provision of nest boxes, modification of natural cavities, supplementary feeding, clutch and brood manipulation and control of nest-fly. Artificial rearing is only used as a last resort.

The echo field monitoring team for the 2002–2003 season consisted of four dedicated staff members that perform possibly, the most physically demanding job in Mauritius wildlife conservation. During the busiest periods of the breeding season, this team works seven days per week for 6–8 weeks straight. Their daily grind involves trekking up and down 6–8 kilometers of the steep mountain terrain in the Black River Gorges National Park. There were 30 nesting sites that were managed in the 2002–2003 breeding season and breeding echos used 23 of these. Depending on the stage of the season, each field worker carries a backpack weighing 10–25kg. Each field worker has to carry all the necessary tools and equipment to deal with any potential problems they might encounter once they arrive at the nest site. Nest access is usually difficult and requires the use of either a metal cable climbing ladder, or the use of jumars to climb up a rope. Every chick is weighed at least every second day, and usually every day during the first 10 days after hatching. Chicks are checked for nest-fly larvae in the nostrils and ear cavities and the nesting substrate is changed when necessary, replacing natural or soiled material with insecticide treated shavings. The field team uses a graph that has the average daily weight of an echo chick as well as a 20% below average weight. As long as a chick is above or on the 20% below average line, it is left in the nest. If a chick starts to fall bellow this line for any longer than a couple of days, it is removed for hand rearing.

Although Echos lay 1–4 eggs per clutch, the wild birds have never been recorded rearing more than 1–2 chicks per clutch. Consequently brood manipulation has proved to be an important management tool in the success of the recovery program. echos with infertile eggs are given fertile eggs from large clutches and some chicks are transferred around, up-sizing and down-sizing broods, so individuals of similar size and age are raised together and are more likely to get an equal share of the food.

The use of artificial propagation has contributed greatly to increasing the population of the echo. Close to 50% of the wild population are birds that were hand reared and released back to the wild. As mentioned earlier, echos are only hand reared when circumstances dictate; many due to several factors. Traditionally there have been problems with captive echo plucking their chicks at the onset of pinfeathers, usually at around 8–12 days of age. These plucked chicks would then be transferred to the nursery for hand rearing. However, during the 2002–2003 breeding season, wire mesh partitions were inserted into the nest boxes to deny the parent birds access to their chicks. The partitions were inserted when the chicks were approximately 16 days old and allowed the parent birds enough access to be able to feed their chicks without being able to pluck them. Although the chicks were young and would normally be brooded at this age, the warm climate of lowland Mauritius meant it was not essential to their development. As the chicks grew, the height of the tray would be raised to ensure that the parent birds could not reach any further than the out stretched beak of their hungry chicks. When the chicks were fully feathered, the trays were removed and the parent birds were no longer interested in plucking the chicks. In 2003–04 heat lamps were used to provide overnight warmth to chicks when the wire mesh partitions were inserted. Last season and for the first two clutches reared this season, the wire mesh partitions were not inserted into the nest box until the chicks were old enough to regulate enough body heat without parental brooding. This is usually at around 16 days of age and requires at least two chicks in the nest. This means the chicks have to endure up to eight days of plucking, which is detrimental to feather growth and undoubtedly stressful, not to mention painful, for the young chicks. With the additional heat provided, the wire mesh partitions were successfully inserted with chicks as young as eight days. The lamp heats up just one wall of each nest box and provides the chicks with a gradient of heat zones to choose which is most comfortable. Once it is noticed that the chicks have moved away from the heated wall, the heat lamp is no longer switched on.

Chicks are brought to the nursery from the wild for a variety of reasons. These include dehydration, poor weight gain, weight loss, cyclones, physical trauma, and hatching difficulties. Sometimes these chicks will be taken to wild cavities, but the majority are hand-reared and released. Efforts are made to ensure these chicks do not imprint on people. Talking in the presence of the chicks is discouraged and actually talking directly to the chicks is very poor practice. Handling and feeding time is kept to an absolute minimum and, when possible, chicks are brooded together to provide socialization opportunity. The young echos are taken to the release site at approximately 45–50 days old and fledge at around 60–65 days old. They are taken prior to fledging in order to allow them to fledge in more natural surroundings and in the presence of wild echos. The fledglings start the release process as soon they are weaned and are totally free by the time they are 65–80 days old.

Hand Rearing Echo Parakeets: Chicks were brooded usually in groups of 2–3 chicks of comparable age and size. Young chicks were fed via a syringe or a plastic pipette until they lost their vigorous feeding response at around 20–25 days. The tips of the feeding syringes were hollowed out so as to fit comfortably between the upper and lower mandible. This hollowing was done with a razor blade and once completed, it was smoothed over by melting it slightly with a lit match. Initially the same was also done with the plastic pipettes. Mid-way through the season, small lengths of silicon tubing were fitted to the ends of the plastic pipettes. The feeding end of the small lengths of silicon were also hollowed and smoothed over. Once the chicks became slow to feed because of reduced feeding response, they were fed directly to the crop via a silicon feeding tube attached to a 20 ml syringe. Avipro was used in alternate feeds, to substitute natural gut flora.

At each feed the following information was

recorded: the date, the chick's age, temperature and humidity of the brooder, the bird's crop status, weights of the bird before and after feeding, the amount of food given, type of formula, and any comments. The growth of each chick was closely monitored, and weights were graphed on Excel. Few echo chicks are raised from the egg and the ones that are raised from the egg are usually problem chicks that needed assistance to hatch. These chicks generally progress slowly for the first 7–10 days, as can clearly be seen on the feeding graph.

Contributions: Ryan Watson of the echo parakeet captive management team who was assisted with information by C.G Jones, the Scientific Director, and J. Malham, the Echo parakeet Coordinator of the Mauritian Wildlife Foundation.

References:

Low, R.1986. Parrots *Their Care and Breeding*. pp 115–152. London: Blandford Press.

Perry, L. 2000. The *Psittacula* of South east Asia and Indonesia. *AFA Watchbird* 4: 35–39.

14.01 Psittacidae Macaws

William Horsfield

Introduction: Macaws belong to the order Psittaciformes, and the family Psittacidae, which now includes all parrot-type birds. Macaws are geographically confined to the Neotropical region, occurring in Mexico and in Central and South America. There are three genera, the monotypic genus *Cyanopsitta,* the genus *Anodorhynchus,* with three species, and the polytypic genus *Ara,* which includes all of the remaining species. The lone species in the first-named genus is the Spix's macaw *Cyanopsitta spixii,* of which there is currently no single known specimen remaining in the wild. One was a male bird paired to a female Illiger's macaw *Ara maracana,* after an attempt to pair him up with a reintroduced captive-bred female of his own species was sadly unsuccessful. This male has unfortunately also disappeared. Three species of blue macaws comprise the second genus, including the hyacinth macaw *Anodoryhnchus hyacinthinus,* the rare and little-known indigo, or Lear's, macaw *A. leari,* and the possibly extinct glaucous macaw *A. glaucus*. Several species of macaws from the genus *Ara* are also now extinct, and the future of others is threatened by loss of habitat.

Macaws have long, graduated tails that are as long as, or longer than, their bodies, and have bare facial patches, which vary in size and color between the species. Those with white facial patches can blush red when excited. Most have tiny tracts of feathers on these bare facial patches. The miniature or dwarf macaws are proportionally scaled down versions of the large macaws, with the smallest being very similar to the *Aratinga* conures. Macaws are monomorphic, with sexes alike, and young birds resembling adults.

Management: Macaws are relatively easy to maintain. They are hardy and require little specialized treatment. Once acclimatized, they can tolerate very cold temperatures and, if housed with adequate shelter, have on many occasions not even needed to be taken indoors after snowfalls. I have seen hyacinth macaws break ice on their water baths and then bathe vigorously! (Sissen, H). They do, however, require increased energy to maintain their body temperature during cold weather, and increased fat is necessary to generate heat (see Diet).

The miniature macaws often roost in their nest boxes and it is therefore advisable not to remove these. All macaws can be noisy, and tolerance depends on the definition of noise. They have very powerful beaks and can certainly be destructive. The preferred housing is a steel-framed structure covered with welded mesh. Aviaries must be designed in such a way as to enable nest inspections without entering the cage (see Breeding). Although macaws will use just about any cage, and any environment, if they decide they are going to breed, a large flighted aviary, perhaps 5 x 2 x 1.8 (16 x 6.5 x 6 ft.) is ideal for the larger species. It must be remembered that they share a strong pair bond and usually fly at the same time. The width of the aviary should therefore be slightly greater than their combined wingspans (as with most parrots). When kept in smaller aviaries they invariably revert to climbing along the wire; this is due to the fact that, when attempting to fly in such facilities, their wings touch the sides of the aviary. The miniature macaws may be housed in smaller suspended aviaries, 2.4 x 1.2 x 1.2 m (8 x 4 x 4 ft.).

Echo Parakeet *Psittacula eques* Feeding Guide

Age	Average weight (g)	Number of Feeds (per day)	Avg. Amount Fed (per feed) (g)	Hand-rearing Formula (% of solids per feed)	Average Brooder Temp. (°C)
0	8.28	10	0.2–0.5	50:50 Ensure/LRS	36.7 (98°F)
1	8.75	8	0.5–1.0	50:50 Ensure/LRS	36 (96.8°F)
2	9.67	8	1.0–1.5	10% Pretty Bird	35.5 (95.9°F)
3	10.3	7	1.0–1.8	12% Pretty Bird	35 (95°F)
4	11.17	7	1.2–2.0	14% Pretty Bird	34.5 (94.1°F)
5	12.08	6	1.5–2.2	16% Pretty Bird	34 (93.2°F)
6	13.9	6	1.8–2.5	18% Pretty Bird	33.5 (92.6°F)
7	14.7	6	2.0–2.8	20% Pretty BIrd	33 (91.4°F)
8	19.24	6	2.2–3.0	20% Pretty Bird	32.5 (90.5)
9	29.54	5	2.5–3.5	20% Pretty Bird	32 (89.6°F)
10	31.14	5	2.7–4.0	20% Pretty Bird	32 (89.6°F)
11	39.42	5	2.7–4.5	20% Pretty Bird	31.5 (88.7°F)
12	43.36	5	3.0–5.0	20% Pretty Bird	31 (87.8°F)
13	42.65	5	4.0–5.5	20% Pretty Bird	30.5 (86.9°F)
14	48.48	4	5.0–7.0	25% Pretty Bird	30 (86°F)
15	64.56	4	6.5–8.0	25% Pretty Bird	29.5 (85.1°F)
16	67.16	4	7.0–10.0	25% Pretty Bird	29.5 (85.1°F)
17	72.95	4	9.0–11.0	25% Pretty BIrd	29 (84.2°F)
18	79.14	4	10.0–12.0	25% Pretty Bird	28 (82.4°)
19	88.67	4	10.0–13.0	25% Pretty Bird	27.5 (81.5°F)
20	93.85	4	11.0–14.0	25% Pretty Bird	27 (80.6°F)
21	101.08	4	12.0–15.0	25% Pretty Bird	26.5 (79.7°F)
22	111.44	4	13.0–16.0	25% Pretty Bird	26 (78.8°F)
23	119.93	4	14.0–17.0	25% Pretty Bird	25.5 (77.9°F)
24	130.11	4	15.0–18.0	25% Pretty Bird	25 (77°F)
25	134.49	3	16.0–19.0	30% Pretty Bird	24.5 (76.1°F)
26	140.15	3	17.0–20.0	30% Pretty Bird	24 (75.2°F)
27	146.5	3	18.0–21.0	30% Pretty Bird	24 (75.2°F)
28	154.78	3	19.0–22.0	30% Pretty Bird	24 (75.2°F)
29	160.97	3	20.0–23.0	30% Pretty Bird	24 (75.2°F)
30	165.79	3	21.0–24.0	30% Pretty Bird	24 (75.2°F)
31	168.17	3	21.0–25.0	30% Pretty Bird	24 (75.2°F)
32	174.72	3	22.0–25.0	30% Pretty Bird	24 (75.2°F)
33	180.62	3	23.0–26.0	30% Pretty Bird	24 (75.2°F)
34	183.3	3	23.0–26.0	30% Pretty Bird	24 (75.2°F)
35	186.5	3	23.0–26.0	30% Pretty Bird	24 (75.2°F)
36	188.6	2	23.0–26.0	30% Pretty Bird	24 (75.2°F)
37	190.35	2	23.0–26.0	30% Pretty Bird	24 (75.2°F)
38	190.94	2	23.0–26.0	30% Pretty Bird	24 (75.2°F)
39	192.45	2	23.0–26.0	30% Pretty Bird	24 (75.2°F)
40	193.04	2	23.0–26.0	30% Pretty Bird	24 (75.2°F)
41	137.37	2	23.0–26.0	30% Pretty Bird	24 (75.2°F)
42	192.17	2	23.0–26.0	30% Pretty Bird	24 (75.2°F)
43	190.91	2	23.0–26.0	30% Pretty Bird	24 (75.2°F)
44	189.23	2	23.0–26.0	30% Pretty Bird	24 (75.2°F)
45	186.63	2	23.0–26.0	30% Pretty Bird	24 (75.2°F)
46	185.72	2	23.0–26.0	30% Pretty Bird	24 (75.2°F)
47	183.9	2	20.0–25.0	30% Pretty Bird	24 (75.2°F)
48	180.84	2	15.0–20.0	30% Pretty Bird	24 (75.2°F)
49	177.05	2	15.0–20.0	30% Pretty Bird	24 (75.2°F)
50	170.91	2	15.0–20.0	30% Pretty Bird	24 (75.2°F)
51	164.13	2	10.0–20.0	30% Pretty Bird	24 (75.2°F)
52	161.95	2	10.0–20.0	30% Pretty Bird	24 (75.2°F)
53	161.4	2	10.0–20.0	30% Pretty Bird	24 (75.2°F)
54	159.41	2	10.0–20.0	30% Pretty Bird	24 (75.2°F)
55	159.91	2	5.0–20.0	30% Pretty Bird	24 (75.2°F)
56	158.16	1	20.0	30% Pretty Bird	24 (75.2°F)
57	157.06	1	20.0	30% Pretty Bird	24 (75.2°F)
58	155.88	1	20.0	30% Pretty Bird	24 (75.2°F)
59	152.29	1	15.0	30% Pretty Bird	24 (75.2°F)
60	155.88	1	10.0	30% Pretty Bird	24 (75.2°F)
61	154.68	1	5.0	30% Pretty Bird	24 (75.2°F)
62	153.5		weaned	30% Pretty Bird	24 (75.2°F)
70	146.66				

Many macaws can also be kept at liberty and, provided the area is safe from predators and overhead power-cables, the non-endangered species may even be allowed to free fly. They should, however, be securely housed at night.

Double wiring to prevent damage to toes and feet through neighborly aggression must be used to separate aviaries spaced next to each other. Newly fledged youngsters will otherwise almost certainly lose toes or feet in their first disorientated days out of the nest. Owners should be particularly familiar with three viral conditions that affect many species of parrots, including macaws. These are Proventricular Dilatation Disease (PDD), which generally affects older birds; Avian Polyomavirus (APV), which affects young birds; and Psittacine Beak and Feather Disease Virus (PBFDV). See Suggested Reading at the end of this section.

As with most intelligent parrots, macaws appreciate routine. They become accustomed to regular activities such as cleaning of aviaries, feeding times, nest inspections, and other management duties. They recognize individual staff members and interact differently with them. Through routine, they are made to feel more secure and less threatened. They are creatures of habit and will behave accordingly at different times of the day. The art of successful management has a lot to do with being able to spot the bird that is acting out of the ordinary, and determining why.

Diet: In essence, largely the same, in terms of fresh fruit and vegetables, as for AMAZONS and COCKATOOS, and the reader is referred to those chapters. There are now a number of formulated diets designed for macaws that are readily available and may be fed exclusively or as part of a varied diet. It is the opinion of many experienced and successful macaw breeders that formulated diets should not be fed exclusively, but rather as part of a varied diet. On the other hand, there are those who sing the praises of these diets. An avian veterinarian or specialist nutritionist should assist in advising on this topic. The diet in the wild is the subject of much ongoing field research and observational studies. As can be imagined, this is no easy task. Certain species are specialist feeders, e.g., hyacinths, which target Acuri and Bocaiuva palm nuts almost exclusively and do not enjoy fruit (Abramson, 1995) while most the others consume a variety of wild fruits, berries, seeds, nuts, and other plant material. Many of these contain animal proteins in the form of insects and their larvae, e.g., fruits of *Ficus* spp., and it is of interest that macaws have even been known to eat freshwater snails and fish trapped in drying ponds (Estudillo Lopez, 1986) as well as lizards and small dead animals (Low, 1990).

Macaws have a keen sense of taste. They relish many bitter, tangy, and spicy-hot items like devil-chillies, pepperdew, jointed cactus (prickly pear), artichoke, onion, garlic, almonds, and orange peel. Macaws need a higher fat, higher energy diet. They should be offered a variety of nuts including pecan, macadamia, peanuts, Brazil, walnut and coconut. Nuts should always be stored in a cool, dry place. They should never be used if there is any sign of mold, as this may cause fatal aflatoxin poisoning or Aspergillosis. A breeding pair of hyacinth macaws on display at Umgeni River Bird Park, RSA, died acutely in 1997 as a result of moldy nuts fed by a well-meaning visitor. For breeding, especially for laying hens, increased protein, energy, and calcium are needed. Birds get more protein by consuming more food, which also increases vitamins and minerals, but fats are needed in yolk formation, and calcium is needed for shell formation.

A variety of legumes, including mung beans, butter beans, broad beans, kidney beans, navy beans, black beans, chick peas, cowpeas, lupins, and lentils, are enjoyed, but should be sprouted and boiled to release trypsin inhibitors. Potatoes and sweet potatoes should be par-cooked in their skins to release the starches in a digestible form. A standard commercial macaw blend of sunflower, maize, sorghum, small seeds, groundnuts, wheat, dried pasta, etc. should be offered in conjunction with all the fresh fruit, vegetables, legumes, nuts, etc. and should not form the bulk of any diet. Animal protein in the form of chicken and chop bones is relished. Individual birds, usually when breeding, may take mealworms. Commercial soft food and vitamin and mineral additives are as for Amazons and cockatoos. Water-soluble vitamins added to drinking water are not generally recommended, as they oxidize quickly and provide an ideal growth medium for bacteria.

Breeding: Most macaws are not difficult to breed, and in captivity do not have a fixed breeding season. In fact, many will go to great lengths to reproduce in what would seem totally inadequate surroundings. Many household pets have decided to take over the laundry basket or the dog-kennel and raise a family. A few rare species have, in fact, been hindered in their attempts to breed by the non-cooperation of their owners with endangered species breeding programs. Macaws will accept almost any kind of dark, defensible retreat, placed at almost any level, and these have included asbestos dog kennels, drain pipes, waste-bins, underground burrows, etc. At Amazona Farm, I use horizontally placed octagonal wooden barrels, made from 50 mm (2 in.) thick pine scaffold planks. The barrels are 1.3 m (4.3 ft.) long, with a diameter of 60 cm (24 in.). For the large macaws, these are placed high in the sheltered area. They have an inspection hatch just off center, and an entrance hole at the other end. Sifted, washed river sand mixed with eucalyptus or pine chips is used as substrate to a depth of 20 cm (8 in.). The pair chew the chips into fine pieces and scratch a hollow in the substrate for the eggs. Those that are particularly destructive are provided with natural syringa logs, capped at both ends.

Macaws are fiercely protective of their breeding area, and nest inspection should be carried out carefully to prevent damage to chicks and eggs. Aviary design should incorporate nest access from outside the cage, with a slide to keep the adults out the nest while inspection takes place. A PVC 110 mm (4½ in.) pipe cut at a 45° angle at the end, of arm's length, may be used to place over the eggs and chicks in cases where adults cannot be lured out of the nest.

Individual macaws may break or even eat eggs, rendering the pair non-productive. Dummy eggs made from hard dental acrylic, of the same color shape, and mass as real eggs may cause the culprit to eventually lose interest in trying to break eggs. A specially designed nest having a concave false bottom with an egg-sized hole in it, through which the freshly laid egg falls onto soft padding, is the ultimate answer for these pairs. Macaws pair for life and will seldom change mates. Scarlets *Ara macao* are an exception, with fierce arguments occurring between pairs. One of my own breeding scarlets "fell in love" with a young hen placed in the adjacent aviary, and spitefully turned on his long-time mate. Within a short time of the young hen being removed, he dutifully returned to his unimpressed hen and they soon produced another round of beautiful youngsters.

The miniature macaws are prolific breeders, with the exception of the red-bellied *A. manilata*. This species is, however, now receiving more attention, and is being bred in increasing numbers. Mini-macaws do very well in vertically positioned wooden boxes or natural logs with an inside diameter of 20–30 cm (8–12 in.). This provides ample space for the 3–5 chicks.

The age of sexual maturity in large macaws ranges from 2–5 years in captive-bred birds. As with the other parrots, an older bird may drive a younger bird to breed earlier than usual. Historically, same-sex pairs have often fooled their owners, even simulating copulation; although such pairs appear totally bonded, they will obviously never breed. With accurate sexing techniques available, this should no longer present a problem. Macaws have also bred in the colony system and in large mixed aviaries. They are, however, most productive if housed in pairs. Macaws that have initially been kept as pets adjust reasonably well to a breeding environment, unlike many cockatoos. The period of their isolation from other macaws seems to be a limiting adjustment factor. Macaws will hybridize readily, and many such progeny are spectacularly colorful. However, the deliberate breeding of hybrids is a controversial issue, with many conservationists and breeders being appalled at the practice.

Artificial Incubation and Hand Rearing: If artificial incubation is only reserved for emergencies, then a single incubator may well suffice. However, if a number of eggs are to be incubated at any one time, then ideally at least three, if not more, incubators are necessary. In order to practice weight-loss techniques, the temperatures in all the incubators will vary by less than 0.5°C (0.9°F), but the various relative humidities may easily vary by 50%. In order to induce a wet egg to lose more weight, it will be moved into a dry machine, and vice versa. Normal eggs are straightforward and follow standard incubation procedures (see AMAZON incubation). Incubation temperature accuracy is critical, and digital temperature, humidity, and egg-turning controllers are now standard features with top brand incubators. These digital read-outs must also be checked against the set point and the controllers re-calibrated if incorrect. The temperature gradient within a single incubator may be surprising, and often quite contrary to the specifications of the manufacturer. This should be determined, from top to bottom and back to front. Incubators are usually most accurate in the area around the center of the machine. Use dummy eggs as spacers to position eggs as close to the center of the unit as possible, if using moving-carpet or roller turning systems, to prevent eggs rolling to the very front or back of the machine. Remember that the reading on the display is taken from wherever the end of the temperature probe is positioned, and this is not necessarily the temperature where the eggs are. Thermometers must never be trusted on their own and should be checked regularly. Those that are inaccurate are often consistently inaccurate and may still be used if this is taken into account. At Amazona Farm I use Grumbach mercury thermometers. There should be no break in the line of mercury if the thermometer is to be accurate to specification.

The use of potassium permanganate (KMNO4) and formalin (Formaldehyde) to fumigate incubators and brooders is no longer recommended, due to its recog-

Scarlet Macaw
Ara macao

nized carcinogenicity (Abramson, 1995). Of all macaw eggs, hyacinth have been the most difficult to hatch when incubated from day one. Many breeders have recorded the death of the chick just before or after internal pip. This seems to have been due to high incubation humidity and, therefore, insufficient weight loss. If incubated at 30–40% relative humidity, results are greatly improved. Eggs removed for artificial incubation should be carefully transported and not jolted, as this increases the incidence of embryo malposition. It is a common error to try and assist the hatching chick in the belief that it is struggling. In most instances, such interference is premature and causes more harm than good. If uncertain, it is preferable to leave the egg or obtain expert advice. After initial vocalization is heard from the egg, there is often a period of quiet from the chick. This is perfectly normal, as the chick rests before continuing with hatching efforts; it is this brief lack of vocalization that is interpreted by the inexperienced ear as a distress sign.

Hand raising is far less problematic with the wide availability of pediatric formulated diets. An avian veterinarian will be able to offer nutritional advice on available products. At Amazona Farm, I use Kaytee (macaw) from hatching for all species except the hyacinth, which requires a higher percentage of fat. At four weeks, we convert to Aviplus Parrot Handraising until weaning (GH: More recently William has used Harrison's diets for all of his birds). Care must be taken when hand feeding young macaws as excessive unilateral pressure, caused by the hand feeder feeding from one side only, may cause mandibular malalignment. This can be corrected by daily direct pressure to the opposite side or in extreme cases by fitting of a prosthetic extension to the lower mandible. This needs to be done as soon as asymmetry is noticed and the beak is still soft and growing. Prevention is, however, infinitely easier than cure.

The feeding response is a vigorous head-pumping movement, often with the whole bird jumping. This is often the cause of syringe and feeding-tube punctures to the pharynx, esophagus, and crop. Chicks will refuse food that is too cold, but will accept food that is too hot, causing crop burns. Care must be taken to check the temperature, making sure the formula is 39°C (102.2°F) when fed. Formula heated in a microwave, which is not recommended, must be stirred well to exclude hot spots. Tube feeding directly into the crop causes a higher incidence of crop burns, due to the fact that the formula has not been correctly temperature tested. Tube feeding is also potentially fatal if formula should accidentally be injected into the trachea instead of down the esophagus. Burns and punctures will generally heal surprisingly well if attended to immediately. Interestingly, birds are far less prone to postoperative infection than cats and dogs. Occasionally a young macaw develops a temporary habit of latching onto and biting siblings in a pumping feeding response manner. The bird should be separated for a few days to prevent it from causing skin and beak trauma to the siblings.

Scarlet macaw*Ara macao* and Blue & Gold macaw *Ara ararauna*

Day	Scarlet Weight (g)	Blue & Gold Weight (g)
1	20	20
5	45	40
10	90	65
15	150	100
20	275	190
25	450	265
30	640	430
35	770	585
40	920	790
45	1000	900

Incubation period for most large macaws is 26 days, with a pip to hatch interval of 24–48 hours, and a desired egg weight loss of 15–19% for the blue-and-gold *Ara ararauna;* 16–20% for the military *A. militaris* and Buffon's *A. ambigua;* 17–20% for blue-throated *A. glaucogularis,* green-winged *A. chloropterus,* and hyacinth; and 14–16% for scarlets.

Only seamless stainless steel bands should be used for the large macaws. They can be fitted 14–21 days. Aluminum rings are too soft and are often crushed onto the leg. Brass rings are unnecessarily heavy. Microchip transponders provide permanent identification and should be implanted on the left side of the neck. They are read with a compatible scanner and are visible on X-ray.

Ring Size:
Hyacinth, Buffon's, green-wing: 15 mm
Blue-and-gold, scarlet, military, blue-throated: 14 mm
Red-fronted *A. rubrogenys:* 12 mm
Severe *A. severa*: 10 mm
Yellow-collared *A. auricollis,* Illiger's, red-bellied: 9.5 mm
Noble *A. nobilis* cumanensis, Hahn's *A. n. nobilis:* 8 mm

Contributors:
John Vermaak, Ara Incubation Company, Durban, RSA.
Sharon Gardner, Birds Galore, Johannesburg, RSA.
John & Eve Daw, Eve Daw Aviaries, Everton, Kwazulu-Natal, RSA.

References Cited and Suggested Reading:

Abramson, J. 1995. *The Large Macaws: their care, breeding and conservation.* Fort Bragg, Raintree Publications., CA:

Berry, R.J. 1998. Breeding cockatoos and macaws in captivity. *AFA Watchbird* 25 (5): 7–14.

Clubb, S. 1998. Psittacine handrearing: problems, causes and solutions. Proceedings IV International Parrot Convention, Loro Parque, Tenerife, Spain.

Forshaw, J.M. 1989. *Parrots of the World.* Third revised edition.

Harvey, R. 1993. *Practical Incubation.* Revised ed. Hancock House. Blaine, WA & Surrey, B.C.:

Jordan, R. 1989. *Parrot Incubation Procedures.*

Low, R. 1990. *Macaws: a complete guide.*

Massie, M.J. & S.L. Clubb. 1998. Utilizing formulated diets in feeding strategies for breeding psittacines. Proceedings IV International Parrot Convention, Loro Parque, Tenerife, Spain.

Phalen, D.N. 1998. Avian Polyomavirus: my thoughts. *AFA Watchbird* 25 (5): 28–39.

Ritchie, B.W. 1995. *Avian Viruses: function and control.*

Ritchie, B.W. et al. 1998. Proventricular Dilatation Disease: a viral epornitic. Proceedings IV International Parrot Convention, Loro Parque, Tenerife, Spain.

Stoodley, J. and P. Stoodley. 1983. *Parrot Production, incorporating incubation.*

14.01 Psittacidae Conures

Terry Irwin

Introduction: The conures comprise one of the largest groups (second only to lories and lorikeets) of all the birds included in the family Psittacidae of the order Psittaciformes. The name conure is derived from the former, now invalid, genus name *Conurus;* species once included in this genus are now classified in the genus *Aratinga*. The term now covers a number of genera of Neotropical parakeets from Central and South America and the Caribbean region. In addition to *Aratinga*, the following genera are all regarded as conures: *Nandayus, Leptositta, Ognorhynchus, Rhynchopsitta, Conuropsis, Cyanoliseus, Pyrrhura, Enicognathus* and *Myiopsitta*.

A number of these genera are not covered here for various reasons. The only species in the monotypic genus *Conuropsis,* the Carolina parakeet *C. carolinensis*, is extinct. Two other species belonging to monotypic genera, the yellow-eared conure *Ognorhynchus icterotis* and the golden-plumed conure *Leptosittaca branickii,* are unknown in aviculture. Also excluded are the two species of *Rhynchopsitta,* the thick-billed parrot *R. pachyrhyncha* and the maroon-fronted parrot *R. terrisi,* which are not generally referred to as conures. They are more closely related to the macaws and are also known as macawlets. Similarly, the monotypic quaker, or monk parakeet *Myiopsitta monachus,* which is very unlike a typical conure in its habits, especially that of building its own nest, is not included here.

Conures range in size from 21 cm (8 in.) in the smallest *Pyrrhura* species, to 45 cm (18 in.) in the Patagonian conure *Cyanoliseus patagonus*. Also known as the "wedge-tailed parakeets," conures are excellent aviary subjects, being active and intelligent. In addition, most species are prolific breeders. Some of the larger conures make endearing pets, learning to mimic the human voice and perform various tricks. They love attention, thrive on human contact, and have been described as "puppy-like." These qualities are well displayed by the blue-crowned conure *Aratinga acuticaudata* that starred as "Paulie" in the movie of the same name.

Management: All of the conure species covered here are cavity nesters and take readily to nest boxes or logs in captivity. In the wild they use natural holes in trees or old woodpecker holes, or burrow into sandbanks or termitaria. Some have been known to nest in cracks or cavities in buildings, or in hollows in the tops of fence posts. Those using arboreal termitaria only nest in those still occupied by termites. The entrance hole is situated near the bottom of the termitarium and the tunnel goes straight up through the hard outer shell, then makes a sharp turn inward and downward into the softer core where the nesting chamber is situated. For about a week after excavating, the birds do not visit the nest. This is possibly to allow the termites to seal off the nesting area. Generally those conures that nest in termitaria in the wild prove to be the most difficult to breed in captivity.

My conures are housed in suspended cages. I have two sizes: 1.8 m (6 ft.) long for the smaller species and 3 m (10 ft.) long for the larger ones. Both sizes are 90 x 90 cm (3 x 3 ft.) in height and width, respectively, and are made from 25 x 25 mm (1 x 1 in.) welded wire mesh. Some breeders prefer the 13 x 25 mm (½ x 1 in.) weldmesh, but I find this difficult to keep clean, and the 25 x 25 mm openings are quite small enough to confine the smallest conures. One end of each cage is covered on top, back and sides with cement ceilingboard to form a shelter area that is 90 cm (3 ft.) or 60 cm (2 ft.) in depth, depending on the size of cage.

At the other end (front of the cage) is a feeding platform 30 cm (1 ft.) wide and 90 cm (3 ft.) long, made from 13 x 25 mm (½ x 1 in.) weldmesh. This platform is fitted a suitable distance down from the top of the cage, depending on the size of bird, and has a 4 cm (1½ in.) lip to prevent the dishes from sliding off into the cage. Both food and water dishes are placed on this platform. It is serviced by a small door, cut out of the front panel, which is hinged from the top and opens inward. The feeding platform is protected from the weather by a 45 cm (18 in.) wide roof of asbestos board covering the whole width of the cage.

A second, larger door is cut into one side, also hinged from the top to open inward; this door allows one to work inside the cage or insert a net to catch the birds. Only two perches are provided in the top line of flight, one placed as high as possible in the shelter, depending on the size of the bird, and the other resting on the lip of the feeding platform. This ensures that the birds have a free flight passage the whole length of the cage. A third perch or branch is placed just above floor level in the center of the cage. For species that are determined destroyers of perches, I have substituted cement perches for the two top perches. To satisfy gnawing needs I supply a piece of a tree stump, or, preferably, dry pine cones.

The cages are placed in rows on timber slats running along the back and front of the cage and mounted on poles about 1½ m (5 ft.) off the ground. Except for the small area of cage resting on the slats at the back and front, the whole area of the base is clear for droppings and waste to fall through onto the ground. The cages are spaced far enough apart to allow for access to the side doors. In addition to the board roof, I have a second roof of corrugated fiberglass sheeting, 1.0 m (3 ft.) wide, running the entire length of each stand of cages. This is fixed about 20 cm (8 in.) above the roof to allow for free circulation of air between the two roofs. This not only provides additional protection against storms, but also protects the nest boxes, which are attached on the sides of the cages in the gap between each cage.

As we have very mild winters my birds are all kept outdoors year round. The whole area containing the cages is covered in a light shade cloth or hail netting to keep out wild birds, especially hawks, and monkeys. This netting is also strong enough to prevent any bird that gets out of its cage from escaping, by holding it long enough to be captured. To prevent grass and vegetation growing up under the cages, I keep a few mallard ducks; they are quiet and non-flyers and are allowed to roam freely under the cages, devouring any greenery.

Diet: In captivity, conures will sample and enjoy a wider variety of fruit and vegetables, in addition to seeds, than most other parakeets. They also enjoy all types of green vegetables. This makes it easier to provide them with a balanced diet. All of my conures receive a seed mixture of millet, canary, oats, and 20% by volume of sunflower seed. This seed mixture I consider a standby, as they are also fed a soft food meal, which they relish, and which constitutes most of their food intake. This soft food consists of various dried peas, corn, and sorghum, which have been previously soaked, then boiled to soften. To this is added soaked oats, grated par-boiled carrot, apple, raw beetroot, and sprouted sunflower seeds. Into this mixture is stirred Avi-plus Parrot, a powdered food supplement containing minerals and vitamins. This is the powdered version of our local parrot pellets and a complete meal in itself. Just before and during the breeding season, a powdered calcium supplement and extra vitamins and minerals are added twice a week.

This meal is offered once a day in the afternoon. I have found that in our hot summers food dries out quickly in the mornings, hardening the peas and corn, which the birds then ignore. By feeding in the cooler afternoon, this does not happen and they have full crops when going to roost. They also find the food still soft in the morning when they start feeding. Therefore, by feeding in the afternoon, they have the advantage of the food still being moist and soft and palatable during these two most important feeding times. In addition, each morning, they also receive any locally grown fruit available, which varies depending on the season. This can include oranges, guavas, papaya, wild figs, banana, mulberries, loquats, etc., but not avocado. At this time they also get chickweed or other favored weeds that-grow in profusion locally. Three days per week we add finely chopped cabbage to their soft food, instead of the chickweed/weeds. Fresh water is supplied daily, bearing in mind that conures are avid bathers and will bathe in their drinking water.

Parents rearing young are given an extra feeding of the soft food mixture in the morning with plenty of Avi-plus Parrot. This can be mixed into a crumbly porridge and may be used on its own as a rearing food. I also feed an egg food of breadcrumbs and hard-boiled eggs, as an alternative. Extra chickweed and fruit is also given to birds rearing young.

Although grit may not be a prerequisite for conures, I provide a grit block, which also serves to keep their beaks and toes in trim. This block, which I make myself, consists of 10 parts river sand, washed and disinfected, plus four parts oyster shell grit and two parts building cement powder. This is mixed with a little water, to give the consistency of a concrete mix, and then poured into plastic containers to set. A short length of wire is coiled and inserted into the wet mixture, leaving two protruding lengths when set. These serve to tie the block to the cage above a perch.

Breeding: I use nest boxes made from 20 mm (0.8 in.) pine or eucalyptus planks, but nest logs are equally suitable. My nest boxes are either the conventional upright or L-shaped type. I prefer the L-shaped, as it gives the chicks more living space when they are old enough to move about. This also reduces the problem of overheating. The size of the box varies depending on the species. For the larger conures, a box measuring 25 x 25 x 40 cm (10 x 10 x 15 in.) high is suitable, while for the small species 18 x 18 x 30 cm (7 x 7 x 12 in.) high suffices. The nest boxes are hung on the sides of the cage in the top rear corner of the sheltered area. A suitably sized hole is cut in the mesh to allow for access into the box. In this position, the nest boxes are protected from rain by the fiberglass roof. All nest boxes are covered on the outside with weldmesh to prevent escapes should the birds chew through the box. There is also an inspection door in each box, which allows nest inspection from the passage behind the line of cages. The boxes have a layer of river sand, topped by pine shavings, as nesting material. As conures also sleep in their nest boxes, the boxes are left in position all year round.

Hand Rearing: Chicks are usually removed for hand rearing at about 15 days old, or just before their eyes start opening. At this time they are also closed-banded. I have found, especially with *Pyrrhura* conures, that the parents will frequently kill their young if returned to the nest box after ringing. One breeder who refuses to ring his birds says this is because he has lost up to 50% by doing so. Chicks removed from the nest are only fed their first meal of hand-rearing formula after their crops have emptied. I never feed formula on top of the parents' food. Usually parent-fed food is dryer and harder, often consisting of whole seed kernels, and takes longer to pass through the crop. Feeding formula on top of this may cause the crop not to empty completely of the harder contents, which could lead to problems.

I use a commercial rearing food and have found that conures thrive on all the different brands I have tried, when fed from 10 days or older. Occasionally chicks have to be hand reared from hatching. To feed from day one, a suitable formula must be used and the manufacturer's instructions should be meticulously followed. For the first few feeds I use pure Lactated Ringers solution, thereafter the feed for at least the first day should be as light as milk.

Conure chicks are generally easy to wean. Feeds are reduced when the chicks are fully feathered and become active. At this time they may also refuse the feed or take much smaller quantities. When this happens I offer my soft food mixture with a larger amount of Avi-plus Parrot. I also offer a dish of sunflower seeds and chickweed. When they start pecking at the food, I reduce the feeds to two per day, and when they are feeding themselves I feed only an evening meal until they are completely independent. The weaning process should not be hurried. The bird will refuse the feeds as soon as it is eating confidently on its own.

Aratinga spp.

Members of this genus are medium-to-large, long-tailed parakeets, ranging in size 24–36 cm (10–14 in.) in overall length. All have naked, or partially feathered, preorbital rings. Although most species of *Aratinga* are generally green with contrasting red or other colors on the head, wings, or breast, some, such as the sun conure *A. solstitialis* and the golden conure *A. guarouba,* are exceptions to this rule. They are yellow or fiery orange with green only in the wings and tail. Most, especially the larger species, have very harsh voices and can be very destructive to woodwork. Despite this, they make very attractive aviary subjects and can be very intelligent and entertaining.

The all-green *Aratinga* species include the Mexican *A. holochlora* and its subspecies, the red-throated *A. h.*

rubritorquis, as well as Finsch's *A. finschi*, Wagler's *A. wagleri*, the mitred *A. mitrata*, the red-masked *A. erythrogenys,* the Hispaniolan *A. chloroptera,* the Cuban *A. euops,* and the white-eyed *A. leucophthalmus*. The sharp-tailed conure *A. acuticaudata* and its better-known subspecies, the blue-crowned *A. a. haemorrhous,* can be listed here, albeit they have varying amounts of blue on the head and brownish-red tail feathers. Except for the last species, all are completely green (Mexican) or with only a red throat (red-throated) or have varying amounts of bright red on the forehead, bend of wing, and under wing coverts. Many are flecked with red on the head, neck, or body. They vary in size from the Cuban at 26 cm (11¼ in.) to the mitred at 38 cm (15 in.). Except for the smaller species, which may have gray eye rings, all have prominent white periopthalmic rings.

These *Aratinga* conures are generally not difficult to breed, although some species are more prolific than others. A vertical or L-shaped box, as described above, or a natural hollowed out log, will be readily accepted. The incubation period varies 24–26 days depending on the size of the species, and the young fledge after 6–8 weeks. Ring sizes vary from 7 mm for the smaller species, to 9 mm for the larger. The diet, as for all conures, is as described above.

The yellow *Aratinga* species include the sun and golden conures, already mentioned, and the jenday conure *A. jandaya.* Also included here is the golden-capped conure *A. auricapilla,* as some authorities regard this as the nominate race, with the sun and jenday conures as subspecies. These are the most brilliant of all the conures, a feature which compensates for their harsh voices! The sun and jenday have proved to be prolific breeders and are being bred in large numbers, especially in the United States, where selective breeding for color enhancement is also practiced. The golden-capped is threatened in the wild, and consequently much rarer in captivity than the sun and jenday. As such, care should be taken to maintain a healthy breeding population in captivity.

Sun Parakeet
Aratinga solstitialis

Under correct conditions of housing and feeding, the golden conure breeds well. One problem that many aviculturists have often noted with this species is its tendency to pluck itself. Opinions on the reason for this are numerous and, once plucking has started, none of the suggested remedies seem to work. Removing the birds to a completely different locality, away from the birds that were housed near them, sometimes alleviates the problem. The incubation period for these conures is around 26 days and they fledge at about eight weeks. The golden conure should be rung with a 10 mm ring.

The small Aratinga are generally green with varying amounts of color browns, yellow, orange or blue on the head, breast and wings. They are generally much quieter than the other Aratinga and also less destructive. They include the Jamaican *A.nana nana* and its subspecies the Aztec or olive-throated *A.n.astec,* St. Thomas *A.pertinax pertinax* and its 13 subspecies collectively known as brown-throated or brown-eared, the cactus *A.cactorum,* the dusky-headed *A.weddellii,* the Petz's *A.canicularis* and golden-crowned or peach-fronted *A.aurea*.

The St.Thomas is endemic to the island of Curacao in the Netherlands Antilles and was introduced to the island of St.Thomas from where it got its name. This is the most colorful of the *pertinax* group, having golden yellow-orange on the forehead, sides of head and chin. The crown is tinged with blue. The most colorful race is *xanthogenia,* known as the Bonaire conure from the adjoining island of Bonaire. Some (about 25% of the wild population) have a slightly larger area of yellow on the head than *pertinax*. However, as the majority of the population resembles *pertinax,* there is some debate as to whether or not it should be classified as a separate subspecies. The remainders of the subspecies differ only in the amount of brown and yellow in the plumage. The St. Thomas and Bonaire are rare in captivity. All the *pertinax* make good pets when hand reared and can learn to mimic a few words.

The Petz's and golden-crowned are very similar but can be distinguished between by their beaks. The Petz's have a horn-colored upper mandible while that of the golden-crowned is black. The Petz is better know in America as the half-moon conure and is very popular as a pet bird. Being a termitaria breeder in the wild, it has proved difficult to breed in captivity. It appears to be reluctant to accept a conventional nest box but I have had success with a horizontal boot type. This has a long tunnel dropping down at the end into a nesting cavity. The nest cavity is lined with damp soil scraped from termite mounds. This is pressed down to form a hollow and allowed to dry. I line this with some fine pine shavings but these are usually discarded and the eggs laid on the bare hardened earth.

The other small Aratinga are good breeders in captivity. They readily accept cockatiel-sized nest boxes lined with wood shavings. Some of the brown-throated and dusky-headed are especially prolific breeders, having as many as four clutches per year. These small conures have an incubation period of about 23 days and fledge at around six weeks. The chicks should be rung with a 6 mm ring at about two weeks old.

Nandayus nenday

This mono-specific genus is represented by the nanday, also known as the black-headed or black-masked conure. It is very similar to the Aratinga conures and its allocation of a separate genus has been questioned.

With regard to its general husbandry and feeding, the nanday should be treated as an Aratinga. It is very common in the wild and in aviculture is one of the best known conures. However, it has a very loud voice, which has made it unpopular. It nests readily and can be bred on the colony system. The incubation period is about 26 days and chicks leave the nest after about eight weeks. Hand-reared chicks make loving pets and enjoy being handled. They are also less noisy. They should be rung with an 8 mm ring.

Cyanoliseus patagonus

This genus is also represented by only one species, the Patagonian conure *C.patagonus patagonus* which has two subspecies, the greater Patagonian *C.p.byroni* and the Andean Patagonian *C.p.andinus*. The nominate race, or lesser Patagonian measures 43–46 cm (17–18 in.) while the greater measures 53 cm (21 in.). The greater Patagonian has more white on the upper breast, which in some specimens forms an almost unbroken band. The yellow and red on the abdomen are more extensive and the yellow is brighter. The Andean conure has little or no yellow and less red on the abdomen and no white on the upper breast.

The greater and Andean Patagonian are rare in aviculture but the lesser is well represented. They are willing to breed in captivity and can be bred on the colony system but do better when pairs are housed individually. They are, however, extremely noisy, and if a number are kept one should have no close neighbors. General husbandry and feeding are the same as for the larger Aratinga conures.

In the wild they nest in the banks of rivers or limestone and sandstone cliffs, excavating a tunnel culminating in a nest chamber. These tunnels, which can be 3 m (10 ft.) in length, may interconnect when several pairs burrow into the same site. In captivity they will accept a vertical box 25 x 25 cm and 45 cm high (10 x 10 x 18 in.) However, more suitable is one with a nesting chamber at the end of a tunnel. We have had success using old sisal poles. These are hollowed almost to the end and then left for the birds to excavate the nesting chamber. Boxes with tunnels leading to the nest chamber also eliminate the problem of eggs being broken by the parents, which is sometimes common with vertical boxes. The incubation period is about 26 days and the chicks fledge at around eight weeks. Hand-reared young make intelligent and affectionate pets. They can learn to talk and are less expensive than, for example, similarly sized macaws with which they have been compared. They should be rung with a 9 mm ring.

Pyrrhura spp.

This is a delightful group of small conures, which are becoming more popular in aviculture. Not all are known in aviculture and the rarer species are much sought after. They are small-to-medium sized birds measuring 22–30 cm (9–12 in.). Their basic coloration is green, with all but two species having edge markings to the feathers of the neck and chest. In all but four species the underside of the tail is reddish-brown or maroon. In the four exceptions this is gray. The sexes in all species are alike.

The maroon-bellied *P. frontalis* and green-cheeked *P. molinae* are now well established in aviculture and may be considered almost domesticated. Mutations of these species are now being bred. Blue mutation green-cheeked are now well established. Hand-reared green-cheeks are becoming very popular as pet birds.

The largest member of the group is the blue-throated (or red-eared) conure *P. cruentata*. This species is endangered in the wild and is the only conure listed in the CITES Schedule I. Once rare in aviculture, cage-bred birds are now readily available. Other species well established in aviculture include the crimson-bellied *P. perlata perlata,* formerly *P. rhodogaster,* and its subspecies, the Pearly *P.p. anerythra, P.p. lepida* and *P.p. coerulescens*. *P.p.anerythra* differs from the other Pearly's in not having any red on the bend of the wing and under wing coverts. Juniper and Parr classify the crimson-bellied as a separate species, *P.perlata,* and the Pearly as *P. lepida lepida,* with the others as subspecies of *lepida*.

The white-eared *P. leucotis leucotis* has four subspecies: *P.l. aucularis, P.l. emma*, *P.l. griseipectus* and *P.l. pfrimer*. The first two subspecies, *aucularis* and *emma,* have the forecrown to nape suffused with bright blue while *griseipectus* has a more extensive and brighter white ear. The race *pfrimer,* which has the sides of the head including ear coverts and forehead chestnut red, is unknown in aviculture.

The painted conure *P. picta picta* is very similar to the white-eared, also having white ear coverts. The two species can be easily distinguished by the scalloping of the breast feathers. While the white-ear's markings are in the shape of half moons, the painted's take the form of an inverted V. It has been found that generally the painted has been the most difficult to breed in captivity. However, it is well established and captive-bred birds are now readily available. Of all the subspecies of the painted, one, the rose-crowned conure *P.p. roseifrons* is the most beautiful. It differs from all the rest in having the crown and facial area bright red. Once rare in aviculture it is a prolific breeder and is now frequently available.

The black-tailed (maroon-tailed) conure *P. melanura melanura* is well established in aviculture. It has four subspecies but probably the most common in aviculture is *P.m. souancei*. The black-tailed is a prolific breeder with clutches of eight eggs being not uncommon. The fiery-shouldered conure *P. egregia egregia,* once rare in aviculture, is now well established due to its prolific breeding. Large clutches of 7–8 eggs are common with this species, and it is usually double brooded. The most beautiful feature of this species is the fiery orange and red shoulder patch and under wing coverts. It also differs from most other *Pyrrhura* conures in having a horn-colored beak. It has one subspecies, *P.e. obscura,* which has darker green upper parts and the maroon belly patch is less prominent.

The black-capped conure *(P. rupicola rupicola),* next to the green-cheeked and maroon-bellied, is probably the most common *Pyrrhura* in aviculture in South Africa. Once regarded as a rarity in captivity, large numbers have been bred and cage-bred birds are readily available. The rose-crowned conure *(P. rhodocephala),* until recently unknown in aviculture, has recently become available. Still rare in captivity, it is reported by those who have pairs, to be breeding well. The Hoffmann's (sulpher-winged) conure *(P. hoffmanni)* is another species that has recently become available to aviculturists. Although still very rare in aviculture, it too is reported to be breeding well. The yellow-sided conure *P. hypoxantha* is now accepted as being a yellow morph of the subspecies *sordida* of the green-cheeked conure. At first only females were known, which, when bred with normal males, gave some normal males and some yellow morph females. Later, yellow morph males were bred and now pairs are available that produce all yellow-sided offspring. It is a beautiful little bird, still rare in aviculture.

Most of the *Pyrrhura* conures are prolific breeders. Many do not take kindly to nest inspection and have been known to destroy eggs and mutilate their chicks after they have been handled. When chicks have been attacked I have found it was after they had been removed from the nest and then returned, still thrashing around. If the parents find them making any unusual vigorous movements they are liable to attack them. This often happens when the chicks are traumatized during ringing and then replaced in the nest. Many will eat their eggs for no apparent reason. This is a trait very common in the blue-throated conure. They have been known to take a few years after they have started laying to settle enough to incubate their eggs and rear chicks. Eggs are either abandoned, eaten, ejected from the nest, or young chicks ignored or killed. The best remedy is to avoid all disturbances until the chicks are ready to be rung, at which time the procedure should be accomplished with as little trauma to the chick as possible, or they should be removed at this time for hand rearing. I have also found that, in the case of large clutches, when the eldest chicks can be 10 days or more older than the last hatched, the parents frequently leave these larger chicks to brood the new ones. If the larger chicks are removed the newly hatched can become chilled. In this condition they do not solicit feeding so are ignored and soon die.

The blue-throated should be rung with a 7 mm ring. The others either 6 mm for the green-cheeked and similar larger species or 5.5 mm for the others.

The incubation period for *Pyrrhura* conures is about 23 days and the chicks fledge at around seven weeks. *Pyrrhura* chicks are very easy to hand rear if pulled at about 15 days. They also wean easily. I have found with all conures that hand-reared babies that have been allowed to socialise with adults straight after weaning soon lose their imprinting on humans. They make steady breeders and often start breeding earlier than parent-reared birds.

Enicognathus spp.

This genus includes the austral *E. ferruginea,* and slender- billed *E.leptorhynchus* conures. The austral has one subspecies, *E.f.minor,* which is a darker bird with less or no red on the belly. The austral is also the most southerly of all parrots in its distribution. The slender-billed has a long, narrow, slightly curved upper mandible, which distinguishes it from all other parrots. They are both very similar in appearance and habits and resemble *Pyrrhura* conures, albeit much larger. Their feeding and husbandry should be as for these conures. They are a lot more vocal than their smaller counterparts, however. Three to six eggs form the normal clutch and the incubation period is about 26 days. The chicks fledge after about eight weeks. They should be rung with 9 mm rings.

References:

Arndt, T. 1993. Atlas of Conures.

Juniper, T & M. Parr, 1998. *Parrots - A Guide to the Parrots of the World.*

Low, R. 1992. *Parrots Their Breeding and Care.*

Weight Gains for 3 Species of Conures:

Day	White-eared	Sun	Golden	Day	White-eared	Sun	Golden
	Weight (g)				Weight (g)		
0	4		9.5	**16**	46	58	116
1	7	7	11	**17**	48	64	127
2	10	8.5	14	**18**	51	69	131
3	12	11	18	**19**	60	73	143
4	14	12	21	**20**	61	78	151
5	17	17	28	**21**	62	82	165
6	19	22	36	**22**	64	87	167
7	24	23	45	**23**	66	92	172
8	26	26	49	**24**	70	96	178
9	28	29	52	**25**	71	100	184
10	32	35	64	**26**	71	101	187
11	38	41	68	**27**	70	103	187
12	31	44	78	**28**	70	105	190
13	37	47	88	**29**	70	106	195
14	40	51	98	**30**	66	109	198
15	43	52	112				

14.01 Psittacidae Amazon Parrots

William Horsfield

Introduction: Amazons belong to the order Psittaciformes, the family Psittacidae and the genus *Amazona.* They are found naturally throughout much of Central and South America including the islands of the Caribbean. Monroe and Sibley (1993) recognize 31 species, while Forshaw (1989) recognizes 28. Much of the debate and research into the classification of Amazons revolves around the relationship of species and subspecies in the *ochrocephala* group, where the former authorities give full specific status to the yellow-headed *A. oratrix* and yellow-naped *A. auropalliata* Amazons, which the latter authority treats as subspecies of the yellow-crowned *A. ochrocephala,* while noting that he agrees that they may prove to be separate species. Monroe and Sibley also recognize the white-faced, or Kawall's, Amazon *A. kawalli,* which was apparently not separated from two similar species, the mealy amazon *A. farinosa* and the scaly-naped Amazon *A. mercenaria,* at the time Forshaw's *Parrots of the World* was published and reissued. Information on the status and distribution of this species can be found in Juniper and Parr (1998).

Amazons vary in size from the small yellow-lored *A. xantholora* of Mexico, measuring 23 in. in length and weighing 200 g (7 oz.), to the magnificent, very large, and highly endangered imperial *A. imperialis,* with a length of 45 cm (17 in.), and the St. Vincent *A. guildingii,* with a weight of 580–700 g (1.3–1.5 lb), from the Caribbean islands of Dominica and St. Vincent, respectively. Amazons are primarily compact green birds with color variations between species mainly on the head, neck, throat, and shoulders. Many species have a speculum of red to orange among the green primary wing feathers and in the base to medial portion of the tail feathers. They have a short, slightly rounded tail and broad, rounded wings. The naked cere is prominent and there is a notch in the upper mandible. Amazons are intelligent, long-lived parrots. Many species have remarkable mimicry talents and enormous pet appeal; however, only non-threatened species should be considered for this purpose.

The St. Lucia amazon *A. versicolor* is a flagship species for Amazon conservation, being the subject of a world-wide educational conservation program co-ordinated by RARE Center for Tropical Conservation, a non-profit, unaffiliated organization based in Philadelphia and presently active in over 20 countries. This program is involved in the preservation of many endangered species of fauna and flora, including the rare Caribbean Amazons. One of the main threats to the survival of these magnificent island species is hurricanes, as seen in September 1998 with the devastating Hurricane George.

Management: With the decline in availability of wild-caught birds and the growing public intolerance of the ruthless practices often associated with wild trapping for the pet bird trade, management requirements are shifting, with a greater emphasis on domestically bred birds. Wild-caught adult Amazons take many years to acclimatize to captivity. Such problems as adapting to an artificial diet, incompatibility, and general wariness toward humans are things of the past with captive-bred stock. Parent-raised progeny may exhibit initial shyness and be inclined to fly full speed into the wire for a brief period after fledging, but soon lose this nervousness. When they are removed from their parents and placed in socializing groups, young birds interact well and settle down quickly. Amazons are very active at sunrise and sunset, when they vocalize loudly. This may be a nuisance to human neighbors, particularly with the larger birds like the mealy and yellow-headed forms, which in terms of volume can be extremely noisy. These periods of activity correspond to their natural behavior in the wild, with the birds keen to start foraging in the morning and making sure they have full crops when they return to roost in the evenings, especially if they have chicks to feed. During the day, captive Amazons are inclined to be lethargic and, unless environmentally stimulated into activity, often become obese. This stimulation can be achieved in part through interesting nutrition (see DIET) and through the enrichment of their immediate surroundings (see PSITTACINE ENRICHMENT).

Amazons are very destructive to vegetation in their cages and, unless housed in a large planted aviaries, generally demolish much if not all plant growth. St. Lucia Amazons kept at Jersey Wildlife Preservation Trust (JWPT) in the Channel Islands are housed in such aviaries, and the heavy vegetation requires annual pruning. An example of excellent management at this institution saved chicks of this rare bird. With the aid of a hidden camera, staff observed that a young hen was brooding, but not feeding, her chicks; they were removed, fed, and replaced until such time as she started to feed them (Hilary French. 1998 pers. comm.).

All of my Amazons are kept in suspended aviaries. These are positioned so that the birds perch above eye level. This reduces stress associated with human approach and territory defense, and the birds remain steady and confident. Only top quality galvanized wire should be used for parrot cages and aviaries. We use U.K. Twilweld welded mesh, which is galvanized after welding. Generally, cheap is cheap, and poor galvanizing is often the cause of lead and zinc toxicosis. Allow shiny new wire to oxidize to a dull color and, if necessary, scrub down with an acetic acid (vinegar) solution. Although Amazons are not destructive to wire, we use 2 mm diameter wire mesh for the suspended aviaries because standard 1.6 mm wire has less tensile strength and the cage tends to lose its shape, unless a framework is used. Preferred aperture is 25 mm^2 (1 sq. in.). The standard 13 x 25 mm (2 x 1 in.) excludes larger rodents but is heavier. An aperture larger than 25 mm^2 allows sparrows, weavers, mannikins, and other feral birds, which are potential parasite and disease vectors, to enter the cage.

New welded mesh eventually oxidizes matt silver-gray, and therefore we do not paint the wire. For those who wish to paint, however, only fast-drying paint that has no toxic base should be used. Slow-drying bitumen should never be used for aviaries, as this soils the feathers, and the resultant damage is long lasting. Our aviaries have an outside section exposed to all elements and a covered and protected inside portion. The cages are 3.6 x 1.2 x 1.2 m (12 x 4 x 4 ft.), with the shelter comprising the first 1.2 m (4 ft.). They are raised 1.2 m (4 ft.) off the ground. This allows easy access under the cage where the door is placed. The large Amazons can have even longer flights. Amazons should never be kept in small battery-style cages, as they will become obese without adequate exercise and their long-term breeding will deteriorate. The nests and feed trays are situated in the sheltered section. There is a wide access passage of 1.2 m (4 ft.) behind the aviaries in this sheltered portion. Hygiene is paramount to successful management. All floors are washed daily using high-pressure hoses, and a disinfectant program is adhered to. Virkon-S and granulated pool chlorine (HTH) are used under supervision. The floor is a sloping concrete screed, which leads to drains. All drainpipes are securely covered to prevent the entry of vermin and to prevent the accidental escape of loose birds. A kea parrot *Nestor notabilis* was tragically lost from a well-known collection through climbing down a drainpipe it had managed to open, then wandering out onto the adjacent road, where it was promptly killed by a passing vehicle.

Our aviaries are separated by a 60 cm (24 in.) gap in which dense vegetation has been established. (Note that the area under the cage is kept clear, otherwise food and feces would soil vegetation.) This vegetation not only provides a partial visual barrier between pairs, but also enables the birds to spend hours chewing at the growing buds that enter the cage. The plants and shrubs used must be carefully chosen to suit the particular climate, and should be species that flourish in spite of heavy pruning. It might occasionally be necessary to enter this area between the cages; if it is solidly blocked by vegetation, this would be difficult. Rodents may find ideal shelter here and this potential problem must be monitored. I use various poison bait-stations, placed safely outside of the aviaries, using primarily liquid poisons, which cannot be removed and carried into cages by the rodents. Poisons which do not cause secondary poisoning of owls and other predators are preferred, e.g., Racumin. Rodent "glue" is also very effective and totally harmless. Smeared on a piece of board and placed where evidence of rats or mice is found, it soon has them all stuck and going nowhere.

The plant species I primarily use between aviaries include hibiscus, Chinese bamboo *Nandina* sp. and other bamboo types, Cape honeysuckle *Tecomaria capensis,* firethorn *Pyracantha* spp., miniature bottle-brush *Callistemon* spp., cigarette bush *Cuphea ignea,* and cotoneaster *Cotoneaster frigada*. All of these have either berries or nectar-rich flowers that keep the birds busy for hours on end. Branches from flowering *Eucalyptus* and *Erythrina* spp. are relished. All our aviaries have a view across the beautiful Assagay valley and the birds are noticeably content. The situation of the aviaries is not critical in our temperate Kwazulu-Natal climate, but in localities where temperatures fall consistently below 10°C (50°F) they should certainly face north and be protected from wind and driving rain. Amazons are extremely hardy if they have a sufficiently protected area into which they can retreat. When properly acclimatized, they can certainly withstand temperatures below the freezing point. A successful breeder in Switzerland uses heated perches in small protected shelters into which the birds retreat during icy weather, with the birds staying outdoors all through the harsh European winter (Kingsley 1998). In these conditions the feathering becomes denser (not more feathers).

In our area, the blazing African heat is far more likely to cause problems than the cold. Adequate ventilation is essential with heat radiating roofing not being directly above cages. Micro-sprayers cool effectively, help keep plumage in peak condition, and raise ambient humidity levels. Swarming African bees can also be calmed down using sprayers. Should a swarm occupy a nest box, immediate action is required before birds start getting attacked and killed. Fumitabs (active ingredient Gamma BHC 440 g/kg) are very effective. When lit and placed in a nest box occupied by bees, the smoke quickly kills them while not being dangerous for the birds. The nest can also be used again, once cleaned properly. An empty hive permanently placed on the aviary roof often attracts passing swarms and can be quickly and safely removed. Household insect aerosols are very poisonous to birds and generally agitate the bees into attack frenzy, as does acetone.

Strategically placed flood-lights can be used to light up aviaries at night in an emergency, e.g., to scare off owls, feral domestic cats, black-spotted genets, raccoons, etc., which have managed to get though the high-voltage perimeter electric fence. If installed correctly, electrified wiring around the aviary buildings is also highly effective against rats and mice. These pests are a serious problem in many aviaries and, even if they cannot actually enter the cage, still pose serious health risks due to urine and fecal contamination of wire, food, and water. All areas where insects and external parasites may hide, e.g., behind nest boxes and at perch ends, should be periodically sprayed to kill and deter them (we use Grenade sheep dip @ 1 ml per liter water). This can be safely sprayed in nest boxes and on birds. They can even be entirely dipped in Grenade if external parasites are observed.

Tapeworm infestation can continue even in birds housed in suspended aviaries. Cockroaches, flies, ants, etc. are suspected intermediate hosts. Bi-annual deworming programs are advised for elimination and control of these and other parasites. Avian malaria is common in RSA. The birds are infected with the *Haemoproteus* parasite by louse flies (a small flat-shaped fly that runs rapidly between the feathers) and this may be confirmed by blood test, although birds often remain otherwise asymptomatic.

In African collections, many rare birds have been killed by snakes, especially spitting cobras, and protective goggles and a snake catching stick can be very useful because the use of firearms in an aviary has, in the past, led to disastrous consequences with birds being shot by accident.

Our aviaries are all fitted with V-shaped perch holders, eliminating the time-consuming need to secure perches with wire. Many innocent birds have hanged themselves on loose bits of wire threaded around their feet, or even managed to wrap wire under their leg-bands along the tarsometatarsus and thereby shut off the blood supply, causing necrosis (tissue death). Oversized leg bands and split rings are also dangerous in this respect. We only use stainless steel feeding dishes and water bowls. They are easily cleaned and are highly durable. They are changed daily and washed in a hypochlorite solution and then rinsed. Anti-pathogen footbaths at the entrances to aviary-blocks help reduce pathogen transmission on shoes. Solutions should be changed regularly. Water is the most common source of contamination in most aviaries and evaluation of water sources is vital (Massie 1998). Our staff receive strict training to prevent any possible contamination through personal hygiene negligence.

Diet: Amazons are not particularly fussy eaters and therefore do not need much coaxing to eat a varied diet. In fact, they tend to over-consume and become obese if balance and volume of feed are incorrect. There are many factors that determine energy levels and therefore influence diet. These include age, reproductive status, environment, and climate. Amazons do well with higher protein, lower energy for maintenance, but breeding birds need 8–12% fat for good production. The advent of complete extruded and pelleted diets presents many options for the aviculturist. Opinion is divided, with many. Exclusive pelleted diets providing little visual stimulation (although some are colored and shaped differently) and have been accused of causing dietary boredom in birds by reducing the need for them to use all their senses of sight, smell, and taste in search of food, as they would naturally. There is, however, less chance of accidental contamination by herbicides, pesticides, or other harmful substances, which is occasionally associated with fresh produce and its storage, as well as a reduced risk of fungal or bacterial poisoning (commonly *Clostridium* and *E. coli*) caused by contaminated seed. Nutritional research and information is nevertheless constantly forthcoming, and will no doubt continue to set new norms and standards.

In RSA we have a few extruded and pelleted options including Kaytee and Harrisons products from the USA and the local Aviplus range. Hagen has also recently entered our market. At Amazona Farm I use a combination of extruded and more traditional recipes. Fresh fruits and vegetables are provided daily, but the high moisture content of this portion must be taken into account in the overall diet analysis. The green and yellow varieties are preferred, such as spinach, green beans, green peas, broccoli, green, yellow, and red chili peppers, beetroot, carrots, patty-pans, grapes, oranges, banana, papaya, guava, melons, and loquat. The fruit and vegetable portion is mixed with Aviplus Parrot soft food. In addition, a vitamin and mineral supplement, also by Aviproducts, is added twice weekly. A formulated mixture of a variety of boiled legumes (see MACAW diet), whole maize, soaked wheat, oats, rice, and striped sunflower are also offered. The food is divided into three portions, with fruit/vegetable soft food being offered early morning, pulses and seed late morning, and then a top-up treat late afternoon, just before lock-up time. Pasta, cheese, chilies, pumpkin seeds, wild figs, berries, grasses, sugar cane, Napier fodder, scrambled egg, and cubes of brown bread with honey and milk are all examples of what I call treats. Pairs that are feeding youngsters are provided with ad-lib portions, while others are rationed according to size and weight, with almost all the food being totally consumed by nightfall.

During the non-breeding season the diet is modified to include mostly extruded Kaytee Maintenance pellets. The birds are best observed at feeding times, when they interact with the keeper. I believe it is important to talk to the birds and make eye contact with them. The blazing eyes of recognition re-enforce the relationship between keeper and bird and strengthen pair bonds. I honestly feel that this communication is where the enjoyment aspect of keeping parrots unfolds to a large extent. This is where valuable behavioral patterns relating to human interaction are observed and where individual personalities display territoriality and recognition, if not acceptance and trust.

Breeding: Most Amazon species are sexually monomorphic. Therefore they should be surgically sexed or DNA-typed to establish true pairs. Males are often larger and bolder across the head, and in some species, slightly more colorful. Exceptions are the white-fronted *A. albifrons* and the yellow-lored, which are sexually dimorphic. Many of the captive-bred birds are breeding at ages previously thought unlikely. Antonie Meiring bred from an 11-month-old vinaceous Amazon *A. vinacea* hen, while David Dennison has bred blue-fronted Amazons *A. aestiva* a number of times from pairs at 2–3 years of age. My own spectacled Amazons *A. pretrei* have bred at 18 months. Generally speaking, most species will start to display breeding behavior between 2–3 years of age. An older bird paired with a younger bird will invariably speed up this process. Young birds that have been parent raised will have an advantage over hand-raised birds, unless the hand-raised birds were well socialized post weaning among members of their same genus. It is an enormous advantage for young birds to be in the vicinity of older birds, preferably breeding pairs, in their formative years, as this learned behavior acts as a back up and stimulus to their innate and instinctive behavior.

Hand-raised Amazons kept isolated as pets for many years will often exhibit behavioral problems when used for breeding purposes, but in most instances they will eventually interact with and accept a mate. However, displaced aggression is often a problem in these pairs. Species in the *ochrocephala* group especially, in which the males even at the best of times are notorious for bullying their females, need to be carefully monitored. The Tucaman *A. tucumana*, red-browed *A. rhodocorytha*, Cuban *A. leucocephala,* and red-lored *A. autumnalis* Amazons are also species that have been known to attack and even kill mates in fits of aggression or jealousy. Potential mates should always be introduced on neutral ground at the beginning of the breeding season. If males are continuously aggressive, one wing may be clipped. I have on a number of occasions seen supposedly incompatible pairs finally settle down and breed. Potential pairs introduced out of the breeding season often only interact favorably once the season commences, whereupon their behavior, and therefore their receptiveness, also changes. Driving males usually "persuade" uninterested females to go to nest, so perseverance is sometimes called for. The synchronized archangel wing-raise, followed or accompanied by loud vocalization, tail fanning, and eye blazing, is typical of bonded pairs. Once a pair has bonded, the two birds will remain paired for life and very seldom switch mates, even if placed in flock situations during the non-breeding season. If space permits, such flocking is a highly recommended, environmentally enriching option, as Amazons naturally congregate in the off season. Aggressive males may be housed in smaller cages adjacent to grouped adult and juvenile birds, in a semi-flock situation, to prevent fighting.

Some of the commonly available but historically difficult to breed species like mealy Amazons are now producing well. Many instances of chronic infertility among supposedly compatible pairs have been solved by allowing the infertile pair to foster fertile eggs from another pair and raise the youngsters to the perch. This strengthens the pair bond. Many Amazon females are excellent foster parents and have been known to feed not only various Amazons, but also cockatoos and macaws. I have a pair of yellow-headed that raise 4–5 red-fronted macaws *Ara rubrogenys* to the perch every year. The parents, right next door, otherwise kill their own young. It is important that these fostered youngsters are socialized with their own species after weaning to prevent behavioral imprinting on the foster parents. Reproductive lifespan is assumed to be approximately 30–40 years, depending on the species. A pair of wild-caught double yellow-headed amazons was known to have bred for 36 years before the male was stolen.

The clutch size is 3–4 eggs in most instances, occasionally five. The incubation period varies between species, and sometimes between individual eggs, ranging from 24 days in the spectacled, 25 days in the Hispaniolan *A. ventralis,* 25–26 days in the red-lored, and 26 days in the blue-fronted and lilac-crowned *A. finschi,* to 26–28 days in both the Cuban and green-cheeked *A. viridigenalis* and 28–29 days in the species and subspecies in the *ochrocephala* group. The pip to hatch interval in all species is 24–48 hours. Incubation is by the female alone, although the male may visit the nest box to feed the female.

Yellow-headed Amazon
Amazona ochrocephala

Nest boxes: A variety of boxes may be offered; if the birds want to breed they will use just about anything available to them. Nest boxes should be removed, cleaned, and stored after the breeding season and re-installed toward the end of winter. Climatic conditions trigger breeding activity. Low-pressure zones (storm formation and rain), seasonal changes (winter to spring), photoperiodic changes (longer days, shorter nights), and a corresponding abundance of food induce reproductive readiness. Partially blocking the entrance hole of the newly returned nest encourages excavation chewing and also acts as a breeding stimulus. Hollowed out syringa logs placed vertically are ideal. These should be capped top and bottom with flat metal sheeting to prevent damage from chewing. The internal dimensions for larger Amazons should be approximately 30–40 cm (12–16 in.) in diameter and 50–60 cm (20–24 in.) in height; for the smaller species, these measurements should be approximately 20–30 cm (8–12 in.) and 40–50 cm (16–20 in.), respectively. Stumps of *Agave* ssp. are readily accepted, but are difficult to inspect and are often chewed through. Smaller boxes may provide more security, but unfortunately also become very hot during intense summer heat. High altitude species like the tucuman Amazon, which live at elevations of up to 2200 m (7200 ft.) are susceptible to heat stress at a young age and nests need to be opened up for better ventilation when ambient temperatures rise above 30°C (86°F). I therefore prefer to have larger boxes for such species so that the whole brood may stretch out in the heat. Conventional plywood, hardwood, or even pine boxes are easier from a management point. For the nesting substrate I use untreated eucalyptus or pine chips, which the birds splinter into smaller pieces, usually along the long axis. This may be sprayed with an iodine-based preparation like Kennel Scrub or Iodet to prevent the growth of fungal spores, which in a warm environment may cause aspergillosis in females that have scratched and chewed at shavings that have become moldy in the nest. This is usually due to infection by *Aspergillus fumigatus*.

Horizontal boxes can also be used. These should measure 60 x 30 x 30 cm (24 x 12 x 12 in.) high for the larger Amazons, divided into 40 x 20 cm (16 x 8 in.) sections inside by a 15 cm (6 in.) partition. If vertical L- or boot-shaped boxes are preferred, they should have an overall height of 60 cm (24 in.) and a base width of 40 cm (16 in.), with the boot being 35 cm (14 in.) deep. This size box is suitable for larger Amazons such as the yellow-naped. For smaller species, such as the white-fronted, the corresponding measurements should be reduced to 30 x 30 x 40 cm high (12 x 12 x 16 in.).

Amazons may be very aggressive when disturbed on the nest, and great caution should be taken to ensure the hen does not suddenly bite a chick that raises its head during the procedure. I prefer to breed these protective hens in a horizontal box, which allows the hen to be pushed away from the chicks and eggs with a towel if necessary. Inspection should preferably be done with the hen off the nest, but this is not always possible. As the youngest chick of a brood must often battle to compete for food, inspection may indicate that a late chick should be pulled for hand raising.

Care should be taken to minimize nest disturbance in order to prevent the formation of dysfunctional incubation habits. Young females in particular may be traumatized by nest inspections, as they are in an incubation trance. Males should stand guard outside the nest box and not enter the incubation chamber except to feed the female and then later, the chicks. Aggressive males may storm into the nest and attack the female when excited by human activities, causing broken eggs or chick mortality, so once again, keep disturbance to a minimum.

Courtship involves mutual archangel wing-raises accompanied or followed by loud calling and eye blazing. The pair may rapidly fly from end to end, and then the male may feed the female and vice versa. The female solicits the male by bending low and forward on the perch and shaking her wings; the male then sidles up to her and they mate crouched side by side with one tail under the other one. Short, rapid, high pitched, yet quiet as well as extended almost "pleasurable whining" vocalizations are uttered during mating.

Egg-laying occurs within 10–21 days of mating. The eggs are normally laid at 2–3 day intervals, but this may extend to 5–7 days in certain hens. Droppings are loose, and greater in volume, just prior to egg laying. Unlike cockatoos and macaws, which may mate at any time with no subsequent egg production, Amazons only mate when intending to breed. Nervous behavior like head shaking or rotating, and lack of coordination at this time or during egg laying, are symptoms of hypocalcemia (low blood calcium). Supplementation of calcium and Vitamin D3, with supportive care, is necessary to elevate calcium serum levels and save the bird. Survivors may exhibit permanent nervous symptoms due to bleeding on the brain.

Chicks are fed initially by the female, who in turn is fed by the male. Later, the male assists in feeding the chicks directly. Depending upon the species, chicks leave the nest anytime from eight weeks of age in the smaller types, to 11 weeks of age in the larger types. They are weaned within 2–3 weeks of fledging, and may be kept with the adults, provided there is no overt aggression, until a few months before the next breeding season.

Artificial Incubation and Hand Rearing: Amazon eggs are fairly straightforward to incubate, generally following the 15% weight-loss guidelines. For accurate weight-loss determination, the mass on lay-date is necessary. At Amazona Farm, I use Grumbach and South African Ara incubators. The Ara can turn the eggs every 15 minutes, enabling up to 96 turns per 24 hour period, giving excellent early vein-growth across the yolk. The sooner the veins have surrounded the yolk, the stronger the embryo. Germicidal ultra-violet lights greatly reduce pathogens circulating in the incubator, and eggs are not dipped or washed prior to being set. Eggs are incubated at a temperature of 37.2–37.5°C (99–99.5°F) with an average relative humidity of 50–55%, depending on the weight loss graph. Eggs are candled and weighed every 2–3 days. I candle on the vertical axis with light source from below through the airspace, rotating egg with left hand in clockwise direction.

Eggs are turned through draw-down till internal pip and then turning is stopped. Only when the embryos are actively moving in the airspace chamber are the eggs moved to the hatcher. By this time they have often made the first external chip. The hatcher runs at 37.5°C (99.5°F) and 70% RH. Eggs are placed, with the chip marks facing up, directly above an accurate thermometer. Newly hatched chicks are weighed and fed, and then placed in brooder at 37.2°C (99°F). If the hatch has been difficult or the chick is dehydrated, Ringers Lactate is fed hourly for 3–4 hours, then normal hand-raising formula is used. I feed all Amazon chicks on Kaytee Exact from the egg. Amazons also do well on Aviplus Parrot Handraising, but this should only be fed to incubator hatched chicks after 2–3 weeks. Leg bands are fitted when the pinfeathers start coming through the skin on the wings.

Chicks are kept singly in Tupperware tubs on soft toilet tissue. Initially, they are fed every two hours between 6:00 a.m. and midnight. The feeding frequency is gradually decreased until the chicks are being fed every six hours at approximately four weeks of age. The brooder temperature is also reduced gradually over this period. At this stage the chicks, which are covered with pinfeathers, are moved out of the brooders and placed in plastic tubs in groups consisting of the same or similar species. The are kept on clean pine shavings, with the room heated to 30°C (86°F). When they are fully feathered and start climbing out of the tubs, they are transferred to small, wire, double-bottomed weaning cages with newspaper in the false bottom to keep them off feces. Here they learn to eat the adult diet. Once they are weaned, they are either sold or placed in outdoor socializing aviaries.

Ring Sizes: (Individuals requirements may vary)
White-fronted, Tucuman: 9 mm
Blue-fronted, orange-wing, green-cheek: 10 mm
Red-lored, lilac-crown, double-yellow, yellow-crowned: 11 mm
Yellow-naped, mealy: 12 mm

Contributors:

J. Vermaak, Ara Incubation Company, Pinetown, Kwazulu-Natal RSA.

J. & E. Daw, Eve Daw Aviaries, Everton, Kwazulu-Natal RSA;

D. Dennison, Shady Streams Birdfarm, Waterfall Kwazulu-Natal RSA

H. French, J.W.P.T., Jersey, Channel Islands, U.K.

A. Meiring, Farmall Parrot Trust, Randburg, Gauteng RSA

References Cited and Suggested Reading:

Arndt, T. 1996. Lexicon of parrots. Vol. 4.

Butler, P. 1998. Promoting protection through pride. In Proceedings IV International Parrot Convention, Loro Parque, Tenerife, Spain.

del Hoyo, J., A. Elliott & J. Sargatal (eds.). 1997. *Handbook of the Birds of the World.* Vol 4. Barcelona: Lynx Edicions.

Jordan, R. 1989. Parrot incubation procedures

Juniper, T. & M. Parr. 1998. *Parrots: a guide to the parrots of the world.* Sussex, UK: Pica Press.

Low, R. 1992. *Parrots in Aviculturė.* Port Perry, ON: Silvio Mattachione & Co.

Murphy, J.J. 1998. Amazon breeding today - blending behavioral biology with environmental enrichment. In Proceedings IV International Parrot Convention, Loro Parque, Tenerife, Spain.

Hummingbirds

15 Trochiliformes

15.01 Trochilidae: Hummingbirds

15.01 Trochilidae Hummingbirds

Karen Krebbs

Introduction: Hummingbirds belong to the family Trochilidae, which is assigned to its own order, Trochiliformes. The family is divided into two subfamilies, Phaethornithinae, the hermits, and Trochilinae, the typical hummingbirds. With a total of more than 320 species in 108 genera, this is the largest non-passerine family of birds. The 29 species of hermits are placed in four genera, while the 293 species of typical hummingbirds occupy 104 genera, many of which are monotypic. Hummingbirds are only found in the New World, from Alaska in the north to Tierra del Fuego in the south. The greatest diversity of species is found in the tropics. Hummingbirds feed on the nectar of flowers and catch small flying insects. These small birds, weighing from 2–20 g (0.07–0.7 oz.) depending upon species, have long slender bills, extensive tongues, iridescent plumage, and highly maneuverable flight. The wings rotate at the shoulder, allowing forward and backward flight, hovering, and the ability to fly upside down. Adult plumage is different for most males and females. Males have highly iridescent feathers and the females are rather drab in comparison. Males are promiscuous, and form no pair bonds with the females. In general, these birds are aggressive, territorial, and unsocial.

Management & General Housing Considerations: Hummingbirds are a very sensitive group of birds to keep in captivity. Care must be taken to allow hummingbirds in captivity to coexist with one another, set up territories, and nest. They must be given the time and attention it requires to keep them healthy and content. I do not recommend mixing North and South American species in the same enclosure. Although this has been successful in a few aviaries, the mixing is usually not a good one. The goal should be to mix those species in captivity that might normally come into contact with one another in the wild. In this way, birds sharing the same facilities are already adapted to the same type of habitat and weather conditions, and are accustomed to confrontations with one another. When putting different species of hummingbirds together, one must also take into account size differences, temperament, sexes, and compatibility. As a rule, larger species are more aggressive than smaller ones; males are more aggressive and territorial than females; and South American species may not get along with North American species. The number of birds, cage size, feeders, roosting plants, nectar producing flowers, nest sites, and local weather conditions should be given careful thought.

If hummingbirds are expected to endure either hot or cold temperatures in a captive situation, the following should apply:

- They should be in excellent physical condition.
- They should have adequate vegetation that provides rest, sleep, shade, and nest areas.
- They should have access to a heat source in the winter such as a heat lamp. If you need to offer supplemental heat, you should consider their territories; it is best to offer one heat source per bird. In this way the birds won't fight over the heat.
- The birds should have access to a waterfall or area where they can bathe.
- The daytime temperature should not stay below freezing.

- The outdoor aviary or holding areas should be covered on the top with tarpaulin or equivalent cover to protect from cold temperatures, wind, snow, ice, hale, and rain when these environmental extremes occur.

Natural Light: Outdoor aviaries are usually more successful when exhibiting or holding hummingbirds. In some of the more northern latitudes this may not be possible due to the weather extremes. A natural environment seems important to the hummingbird's health, well-being, and nesting success. Indoor aviaries or holding areas should use natural light whenever possible in the form of skylights, windows, etc.

Water Quality: For food preparation and bathing, the water should be void of chemicals. The water should be tested for impurities. Even small amounts of impurities can mean illness or mortality to a hummingbird. A water filtration system is recommended.

Air Filtration: Indoor aviaries and holding areas should have good quality air and be free of pollutants. Outdoor aviaries should not have a problem with this.

Waste Disposal: Indoor aviaries should be hosed daily to remove nectar drippings on floors. Unclean floors can attract ants and other insects. The floors should have good drainage so that water does not pool in these areas. Outdoor aviaries or holding areas should have a good soil substrate and any hard surfaces like concrete should also have drainage.

Insects: Indoor holding cages should have "ant guards" to protect birds and feeders from ants. Legs of the holding cages can be set in containers filled with water to prevent ants from crawling up to the feeders or perches. Vaseline rubbed on feeder hooks or wires will also discourage ants. Insecticides should never be used to control ants or insects around hummingbirds.

Glass Enclosures: A glass enclosure can be used if precautions are taken in preparation to exhibit birds. The greatest task is to prevent birds from flying into the glass and injuring themselves. Before releasing the birds into the enclosure, the glass should receive a thick coating of white or green water-soluble paint. The thick coating is placed on the outside of the enclosure so the hummingbirds cannot get to it. The coating still allows light penetration into the enclosure. When the birds are released into the enclosure, they should have time to adjust to the boundaries. Generally, after a week a layer of the coating is removed, and then another a week later. If all the birds are adjusting to the situation, then the remaining layer of coating is removed from the glass. Hummingbird fledglings seem to be aware of the boundaries and usually have no problem with the glass after they fledge.

Recommended Holding Pens and Aviary Sizes: Outdoor holding pens for single hummingbirds should be at least 0.9 x 3.6 x 2.1 m high (3 x 12 x 7 ft.). The holding pen should have double doors to guard against escapes. An alternative could be a pen 1.8 m^2 (6 sq. ft.) and the same height. Holding pens for several birds should be larger. For 2–3 hummingbirds, an enclosure measuring 1.8 x 4.3 x 2.1 m (6 x 14 x 7 ft.), also with double doors for escape proofing, should suffice. Holding pens for larger groups of up to 8–10 hummingbirds should be around 3.6 x 5.5 x 2.4 m (12 x 18 x 8 ft.), with safety doors. Larger collections should be housed in correspondingly larger aviaries, or divided into smaller groups according to aviary size.

Temporary indoor holding facilities can be smaller. For a single hummingbird, a cage 45 x 75 x 45 cm high (18 x 30 x 18 in.) is sufficient. This cage would sit on a table or shelf inside a room. Hummingbirds should not be kept in this size cage for more than 60 days.

An outdoor public viewing aviary for hummingbirds could be at least 167 m^2 (1,800 sq. ft.) in area, and 3 m high (10 ft.). This size aviary can hold up to 20 birds. An aviary with an area of 512 m^2 (5,500 sq. ft.) and 3 m high (10 ft.) can accommodate up to 30 birds. I would suggest that all outdoor holding pens and aviaries be constructed of tubular steel frames or equivalent, with 12.7 x 12.7 mm (0.5 x 0.5 in.) wire mesh for the top, sides, and bottom. Holding pens should have adequate shade in the summer, either natural shade from plants or some type of covering to provide a shaded area. Heat lamps should be provided for warmth during winter. If winter temperatures are extremely low (below

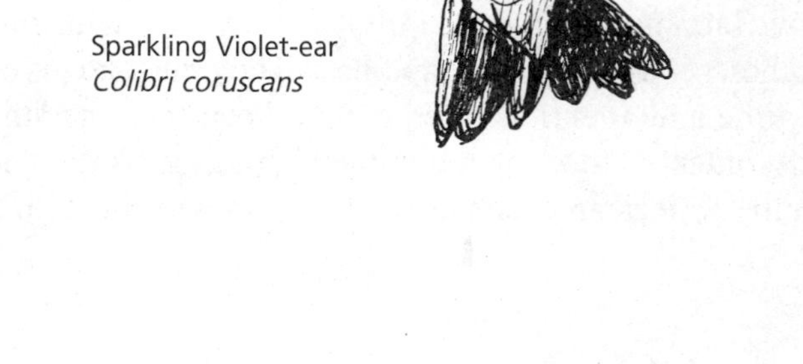

Sparkling Violet-ear
Colibri coruscans

freezing) for long periods of time, hummingbirds should be housed indoors. Indoor facilities should be of sufficient size to accommodate the number of birds held. The hummingbird aviary, whether indoors or outdoors, should feature trees, shrubs, and herbaceous plants to provide shade, roost areas, bathing platforms, nesting sites, territorial separations, and feeder sites. A succession of flowering nectar plants is desirable. Hummingbirds love to bathe, and waterfalls or small pools of water are necessary for their hygiene and good health. Aviaries and holding pens should be misted daily with a hose or sprinkler to allow the occupants to "rain" bathe.

Diet: Hummingbirds in captivity should be given a choice of food. They should be offered feeders with nectar, and have access to nectar flowers and small flying insects. Their main diet should be a nutritionally complete and balanced diet, which contains carbohydrates, protein, fat, vitamins, and minerals. Care should be taken to use chemical free water to mix the food. Nectar feeders should be changed twice daily, and even more often when daily temperatures exceed 38°C (100°F). Two-ounce (62.5-ml) feeders are easy to clean and work well for feeding hummingbirds. These feeders should be disinfected with a bleach solution at a ratio of 10:1 (10 parts water to one part bleach) each time they are changed. The feeders should be placed all over the aviary at various heights to allow the birds to set up different territories. Different species of hummingbirds will use portions of the aviary differently. The feeders should accommodate these differences among species. The more feeders you have, the less chance of fighting.

A variety of nectar producing flowers will provide a natural food source for the hummingbirds. The flowers will be defended by the dominant birds, but all of the birds will have the opportunity to feed on them. They should be planted throughout the aviary. Hummingbirds should be provided with fruit flies *Drosophila* throughout the year. During times of molt or nesting the flies provide an extra source of protein for the birds. Fruit flies can be bred in bottles and kept in an incubator. Each day the fly jars can be taken to the aviary and the flies released for the birds. Fly jars can be placed in the enclosure for the entire day and picked up before dark. Fruit flies can be obtained from local universities or purchased from biological supply companies. If the aviary is outdoors, old fruit can be placed in the area to attract wild flies and insects. This will provide a source of protein for the birds too. Care must be taken to avoid attracting ants to the fly jars, fruit, or nectar feeders. Fly jars and the fruit can be placed in the center of a pan that has a small amount of water in it to prevent the ants from getting into the fly jars or on the fruit.

Breeding: It is difficult to breed hummingbirds in captivity. The birds must feel comfortable with their surroundings and the aviary should resemble their natural habitat as much as possible. Hummingbirds do not pair off or form a pair bond during the breeding and nesting cycle. Separating pairs of hummingbirds for breeding purposes usually does not work. The males can be very aggressive and territorial, and if separated with a single female, the two birds spend most of their time fighting. It is better to keep potential breeding birds in an aviary large enough to hold all of the birds. In this way, the female can select her nesting site and mate with the male of her choice. Competition from other males resembles a natural situation in the wild. If a male and female are separated for breeding purposes, it is best to introduce the male to the female's enclosure for several hours a day when the female is building a nest. This way the pair only comes into contact with each other for a short period of time each day. The male and female could be housed side by side during the times they are not in the same cage so that the birds could see each other. This method of breeding is widely practiced in Europe and has been successful. With fewer and fewer non-native hummingbirds available in the United States, this type of breeding arrangement may be useful in the future. When female hummingbirds want to nest, they will hover around trees, bushes, plants, and people, looking for nest sites and nesting material. The birds will "try out" nest spots by sitting on objects before they start the nest construction. They may approach people searching for hair, strings, and sweater yarn to use in their nests. Such material as animal hair, plant down, clothes dryer lint, cotton, lamb's wool, feathers, bark, moss, lichens, and spider webbing should be supplied. Artificial spider webbing is not recommended because birds can get tangled up in it. There should be a healthy population of spiders living in your enclosure to provide the spider webbing these birds need to nest successfully. The nesting material can be hung on the vegetation and placed on the ground for the birds to use. Hummingbirds will construct a small, cup-shaped nest from some of these materials. Hummingbirds are very tolerant of humans around their nests, but will get stressed if the eggs or nest are approached too closely. Efforts should be made to give the birds the privacy they need to nest successfully.

It is recommended to have at least one male for every female in your aviary. If the female decides to nest and a male of the same species is not available she may breed with other males. This will encourage hybridization and can be a problem with captive hummingbirds. If the female builds a nest and lays eggs and there is no male of the same species in the aviary, you should assume that she bred with another male. You can remove the eggs or substitute non-fertile eggs in their place. If you have collected eggs that haven't hatched in the past you can reuse them for this type of situation. This way the female can incubate for a period of time before she abandons the nest. Otherwise she will continue to lay eggs to replace the removed ones and this can cause a drain on her health.

Normally two eggs are laid 48 hours apart and the female usually starts incubating with the laying of the first egg. The eggs will hatch after 15–17 days. At hatching time it is important to supply the female with additional insects to feed to her young. Common predators on hummingbird nestlings are snakes, lizards, ants, rodents, and other types of birds. All efforts should be made to eliminate these types of predators from your bird enclosures. Nestlings can fledge at 21 days or longer after they hatch. The female will continue to feed the fledglings another 7–21 days until they are weaned. Fledglings are usually not bothered by the other adult birds in the enclosure as long as they are under the care of their mother. But once the young are weaned, they will be harassed by other adult birds. Young hummingbirds can breed at one year of age and it is advisable to remove them before that time to avoid inbreeding.

Behavior & Social Management: Since hummingbirds are very territorial any newcomers are aggressively chased and harassed by the resident birds. It is recommended that several birds be introduced at a time in an exhibit or aviary. This way the chasing and aggression is dispersed among the new birds. New birds should be observed closely to assure that they are getting to feeders or flowers. Adding extra feeders on the day of the introduction gives the new birds a better chance at getting to the feeders and the resident birds have not had the chance to claim the new feeders. If one of the resident birds is too aggressive in the enclosure, it may help to remove this bird on the day of the introduction. This gives the new birds an opportunity to settle into the exhibit and claim feeders before the removed bird is returned to the enclosure. Even resident birds have a disadvantage when reintroduced into the enclosure after several days. It is also recommended that the new birds be placed in "howdy" cages 1–3 days before the introduction. This allows the new birds to locate landmarks, nectar, and flower sources. In addition it allows the new birds to meet the established birds in a controlled situation.

Capture & Handling: Hummingbirds in small aviaries or enclosures can be caught up in the early morning before light. You must know where the bird is roosting and quietly approach the sleeping bird. If you gently place your hand or net over the bird there is little struggle. In larger aviaries, a catch cage will have to be used to trap the bird. A cage made of welded wire and a door that can be held open with rope or wire is an adequate trap. A feeder is hung inside the cage and when the bird enters to feed, the door is carefully closed to trap the bird. The cage door must never be allowed to drop quickly or the bird may be harmed. Once the bird is in hand, care must be taken so the bird is not injured. Wings are easily damaged and the bird should be held gently but firmly to avoid injury. Wings should be against the bird's body and the bird cupped within the hand with the head protruding out. The bird should then be transferred to a soft mesh bag or cloth bag and placed in a box for transport. If at all possible, the bird should not be left in this bag for more than a few minutes before placing it in another cage. If the bird will be in the bag for more than five or ten minutes the bird should be fed by hand. Hummingbirds should only be trapped and handled when absolutely necessary.

Removing Members of the Group: Hummingbird will defend as large a territory as they can. Removing birds from an enclosure allows others to expand their own territories. If one bird is causing problems in an enclosure it is best to ride out the bad period rather than remove the bird. Exceptions would be new introductions where the new birds might benefit initially from the other bird's temporary removal. If conditions are crowded in the enclosure, removing birds will probably benefit the remaining birds. Young birds that have hatched in the exhibit should be removed before the next breeding season. Male fledglings can be very aggressive to the older birds and the enclosure can benefit if the young birds are removed. Removing the offspring also prevents inbreeding.

Indicators of Stress, Harassment, or Social Changes: Torpor is a form of hibernation that hummingbirds use when stressed, during cold periods, and when they have not obtained the necessary calories they need daily. During torpor the birds' metabolism, breathing, body temperature, and heart rate are reduced. The birds do not respond to touch, light, or sound while in torpor. Torpid birds should be allowed to wake on their own and in a reasonable time (30 minutes or less) in the morning hours. If the birds do not wake normally, they should be placed in a soft cloth bag and placed in an incubator until they come out of the torpor. Once awake they should be hand fed and placed in a flight cage for observation. When the bird is flying and feeding normally, it can be returned to the aviary.

It is normal for birds confronted with strong winds or cold temperatures to fluff up their feathers. If a bird consistently fluffs up when other birds are not doing this, the fluffed bird may be ill or in declining health. If you suspect that the bird is not well you should remove it from the aviary for further observation.

When a bird is being harassed, it will be chased around the enclosure. If the harassed bird is no longer defending a territory, it may have problems getting to feeders. The bird may perch with bill open, wings drooping, or try to hide from the other birds. If the bird does not obtain the food it needs it will die. Efforts should be made to accommodate the bird in the aviary by setting up hideouts. Extra vegetation (can be dead limbs or cuttings) can be placed around the exhibit for the harassed bird to hide in. A feeder can be placed in the temporary hideouts. Harassed birds may go to the same place to perch or rest and these areas are ideal for temporary hideouts. If this does not improve the situation for the harassed bird, it should be trapped and removed from the exhibit. Since hummingbirds have a high metabolism, it does not take them long to decline when they are not able to feed normally.

When a hummingbird consistently hangs on to the cage wire it is probably a response to stress. The bird is trying to escape the unpleasant situation. Efforts should be made to improve the conditions for the stressed bird. The bird should be watched closely to assure it is feeding properly. Stressed or harassed birds will spend more time around entrance and exit doors of the enclosure. If given the opportunity, these birds will get through the open doors and escape. Unfortunately many hummingbirds do not show any physical signs of health or social problems and will die or disappear unexpectedly. Good observation skills and familiarity with each individual bird will allow you to recognize problems when they develop.

Display Considerations: Outdoor exhibits should take advantage of the natural light when viewing hummingbirds. Iridescence on hummingbirds is best viewed with the sun to your back. In this way you are not looking into the sun and you have the advantage of the light behind you as you view the bird. The exhibit should also be located in an area where there are fewer people and cart traffic for evening events. This way the birds are not subject to unnecessary noise or light at night when they are sleeping.

If hummingbirds are provided with enough space to co-exist with other hummingbirds and their other basic needs are met, they should remain disease free. Stress is a major cause of weakness and disease in these birds. Larger birds are more aggressive than smaller ones, males are more aggressive and territorial than females, and some species may not get along with other species. You should choose the number and species carefully, provide numerous feeders (two feeders per bird), allow the birds to set up territories with numerous plants and barriers, and supply the necessities for nesting.

Mixed Species Exhibits: Hummingbirds can be exhibited with other species of birds but care should be taken not to place large predatory birds such as cowbirds or roadrunners within the enclosure. Small species like tanagers, honeycreepers, manakins, cotingas, small doves, and waterfowl usually work well with hummingbirds. Aviaries that exhibit both hummingbirds and butterflies can be very successful.

Transport & Handling: There are several airlines that regularly ship birds. You should check with the individual airlines since regulations or shipping requirements can change daily. Hummingbirds can be shipped in two different ways. Some of the airlines will allow you to carry the birds with you on the plane if the birds are contained in a carrier that will fit underneath the seat. You will have to purchase a ticket for the birds and the airline usually limits the number of birds you can carry on board. The other method of shipping the birds is by sending them by freight as you would domestic pets. You can usually ship counter to counter and the birds can be picked up at the airline freight counter in the receiving city. These flights should be non-stop, if at all possible, to minimize the time the birds will be on the flight.

A wooden box 36 x 30 x 17 cm deep (14 x 12 x 6.5 in.) with a handle and several small windows covered with mesh for air can be used to carry the birds on board. The top of the box should open and the inside of the box should be lined with a soft material on which to lay the birds during shipment. Before building the carry container you should check with the airlines for size specifications. The birds are kept in a soft mesh bag so the bill is readily available for hand feeding during the flight. The birds will have to be fed every 45 minutes throughout the flight. Do not take the birds through the X-ray machines at checkpoints in the airport. Ask airport officials to hand check the birds. Hummingbirds that are subjected to large amounts of radiation could be harmed or killed.

Small or medium sky kennels can also be modified

to ship hummingbirds by air. The windows must be covered with mesh and small perches wired to the inside of the kennel so they do not come loose during shipment. Feeders can be tied to the inside of the kennel and it is recommended to have several large flat saucer-type feeders at the bottom of the kennel for the birds to use as well. This way the birds have a food source on the bottom of the kennel in case some of the nectar spills out of the feeders affixed to the sides of the kennel. The birds can fly free inside the kennel. Do not put more than 2–3 birds in the same kennel.

If birds are hand carried onto the plane you will need a container of mixed nectar and feeders (2–3) for the feeding. Bring paper towels, water, and extra dry food in your carry-on luggage. You should check all other luggage so you can focus on the care and handling of the birds. Birds can be removed from the carrier for feeding but do not take them out of the mesh bag. The birds' bills can be easily manipulated for hand feeding. You may have to expose the birds to some sunlight if they refuse to feed and it may help to sit by the window on the plane. Most birds feed easily, but it may take longer for others. Be patient when a bird does not feed, and return it to the box for later feeding. Handling the birds as little as possible will reduce the stress. A stressed bird may go into torpor. Holding them on your lap while keeping them in the mesh bag may help, but you should keep them as warm as possible. You can gently blow on them or use your voice to waken them. If these attempts do not work to waken the birds it is best to keep the birds in the box and quiet.

Once you have reached your destination, you can continue to work on getting the birds to wake. Do not try to force feed a torpid bird. Once the bird is awake, you should feed the bird and place it in a flight cage after you reach your destination. It is important that the birds be allowed to fly and feed before they go to roost. This may require that the lights be kept on for a short period (20–30 minutes) before the birds go to roost. Birds that have traveled in the sky kennel should be removed from their carrier and placed in flight cages at the final destination. You should ask the receiving institution to call you when the birds arrive. This way you know the birds have arrived safely and have not gotten lost during the flight.

Species Requiring Similar Management: Hummingbirds are in a special class by themselves, and their captive management is unique. The SUNBIRDS of Africa and Asia, treated separately, require similar management.

Band Size: Hummingbird bands come in two different sizes. Normally, for most North American species weighing 2–6 g, a band with a diameter of 2 mm (0.08 in.) is used. For the larger North American species, weighing 7–12 g, the band should be 2.5 mm (0.1 in.) in diameter. Extreme care should be used when banding hummingbirds so as not to harm the bird or injure legs. In the United States, the Bird Banding Laboratory in Laurel, Maryland can be contacted for more information about banding hummingbirds. A hummingbird manual can be obtained from the Bird Banding Laboratory that details the banding technique.

Author: Karen Krebbs, Conservation Biologist in Conservation & Science, Arizona-Sonora Desert Museum, Tucson, Arizona, USA.

References:

Brice, A.T. 1991. Protein Nutrition of captive Hummingbirds, Procedings. of the 9th Dr. Scholl Conference. on the Nutrition of Captive Wild Animals.

Johnsgard, P.A. 1983. *The Hummingbirds of North America*. WA, D.C: Smithsonian Institution Press.

Krebbs, K. 1992. Hummingbird Breeding and Nesting at the Arizona-Sonora Desert Museum. AZA Western Regional Conference, Tucson, Arizona, USA.

Krebbs, K. 1997. The "Hummingbirds of the Sonoran Desert Region" Exhibit. *Journal of the American Association of Botanical Gardens and Arboreta Public Garden, P.A.*

Krebbs, K. 1999. The Arizona-Sonora Desert Museum's Hummingbird Propagation Program. 22nd Annual International Wildlife Rehabilitation Council Conference Proceedings, Tucson, Arizona.

Krebbs, K. 2002. *Hummingbirds: Beauty in Flight*. IL: Publications International, Ltd. Lincolnwood.

Palmer, B., K. Krebbs & S. McLalan. 1989. The Creation and Operation of a Native Hummingbird Exhibit. Proceedings of the 15th National Conference of the AAZK. Syracuse, N.Y.

Scamell, K.M. 1966. Breeding the Violet-eared hummingbird (*Colibri coruscans*). *Avicultural Magazine*. 73.

Skutch, A.F. 1973. *The Life of the Hummingbird*. NY: Crown Publishers.

Stoppelmoor, G. 2000. Captive breeding of the Sparkling violet-eared Hummingbird. *AFA Watchbird*, January/February: 52–54.

Turacos

16 Musophagiformes

16.01 Musophagidae Turacos

Glen Holland

Introduction: The African continent is the exclusive home of the family Musophagidae, the turacos and plantain-eaters. Formerly aligned with the cuckoos, they are now separated, on the basis of DNA evidence, into their own order, the Musophagiformes. Although the classification of turacos is still the subject of much controversy, under the Sibley-Ahlquist-Monroe system followed here, the family is divided into two subfamilies. The subfamily Musophaginae includes 17 species, 13 in the genus *Tauraco* and four in the genus *Musophaga;* the subfamily Criniferinae includes six species, three in the genus *Corythaixoides,* known as go-away birds, two in the genus *Crinifer,* the plantain-eaters, and the monotypic great blue touraco *Corythaeola cristata.* Apart from the great blue, all species are much the same size, weighing 250–350 g (8.8–12.3 oz.) and measuring about 45 cm (18 in.) in total length, of which roughly half is tail. All species except one have erectile crests. The sexes are alike, except in one species, where they differ only in bill color. The majority of the species occur in forest habitats, with a few inhabiting dry woodlands. Turacos are canopy dwellers, where they run and bound among branches. They are weak flyers but glide well. They feed on a variety of fruits and leaves. Wild figs in particular are an important food source for these birds. Although territorial and usually found in pairs, they also occur as family groups, and out of the breeding season, in small, loose groups. For the purposes of describing the captive management required for this group, I have concentrated on the gray go-away bird *Corythaixoides concolor,* representing the subfamily Criniferinae and the green-crested turaco *Tauraco persa,* representing the subfamily Musophaginae, with additional notes on other species supplied by David Bender of Hancock Wildlife Research Center in British Columbia, Canada (HWRC).

16.01a True Turacos

Musophaginae

• Glen Holland and David Bender

Natural Habitats: Green-crested turacos inhabit evergreen climax forest where they are usually spotted in flight, as the crimson of the primaries flash through the forest. They are typical of the essentially canopy dwellers, bouncing from branch to branch with amazing agility. They will descend to feed on the fruits of subcanopy trees and shrubs. A strictly territorial species, their calls consist of a series of deep growls.

Much of the controversy surrounding the classification of turacos revolves around the question of whether the green-crested forms represent subspecies of a single species, or a number of full species. The S-A-M system, followed here, recognizes the Guinea, or Gold Coast turaco, *Tauraco persa,* Schalow's turaco *T. schalowi,* Livingstone's turaco *T. livingstonii,* and the Knysna turaco *T. corythaix* as separate species. This view is confirmed by comparative vocal studies of these birds in the wild (Dowsett-Lemaire & Dowsett, 1988), as well as by behavioral observations of captive birds. The staff at HWRC — where three of these four species, along with Buffon's turaco *T. p. buffoni,* a subspecies of the Gold Coast — have been reared in considerable numbers for several years, strongly supports the recognition of full species.

In addition, a study of the comparative chromosome banding analysis of nine forms of turacos, though considered inconclusive by the authors, strongly suggests that these forms should be accorded full specific status (Van Tuinen & Valentine 1986).

There is also a difference in the size and shape of eggs in the green turacos. The Gold Coast and Buffon's lay rather round to broadly oval eggs, while those of Livingstone's are smaller and more elongated, even though the species itself is slightly larger. The eggs from two pairs of Schalow's were more oval than those of the Gold Coast, but not as elongated as those of Livingstone's. These differences were consistent with a number of pairs of birds of several unrelated bloodlines. Turaco eggs are seldom egg-shaped, i.e., having a large end and a small end, and should, therefore, be candled before being placed in an incubator to determine the position of the air sac.

Management: These active birds require a large planted aviary. A large flight of 6 x 2 x 2.5 m (20 x 6.5 x 8 ft.) high is the minimum requirement. They are aggressive toward many other species and will dominate the feeding area, often out of jealousy if not hunger. They can be mixed with other softbills and even finches, but never attempt to introduce a second pair or another species of turaco as this will result in casualties. Aggression has also been noted toward doves and green fruit pigeons. They have even been recorded attacking ground birds placed in the same aviary. Visual barriers between aviaries are recommended to stop adjacent pairs of turacos from fighting.

In common with the other turacos, the most common cause of death is due to persecution from a mate, usually the male. A well-planted aviary will allow the persecuted bird some place to hide and hopefully this is noticed by the keeper before death results. Birds that turn aggressive emit a growling call, which is the first sign of a problem. When introducing new partners to each other, always introduce the male to the female's aviary, and if necessary one wing of the more aggressive bird can be partially clipped.

At the Hancock Wildlife Research Center, where multiple pairs of four species of green-crested turacos were kept, this tendency was experienced on several occasions. When the turaco breeding program at HWRC was inaugurated in 1992, owner David Hancock and aviculturist David Bender were advised by established turaco breeders that captive turacos are noted for suddenly turning upon, and even killing, mates with whom they have lived in harmony for years. Some species were said to be more aggressive than others, the general rule being the greener the meaner. Indeed, the "green" turacos did prove more aggressive and murderous than other species. However, it was noted that the nominate race of the Gold Coast turaco was more likely to attack its mate than was its subspecies, the Buffon's turaco. The Livingstone's turacos proved to be even more aggressive. The male of the original pair killed his mate during the first breeding season, and a clutch of fertile eggs was lost in the process by becoming chilled before the death of the hen was discovered. They were duly transferred to an incubator, but the embryos were already dead in the shell. The following year a replacement hen was also killed, again during the incubation period, and it was speculated that the male was "driving" the hen whenever she left the nest to eat, exercise, or defecate, in an attempt to force her back onto the eggs, rather than taking a turn on the nest himself. Fortunately, the eggs were salvaged this time, and placed in an incubator, where a single chick hatched and was hand reared, proving eventually, through DNA-sexing, to be a male. Meanwhile, a third hen had been obtained, along with a second pair, arriving just in time for the next breeding season. It was decided to remove the aggressive male as soon as his

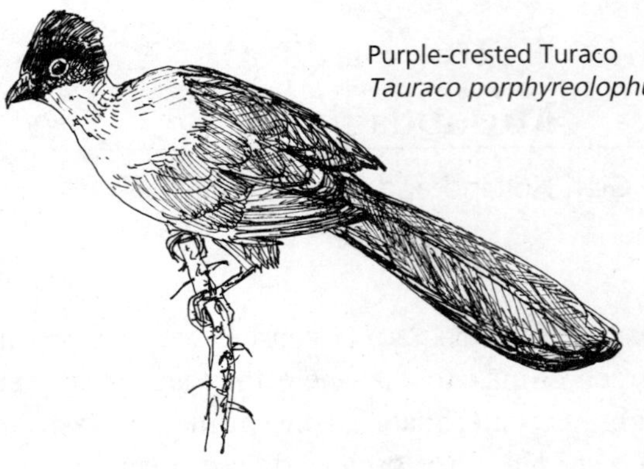

Purple-crested Turaco
Tauraco porphyreolophus

new mate had laid a clutch of eggs, allowing her to incubate them unmolested. This was done, but the eggs, though fertile, apparently became chilled sometime during incubation, and failed to hatch, perhaps because the male was not around to tend the nest while the female was off eating. The male was then returned to the aviary in the hope of producing a second clutch of fertile eggs, which would be pulled immediately for artificial incubation. However, several weeks later, about the time eggs were expected, the hen killed the male. This hen was later paired with the young male reared the previous year, and they bred successfully for a number of years with no signs of aggression from either bird (except toward the keeper during nest inspections). Others pairs, however, have exhibited various degrees of aggression, usually by the male toward the hen. A period of separation, or a change of mates, has usually solved the problem. Interestingly, none of the closely-related Schalow's turacos kept by HWRC over the years showed even the slightest sign of aggression or mate-battering. This included two long-established, wild-caught pairs and a number of their offspring.

The other species maintained in the Hancock collection exhibited this tendency relatively infrequently, although occasional problems arose with breeding pairs of white-cheeked turacos *T. leucotis* and red-crested turacos *T. erythrolophus*. No deaths could be definitely attributed to this cause among the Hartlaub's *T. hartlaubi* or white-crested *T. leucolophus* turacos, but the loss of a Fischer's turaco *T. fischeri* hen may have been caused by mate aggression. It should be noted, however, that these species were held in smaller numbers in the collection, in the case of the white-crested, only a single pair.

Over a ten-year period several female violaceous turacos *Musophaga violacea* fell victim to mate aggression. A pair of this species, or at least one member of the pair, was also responsible for killing, in a single afternoon, three of four young Edwards's pheasants *Lophura edwardsi* sharing the same aviary. They had been housed together for a number of weeks with no observed interactions before the sudden and vicious attacks occurred. Of all species of turacos bred at HWRC, the violaceous were also the most aggressive in defending the nest. While most species were extremely tolerant of nest inspection, generally remaining on the nest during inspections and offering only a few token pecks at the intruding hand in protest, the violaceous hen would literally attack the keeper during nest inspection, often coming off the nest with flailing wings and claws in an assault directed at the exposed skin of the keeper's hands and face. When chicks were in the nest, the male would join in the attack, making nest inspection difficult with-

Livingstone Turaco
Tauraco livingstonii

Lady Ross Turaco
Musophaga rossae

Knysna Turaco
Tauraco corythaix

Hartlaub's Turaco
Tauraco hartlaubi

Violaceous Turaco
Musophaga violacea

out some kind of protection. This aggressive nest protection seems characteristic of the species, having been encountered in every breeding pair kept at this institution. Occasionally, individuals of other species, particularly Livingstone's, and to a lesser degree, white-cheeked and red-crested, would be more than normally aggressive in nest defense, sometimes with the male joining in, but never as consistently, nor to the same extent, as the violaceous. The congeneric Lady Ross turaco *M. rossae* was only represented in the collection by a single breeding pair, of which neither the female nor the male has displayed such aggression. However, several two-year-old offspring of this pair, kept in a huge flight with a younger sibling and group of other young turacos, suddenly became aggressive, killing a red-crested and harassing several others. When the four Rosses were removed to a separate aviary, they eventually turned upon, and killed, one of their own.

Threat displays involving calls, ruffled plumage, raised crests, and outspread wings are typical of rival birds in visual contact. The aviary should be planted with some dense shrubs and trees as described for Habitat A in the chapter on LANDSCAPE. Turacos are fond of bathing and enjoy a mist spray hung over a shrub in which they will dash about until thoroughly soaked. This will help maintain the feathers in good condition when they are housed in a dry climate. A pair I possess fly to a favorite low perch immediately when they see me approaching with a hose in my hand.

Juveniles can be housed together until they become sexually active, at around one and a half years of age in most species, although some pairs have been known to breed at a younger age. At the HWRC, a mixed species group of up to 30 young turacos ranging in age from 1–3 years has been kept in a covered safety corridor between aviaries. This corridor is about 3.6 m wide, 3.6 m high (12 x 12 ft.), and approximately 60 m (200 ft.) in length, with three side alleys, each measuring roughly 2 m x 7.5 m (6.5 x 25 ft.). In this large flight, little aggression has been noted, with the exception of the Lady Ross touracos mentioned earlier. On several occasions, birds in this group have paired up and nested, although the successful hatching and rearing of young under these circumstances has been rare.

Diet: These turacos will feed on most fruits, the large gape enabling them to swallow large fruits whole. Figs, diced apple, paw-paw, grapes, guavas, par-boiled grated

Growth Chart for Red-crested Turaco *Tauraco erythrolophus*

Date	Age (days)	Hand-reared Weight (g)	Parent-reared Weight (g)	Notes
April 2nd, 2006	0	N/A		
April 9th	8	45.3	pulled for hand rearing	
April 10th	9	39.6		
April 11th	10	45		
April 12th	11	51	56.6	
April 13th	12	53.9		
April 14th	13	59.5		
April 15th	14	68	76.5	
April 16th	15	73.7		
April 17th	16	82.2		
April 18th	17	87.8		
April 19th	18	90.7		
April 21st	20	99.2	110.5	parent reared fledged
April 22nd	21	103.5		
April 24th	23	107.7		
April 29th	28	119		
May 2nd	31	136		
May 6th	35	167.2		
May 8th	37	175.7		
May 11th	40	187.1		
May 15th	44	198.4		moved hand reared into larger pen
May 17th	46	200.3		
May 21st	50	204.1		
May 27th	56	200.4		
May 28th	57	202.3		
June 19th	79	209		moved outside
June 27th	85	210		

carrots and soaked raisins are among the favored foods. Squares of moistened brown bread, and commercial softbill mixtures are also taken. Twice a week I also add a grated boiled egg to the soft food. If indigestible articles such as large seeds are swallowed, they are later regurgitated. At the HWRC, the diet includes not only such staples as apples, bananas, grapes, pears, tomatoes, papayas, and cooked carrots, but also generous amounts of broccoli, cauliflower, lettuce, and/or other greens, and lesser amounts of zucchini, eggplant, canteloupe, honeydew and, occasionally, radishes, onions, and other garden vegetables.

Other fruits and berries are incorporated when available, such as peaches, plums, apricots, mangoes, blueberries, raspberries, blackberries and strawberries. The fruit and vegetables for this large collection are run through a commercial food processor and mixed together to prevent the birds from picking out favorite items at the expense of others. Steamed bone meal, and sometimes vitamin supplements, are mixed in. Soaked Kaytee softbill pellets are placed on top of the fruit mix. These birds also like watermelon, which is fed separately when available. Avocado and citrus fruits are avoided, the former being too high in lipids and the latter being too high in sucrose. The few studies available on the composition of the wild diet of turacos indicates that they strongly prefer foods that are high in the simple sugars, fructose and glucose, and relatively low in the compound sugar sucrose, as well as in lipid content.

Breeding: In courtship, the male feeds the female while she flutters her wings as would a chick. Allopreening and carrying of nest material also occur. Gaping, with the pair facing each other and shaking their heads from side to side, is another part of their courtship display. The clutch is 1–3 eggs, but usually two, as with most species of *Tauraco* and *Musophaga*. The eggs are creamy white in color. Incubation is by both sexes. The nest is a flimsy platform of interlaced twigs, often built on a wire platform or on top of a dense dry thicket. A dense shrub growing in the aviary also makes a good nest site. Fortunately in an aviary they will use open boxes, baskets, and wire platforms. This stability adds to the chances of successful incubation and rearing. Nest construction takes 4–5 days and twigs are collected from surrounding branches. When building the nest, one bird fetches the material and passes it to the mate, which remains on the nest and positions the material. A bird arriving to relieve a bird on the nest often brings a twig in its beak and calls softly to the sitting bird.

The eggs are usually laid on consecutive days, particularly in the smaller species such as Gold Coast, red-crested, and Hartlaub's. This is sometimes true with the larger species, but often, particularly in the white-cheeked, violaceous, and Ross's, the eggs appear at a two-day interval. Some turacos begin incubation with the first egg, with the chicks hatching asynchronously; others begin with the second egg, with the chicks hatching simultaneously. This is usually the case with the Gold Coast and Buffon's, while Livingstone's and Schalow's are more likely to hatch on consecutive days. Three-egg clutches are more frequent in the Livingstone's than in other members of the genus, and when this occurs, the first two eggs usually hatch on the same day, the third a day later. Whether incubation begins with the first or second egg does not seem to be species related so much as an individual choice, and is not always consistently practiced by the same pairs.

The incubation period is variable in most turacos, probably due to weather conditions and other external circumstances, such as the nest attentiveness of a given pair. The incubation period also varies between species. The data below have been compiled from the literature and from records kept at HWRC over a decade. The

Growth Chart for White-cheeked Turaco *Tauraco leucotis*

Date	Age (days)	Hand-reared Weight (g)	Parent-reared Weight (g)	Notes
June 29th	6	88.85	92.05	Pulled 88.85g chick
June 30th	7	80.7		Ate 8.65g for first feeding
July 1st	8	91.15		Ate 14.95g for first feeding
July 3rd	10	115.1		
July 5th	12	129.7		
July 7th	14	139		climbing out of dish
July 8th	15	133.2	141.45	parent reared fledged
July 10th	17	130.1		
July 11th	18	148.05		hand reared chick close banded
July 17th	24	173.3		moved to larger pen
July 23rd	30	202.6		
August 25th	63	247	290	vocalizing like adults, moved both into their own pen
October 30th	99	323	281.4	moved in with sibling

incubation period repeated in many African bird books for Hartlaub's turaco, 16–18 days, was based on V.G.L. Van Someren's account of this species in *Days with birds: studies of habits of some East African species*.

Although there have been reports of shorter incubation periods in the wild than in captivity, these may be due to difficulty of determining the actual laying and hatching dates of eggs in nests high in trees and largely hidden from sight by foliage. Even given the variation of the incubation periods in captive birds, an incubation period of only 16–18 days seems implausible for any turaco, and indeed, what most subsequent authors have failed to note is that Van Someren qualified his statement by adding "so far as I have been able to estimate" (Van Someren 1956); perhaps his estimate was even influenced by some of the inaccurate earlier reports of incubation periods for other species of turacos.

The young are only brooded closely for the first week. They are fed by regurgitation and, after feeding, the chick presents the anus to the adult, which then swallows the loose feces. If chicks on the nest are threatened, they hiss and raise their wings. The chicks fledge at about 25 days but only fly after five weeks. It is therefore advisable to provide plenty of loose branches in the area surrounding the nest. The chicks should be removed at about 10 weeks of age. By three months, the chicks resemble the adults but for the beak and bare skin around the eye, which is dark. If left together too long, the parents will become aggressive toward the chicks. Also, chicks left in the aviary are likely to interfere with the flimsy nest of the second clutch. The adults will often lay almost immediately the first brood fledges, and sometimes even earlier. Red-crested turacos at HWRC, for instance, have produced a second clutch of eggs while chicks were still in the nest, and if the first brood of chicks is removed at 10 days of age for hand-feeding, this species will almost invariably lay again about 10 days later. Three broods in a season are not uncommon.

Chicks of most, if not all, *Tauraco* and *Musophaga* species resemble each other closely enough that cross fostering may be used to increase the production of a rare species such as Ross's. Cross fostering between species, or fostering within a species, can also be practiced when eggs or chicks are for one reason or another abandoned, which sometimes occurs.

I have seen a lone green-crested adopt and raise an orphaned gray go-away chick. An incubating pair will usually adopt a hatching egg, the parents eating the shell of the newly hatched egg being an indication of their acceptance. Pairs that break their own eggs can be given a clutch of dummy eggs, and their chicks given back to them on hatching.

The transfer of eggs or chicks from one nest to another is facilitated in turacos by the fact that, when several breeding pairs of a species are kept, they will often nest synchronously, particularly at the beginning of the breeding season. In the Hancock aviaries, up to four pairs of red-crested turacos have laid within a day of each other on several occasions, and in 2002 two pairs laid on the same day, with a third pair in another breeder's aviary several miles away also laying that same day. This phenomenon has been noted in other species as well, particularly Gold Coast, Buffon's, and white-cheeked, and to a lesser degree, in Livingstone's, while other species seem more individualistic in this respect. This cannot be attributed to photoperiod, at least not to photoperiod alone, because some of the pairs were housed indoors under 12-hour lighting at the time, while other pairs had spent some

Incubation Periods for Turacos

Species	Incubation period
Gold Coast *T. persa persa*	20–21 days (Roles 1971)
Buffon's *T. p. buffoni*	21–22 days
Livingstone's *T. livingstonii*	23 days
Schalow's *T. schalowi*	20–22 days (Roles 1973)
Knysna *T. corythaix*	23–24 days (Jarvis & Currie 1978; Rowan 1983)
Red-crested *T. erythrolophus*	22–24, usually 23
White-crested T. leucolophus	
Fischer's *T. fischeri*	3 days (Milne 1991)
Hartlaub's *T. hartlaubi*	20–22 days
White-cheeked *T. leucotis*	21–23 days (Foxall & Burton 1975; Evans 1984)
Lady Ross *M. rossae*	24–28 days (Steel 1973; Candy 1984; Milne 1990)
Violaceous *M. violacea*	23–24 days (Bent & Corbett 1993)
Purple-crested *M. porphyreolopha*	22–23 days (Jarvis & Currie 1979; Raison 1992)
Gray go-away *C. concolor*	26–28 days (Roles 1970; Rowan 1983)
White-bellied go-away *C. leucogaster*	25 days (Todd et al. 1985)
Western gray *C. piscator*	
Great blue *C. cristata*	29–31 days (Candy 1984)

time outdoors under natural lighting. When eggs are fostered under the same species, they should be marked, and the chicks removed at hatching for hand rearing, or somehow marked, in order to keep accurate pedigree records. For this reason, cross fostering between species is actually preferred, whenever possible.

During the past 20 years, 400 turacos have been bred by Houston Zoo, a truly remarkable feat for which they have received a significant achievement award from the American Association of Zoological Parks and Aquaria. The oldest breeding turaco at Houston Zoo is a Ross's, which was wild-caught 24 years ago. Gold Coast, Buffon's, and Schalow's turacos exceeding 20 years of age have bred in the Hancock collection, where several hundred turacos have been reared since 1992. The International Turaco Society, based in England, was formed to promote captive breeding and long-term continuity of this fascinating group of birds. Members receive an informative newsletter covering current turaco research, diet, breeding, and rearing techniques. Every aviculturist with an interest in turacos should join this organization.

Incubation and Hand Rearing: Artificial incubation and hand rearing are considered for two reasons. Firstly, it is not unusual for a pair to lay a clutch during the cold months, and fail to incubate. Secondly, the removal of a clutch is likely to induce the pair to produce a second clutch. The eggs are almost uniformly round in most species and they should be incubated at 37.2°C (99°F) with a relative humidity of 62%, for 4–5 days to determine the location of the air space. This can be marked with a spot on the shell, and the egg then set for turning accordingly. Newly hatched chicks are initially maintained in the incubator at 36°C (96.8°F) for a few days. The temperature must be dropped gradually. They are best kept in a small bowl lined with absorbent paper towel with a few twigs to provide a grip and reduce the risk of the young developing leg problems.

Mossie Weber describes his hand-rearing protocol as follows:

> "The first feed is given after 6–12 hours and consists solely of ripe paw-paw, mashed by a fork with a little water to give a liquid consistency, and warmed to body temperature in a microwave oven. All feeding is done with a syringe. It is important to start the chicks on paw-paw because it contains special enzymes. Approximately 2 ml of liquidised paw-paw is fed every two hours. The amount is gradually increased as indicated by the chicks, which refuse to take any more food when they have had enough. On the third day a proprietary parrot hand-rearing food is added to the liquidised paw-paw in a proportion of one-third to two-thirds paw-paw, and considerable experimentation with different brands took place before I hit upon a successful formula. I started with Nutribird 19, which is a hand-rearing formula used mainly on macaws and African grays, initially in the proportion of one-third to two-thirds of liquidised paw-paw. I am now using Aviplus Premium, which is a formula for newly hatched parrots, instead of the Nutribird 19. Additional water also has to be added to maintain a fluid consistency. On the fifth day the brooder temperature is lowered to 34°C (93.2°F). The proportion of hand-rearing powder is increased to half of the diet. I raised 34 turacos in the 2002 season and aimed for 50 in 2003."

They have an extremely large gape and as food is offered, they strike at it like a snake, with the mouth open and the throat expanded as well. Newly hatched chicks are induced to defecate by gently massaging the cloaca with a cotton swab dipped in luke warm water. Initially they require two-hourly feeds from 6:00 a.m. to 8:00 p.m. They will continue to beg, even when they have had enough, and care must be taken not to overfeed them. As they begin to feather, feedings can be reduced to five per day from 6:00 a.m. to 6:00 p.m. By day 21 they are fed on parrot hand-rearing formula alone. The chicks have by now developed their flight feathers. The chicks should be in an indoor cage with a source of warmth at night. The cage should measure 750 x 450 x 700 mm (2.5 x 1.5 x 2.3 in.) with a plastic base covered by a double sheet of newspaper. The chick sits comfortably on one or other of the two perches fixed about 75 mm (3 in.) above the base of the cage and from which feeding dishes are within easy reach. The chicks enjoy jumping from one perch to the other. In this cage the babies are introduced to fresh fruit, mainly paw-paw and banana diced into 0.5 cm cubes and placed in the feeding dishes, while they continue to be fed with the parrot formula three times per day. Even when turacos have been weaned onto the

Turaco chick

adult diet, they do not forget the syringe and will move over to be fed when they see a syringe. Mr. Weber has a three-year-old female of three years old that was easily medicated by mixing the medication into some parrot formula. Chicks should never be left in direct sunlight for long periods as, due to their black plumage, they absorb the heat and soon start panting. When fledged, their diet should be kept varied and the fruit chopped finely and mixed together. The soft food is added to the mix. Mixing the food prevents the chick from eating only the favored items. A calcium-phosphorous supplement and a multivitamin should be added to the hand-rearing diet. Large nestlings and fledglings imitate the calls of the adults.

Hand Raised Lady Ross Turaco
Musophaga rossae

Day	Weight (g)	Day	Weight (g)	Day	Weight (g)	Day	Weight (g)
1	35	**11**	87	**21**	165	**31**	219
2	33.5	**12**	93	**22**	170	**32**	225
3	39.5	**13**	99	**23**	183	**33**	234
4	35	**14**	115	**24**	187	**34**	240
5	34	**15**	117.75	**25**	190	**35**	255
6	36	**16**	132	**26**	197	**36**	261
7	42.5	**17**	141.5	**27**	203	**37**	268
8	49.5	**18**	149	**28**	209	**38**	275
9	58.5	**19**	157	**29**	211	**39**	279
10	71.5	**20**	163.5	**30**	215	**40**	284.5

16.01b Gray Go Away Bird
Criniferinae

• Glen Holland

Introduction: The Afrikaans name "Kwevoel" aptly describes the bird's *kweee* type call. This alarm call causes a response in most wildlife, both birds and mammals. When greeting a mate that comes down to perch, or when chicks beg for food, a *wuk, wuk* type call is made. Hand-raised birds also greet their keeper with this call. Go-away birds favor drier, open woodland and thornveld thickets. In recent years, largely due to droughts, these birds have moved into many suburbs and are common on garden bird-feeding tables. Most of their feeding is done while perched in a tree, but they are quite comfortable on the ground.

Management: The rather drab appearance of this species has resulted in relatively little interest from aviculturists. However, what these birds lack in coloration, they make up for in character. They tame quickly and will learn to take food from the hand. Habitat B will suit go-away birds. In a large planted aviary they can be mixed with a turaco species, but in a single-flight situation they are best kept as sexed pairs. Naturally, and in large aviaries, they often occur in loose groups of 3–5 birds which share the nesting and raising of the chicks. When confined in a flight aviary, the dominant bird inevitably tends to turn on a submissive bird. They can be mixed with doves, pigeons, softbills, and large finches.

Diet: The feeding of gray go-away birds is as described for the green-crested turaco, but they are also fond of leaves and will take diced spinach, the leaves of aviary plants, and occasional insects such as winged termites. Flowers, particularly those containing some nectar, are also eaten. These birds are big eaters and this has made them unpopular in gardens where they descend and finish everything put out on bird tables. I have found them to be fond of small squares of whole-wheat bread that have been slightly moistened.

Breeding: They are willing to nest in captivity and will produce 2–3 broods each season. Courtship consists of one bird soliciting with the wings held open and quivering. After copulation, the members of the pair sit quietly together for some time. Nests are built in a wire basket or on a thick, dry, preferably thorny bush hung high inside the shelter. The nest is lined with twigs, which one bird, presumably the male, brings to the nest for the mate to put in place.

The 2–4 rounded, glossy eggs are grayish-white or faint bluish-white. The eggs are laid on alternate days and incubation is by both sexes for 27 days. Chicks fledge at 21 days and fly well by 40 days. When either bird replaces its mate on the nest, they both greet with vigorous shaking of the head. The chicks are fed by regurgitation and often, as already noted, there are three or more birds in attendance. Gray go-away birds have a rather endearing feature in that they can seldom resist feeding another gray go-away chick even if it is large and not their own. While working on the Pilanesberg project, we regularly had chicks brought to us and our birds adopted most of these. The chicks are independent at seven weeks and, although aggression from the adults is not a problem, they should be removed to avoid interference with the second clutch.

Hand-rearing: The hand-rearing diet described for green-crested turacos should be used, but these birds require more roughage, such as diced apple, including the skin with the fruit. As they naturally feed on leaves, diced spinach should also be added.

Ring Size: 6 mm

Species Requiring Similar Management:

Bare-faced go-away bird *Corythaixoides personatus*
White-bellied go-away bird *C. leucogaster*
Eastern gray plantain-eater *Crinifer zonurus*
Western gray plantain-eater *C. piscator*

All of the previously mentioned species have been kept in aviculture, although in lesser numbers than the more colorful members of the family. The white-bellied go-away bird, while not common, has been bred in several American collections over the years. The bare-faced has been available from US bird dealers in recent years, and will hopefully become established in American aviaries. The eastern gray plantain-eater has been particularly rare in collections. Although the western gray has been bred on a number of occasions, the chicks have proved more difficult to rear to maturity than those of the forest-dwelling turacos. The experience of Cotswold Wildlife Park in England is typical. During 1996 a total of nine chicks were hatched by one pair, but none survived beyond 12 days. The following year, when two chicks survived the critical 12 day period, it was noted that the parent birds were eating, and presumably feeding the chicks, various plants growing in the aviary, particularly the rose-bay willow-herb, or great herb willow, *Epilobium angustifolium*. Whole lettuces were spiked on branches, and consumed by the birds, which also continued to feed on the aviary plants. One chick was reared to independence. When a subsequent clutch of chicks failed to survive in an aviary containing no rose-bay willow-herb, it was speculated that this plant had been a vital ingredient in the earlier success. Since that time over a dozen chicks have been reared, always with a generous supply of rose-bay willow-herb in the aviary, supplemented with other greens such as lettuce (Crockford 2002).

A breeding pair at HWRC followed a similar pattern. Although this species was routinely fed a greater proportion of roughage than other turacos, parent-fed chicks did not survive beyond two weeks. A later clutch, pulled at six days of age for hand feeding, failed to thrive on the rearing diet used successfully for 10 other turaco species and subspecies. Although the chicks survived for over three weeks, they stopped growing after about a week of hand feeding. Since the rearing diet of this species at the San Diego Zoo, where one of the parents had originated, included pinkie mice, the protein level of the diet was increased, but did not solve the problem. Before the beginning of the next breeding season, it was decided to increase the percentage of green food even further, and the adults were also transferred to a larger aviary containing more deciduous plant growth; chicks were finally reared successfully by the parents that summer.

The Houston Zoo had a similar experience with the white-bellied go-away bird. After an unsuccessful rearing attempt in which the chick was lost to impaction of leaves from a crepe myrtle *Lagerstroemia indica* fed by the parents, a pair was set up in a flight containing a small acacia *Acacia radii,* the leaves of which were also readily eaten by the adults. Due to the strong inclination of the parents to feed green food to the young, an entire head of romaine lettuce was offered daily, along with honeysuckle *Lonicera japonica* stems with leaves and blossoms. All plant material provided was consumed, and the staff at Houston Zoo felt that this was "crucial to their successfully rearing chicks" (Todd et al. 1985). The importance of foliage in the diets of all the members of the subfamily Criniferinae cannot be overstressed.

Ring Size: 6 mm

Species Requiring Similar Management:

Other turacos *Tauraco* and *Musophaga* spp.
Fruit pigeons *Ptilinopus* spp.

Contributors:

David Bender, Hancock Wildlife Research Center, British Columbia

Rochelle Plasse, curator of birds, Disney Wild Animal Park, Florida.

William "Trey" Todd, senior zoo keeper, Houston Zoo, Texas.

References:

Bent, N. & F. Corbett. 1993. Helping behaviour and other

observations on nesting in the violet turaco. *Avicultural Magazine* 99 (3): 132–135.

Candy, M. 1984. Habits and breeding biology of the great blue turaco. J. East Africa *Natural History Society and National Museum* 190:1–19.

Crockford, N. 2002. Cotswold Wildlife Park, Burford, UK. *International Zoo News* 49(2): 106–107.

Dowsett-Lemaire, F. & R.J. Dowsett. 1988. Vocalizations of the green turacos (Turaco species) and their systematic status. *Turaco* 1: 64–71.

Evans, K. 1984. Hand-rearing the white-cheeked turaco at the Padstow Bird Gardens. *Avicultural Magazine* 90 (1): 32–34.

Foxall, I. & P.J.K. Burton. 1975. Notes on the nesting of the white-cheeked turaco. *Bulletin of the British Ornithologists' Club* 95 (1): 25–29.

Jarvis, M.J.F. & M.H. Currie. 1974. Breeding the purple-crested turaco. *Avicultural Magazine* 80 (6): 217–219.

Milne, L. 1990. Breeding the Lady Ross's turaco. *AFA Watchbird* 17 (4): 39–40.

Milne, S. 1991. First breeding of the Fischer's touraco. *AFA Watchbird* 18 (5): 41–42.

Raison, F. 1992. Captive breeding of the purple-crested turaco. *Avicultural Magazine* 98(3): 111–114.

Roles, D.G. 1971. Breeding the Gold Coast turaco at the Jersey Zoo-Park. *Avicultural Magazine* 77(5):163–165.

Roles, D.G. 1973. Breeding Schalow's turacos at the Jersey Wildlife Preservation Trust. *Avicultural Magazine* 79(3):75–78.

Roles, G. 1970. Breeding the grey turaco at the Jersey Zoological Park. *Avicultural Magazine* 76(6):232–233.

Rowan, M.K. 1983. *The Doves, Parrots, Louries and Cuckoos of Southern Africa*. The John Voelcker Bird Book Fund and David Philip, Cape Town and Johannesburg.

Todd, W., D.C. Grubbs & H. Lahera. 1985. A successful breeding of the white-bellied turaco at the Houston Zoo. *Avicultural Magazine* 91(3):121–123.

Van Someren, V.G.L. 1956. Days with birds: studies of habits of some East African species. *Fieldiana: Zoology* 38:1–520.

Van Tuinen, P. & M. Valentine. 1986. Phylogenetic relationships of turacos (Musophagidae; Cuculiformes) based on comparative chromosome banding analysis. *Ibis* 128(2):364–381.

Weber, M. 2001. Hand Rearing Turacos. *Avizandum* 13(4):33.

Spectacled Owl *Pulsatrix perspicillata*

Day	Weight (g)	Day	Weight (g)
1	44	15	283
2	49	16	315
3	62	17	326
4	73	18	334
5	94	19	379
6	103	20	388
7	119	21	402
8	133	22	449
9	146	23	456
10	167	24	461
11	189	25	504
12	208	26	521
13	229	27	529
14	254	28	578

Ring Size: 13 mm.

Species Requiring Similar Management:

Band-bellied owl *Pulsatrix melanota*
Tawny-browed owl *P. koeniswaldiana*
Brown wood-owl *Strix leptogrammica*
Eagle-owls *Bubo* species
Barn owl *Tyto alba*

Northern Pygmy-Owl *Glaucidium californicum*

Introduction: The Northern pygmy-owl *Glaucidium californicum* is one of the smallest owls in North America, measuring 170 mm (6.7 in.) in length and weighing 58–72 g (3.6–4.5 oz.), and was just recently accorded full specific status, having formerly been regarded as a subspecies of the mountain pygmy-owl *G. gnoma.* Its distributional range is limited to the mountainous areas of the west coast of North America. Northern pygmy-owls prefer the higher altitudes of the mountains, ranging from mixed mountain coniferous forests with openings for hunting purposes, to scattered conifer trees at the tree line, as their breeding biotope. In the autumn, one can observe them in mountain valleys as well as in village parks and gardens and in villages surrounding open farmland. This owl is clearly a diurnal bird, although it is most active in the twilight of morning and evening. Sitting in bright sunshine in the dense top of a spruce tree, it can be located easily by its call in the spring and fall. The monotonous territorial song catches the attention of even a novice birder. A repetitive, well-spaced series of *fhoot* or *fhoot-fhot* notes are easy to imitate. This little gnome lacks the silent flight characteristic of most owls; the whirring of its wings can be heard even at a short distance.

The northern pygmy owl occurs in three distinct, genetically determined color-morphs (dichromatism). There are birds that in general have gray, brown, or reddish plumage. The lower hindneck is dominated by two oval, black spots, which give the illusion of two dark eyes (occipital face). The whole back is speckled with small buffy spots, and the under parts are white, streaked with dark brown. Sexes are alike, except that females are bigger and heavier than males.

Management: Northern pygmy owls are very rare in captivity. To my knowledge, there are only two places where this owl is being aviculturally bred. However, their requirements under human care seem to be no different from those of other pygmy-owl species. In spite of their size, they need a relatively large aviary, at least 2 x 3.5 x 3 m (6.5 x 11.5 x 10 ft.), to satisfy the need for their flying activities. Their aggressiveness toward others of their own kind is one reason why not too many aviculturists can afford to keep them, because it is wise to keep the male and female in separate aviaries during the non-breeding season. So far, I have been lucky in being able to keep whole families together for months without any incidents, but I have heard different stories from other people.

It seems almost unbelievable, but these little owls can withstand sub-zero temperatures as low as –40° and perhaps even colder. For this reason, it is not necessary to provide them with a heated aviary; on the contrary, this would prevent them from displaying the behavioral patterns that enable them to survive in a bitter cold environment. In any case, the aviary should be designed with a wind and draft free section topped with a roof. This offers the birds the possibility of avoiding hot sunshine and heavy rainfall. An aviary for this species can

Oriental Bay-Owl
Phodilus badius

easily be created to reflect its natural, mountainous environment by planting two or three small conifers. A few really big rocks and a bizarre looking dead tree, which should be the highest point in the aviary to serve as a lookout for the owls, completes the landscaping. It becomes even more effective when this same biotope is continued outside the aviary as well. A properly landscaped pygmy-owl aviary is very rewarding. Not only does the vegetation stay nice and healthy, but being able to observe the owls' activities in broad daylight in a natural surrounding is very enjoyable. Aviary maintenance is simple. One must merely prune the vegetation periodically so it doesn't become too dense and so the flight corridors stay open.

Diet: The summer diet of these owls consists mainly of mice and voles; in the winter their natural diet in the wild is mainly small birds. In captivity, they show a clear preference for day-old chicks. Only during breeding time do they develop some liking for mice and mouse-sized rats. One will notice that shortly before winter arrives, they will start caching food in nest boxes as a precaution for times of food shortage. During the winter, when the stored food items are frozen rock-solid, the owls will spend considerable time in the nest boxes to warm up pieces of food by brooding on them. Of course there should always be enough fresh food available, and during heavy frost periods, they should be fed twice daily. Northern pygmy-owls have developed another strategy to keep fresh food items from freezing; they will carry food around for hours, thereby holding it close to their bodies. This is referred to as "thermo-storaging." During cold weather and at molting time, these owls eat more, consuming about 50–60 g (1.7–2.1 oz.) during a 24 hour period, which amounts to approximately 1½ day-old chicks or three mice. All of our owl aviaries are constructed in such a way to allow access to small mammals and reptiles, as well as insects, of course. The owls are eager to hunt, and in the case of the northern pygmy owls, kill the incoming mice, remove their heads, but do not pay their beheaded prey any further attention. I have, however, seen them eating some larger insects from the wild, as well as the locusts we provided.

Breeding: Of the northern owl species, the northern pygmy-owl is the last to start breeding, never before the beginning of May. During the mating time, a period lasting a month and a half prior to the breeding season, the pair becomes very fond of each other. They call each other in a soft, low voice, and the male is very active in showing the female different nest boxes from which she will finally accept one. I provided them with two different kinds of nest boxes, one large, 20 x 25 x 40 cm (8 x 10 x 16 in.), and one small, 15 x 30 x 25 cm (6 x 12 x 10 in.), and with one exception, they always used the smaller nest box. Contrary to most North American owls, in which incubation commenced with the laying of the first or second egg, this small owl does not start before laying the fourth egg, with a normal clutch of 5–6 eggs that average 29.05 x 23.32 mm (1.14 x 0.9 in.) in size and 7.7 g (0.27 oz.) in weight. As an adaptation to the sub-zero temperatures of May, they covered their first-laid eggs, which were otherwise unprotected, with a mixture of peat moss and wood shavings that I provided for them. The interval of egg laying is normally 56 hours (two and a third days) but it can be even longer. During incubation, which is done only by the female, the male provides his mate with food. The incubation time is around 28–30 days. Out of a clutch of five eggs, four will hatch within a span of 24 hours while the fifth will hatch three days later. With a clutch of six, four will hatch within 24 hours, the fifth one day later and the sixth three days later. In both cases, the last one will not survive. During the first few days, the hatchlings are covered with white downs, but with the juvenile plumage developing quickly they appear more grayish. The young owls stay in the nest for 25–28 days before fledging. Once they have left the nest they need three days until they are able to reach every point in the aviary. At this time, the male takes over completely and provides the youngsters with food. The young pygmy-owls do not go back into the nest box. At night, they sit together with their parents somewhere in the aviary where they are sheltered by a roof.

Hand Rearing: So far, artificial incubation and hand rearing has not been practiced with this species. If hand rearing becomes necessary, there should be no problems encountered, as several related species, such as the European pygmy-owl *G. passerinum,* have been hand reared without any difficulties. Due to the northern pygmy-owl's preference for higher altitudes, it is greatly threatened by logging activities. Therefore no conservation-breeding program is currently in place.

Ring Size: 8mm

Species Requiring Similar Management:

Austral pygmy-owl *Glaucidium nanum*
Eurasian pygmy-owl *G. passerinum*
Pearl-spotted owlet *G. perlatum*
White-faced owl *Sceloglaux albifacies*
Scops-owls *Otus* species

17.02 Aegothelidae Owlet-Nightjars

Trena Carney

Introduction: The family Aegothelidae, owlet-nightjars is composed of one genus, nine species and 18 taxa, which range in size from 18–30 cm (7–11.8 in.) (Holyoak 1999).

The family is endemic to New Guinea, Australia, New Caledonia and the Mollucas, although an extinct owlet-nightjar Megaegotheles novaezealandiae is known from fossil evidence in New Zealand, and the Quercy Aegothelid from fossil deposits in France (Holyoak 1999).

Owlet-nightjars are shy, nocturnal birds that are rarely seen during daylight unless flushed from their roosts. They are all very similar morphologically, with forward facing eyes, well developed facial bristles, a wide, bright gape and rounded wings suitable for short flights through their forest habitats (Holyoak 1999). The legs and feet are small and weak in most species, however, they are slightly larger in *A. cristatus* and considerably longer and stronger in *A. savesi,* suggesting that these species spend a greater proportion of their time on the ground (Holyoak 1999). Plumage coloration ranges from brown, rufous, chestnut and grey. The smaller species are mainly brown with grey or blackish streaks and a barred pattern. The two largest species *A. insignis* and *A. tatei* have large areas of rufous to chestnut plumage with bold markings in white and black (Holyoak 1999). Both sexes of *A. insignis, A. archiboldi* and *A. albertisi* and female *A. cristatus* have a rufous morph (Holyoak 1999). There is very little information available for most species and in some cases the only records are those obtained when original type specimens were collected. The Australian owlet-nightjar is the only species that has been studied in any detail and is also the only species that has been recorded to have been kept and bred in captivity.

Owlet-nightjars are sedentary and generally inhabit forests, forest edges and woodlands (Holyoak 1999). *A. crinifrons* and *A. cristatus* inhabit a wider range of environments and are also found in cultivated woodlands and farmland with *A. cristatus* being the only species to be found in arid and semi-arid habitats (Holyoak 1999). They have been recorded at elevations ranging from sea level to 3,700 m (12,140 ft.) (Holyoak 1999).

As there is so little information available on most species of owlet-nightjar, it is difficult if not impossible to determine the conservation status for most of them. *A. savesi* was thought to be extinct and known only from fossil evidence and a single type specimen until a single bird was sighted in 1998 (Holyoak 1999). A detailed survey needs to be completed to establish the validity of this sighting and the status of the species. There is a subspecies recovery plan for *A. c. tasmanicus* and this subspecies is listed as lower risk (near threatened): d (Garnett 1999).

Owlet nightjars are not commonly kept in captivity. Difficulties in displaying and feeding this species are thought to be the main factors that to date have restricted their popularity. Muller & Clayton (1972) state that wild-caught adult birds have never survived for more than six weeks in captivity, with only birds introduced to captivity as fledglings or chicks surviving for any length of time. The first captive breeding occurred at Taronga Zoo in 1970 with the same pair breeding again in 1971. Due to the low numbers of birds in captivity, no further breeding occurred until a pair were introduced at Taronga Zoo in 1999. Advancements in husbandry and exhibitory techniques, and increased availability of live food at more affordable prices, have enabled this species to be maintained in a wider range of captive situations. Several specimens acquired as adults not suitable for release from wildlife rehabilitation clinics have survived to breed in captivity.

Management: In general the Australian owlet-nightjar is relatively easy to manage in captivity. They are a shy, unobtrusive species and their nocturnal activities generally have no effect on other aviary inhabitants. The most important aspect of their husbandry is to supply sufficient roost sites to enable each bird to have a territory that includes several logs or boxes to ensure that they are not forced to roost together. They also require multiple roost sites to ensure that they have a "bolt hole" to retreat to when flushed from their log or box. In general it has been found that the Australian owlet-nightjar is compatible with most small passerines and pigeons. However, housing with other nocturnal species such as the tawny frogmouth, *Podargus strigoides* is not advised.

While they will defend nest and roost sites, they are easily driven out of nest boxes and logs by other competitive species such as small parrots and finches. In several cases at Taronga they have been found sharing nest boxes with nesting zebra finches, *Taeniopygia gut-*

tata, although the finches usually succeed in evicting the owlet-nightjars from the nest. Owlet-nightjars always roost in a log or box. If they are found roosting constantly in the open or in a dark corner of the aviary it is generally due to being out competed, and more boxes should be introduced to the aviary or the inhabitants of the aviary should be reconsidered. Although the Australian owlet-nightjar does not impact greatly on cage mates, some of the larger species of owlet-nightjar may be more aggressive and capable of inflicting damage. There is one suspected case of aggression from a pair of Australian owlet-nightjars where a pair of superb fruit-doves *Ptilinopus superbus* were severely plucked (pers. obs.). It is suspected that the fruit doves were roosting in the feed station at night and the owlet-nightjars chased them from the area, as they were between the owlet-nightjars and their food source.

Exhibits with wire mesh flights and solid shelters have been used to house breeding pairs of the Australian owlet-nightjar at Taronga Zoo. In 1970 and 1971 a breeding pair of the owlet-nightjars were housed in an exhibit measuring 2.5 x 8.5 x 2 m high (8.2 x 28 x 6.5 ft.) at the front and 2.3 m (7.5 ft.) high at the rear with the last 2 m (6.5 ft.) consisting of an openfronted concrete shed (Muller & Clayton 1972). In 1999 a similar exhibit was used measuring 9.7 x 8.5 x 2 (32 x 28 x 6.5 ft.) at the front and sloping up to 2.3 m (6.5 ft.) high at the rear. The last 2m (6 ft. 8 in.) consisted of an open-fronted concrete shed. The flight area was well planted with native plants. Webb (1989) found that owlet-nightjars spend a large proportion of their time when hunting foraging on the ground. Therefore, areas of open ground are provided in all aviaries to allow room for pouncing and foraging. Roosting logs and nest boxes are mounted high on walls in sheltered areas of exhibits. At least one natural log and one artificial nest box are provided for each bird. These birds prefer to roost alone and also like to have a "bolt hole" to fly to when disturbed during the day (Higgins 1999; Strahan 1994). The birds do appreciate having the entrance to their roost log or box facing the sun for a period of the day to allow them to sunbathe and receive sufficient ultraviolet radiation. Ultraviolet lights operated on timers are installed in most exhibits to attract nocturnal insects.

The only species for which information is recorded on social organization is the Australian owlet-nightjar, which is permanently territorial (Higgins 1999). However there are several references to *A. c. cristatus* forming loose colonies in the wild (Higgins 1999). This species has been successfully housed in captivity in single-sex groups of 3–4 individuals, providing that there are sufficient roosting and feeding sites within the aviary. With older birds these

Owlet-Nightjar
Aegotheles cristatus

groups have also generally been comprised of individuals of the same sex. However, they have never been successfully bred in a colony situation or when young from a previous clutch remain in the same aviary. Schodde & Mason (1997) state that the pair bond is lifelong and is maintained throughout the year; however, Bringham *et al* (1997) observed that birds will re-pair if their mate dies during the breeding season. Pairs have been successfully rearranged for genetic management without undue effects on the birds, resulting in successful breeding of new pairings even when the mates from previous seasons were housed within visual and auditory contact.

The Australian owlet-nightjar is widespread throughout Australia (Higgins 1999), therefore, ambient temperature, humidity and normal daylight cycle are suitable if housed in outdoor aviaries in areas with a similar climate to most parts of Australia. If housed in a nocturnal house, provisions for bathing and UV light need to be considered. Some species are found in colder areas of montane forests at high altitudes; *A. insignis, A. archibold,* and *A. albertisi* inhabit altitudes of up to 3,000 m (9,840 ft.) and *A. crinifrons* can be found at altitudes of 1,800 m (5,900 ft.) (Holyoak 1999). High temperatures may affect the health of these high altitude species. The Australian owlet-nightjar copes with low temperatures during the day and especially at dawn by entering torpor, which can last for up to four hours (Brigham *et al* 2000). It is highly likely that the montane species of owlet-nightjar also make use of torpor to conserve energy during the day.

Diet: There is very little information available on the foods and feeding habits of most species of Owlet-nightjars. There is no information available for *A. tatei, A. wallacii, A. archiboldi* and *A. benettii,* with the diet of *A. savesi* being recorded as beetles from the stomach contents of the single type specimen (Holyoak 1999). *A. crinifrons* feeds mainly on small insects that are captured in short sallying flights in the middle storey of the forest, or gleaning from leaves while hovering (Holyoak 1999). *A. albertisi* is recorded as feeding mainly on beetles with some moths, flies and grasshoppers, and infrequently earthworms, suggesting that they do sometimes feed on the ground (Holyoak 1999). Dietary information and foraging habits are more completely documented for the Australian owlet-nightjar, which obtains the majority of its food by catching insects in its bill after short sallying flights or by swooping to snatch insects from the ground, tree trunks or foliage (Higgins 1999). The Australian owlet-nightjar also spends a considerable amount of time foraging on the ground in short hops and fluttering flights (pers. obs.). This species feeds on beetles (larvae and adults), grasshoppers, ants, moths (larvae and adults), bugs, cockroaches, and earwigs (Barker & Vestigens 1984).

Although there is a lack of information in regard to the wild diet and feeding strategies, for most species the general body plan and habitat for all species is very similar. It is, therefore, possibly safe to assume that the diet used to maintain and breed *A. cristatus* in captivity, with appropriate adjustments for size variations, would also be suitable for other species. In captivity, both adults and young need a copious supply of live food. Artificially reared birds will eat small quantities of artificial diets but the majority of their diet is live food.

Taronga Zoo maintenance diet for the Australian owlet-nightjar:

- Approximately 10 adult crickets *Archeta domestica* per bird daily (preferably over 10 mm long.)
- *Ad lib* Mealworms.
- Maggots *Mus domesticus,* several times per week.
- Nocturnal flying insects.
- Native wood cockroaches 5–10 per bird twice weekly
- ¼ cup insectivore diet per exhibit
- ¼ cup carnivore diet per exhibit

Taronga Zoo breeding diet for the Australian owlet-nightjar:

- 80 small crickets daily for the first four days after chicks hatch (5–10 mm long).
- 40 adult crickets (over 10 mm) daily for each chick in the nest from day four.
- Mealworms *ad lib*
- Maggots *Mus domesticus, ad lib* daily
- Native wood cockroaches 5–10 per bird twice weekly
- Nocturnal flying insects
- ¼ cup insectivore diet per exhibit
- ¼ cup carnivore diet per exhibit

If there are large numbers of nocturnal insects available, the number of crickets fed are adjusted accordingly, however, there should always be live food left the following morning.

The number of crickets is not reduced until about two weeks after the chicks fledge. It is then reduced to 10–15 adult crickets per bird per day until the juveniles are separated into separate exhibits when they are then placed on the maintenance diet.

As owlet-nightjars are nocturnal, all food is offered as late in the afternoon as possible. This also reduces the time available for other exhibit inhabitants to consume the live food intended for the owlet-nightjars, and also prevents deterioration in the quality of the food.

All live food produced in captivity is fed out to the birds with food in the gut to increase nutritional value of the food. Crickets are maintained on Zeigler Cricket Food™, grated carrot and lettuce. Mealworms are maintained in wheat bran and high protein baby cereal and cockroaches are maintained on carrot and dog kibble. Mealworms are presented in trays containing carnivore and insectivore diets. The trays are prepared several hours before being fed to enable the mealworms to eat some of the prepared diets. These trays are placed on raised wire racks to prevent ground-feeding birds from eating the mealworms. Crickets, maggots and cockroaches are presented in large plastic tubs on the ground. Some Zeigler Cricket Food™ is placed in the tub to enable the insects to continue feeding. Nocturnal flying insects are attracted directly to the exhibits using fluorescent ultra violet lights. An insect trap is also used to catch insects, which are then later fed to the birds.

The main dietary change required in the breeding season is an increase in the quantity and variety of live food offered to the birds. In spring the quantity of nocturnal insects increases naturally. However at Taronga Zoo owlet-nightjars have started breeding before this natural flush of insects occurs. In this case the quantity of crickets and mealworms is increased.

Breeding: The breeding biology of most species of Owlet-nightjars is completely unknown, with the nests of *A. tatei, A. insignis, A. crinifrons, A. wallacii, A. archiboldi* and *A. savesi* completely undescribed

(Holyoak 1999). Eggs are only known for *A. benettii, A. albertisi* and the Australian owlet-nightjar, and are white and unmarked and laid in hollows in dead tree branches (Holyoak 1999). The breeding season of most species is completely unknown, however, in several species the size of gonads of collected specimens has given some indication of when breeding may occur.

The Australian owlet-nightjar is the only species for which there is a detailed account of the breeding biology. In the wild, breeding season varies according to geographical location and has been recorded from July through to April in Australia (Higgins 1999). Captive birds in Sydney have started egg laying as early as May and continued through until late January. Holyoak (1999) states that reports of breeding in any month in arid regions, especially after rain, to be incorrect. However it appears that captive birds can be stimulated to breed early in the year, possibly due to an abundance of live food provided.

The nests of this species are generally in holes in trees but have also been recorded in fence posts, fallen logs, crevices in buildings and burrows in riverbanks usually 1–5 m (3 ft.) above the ground (Holyoak 1999). The first captive breeding of this species occurred in a hollow log hung horizontally in an exhibit at Taronga Zoo (Muller & Clayton 1972). Subsequent to this, all nesting has occurred in artificial nest boxes that are generally used for breeding small parrots such as peach–faced lovebirds *Agapornis roseicapillus*. Natural logs were always available, however the birds prefer to build and lay in the artificial boxes. Larger parrot nest boxes have been made available, and although they will roost in these boxes they still prefer to nest in the smaller boxes. Wild birds have also been known to nest in man-made boxes. A substrate of equal parts of fine sawdust and peat is placed in the bottom of the nest boxes to absorb moisture, as parents do not remove feces from the nest.

The nest is constructed by both sexes and consists of fresh Eucalyptus or Acacia leaves (Higgins 1999). Captive birds however have used a majority of dried leaf litter from the floor of the aviary even when fresh leaves were available. Clutch sizes are 2–5 eggs but generally 3–4 are laid one to two days apart, with incubation starting with the laying of the first egg (Higgins 1999). Eggs are white and slightly glossy, rounded-oval to elliptical in shape (Holyoak 1999; Higgins 1999). Eggs of captive birds range from 27–28.5 x 22.1–23.7 mm (1–1.12 x 0.87–0.93 in.) (n=9). Both sexes share incubation, however only one bird will be in the nest at any one time (Higgins 1999; Holyoak 1999). The incubation period is recorded as 25–27 days (Strahan 1994), however incubation in captivity has ranged from 22–25 days with an average of 23 days (n=18). Adults often do not flush from the nest, and defend the eggs with a defensive posture with the mouth open and body swaying from side to side. Chicks hatch with a covering of long, white down and with the eyes closed. A vigorous feeding response is elicited when the chicks are touched, and they respond by lifting the head and opening the mouth to reveal a large red gape. Chicks are brooded constantly throughout the day until they are about five days old and pin feathers start to emerge on the wings and body. At this stage they are often left alone during warmer afternoon hours. The eyes begin to open at around three days old but are not fully open until 10–11 days old. At this stage the body is covered in pin feathers and the flight feathers and tail feathers are beginning to emerge from sheaths carrying the white down on the tip of each feather. At two weeks old contour feathers are emerging from pins and the rictal bristles are fully developed. The parents generally only brood the chicks in the early morning but this is dependent on the ambient temperature. When disturbed, the chicks exhibit a defensive posture very similar to the parents' defensive behavior in the nest. Brooding is discontinued by the time chicks are 18 days old, when more than 75 % of the body is covered in contour feathers. Age at fledging has been recorded 21–29 days (Higgins 1999) and 35 days (Muller & Clayton 1972) however, recent captive breeding has established the age of chicks at fledging to range from 21–26 days with an average of 24 days, (n=15). Chicks fledge with some down still clinging to the ends of feathers, especially around the face and neck, and the tail is generally not completely grown. Holyoak (1999) states that the chicks do not return to the nest in the wild, however in captivity fledglings are often found roosting in the nest box where they were reared. For the first few days after fledging the chicks are generally chaperoned by one of the parents in a roost log or nest box. The age of independence has not been documented and it has been reported that wild fledglings remain with their parents for several months (Holyoak 1999). At Taronga Zoo three juveniles were separated from their parents 27 days after fledging with no ill effects. Juveniles have also successfully been left in the same aviary as the parents until the following breeding season.

Higgins (1999) states that there have been several observations of birds re-nesting in the wild if the first clutch of eggs fails. Observations of captive birds reveals that birds will re-nest if a clutch fails and will also produce multiple clutches in a season. The interval between clutches of captive birds at Taronga Zoo was 9–20 days, with an average of 12 days (n=10). Multiple clutching may be a result of the continuos supply of live food in captivity. Subsequent clutches of eggs are generally not completed and usually do not hatch if juveniles from the first breeding are left in the aviary, as the young birds disrupt the nest by returning to it to roost. Females housed alone or in groups will lay multiple clutches of eggs, however they generally do not incubate, and males housed singly or in groups will nest build. There is no information available for age of first breeding and longevity or wild owlet-nightjars. A. c. cristatus has been bred in captivity at 12 months old and one male at Taronga Zoo is still breeding at 12 years of age.

Parent Reared
Australian Owlet-nightjar *A. cristaus*

Age (Days)	Weight (g)	Age (Days)	Weight (g)
1	4.46	**13**	48.9
3	8.79	**15**	50.7
5	15.4	**17**	51.0
7	27.3	**19**	48.3
9	31.7	**21**	47.1
11	44.1	**30**	51.8

Artificial Incubation: There is no information available for artificial incubation of any owlet-nightjar species. However eggs of most species of birds can be successfully incubated at a temperature of 37.2–37.4°C (99–99.4°F). The relative humidity in the incubator should initially be maintained at 50–60% and adjusted to produce a weight loss of approximately 15% of the fresh egg weight at internal pip. Although there is no information on the frequency of turning required, it should be safe to turn the eggs though 180 degrees up to 12 times per day until 100% veination is reached when the turning should be reduced to 100–120 degrees per turn. Internal pip generally occurs approximately 48 hours before hatching. Therefore, based on the range of the incubation period, eggs should be examined closely for signs of drawdown as early as 20 days into incubation. After internal pip has occurred turning should cease, to allow the chick to position itself correctly for hatching. External pip should occur approximately 24 hours after internal pip, and at this stage the relative humidity should be increased to 75–80% and temperature reduced by 1°C to prevent desiccation of the shell membranes. Hatching should be completed 24–48 hours after external pip occurs.

Artificial Rearing: There is no record of captiv-bred chicks being artificially reared, however chicks at various stages of development are reared by wildlife carers on a regular basis. Unfortunately detailed records are not available. Immediately after hatching the umbilicus of the chick should be swabbed with a 1% solution of providone iodine. Newly hatched chicks should be maintained at a temperature of 35°C (95°F) for the first 24–48 hours until the down is dry and fluffy. However the behavior of chicks should always be used as a guide to the temperature at which they are maintained. The temperature can be reduced gradually until the chicks are approximately 14 days old when they should be well feathered enough to be maintained at room temperature or 25°C (77°F). The parent birds do not remove feces from the nest, and it is not contained within a fecal sac and is very messy. The substrate that the chicks are kept on must, therefore, be changed regularly to maintain a hygienic environment. It is also important to provide the chicks with sufficient traction to prevent splayed legs and abrasions to the feet. Rubber non-slip matting placed over kitchen towel is an excellent combination of substrates as the towel absorbs moisture and the perforations in the rubber matting allow the feces to pass through, keeping the chicks clean as well as providing traction. Once the chicks are partially feathered with well-developed feet they can be housed in containers with wire mesh bottoms, with a lining of paper towel or wood shavings underneath.

As owlet-nightjars are insectivores, the rearing diet should be predominately insect based with insects such as crickets, cockroaches, maggots, mealworms, moths and grasshoppers. When the chicks are young the insects should be cut into appropriate-sized pieces and tenderised slightly before being offered. To ensure that chicks are well hydrated and to help balance the diet, the insects can be fed in slurry of Wombaroo Insectivore Rearing Powder or an equivalent product. Soaked dog kibble is also often used as a part of the diet when rearing insectivorous and omnivorous birds, as it is generally high in calcium and protein and the extra moisture also helps to maintain hydration. A feeding response can easily be elicited from young chicks by touching the side of the beak or tapping the side of the nest. Older chicks with open eyes are extremely visually oriented and will respond by gaping at the slightest movement in their field of vision.

Parent-reared owlet-nightjars are only fed at night, however it is not very often that we can mirror the frequency at which chicks are fed in the nest by their parents. Rather than setting out rigid feeding guidelines it is probably far more effective to judge the frequency of feeds and the quantity given by looking at the growth rate and appetite of each individual bird. Six parent-reared chicks at Taronga had an average growth rate of 10.24% of their body weight each day from day one to 14. The growth rate dropped off rapidly at this point to an average of only 0.7% per day. Weaning should be relatively simple as the young birds should be stimulated to feed on moving insects by instinct. A shallow tray of crickets and mealworms placed in the bottom of a weaning cage should have fledglings feeding on their own quite rapidly.

Recipes for diets used at Taronga Zoo:

(All diets are prepared the day before they are used unless otherwise stated.)

Carnivore Diet

- 3.5 lt Pedigree Pal Meaty Bites Minis soaked in 1 lt warm water
- 1.5 lt insectivore Rearing powder
- 1 kg kangaroo mince
- Mix together and add small amount of water if required to moisten end product. Refrigerate overnight and add the kangaroo mince the following morning.

Insectivore Diet

- 34 finely mashed boiled 60 g eggs
- 1.5 kg Vetafarm insectivore diet 1
- 500 ml fly pupae
- Mix together and add small amount of warm water if required to moisten end product.

References:

Barker, R. D. & W.J.M. Vestigens.1984. *The food of Australian Birds 1 Non-passerines*. Melbourne: CSIRO.

Bringham, R. M., G. Kortner, T.A. Maddocks, F. Geiser. 2000. Seasonal use of torpor in free-ranging Australian Owlet-nightjars *(Aegotheles cristatus). Physiological & Biochemical Zoology* 73(5): 613–620.

Higgins, P.J. (Ed.) 1999. *Handbook of Australian New Zealand & Antarctic Birds. Volume 4: Parrots to Dollarbirds*. Melbourne: Oxford University Press.

Holyoak, D. T. 1999. Family *Aegothelidae* (Owlet-nightjars) In *Handbook of the Birds of the World. Volume 5 Barn-owls to Hummingbirds.* Barcelona: Lynx Edicions.

Muller, K. A., & L. J., Clayton. 1972. Maintaining and breeding Australian owlet-nightjars *Aegotheles cristatus* at Sydney Zoo. *International Zoo Yearbook* 12:138–140.

Schodde, R. & I.J., Mason. 1997. *Zoological Catalogue of Aust.* 37.2 Aves. Melbourne: CSIRO.

Strahan, R. 1994. *Cuckoos, Nightbirds & Kingfishers of Australia.* Sydney: Angus & Robertson.

Webb, G., A. 1989. A note on the diet of the Australian Owlet-nightjar *Aegotheles cristatus. Corella,* 13(3): 90–91.

17.03 Podargidae Frogmouths

Liz Romer

Introduction and Natural Habits: Frogmouths are now included in the order Strigiformes along with the owls, nightjars, potoos, and the oilbird. The Australian frogmouths are assigned to the family Podargidae, which contains three species belonging to a single genus. The tawny frogmouth *Podargus strigoides* is endemic to Australia; the Papuan frogmouth *P. papuensis* ranges from the Aru Islands to New Guinea and northeastern Australia; the marbled frogmouth *P. ocellatus* is found from the Aru to the Solomon Islands, as well as eastern Australia, where the nominate race occupies the northern portion of its range and a subspecies, the plumed frogmouth *P. o. plumiferus,* occurs further south, in the central part of eastern Australia.

The Asian frogmouths are now placed in a separate family, Batrachostomidae, which includes 11 species, also assigned to a single genus. One species, Blyth's frogmouth *Batrachostomus affinis,* occurs in Malaya and the Greater Sundas, with a subspecies *B. a. continentalis* in Indochina. Two species, the Dulit frogmouth *B. harterti* and the Bornean frogmouth *B. mixtus,* are found in Borneo. The Sunda frogmouth *B. cornutus,* is confined to the Greater Sunda Islands, and the Philippine *B. septimus,* to the Philippines. The Javan frogmouth *B. javensis* occurs only in western and central Java; the short-tailed *B. poliolophus,* only in Sumatra. The Ceylon frogmouth *B. moniliger* is found in Sri Lanka and in southwestern India. The large frogmouth *B. auritus,* Gould's frogmouth *B. stellatus,* and Hodgson's frogmouth *B. hodgsoni* are all found in Southeast Asia, with the geographical range of the latter species extending westward through southwestern China to northeastern India.

The tawny frogmouth is the most common species in aviculture, particularly in zoos. This species is widespread in Australia. It is a medium-to-large bird weighing 400–420 g (14–14.8 oz.). The plumage is mottled gray and charcoal gray. This species is characterized by a large head with a huge yellow-pink gape and a large, flat, horny, triangular, sharply hooked bill, with the eyes facing sideways. At rest, this frogmouth appears much larger than it really is, but when alarmed it freezes, compacting its plumage and closing its eyes to slits. When this occurs, the bird is camouflaged to look like a broken branch. The typical call is a deep, resonant *oom-oom-oom-oom,* and when threatened, they clap their bills loudly.

Management: This frogmouth should be kept in separate pairs otherwise fighting will occur, mainly between pairs during the breeding season. The birds should be housed in an aviary of at least 3 m (10 ft.) square and 2.5 m (8 ft.) high. Habitat B is suitable for frogmouths, but I have seen them displayed and bred in aviaries containing a minimal amount of plant growth. Perches of varying thickness and heights, plus a small pool for drinking and bathing, should be features of their aviary. They also adore showers from a hose, or a mist spray positioned above a favorite perch. One must be careful on hot days, however, that water does not condense on the inside of the tin roof and fall, boiling, onto the birds. Showering is an important aid in waterproofing the feathers and will also help to stimulate breeding. The aviary floor is covered in mulch and leaf litter where the birds will pick up any insects that visit the aviary. As all frogmouths naturally inhabit warm areas, in severe climates where temperatures remain below 7°C (44.6°F) for extended periods, they should be housed indoors with some supplementary source of warmth.

Frogmouths should never be taken for granted. They are usually docile, but they are capable of removing a fair lump of skin with their beaks, should the keeper become careless. It is also a good idea to always wear safety glasses when handling such birds. They become more active and aggressive at dusk, and are therefore best handled during the day. Frogmouths kept in close confinement, such as injured birds, will often refuse to eat and should be force-fed small mice in the morning and evening. They will resume feeding themselves once released. When force-feeding is necessary, the food should always be placed carefully over the tongue to prevent damage. Tawnys can change color when they have been housed inside for a period of time and are then given access to an outside aviary. They generally turn a brownish color when indoors for an extended period, and then turn dark gray when put outside. The plumage color does, however, vary from bird to bird and from area to area.

Coccidia is present in most adult tawny frogmouths, and it is important that young birds, which may be killed by coccidiosis, have fecal samples checked and undergo treatment when necessary. Tawnys frequently damage their tongues as they bash their catch on a branch to kill it. Sometimes the tongue will be pushed back as the catch is swallowed, forming a crease across the tongue, or it will be pushed to the side and remain there. The tongue will eventually drop off at the damaged crease with no discomfort to the bird. The keeper need not panic when this occurs, as a new section of tongue will eventually grow in a way similar to the re-growth of the tail of a lizard. When tawnys fight, both combatants open their beaks and whichever bird is lucky enough to get the hook of its top beak into its opponents mouth first is the one who does the damage. The hook will pierce and rip the inside of the mouth of the victim and the battle is won. Again, the tongues of frogmouths are frequently damaged, and the sides of their beaks can be broken.

A captive bird was known to be at least thirteen years old at death.

Diet: Frogmouths naturally take most of their food from the ground. They flit from perch to perch silently and sit and watch for the movement of prey, then swoop down with beaks open and scoop up their prey without actually touching the ground. Most of their hunting is done at dawn and dusk. Large prey, such as frogs and mice, is held in the bill and beaten against a branch before being swallowed.

Food is offered on two feeding stations at different heights in the aviary. Small food is placed in a dish fixed to the lower feeding platform; mice and larger food items

Tawny Frogmouth
Podargus strigoides

are laid on the upper feeding platform. The staple diet consists of mice of approximately 2–3 weeks of age. These "fuzzies" are far more nutritious than younger "pinkie" mice. Day old chickens are also consumed. If the food being fed has been frozen prior to use, a multivitamin should be sprinkled over the food daily. Food items that are too large or heavy are refused. Lizards, skinks, grasshoppers, mealworms, cockroaches, fly larvae, earthworms, and other insects are also offered. An ultraviolet light in the aviary, which is turned on at night, will attract a variety of insects that the birds will readily consume. Frogmouths detest any food that sticks to their beaks. Even chopped mice and meat are rejected for this reason. Like owls, frogmouths regurgitate pellets consisting of fur, bone, invertebrate exoskeletons, etc. If a bird refuses to feed for a while, check it later on. Once the pellet has been regurgitated, it should resume feeding as normal.

One must never force water or other liquids into a tawny frogmouth; they seldom drink water and can develop pneumonia if liquid is inhaled. Tawnys obtain ample liquid from their prey. In extremely hot weather, however, they have been observed dipping their lower mandible into water, scooping it up, and throwing the head back to swallow it. Therefore broad dish of water at the level of their normal perch should be provided.

Breeding: Wire platforms are usually accepted as nest sites, and materials such as twigs (including some with leaves) and straw should be available for nest construction. In nature, the nest is a flimsy platform of twigs. Prior to egg laying the female occupies the nest for about a week. Two white eggs are laid, three days apart. The incubation period is 28 days. The chicks appear as white balls of fluff with bald patches on the tops of their heads. They have ferocious appetites and both parents participate in feeding. Chicks solicit food with plaintive wheezing calls. The parents go through an elaborate ritual before feeding the chicks. The chicks fledge at 36–43 days of age and are weaned at approximately 78 days. The young sometimes disappear and are presumed to have been eaten by the parents.

Hand Rearing: Frogmouth chicks, like owls, should not be allowed to become chilled, but also should not be heated, as one would do for most chicks. Rather, they should be kept in a draft-free environment at an ambient temperature of about 26°C (78.8°F). Bright lights for heating or any other purpose should be avoided, as these can cause retinal damage to these nocturnal birds. In the natural state, they begin to perch at a very young age, so some suitable twigs should be available to hand-reared chicks for this purpose.

The hand rearing diet should consist of a quarter of each of the following:

- Ox-heart strips
- Diced day-old chickens
- Pinkies and pieces of young mice
- Insects such as grasshoppers, cockroaches etc.

Frogmouths hand raised at the Bronx Zoo were started on two pieces of pinkie and were carefully watched to make sure they passed their first fecal sac before reaching 25 g (0.88 oz.). On day three, half pinkies were introduced. The chicks should produce their second fecal sac before reaching 30 g (1.05 oz.) and their third before 36 g (1.27 oz.). On day seven, whole pinkies were introduced, and by day 13, small mice were fed.

Tawny Frogmouth *Podargus strigoides*

Age	Weight (g)	Feeds/day	Age	Weight	Feeds/day
Hatch	17–20	3–4	**10**	70–80	3–4
1	18–20	4–5	**11**	80–85	3–4
2	19–23	4–5	**12**	90–105	2–3
3	21–26	4–5	**13**	100–110	2–3
4	25–30	4–5	**14**	110–120	2–3
5	29–36	4–5	**15**	120–130	2–3
6	34–45	4–5	**16**	130–140	2–3
7	40–50	3–4	**17**	140–155	2–3
8	50–62	3–4	**18**	150–170	2–3
9	58–70	3–4	**19**	160–180	2–3

Food is presented either by hand or with a blunt forceps. Initially, only the soft bodies of insects are fed. Prior to feeding, insects are killed and slightly pulverized, as would be done by the adult bird. Fur and feathers are left on the food, but sharp pieces of bone and the legs of grasshoppers, etc. are removed before feeding. Multivitamin and calcium supplements are added to the hand-rearing diet daily. After fledging, juveniles are compatible and up to six can be kept in an aviary measuring 2 x 2 x 2 m (6.5 x 6.5 x 6.5 ft.). Even a newcomer can be released directly into the aviary. Tawnys do not fight to the death, and in the unlikely event of fighting, only minor beak and tongue damage may occur. Young birds often have a milky color across the eyes. This is normal. As they get older, the eye will change from a milky blue to white with a dark pupil. Their feces are very sloppy and have a high odor which sometimes makes people think they are sick, but this is normal.

References:

Wakelam, V. 1983. Tawny Frogmouths - Accidental Breeding 1983.

Henderson, N. 1997. Australian Bird Rehabilitation Manual.

17.04 Caprimulgidae Nightjars and Nighthawks

Joseph Lindholm III

Introduction: The roughly 70 members of the family Caprimulgidae are distributed worldwide, and several are common and well-known birds. While these birds are primarily nocturnal, the unique calls and distinctive flight of some species make them conspicuous. All are strong flyers and consume only living animals, almost entirely flying insects. Nightjars have traditionally been considered a fairly homogeneous group. The protein research of Charles Sibley and others has more recently lead to the segregation of the eight nighthawks, found only in the Western Hemisphere, as a subfamily, and the removal of the seven members of the genus *Eurostopodus,* the eared nightjars of the Oriental and Australian Zoogeographical Realms, as a separate family Eurostopodidae. As I am not aware of any recorded aviculture of eared nightjars, they will not be further discussed here.

Management: In nature, nightjars and nighthawks eat only what they catch while in flight. As with insectivorous bats, they have thus proved rather difficult captive subjects. Only a handful of species have been maintained in captivity for significant lengths of time. They are generally not considered suitable subjects for aviculture. There have, however, been a small number of cases where fairly satisfactory results were achieved.

At present, long-term private aviculture of nightjars or nighthawks in Europe or North America is unlikely, as native species are protected, and animal dealers never offer foreign ones. In the earlier part of the twentieth century, several private aviculturists published their experiences with the European nightjar *Caprimulgus europeaus*.

Unique results were achieved by one of Europe's leading ornithologists, Oskar Heinroth (1871–1945). Dr. Heinroth worked and lived at Zoologischer Garten Berlin, from 1904 to 1945, initially as Director's Assistant, then as Aquarium Director from 1913 to his death in 1945. In the first decades of the twentieth century, he and his wife Magdalena Heinroth engaged in a 20-year study of the development of European birds from the egg, culminating in the publication of the massive *Die Vogel Mitteleuropas* from 1924 through 1932. As will be described below, the Heinroth's achieved the only captive breedings of European nightjars in their residence at the zoo. The birds were at liberty in the study and dining room of this house. A male was received as a juvenile in 1906, followed by a female in 1907 (Heinroth, 1909). As discussed below, two broods were parent reared in 1908. No further reproduction is recorded. The female flew off in the winter of 1911, while the male died in 1912, after being stepped on by a "new housemaid" (Heinroth & Heinroth, 1926).

The British aviculturist Hugh Wormald (1915) discussed a bird that enjoyed liberty in a friend's schoolroom, where it became extraordinarily tame, spending much time on a log by the fireside. He noted that, after several months in captivity, it had "several narrow escapes from being trodden on." Wormald had collected this bird in August of 1914 as a nearly full-grown young one, capable of flying about 20 yards at a time. It was kept in a cartridge bag all day after capture, then was force-fed that evening.

Another British aviculturist P.F.M. Galloway (1915a) maintained one that lived in perfect condition just over five years. Hand reared as a chick, this bird could be freely handled. It was kept in a cage, but given some time at liberty in a room each evening. Galloway noted that this bird enjoyed both sun-bathing and dust-bathing in sand. As noted below, he devised a method by which this bird fed itself.

The French aviculturist Marcel Legendre (1926) twice reared nightjars from the nest. His birds were maintained in a large aviary where, at night, they "flit about on their great wings, velvety and beautiful, like those of a large nocturnal moth." His birds were kept in large indoor cages in the winter. He specified that the few perches should be "made from branches as thick as one's wrist with the bark left on." Legendre's birds squatted lengthways on these. They were very confiding and would take food from the hand and very gentle and would allow themselves to be caressed.

I am aware of a single instance where a private aviculturist maintained a caprimulgid outside its native continent. The famed Sydney Porter (1941) kept a Pauraque *Nyctidromus albicollis* in England. Unfortunately he did not document how long it lived or when it was obtained, instead writing in detail about its feeding habits, which will be summarized below. Pauraques are found from Southeastern Texas through Paraguay. Despite this enormous range, I am aware of no other documented captive specimens. Porter's specimen, which was given to him by another imminent British aviculturist the Hon. Anthony Chaplin, was collected as a nestling in South America, arriving in England while still in juvenile plumage. It initially spent a week in an aviary at the London Zoo. The only other caprimulgid I am aware of having been exported alive from its native continent is a Nacunda nighthawk *Podager nacunda,* also kept at the London Zoo. Received in January 1911, from a sailor who captured it at sea off the coast of Brazil, it lived for nearly three weeks, apparently injured on arrival (Seth-Smith, 1911).

In general, Caprimulgids have been infrequent and ultimately unsatisfactory zoo animals. From 1870 to 1902, the London Zoo held only four European nightjars that lived longer than one month (Mitchell, 1911). The longest-lived of these birds died after seven months at the zoo, and the average longevity among the four was four months. On the other hand, Oskar Heinroth (1926) wrote that the birds hatched at his house in 1908 lived a "long time" at the Frankfurt and Breslau (Wroclaw) Zoos. Earlier, Heinroth (1909) wrote that in 1908, the Breslau and Cologne Zoos were each given a pair of captive-bred nightjars, so one wonders if birds were subsequently transferred from Cologne to Frankfurt, or if Frankfurt was listed in error, nearly 20 years after the donation.

More recently, Alpenzoo, at Innsbruck, Austria, which maintains only animals of the European Alps, successfully exhibited European nightjars. In spring, 1989, I was informed by Dr. Ellen Thaler-Kottek, then Curator of Birds that one of the Alpenzoo's two nonreleasable males had been there six years, the other four years. They shared an outdoor aviary with pairs of nightingales, white wagtails, coal tits, and hoopoes (Thaler-Kotek, 1988). This exhibit is one of a series of aviaries, discussed in detail by Dr. Thaler-Kotek (1988) in a *Gefiederte Welt* article. As described below, they were generally hand-fed five times a day. I am not aware how long these birds ultimately lived. At any rate, as of December 2003, the International Species Inventory System (ISIS) listed three collections holding European nightjars. The Tallinn Zoo, in Estonia, held one. The others zoos were in Poland: Warsaw with one, and Wroclaw with two.

Writing for the guidebook to the New York Zoological Park, Lee Crandall, the great aviculturist and zoo historian, noted, regarding caprimulgids: "Unfortunately, the conditions required by these birds are not easily complied with in captivity and we make no effort to keep them regularly. However, we may occasionally have specimens of the… local species…" (Ditmars & Crandall, 1939). While the whip-poor-will *Caprimulgus voceriferus* is widespread, and very familiar by name and call, the only definite captive record I have found concerns the New York Zoological Park (Bronx Zoo) (Lindholm, 1989). This species is included in an extensive list of birds new to this collection obtained in 1911 (Anon. 1912). From Crandall's above-noted statement, it can be gathered this specimen was short-lived. Another well-known North American nightjar, the Chuck-will's-widow *Carprimulgus carolinensis* has a similarly sparse captive history (Lindholm, 1989). The Bronx Zoo did not obtain one until 1926 (Crandall, et al, 1927). The only caprimulgid included in a list of all animals received at the National Zoological Park in its first forty years (through 1928), is a Chuck-will's-widow that survived only half a month (Mann, 1930). To my knowledge, this is the captive longevity record for the species (Lindholm, 1978). Wanda Weaver, Registrar at the Dallas Zoo, informed me that a Chuck-will's-widow received there May 1981, lived a week, and another donated May 16, 1982, was released the same day.

I have found no documentation that whip-poor-wills or Chuck-will's-widows were ever actually placed on exhibit. On the other hand, I have seen the much smaller poorwill *Phalaenoptilus nuttallii* displayed at two institutions. No records currently exist for a poorwill I saw at the Woodland Park Zoo in Seattle on November 1, 1971. Exhibited in a long-since demolished bird house, it shared a tall, narrow, glass-fronted exhibit with a single calliope hummingbird. It sat on the mossy floor of this enclosure, and although displayed in bright light, was interested in visitors' fingers on the other side of the glass, opening and shutting its beak rapidly. Obviously it associated hands with food. I presume this bird did not live long, as the master aviculturist Jan Van Oosten, who was closely associated with this zoo since the 1960s, and served as its director from 1972 through 1974, has no recollection of it.

A poorwill was exhibited for more than four years at the Living Desert, near Palm Springs, California. When I saw it in 1988, it occupied an exhibit with a glass front and sides, 90 cm^2 (3 sq. ft.), and 1.5 m (5 ft) high. On either side were identical enclosures, one with a sidewinder, the other with a kingsnake. Along with the reptiles and small mammals in the "Desert after Sundown Room" it was displayed under blue light, produced by "Vita Light" bulbs. I found it awake, mostly sitting still amidst the rocks and shrubs on the floor of its exhibit, but occasionally shuffling around. Terrie Correll, Curator of Animals, informed me it was exercised with flights up and down the service corridor behind its enclosure. This bird was a public donation, received November 6, 1984, and died February 7, 1989. This, to my knowledge, remains the captive longevity record for its species.

From Terrie Corell, and Anne Fletcher-Jones, Registrar, I received data on seven other poorwills maintained at the Living Desert. All were received either from members of the public or the California Department of Fish and Game. One received May 2, 1985, died February 10, 1987, being exhibited with the 1984 bird. Another was maintained from October 20, 1986, until its death November 5, 1987. Records do not indicate where this specimen was kept, but one that arrived November 4, 1988, and died July 25, 1990, is recorded as having lived in an outdoor walk-through aviary. Two other specimens were comparatively short-lived. One received May 30, 1987, died June 12, 1987, and one obtained August 3, 1993, died April 8, 1994. The final specimen documented in the Living Desert's records was one received May 6, 1994, and turned over to a private party, who I assume was a rehabber, four days later.

Most of these specimens are indicated to have arrived at the Living Desert as adults. While I have no details, I imagine some of these accessions may be a result of a well known feature unique to poorwills. During the winter, this species may maintain a state of torpor for three months, during which digestion ceases, respiration is reduced dramatically, and body temperature falls more than 30 degrees. It is not surprising that birds found in this state should be presumed to be ill or injured.

The natural history of the common nighthawk *Chordeiles minor* definitely makes it liable to being brought to zoos or rehabilitators. Nighthawks spend far more time in flight than nightjars. Their habit of feeding in numbers around lights makes them especially accident-prone. The ornithologist George Sutton (1977) noted that every fall, without fail broken-winged nighthawks were brought to him at the University of Oklahoma. He attributed these injuries to utility wires in the vicinity of lights. This species' nesting habits also contribute to its frequent appearance as a public donation. Eggs are laid and chicks are raised directly on such substrates as bare cement (Sutton, 1977), on the ground or on top of buildings, often without any nest materials. It is thus not surprising that persons discovering these sites should presume they have found a bird that fell out of its nest.

Over the last half of the 1970s, the Lincoln Park Zoo, in Chicago, received large numbers of young nighthawks. In 1976 alone, 16 were brought in by the public (Meritt & Meritt, 1979). Their weights ranged from 24–64 g (0.8–2.3 oz.). Developmental stages ranged from downy chicks to almost fully feathered fledglings. Dennis Meritt, Lincoln Park's Assistant Director, and his wife Gail F. Meritt, found that while most of these birds were alert, responsive and hungry, they fell into two distinct categories of behavior. While some were easily handled and hand-fed, others attempted to struggle, hiss, and bite, and proved difficult to feed. It was found that these behaviors were not altered by captivity, with docile birds, and excitable and hostile ones, all remaining so throughout the time they were held. Following the remarkably simple feeding regime described below, all 16 of the 1976 specimens were reared and released when they achieved flight capacity. Their time in captivity varied ranged 4–7 weeks. They were liberated at early evening in suitable habitat, after being fasted for six hours. All were banded, and subsequent sightings and recaptures of some of these birds confirmed the success of Lincoln Park's program, described in detail by Meritt and Meritt (1979).

Aside from my own review of caprimulgid aviculture (Lindholm, 1989), the only detailed published account of the public exhibition of a common nighthawk I have found concerns what I believe to be the first specimen displayed in a zoo (Anon. 1909). In 1908 the New York Zoological Park (Bronx Zoo) received a bird slightly injured by flying against a telegraph pole. It is recorded that its injury soon healed. Following a week of force-feeding, the bird progressed to hand-feeding. In an account published in January, 1909, written after the bird had been in captivity over a month, its exhibit and feeding routine are described:

> "Its favorite position, true to the custom of its family, is lengthwise upon a small prostrate tree-trunk. At the approach of a keeper with food, the bird flies down to the door and greets the man with its great mouth wide open and sings quivering with eagerness. Pellet after pellet of meat, egg, and mealworms is caught and swallowed, until the bird signifies its satiety by flying back to its perch. Few people have seen a nighthawk or a whippoorwill alive, and this bird attracts a great deal of attention."

(Anon. 1909). I am unaware of the subsequent history of this specimen.

The most remarkable records for captive common nighthawks have been achieved by the Assiniboine Park Zoo in Winnipeg, Manitoba, Canada Province (Lindholm, 1989). Having noted listings for nighthawks at Winnipeg in 1980s ISIS records, I contacted Rolf Henning, animal supervisor, and was given the following information:

On 21 July, 1976, a male and female were brought in injured. Both were presumed to be less than two months old, though fully fledged. After a stay in the zoo hospital, where they progressed from force feeding, to hand feeding, to eating on their own, they were exhibited in an outdoor aviary for native birds, 16 x 9 m (50 x 30 ft.) deep. Cagemates included everything from bitterns to mourning doves. The exhibit was thickly planted, especially with Junipers. During Winnipeg's fierce winters, all birds were transferred together to an indoor, planted, public exhibit. A third nighthawk, received July 9, 1980 was added to the same aviary. This bird was discovered missing at inventory December 31, 1983. The male received in 1976 died October 15, 1983. The female that arrived with it died September 29, 1984, having achieved a captive longevity of eight years, two months, and eight days, a record, so far as I know, for any member of the family Caprimulgidae.

I am aware of two other institutions that have maintained common nighthawks for any length of time, in both cases as education animals. When I visited the Tracy Aviary, in Salt Lake City, Utah, in October 1988, I found two birds were included in the education program collection. A broken-winged adult was received August 8, 1986. A fledgling arrived in 1987. They were maintained a large off-exhibit room, with a rather limited amount of natural light, sharing this facility with rehabilitated barn swallows and local hummingbirds. In rotation, one bird or the other was taken out and shown in the aviary's amphitheater or transported to local schools. Children were especially intrigued by their enormous gape as they were being handfed. I found both birds wonderfully tame. When one was taken out of its building and set on a lawn, it made no attempt to escape, preferring to investigate the edible potential of proffered fingers. Both birds were only hand-fed.

Liz Rank, registrar at Tracy Aviary, found no exact dates for the demise of these birds, but found that while two were still inventoried 1 January, 1989, only one was present a year later. The remaining specimen died around September 1990, but whether it was the 1986 or 1987 specimen is not recorded. I was also informed that the education department held two on January 1, 1994, but no further data is available. The final nighthawk recorded for Tracy Aviary was received as a juvenile in September 1995, and was euthanized due to an injury on April 8, 1999.

As of December 2003, ISIS documents two common nighthawks. I have no information on the one held by the Guadalajara Zoo in Mexico. On the other hand, the one at the Virginia Marine Science Museum, at Virginia Beach, was acquired from a rehabilitator in November 1999. Charlotte Allred, a keeper at the museum, informed me that this bird was received as an adult, and though capable of flight, had become so dependent on hand feeding while being rehabilitated that it was considered nonreleasable. Like the ones at Tracy Aviary, it is brought out for educational programs. It is otherwise kept in an off-exhibit large enclosure, shared with an eastern kingbird. While the nighthawks at Tracy Aviary appeared to spend all their time on the floor of their enclosure, the Virginia Beach specimen roosts on branches.

Diet: As all nightjars and nighthawks capture insects in flight, adjusting captive specimens to an alternate diet has always been an essential element in their maintenance. Reviewing the long-term captive records discussed above, a variety of methods have been devised to feed caprimulgids in captivity, producing varying results.

The diet of Oskar Heinroth's breeding pair of European nightjars consisted largely of mealworms, supplemented with two species of cockroaches, two species of beetles, moths, meat, and egg conserve (Heinroth, 1909). I imagine egg conserve was Abraham's Food or some other commercial dried-egg preparation.

I am not certain if Heinroth's birds were self-feeding. P.F.M Galloway (1915), who maintained a European nightjar for five years, devised a method by which his bird fed itself. He created a miniature shelf (of which he provided a sketch (Galloway, 1915) studded with headless pins, 7.6 cm (3 in.). This was affixed inside the cage along the floor. Moths, mealworms, cock-chaffer beetles, and pellets of "Life" (a commercial insectivorous food) were impaled on the pins. Galloway noted that when the nightjar was hungry he would run forward and pull the food off the pins and bolt it. Galloway did not have the slightest fear of the bird hurting his mouth with the headless pins, because he always ran up to them and took hold of the food only, pulling it off.

Hugh Wormald (1915) who documented a European Nightjar kept at liberty in a schoolroom, noted this bird (collected nearly fully grown) initially "made a few half-hearted efforts to catch moths thrown up for it in the summer, but now it won't try at all!" It accepted hand feedings readily. During its first two weeks in captivity it was fed almost entirely on grasshoppers, with some moths and butterflies. It was later maintained on maggots, mealworms, and raw meat, with some hard-boiled egg, but refused any generic soft food.

Marcel Legendre (1926) not only found his European nightjars accepted the usual insectivorous food, but would take it from a food vessel, to which they were easily accustomed. During the winter, he made sure the food was rich with ephimera and dried flies. In months that moths and butterflies were available, as many as possible were provided live. Both of Legendre's birds were taken from the nest, and also took food from the hand.

At the Alpenzoo, Innsbruck, where European nightjars had been kept in a planted, multispecies aviary for four and six years as of 1989 (Lindholm, 1989) the diet was insects and [horse?] heart, hand-fed five times a day (Thaler-Kottek, 1988b). Alpenzoo's Curator of Birds, Dr. Ellen Thaler-Kottek, informed me that, on occasion, one of these birds would refuse to eat for more than a week, becoming very thin before resuming its feedings. I wonder if this ability of a highly insectivorous bird to survive so long a fast might not bear some relation to the poorwill's previously mentioned capacity for assuming a torpid state?

In contrast to the five daily feedings of the Innsbruck European nightjars, only three daily hand feedings were accepted by Sydney Porter's Pauraque (Porter, 1941, Lindholm, 1989). During English warm weather, moths were fed, but otherwise, the diet consisted of minced meat made into small pellets, dipped into water and then rolled in dry insectivorous food. Porter (1941), a master aviculturist, marveled at this bird's small food intake:

> "It was remarkable that the small amount of food the bird required, two pellets (about the size of one's thumb) given in the evening, one in the morning, and one at lunch-time was all that was required. If not hungry the bird would refuse to swallow the food and being put on the ground, it would throw it out of its mouth." Porter also noted that "at first the bird would readily run towards one with out-stretched wings, seize the meat, and swallow it, but later on it became more nervous and had to be hand-fed".

Terrie Correll, Curator of Animals at the Living Desert, informed me the poorwills kept there were voluntarily hand-fed in the morning and the evening. Their diet consisted primarily of Nebraska Brand Bird of Prey Diet supplemented with waxworms, mealworms, and frozen crickets (Lindholm, 1989). As note above, one specimen lived more than four years.

The diet by which 16 young common nighthawks at the Lincoln Park Zoo were reared to release in 1976 is startling in its simplicity. After disappointing results in previous years that resulted from diets of dog meal, dog biscuit, horse meat, canned dog food, or combinations thereof, the diet was restricted entirely to Nebraska Brand Bird of Prey Diet, fed without any additional vitamins, minerals, or water (Meritt & Meritt, 1979). Depending on their age, the nighthawks received 3–8 balls of food at each feeding. The smallest balls were pea-sized. The balls were formed previously, and allowed to desiccate slightly so they did not fall apart when fed. Feedings were administered four times a day, in daylight only.

As noted above, the common nighthawk received as an adult at the Bronx Zoo in 1908 enthusiastically solicited hand feedings, which were apparently tossed to it by a keeper who was greeted with open beak and wing fluttering. The bird would fly to the door of its exhibit when the keeper approached and was fed until it flew back to its perch. At arrival, a week of force-feeding had been required. The diet was pellets of "meat, egg, and mealworms."

The common nighthawk that attained a longevity of more than eight years at the Assiniboine Park Zoo, and the other two kept with it, fed on their own in large mixed-specie aviaries. Arriving as adults, the more long-lived specimens progressed from force-feeding to hand feeding to self feeding, during their stay at the zoo's hospital, under the care of keeper Nancy King (Lindholm, 1989). Along with mealworms, their diet was a horse-hamburger concocted by Clive Roots, director at Assiniboine, and an internationally recognized softbill aviculturist. Aside from horse meat, this included bone meal, shrimp meal, and alfalfa meal, and other ingredients.

The common nighthawks maintained at the Tracy Aviary in the late 1980s were hand-fed small balls composed of soaked dog food and Bird of Prey Diet, as well as crickets and mealworms (Lindholm, 1989). In interesting contrast to the previous diets, the common nighthawk that was still alive as of December 2003, after four years at the Virginia Marine Science Museum received nothing but insects. In the morning this bird receives 40 mealworms and five waxworms, all dusted with vionate powder. In the after noon it received 10 large crickets. All items were hand-fed. Thus, positive results with nighthawks may be achieved through a variety of diets.

Breeding: The aforementioned repeated breedings of European nightjars by the Heinroth's in 1908 remain the soul documented captive propagation of any member of the Caprimulgidae. Highly detailed accounts with photographs were published in German (Heinroth, 1909, Heinroths & Heinroth, 1926). Rutgers (1972) provided a brief English summary, and I published another, with further translation from the German (Lindholm, 1989).

> "While the male was held from 1906 through 1912, and the female from 1907 through 1911, and both were received as juveniles (Heinroth & Heinroth, 1926, Lindholm, 1989), 1908 appears to be the only year reproduction occurred. At the end of May 1908, the male was noted twisting round and round and scraping with its feet on a peccary skin rug in the dining room (where both birds had liberty , as well as in the study). On June 2 and 4, eggs were laid on the site thus prepared. The female was the primary

brooder, but was relieved by the male when she left to eat. The incubation period was 16½ days, both eggs hatching. These chicks were parent reared, and present when a second clutch of two hatched the same season. Heinroth and Heinroth (1926) published a photo of six birds together, two decidedly smaller. As mentioned earlier, all four offspring were given to zoos in September 1908."

References:

Anon. 1909. The Nighthawk. *Zoological Society Bulletin*. 2 (No.32):478.

_____. 1912. Recent arrivals. *Zoological Society Bulletin*. 2 (No. 49):836–837.

Crandall, L.S., W. Beebe &. S. Stacey. 1927. Department of birds. *New York Zoological Society Annual Report* 31.

Ditmars, R.L. & L.S. Crandall. 1939. *Guide to the New York Zoological Park*. New York Zoological Society.

Galloway, P.F. M. 1915. The Nightjar *(Caprimulgis europaeus)*. *Avicultural Magazine*. (Series 3) 6:137–139.

_____. 1915. Feeding a nightjar. *Avicultural Magazine*. (Series 3) 6:231.

Heinroth, O. 1909. Beobachtungen bei der zucht des zeigenmelkers. *Journal fur ornithology*. 57:56–83.

Heinroth, O. & M. Heinroth. 1926. *Die vogel Mitteleuropas* (Vol.1). Hugo Bermuhler Verlag.

Legendre, M. 1926. Nightjars. *Avicultural Magazine* (Series 4) 4:30–32.

Lindholm, J.H. 1989. Caprimulgiform Birds in Captivity (Part II.) *Honeycreeper* 4(Nos.1,2):24–38.

Mann, L.Q. 1930. List of animals exhibited in the National Zoological Park during its first forty years. IN Mann, W.M. 1930. *Wild animals in and out of the zoo*. Smithsonian Institution Series, Inc.

Meritt, G. F. & D.A. Meritt. 1979. Hand-rearing and nutrition of nestling nighthawks *(Chordeiles minor)* at Lincoln Park Zoo, Chicago. *International Zoo Yearbook*. 19:256–257.

Mitchell, P. C. 1911. Longevity and relative viability in mammals and birds. *Proceedings of the Zoological Society of London*. (1911, Part II):425–555.

Porter, S. 1941. Notes on the White-necked Nightjar *(Nyctidromus albicollis)* in captivity. *Avicultural Magazine*. (Series 5)6:218–220.

Rutgers, A. 1972. Caprimulgifornes IN Rutgers, A. & K.A. Norris, 1972 *Encyclopedia of aviculture* Vol.2:259–264., Blandford Press.

Seth-Smith, D. 1911. Bird notes from the Zoological Gardens. Avicultural Magazine. (Series 3)2:128–129.

Sutton, G. M. 1977. *Fifty common birds of Oklahoma and the southern Great Plains*. Univeristy of Oklahoma Press.

Thaler-Kotek, E. 1988. Kleinvogelhaltung im Alpenzoo Innsbruck (1. Teil). *Die Gefiederte Welt*. 112:220–223.

_____. 1988. Kleinvogelhaltung im Alpenzoo Innsbruck (Schluss). *Die Gefierderte Welt*. 112:247–249.

Wormald, H., 1915. Nightjar, snipe and phalarope. *Avicultural Magazine*. (Series 3) 6:195–197.

Pigeons

18 Columbiformes

18.01 Columbidae: Pigeons & Doves

18.01 Columbidae Pigeons and Doves

Glen Holland

Introduction: Pigeons and doves belong to the order Columbiformes and the family Columbidae. The order Columbiformes also includes the family Raphidae, which consists of three species, all extinct, including the dodo *Raphus cucullatus*. The sandgrouse, family Pteroclidae, formerly included in Columbiformes, has now been transferred to the order Ciconiiformes. The number of extant species of Columbidae has been assessed at 310, with an additional three species, including the legendary passenger pigeon *Ectopistes migratorius*, now extinct. The family is probably best known internationally for the city pigeon, which originated from the feral rock dove *Columba livia,* and the barbary dove *Streptopelia risoria,* both of which have colonized many regions of the world. The diamond dove *Geopelia cuneata* of Australia, one of the smallest doves, is an international favorite, long considered an ideal bird for beginners in aviculture. As pigeons and doves vary widely in terms of habitat and dietary requirements, I have chosen to cover a small number of species and those requiring similar management have been recorded. Pigeon and dove species adapt well to aviculture, the most challenging being the fruit pigeons. Most sustain themselves largely or entirely on vegetable foods, some being predominantly seedeaters, others frugivorous. Another feature of this family is their method of drinking by suction rather than scooping water with their beaks. In all species that have been studied, the female constructs the nest with material brought to her by the male. Generally, they can be housed with other birds of a similar size, including other pigeons and doves, but in many cases one pair per species is recommended.

Pink Pigeon *Columba mayeri*

The information in this section is taken directly from the husbandry guidelines supplied by Durrell Wildlife Conservation Trust.

The pink pigeon is endemic to Mauritius in the Indian Ocean (Collar and Stuart, 1985) and was once the world's most endangered pigeon (King, 1981). *Columba mayeri* measures 32 cm (12.5 in.) in length and weighs approx. 0.35 kg (0.75 lb). Pink pigeons feed and roost in small flocks. Being an arboreal species, they are very agile when foraging in the canopy. During the breeding season monogamous pairs defend small territories. Both sexes are involved in nest building, incubation and rearing of the young. Conservation Status: IUCN category: Critically Endangered (IUCN, 1996).

By the late nineteenth century the pink pigeon was regarded as rare, its population was estimated to be in the low hundreds. Only 40–60 birds remained by the late 1950s and the population declined further during the 1960s, stabilizing at approximately 20 birds (Jones and Owadally, 1988). Today, numbers are increasing due to the release of captive-bred birds and habitat protection. Population estimates are now approximately 330 birds. The main threats are thought to be habitat degradation and destruction, and predation, especially by introduced mammals such as the black rat *Rattus rattus,* the long-tailed macaque *Macaca fascicularis* and feral cats, the latter taking newly fledged young and adult birds. Human persecution plays a part too (Jones and Owadally, 1988).

A conservation project for this species was initiated

in the mid-1970s and the Durrell Wildlife Conservation Trust (formerly Jersey Wildlife Preservation Trust) became involved in 1977 when it set up a captive breeding project in co-operation with the Ministry of Agriculture, Fisheries and Natural Resources (Jeggo, 1977, 1978, 1979; Hartley, 1977; Durrell and Durrell, 1980). Since 1984, a total of 63 captive-bred birds have been released in Mauritius, including a trial release of 22 pigeons in the Pamplemosses Bontanic Gardens, Mauritius and 42 birds (19.13.10) into native forest at Brise Fer and Macchabe. Of these 42 juvenile and adult birds, 40 were bred on Mauritius and two at Jersey Zoo. One bird released in Pamplemousse was re-released into native forest.

Age at release varied from 72–385 days but 35 (83%) of the birds were <1 year (Jones et al, 1992). Only healthy birds were set free. Prior to release, the pigeons were held in an aviary at Plaine Lievre for 24–133 days. This acclimatized the birds to the new area and the supplementary feeding point. After 30 days of liberty, 27 (64%) of 42 pigeons continued to return for food, and a year later 13 (36%) were known to be alive. Supplementary feeding and rat control (intensive poisoning) around both the wild and introduced populations have much improved the species' status. The captive-bred birds have adapted well to their native environment and are known to have fed on at least 42 plant species, many of them wild pigeon food plants. By May 1994 the introduced population numbered 52 birds. The release program is being extended. The *Cryptomeria* grove used by the wild population is now within the Black River National Park.

Green Imperial Pigeon
Ducula aenea

Captive management: A captive breeding program was launched for the pink pigeon in 1976, with colonies established at the government aviaries in Mauritius and in 1977 at the Jersey Wildlife Preservation Trust (now Durrell Wildlife Conservation Trust) (Jones et al, 1983). There were 16 founders of which 11 (8.3.0) have descendants represented in the current captive population. In the early years of the breeding program, the small size of the founding population created a genetic bottleneck. A comprehensive management strategy was therefore needed (Lind, 1994). Currently the captive population is divided into two regional sub-populations, both of which are co-ordinated through regional management plans. Collections in North America have been consolidated under a Species Survival Plan (SSP) and those in Europe under the European Endangered Species Programmed (EEP).

The sex ratio in the latter population is almost equal, with males comprising 53% of the population. The EEP program hopes to regain 95% of the current genetic diversity over the next 100 years (Lind, 1994). For identification, numbered metal rings are attached after fledging (2–3 months of age).

At Jersey Zoo the pigeons have been housed in a variety of aviaries. Experience has shown that, if a few basic criteria are met, this species will breed in almost any aviary that provides an outside flight and a sheltered inside area. The latter should be heated if the birds are to breed throughout the year. Ideally, the aviary should be 3m (10 ft.) or more in height. The nesting sites should be located at high points in the inside area, which give the birds a greater feeling of security (Lind, 1991). Adjacent aviaries housing pink pigeons at the zoo are constructed with adjoining doors that can be opened to allow free movement between the aviaries. Normally, one pair is housed in two adjoining aviaries; this allows separation if the male becomes overly aggressive towards the female. Males in adjacent aviaries may display aggression towards each other; the involved pairs usually build poor quality nests and fail to brood the eggs properly. Pairs left to parent rear should therefore be isolated (Lind, 1991).

Imaginative planting in the aviaries enriches the birds' environment. Perching and appropriate leaves to feed on should be available. Extra heating or lighting is not entirely necessary in Jersey, but it is recommended and would obviously be important in colder climates.

The provision of nesting material is one of the most important aspects of successful breeding in this species. A constant supply of fresh nesting material is crucial to promoting proper breeding behavior. Initiation of broodiness in the female is thought to be brought on by the male carrying nesting material. At Jersey Zoo fresh twigs or small sticks are given almost daily when the birds are nesting (Lind, 1991).

Life span is estimated to being the late teens (in a few cases even in the early twenties); males live longer than females. In the past many squabs developed "inclined feet" (Jones et al, 1983), a rotation of the distal tarsometatarsus which is not, as originally thought, a consequence of inbreeding but of improper substrate in the nest box. Therefore nest boxes now consist of an open fronted wooden box 20 x 20 x 20 cm (8 x 8 x 8 in.) with a 5 cm (2 in.) lip on the open front with a heavy gauge wire mesh 1 cm^2 (0.5 sq. in.) insert in the base of the box with the edges bent down to create an elevated platform approximately 4 cm (0.75 in.) above the bottom of the box. A nest of straw is added on top of the wire mesh insert. The straw provides cushioning for the eggs while brooding. Both wire mesh and the straw provide a substrate for the chicks to grip with their feet (Lind, 1991).

Diet: The diet of the pink pigeon in the wild is comprised predominantly of the fruit, flowers, and leaves (Mc Kelvey, 1976; Jones et al, 1983) of both native and exotic plants. Jones et al, (1992) observed released birds feeding on 37 identified species, 27 of which were native. Thirty-four percent of the feeding observations were on four native plants: *Premna corymbosa, Homalium integrifolium, Securinega durissima* and *Ochna mauritiana*. Pink pigeons proved to be selective feeders in captivity. To encourage acceptance of a balanced diet, the different food items are presented as a mixture.

Nutritional Requirements in Captivity: The main diet consists of a mixture of the following:

- 1 part mixed grains (wheat, cut maize, whole maize)
- ½ part SDS Pheasant Breeder pellets
- ½ part Universal food
- ½ part apple
- ½ part diced cucumber
- ½ part diced celery
- 1 part leaves (lettuce, water cress, celery)
- 1 part brown bread crumbs
- ½ part grated carrot
- 1/10 part grated egg

The fruit is cut in small pieces, approx. 0.5 cm cubed (0.2 in.), and the lettuce cut to roughly the same size. The egg is finely grated. One feed per day is prepared for each pigeon and is presented in a 11.5 cm (4.5 in.) diameter crockery bowl that is placed individually in the aviary. As a supplement, Nutrobal is sprinkled on top of the feeds every day (0.1 g each). The pigeons also feed on the shrubs that are planted in the aviaries. Water is provided ad lib in water bowls and changed daily.

Breeding: Males and females have different reproductive life spans. Males are known to breed through the 17–18 year class, whereas females do not reproduce beyond 10–11 years, although they may live longer than this (Lind, 1994). Young pigeons kept in a group can be sexed once they reach sexual maturity, at about seven months, when they will display courtship behavior; separation of the individuals will then be necessary (Lind, 1991). However, the preferred sexing technique is DNA analysis of either blood or feathers. Feather sexing is done at an early age when the birds are ringed, blood sexing at the age of 4–5 months.

In the wild, pink pigeons show breeding seasonality, but at Jersey Zoo they breed throughout the year. Problems may arise when temperatures fall below 10°C (50.0°F). Without heated accommodation, females may then lose interest in breeding and males, still in breeding condition, may become aggressive. Therefore, it is important to monitor breeding pairs that stay together during winter (Lind, 1991). Pink pigeons show no obvious changes in physical appearance or body weight that might be used as indicators of optimal breeding condition (Lind, 1991). The female will begin to sit tightly on the nest two or three days before laying the first egg of the clutch. During this period the male will also sit on the nest with the female nearby, spending more and more time on the nest until the laying of the second egg. Both male and female incubate the eggs (Lind, 1991).

Pairing must be individually tailored to suit the birds' temperaments. However, the usual method involves placing a new male into an aviary with a previously established female. Males are usually more aggressive than females. If a new female is placed into an aviary with an established male, the male often becomes aggressive and this results in the birds being separated (Lind, 1991). At Jersey Zoo the pairs usually remain compatible for 3–4 months. This is independent of both the number of eggs laid and whether or not the eggs are removed. Pairs are then separated and the female rested for at least 6–8 weeks before being paired up again (sometimes with a different partner, depending on studbook recommendation).

The number of clutches per year varies: mean of 5.4 (SD=3.3) with a range of 1–14. The majority of clutches (77%) contain two eggs, laid on the first and third days. Initial egg weights also vary greatly, with a mean of 14.7 g (SD=1.3) and a range from 8.6g to 17.6g (Lind, 1989). Eggs are biconical, white and smooth (Cheeke and Jones, 1987). At hatching, chicks weigh 12–17g (0.4–0.6 oz.). Chicks left with their parents will normally be brooded continually for seven days; afterwards they will be left unbrooded for increasingly longer periods until fledging (Bell et al, 1988). Particular attention should be paid at this time because very often the first hatched chick falls, or is pushed off the edge of the nest. By this time the parents will be brooding a second egg, and so it is often difficult to notice that the first chick is not in the nest. Squabs should be removed from the parent birds shortly after fledging (Lind, 1991).

Pigeons reared under foster doves, are socialized in groups after fledging (Jones et al, 1983), which takes place at approximately 26 days (Lind, 1991). Growth and development of the foster-reared squabs is monitored regularly, resulting in a very low mortality rate. The best results with mixing young pigeons have been achieved by taking individuals straight from the foster birds and immediately mixing them. Aggression may arise among young pigeons if this strategy is not followed.

Pink Pigeon *Columba mayeri*

Day	Weight (g)	Day	Weight (g)
1	13.4	12	145.3
2	22.2	13	155.6
3	32.9	14	165.5
4	45.1	15	171.4
5	59.1	16	176.3
6	73.4	17	179.8
7	86.3	18	181.7
8	99	19	187.2
9	110.1	20	191.7
10	124	21	193
11	135.6		

Jersey Zoo (Lind, 1991).

The incubation period is approximately 14 days, with a range of 13–16 days. The change in brooding behavior is an indicator of egg hatching. Chipping occurs 24 hours prior to hatching and often beings early in the morning (Lind, 1991). The greatest breeding success in pink pigeons has been achieved through the use of foster Barbary doves *Streptopelia risoria* to hatch and rear pigeon squabs (Jones et al, 1983). Barbary doves have the same incubation period as pink pigeons (14 days), as opposed to domestic pigeons, for example, which incubate for 18 days (Lind, 1991). Pink pigeons' eggs are removed from the parent birds within a few hours of laying. This strategy reduces chick mortality and improves productivity by allowing multiple clutching. However, this can create unnatural stress for the pigeons who have then laid up to 21 clutches per year (Lind, 1989). This practice is now restricted.

Lind (1989) found a trend towards decreasing "hatchability" with an increase in the number of clutches laid. It even decreased sharply after the second clutch. This resulted in a management change with more pairs of pigeons being bred and fewer eggs being taken from each pair. Additionally, pigeons are only bred once and then exported. The former strategy, with a large number of eggs being taken, is applied only in the case of genetically underrepresented individuals (Lind, 1989). Artificial incubation and hand rearing should be avoided with pink pigeons as it proved difficult to produce a suitable crop milk to feed the squabs and there is a problem of imprinting (Lind, 1991).

Management of Foster Birds: The foster doves are housed in stacked cages 86.4 x 78.4 x 61.0 cm (34.5 x 31.5 x 24.5 in.). Each cage is provided with two perches: one spanning the length of the cage and one the width. A wooden nest box is also provided. The bottom of the cage is lined with newspaper (changed daily), which is used as nesting material. The birds are fed a mixture of wheat and maize and are provided with a commercial pigeon grit. A general avian multi-vitamin, Vetark, is given daily in the drinking water. A photoperiod of 12 hours light/12 hours dark is used to optimize the egg output.

It is important to have a sufficient number of breeding pairs of foster doves so that any pink pigeon eggs can be set under doves whose own eggs are of the same age (their own eggs are then discarded). This ensures that the doves have the required amount of crop milk (a secretion from the lining of the crop) to successfully rear the pigeon squab as one of their own. Crop milk is produced after 14 days of incubation, but its production ceases in the absence of a chick, so the synchronization of pigeon and dove reproductive cycles is critical. Generally, a minimum of three pairs of foster doves should be kept per pair of pink pigeons (Stewart, 1999).

Nicobar Pigeon *Caloenas nicobarica*

Introduction: A distinctive feature of the Nicobar pigeon *Caloenas nicobarica* is the long hackle feathers, which reminds one of the long hackle feathers on the

Nicobar Pigeon
Caloenas nicobarica

game bantams. They hang down from the body and tend to be longer on the males than on the females. There is a distinct size difference between sexes, with the males being larger; they also have a small black knob at the base of the bill near the forehead, which is smaller in females. Nicobar pigeons live on small uninhabited islets off the coast of the larger islands. At night they return to these uninhabited islets where they roost and breed. This gives them protection from human and animal predators. During the day they fly to the larger islands, sometimes over considerable distances, in order to feed. They are found over a wide area, including the Philippines, New Guinea, and the Solomon Islands, as well as the Nicobar Islands from which their name is derived. In some areas they live in large numbers, but only if not disturbed by animals and man.

Management: Most of the time these birds seek their food on the ground where they turn the leaves over and dig with their bills looking for food items. They are large birds, and powerful flyers, and therefore need a spacious aviary if they are to be comfortable. They normally live and breed quite satisfactorily in captivity. Aviaries should measure about 25 x 10 x 3 m high (82 x 33 x 10 ft.), with an earth floor planted to trees. Habitats A or B will prove suitable for this species. Up to a dozen pairs of Nicobars can be housed in an aviary of this size. Although there is a certain amount of bickering among the birds in such a group, there is no actual fighting and no damage by one bird to another. The Nicobar pigeon is very amicable toward other birds and can be housed with such species as golden-heart doves *Gallicolumba rufigula,* Celebes doves *G. tristigmata,* superb starlings *Lamprotornis (Spreo) superbus*, fairy bluebirds *Irena puella*, and red wood quails. Toward these birds they are very docile, tolerating them all without any interference.

Diet: In captivity the Nicobar is normally quite easy to care for. They require a good mixed pigeon diet, that contains corn and various seeds, including peas. They will also take commercial parrot food and gamebird pellets for extra proteins, vitamins, and minerals, as well as fruit such as grated apple and chopped pawpaw.

Breeding: The Nicobar pigeon only breeds really well in colonies. In some instances they are bred in aviaries of 10 x 4 x 3 m high (33 x 13 x 10 ft.). These smaller aviaries can accommodate three pairs, but usually only two pairs will breed. In the larger aviary described earlier, not all the pairs breed at one time; in one instance, there were three females with chicks, three females incubating, and two building nests.

Another interesting point about these birds is that they breed more successfully in shady aviaries. These aviaries are either situated among large trees with overhanging branches or have trees planted within, or sometimes a combination of the two. This seems to give the birds an extra sense of security, which seems to be a positive requirement for breeding. In enclosures that do not provide the shade of the extra trees in the flight, breeding has not been successful. When the trees were full of leaves the Nicobars bred well, but when the trees had lost their leaves, for instance during the cooler part of the year, the birds still laid eggs but did not hatch them. Possibly they just do not incubate properly due to the lack of a sense of security.

Another interesting observation is that the birds that bred were all housed in aviaries with natural sand or earth floors. It appears that this feature is important. Having normal earth floors, however, means that the birds must be periodically dosed for worms. We do this once a year. A feature of incubation is that the female incubates during the day and the male at night. Once the babies hatch, they grow very rapidly in the beginning, then more slowly. They have a very long nestling period; I estimate it to be about eight weeks and possibly longer before the young are fully fledged. In those aviaries that are planted with trees, the Nicobars prefer to build their own nests. Others will also take an open nest box about 30 cm^2 (12 sq. in.) with either a pine or wire bottom. When these birds build their own nests, particularly in the trees, they make a very substantial nest at least 75 mm (3 in.) high, of sticks about as thick as a pencil. Thinner sticks are then placed on top of this, and finally, fine grasses and twigs on the very top. While the adult's tail is white, in the youngsters under the age of one year the tail is gray; youngsters also lack the long glossy hackle feathers.

While at times it is difficult to get Nicobars to start breeding, once they do get going they seem to breed fairly continuously. It is important to have suitable aviary. One must also remember that these birds are now listed on CITES 1, which means that they may no longer be traded internationally unless they are captive-bred. Those who have Nicobars and are successful in breeding them are advised to close ring young birds to prove their captive origin.

Nicobar *Caloenas nicobarica*

Day	Weight (g)	Food (g)	Day	Weight (g)	Food (g)
1	21	8	18	112	
2	22		20	167	
4	23		22	193	60
6	25		24	242	
8	26	14	26	270	
10	35		28	290	85
12	48		30	315	
14	63	35	32	330	
16	85		34	355	

Speckled Pigeon *Columba guinea*

Natural Habits: The speckled pigeon *Columba guinea* is also known as the triangular spotted dove or rock pigeon. These birds are found on rocky outcrops and buildings, moving on to open flats and cultivated fields to feed. Flocks in excess of 500 birds have been recorded on maize and sunflower fields. Their calls consist of a series of deep growling *coos* made in quick succession. During displays or aggression, a series of three or four deep loud *coos* is made.

Management: These pigeons prefer an aviary with sparse vegetation and a few rocks and stumps on which they can perch. A mixture of Habitats B and D will suit them well. As much of their day is spent on the ground, open sandy patches must be available. It is here that the birds will often sunbathe by lying on one side with the exposed wing outstretched in the sun. They also enjoy a mist spray, which, if turned on in warm weather, will attract the birds. In large aviaries a few pairs can be kept together, provided enough nests are supplied for all the pairs. Their large size and boisterous nature make them unsuitable to mix with small birds or other pigeons and doves. If hand reared, they become very tame; a male that I had regularly flew down to attack the feet of visitors in the aviary. During these "attacks" he walked around in circles making his growl-like call.

Diet: Speckled pigeons should be fed on crushed corn, sorghum, wheat, buckwheat, and sunflowers. A slice of moistened whole wheat bread is a favorite food of this species. In addition, they will eat cereal based soft food such as prepared for softbills or parrots.

Breeding: Sexes look alike, and although the gap in the pelvic bone is a relatively reliable method of sexing them, surgical or DNA sexing is advised for certainty. Courtship includes a display flight with noisy wingbeats, as well as parading and bowing displays with rumbling *coos*. The male usually feeds the female by regurgitation before copulation takes place, and allopreening often takes place following copulation. They are reliable breeders and will build a loose structure of sticks on any suitable platform, or in a box or basket. The two white eggs are laid at an interval of approximately 24 hours. Proper incubation begins once the clutch is laid, and the eggs hatch after 16 days. Both chicks normally hatch on the same day and fledge after 25 days. Up to six broods have been recorded in a season. When very young, chicks bob their heads weakly and "pout" their crops when disturbed. At a later stage they will peck, snap their beaks, and beat their wings when disturbed on the nest. The chicks are very noisy when begging for food.

As with all doves and pigeons, both sexes feed the offspring on pigeon's milk, which has been described as "a remarkable secretion produced in the bilobed crop, in which the lining epithelium thickens during the last few days of incubation, and its cells proliferate. When the chicks hatch, these cells start sloughing off to form the "milk," which looks and smells rather like soft cheese, and which has much the same nutritional value as mammalian milk, being rich in protein (13–18%) and fat (7–13%). It is regurgitated by the adults and in most species evidently forms the chicks' sole nourishment during the first few days of life" (Rowan 1981).

In the natural state, speckled pigeons have learned to nest in buildings, under bridges, on window ledges, etc. They are multiple brooded, and will start a second clutch before the first clutch is independent. The same nest is used repeatedly. Immatures are pale versions of the adults, with less prominent spots on the wing, and the bare skin around the eye is a grayish color. White and pied mutations have been established in captivity. These birds have also been crossed with homing pigeons to increase the power of flight in these domestic birds, but this has resulted in a reduced homing instinct in subsequent generations.

See the section at the end of the chapter for advice on HAND REARING Doves and Pigeons.

Ring Size: 8 mm

Species Requiring Similar Management:

Feral pigeon *Columba livia*
Bare-eyed pigeon *C. corensis*
Picazuro pigeon *C. picazuro*
Other *Columba* species

African Olive or Rameron Pigeon

Columba arquatrix

Natural Habits: Also known as the African olive pigeon and speckled forest pigeon, the rameron pigeon *Columba arquatrix* is a forest species that is particularly fond of wooded gorges in montane areas. Out of the breeding season, they occur in flocks of 6–15 birds, but will also congregate in numbers of 40–60 birds in fruiting trees. They feed on the fruits and seeds of forest trees and shrubs. Their calls consist of soft rumbling *coos*.

Management: Rameron pigeons tend to be rather nervous in captivity, and a well-planted aviary will suit this species best. Newly introduced birds must be given time to settle into their new surroundings. They are heavy flyers and will damage themselves against the aviary wire if they are panicked. Being a forest species, they require plenty of shade in their aviary; they are not fond of too much exposure to the sun. The sexes are alike in appearance, but females tend to be smaller. However, it is best to have the birds surgically or DNA sexed. In large aviaries, it is possible to keep more than one pair together, provided the aviary is well planted. At least a third of the aviary should be planted with tall shrubs and small trees in which the birds will spend most of their day, and also nest. Habitat A will meet the basic requirements of these birds. I have not noticed any serious aggression either toward their own or different species. In severe climates, where daytime temperatures regularly remain below 10°C (50°F), I would recommend that these birds be housed indoors during the cold months.

Although I have observed this species breeding in relatively small aviaries, I would recommend an enclosure at least 8 x 5 x 2.4 m (26 x 16 x 8 ft.) high. In larger aviaries, it is possible to house more than one pair together. This beautiful pigeon deserves more effort from aviculturists to produce viable, close rung, domestic strains in order to ensure that the species flourishes in captivity.

In common with all members of this family, they tend to be rather nervous birds, and any sudden movements should be avoided in their aviary. Cats and predators on the aviary roof, either by day or night, will panic the birds, resulting in damaged foreheads and ceres from bashing into the netting. Regular disturbance of this type will result in pigeons and doves either failing to breed or deserting their nests. When transporting pigeons and doves, it is important that the transport box has a foam-lined roof to protect the birds' heads when they jump upward, as they tend to do.

Diet: In the natural state, Rameron pigeons will occasionally go to the ground to feed on fallen seeds and berries and on weeds, but they feed primarily from perches in fruiting and budding trees. In captivity, they are best fed on crushed corn, sunflower, buckwheat, sorghum, lentils, groundnuts, and fruits such as pawpaw (with seeds) and soaked raisins. They are fond of small berries, but I have found them reluctant to take diced fruits such as apple. A multivitamin/trace element powder should be included in the diet twice a week.

Breeding: In display, the male will undertake display flights and also perch and bow toward the female while repeatedly uttering his very soft call. This species requires time to settle in an aviary, but, provided they are not excessively disturbed, they are willing breeders and will nest 2–3 times per season. The nest is a platform of twigs built in the fork of a tree or on a wire platform or basket, which is best placed in the canopy of a tree. Nesting material is not picked up from the ground, but snapped from dry branches, which should be made available to accommodate this behavior. Pairs will nest within a few meters of each other. The clutch consists of one creamy white egg that hatches after an incubation period of 17–19 days. The chick fledges at 21 days and is a dull, browner bird than the adult, with the feathers on its chest and belly having white edges.

See the section at the end of the chapter for advice on HAND REARING as per fruit pigeons.

Ring Size: 8 mm

Species Requiring Similar Management:

Wood pigeon *Columba palumbus*
Western bronze-naped pigeon *C. iriditorques*
Eastern bronze-naped pigeon *C. delegorguei*

Fruit Doves and Pigeons

Rose-crowned Fruit-Dove
Ptilinopus regina

Description and Natural Habits: There are three types of fruit pigeons. The imperial pigeons are large fruit-eating pigeons comprising the genus *Ducula,* with 38 species. The green pigeons of the genus *Treron,* with 23 species, and the brown doves of the genus *Phapitreron,* with three species, are another group. The genus *Ptilonopus,* the fruit doves, is the second largest genus in the family Columbidae, with 51 species. These birds are all largely arboreal, seldom descending to the ground. The imperial pigeons are large, heavily built pigeons with big heads and purplish red legs and feet. The sexes, with one exception, are alike. Fruit doves vary in size from some species hardly bigger than a sparrow to the majority, which are the size of domestic doves. An almost incredible variety of bright colors mark their plumage, with some species sexually dimorphic. They have short, small bills and legs, and lay only one egg per clutch. The green pigeons favor habitats varying from evergreen forest to open woodland and coastal dune forest. They are particularly fond of wild figs on which they feed, and in which they often breed. Most are fast-flying species with a characteristic short tail. In Afrikaans these birds are aptly known as the *papegaaiduif* which, when translated, means the parrot pigeon. Their greenish-yellow coloration and habit of clambering about in trees has resulted in this rather descriptive Afrikaans name. In the non-breeding season they congregate in large numbers to feed. When disturbed they sit still, and only after a while will they begin clambering about in the trees again. The unusual call starts with a series of clicks, moving into a variety of flute-like whistles, and ending with a few deep growling sounds.

Management: Wild-caught birds are difficult to establish and post-capture mortalities are high. Newly acquired birds are extremely nervous and often refuse to feed. These stubborn birds can be force fed twice per day as described for hand rearing, and after a few days they will begin to feed independently. They can be rather aggressive in a mixed collection, particularly with other doves, and no more than one pair of a species should be kept per aviary. They will dominate a food table and finish preferred foods, such as soaked raisins.

A large planted aviary, with a minimum size of 15 m^3 (530 cu. ft.), is ideal for these birds. A well-planted aviary of 1.8 x 3.6 x 2.4 m (6 x 12 x 8 ft.) high could adequately house a pair each of fruit doves, quail doves, and smaller doves. A large aviary of 9 x 22 x 3.6 m (30 x 72 x 12 ft.) high can house five pairs of fruit doves and 5–6 pairs of other species of pigeons and doves. A mixture of Habitats A and B will suit fruit pigeons and doves, and the aviary should be situated in a sunny position. A mist spray over a shrub is enjoyed on a warm day. These doves will lean to one side and lift the opposite exposed wing, and then change sides, while enjoying the rain from the mister. The most important part of the management of these birds, however, is to ensure that all feeding dishes, platforms, etc. are kept spotlessly clean. With the soft foods on which these birds feed, they are susceptible to gastric disorders, which soon results in death. As these birds originate from subtropical areas, they are not comfortable in temperatures that remain below 15°C (59°F) for extended periods, or in combined cold and wet climates. In such climates the birds should be brought indoors during the severe months. In cool, damp climates where temperatures are not too low, ensure that plenty of good shelter is available.

Diet: All food given to these birds should be fresh and fit for human consumption. The inclusion of dog food in the diets is not recommended, as many of the birds on this diet are lost due to fatty liver disease. It is possible that the fats and formulation of the dog biscuits do not suit these birds, or maybe there is not enough biotin in this diet.

In captivity they are best fed on a firm porridge-textured softbill food with grated apple, diced fruits such as grapes, figs, guava, pear, soaked raisins, and papaya. I have found they enjoy the seeds of papaya as well. A multi vitamin/trace element powder is added to the diet twice a week. A recipe that proved successful includes fruits as described above, plus sweet potatoes, pomegranates, kiwis, etc.:

- 3 parts fruit cut into small cubes
- 1 part soaked raisins
- 1 part cooked brown rice

Grind to a fine powder:

- 1 cup racing pigeon pellets
- ½ cup dry parrot pellets
- 2 tsp. vitamins
- ½ tsp. vitamin E
- ¼ tsp. Calcium

Sprinkle enough of the powder over the fruit mix to make it dry. Feed in late afternoon for eating early evening and early morning. Fruit pigeons do not require a high protein diet. During rearing, however, an insect- or egg-based food can be sprinkled over the fruit.

The diets fed to the Disney's Animal Kingdom aviary birds are complete and varied but relatively simple to prepare. The majority of the diet bowls include Mazuri brand parrot pellets and Mazuri small bird breeder pellets (both soaked until moistened through) a 50/50 mix of finch seed and dry small bird breeder pellets, freshly chopped fruit mix containing grapes, cooked sweet potato, apple, raisins, pear, and pineapple. Shredded carrots and dark leafy greens, along with a commercially prepared insectivore mixture are also sprinkled in the mix. Soaked parrot pellets seem to be the pink-necked's favorite food item. Several of these diet dishes are placed throughout the aviary in an attempt to dissuade displacement and aggression among species. The diets are prepared and offered to the birds in the morning, and fresh diets are prepared and fed at midday. Diets are pulled overnight unless chicks are being reared. Large fig trees *Ficus microcarpa* also offer the birds fresh figs, buds, and flowers that they relish.

Breeding: Recently, fruit pigeons have been bred on a number of occasions, with some pairs proving to be quite prolific. The nest is typical of pigeons, consisting of a small platform of twigs placed in a bush or on a wire basket placed 1–2 m (3–6.5 ft.) off the ground. One breeder described how one pair continually laid in a food bowl. These birds will also nest on thick, dry, matted vegetation packed into a corner of the aviary. The green pigeons and imperial pigeons generally lay two pure white eggs; the fruit doves lay one, sometimes two. Young *Ptilinopus* doves are very small and sometimes leave the nest at nine days, with only wing feathers and very little other covering. They chill very easily, especially if they get wet. Green pigeons leave the nest when they are older and better developed, and are therefore less likely to succumb so easily to cold and wet conditions. The incubation periods vary from 12–16 days, and the chicks fledge at 15–18 days. The newly fledged chicks remain close to the nest for a few days. The immature birds are generally a dull version of the adults. One of the problems experienced with these birds is night frights, which may result in the adults deserting the clutch or chicks dying from exposure because the adults have moved off the nest during the night.

Staff at the Disney's Animal Kingdom give an excellent account of their success with breeding the pink-necked fruit dove *Ptilinopus porphyrea:*

> "Disney's Animal Kingdom acquired a male in early 1998 followed by a female in 1999. The pair appeared to bond quite quickly and after the required quarantine period was "howdied" and then released into a large mixed species Asian aviary. The aviary measured 30 x 17 x 12 m (100 x 55 x 40 ft.) tall. Thirty-eight bird species inhabit the Asian aviary, from tri-colored flycatchers *Picedula zanthopygia* and Dhyal thrushes *Copsychus saularis* to azure-winged magpies *Cyanopica cyana* and Amboina king parrots *Alisterus amboinensis*.
>
> The male pink-necked had been an aviary resident for six months prior to the release of the female. The male took an immediate interest in the female when she was placed in the aviary "howdy" or introduction cage and was observed strutting across the top of the cage vocalizing. After a brief introduction she was released and the male commenced to courting her. The pair formed a strong bond within two weeks. Three weeks later the male was seen bringing short twigs and dried laurel oak leaves *Querus virgintana,* which has small rounded leaves 1–25 inches in length, to a site 28 feet overhead in a large *Ficus* tree. Precopulatory behavior was observed to consist of brief bouts of allopreening preceeded by, or during which, the male inflates his chest and vocalizes while bowing. The female was then observed to stoop forward and spread her tail at which time the male mounted. Actual copulation lasts 1.5–3 seconds.
>
> By the end of their first month together the birds were incubating an egg. From the time of the female's introduction into the Asian aviary the pair was constantly harassed and displaced by a pair of Wompoo fruit doves. The male Wompoo and the male Temminck's pursued each other relentlessly whenever either came into view of each other. Scuffles and displacement between these two species sometimes occurred for several hours throughout the day and frequently occurred near feeding stations and in the ficus tree near to where the Temminck's

nesting site was located. On one occasion the two males were observed to lock feet and tumble to the ground. The male Wompoo even invaded the Temminck's nest site causing the birds to desert the site, always rebuilding nearby. The Wompoo fruit doves were determined to be the more aggressive of the two species and after six months, and five nesting attempts, the Temminck's pair was moved to an off-exhibit bird research and holding facility.

The new aviary in bird holding was 1.5 x 2.85 high x 3.6 m deep (5 x 9.5 x 12 ft) made of 1.27 x 7.62 cm (0.5 x 3 in.) welded wire, the floor made of soil covered to a depth of 10 cm (4 in.) with 3 mm (1/8-in.) pebbles, allowing for good drainage, raking and sifting to remove debris. The pair's only neighbor was a single male brimstone canary *Serinus sulphurathus* housed to one side. As a newly constructed facility, the planting of trees and shrubs was not yet completed. The pair quickly settled and the male began courting the female. Copulation was observed the next day. Nest platforms measuring 15 x 20 cm (6 x 8 in.) and 15 x 15 cm (6 x 6 in.) made from 0.4 x 1 cm (1 x 3 in) boards were placed in their enclosure the following day. On the fourth day the pair was observed to fly to the floor of the enclosure and investigate the far corner. They spent several minutes allopreening on the ground in this corner, the female was observed walking in tight circles in the area.

On the same day she began to pull small leaves toward her while sitting near the corner. A macho fern *Nepbrolepis bifurcata* was immediately planted in the corner in an attempt to give the birds something to nest on, or to dissuade them from nesting on the ground all together. The pair investigated the fern and began to nest build on the ground in front of it. The male was soon breaking small twigs and leaves from the dead oak tree branches and bringing them to the female. Dense, dry and fibrous rootlets from a nearby orange tree were added to the nest site by a keeper, providing a thicker barrier between the bird and the pebble floor. Two days later the pair laid its single white egg atop five small twigs, the size of wooden matches, and the rootlets. The average measurements for the two eggs laid to date are 31.20 mm x 22.90 mm. The pair did not begin to incubate the egg continually until the third day. Prior to this, both the male and the female took long breaks from incubation, lasting from 15 minutes to three hours. The pink-necked studbook (Morris, 1998) describes the average length of incubation, an average of three nesting instances, as being 18.5 days. In our case incubation lasted 20.5 days. This may be due to our unusual case in which the eggs were laid in a nest on the ground. (The ground temperature being cooler than the 26.6–29.4°C (80–85°F) daytime air temperature may have caused the embryo to develop more slowly; although a second egg laid by the pair in a tree nest site hatched after 20 days of incubation.)

The pair shared the incubation, with the female performing the majority of the duty. The female was observed to incubate from late afternoon, beginning at 2:30 p.m. and 4:15 p.m. until early morning ending between 7–9:00 a.m., with the male replacing the female as soon as she was off. The pair appeared to become anxious whenever the incubating member left the nest. At these times the birds would pace in the branches above the nest while peering down at the egg, within 60 seconds the next bird would fly to the ground and take its turn. On three occasions both male and female were observed to incubate, or tend to the egg, by resting to its side (not incubating by sitting directly over the egg).

Parent-raised Pink-necked Fruit Dove Chick *Ptilinopus porphyreus*

Age/days	Weight/grams
2	11.3
4	18.8
7	33.7
9	40.3
11	47.5
14	50.8
31	113.3
52	136.2

Once the chick hatched it was dry within 24 hours and was covered in a dense white down. It was brooded by both parents in nearly the same turns they had taken when incubating. The chick grew very quickly and although the parents were not seen to feed it, it always seemed to have a very full crop. At three days of age the chick was alert and could shuffle in the nest to turn itself around. The parent birds occasionally brooded to the side of the chick rather than sitting directly over it. The chick was soon covered in dark, white tipped pinfeathers with the secondary and primary feathers beginning to appear at five and six days of age. The parent birds began to brood the chick less during the day from day eight. From day 10 the chick began performing threat displays, puffing up and raising its wings when approached to be handled. At day 12, the chick weighed less than 50 g (1.7 oz.) and was a third the size of the parents but was already walking around the edge of the nest area. In a tree nest the chick would have undoubtedly been perched on the nest edge. At 16 days the chick would fly when frightened. It was observed eating from a food bowl at 26 days but failed to feed itself when isolated from its parents at 34 days. The chick was then given access to the parent's aviary where the parents fed it and it did not interfere with a second nest that already contained eggs."

See the section at the end of the chapter for advice on HAND REARING fruit pigeons.

Ring Size: 8 mm for green pigeons and fruit doves.

Species Requiring Similar Management:
Fruit-doves *Ptilinopus* species
Green pigeons *Treron* species
Brown-doves *Phapitreron* species
Imperial pigeons *Ducula* species
New Zealand pigeon *Hemiphaga novaeseelandiae*

Seminivorous Doves

As there is a wide variety of doves, many of which require similar captive management, I will discuss four representative species and list other species that require similar management. The divisions have been made between smaller species from woodland habitats, forest-related species, and larger species from woodlands and towns.

Cape Dove *Oena capensis*

- Glen Holland

Natural Habits: The Namaqua, or Cape, dove *Oena capensis* occurs naturally in hot, dry thornveld country, particularly in areas with short thorn scrub, where they feed from the ground on the seeds of various plants, weeds, and grasses. When disturbed, they take off quickly, showing two bands over the rump and the long distinctive tail. They are extremely fast flyers, and I have seen them hit a mist net with such speed that they were catapulted out again. They are regular visitors to waterholes, and their arrival can be heard by their hard, noisy wing beats. Their call is a soft deep *doooooo,* which rises toward the end and carries for some distance. The Cape doves' closest affinities are with the members of the genus *Turtur*. All are characterized by metallic wing spots, a double band on the back, cinnamon coloration on the wing, multi-colored bills, cream-colored eggs, and barred juvenile plumage (Rowan, 1983).

Management: The female lacks the black facial mask of the male, which has given the species the alternate name masked dove. Immatures are grayish in color with dark wavy barring over most of the plumage. These small doves are gentle, quiet aviary inhabitants, and in a large aviary, more than one pair can be kept together. In smaller flights, to avoid bickering, they must be housed in pairs. Young males start to show the black feathers on the chest by three months. They are often kept with small waxbills and finches. Cape doves tend to sleep outside of sheltered areas, and are particularly susceptible to disturbances at night. Disturbance by cats will soon be obvious, as the doves will show damaged feathers on their heads; if the problem is not solved, it may result in death due to a damaged skull.

The aviary should be lightly planted, as described for Habitat B in the LANDSCAPE chapter. The aviary floor should be covered in coarse sand, with a few scattered rocks and stumps upon which the birds will often perch. As with most doves, they spend much of their time on the ground, and regularly sunbathe in exposed sandy areas. Cape doves are rather delicate in cold, wet climates. In some areas of their natural distribution the cape dove is common in the dry season and disappears during the wet season. In their natural habitat, overnight temperatures will drop below zero, but the winter days are warm and dry. In severe climates they should be sheltered from the cold and wet, and when low temperatures continue for prolonged periods, additional warmth should be provided.

Diet: These doves feed on the ground on finely crushed maize, wheat, and a mixture of millets. Due to their habit

Namaqua Dove
Oena capensis

of feeding on the ground, doves in general tend to become infected by worms, and a de-worming program should be implemented every four months. Regular fecal checks will highlight any persistent parasite loads. Care should be taken if the birds are being moved from an aviary where the food dish is on the ground to one where the food dish is elevated. Cape doves need some time to adjust to this.

Breeding: During courtship, the male will sit on the nest calling, flitting his wings, bowing his head, and at the same time raising his tail to attract his mate. A form of chasing also occurs, with the female walking swiftly on the ground and the male following her with a pounce and landing slightly behind her. This chase usually ends with the female flying off. Under the correct conditions, a pair can prove to be prolific breeders. The nest is a tiny structure placed close to the ground in a dense shrub, or a thornbush, and is constructed of a few small twigs with a few rootlets as a lining. Both sexes build the nest, which measures about 7–8 cm (3 in.) in diameter. In aviculture, they will also build a flimsy nest in a basket, or on a wire platform. The nest is used for several consecutive broods, sometimes with gaps of many months between some of the broods.

In the natural state, in suitable climates, they breed throughout the year, with a peak between September and December. In cold climates it is recommended that the sexes be separated during winter, as few chicks will survive, and the birds' efforts, and energy, will be wasted. A clutch consists of two cream-colored eggs that hatch in 14 days. Incubation is shared between the sexes and starts with the first egg laid, resulting in the chicks hatching on successive days. The chicks fledge after 15–16 days. They are often multiple brooded, raising 4–5 broods in succession before taking a break. The skin of newly hatched chicks is dark purplish gray to black, sparsely covered with white or ashy colored down (Rowan, 1983). Many of the young that fledge in the cold or wet are lost.

Hand Rearing: As described at the end of the section under HAND REARING, but these small doves require more frequent feeding and more warmth than the larger species. As syringes can prove too big for feeding, a section of thin surgical drip tube can be fixed to the end of a syringe. The food in this case must be finely ground to avoid the syringe becoming blocked. It should be mentioned that these small species are very difficult to raise and this should only be attempted in emergency situations.

Ring size: 4.5 mm

Species Requiring Similar Management:

Emerald-spotted wood-dove *Turtur chalcospilos*
Diamond dove *Geopelia cuneata*
Zebra doves *G. striata, G. maugeus*
Inca dove *Columbina inca*
Scaly dove *C. squammata*

It must be remembered, however, that due to aggression, for most dove species no more than one pair should be kept in an aviary unless the aviary is large.

Emerald Dove *chalcophaps indica*

Natural Habits: In its native habitat, the green-wing, or emerald, dove *Chacophaps indica* occurs, in numerous subspecies, throughout southern Asia, the Philippines, New Guinea, coastal northern and eastern Australia, as well as many of the islands surrounding these areas. These doves inhabit areas of rainforest, wet eucalypt forests, heathlands, and thickets of disturbed vegetation such as lantana. This preferred habitat should be taken into account when housing this species. They spend a large proportion of their time on the ground, feeding on seeds, berries, and insects.

Management: Both sexes have a light brown body cover, but differ as follows:

Coloration	Male	Female
Neck, breast & back	Purple-brown	Brown
Rump and upperside of tail	Black	Brown
Shoulder patch	White	Gray
Bars over back	Distinct	Indistinct

*Table from Brown, D. 1993. The green-wing. Australian Birdkeeper June-July 1993.

Juveniles are of the same light brown body color as the female, but have bars and scalloping of chocolate brown and emerald green on the back, wings, and breast. Approximately 4–6 weeks after fledging, the color pattern becomes more like that of the adult female, with the young males starting show a few purplish-brown breast feathers.

Green-wings should be housed in a medium to large planted aviary, 2.5 x 4 m (8 x 13 ft) to 4 x 4 m (13 x 13 ft), landscaped to Habitats A and B, and can be kept in mixed collections of finches, quail, and other pigeons and doves. In such collections they are very docile, causing minimal problems. As with all aviaries containing pigeons or doves, any movement within the aviary should be slow and cautious so as not to panic the birds, as they can be rather clumsy when flying. Once settled, however, they become quite tame and confiding. In holding cages or extremely large aviaries, more than one pair can be housed together, but in aviaries of the size described, to avoid bickering, only one pair should be housed per aviary. They appreciate an aviary with open patches on the floor, where they will spend most of their day. They enjoy sunbathing, and a water bath from an overhead sprinkler, turned on during hot days, will also be thoroughly enjoyed by these birds. They can cope with temperatures ranging from 0–40°C (32–104°F) but daytime temperatures should be in excess of 12°C (53.6°F).

Diet: Although in the natural state these birds subsist on a diet of seeds, berries, fruits, and some insects, they will thrive in an aviary on a basic diet of dry seed, supplemented with seeding grasses, green food, and occasionally a little fruit, such as apple or grapes. It is important to provide insoluble grit for these birds to allow normal gizzard function, as, like all pigeons, the seed is not hulled before it is eaten. They show no particular preference for the smaller or larger seed types in the finch mix. They may also occasionally take turkey starter crumbs and coarse layer mash. Hulled sunflower is a favorite but, due to the high fat content, it should only be offered in limited quantities.

Breeding: Green-wings generally nest at any time of the year and can be most prolific. Over a 15-month period, one pair owned by Danny Brown hatched 17 chicks from nine nests, of which 16 survived. Two eggs form the usual clutch, with incubation, which commences after the laying of the second egg, lasting 14–16 days. The nest, if built in a natural bush, is a typical flimsy structure of fine twigs and stiff grasses. They are best encouraged to nest in wire trays, filled with grass, placed at a height of 1.2–1.8 m (4–6 ft.) above the ground. Some pairs cannot be restricted from breeding, and if the nest platform is removed, they simply nest on the ground.

The parents are usually tight sitters, although care should be taken not to disturb them during the first two days of incubation, as desertion may result. After this, the nest may be checked regularly and young may be handled, although this should be avoided in the first two days after hatching. The parents share incubation duties, with the male usually incubating during the day and the female at night and occasionally during midday. Once the young have hatched, they are fed by both parents.

The squabs feather quickly, with pin feathers appearing at approximately 5–6 days of age. The eyes also open at about this time. The parents brood the young for approximately 8–10 days after hatching, after which the young are left uncovered. One parent, usually the female, will still roost beside the young at night until they are fledged. Fledging occurs approximately 2½– 4 weeks after hatching, depending on the weather. During the hotter months, the young may leave the nest early and roost on the floor for sometime before being able to fly. On the floor they are surprisingly well camouflaged, especially among foliage. Once able to fly, the young only occasionally return to the nest.

A good pair will usually embark on a new clutch within days of the previous young leaving the nest, or occasionally, even before they have left. The young are independent approximately 3–4 weeks after fledging. Occasionally, a successful pair may have a period of failed nests or inactivity, following a period of successful nests in quick succession. This appears to be the way of self-regulating themselves and having a well-earned break. Normal breeding resumes after 4–6 months.

See the section at the end of the chapter for advice on HAND REARING Pigeons and Doves.

Ring size: 5 mm

Species Requiring Similar Management:

Tambourine dove *Turtur tympanistria*
Cuckoo-doves *Macropygia species*
Crested pigeon *Geophaps lophotes*
Ruddy quail dove *Geotrygon montana*
Bleeding-heart *Gallicolumba luzonica*
African cinnamon *Alopelia larvata*

Senegal Dove *Streptopelia senegalensis*

The genus *Streptopelia* includes several favorite avicultural birds, collectively referred to as turtle doves. The larger species, such as the laughing, or Senegal, dove *S. senegalensis,* the red-eyed dove *S. semitorquata,* and the ring-necked dove *S. capicola,* are among the most hardy and easiest to maintain avicultural subjects of the dove family. While the husbandry described in this section focuses on the Senegal dove, the other members of this genus require similar captive husbandry.

Management: In a large aviary, a few pairs can be kept together with little serious aggression. Males will squabble over nest sites, and some chasing and beating

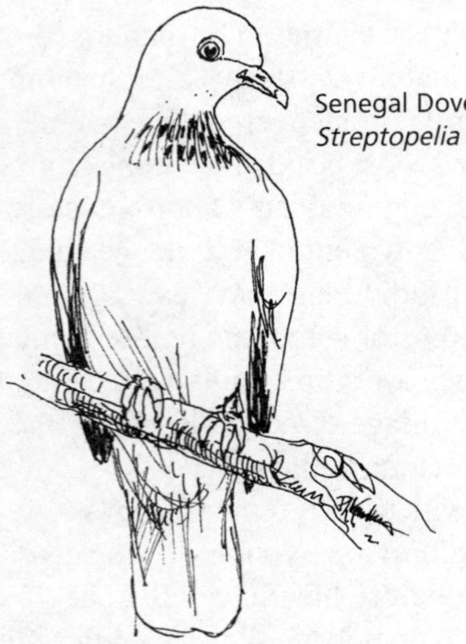

Senegal Dove
Streptopelia senegalensis

with wings will occur. In smaller flights, they are best kept as sexed pairs. Apart from some inter-specific chasing, these doves are not aggressive toward other species. Their size, however, makes them unsuitable to mix with small birds such as finches, which would be intimidated by these large birds.

Aviaries should be sparsely planted with small trees and shrubs, which will provide perches and nesting sites. Habitat B as described in the LANDSCAPE chapter is suitable for turtle doves. The ground should have plenty of open sandy patches where the birds will spend much of their time. On a hot day, these large doves are fond of bathing under a mist spray positioned above a perch, and of sunbathing, lying on one side with the upper wing stretched out in the sun. Domestic mutations of the Senegal dove, which have fawn and white feathers, are now well established.

Diet: In captivity they will feed on mixed finch seed, crushed corn, whole sunflower, wheat, sorghum, and shelled peanuts. They will devour a slice of whole-wheat bread that has been slightly moistened with water. I have also seen them take moistened insectivore food and, in the natural state, the Senegal dove has also been recorded eating insects and fruits such as small figs.

Breeding: The sexes are alike. The Senegal dove naturally has a towering display flight like most other members of its genus. The bird leaves its perch with noisy wingbeats and rises to a considerable height before checking and sailing down on outspread wings and tail. In the bowing display the male pursues another bird on foot, puffing out its neck and bobbing the forepart of the body to the accompaniment of a series of laughing *coos*. This display is not known to occur in association with copulation but is often seen in aggressive situations. Pre-copulation, the male usually feeds the female once she has placed her beak into his. Allopreening usually occurs post-copulation. (Rowen, 1983).

They nest readily and will build a flimsy nest of twigs in a basket, on a wire platform, or in a dense shrub. The male collects and supplies the nest material to the female who then places it in the nest. The nest usually takes about 2–3 days to construct and is used repeatedly, 2–8 times, and with re-use, the accumulated droppings and added material may form a platform of up to 12 cm (5 in.) thick. Incubation starts either before the first egg is laid, once the first egg is laid, or sometimes once the second egg has been laid. One of the problems that occurs with these prolific species is that aviaries soon become overpopulated. In large, public display-type aviaries, these species will take over the aviary if the fledglings are not removed regularly. Incubation of the two white eggs is shared between the sexes with the female incubating at night and the male during the middle of the day. Usually, but not always, the switching of incubation duties is accompanied by cooing from both birds.

The first external pip on the eggshell will often

Descriptive Notes of Senegal Dove *Streptopelia senegalensis*

Plumage	Male	Female
Overall plumage	Dark gray/brown with pinkish tinge	Light gray/brown
Wing spots	Metallic blue	Metallic mauve
Nape	Silver/gray	Gray/brown
Forehead	Pinkish	Brown
Cheeks/throat	Pink/gray	Light brown
Chest	Pink/gray / scaly	Light brown / scaly
Belly	Pink/gray to white	Gray/brown to white
Beak	Scarlet with black tip	Red with black tip
Iris	Scarlet	Red

Immatures are initially a dull grayish brown color, but at two months old, both sexes show considerable scaly markings on the chest, and males begin to develop the pink tinge on the neck and chest. They are fully colored at three months.

occur a day prior to hatch. Eggshells are removed from the nest by either sex. The newly hatched chick has a dark purplish gray skin and is covered with yellow down. It weighs about 5 g at hatch and its bill is white-tipped, as is usual in Columbidae (Rowen, 1983).

See the section at the end of the chapter for advice on HAND REARING Pigeons and Doves.

Species Requiring Similar Management:

Lace-neck dove *Streptopelia chinensis*
Cape turtle dove *Streptopelia capicola*
Madagascar turtle dove *S. picturata*
Mourning collared dove *S. decipiens*

Ring size: Senegal: 5.5 mm; turtle dove: 6 mm.

Passerina Dove *Columbina passerina*

Natural Habits: The passerina dove, or common ground dove, *Columbina passerina,* has a widespread distribution that includes the southern United States, the Bahamas, Jamaica, Guatemala, Columbia, Ecuador, and a number of the surrounding islands in these regions. To date, at least 19 subspecies have been recorded. They spend much of their day on the ground where they feed on small seeds. They seldom venture far from cover and prefer to feed under a canopy of low shrubs.

Management: These tiny doves are sexually dimorphic.

I have bred this species in flights measuring 6 x 1.2 x 1.8 m (20 x 4 x 6 ft.) high. It is possible to house them in smaller flights, and I have even bred them in a suspended weldmesh cage but, since they are rather nervous birds, larger flights help to ensure that they do not panic if approached. Habitat B is ideal for these doves. The aviary should not be too thickly planted. It should have plenty of open, dry, sandy spots on which an occasional bare rock or stump should be positioned, as passerina doves spend a considerable amount of time on the ground, where they prefer to sit on a rock or stump slightly above ground level. Pairs usually sit close together and often lean to one side, stretching out the opposite wing and sunning themselves. I have also often found them resting in their food bowls.

As with most of the small doves, passserinas are aggressive to others of their own or a similar kind, such as pygmy doves *C. minuta.* The seriousness of this aggression and the negative impact it may have depends on the size of the aviary. I know of one breeder with a large aviary in which bleeding-hearts *Gallicolumba* spp. breed. He has a number of pairs of passerina doves that, as he puts it, "breed in between the bleeding-hearts" with little sign of aggression. In flights such as mine, however, the aggression would disrupt breeding, and even result in injured or badly plucked birds. They can, however, be mixed with finches, canaries, etc. with no problems. To date, I have not lost birds to worms. Passerinas suffer over long periods of cold and wet weather, and in harsh climates I would recommend that they be housed indoors during the cold months.

Diet: My birds feed from the same dish, (containing a mixture of millets) as the finches but, due to their presence, I add some finely crushed corn, which they enjoy. In addition, cuttlebone, eggshells, and a bowl of coarse river sand for grit are always available. One problem I have experienced with most doves is the way they scatter seed out of the bowl while feeding. One solution to having seed soiled or wasted in this manner is to place the seed bowl in a larger flat tray from which all the birds continue to feed on the spilled seed. I have also seen my birds feeding on the sprouted Japanese millet/soft food mixture provided for the finches, particularly when feeding chicks in the nest.

Breeding: Males call continually, particularly if housed in adjacent aviaries. This call is a mournful *cooo* made repetitively. While the pair is on the nest together, males also make a soft growling sound as part of courtship. While males call and both birds flit their wings excitedly, the females are more aggressive during courtship. They will chase after males and aggressively preen the male's nape. The two matings I have observed took place after a few minutes of calling and the male chasing the female on the ground.

The nest, as with most doves, consists of a small platform of grass, thin twigs, and rootlets. Fortunately, however, it is a reasonably firm nest, generally placed high in the aviary, with a slight cup in the center, which results in few clutches lost even in bad weather. The nest is used up to four times consecutively, with a new top lining added with each new attempt. Passerina doves are not fussy with regard to nest sites, and will build on a dry thicket hung in the aviary, in a dense shrub, or in an artificial nest, such as a canary cup.

Two white eggs are laid on consecutive days and incubated by both sexes for 13 days. After the first 4–5 days of incubation, they sit tight. If made to fly off the nest, the birds return immediately after the nest inspection is complete. Incubation begins with the first egg laid, and the chicks hatch on alternate days. Telltale shells dropped some distance from the nest on the aviary floor signify that the chicks have hatched. If a nest containing newly hatched chicks is inspected, the adults will sometimes run about on the ground feigning injury.

Both sexes feed the chicks with a pumping action lasting about 30 seconds at each feeding. For the first 7–9 days, the chicks are brooded constantly by one of the parents. Chicks develop rapidly and fledge at 16 days. The fledglings spend most of their first two weeks on the aviary floor where they pester their parents for food at every opportunity, with flitting wings. Although they begin to peck at food after a week, I remove the chicks at about a month of age. I have found that chicks left longer than this in the flight situation will begin to interfere with their parents on the nest. Chicks that fledge in extended cold, wet periods are often lost. In harsh climates, it would perhaps be better to discourage breeding or ensure that sufficient shelter is available where the chicks can avoid severe weather. These doves will breed almost continually, if allowed to, with a new clutch usually laid within three weeks of the previous clutch fledging. Some pairs occasionally take a break for a few months before resuming their breeding.

Ring size: 4 mm

Species Requiring Similar Management:

Ground-doves *Columbina, Claravis, Metriopelia* and *Uropelia* species
Pygmy dove *Columbina minuta*
Diamond dove *Geopelia cuneata*
Bar-shouldered dove *G. humeralis*

HAND REARING PIGEONS, DOVES AND FRUIT PIGEONS

Allison Beutel of the San Diego Avian Propagation Unit provides the following advice on hand rearing doves and pigeons:

> "Initially the hand rearing formula must mimic pigeon milk. Pigeon milk (or crop milk) is a holocrine prolactin-mediated secretion produced by both male and female parent pigeons. The milk is fed exclusively to squabs during the first 3–10 days of life. Crop milk is high in protein (11–23%), high in fat (6–12%) and high in water content (65–82%). Carbohydrates are virtually absent from this mixture. The young chicks will drink this mixture from the oral cavity of the parents in much the same way that parent birds drink. The diet should be gradually changed to become more similar to the adult diet as the chick ages. This gradual change should slowly decrease protein and fat content while raising solid matter and carbohydrate content. Supplying the diet to the chicks is easiest if an artificial 'pigeon beak' is constructed. These can be fashioned from pipette bulbs, eyedroppers, rubber tubing and modified syringes. My preference is a syringe that has had the end cut/drilled out leaving a

Superb Fruit Dove *Ptilinopus superbus* Hand Rearing Diet

Basic Mix
250 g quark
250 g tofu
90 g hard-boiled egg
1 level teaspoon SA37
1 level teaspoon calcium supplement

Day 1–4
Basic Mix
100 ml saline

Day 5–9
Basic Mix
28 g fruit puree
70 ml saline

Day 10–14
Basic Mix
28 g fruit puree
92 g wheatbread
30 ml saline

Day 15–19
Basic mix
28 g fruit puree
50 g wheatbread
40 ml saline

Day 20–43
Basic mix
28 g fruit puree
50 g wheatbread
10 ml saline
100 g sprouted pulses

Day 44–80
Basic mix
28 g fruit puree
50 g wheatbread
100 g sprouted pulses
100 g banana (fresh)
263 g tinned guava
peanut butter

From Avicultural Magazine 105(2):60.

1.0 mm margin around the inside edge. The syringe size should be increased, as the chick grows older. The diameter of the syringe (or alternative feeding tube) should be such that the chick's beak is able to fit inside the tube opening and open up inside the tube body. The chick will then drink the hand-rearing formula in which its beak is submerged.

The feeding regime will depend on the species but the following routine is a general guide. The feeding period is roughly divided into four stages. The first is the newly hatched stage, and in a chick with a 30-day fledging period this period lasts approximately until four days old. The second stage is the early growth stage, and in the same chick lasts until 5–7 days of age. The third stage is the late growth stage, which will last until 8–14 days of age in the above chick. The final stage is the fledgling stage, which lasts from 15 days to weaning. From stages 1–3 the chick should be maintained on a 'crop milk' substitute and this mix should become progressively thicker by the end of stage three. At the start of stage four, the diet should be changed to a formula similar to that used in the normal raising of a granivorous species such as parrots. From stage four until weaning the proportion of adult diet fed should increase. Feeding rates should initially be 3–4 times a day (depending on crop emptying rates) and should be reduced to one feed a day at weaning age. Actual weaning age will depend on the acceptance of the adult diet by the chick, so adult diet should be offered as soon as practicable (usually around normal fledging age).

The following stages are those recommended for a species with a 30-day fledging period, and the starting point for an equivalent stage in a rapidly fledging species needs to be calculated. The equivalent periods correspond to approximately 14% of the fledging period, 15–24% of the fledging period, 25–46% of the fledging period and 47–100% of the fledging period. If a delay in dilution change occurs this will usually result in slow or stunted growth, particularly with commercial diets. The following products have been successfully used for hand rearing of various pigeon and dove species or for supplementing chicks in the nest.

I have used these formulas — Roudybush Squab Hand-feeding Formula and Roudybush Formula — successfully on ground-dwelling granivorous pigeon species from day old chicks:

Stage 1: Squab Handfeed
- 6 parts water
- 1 part formula by weight, OR 2.2 parts water
- 1 part formula by volume.

Stage 2: Squab Handfeed
- 4 parts water
- 1 part formula by weight, OR 1.5 parts water
- 1 part formula by volume.

Stage 3: Squab Handfeed
- 3 parts water
- 1 part formula by weight, OR 1 part water
- 1 part formula by volume.

Stage 4: Formula 3
- initially feed at a dilution of 3 parts water
- 1 part formula by weight, OR 1.2 parts water
- 1 part formula by volume and reduce the proportion of water as the chick approaches weaning age.

For fruit-eating species, stewed or pureed apple or pear may be used to replace water portion in stage four. As the chick approaches weaning, the formula may be mixed with apple or pear to a putty-like consistency and rolled into fruit sized bits. The chicks are able to recognize the smell of the Roudybush balls and will usually accept them readily. These balls are then gradually replaced with fruit pieces."

Paul Harrington, Amada Ferguson, and Patsy Josef give an excellent account of the hand rearing of a superb fruit dove *Ptilinopus superbus* at London Zoo:

"A hand rearing diet to stimulate a natural crop milk was formulated using the following ingredients: 250 g quark (a soft cheese product), 250 g tofu (soy bean curd), 90 g hard-boiled egg, 1 level teaspoon SA37, and 1 level teaspoon calcium supplement.

The protein and fat percentages in the mix were 11% and 4% respectively. These were at the lower end of the scale for the known fat and protein levels in natural crop milk. It was thought that higher percentages of these two dietary components might lead to too viscous a mix, which would reduce the motility of food through the crop. It was also thought that too high levels of these components might compromise liver function. Animal and vegetable proteins were used in the formula to cover the full spectrum of essential amino acids. Carbohydrate in the form of starches and sugars were close to zero, which was in line with the levels found in natural crop milk. Carbohydrate content was increased as the chick matured. The consistency of the mixture needed to be such that it was neither too liquified, which could result in aspiration of food, or too thick, which could result in crop stasis and impaction. The chick was fed using a syringe.

In the absence of the benefits derived from crop milk to promote rapid development in the chick, it was considered essential, when hand rearing with an artificial diet, to maximize the chick's food intake. This was achieved by:

- Incorporating a high number of feeds. The number of feeds over the first ten days averaged eight per day over a 16-hour day.
- Filing the crop to capacity at each feed. Care was taken not to overfill or overstretch the crop.
- Not allowing the crop to empty between feeds over the 16-hour day. The crop was allowed to empty overnight to reduce the risk of bacterial infection.

Weaning was encouraged by:

- Including chopped fruit and pulses in the rearing mix. This encouraged the chick to madibulate the food particles before it swallowed them.
- Offering the food from a pair of forceps, particularly favorite food items such as tinned guava.
- Placing a female congener in its enclosure to stimulate copycat feeding.
- Reducing the amount of hand-rearing mixture fed to the chick.

To avoid imprinting, which occurs at an early stage in a pigeon's life, a puppet modeled on a female superb fruit dove was introduced to the chick on the sixth day, three days after the chick's eyes had begun to open. Ideally the puppet would have been introduced earlier. The chick's response to the puppet was to beg for food. This it did by calling and biting the bill of the puppet. In experiments presenting the syringe and the puppet simultaneously to the chick, the puppet would always be selected in preference to the syringe, while only the former provided food. The chick would also attempt to be brooded by the puppet. As it was unable to get underneath the puppet it would snuggle as close as possible beside it. At a later developmental stage the chick would perch leaning against the puppet, which was attached to a perch, thus adopting the same behavior as in parent-reared chicks.

Hand-rearing diet for Torres Strait pigeons and rose-crowned fruit doves (Gibson, 1981):

Stage 1–3:

- 1 part finely crushed Arrowroot biscuits
- 1 part finely blended turkey starter crumbles
- 2 parts Heinz high protein baby cereal
- 2 parts Farex
- Mix with enough water to produce a runny consis tency and feed by syringe or rubber tube.

Stage 4:

- Add mashed fruit (baby food) to make approximately one third of the daily food intake.

At weaning age the young bird is hand fed diced fruit until it is able to pick it up itself."

Glen Holland's personal formula for hand rearing doves and pigeons from about a week old consists of the following:

"The chicks are fed on a mixture of equal portions of commercial psittacine hand-rearing formula moistened with water, and millet seed which has been soaked in hot water. A multi-vitamin and a calcium supplement should be added to this as well. With small chicks or smaller species, a 1.0 ml syringe is used, and with large chicks a 2.5 ml syringe. The entire end of the syringe is cut off, leaving a hollow tube. Should the syringe not draw with ease, a little vegetable cooking oil can be placed on the stopper. The chick should be able to swallow the tip of the syringe with ease. This is placed well into the back of the throat before feeding.

The food mixture should be just liquid enough to be able to draw it into the syringe. Chicks over a week old are able to digest more coarse food than newly hatched chicks. Food can be prepared in bulk for freezing and daily portions defrosted as required. I have found that if the rearing food is fed without the addition of the soaked millet seed, the chicks develop diarrhea. If the chicks are large when first obtained, they are best wrapped in a towel to avoid struggling while being fed. The beak is held open with a finger on each side at the base of the beak and the syringe is then inserted well into the throat.

After 2–3 days, the chick will begin pushing its way between the attendant's fingers and at the same time opening its mouth. To judge if enough food has been given, feel the crop, which should be bulging slightly. Additional warmth should be supplied until the chicks are well feathered. From then on, supplemental heat is only required during cold weather. If feathers have not yet appeared, the chicks should be kept in a brooder at 30°C (86°F); by the time the chicks are half-feathered, the temperature can be reduced to 25°C (77°F). As soon as the chicks fledge, a sprinkling of millet placed on the cage floor will usually stimulate them to begin pecking almost immediately. At this stage, to check if the chick still requires hand feeding, one should feel the crop; if food is present continuously, hand feeding can be stopped. Depending on the size and age of the bird, young doves will take 5–15 ml of formula, fed four times per day to chicks less than 10 days old and three times per day from then on.

I have personally hand raised fruit pigeons from about a week old using the following protocol: the chicks may be easily hand reared using a commercial psittacine hand-rearing formula with enough water added to give a porridge consistency, mixed with grated apple and mashed banana. This is fed using a syringe of +/- 8 mm in diameter, with the fine tip cut off. As with all pigeons and doves, the birds will resist feeding if the food has not been warmed to body temperature. Initially the beak is forced open, but within a few days the chicks will be only too eager to be fed. Chicks that are covered with down are fed four times per day.

This is decreased to three feeds by the time half the body is covered by feathers, then to two feeds when fully feathered. Thereafter, the amount fed is gradually reduced, starting about 10 days after fledging. The chicks should be independent by 18 days post fledging. These chicks do not wean as fast as do other pigeons. Soaked raisins sprinkled over the usual diet will entice the chicks to start pecking. The hand-reared birds can imprint and remain very tame;

Superb Fruit Dove *Ptilinopus superbus* hand-rearing diet

Day	Weight (grams)	Total food intake (grams per day)	Number of feeds / day	Brooder temp	Comments
0	5.7	0.1	3	36°C	
1	5.4	1.4	10	36°C.	Wild reared weight = 8.3 g
2	5.6	3.2	8	36°C	Wild reared weight = 11.2 g. Wing pin feather showing.
3	7.5	4.2	9	35°C	Wild reared weight = 18.3 g. Eyes opening.
4	9.3	6.7	8	35°C	Wild reared weight = 24.1 g
5	12.5	9.9	9	33.5°C	Wild reared weight = 28.3 g
6	15.5	9.4	8	33.5°C	Wild reared weight = 31.6.
					Puppet introduced.
7	18	13.1	7	34.5°C	Wild reared weight = 32.7 g
8	21	13.7	7	34°C	
9	23.8	16.5	7	33°C	
10	25	16.8	7	32.5°C	
12	27.9	24.4	7	32°C	
15	42	26.2	6	30°C	Moved from brooder to small cage
17	43.6	34.3	4		Perching
19	56	46.7	4		
21	60.4	45	3		First drink on day 20
23	67	45.9	3		
25	67	56.6	3		Attempted to fly
28	74.5	45.7	3		
32	78	44.3	3		
35	75.8	61.2	4		
38	78.1	48.8	5		
43	83.7	49.3	4		
47	92	46.8	4		
51	96.7	14.9	3		Self feeding started.

From: Avicultural Magazine 105(2):61–62.

I have one African green pigeon that puffs itself up, gives his clicking and whistling call, and pounces around whenever he sees me. Another that I raised was housed with a fledgling wagtail; after fledging, these two remained companions for many months and would often huddle up together on a perch. In New Zealand, hand-raised New Zealand wood pigeons *Hemiphaga novaeseelandiae,* which are fruit pigeons, although undoubtedly slightly imprinted, make a fascinating addition to a free-flight native species display. These hand-raised birds do also provide good breeding stock but should be housed with a conspecific, ideally as a chick before fledging, as if they are handled too much they can imprint and then fail to breed."

References:

Bester, M. 1991. The Nicobar pigeon. *Avizandum*. January: 21–23.

Bockheim,G., J. Daugherty & C. Sky. 2000. Breeding the Pink-necked Fruit Dove at Disney's Animal Kingdom. *AFAWatchbird* 27(4): 20–22.

Brown, D. 1993. The green-wing. *Australian Birdkeeper* June-July.

Harrington, P., A. Ferguson & J. Patsy. 1998. Hand rearing a Superb Fruit Dove at London Zoo. *Avicultural Magazine* 105(2): 57–64.

Jeggo, D. 1977. Preliminary notes on the Mauritius pink pigeon at Jersey Zoological Park. Dodo, *Jersey Wildlife Preservation Trust* 14:26–30.

Jeggo, D. 1979. Breeding the Mauritius pink pigeon at Jersey Zoological Park. *Avicultural Magazine* 85(1): 1–2.

Jones, C.G, D. Jeggo & J. Hartley. 1983. The maintenance and captive breeding of the pink pigeon. Dodo, *Jersey Wildlife Presrevation Trust* 20: 16–26.

Lind, C.R. 1989. The effects of multiple clutching on the size, fertility and hatchability of the eggs of the pink pigeon at the Jersey Wildlife Preservation Trust. Dodo, *Jersey Wildlife Presrevation Trust* 26: 93–98.

Lind, C.R. 1989. Management of the EEP pink pigeon population. Dodo, *Jersey Wildlife Preservation Trust* 30: 106–113.

Rowan, M.K. 1981. *The Doves, Parrots, Louries & Cuckoos of Southern Africa*. Cape Town, SA: The John Voelcker Bird Book Fund and David Philip.

Bustards, Cranes & Trumpeters

19 Gruiformes

19.01 Otididae Bustards

Charles E. Siegel, Sara Halliger & Tom Bailly

Introduction: The bustards comprise the Family Otididae in the Order Gruiformes. Twenty-two species from six genera are currently recognized. An excellent source of information is Paul A. Johnsgards book, *Bustards, Hemipodes, and Sandgrouse: Birds of Dry Places* (1991, Oxford University Press). The group is concentrated in sub-Saharan Africa, but also occurs in all other major arid lands of the Old World, from Europe to eastern Siberia, and discontinuously in Australia and in India. All bustards have cryptically patterned upperparts with black on buff, rufous, or brown. The feathers show a pink tinge at their base, the result of light-sensitive porphyrins. The males have fascinating courtship displays and, depending upon species, use combinations of elaborate plumage displays and postures, distinctive vocalizations, or flight displays. They lack a preen gland, but produce powder down to waterproof themselves. All bustards lack a hind toe, and the feet have no grasping ability. First-time observers are often struck by the relatively small size of the feet compared to the body size.

Bustards are completely terrestrial, but are strong, if reluctant, fliers. Adult male kori bustards, *Ardeotis kori,* may weigh up to 19 kg (42 lb.) and are among the worlds heaviest flying birds, while the little bustard, *Tetrax tetrax,* is a more modest 0.55 kg (1.2 lb.). Bustards typically avoid heavily wooded and rocky, precipitous terrain, and prefer open habitats. On the ground, they are typically nervous and alert, and move into cover at the first sign of danger. When foraging, these omnivores are strong but slow walkers, taking invertebrates, small vertebrates, and vegetable matter from scrub or grassland. Bustards do not have a crop, but consume large quantities of grit, have a powerful gizzard, and have a highly developed cecum to aid in digestion. Two kinds of well-formed feces are excreted and considered normal — one is light brown, the other is black and tarry. Bustards suck up water, rather than scoop it with their bills (Hallager 1994).

Kori Bustard
Ardeotis kori

The mating system is still unclear for many bustard species. They have variously been described as probably monogamous or polygynous or promiscuous, although present evidence indicates the majority is polygynous. In general, bustards are K-selected species-long-lived with a low reproductive output (Collar 1996). In most species, males and females do not establish a true pair bond, and depend instead on a dispersed lek breeding system in which the males advertise themselves in traditional areas. However, the great bustard, *Otis tarda,* forms true leks. Male bustards perform elaborate displays to attract females and maintain a dominance hierarchy. In polygynous species, the females are left alone to undertake nesting, incubation, and rearing of the young.

Not all bustards are polygynous, and in the genus *Eupodotis* there are at least three and perhaps five monogamous species. The black-throated bustard, *Eupodotis vigorsii,* blue bustard *Eupodotis caerulescens,* and white-bellied bustard *Eupodotis senegalensis* are monogamous. It is speculated that monogamy occurs in Ruppell's bustard *Eupodotis rueppellii* and little brown bustard *Eupodotis humilis* (Johnsgard, 1991), although data on these species is lacking and anecdotal. What little is known of the mating strategies of any of these species contrasts with that of polygynous bustards in that these species form monogamous pair bonds throughout the year, often calling together to maintain the pair bond (Johnsgard 1991). In contrast to polygynous bustards, there are no known male self-advertisement displays and both sexes are monomorphic in size and plumage. The extent of paternal care in many monogamous species is unknown.

Some species of bustard are highly gregarious during the non-breeding season. Even during the breeding season the sexes tend to stay apart, and the bustards probably exhibit male dominance rather than resource defense polygyny. In nature, larger species become sexually mature at 4–6 years, females at 3–4 years, although this may be reduced in captivity. Most bustards lay 1–2 (larger species) or 3–4 (smaller species) eggs in a simple ground scrape at 2–3 day intervals. If the eggs are removed, the females will lay replacement clutches. Incubation in most species lasts 22–24 days. Hatching is staggered. The chicks are semi-precocial and are initially fed bill-to-bill by the mother.

Management strategies for breeding vary according to the species but there are basically three regimes (Table 1). In captivity, houbara bustards *Chlamydotis undulate.* managed as pairs or as mixed-sex groups rarely, if ever, breed. Similarly, houbara bustards caught as adults remain sensitive to stress and in most cases fail to breed. Successful houbara bustard breeding projects use hand-reared birds, which are more suitable for intensive production efforts utilizing artificial insemination. White-bellied bustards *Eupodotis senegalensis,* are monogamous and adults are maintained in pairs. Buff-crested bustards *Eupodotis ruficrista,* tend to be bred in captivity as pairs, but in the 1980s and early 1990s they were successfully bred at Al Ain Zoo (United Arab Emirates) in polygynous breeding groups of three females and a number of males in a large aviary. Kori bustards are managed in mixed-sex flocks, but can be managed as trios. If managed as mixed-sex flocks, provisions must be made for the males to be kept physically separate due to extreme aggression during the breeding season.

Rainfall plays a strong role in the breeding success of free-living kori bustards, with breeding greatly reduced during time of drought (Osborne 2000). At the National Wildlife Research Center (NWRC) Heezik et al (1999) considered that low fecundity in the captive houbara bustard flock in 2000 was related to a warm winter. Cooler winters were correlated with high proportions of females laying and more eggs being laid, and the timing of the first rainfall of the season also stimulated some birds to breed (Heezik et al 1999).

In captivity, recruitment of female houbara bustards into the breeding population takes place over five years and the median age at which females breed is

Table 1. Examples of Breeding Strategies of Bustards

Strategy	Species	Age	Social Grouping
Isolated for artificial insemination	Houbara *chlamydotis undulata*	AdultJuvenile	Singles Small groups (=5)
Pairs	White-bellied and Buff-crested *Eupodotis senegalensis* & *Eupodotis ruficrista*	AdultJuvenile	Pairs Small groups (=5) Heterosexual groups
Mixed-sex flocks	Kori *Ardeotis kori*	Adult AdultJuvenile	Large group (~25) Trio or 2.3 Large group (~15)

Bailey and Hallager, 2003

three years (Heezik 2000). Females lay 6–9 eggs per season (Heezik et al. 1999). At NWRC avicultural research over a six-year period from 1993 to 1999 has shown that less than 5% of females breed for the first time at one year, 21% at two years, 18% at three years, 13% at four years and 6% at five years (Heezik et al 1999). There was variation between cohorts in the proportions of females recruiting at different ages. Research to increase the proportions of females that breed each year would have a great impact on productivity of houbara projects.

Gonadotrophins have been used to try to stimulate reproductive development in houbara bustards at National Avian Research Center (NARC), but preliminary trials were of limited success and the work has been discontinued (Jacquet 1996).

Diet of Wild Bustards: Emperor Frederick II was probably the first person to describe the feeding habits of wild bustards (1248 AD, translated by Wood and Fyfe 1943). Frederick II commented that bustards, when seeking their food, (herbs, grains and worms) grasp it with their mandibles, taking it as it is found on the surface of the ground.

Most bustards, including houbara bustards, are opportunistic omnivores and in the wild their diet reflects the local and seasonal abundance of plants and small animals (Schulz and Seddon 1996; Tigar and Osborne 2000). Kori and black-bellied bustards *Lissotis malanogastor* will even eat carrion and have been found at the road side eating corpses of birds (Collar 1996). Plants appear to be a more important source of food for houbara bustards during winter and early spring while animals, mostly invertebrates and small vertebrates, are more likely to be consumed in the late spring and summer (Cramp and Simmons 1987; Johnsgard 1991). In the wild, houbara bustard chicks are fed mainly on insects and small reptiles (Schulz and Seddon 1996).

Field studies of the gut contents of bustards have shown that vegetation constitutes a very significant proportion of their diet (Collar 1996). In the case of the Vigor's bustard *Eupodotis vigorsii,* animal material only makes up 6% of the aggregate volume and 12% of the aggregate mass of the intestinal contents (Collar 1996). Table 2 summarizes what is known about the diets in the wild of the species most commonly kept in captivity.

Lifespan: Longevity in the wild is largely unknown, but can approach 20 years or more in captivity in the smaller species, and 30 years or more in larger species. The average lifespan of great bustards is at least 12 years in the wild and 28–50 years in captivity (Carey and Judge 2000, Osborne and Martin 2001). Osborne and Martin (2001) comment that most animals have a shortened life expectancy in the wild and cease breeding before reaching the maximum age reached in captivity. The maximum reproductive age of great bustards is considered to be 20 years. Flower (1925) reported examples of a great, a Nubian *Neotis nuba,* and a white-bellied bustard living for at least 27, 10 and 9 years respec-

Buff-crested and Kori Bustards (includes food consumption of buff-crested)

Day	Buff-crested Weight (g)	B-c Food (g/day)	Kori Weight (g)	Day	Buff-crested Weight (g)	B-c Food (g/day	Kori Weight (g)
1	28	3.7	103	**42**	254	38.2	
2	29	6.2	98	**44**	267		1386
3	30	7.6		**45**	280	40	
4	35		115	**48**	302	38.4	
6	40	10.1		**51**	325	39	
8	45		222	**52**	339		1635
9	51	12.8		**54**	357	60.1	
12	64	16	354	**57**	384		
15	77	19.4		**60**	423		
16	90		472	**61**	435		1950
18	103	20.5	542	**63**	448		
20	112		630	**66**	472		
21	121	20		**67**	483		2140
24	133	14.6	756	**69**	494		
27	143	11.2	812	**72**	503		
30	161	23.1		**75**	521		
33	177	25.4	1074	**78**	527		
36	201	24.8		**79**	544		2460
39	231	30.1		**81**	561		

Table 2. Bustard Diets in the Wild

Arabian bustard *Ardeotis arabs*	Chiefly grasshoppers, but also locusts when swarming. Other invertebrates include beetles, crickets, and caterpillars. Small mammals, mice, nestling birds, grass, seeds, fruits of trees, succulent parts of plants, and Acacia gum are also taken.
Australian bustard Ardeotis australis	A wide range of vegetable material and animals including shoots, roots, leaves, flower heads, seeds, berries, molluscs, myriapods, arachnids, insects (grasshoppers, beetles, caterpillars), reptiles, young birds, and rodents.
Buff-crested bustard ***Eupodotis ruficrista***	Diet is poorly studied, but it is known to eat seeds, green herbage, berries, Acacia gum, and insects (Tenebrionidae, scarabs, beetle larvae, grasshoppers).
Black bustard	Diet is poorly studied, but it is known to eat vegetable matter, seeds, roots, shoots, and insects.
Heuglin's bustard ***Neotis heuglinii***	Diet is poorly studied, but known to eat Orthoptera, small vertebrates, berries, and vegetable matter.
Houbara bustard ***Chlamydotis undulata***	Vegetable matter includes fruits, seeds, shoots, leaves, flowers. Animals eaten include mainly Orthoptera, Coleoptera, Tenebrionidae, as well as other invertebrates, small snakes, lizards, and hatchlings of ground-nesting birds such as sand grouse and larks.
Great bustard ***Otis tarda***	Mainly plant material and invertebrates, although small mammals, amphibians, and nestling birds are sometimes taken.
Great Indian bustard	Mainly insects including grasshoppers and beetles, but also scorpions, reptiles, bird eggs, and small mammals. Favored crop plants include groundnut, rocket salad, millet, and Bengal gram.
Kori bustard ***Ardeotis kori***	A wide range of vegetable material and animals including seeds, berries, bulbs, Acacia gum, snails, insects, locusts, grasshoppers, dung beetles, rodents, lizards, snakes and birds (eggs, nestlings and road kills). Studies of the gut contents of kori bustards in Namibia have revealed mostly insects and very little plant material (T. Osborne, pers. comm.).
Little bustard ***Tetrax tetrax***	Animals eaten include beetles, grasshoppers and other terrestrial invertebrates. Plants eaten include shoots, clover, grain, leaves, flowers and seeds.
Nubian bustard ***Neotis nuba***	Invertebrates such as ants, locusts, beetles, caterpillars, grass seeds and vegetable matter, leaves, fruit and berries of desert plants, and Acacia gum.
White-bellied bustard ***Eupodotis senegalensis***	Animals eaten include termites, locusts, caterpillars, beetles, spiders, scorpions, snails, snakes and lizards. Vegetable material includes green herbage, grass seeds, bulbs, berries, and flowers.

Summary of the wild dietary requirements of some bustard species that are kept in captivity (Rutgers and Norris 1970; Rahmani 1987; Johnsgard 1991; Bhushan and Rahmani 1992; Collar 1996; Cornes 1997; Snow and Perrins 1998; Tigar and Osborne 2000; Tourenq et al 2003.) Reproduced with kind permission of the National Avian Research Center (Bailey, in press).

tively. To date, the record for longevity in kori bustards is a 30-year-old female and a 28-year-old male. Both birds are currently alive in the United States (Hallager 2004) and are reproductively active.

Conservation: Bustards are in need of conservation. The 1996 IUCN Red List and BirdLife International place several species of bustard in population risk categories:

- lesser florican *Eupodotis indica,* CR A1a (critically endangered) habitat loss, hunting, and drought
- great Indian bustard *Ardeotis nigriceps,* EN C2b (endangered) due to habitat loss and disturbance, and hunting of birds and eggs for food and sport
- Bengal florican *Eupodotis bengalensis,* EN C2a (endangered) habitat loss
- great bustard *Otis tarda,* VU A2c (vulnerable) habitat loss and hunting
- blue bustard *Eupodotis caerulescens,* LR/nt (near threatened)
- little brown bustard *Eupodotis humilis,* LR/nt (near threatened)
- Nubian bustard *Neotis nuba,* LR/nt (near threatened)
- little bustard *Tetrax tetrax,* LR/nt (near threatened).

Some species have scarcely been studied and so their true conservation status is unknown. Agricultural practices, including overgrazing, hunting, trapping, habitat loss, droughts and wars are the foremost threats facing all bustard species.

Management of Bustards: Bustards have been managed in menageries from historical times. In the nineteenth century great bustard eggs were frequently taken from the wild and hatched under domestic hens, so it is likely that this species was often found in captivity. Hopkinson (1926) noted that the first documented breeding of great bustards in captivity was in the Tyrol in 1858–1860. In 1889, a letter from Edward Meade-Waldo, an eminent ornithologist, to Herbert St. Quintin, a well-known falconer, describes the diet of captive houbara bustards kept in Tenerife (cited by Upton 2002). The earliest records of captive bustards being added to the collection at the Zoological Society of London (ZSL) are of houbara bustards caught in the Red Sea in December 1908 and in North Africa in May 1915 (Bradford 1908; Woodward 1915). However, it is likely that bustards were maintained at the ZSL before this, as Beddard gave descriptions of the alimentary tract of dead bustards in 1898.

Sexing Bustards: Bustards show sexual dimorphism. Some smaller species such as white-bellied bustards can be sexed by differences in head and throat plumage. Larger species, for example, kori, are easily sexed at one year of age, as males are considerably larger. Juvenile bustards can also be sexed using endoscopy after about six months of age.

Jarrett and Warren (1999) presented morphometric data to guide age and sex determination in houbara bustards. It is important to remember that there is a seasonal variation in body weight and food intake in some species, including houbara, kori, and Australian bustards. Houbara bustards are heaviest in winter, progressively losing weight towards summer (Jacquet 1997). Male Australian bustards have a 50% increase in bodyweight at the end of winter before they begin to display (Fitzherbert 1983). Kori bustard males can increase by 4 kg (9 lb.) during the breeding season (Hallager 2005).

Molecular sexing offers many advantages, being non-invasive and, if done from feathers or blood collected from freshly hatched chicks, it can allow different sexes to be reared under different protocols (D'Aloia and Griffiths 1999).

Large Bustards (kori bustard, great bustard): It is most convenient to consider the management of bustard species by dividing them according to size. Wild-caught birds are shy of humans, as are parent-reared captive offspring. Hand-reared birds are not as nervous around humans, and imprinted males have been known to display to their keepers. After shipping, newly arrived bustards may refuse to eat, although they can be encouraged to do so when provided with privacy and an abundance of live insects or baby mice provided continuously. The larger bustards are easily startled by people, noises, quick movements, ungulates, feral dogs and cats, and so on. When nervous, they bark a rough-sounding *Ahuff.* Large adult bustards appear to be susceptible to a variety of disorders similar to those found in ratites. Most of these problems are associated with improper handling and management (Bailey et al 1997).

Koris, especially adult males, are very powerful birds and require somewhat different handling skills from other long-legged birds of similar size. Historically, this species has been prone to leg and wing fractures at the time of capture. Capture-related miopathy following episodes of capture and handling has been observed in houbara and kori bustards. Early treatment with fluids, vitamin E/selenium, and physiotherapy are essential to recovery, as well as reducing the conditions that lead to this condition (extended pursuit times, transport, excessive handling and over-exertion during times of high heat).

Bustards are highly sensitive to stress and incorrect handling can cause:

- temporary or permanent neural damage
- hyperthermia
- fractures of legs or wings
- skin lacerations, bruising and feather loss
- luxation of the tibiotarsal bones
- dislocation of the cervical vertebrae
- compression of the trachea and internal organs
- stress and the progression of a disease process and even death

Handling: There are three aspects of large bustards that demand different handling.

1) The physical strength of the birds, especially adult male koris, which can weigh as much as 19 kg (42 lb.) during breeding season.

2) The thinness of their skin. Bustard skin is very thin and rips easily when a bird is struggling. Torn skin will result in further handling for sutures, etc., which can lead to maggot infestation.

3) Their inherent nature to shed feathers when captured. Bustard feathers are very loosely embedded in the follicles and even the best handling will result in lost feathers. Improper handling can cause major feather loss (Hallager and Boylan 2004).

Two people are needed to capture and handle a bird and, in the case of males, it is absolutely essential to avoid injuries. The initial catch is made single-handed and the second person is there to help bring the legs and wings under control. Adult males require that a second person be there to take additional control of the legs and wings. A person of small stature (male or female) may be incapable of adequately restraining an adult male kori, regardless of their skill and experience.

The following capture protocol is from the Kori Bustard SSP Husbandry Manual (Hallager and Boylan 2004):

- Captures that can be done in the early morning hours (especially in southern zones) are recommended to reduce the risk of heat stress to the birds.

- Equipment needed for handling birds consists of a dark hood or towel for the bird to reduce stress and struggling (note: hand-reared birds may prefer to remain unhooded), protective eyewear (if desired or an institutional requirement), gloves (if desired) and a chair for the handler (highly recommended if the bird is to be restrained in the same place for several minutes).

- If birds are to be bled (from the leg) when the temperature is below 4°C (39ºF) it is recommended that birds first be housed in a warm area for a minimum of 30 minutes to allow their legs to warm up. This will allow their blood to clot much quicker following blood collection than if the leg is cold, and will reduce the amount of time the bird must be restrained by as much as 10 minutes.

Safety Concerns for the Birds: Choose the preferred capture site. The capture is preferably done in a small, darkened and enclosed area such as a shed or stall with solid walls so that the chance of evasion is greatly reduced. Whenever avoidable, do not corner a bird against chain-link or welded-mesh fences as the birds are more likely to receive trauma to their beaks, heads, carpal joints and feet because they tend to try to "climb" a see-through barrier. Nets should not be used on large bustards. If catching a bird in an enclosed area, such as shelter or stall, you may wish to use the shift stall doors to your advantage and capture the bird as it walks through the door into the adjacent area. Again this is potentially dangerous to the bird and will require rapid and firm control of the bird as quickly as possible. The bird should be moved away from the walls to reduce injury. Padded walls may be beneficial in a capture area to reduce the chance of trauma. Even fixing corrugated cardboard to the walls can be beneficial although there are plastic alternatives for more permanent use. If birds are to be captured in their yard (not recommended), make certain that the fence line at the capture area is at least 2.1 m (7 ft.) high, preferably more. Even pinioned koris can jump high.

Safety Concerns for the Handler: The main cause of injury to a keeper from a large bustard will most likely come from the pushing out of the feet during restraint. They are very strong-legged birds and to tuck them into a position to carry them can be rather difficult. This is why it is necessary to have two keepers present to insure proper safety of the bird as well as the keepers involved. Goggles may be worn as an added safety measure to protect the eyes of the restrainers. It is unusual, but not unheard of, for kori bustards to poke at their holder's face.

Capture: The ease of capture will vary widely, depending on exhibit design. If the birds are accustomed to coming into a shed or stall daily for their food, there may be less stress in the capture because the birds can be enticed in with food. In the absence of an indoor capture area, use a small area with either trees or brush to assist in directing the animals into the catch area. This can also be created using a funneling system. If there is not an opportunity to funnel, use the boundaries of the exhibit (such as fences or walls) to direct the animal to your catch area. The catch area needs to be such that you can quickly gain control of the bird.

Once in the containment area it is very important to catch the bird quickly. The less the bird crashes around to avoid capture, the less it will be stressed and the less the chance of resulting injury. When cornered against a solid wall, most kori bustards will generally attempt to leap into the air and even a pinioned bird can jump at least 2 m (6.5 ft.) into the air. Ideally the bird will be caught before it actually takes the leap. If it does leap upwards try to catch it just as its feet get back onto the ground and before it can leap again. Never grab for the tail or wings, as you will be left with a handful of feathers and an even more panicked bird.

It is strongly recommended whenever possible not to catch a large bustard with a net as the risk of injury to the bird escalates considerably with this capture technique. Injuries will be greatly reduced if birds are guided into darkened areas and then caught by hand.

Restraint: The preferred method of restraint is fairly standard — tuck the body of the bird under one arm with the head at your back. The weight of the bird rests primarily on the holder's forearm while the other hand is used to restrain the legs. Legs are generally not tucked up under the body, as it is possible for a bird to break its leg if is too tightly restrained this way when it struggles. Restrain the legs at the tarsal joint with at least one finger keeping the legs apart so that they do not abrade the joints against each other. The second handler should immediately cover the eyes of the bird without blocking the nares or holding the beak shut so that the bird can breathe and pant.

Once in hand, it is best not to apply pressure to the body from above to make the bird "sit" as this may cause the bird to resist, as well as put unnecessary pressure on the legs. Birds will flap their wings and, since they are large birds, it is important to make sure the wings are tucked firmly against the body of the bird. It is not necessary to restrain the head (bustards do not poke like cranes), however, it may be in the best interest of the restrainer to have the head under control. When doing this, make sure to not block off the airway for breathing. A towel may be used to cover the eyes in order to calm the bird.

Transporting: For internal transfers on property, it is better if the bird can be hand carried, as this will reduce the problems associated with recapture once the bird is released from its crate. If there are more birds being caught than can be carried, the use of crates will be necessary. Use a crate that is padded on top (to protect the head), tall enough for the bird to stand upright and just narrow enough to restrict a large amount of movement. Crates work very well for transport, but removing a large bustard from a crate so that it can be handled is tricky. If the bird is to be re-captured once released from a crate, it is recommended to release the bird into a small area to let it calm down, and then recapture it again using the technique described above. If a bird is being caught for a procedure that will involve anesthesia, it can be helpful to begin the anesthesia at the exhibit so that the bird is already unconscious when it arrives at the clinic. Use of this technique will of course be up to the discretion of the veterinarian. However, it can greatly reduce the stress on the bird.

Release/Recovery: For the release of a manually restrained bird it is important to be cautious in regard to stabilizing the bird before release. Always release the bird in the direction of a clearing that allows it time to gain its balance and get its bearings. Before releasing the bird, position yourself so that the legs of the bird are touching the ground, keeping in mind they can jump high. Direct the bird towards your clear area. Once the bird feels stabilized, simply release the bird from your control and slowly step back from the bird. The bird will most likely move away from you, but they can just as easily stand there looking at you. In either case, moving towards the exit in a slow and deliberate manner is the best practice.

When recovering a bird from anesthesia, it will be necessary to restrain for a longer period of time. As expected, individual birds can recover at different rates and some discussion should occur as to what will work best for your team. If you are manually restraining the bird it will be important to maintain firm and constant pressure/control of the body and legs to restrict any sudden outburst of energy. This is common and using an eye guard (towel) will help in reducing this reaction. Once the animal is fully recovered from anesthesia you may release the bird as described above for a manual release, but be sure that the bird has regained sufficient control of its legs to be released.

When recovering in a crate, it is best to make sure the bird is safe from injury and has recovered completely before release should occur. If you move the crate during recovery, take the time to do this slowly so as not to startle the bird. Release is similar to the manual release, with the direction of the door to the crate positioned to a clearing that the bird can see. Open the door and step back so that the bird can step out of its own accord. Once the bird has moved away from the crate, simply move the crate out of the area while watching the bird for any negative reaction.

Medical Concerns: Although bustards do not have many unique infectious disease conditions, a wide variety of diseases have been reported in both wild and captive species (Bailey, in press). The diseases of houbara bustards have been best described, owing to the proliferation of well-funded breeding projects in the Middle East over the last 15 years. Although generally hardy birds, most large bustards in captivity die as a result of complications from puncture wounds or compound fracture of legs or wings, in addition to complications caused when housed with incompatible exhibit mates. Nervous kori bustards will run into, or pace in contact with, their perimeter fence or pen walls. It is important to use fencing materials that are smooth to avoid abrasions to faces, the bends of the wings, or pinion sites. They may compact the soil by chronic pacing, and may become prone to lameness and pododermatitis.

In our experience, kori bustards do not preen fly eggs or maggots off their wounds, and even superficial abrasions can open the door for lethal invasions. Despite the added stress to an injured bird, it is important to capture the bird and either place it in a fly-free room, or smear the area around the wound with a generous supply of fly-repellant gel such as that used for horses. Kori bustards do not thrive in damp environments and this should be taken into consideration before acquiring them (crane species make a wonderful substitute under these circumstances).

Significant viral diseases of bustards include Newcastle disease, avipox and avian influenza. Reovirus has been implicated in losses in young Heuglin's bustards *Neotis heuglinii,* in the United Arab Emirates. Other viruses isolated from bustards include adenovirus, Gumboro disease virus, herpes virus and PMV-2, but the significance of these findings is as yet unknown (Bailey, in press). Lymphoid leukosis has been reported in bustards. Serological surveys of houbara bustards in the Middle East have shown antibodies against Gumboro disease, pigeon herpes, infectious bursal disease type 2 and Marek's disease viruses (Bailey, in press).

Studies on captive and free-living bustards have shown that healthy bustards are colonized by a wide range of bacteria. *Proteus* spp., *Enterobacter* spp., *E. coli, Klebsiella* spp., *Aerococcus* spp., and *Enterococcus* spp. should be considered part of the normal aerobic intestinal flora of captive bustards (Silvanose et al. 1997). Gram-negative bacilli are the predominant aerobic microflora of developing bustard chicks (Naldo et al 1998). The most frequently reported bacterial diseases described in bustards include colisepticaemia, mycobacteriosis, clostridiosis, salmonellosis, yersiniosis, pseudomoniasis, and erysipelothrix. Chlamydophilosis is a common finding in houbara bustard rehabilitation projects in the Middle East.

Aspergillosis was one of the first mycotic diseases to be described in animals and *Aspergillus fumigatus* was first found in the lungs of a great bustard in 1863 (Fresenius 1863). Aspergillosis is common in smuggled bustards in the Middle East, but does occasionally occur in captive-reared chicks. Aspergillosis has also been reported in captive and free-living great bustards (Garcia-Montijano et al 2002). Infection of the oropharyngeal cavity and esophagus in bustards by the yeast *Candida albicans* often occurs as a consequence of antibiotic treatment. Infection occasionally occurs in juvenile bustards with an immature immune system, but is more common in adult birds housed in unsanitary conditions.

Helminth infections are some of the earliest diseases of bustards to be described (Beddard 1912, 1914; Dollfus 1957). Cestode infestations were found in 26% of bustards examined in one survey (Jones et al 1996) in the United Arab Emirates and heavy burdens are suspected to have contributed to morbidity in some birds. Ostrowski et al (1996) considered that heavy cestode infestations in houbara bustard chicks could cause weakness and nutritional deficiencies.

Furley worked at al Ain Zoo in the 1980s when large numbers of bustards were being brought into the United Arab Emirates for falcon training. He commented that in all species of bustards, newly arrived birds appeared to suffer an extraordinary degree from internal parasites and considered that all newly arrived bustards originating from the wild should be given antiparasitic medication (Furley pers. comm.). This advice holds good today and a number of subsequent authors have also made similar comments (Jones et al 1996; Bailey et al 2000; Schuster and Kinne 2003).

In the Middle East, trichomonosis is probably the most important protozoal disease of captive bustards. However, other protozoa, including *Eimeria* spp., *Giardia* spp., *Histomonas* spp., and *Entamoeba anatis,* have also been reported as causative agents of digestive tract disease (Silvanose et al 1997).

Preventive medicine programs in the Middle East include annual vaccination with inactivated Newcastle disease vaccine, live canary pox vaccine and regular (2–3 x year dependant on risk) anthelmintic and antiprotozoal medication given in the water or food. Vaccines are given during health-assessment catches held 1–2 months before the breeding season and during the cool season in the Middle East (Bailey and Hallager 2003). Kori bustards in zoo settings need to be carefully monitored for signs of impaction and zinc toxicity as they have a particular tendency to consume items thrown into their enclosures (e.g., pennies, camera batteries, nails, etc.) by visitors. Daily inspection of pens and the removal of foreign material are very important.

An excellent source of information concerning medical conditions, parasites, supportive care and therapy, anesthesia, and medication dosages is Bailey, T.A. (in press). Diseases and medical management of houbara bustards and other Otididae and Propagation of the Houbara Bustard. Recently, some zoos have been providing glucosamine to older kori bustards as means of slowing the effect of degenerative joint disease. There is evidence that this is effective in some birds.

Enclosure: Pen size must be large enough for bustards to escape keeper intrusion. The suggested minimum pen size is 13 x 20 m (42.5 x 65.5 ft.) per bird, and birds will be much better off if provided with a few hectares of space. The preferred boundary for large bustard enclosures is 2.5 cm (1 in.) chain link mesh. This size mesh reduces the chances of chicks getting out and large rodents getting in. The smaller welded wire also eliminates any chance of a bird getting a leg caught in the fence during a capture. Even though some birds are successfully maintained in exhibits with a 2.4–m (8 ft.) high fence, the preferable height is 3–3.7 m (10–12 ft.). Kori bustards are powerful flyers and even flight-restrained birds can escape a 2.4 m (8 ft.) fence when startled or on a windy day. Site barriers, such as thick shrubs placed 1.3–2.5 m (4–8 ft.) in from the perimeter walls, give the birds a sense of security in pens of minimum size.

Native/feral predators are dangerous for both adult and young bustards. Foxes and/or raccoons have attacked adult kori bustards, sometimes fatally, and small chicks left on exhibit usually disappear. Thus, enclosures must be built to minimize predator access. Digging predators (e.g., dogs, foxes) can be excluded by burying the base of the boundary 0.3 m (1 ft.) in the ground. Surrounding the enclosure with electrical wire can deter climbing predators, such as raccoons. Covered pens are necessary if hens are allowed to raise chicks. In areas where large predators (e.g., coyote, bobcat, cougar) are common, birds may need to be housed indoors at night.

Housing requirements: Facilities in zones where temperatures fall below 0°C (32°F) must have winter holding facilities available for housing birds (especially kori bustards) during inclement weather. Bustards are susceptible to frostbite and must not be left outside during periods of freezing rain or snow. Even facilities that do not experience extremely cold weather should have a shelter available both for koris and great bustards. The shelter can be used for times when pen repairs are needed, for medical confinement, to minimize food loss from wild birds, or when birds must be caught.

Kori bustards do not like to be confined, but the birds should be locked into a heated stall if the windchill temperature is forecasted to be –5°C (23°F) or below, especially if rain or sleet is expected, and always during snowy periods. For winter holding of two compatible kori bustards, the minimum recommended space per bird is:

- 2.4 x 3 m (7–10 ft.) for overnight holding
- 2.4 x 4.9m (7–16 ft.) for housing up to seven days
- 3 x 6.1 m (10–20 ft.) for housing longer than seven days

Outdoor holding yards should exist so that birds can be allowed outside access.

The recommended oblong dimensions allow for some exercise space, as well as allow the birds to distance themselves from keepers during shelter cleaning. These recommendations are highly dependent on the compatibility of cage mates. Larger dimensions or separate shelters may be required for birds that are have a lower degree of compatibility. Sheds should have the capability to be divided, in the event that birds are not compatible in small areas. Sheds should be heated to 10–15.5°C (50-60°F). Heat bulbs, if used, should be encased in protective wiring to prevent bulb breakage in case a bird makes contact with the bulb. Lights should be installed that can be turned on during the day and off at night. Skylights, which allow for natural light in the stalls, may be advantageous for birds that must be housed for extended periods. A small night-light will provide some degree of light for the birds during the overnight hours. The floor of the shed can be dirt or sand that can be covered with bedding hay (straw). Concrete is not recommended as birds can slip if startled. Sliding doors, operable by ropes outside the shed, are useful for controlling access without having to enter the shelter (Hallager and Boylan 2004).

Although zoos can often provide spacious pens, most do a great disservice to large bustards, especially the kori, by housing them in mixed-species exhibits with hoofed mammals and ostrich. We strongly recommend against housing this species together with known aggressive species based on data that incriminates this practice (Hallager and Boylan 2004). For optimal breeding and management, bustards are best housed by themselves.

Kori bustards do not thrive in climates that are consistently wet, rainy and damp. These conditions lead to poor feather condition and unhealthy birds. Likewise, exhibits that offer no areas of sun are also detrimental to bustard health. Kori bustards need areas of full sun to allow them to dry off damp. Additionally, areas of shade should be available, especially in hotter climates.

Social Organization: Captive female kori bustards develop a social hierarchy among themselves, and subordinate females may suffer attacks about the face and head. If given sufficient room, approximately 250 m^2 (820 sq. ft.) per bird, and a few well-placed thick shrubs that can act as visual barriers, females may be housed together. Males should be housed separately from each other. Dominance interactions come in both subtle and overt forms. In captivity, dominant males are known to attack and even kill subordinate males. More commonly, the subordinate male is chased away from food sources and is relegated to marginal areas of the enclosure and suffers chronic stress.

Water: Bustards require only small areas of water from which to drink. They are not heavy drinkers but do drink on a daily basis. Heated water dispensers for northern zones are recommended. Bustards do not bathe in water (they dust bathe) so pools are not needed in exhibits, except for aesthetic reasons. If pools are present in enclosures, they should be shallow enough that a bird can walk through the water, and the sides should gradually slope to the deepest portion. Pools deeper then 0.6 m (2 ft.) are not recommended.

Diet: Bustards are omnivorous, consuming mostly insect and plant material in their native habitat. Small vertebrates are consumed when available. A complete nutrient analysis of these food items and their contribution to the overall diet has not been determined. The gastrointestinal tract of the kori bustard is typical of an insectivorous bird, but to this point they have been fed in captivity as "primarily carnivorous" omnivores.

Suggested Diet Composition: Based on free-ranging ecology and gastrointestinal morphology, the recommended diet is based on nutritionally complete feeds, whole prey (vertebrate and invertebrate) and produce. Recommended crude protein in diets offered to captive koris should be 16.5–30.0% on dry matter basis. Based on the reported foraging strategy of free-ranging kori bustards, proposed diet proportion guidelines are presented in Table 3. These guidelines assist with diet formulation by proportion in order to insure that nutrient needs are met (and levels of specific nutrients are not grossly exceeded, i.e. protein). When ingredients are combined according to the proportions recommended in Table 3, diets can be formulated to meet the proposed nutrient guidelines for koris. (Maslanka and Ward 2004).

The diets listed in Table 4 are not recommended diets, but examples of how the proportions listed in Table 3 can be used to formulate diets that meet nutrient guidelines. A variety of ingredients can be chosen based upon availability, palpability, and management needs.

Feeding Schedule: Bustards should be fed at least once per day (many facilities feed twice a day), with additional feedings to allow for management and behav-

Table 3. Kori Bustard Recommended Diet Proportion Guidelines (as fed basis)

Item	Minimum, Percent of Diet	Maximum, Percent of Diet
Vertebrate prey	0	25
Invertebrate prey	5	30
Nutritionally complete feeds*	40	55**
Produce	10	20

* Nutritionally complete feeds are those designed to meet specific recommended nutrient levels.

** Diets that exceed 55% complete feeds can be considered. A diet comprised of 75% complete feed has maintained captive koris (Anderson 1995).

ioral needs. As long as birds have adequate time for food consumption, the period of access to the diet can range from several hours to all day. Minimizing the presence of pest species who will devour the food may shorten the periods of offering the diet. This should be considered when determining the period of time the birds have access to the food.

Food should be offered in such a way as to minimize depletion by pest species and insure complete consumption of the balanced diet by the birds. Diet can be offered in pans, tubs, buckets, platforms, etc., or hand-fed to individual birds in a group. Bustards will consume pelleted foods, so pellet dispensers should be placed in pens to encourage the consumption of dry, nutritionally complete feeds.

In the 1990s a flock of great bustards was kept at Whipsnade Wild Animal Park in the UK. These birds were fed at various times on proprietary concentrate pellets, Zoo A diet (Mazuri Zoo Foods, UK), a proprietary crane diet, and a formulated bustard diet (Special Diet Services) while fresh cabbage and cauliflower were fed ad libitum (Flach 1995). Locusts, crickets, mice, dog food, insectivorous bird food, mealworms, spinach, cress, spring greens and sprouting broccoli were also fed in order to increase the condition of the birds (Bailey and Flach 2003). Great bustards maintained at the breeding centre in the University of Poznan in Poland are fed a diet consisting of ground horse or beef meat, ground pigeon (with feathers and bones), boiled hens eggs, barley gruel, barley grain, wheat grain, wheat bran, oat flakes, dry green forage, ground carrot and cut onion (Bereszynski and Graczyk 2000). Bereszynski and Graczyk (2000) estimate that adult great bustards consume 5–10% of their body mass a day.

Great Indian bustards at Jodhpur zoo have been provided with a diet comprised of meat (liver, species not specified), wheat flour, spinach, alfalfa and bread (Makwana et al 1980). It was also observed that bustards

Table 4. Sample Diets that Meet Proposed Nutrient Guidelines for Kori Bustards (analysis on a dry matter basis).

Nutrient	Diet 1	Diet 2	Diet 3	Diet 4	Diet 5	Proposed Nutrient Guidlines
Protein, %	29.1	27.9	24.3	21.2	23.6	-
Fat, %	7.5	7.6	5.5	4.7	2.8	-
Crude Fiber, %	5.3	5.0	14.4	5.3	4.3	-
Ca, %	1.1	1.2	0.9	0.8	3.6	0.66–2.75*
P, %	0.9	0.8	0.9	0.7	1.8	0.33–1.0*
Ca:P	1.2	1.5	1.0	1.1	2.0	-
K, %	0.6	0.7	1.4	0.7	1.1	0.44-0.72
Na, %	0.1	0.1	0.2	0.1	0.5	0.13–0.18
Mg, %	0.1	0.2	0.3	0.2	0.4	0.05-0.06
Cu, mg/kg	13.1	13.8	25.6	14.2	8.9	5.5-8.8
Fe, mg/kg	187.4	1.7.7	291.0	264.4	182.0	55–77
Zn, mg/kg	93.2	121.6	194.9	114.7	158.5	55.5–70
Mn, mg/kg	42.7	83.4	183.1	5.3	127.1	66–72
Se, mg/kg	0.3	0.3	0.8	0.4	0.2	0.2
Vitamin A, IU/g	7.9	11.4	5.5	8.8	7.8	1.65–5.5
Vitamin D3, IU/g	0.5	1.8	1.4	2.1	2.7	0.22–1.2
Vitamin E, IU/kg	105.8	96.1	162.3	118.0	203.8	11.1–50
Thiamin, mg/kg	5.1	5.1	11.5	8.5	15.1	2.2
Riboflavin, mg/kg	4.4	2.7	8.9	5.3	13.0	2.75–4.4
Pyridoxine, mg/kg	7.6	4.0	11.9	5.5	12.8	3.3-5.0
Vitamin B12, mg/kg	0.01	0.01	0.02	0.01	0.003	0.003–0.01
Biotin, mg/kg	0.3	0.2	0.4	0.4	1.1	0.11–0.25
Folacin, mg/kg	1.6	1.5	5.4	2.8	0.1	0.8-1.1
Niacin, mg/kg	1.6	1.5	5.4	2.8	0.1	0.8-1.1
Pantothenic Acid, mg/kg	22.1	14.8	38.8	9.8^	47.3	10.5–17.6

Proportions listed by weight:

Diet 1: 25% mice, 25% crickets, 35% Zeigler Avian Maintenance (Zeigler Brothers, Gardners, PA 17324), 15% produce.

Diet 2: 15% mice, 25% crickets, 40% Mazuri Waterfowl Maintenance (PMI International, LLC, Brentwood, MO), 20% produce.

Diet 3: 0% mice, 25% crickets, 55% Zeigler Ratite Grower/Maintenance (Zeigler Brothers, Gardners, PA 17324), 20% produce.

Diet 4: 20% mice, 25% crickets, 55% Mazuri Exotic Gamebird Maintenance (PMI International, LLC, Brentwood, MO).

Diet 5: 10% beef, 75% Kock Kori Production Pellets (Breeder; Kock 1990), 15% produce.

Target nutrient levels established using ranges from domestic poultry (NRC 1994), and Anderson (1995). Breeding values are at the high end of the range and are indicated with an asterisk (*). ^ Values generated as a result of missing values in database.

caught and ate small birds, squirrels, lizards and insects in their enclosure.

At NARC and NWRC maintenance pellets (14% protein) are fed to houbara bustards outside the breeding season and productioner pellets (22% protein) are fed during the breeding season (Paillat and Gaucher 1996; Sleigh and Samour 1996). Additional live food such as mealworms or crickets is supplemented as part of taming protocols to reduce keeper-induced stress.

Captive bustards will also eat invertebrates attracted to vegetation in naturalistic aviaries. Beef mince has been used to replace mice or mealworms if either component is unavailable. It should be noted that meat is perishable and, particularly in hot climates, needs to be handled carefully to prevent spoilage (Hallager and Boylan 2004).

Breeding: The presence of more than one male may be crucial to stimulating successful reproduction in certain species. Several zoos have had success by housing female kori bustards in an enclosure that is adjoined by two or more enclosures that each contains a single male. The males are stimulated by hearing the vocalizations and seeing the postural displays of their competitors. The females are presumably stimulated by the opportunity to choose between males. Females gain access to the males by means of openings in the fence, 75 cm (29.5 in.) wide x 57 cm (22 in.) high, located midway along the dividing fences. The openings are large enough for the females, but too small to allow the males passage. It should be pointed out, however, that more than one zoo, including the Audubon Zoo, Cameron Park Zoo, Dallas Zoo and Smithsonian National Zoological Park (all in the United States), have also had successful breeding when housing only one male in the collection. Copulation lasts only a few seconds, and is performed on the ground. It is preceded by up to 20 minutes of head pecking by the male to the recumbent female. Once over, the male leaves to resume displaying. He plays no part in incubation or chick rearing (Hallager 2003).

A female entering reproductive condition creates a nest scrape, usually placed next to a shrub or large rock for camouflage. She sits tight on the nest scrape around 2–4 days before laying. All other females should be separated from her at this time. The 1–2 pale-to-dark olive eggs with brownish markings are laid at two-day intervals. Some females begin incubation with the first egg, while others begin after the arrival of the second egg. Several eggs may be produced by immediately pulling newly laid eggs. Clutch recycling occurs in about 14 days. Egg dimensions in kori bustards in the United States average 82 x 57 mm (3.2 x 2.2 in.) and average weight is 149 g (5.2 oz.). Incubation is 22–24 days. The female alone incubates the eggs, leaving occasionally to eat. She returns quickly to the nest. Preliminary data using a telemetric egg (Advanced Telemetry Systems, Inc. model R2100) show that female kori bustards at the Smithsonian National Zoological Park rotate their eggs 1–26 times per day, with a median turning rate of eight times per day. Eggs are also turned at night (Hallager, unpublished data).

During incubation, the food bowl should be placed as close to the incubating hen as possible without causing a disturbance. She may defend her nest. Females have been known to leap up from their nest, roaring and kicking at the keeper who attempted to observe an egg. To get hold of the egg it is best to reach under the female from the rear. The presence of a second person is recommended in case the female becomes aggressive.

Bustard chicks begin vocalizations at internal pip. Hatch occurs about 48 hours after internal pip, 12–24 hours after external pip. Chicks are precocial and nidifugous at birth. Following birth, chicks are usually dry within an hour or two. Eyes are open upon hatch, although they have an opaque appearance to them for the first several days. Chicks are somewhat mobile approximately 4–6 hours after hatch. They are not fully mobile until they are 24 hours old.

Artificial Incubation and Rearing: For artificial incubation, the incubator dry bulb temperature is 37.5°C (99.5°F) using 70–75% humidity. Target weight loss for three bustard species is presented in Table 6. Adjustments are made to the humidity to achieve the targeted weight loss. Eggs are turned every hour. The thickness and pigmentation of the egg makes candling

Table 5. Average Weight Loss During Artificial Incubation of Different Bustard Species

Bustard species	Weight loss (%)
Kori	14.9
Buff-crested	12.3
White-bellied	16.0

(Source: Anderson, 1998a; 1998b, 1998c)

Table 6. Incubation Parameters (mean, range) in Bustard Species.

Bustard Species	Start of Incubation to Internal Pip (days)	Time to External Pip (hrs)	Time to Hatch from External Pip (hrs)	Source
Houbara	22 (21-22) n=8	9 (2-22) n=8	23 (12-31) n=13	Bailey and Anderson 2000
Kori	21 (20-22) n=10	13 (3-24) n=10	26 (6-39) n=20	Bailey and Anderson 2000
Buff-crested	19 (17-24) n=15	13 (4-24) n=14	18 (6-44) n=34	Bailey and Anderson 2000
White-belly	19 (19-19) n=3	19 (14-24) n=2	26 (8-38) n=5	Bailey and Anderson 2000
Great	nd	nd	23	Seidal 1995
Little	nd	nd	22	Seidal 1995

difficult, but candle every five days until the air cell begins to drop down. At that point, increase humidity to 90% and candle the egg daily. At internal pip, increase the humidity to 95–98%. Hatching should occur within 24 hours after external pip.

Incubation Parameters for Bustards: The incubation parameters for houbara, buff-crested, kori, and white-bellied bustards are listed in Table 5. If bustard chicks greatly exceed these pipping intervals, or are weak or not progressing properly, radiography, endoscopy and ovotomy should be used to determine the position and status of the embryo. These assisted hatch techniques are described in more detail by Bailey and Anderson (2000). Experience shows that most of the chicks that have assisted hatches have a lower chance of survival.

Parent Rearing of Kori Bustards: If the egg has been artificially incubated, pull the dummy and replace the real egg at internal pip. Leave the chick and mother alone for the first 24 hours to allow bonding to occur. The next day, briefly examine the chick, weigh it, swab its umbilicus with betadine, and check the status of yolk absorption before returning it to its mother. Weigh the chick on the second, fourth and sixth days after hatching, then every four days after until the chick is too big for the scale. Young chicks will sit nicely in a large, towel-lined crock, but become more restless with age. They will softly vocalize to the mother while being handled.

The chicks may lose 10% of their body weight the first two days, but then should gain from 2–18% body weight per day thereafter. Continued weight loss may indicate a need for supplemental feedings. Chicks sometimes develop slipped wing due to the high protein diets provided, but taping the wing will correct the problem within a few days if started immediately. Initially, the female offers food items to the offspring bill-to-bill. The chick begins to peck at food on its own after four or more days. Until the chick weighs 1 kg (2.2 lb.), pinkies should be substituted for full-grown mice in the mother's diet; mouse fur may cause impaction. Similarly, giant mealworms should be left out of the diet and the hind legs from crickets removed. Chicks may stay with the mother up to the next breeding season.

Hand Rearing of Kori Bustards: Hand rearing is the recommended method for raising kori bustards that will be tolerant of human disturbance and handling. Future reproduction has been shown not to be compromised in kori bustards. Once the chick is hatched, clean the umbilicus with Betadine (100%) solution. If the chick has had difficulty hatching, 2 cc of half-strength Lactated Ringers and 2.5% dextrose solution should be provided subcutaneously. After being allowed to rest for several hours in the hatcher, move the chick to an Animal intensive care unit, 69.85 x 33 x 35.5 cm deep (27.5 x 13 x 14 in.) at 36.1°C (97°F). The floor of the brooder should be carpeted and a feather duster hung in one corner to simulate the mother. A small mirror affixed in the brooder acts as a visual stimulus for the chick. Line the floor with towels for frequent cleaning.

Growth Rates in US Kori Bustards: Weights from 30 hand-reared males, 20 hand-reared females, six parent-reared males, and three parent-reared females born in US zoos were used to produce graphs of weight gain and growth rate in kori bustards (Hallager and Boylan 2004). It is evident that males gain weight faster than females. Though parent-reared males gain weight faster than hand-reared males, parent-reared and hand-reared females gain weight at similar rates (Figure 1).

At hatch, males are slightly larger than females ($p=0.05$) but much overlap exists. At day 30, males are larger than females ($p=0.003$) but some overlap exists. By day 60, males are much larger than females ($p=0.000$) and almost no overlap exists in weights (Figure 2).

Parent-reared chicks seem to have higher growth rates than hand-reared chicks during the first week of life. The sample size is small, so conclusions are not robust. By 30 days of age though, all chicks have similar growth rates.

Captive-reared Bustard Chicks at NARC: Growth rates of captive-reared chicks at NARC are pre-

Hand Rearing Protocol for Kori Bustard Chicks

Day 1

Brooder temp: 36.1°C (97°F)

Brooder: 69.85 x 33 x 35.5 cm deep (27.5 x 13 x 14 in.)

Floor is carpeted and a feather duster is hung in a corner to simulate the mother. Single chicks are given a mirror.

Number of Feedings: 11

Diet: Each feeding:

1–2 cricket abdomens (from freshly killed crickets)

1–2 small pieces watermelon

1 green bean

Later feedings on day 1:

- Add soft fruit mix with apple paradise (offer both fruit and pellets). Do not feed more than five apple paradise-sized pellets.
- To ensure proper hydration, water is given via a syringe (no needle) at every feeding in addition to the small pieces of watermelon.
- Do not feed for the first 12–18 hrs after hatch.
- Feed every 60–90 minutes, 0600–1730 h (6 a.m.–5:30 p.m.).
- Puffy legs are a good sign of hydration, if skin on the legs appears tight, hydration is poor and chicks must be given SQ fluids.
- Lightly dust crickets with dicalcium phosphate.

Comments: Healthy chicks will attempt to grab offered food items. Coordination is not always well developed, however, and some chicks may need to be force fed for one or two days. Generally, if chicks are not attempting to grab food items, this is a sign they may be dehydrated. Chicks are fed every 60–90 minutes the first day beginning at 6:00 a.m. and ending at 5:30 p.m. Chick weights are recorded before the first feeding of the day from day 1 to day 30. Because hand-reared chicks do not have the constant presence of a mother, they will cry persistently if left alone. To reduce stress and provide comfort, keepers should brood chicks at every feeding and as often as possible during the first seven days.

Day 2–3

Brooder temp: 36.1°C (97°F)

Brooder: 69.85 x 33 x 35.5 cm deep (27.5 x 13 x 14 in.)

Floor is carpeted and a feather duster is hung in a corner to simulate the mother. Single chicks are given a mirror.

Number of Feedings: 7

Diet: Total diet for days 2 and 3 (each):

1 pinkie mouse (1–1.5 g). Chicks can take whole pinkies.

12–15 cricket abdomens

15–20 small watermelon pieces

Soft fruit mix (including ~ 12 apple paradise pellets)

Chopped greens

3 green beans

If chicks are well hydrated, they may be offered soaked pheasant starter and soaked dog food on some of the later feedings on day 2 or day 3.

Comments: It is important to keep the chick hydrated. Chicks are pinioned on day 3 depending upon their weight, activity, and health status. Chicks are fed at two-hour intervals from 6:30 a.m.–5:30 p.m. Healthy chicks normally lose 3–5g on day 2, but their weights stabilize on day 3 and increase thereafter. Chicks can stand and walk on day 2, and readily accept food items presented by hand. Water is given using a syringe (do not use a needle) by drops. Try to train them to drink from dish on their own by using the syringe to lead them to water dish and to get them to peck at the water. Do not offer mealworms until chicks are seven days old. Continue to brood chicks.

Day 4

Brooder temp: 35°C (95°F)

Brooder: Carpeted nursery area, 1.8 x 3.7 m (6 x 12 ft.)

Three heat bulbs suspended from the ceiling so that three separate sections of the floor are kept at 35°C (95°F).

Several feather dusters are suspended near the heat bulbs.

Single chicks are given a mirror. Mirrors can be used with multiple chicks but this may cause some chicks to be agitated. If so, remove the mirror.

Number of Feedings: 7

Diet: Total diet for day 4:

1 pinkie mouse (1–1.5 g each) dusted with Ca2P.

1–1.5 g carnivore meat (rolled into balls). Do not put meat in tub with bird; ration out throughout day.

Offer daily (proportions will vary with chick). Marion scenic pellets, Zeigler Bird of Paradise pellets, green beans, peas, crickets, watermelon, apple, kale, and banana, crickets, waxworms, small watermelon pieces, soaked pheasant starter, soaked dog food, chopped greens.

Comments: Feed every two hours, 6:30 a.m.–6:30 p.m. Chicks must be encouraged to exercise in the pen following each feeding to avoid problems such as slipped tendons. Chicks should be trained to drink from a dish on their own. Use a syringe or a finger to lead them to water dish and to get them to peck at the water. Floating pieces of greens on the water will encourage them to drink. Use a shallow water bowl and a rock in the bowl to prevent chicks from accidentally falling in and becoming wet. Do not overfeed chicks with favored food items such as crickets. Emphasis should be on the feeding of fruits and vegetables at each feeding. Continue to brood chicks.

Day 5–6

Brooder temp: 35°C (95°F)

Brooder: Carpeted nursery area, 1.8 m x 3.7 m (6 x 12 ft.)

three heat bulbs suspended from the ceiling so that three separate sections of the floor are kept at 35°C (95°F).

Several feather dusters are suspended near the heat bulbs.

Single chicks are given a mirror. Mirrors can be used with multiple chicks but this may cause some chicks to be agitated. If so, remove the mirror.

Number of Feedings: 7

Diet: Total diet for day 5 and 6 (per day):

1 pinkie mouse (1–1.5 g each) dusted with Ca2P.

1–1.5 g carnivore meat (rolled into balls). Do not put meat in tub with the bird, ration out throughout day.

Offer daily (proportions will vary with chick). Marion scenic pellets, Zeigler Bird of Paradise pellets, green beans, peas, crickets, watermelon, apple, kale, and banana, crickets, waxworms, small watermelon pieces, soaked pheasant starter, soaked dog food, chopped greens.

Comments: Feed every two hours, 6:30 a.m.–6:30 p.m. Chicks must be encouraged to exercise in the pen following each feeding to avoid problems such as slipped tendons. Chicks should be trained to drink from a dish on their own. Use a syringe to lead them to water dish and to get them to peck at the water. Floating pieces of greens on the water will encourage them to drink. Use a shallow water bowl and a rock in the bowl to prevent chicks from accidentally falling in and becoming wet. By day 5, chicks are keen to pick food floating in water, so to encourage self feeding, some food is left in shallow water bowls during the day. Emphasis should be on the feeding of fruits and vegetables at each feeding. The need for brooding diminishes around day 6. Chicks will now object strongly when held.

Day 7–13

Brooder temp: 35°C (95°F)

Brooder: Carpeted nursery area, 1.8 m x 3.7 m (6 x 12 ft.)

Three heat bulbs suspended from the ceiling so that three separate sections of the floor are kept at 35°C (95°F).

Several feather dusters are suspended near the heat bulbs.

Single chicks are given a mirror. Mirrors can be used with multiple chicks but this may cause some chicks to be agitated. If so, remove the mirror.

Number of Feedings: 7

Diet: Increase to: 2 pinkie mice (1–1.5 g) daily

2–3 g carnivore meat (rolled into balls)

On day 7, the amount of pelleted food is increased depending on appetite, with a continued emphasis on fruit and vegetables.
On day 13, feedings are reduced to six per day.

Comments: Feed every two hours, 6:30 a.m.–6:30 p.m. If mealworms are fed, limit the total number fed each day to 10, to reduce the risk of impaction. Hand-reared chicks may develop slipped wings (i.e., an outward turning of the manus) anywhere from day 7–11. Slipped wing is easily and permanently corrected if the primaries of the affected wing(s) are taped to the body in a natural position for 7–10 days at the first sign of the problem.

Day 14

Brooder temp: 35°C (95°F)

Brooder: Carpeted nursery area, 1.8 m x 3.7 m (6 x 12 ft.)

Three heat bulbs suspended from the ceiling so that three separate sections of the floor are kept at 35°C (95°F).
Several feather dusters are suspended near the heat bulbs.
Single chicks are given a mirror. Mirrors can be used with multiple chicks but this may cause some chicks to be agitated. If so, remove the mirror.

Number of Feedings: 6

Diet: Same as above, but the number of mouse pups (5–6 g) is increased to four per day.

Comments: Feedings are reduced to six per day, 6:30 a.m.–6:30 p.m.). Chicks are taken outside for the day if the temperature is above 24°C (75°F). Once chicks have been given access to outside yards, they are carefully monitored for the ingestion of foreign material that could result in impaction. Watch chicks to be sure they do not consume too many small pebbles. Chicks normally will seek out tiny pebbles as a source of grit. While a few pebbles are acceptable, ensure that chicks do not try to consume stones, which are too large for them. Ensuring that chicks continue to defecate normally is important.

Day 14–30

Brooder temp: Same as above

Number of Feedings: 6

Diet: Same as above

Comments: Move the heat lamps up as the chicks grow so that they do not burn the top of their head. Also, beware of chicks jumping- they can jump very high and if heat lamps are too low, they will hit the bulb and break it.

Day 30+

Brooder/Temp: When chicks are 30–40 days old, they are moved outside to a covered yard measuring 5 x 15 m (16 x 49 ft.). The young birds spend the day outside and are housed in a heated shed at night.

Number of Feedings: 5

Diet: The number of mouse pups (5–6 g) is increased to six per day. The diet at this point is similar to the adult diet — mainly small mice, carnivore meat (mixed with crane and ratite pellets), Marion scenic jungle pellets, chopped kale, eggs, mixed vegetables and fruit.

Comments: If chicks are scale trained, weights may be continued. Otherwise, daily weights are discontinued around day 30 to minimize the risks associated with repeated handling.

sented in Table 9. In a review of a large sample of captive-bred houbara bustards at NWRC several factors including: sex, year of hatch (i.e. different management practices) and size of group, were found to cause variability in growth rates between years (Heezik and Ostrowsk, 2001). Houbara chicks kept in groups of eight had slower growth rates until fledging, compared with chicks reared as groups of five (Heezik and Ostrowsk, 2001). In the early days of NARC bustard chicks were reared singly and may have been an important stress factor for these birds (Deeming 1995). This may have contributed to the poor growth rates and susceptibility to diseases seen in these birds. It is thought that early differences in growth rate may have long term effects on reproductive performance, although further studies are required to demonstrate this. Interestingly, houbara chicks that were regularly weighed had a reduced risk of mortality during their time in the rearing unit (Heezik and Ostrowsk 2001). Regular weighing is thought to accustom birds to handling and human presence and may make birds less flighty and prone to trauma. Anderson (1998c) reports that musculoskeletal problems were seen in buff-crested bustards that had growth rates in excess of 10% during one season and, as a consequence, target growth rates for this species are recommended at less than 10%. Rearing regimens may also influence long term reproductive performance, and this is another important area for consideration by bustard projects aiming to improve productivity and lower costs.

Small Bustards (buff-crested bustard, white-bellied bustard, houbara bustard).

Capture: Correct methods must be used to catch and handle bustards to avoid injuries. Frederick II described the escape responses of bustards in his chapter on avian means of offence and defence in *Arte Venandi cum Avibus* (1248, translated by Wood and Fyfe 1943):

> "There are birds that springing up from the ground, attempt to throw an attacking bird of prey beneath them, among these are the large bustards and the lesser bustard."

Moody (1932) commented that little bustards damage their heads by their habit of springing up in the air when alarmed. These predator-escape responses make bustards vulnerable to traumatic injuries in captivity. Male buff-crested bustards also have spectacular aerial displays (Johnsgard 1991) and trauma has occurred in these birds during the breeding season, presumably by damaging themselves against the roof of their pens. Captive bustards, particularly birds originating from the wild as adults, are highly susceptible to trauma-related death or injury. This can be minimized through corrective husbandry, well-considered design of aviaries, minimization of unnecessary disturbance, and the implementation of protocols to diminish the sensitivity of birds to stress.

In a study at NARC, traumatic causes of death (intra-specific trauma, fractured neck, euthanasia due to fresh fractures, and dislocations and aviary-related trauma) were responsible for 23.5% of the deaths of captive adult houbara bustards, and 57.1% of the deaths of captive adult buff-crested bustards (Bailey et al 1996a, 1996b). Some 60% of deaths of captive adult houbara bustards at the NWRC were caused by trauma (NWRC 1993; Ostrowski and Combreau 1995). Stressed birds and birds that have not been pinioned or had their primary flight feathers clipped appear to be particularly susceptible to trauma. Disturbed or frightened buff-crested bustards have been observed to fly vertically and are prone to breaking their necks. Consequently, attention to the design of facilities and the behavior of staff working with the birds is important to minimize trauma-related problems. Trauma can be reduced by:

- Using plastic coated foam padding to surround the sides of cages or pens, for example in areas where birds are regularly caught, such as in hospital or quarantine pens. This minimizes abrasion injuries to wingtips.
- Using shade-cloth or tension netting on the roof and sides of aviaries to cushion the impact resulting from birds flying inside a pen.
- Modifying behavior by taming nervous individuals or housing such birds in naturalistic pens with cover.
- Ensuring that stock that has not been pinioned is regularly feather cut.
- Preventing stress by reducing the number of non-essential people who visit the birds.

Additionally, it is thought that houbara bustards derived from migratory populations, such as the captive stock originating from Kazakhstan at NARC, may be more restless during the migratory season and more vulnerable to trauma-related deaths (Combreau pers. comm.). In the future, chemical modification of behavior, a technique widely used in mammalian wildlife species, may be an option for managing confiscated wild-caught bustards that are being rehabilitated. However, in our experience wild-caught houbara bustards appear to be refractory to even high doses of tranquillizers such as haloperidol, zuclopenthixol acetate and perphenazine that have been successfully used to modify behavior in psittacines (Chitty 2003, haloperidol) and game species (Swan 1993, haloperidol, zuclopenthixol acetate and perpehazine). It remains to be seen if tamer captive-reared houbara bustards are susceptible to the effects of behavior-modifying tranquilizers.

For birds maintained in outside aviaries in the Middle East, as well as North American zoos in the southwest, catches should be carried out early in the morning to reduce the risk of heat stress. In the Middle East, multivitamin solutions are given in the water before large catches to reduce the potential for capture-related myopathies. Single small- to medium-sized bustards in small outside aviaries are caught with nets. Single tame bustards in small indoor aviaries are caught by hand, after herding the bird into a corner. Flocks of birds in large aviaries can be captured using a catching corral. This is a blind-ended funnel with a wide mouth and a circular catching area at the blind end. Small and medium-sized bustards can be caught using nets in the blind end (Bailey and Hallager 2003).

Restraint: Birds should be held firmly against the handler's body, the wings should be fully closed and the legs held together, but prevented from rubbing against each other by positioning fingers between them. Legs can be held so that the legs are bent at the tibiotarsal joint. Some movement of the leg is allowed without letting the bird kick uncontrollably. Hoods or cloth bags with holes for the nares can be used to calm birds down although hand-reared birds may be more comfortable without hoods. Birds should be kept as sternal as possible while being held, as this reduces stress levels (Bailey and Hallager 2003).

Transporting: Small bustards up to 2 kg (4.4 lb.) can be transported in commercial pet carriers or cardboard boxes. Crate requirements for all bustard species can be found in the International Air Transport Association manual (IATA 1998).

Medical Concerns: Unlike larger bustards, trauma to wings and legs is less of a problem in the small species of bustards. One of the main concerns is scalping and trauma to females by overly aggressive or hand-reared males. A nervous bird may incur injury if flushed upward into the ceiling; also males of some species fly directly upward as part of their courtship display. Wounds are also apt to get fly blown in the smaller species. It is safest to house male and female bustards separately, in adjacent pens which allow visual access.

Enclosure: The enclosure may be as small as 5 x 3 x 2.5 m (16 x 9.8 x 8 ft.) high. A small gate, which can be fastened open, is built into the dividing wall. The gate opening should be at ground level with dimensions 13 cm (5 in.) wide x 20 cm (7.8) high. Plant grass and shrubs as hiding places, but leave some areas and trails with bare dirt. A nervous bird is less likely to fly up to the ceiling if it can hide in vegetation during servicing of the enclosure. Outdoor aviaries may have alfalfa beds in addition to natural vegetation. This provides birds with cover to hide in and also encourages foraging for invertebrates attracted to the vegetation. Smaller bustard species can be exhibited in mixed aviaries (Bailey and Hallager 2003) but caretakers should be aware of the risk to small fledglings of other species, as all bustards are opportunistic carnivores.

Housing Requirements: In temperate climates, small bustards should have continuous access to a heated stall and should be shut inside if the wind-chill temperature is forecasted to be –1°C (30°F). Differences exist in the housing of bustards both between and within species and according to the region in which the birds are maintained in captivity. In the Middle East, where captive

Table 7. Food Consumption and Growth Rate (Day 1-14) of a Male Kori Bustard at the Smithsonian National Zoological Park.

Day	Food (g/day)	Weight (g)	Day	Food (g/day)	Weight (g)
1	6.2	99.2	**8**	49.1	171
2	16.38	100.3	**9**	53.4	190
3	37	103.6	**10**	71.4	220
4	30.6	110.3	**11**	56.6	251
5	24.6	122	**12**	73.3	278
6	34.8	138	**13**	53.2	314
7	44.6	155	**14**	51.5	368

Table 8. Commonly Encountered Problems of Hand-reared Chicks

Problem	Recommendation
Day 1-13	
Dehydration	Newly hatched chicks can be prone to dehydration for the first 2–3 days of life. Ensure that chicks are properly hydrated by feeding watermelon or dipping food items in water immediately prior to feeding.
Curled toes	Chicks may hatch with inward pointing toes. This condition typically corrects itself. Taping is only necessary in cases where the condition is severe or worsens with time.
Pinioning	Pinioning is recommended when chicks are 2–3 days old. Future fertility will not be compromised as copulation in the species occurs on the ground. Additionally, birds will not need to be routinely captured for feather clipping, reducing capture-related injuries.
Exercise	Chicks should be exercised following each feeding session and as often as possible throughout the day. This will minimize problems like slipped tendon. Additionally, chicks should be allowed outside after day six for exposure to sunlight.
Housing	If they differ in age by less than 14 days, chicks benefit from being housed together. If they differ in age by more than 14 days, chicks must be housed apart because of aggression from the older chick. Single chicks benefit from a mirror in their pen, or a companion (e.g. quail or guinea fowl chick).
Handling	At a young age, chicks are less stressed if handling simulates brooding, such as cupping in one hand and covering with the second hand, keeping it in a "nest" with a feather duster on top, or covering it with a breathable cloth.
Angel wing	This condition typically begins at day 11 and as early as day seven. Taping the affected primaries in a natural position at the first sign of outward turning will permanently correct the deformity. A definitive cause of the problem has not been determined.
Day 14-29	
Ingestion of foreign material	Chicks normally eat small pebbles to aid in digestion. When exercising on natural substrates, monitor chicks closely to ensure that they do not consume too many or any large pebbles, which can lead to impaction. Additionally, closely monitor defecation.
Handling	Bustard chicks have very sharp nails. When handling chicks, keep the feet away from the body to avoid injuries (e.g. a ripped neck) to the chick caused by the bird kicking.
Day 30+	
Eye pecking	One bird may peck at the eyes of another bird. The aggressor must be removed until the eyelid of the injured bird has healed. Providing the birds with food items to pick at (e.g. slices of watermelon, tomatoes etc.) may lessen the behavior.
Weighing	Do not weigh healthy chicks daily. This will reduce injuries caused by frequent handling. If chicks are trained to stand on a scale, then daily weights may be continued.
Housing	Avoid housing in enclosures with barriers (e.g. trees) that the birds may run into, or large-mesh chain-link. These may lead to injured or broken wings. When introducing to a new enclosure, have a familiar keeper acquaint chicks to a new enclosure immediately after releasing them. This entails walking around with the birds and showing them plants, fences, walls, dishes, etc
Capture	Avoid catching birds against a fence. Rather, walk the bird into a dark stall or sectioned-off area of the enclosure. Firmly, but loosely, grab the bird around the body and tighten in when the wings are properly secured. Another person needs to help secure the legs and head. Handreared birds often prefer to remain unhooded, whereas others are calmer if a breathable cloth is loosely draped over their heads. Some birds are calmer with legs stretched and feet on the ground (still being held securely), whereas, others are calmer sitting on the ground with legs folded. A keeper must constantly monitor and restrain the bird if it jumps up. Most birds are least comfortable when held up off the ground.

houbara are managed in an extreme climate outside of their natural breeding range, they are maintained in environmentally controlled buildings. In North Africa, where captive houbara are bred within their natural breeding range, birds are maintained in outdoor aviaries. In the United States, smaller species such as the buff-crested bustard and white-bellied bustard are kept in both indoor and outdoor aviaries, and provided with supplemental heat during the winter months if housed outdoors and/or moved into winter holding facilities. The provision of shade and shelter are important for birds managed outdoors in tropical and temperate climates respectively, as is the use of predator-proof fencing material. Supplemental heat must be supplied when temperatures drop below 4ºC (39°F) (Bailey and Hallager 2003).

Diet: All small bustards are omnivorous with a well-developed caeca, indicating their dependence on plant material. Insects form a large part of their diet and include beetles, beetle larva, grasshoppers, ants, and centipedes. They also eat berries, fruits, and seeds. Small vertebrates are consumed when available.
Daily diet for one adult buff-crested bustard (Dallas Zoo):

- 25 g softbill pellet or standard dog food, soaked
- 6 juvenile mice (fuzzies)
- 20 mealworms or waxworms
- 10 g chopped fruit (apple, melon, papaya, raisin)
- 10 g chopped vegetables (carrot, yam, kale, peas)
- Supplement with Vitamin/Calcium powder.
- Place food items in a shallow pan set on the ground.
- Place water in a bowl or tub.

Stiévenart (2002) described the rearing procedure of houbara bustard chicks at NWRC. The diet for the first 45 days consists of:

- Dry pellets (dry matter (DM) 88%, crude protein= 22% DM, calcium=1.5% DM, phosphorus= 0.8% DM).
- Fresh alfalfa (dry matter 24%, crude protein= 4.8% DM (Anderson 1995).
- Mealworms *(Tenebrio molitor)* (dry matter=42.2%, crude protein=22.3% DM (Jones et al 1972).
- "Humid pellets" SDS (Special Diet Services) pellets (crude protein=23.5% of dry matter) swollen by immersion in water enriched with a vitamin B complex solution and containing about 75% of water.
- Crickets *(Acheta domestica)* (dry matter 31.7%, crude protein=17.8% DM) (Anderson 1995).

Chicks are given mealworms (one meal every three hours from sunrise to sunset) upon entry to the nursery in order to stimulate pecking behavior (Stiévenart 2002). This period of limited food supply ends after the first night. Afterwards, chicks are hand fed with small pieces of humid pellets, mealworms, alfalfa and dry pellets. From one to 10 days, chicks are given six meals per day, including these four food items, and are progressively accustomed to feed independently. From 10 days of age, chicks are able to pick up food items independently. They are stimulated to eat dry pellets, while fresh food is offered during taming activities at least four times a day. Fresh alfalfa, dry pellets and water should be available to the chicks ad libitum four days after rearing groups of chicks are formed.

Mealworms and crickets contain levels of calcium and phosphorus that are inappropriate for normal bone development in young animals (Allen and Oftedal 1989). Ca:P imbalances in the diets of growing animals result in metabolic bone disease. Anderson (1997) demonstrated that it is possible to improve the calcium status of mealworms and crickets by maintaining them in vitamin/mineral supplements for as little as 24 hours before they are fed to bustard chicks.

Breeding: Sexual maturity in small bustards can occur as early as two years. The presence of more than one male may be crucial to stimulating successful reproduction in buff-crested bustards. Several zoos have had success by housing females in an enclosure

Table 9. Growth Rates of Captive Reared Bustards

Species	Growth rates	Source
White-bellied	7.5–10% for the first month; adult bodyweight by six months.	Anderson 1998b
Great	5.3–6.5% for the first month; 1.9% for the second month. 1.3% for the third month; 0.2% from 6 months of age.	Zubko 1992
Kori	7.5–10% for the first month; adult bodyweight not achieved in males by 330 days. Adult bodyweight achieved in females at 300-450 days.	Anderson 1998a
Buff-crested	5–7.5% for the first month.	Anderson 1998c

adjoined by two other enclosures; each contains a single male. The males are stimulated by hearing the vocalizations of their competitors. The females are presumably stimulated by the opportunity to choose between males. It should be noted that the Dallas Zoo has bred buff-crested bustard using one compatible pair that was housed together in a small, 3 m (9.8 ft.) in diameter public exhibit. In captivity, houbara bustards managed as pairs or as mixed-sex groups rarely, if ever, breed. Similarly, houbara bustards caught as adults remain sensitive to stress and in most cases fail to breed. Successful houbara bustard breeding projects use hand-reared birds. White-bellied bustards are monogamous and adults are maintained in pairs (Bailey and Hallager 2003).

Females entering reproduction condition create a nest scrape, usually placed next to a shrub or large rock for camouflage. In buff-crested bustards, the first infertile egg of the season is removed and the female given access to the male's enclosure. Usually, the male will enter the female's enclosure. Birds are separated after 30 minutes to three hours. This technique is repeated until the female lays her next egg at which point introductions are stopped.

Females sit tight on the nest scrape 1–2 days before laying. The two eggs are olive with brownish markings, and are laid at two-day intervals. Several eggs may be produced by immediately pulling newly laid eggs. Clutch recycling occurs in about 9–14 days depending on the species. Egg weights average approximately 45 grams. Incubation is 19–22 days. Females do all the incubation, leaving the nest once or twice per day to eat, preen and conduct other activities.

Incubation and Rearing: For artificial incubation, the incubator dry bulb temperature is 37.5°C (99.5°F) using 70–75% humidity. Target weight loss is 12%, and adjustments are made to the humidity accordingly. Eggs are turned every hour. Candle every five days until the air cell begins to drop down. At that point, increase humidity to 90% and can-

Table 10. Food Consumption and Growth Chart of a Buff-crested Bustard at the Dallas Zoo

	Food (g/day)	Weight (g)		Food (g/day)	Weight (g)		Food (g/day)	Weight (g)
Day 1	3.7	28	**Day 31**	n/a	n/a	**Day 61**	n/a	435
Day 2	6.2	29	**Day 32**	n/a	n/a	**Day 62**	n/a	n/a
Day 3	7.6	30	**Day 33**	25.4	177	**Day 63**	n/a	448
Day 4	n/a	35	**Day 34**	n/a	n/a	**Day 64**	n/a	n/a
Day 5	10.1	n/a	**Day 35**	n/a	n/a	**Day 65**	n/a	n/a
Day 6	n/a	40	**Day 36**	24.8	201	**Day 66**	n/a	472
Day 7	n/a	n/a	**Day 37**	n/a	n/a	**Day 67**	n/a	483
Day 8	n/a	45	**Day 38**	n/a	n/a	**Day 68**	n/a	n/a
Day 9	12.8	51	**Day 39**	30.1	231	**Day 69**	n/a	494
Day 10	n/a	n/a	**Day 40**	n/a	n/a	**Day 70**	n/a	n/a
Day 11	n/a	n/a	**Day 41**	n/a	n/a	**Day 71**	n/a	n/a
Day 12	16	64	**Day 42**	38.2	254	**Day 72**	n/a	503
Day 13	n/a	n/a	**Day 43**	n/a	n/a	**Day 73**	n/a	n/a
Day 14	n/a	n/a	**Day 44**	n/a	267	**Day 74**	n/a	n/a
Day 15	19.4	77	**Day 45**	40	280	**Day 75**	n/a	521
Day 16	n/a	90	**Day 46**	n/a	n/a	**Day 76**	n/a	n/a
Day 17	n/a	n/a	**Day 47**	n/a	n/a	**Day 77**	n/a	n/a
Day 18	20.5	103	**Day 48**	38.4	302	**Day 78**	n/a	527
Day 19	n/a	n/a	**Day 49**	n/a	n/a	**Day 79**	n/a	544
Day 20	n/a	112	**Day 50**	n/a	n/a	**Day 80**	n/a	n/a
Day 21	20	121	**Day 51**	39	325	**Day 81**	n/a	561
Day 22	n/a	n/a	**Day 52**	n/a	339			
Day 23	n/a	n/a	**Day 53**	n/a	n/a			
Day 24	14.6	133	**Day 54**	60.1	357			
Day 25	n/a	n/a	**Day 55**	n/a	n/a			
Day 26	n/a	n/a	**Day 56**	n/a	n/a			
Day 27	11.2	143	**Day 57**	n/a	384			
Day 28	n/a	n/a	**Day 58**	n/a	n/a			
Day 29	n/a	n/a	**Day 59**	n/a	n/a			
Day 30	23.1	161	**Day 60**	n/a	423			

n/a= information not available

dle the egg daily. At internal pip, increase the humidity to 95–98%. Hatching should occur within 8–24 hours after external pip. The incubation period is 19–22 days.

Parent Rearing: If the egg has been artificially incubated, pull the dummy and replace the real egg at external pip. The egg should hatch within 8–24 hours. Leave the chick and mother alone for the first 24 hours to allow bonding to occur. The next day, briefly examine the chick, weigh it, swab its umbilicus with betadine, and check the status of yolk resorption before returning the chick to its mother. Chicks may lose 10% of their body weight the first two days, but then should gain 2–18% body weight per day thereafter. Continued weight loss may indicate a need for supplemental feedings of mealworms, waxworms, and Startena. Chicks sometimes develop airplane wing (slipped tendon) due to the high protein diets provided, but taping the wing can fix the problem within a few days if started immediately. Females alone care for the semi-precocial chick. Initially, the female offers food items to the offspring bill-to-bill, but chicks begin pecking at food on its own after four or more days. If grass with seed heads is growing in the pen, the mother may strip the seeds from the stem and feed it to the chick. Buff-crested bustards carry young chicks under their wings in a manner similar to jacanas (Falzone 1992). Chicks may stay with the mother up to the next breeding season.

Hand Rearing: Hand rearing is not recommended for buff-crested bustards due to its probability of reducing breeding success due to imprinting. Imprinted male buff-crested bustards will display to humans, but have been known to scalp or kill female buff-crested. If the chick must be hand reared, the brooder should initially be 32.2–35°C (89–95°F) and lower the temperature gradually to 29.4°C (84.9°F) by the end of the second week. A large, sterilized feather duster will provide the chick with a surrogate mother. A mirror or other hatchling bustard will decrease imprinting. Do not feed the first 24 hours after hatching. Then use tweezers to offer bits of food. Food items with indigestible components such as fur and excessive chitin may cause impaction. If too many high protein items are offered, angel wing or slipped leg tendons may result, so equal amounts of the fruits and vegetables should be fed. Weigh the bird daily. The chick should gain 2–18% body weight per day, with 10–12% the most desirable. Water may be offered via droplets from an eyedropper, or a small, very shallow dish with clean medium-size pebbles. For the latter method, use the tweezers to tap at the pebbles and water to stimulate the bird to drink. Keep the water depth no more than 1.0 cm (0.4 in.) to prevent the chick from drowning.

Hand rearing diet for buff-crested bustard (Dallas Zoo):

- Initially, offer food items with tweezers until chick begins eating on its own (4–14 days of age). Dip each food item in water before offering it to the chick. The first few wet items should also be lightly dipped in a vitamin powder. A sample food consumption and growth chart of a buff-crested bustard from Dallas Zoo is presented in Table 10.

- Day 2–7: Five feedings per day. Quantity 25% per day of chicks daily body weight (5% per feeding). Offer tiny pieces from each of the following: diced fruit (apple or papaya), diced vegetable (carrot, spinach, cooked yam, fresh peas), hairless baby mice, cricket body (wings and legs removed), soaked softbill pellet, mealworms and waxworms.

- Day 8–14: Four feedings per day. Quantity 25% per day of chicks daily body weight (6.25% per feeding). Diet as described for day 2–7.

- Day 15–21: Three feedings per day. Quantity 25% per day of chicks daily body weight (8.3% per feeding). Offer small pieces from each of the following: diced fruit (apple or papaya), diced vegetable (carrot, spinach, cooked yam, fresh peas), juvenile mice, crickets, soaked softbill pellet, mealworms and waxworms, Layena (poultry layer mix).

- Day 22–28: Three feedings per day. Quantity 20% per day of chicks daily body weight (6.6% per feeding). Diet as for day 8–14.

- Day 29–60: Three feedings per day. Quantity 15% per day of chicks daily body weight (5% per feeding). Diet as for day 8–14.

References:

Allen, M.E. & O.T. Oftedal. 1989. Dietary manipulation of the calcium content of feed crickets. Journal of Zoo and Wildlife Medicine. 20: 26-33.

Anderson, S. 1995. Bustard micronutrient review. National Avian Research Center, Abu Dhabi, External Report No. 4.

Anderson, S. 1997. Improving the calcium content of cultured insects for rearing bustard chicks. National Avian Research Center, Abu Dhabi, Internal Report No. 57. 19 pp.

Anderson, S. 1998. Captive management and breeding of the kori bustard (Ardeotis kori) at the National Avian Research Center. National Avian Research Center, Abu Dhabi, External Report.

Anderson, S. 1998. Captive management and breeding of the white-bellied bustard Eupodotis senagelensis at the National Avian Research Center. National Avian Research Center, Abu Dhabi, External Report.

Anderson, S. 1998. Captive management and breeding of the rufous-crested bustard Eupodotis ruficrista at the National Avian Research Center. National Avian Research Center, Abu Dhabi, External Report.

Bailey, T.A., J.H. Samour, J. Naldo, J. Howlett & M. Tarik. 1996. Causes of morbidity in bustards in the United Arab Emirates. Avian Diseases. 40: 121-129.

Bailey, T.A., P.K. Nicholls, J.H. Samour, J. Naldo, U. Wernery & J. Howlett. 1996. Post-mortem findings of bustards in the United Arab Emirates. Avian Diseases. 40: 296-305.

Bailey, T., J. Naldo, J. Samour & J. Howlett. 1997. Bustard Therapeutics: Techniques for the medical management and care of captive bustards. External Report No. 5, Wildlife Veterinary Research Institute, Environmental Research and Wildlife Development Agency, Abu Dhabi, UAE.

Bailey, T.A. & S.J. Anderson. 2000. Commonly encountered hatching and post-hatching problems in bustards: radiographic imaging, endoscopy, surgical intervention and post-hatching care. The Avicultural Magazine. 105 (3): 114-126.

Bailey, T.A. & E. Flach. 2003. A review of morbidity and mortality of the Great bustard (Otis tarda) flock at Whipsnade Wild Animal Park 1989-1999. Veterinary Record. 153: 397-400.

Bailey, T. & S. Hallager. 2003. Management of Bustards in Captivity. Avicultural Magazine. 109 (1): 1-8.

Bailey, T.A. (in press) Diseases and medical management of houbara bustards and other Otididae. Published by the National Avian Research Center, Abu Dhabi.

Bailey, T.A., C.D. Silvanose, J. Naldo, O. Combreau, F. Launay, U. Wernery, J. Kinne, R. Gough & R. Manvell. 2000. Health considerations of the rehabilitation of illegally traded houbara bustards Chlamydotis undulata macqueenii in the Middle East. Oryx. 34: 325-334.

Beddard, F.E. 1898. The Structure and Classification of Birds. London: Longmans, Green and Company. Pp 331-336.

Beddard, F.E. 1912. Contributions to the anatomy and systemic arrangement of the Cestoidea III. On a new genus of tapeworms (Otiditaenia) from the bustard (Eupodotis kori). Proceedings of the Zoological Society of London. XIII-XIV: 194-221.

Beddard, F.E. 1914. Contributions to the anatomy and systemic arrangement of the Cestoidea XIV. On a new species of Rhabdometra and on the paruterine organ in Otiditaenia. Proceedings of the Zoological Society of London. LVIII-LIX: 859-887.

Bereszynski, A. & R. Graczyk. 2000. Results of studies on the breeding of bustard, Otis tarda, in Poland. Animal Science. 2: 11-30.

Bhushan, B. & A.R. Rahmani. 1992. Food and feeding behaviour of the great Indian bustard (Ardeotis nigriceps). Journal of the Bombay Natural History Society. 89: 27-40.

Bradford, J.R. 1908. General meetings for scientific business. Proceedings of the Zoological Society of London. Pp 1.

Carey, J.R. & D.S. Judge. 2000. Longevity Records: Life Spans of Mammals, Birds, Amphibians, Reptiles and Fish. Pp 106. Denmark: Odense University Press.

Chitty, J.R. 2003. Feather plucking in psittacine birds 2. Social, environmental and behavioural considerations. In Practice. 25: 550-555.

Collar, N.J. 1996. Otididae (Bustards). In: del Hoyo, J., Elliott, A. and Sargatal, J. Handbook of Birds of the World. Hoatzin to Auks. Barcelona: Lynx Edicions. Pp 240-275.

Cornes, S. 1997. Dietary analysis of the houbara bustard (Chlamydotis undulate macqueenii). ERWDA External Report 10. National Avian Research Center, Abu Dhabi. 22 pp.

Cramp, I.S. & K.E.L. Simmons. 1987. Handbook of the Birds of Europe, the Middle East and North Africa, The Birds of the Western Palaearctic, Vol 2: Hawks to Bustards. UK: Oxford University Press. Pp 636-668.

D'Aloia, M.A. & R. Griffiths. 1999. Molecular sexing of the houbara bustard Chlamydotis undulata, and other arid-land species. Zoology in the Middle East. 18: 33-40.

Deeming, C. 1995. Report on consultancy visit to National Avian Research Center, Abu Dhabi. Unpublished Report, National Avian Research Center. 9 pp.

Dollfus, R. 1957. Miscellanea helminthologica Maroccana XXI. Archives de Línstitut du Maroc. 7: 329-393.

Falzone, Celia, K. 1992. First Observations of Chick Carrying Behavior in the Buff-crested Bustard. Wilson Bulletin 104(1): 190-192.

Fitzherbert, K. 1983. Seasonal weight changes and display in captive Australian bustards Ardeotis australis. In: Goriup, P and Vardhan, H. (eds), Bustards in Decline. Proceedings of the International Symposium on Bustards, Jaipur, India. Pp 210-225.

Flach, E.J. 1995. Seasonal changes in haematological parameters in the great bustard Otis tarda. Verhandlungsbericht ber Erkrankungen der Zootiere 37: 351-356.

Flower, S.S. 1925. Contributions to our knowledge of the duration of life in vertebrate animals - IV Birds. Proceedings of the Zoological Society of London. Pp 1365-1413.

Fresenius, G. 1863. Schimmel in den lungen der trappe. (Mould fungus in the lungs of the bustard.) Zool. Garten. 4: 89.

Garcia-Montijano, M., A.M. Tebar, B. Barreiro, P. Rodriguez, J.C. Alonso, C. Martin, M. Magana, C. Palacin, J. Alonso, A. Montesinos & I. Luaces. 2002. Postmortem findings in wild great bustards Otis tarda from Spain: a clinical approach. European Association of Zoo and Wildlife Veterinarians Conference, Heidelberg. 8-12 May. Pp 41-46.

Hallager, S. 1994. Drinking Methods in Two Species of Bustards. Wilson Bulletin. Vol. 106 (4): 763-764.

Hallager, S. 2003. A description of copulation in the Kori Bustard Ardeotis kori struthiunculus. Bulletin British Ornithologist's Club. 123(2): 125-129.

Hallager, S. 2004. International Studbook for Kori bustards. Washington, DC: Smithsonian National Zoological Park.

Hallager S. & J. Boylan. 2004. Kori bustard SSP husbandry manual. Washington, DC: Smithsonian National Zoological Park.

Hallager, S. 2005. Breeding and hand-rearing the Kori bustard Ardeotis kori at the Smithsonian National Zoological Park, Washington, DC. International Zoo Yearbook 39: 143-161.

Hopkinson, E. 1926. Records of Birds Bred in Captivity. Holborn, London: H.F. and G. Witherby. Pp 138-139.

IATA. 1998. Live animal regulations, 25th Edition, International Air Transport Association, Geneva. www.iata.org/index.asp

Jacquet, J.M. 1996. Use of gonadotrophins to stimulate reproductive development in the houbara bustard 1995: preliminary trial. Internal Research Report, National Avian Research Center, Abu Dhabi. 11 pp.

Jacquet, J-M. 1997. Reproduction in houbara bustards kept under natural conditions at the Sweihan Research Station 1994/95 season. Internal Research Report, National Avian Research Center, Abu Dhabi. 12 pp.

Jarrett, N.S. & S.M. Warren. 1999. A preliminary guide for age and sex determination of the houbara bustard Chlamydotis undulate macqueenii. Journal of the Bombay Natural History Society. 96: 28-41.

Johnsgard, P. 1991. Bustards, Hemipodes, and Sandgrouse: Birds of Dry Places. UK: Oxford University Press.

Jones, A., T.A. Bailey, P.K. Nicholls, J.H. Samour & J. Naldo. 1996. Cestode and acanthalocephalan infestations in captive bustards: new host and location records, pathology, control and preventive medicine. Journal of Zoo and Wildlife Medicine. 27: 201-208.

Jones, L.D., R.W. Cooper & R.S. Hardin. 1972. Composition of mealworm Tenebrio molitor larvae. Journal of Zoo and Wildlife Medicine. 3: 34-41.

Kock, Richard A. 1990. National Avian Research Center Internal Research Report. Bustard Diet Project, No. 1. Abu Dhabi, UAB.

Makwana, S.C., A. Wason & A. Parihar. 1980. Some observation on the great Indian bustard in captivity.

Goriup, P.D. & H. Vardhan (eds). 1980. Bustards in Decline. Proceedings of the International Symposium on Bustards, Jaipur, India, 1-4 November. Tourism and Wildlife Society of India, Jaipur. Pp 152-154.

Maslanka, M. & A. Ward. 2004. In: Hallager S. & J. Boylan. 2004. Kori bustard SSP husbandry manual. Washington, DC: Smithsonian National Zoological Park.

Naldo, J.L., C.D. Silvanose, J.H. Samour & T.A. Bailey. 1998. Developmental intestinal aerobic microflora in the kori bustard Ardeotis kori. Avian Pathology. 27: 359-365.

National Research Council (NRC). 1994. Nutrient Requirements of Poultry, Ninth Rev Ed. Washington, DC: The National Academies Press.

NWRC. 1993. Annual Report of the National Commission of Wildlife and Development. In: Schwede, G. (ed), National Commission of Wildlife Conservation and Development, Saudi Arabia. Pp 39-50.

Osborne, P.E. & A. Martin. 2001. Reintroduction of the great bustard to Britain: a discussion paper. Consultancy Report to the Great Bustard Group. Department of Environmental Science, University of Stirling, Scotland. 65 pp.

Osborne, T. & L. Osborne. 2000. Ecology of the Kori bustard in Namibia. Annual Report of the Ministry of Environment and Tourism Permit Office, Namibia. February.

Ostrowski, S. & O. Combreau. 1995. A survey of the causes of mortality in captive, captive bred released, and wild born houbara bustards Chlamydotis undulata in Saudi Arabia. Proceedings of the Joint Conference of the American Association of Zoo Veterinarians and the Wildlife Disease Association and the American Association of Wildlife Veterinarians. East Lansing, Michigan, USA. Pp 87-91.

Ostrowski, S., A. Greth & I. Mikaelian. 1996. Pathology and veterinary care. In: Saint Jalme M. & Y. van Heezik (eds). Propagation of the Houbara Bustard. London: Kegan Paul International. Pp 70-96.

Paillat, P. & P. Gaucher. 1996. Management and facilities In: Saint Jalme, M. & Y. van Heezik (eds). Propagation of the Houbara Bustard. London: Kegan Paul International. Pp 15-31.

Rahmani, A. 1987. The Great Indian Bustard. Bombay Natural History Society. St Francis ITI, Bombay. 24 pp.

Rutgers, A. & K.A. Norris. 1970. Encyclopaedia of Aviculture. London: Blandford Press. Pp 266-269.

Saint Jalme, M., & Y. van Heezik. (eds.). 1996. Propagation of the Houbara Bustard. London: Kegan Paul International.

Schulz, H. & P.J. Seddon. 1996. Biology and status of the houbara bustard. In: Saint Jalme M. & Y. van Heezik (eds). Propagation of the Houbara Bustard. London: Kegan Paul International. Pp 3-14.

Schuster, R.K. & J. Kinne. 2003. Acanthocephalans and other helminths of the Asian houbara bustard Chlamydotis undulata macqueenii. Parasitological Institute of the Slovak Academy of Sciences. Pp 36.

Silvanose, C., J. Samour, J. Naldo & T.A. Bailey. 1999. Intestinal protozoa and associated bacteria in captive bustards in the United Arab Emirates. Avian Pathology. 28: 94-97.

Sleigh, I. & J. Samour. 1996. The National Avian Research Center Bird Care Manual: management techniques for a collection of bustards. Internal Research Report No. 42. National Avian Research Center, Abu Dhabi.

Snow, D.W. & C.M. Perrins. 1998. The Birds of the Western Palaearctic. UK: Oxford University Press. Pp 519-531.

Stiévenart, C. 2002. Monitoring the growth in houbara bustard Chlamydotis [undulate] macqueenii chicks. In: Scullion, F.T. & T.A. Bailey (eds). Proceedings of the World Association of Wildlife Veterinarians. 27th World Veterinary Congress, September 2002, Tunis, Tunisia. Pp 80-84.

Swan, G.E. 1993. Drugs used for the immobilisation, capture and translocation of wild animals. In: McKenzie, A.A. (ed). The Capture and Care Manual. South Africa: Wildlife Decision Support Services. Pp 1-115.

Tigar, B.J. & P.E. Osborne. 2000. Invertebrate diet of the houbara bustard Chlamydotis undulata macqueenii in Abu Dhabi from calibrated faecal analysis. Ibis. 142: 466-475.

Tourenq, D., O. Combreau, M. Lawrence & F. Launay. 2004. Migration patterns of four Asian houbara Chlamydotis (undulata) macqueenii wintering in the Cholistan Desert, Punjab, Pakistan. Bird Conservation International. 14: 1-10.

Upton, R. 2002. Arab Falconry. History of a Way of Life. Surrey, Canada: Hancock House Publishers.

Van Heezik, Y. & S. Ostrowski. 2001. Conservation breeding for reintroductions: assessing survival in a captive flock of houbara bustards. Animal Conservation. 4: 195-201.

van Heezik, Y., S. Hemon & P. Paillat. 1999. Profile of the captive houbara bustard flock 1993-1999. National Wildlife Research Center, Taif, Saudi Arabia, Internal Report. 12 pp.

van Heezik, Y., S. Hemon & M. Saint Jalme. 1999. Influence of breeding experience, age and climate on the breeding performance of captive houbara bustards. National Wildlife Research Center, Taif, Saudi Arabia, Internal Report. 18 pp.

van Heezik, Y. 2000. Population parameters of the captive houbara bustard flock. National Wildlife Research Center, Taif, Saudi Arabia, Internal Report. 10 pp.

van Heezik, Y. & S. Ostrowski. 2001. Conservation breeding for reintroductions: assessing survival in a captive flock of houbara bustards. Animal Conservation. 4: 195-201.

Wood, C.A. & F.M. Fyfe. 1943. The Art of Falconry by Frederick II of Hohenstaufen (1248). Translated and edited by Wood, C.A. & F.M. Fyfe. CA: Stanford University Press. Pp 63.

Woodward, A.S. 1915. Exhibitions and notices. Proceedings of the Zoological Society of London. Pp 529.

Zubko, V.N. 1992. Experience from incubating eggs and rearing chicks of great bustards in Askania-Nova. Bustard Studies. 5: 139-149.

19.02 Gruidae Cranes

Glen Holland

Introduction: The 15 species of crane are in one of two subfamilies, Balearicinae (two species of crowned cranes) and Gruinae (the other 13 species), in the family Gruidae. Seven of the 15 species are considered threatened. The most popular of the African cranes in aviculture, which is the species with which I have the most experience, is the blue or Stanley's crane *Anthropoides paradisea,* which is endemic to South Africa. The numbers of this species in the natural state have decreased alarmingly due to poisoning by grain farmers and the loss of their grassland habitat due to afforestation. Recent community conservation initiatives are seeing a recovery in numbers of this species. Twelve of the 15 crane species are reliant on wetland habitats for their survival and these same habitats are being drained, burned and planted by man. Species such as wattled crane however do make good "flagship species" with which to advocate the conservation of their habitats. I found farmers generally were keen to implement conservation strategies, once their role in the plight of the species had been acknowledged. Aviculturists can, therefore, play an important role in the conservation of this and other species. In the natural state they occur as pairs in the breeding season and flocks in the non-breeding season. Flocks of 300 and more birds have been recorded. They inhabit large areas of open space where short grasslands occur, and are also often found in rather arid areas. Blue cranes now regularly breed in the harvested wheat fields of the Cape Province.

Management: Although not common in private collections, cranes are now well represented in the zoological collections around the globe. They are noisy birds and require large enclosures, which should include open sandy patches where the birds can dig, interspersed with clumps of grass and a few trees for shade. A pool of water measuring a meter (3 ft.) at its deepest should be provided. At present, cranes are being successfully bred in enclosures of 150 m^2 (179 sq. yd.), but I would recommend a minimum of 400 m^2 (478 sq. yd.). Sexes are alike and best DNA-sexed (blood or feather). Surgical sexing poses a risk to such large birds.

Blue cranes are a hardy species that can withstand extremes in temperature from extreme heat to walking in snow. Temperate and subarctic cranes can tolerate temperatures of –30°C (–22°F) or colder as long as they have shelter and food. All of these species require only a three-sided shelter (windbreak). Subtropical cranes need supplemental heat or should be locked indoors when the temperature is below 0°C (32°F). Other species such as geese, storks, pelicans, and ostrich can be mixed with the cranes, but smaller birds will be intimidated by them. Due to aggression, it is advisable to house pairs alone, but in an open park-type situation where all the birds are introduced simultaneously, the pairs will separate into their own territories. Fighting with pairs in adjacent enclosures can be a problem and a visual barrier is advisable between the enclosures. They are excitable and the pair often suddenly bursts into a dance, which includes chasing, jumping, and throwing twigs into the air. Due to their bold and excitable nature they are rather dangerous to both keepers and the public. As with most of the larger crane species, it is advisable to ensure that the birds are not able to peck at the public. I have observed one of these birds peck a small child just below the eye, fortunately not resulting in much more than a bruise.

With some pairs it is necessary to carry a branch or broom for defense when entering their enclosure while breeding is in progress. When attacking, they use the beak and also kick, both of which can cause serious injuries. They are strong flyers and should ideally be kept in flight-netted pens. Alternatively, but less desirably, they can be kept pinioned or wing clipped. Personally I prefer to clip a wing, but it is necessary to remember the dates to clip or else the bird will take off with no warning. It has been suggested that pinioning inhibits successful mating. The primaries and secondaries are molted simultaneously once in two years. Cranes are prone to leg, neck, and wing injuries. The recommended method of capture is to herd the birds into capture pens covered with soft netting. Serious injuries from cranes have been recorded and, when capturing cranes, keepers are advised to wear eye protection and work with 2–3 people assisting, one with a broom for protection and the other people capturing the birds. Use elevated arms to guide the birds into the pen and as the bird is about to dash to escape, grab both wings close to the body (bustle). They can also be held between the catcher's knees. The bird is then held close to the catcher's body while a second person holds the neck. Once captured the bird should be hooded, which will calm it down. Avoid capture/transport on extremely hot days as heat stress combined with the stress of capture could result in dead birds. See the section on capture and restraint of large BUSTARDS for additional relative advice.

A longevity record for blue cranes is of at least 22 years and 10 months. A wattled *Bugeranus carunculatus* of 33, Siberian crane *Grus leucogeranus* in his 70s, and white-naped cranes *G. vipio* of more than 67 years have been recorded. Demoiselles *Anthropoides virgo* have bred when both birds of the pair were over 60 years.

The crane husbandry manual from the International Crane Foundation is a must for anybody managing cranes and was used a refrence for this chapter.

Diet: The recommended diets from the International Crane Foundation have 21% protein, which is started a month before the breeding season, and 19% protein maintenance diet out of breeding. Their diet should consist of a mixture of grains such as wheat, maize, sunflower seed, and an equal portion of poultry breeder pellets. In addition to this, insects such as crickets and grasshoppers are enjoyed and some plant material is eaten. I would suggest that edible grasses such as Kikuyu and rye be planted in their enclosures. In bare enclosures there seems to be a high incidence of infertility and, although not proven, this may well be due to a lack of live and green food. Planted enclosures will also supply a quantity of insects that the birds will catch.

Adult brolga diet:

- Feed two large cups per bird, twice daily.
- 60 g wheat
- 60 g maize
- 60 g oats
- 40 g sunflower seeds
- 40 g peanuts (in shell)
- 60 g peanuts (husked)
- 60 g Peck 'n' Lay poultry pellets
- 60 g cat biscuits (Science Diet brand)
- 3–4 slices wholemeal bread cut into small squares
- 1 apple, diced
- 2 raw sprats, chopped into chunks
- raw meat, chopped into chunks
- 2 boiled eggs, whole or cut into quarters

Extras for variety:

- Boiled root vegetables (potato, kumera)
- freshly killed mice
- husked nuts
- diced spinach
- live invertebrates (mealworms, locusts)
- diced cheese
- fruit cut into small squares

In captivity there is often a lot of food wastage due to loss to doves, sparrows, and other feral birds. In order to prevent this, at Jersey Zoo in the Channel Isles they used vertically placed 110 mm (4.25 in.) PVC pipes that were approximately 500 mm (20 in.) high as feeding stations. The cranes soon learned to reach down into these to retrieve their food.

Breeding: Cranes have a strong pair bond and, ideally, birds should be able to choose their own mates in a flock situation. Flocks must, however, be carefully observed for aggression that could result in injured birds. Birds from about two years old onwards will choose mates in these situations. Mature birds should be introduced by placing them in separate, adjacent pens, and only once a positive acknowledgment of the other bird has been observed are the two carefully guided together. Once again, the birds have to be carefully observed at all times. Well-paired cranes synchronize their activities such as feeding, resting, and dancing. Blue cranes are sexually mature at four years under ideal conditions, but otherwise take up to seven years to breed. Some species start as early as two years (wattled, red-crowned, eastern sarus, and sandhill) but these are the exceptions and not the rule. Their first few clutches are often infertile. A female blue of four years that was placed with an older male produced fertile eggs. Some pairs can be induced to incubate eggs and adopt chicks. Patuxent has introduced dummy eggs to eight pairs of whooping cranes that had never before laid, and two of these pairs accepted and incubated the eggs. One of these pairs hatched a chick and the other adopted a chick. Pairs for possible egg adoption would be those pairs that fail to lay after showing all the other signs of breeding behavior, courtship, nest building etc. Courtship in the blue includes the dance described earlier and duet calling.

Pairs display territorially by bowing and lowering their heads towards the ground while simultaneously and vigorously shuffling their wings, and then lifting their heads back up in a forward scooping motion. When dueting the male throws his head further back than the female and he also lifts the humeri above his back. Other social displays include rigid threat posturing, rigid strutting, ritualized preening of the back or thigh, feather ruffling, tail fluttering, crouching, growling, and hissing. Interestingly, the form of these complex behaviors is apparently independent of learning, with blind cranes being able to perform a full complement of behaviors. However, the object towards which this behavior is directed is learned. During their excited dance, and while running around, other birds in the enclosure will scatter, and the cranes will often chase them with the beak open. The female will then invite copulation by crouching low, spreading the wings, holding the tail up and bending the neck forwards (horizontal position). The male then mounts while flapping his wings (helicopters). The female can actually be pinioned on both wings and this will not affect successful mating, so long as she can form this small platform. The male must be able to hop onto the back of the crouched female and copulate with her. For this he needs to balance himself, as he maneuvers his tail beneath the raised tail of the female to make cloacal contact and he uses his extended wings to do this. After copulation the pair call in unison. Fully-flighted male birds (or properly wing-clipped birds) are therefore at a seeming advantage at this point. Mainly for this reason, the last three primaries (long, outermost wing feathers) must never be cut in wing-clipped birds, especially males. Another important reason is because these feathers protect the blood-filled quills that develop when the bird is molting. Without the old feather, the fragile new one has no protection and is easily damaged, being at the tip of the wing. Although this may seem unimportant, a broken blood-feather may cause excessive blood loss through continuous bleeding in a long feather such as the primary. Should a blood-quill become damaged it must be pulled out from the base to stem the bleeding.

The nest of the blue is more the exception than the rule among cranes, and is often built among stones and lined with a few pebbles and sometimes vegetation. Sometimes there is no nest at all, the clutch being laid on the bare soil. The clutch consists of 1–2 eggs that are buff brown in color with dark brown and rufous brown markings. Many crane species such as the wattle, brolga, and crowned species nest on mounds of vegetation, which form islands in aquatic habitats. Some of these birds will nest on dry ground in captivity but a shallow pond with an island mound of soil, slightly above the water level, will provide a good nest site. Placing some vegetation such as sedges and bulrush *Typha* spp. will often be accepted by the birds as the start of a nest. By pulling eggs after two weeks of incubation, adults can be stimulated into laying 2–3 clutches, but personally I feel that they should be left to raise the second clutch. Single-clutch eggs tend to be larger than the average egg. Clutches of three eggs can be induced by lifting the first egg immediately it is laid (prior to incubation). By removing eggs as laid, one captive Florida sandhill crane at ICF laid 19 eggs. Hatchability of the eggs for some species can be improved if at least the first 7–10 days of incubation are natural. Incubation for the blue starts with the first egg and is shared by both sexes, but largely by the female, and lasts 30 days. This is 27 days incubation and the chicks take three days to emerge after the first sign of hatching. While incubating, the male is particularly aggressive, but once the chicks have hatched both birds are equally aggressive.

The chicks are usually led from the nest the day after hatching and are fed by both birds. The parents present food in the tip of the bill and at the same time give a soft purring call. After 10 days the chicks pick up their own food. In the natural state, hatching success rates of 78% and 90% have been recorded but in captivity this rate tends to be lower. There is some aggression between the chicks. When hatched under the parents each parent takes a chick and initially they remain a few meters from each other. When the chicks hatch, a bowl containing small strips of heart, crickets, mealworms, earthworms, grasshoppers, and a rearing food that can be mixed with grated boiled egg, should be placed near the adults. They will soon begin feeding this to the chicks. Chicks reared by the parents show a degree of tolerance towards human presence, but here the relationship ends.

Artificial Incubation: Eggs removed for artificial incubation have a far higher success rate if they are first allowed to be incubated with the parents for two weeks. Eggs are then transferred to an incubator set at 37.6°C (99.5°F) and a relative humidity of 60%. Due to the dark coloration, it is hard to candle the egg to determine fer-

Demoiselle Crane
Anthropoides virgo

tility. Egg flotation may be used with eggs older than 21 days. The fertile egg will float nearly vertically with the large end up. Use a mild iodine-based disinfectant solution at 43°C (109°F) for this technique and float for less than one minute.

Optimally, eggs lose 15% of their fresh weight over the incubation period. In the hatcher the temperature is dropped to 37.2°C (99°F) and the humidity increased to 75%. Eggs are placed in the hatcher once the chick has pipped and should be positioned with the pip up. Chicks are often heard peeping two days before they hatch so do not be too hasty to assist the hatching process. From the first pip to successful hatching usually takes 24–36 hours and if not yet hatched thereafter assistance can carefully be given. I usually make a rasping purr type sound to a chick that has become vocal in the egg. This call stimulates activity from the chick and is usually performed by the adults. If a chick's cheeping becomes less audible or its movements less vigorous, the chick may be weakening and unable to hatch. The chicks are left in the hatcher for the first 18–24 hours. Once hatched, chicks must be shielded from seeing humans to prevent imprinting, until they are dry and can be returned to their parents, foster parents, or be placed in an isolation-rearing facility where crane look-alike puppets are used. It is important, where possible, that incubator-hatched chicks are not reared in isolation from other cranes as this has a detrimental effect on their future breeding capacity. Bantams make good initial foster parents but, once again, it is much better to have at least two crane chicks growing up together than singles. As soon as they are eating on their own, they should be placed in view of the adult birds for correct behavioral development. Isolation rearing using puppets is the best form of artificial rearing. The first to hatch is usually the larger chick and initially, due to intimidation of the smaller chick, they should be placed in separate brooders. Average incubation periods are as follows:

Black and gray crowned, and demoiselle.	28
Siberian, hooded and whooping	29
Blue, sandhill, white-naped, brolga, Eurasian and black-necked	30
Sarus	31
Red-crowned	32
Wattled	33

Hand Rearing: Recently hatched chicks must be carefully handled, as improper handling can cause damage to the limbs or ruptured yolk sacs, which can end in death. Hold the chicks on the palm of the hand with the legs hanging free between the fingers. The indoor brooder box is fitted with a brooder lamp suspended over one side and should have a bed of river sand on the floor. For the first week the brooder is maintained at a temperature of 37°C (98.6°F) below the brooder lamp. The chick then has the option to choose its preferred temperature.

It is important that artificially raised chicks are seen to be drinking water on their own soon after hatching. Dehydration accounts for many deaths and the chicks don't always know what water is, even if they are standing in it. They can be coaxed by putting floating insects on the water to peck at or finally by dipping their beaks into the water. Crane chicks are excellent swimmers and can paddle about like ducklings. Swimming therapy is useful in correction of leg problems if detected early enough. They must be able to easily get out of drinking water containers, otherwise they may become exhausted and drown. A 10–15% weight loss in the first few days is normal as the chick absorbs the yolk sac. Chicks begin feeding the day after hatching, and small items such as insects and pellets can then be offered in a forceps tweezer or a spoon with a puppet glove resembling the adult's head and beak. Tests have shown that chicks respond best to a red tweezer or spoon. The starter diet should have 24% protein. A higher percentage protein will result in leg problems. The puppet glove is used to avoid imprinting. They begin feeding themselves at 5–10 days. A shallow bowl of water with pebbles in the base must always be available. Initially the chicks' pecks at the food are largely unsuccessful, but this improves within a few days. Naturally the chicks follow the adults as they walk, and it is vital to ensure that the chicks have regular walks where they will follow the keeper, thus ensuring proper development of their legs. Swimming is another option for exercise. Provided it is done on a warm day, and only for a few minutes, it does the chicks no harm.

After a few days it is possible to place groups of 5–8 chicks of similar ages together in indoor (night) pens and outdoor (day) pens. If a chick shows dominance and aggressive behavior it is moved up to a more advanced age group. This group rearing assists to reduce imprinting on their human keepers. Young birds are initially placed in outdoor pens with a few other chicks during the day. Here they must be monitored for signs of inter-chick aggression. At night each chick is placed in a separate adjacent run. They are watched for about two months, after which they are placed permanently in outdoor enclosures with other juveniles. Groups of 5–8 chicks are ideal.

By 10 days the chicks will feed themselves, and at this stage they should be placed in an outdoor pen during the day in warm weather. Without sufficient exercise and a correct diet the chicks may gain too much weight, which causes deformities in the legs. Should bent toes occur, they are easily rectified by splinting with a wooden or metal splint to the toe for 2–3 days.

Stanley Crane
Anthropoides paradisea

Day	Weight (g)	Day	Weight (g)
0	115.1	**37.5**	1960
2.5	126.5	**40**	2200
5	163	**42.5**	2400
7.5	209	**45**	2620
10	292	**47.5**	2810
12.5	339.5	**50**	2910
15	462	**52.5**	3090
17.5	554	**55**	3210
20	757.5	**57.5**	3400
22.5	850	**60**	3540
25	1091	**62.5**	3660
27.5	1222	**65**	3800
30	1510	**67.5**	3880
32.5	1605	**70**	3955
35	1835	**72.5**	4000

The recommended diet for hand rearing chicks has a protein content of not more than 24% and 0.73% sulfur amino-acids to slow growth and thereby reduce leg and wing abnormalities associated with higher protein levels. Chicks in the wild have a fairly harsh time and walk long distances with their parents while foraging, and seldom have leg problems. This is often not the case in captivity, and in smaller camps with too rich a diet the legs often grow askew, with ultimate fatal consequences. It would seem therefore that lack of exercise and overfeeding will likely cause leg problems, so chicks should be kept on the lean side and be allowed to exercise as much as possible. Daily weight gains should not exceed 10–15%. Once the chicks are feeding themselves the diet should consist of turkey starter crumbles (24% protein), insects, grated boiled egg, and diced greens such as spinach. A multivitamin and calcium/phosphorous/vitamin D (ICD granules) supplement is added to the food from the initial feed and continued until the chick is fledged. At a month old the diet is gradually changed to turkey grower pellets (22% protein). At six weeks of age chicks may be carefully introduced to handicapped adults, if available, who will usually willingly act as nursemaids. At two and a half months the chicks are put on to maintenance pellets, which have a protein content of 18%. By three months the chicks should be living with others in an outdoor enclosure, and the pairs are separated at 18 months to two years once they have bonded.

A report by Chantal Winton, a keeper from Auckland Zoo, gives a good description of the incubation and hand raising trials of Australian brolga *Grus rubicundus* chicks:

> "On September 1, at 29 days, the chick could be heard moving and calling from inside the egg. The chick also could be seen moving while candling the egg. The humidity was increased to around 75%. At 11:30 a.m. the next morning (September 2), the chick had externally pipped and was very vocal throughout the day. At 8.00 a.m. the following day (September 3) the larger end of the egg had been broken off and the chick was halfway out. An hour and a half later we had a 121 g (4.3 oz.) brolga chick. Using a cotton bud, a biocil solution was applied to the umbilicus. The chick was then moved to the brooder at a temperature of 37°C (98.6°F) degrees and with a lower humidity to allow the chick to dry out.

Hand Rearing: The chick then slept until 3:00 p.m. and when it awoke, began looking around. Unsure if the chick was ready to take food, a small amount was prepared using a chick starter crumble, wombaroo insectivore mix (high in protein) and boiled egg (both the egg yolk and white) blended together with a small amount of water. The food was offered to the chick using a specially built 'brolga head puppet.'

This very realistic looking puppet was designed and made by Michael Batty, team leader of the exotic bird section. The objective of using the puppet was to ensure that the chick did not become imprinted. Along with the puppet, sheets were also put up around the brooder so the chick could be fed without it seeing any people. As a further insurance, a 'brolga outfit' was made to mask the shape of the keepers, who appeared as just a sheet with eyeholes with a brolga puppet on their hand.

Using the tweezers built into the head of the puppet and protruding approximately 1.0 cm (0.4 in.) from the 'beak', the chick was offered a mealworm and a small amount of

Puppet-rearing a crane chick.

the food mix. Both food offerings were readily accepted. An hour later the chick was observed trying to pick up more food but was still too uncoordinated to be successful.

Day 2: The temperature was turned down to 36.8°C (98.2°F). The chick's weight had dropped by 8 to 113 g (3.9 oz.) (7%). Using the brolga puppet, the chick was fed with the food mixture, mealworms and waxmoth larvae Galleria six times during the day.

Day 4: The chick's weight had increased and continued to increase everyday from this day forth. Calcium and spinach were added to the food and small crushed locusts were also provided, which were eaten straight away.

Day 5: The tweezers in the puppet proved to be very difficult to manipulate. This was further compounded by the chick's lack of co-ordination as it attempted to aim for the small area at the end of the tweezers. The solution was to attach a teaspoon to the lower beak of the puppet. This way it was much easier on our hands and the chick usually found food because of the larger area to aim for. Grated cheese (one tablespoon) was added to the food mixture along with earthworms, which the chick seemed to enjoy. Cheese is high in protein and fat, so it's a fine line between putting weight on the chick and putting on too much weight, which can cause leg problems.

Day 7: As it is well documented that cranes can develop leg problems if not well exercised, the chick was taken outside for exercise. This problem has been attributed to the rapid weight gain of the chicks if not exercised. It is important for the chicks to get natural sunlight for vitamin D, which aids calcium absorption.

Despite the keeper wearing the sheet to disguise the body shape and using the puppet, the chick showed no recognition, became frightened and ran off. To avoid a repeat of this reaction, the following day the keeper began the session from a crouching position and each day moving a little more till we were standing. This was continued until the chick felt comfortable with its 'parent.'

Day 14: At 339 g (11.9 oz.) the chick starter crumble was changed to a turkey pellet (high protein level of 25%). The chick showed a definite preference for the silverbeet and one teaspoon of cheese and of course live food, which was always eaten first.

At three weeks the chick was walked down to the wallaby paddock to a temporary enclosure made of a wire frame covered with hessian (burlap). On the first day the chick was left in for two hours, with a volunteer observing the chick for safety reasons. The chick had been pacing up and down the fence and a small patch on her neck had been rubbed into a small graze. Unfortunately the volunteer didn't tell us the chick was pacing until the graze was already there. It took about a week for the chick to gradually get used to the enclosure. The chick continued to be fed several times a day by the keepers wearing their disguise and using the puppet. Live food was also given to keep the chick occupied, which helped to stop the pacing.

Every morning the brolga chick (now named Bimbimbie — aboriginal for "place of many birds") was walked down to the wallabies. In the late afternoon, she was walked back to an inside enclosure, much to the amusement of both visitors and staff.

Meanwhile in the main brolga enclosure the female was sitting on another two eggs. The first egg externally pipped on October 14 2002. The chick took two days to break out of the egg. The second egg pipped the morning of October 16. The parents took the first chick away from the nest to feed, leaving the second egg unattended. The decision was made to place the egg in the incubator to hatch. By 1:30 p.m. the next day (October 17) the chick had not made any more progress and was becoming less vocal. At this stage the decision was made to assist the hatch of the chick. This involved carefully removing small amounts of shell using tweezers, being careful not to rupture any blood vessels. The chick, weighing 122 g (4.3 oz.), was placed under a heat lamp at around 36°C (96.8°F).

The next morning the chick was placed back in the nest along with the remaining eggshell, in the hope the parents would raise both chicks. Brolga can raise two chicks, although normally one. When food is scarce usually only one chick survives. The parents came over to investigate and ate the eggshell but ignored the chick. The chick moved into longer grass and stayed there for around 45 minutes. The female then came over to the chick and started to peck it on the head and neck. The first chick also came over and pecked at the chick, which is a natural behavior. It was decided that the chance of the parents raising both chicks was extremely low so the second chick was returned to the brooder room for hand rearing.

Over the next nine days the chick was fed using the puppet as often as possible. It took this long for the chick to start to put on weight, which had decreased to 88 g (3.1 oz.) losing 27%. The chick was very uncoordinated and would often peck at the spoon and fall over backwards. It may have been from the peck on the chick's head from the adult female. To this day she has a bald patch on the top of her head.

On day 14 chick #3 weighed 161 g (5.6 oz.), compared with chick #1 that weighed 339 g (11.9 oz.) at the same age. The chicks were housed beside each other in the brooder room. They could see and hear each other, which made it easier to introduce them to one another when chick #3 was three weeks old. Both chicks were walked down to the wallaby enclosure where they remained, unless it was raining, and they were then returned to their house at night. Bimbimbie was put into an off-display outside enclosure on day 72, where chick #3 joined her during the day until day 45 after which they both stayed outside together overnight. During feeding, the younger chick would quite often take food from the older bird's beak. This behavior (similar to what would be expected if the chick was with the parents) didn't seem to bother her too much, she simply got more food from the food dish.

Chick diet, first two weeks:

- 2 tbsp. turkey starter crumble
- ½ tsp. Womabaroo Insectivore mix
- ½ shelled boiled egg, white and yolk
- 1 tsp. grated cheese
- spinach, finely sliced
- live food
- mealworms, galleria, small locust crushed and small earthworms
- first two days, small or white mealworms
- feed three times a day
- live food ab lib

Chick diet, after 14 days:

- ½ insect container turkey "grower" pellets
- 100 grams Womabaroo Insectivore mix
- 1 boiled egg, chopped finely
- 2–3 tsp grated cheese
- spinach, sliced finely
- feed three times a day
- live food five times a day"

Ring Size: 16–18 mm. A permanent metal band with the ID number is placed above one hock. A larger color plastic wrap around band is placed above the other hock. This can also carry ID numbers, year, etc.

Contributions:

Ann Burke, aviculturist at the International Crane Foundation.

Walter Mangold, World of Birds, Cape Town, SA.

References:

Ellis, D.H., F.G. Gee & C.M. Mirande. 1996. *Cranes: Their Biology, Husbandry, and Conservation*. Surrey, Canada: Hancock House Publishers.

Horsfield, W. 2000. Blue Cranes *(Anthropoides paradiseus)* A personal perspective.

Johnsgard, P.A. 1983. *Cranes of the World*. Indiana Unviversity Press.

Winton, C. 2002. Hand Rearing Brolga at Auckland Zoo. (unpbl)

www.savingcranes.org. The crane husbandry manual. International Crane Foundation.

19.03 Psophiidae Trumpeters *Psophia* spp.

Piet and Trijn Kreeft

Introduction: Trumpeters belong to the order Gruiformes and form the family Psophiidae. There are three species, the common *Psophia crepitans,* white-winged *P. leucoptera* and the green-winged *P. viridis*. They occur in the forests of Brazil, N. Bolivia, E. Peru, S. Venezuela, the Guianas and N. Upper Amazonian. Their name trumpeter originates from their calls, which sound like a loud musical instrument. Naturally they occur in groups of 5–15 and occasionally in groups of up to 200 birds. They feed on the ground, consuming fallen leaves, fruits and insects. Their behavior is similar to that of both cranes (Gruidae) and rails (Rallidae). They are co-operative, breeders with the entire family assisting to incubate and care for the young. The native people in the birds' range often keep these birds as pets. They tame the young birds and let them mingle with their children whom they warn if snakes or other dangers are present. Adult birds weigh 900–1100 grams. Sexes are alike.

Management: Trumpeters are tame and confiding birds, which become very attached to their owners. They enjoy being stroked, during which they make a deep growling sound. Once a pair/group is settled in their aviary they will not tolerate any other trumpeters being introduced into the aviary. They will, however, ignore other birds, even as small as finches. The ideal aviary for a pair in warm climates is outdoors with floor space of 30 m^2 (98.4 sq. ft.). The base should be kept moist and have a covering of grass and leaf litter. If left on a dry substrate the birds are prone to split skin on the toes, which can lead to secondary infections such as bumblefoot. Humid conditions, however, create ideal conditions for fungal (aspergillosus) infections to which these birds are prone. The aviary should include a bath and shrubs to provide shade. They are prone to frostbite on their toes, and in climates where temperatures drop below 5°C (41°F) they should be provided with a frost-free shelter. They roost on perches near the ground. A few branches of 5–10 cm (2–4 in.) thick should be provided at a hight of 1.5–2m (5–6.5 ft.)above the ground.

Ideally the birds should be kept in pairs, but one can also keep a group together. There is usually one dominant female that lays eggs, and if a second female begins to nest, fighting will result. The second female should then be moved to another aviary. Males are generally more tolerant of each other, but even after years of being in the same aviary they may turn on each other. Males are sexually mature at one year and females begin to lay at 2–3 years. Trumpeters have been known to reach an age of 15 years.

Diet: Trumpeters are easy to cater for in captivity. They are fed primarily poultry pellets ad lib but also get some soaked puppy food, 20 mealworms and half a piece of fruit like apple, banana, pear, melon per bird/day. Oyster shell grit is always available.

Breeding: Trumpeters dance in the same manner as do cranes; this is part of pair bonding but is not part of courtship. When a female is ready to copulate she will stand with her head down and her tail end turned towards the male. Copulation will ensue without any dancing. A few days before the female begins to lay she will start to call. The clutch is usually 3–5 white eggs, which are laid at 48-hour intervals. The eggs weigh 70–80 g (2.5–2.8 oz.) and measure about 45 x 60 mm (1.8–2.4 m) and are rather blunt.

They like a nestbox with an opening measuring 20 cm (8 in.) at the front. The box should measure 60 x 60 x 60 cm (24 x 24 x 24 in.) and should be placed on the ground next to a shrub or grass tussock. The female sometimes scratches a nest scrape somewhere under a bush or plant. They breed easily and, if in a group, they all share the responsibility of incubation and rearing (co-operative breeders). Aviaries must be kept moist; dry conditions have resulted in poor hatchability of eggs. In dry climates artificial incubation, where humidity is easily controlled, is recommended.

Pale-winged Trumpeter
Psophia leucoptera

Common Trumpeter *Psophia crepitans*

Day	Weight (g)	Day	Weight (g)
0	60	8	550
1	100	9	600
2	135	10	680
3	200	11	740
4	285	12	810
5	340	13	910
6	395	14	1000
7	490		

Artificial Incubation and Rearing: Trumpeters' eggs must be hatched in a moist environment with a relative humidity of about 70%. The incubation temperature should be 37.8°C (100°F) at the top of the egg, in a still-air incubator. The eggs are turned over the point three times per day. Incubation lasts 27–28 days. Two days prior to hatching, all turning is stopped. As soon as the chick has entered the air cell and begins the external pip, the shell fracture is turned to the top. Once hatched, the chick remains in the hatcher for a further 24 hours. There after the chicks are removed to a brooder measuring 50 x 100 x 40 cm (19.6 x 39 x 15.7 in.) which has a heat source suspended above it. The initial temperature on the brooder floor, directly under the heat source, should be 37°C (98.6°F) for the first week. After the first week the temperature is reduced by 2°C (0.9°F) per day. For the first week the brooder base is covered with paper towel on which some chick rearing food (poultry mash) is sprinkled. Mealworms, which have new soft skins, are also offered. Small crickets are also fed from the tip of tweezers. Once they begin to feed, some moistened rearing mash is fed with the live food. They are fed from the tweezers every two hours from 7 a.m. to 8 p.m. On day three a shallow dish with moistened chick meal and some mealworms is offered. They are encouraged to begin feeding from the dish by offering food from the tweezers while digging into the dish. As soon as they begin to feed themselves, after about seven days, the paper towel is removed and the chicks left on a bed of wood shavings. At this stage a little moistened puppy food is added. By 10 weeks they receive the same diet as described for the adults.

Newly hatched trumpeters weigh 40–60 g (0.09–0.13 oz.) and double this weight within 12 days. At seven weeks they have reached about half their adult weight, which they achieve at about 14 weeks. Young trumpeters can be very aggressive towards each other for the first week after which the aggressive tendency disappears. Young birds are prone to aspergillosus infections and therefore it is important to disinfect the incubator and brooder with formaldehyde. When chicks grow too fast or fail to exercise sufficiently, leg problems will result. At three weeks the chicks are removed to a warm 24°C (75.2°F) environment. Wood shavings are used as a base and a perch is placed below the brooder lamp.

19.04 Cariamidae Seriemas

Sara Hallager

Introduction: The seriemas comprise the family Cariamidae in the order Gruiformes. There are two species, both of which are found in South America — the slightly smaller black-legged seriema or Burmeister's seriema, *Chunga burmeisteri,* and the larger red-legged seriema or crested seriema, *Cariama cristata.* The placement of seriemas within the order Gruiformes has been disputed over the years. They are generally placed closest to bustards, with which they share similar cestode parasites.

Fossilized predecessors of modern seriemas have been found in North and South America and Europe. These ancestors were 1–3 m (3–10 ft.) high. They had huge, sharp beaks making them significant predators. And, like seriemas today, they could run extremely fast. The two extant species of seriemas are both found in Central South America where they inhabit the grasslands and wooded savannas. Red-legged seriemas are generally found in open habitats while black-legged seriemas prefer more wooded areas. The range of black-legged seriemas is considerably smaller than that of the red-legged seriema (del Hoyo 1996).

Sexes are similar, with males being slightly larger than females. Depending on the species, birds weigh approximately 1.5–2.5 kg (3.3–5.5 lb.) and stand about 75–90 cm (29.5–35 in.) tall. Red-legged seriemas have a crest. Unique among South American birds, the crest is comprised of tufts of permanently raised, slightly stiff feathers at the base of the bill, and can reach 10 cm (4 in.) in length. Males have longer crest feathers. Seriemas have loose grey-colored plumage, especially on the neck and under parts. The hooked beak is often compared to that of a raptor. Mostly ground dwelling birds, seriemas have three short, very sharp front toes, and a raised smaller hind toe. This anatomical arrangement prevents them from catching prey with their feet. Instead, the sharp claws on the feet are used to hold prey in place while the sharp bill tears the flesh.

Although seriemas are omnivorous, they are highly carnivorous. Their diet includes insects, small rodents, lizards, frogs, birds, and occasionally snakes, although their reputation of consuming large numbers of snakes is exaggerated. Leaf material, seeds, fruits, and crops cultivated by humans, such as corn, beans, and grains, are also eaten. Small prey items are swallowed whole and head first. Larger animals are held between their sharp claws, and torn apart. Killing their prey may involve beating the animal on the ground, or throwing it against a hard surface such as a rock.

They are quick runners, generally preferring to run from danger than fly. They have been clocked running upwards of 70 km/hour (43 mph) (del Hoyo 1996). When they do fly, it is generally only when they cannot out run their pursuer, or when they must reach an elevated perch.

Neither species of seriema is considered threatened. However, exact counts of both species have not been conducted, so their true status is not really known. Further complicating the matter is their far-reaching call, which may yield a false count of the actual numbers of birds in a particular area.

Central Brazil is the stronghold of the red-legged

seriema. Here, the species favors the grassy, savanna areas dominated by termite mounds. These favored lookout perches provide the red-legged seriema with elevated roosts from which to call and advertise their territory. Farther south, the species is found in the grasslands of southern Brazil, Uruguay and Argentina. In Argentina, numbers range from abundant to scarce. It is considered a vanishing species in Uruguay. In northeastern and southeastern Brazil, illegal traders of live animals occasionally offer red-legged seriemas for sale (del Hoyo 1996).

Black-legged seriemas are found in the dry woodlands of Bolivia through western Paraguay and central Argentina. The species is sympatric with red-legged seriemas from southern Bolivia through the Chaco woodlands of Paraguay to Argentina. It may be segregated by temperature with the red-legged seriema being less common during the warmer season and the black-legged less common during the cooler season (del Hoyo 1996).

Threats to both species' habitat are in the form of agricultural development and hunting. Red-legged seriemas seem to have adapted and in fact have colonized certain deforested areas in Brazil (del Hoyo 1996). Populations of both species should continue to be monitored given their precarious status in certain parts of their range. The mating system of both species is monogamous. In central South America, the breeding season corresponds to the rainy months of September to May.

Management: Red-legged seriemas are more commonly kept in captivity than black-legged seriemas. Red–legged seriemas have been kept in captivity since the late 1800s. They are generally long-lived birds, with life spans of 25 years not uncommon in captive situations and some birds reaching 30 years (Hallager 2004). Management of both species is similar.

Seriemas are easily caught using a large net. Care must be taken when extracting the bird from the net so that the sharp talon of the bird's inner toe does not pierce the handler. Once in hand, the greatest threat is the razor-sharp nails, so handlers must immediately secure the feet. The bird can then be held much like a crane. Once the feet are secure, the handler should immediately pull the seriema's body against his/her own. A second person should hold the head, although the bill of the seriema is not particularly dangerous. Seriemas do not use their bill as a weapon although they will hold it open in an attempt to appear more ferocious. Restrain the head by encircling the bird's upper neck with one hand without restricting the airway; do not cover the nares. Hood the bird if necessary to induce calmness.

Another method is to hold the wings and body of the seriema with one arm, and the legs with the other. Grasp the legs just above the hocks and place one finger between the hocks to prevent the legs from abrading one another. Make sure the tarsi cannot contact the seriema's neck or the person holding the bird in order to prevent the bird from ripping its neck open. If the seriema's legs must be folded, gently force the tarsi around, but if the bird locks its hocks rigidly, do not force the legs to fold. Instead, maintain steady pressure on the tarsi until the bird allows you to fold the legs. When holding a seriema with folded legs, support the bird's weight with the arm holding its body. Do not keep a seriema's legs folded for more than 30 minutes. These precautions will help prevent slipped tendons, capture miopathy, and in some cases leg fractures. When releasing a seriema, allow its legs to touch the ground before letting go of the body or wings.

They are best transported in medium-sized commercial pet carriers. Crate requirements for seriemas can be found in the International Air Transport Association manual (IATA, 1998).

Growth of a Red-legged Seriema
Cariama cristata

Day	Weight (g)
0	45
5	83
10	200
15	310
20	416
25	515
30	650
35	795
40	875
45	950
50	1045
60	1240
70	1280
80	1385

Seriemas are generally hardy long-lived birds with few medical complications arising during their life. Adult mortality is often the outcome of complications from puncture wounds or compound fracture of legs or wings or from complications caused when housed with incompatible exhibit mates. Seriemas have also succumbed to frostbite and must be housed in heated shelters when temperatures are below 0ºC (32ºF) or colder. Birds should be routinely screened for internal parasites and dewormed if necessary.

Pen dimensions should be longer than they are wide. The suggested minimum pen size is 7.5 x 5.0 x 2.5 m (24.5 x 16 x 8 ft.) for a pair of birds (Collins 1998).

Fowler (2001) recommends a space of 10 m2 per bird and a minimum aviary height of 2 m (6.5ft.). Seriemas like to roost at night as well as nest above 5 m (16 ft.) so elevated perches should be provided for them, especially for non-flighted birds. They like to run and need long exhibits in which to exercise. Long runs are also important for chicks, which need plenty of exercise to avoid developmental problems. A suggested boundary for enclosures is 2.5 cm (1.0 in.) chain link mesh. This size mesh reduces the chances of chicks getting out and large rodents getting in. Site barriers, such as thick shrubs placed 1.3–2.5 m (4–8 ft.) in from the perimeter walls, give the birds a sense of security in pens of minimum size. Areas of full sun should be provided, as seriemas are heavy sunbathers. Additionally, areas of shade should be available, especially in hotter climates. Seriemas dust bathe so areas of sand should also be provided.

Native/feral predators are dangerous for both adult and young seriemas. Foxes and/or raccoons can fatally wound adult birds. Thus, enclosures must be built to minimize predator access. Digging predators (for example, dogs and foxes) can be excluded by burying the base of the boundary 0.3 m (11 in.) in the ground. Surrounding the enclosure with electrical wire deters climbing predators such as raccoons. In areas where large predators such as coyote, bobcat, or cougar are common, birds may need to be housed indoors at night.

Zoos in zones where temperatures fall below 0ºC (32ºF) must have winter holding facilities available for housing birds during inclement weather. Seriemas are susceptible to frost bite and must not be left outside when temperatures are below 0ºC (32ºF). They should also be housed during periods of freezing rain or snow. Sheds should be heated to 10–15.5°C (50–60ºF). Heat bulbs, if used, should be encased in protective wiring to prevent bulb breakage in case a bird makes contact with the bulb. Lights should be installed that can be turned on during the day and off at night. Skylights, which allow for natural light in the stalls, may be advantageous for birds that must be housed for extended periods. A small night-light turned on at night will provide some degree of light for the birds during the overnight hours.

The floor of the shed can be dirt or sand that can be covered with bedding hay (straw). Perches should be provided in sheds so that birds can roost off the ground. Even zoos that do not experience extremely cold weather should have a shelter available. The shelter can be used for times when pen repairs are needed, for medical confinement, to minimize food loss from wild birds, or when birds must be caught.

Seriemas can be exhibited with a wide variety of bird and mammal species. Seriemas are hunters though, and small fledglings of any species will be consumed if the opportunity presents itself. Birds that have been successfully exhibited with seriemas include:

cinnamon teal *Anas cyanoptera*
ringed teal *Callonetta leucophrys*
Egyptian goose *Alopochen aegyptiacus*
black-necked swan *Cygnus melanocoryphus*
paradise shelduck *Tadorna variegate*
common shelduck *Tadorna tadorna*
Andean goose *Chloephaga melanoptera*
white ibis *Eudocimus albus*
kookaburra *Dacelo novaeguineae*
hyacinth macaw *Anodorhynchus hyacinthinus*
scarlet ibis *Eudocimus rubber*
boat-billed heron *Cochlearius cochlearius*
king vulture *Sarcorhamphus papa*
grey-winged trumpeter *Psophia crepitans*
red-billed hornbill *Tockus erythrorhynchus*
tawny frogmouth *Podargus strigoides*
common peafowl *Pavo cristatus*

As well, various species of the following birds have all been exhibited with seriemas at zoos:

guan *Penelope sp.*
dove *Columba sp.*
turaco *Tauraco sp.*
thrush *Garrulax sp.*
stork *Ciconia sp.*
crow *Corvus sp.*
conure *Aratinga sp.*

Mammals species that have been successfully exhibited with seriemas include:

collared peccary *Pecari tajacu*
Brazilian tapir *Tapirus terrestris*
squirrel monkey *Saimiri ustus*
capybara *Hydrochaeris hydrochaeris*
guanaco *Lama guanicoe*
giant anteater *Myrmecophaga tridactyla*

Seriemas are most often seen alone or in pairs. Groups of three or more birds are most likely parents and offspring. They are a diurnal species and roost in trees at night. More than one pair of seriemas cannot be exhibited in the same enclosure, as they are extremely territorial. Seriemas should not be introduced to new mates during the breeding season. Males can be overly amorous in their efforts to breed, causing undue stress upon a newly introduced female.

Seriemas require only small areas of water from which to drink. They are not heavy drinkers but do drink on a daily basis. Heated water dispensers for northern zones are recommended. Seriemas do not bathe in water so pools are not needed in exhibits (unless for aesthetic reasons). If pools are present in enclosures, they should be shallow enough that a bird can walk through the water and the sides should gradually slope to the deepest portion. Pools deeper then 0.6 m (24 in.) are not recommended. To prevent accidental drowning, pools should be lowered once chicks leave the nest.

Diet: Diets for captive seriemas are varied, but most zoos feed a combination of small and large mice, mealworms, crickets, canned dog or cat food, avian pellets, and either horsemeat or beef. Fruit may be occasionally consumed. Food should be offered twice daily and on the ground in a pan. When chicks are present, hard-boiled egg and pinkies should be given and all amounts of food should be increased. Food should be offered three times per day when chicks are present.

Breeding: In captivity, breeding can occur at any time of the year especially in southern zones. In colder climates of the United States, breeding typically commences in early March. Males as young as eight months have bred (Collins 1998); females have bred at one year (Hallager 2004). The male initiates the breeding season by showing off his flight feathers, stretching them to one side. The display is also accompanied by strutting in front of the female, with head pointed down and crest raised. The male often presents food to the female. Both birds call, thereby strengthening the bond between them as well as defining the territory of the pair (Redford and Peters 1986). Copulation occurs on the ground (de Almeida 1994).

Once a pair shows signs of nesting (carrying sticks), keepers should provide extra nesting material such as twigs, dried grasses and small branches. Nest building is done by both sexes, although the male does more of the construction. Nests are usually in a tree 1–5 m (3–16 ft.) off the ground and construction generally takes one month. Pairs will use the same nest site from year to year. The nest is assembled of twigs and branches, and is lined with mud and leaves. It is usually placed so that the birds can jump up to it in one or two leaps. Commonly, two (or three) white, slightly spotted eggs are laid 57–71 x 46–50 mm (2–2.8 x 2 in.) in *Cariama cristata* and 56–61 x 42–46 mm (2–2.4 x 1.8 in.) in *Chunga burmeisteri* (del Hoyo 1996)). Incubation is carried out by both sexes. Captive pairs sometimes eject the first clutch of the season and some pairs are notorious for repeatedly ejecting eggs. It may often take several years for the pair to stop this behavior or even re-pairing of birds to correct this problem. Following a 27–28 day incubation period, the chicks hatch.

Chicks are nidicolous and helpless when they hatch, but only for a short period as their growth rate is very fast. At birth, the eyes are open but chicks cannot stand. Chicks are born with brown-colored down and weigh about 40–60 g (1.4–2 oz.). At one week, pinfeathers begin to emerge and chicks attempt to stand. Young birds may utter the species characteristic vocalization at this age. As chicks approach 10 days, they are able to stand, and by 14 days may even leave the nest, although most chicks remain in the nest for 3–4 weeks.

Chicks are brooded by their parents for the first three weeks of life. After three weeks, the juvenile birds roost in trees at night. Because they are unable to fly at this age, they must have perching positioned so that they can reach high roost sites at night. Fledging occurs at about 30 days. At three months, the parents will start to wean the chick by encouraging them to pick up food for themselves. By three and half months, chicks are nearly full-grown although their plumage is still in juvenile form. At four to five months of age, chicks can be moved to their own enclosure or sent to other zoos.

Parent Rearing: If the egg has been artificially incubated, pull the dummy and replace the real egg at external pip. Leave the chick and parents alone for the first 24 hours to allow bonding to occur. Seriema parents are normally very attentive, but sometimes chicks need some extra help if they are not growing at the proper rate or they appear weak. Younger siblings can get pushed out or dominated by older chicks and may require supplemental feedings from keepers. The parents usually accept these extra feedings but keepers should be mindful of the sharp talons of adult seriemas when providing supplemental feedings to chicks. Occasionally if two or more chicks are hatched, one chick may die or be killed by the parent (likely the weakest chick) and fed to the remaining chick. Seriemas in the wild are reported to sometimes only raise one chick (del Hoyo 1996).

Artificial Incubation: Eggs are best left under the parents, but this is not always possible as some pairs routinely eject eggs from the nest. Should artificial incubation be necessary, an incubation temperature of 37.5ºC (99.5ºF) and a relative humidity of 55% is recommended. Eggs should be automatically turned in the incubator every two hours. The humidity should be increased to 75% once the chick enters internal pip. Eggs hatch after 26–27 days of incubation.

Hand Rearing Protocol for Seriema Chicks

Day 1–10

Offer first feeding at 12–18 hours post hatch. Healthy chicks should attempt to grab offered food items. Generally, if chicks are not attempting to eat it may be a sign they are dehydrated. Chick weights are recorded before the first feeding of each day from day 1–30.

Diet: Pinkies, soaked dog chow, cricket abdomens. To ensure proper hydration, water is given via a syringe (no needle) at every feeding and all food is dipped in water prior to feeding. Feed chicks every 90 minutes, 0600h–1730h (6 a.m. to 5:30 p.m.). Night feedings are not necessary unless chicks are not growing properly.

Brooder/Temperature:

- 36.1°C (97°F). Temperature may be lowered to 35°C (95°F) on day 5.
- Brooder 69.85 x 33.0 x 35.5 cm (27.5 x 13 x 14 in.) deep.
- Floor is carpeted and a feather duster is hung in a corner to simulate the mother. Single chicks are given a mirror. A towel formed in the shape of a nest will help the chick feel secure when placed in it.

Day 11–15

Exercise: Chicks must continue to be encouraged to move about throughout the day. If outdoor temperatures are 24°C (75°F), chicks may be taken outside for exercise. Once chicks have been given access to outside yards, they are carefully monitored for the ingestion of foreign material that could result in impaction. Watch chicks to be sure they do not consume too many small pebbles. Chicks normally will seek out tiny pebbles as a source of grit. While a few pebbles are acceptable, ensure that chicks do not try and consume stones, which are too large for them. Ensuring that chicks continue to defecate normally is important.

Diet: Feed every three hours, 0600h–1600h (6 a.m. to 4 p.m.). Encourage chicks to drink out of a water bowl at each feeding. Diet same as above, but replace pinkies with small mice.

Brooder/Temperature

- Carpeted nursery area, 0.5 x 1.0 m (19.5 x 39 in.).
- Small-medium towel lined bowl or artificial nest.
- Three heat bulbs suspended from the ceiling so that three separate sections of the floor are kept at 35°C (95°F).
- Several feather dusters are suspended near the heat bulbs.
- Single chicks are given a mirror.

Day 31+

Diet: Feed three times per day, 0600h–1600h (6 a.m. to 4 p.m.). Eliminate egg from the diet. Chicks should be on the adult diet after 31 days.

Enclosure:

Chicks should be moved outside to a covered yard measuring 5 x 15 m (16 x 49 ft.) and the pen set-up should be similar to that of an adult (see above). They should be housed in a heated shed at night.

Hand Rearing: Once the chick is hatched, clean the umbilicus with Betadine (100%) solution. If the chick has had difficulty hatching, 2 cc of half-strength Lactated Ringer's and 2.5% dextrose solution should be provided subcutaneously. After being allowed to rest for several hours in the hatcher, move the chick to an Animal Intensive Care Unit, 69.85 x 33 x 35.5 cm deep, (27.5 x 13 x 14 in.) at 36.1ºC (97ºF). The floor of the brooder should be carpeted and a feather duster hung in one corner to simulate the mother. A small mirror affixed in the brooder acts as a visual stimulus for the chick. Line the floor with towels or carpet for frequent cleaning.

References:

Collins, S. 1998. Breeding the Red-legged Seriema. *AFA Watchbird* 25(6): 50–51.

de Almeida, A.C. 1994. Notas Sobre a Biologia Reprodutiva da Seriama *Cariama cristata* (Linnaeus, 1766) (Gruiformes-Cariamidae). Revista Nordestina de Biologia 9(1): 49–59.

del Hoyo, J., Elliot, A. & Sargatal, J. (eds). 1996. *Handbook of the Birds of the World.* Vol. 3. Hoatzin to Auks. Lynx Edicions, Barcelona. pgs. 234–239.

Fowler, M.E. and Cubas, Z.S. (eds). 2001. *Biology, medicine, and surgery of South American wild animals.* Iowa State University Press.

Hallager 2004. North American Studbook for the red-legged seriema. Smithsonian National Zoological Park.

Heinroth, O. 1924. Die Jugendentwicklung von *Cariama cristata.* Journal fur Ornithologie 72: 119–124.

International Air Transport Association manual 1998. Live animal regulations, 25th Edition, International Air Transport Association, Geneva. www.iata.org/index.asp.

Redford, K.H., and Peters, G. 1986. Notes on the Biology and Song of the Red-legged Seriema *(Cariama cristata).* Journal of Field Ornithology 57(4): 261–269.

19.05 Rallidae Gallinules, Rails & Crakes

Glen Holland & Myles Lamont

Thirty-four species exist in the family Rallidae, according to Sibley & Monroe. Originating primarily from Africa, South America, Australia and the eastern islands, the rails have one of the largest natural distributions of any terrestrial vertebrate. Besides rails, the family includes coots, gallinules, flufftails and crakes.

Associated with aquatic habitats, they are found either singly or as pairs. Usually located by their rattling duets, they are rather secretive and found on the vegetated fringes of pans of water and slow-moving rivers. They will venture out onto floating vegetation or open marsh, but disappear into thickets immediately when threatened. Most species in the family are rather vocal birds, and pairs regularly call and answer each other in duets.

Moorhens & Gallinules

The common moorhen *Gallinula chloropus* belongs to the rail family, or Rallidae, and is closely related to other species of moorhens, gallinules, and native-hens in the same genus, as well as to the gallinules and swamphens in the genus *Porphyrio*. Common moorhens are widespread, having been introduced to many regions of the world. They are abundant wherever there is suitable habitat. They occur naturally in reedbeds and vegetation on the edges of pans of water, and dislike open expanses of water with no refuge. Moorhens are strictly territorial and occur as pairs on small bodies of water.

Management: Most of the larger gallinule species inhabit similar areas as described for moorhen; but the endangered (less than 200 remain), and once thought extinct, takahe *P. mantelli,* a large, flightless gallinule of New Zealand, was rediscovered in 1944 in sub-alpine grasslands of New Zealand Fiordland. Most gallinules are rather vocal birds and pairs regularly call and answer each other (duet). I will concentrate on my experience with moorhens and gallinules *Porphyrio porphyrio,* and also mention husbandry techniques for the takahe.

Moorhens and gallinules are common species in most zoological parks and gardens. They are not as nervous as are other rails and adapt well to captivity. It is possible to mix moorhens, but not gallinules, with other waterfowl and larger species, such as hornbills and starlings, but they are not suited to mix with small birds as they will eat their eggs, and their mere size will intimidate the other birds. Gallinules have even been recorded stealing the eggs out of the nests of egrets.

In a large planted aviary it is possible to keep more than one pair of moorhens or gallinules. Their aviary should be landscaped according to Habitat F and include a pond or, preferably, a slow-flowing water feature surrounded by thickets of cypress, sedges, reeds and shrubs. Gallinules are fond of bathing and stand in shallow water making ducking movements of the head in bursts, alternating with bursts of flapping partly opened wings. Coots *Fulica atra* and moorhens bathe when standing with their feet on the bottom of breast deep water, resembling the less aquatic water rail *Rallus aquaticus* and black crake *Porzana flavirostris*. Sunbathing is commonly performed, particularly in warm weather, especially after bathing and the subsequent bout of oiling and preening. Occasionally gallinules will bathe while sitting on the ground, with their legs projecting forwards under the body, all four toes held apart and pointing forward, and the wings closed.

It is necessary with moorhens, if keeping more than one pair in an aviary, to ensure that each territorial pair have their own stretch of water to use. Pairs waste no time in building a nest of vegetation, which is positioned in a thicket. Moorhens are extremely aggressive when breeding or with newly hatched chicks, and will not hesitate to attack their keeper if approached too closely. I always found it fascinating how in a public walk-through aviary they are able to identify and attack only their keeper. When attacking, they charge with horizontal necks or rush across the water, deliberately splashing the target.

Gallinules and moorhens have white under-tail coverts, which contrast sharply with the darker plumage surrounding them. The patch formed in this way is used as a signal to warn other bids of the appearance of a potential predator and is emphasized in the gallinule by tail flicking movements. The moorhen spreads the white patch on its under-tail coverts in territorial threat, but does not flick its tail. Body language, including head and neck stretched forward either horizontally or downwards, and also an upright posture with tail flicking, is common during aggression. From the evidence from known-sex birds, it would appear that the calls of the females are softer and less harsh than those of males.

Diet: Swamphens and gallinules, such as takahe, are largely vegetarian and require reeds *Cyprus, Hypolepis* rhizomes, and grass tussocks on which they feed, and in which they nest. Gallinules and moorhens are more omnivorous and, apart from vegetation, will include insects, leeches, fish and carrion in their diets. With gallinules, food is often picked up in the bill and then transferred to feet and manipulated with them. Food is held in the feet by apposition of the hind toe with the three closed fore toes, and the feet are sometimes used to hold down large food items while they

Common Moorhen
Gallinula chloropus

Water Rail
Rallus aquaticus

Purple Swamphen
Porphyrio porphyrio

are pecked. Adult gallinule will also pick insects off the surface, and shovel aside gravel with the closed bill to reach earthworms and snails. Food is very often carried to the water and washed before it is swallowed. Moorhens and gallinules feed on the same grain and pellet diet as described for waterfowl and, in addition, they are fond of greens such as duckweed, insects and soft food. Soaked dog cubes or a slice of moistened whole wheat bread are also relished by moorhens. In addition to their natural forage, takahe are fed approximately 50 g (1.7 oz) of high-energy poultry pellets daily.

Breeding: Aggressive behavior often merges into sexual behavior. If two birds approach each other threat posturing, they usually follow with allopreening (male preens the female) or, less often, pass a small food item between their bills. In contrast, allopreening at the gallinule nest is most often mutual and other parts of plumage are usually preened in addition to the head and neck. Sometime the oil gland of the other bird is preened and the oil wiped on the feathers of other birds. A typical behavior pattern preceding copulation in gallinules is for the two birds to approach each other giving slight threat display; then the female solicits allopreening and is preened on her head and neck by the male. At this stage either or both birds may start "walking time," a peculiar action where the bird walks on the spot, raising its feet without closing the toes as they usually do when the foot is lifted in walking. The allopreening solicitation posture of the female may then be exaggerated into the sexual solicitation posture and the male may step or jump onto her back and copulate. In moorhen courtship the male moorhen will chase the female around the aviary. Prior to copulation, to indicate she is receptive, the female will squat with her neck arched.

Both male and female birds share in bringing material to the nest site, but much of the material seems to be incorporated into the nest by the female bird that receives it from the bill of the male, and sometimes other females. Nests are generally built on the ground in dense vegetation, and nest building continues throughout the incubation period. Coots, in contrast, tend to construct their nests on platforms of vegetation that float on water. Among a group of gallinules, a number of birds may assist the breeding pair by bringing nest material to the nest site, but only the breeding female actually places the material in the nest. During incubation the bird, which is sitting, spends a considerable amount of time plucking material from within bill reach of the nest and adding this to the nest. It is not uncommon for more than one gallinule to sit in the nest at one time but only one bird will be in contact with the eggs. Often a bird approaching, to relieve another that is busy incubating, will bring a strip of vegetation to the nest. Apart from the breeding nests, both gallinules and moorhens will also construct platforms on which they will stand, feed and roost.

The sexes are alike. Moorhens are extremely productive and it is often necessary to thin out their numbers. If the population is allowed to expand unchecked, the birds begin to fight among themselves, and with other species. The larger gallinules, however, are less prolific, more seasonal, and more prone to desert nests if disturbed, than moorhen. The 5–8 eggs are off white, covered in rust- and slate-colored spots. The incubation period is 21 days for

Takahe
Porphyrio mantelli

moorhens and 23–24 days for gallinules. With moorhens, the male does most of the incubation.

For the first day the newly hatched chick remains on the nest and receives little food, if any, but it is brooded almost continually. After this, chicks begin to wander from the nest and are brooded and fed either by a parent or by other bids of their particular group. The section on black crakes describes the co-operative breeding strategies that are similar in the moorhen. At first the food is invariably vegetation, but as time progresses the adults begin to feed the chicks artificial diets as well. The food is fed from the tip of the bill.

Moorhens and the larger gallinules are multiple clutched and usually lay within 25 days of the eggs being lifted for artificial incubation. The chicks are unusual little creatures, looking like black balls of wool with a red and yellow bill and a small red facial shield. The legs, and particularly the toes, are long. They are happy to walk on land, but often follow the adults into the water, waving their thin wings and head while begging to be fed. They are independent at two months, by which time the parents are usually sitting on a second clutch. A pair of moorhens with their unusual chicks makes a very interesting viewing in a public display. I have observed visitors sitting for long periods watching the antics of a pair of moorhens with small chicks. For the purposes of environmental education, a few mealworms will cause a rush of activity as the parents dash to and fro feeding the chicks. Should you decide to hand raise a clutch, do not attempt to rear young that have hatched under the mother. They are extremely nervous and usually refuse to feed, and soon die.

Takahe are rather unsuccessful parents and are left with only one chick to raise at a time. While raising chicks, they are fed daily for the first two weeks with 400 g (14 oz.) high-energy poultry pellets and 100 g (3.5 oz.) of chick food (see under hand rearing), per day.

Incubation and Rearing: Artificially incubated eggs produce tame chicks, which, if raised alone, imprint on the keeper. The eggs are incubated at 37.2–37.3°C (99°F) and at a relative humidity of 55–60%. At hatching, the humidity is increased to 75%. The chicks are left in the incubator for the first 24 hours, during which time they are fed a few tiny morsels with tweezers. When hungry, the chicks often peck at each other's wings and feet. Chicks lift their wings while they feed. When moved to a brooder box they are initially kept on a soft material, such as rubberized brooder mat, which is easily washed and sterilized for re-use.

Moorhen chicks are fed on duckweed and diced greens, moistened tinned dog food, minced ox heart, diced mealworms, flies, and commercial insectivore soft food. A multi-vitamin and calcium supplement is added to the food. Water is supplied in a shallow dish, with some pebbles to ensure the birds have a grip and do not struggle to get out of the water. A little duckweed is sprinkled on the surface of the water. After four days the chicks begin to pick at food in a bowl, particularly if a few small, live mealworms are added to the food. Hand feeding can be slowly reduced, and by 10–12 days stopped altogether. Water should be available at all times. From day one, once a day, the chicks are placed in a deep bowl of water for a few minutes to allow them to swim. The swimming stimulates the chicks to defecate.

By two weeks feathers begin to appear and the chicks begin to flit their tails in typical moorhen style. Provided there is enough shelter and a suitable climate, three-week-old chicks can be placed in an outdoor flight. Immatures are dark on the back, grayish below, the bill and shield greenish-brown.

Takahe chicks imprint easily; adult vocalizations are broadcast over speakers until the chicks are a month old. Initially fed every one-half hour with a puppet glove, chicks receive a vegetable mix containing finely diced carrot, potato, cabbage, spinach, clover, and a few drops of vegetable oil. To 30 g (1.0 oz.) of this mixture, 10 g (0.3 oz.) of Farex (high protein baby food) and some water is added to a consistency of moist porridge. After day four, some rearing mash is added to the diet and diced clover sprinkled over the water dish. By the second week chicks are fed every 1–1½ hours. During the third week vegetation is held in the glove while the chicks break off pieces of food.

I once worked with a hand-raised, deliberately imprinted, purple gallinule, which was a fantastic bird for environmental education and enjoyed people's company. This bird would take a piece of wholemeal bread, hold it in one foot and devour it piece by piece, much to the amusement of park visitors.

Ring Size: Moorhen: 8 mm, Gallinule & Coot: 11 mm

Crakes & Rails

Crakes and rails are classified in the family Rallidae. There seems to be no consistent criteria for determining whether a given species should be referred to as a crake or as a rail, and one suspects that the common names were bestowed rather arbitrarily. In many instances, species called crakes and species called rails can be found within

Spotted Crake
Porzana porzana

the same genus. The genus *Laterallus,* for instance, contains several species called crakes and several called rails. All species in the genus *Porzana* are called crakes except the North American sora rail *Porzana carolina*. The genus *Amaurornis* includes different species called crakes, rails, bush-hens, and waterhens. While this can be confusing, to use the terms interchangeably would result in further confusion. For example, the black crake *Amaurornis flavirostris* should not be referred to as the black rail simply because there is already a species called the black rail in the genus *Laterallus*.

Rails and crakes are distributed worldwide in suitable habitat. They are associated with aquatic environments, particularly marshes, but may also be found on the vegetated fringes of more open bodies of water, or even slow-moving rivers. Some species can be found inhabiting forested or brushy areas, or even fairly dry fields. They are found either singly or in pairs, and can often be located by their rattling duets. They are rather secretive, sometimes venturing out onto floating vegetation, but immediately disappearing into thickets when threatened. Their laterally compressed bodies enable them to escape through dense vegetation with ease. It is interesting to note that the words "rail" and "crake" both have their origin in attempts at describing the noise made by the bird. Almost certainly the original bird described was the European corncrake *Crex crex,* which resembles the bird's call.

Management: Black crakes *Porzana flavirostra* adapt well to aviary life; the secret is to provide the correct habitat. A small, shallow pond with some adjacent tall grass thickets, miniature bamboo, short reeds such as Cyprus, and sedge thickets will provide the cover these secretive birds require. Habitat F as described in the LANDSCAPE chapter will suit crakes. After a period they do tame to an extent, particularly aviary-bred birds, which soon learn to come out from hiding to fetch insects that are tossed to them. They can be mixed with other birds, but the one drawback with this specie is that they are partial to other birds' eggs. I know of one case where this species even ate the egg of a Victoria crowned pigeon. Black crakes kept in Colin Wintle's aviaries in Zimbabwe have also killed small waders, such as rails and snipe. They are best mixed with species that are either cavity nesters or those species that are capable of defending their nests successfully. Immatures are grayish-brown in color and have a whitish patch on the throat. They lack the bright red eyes of the adults.

The white-breasted or red and white crake *Laterallus leucopyrrhus* is unusual in the group in that it spends much of its time in bushes, trees or other high spots in the aviary. On the ground, the bird likes to be within sprinting distance of cover. They are quiet birds and communicate with a low whistle. A few extra calls are added when chicks are around. They are peaceful and, unlike many members of their family, they do not molest small birds, even when breeding. The one drawback is that they find nests irresistible and like to see how many of their group can pile into one nest. Four have been noted squeezing into a canary nesting basket. They roost in a nest all year and will use any suitable nest, or use an open-topped box or basket to which some leaves have been added.

Diet: Crakes are fed on a mixed millet, crushed maize, and sorghum diet. In addition they will pick at fruit, insectivore soft food, minced heart, soaked dog or cat biscuits, and are fond of insects. A grated boiled egg and insectivore mix is offered to the adults while they are feeding chicks. A variety of insects should be offered 2–3 times per day; the adults immediately carry these to the chicks. *Laterallus* are fond of earthworms.

Breeding: Black crakes are strictly territorial and only one pair should be kept per aviary. The nest is a small, deep cup made of plant material, and usually placed low in vegetation in close proximity to their aquatic habitat. The 2–4 buff-colored eggs are covered in gray- and rufous-colored spots. Both sexes incubate the eggs, which hatch after 15–16 days. The chicks leave the nest to follow the parents soon after hatching. The downy chicks are black in color with a pink bill with a vertical black band in the centre. The legs are grayish. Both parents care for the chicks for the first 10 days, offering them food from the beak. By three weeks the pink tip of the bill has disappeared and flight feathers start to show. They are fledged by five weeks but are tolerated near the new nest, which they may help build, and even assist in feeding the next brood. By two months old the chicks must be removed, as the adults become aggressive towards them and will kill them.

Notes from Greg Bockheim and Shannon Mezzell at Disney's Animal Kingdom on cooperative breeding in black crakes provide some interesting detail that needs to be considered in the pursuit of success with this species:

> "Single chicks were produced from all three of the crakes' first three nesting attempts. Aviary staff theorized that the crakes would rear larger broods of chicks once they had a support group that would assist in raising more chicks. After raising several clutches of chicks, the crake breeding results supported the hypothesis that a larger number of participants, supporting the cooperative breeding effort, results in more chicks being raised. The data collected also supports the idea that nestlings and juveniles take a more active role in the cooperative breeding association at a younger age when they have a larger peer group from which to learn.
>
> Eggs in our broods have taken 24–72 hours to hatch. With earlier nests our adult birds left the nest within 24 hours of the first chick hatching. In successive nests, our female remained brooding the chicks for up to four days, giving the eggs time to hatch. While remaining at the nest for this length of time the chicks that hatched earlier have gone for "walks" within close proximity to the nest, all the while being brooded and fed by older nestlings, juveniles and adult birds. After the female left the nest, much of her time was spent brooding the newly hatched chicks, but she soon left the chick-raising responsibilities to other group members and selected a new nest site in which to lay her next clutch of eggs. During the incubation process the alpha male built one or two brood nests near to where the clutch was being incubated.
>
> Frequent exchanges of food between birds, chick brooding and nest sharing (behavior in which juveniles and/or adult birds sit side by side in a nest) are the most frequent means of social contact and bonding type behavior that we have observed among our crates. Exchanging food occurs between adult and chick, adult and juvenile and fledgling, juvenile and juvenile, and juvenile and fledgling. All mature crakes, juveniles and fledglings have been observed feeding nestlings from two days of age. Our youngest fledgling recorded feeding another group member, nestling or juvenile, was aged 27 days. Thus the period in development, from fledgling (still dependant on its parents) to juvenile must be a relatively short stage. It is likely that a 7–14 day old chick can feed itself and begin to feed its younger siblings at around 20–27 days old, but is certainly not capable of completely fending for itself, given that it has yet to learn all the skills necessary for survival. Juveniles and fledglings have been observed feeding adults,

> then the adults returning the same morsel to the same bird, or feeding it to another group member. Fledgling and juvenile crakes have been observed participating in incubating eggs only for brief periods. From the age of 37 days, fledglings and juveniles have also been observed brooding nestlings. Juvenile and fledgling crakes have been observed brooding younger nestlings more frequently than they have been observed incubating eggs."

White-breasted rails construct a substantial nest for breeding and will add a thick lining of leaves to the roosting nest. Large or long leaves, such as bamboo, are used and pasted on when wet to give a smooth interior. Leaves are piled into the water dish for several days until they are soft and pliable. Leaves in the water dish are a good indication that breeding is underway, providing of course that there is no supply of wet leaves. Two eggs, white and sharply tapered (and huge for the size of the bird) are laid.

The chicks are fascinating to watch, jet black all over save for a white tip on the upper mandible. The feet are huge and the chicks resemble a stuffed toy. Chicks leave then nest within two hours of hatching and follow the adults in an unsteady fashion. By the second day they are quite steady and much faster. They tend to follow whatever adult happens to pass, which does not seem to matter as all the adults fuss over the chicks. For the first week, until the chicks are able to gain access to the arboreal nest, they are brooded on the ground. While being brooded the chick jumps onto the back of the adult and is completely covered by the adult's wings. The chicks are fed and preened by all the adults in the aviary. In the early stages the brooding bird feeds the chick, and is in turn fed by the other adults. An approaching bird runs up to the brooding adult and the food exchange is accompanied by a few clucks. The brooder in turn clucks to the chick, which pops its head out and takes the food from the adult's beak. By 30 days the chicks roost in the highest nest with the adults, and by six weeks of age they resemble the adults closely. Females of this species tend to be larger and heavier than males.

Unlike the white-breasted, Virginia *Rallus limicola* and Sora *Porzana carolina,* rails are strictly territorial and pairs defend strict borders. Insufficient space between pairs results in high levels of aggression and suppressed breeding. It was also found with both of these species that wild-caught specimens required a period of training onto artificial diets and initially would take only live food. The captive specimens of these two species did not build a complete ramp from the water to their nest, as did wild birds. When their chicks fell into the water the Soras never appeared to be concerned at the plaintive peeping of their young and merely watched them drown. The Virginia rails however have been observed rescuing their young and then brooding them.

The recommendation for breeding these two species is as follows: Place them in a simulated marsh environment in spring with a minimum area of 5 x 9 m (16 x 29.5 ft) per pair. The water level of the pool must be stable. Profuse numbers of aquatic invertebrates and duckweed should be present in the pool. When young hatch, make certain the ramps between the water and the Soras' nests are sloping inclines to avoid the cxhicks drowning. Supply mealworms and maggots in profusion.

Hand Rearing: Hand rearing crakes is not difficult but, being very small at hatching, they are rather delicate initially and require regular attention for the first week. Rearing is as described for the moorhen and fluff-tails. They also require abdominal contact with water to stimulate defecation.

Ring Size: African rail, corncrake, black crake: 6 mm.

Giant Wood Rail *Aramides ypecaha*

• Myles Lamont

Management: The giant wood-rail *Aramides ypecaha* is one of the largest species in the family Rallidae and lives in the marsh and forested areas of Central South America. This species, although taking advantage of a large body of water, does not require it, and is similar in respect to the nene goose, which prefers smaller pools of water in which to preen and bathe. This species needs to be watched closely if it is housed with other birds, as eggs and chicks will be readily consumed if given the opportunity. Pairs are best placed in aviaries to themselves, but can be placed adjacent to others, as this seems to help stimulate stronger pair formation and helps induce breeding activity. Sexes are most easily identified via DNA analysis, as they are monomorphic. The only behavioral difference is males being slightly more aggressive and having a larger role in the rearing and incubation of young, but this method of visual identification is not reliable. From a management perspcective, this species, and the family in general, can usually be treated in a very similar fashion to their larger cousins, cranes.

Diet: The natural diet of this species is quite diverse, and they are quite obviously opportunistic feeders. Fish, frogs, rodents, and other small vertebrates are readily taken, and this species has been observed cooperatively hunting down small prey, including small Passerines (Sheila.. McKay, 2005 Pers. comm.). Larger invertebrates such as beetles, mollusks, earthworms, grubs and caterpillars are also consumed. The wood-rails are not picky eaters and take well to a captive diet.

The captive diet used for this species at Hancock Wildlife Research Center (HWRS) is as follows:

- 3 parts Mazuri Crane Diet
- 2 parts Mazuri Waterfowl Maintenance Diet
- 2 parts Mazuri Sea Duck Diet
- 2 parts dog kibble
- 1 part Mazuri Parrot Maintenance
- 1 part Mazuri Zulife Softbill
- 1 part Kaytee Softbill Pellet
- 1 part Trout chow

A regular frugivorous diet was also offered and readily taken, but the species seems to enjoy feeding on halved oranges and apples. Mice, fish and mealworms can be offered on occasion as well. The pelleted portion is soaked overnight and fed in the morning

Breeding: Breeding for this species occurs all summer, but can commence as early as late March in the northern hemisphere. Baskets or platforms can be placed in the middle of a tree or shrub. Moss, peat and leaves can be used as a substrate, however the birds themselves will add some material to the nest, mainly grass, lichens and other soft debris found on the aviary floor. Females kept individually and not offered a nesting platform have taken to nesting on the ground against the base of a tree. A scrape in the sand approximately 5 cm (2 in.) deep was made and small leaves were added. Eight eggs in total were laid between two sibling hens in this nest, accounting for the normal clutch size of four eggs that has been observed at HWRC. Both females took part in incubation duties and acted as a pair once egg laying began; both were approximately 10 months of age when they laid, showing the early reproductive age for this species and likely the genus in general.

Birds are often quite nervous when caught up and introduced into new surroundings. They usually calm down within 2–3 weeks, sometimes sooner if unhandled. Territorial calling usually begins once a pair bond is established, and aggressiveness can be seen and often heard between adjacent pairs. Individuals may be seen carrying around nesting material within their aviary, usually a good indicator of breeding behavior. A female kept by herself attempted to build an elevated nest on her own since no platform was offered. Large amounts of vegetation could be found between the branches of a short tree, but no nest was actually constructed.

Sudden behavioral changes are often the best indication that laying has begun, however this usually begins after the clutch is nearly complete. The average egg-laying interval is two days, however three or four has been observed in younger hens. Wood-rails have been found to be excellent parents and both sexes are extremely protective and attentive to their offspring, males in particular. Brooding duties are shared, and while one is caring for the offspring, the other is often gathering food or acting as a sentry. Food is carried by one parent to the nest site, where it is not uncommon for the chicks to remain for 48–72 hours, after which the nest is vacated and brooding occurs on the ground, often in thick vegetation. An interesting behavior of constructing a "roost site" has been observed in the most experienced breeding pair at HWRC. A collection of material similar to that used in the actual nest was placed on the ground amongst English ivy where the pair would brood the chicks overnight.

Giant Wood-Rail *Aramides ypecaha*

Age	Chick #1 weight (g)	Chick #2 weight (g)
Day 2	28.4	22.7
Day 3	28.4	28.1
Day 4	29	28.5
Day 5	39.8	36.9
Day 7	42.5	37
Day 8	45.3	39.7
Day 9	48.2	42.5
Day 10	51	49
Day 12	59.5	56.7
Day 14	65.2	59.5
Day 15	62.3	56.7
Day 17	68	59.5
Day 18	76.5	70.8
Day 21	96.4	87.9
Day 29	164.4	153
Day 37	252.3	195.6
Day 46	371	277.8
Day 51	442	328.9

Born Sept. 15, 2005

Hand Rearing and Incubation: Although considered precocial in nature, a more accurate description would be semi-precocial, as chicks require regular hand feeding and do not show much interest in moving live food placed in front of them. Hand rearing should be attempted at a very early age or after 4–5 days; otherwise chicks can be difficult to convince to eat, as they do not recognize their keeper as a food source. Chicks can be treated similar to young cranes, needing to be fed from tweezers or fingertips in order to eat. Food can be placed in front of them, however it is not usually recognized until at least a week of age. Some clutches begin eating on their own as early as four days; others have taken up to two weeks.

Gerald Kaufmann briefly describes hand rearing Sora *Porzana carolina* and Virginia rails *Rallus limicola:*

> "Chicks were kept in a box along with a saucer of water and a light bulb, and were hand fed every two hours from 4:00 a.m. until midnight for about one week after hatching, at which time they began feeding for themselves. I mixed two hardboiled egg yolks, one can of horsemeat dog food, one can of nearly all vegetable dog food, and several vitamin drops for rail food."

Hand rearing diet at HWRC was composed of the following ingredients:

- 2 parts Mazuri waterfowl starter
- 2 parts Mazuri crane diet
- 1 part Mazuri waterfowl maintenance
- 1 part Kaytee softbill
- 1 part Mazuri Zulife softbill diet
- 1 part Mazuri insectivore diet
- 1 part Mazuri parrot maintenance diet

All was soaked prior to feeding, with Mazuri Zulife softbill being a favorite item. Chicks that are difficult to get to eat are offered live insects such as waxworms and/or mealworms.

The captive diet previously used for the Lord Howe Island rail was:

- Poultry layer pellets
- Whiskas pelleted cat food
- Grated cheddar cheese
- Cottage cheese
- Madeira cake
- An occasional complete raw egg
- Newborn lab mouse
- Mealworms
- Woodgrubs (Cerambycidae family)
- Other invertebrates

During the non-breeding season, diet consisted of:

- 5 g poultry pellet
- 5 g grated cheese
- 5 g mealworms
- 20 g Whiskettes cat food

Breeding season diet consisted of:

- 5 g poultry pellet
- 10 g grated cheese
- 20 g Whiskettes cat food
- 10 g cottage cheese
- Raw egg once weekly per pair
- Baby mice every 2–3 days

Chicks are usually pulled in pairs, so imprinting is reduced. To date, hand-reared birds display no unusual behavior, reproductive or otherwise. Both hand- and parent-reared females have been observed making a submissive copulation posture towards their keepers and aggression levels are to the same degree for birds reared both methods.

Day-old chicks are placed in stainless steel brooders with a 12.7-mm (0.5-in) wire bottom, covered in a rubber drawer liner to protect the toes of the young birds. An infrared heat lamp and ceramic heater are used as a heat source with an ostrich feather duster placed in the corner to give the chicks a location in which to hide when intimidated. Chicks that are pulled from their parents can be lured out of their hiding by replicating an easily imitable cooing noise that the parents make.

Chicks are moved to either a larger brooder or a small brooding room with a sand substrate at 2–3 weeks. In good weather, chicks can be moved outdoors as soon as feathers begin to emerge (approx. 4–5 weeks). It is not wise to introduce different clutches unless at a very early age, as there is likely to be some strong aggression between individuals. Sibling rivalry is not uncommon, but has never been observed to the extent of that of cranes. Some breeders have witnessed stronger aggression, and separation may be necessary. Some chicks have proven to be fatally aggressive to their siblings without any prior indication of aggressiveness, therefore it is recommended that close observation be maintained, and possible separation, as the chicks increase in age (McKay, S. 2005. Pers. comm.).

Parent Rearing: As previously mentioned, this is normally a very successful method of reproduction as parents are quite dedicated, despite prior parental experience. If pairs are adjacent to one another, care must be made to ensure there is no possibility of jealous pairs pulling and killing neighboring chicks through the wire. A solid partition between pairs will help remedy this, however they will often dig to some extent below the barrier if it is not placed deep enough. The lower half of the perimeter must be tight and secure, as when chicks leave the nest they search immediately for cover and will attempt to escape if there are any holes in the enclosure. Chicks often select a small, dark cavity of some sort within the aviary to spend the majority of their time until they are old enough to fend for themselves. The placement of an oak barrel or large plant pot covered in vegetation will be warmly accepted. The pair will exchange parental duties throughout the day, but both will greet their keeper with beak and claws if they attempt to even enter the aviary for anything other than daily feeding.

Chicks should be weaned by four weeks of age and will begin to follow their parents around the aviary. Around this age they need to be close banded, normally at 4–5 weeks, with a 14 mm inside diameter band above the knee joint. Plasticine or a similar material may be needed to keep the band above the knee for a few days until they grow into it.

Giant wood-rails have an incubation period of 24 days. Eggs are a cream to off-white color with brown spotting over the entire egg, with a greater concentration of pigmentation towards the rounded end of the egg. A lighter gray spotting is also visible in a much lower opacity, as if the coloration occurs on two separate layers with the brown speckling being found above the grey.

Band Size: 14 mm

Flufftails *Sarothura* spp.

• Glen Holland

Introduction: Named as crakes by Howard & Moore, but known as flufftails in Africa, the sub-family Coturnicops that includes twelve species. They are part of the family Rallidae and the order Gruiformes. Flufftails inhabit wetlands and densely vegetated river and pond banks. Associated with rank grass, reeds and sedges where they are found in pairs that are strictly territorial. They are extremely active and fast moving. Flufftails are usually located by their call, which is a mournful *wooah-hoo* sound often made in wet or misty weather. A variety of other calls exist that include duets between a pair. They call by day and night, largely in the summer breeding season.

Management: Flufftails are not well established in captivity. The avicultural experience described here is based on work with the red-chested flufftail *Coturnicops rufa.* Although common in some locations, they are secretive and easily overlooked. Wild birds brought in after being caught by cats have been extremely nervous, and initially they have refused to feed.

A shallow pond surrounded by grass and sedge thickets will provide both cover and nesting sites for the birds. They are extremely territorial and only one pair should be kept per aviary. In a large aviary it is possible to keep two pairs but only one pair will breed successfully. Dominant males will also "rape" other females in the aviary. Females tend to be the more aggressive birds. Species such as pygmy geese, waxbills and doves can be safely mixed with flufftails.

Flufftails have a recorded longevity of over 10 years. To establish wild birds, they should be confined to a small box where they are not disturbed and an ample supply of mealworms and termites made available.

Flufftails are relatively hardy and are able to withstand overnight temperatures to freezing. They do however require daytime temperatures in excess of 14°C (57.2°F). In countries with more severe climates they are best housed indoors during the winter months. Even when housed indoors they require a thicket into which they can retreat if threatened.

Diet: These birds naturally feed largely on insects with a small quantity of seed. In an aviary they will take some millet and sunflower, but a commercial softbill mix with some minced ox heart and a good supply of insects is essential. They will learn to come out of hiding to fetch their live food, which can be tossed to them at close range.

Breeding: A male in breeding condition will strut about while picking his feet up high. They call regularly in the breeding season, the call of one bird being taken up by another as it ends. The nest is a shallow cup made of vegetation, placed 20–30 cm (7.9–11.8 in.) above the ground in a grass thicket. Both sexes, the female by

Red-chested Flufftail
Sarothrura rufa

night and the male by day incubate the 2–6 white eggs by day, for 14–16 days. Incubating birds leave the nest as soon as the aviary is approached.

Artificially incubated eggs should be placed in a bowl lined with grass when they begin to pip. Hatching will be complete in 12 hours. The chicks remain in the nest for the first 3–4 days and, as their legs are initially rather weak, the grass-lined bowl will avoid spraddled legs. Initially, the chicks are fed from the parent's beak. The chicks are black, with the bill pink at the base, centre black, tip white, and legs purplish black. The chicks *cheep* while following their parents, and the moment the parent catches an insect they wave their heads from side to side begging to be fed. Males construct a number of roosting nests to which they return at night. Chicks are fully feathered black by 42 days and then immediately begin to molt. The sexes can be differentiated by two months. As described for black crakes, the first brood will help to care for and feed the second brood and so on, up to four broods or more. To succeed with these birds, the correct habitat must be provided.

Hand Rearing: The chicks first feed at 36 hours and have to be fed from tweezers five times a day for the first four days, or until they begin to feed themselves. The hand-rearing diet is as described for moorhens and fluff-tail chicks; they also require placement in shallow water to stimulate defecation. As they have soft feet, the base of the brooder box should be covered with a cloth or sacking. While in a brooder box with a light, if a window is left open it will allow insects to be attracted to the light, ensuring a natural food supply for the chicks. At a month old, in reasonable weather, the chicks can be placed in an outdoor aviary. It is important to maintain regular contact with the chicks to ensure they grow tame. The most difficult stage in artificial rearing is when the chicks are moved. They are highly strung and chicks have been known to die from stress within minutes of being moved. Adults may also go into shock when caught.

Ring Size: 4 mm

Species Requiring Similar Management:

Eulabeornis
Rallus
Rallina
Aramides
Coturnicops
Porzana
Laterallus
Gallinula
Porphyrio and *Fulica* spp
Jacanas

Contribution: Mr. Colin Wintle, Bromley, Zimbabwe.

References:

Gibson, L. 1979. Breeding Laterallus leucopyrrhus. *Avicultural Magazine*. Vol. 85.

Bockheim, G & S. Mezell, 1999. The Black Crake. An Effective Cooperative Breeder at Disney's Animal Kingdom. *Avicultural Magazine*. 105(1): 12–21.

Holyoak, D.T.1969. The behaviour of captive Purple Gallinules. *Avicultural Magazine*. Vol. 76.

Kaufmann, G.W. 1987. Growth and development of Sora and Virginia rails. *Wilson Bulletin* 99(3): 432–440.

Lamont, M.M. 2006. Breeding the Giant Wood-Rail. American Federation of Aviculture Annual Proceedings Dallas, Texas. Pp. 63–67.

Lourie-Fraser, G. 1983. Captive Breeding of the Lord Howe Island Woodhen. *Watchbird* 10(2): 30–44.

Urban, E.K., C.H. Fry, & S. Keith, Eds. 1986. *The Birds of Africa*. Vol. 2: *Game Birds to Pigeons*. London & New York: Academic Press.

Taylor, Barry. 1998. Rails: *A Guide to the Rails, Crakes, Gallinules, and Coots of the World*. New Haven, CT: Yale University Press.

Egg Weights for Giant Woodrail *Aramides ypecaha* for 2006

Clutch #1	Fresh Egg Weight	after 10 days	after 17 days	Weight at Hatching	Egg Size
Egg #1	45.4g	42.5g	39.6g	parent reared	5.1cm x 3.6cm
Egg #2	42.5g	39.6g	36.8g	"	5.3cm x 3.4cm
Egg #3	45.4g	42.5g	41.0g	"	3.2cm x 3.5cm
Egg #4	41.4g	36.8g	34.1g*	cracked shell	5.2cm x 3.3cm
Egg #5	44.5g	42.5g	39.5g*	31.2g	5.1cm x 3.5cm

** pulled for AI*

Clutch #2	Egg Weight after 8 days	after 17 days	after 22 days	Weight at Hatching	Egg Size
Egg #1	42.5g	36.8g	34.0g	25.5g	5.2cm x 3.0cm
Egg #2	48.2g	45.3g	42.4g	31.2g	5.9cm x 3.1cm
Egg #3	48.1g	Not pulled for AI	N/A	N/A	5.75cm x 3.05cm
Egg #4	48.4g	Not pulled for AI	N/A	N/A	6.0cm x 3.5cm

Sandgrouse, Jacanas, Plovers, Gulls *et alia*

20 Ciconiiformes

20.01 Pteroclidae Sandgrouse

Glen Holland

Introduction: The 16 species of sandgrouse occur primarily in Africa and Asia, with a single species found on the island of Madagascar. Formerly placed in their own order and considered to be related to the pigeons, which they resemble superficially in appearance, they are in fact more closely related to the woodcocks, snipes, seed-snipes, plovers, and other waders, as evidenced by their nesting habits and confirmed by DNA studies. They currently occupy their own family, Pteroclidae, in the order Ciconiiformes. The family includes only two genera, with the two species in the genus *Syrrhaptes,* the Tibetan sandgrouse *S. tibetanus* and Pallas's sandgrouse *S. paradoxus,* confined to the mountains of central Asia and western Europe. The genus *Pterocles* has a wider distribution, with most of the 14 species scattered throughout Africa. Of these, the chestnut-bellied sandgrouse *P. exustus* also occurs in southern Asia, as does the crowned sandgrouse *P. coronatus*. The spotted sandgrouse *P. senegallus* is found in Africa and in southwestern Asia. The nominate race of Lichtenstein's sandgrouse *P. lichtensteinii* occurs in north central Africa, with a subspecies, the close-barred sandgrouse *P. l. arabicus,* extending to south central Asia. The black-bellied sandgrouse *P. orientalis* and the pin-tailed sandgrouse *P. alchata* are of southern Palearctic distribution, the latter wintering to southern Asia. The subspecies *P. a. caudacutus* occurs in extreme northern Africa, in Morocco, Algeria, Tunisia, and Libya. The painted sandgrouse *P. indicus* is found in central and southern Asia. The strictly African species include Burchell's *P. burchelli,* the four-banded *P. quadricinctus,* the double-banded *P. bicinctus,* the black-faced *P. decoratus*, the yellow-throated *P. gutteralis,* and the Namaqua *P. namaqua*. The Madagascar sandgrouse *P. personatus* is found only on that island.

Sandgrouse prefer open arid country where the soil is sparsely covered by pebbles, grass, and short succulent shrubs. Nomadic by nature, they follow supplies of both food and water. Their flight is powerful and fast. Once a day they visit waterholes and, outside of the breeding season, flocks may number thousands of birds that arrive at the waterholes, drink, and disappear, all within a short space of time. Different species drink at different times of the day. In the natural habitat of most species, the winter nighttime temperature often drops to freezing, but the days are generally warm and dry. Some of the Asian species are, however, completely winter-resistant.

Management: Sandgrouse are rather unique birds that are naturally adapted to harsh conditions. As a survival measure, the birds lead a nomadic life. Of the 12 species occurring on the African continent, the best known is undoubtedly the Namaqua. It is with this species that most of my avicultural experience is associated. Their calls are best described by the descriptive Afrikaans name: *Kelkiewyn*. When newly captured or imported, these birds are prone to stress. This problem needs to be dealt with to avoid a high of mortality rate. The first step is to clip one wing so that the birds are unable to fly up and knock their heads on the top of the aviary or cage. Some breeders keep one wing of their birds permanently clipped. The second step is to provide additional warmth, particularly at night and in cold weather. This supplemental heat can be gradually reduced after a week. The third step is to add an electrolyte/anti-stress formula to the drinking water. Additionally, the food should be scattered liberally on the cage floor. Initially it is advisable to confine the birds to a cage, the front of which has been covered in soft nylon gauze, against which the birds are unable to damage themselves. They should be left in a quiet spot and fed once per day. When moving sandgrouse in a carrying box, the box should have a foam layer on the roof, and gauze over the exposed areas, again to keep the birds from damaging themselves.

Sandgrouse chicks drinking from male's water-soaked breast.

An aviary for sandgrouse should be situated in a dry, sunny position. It should measure at least 12 m² (130 sq. ft.) and be sparsely planted, as described for Habitat E, with clumps of grasses interspersed with open patches of sandy soil and scattered pebbles. Hinsley and Hockey (1989) used an aviary in a greenhouse at Cardiff University to house seven male, two adult female, and one juvenile female double-banded sandgrouse. This aviary measured 12 x 6 x 2 m (40 x 20 x 62 ft.) and was constructed of soft nylon mesh held under tension by a string framework. These birds are unable to withstand long periods of cold wet conditions, but are naturally accustomed to cold nights in dry conditions. The secret is to keep them, and their aviary, dry. The greenhouse aviary at Cardiff was furnished with heat lamps (a combination of 150 W spot lamps and 250 W infrared lamps). Although these lamps were constantly available, the birds were observed roosting away from them even with night time temperatures as low as 0–5°C (32–41°F). They are not aggressive birds, and in large aviaries a few pairs can be housed together.

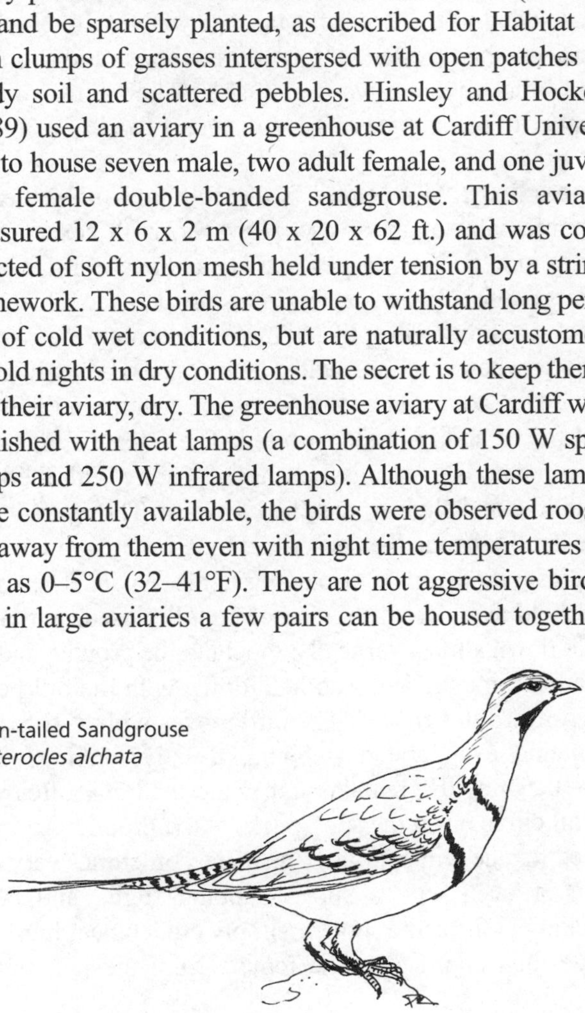

Pin-tailed Sandgrouse
Pterocles alchata

Each pair should have approximately 20 m^2 (215 sq. ft.) of ground space. After cleaning the aviary, which should be performed regularly to prevent infections that these birds are not exposed to in their natural, rather sterile environment, a little shell grit should be scattered over the sandy areas. A single longevity record is of 17 years.

Diet: In the natural state, sandgrouse eat mainly protein-rich legume seed, so the inclusion of legume seed, e.g., lentils, in the captive diet is encouraged. Sandgrouse in captivity feed on a mixed seed diet of finely crushed maize and sorghum, millet, game bird crumbles, and a small amount of sunflower. If not available in the aviary, a small quantity of diced green food should be included in the diet. These birds enjoy pecking the seeds off the heads of seeding grasses.

Hinsley and Hockey (1989) describe the diet used at Cardiff University as "three parts mixed millets, two parts chick crumbs, two parts red milo, two parts groats, two parts whole hemp and one part linseed." Approximately twice a week their birds were also given Haith's Rearing and Condition Food, provided dry. The food was placed in a trays at various sites in the aviary and scattered on the ground. A soluble vitamin supplement was added to the water for three days every two weeks. Grit, crushed oystershell, cuttlefish bone, and salt/mineral blocks were always available. Their birds did not appear to eat any green food or insect food.

Breeding: Monogamy is the rule and males defend their females aggressively. However, both sexes sometimes indulge in extra-pair activity and may form trios. In a large aviary where a few pairs are kept together in a flock, the pairs will split and move to separate areas in the aviary when they are ready to breed. In courtship, the male Namaqua chases the female on the ground with his tail raised and his head lowered. The female's invitation to copulate is a submissive sitting posture with the head held low and slightly stretched forward. The courtship of the double-banded, as described by Hinsley and Hockey, begins with the male displaying to the female. The wings are held away from the body but still closed, the tail is held down almost vertically, touching the ground, and the head is retracted and also held down with the bill pointing toward the ground. The bird moves with a stiff, slow gait and exaggerated stepping, usually within 1–3 m (3–10 ft.) of the female, and while walking, often in a small circle or on the spot, pecks at the ground. An interested female will approach the male, sometimes very rapidly or even with a short, flapping flight, and begin pecking at the same spot, or simply crouch near him. The male then approaches the female from the rear with a slow, deliberate gait. During copulation, the male slowly opens his wings. By the time he dismounts over the female's head, his wings are spread totally laterally.

The nest is a small scrape in the sand, about 120 mm (5 in.) across, often among pebbles, and is sometimes sparsely lined with grass. The male generally excavates a number of scrapes, from which the female then selects one, usually situated on an exposed stony site. The chosen site is defended against any strange bird approaching within a distance of approximately 2 m (7 ft.). Females are more aggressive in this behavior than males. Birds may even peck at an approaching hand or, if disturbed near the nest, perform a distraction display. The clutch consists of three, rarely two, eggs which are glossy, creamy-white to pinkish or buff and speckled with dark brown and gray. The eggs are laid 48 hours apart. Both sexes incubate for 21 days from the day the third egg is laid. This is the shortest incubation period of all sandgrouse. The female incubates during the day and the male at night. The double-banded sandgrouse sexes do their incubation in the reverse of observations of other sandgrouse species in the wild.

When the aviary is entered, the male will position himself between the intruder and the female. Parent birds remove the eggshells from the nest. The precocial young hatch covered with down and are able to feed themselves soon after leaving the nest. After hatching, the first chick follows the female away from the nest, while the male incubates and hatches the second egg. The male brings chicks water in his tubular shaped belly feathers, which are adapted for this purpose. Each day when he flies back to his brood from the waterhole, he returns with a load of water which he has acquired by soaking these specialized feathers.

The young feed on very small items, particularly seeds and green food such as clover and other legumes. Young are able to fly by three weeks of age; at this stage they are half the size of their parents. By two months of age they resemble the adults. Sandgrouse are usually single brooded and remain as a family group for many months. However, some captive pairs have been reported to produce up to three clutches in a season, with one or two unsuccessful clutches. Hinsley and Hockey (1989) describe their chick diet as a mixture consisting of three parts chick crumbs, two parts red lentils, two parts green lentils, two parts Haith's Rearing and Condition Food, one part hemp seed, and one part linseed. All of the ingredients, except the rearing food, are finely ground in a blender before being mixed together. As the chicks grow, the food is less finely ground and whole seed is gradually introduced, starting with millet. When about two months old, the chicks are eating the adult diet supplemented with lentils and rearing food. Newly hatched chicks and parents are supplied with a shallow, sloping water tray with a gravel base. As is usual with precocial young, the chicks are self-feeding but also receive encouragement to feed, and some actual food items, from the parents that drop these items in front of them.

Incubation and Rearing: As sandgrouse are rather highly strung birds, it is not uncommon for them to refuse to incubate in captivity. By removing clutches for artificial incubation, the birds will lay up to four clutches in a season. Eggs are incubated at 37.5°C (99.5°F) with a relative humidity of 55%. Chicks are left in the incubator for the first 16–18 hours after hatching.

Pin-tailed *Pterocles alchata*
Namaqua *Pterocles namaqua*

Day	Weight (g)		Day	Weight (g)	
	Pin-tailed	Namaqua		Pin-tailed	Namaqua
1	13.8	5.43	**20**	90.7	55.83
3	18.6	8.15	**22**	99.98	62.58
6	29.15	13.63	**24**	109.33	69.15
8	35.3	18.21	**26**	115.68	75.47
10	44.4	23.45	**28**	129.55	81.46
12	60.55	29.32	**30**	137.73	87.11
14	66.25	35.62	**36**	154.5	101.68
16	70.8	42.23	**42**	180.5	112.8
18	84.25	49.01	**53**	214	125.84

Once they appear dry, they are placed in a brooder box that is warmed, as described under the section on the CALIFORNIAN QUAIL, by a brooder lamp. The floor of the brooder is covered with paper towel upon which small seeds and finely crushed turkey starter crumbles, rearing mash, small millets, and some finely diced greens are sprinkled. Aviculturists have reported using young quail to teach the sandgrouse to feed, but Etienne Dierick found that young Pallas's sandgrouse begin feeding without any coaching. Penn Lloyd, however, describes having to coach young Namaquas to feed. Some breeders use pipettes or soaked cotton wool to give the chicks water, but by gently pushing the chick's bill into a shallow dish of water they soon learn to drink. The latter routine should be performed twice daily for 3–4 days, until they begin to drink independently. Recently dried chicks respond to a touch or nearby movement by stretching up and making searching movements with the head and bill while calling. They should remain in the brooder, at gradually reduced temperatures, until at a month to six weeks, in favorable weather, they can be placed into an outdoor flight.

The growth chart was compiled for pin-tailed sandgrouse by Shelley Hinsley (UK) and the weight gain for Namaqua sandgrouse extrapolated from this.

Ring Size: 6 mm

Species Requiring Similar Management:
Sandgrouse *Syrrhaptes* and *Pterocles* spp.
Quail
Partridge

Contributor: Penn Lloyd, Percy FitzPatrick Institute for African Ornithology, Cape Town, SA.

References:

Dierick, E. Sandgrouse, a review. *AFA Watchbird* 17 (5): 22–23

Hinsley, S.A & D.J. Hockey. 1989. Breeding and breeding behaviour of double-banded sandgrouse. *Avicultural Magazine* 95 (2): 62–73

Maclean, G.L. & C.H. Fry. 1986. Pteroclidae, sandgrouse. Pp. 422–441 in E.K. Urban, C.H. Fry & S. Keith (eds.), *The Birds of Africa*, vol. 2. London and Orlando: Academic Press.

20.02 Scolopacidae Waders

Achim Johann

Introduction: The suborder charadrii includes about 365 species and has members in all parts of the world. Most live near water and eat invertebrates or other small animals; however some are pelagic, some occupy deserts and a few are found in thick forest. The order is usually divided into three suborders:

- The waders, or Charadrii, that are typical wading shorebirds in both coastal and freshwater environments.
- The auks, coastal species that nest on sea cliffs and "fly" underwater to catch fish.
- The suborder Lari, which includes the gulls and skuas and their allies.

This section is devoted to two wading species in that family Scolopacidae but should be read in conjunction with the chapters on STILTS and PLOVERS as all three chapters provide information relevant to each other. The wading species covered in this chapter are the ruff *Philomachus pugnax* and the redshank *Tringa totanus*. It should be noted, however, that although dealt with in separate chapters, much of the information on stilts, plovers and waders is relevant to each of the other groups. The ruff occurs in NW Europe through Siberia to Chukotskiy Peninsula and Sea of Okhotsk. They winter from Mediterranean and Sub-Saharan Africa through the Middle East to the Indian Subcontinent. The redshank has six subspecies and occurs from NW Europe through Siberia, Mongolia, Russian Far East, and NW China. They winter from Mediterranean and Sub-Saharan Africa through the Middle East to the Indian Subcontinent and South East Asia. They are chiefly migratory, although some birds in Iceland and W Europe are virtually resident.

Management: There is a great interest among zoos in keeping ruffs. However, there are very few prolific breeding colonies in captivity to fill the demand. This is typical for animal species for which the answers to husbandry issues are still being discovered. Ruffs are listed on the EEP Charadriifomes Regional Collection Plan as a species that is recommended to be kept because of their high educational value (behavior, flagship species for wetland protection). In spite of the general great interest and evident husbandry problems reports on the maintenance of ruffs are very scarce. Some collection of data and special husbandry research has to be done to fill gaps in our knowledge or to prove hypothetical thoughts. Redshanks are listed on the EEP Charadriiformes Regional Collection Plan as a species recommended to keep because of its educational value and as a model species. There is limited space for the different species of the genus *Tringa* in captivity. Redshanks are given preference as they are currently bred quite successfully and, therefore there is a founder stock for long-time propagation. Furthermore, they add attractively to mixed-species enclosures because of their unique vocalization. Husbandry reports are scarce. Experiences at Rheine Zoo, Germany proved that this species is ideal keep and propagate in captivity and that breeding and rearing do not pose great problems.

Housing should allow maintenance of at least 10 ruffs. However, it is currently not possible to determine the minimum group size for a breeding unit of ruffs, nor can be said what are the minimum standards to maintain and breed ruffs. To meet the needs for fully developed courtship display, large landscaped enclosures and a group size of at least 10 (5.5) birds seems to be appropriate. Systematic collection of data on group composition related to breeding results and enclosure size is required. Landscaped outside aviaries seem to be ideal. There are reports (pers. comm.) that redshanks might be highly aggressive among each other if kept in smaller groups, or even as pairs. This might be seen in relation to aviary size. Small sized aviaries of ~10 m² (32.8 sq. ft.) might be a minimum. There are no minimum standards for aviaries to successfully maintain redshanks available. Most ruffs are kept in mixed-species wader exhibits, some laid out as walk-through exhibits. At Rheine Zoo a mixed species wader-enclosure (walk-through) for ~80 waders of six species (including ~30 ruffs) has a floor area of 370 m² (3982 sq. ft.) and height of up to 8 m (26.3 ft.). These dimensions are well suited to house this number of birds. Breeding results (including natural brooding) are good. However, success by natural rearing of both ruffs and redshanks is unlikely because of the high density and the territorial behavior of the birds. In the 1999 breeding season, four chicks had been left for natural rearing, as they had been the last hatchlings in the aviary. After one day, three had been found dead, the last one was not seen after four days. Competition with other ruffs or other wader-species has been very low. The redshanks started with five birds kept in a mixed-species exhibit. The number had been enlarged to seven by retaining two offspring.

Soft netting on the roof is required as ruffs fly straight up when disturbed or in panic. Nylon netting with small mesh of 2 cm² (0.8 sq. in.) is recommend to prevent harmful accidents with skull or wing damage; small mesh prevents accidental hanging. Adult ruffs are hardy birds and can easily withstand freezing temperatures. An indoor room with a temperature of about 5°–10°C (41–50°F) is fully sufficient to allow permanent access to water and food in winter. The ruffs should also have access to outside enclosures in winter.

A varied landscaped enclosure is essential for both ruffs and redshanks. Short-cut meadow will allow for courtship display and foraging behavior; clumps of grass and weeds as well as small scrubby places to hide and for nesting. Sandy areas and places covered with pebbles of different sizes will provide for roosting. A very shallow pool, preferably with a sandy or muddy floor, will provide for foraging behavior, roosting and bathing. Sandbanks are favored places to roost. A slowly running brook stimulates natural foraging behavior. Floors of indoor rooms should be solid and easy to clean. Coverage with straw or hay and also straw bales should be supplied; ruffs like this bedding for roosting. Bales are used as elevated outlooks, and foraging behavior is stimulated when mealworms are scattered in the straw-bedding. Food dishes are placed on the open floor and food places have to be cleaned thoroughly each day.

The life cycle of ruffs shows a high degree of seasonal variance. Influence of light seems to be crucial to maintain the life cycle and for synchronizing behavior. At Rheine Zoo, Northwestern Germany, ruffs and redshanks are kept on a light cycle in winter of 12 hours day/12 hours night (7:00 a.m. to 7:00 p.m. light). Increasing daylight in spring happens by natural light only, as the animals have all year access to the outdoor aviary; there are also large skylights covering the indoor rooms.

Ruffs are peaceful with any other inhabitants in a mixed-species enclosure. At Rheine Zoo, there is no indication that the ruffs or redshanks suffer in any way from other waders co-inhabiting the enclosure (Avocets, stilts, redshanks, ringed plovers and lapwings). Oystercatchers had to be removed, as there were hints that they damaged eggs laid by the other waders. The most common method for identification is ringing. Authorities recommend closed aluminum rings of 5.0 mm for male ruffs and rings of 4.0 mm for females. However, experience shows that these recommended sizes might be too tight. Currently closed aluminum rings of 5.5 mm are used for both sexes and are found to be appropriate. Open aluminum rings are preferable if allowed by local authorities. Additionally colored plastic rings allow individual identification at a distance.

Adult ruffs are hardy birds. Causes of death are mostly related to accidents (drowning, fractures, becoming tangled in scrub or bamboo). Older birds show symptoms of stiff legs and lameness (with thickened tibiotarsus). Older males (~10 years) especially show symptoms of dropped wings, in most cases the same animals with stiff legs. They can live well with these, and are also able to fly for one or two more years. However, over time they show more signs of old age (imperfect molting, no participation in courtship display).

Diet: In the wild during the breeding season they feed mainly on terrestrial and aquatic insects; outside breeding season the diet is more varied, including caddisflies, water bugs, mayflies and grasshoppers, small crustaceans, spiders, small molluscs and annelid worms. Birds wintering in Sahel often feed on rice seeds and other cereals, grasses and aquatic plants. They are both nocturnal and diurnal feeders.

Ruffs adapt to diets offered in captivity. A softbill mixture enriched with small amounts of meat and shrimp can be used as a basic food. At Rheine Zoo the year-round diet consists of a softbill mixture (enriched with mashed carrots and other vegetable, curd, cooked rice), dried and frozen shrimp, dog chow, seed mixture, and small cut fruit. There may possibly be a need for higher level of vitamin E in winter and early spring to influence breeding behavior. In spite of the sesonal lifestyle and migratory behavior, there is no indication that year-round provision of high quality food, as is the case in captivity, has a negative influence on the health of ruffs. Also, in the wild ruffs obviously have a well balanced intake of food even when migrating, which is energy-consuming.

Breeding: Ruffs breed naturally in the Arctic and sub-Arctic, and south to temperate zones in Europe. They require habitat with adjacent feeding, lekking and nesting areas. This occurs naturally in the coastal tundra to forest tundra, near small lakes, in marshes and deltas with shallow water margins, variably covered with vegetation, and dry mounds and slopes with some low scrub for leks. In Western Europe, they are found in damp to swampy meadows, often with shallow pools or ditches, and patches of birch and willow scrub and short grass or bare ground. Outside breeding season, they prefer muddy margins of lakes, rivers or flooded areas, including brackish, saline or alkaline waters. Ruffs are sexually dimorphic. The male head to body length is 26–32 cm (10–12.5 in.), wingspan 54–58 cm (21–22.8 in.), weight 130–254 g (4.6–9 oz). The females head-body-length 20–25 cm (8–10 in.), wingspan 48–52 cm (19–20 in.), weight 70–170 g (2.5–6 oz.). Sexual identification by size and weight is possible as early as the first few weeks after hatching. The males have individually unique courtship plumage. This is extremely variable and a useful means to identify individuals.

Ruffs are polygynous, with no true pair bond. Mating generally takes place at traditional leks where males gather to display and females visit. Three types of males appear on arena:

- residents, possessing small territories,
- marginals, without territories,
- satellites appearing on periphery and mating opportunistically.

There is no site fidelity and natal philopatry. Eggs are laid from early May to mid June. They nest solitary or semi-colonial, occasionally only a few metres apart. Nests are well concealed in marsh plants or meadow grass; shallow scrape lined with grass, leaves and stems.

Redshanks are monogamous and nest either solitary or in loose colonies. The nest is typically at the base of a tall clump of grass, with leaves covering overhead.

Whereas the courtship display of the male ruff is a spectacle that cannot be missed, copulations are rarely observed. Nest building is by the female alone and brooding happens very secretively. In landscaped aviaries it is difficult to find nests. There is an indication that females leave nests and eggs when disturbed, or even when they realize that their nests are detected (by keepers). At Rheine Zoo, nest sites are marked or noted on a map when detected. Clutches of eggs are transferred to the incubator as soon as the full clutch of four eggs (or three) is completed. Multiple clutching is possible. Observations from the wild, however, indicate that ruffs do not normally replace lost clutches as their season for brooding is very limited. At Rheine, natural incubation in the mixed-species exhibit with a large population of different waders can be successful, as has happened several times with non-detected clutches. Chicks had to be transferred for hand rearing, as conditions in the enclosure are not suitable for natural rearing. There are observations of female ruffs destroying their own eggs or eggs laid by other females. This may be caused by disturbances or lack of suitable places to hide their nests.

For ruffs the 3–4 eggs are laid at 24–36 hour intervals with the incubation lasting 20–23 days, starting with the last egg. Incubation and brood rearing is by the female alone. The chicks fledge at 25–28 days. The usual age of the first breeding is two years but they have been known to breed at a year in captivity. The oldest ringed wild bird was 10 years, 11 months but the oldest bird in captivity was 14 years, 3 months, 11 days.

With redshanks, which are monogamous, the four eggs (3–5) are laid at intervals of 35–43 hours. Replacement clutches laid when clutches are lost in the early part of the season. Incubation lasts 23–24 days and is by both sexes with both parents initially tending to the chicks. The hatching success in the wild has been found to be a rather low 14%, and from hatching to fledging, 20–50% success. The age of first breeding is 1–2 years. The oldest ringed bird was 17 years. Successful breeding might occur from one pair. However, there are hints that partners of a pair might be highly aggressive towards each other, at least in smaller aviaries. Keeping a group of ~6 redshanks might be more promising for successful maintenance and reproduction (depending on the size of the aviary). At Rheine, keeping redshanks started with five birds kept in a mixed-species exhibit. Retaining two offspring had enlarged the number to seven. Sexes are not identified but are suspected to be 4.3. From these seven birds up to 30 eggs had been produced in 1998 and 1999.

Redshanks courtship and territoriality displays are easy to observe. However, it is difficult to detect nests in landscaped aviaries. Warning calls of the parent-birds might give hints if approaching a nest site. At Rheine, nest sites are marked or noted in a map when detected. Clutches of eggs are transferred to the incubator as soon as four eggs (or three) are laid. Replacement egg clutches do occur. Natural incubation in the mixed-species exhibit with a large population of different waders can be successful, as happened several times with non-detected clutches. Chicks had transferred for hand rearing, as conditions in the enclosure are not suitable for natural rearing. It once occurred that one chick was parent reared, as this hatchling was the last one of the breeding season in the aviary.

Artificial Incubation and Rearing: As previously mentioned, success with parent rearing is unlikely in a mixed-species exhibit because of the high density and the territorial behavior of the birds. Eggs are removed for artificial incubation and rearing. Incubation of ruff eggs is done at 37.5 °C (99.5°F) and ~60% relative humidity and hatching at 37.5 °C (99.5°F) and 60–80% humidity. Time from first pipping to hatching might be as long as 3–4 days but hatching success under these conditions are generally 70–80%. There is an indication that ruff chicks are highly sensitive and easy to stress already at the stage of hatching. There is, therefore, strict regulation that incubators be checked only twice daily to avoid any disturbances that might prolong the hatching-period.

Redshanks pose the least difficulties to hand rear compared to the other wader species kept at Rheine. Incubation and rearing parameters are as described for ruffs, but hatching success under these conditions is at nearly 100%.

Rearing of ruff chicks naturally or artificially is the most critical task in the successful husbandry of this species. At Rheine our impression is that the chicks are very easy to stress, which causes sudden death or "twisted-neck" syndrome, with consequent death. Current management works towards minimizing disturbance. Boxes 50 x 60 x 25 cm (22 x 24 x 11 in.) to 140 x 60 x 25 cm (56 x 24 x 11 in.) to rear ruff chicks are placed in a separate room and are visited only twice daily for feeding and cleaning. Boxes are decorated with grass clumps and twigs with leaves to allow the chicks to hide. The floor is covered with green towels and boxes are made of plywood to keep the surroundings quite dark. Ambient temperature lies around 15°C (59°F) to more than 20°C (68°F) and there is an infrared light, for local warming, fixed over the box. The water dishes are small and relatively flat. A large pebble is placed in each dish as chicks have drowned in even very flat small dishes. Rearing food is the softbill mixture (and all the other items mentioned above) for adults, with added crumbled egg yolk. Some freshly molted mealworms are offered daily. More live food (young crickets, *Drosophila*) might be favorable. (GH: I would recommend the artificial rearing diet as described for SHORE PLOVER be read with this. I had the plover diet analyzed with very good results. Food does not however seem to be the crucial point in rearing ruffs).

Food is offered in dishes and is also scattered around in the rearing box. There is indication that crowding causes stress. Therefore, a maximum of four chicks are placed together from hatching to two weeks of age. After this they can be combined with others. Current management precludes mixing young ruffs with other wader chicks. As they grow, the chicks are moved to larger boxes and finally to indoor aviaries with connected outdoor aviaries, all in someway landscaped with grass and weed clumps, scrub. The critical phase of rearing seems to be from hatching to the end of the second week. Sudden death and twisted neck-syndrome are the main causes of chick mortality. Current rearing rate at Rheine zoo is about 50 % (which is low compared with species such as Redshanks — nearly 95 %, Avocets, lapwings and ringed plovers).

Redshank chicks are raised under the same conditions but tend to spoil a lot of water and this has to be replaced repeatedly. We have made no negative experiences with "crowding-conditions" and about 10 hatchlings had been kept together at one time. Single hatchlings are preferably mixed with other species (lapwings, very young avocets). Young redshanks are very lively and active and therefor mixing with hatchlings of calmer species is not recommended in favor of the latter ones. Current rearing rate at Rheine is nearly 95%. In fact, Redshanks proved to be the species of the waders kept which is the easiest to rear.

References:

Johann, A. 1999. *Husbandry Guidelines for the Ruff* (*Philomachus pugnax*).
Johann, A. 1999. *Husbandry Guidelines for the Redshank* (*Tringa totanus*).

20.03 Jacanidae Jacanas

Glen Holland

African Jacana *Actophilornis africanus*

Introduction: The eight species of jacanas belong to the family Jacanidae in the order Ciconiiformes. Despite the limited number of species, jacanas enjoy a wide distribution, with two species, the lesser jacana *Microparra capensis* and the African jacana *Actophilornis africanus*, in Africa; a single species, the Madasgascar jacana *A. albinucha,* on the island of Madagascar; two species in Asia, the range of one, the bronze-winged jacana *Metopidius indicus,* extending from southern through southeastern Asia, and the other, the pheasant-tailed jacana *Hydrophasianus chirurgus,* occurring in southern Asia and the Philippines; another species, the comb-crested jacana *Irediparra gallinacea,* inhabiting Indonesia as well as northern and eastern Australia; and the final two species, the northern jacana *Jacana spinosa* and the wattled jacana *J. jacana,* found in the New World. The northern ranges from southern Texas through Mexico to Middle America and the Greater Antilles, while the wattled occurs from Panama southward through South America.

Jacanas avoid dry areas, choosing areas of open water fringed with, and preferably including, floating vegetation. Jacanas walk, with the assistance of their long toes, over floating vegetation, and this has resulted in them commonly being referred to as lilytrotters. They are also good swimmers and divers. The avicultural experience described here is from work done on the African jacana. This species favors water lilies, and occurrs in pairs, or loose groups with constant territorial chases accompanied by a variety of screeches and the common contact call *kree-kreep-kreep*. On small stretches of water only a single pair will be found, due to territorial aggression.

Management: The sexes are alike in plumage, but can be differentiated by size. Females are noticeably

Pheasant-tailed Jacana
Hydrophasianus chirurgus

larger than males, with the average weight of females being 261 g (9.2 oz.) while that of males is 137 g (4.8 oz.). Due to space limitations, an aviary will only accommodate one territorial pair that should be accurately sexed prior to introduction. At the San Diego Zoo these birds were not found to be aggressive toward any other species, and in fact were easily displaced by most other species. However, when a pair formed, they became very aggressive toward any other jacanas in the enclosure, to the point of killing any individual that could not get completely out of sight. This has happened in enclosures as large as 60 m x 50 m x 15 m (200 x 164 x 49 ft.) high.

Jacanas are not easy to establish or maintain in captivity. They are rather nervous and, when approached in an aviary, will dash for the closest available cover. The secret of success with jacanas lies in creating the correct habitat for the birds. The ideal aviary for this species should include a shallow pond of about 60 cm (24 in.) at the deepest point. The pond should be fringed on one side with an area that is slightly flooded when the pond is full, planted with reeds and sedges. Floating vegetation such as waterlilies and water hyacinth should be placed in the pond. These birds also appreciate a large dead tree stump protruding from the pond. Habitat F, as described in the chapter on LANDSCAPE, provides examples of the plants that should be used for a jacana exhibit.

Jacanas are compatible with waterfowl, but the latter tend to ruin the specialized landscape, which is required for success with this species. Considering the beauty and unique features of these birds, I feel that they deserve to have an aviary to themselves. At San Diego, they have been mixed with a variety of birds from pygmy geese to storks, weavers, and parrots. The biggest problem in a mixed collection is protecting the eggs and chicks from predation by the other occupants, which generally necessitates removing the eggs or chicks for hand rearing. The aviary should cover a minimum area of about 50 m^2 (538 sq. ft.). Although I have not experienced this, I have no doubt that they, like crakes and moorhens, will rob the nests of small birds such as quails. Juveniles must be removed when they become independent.

Diet: The natural diet of jacanas consists of insects, mollusks, crustaceans, seeds, and waterlily bulbs, but they are largely insectivorous. In captivity, they will feed on moistened commercial softbill mixtures, grated boiled egg, millet (particularly sprouted), soaked dog cubes, and a daily ration of insects. Bird of prey meat and chopped smelt are also added to the diet.

Breeding: During courtship, the male calls to the female from a basic nest platform with a repetitive *chuk-chuk*. The female then approaches, and copulation follows. The nest is built on floating vegetation and constructed with aquatic vegetation. The clutch consists of 2–5 eggs, which are of the most beautiful of any African bird. They are bluntly oval in shape with a smooth, polished looking shell that is a light brown-mustard color and thickly marked with black or brown scrolls.

Incubation in the natural state, where a number of birds are present, is by the male alone. It has been found that a single female will protect a harem of up to four males for whom they lay numerous clutches. This leads to the females becoming the dominant, aggressive birds. In aviculture, as well as on small natural bodies of water where only a pair occur, both sexes incubate the eggs and raise the chicks. During incubation, the feet are placed wide apart and the wrists of the wings are placed under the clutch. This serves to keep the eggs dry. In between periods of incubation, the birds leave the nest regularly for short periods. The eggs hatch in 22 days and the male removes the shells after hatching. The chicks are unstable for the first 24 hours, but begin to follow the parent by 48 hours. The male is very attentive of the chicks and will sometimes carry them under his wings with their legs dangling. Chicks feed themselves and are independent by 40 days of age.

Artificial Incubation & Hand rearing: Eggs weigh 8.4–8.77 g (0.29–0.31 oz.) and are incubated at 37.5°C (99.5°F) with a relative humidity of 88%. Jacana chicks are semi-precocial and must be taught to eat. The first few days are extremely important for success. A box brooder 45 x 60 cm (18 x 24 in.) with a rubber mesh floor must be maintained initially at 36.7°C (98°F). At lower temperatures the chicks become extremely lethargic. A feather duster, into which the chicks can snuggle, should be provided in one corner of the brooder. As chicks are visually stimulated to feed by seeing movement, it is important to have the brooder well lighted by a fluorescent light. Chicks should be stimulated to feed, with food held in a forceps, every 1–2 hours for the first few days, or until they are eating well on their own and steady weight gains are observed. Both wet and dry food should be changed every 4–5 hours. When the chicks are observed self-feeding, and are gaining weight, the assisted feedings can be reduced. By Day 30 the temperature can gradually be lowered, but if the chicks appear lethargic, it should be increased slightly and maintained at that level for a few days before attempting to lower it again. As soon as lower temperatures are tolerated well, the birds can be moved into larger outdoor pens.

A shallow pan of 25 cm (10 in.) in diameter and 12 mm (½ in.) deep is filled with water and placed on the floor of the brooder. The chicks need to be able to step in and out of this pan with ease. The chicks will soon begin to forage around a few small pieces of lettuce or waterweed and a few pebbles placed into the water. The following should also be placed in the water:

- 2 tablespoons live brine shrimp
- 2 teaspoons tubifex worms
- 10–15 very small mosquito fish or guppies
- 2 teaspoons dried daphnia
- 2 teaspoons live fruit flies, chilled
- 1 cube frozen glassworms
- 1 cube frozen discuss formula fish food, thawed
- 10 pinhead crickets, slightly chilled.

These items are to offer as much variety as possible, visually as well as nutritionally, and to stimulate self-feeding. Chicks are visually stimulated not only by movement of food items, but by color contrast of food items in the water. In addition, offer two teaspoons of each of the following in one or more mounds on the carpet of the brooder floor, or on a smaller piece of carpet placed on top of the brooder floor mat:

- dried daphnia
- guppy food flakes
- high pro baby cereal
- spiralina flakes
- soaked ground dog food

Place these items so that the chicks will walk through them. When walking through these items, they tend to pick off and eat the food that has adhered to their feet. The object is to get the chicks to forage on their own as soon as possible, in order for them to achieve sufficient weight gains (see graph). Until this has been accomplished, it is necessary to stimulate the chicks to feed by using long forceps to focus the attention of the chicks on the various food items. This can be challenging and may need a little creativity in technique. At the following ages, new foods are introduced:

Day 13 – small mealworms and powdered bonemeal
Day 30 – bird of prey meat
Day 38 – adult crickets, waxworms, and mealworms
Day 59 – chopped smelt

As the chicks are moved to outdoor pens and are eating more of the insects and bird of prey meat, the wet pan items can be gradually phased out as long as the weight gains continue.

African Jacana
Actophilornis africanus

Day	Weight (g)	Day	Weight (g)
1	4.95	30	48.5
5	5.72	35	62
10	8.82	40	75
15	14.36	45	92
20	24.85	50	102
25	34.5		

Supplied by San Diego Zoo

Ring Size: 8 mm

Species Requiring Similar Management:
Rails, all species
Crakes, all species
Swamphens *Porphyrio* spp.
Plovers, aquatic species

Contributor: Ed Lewins, Assistant Curator of Birds, San Diego Zoo

20.04 Burhinidae Thick-Knees

Glen Holland

Natural Habits: Stone-curlews belong to the order Ciconiiformes and the family Burhinidae. There are nine species within one genus. Three species are African, two South American, two Australian, and one found in north Africa, southwest Asia and Europe. They occur primarily in arid regions with scattered sparse vegetation and a mixed stone and sand substrate. However, three species are associated with the fringes of aquatic habitats. The nominate race of stone curlew *Burhinus o. oedicnemus* has become one of the most endangered species of birds in Europe (Cwiertnia 1999). The voice is a repeated whistle, which rises in volume and then fades away towards the end. Adults give a three-note call to warn chicks to hide when threatened. They also give a harsh grating call if the chicks or eggs are under direct threat.

Management: Stone-curlews are also known as "thick-knees" in Australia and *dikkops* (an Afrikaans name referring to "thick-head") in Africa. Both terrestrial and aquatic related species occur and size varies from species of about 30 cm (12 in.) in height to others double that height. Most species feed on both small vertebrates and invertebrates. They usually occur in pairs but will also form small groups out of the breeding season. Stone-curlews are generally nocturnal in habits. The sexes are alike. Most of my experience has been with the Cape dikkop *Burhinus capensis,* which is a common terrestrial species in southern Africa.

They are best suited to public displays where they are kept in large aviaries of at least 10 x 3 x 2 m (33 x 10 x 6.5 ft.). They are not difficult avicultural subjects but require a diet that the average aviculturist does not stock. The aviary should be planted with a few sparse shrubs between which open sandy and pebble-strewn areas are left. Habitats varying between B and E will be suitable, provided that sufficient open areas remain. In wet climates sufficient shelter must be provided to ensure dry areas where the birds will rest during the day. As dikkops are largely nocturnal they should be mixed with larger species such as coucals, owls, and hornbills, which are unlikely to be disturbed by their nocturnal activities. Although it is possible to keep a number together outside the breeding season, due to their territorial nature I would recommend a maximum of one sexed pair per aviary. If two pairs are housed together, only the dominant pair will breed. Although relatively hardy birds, I would recommend that in extreme climates, the African species be housed indoors in cold and wet conditions, particularly where temperatures remain below 10°C (50°F) for extended periods.

Diet: They feed on minced heart and bird of prey diet, which in turn is mixed with equal portions of commercial softbill food and a few cat cubes that have been moistened. A multi-vitamin supplement is added twice a week. In addition, they will take small mice, crickets, grasshoppers, and mealworms. At least one form of live food is offered daily.

At Poznan Zoo the birds are fed once per day, five days per week (G.H.: Not a policy I would recommend for *Burhinus*). The food consists of boiled egg, cooked rice, white cheese, ground meat, and minced fish. In addition, from time to time the birds also receive small laboratory mice *Mus musculus*. The preferred food is mice, followed by boiled egg and ground meat (Cwiertnia 1999).

Cape Thicknee
Burhinus capensis

Breeding: At Poznan Zoo the paired/breeding birds are moved to off-display breeding aviaries measuring 2 x 3 x 2 m high (6.5 x 10 x 6.5 ft.). Each aviary has a den that is slightly heated in winter to ensure the temperature does not drop below freezing. The bank of five aviaries are adjacent and all occupied by stone curlews. All unpaired and young birds are kept together in a communal aviary. Birds aged three years, which is when breeding activity usually begins, are allowed to select a partner of their own choosing. A mated pair soon begin to chase away other birds and are removed to their own breeding aviary. In 1997 we tried to leave the eggs with this pair for them to incubate and hatch, but this ended in failure. The eggs of the fifth pair were destroyed by the birds themselves immediately after being laid (Cwiertnia, 1999).

The season in the northern hemisphere starts in March or April, with the earliest clutch laid on March 10. The second egg in a clutch is laid 1–4 days after the first, with two days being the usual interval. If the clutch is removed or lost, the replacement clutch is laid 9–57 days later. Pairs usually lay two clutches per season, but one pair at Poznan laid 10 eggs in 1996 and 20 in 1997.

The nest is a shallow scrape in the earth lined with a few pebbles and pieces of vegetation and often situated next to a rock or stump. The eggs are blotched with brown and gray markings. Incubation starts with the second egg and it may be advisable to remove the first, replace it with an artificial egg and then replace it again when the second egg is laid, thus avoiding predation from other birds. When the nest is approached both birds put on a threat display with open wings while making harsh calls.

Incubation lasts 24–25 days and the chicks are led from the nest within a day of hatching. The adults feed the chicks largely on live food, which is either offered in the tip of the bill or placed next to the chick. Chicks fledge at eight weeks. Unfortunately they are not particularly successful parents in captivity and I would recommend hand rearing. It is better to remove the eggs for artificial incubation, inducing the birds to lay a second and sometimes third clutch. Chicks that are removed from the parents after hatching, are wild and more difficult to raise.

Hand Rearing and Incubation: The incubator is set at 37.2°C (99°F) and a relative humidity of 55% and adjusted to 36.9°C (98.2°F) and 66% humidity at hatching. Eggs take 25–28 days to hatch and chicks are left in the incubator for the first 18–24 hours. They are then removed to a brooder box with a brooder lamp over one end and a river-sand base. This sand provides both grip and grit. Mealworms, small crickets, waxmoth larvae, grasshoppers, and small strips of ox heart are dipped in commercial rearing food, such as the artificial diet as described for SHORE PLOVER, which has been moistened with water. This is fed eight times per day with a long forceps tweezer. It is important that the chicks are not overfed or fed too rich a diet, and they must obtain sufficient exercise. Overweight chicks will develop deformed legs. Rearing stone curlews is very similar to the methods described for PLOVERS.

The chicks soon begin to peck at the food offered in the tweezers, and by day four the chicks begin to take these items from the floor. They should then be offered food in a shallow dish, mixed with rearing food as described for SHORE PLOVER. A powdered multivitamin and calcium supplement is sprinkled over the food. By day seven the chicks feed themselves with fresh food offered three times per day. At two weeks they should be placed in an outdoor pen, in favorable weather, and after a month they can remain outdoors if temperatures are above 18°C (64.4°F) at night.

At Poznan Zoo where all the eggs were removed for artificial incubation and rearing, the results between 1989 and 1997 were as follows: eggs laid (69), fertile (43), hatched (30), reared (19).

The chicks have a strange habit of lying on their side with the legs stretched out straight. Different size chicks can be kept together without problems except competition for food. This problem is easily avoided with extra food bowls. Chicks, that are received after hatching under the adults are stubborn and will instinctively lie flat on the ground, refusing food. These chicks are best left alone with a bowl of live food, which they will take unobserved; however, sometimes they will have to be force-fed. Birds that are hand raised alone and given too much attention tend to imprint on humans, later refusing to breed with their own kind.

Ring Size: 8 mm

Species Requiring Similar Management:

Coursers
Plovers

Contributions:

Walter Mangold , Director, World of Birds, Cape Town, SA.

References:

Cwiertnia, P. 1999. Multi-generation Breeding of Stone Curlew. *Avicultural Magazine* 105(1): 6–11.

Urban, E.K., C.H. Fry, & S. Keith. 1986. *The Birds of Africa* Vol 2. London, UK: Academic Press, London.

25.05 Charadriidae Stilts & Plovers

25.05a Stilts

Recurvirostridae

• Glen Holland

Introduction: Stilts belong to the order Charadriiformes and the family Recurvirostridae. There are nine species worldwide. They are found in pairs or loose congregations, primarily on mud flats, which include areas of shallow water, both fresh and saline, where they feed. Stilts walk with long strides often bobbing the head. They catch most of their food items on, or in, the mud and within the water column. They will sometimes wade to where the water touches the belly. They are usually located by their calls, which consist of a sharp *kwik-kwik-kwik* or *yap* contact call, and some species have a loud *kik-arik* call given in flight.

Management: Stilts are rather sensitive birds requiring specialized management. They are best kept only by those who can afford the time and effort. Although most species require similar management, I have concentrated on the black-winged stilt *Himantopus meridionalis* with which I have had most of my experience. Additional contributions were received from Taronga Zoo, a very successful breeder of the pied stilt *H. himantopus leucocephalus,* and the New Zealand Department of Conservation that has an entire breeding center dedicated to the endangered black stilt *H. novaezelandiae.*

Aviary sizes are recommended as follows:

- Public viewing and holding: 2 birds, 80 m^2 (860 sq. ft.)
- Flock mating: 400 m^2 (4300 sq. ft.)
- Breeding: 140 m^2 (1500 sq. ft.)
- Holding/hand rearing juveniles (up to 4): 12 m^2 (130 sq. ft.)

Stilts are a valuable species in a public display, where they add color and character to a shallow water feature. Black-winged stilts have successfully bred and raised young when supplied only with a water dish of 40 cm in diameter and 5 cm depth (16 x 2 in.), but a shallow pond of approximately 2 m^2 (21.5 sq. ft.) and 10–15 cm (4–6 in.) deep is preferable. The pond should have a natural silt/gravel lining, which will provide the birds with some natural feeding. The pond should have gently tapered sides. To bathe, the stilt will dip its head into the water and then shake the water through the feathers.

Stilts do not require much protection from the elements, provided they are not too extreme. Tall vegetation such as reeds should be avoided in the aviary, as these will entangle birds when descending to land. Short, swamp-type vegetation is more suitable for them. Weed and grass areas are best either kept very short or killed with a chemical spray that is non-toxic to birds, at which time the birds are best removed from the aviary for 10 days. Damage to feet, legs, and bills are the most common injuries. Cut or cracked feet can be avoided if dry areas are made available to the birds and pebbles or river sand covers a large portion of the aviary floor. They should not be kept on a concrete base as their toes are covered with a very thin skin and they are prone to foot problems. Earth substrate should also be avoided. A mix of sand and fine pebbles is ideal. Stilts are aggressive when breeding and they are best housed as single pairs in an aviary. Young birds of six weeks or older and adults, out of the breeding season, can be housed together with numerous feeding stations (black stilt juveniles are an exception unless reared together).

The least stressful capture technique in aviaries is the use of a drop trap over a food dish. This is triggered from outside the aviary. Once caught, the birds should be held from above by placing both hands over the wings, leaving the legs and neck/head free. Each compartment of the transfer box should be at least 35 cm (14 in.) high with a padded roof. The internal diameter should be sufficient to allow the bird to turn with ease but not to pace around the box.

Diet: Stilts are fed on a mixture of minced heart and fish, hard-boiled egg, commercial insectivorous food, and insects such as mealworms. Water is added to this mixture to give it a soup-like consistency, and is fed from a clean dish morning and afternoon. Each bird consumes 30–50 g (1–1.75 oz.) per day. A soluble multi-vitamin is added to the mixture twice a week. The diet presently used at Taronga Zoo, Sydney, Australia is as follows:

- 3 parts French millet
- 3 parts Panicum
- 2 parts turkey starter crumble
- 4 parts egg cake mixture (wholemeal cake, egg, cheese, and pupae)
- 3 parts minced heart
- 1.5 parts commercial insectivorous mixture
- 1.5 parts fly pupae
- 0.5 tbsp calcium carbonate
- 1 tbsp Petvite (multi-vitamin)

In addition, a handful of fly pupae and mealworms (5–10 per bird) are placed in a water dish or pond. The birds should be fed twice daily.

The diet used in New Zealand for adults consists of 1 kg (2.2 lb) minced heart to which a tablespoon of a supplement premix is added. I have also added finely flaked pet's fish to a similar diet. However this diet lacks iodine and is not ideal. Chicks are reared on insectivore mix and beef heart. A shore plover diet (blended cat kibble [1/3], minced heart [1/3], insectivorous mixture [1/3], and about 20 mealworms per bird/day) has been analyzed and found to have excellent nutritional content and absorption rates. Research and breeding results to date indicate this diet to be more suitable for stilts than the meat diet previously used.

Recent research into the deaths of released juvenile black stilts has shown that goiter is a prevalent cause of death. Goiter is an enlargement of the thyroid glands usually caused by an iodine deficiency (or, less commonly, excess). Iodine is an integral part of the thyroid hormones in birds. Insufficient dietary iodine inhibits thyroid hormone biosynthesis. When severely enlarged, the thyroid glands can press on the trachea and syrinx causing labored breathing and difficulty in swallowing and regurgitation. This condition can be fatal. An iodine deficiency, by reducing the level of circulating thyroid hormones, lowers the metabolic rate, which is naturally high in birds. This reduces feeding efficiency, heat generating ability, slows the reaction to predators, and increases susceptibility to infection.

In the 1997/98 season 80% of wild collected black stilt eggs hatched; however, only 28% of fertile captive eggs hatched. The effects of an iodine deficiency on embryo development and hatchability are well documented. Generally the chicks develop fully, move into the correct hatching position, internally pip, vocalize, star crack, break a small hole, then instead of rotating inside the shell, "unzipping" a section while absorbing the remaining yolk, they form a large hole through which the beak protrudes and then after a few days they die. Research on captive and wild-origin hatchlings showed the wild birds had 16 times more iodine in the blood than the captive-origin chicks that died at hatching. We can very easily correct the problem by adding iodized salt (2.5 g per kg [0.09 oz. per 2.2 lb] of diet); thyroid levels respond quickly (within a month). Food placed in a saltwater tray (iodized salt), as described for shore plover is another method of achieving this. Some live food can even be sprinkled in the saltwater tray.

Wild birds and parent-reared young are trained to substitute diets using live food mixed into the substitute food. Conversion to a substitute diet may take up to six weeks.

To avoid unhygienic conditions developing, feeding stations should be moved slightly on a regular basis. They are rather messy feeders and food bowls can be placed on a rubber, polythene or plastic mat, which is replaced and cleaned daily. It must be remembered that clean water for drinking and cleaning the beak should always be available. To save preparing the food on a daily basis, the solid mixture can be made in quantity and then frozen in smaller portions as required daily. This is then defrosted daily and water added.

Black-winged Stilt
Himantopus himantopus

Breeding: Flock mating is the most successful method of establishing pairs. This generally occurs in late winter. All birds being placed in the flock should first be reliably sexed and color bands fitted. Two males and two females should preferably be housed together for 2–3 hours on a "neutral" territory, which has not been occupied by any of the four prior to flock mating. The aviary should be at least 400 m^2 (478 sq. yd.). A close watch must be kept on the birds and the dominant pair will usually be obvious within hours, at which stage they are removed to their breeding aviary. Under no circumstances should flock mating or pair stimulation continue overnight unless there are absolutely no signs of aggression. Visual barriers to avoid birds wasting their energy on aggression should separate pairs placed in adjacent aviaries for breeding. The vocalizations between pairs will, however, assist to stimulate breeding. Courtship includes the birds flying high over the mate with the legs and beak pointing downwards (although I haven't seen this in black stilts) and also walking on the ground and tossing small twigs over the back.

The nest is either a shallow scrape in the sand or built on exposed aquatic vegetation. Islands are a favorite nesting site. The nest is lined with shells, feathers, vegetation, or dry mud, and built in an exposed position, which will allow 360° visibility. Captive black stilts prefer sites close to fences and behind tussocks. The 2–6 eggs are cream- or green-colored, mottled with brown and black. If a clutch is lost, another clutch may be started within four days. Incubation starts when the third or fourth egg is laid and is shared by both sexes for 23–25 days. During incubation the adults sometimes soak the belly feathers in water before returning to the nest to incubate. In one case at Copenhagen Zoo another bird in the aviary killed the first chick, and the entire nest with the adults was moved to a new aviary. After a while the adults continued to incubate the eggs and all four were hatched successfully. Male black stilts will also sometimes become aggressive to their chicks, particularly male chicks. Adults are, however, usually model parents raising 80–90% of their chicks on a diet consisting of ox heart and commercial insectivorous food in a ratio of 2:1. Live invertebrates are important for at least the first week of the chicks' lives. Adults with chicks should not be removed to brooder boxes as the adults suffer from stress under these conditions. Chicks require brooding by their parents until they are three weeks old. A careful watch should be kept to ensure that adults are brooding successfully and if not, the chicks are removed for hand rearing. The removal, and artificial incubation of clutches, will induce pairs to lay second and third clutches. One black stilt pair has even been known to double brood after successfully raising their first clutch of the season. All parent-reared chicks should be provided with a tray of stream or pond insects twice a day and 30–50 mealworms added to the supplementary food twice a day. These insects are gradually removed from the diet after two weeks. At this stage chicks should carefully be observed at the feed tray to ensure they are eating supplementary food.

Parents do not feed newly hatched chicks. On the first day chicks forage within a few meters of the nest then return to the parent to brood. They do not take much substitute food in the first few days and a plentiful supply of natural invertebrate life is essential. When chicks are cold they cheep loudly and the adults immediately sit down to allow the chicks to retreat under their belly feathers. The first brood will interfere with the second and should, therefore, be removed. Naturally a brood will often remain with the parents until the following spring, at which stage they will be driven off. Chicks are independent by two and a half months, resemble the adults at a year, and are sexually mature when two years.

Hand Rearing and Incubation: Eggs are artificially incubated at 37.2–37.5°C (99–99.5°F) and 55% relative humidity. Survival of black stilt eggs incubated at 37.5°C is significantly higher than at 37.2°C. At first pip the eggs should be transferred to a hatcher at the same temperature and 80% relative humidity (the floor of the hatcher should have a full-sized tray of water). Late stage eggs can be checked for signs of life by placing a thin straw horizontally across the egg, attached at the large end with a non-toxic tacky substance. Any movement in the egg will be accentuated at the tip of the straw.

- Chicks aged 0–7 days: Each newly hatched chick is marked with a temporary band (ring). They are moved into a brooder and placed below an artificial stilt model brooder or else placed below a brooder lamp. The entire brooder and brooder room should be maintained at about 20°C (68°F). The temperature on the ground below the heat source should be as for incubation. For the first 3–4 days chicks should be checked regularly at night to ensure they are using the source of warmth. The entire floor of the brooder is covered with artificial grass matting (including the water bath at the end of the brooder, which is empty). At this stage chicks remain indoors. Live insects should be provided to the chicks from hatching until they begin to take supplementary food, which is usually at about 10 days. Hand-reared chicks require sufficient insects 3–4 times daily such that their plates are continuously full. Thousands of aquatics such as mayflies, caddis or damselflies, and brine shrimp are suitable for young chicks. Small mealworms should be placed in a bowl containing a moistened mixture of finely minced heart and a ground commercial insectivorous mix. Once chicks are taking the artificial diet well, the live food is gradually phased out altogether.

- Chicks aged 7–14 days: As above, except that the grass matting is removed from the water bath, and provided the chicks are healthy, the bath is filled with water. The water bath should be 6–7 cm (2.4–2.8 in.) deep with tapered sides. On warm days chicks are allowed out into an adjoining small covered 1 x 2 m (3 x 6.5 ft.) aviary. They should be returned to the brooder before the day cools. Ideally the chicks should not be caught to do this, but can be enticed back inside with live food or the light of a brooder as the sun sets. During wet and/or cold days these chicks should be left inside. The brooders should have good ventilation, and one-way glass is ideal as a front door so that their normal behavior can be observed without disturbance.

- Chicks aged 15–30 days: Except in severe weather, the chicks are allowed into the outdoor aviary. They are still fed in the brooder to encourage them to move in and out freely. After 30 days they should be left outside day and night, with a door open to the brooder in case of bad weather. At 30 days the chicks are fitted with permanent bands. In extreme cases, where chicks do not learn to eat substitute foods, they have to be force-fed the food four times per day until they are feeding themselves. This stress is, however, dangerous and may negatively affect the birds' health. It is a last resort.

Chicks should not be raised alone and, once banded, broods can be mixed if they are less than a week old.

With older chicks it is advisable to introduce them to each other on neutral ground.

Chicks are susceptible to bacterial infections and aspergillosis, both of which arise when chicks are stressed and/or housed in unhygienic conditions with poor airflow. Chicks that become lethargic should be isolated from others and, if not feeding, some force-feeding can be done. A 100 g (3.5 oz.) chick should be fed 4 g (0.14 oz.) of food four times per day. Fluids are vital and a clean source of water should be maintained. Bacterial infections should be treated with antibiotics, under the direction of a veterinarian. Should impaction occur, a little vegetable oil can be added to the diet. Bumblefoot, which becomes apparent by the swelling of a knee or ankle joint with a secondary invasion of *Staphylococcus aureus,* is a common problem with captive birds. It can be treated in the early stages with antibiotics with only moderate success. Holding stilts on concrete or damp ground should be avoided, as this tends to aggravate the condition. Tapeworms, cestodes, ascarids, coccidian, and capillaria are all known to infect stilts. Should birds break a section of either mandible this is best splinted. A feather shaft split lengthwise and glued to the bill is useful, although it will tend to be shed reasonably often. Do not trim either mandible, as this stress has been known to cause death. Should a stilt break its tarsus or tibia the only option is to euthanize the bird. Dislocations of the knee joint can be splinted. Splints and slings have been used to treat complete bone breaks but birds have always faded and died anyway.

The various stilt species will hybridize and at Copenhagen Zoo the black-winged crossed with a black-necked stilt. In the natural state in New Zealand the hybridization of black and pied stilts is a cause of concern, but is being carefully managed by conservationists to safeguard the future of pure black stilts.

This chapter should be read in conjunction with the chapter on SHORE PLOVER.

Ring size: 5.5 mm

Black Stilt *Himantopus meridionalio*

Day	Weight (g)	Day	Weight (g)	Day	Weight (g)
1	17.7	**15**	75.3	**28**	144.5
2	17	**16**	56.8	**29**	148.2
3	17	**17**	68.9	**30**	153.6
4	18	**18**	85.2	**31**	164.4
5	20.5	**19**	87.8	**32**	158.5
6	24.2	**20**	92	**33**	161
7	24.8	**21**	105	**34**	165.5
8	35.7	**22**	104.7	**35**	167.7
9	29	**23**	107.2	**36**	165.6
10	38.5	**24**	112.7	**37**	167.2
11	45.7	**25**	125.7	**38**	168.5
12	47.5	**26**	127.7	**39**	175.1
13	47	**27**	146.1	**40**	178.3
14	59.2				

Species Requiring Similar Management:

Stilt species
Avocets
Snipe
Sandpipers
Aquatic plovers

Contributions:

Emily Sancha, New Zealand Department of Conservation

Phillip Zammit, Keeper, Taronga Zoo, NSW, Australia.

References:

Poulsen, H. 1965. Breeding Stilts at Copenhagen Zoo. *International Zoo Yearbook* No 5: 130–131.

Reed, C., & I. Fraser. 1998. Husbandry Manual for Black Stilts. New Zealand Department of Conservation.

Urban, E. K., C. H. Fry & S. Keith. 1986. *The Birds of Africa* Volume 2. London, UK: Academic Press.

20.05b Plovers

Charadriinae

• Glen Holland and Dick Schroeder

Introduction: The plovers belong to the order Ciconiiformes, the family Charadriidae, and the subfamily Charadriinae, which encompasses 66 species in 11 genera including many species commonly known as lapwings and dotterels. Plovers are generally found in pairs while breeding, but may occur in flocks of up to 200 birds outside the breeding season. Inhabiting open country ranging from wetlands, seashores, and salt pans to grassland, old pastures and sports fields, they avoid tall vegetation. They rarely wade to forage, but feed mainly on insects and their larvae found in grass. Some vegetable matter is eaten as well. Pairs are highly vocal and territorial, and are well known for their brave aerial attacks on anything that threatens their nest. Plovers keep in regular contact through their calls, which include alarm calls to warn others of approaching danger.

Management: My recommendations for managing plovers in captivity is based on experience with the crowned lapwing *Vanellus coronatus* and the blacksmith plover *V. melanopterus* of Africa and the endangered shore plover *Charadrius novaeseelandiae* of New Zealand. In common with many other plovers, the crowned is a bird of purely terrestrial habitats; the blacksmith is usually associated with fresh water and estuarine habitats; and the shore plover is found entirely on coastal rocky and sandy beaches. These habitat variations must be catered to in aviculture. If an appropriate habitat for the particular species kept is not provided, the chances of success are drastically reduced.

Plovers adapt very well to captivity, making the common species ideal for public displays. The aviary landscape for terrestrial species should correspond to either Habitat B or Habitat E, with plenty of open sandy areas combined with mowed lawns. A few shrubs will provide shelter, shade, and a place to hide from raptors flying above the aviary, which plovers are usually the first birds to spot. Species such as blacksmith and three-banded plovers *C. tricollaris* require the aquatic Habitat F, which is easily created. A shallow pond, preferably with a natural mud base, surrounded by a few sedges, will provide a perfect environment. Shore plovers require a relatively sterile beach-type environment. Our aviaries for this species have a pebble base with a few scattered grass tussocks and a continual stream of clean water flowing through the center. Aviaries for plovers and stilts should not include tall dense thickets of vegetation in which the birds may become entrapped, particularly when they are panicked or landing in a stressed state.

In extremely large aviaries, several pairs of some species can be kept together, as long as a sexual balance is maintained. All the birds should be placed in the aviary at the same time, as individuals introduced into an established group will be persecuted. Sometimes several species can share a large aviary; I have mixed crowned and blacksmith with no serious problems. Ideally, however, plover species should not be mixed, as most aviaries are not large enough to avoid fighting over territories. As they are not aggressive toward other birds plovers can be kept with unrelated species.

Shore plovers are highly territorial. Because they are endangered in the wild, the captive population plays an important role in re-establishing the species in areas of its former habitat where it has been extirpated or reduced in numbers. For this reason, each pair should be afforded their own large aviary — perhaps 40 x 15 m (30 x 50 ft.) off public display. Territorial aggression is avoided by visual barriers between the pairs. Note the chart, in the chapter on STRESS, indicating the extent to which this species is affected by stress.

Foot problems such as bumblefoot, or damaged feet in general, can be averted by placing the food bowl in the center of a shallow footbath containing a saline solution, which is easily prepared by mixing 350 g (12.3 oz.) salt in each 10 L (10.5 qt.) of water (3.5% saline solution). This strengthens the skin on the feet, making the birds less susceptible to cuts and infections. The saline water should be renewed every second day.

Diet: Unlike hand-raised birds, wild-caught birds will initially take only live food. By mixing live food with other food items, the birds will gradually begin to accept an artificial diet. The mixture should consist of equal portions of soaked cat kibble, insectivore mix, minced heart, and insects. Mealworms, bloodworms, and other live food can be used to tame these birds to the point where they will feed from the hand. Shore plovers should receive around 20 mealworms per bird once a day. During the breeding season, the amount should be increased to 30 mealworms per bird twice each day. At this time, the mealworms should be dipped in a calcium supplement before being fed. This entire diet was thoroughly investigated for nutritional values and rates of absorption and proved to be excellent on both accounts. Plovers spend much of their time hunting for insects in open areas. An aviary of sufficient size, with the proper habitat, will provide a good source of natural food. Water levels can be raised and lowered periodically to expose live food on the pond margins.

Breeding: Provided they have a secluded spot, most plovers are willing to nest in captivity. Courtship display is rather meager; following copulation, one or both of the birds may raise either one or both wings over the back. The nest is a shallow scrape in the ground lined with a few pebbles and rootlets. The 2–3 and sometimes, in the case of shore plovers, four eggs are laid 24 hours apart. They are buff colored with numerous black, brown, and gray markings over the entire surface. Incubation is shared by both sexes. The incubation period is 28–32 days. If disturbed during incubation, the parent leaving the nest will give a distraction display with the wings and tail spread while either standing or crouching. As the chicks hatch, they join the parent that is not incubating. One bird remains on the nest until the entire clutch has hatched.

The adults feed the chicks by offering insects from the tip of the bill. Chicks grow rapidly, reaching adult size at only a month of age. They are sexually mature at 18 months. The removal of a clutch of eggs for artificial incubation will entice the adults to lay a second clutch soon after. In fact, shore plovers have laid up to four clutches in a season when all the eggs were removed for artificial incubation. As the chicks are easy to raise artificially and do not imprint on their keeper if raised in groups of two or more, I would recommend this practice, particularly for rare species.

Incubation & Hand Rearing: Plover eggs should be incubated at a temperature of 37.2°C (99°F), with a relative humidity of 60%. At hatching time, the temperature should be reduced to 36.5°C (97.7°F) and the relative humidity increased to 66–70%. Squeaking and movement inside the egg are detectable up to two days before the egg hatches. Chicks should be left in the incubator for the first 8–12 hours, until their down is dry. After that, they can be removed to a brooder box with a temperature of 32–34°C (89.6–93.2°F). Plover chicks are often inactive in the brooder for up to 36 hours before they finally begin to move around and feed. The temperature can be gradually reduced until day 10, when the chicks will be comfortable at a room temperature of about 25°C (77°F). For shore plovers, which originate from a cooler climate, the temperature can be reduced to 22°C (71.6°F) at this stage. However, some heat should still be supplied over one end of the brooder. A substrate of river sand is ideal for plover chicks as it provides both good footing and a source of grit, which these birds require. Alternatively, rubber brooder matting or towels, which are easy to clean in a washing machine, can be used. A cardboard or wooden box with three open sides and a few grass tussocks can be provided as refuge for the chicks.

Northern Lapwing
Vanellus vanellus

Young chicks soon begin to peck at anything that moves. Interaction with slightly older chicks will enhance this feeding behavior, as they quickly learn from their older siblings. Shore and blacksmith plovers will start feeding on live aquatic insects within two days of hatching. Live food such as mealworms, white ants, etc., sprinkled on artificial food will assist in getting chicks onto an artificial diet. With the larger plovers, food items such as thin heart strips and insects can be offered with a forceps, from which the chicks soon learn to take food. The heart should be moistened with a little water, and once a day a multivitamin and calcium supplement should be added. A shallow bowl of water, the base of which is filled with small pebbles, should be available from day one.

By the fourth day, the chicks will pick up food that is dropped in front of them. From day five, a bowl of food consisting of a mixture of moistened and ground cat kibble, softbill mix, minced heart and live insects should be offered. Plover chicks require 24% protein in their diet. A diet that is too high in protein will result in leg problems. Chicks removed from the parents for hand-rearing will initially accept only live food; they need to be trained onto the substitute diet. In warm weather, chicks of the larger species can be transferred to outdoor rearing pens during the day when they are a week or more old. Such pens should have mottled sunlight and a coarse sand base. This will allow the chicks the exercise that is necessary for the development of healthy legs.

The following hand-rearing procedure has been successfully used to rear shore plovers.

- Day 1–10: Fresh water is provided in a shallow tray 15 mm (2/3 in.) deep. This is replaced five times daily, and 200 or more small invertebrates such as boatmen, daphnia, etc. added each time. Insects must be collected from a clean source and, if stored before feeding, have their water changed daily. The artificial diet is offered ad-lib and replaced twice daily. It consists of a fine blend of the diet described for adults. To this, 50 small mealworms, 70 whiteworms and 20 small waxmoth larvae are added. As described for adults, the feed dish containing this mix is placed in a shallow tray of saline solution.

- Day 10–20: The tray containing live aquatic invertebrates is reduced to three changes daily, whiteworms are removed from the diet, and the size of mealworms and waxmoth larvae offered increased to medium.

- Day 20–30: The shallow food and water trays are replaced by trays that are 25 mm (1 in.) deep. Live aquatic food is given only once a day. The size of mealworm and waxmoth larvae is increased to large. Chick starter crumbles are removed from the diet.

Strict hygiene is enforced by moving the chicks to a clean brooder every second day. To ensure that they are gaining weight properly, chicks are weighed every second day for the first week and twice daily thereafter. Dehydration from overheating should be monitored by observing behavior; signs to watch for include panting, outstretched wings, and general lack of vitality. Legs and feet are checked to avoid problems. Healthy legs, feet, and toes are plump in appearance, as opposed to a shriveled appearance with the skin stretched tightly over the limbs. Aggression at food sources, particularly among chicks of mixed sizes, must be carefully monitored. At 30 days of age, fledglings can be placed in a flock, but color bands should be fitted for the identification and possible removal of aggressive or submissive birds.

Shore *Charadrius novaeseelandiae* and Masked Plover *Vanellus miles*

Day	Weight (g) NZ Shore	Weight (g) Masked
0	9.1	18
5	17.6	25
10	29.7	30
15	43.7	47
20	48	65
25	50.8	100
30	51	130
35		180
40		225
45		280

Spur-winged Plover *Anseranas Semipalmata*

- Dick Schroeder

Introduction: The spur-winged plover *Vanellus spinosus*, also known as the spur-winged lapwing, ranges across central Africa. Sexes are alike in these 35 cm (14 in.) birds. The crown, forehead, lores, and nape are black, as are the throat, breast, and belly. The sides of the face and neck are white and the upper body is grayish brown. The rump is white and the tail is black. These birds congregate in large flocks during the non-breeding season, but split off into pairs when ready to nest. They are very territorial when nesting and will defend their nest against any intruder, be it man or beast.

Management: I enjoy ground birds in my aviaries, and the spur-winged plover is among my favorite species. Only one pair can be kept in a flight, and it is a good idea not to mix other species of plovers or jacanas in with them. They are fine with any of the passerines. During the winter I have housed them with blacksmith plovers *V. armatus* without a problem, but I was quick to separate them before summer breeding season. They do well on grass or soil, but prefer a sandy or gravelly area for nesting. They also enjoy wading in a shallow pond, however, this should have fresh, flowing water to eliminate the chance of botulism, to which many of these birds are susceptible. An alternative would be to have a pool that is shallow enough to flush with the hose daily. Plovers spend their time wandering about the ground, picking through the litter for any insects or larvae. I have not seen them attempt to eat the lizards that can be found in their aviary, and I imagine these are too large for their fine beaks. Many of the plovers and lapwings can be very noisy at night, but this species will not rob you or your neighbors of any sleep. I only hear them at night if there is an intruder, such as a raccoon, near the aviary.

Diet: My plover diet is quite simple. I soak Kaytee Exact softbill pellets in warm water until soft, then add some game bird crumbles, a few mealworms and crickets, and perhaps a scoop of the chopped fruit and vegetable mix I use for softbills. The plovers seem to pick the corn from the softbill mix.

Breeding: The nest of these plovers is nothing more than a shallow scrape, sometimes with a few pebbles added, in a sandy or gravelly area. The 2–4 eggs are an olive-clay color, heavily mottled with brown. In their nat-

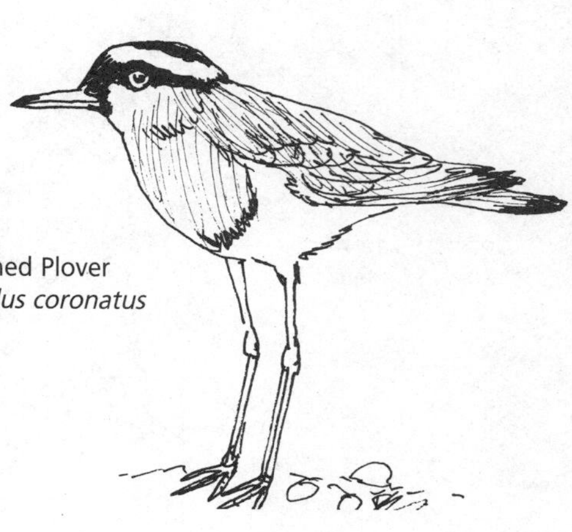

Crowned Plover
Vanellus coronatus

ural setting, they are nearly impossible to see. Incubation is 28 days. I have read in the literature that the incubation period for this species is 24 days, but this is not the case, at least in my experience. Unfortunately, I learned this the hard way, by opening an egg that had been incubated for 27 days, thinking it was overdue, only to find a chick that was ready to hatch. Of course, it was the only fertile egg in a clutch of four. The parents share incubation and defense of the nest. If approached too closely, they will immediately fly at the intruder's head and face. One soon learns why they are called "spur-winged"; they carry a tiny dagger in the bend of their wing! Other birds are kept a distance by the non-incubating adult, which lowers its head and charges the trespasser.

Newly hatched chicks are able to run around almost immediately. They are a downy buff color with black streaks over most of their body. They learn to eat quickly and soon are following the parents around the enclosure. If there is a pond in the aviary, it is a good idea to drain it at this time, as I have had chicks a few days old fall in and drown. A shallow dish of water can be provided for drinking. I usually add black or tubifex worms, which are available in most aquarium stores as fish food, to this shallow bowl. The chicks are soon drinking and feeding there. They will eat the same diet as the adults, however the mealworms must be very small at first. Some finely choppped, hardboiled egg will be eaten as well during this time. They can be separated from the parents at eight weeks of age, or can remain until next breeding season, at which time they will be treated by their parents as just another intruder. I have left a first brood in the aviary when the parents have double clutched, and they have been fine throughout the rearing of the second brood.

Ring Size:

Crowned, blacksmith and spur-winged: 6 mm.
Three-banded and shore plover: 3 mm.

Species Requiring Similar Management:

Plovers	Stilts
Lapwings	Greenshanks
Sandpipers	Ruffs

References:

O'Connor, S. 1997. Incubation & hand rearing protocols for New Zealand shore plover. New Zealand Department of Conservation.

Urban, E., C.H. Fry & S. Keith. 1986. *The Birds of Africa*, Vol. II

20.06 Laridae Gulls & Alcids

20.06a Gulls

Larinae

• Joseph Lindholm III

Introduction: Gulls belong to the order Charadriiformes and the family *Laridae*. The taxonomy of gulls is still hotly debated by ornithologists, with some species recently being split into numerous sub-species. While numbering only about 50 species, gulls are one of the most familiar and successful groups of birds, found in shorelines and other suitable aquatic habitats across the world, with the exception of some of the Pacific islands. While many species are of distinctive appearance and easily identified, all gulls share a very similar basic morphology, and are not likely to be mistaken for anything else. A number of natural hybrids have been identified.

Management: Gulls have always held a limited attraction for private aviculturists, though individuals may make engaging pets. On the other hand, they have been widely kept by zoos, although in recent years there has been a decline in their popularity as zoo animals. In days when there was a far greater emphasis among zoos on exhibiting large numbers of species and specimens, and propagating anything was regarded as a worthy achievement, gulls made ideal exhibits, being generally hardy, easily fed, and able to withstand crowded and otherwise less than ideal captive environments. In addition, there was a tendency to collect the various forms like stamps, often exhibiting together a variety of species from around the world. Of the nine species of gulls documented by David Coles (2003) to have been bred in the United Kingdom, all but one were first bred by the London Zoo, and seven of those records were achieved no later than 1912, the first being for herring gulls *Larus agentatus* in 1850, followed by lesser black-backed gulls *L. fuscus* in 1872, and silver gulls *L. novaehollandiae* in 1879. Through its first century (1828–1927), the London Zoo procured 22 gull species (Low 1929). By 1939, the New York Zoological Park (Bronx Zoo), opened in 1899, had exhibited 26 forms (Ditmars & Crandall 1939).

Gulls have historically proven to be ideal inhabitants of outdoor exhibits due to their ability to tolerate practically all weather conditions, as long as open water is available. This hardiness is not confined to birds from typically harsh climates. The silver gull is a very familiar bird in Australia, and has proven particularly weather resistant. A breeding group at the St. Louis Zoo withstood a range of temperatures ranging from –24°C to 38°C (–18°F–100°F) (Bohmke & Fisher 1992). Their exhibit did provide ample shade, open water, shelter, and heat sources. In part because of this sturdiness, gulls were formerly standard inhabitants of large flight cages, and were often the most conspicuous features of such exhibits. The walk-through aviary at the National Zoological Park in Washington DC was informally known to its keepers as the "Silver Gull Cage". According to the *International Zoo Yearbook* (Zoological Society of London, 1960–1998) 41 were hatched there, and 34 reared, from 1971 through 1994 (Lindholm 2004). Over the years covered by the breeding records of the *International Zoo Yearbook* (1959–96), well over 150 gulls of five species were hatched in the Bronx Zoo's flight cage, which was 22.8 x 46.3 x 16.8 m (152 x 55 x 75 ft.) high. (This aviary, opened in 1900, collapsed under the weight of snow in early 1995). In London Zoo's famed Snowden Aviary, 108 African grey-headed gulls *Larus cirrocephalus poiocephalus* hatched from 1975 through 1996, only two being noted to have died prior to fledging (Zoological Society of London, 1960–1998, Lindholm 2004).

While it might be imagined that the presence of gulls in a community aviary would have a deleterious effect on other inhabitants, this is has not always been the case. Certainly, such robust and omnivorous species as herring gulls or greater black-backed gulls *Larus marinus* can be a positive menace to other birds, but quite satisfactory results have been achieved with other smaller and less aggressive gulls. From 1966 through 1992, 114 laughing gulls *Larus atricilla* were hatched at the Fort Worth Zoo, and 72 are recorded as being reared (Zoological Society of London 1960–1998). This constitutes the record for the most specimens of a gull species hatched in a single collection recorded by the *International Zoo Yearbook* (Lindholm 2004). These hatchings took place in the Pagoda, an aviary built over the framework of a fete pavilion, with a diameter of 12.2 x 7.6 m high (40 x 25 ft.), and an additional section 5 x 5 m (16 x 16 ft.) at one end. In 1977, the year that twenty laughing gulls hatched in this exhibit, and 19 were fully raised, two cattle egrets, four American green herons, eight snowy egrets, and one roseate spoonbill were also successfully reared there (Zoological Society of London, 1979).

From 1987–1992, 20 silver gulls were hatched and 15 were reared at the St. Louis Zoo's 1904 flight cage, 69.2 x 25.6 x 15.3 m (227 x 84 x x 50 feet) high. Approximately thirty species of primarily aquatic birds were maintained at the same time in this aviary (Bohmke & Fisher 1992). Tim Snyder, now Curator of Birds at the Birmingham Zoo, who worked with this exhibit, informs me that the only interspecific aggression by the gulls that he witnessed was in defense of their immediate nest areas when these were encroached upon. He never saw any aggression against Inca terns or their offspring.

A specimen of the small Bonaparte's gull *Larus philadelphia* was maintained for some years in the San Diego Zoo's Hummingbird Aviary, in the company of a diverse collection of shorebirds, roulroul partridges, lesser green broadbills, a number of species of hummingbirds and tanagers, and a variety of other small softbills (Lint 1971).

The maintenance of gulls in large community flights has declined as these exhibits, often built decades ago, have been "themed" to represent a particular zoogeographical area. The San Diego Zoo's Scripps Cage, built in 1922, where such species as the band-tailed gull

Black-backed gull
Larus fuscus

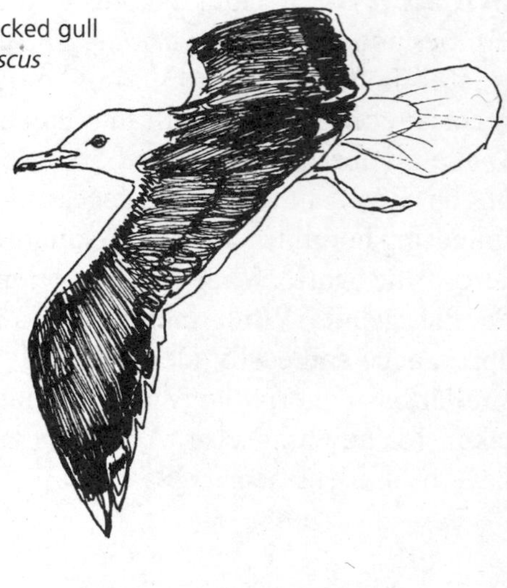

Larus belcheri from the Pacific coast of South America, the galapagos lava gull *L. fuliginosus,* and the black-headed gull *L. ridibundus* of the paleoarctic were exhibited over the years, was converted to a tropical South American display in the 1980s, and then an African aviary, forming part of San Diego's gorilla tropics, in 1991. The aforementioned silver gull cage at the National Zoo has recently been restricted to temperate climate Asian birds, while the St. Louis Zoo's 1904 flight cage now displays only North American species.

Another widespread traditional method of gull exhibition in zoos has been the maintenance of non-flighted birds in open enclosures. This has been especially the case with larger, more aggressive species such as herring gulls and great black-backed gulls, which easily breed in these circumstances, but also work well with other species such as laughing gulls. Despite the apparent impression of indestructibility presented by the larger gulls, care must still be taken against predators. The colony of herring gulls thus maintained at the New York Zoological Park (Hornaday 1904), which commenced rearing chicks in 1901 (Hornaday 1901) was exterminated by wild minks (Beebe &Crandall 1909). Attempts to establish a resident feral colony at the New York Zoological Park were not successful, as any unclipped birds, including ones hatched there, flew off (Hornaday 1904). In light of the nuisances that "fly-in" gulls have become in many American zoos, such experiment may seem most odd, but in the early twentieth century, the Atlantic coast population of herring gulls was in decline due to massive exploitation of their eggs for bakeries.

In marked contrast to the New York Zoological Park, Tiergarten der Stadt Nurnberg (Nuremberg Zoo) in Germany has been very successful in maintaining a liberty flock of herring gulls. Having noted that the *International Zoo Yearbook* documents more than 75 hatched there from 1959 through 1996, in 2004 I contacted Dr. Bernhard Neurohr, head veterinarian, who kindly provided the following information:

For at least four decades herring gulls have bred on a small island in a pond of over 4,000 square meters (43,000 sq. feet). In previous years they have also nested in the gorilla and polar bear exhibits. In all cases, the birds were free-flying. The zoo's lakes are also host to over 1,000 mallards, 500 great cormorants, and 60 grey herons, all breeding. The gulls prey extensively on mallard ducklings, as well as those of mandarin ducks. Occasional gulls have been killed and eaten by polar bears. At any time, there are usually only three breeding pairs of herring gulls on the grounds. The chicks are tagged and usually allowed to fly free. In 2002, three young birds were placed in the enclosure housing a breeding herd of California sea lions. While the sea lions "sometimes play with the gulls", they have never attacked or injured them. (The tradition of maintaining pinnipeds with either non-flighted or netted in gulls is a fairly old one. I believe California sealions have proved unpredictable exhibit mates, and such exhibits have fallen into decline, but common (or harbor) seals are apparently less problematic, and several European collections display them with gulls. [Harry Schramm, pers.com.])

In general, and in the United States in particular, the display of gulls in open enclosures has declined. This is in part due to the establishment of self-sustaining populations of *Spheniscus* penguins, now maintained under careful guidelines. Another factor is the availability of a much wider array of ducks, geese, and swans than in previous decades.

Gulls continue to find a place in zoo collections, but more often as deliberately chosen components of carefully designed exhibits. The Living Shores at the Chicago Zoological Park (Brookfield Zoo) was opened in 1996 and highlights the ecosystems of the arid coast of Peru and Chile. Along with a series of aquaria and displays for small mammals and reptiles, is a spacious, sunlit, and entirely indoor walk-through habitat, dramatically representing an abandoned guano mine on a sea cliff. Around 10 grey gulls *Larus modestus* are maintained with roughly 25 Inca terns and around 30 Humboldt's penguins. All three species reproduce successfully. The nucleus of the grey gull colony was collected as eggs by a special expedition. (*L. modestus* is famed for nesting in the high Atacama Desert of Chile, from where birds may fly 60 miles to the coast to feed).

Administered by the Paignton Zoo, the Living Coasts is located on the shore of Torbay in Torquay, Devon. Opened in 2003, this 1.4 acre facility features aquatic birds, South American fur seals, with black rats being the only other animals displayed (Ashby 2003). This complex is enclosed in 5,500 square meters (59,200 sq. ft.) of netting at a maximum height of 19 m (64 ft.). Grey gulls are at large here, sharing this space with Inca, common and sandwich terns, European avocets, ruffs, redshanks, black-tailed godwits, black-necked stilts, black scoters, spectacled and common eiders, barrow's goldeneyes, and gentoo and South African penguins. In a separate outdoor exhibit, tufted puffins and pigeon guillemots are exhibited with red-legged kittiwakes *Rissa brevirostris,* known to breed only in the Pribilof and Aleutian Islands and, so far as I know, not previously maintained in captivity.

The far more widespread black-legged kittiwake

Rissa tridactyla, a specialized cliff-nester like the red-legged kittiwake, has a long captive history, but has never been especially common in collections. It does not appear to have hatched in captivity until 1995, when success was achieved at Biodome Montreal in Quebec. Serge Pepin, curator of aquatic animal collections, at the Biodome has provided me with the history of black-legged kittiwakes there. This species is exhibited in the St. Lawrence Marine Ecosystem, an indoor walk-through exhibit. Through the authority of the Canadian Wildlife Service, during the summers of 1990 and 1991, 17 unfledged chicks were removed from as many nests, all containing more than one offspring. The collection sites were in Northern Quebec seabird refuges. The kittwakes share their exhibit with a breeding colony of common terns, as well as black guillemots, common eiders, pintails, green-winged teal, greater and lesser yellowlegs, short-billed dowitchers, ruddy turnstones, a breeding pair of black-crowned night herons, and some passerines. There are also a variety of northwestern Atlantic fishes and marine invertebrates. More than thirty kittiwakes have hatched in this exhibit since 1995. On the whole, these gulls have proved highly satisfactory inhabitants of the Biodome. The tidepools are shrouded with netting at night to prevent them from eating whelks and sea-urchins, and they are fond of stealing food brought up from the bottom of the aquarium by the eiders.

Black-legged kittiwakes are also a feature of an outdoor exhibit, Bass Rock, at the Royal Rotterdam Zoological Gardens, shared with common murres and Atlantic puffins. While they have done well and bred there, a group sent from Rotterdam to Tierpark Berlin, succumbed one by one to aspergillosis (Cathy King, pers. com.). As the Tierpark facility is an outdoor aviary devoted entirely to gulls of the world, where a number of species have bred over the years, it may be concluded that kittiwakes do not share the robustness of other gulls, at least when kept with them.

In general, gulls are fairly long-lived birds. The eminent zoo historian, Marvin Jones, Registrar Emeritus of the Zoological Society of San Diego, has provided me with the following significant longevities. A herring gull lived from June 17, 1921 to May 14, 1941 at the Philadelphia Zoological Garden, while a Pacific gull *Larus pacificus* received there November 10, 1920, died October 19, 1940. A great black-backed gull received at Philadelphia 29 September, 1924 was sent to the San Diego Zoo August 7, 1951. Two others, hatched at Philadelphia in 1948, were sent to San Diego at the same time. Of these three birds, one died at San Diego January 1, 1952, another December 1, 1953, and the last February 1, 1954. (I suspect they were kept with sealions). Unfortunately no record was made of which individual specimens died, so the death date of the 1924 bird is uncertain, but certainly no later than 1954. Significant longevities are not confined to the largest and most predatory species of gulls. A wild-caught silver gull, received at the St. Louis Zoo in 1969, was still alive in 1992 (Bohmke & Fisher, 1992). Of two female African grey-headed gulls that arrived June 2, 1965 at the London Zoo (where 108 were hatched from 1975 through 1996), one died June 14, 1993, the other October 2, 1993.

Diet: To inhabitants of densely populated areas at least, gulls are notorious for devouring garbage and other refuse. K.C. Lint, longtime curator of birds at the San Diego Zoo, considered them "the easiest of all birds to feed in captivity" (Lint & Lint 1981). Many species will readily consume anything they can swallow that contains protein or starch. In enclosures containing a variety of animals, they are likely to sample, if not steal, food intended for other species, and, as previously noted, some can be a definite menace to eggs, nestlings, or any small defenseless bird.

Traditionally, gulls in zoos have been fed whole or chopped fish, chopped raw meat, processed meat for zoo animals, or simple combinations thereof. As examples of current practice, I present diets from three institutions. The grey gulls and Inca terns in the afore-mentioned Living Shores exhibit at the Brookfield Zoo receive three species of whole small fish daily: 44% silversides, 36% capelin, and 20% marine smelt. These are supplemented with about 0.1% total weight thiamin-E paste, and about 0.02% multi-vitamin/mineral powder (Anne Oiler, pers. com.). The black-legged kittiwakes and common terns at Biodome Montreal's St. Lawrence Marine Ecosystem are likewise fed silversides, capelin, and smelt, as well as krill, with a daily vitamin supplement (Serge Pepin, pers.com). The silver gull colony at the St. Louis Zoo receives capelin, krill, and a commercial bird of prey diet, consisting largely of horse meat, supplemented with vitamins (Bohmke & Fisher 1992).

Breeding: According to the breeding records of the *International Zoo Yearbook,* 2,413 gulls of 28 taxa, as well as a further 88 hybrid specimens, were hatched among 138 collections representing 40 countries, from 1959 through 1996 (Lindholm 2004). Remarkably little in the way of published information (in English at least) of gull captive breeding has been produced.

1,316 herring gulls were hatched among 78 collections representing 27 countries (with a concentration in Central and Eastern Europe) from 1959 through 1996

(Lindholm 2004). The only detailed published accounts of captive breedings of this species of which I am aware are those of Hornaday (1901 & 1904) and Beebe & Crandall (1909), all concerning the above-noted colony maintained in an open enclosure at the New York Zoological Park (Bronx Zoo). William Hornaday (1904), first director of the zoological park provides a typically vivid description of the first success:

> "In an enclosure that embraced a pond and an island inhabited by about 20 gulls, 12 Canada geese, and a few other birds, two enterprising gulls decided to nest and rear a family. Accordingly they built a nest under a bush that stood on a point of the island, in a position that strategically was well chosen for purposes of defense. The two birds made a very wise division of the labor. The female built the nest, laid the eggs and hatched them, and the male did the screaming and fighting that was necessary to protect the family from molestation. Never was there a more bonnie fighter than that male bird. During that whole nesting period, lasting from April–May 15, he either bluffed or fought to a stand-still everything that came within 10 feet of that nest. Before his defiant and terrifying screams, and his threatening beak and wings, no other gull could stand for a moment. When a Canada goose crossed his line, the gull would rush at him, seize him by the nearest wing, wing-beat him, and hang on like a bulldog, regardless of being dragged about by the stronger bird, until the goose was glad to purchase peace by retreating. During all of these battles, the female sat firmly on her eggs, but pointed her bill at the sky, and screamed encouragement with all the power of her vocal machinery. Eventually the three eggs were hatched, and the young were reared successfully."

The most comprehensive published account of gull aviculture I am aware of concerns silver gulls at the St. Louis Zoo (Bohmke & Fisher 1992). A total of nine birds were obtained from 1969 into the mid 1980s, and 15 fully reared through 1992. As of 1992, the colony (since disbursed) consisted of four each of males, females, and unsexed birds. As previously mentioned, it was maintained in a walk-through aviary 69.2 x 25.6 x 15.3 m (227 x 84 x 50 ft.) high, with approximately thirty species of other birds. Four wooden A-frame shelters measuring 40 x 60 x 40 cm (16 x 24 x 16 in.) high were placed directly on the gravel substrate along the perimeter of the aviary. The gulls did not nest inside these structures, preferring to nest alongside them, or alongside a shrub or log. The nesting pairs bred in proximity to each other, sometimes as close together as 1.5 m (5 ft.). Only the immediate nest area was defended from other birds (Tim Snyder, pers.com.). Clutch size varied from 1–3 eggs, and an average of 20 nesting attempts was 1.85 eggs (Bohmke & Fisher 1992). Parental incubation ranged from 23 to 26 days. Chicks left the nest at 3–4 weeks of age, but continued to be fed by parents as late as 12 weeks.

Tim Snyder, now curator of birds at the Birmingham Zoo, provided me with further notes on this colony:

> "Our silver gulls in Saint Louis would nest just about anywhere but preferred to nest along the walls hidden behind low growing juniper bushes... Being against the wall and behind the bushes allowed them some sense of security. They defended their little territories against everything. They built their nest out of leaves, sticks, twigs, bits of hay, and of course, some gravel. [I] never saw a gull go through so much trouble building a nest before. I just thought they used a little scrape and some pebbles".

The tendency for captive gulls to nest in a different fashion from that in the wild has been observed elsewhere. Christine Shepherd, curator of birds at the Bronx Zoo, informed me that the prolific silver and Andean gulls *Larus serranus* kept prior to 1995 in the 1900 flight cage nested in hollows in the rockwork, using grasses and weeds from the exhibit. Christopher Davis, a keeper at the Fort Worth Zoo, where 114 laughing gulls hatched from 1966 through 1992 in the aforementioned community aviary, informed me the breeding pairs generally nested in the rockwork of the exhibit, selecting crevices between pieces of concrete rubble, definitely a departure from the usual depression in the sand. This parallels a similar situation at the Bronx Zoo decades previous, described by L.S. Crandall (1919):

> "The laughing gulls in the flying cage are among our most satisfactory exhibits. They are always immaculate and their raucous voices are never still. Some of the older individuals have nested for several years but with little success. Pelicans and herons developed a great fondness for young gulls, and hustling the harassed parents from the nests became a favorite sport with others. We finally adopted the expedient of surrounding a labyrinth of paving stones with a ring of stout wire of small mesh, with openings cut at the bottom for the passage of the gulls. The latter birds took to it at once and several pairs nested during the first season. Still the babies mysteriously disappeared, and after much watching we found that great

> herons and ibises were forcing themselves through the entrances, apparently far too small for birds of such size. Last spring we remedied this defect and two young gulls were added to the collection."

On the other hand, laughing gulls have also nested in zoos in a more typical fashion. Johnny Binder, general curator of the Cameron Park Zoo, as well as its previous incarnation, the Central Texas Zoo, also in Waco, Texas, told me that laughing gulls, which nested on and off at the Central Texas Zoo from the 1960s through the 1980s, nested directly on the gravel substrate of their aviary, without any sort of shelter. This exhibit was shared with herring gulls, crowned and sandhill cranes, ruddy shelducks, cattle egrets, and Lousiana and black-crowned night herons. Laurie McGivern, lead bird keeper at the Dallas Zoo provided me with the following information:

> "For the past two years [2002 and 2003] we have had laughing gulls nesting in two different locations. Due to difficulty with placement of chicks or permitting for release, we chose to substitute with dummies and not raise chicks. The first year, in our naturally planted flight cages, two pairs nested about 15 feet apart. The dominant pair had a scrape under a small bush and the second pair's was on open ground. The dominant pair incubated until term and the second pair's nest was predated. We have a problem with rat snakes in the area, as well as two male hyacinth macaws who routinely steal from the duck nest boxes. The dominant gulls were actually able to fend off the macaws. It was… quite funny to watch the macaws march toward the gulls on the ground and be driven off. Last year one pair nested in our old hippo enclosure. A weed had grown through the concrete curb near the edge of the exhibit and a small bit of straw lodged as well. They fiercely guarded about a 4.5 m (15 ft.) span from ducks, cattle egrets, pelicans, other gulls and keeper staff for the full incubation period. The two male and a female gulls are getting along and socializing [as of January] but I expect the pair will soon start to shun the lone male, as they come into breeding plumage."

Several species of gulls have specialized nesting requirements not met with in most zoo settings. The fairly common Bonaparte's gull has been exhibited by a number of North American zoos over the years but, to my knowledge, has never bred in captivity. This may be due to the fact that it most atypically builds a stick nest in small conifers (Gilliard 1958). The black-legged kittiwake is a specialized cliff nester and, while it has had a fairly long captive history, does not appear to have bred in captivity until 1995, in the above noted St. Lawrence Marine Ecosystem exhibit at Biodome Montreal, where breeding has since occurred every year, with more than 30 hatched. Nesting occurs on "artificial rocky cliffs surrounding a large saltwater basin". Nests are composed of algae, plant debris, soil, feathers, and guano (Serge Pepin, pers.com.).

As previously noted, many gull species have a potential for sustained propagation in zoos, and multiple generational breedings have occurred in a number of collections. Silver gulls received by the National Zoological Park in 1922 commenced breeding in 1925 (Tongren 1989), producing 16 chicks through 1928 (Mann 1930), and for some years afterwards. Five specimens were sent from the National Zoo to the Brookfield Zoo in 1933 (Plath 1946). Hatchings commenced at Brookfield in 1937 (Plath 1937), and occurred each year at least through 1945 (Plath 1946). All breedings took place in a 30.5 m (100 ft.) long indoor flight cage that, throughout this period was densely populated with herons (including great blue herons), storks, spoonbills, and pelicans, as well as herring, California and western gulls (none of which appear to have bred).

While the colony of African grey-headed gulls that produced 108 chicks at the London Zoo from 1975 through 1996 has declined to three non-reproductive birds as of 2004, the colony at Tierpark Berlin is entirely derived from it (Rudolf Reinhard, pers com.). Breeding at Berlin commenced in 1986 and, through 1996, 72 had hatched there, with 57 surviving to fledging (Zoological Society of London 1960–1998).

Tierpark Berlin maintains this species in the company of several others in an aviary constructed over a natural rock formation in 1984, expressly for gulls of the world. (Blaszkiewitz 2003). Along with the above-noted 72 grey-headed gulls, the following gulls were hatched in this exhibit over the period covered by the *International Zoo Yearbook:* 29 black-tailed gulls *Larus crassirostris* (26 reared) 1988–96, nine laughing gulls (five reared) 1985–96, 13 Mediterranean gulls *L. melanocephalus* (nine reared) 1990–95, and two black-headed gulls (1985 & 1987). As of 31 December, 1999, the aviary held 25 black-tailed gulls, eight laughing gulls, 39 grey-headed gulls, and 11 Mediterranean gulls.

As previously mentioned, the *International Zoo Yearbook* documents the Bronx Zoo as hatching over 150 gulls of five species from 1960 through 1992, in a walk-through aviary inhabited by a wide variety of other birds. I do not have exact figures, as the number of specimens was not specified before 1970, but over 60

Andean gulls (with at least 41 fledging) were hatched from 1967 through 1992. (The flock escaped when the aviary collapsed due to heavy snow drifts in 1995.) More than 60 silver gulls hatched from 1964 through 1988, with at least 50 fledging. Eight grey gulls hatched from 1970 through 1990, with five reared. From 1960 though 1977, more than 10 laughing gulls hatched, with at least nine reared. An unspecified number of ring-billed gulls *Larus delawarensis* were hatched over four years from 1960 through 1968, and in three of those years, laughing gulls were also hatched, while Andean gulls hatched during two of those years (Zoological Society of London 1960–1998).

The *International Zoo Yearbook* lists no hybrid gulls hatched at the Bronx Zoo, but does list a single hybrid for Tierpark Berlin: a common (mew) gull *Larus canus* X laughing gull hatched in 1985. The St. Louis Zoo, on the other hand, has had several hybrids hatch in the 1904 flight cage. The *International Zoo Yearbook* lists laughing gull X ring-billed gull hybrids hatched in 1967 and 1968. Not listed by the IZY, but documented by Bohmke & Fisher (1992), are two chicks reared by a male band-tailed gull and a female silver gull in 1985. Eighty-eight hybrid gulls are documented by the *International Zoo Yearbook's* breeding records. These were hatched among 13 zoos and involve 13 species (Lindholm, 2004). On the whole, however, it appears that if appropriate mates are available, hybridization is less likely to occur.

Artificial Incubation & Hand Rearing: As with most of the captive management requirements, gulls are not difficult to incubate or raise. I would recommend the section on PENGUINS and/or IBIS as a reference. Two eggs from the St. Louis colony were hatched in an incubator and hand reared. Bohmke and Fisher (1992) provide a table of one specimen's daily feedings, intake in grams per day, and chick weights from the age of two to 22 days. The number of daily feedings declined from six to three, while daily consumption rose from 9.6 grams to 118 grams. The chick was initially offered food at the end of a forceps, but fed on its own by the eighth day. The diet consisted of "soaked trout chow, krill, capelin, and Bird of Prey diet" (Bohmke & Fisher 1992).

Serge Pepin, curator of aquatic animal collections at Biodome Montreal provided me with the following notes on hand feeding the Kittiwake chicks collected from the wild in 1990 and 1991:

> "From capture to 30 days old, chicks are fed ad libitum on a mixture of silversides and bait smelts 7.6–10 cm (3–4 in.) long (about 80% of the diet), superba Japanese krill (about 5% of the diet) and squid pieces (about 5% of the diet). Multivitamins, thiamine, and calcium supplements are added to fish. Until they reach 30 days old, chicks receive food three times a day (7 a.m., 3 p.m., and 8 p.m.). After this period, food is given two times a day (7 a.m. and 3 p.m.)."

References:

Ashby, A. 2003. The Living Coasts. *Zoo!* No.22: 39–41.

Beebe, W. & L.S. Crandall 1909. Wild birds bred in captivity in the Eastern United States. *[New York] Zoological Society Bulletin* No.36: 580–583.

Blaszkiewitz, B. 2003 *Tierpark Berlin-Friedrichsfelde – Wegweiser durch den Tierpark*. Tierpark Berlin –Friedrichsfelde.

Bohmke, B.W. & M.T. Fisher 1992. Management of a Silver Gull Colony at St. Louis Zoological Park. *Avicultural Magazine* 98:173–177.

Brouwer, K., F. Rietkerk, S. Smits, B. Higginga, & M. Kurtz 1999. EAZA Charadriiformes TAG Survey: Update. *The EAZA TAG SURVEY*. Eighth Series: 29–40.

Coles, D. 2003. *First breeding records for birds reared to independence under controlled conditions in the United Kingdom*. DC Books Co.

Crandall, L.S. 1919. Rare birds bred in the Zoological Park. *[New York] Zoological Society Bulletin*. 22: 16–18.

Ditmars, R.L. & L.S. Crandall 1939. *Guide to the New York Zoological Park*. New York Zoological Society.

Gilliard, E.T. 1958. *Living birds of the world*. NY: Doubleday & Co.

Hornaday, W.T. 1901. Young birds and mammals. *[New York] Zoological Society* Bulletin No.5:15–16.

_____. 1904 *Hornaday's American Natural History*. N.Y.: Charles Scribner's Sons.

Lindholm, J.H. 2004. The *International Zoo Yearbook* breeding records, 1959–96: Gulls. *Avicultural Magazine* In Litt.

Lint, K.C. 1971. Shore and wading birds in the hummingbird aviary. *Zoonooz* 44 (No.6): 8–20.

Lint, K.C. & A.M. Lint 1981. *Feeding cage birds. A manual of diets for aviculture*. Blandford Press.

Low, G.C. 1929. List of the vertebrated animals exhibited in the Gardens of the Zoological Soeciety of London, 1828–1927. Zoological Society of London.

Mann, L.Q. 1930. Appendix II: List of animals born in the National Zoological Park during its first forty years. IN Mann, W.M. 1930. *Wild animals in and out of the zoo*. Washington, DC: Smithsonian Institution Series, Inc.

Plath, K. 1937. The birds at the Chicago Zoological Park (Part 3. The aquatic bird house; pheasants; cranes. *Aviculture* 7: 112–114.

_____. 1946. Avian oldtimers at Brookfield. *Aviculture* 16: 1–5

Tongren, S. 1989. *The birds of the National Zoological Park*. National Zoological Park

Zoological Society of London 1960–1998. Species of birds bred in zoos and other institutions 1959–1996. *International Zoo Yearbook*. pp. 1–36.

20.06b Alcids

Alcinae

• Heidi Weingartz

Introduction: The family Alcidae includes the little auk *Alle alle,* razorbill *Alca torda,* guillemots *Uria* and *Cepphus* spp., murrulets *Brachyramphus* and *Synthliboramphus* spp., auklets *Ptychoramphus, Cyclorrhynchus, Aethia* and *Cerorhinca* spp. and puffins *Fratercula* and *Lunda* species. The family falls under the order Charadriiformes. Alcids are pelagic seabirds of the colder waters in the northern hemisphere, spending most of the year living on the open ocean, only coming to land to breed. These characteristics make for some unique challenges when housing and breeding these birds in captivity, and these will be addressed in this chapter. They occur predominantly in the Bering Sea and the Sea of Okhotsk. There are also colonies off the coast of Alaska and in Britain and Newfoundland.

They are one of few groups of birds that are adapted for both air and sea. The only other group of birds that has adapted so well to their marine environment is the penguins, but they have forfeited flight in exchange for their adaptations. The alcids can both fly in the air and, literally, use their wings to fly underwater. There are 21 species of alcids in 10 genera. The group includes the well-known puffins, the murres and razorbill, the murrelets, auklets, guillemots and the Dovekie.

Management: Most of the alcid species are gregarious in the wild. It is therefore advisable to keep large groups of each species, 20 individuals or more, to maintain their natural social behavior in captivity. It is also important to mimic their natural surroundings as much as possible to facilitate breeding. For example, those species that commonly nest on rocky coastal cliffs would need a similar environment in captivity to breed. For the crevasse-nesting species, this is simple to achieve with the various artificial rock constructions that are widely available. For those species that use burrows, this is achieved by embedding flexible PVC piping 60–120 cm long (24–48 in.) into the artificial rockwork, which leads to easily accessible nest boxes. Be sure not to make the tunnels too long as this makes egg/chick collection and cleaning difficult. The nesting boxes can be attached to the wall at the end of the tunnel and should measure 30 cm^3 (12 cu. in.). Nesting material should be placed in the boxes and should also be provided in the enclosure so the birds can partake in nest building. Moss (growing or collected) and grass (growing or collected) are the safest options for nesting material.

Two of the most important stimuli for breeding these birds are day length and an abundance of food. Day length can easily be achieved with artificial lights and should be scheduled to fit closely to the natural summer daylight hours of the North Pacific coast. An increase in food supply is also simple to mimic. It is important to keep the swimming water and the air temperature at low levels. Ideal water temperature would be 4–7°C (39–45°F) and air temperature 4–16°C (39–61°F). Keeping these birds at these temperatures will lesson the risk of fungal infections and also reduce the risk of heat stress. Although many of these birds in captivity are kept in shallow water, it would be advisable to give them water that is as deep as possible to enable them to forage in a natural manner. Salt water is preferable to fresh since this will keep their salt glands active, thus preventing any congestion in the glands or nasal passages that could otherwise result in a susceptibility to respiratory disease. Due to the pelagic nature of these birds, it is important to ensure they have a large surface area of water to use relative to the land surface area. The water surface area should be at least three times as much as the land area. This will reduce the amount of time spent on the land and minimize bumble foot.

Alcids are monomorphic and therefore need to be sexed either with DNA or surgical sexing, to ensure reasonably even sexes in a collection. Large deviations from this can result in aggression. Also, the larger species, such as murres, have been known to be aggressive to smaller species, so a limited combination of species can be kept together.

Diet: A variety of fish species forms the basis of the alcid diet. Fish fed should include capelin, herring, silversides, salmon and sandlance. Most fish should be up to 15 cm (6 in.) in length. Great care should be taken to ensure safe handling of all fish products including storage in freezers and thawing in compliance with relevant governmental regulations in the applicable location. A number of other seasonal foods can be fed as enrichment and as a breeding stimulus, such as krill, squid and fish roe. Prior to and during the breeding season, plumage hues can be boosted with red Cyclopes (copepods) that are mixed in with the other food. Throughout the year, these birds should have vitamin supplements to make up for those lost in the freezing process. These can easily be administered by putting vitamin tablets under the operculum of the fish fed. If these "vita-fish" are fed first thing in the morning, there is a good chance that most birds will be hungry and will get a fish with vitamins. A maintenance vitamin supplement should be fed during the non-breeding season and a breeding vitamin should be fed during the spring and summer. Both are freely available commercially. It is important to feed the birds a number of times during the day; four times is recommended. A good method is to feed by broadcasting, whereby handfuls of fish are thrown over the water. This allows alcids of different species to fish at different depths.

Breeding: All eggs and chicks are handled using separate pairs of disposable gloves.

Nest boxes are checked every second day. As soon as a new egg is laid/detected, an initial mass (g) is taken, together with measurements (mm). After 7–10 days, the egg is weighed again and candled. If eggs appear infertile, wait another week and candle again to verify fertility. If the eggs are fertile at first candling, they are candled once a week (maximum) to monitor development. It is important to maintain a high humidity in the nesting boxes or cavities since these birds inhabit a moist environment. Ambient humidity in their natural habitat averages around 80% RH and nest boxes can be maintained close to this by spraying the nest boxes with cool mist/water once a day (or by the setting on the incubator if the eggs are being artificially hatched.)

Parent rearing until the age of 14 days is preferred. Upon initial inspection, a chick should appear dry, fluffy, alert, responsive, the eyes should be bright and moist, the cloaca clear, and the yolk sac properly absorbed. The umbilicus should be swabbed with a betadine solution. Chicks are weighed at the first opportunity and will be weighed every day thereafter and every time the chick is handled.

Supplemental feeding of chicks in burrows:

Starting on day two, chicks are hand-fed thin slivers of silversides. For the first two days of feeding the chicks are fed three times per day. The number of slivers fed per feed depends on the chick's appetite and mass. Chicks will usually hand feed readily if the fish is small enough. Once the chick is eating whole fish (at about day five), fish should be offered head first so they go down smoothly and there may be as many as 4–5 feeds per day. It is important to condition chicks to hand feeding for vitamin supplementation later. Alcid parents are usually very tolerant of these nest disruptions.

Chicks should gain about 10–13% of their body

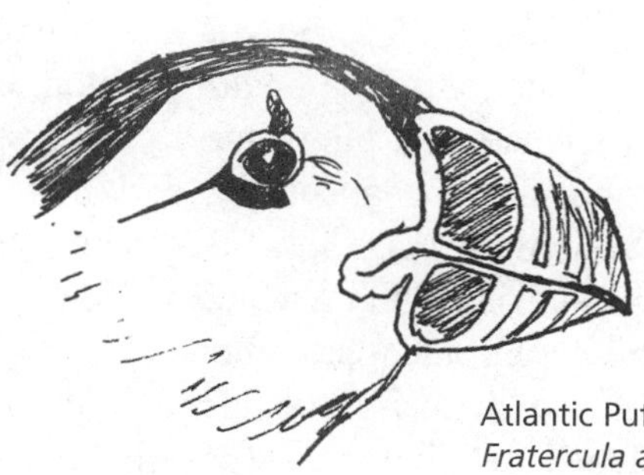

Atlantic Puffin
Fratercula arctica

Razorbill
Alca torda

Crested Auklet
Aethia cristatella

Rhinoceros Auklet
Cerorhinca monocerata

Dovekie
Alle alle

mass on a daily basis. If this growth rate cannot be maintained, it is presumed that a health problem exists or parental inexperience is at play. During the first few days, the yolk will still be absorbing and there will likely be a post-hatching mass loss. Beginning at day four or five, chicks have usually reached sufficient body mass and will be eating adequate amounts for vitamin supplementation. Use Mazuri Vita-Zu Bird 7537 tablets. Vitamins are placed in silversides (usually under the operculum) and fed based on Mazuri's specifications (One tablet per 250 g (8.8 oz.) of fish fed). Chicks are started on one-quarter of a vitamin tablet every other day around day five. Based on their food consumption, they can be increased to one-half of a tablet every other day. This usually occurs around day 13. By about day 24, the chicks can receive a full tablet every other day.

At age 14 days, remove the chicks before fledging and continue hand rearing in a holding area. Pulling chicks at two weeks after hatching will ensure well adjusted adults that are also quite approachable by keepers. Supplemental heat in the absence of the parents will be necessary by means of infrared lamps. 3M Nomad matting should be used to cover the kennel floor and cotton terry towels should be rolled lengthwise, formed into a semi-circle, and placed in the back of the kennel for the chick to nestle against. Keep the kennel quiet and dark. The hand-rearing room should be no warmer than 20 °C (68°F). On average, chicks at day 15 should be ready to move into a cooler holding room, around 13°C (55°F). At this age, contour feathers should be observed growing under down and will help to further insulate the chick. It is recommended that growth curves are kept on each bird that is being reared.

Hand Rearing: Eggs should be artificially incubated at no warmer than 37°C (99°F). Humidity should be set at 80% RH. The eggs should be turned every two hours. The plastic containers should measure 31 x 25 x 17 cm (12.5 x 10 x 8 in.) high. The lids must snap tightly onto the containers and holes should be drilled in each lid for ventilation. 3M Nomad matting is used to line the floor of each container, smooth side up, to prevent nails from becoming entangled. Washing in diluted bleach is the method used to clean the matting. Keepers should be very careful to rinse the matting well and let it dry completely before replacing it in the bins. If possible, a defecating hole is provided on one side of the container for the chick to use.

To monitor the temperatures inside the container, the sensor wire of a digital thermometer is placed under the lid. Temperatures are maintained at 27°C (81°F) for the first few days. To maintain humidity, a plastic container of warm water with small holes in the lid can be placed in the kennel. Newly hatched chicks are hand-fed small silversides or sand lance with the heads, tails, and fins removed. To insure chicks are hydrated, the fish are injected with water or a 5% Dextrose/Lactated Ringers solution for chicks with compromised health. Chicks are weighed first thing in the morning. The goal is to feed 40% of their body mass daily. This amount usually yields a 10–13% mass gain per day. Once the total daily percentage to be fed is decided, it can be divided over the number of feeds for that day. Daily charts should be kept on each chick to monitor feeding etc. Chicks should be fed 4–6 times a day initially. As the volume per feed increases, the feeds become less frequent.

Once chicks have reached 250–300 g (8.8–10.6 oz.), the chick can be fed three times a day.

Before each feeding, the keeper should palpate the chick's crop for any undigested fish. If leftover food is felt in the crop, no additional fish are fed, and keepers should wait one-half hour to see if the food passes. If there is a persistent sour smell to the chick's breath, it may become necessary to flush the chick's crop with water. If chicks are too weak to self-feed, they can be hydrated by tube feeding a mixture of water and a caloric supplement. A fish and water puree can be added as a step to wean from the caloric supplement to solid food items.

Horned Puffin *Fratercula corniculata*

Day	Weight (g)	Day	Weight (g)
1	46	**22**	281
2	44.4	**23**	305
3	45.9	**24**	324
4	52	**25**	345
5	58	**26**	345
6	65.8	**27**	357
7	76.1	**28**	381
8	88.6	**29**	393
9	94.1	**30**	400
10	106.5	**31**	402
11	122	**32**	402
12	137	**33**	406
13	147	**34**	407
14	160	**35**	421
15	168	**36**	406
16	185	**37**	400
17	199	**38**	398
18	219	**39**	390
19	237	**40**	385
20	251	**41**	388
21	263	**42**	

Usually between days 7–10, chicks are ready to begin self-feeding from a food dish. Fish are offered on a small dish, about 15 cm (5.9 in.) in diameter. Keepers

Puffin chick

should be careful not to leave fish in kennels with chicks for more than 15 minutes. This is to avoid the nutrient breakdown as fish thaw, as well as to prevent soiled plumage as a result of chicks lying on the fish. To maintain moisture, the fish can be misted with a spray bottle filled with water or simply dipped in a shallow dish of water before feeding. As chicks become more comfortable with self-feeding, dishes can be left in kennels for a longer time frame on a bed of crushed ice.

Usually by 15–25 days of age, chicks can be moved into the cooler room at about 13°C (55°F). By then, they should have a sufficient layer of contour feathers developing to thermo-regulate on their own. When the majority of the contour feathers have developed and most of the down has been lost, keepers should begin lightly misting alcid chicks to stimulate bathing behaviors and preening. At this age a mixture of food is offered on a bowl: krill, whole and cut up capelin, cut up squid, herring and silversides.

Chicks can be introduced slowly at this age to water in a tote. Initially they should only be swimming under supervision until it can be determined that they are completely waterproofed. Ensure that the birds have a ramp no steeper than 45 degrees leading out of the water onto a haul-out area. The haul-out area should be covered in nomad matting, which drains well and is soft enough to prevent bumble foot.

As can be seen from the graph below, these birds reach a peak mass at an age where they would fledge from the nest in the wild. At this age the wild fledglings need some extra reserves while they take a few days to learn how to catch fish by themselves. The horned puffin chick below is used as an example of a medium-sized alcid that was reared at the Alaska SeaLife Center in Seward, Alaska, USA in July/August 2003. This chick was at optimal health at all stages of growth. Average daily weight gain was 11.8% for the first 30 days.

20.07 Accipitridae Eagles

Glen Holland

Introduction: The order Falconiformes includes 292–304 species of Falconiformes split into five groups. Eagles fall into the group Acciptridae. This group consists of 223–232 species, including approximately 190 species of hawks and relations, including eagles and buzzards; around 30 species of kites; 13 species of harriers; and 15 species of Old World vultures.

Most of my personal experience with eagles is based on four African species: the crowned hawk-eagle *Stephanoaetus coronatus;* the black, or Verreaux's, eagle *Aquila verreauxii;* the martial eagle *Polemaetus bellicosus;* and the booted eagle *Hieraaetus pennatus*. In particular, I have been intensely involved with the crowned hawk-eagle, both in captivity and in the wild. This species is widespread in Africa from Guinea and Somalia southward to the Cape Province, but absent from dry areas and other areas where no large trees can be found. Known largely as a forest species, they also inhabit open woodland areas, particularly in densely vegetated valleys. In many areas, particularly where leopards are absent, crowned hawk-eagles are the largest predators, and therefore positioned at the top of the food chain. They are easily located by their aerial display flights, which are accompanied by loud, high-pitched, undulating calls. These birds perform an important role in controlling monkeys. When monkey populations are left to expand unchecked, they have a devastating effect on nesting birds. Relatively common, their numbers are limited largely by suitable habitat and size of territories.

Management: My experience with these birds started while I was employed by the Natal Parks Board in Zululand. I was at that time coordinating a raptor and specially-protected bird survey, and applied to the board to hand raise a single crowned hawk-eagle, from a second egg, to be used as a tool to promote raptor conservation in talks to farmers, schools, and other groups. At the same time, unknown to me, the San Diego Zoo had approached the parks board with a request to obtain and raise two birds in the hope that they would obtain a male to pair up with their single bird that was part of a new

African forest exhibit. Permission was granted, and I raised two birds, which fortunately turned out to be a pair. Following the advice of falconers, I chose the smaller male to use for the talks. My first trip out was to a farmers' meeting, and I began the talk outdoors with the bird perched on my arm. All went well until a farmer's dog walked out across the lawn. The eagle raised his crown and began to call and then suddenly clenched his talons which went straight through the double set of welding gloves on my fist and lower arm. I continued to smile and talk but soon had the bird tethered to the log on which he traveled. This was purely due to my inexperience as a falconer but from that day on he remained on his log during the talks.

These majestic raptors require a large, sturdy aviary covered with weld mesh of approximately 10 x 10 cm (3 x 3 in.). The aviary at the San Diego Zoo in which these birds have successfully bred is described as being 10 x 20 x 6 m (25 x 50 x 30 ft.) high, covered by 5 x 10 cm (2 x 4 in.) weld mesh, with two 3 m^3 (106 cu. ft.) holding cages adjacent to the large enclosure. These holding cages are necessary for management purposes, as the adults are very aggressive, especially the male, and must be locked out when keepers enter the enclosure. Crowned hawk-eagles are not compatible with any other species, either bird or mammal. If keepers are required to enter an enclosure with these birds, they must be aware of the danger and have access to easy escape routes should a problem arise. In addition, food should be delivered through a hatch or gate system that eliminates the necessity of entering the enclosure. San Diego feed in the holding cages so that birds are used to going in and can be trapped fairly easily. As mentioned, when hungry, as with any predator, they switch to a killing mode and are extremely dangerous. Hand-raised, imprinted birds in particular lack all fear of humans; when the birds are hungry, even a keeper falls into the category of prey. Fortunately, with experience, a keeper can learn to read an eagle's body language. For example, a raised crown, wings held slightly away from the body, or mantling over food are all warning signals. The

Harpy Eagle
Harpia harpyja

bird's calls also vary according to its mood. At San Diego, the crowned hawk-eagles are housed in the African section, which includes colobus monkeys and pygmy chimps. Similarly, at the Los Angeles Zoo a pair of harpy eagles *Harpia harpyja* was housed adjacent to spider monkeys. In neither instance did the close proximity of their natural prey seem to concern the eagles.

The aviary should be landscaped as described for forest Habitat A. Large *Ficus* trees will provide excellent natural habitat and perches. Ideally, large perches, 10–15 cm (4–6 in.) in diameter, should be placed high at both ends of the aviary to ensure that the birds obtain sufficient exercise. These hawk-eagles also enjoy bathing in a mist spray, which should be situated above a favorite perch and turned on for 20 minutes on warm days. In temperate climates, where freezing temperatures and extended periods of cold and wet weather occur, these birds will require indoor facilities with some additional warmth. As large as they are, cases of frostbite on toes have been recorded in these large raptors.

Claus Fentloff describes the aviaries used for breeding white-tailed sea eagles as follows:

> "The aviaries vary in size from 7 x 8 to 9 x 13 m (23.3 x 26.6" x 43.3 ft.) and are 5–6 m (16.6–20 ft.) high; eagles have bred successfully in much smaller cages measuring, for example, 4 x 5 x 3.2 m high (13.25 x 16.5 x 10.5 ft.). Since it is important that the birds are protected from noise and other disturbances, all the walls of the aviaries are solid up to 3 m (9.8 ft.) beyond which lattice extends upwards to allow light to penetrate even when the sun is low in winter. Apart from a small solid roof section, the top of each aviary is covered with galvanized 10 x 20 cm (4–8 in.) mesh chainlink fencing. The birds are kept individually or in pairs only, since the dangers of communal flight cages are known to be considerable.
>
> Dark elevated observation cabins have been installed between the eagle cages, and viewing windows made of one-way mirrors allow the breeding compartments to be observed without detection. Video apparatus has been set up in a number of aviaries containing breeding pairs, and the day's events can be played back on monitors in a central studio. At any time the apparatus can be set to record pair behavior, incubation and rearing of young without risk of the birds being disturbed by the sound of an observer. Slow-speed recordings of activity phases lasting several hours can be run at a faster speed and thus save valuable time."

Diet: The natural diet of crowned hawk-eagles consists

of monkeys, antelope, monitor lizards, mongoose, rock hyrax, guineafowl, and some other avian prey. Often the prey is caught co-operatively by the pair. Antelope of up to 20 kg (44 lb.) are a regular feature in the diet, and one record is of a bushbuck that was estimated to have weighed 30 kg (66 lb.). Once killed, the prey is dismembered; if too large to carry, various parts are torn off and cached in trees, away from other predators. They may survive for a few weeks on only 2–3% of their body weight but, when able, they will gorge up to 15% of their body weight in one feed. Following is the weekly captive diet for a single bird at the San Diego Zoo.

Sunday	medium rabbit	850 g (30 oz.)
Tuesday	bird of prey diet	680 g (24 oz.)
Wednesday	3 adult rats, whole	950 g (33.5 oz.)
Friday	2 adult rats, whole	630 g (22 oz.)

Monday, Thursday, and Saturday are fast days unless a chick is present, in which case the birds are fed daily. They are best fed in the early morning. Very little is wasted, with most of the bones being swallowed and, for example, in the case of a small antelope, only the skin and skull being discarded. Falconers have used crowned hawk-eagles to hunt jackal with the aid of a spotlight at night, and in cities such as Pietermaritzburg in KwaZulu-Natal, these birds are often reported taking pet dogs and cats caught on open lawns in gardens.

Breeding: In the natural state, due to the length of time required to complete the breeding cycle, and depending upon the availability of food, these birds tend to breed two out of every three years. In captivity, however, where immatures do not really have to learn to hunt and can therefore be removed from the parents about six months after fledging, it is possible that pairs could reproduce more often, although I would hesitate to say annually. Regarding the frequency of breeding at San Diego Zoo, each year the chick is moved into a holding facility at approximately six months of age. At this point, the parents are already showing signs of nest building. The pair at San Diego Zoo, of which the male was wild caught in the mid 1970s and the female hand raised from a young chick that was partially parent raised by a wild bird, were introduced to each other in 1995 and produced their first egg, which resulted in a chick, in 1996. In 1997, a two-egg clutch was laid. Both eggs hatched, one on March 30 and the second on April 1. The younger chick died in the nest on April 9, but the older chick was successfully reared. This pair continues to do well and have raised five chicks up to 2003.

African Fish-Eagle
Haliaeetus vocifer

Courtship of the crowned hawk-eagle in the natural state includes terrific aerial display flights, but in captivity it is limited to some chasing; the male running around his mate with his wings raised to display his chestnut underwing coverts; and much excited calling between the pair. The large, 2.5 m2 (27 sq. ft.) nest, usually placed in the fork of a tree and concealed below the canopy, is used repeatedly. A plentiful supply of large twigs, up to 8 cm (3 in.) in diameter and 1.2 m (4 ft.) in length, should be supplied. Naturally, many of these are broken from live trees while the birds are in flight. With most of the large eagles, the male supplies the building material that the female then positions in the nest. Copulation most often occurs on the nest, and with some species, such as the golden eagle Aquila chrysaetos, copulations continue during incubation and even after the chicks have hatched.

A number of publications have recommended the continual supply of fresh nesting material as an important factor in maintaining nest hygiene and the pair bond. At the San Diego Zoo the nesting birds were offered green sprays of foliage that they took through the wire and then carried along a perch leading to the nest platform that was located on a pole at a height of approximately 6 m (20 ft.). It is advisable that provision be made for examination of the nest site and its contents at any time of the year. A ladder positioned outside the aviary, near the nest, can be left in place year round for this purpose (the San Diego Zoo has a video camera positioned so that staff can see inside the nests for the crowned and harpy eagles). To support my warnings on the dangers of captive crowned hawk-eagles, similar experiences have been described with captive harpy and golden eagles. While collecting the eggs from the two wild crowned hawk-eagle nests that I had been monitoring, the adults remained close by, calling loudly, but did not attempt to attack us.

Crowned Hawk-eagle
Stephanoaetus coronatus

Day	Male	Female	Day	Male	Female
1	89.04	94	**38**	2160.00	1593
2	85.68	90	**39**	2360.00	1684
3	90.51	98.4	**40**	2360.00	1755.5
4	101.98	111.3	**41**	2460.00	1805
5	113.00	118.4	**42**	2530.00	
6	129.60	141.2	**43**	2520.00	1945.5
7	149.40	166.2	**44**	2700.00	2025.5
8	170.50	186.1	**45**	2720.00	2131
9	203.10	202.1	**46**	2810.00	2217.5
10	246.40	215.3	**47**	2850.00	2243
11	263.40	228	**48**	2860.00	2320
12	289.90	261.4	**49**	3040.00	2227
13	327.70	301.2	**50**	3160.00	2285
14	365.70	326.4	**51**	3100.00	2400.5
15	412.00	360	**52**	3240.00	2512
16	469.00	387.4	**53**	3360.00	2583.5
17	502.10	441.1	**54**	3420.00	2582.5
18	570.00	432.2	**55**	3520.00	2617.5
19	637.10	425.5	**56**	3340.00	2728
20	703.10	464.7	**57**	3680.00	2765
21	778.80	525	**58**		2811.5
22	861.40	599	**59**	3760.00	2939
23	945.90	664	**60**		2954.5
24	1010.00	733.2	**61**	3940.00	
25	1030.00	811	**62**		2891.5
26	1110.00	884.1	**63**	3840.00	
27	1190.00	946	**64**		3006
28	1250.00	1015	**65**	4000.00	
29	1350.00	1103	**66**		3177
30	1460.00	1119.3	**67**	4100.00	
31	1565.00	1165	**68**		3200
32	1603.00	1251	**69**	4220.00	
33	1700.00	1340	**70**		3254
34	1820.00	1423.5	**71**	4400.00	
35	1860.00	1459	**72**		3352
36	1940.00	1499	**73**		
37	2060.00	1548.5	**75**		3394

The usual clutch is 1–2 chalky white eggs, but sometimes only a single egg is laid; however, in some species, three eggs are not uncommon. The normal laying season in the southern hemisphere is from August to November. Most of the incubation in crowned hawk-eagles and harpy eagles is carried out by the female, with some assistance, usually about an hour a day, from the male. An interesting swap-over at a golden eagle nest has been described. The male landed on the nest and walked on the female's tail, deliberately touching her tail with his foot. The female stood up, moved off the nest, and was immediately replaced by the male. The interval between the laying of the first and second egg in most large species, including golden and bald eagles *Haliaeetus leucocephalus,* can be 2–6 days, but is usually 3–5. (Harpy eagles can lay a second egg as much as nine days after the first.) The interval may even vary in the same pair, between clutches.

Claus Fentloff advises on the formation of captive pairs of white-tailed sea-eagles as follows:

> "It is becoming apparent that, depending largely on climatic variations, there are marked differences in the courtship cycle of birds originating from different geographical areas. This means that a pair consisting of birds from areas perhaps only 6–7 degrees latitude apart is unable to synchronize their physiological and behavioral mating cycles to produce fertile eggs. Furthermore, observations to date indicate that the cycle of a bird from one cline does not become adapted to the conditions of another even after 6–10 years. Thus, a pair may remain together for years, perform courtship displays and copulate but never lay fertile eggs. It is therefore of vital importance in a breeding program to discover the geographical origins of captive pairs.
>
> When planning the formation of captive pairs of white-tailed sea eagles, where the birds are kept in a comparatively confined space, the 'concept of territory' cannot be ignored even though territorial behavior is subject to seasonal fluctuations, and other *Haliaeetus* species show various transitional forms between individual and group territories. In the wild the male offers a territory to the female, advertising it with display flights and calls. This factor has to be taken into account with captive birds, so a female is always allowed to become habituated to an aviary before a proposed mate is introduced. With conditions in captivity making it necessary to provide a partner instead of allowing free choice as exists in the wild, this is an essential precautionary measure. If there is any mutual aggression, a change of partners must be made immediately, but any male to which the female is later introduced must again have been in possession of its enclosure for some time. The addition of calcium preparations (a powder mixture is used made up of phospho-calcium salts and trace elements with vitamins A and D3 added) has resulted in an increase in egg-shell thickness from an initial 0.35–0.47 mm to 0.49–0.56 mm.
>
> In captive white-tailed sea eagles preliminary or autumn courtship, which begins in October, takes the form of frequent flights to and from and much calling; the birds then start working on the nest. A regular part of the nest-building routine is bark peeling as observed in the Steller's sea eagle *H. pelagicus* and the American bald eagle *H. leucocephalus* (pers. obs.). The earliest observed copulation was seen 11

days before laying and the last occurred 26–20 hours before laying. Ovulation, which is indicated by the female making quiet plangent sounds with wings and the head drooped, usually takes place around the latter time and the female will often interrupt the final copulation by throwing off the male. During this phase, however, the male makes more frequent copulation attempts than before but is vigorously thwarted by the female who beats her wings and lunges with her beak.

Laying takes place about 24 hours after ovulation. Two days before this the female will have adopted a sitting position between bouts of nest building; she now lies down in the nest supported on her wings and, with beak open, makes quiet cheeping sounds. Displacement activity by the male taking the form of vigorous attacks on the female so that she is driven from the nest, often occurs only 15 minutes before laying; nevertheless, all invariably goes well. It is possible that the male's aggression arises because the female rejects his mating attempts.

"We have discovered that individual white-tailed laying for the first time produce eggs 10–14 days later than in subsequent years when there appears to be a fairly standard period for laying. Consideration should be given to this in the preparations for captive breeding (e.g., provision of nesting material, food additives). We have found that such 'virginal clutches' are mostly infertile and other breeders have confirmed this finding. The laying date is not apparently affected by climate. First clutches usually consist of 2–3 eggs laid at intervals of 2–4 days. Replacement clutches are of 1–2 eggs and are laid three weeks after the last egg of the first clutch has been removed from the nest. When their eggs were taken immediately, one pair produced as many as eight eggs in a breeding season. Systematic observations on one pair over four days (days 2–6 of incubation) showed the proportion of the incubation time taken by the female was 59%, with an average shift of 73.8 minutes.

Throughout the observation period the eggs were left uncovered from only a few seconds to a maximum of seven minutes. The female was reluctant to leave the eggs at each change-over and frequently rejected her mate's persistent efforts to take his turn; she was also quite prepared to force the male off the eggs if he, atypically, did not immediately rise to let her take over. The incubation time for eggs for four of our pairs was 38 days; 21 artificially incubated eggs took 36–38 days to hatch. The hatchlings emerge during the days, with their eyes open, and from the first pipping of the egg to hatching takes 27–30 hours, activity being suspended at night. For the first few days the adult eagles are given freshly killed rabbit and duck, later also carp and trout, as food for their young. Although the male flies to the nest with the well-plucked prey and helps to feed the young, his share in feeding amounts to only 30%. During the early stages feeds take place hourly, each lasting 8–12 minutes. Intervals between feeds become longer as the young grow and may last up to two hours on day six."

Clutch and incubation details for some larger species

Species	Clutch	Incubation (days)
Crowned	1–2	49–51
Black (Verreaux)	1–3	45–46
Martial	1	47–50
Bald	1–2	35
Harpy	1–2	55
White-tailed sea	2–3	36–38
Golden	2–5	42–45

In many species, if both eggs, or all three in the case of three-egg clutches, hatch, the second, and third, chick is eliminated as a result of canism. Some species, such as the golden eagle and the tawny eagle *A. rapax,* have been known to raise more than one chick on occasion. At hatching, the chick is covered with white down, the first quills only appearing at about five weeks of age. As soon as the chick hatches, the role of the male changes largely to one of supplying food to the family. If the male crowned or harpy remains at the nest too long, he gets a good dose of vocal and body language from the female, who raises her crest and mantles over the food that she has snatched from him. At a harpy eagle nest, muscle from the fore — and hind — limbs of prey animals was fed most often, in pieces of about 7 mm^2 (.25 sq. in.), offered from the tip of the female's bill.

To facilitate feeding, her head was turned nearly upside down. The young eagle ate voraciously, even at five days, eating nearly 30 pieces at one feeding session.

David Rimlinger provided the following information on the breeding of harpy eagles at San Diego Zoo:

"Our harpy pair consists of a male, captive hatched at Tierpark Berlin in 1985 and received at SDZ in 1987. The female was wild origin in Columbia and came to SDZ in 1991. The pair produced their first fertile egg in 1992, which they hatched, but the chick died at 10 days of age after apparently being accidently grabbed by the female when reaching for food. By reviewing the video it appeared that the male had just dropped off a rat and she caught a talon on the chick by mistake when reaching for it.

Since that first clutch, whenever we have left an egg it gets broken during the process of incubation. Our routine has been to pull the egg after 10 days of natural incubation (we feel they hatch much better this way) if it was a single egg clutch or to split the clutch if two were laid. We have raised eight of 10 chicks since 1992, all puppet reared in isolation. The growth chart and rearing protocol is based on those hand-reared (puppet) birds. See Table 1 for the weight gains of these eight hand-raised chicks. Of the eight chicks, two were released to the wild in Panama (by the Peregrine Fund) and made it to independence catching monkeys and sloths, etc. A female that we raised in 1995 produced her first chick in 2000."

With the golden eagle, the male regularly participates in feeding the young. A summary of feedings recorded at a golden eagle nest indicates that 45% of the feedings were given by the male, 41% by the female, and the remaining 14% by both parents. Generally a carcass is carried to the nest, from which the parents tear pieces to feed the chicks. Using a commercial bird of prey diet for captive birds may inhibit the adult's ability to carry the food, already in small pieces, to the nest. Larger, chunky sources of food should therefore be supplied to nesting birds to allow for normal feeding behavior. It is quite natural for any chick that dies on the nest to be consumed, although often only in part."

Brian Jones of the Moholoholo wildlife rehabilitation center describes a rather unique breeding regime. He has an imprinted crowned hawk-eagle female, called Queen, which has accepted him as her mate. Brian describes how he and his bird have successfully bred six times. He assists by supplying branches and leaves for the nest, and he also shares in incubation. When relieving her on the nest, he carries a spray of green leaves to the nest. He then sits on the nest while Queen bathes and preens. She lays one egg and two egg clutches on alternate years. Once a wild nest has been located and the "Cain" egg has hatched, the "Abel" egg is then removed and replaces Queen's eggs. She then hatches and raises the chick, with Brian delivering food daily to his "mate" and her foster chick.

Hand Rearing: The eggs I collected were placed in an incubator at 37.1–37.5°C (98.8–99.5°F) with a relative humidity of 68%. Four days after the first egg was collected, calls and vigorous movements were detected within the egg. Within 36 hours the chick had pipped the shell, and after a further 40 hours it finally hatched. At hatching, this female chick weighed 71 g (2.5 oz.) and was extremely vocal, but rather unsteady. The feet of the newly hatched chick were small, soft, and limp, with no ability to grasp or support its body weight. For several weeks it either rested on its abdomen or sat on its hocks. At about a month of age the legs are strong enough to permit standing for a minute or two at a time. Behavior associated with preening is usually observed soon after hatching, with the bird rising on its hocks, stretching its body, and flapping its wings two or three times. By day eight the chick was able to sit for brief periods while preening a wing.

Claus Fentloff describes the incubation and rearing for white-tailed sea eagles as follows:

"White-tailed sea eagle eggs are incubated at a temperature of 37°C (98.6°F) until six days before hatching when the eggs are moved to a hatching compartment where the temperature is kept at 36.5°C (97.7°F). Relative air humidity is 64% in the incubator and 85–90% in the hatcher. We have found that the best hatching results using artificial incubation are achieved at a weight loss through evaporations of 13–14% over the whole incubation period. The fact that brooding pairs have never been seen to leave the eggs uncovered for more than a few minutes has led us to stop the practice of cooling eggs during artificial incubation. In captivity adults bathe readily after feeding and often come to nest wet. To simulate this moistening effect the eggs in the incubator are sprayed twice a day with water that has been boiled and allowed to cool to hand temperature.

Depending on whether they hatch in the morning or evening the chicks remain for 7–12 hours in the hatching compartment where the first defecation takes place after 20–30 minutes. The young are then transferred to the brooder, maintained at a temperature of 30°C (86°F) with a relative humidity of 50%, and fed. The first feeds given to the chicks are of chick or mouse heart and tender muscle flesh. From day six minced whole chicks or mice (without feathers/fur but including bone) are given. From day seven potentially pellet-forming material such as down and mouse fur may be added, but this is not essential. A calcium/phosphorous and a vitamin supplement is given from day one. Our observations of the extremely marked salivation by the adults while rearing young may indicate that calcium and enzymes are transferred in the saliva. Birds that were hand raised and given the same diet as was given to the parent birds raising chicks, showed signs of rickets.

We have noticed a marked improvement in development, with a disappearance of symptoms of nutrient deficiency, since we adopted the practice of immersing the pieces of meat in enzyme solution (One pellet of Enynorm dissolved in 130 ml of luke warm water) before feeding the young. Feeds are given every two to three hours, the intervals increasing with the growth of the chick."

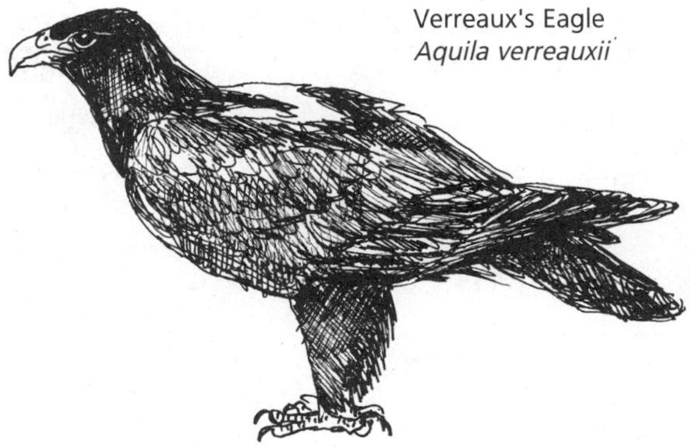

Verreaux's Eagle
Aquila verreauxii

The first feed for my crowned eagles was offered 18 hours after hatching, at which time two small pieces of partially plucked, minced bird, including all bones, were taken. Five feeds were given per day, and at the end of the first week, when the chick weighed 144 g (5.1 oz.), up to 10 pieces of food were taken per feed. The nest consisted of a large plastic bowl with a layer of branches and sticks in the base. Green sprays of vegetation were regularly placed in the nest. At two weeks, the chick weighed 316 g (11.1 oz.) and was feeding voraciously, with loud calls, whenever movement was detected near the nest. Any food dropped in the nest was picked up by the chick. At this stage, supplementary heat was removed during the day and the chick was regularly placed outdoors in dappled sunlight. This forest dwelling species soon began panting if placed in direct sunlight. Food consisted mainly of vervet monkey (from problem animal control) and rock hyrax meat. Many of the smaller bones were fed whole and the chick swallowed these with no problems. A two-month-old chick in the wild was observed swallowing an entire bushbuck leg bone, which took a few hours to finally disappear down the throat. Including bones in the diet is important, as powdered calcium supplements are insufficient to guard against bone growth problems when raising such large raptors.

At a month of age the chick, which weighed 830 g (29.3 oz.), was standing strongly and regularly exercising its wings, which were starting to show the first signs of quills. At six weeks, with the chick weighing 1,562 g (3 lb. 7 oz.), feeds were reduced to three per day. By this time dark quills had appeared over the back, wings and tail. The chick spent increasing amounts of time standing, but while sleeping it stretched out flat in the nest with the limbs extended in all directions. At two months the chick, now weighing 2,985 g (6 lb. 9 oz.), was feeding so aggressively that all food was offered with forceps and fingers kept well clear of the beak. Even at this age, only relatively small pieces of meat were swallowed, and feeds were reduced to twice daily. At this age a piece of a carcass can be left on the nest and the chick will begin to help itself. The chick's crest is visible by 60 days, and it is fully feathered at 11 weeks. Chicks fledge at 110 days, remaining in the vicinity of the nest and calling continually to the parents. Once fledged, they were fed about 700 g (24.7 oz.) of meat with some bone and fur, five times per week. Immatures are light colored with the head and chest white, and attain full adult plumage at four years. The information on hand rearing falcons in the FALCON section, which was obtained from the Peregrine Fund, provides valuable tips on hand raising raptors.

Weight chart for three eagle species, includes approximate food consumption for two species:

Week		Food /day	Weight (g)	Food /day	
	Bald Eagle		White-tailed Sea		Crowned
0	71	20	87.8	8	70
1	144	60	156	67	160
2	316	90	426	117	280
3	580	200	1030	331	600
4	830	300	1750	320	1450
5	1300	460			2300
6	1562	540			3120
7	2350	700			3850
8	2985	600			4030

Wedge-tailed Eagle *Aquila audax*

Day	Food/day (g)	Weight (g)	Day	Food/day (g)	Weight (g)
1	4	72	7	47	150
2	14	80	8	57	170
3	19	87	9	73	200
4	33	100	10	68	230
5	34	115	11	70	250
6	41	120	12	50	280

Table 1. Weight Gains for hand raised Harpy Eagle chicks

The New York Zoo described their hand-rearing diet for bald eagles, which includes a high percentage of fish in their natural diet, as chicken breast initially, being changed at day 10 to trout fillets, with mice being added on day 15, along with cod-liver oil, wheat germ oil, Calphos, and vionate powder.

Species Requiring Similar Management:

Harpy eagle *Harpia harpyja*
Hawk-eagles *Spizaetus* species
Golden eagle *Aquila chrysaetos*
Verreaux's eagle *A. verreauxii*
Martial eagle *Polemaetus bellicosus*

Contributors:

Brian Jones, Curator, Moholoholo Wildlife Rehabilitation Centre, Hoedspruit, SA.

David Rimlinger, Curator of Birds, San Diego Zoological Gardens.

Ed Lewins, Assistant Curator of Birds, San Diego Zoological Gardens.

References:

Fentloff, C. 1984. Breeding, artificial incubation and release of White-tailed sea eagles. *International Zoo Yearbook,* no. 23: 18–35.

Gilbert, S., P. Tomassoni & P.A. Kramer. 1980. History of captive management and breeding of bald eagles. *International Zoo Yearbook*, no. 21: 101–109.

Minnemann, D., H. Busse & B. Gast. 1987. Bemerkungen zur Haltung und Zucht sowie Beobachtungen zur Biologie und zum Verhalten der Harpyie *(Harpia harpyja)* im Tierpark Berlin (Notes on the Husbandry and Propagation as well as Observations on the Biology and Behavior of the Harpy Eagle *(Harpia harpyja)* at Tierpark Berlin.) Zool Garten N. F. 57 2/3: 69–115.

Steenberg, J. 1980. Captive breeding of American golden eagles. *International Zoo Yearbook*, no. 21: 109–115.

Steyn, P. 1985. *Birds of Prey of Southern Africa.*

Todd, F.S. & T. Meachan. 1974. Breeding the harpy eagle at the Los Angeles Zoo. *International Zoo Yearbook* 14: 90–94.

Old World Vultures

Introduction: Vultures belong to the order Ciconiiformes. The three species on which this chapter is largely based are all Old World vultures, which are classified along with hawks and eagles in the subfamily Accipitrinae of the family Accipitridae. There are 15 species of Old World vultures, seven in the genus Gyps, which are called griffons, and eight others in monotypic genera. The New World vultures are assigned to their own subfamily, Cathartinae, in the family Ciconiidae. The seven species fall into five genera, four of which are monotypic and the fifth, *Cathartes,* containing three species, the turkey vulture *C. aura*, and the lesser and greater yellow-headed vultures *C. burrovianus* and *C. melambrotus*. Other New World vultures include the black vulture *Coragyps atratus,* the king vulture *Sarcoramphus papa,* and the California and Andean condors, *Gymnogyps californianus* and *Vultur gryphus,* respectively.

The bearded vulture or lammergeyer *Gypaetus barbatus* is confined to mountainous terrain. Four geographically separated subspecies are recognized, two occurring in Africa, one in southern Europe and the Middle East, and another in southern Asia. Nowhere are these birds common, and they are now absent from much of their previous range in both hemispheres. Bearded vultures are adapted to their cold alpine environment by having broad, stiff, overlapping contour feathers that prevent cold winds from penetrating the downy insulating layer beneath. At night they roost in caves or on their wool-lined nests.

The seven species of griffons are found through various portions of Africa, India and central Asia. Vultures are destructive birds, and the griffons in particular will destroy soft plants. I vividly recall how one imprinted griffon aggressively chased me around the aviary, bouncing from perch to perch and running along the ground with his wings outstretched. A pair of griffins once managed to pull a large jacket of mine through the wire, tearing it to shreds before I could save it.

Management: Vultures are robust birds that require a large aviary with a length of at least 20 m (66 ft.) and a minimum width and height of 8 m (26 ft.). Their aviary, ideally should have a grass floor with a few bare sandy patches, large boulders and dead tree stumps on which the birds can perch. In addition, a pool of water with gently sloping sides and a maximum depth of 40 cm (16 in.) should be available, as these birds will bathe regularly. Some species of vultures, such as the bearded and the Egyptian *Neophron percnopterus,* occur in pairs; to avoid aggression they should be housed as such. Others, like the griffons, breed in loose colonies, and it is possible to keep a number of these birds together, depending on the size of the enclosure. It is, however, advisable to maintain an equal sex ratio in the group, to avoid excessive aggression among the birds. In cold climates, although the birds can live outdoors throughout the year, shelter is essential to protect the birds from prolonged rain, snow, and cold winds. This requirement varies between species, with the birds of alpine origin particularly resistant to severe weather.

Although I have seen griffons housed with other large raptors, such as black eagles *Ictinaetus malayensis,* this is not recommended in a typical aviary if breeding is to be encouraged in either species. Griffons are inquisitive and mischievous birds that will disturb any other birds in the enclosure. An interesting exhibit is found at the San Diego Wild Animal Park, where six Ruppell's griffons *Gyps rueppellii* are housed in a 36 hectare (89 acre) enclosure with 200 mammals of 12 species, including rhinos, antelope, and zebras, as well as cranes and spur-winged geese. The Ruppell's are kept pinioned by the removal of the distal portion of the wing.

Bearded vultures have been the focus of a coordinated captive breeding program to provide birds for release in the European Alps. The Innsbruck-Alpenzoo and Wassenaar Zoo have both been particularly successful in breeding bearded vultures. The release of captive-bred birds started in 1986; by 1996, 68 birds had been released in Austria, Switzerland, France and Italy. Of these, eight birds subsequently died and three had to be recaptured. In 1996, the first chick was successfully raised in the wild by a pair of captive-bred birds that had been released through this program.

Diet: Vultures are not difficult to feed in captivity, and will consume a variety of meats as well as the intestines of animals. However, their diet should always include some bone as well. Fresh bones that have been crushed with a hammer should be offered on a weekly basis. The birds will consume some of these bone fragments. Bearded vultures in particular specialize in feeding on bone marrow, which they obtain naturally by dropping bones from the air onto rocks in order to break them. A multivitamin and trace element powder should be sprinkled over the food twice per week. Adult griffons have been recorded in the natural state with as much as 1,000 g (2.2 lb.) of food in the crop, but such gorging is unnecessary in captivity, where they require about 300 g (0.66 lb.) per bird per day. The Ruppell's griffons at the San Diego Wild Animal Park are fed daily on a diet that includes varying proportions of rabbit, beef, and trout. Condors at the New York Zoo are fed a diet of rats on three days a week, bird of prey diet two days a week, and rib bones one day a week, with one fast day per week. King vultures at the Paris Menagerie are fed pieces of beef or sheep carcasses sprinkled with Tonacanin, a vitamin and mineral supplement. Throughout the breeding period, rats, guinea pigs, mice, and day-old chicks are regularly provided. Rabbit meat was refused whether or not the carcass had been skinned.

Breeding: Prolonged or repeated disruption near potential nest sites inhibits reproduction. It has been suggested that disturbance in general is the main factor limiting successful breeding in zoological parks. Female king vultures have been witnessed standing for long periods at potential nest sites, looking around and occasionally raising and lowering one foot after the other and pivoting, and it is believed that the purpose of this behavior is to test the amount of disturbance. A few loose items may be gathered together at a prospective site and investigated periodically. As egg-laying approaches, either sex may sit for long periods at potential nest sites. In species such as king vultures, which do not nest colonially, breeding can be inhibited by the presence of a juvenile that fledged the previous season.

In captivity, tree-nesting vulture species will build their nests on flat platforms, while cliff-nesting species,

Griffon Vulture
Gyps fulvus

including condors and bearded vultures, will use a nest cave. I have also seen the Cape griffon *Gyps coprotheres,* a cliff nester, breeding on an artificial cliff constructed for them. Pairs normally roost on the nest for a few weeks before laying begins, and the female usually refuses to eat for a few days immediately prior to laying. Generally, the griffons lay only one egg and incubation, although shared by both sexes, is predominantly by the female, and lasts about 55–59 days. The chick can be heard in the egg two days before hatching, and may take three days to finally hatch. The chick is entirely downy for the first month; quills start to erupt at day 35. The nesting period is long, with the chick fledging at 130 days. Both sexes feed the chick by regurgitation. Young are dependent on the adults for a month after fledging, and then join the adults when feeding.

The Ruppell's griffons at San Diego nested on a rocky, elevated hillside. Wooden nest platforms, which were inaccessible to the mammals, were placed in such areas, and nesting material in the form of dry twigs and branches were available ad lib. The birds have nested the past two seasons and produced three chicks in two years. Due to the presence of native predatory birds, the eggs are removed for artificial incubation. The real eggs are replaced with dummy eggs to give the vultures a chance to incubate and to provide the zoo staff with an opportunity to assess how successful the vultures are at defending the egg. The objective of this procedure is to return the egg to the parents at hatching time.

Cathartids such as the king vulture do not build actual nest structures. Instead, eggs are deposited in hollow stumps, crevices, caves, or on the sides of cliffs. Many of the sites are naturally covered with leaves, humus, or decaying wood, and such loose material is usually manipulated and rearranged. Kings, in common with other cathartids, prefer dark corners such as lofts; they also show a preference for sites with just enough space to squeeze into. The female carries out most of the nest building, with some assistance from the male. Typical nest building behavior includes picking up and nibbling objects in the immediate vicinity of the nest area, particularly when egg-laying is imminent, tossing material sideways or backward, picking up material in the bill and dropping it at the toes, dragging material and pieces of carcass to the nest site, and holding down potential nest material with the feet and tearing it apart. The substrate upon which the egg is to be laid is often loosened by probing with the bill, and a scrape then made. The scrape may be made by pushing the chest into the substrate and alternately thrusting the legs backward to push the nest material into place. This behavior is common in griffon species as well. The junction where a natural log has been placed at an angle to a wall is often used as a nest site.

Incubation, which in the king vulture is shared by both sexes, normally lasts 54–56 days, however, a two-day margin either way is common when eggs are artificially incubated. During changeovers, the relieving bird approaches in a hunched posture, walking slowly and carefully following a stereotyped path along one of the supports. The change is accompanied by a grunting sound emitted by both birds, followed by billing, the touching of wattles, or nibbling at the fleshy base of the bill and exchanging food. Egg shifting usually occurs as the relieving bird settles on the nest. The distance the egg is shifted varies among species and substrates, with king vultures moving their egg 10–15 cm (4–6 in.). Black vultures, which nested on the floor, moved their egg up to 2 m (62 ft.) in a 24-hour period. Broken eggs, or eggs removed for artificial incubation, are usually replaced 28–47 days later. Second clutches are not always laid, but are more common when birds lose a fresh rather than full-term egg.

As described by Marsha Schlee (1994), the hatching of a king vulture egg involves a high degree of assistance and interest by the adult birds. This natural parental participation in the hatching process supports the conclusions of a number of institutions that have artificially incubated the eggs of these birds, that vulture chicks require assistance in hatching. On the morning the egg hatched, the frequency of changeover increased from once in six hours to hourly. By afternoon, the birds were changing up to three times per hour. Both adults rolled the pipped egg at a highly accelerated rate, and sometimes flipped it over by seizing the shell at the area of breakage. The kings were not however seen nibbling the shell to enlarge the hole, as has been witnessed in the griffon vultures. Less than a minute and a half prior to actual hatching, the male stepped on the egg with one foot, providing the final shock necessary to crack it open.

Immediately after hatching, both parents tried to take the chick for brooding; the male pulled it by one leg but the female recovered it by its wing. In spite of this rough treatment the chick survived. However, this type of behavior may account for much of the mortality of eggs and young nestlings in captivity. The chick usually sits on the parent's foot while being brooded. By day eight the parents begin to leave the chick alone for short periods; they stop brooding altogether by day 25. The chick first begins to take food when about 18 hours old. On day 24, a parent-reared nestling was seen nibbling on a bone left at the nest; by day 45 it was actively standing on top of and feeding upon a piece of ribbed carcass. Similar behavior has been noted in hand-reared birds. King vulture chicks fledge at 115 days of age.

At the European zoos, the bearded vultures were supplied with branches and large masses of sheep wool. They spent about a month arranging this material before laying, which occurred in December. As problems were initially experienced with adults eating their offspring, subsequent eggs were removed about a week before hatching to be incubator hatched and hand-raised. The bearded vulture, unlike the griffons, usually lays two eggs, the second egg following the first after an interval of five days, and consequently hatching five days later. When both eggs hatch, the younger chick is invariably lost through cainism. Observations of a nest at Alpenzoo revealed that the parents fed both chicks, but the elder pecked almost continually, even when fed, at its sibling, grasping its wing, foot or head and apparently trying to swallow it. After seven days, the younger bird was removed for hand-rearing. Despite the injuries it had suffered, the chick recovered and was successfully reared. Parents feed the chicks small items together with much saliva and small bone fragments, and this procedure was duplicated as nearly as possible during hand-feeding.

The breeding of condors has been well researched, with the first Andean condor *Vultur gryphus* egg being incubated under a chicken as far back as 1846, at London Zoo (H.J. Samour, P.J.S. Olney, D. Herbert, F. Smith, J. White, & D. Wood 1982). Numerous zoos have now bred Andean condors to the point where no more birds are required, and many eggs are now discarded annually. Condors are sexually mature at six years of age. Courtship displays begin at the end of December or early January. The breeding techniques developed by working with the Andean condor were used as an analogue study in preparation for the Californian condor *Gymnogyps californianus* breeding project undertaken by the San Diego Zoological Society. When I visited this project, the first question which came to mind was why the "Fort Knox" appearance of the enclosures. It was explained to me that when the California condors were brought into captivity, armed guards were employed to protect them from the "greenies" that threatened to kill them rather than see them placed in captivity. This leads to the story of a long, controversial, interesting, and expensive breeding project which ultimately succeeded, with captive-bred birds being re-established in the wild. Most pairs of condors can be housed together through the year, but there have been some cases of extremely aggressive males that have had to be held separately from the females, and only introduced to them for short periods to allow copulation to take place. In the wild, condors only breed every second year, but through captive manipulation, such as the removal of eggs and chicks, they can be managed to breed every year.

Hand Rearing & Incubation: The removal of the first egg will usually result in a second egg being laid about a month later. The egg is collected as soon as possible after it is laid. To minimize human contact, eggs should only be handled using paper towels or wearing surgical gloves. Vulture eggs are artificially incubated at a temperature of 36–36.5°C (97–98°F), with a relative humidity of 55%. The natural incubation temperature under the European griffon *Gyps fulvus* has been recorded as 34.5°C (94.1°F) and, based on this, lappet-faced vultures *Torgos tracheliotus* have been successfully hatched at this temperature. The egg should be automatically turned a minimum of eight times daily during incubation. If the turner only rotates the egg 90–100°, it should be turned manually 180° at least once per day. After about 56–60 days of incubation, in the case of condors, or 52–56 days for griffons and bearded vultures, the egg should be examined daily. At the first sign of pipping, the egg should be transferred to a hatcher set at a temperature of 37.2°C (99°F) and a relative humidity of 80%. The egg should be placed with the pip facing upward. At this point, the egg must be sprayed at regular intervals with a fine mist of distilled water to prevent the membranes from drying and sticking to the chick. This is particularly important during the final stages of hatching.

D.F. Bruning, (1982) describes that there is usually a period of inactivity lasting12–48 hours after pipping, but once the chick begins to pip, it usually manages to break free of the egg within a few hours. A long delay at this stage may indicate that manual assistance is necessary, but great care must be taken to avoid rupturing the allantoic blood vessels in the chorioallantoic membranes, as severe bleeding and death can result. After hatching, most chicks appear exhausted and will lie quietly for 1–2 hours. When the chick is dry, it is placed in an infant incubator or a brooder. Most newly hatched chicks dehydrate easily in their first week. It is therefore advisable to include an electrolyte solution such as Pedialyte, Ringers Solution, dextrose, or glucose as part of the initial diet. The first feed is offered about 15–20 hours after hatching. The use of an enzyme such as Viokase or Tryplase to partially digest the meat to be fed to young vultures has proved useful. Ruppell's griffons, lappet-faced vultures, and a hooded vulture *Necrosyrtes monachus* have, however, been successfully reared without such additions, developing normally without any problems. The basic rearing diet used at New York Zoo consists of 45 g (1.58 oz.) horsemeat, 45 g (1.58 oz.) newborn mice, 50 ml (1.69 fl oz.) water, a drop of liquid multivitamin, a crushed digestive enzyme tablet, and 2 teaspoon of calcium lactate. Food is fed warm, and a new mixture prepared after every second feed. Until day 30, the chick is fed three times daily between 6:00 a.m. and 6:00 p.m. A steady weight gain is a good indication of adequate feeding. When the chick's food intake increases sharply, the diet is modified by increasing the amount of baby mice, while the horsemeat portion remains the same. By 40–60 days of age, the chick is offered skinned rats, and by 70–90 days, it will take whole dead rats. Over the next two months the diet is modified to conform to the adult diet.

Eight Hand-reared Andean Condors

Age (days)	Food intake (g/day)	Chick weight (g)
1	20	163–200
2	25–35	160–198
3	25–45	163–199
4	30–50	185–212
5	40–60	214–240
6	60–72	239–260
7	65–75	263–278
8	70–85	276–295
9	75–90	298–323
10	max 100	320–378
15	120	417–467
20	200	640–747
25	315	1000–1200
30	385	1500–1750
35	360	1950–2350
40	400	2500–2850
45	470	2850–3200
50	480	3200–3750
60	gradual	3400–5000
99	change to	5616
114	adult	5990
136	diet	7862

Table copied from Bruning, D.F. International Zoo Yearbook 23. 1982

The chicks react to high temperatures by hyperventilating, and should be cooled slightly if this occurs. Contact with the keeper should be kept to a minimum. The chick is encouraged to feed itself from a bowl as early as possible, usually around 3–4 days of age. When there is more than one hatchling, the young are placed together as soon as possible. Chicks can first be introduced with a glass barrier between them; later, at about two weeks of age, a mesh barrier can be used to replace the glass. This is, however, not possible with species such as the bearded, where cainism occurs. When chicks are about 12 days old, they are allowed outdoors in dappled sunlight for fresh air and exercise. Chicks placed together should always be under supervision. Lappet-faced vulture chicks differ from most other vultures in that they have strong feet that are able to grip food and on which the claw of the second toe is large and strongly curved, as in non-vulturine raptors.

Bearded *Gypaetus barbatus* and W. Ruppell's *Gyps rueppellii* Vultures

(with food intake for bearded)

Day	Bearded Vulture Weight (g)	Bearded food Weight (g)/day	Western Rupells Weight (g)
0	142	20	145
5	175	30	153.2
10	195	40	277.87
15	270	80	478.47
20	640	80	858.37
25	1090	110	1305.47
30	1800	210	1787.33
35	2800	625	2527
40	3300	870	
45	4270	1000	

One-way glass barriers have proved useful in reducing keeper contact with the birds. Marsha Schlee (1998) gives examples of how imprinting can be reduced or avoided. To minimize the chances of imprinting on humans, keepers at Milwaukee Zoo wear a mask and use a puppet glove when feeding chicks. The puppet is also used to groom the chicks and to manipulate food items in order to stimulate interest in eating. As soon as a chick is eating without assistance from the puppet and moving about on its own, it is placed in visual contact with conspecifics. At Arnhem and at the Paris Menagerie, mirrors are installed near the chicks to provide social stimulation as well as a non-human image

Bearded Vulture
Gypaetus barbatus

on which the nestling can develop a social attachment. The nestling at Paris spent much of its time next to the mirror, occasionally looking at it, vocalizing, and pecking or touching the image with its beak, but at Arnhem, no direct interactions were observed. Similar methods have been used with incubator-hatched lappet-faced and griffon vultures, where nestlings, as soon as the feathers started to grow, had visual contact with conspecifics throughout the day. Human contact was restricted to placing food in the enclosures. The individuals of these species reared in this way subsequently exhibited normal courting behavior and took no social interest in humans.

The daily food allowance for a hand-reared bearded vulture chick was about 20–30% of its body weight. A chick hand raised at Wassenaar was initially fed small pieces of rabbit and mouse, with some of the intestines included. This particular bird was incubated in a still-air incubator at 38°C (100.4°F) and a relative humidity of 60%. It was fed six times per day, but the crop was allowed to empty before the next feeding commenced. From day 13 onwards, entire minced guinea pig, without the fur, but with all the bones and intestines, was fed. After this more solid food was introduced, it was noticed that the chick's feces were hard and dry, but the bird was apparently thriving. However, due to the lean meat on guinea pigs, the chick did not show sufficient weight gain, and therefore a meat based, tinned commercial cat food was introduced as 50% of the diet. At day 19 the hind toe was occasionally remaining in the hind position and no longer tucked under the foot. Between days 10–30, it exhibited the typical cainism response, pecking at and trying to swallow grayish objects — a woolen sock, a model of a young vulture's head, etc. On day 35, an x-ray revealed that the chick was suffering from compaction. Liquid paraffin, administered at the rate of 4–10 ml per day, brought relief. To avoid imprinting, the chick was returned to the parents' nest at 47 days of age. The nest was divided by installing a barrier between the two chicks. On day 67, when the barrier was accidentally removed, both chicks ended up with bloodied heads! From day 110 on, however, the chicks were happy to sit near each other with slight aggression being shown only when they were extremely hungry. They fledged at approximately four months.

Andean Condor
Vultur gryphus

Ring Size: Griffons: 26 mm

Contributors: Dr. Christiane Bohm, Innsbruck Alpenzoo — bearded vulture; Mike Mace, Curator of Birds, San Diego Wild Animal Park — griffons.

References:

Bruning, D.F. 1982. Breeding the Andean condor at the New York Zoological Park. *International Zoo Yearbook* 23: 11–14.

Samour, H.J., P.J.S. Olney, D. Herbert, F. Smith, J. White, & D. Wood. 1982. Breeding and hand-rearing the Andean condor at London Zoo. *International Zoo Yearbook* 23: 7–11.

Schlee, M.A. 1994. Reproductive biology in king vultures at the Paris Menagerie. *International Zoo Yearbook* 33: 159–175.

Schlee, M.A. 1998. Hand rearing of Ruppell's griffon vulture. *International Zoo Yearbook* 36: 179–194.

Thaler, E. & H. Pechlaner. 1980. Cainism in the lammergeier or bearded vulture. *International Zoo Yearbook* 20: 278–280.

20.08 Falconidae Falcons and Kestrels

Glen Holland

Introduction: The Falconidae includes the falcons of which 13 species are called kestrels. All of these are still retained in the genus *Falco* that includes all the large hunting falcons. In total the family Falconidae includes 63 species. Together the genus *Falco* comprises 38 species, or 60% of the Falconidae. The other species within the Falconidae are scattered among the caracaras, long-legged neotropical birds, some of which include carrion in their diet; the unique laughing falcon *Herpetotheres cachinnans;* and the forest-falcons of the neotropics (many of them secretive). Also included are the pygmy falcons and falconets of the Old World tropics, including the small falconets that hunt the forests of southeast Asia, Borneo, and the Philippines. A section on captive husbandry of pygmy falcons and the Mauritius kestrel is included after the section on larger falcons.

The prime focus of the captive husbandry section is on the peregrine falcon *Falco peregrinus*. This species is distributed through most of the world, with 18 recognized subspecies. Peregrines are extremely able hunters, known for their speed, which has been estimated to reach up to 240 km/h (144 mph) during a stoop. In both the USA and Britain the peregrine was nearly exterminated. In the eastern United States, this was attributed to persistent organochlorine pesticides, which caused eggshell thinning and abnormal adult brooding behavior, resulting in breeding failure. Once considered the poster child of endangered species legislation, the peregrine falcon now represents an underpublicized triumph for the concept of captive breeding as a tool for conservation. In the USA, the peregrine has been captive bred and re-introduced into areas where it formerly occurred. During the 1970s, the Peregrine Fund in Idaho began to break new ground, achieving astounding success with the captive breeding of the species. In fact, this entire chapter is based largely on *Falcon Propagation,* a manual on captive breeding produced by the Peregrine Fund, Inc., in 1991.

Management: The peregrine breeding facilities located at Cornell University in Ithaca, New York, consist of a large barn divided into 40 rooms, each measuring 3 x 6 m (10 x 20 ft.). The roof of each room offers complete cover and includes a single 0.9 x 3 m (3 x 10 ft.) translucent fiberglass panel to allow more light to enter the room. In suitable climates, such a panel can be opened in good weather to provide increased ventilation. Rooms with solid roofs need some source of ventilation to prevent heat build up. The front wall is open to the weather, with vertical bars 6.2 cm (2.5 in.) apart and separated from the external wire mesh by 15 cm (6 in.). In order to reduce the amount of snow entering the rooms, the lower 0.9 m (3 ft.) of the front is paneled. A wire mesh predator barrier is buried all around the building. The roof is 6 m (20 ft.) high on the interior and 4.2 m (13.75 ft.) at the front wall. Each room receives either morning or afternoon sunlight. Birds are observed through one-way glass windows from both the upper and lower levels of the central hallway, to which all rooms are connected. There is also an upper and lower feeding port and a small access door to the corner-nesting ledge in each pen. Noise is minimized by playing music in the central passage, and through the use of a television monitoring system.

The New Zealand Falcon Husbandry Manual recommends wide ledges as resting areas for falcons. An easily cleaned ledge with a food chute above it provides a means to feed the birds from outside the aviary during the breeding season and should be inside the double door system. A large plastic or concrete bath of at least 50 cm (20 in.) in diameter should be provided for the birds to bathe.

Viewing access should be incorporated in the design of the aviary either using one-way glass or peepholes. Additional secluded areas or screens where birds can remain out of sight of each other should be provided to lessen the risk of females killing males. The aviary should be positioned where the birds have access to morning sun.

Perches should be at least 5 cm (2 in.) in diameter and arranged to allow a minimum 1.2 m (4 ft.) of headroom, which is enough to facilitate copulations and limited flight displays. The highest perches in the room should be natural branches, and large smooth rocks should be placed both on the floor and on the edge of each nest ledge. Good quality coco fiber mats are recommended for the other perches in the room. Floors should be covered with a minimum of 7.5 cm (3 in.) of coarse, round gravel, which is not dusty and does not compact. Each room should be fitted with a light source that allows for photoperiod manipulation and lighting for emergencies, such as a brooding bird off the eggs in the dark.

In order to maintain a respectable level of cleanliness and at the same time minimize stress, the chambers are ideally cleaned and reconditioned twice a year. In addition to the biannual general cleaning, some food remains should be removed any time a chamber is

Bat Falcon
Falco rufigularis

entered. The New Zealand Falcon Husbandry Manual recommends that aviaries should be cleaned by a weekly sweeping up of fecal matter, castings, removal of uneaten food, and scrubbing out of water dishes and food ledges. Aspergillosus is a known health problem of falcons, especially in damp conditions, but the removal of substrates for this fungal disease, the use of gravel floors, minimization of stress, and good ventilation largely prevents this often fatal disease. Any capture and subsequent handling should be accomplished as quickly as possible and with a minimum of stress. Falcons too wild to be taken up willingly are quickly caught with a long handled net. This method is far superior to a protracted chase around the chamber. Prior to capture, keepers make some noise to warn the birds of their presence and thus avoid a panic flush from the bird. An assistant usually allows the falcon to grip his gloved hands to prevent self-inflicted foot punctures. The bird may be hooded if the operation is to take more than a few moments.

Diet: The diet of wild New Zealand falcons consists of 61% birds, 38% mammals, with 1% reptiles and insects. The captive diet should consist of 40% chickens raised to the weight of 284 g (10 oz.) and 60% coturnix quail raised to 60 days, both of which are easily and economically reared in large numbers to provide the bulk of the whole carcass diet for falcons. Domestic pigeons should be avoided, as they are more likely to carry diseases.

The New Zealand Falcon Husbandry Manual mentions that some breeders do, however, use pigeons and, if used, it is recommended that they be frozen for at least two weeks at –20°C (–4°F) to kill the protozoan that causes "Frounce" or trichomoniasis. The staple diet for New Zealand falcons is recommended as 60% day-old chicks, 20% mice, 20% coturnix quail with an added vitamin/mineral supplement. Large prey should be cut into at least two pieces to prevent fighting between a pair. Females, which are 30% larger than the males, have been known to kill their mates. During the breeding season from August (in Southern Hemisphere), food should be given through the food chute at least three times daily in small portions to encourage "prey" transfers from male to female as part of courtship. During the breeding season the proportion of quail should be increased to 60%.

It pays to insure that these food animals are maintained on the highest quality feed and under the most sanitary conditions practicable. After 24 hours of fasting to allow clearance of materials from the digestive tract, food animals are killed and quick-frozen. Daily rations are thawed as needed and fed immediately. The chickens are cut in half, the two halves providing an ample meal for a pair of peregrines. Birds are fed once per day. In order to deter any food-related aggression, pairs are always fed with two or more food items. If excessive food remains are observed, the amount being fed may be reduced. A well-fed pair will leave chicken feet and wings uneaten. Untouched food is an indication that something may be wrong with one or both of the falcons. As a rule of thumb, it is always better to overfeed slightly than not feed enough.

A vitamin supplement can be injected into the major food items throughout the breeding season. This is in addition to, not as a substitute for, good basic nutrition. Vitamin E has been found to be eight times lower in captive falcons than in their wild counterparts. This should be supplied to captive birds through one of two methods. In small-scale breeding programs, a supplement is injected into the breast muscle of the quail at the rate of 44 IU per 100 g (3.5 oz.) quail; in large-scale operations, the supplement is added to the quail feed at a concentration of approximately 200 IU/kg (200 IU/2.2 lb.) dry matter. Biotin deficiency causes a reduction in hatchability without adversely affecting egg production. Quail diets should contain 0.2–0.3 mg/kg of biotin.

Breeding: The sexes are alike in plumage, but the male is about a third smaller than the female. Falcons begin breeding from two years of age, but pairs can take several years before successful breeding takes place. Each pair of falcons represents a challenge to the aviculturist, requiring much experimentation and intuitive insight to bring the mates into reproductive condition. These procedures are both time consuming and laborious. Birds trapped as migrants show the least potential as breeders in the captive situation. The hand rearing of potential breeders in group situations is a favored means of insuring the maximum percentage of successful natural pairs. However, it is suggested that actual hand-feeding be kept to a minimum after 14 days of age. The group should be allowed to feed from the same bowl in the presence of the handler, and be fed often enough so that there is not a clamoring for the food when it is presented. It is also important to spend some time with these youngsters that is not food related.

Oscar Beingolea provides the following comment on conditioning raptors to breed:

> "Photoperiod is said to be the main environmental component to which the breeding cycle of most animal species is adjusted, but little is known about its influence in many subtropical animal species, including avian raptors. Around the Equator the photoperiod remains almost constant, so it is believed that the strongest element stimulating breeding is food supply.

Thus the breeding season is linked to the time when food supply turns abundant. Our captive experiences with South American falcons (genus Falco) and hawk species (Accipitridae) around 12°S is evidence that the falcon breeding season is influenced by photoperiod while some hawk species are not. In fact, with a different food spectrum and aerial needs, it seems that falcons are more sensitive and that they have adjusted to the photoperiod, in anticipation or during food abundance and avoid any possible interference induced by the main rainy season. In this way each falcon species is averaging a five-months laying span. Falcons have also adjusted their size to the food spectrum. The peregrine falcon is the largest falcon and a bird-eating specialist, so that its laying begins earlier (late April). The peregrine falcon begins when the photoperiod gets increasingly reduced and some two months earlier than the rest of smaller and insect-eater falcon species, such as the Aplomado falcon *Falco femoralis* and Bat falcon *Falco rufigularis*. These two species begin laying after the shortest day of the year (June 23) and peak when the photoperiod announces the oncoming spring. A similar occurrence happens with the Eurasian kestrel and the European hobby; it is interesting to note that our American kestrel *Falco sparverius* begins earlier than the rest of the falcons but not earlier than the peregrine falcon. We have found that peregrines are finishing in mid September but Aplomado falcons and Bat falcons lay their last clutches around late November, so the pairs of these species breeding this late are also benefiting from the surplus of insects emerging during the beginning of the next rainy season. The number of eggs and fertility decreases towards the end of the season. As described, the late breeding wild hawks have lower reproductive success. The lower fertility of our captive falcons is probably explained by the switching off of the reproductive system in both male and female, induced by photoperiod changes. These changes in the hormonal system probably include a reduction in the impulse to rear young, further decreasing the success of late breeders.

Perhaps due to wider food sources and different aerial needs, some hawk species such as Harris hawks *Parabuteo unicinctus* and bi-colored hawk *Accipiter bicolor* display a year-round laying span, suggesting that they are not stimulated by photoperiod. Perhaps the more interesting finding in our captive experience has been our female bi-colored hawk that laid 25 eggs in eight clutches within a 13 month period. Then she rested seven months before beginning again. Seven months is about the longest resting time, between two clutches, that our female ornate hawk-eagle *Spizaetus ornatus* took after raising a chick up to three months and 30 days. Since the laying in some hawk species is not influenced by the photoperiod at this latitude, the females of these hawks have a greater endurance, enhancing their chances of completing a breeding cycle. We have not discovered as yet if the fertility of a male, either the male bi-colored that began copulating during and after the clutch number six, or the ornate hawk-eagle, could match the female's ability to produce eggs. Although the ornates were copulating, the male did not fertilize any of the eggs laid between seven months and up to 20 months after the hatching of an egg. In the wild, the flying young of this species remain dependent up to one year after hatching and are mostly fed by the male. We don't know if infertility in the male really occurs (we have not performed a spermatic count) as an evolutionary development preventing him from breeding too soon, ensuring the survival of the dependent young.

Some of our captive females could respond to our artificial overfeeding rather than to any male in courtship. Does this mean that the wild females of some hawk species could respond to the courtship and overfeeding from a new and fresh mate. This would mean that she could achieve another breeding cycle and earlier than with the previous male? This has not been ascertained as yet in wild pairs of the ornate hawk-eagle, nor have I heard about similar findings in similar sized hawks, but I believe it is possible. While any individual breeds to ensure continuity of the species, it also works to ensure continuity of its own genes."

With the onset of the breeding season in early spring the female peregrine falcon begins to increase her vocalizations and the feeding regime is altered somewhat. To encourage food transfers, smaller amounts of food are fed several times per day. Quail culled at a young age are useful to encourage this behavior and are particularly attractive to males. One or two nest ledges, larger than the area required simply for nesting, are located near portholes, which allow for easy access. New Zealand falcons regularly nest on the floor, and patches of fine gravel are placed near cover such as a dense shrub or a log. These are covered with round gravel and the leading edges covered with synthetic turf or open loop carpeting. Courtship begins with fly-bys by the male, scraping by one or both birds, and the female waiting longer before feeding and pushing or driving the male. Either falcon, but usually the male, may stand in the scrape and call to its chamber mate. As the pairing behavior progresses, food transfers increase in frequency. Before egg laying, the female should begin to solicit copulation by holding her head low while raising her tail. A whining vocalization may accompany this posture. Many pairs require more than one season to learn to copulate well enough to fertilize all or most of the eggs. As egg laying becomes imminent, the female will look sick and become reluctant to leave the ledge area. This condition is called egg lethargy and is quite normal.

On occasion, young are fostered to pairs. When young are exchanged for eggs, the chicks are grouped on the ledge against the wall near the scrape. A few dead quail are also placed on the ledge. This can be a risky process and careful observation is required to ensure that the female, in particular, accepts and broods the young. Tapping on the wall of the chamber can halt any slight aggression. Once the pair has accepted the young, fresh food is provided three times each day. The young are carefully observed to make sure that all are being fed. Peregrines should not be supplied with more than four chicks, but will raise two successive broods. However, caution is advised, as the adults will tire of raising young and become less attentive. The majority of imprinted birds, both male and female, have proven to be good parents for young that require no brooding.

First time breeders should not be subjected to stress, but pairs that fail to breed after two or three years may benefit from separation and pairing with new mates. Some pairs will produce no more than four eggs in a clutch, whereas others will produce 12–16 eggs when eggs are removed sequentially. When entering the chamber during the breeding season, some noise should be made to warn the birds of human presence. This will help to prevent broken eggs or ejected chicks as a startled adult bolts the nest. Some females are particularly aggressive; in such cases a stuffed glove on a pole is placed near the nest for the female to latch onto. She is then gently lowered to the ground.

Artificial insemination and the training of imprinted birds to be semen donors are procedures that the Peregrine Fund has perfected. Using these techniques, they have produced remarkable numbers of falcons. Artificial insemination is, however, an involved procedure that would deserve a chapter in itself; as such, I will not cover the topic in this book.

Incubation and Hand Rearing: The incubation room should have a regulated ambient temperature of 21–26°C (70–80°F). Extremes in room temperature make it difficult for the incubator to maintain a consistent internal temperature. In areas where ambient humidity is high, especially in the warmer months, a dehumidifier may be necessary to control relative humidity in the incubator room. The room should not allow sunlight to enter directly, resulting in raised temperatures, and should be sufficiently dark to allow for eggs to be candled. Prior to the beginning of incubation, the incubator should be cleaned with a bacterial and fungicidal disinfectant. Although generally, due to health hazards, the practice is no longer used, some breeders use fumigation with formaldehyde gas. This gas is highly noxious, so the incubator should be placed in a well-ventilated area. The gas is easily produced by mixing 0.4 g of potassium permanganate with 0.8 cc of formalin (37.5%) for each cubic foot of incubator space. The gas should be allowed to circulate for at least 20 minutes. If necessary, eggs can be stored for up to five days at a temperature of 15°C (60°F) and a relative humidity of 75–80%.

Natural incubation, by either a raptor or a broody hen, provides the best start for an egg. Eggs that were placed in an incubator after receiving less than 3–4 days of natural incubation hatched at a rate of 50–60%, while eggs that were naturally incubated for the first 5–10 days hatched at a rate of 75–85% after being placed in an incubator. Lifting a clutch after 7–10 days allows the eggs a good start. At this stage, the parents have not incubated so long that they will not recycle and lay a new clutch after the first is removed.

For optimal hatching success, the eggs should be turned at least eight times every 24 hours with the incubators set 37.5°C (100°F). Regardless of the number of times an egg is turned each day, the intervals between turning should be evenly spaced throughout the 24-hour period. Furthermore, the eggs should be turned in alternate directions, as turning in only one direction will increase embryo mortality. A scale that weighs accurately to 0.1 gram is essential to weigh the eggs and ensure a proper loss of egg weight, which is an important factor for successful incubation. An egg weight loss of 15% from beginning of incubation to pipping is a crucial element in ensuring a good hatch rate. This is best controlled by determining the fresh egg weight and then plotting a 15% weight loss on a graph. The weight loss can then be controlled through adjustments in relative humidity. If, for example, weight loss falls below the 15% line, the egg is losing too much weight and the rate of weight loss is then decreased by slightly increasing the humidity in the incubator.

Approximately 24–48 hours before the egg pips, candling reveals that the air cell expands and gradually extends down one side of the egg. This process is called draw down; once it begins, turning the egg is no longer necessary. Upon pipping, the egg is transferred to the hatcher, which has the relative humidity set at 55–60% and the temperature set about 0.55°C (1°F) less than the incubator. At this hatcher temperature, the pip to hatch interval averages 50 hours, with a range of 24–72 hours. When hatching is well underway, chicks become vocal; if more than one egg is hatching, they should be placed close together to stimulate each other. Once the chick has emerged, the navel is swabbed with a mild antibiotic ointment that is known to be safe for avian use or 1% iodine. A problem is usually indicated when a chick in the shell stops for an extended period or when it reverses direction back to the pip site. Improper action at this stage can be severely debilitating or fatal to the embryo. The greatest danger in assisting a problem hatch is the risk of severing an active blood vessel, which could result in the chick bleeding to death. An unretracted yolk sac from a chick that has been assisted will require a surgical procedure. Anyone not experienced at dealing with problem hatching should contact an experienced person prior to the expected hatch date. In addition, there are several good incubation books available which cover the topic of problem hatching.

After the chick has hatched, it should be placed in a still air brooder on a piece of gauze, which will prevent litter from sticking to it. Initially, up to four chicks can be kept in an aluminum cake pan of 23 cm (9 in.) diameter, lined with a bed of corncob litter. This is surrounded by an aluminum ring of 25 cm (9.8 in.) diameter, which is 12 cm (4.7 in.) high and elevated slightly above the newspaper on the brooder base. When the chicks defecate, their feces hit the outer ring and slide down onto the newspaper. This system is easy to clean. The thermometer used to monitor temperature must be placed at a level near the chicks. The general rule of thumb about brooding is to reduce the brooder temperature by 0.55°C (1°F) per day until the chicks can be moved from the brooder to room temperature.

The most important factor, however, is observation of the chick's posture and behavior. A comfortable chick will lie quietly sleeping with its wings loosely folded at its sides and its legs and feet tucked under its body. Occasionally, the sleeping chick will flutter its wings as if flying in its sleep. Two or more chicks will sleep quietly touching each other, frequently in a heap, one draped over the other in a loose bunch. A hot chick will lie with its wings stretched out and one or both legs extended to the rear, sometimes panting and frequently complaining vocally. A cold chick may move around the pan with its neck arched trying to locate its mother's breast while protesting to a great degree.

It is very important to feed young raptors the highest quality diet possible. Young chicks grow very rapidly; therefore, nutritional deficiencies can arise quickly if diet quality is not maintained. Freshly killed, mature coturnix quail are the ideal food for the first 10 days. The use of fresh-killed chicken, or previously frozen chicken and quail, has in every case resulted in poor growth or other problems affecting the health of the young chicks. Chicks older than 10 days can be successfully raised on other diets if the availability of quail is limited. One such diet consists of 50% whole, ground, six-week old chickens with only the large feathers excluded and 50% lean, ground horsemeat with a vitamin/mineral supplement. All carcasses should be skinned and the head, neck, crop, wings, feet, tail, and digestive tract removed. All other organs should remain. The carcasses should then be ground thoroughly in a meat grinder. Immediately after grinding, the meat should be packaged in plastic bags, covered dishes, or other sealed containers, and refrigerated. When feeding time arrives, only the quantity required for a feed is defrosted and warmed to room temperature. Supplemental vitamins can be added to the ground quail if desired. Although this has not been found to be necessary, it will do no harm if administered properly.

The young falcons are fed using a pair of blunt forceps. The newly hatched chick is not fed its first meal until it is 8–12 hours old. It should at this stage be able to sit upright. A begging response can be elicited by imitating the *chup* call a female falcon uses when feeding her young. Feed the begging chick 2–3 BB-size bites of mashed breast meat. Gradually increase the meal size for newly hatched chicks; at three days of age they should be able to consume a full crop at each feeding. Up to this point the young birds have been fed strictly the breast meat of the quail. When the chick begins to consume a full crop at each feeding, it should be fed the complete ground quail mixture without selectively choosing bits from it. Common sense should be used, however, and any large chunks of meat or chunky, sharp bones, which could cause problems, should be removed. Under no circumstances should newly hatched chicks be force-fed. A chick that does not beg on its own is not ready to eat. If, however, after 12 hours of age, the chick has not begged for food, it may be generally weak and would probably benefit from a subcutaneous injection of Lactated Ringers. Repeat the injection every 3–4 hours as necessary. A positive response is usually evident after one or two treatments, and frequently the chick is soon begging and eating normally.

While chicks are under 10 days old it is important to check their crops and stomachs to determine that the crop is empty before offering more food. The stomach is positioned in the lower abdomen, just to the right of center. If that area appears dark in color or feeds very hard when touched by a finger, then a large amount of food remains in the stomach and feeding should be delayed until the stomach is soft and flaccid and has returned to its normal light pinkish color. Once the yolk has been cycled through the gut, which is indicated by yellow colored feces, the feces should appear well formed and black in color surrounded by white. This indicates that the chick is digesting its food properly and that the gut is functioning as it should. Young raptors are normally fed every 3–5 hours, except during an eight-hour interval overnight. It may be beneficial to allow them this long interval to empty their digestive tracts more or less completely. When chicks begin self-feeding at just over 10 days, providing some whole ground food encourages casting. Five-week-old chickens with only the large feathers and crops removed are ground daily.

Occasionally a chick will exhibit splaying of the legs. This is not a serious problem, and is easily corrected if immediate action is taken. Either place the chick in a small bowl or cup, in which the legs will be pushed under the body where they belong, or tie the legs together using two small bracelets and a string or rubber band. Leave about 2 cm (1 in.) between the legs. The bracelets are to prevent constricting the blood flow to the feet. Remove the fixture after about five days. This condition indicates a problem with the surface on which the chicks are being kept. One other thing with potentially serious consequences is food or fluid getting into the eye of a chick as it weaves back and forth at feeding time. If this happens, rinse the eye immediately with eyewash and then administer a small amount of ophthalmic ointment. If untreated, there is a possibility of infection and blindness.

Ring Size: 11 mm

Species Requiring Similar Management:

- Saker falcon *Falco cherrug*
- Lanner falcon *F. biarmicus*
- Prairie falcon *F. mexicanus*
- Hobbies *F. cuvieri, F. subbuteo, F. severus, F. longipennis*

Pygmy Falcon *Polihierax semitorquatus*

Introduction: Two races occur, one in northeast Africa *Polihierax castanotus* and another in southwestern Africa *Polihierax semitorquatus*. Preferring dry acacia savanna to semi-desert habitats they are strictly territorial and found in pairs. They tend to perch high and drop to the ground to catch prey but they also catch insects in flight. Pygmy falcons are extremely vocal and a duet between a pair is accompanied by head bobbing and tail wagging. In the southwest their range overlaps with that of the sociable weaver in which nests the pygmy falcon roosts and breeds. In northeast Africa they use the nests of the white-headed buffalo weaver. A coating of "whitewash" around the entrance easily identifies a nest chamber, which is occupied by these birds.

Management: Females resemble the males but are distinguished by a dark chestnut patch on their upper back. They are strictly territorial and best housed in pairs. It is possible to house them with larger birds than themselves, such as turacos and starlings, but always remember that even in the natural state these birds occasionally prey on their host species of weaver bird. Landscape is not particularly important but moist forest-type habitats should be avoided. Landscape varying between woodland Habitat B and semi-arid Habitat E as described in the chapter on LANDSCAPE will be most suitable for these birds. At the San Diego Zoo they have been bred in aviaries as small as 2 x 2.5 x 3 m (6.6 x 6.6 x 10 ft.) high which are covered by weldmesh of 13 x 75 mm (0.5 x 3 in.). A nest box should be available year round, as they use these for roosting in the non-breeding season. As the birds are very sensitive to cold, they have at times been supplied with heated nest boxes. In cold, severe climates where day time temperatures remain low, and are combined with wet conditions, the birds should be housed indoors with additional heat during the cold months.

Diet: Naturally they feed largely on lizards and insects with some avian and small rodent prey. The captive diet at the San Diego Zoo consists of the following per bird, per day:

1 adult, skinned mouse	20 g
1 lizard/skink	2 g
Bird of Prey Diet	10 g
mealworm larvae	5 whole
crickets	5 whole

When chicks are present:

fuzzy mouse	2 whole

Problems have been encountered with more than one bird becoming impacted on mouse hair. It is for this reason that mice are skinned prior to feeding to the birds.

Breeding: Courtship includes the tail wagging ceremony during which the female adopts a submissive posture while also exposing her chestnut back. Much head bobbing occurs after copulation and the pair then usually fly to the nest. Naturally they breed irregularly, depending on the availability of food but in aviculture they can double clutch. Males roost with females until the first egg is laid and then move to an adjacent cavity. Some pairs, particularly when breeding for the first time, are rather sensitive and prone to desert their nests if disturbed. In captivity they nest in a nest box which should measure 25 x 35 x 27 cm high (10 x 14 x 10.5 in.) with wood shavings on the base. At the San Diego Zoo the birds even used a white-headed buffalo weaver nest, which is the natural host of the east African species.

Pygmy Falcon
Polihierax semitorquatus

Day	Weight (g)	Day	Weight (g)	Day	Weight (g)
1	6.34	11	22.15	21	54.15
2	6.31	12	25.54	22	56.22
3	6.41	13	28.27	23	58.79
4	7.05	14	30.02	24	59.71
5	8.11	15	35.3	25	60.84
6	9.76	16	39	26	63.59
7	10.16	17	42.39	27	64
8	13.39	18	43.69	28	65
9	15.62	19	44.6	29	65
10	18.94	20	51.61	30	66.65

The clutch numbers 2–4 white eggs that are incubated for approximately 28–30 days. Both sexes incubate; the female, however, does the majority and is fed by the male. He brings prey to a perch near the nest and then calls her out to where she will feed before returning to the nest. Newly hatched chicks are bright pink and covered with white down, but quills appear within a week and by three weeks they are fully feathered except for some down on the head and a short tail. Chicks fledge at about 30 days and usually return to the nest at night. They are independent at 6–8 weeks. Pygmy falcons have been recorded hitting the heads of keepers while nest inspections were in progress.

Incubation & Hand Rearing: This species has been bred regularly at the San Diego Zoo and Wild Animal Park and, to date, over 70 pygmy falcons have been hatched since the late 1980s, but many of these had hatching problems and did not survive. Hand-reared falcons become very imprinted and eggs are only pulled for hand rearing as a last resort to increase hatching success. When hand rearing, puppet gloves are used and the chicks kept isolated from people to minimize imprinting. San Diego has supplied the following advice on hand-rearing techniques and diets:

- Day 1: 33–34.5°C (92–94°F). Feed every two hours (seven feeds/day) chopped pinkie mice and added calcium/multivitamin. The goal is to feed enough for each chick to gain 15–20% daily. Intake 20–25% of body weight.
- Day 3: Eyes opening; start using puppet glove. Intake 30–35%.
- Day 4: Feed every three hours (five feeds). Intake 35–40%.
- Day 5: Add twigs to nest cup.
- Day 6: Add young crickets, ½ size mealworms and chopped fuzzies. Intake 50%.
- Day 10: Add chopped lizards. Feed every four hours (four feeds).
- Day 11: Place in box brooder. Add adult cricket, without head or legs, and whole mealworms.
- Day 18: Begin sunning.
- Day 24: Place in cage with some heat. Feed every five hours (three feeds).
- Day 29: Outside during the day with heat.
- Day 31: Outside permanently with heat. Begin eating on their own. Feed every six hours (two feeds).
- Day 44: Remove heat in favorable climates. One feed per day.

Percentage food intake based on daily percent weight gain. If gain is 15–20%, then increase food intake by 5%. If gain too high, hold at previous day's percentage or decrease by 5%.

Ring Size: 6 mm.

Species Requiring Similar Management:
Falconet
Forest Falcon

Contributions:
Ed Lewins, Assistant Curator of Birds, and Pat Witman, mnager of the Avian Propagation Center, San Diego Zoo.

Mauritius Kestrel *Falco punctatus*

This section is based on the Mauritius kestrel *Falco punctatus,* which is an endemic on the island of Mauritius. The Mauritius Kestrel Management Guidelines were compiled by Hilary French and published by the Durrell Wildlife Conservation Trust. The information in this chapter is directly from these guidelines. The Mauritius kestrel was threatened with extinction in 1974 with numbers down to four known individuals in the wild. The species declined due to pesticide contamination, loss of suitable habitat and subsequent decline in prey availability. Extremely successful conservation measures including the release of captive-bred birds and supplementary feeding of wild pairs has resulted in a remarkable recovery in the wild population. The population has risen to an estimated 400 birds at present. Birds have been established at institutions outside of Mauritius at institutions such as the World Center for Birds of Prey, Boise, USA and the Jersey Wildlife Preservation Trust. It was, however, never the intention to use birds from these sources to bolster wild populations. Adenovirus was the cause of a number of deaths at institutions outside Mauritius.

The Mauritius kestrel, with its short rounded wings, is adapted to hunting in forests. The tarsus is long, 32% longer than in the European kestrel, and only 6% shorter than the tarsus of a sparrowhawk. This is probably an adaptation for hunting in vegetation. Reversed sexual dimorphism is the most distinctive feature between the sexes, with females being larger than males.

Management: At the Trust the kestrels are housed in aviaries consisting of a metal framework outside flight covered in weld mesh, with a "house" built of cement blocks providing an indoor area. The outside flights measure approximately 6.5 x 4 x 3.5 m (12.5 x 13.1 x 11.5 in.) high. This is recommended as the minimum size to house a breeding pair. Much less than this and there is insufficient space to allow for any chasing or aggression, particularly when the birds are first introduced. Because the female has the potential to kill the male, there must be plenty of room for the male to escape. This can not be overemphasized, and should be taken into account when either building new accommodation for the species, or else modifying existing aviary space.

Bark chips are used as a substrate and are replenished annually. A variety of shrubs help provide a more natural and attractive environment for the birds, but should not be so dense when mature that they restrict the birds' movements. The plantings should also allow for patches of natural sunlight, which the birds enjoy. A few scattered large boulders should be placed on the aviary floor. The flight is perched with branches of various sizes. Upright tree trunks are also planted in the ground, as the birds like to perch at height. Access to indoor heated area is via two popholes 1.7 m (5.6 ft.) above the ground. It is also useful to have popholes at ground level as an extra mechanism of escape for the male should an aggressive female ground him.

One of the most important features of the aviary that needs to be considered is the indoor area and the facility to separate birds. Even if housing an established pair, a separation facility is important. This is best incorporated into the indoor area. At the Trust these areas are 1.8 x 3.7 m (6 x 12 ft.) and are divided into two by wire mesh mounted on a wooden framework so that the birds can be separated but still have visual/auditory contact. This is particularly useful as a starting point for introducing two birds. Prior to introductions outdoor access can be alternated using the popholes.

A thin layer of sand is used as a substrate to cover the floor of the indoor area. This can be sieved weekly to remove pellets and food remains. Sand is replenished with fresh as needed and before it becomes too dusty. They do not like very cold, windy weather or heavy rain and will do better in winter with good quality heated indoor accommodation. Kestrels are shut indoors at night over the winter months, November to March, at about 2:30 p.m. The provision of food indoors will usually facilitate this. They are usually housed as pairs in the same aviary throughout the year, or as single-sex siblings in holding aviaries until 12 months old. Holding aviaries for single-sex siblings are similar but a smaller and more basic structure. It is best to avoid moving established breeding pairs, as cases have been recorded where the female of a well established pair has turned on her mate after being moved to a new aviary. Clipping her flight feathers twice did not stop the problem and eventually the pair had to be split.

Early treatment of bumblefoot is important. Whenever birds are in the hand a routine medical examination should be carried out, which includes the feet. In all cases of bumblefoot, irrespective of the state of development, include the following: a good diet with vitamin and mineral supplements, all perching covered with Astroturf, moving birds off concrete/gravel surfaces, and increased exercise. Stress must be considered as a factor in the manifestation of respiratory disease. Ensure suitable substrates in the indoor area, sawdust can result in sinusitis issues, and adequate ventilation is important.

Diet: In the wild the Mauritius kestrel feeds mainly on lizards, in particular day geckos, and insects, especially locusts, plus dragonflies and grasshoppers.

Proximate Chemical Composition of Whole Mice

Mice Size	Water %	Ash %	Crude fat %	Protein %
Pinkie	80	10	15	65
Fuzzy	70	8.5	45	40
Crawler	70	8.5	45	45
Small	70	13	23	55
Medium	70	17	20	58
Large	62	13	30	45

Different size classes (dry matter basis).

In captivity at the Trust all adult Mauritius kestrels are fed once daily with a fresh laboratory mouse. One mouse per adult bird weighing approximately 25 g (1 oz.) is offered in the afternoon. During the breeding season extra food is available to the bird since courtship feeding is a very important part of the reproductive cycle. The males will also cache food and retrieve this later to feed to the female. In addition, during spells of very cold weather, extra food is provided as necessary. Frozen mice that have been defrosted naturally are adequate for the birds although, especially during the breeding season, the provision of freshly killed mice is a preferable option. If frozen mice are used, vitamin and trace elements supplements are included. A change in frozen food to freshly killed food in the breeding season produced a dramatic improvement in breeding success.

zFor the feeding of newly hatched chicks, it is recommended that only freshly killed mice be used and that, only if absolutely essential, frozen mice be used at a later stage of the rearing process. Mice that are considered soggy when defrosted are definitely not suitable for rearing and should not be fed to adult birds either. Adults will usually reject or only partly consume poor quality food. Mauritius kestrels, in common with other small falcons kept in captivity, are fed daily. The birds should be keen to come down for their food as it is placed in the aviary. It is only when birds are slow to come for food, or else show no interest at all that overfeeding becomes an issue. At the Trust, no signs of overfeeding are apparent with the diet as given here. Locusts are offered a minimum of four times per week. They can either supplement the daily staple of mice (add 4–5 locusts per bird) or, alternatively, fed on their own as a small feed (about eight locusts per bird). During the warm months the locusts are fed slightly chilled to allow the kestrels to catch most of what is offered.

Concern over post mortem results that showed considerable fat deposits led to research into the fat content in mice of different ages.

It is recommended that the birds be fed mice 4–6 weeks old (medium sized).

Breeding: Factors that will assist successful pairing are, ensuring the aviary is large enough to allow for some chasing/aggression, and that sufficient escape routes exist. The birds to be paired are usually placed in the separate holding quarters previously described for 4–5 weeks. Observations of the behavior between the birds, through the wire, can give a good indication as to how the pairing may develop. Once confident that the pair are ready to be introduced, the female is caught up and numbers 5–6 primaries trimmed almost back to the base on one wing. Marginally decreasing the female's maneuverability will greatly reduce the risk of males being killed by females. It is recommended that if established pairs are moved to new aviaries, that the female's wing is trimmed as described. Never assume that a stable pair situation will always remain so. Aggression between pairs that have bred successfully can occur, and action must be taken promptly to avoid injuries or stress related illness.

An introduction of new pairs is a gradual process and

Hand-reared Mauritius Kestrel *Falco punctatus* Chicks.

Age	Weight of Food/Feed (g)	Number of Feeds	Average Chick Weight (g)
0	0.3–0.5 internal organs	Five times per day 07h00 – 20h00	
1	0.5 max internal organs	Five times per day 07h00 – 20h00	12
2	0.5–1 internal organs	Five times per day 07h00 – 20h00	13
3	1 organs + flesh/bone	Five times per day 07h00 – 20h00	14
4	1.4	Five times per day 07h00 – 20h00	16
5	1.7	Five times per day 07h00 – 20h00	19
6	2	Five times per day 07h00 – 20h00	23
7	3–4	Five times per day 07h00 – 20h00	27
8	4–5	Four times per day 07h00 – 20h00	34
9	Up to 5	Four times per day 07h00 – 20h00	40
10	5–6	Four times per day 07h00 – 20h00	46
11	6–8	Four times per day 07h00 – 20h00	52
12	8	Four times per day 07h00 – 20h00	60

best done with all escape routes open for the male. The birds are mixed for about 20 minutes to start. Provided all goes well, the length of time together is gradually increased, but at all times they are watched. When it is considered safe to leave them together overnight, again all escape routes must be left open. It can be considered going well when the male starts to feed the female. In the early stages this may consist of the female snatching the food from the male rather than actually being offered it. Eventually beak-to-beak food passing will occur.

Nest boxes should be situated inside the heated area and fixed at a height of 2.5 m (8 ft.) above floor level. At the Trust the breeding season starts in March with aerial display flights and courtship feeding. Nest boxes measure 61 x 25 x 25 cm (24 x 10 x 10 in.) and are divide in two sections by a shallow sill of 6 cm (2.4 in.). The rear half is filled with disinfected pea-sized gravel to the level of the sill. This allows sufficient depth for the female to make a scrape and lay her eggs. An inspection door opens into the nesting chamber.

Eggs are usually laid in April. The clutch consists of 3–5 eggs with egg laying intervals of 3–5 days. Females are generally quite lethargic while laying. Once copulations have been observed, nest boxes are checked on a daily basis, which allows birds to become conditioned to this intrusion. Tap gently on the box before an inspection, as this avoids startling the birds when the door is opened. Approximately 10 days after the last egg has been laid the clutch is candled. If the eggs are infertile or a double clutch is desired, they are not returned to the female. Incubation by the female commences before the clutch is complete with the result that hatching is asynchronous. The incubation period is 28–35 days. It is worth noting that adults will attack keepers while breeding.

Artificial Incubation & Hand Rearing: All eggs at the Trust are left with the female until several days before pipping or hatching. Rearing from hatching by females has not been successful, so chicks are hand raised for 10–12 days before being returned to their parents to fledge from the nest box. Eggs are not cleaned before being placed in a forced-air incubator that is set at 37.5°C (100°F). Eggs are turned by hand seven times per day between 8:00 a.m. and 8:00 p.m. Eggs removed at an earlier stage are weighed daily to ensure a 15% weight loss. Humidity is adjusted to compensate for excessive or insufficient weight loss.

At external pip the eggs are placed in a hatcher set at the same temperature as the incubator. Humidity is increased and each egg placed in a wire basket which is lined with kitchen roll (paper towel) secured in place with paper clips. Hatching times vary from 20 hours, exceptional, to 70 hours which is more usual. Only very few small or weak chicks are assisted to hatch. After hatching chicks are left to dry and rest before being moved to a brooder, which is set at 36°C (97°F) for rearing. The first feed is offered 4–5 hours after hatching.

Hand-reared Common Kestrels
Falco tinnunculus

Day	Weight (g)
1	18
2	19
3	22,5
4	30
5	37.5
6	49
7	64

As given in the Zoo Yearbook No 23, page 75.

It must be remembered that young chicks are not capable of digesting mice fur. Mice are therefore skinned and prepared for feeding. Temperatures in the brooder are gradually reduced by 0.55°C (1°F) per day until an ambient temperature is reached. This is important for chicks that are to be placed back in an outdoor nest box for rearing by parent birds. A water dish in the brooder supplies humidity and the chicks are placed in bowls lined with tissue that can be changed at each feed. For older chicks this can be changed to fine grade disinfectant gravel.

Overfeeding is the greatest risk to the chicks. The feces of newly hatched chicks will be yellow in color indicating the digestion of yolk material. Falcons have a small nasal gland that excretes salt from the body and, when feeding chicks, a clear salty-tasting fluid collects in droplets at the tip of the bill and falls into the food before being fed to the chicks. A pinch of salt is therefore added to the boiled water in which food is dipped before feeding. Older chicks will produce darker feces and the stomach will appear a dark green color when food remains. It may also feel hard to the touch. If either or both of these conditions apply, then the chick should not be fed again until the stomach is flaccid.

From days two and three, chicks should be offered tiny pieces of bone in the diet and, in addition, supplements are used: calcium lactate and Vionate, with a pinch added once daily. The digestive tract of the mice is never used for feeding while the chicks are being hand reared.

At 10–12 days old chicks can be returned to their parents for rearing and fledging. Problems with human imprinting occur from day 14 onwards. Because of the nature of hatching, the return of the chicks is spread over a period of days as each chick reaches the desired age. Usually chicks are returned after the first morning feed. The chick is placed under the female and at the same time the dummy eggs are removed. Fresh food is left for the male to take to the female and chick. A close watch should be kept to ensure all is well. If problems occur prompt intervention is required. If all goes well fresh mice are fed throughout the day starting at 7:30 a.m. The chick must be checked later in the afternoon and thereafter chicks should be checked once per day to ensure that all is well. A little blood around the bill will confirm feeding by the female. If all progresses well, after 24 hours more of the brood can be returned. Fledging occurs at 36–38 days and for the last week the daily nest inspection should be stopped. At this stage chicks spend most of their day peering from the front of the nest box. Adults continue to feed the chicks for about 70 days. Longevity is recorded as 14+ years. Older infertile pairs can be extremely valuable as foster pairs.

References:

French, H.J. 1999. *Mauritius Kestrel Management Guidelines*. Durrell Wildlife Conservation Trust.

Lawrence, S. 2002. New Zealand Falcon Husbandry Manual.

Weaver, J.D. & T.J. Cade. 1991. *Falcon Propagation — a manual on captive breeding.* The Peregrine Fund, Inc., World Center for Birds of Prey, Boise, Idaho USA.

20.09 Ardeidae Herons, Bitterns and Egrets

Glen Holland

Introduction: Herons, bitterns, and egrets belong to the order Ciconiiformes and the family Ardeidae, with 65 species in 21 genera. Traditionally, this family has been divided into four groups, the day herons and egrets, with 38 species in 10 genera; the night herons and boat-billed heron, with eight species in four genera; the tiger herons, with 3 species in a single genus; and the bitterns, with 16 species in six genera. While species in the three larger groups are distributed worldwide, the tiger herons are confined to Central and South America. Herons occur primarily on inland waters, with a few species favoring marine environments, particularly estuaries. The sexes are similar in most species. These birds are generally quite vocal, particularly when breeding, which most do colonially. The typical nest is a platform of sticks placed in vegetation. The sparsely downy, altricial chicks are cared for by both sexes. Most species of herons and egrets feed by wading in shallows. They are popular birds in modern public displays. In this chapter I will discuss three representative species with which I have had experience: the gray heron *Ardea cineria,* the little egret *Egretta garzetta,* and the black-crowned night heron *Nycticorax nycticorax*.

Management: Herons, egrets, and bitterns are usually displayed in walk-through aviaries. The four key rules with this family are to:

- construct a large aviary
- provide the correct habitat
- ensure compatible species
- avoid overcrowding

Most of these birds enjoy perching on dry branches placed high in the aviary. Night herons are the exception, preferring dense vegetation, such as a weeping willow overhanging water, which gives them a sense of security. I have seen them housed and bred in aviaries measuring only 3 m (10 ft.) high with a floor space of about 20 m^2 (215 sq. ft.), but such a small aviary is not conducive to their welfare, and those who decide to keep this family of birds should not be aiming for the minimum. In a larger aviary, 5–6 m (17–20 ft.) high and about 30 x 50 m (100 x 165 ft.) in area, a nesting colony can be established. These colonies are a little smelly, but they are noisy and exciting to watch, generally active for many months of the year, with vocal and visual displays often still occurring among some pairs while others are already feeding their noisy chicks on the nest.

It should be remembered, however, that herons are predators. While they feed primarily on fish, frogs, and aquatic invertebrates, they are opportunistic, and not opposed to consuming chicks from a neighboring nest. This has been recorded for many species, including some as small as the little egret. I have personally witnessed a gray heron consuming an entire brood of ducklings, purple herons *Ardea purpurea* catching weaver birds in a reed bed, and black-headed herons *A. melanocephala* catching and consuming rats. I have, for display purposes only, housed a single little egret in a mixed walk-through aviary containing starlings, hornbills, etc., and never noticed any problems. However, the pair of resident moorhens, which always had chicks of some size, never ventured into the egret's territory. One reason for not overcrowding these birds is that to watch a single heron or egret fishing patiently is a fantastic experience, with far more aesthetic and educational value than a mass of noisy birds. A gently flowing stream between deeper pools, planted along the edges with patches of reeds and sedges, provides ideal public viewing and photographic opportunities. If the stream is stocked periodically with live fish, the public will have the opportunity to observe the "master at work."

Members of this family have a variety of interesting postures, vocalizations, displays, and hunting techniques. While feeding, these birds often assume an erect posture with the head and neck tilted to one side, which seems to assist in reducing glare and distortion as they peer into the water. This characteristic stance provides a binocular view of the prey beneath. The crouched posture is used by some species almost as a position from which to pounce. Egrets in particular use this posture when hunting small fish or flying insects. Some species also stalk their prey. They may walk either slowly and deliberately or quite rapidly, or actually run when pursuing prey. Some species hold their wings out while chasing prey; others, such as the black heron *E. ardesiaca,* form a full canopy with the head held under the umbrella. I have hand raised black egrets which, even in ponds containing no natural prey, would regularly disappear below their umbrella, to the delight of the visitors. The use of feet while feeding is common in some species. The foot is used to stir prey from within the water or below the substrate. Head swaying is another popular activity with cattle egrets *Bubulcus ibis* and black-headed herons, and serves to more accurately determine the location and distance of prey items. Cattle egrets often employ this technique to catch flies in an aviary. All of these displays and postures offer ideal focal points to use in educating visitors.

Most heron species are hardy, and are quite happy feeding in ponds that have frozen edges. It would, however, be advisable to carefully monitor those species from the tropics during extreme weather conditions. Most of the smaller species such as black herons, little egrets, and night herons can be housed together. They will breed in a colony, along with other species such as reed cormorants *Phalacrocorax africanus*. Other birds that require a similar diet, such as hammerkops, can also be housed with herons and egrets. As mentioned earlier, however, herons are predators and will not hesitate to consume chicks from neighboring nests. The larger species, such as gray herons, are best housed with other large birds, such as

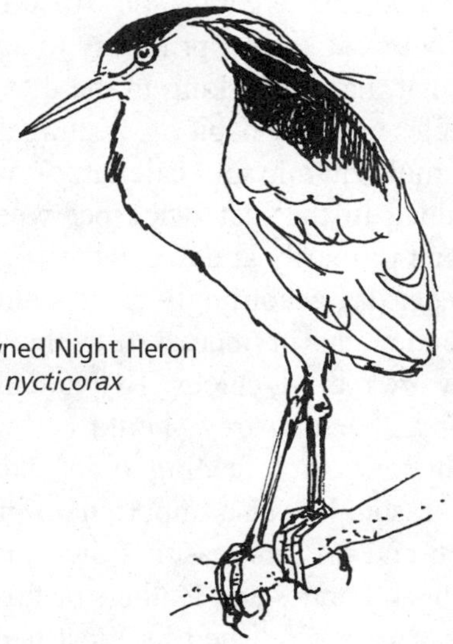

Black-crowned Night Heron
Nycticorax nycticorax

sacred ibis *Threskiornis aethiopicus* and pelicans *Pelecanus* spp. They will consume any precocial chicks, e.g. waterfowl, even if defended by the parent.

When catching and transporting herons and egrets, the most common problem is damage to the legs. The proper procedure involves securing the head by holding the bird with one hand behind the head while the second hand reaches over the body, with the arm securing the wings, to grasp the legs near the hock. The birds' first weapons are their bills, and they know instinctively that attackers' eyes are their weak points. This should always be kept in mind when handling these birds. Larger species, such as gray herons and Goliath herons *Ardea goliath,* are best handled by two people. When being transported, adults should always be crated individually. Youngsters, however, can be transported 2–3 per box. A firm bedding of grass or twigs will prevent slipping, which can cause leg injuries during transportation. Air vents should be arranged to preclude the possibility of visual or bill contact with the person handling the crate.

Diet: Herons and egrets feed primarily in aquatic habitats. Their natural diet includes items such as fish, crabs, eels, crustaceans, frogs, invertebrates, and small vertebrates. Only three species, the cattle egret, the black-headed heron and the whistling heron *Syrigma sibilatrix,* regularly feed on dry land. Some species, particularly the larger herons, forage singularly, while others feed in groups, which at times become large aggregations. One species often benefits by the feeding strategy of another; for example, egrets *Egretta* spp. often follow spoonbills *Platalea* spp., which chase fish about when feeding. The green-backed, or striated, heron *Butorides striatus* has been recorded baiting prey by placing a floating lure on the water.

The captive diet consists basically of a variety of

fish, meat, and chicken, with some institutions including soaked dog or cat kibble, primarily to add vitamins and minerals that may be lacking in the diet. Unfortunately, many species have the habit of washing their prey before eating. A multivitamin and calcium powder supplement can be added to the diet twice per week. Since many species hunt primarily at dawn and dusk, they should be fed twice per day accordingly, from stainless steel trays. The frequency can be increased to three times per day when they are rearing chicks. To avoid spoiled food and overfeeding, the food tray should be emptied approximately an hour after feeding time, except during the breeding season. It is also important to include roughage in the diet, and for this reason scales and bones are left on the fish, and bones and feathers or fur left on some of the meat. Mice are relished by most herons and egrets, and can be chopped up for the smaller species. Live food, in the form of small fish, if available, can be stocked into ponds. This will not only supply an ideal food source, allowing the birds to forage in a natural and instinctive manner, but their hunting behavior also provides an exciting display for visitors.

Great Blue Heron
Ardea herodias

Breeding: Herons form pair bonds through a series of vocal and visual courtship displays. The males select potential nest sites from which they solicit, and females then choose the male with which they wish to engage in courtship. Males of the communal species, which nest in close proximity to each other, require more elaborate and energetic vocal and visual displays to attract the females. The stretch display, which involves vertically lifting the head and closed bill, is the most common display among these birds. At the peak of the stretch, a call may be uttered. As the head is lowered, the legs are bent and the feathers of the nape, crest, and lower neck are usually raised. In some displays the bill is snapped shut. Bill clappering is a common occurrence among pairs of most species, often as a prelude to copulation. Some species grasp a twig on the nest and shake it from side to side, in a manner similar to what they might do when nest building. Aerial display flights are performed by some species, but this behavior is limited by the captive environment. Black-crowned night herons have an unusual display which involves standing erect, swaying from one foot to the other, then finishing in a hunched position with its bill down while mock preening its belly feathers. During the display, the male gives a hissing call or sometimes performs bill clappering. Little egrets develop characteristic display plumes during the breeding season. The skin, bill, and plumage color changes in some species during the breeding season. For example, in the gray heron the iris, bill, and legs flush deep orange. In little egrets the lores turn red; in the yellow-footed races, the feet turn red as well.

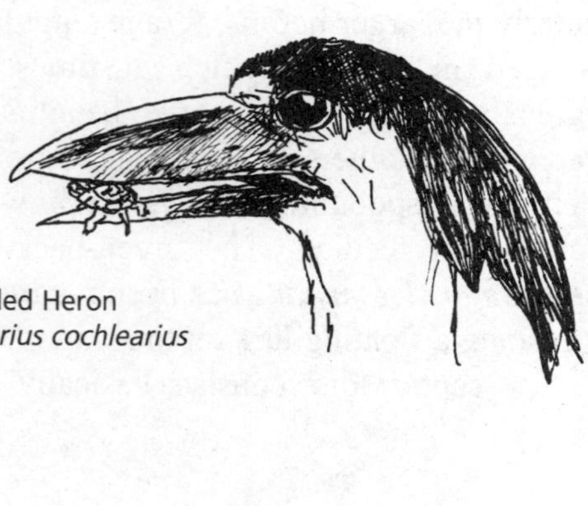

Boat-billed Heron
Cochlearius cochlearius

In general, the sexes are alike, but on average females are smaller. The gray heron is primarily a colonial breeder. Black-crowned night heron colonies in excess of 1,500 pairs have been recorded. The nest is constructed from the most easily available material. A plentiful supply of twigs, some dry and others with a few leaves, should be supplied. New nests are often so flimsy that the eggs can be seen through them. A platform, such as a wire mesh hoop, a slatted wooden box, or a cane dog basket may be accepted as a base for a nest. Placing material into such artificial platforms may stimulate birds to use them. The females complete the nest, which is initiated by the male. During the completion stages, the male supplies building material to the female, which remains on the nest. In captivity, nest structures can be attached to a variety of natural or artificial supports, such as trees or wooden posts. Nests should be spaced 2–2.5 m (6.6–8 ft.) apart. Incubation of eggs, and feeding the young, is shared by both sexes. The food is initially fed only from the adult's bill. Soon afterward, however, food is regurgitated onto the nest and the young squabble noisily over it. Most young only leave the nest area a few weeks after fledging. In large broods, the youngest bird often does not survive. Such birds can be either hand raised or cross fostered to a suitable nest.

Juveniles are generally duller versions of the adults, but in the case of night herons, juveniles have a streaky brownish plumage, very different from that of the adults. By the time they reach a year of age, they closely resemble the adults. Goliath heron chicks may differ in coloring, even within the same nest. Susan Fowkes reports a nest of two Goliath chicks in which one had a speckled, but mostly white, belly from the base of the neck to the legs, while the other was completely streaky with very little white showing through.

Many species have been recorded naturally breeding on the ground when insufficient nest sites were available. In captivity, a number of species have been pinioned and bred successfully. Disturbance in the colony should be kept to a minimum; large chicks that are disturbed may leave the nest prematurely, resulting in many anxious hours for the parent birds and the keeper. With most species, fledging does not indicate that the chicks have reached independence; fledglings remain reliant on the adults for some time.

Little Egret
Egretta garzetta

Artificial Incubation & Rearing: Heron eggs should be incubated at a temperature of 37.2 –37.5°C (99–99.5°F) with a relative humidity of 63–65%. The usual egg weight loss of 15% at hatching should be the goal. In his indispensible book, *Practical Incubation,* Rob Harvey describes an interesting experience with the eggs of the black-crowned night heron. He found that eggs that were incubated at 37.5°C (99.5°F), with the normal 15% weight loss, in an incubator which automatically turned them 24 times a day by rocking them 45 degrees from the vertical, never achieved complete vein growth; all ended up dead in shell. The next group of night heron eggs was incubated at the same temperature and humidity, but on their side on rollers, which also turned them 24 times a day. On rollers, the eggs often ended up in strange positions, such as pointing upward at an angle. The hatching results were better, but still not adequate. A third group of night heron eggs was incubated in the same manner, but hand turned seven times a day in addition to 24 times a day on the rollers. All of these eggs hatched successfully.

Eggs collected late in the incubation period do not require such frequent turning, and I have hatched a variety of late term egret and heron eggs by hand turning three times per day. The incubation and hand-rearing section on SPOONBILLS outlines similar protocols to those required for the heron family.

Newly hatched heron or egret chicks are naturally reliant on their parents to feed them directly from the bill.

Breeding Details

Species	Egg Color	Clutch Size	Incubation	Fledging
Goliath heron	pale blue	2–4	28–30 days	40 days
Black-headed heron	blue/greenish blue	2–4	23–26 days	40–55 days
Gray heron	pale greenish-blue	3–5	5–26 days	50 days
Black-crowned night heron	as above	3–5	21–22 days	45–50 days
Green-backed heron	as above	2–5	21–25 days	21–25 days
Cattle egret	as above	3–7	24 days	20–25 days
Great white egret	pale blue	3–5	25–27 days	40 days
Little egret	as above to off-white	3–6	21–25 days	45–50 days

Goliath heron - *Ardea goliath*
Black-headed heron - *Ardea melanocephala*
Gray heron - *Ardea cinerea*
Black-crowned heron - *Nycticorax nycticorax*
Green-backed heron - *Butorides striatus*
Cattle egret - *Bubulcus ibis*
Great white egret - *Casmerodius albus*
Little egret - *Egretta garzetta*

Because of this, it is necessary to hand feed incubator-hatched chicks. I prefer to use a 10-ml syringe, depending on the gape and bill size of the species in question, with the entire tip section removed, leaving a tube with a plunger. The initial diet should be of rather watery porridge texture, and consist of equal portions of finely minced ox heart, beef- and fish-flavored tinned dog food, and soaked trout pellets, as well as extra calcium. Water should be added to attain the correct consistency. The formula should be warm when fed to the chicks. Initially, it is necessary to support the chicks with one hand while feeding them, but they are soon able to support themselves. At day 4–5, the texture of the food can be altered slightly by adding less water. At 7–10 days, a slightly coarser textured food is prepared. Feeding frequency and correct brooder temperatures are shown in the following chart.

Age (days)	Feeds Per Day	Brooder Temperature
1–10	5	35°C (95°F), decreasing to 30°C (86°F)
11–21	4	30°C (86°F), decreasing to ambient, at about 25°C (77°F)
22–35	3	outdoors in day, indoors at night
	2	thereafter outdoors permanently, if ambient temperature is suitable

The texture of the food should be gradually changed. At a month of age the chicks can be given diced, rather than minced, food, with a small amount of moisture added. During all stages of growth, the internal organs of the fish, as well as some bone and scales, are included in the diet to supply both roughage and calcium. In addition, a multivitamin and trace element supplement and a powdered calcium supplement should be added to the food three times per day until the chicks are fledged. Herons grow slowly and consume large quantities of food while in the nest. They have a long nestling period; Goliaths, for instance, are unsteady on their long legs for at least three months. The chicks of many species also rearrange nest material and carry sticks around on the nest, as well as throughout their enclosure when they are able to stand and move about. At 2–3 weeks, if ambient temperatures are sufficient, the chicks can be placed, in their nest, in an outdoor enclosure with dappled sunlight during the day. The sun and fresh air are vital for the good health of the chicks and also result in less cleaning for staff. The diet, management, bedding, and housing protocols for this family as for those described for SPOONBILLS.

Goliath Herons *Ardea goliath*

Day	Weight (g)	Weight (g)	Weight (g)	Weight (g)	Weight (g)
0	85.0	75.1	72.2		74.1
1	70.4	73.8	74.1	64.6	68.8
2	79.8	80.2	80.9	67.5	74.5
3	87.7	94.8	101.4	78.5	86.0
4	105.2	114.6	119.9	92.3	99.8
5	135.1	140.5	153.6	114.1	121.7
6	162.5	169.9	170.1	145.0	145.8
7	199.1	209.6		191.0	167.1
8	242.3	244	282	186.6	201.2
9	283.5		347	218.8	253.3
10	355.4	374	404	266.0	305.0
11	431.8	451	485	334	347.7
12	494.1	531	545		397.4
13		608	667	452	485.0
14	649	705	761	555	547.8
15	738	814	884	631	643.9
16	794	877	1000	691	731.1
17	891	1026	1121	684	818.7
18	984	1104	1256	664	900.4
19	1134	1205	1363	702	996.3
20	1217	1304	1515	810	1119.2
21	1377	1420	1491	941	1196
22	1514	1573	1511	1106	1348
23	1458	1710	1675	1219	1443
24	1623	1730	1836	1384	1500
25	1753	1877	2028	1546	1727
26	1928	2019	2186	1690	1861
27	2016	2134	2362	1792	2025
28	2035	2253	2475	1882	2062
29	2219	2371	2627	1930	2273
30	2308	2507	2783	2179	2315
31	2396	2666	2877	2179	2370
32	2519	2743	2842	2349	2433
33	2639	2768	2955	2483	2576
34	2644	2878	3094	2420	2605
35	2876	2977		2693	2832
36	2888			2809	2759
37	2952			2901	3023
38	3012			2962	3060
39	2964			3160	3140
40	3078				3090
41					3259
42	3420	3540	3620	3232	3232
43					3470
44					3449
45					3511
46					3797
47					3627
48					3829
49	3940	3870	4060	3552	3764

*Growth chart for five Goliath herons hand raised at the Dallas Zoo

Species Requiring Similar Management:

Whistling heron *Syrigma sibilatrix*
Herons and egrets *Ardea* and *Egretta* spp.
Capped heron *Pilherodius pileatus*

Great egret *Casmerodius albus*
Intermediate egret *Mesophoyx intermedia*
Cattle egret *Bubulcus ibis*
Squacco and pond herons *Ardeola* spp.
Herons *Butorides* spp.
Agami heron *Agamia agami*
Night herons *Nyctanassa, Nycticorax*, and *Gorachius* spp.
Boat-billed heron *Cochlearius cochlearia*
Tiger herons *Tigrisoma* spp.
Zigzag heron *Zebrilus undulates*
Bitterns *Zonerodius, Ixobrychus, Dupetor,* and *Botaurus* spp.

Contributors:

Walter Mangold, Director, World of Birds, Cape Town, SA.

Susan Fowkes, Curator, Kuimba Shiri Wildlife Park, Harare, Zimbabwe

References:

Hancock, J. & J. Kushlan. 1984. *The Herons Handbook.*

Harvey, R. 1990. *Practical Incubation.* Surrey, Canada Hancock House Publishers.

Lyles, A.M. (Ed.), AZA Heron, Ibis and Hammerhead Advisory Group. Husbandry Notebook. American Zoological Association.

20.10 Scopidae Hammerkop

Glen Holland

Introduction: The hammerkop *Scopus umbretta,* like the shoebill, has long been associated with the storks. It is the only species in the monotypic family Scopidae. The name hammerkop is Afrikaans for hammerhead, and was bestowed upon this species due to the strange shape of the bird's head, which has a long bill that is balanced with a backward pointing crest. The range of the hammerkop is sub-Saharan Africa, in areas varying from open woodland and savanna to wetlands, where areas of shallow water occur. These birds are usually found in pairs, but will sometimes congregate in larger numbers to roost. Their diet consists predominantly of amphibians and small fish such as *Tilapia,* catfish, and minnows. They also feed on insects, worms, shrimp, and occasionally, small mammals. These birds also have a feeding behavior unique in Ciconiiformes. Hammerkops will hover over water while flying directly into the wind, capturing small fish and tadpoles while in flight. At one time hammerkops were thought to nest in colonies because of the number of nests in a particular area. The fact is, a single pair of hammerkops will build many nests in a single season. The nest building activities are believed to reinforce the pair bond in a manner similar to the pseudo-mounting behavior which is also exhibited by this species.

Management: Hammerkops are unique in Ciconiiformes because of their nest strategy. They are the only stork-like bird to build a domed structure and nest in a chamber. To propagate this species in captivity, enormous amounts of nesting material must be made available. The nest itself can weigh 100–200 pounds, making it one of the largest structures in the avian world. A hammerkop nest can also support the weight of an adult human. The birds will build a number of nests during the year. The location is usually situated in the fork of a large tree, or on the top of artificial rock in the aviary. It is often built close to water. The hammerkops mix well with other bird species ranging in size from larger stork and crane species to birds much smaller than themselves. Care must be taken, however, as hammerkops have been recorded attacking young or debilitated birds in the same aviary. They are a good species to exhibit in a zoo because of their unique behavior and the African folklore associated with these birds. A flowing stream with a few small waterfalls will provide the ideal habitat for this species, which likes to stand in the water while fishing. Live frogs and small fish can be stocked weekly to provide natural food and give visitors a chance to observe them catching their prey.

Diet: The daily diet for each individual hammerkop consists of one adult mouse, one small trout, one smelt, and 112 g bird of prey meat supplemented with bonemeal.

Breeding: As far as providing nests is concerned, hammerkops are persistent at building their own. The most effective method is to provide a large tree and an enormous amount of nesting material. Predictably, the birds will build their own nest structure. Historically, hammerkops have been inconsistent in the production of offspring. Many times storks and similar species will use the abandoned hammerkop nest on which to nest. Platform nesters such as spoonbill, ibis, and Egyptian

Hamerkop
Scopus umbretta

geese *Alopochen aegyptiacus* will also nest on the top of old hammerkop nests. Rufous-crowned rollers *Coracias naevia* and barbet species have been observed tunneling into the side of the hammerkop nest to build their own. Glen Holland reports that he once climbed a tree to check the contents of a hammerkop nest, only to be bowled out of the tree by an African civet cat that dashed out of the nest as he peeped into the entrance. He has also regularly seen barn owls use these nests. In addition to all of these interesting activities in and around the nest, from an educational standpoint, the African folklore regarding hammerkops makes them very interesting.

Hand Rearing: Glen Holland notes that many years ago, in his youth, he was asked to raise four hammerkop chicks that had been turned in to the local animal welfare society. Unfortunately, he did not keep records of the exact details, but he does recall feeding the birds equal portions of diced sardine, chicken, and ground dog biscuits. Calcium and multivitamin supplements were added. Although by today's standards this is hardly an ideal diet, all four chicks were successfully fledged.

20.11 Phoenicopteridae Flamingoes

Glen Holland

I must start this chapter by thanking Christopher Brown, AZA Ciconiiformes TAG chair, for allowing me to use the Flamingo Husbandry Guidelines as a reference for this chapter. Until this document was made available I found much of the literature on flamingos in aviculture either outdated or contradictory. I take no credit for this work, and the entire chapter, with the exception of the section on artificial incubation and rearing, is taken directly from this excellent husbandry manual. I would recommend that anybody who manages flamingos obtain a copy of it.

> "Due to the unreliable reproduction of flamingos in captivity, aging population, and the responsibility of captive managers to no longer depend on collecting wild birds, members of the AZA Ciconiiformes TAG, EAZA Ciconiiformes/Phoenicopteriformes EEP, and Joint Management of Species (JMS) collectively agreed on the need for a guide to assist captive managers and keepers with a collaborative approach to methods for managing flamingos in captivity. The AZA Ciconiiformes TAG facilitated the beginning of a workshop that would include zoo and aquarium professionals from North America and Europe to come together to produce husbandry guidelines for flamingos in captivity. Christopher Brown, AZA Ciconiiformes TAG chair (Dallas Zoo, USA), Catherine E. King, EAZA Ciconiiformes/Phoenicopteriformes EEP chair (Rotterdam Zoo, The Netherlands), and Tony Richardson, then of the Wildfowl and Wetlands Trust (Slimbridge, UK), worked cooperatively to prepare for the process. Disney's Animal Kingdom provided the facilities, equipment, and personnel in order to make the husbandry workshop possible."

Introduction: In general, most taxonomists seem to agree that the flamingos belong somewhere between the ciconiiformes and the anseriformes with affinities to the charadriiformes. Some place them in their own order as a compromise. Despite taxonomic issues, it remains obvious that there are six different types of flamingo. Whereas Sibley and Ahlquist consider them closely related and place all six into a single genus *Phoenicopterus,* others continue to recognize the three traditional genera *Phoenicopterus, Phoeniconaias,* and *Phoenicoparrus*. Flamingos appear to be among the longest-lived birds. Many captive birds are of unknown age but there are records of a female greater flamingo as not only alive but breeding successfully at 53, and eight birds in the same flock have lived to at least 44.

All flamingo species are highly nomadic, with their movements apparently driven by food availability as opposed to the traditional notions of migration based on climatic changes.

Flamingos are "confined to shallow soda lakes and salt lagoons with a high pH (up to 10.5), usually barren of vegetation and surrounded by almost desert-like wastes." Prey availability seems to be the main factor affecting the occurrence of flamingos within a habitat. Flamingos can be divided into two groups according to their diets: the *Phoenicopterus* species feed primarily on arthropods and mollusks, while the other species take algae and diatoms. All flamingos feed mainly by filtering large numbers of their small food organisms from water or mud using their bills, which are highly adapted for filter feeding. The *Phoenicopterus* species have "shallow-keeled" bills with which they take their animal prey from at, or just above or below, the surface of the substrate, either below water or in soft exposed mud. Of the "deep-keeled" species, *Phoenicoparrus* also feed largely at or near the mud surface, while *Phoeniconaias* feeds primarily near the water surface.

Management: Unfortunately, most current exhibits are open-air enclosures creating the need for the birds to be rendered flightless to prevent escape. Most common methods for restricting flight are pinioning and wing feather clipping. It has been shown that reproduction is severely hindered by pinioning, due to the male's difficulty in balancing himself during copulation. The recommendation is to keep flamingos full-winged and design new exhibits to accommodate a full-winged flock. Sadly, there are very few exhibits that are designed for full-winged flocks. There are advantages to keeping full winged or pinioned birds in an aviary. These include minimal threat of predation, reduced food stealing by gulls and mallards and increased fertility. Flamingos are colonial and are best kept in large flocks of at least 20 birds. Reliable breeding occurs with flocks of 40+ birds. A facility should be designed to hold the future desired number of birds and be able to accommodate potential offspring. "Display only" birds are not recommended. In creating an enclosure for flamingos, the prime consideration must be to ensure the best possible conditions for the birds while taking into account the impact of the exhibit on the viewing public. The nesting area should be visible to the public and, where possible, designed in such a way that the public feels they are not just viewing a stage set but approaching a flamingo colony.

Outdoor exhibits should be built with the maximum desired flock size in mind versus the current flock size. There should be enough land space to allow a flamingo flock to engage in courtship displays and breeding activities. Areas for sunning and nesting should be taken into account. The recommended size for an outdoor enclosure is 1.4 m^2 (15 sq. ft.) per bird, and hard abrasive substrates should be avoided. The pool should be large enough for the entire flock to occupy the area at the same time. Water should be able to flow freely in all areas to avoid botulism. A nesting area should be large enough to accommodate the entire flock, with room to allow for potential

Natural Distribution of Flamingo

Species	Distribution
Caribbean *Phoenicopterus ruber ruber*	Bahamas, Cuba, Yucatan, Bonaire, Venezuela, Colombia. Feral in Florida. Apparently isolated population on Galapagos Islands.
Greater *Phoenicopterus ruber roseus*	The most widely distributed flamingo, ranging from the western Mediterranean, where it occurs in Spain and southern France as well as in North Africa (principally Tunisia, Algeria and Morocco), into West Africa, and in the eastern Mediterranean and eastward to Iran and the former USSR, India and Sri Lanka, and also in southern and east Africa.
Chilean *Phoenicopterus chilensis*	The most widespread of the three South American flamingos, ranging from the highlands of central Peru to Tierra del Fuego.
Andean *Phoenicoparrus andinus*	The Altiplano of southern Peru, northern Chile, western Bolivia and north west Argentina.
James *Phoenicoparrus jamesi*	Restricted to altiplano salt lakes, in southern Peru, western Bolivia and the north of Chile and Argentina.
Lesser *Phoeniconaias minor*	Sub-Saharan Africa and India, often with greater flamingos. The largest flocks are found in the East African Rift Valley.

aggression. A recommended size for a nesting area is 0.9m^2 (10 sq. ft.) per bird. Always allow enough space to accommodate offspring and any future acquisitions. Enclosure orientation should also be looked at in these cases to eliminate long runs into the wind. If possible, also include a built-in catching and holding area. Outdoor exhibits usually have natural plantings to help enhance the esthetics of the enclosure. The plant density should be kept to a minimum to allow the birds to move unobstructed through the area. Flamingos are very prone to injuring themselves on bushes and trees.

Combining bird and mammal species to show the diversity in a specific habitat is not a good idea for flamingos. Some flamingo exhibits do, however, mix other species of birds with their flamingos. Swans and geese should be avoided. These types of birds have a tendency to be aggressive and can cause injury to the flock and interfere with breeding activity. Some suitable species of birds to mix with flamingos are non-aggressive waterfowl species and ibis. Avoid isolating individual flamingos by themselves. Sick flamingos should be housed with at least one other bird. If this is inappropriate for the specific medical problem (e.g. infectious disease issues), then mirrors or a tape recording of vocalizations can be used as alternatives. Another flamingo should be placed with the ill flamingo as soon as is appropriate (all infectious disease issues resolved).

Any fences should be constructed with the posts on the outside of the enclosure. Once settled in an enclosure, pinioned flamingos will not normally cross fences 1 m (3 ft) high although they are quite capable of jumping over such fences. In strong winds even pinioned flamingos may be able to fly/jump considerable distances. Banks, fences or trees can be used to form wind breaks, and enclosures should not be oriented to allow long run ups into the wind. A factor that should be considered when choosing an effective outdoor containment barrier is predator control. Several institutions have experienced losses within their flocks due to predation by feral dogs and, in Europe, foxes. Some alternatives that have been used in some zoos are electric fences, routine trapping of predators, and enclosed areas that have roofs or meshed ceilings that the flamingos are shifted into at night.

Construction of the pool should allow water to flow freely through every part with no dead ends where sediments or concentrations of botulism can build up. Water quality and filtration can be important issues when managing a flock of flamingos. There have been reported die-offs due to contaminants in the birds' water source. Flamingos rely heavily on the use of bodies of water for various activities, such as courting, copulating and sleeping. The depth of the pool should be 46–70 cm (1.5–2.3 ft). Some facilities provide deeper pools that allow birds to swim. Flamingo pools can be a natural stream/lake or a man-made pool. Natural streams or lakes should be tested periodically for contaminants. The pool should allow the birds to easily access the water. Special attention should be paid to the slope, which should be gentle. A steep slope will pose a trip and slip hazard for the birds. The ability to raise and lower the water level around the nest site would assist with the management of a breeding flock and diminish the amount of keeper intrusions. Several different types of pool substrates that have been used in zoos are outlined below:

a) Natural bottom: has clay, sand, or mud. Sand is not recommended for lesser flamingos due to risk of ingestion and impaction. Pools with this type of substrate allow natural behaviors and are easy on bird feet. Clay, mud, or sand substrates do, however, pose a potential disease problem and need to be changed periodically. A natural bottom pool is recommended only if there is good water flow.

b) Concrete: This type of substrate can be inexpensive to install and is easy to clean. It can, however, be abrasive to bird feet and slippery if not textured.

c) Fiberglass.

d) Combination of rock and concrete: This substrate can be attractive and naturalistic. Depending on the texture, it can be abrasive to bird feet.

e) Gunite-formed concrete: This type of surface can also be made to look natural. It also can be abrasive to bird feet and is expensive to install.

f) Plastic liner: A substrate of this nature often occurs in recirculation pools and is designed to operate without routine cleaning. It can also be easy on bird feet. The surface can become slick and could pose a potential fall hazard for keepers and birds alike.

When designing a new exhibit, separate feeding areas should be set aside within the enclosure. These areas should be separate from the pool or main water source, with their own drainage system and water source.

Outdoors the following substrates have been used:

a) Grass: This substrate is recommended for use in outdoor enclosures.

b) Soil/clay: Although this substrate does not require much maintenance, it has been attributed to numerous foot problems, and lesser flamingos are prone to ingesting it.

c) Mulch: This substrate requires frequent replacement and can be quite expensive. It has also been attributed to numerous foot and leg problems.

d) Sand and decomposed granite: These two substrates are often attractive and easy to maintain but ingestion has caused impaction and numerous foot problems.

Two factors usually determine when flamingos need to be brought indoors. They are predators and weather. Northern zoos are faced with cold weather during the winter months, which necessitates the birds coming indoors for a portion of the year. These indoor facilities should be built adjacent to the exhibit to allow the birds to move into winter holding without capture. This building does not need to be as large as the outdoor exhibit space, but should be large enough to allow the birds to move freely and to separate from aggressive members of the flock. A recommended size (land and pool space combined) is 1.4 m^2 (15 sq. ft.) per bird. Ideally, a public viewing area should be designed into any new winter holding structure, with viewing limited to one side. Since most indoor holding facilities' floors can remain damp for long periods of time and are usually constructed of concrete, it is essential that rubber, vinyl, or other easy-to-clean, fast-drying matting material be used to separate the birds' feet from the caustic and abrasive properties of concrete. Long-term exposure to damp concrete causes the birds' feet to become receptive to abrasions, thus opening up the opportunities for infection and the debilitating condition known as bumblefoot. The building should have adequate ventilation, and there should be at least ten air changes per hour. Access to the outdoors during the winter months is essential for birds that are housed in winter holding facilities. This practice allows the birds to go outside on nice days and keeps the flock acclimated to outdoor temperatures.

The three species most often managed in zoo and aquarium collections are surprisingly hardy and can benefit from some limited exposure to mild low temperatures. A suggested general minimum ambient cold temperature for the birds to remain outside is –6°C (24°F). However, wind chill will more often be the determining factor. In some areas summer temperatures can be extremely hot. Misters and sprinklers are used to provide relief for the birds. Shade from natural vegetation can also provide relief from the sun.

It is particularly important to keep accurate records on long-lived birds such as flamingos, and it is vital to ensure all captive flamingos are marked so that they can be individually identified. There are two main types of ring or band available. These are metal rings/bands or large plastic rings/bands. Metal rings/bands probably last longer but it is necessary to catch birds to identify individuals. Engraved plastic rings/bands have the advantage that they can be read easily without the need to catch birds to identify them. Because flamingos spend a large proportion of their time standing in water, it is advisable to apply the ring/band on the upper leg above the tarsi-tibia joint. Plastic rings/bands are the recommended form of physical identification. They are relatively inexpensive, providing numerous color variations, numbers, lettering and stripes. Secondary identification can be plastic or metal ring/band, or transponders. The transponder should be placed dorsally at the juncture of the neck and the body on the left side.

Plastic cable ties, such as those used by electricians to bind a group of cables or electric wires together, are not recommended except as a temporary form of identification. Wing Web Clip/Patagial Flags and tattoo/dyes are not recommended.

Avoid sudden movements and loud unexpected noises that could frighten the flock. Do not leave objects (e.g., hoses, shovels, rakes) unattended in the exhibit. Care should be taken to protect birds from potential hazards in the exhibit, such as rocks, branches, pipes, slopes, holes, etc.

Many exhibits have attached holding areas that the birds can be herded into. This type of area can be very useful when dealing with medical procedures, exhibit maintenance, and flock management activities, such as ringing/banding. The area should be large enough to hold the flock comfortably for an extended period of time and be predator proof. For capture, flamingos can be herded into primary, and even secondary, corral systems when it is necessary to catch them up. Some commonly used barriers include fencing, plywood, burlap, regular or closed-cell foam, and black or clear plastic sheeting. Mesh barriers are not recommended because of the high risk of injury due to entanglement. Care must be taken when using fencing to ensure the holes are small enough so wings or feet do not penetrate. Fencing not greater than 2.5 x 2.5 cm (1 x 1 in.) is recommended.

It is recommended that flamingos be caught by hand; it is not recommended to use nets of any type. Once captured, extra persons may be required to hold birds. Planning and discussion of capture and handling procedures should be undertaken prior to the event. Whenever possible, birds' legs should remain unfolded. Folding can increase the risk of capture miopathy and leg paralysis or injury. If folding of the legs is to be undertaken, it must be done with caution. To fold, grasp the legs below the ankle joint and apply gentle pressure until the bird relaxes. Legs should never be forced into a folded position. Additional care should be taken to avoid crossing the legs at the ankle.

a) Hand-held restraint methods are:

- Body close to holder, head facing backward; legs may be down or folded.
- Body close to holder, head facing forward; legs may be down or folded.
- Wing hold at shoulders with one hand, grasping neck gently with the other, legs down allowing bird to be guided along in a semi-walking fashion. It is recommended that only handlers with experience use this method.

b) Mechanical restraint recommendations are:

- On occasion it may be necessary to restrain a bird in a sling. Slings are made from fabric, foam and/or plastic.
- Towels can be used as a temporary form of restraint for weighing. The flamingo is wrapped snugly with legs folded. Another towel can be draped over the head (as it lays on the back) to reduce external stimuli.

During crating and transportation it is recommended that flamingos be shipped individually and, when possible are free standing. This method eliminates folding and may reduce the incidence of capture miopathy, leg paralysis and overheating. It is not recommended to ship freestanding birds grouped together due to reported wing and leg injuries. International Air Transportation Association (IATA) regulations apply. Padding, made from a soft non-destructible material, is required under the top of the roof. The 2-cm (3/4-in.) wood frame is covered with lightweight plywood or similar material. The floor must be covered with a solid non-slip, leak-proof absorbent material (non-agricultural) and securely attached to the floor to give the bird a solid foothold. Adequate ventilation must be provided in at least three sides of the container. Although IATA includes slings within the shipping specifications as optional, flamingo managers should seriously consider the hazards associated with slinging birds unmonitored during shipping.

A variety of methods have been used to ship flamingos. Among them are the following:

- Freestanding, individually crated.
- Standing, crate height to back (keeps bird's head down). Construct a crate of a height no higher than the back of the bird.
- Folded, wrapped, and crated.
- Folded, wrapped, in back of truck. Flamingos have been transported as a group in the back of a truck lined with clean hay. Each bird is individually wrapped in a pillowcase then nestled in the hay, which prevents the birds from rolling around.
- Freestanding in back of truck/van. This method involves putting freestanding birds together in the back of a truck, commercial van or trailer. The vehicle should be padded and air ventilation assured. The substrate should be non-slip and securely affixed to the floor. The disadvantages of this method include the high propensity for a bird to be injured from falling and being trampled. Driving must be slow with careful starts, stops and turns.
- If the birds are wrapped, a pillowcase of 100% cotton or gauze-like material is used to enclose and immobilize folded birds for transport. Nylon or body stockings are not recommended. The pillowcase is affixed with masking or duct tape to secure the legs and body. Caution must be taken to ensure the legs are not crossed inside the case. It is critical that this material allow heat exchange to prevent overheating.

Two examples of birds that were shipped are:

1) Sixty-day-old greater flamingo chicks, hand reared in isolation at Slimbridge, special crates, each with four compartments and one bird per compartment, over 11,000 miles to New

Zealand. In transit (over 24 hours) they were tube fed and attended by Michael Batty, a senior keeper at the Auckland Zoo. The five transport crates were made from untreated 12 mm (0.5 in) plywood and 10 x 10 mm (0.4 x 0.4 in) galvanized weld mesh at Auckland Zoo, and the crates were shipped to WWT Slimbridge disassembled. Assembled, the crates measured 1.18 x 0.88 x 1.11 m (3.9 x 2.9 x 3.6 ft). The bases were lined with heavy-duty polythene to a height of 0.15 m (5.9 in) to prevent any fluids leaking from them in transit (as required by IATA). Each crate comprised four compartments measuring approximately 0.44 x 0.42 x 1.11 m (1.4 x 1.3 x 3.6 ft). Internally, each compartment was lined with Hessian (burlap). The floor of each compartment was covered with a woven nylon mat. Almost half of the wall panel areas were made from weld mesh to allow airflow through compartments. Each compartment had an access door measuring 0.3 x 0.15 m (11.8 x 5.9 in) and positioned 0.22 m (8.6 in) from the top of the crate to allow hand feeding during transit. Translocation began on July 27, 2001 when the 20 chicks were aged, on average, 55 days (range 33–71 days). Examination of the birds at Prestwick airport revealed that four birds had superficial wounds on the dorsal surface of their upper mandibles caused by rubbing against the Hessian lining inside crates. To protect beaks against further abrasion, 1–3 layers of surgical tape were positioned along the length of the mandible of all birds. Tape was replaced three times during the flight.

2) Disney's Animal Kingdom shipped 32 Chilean flamingos and 65 Caribbean flamingos in three separate shipments. Two shipments were by ground vehicle, and one shipment of 32 Chileans flamingos was by air. For these shipments, birds were in the crates from 8–14 hours. All flamingos were shipped in crates that allowed the birds to stand but did not have any sling device. For future shipments staff would consider adding padding to help with abrasions that occurred on the wings.

Sea World has designed a very functional sling. This sling design is suitable for long-term slinging of long-legged birds for recovery of leg injuries or for short-term slinging required during treatment of miopathy. Flamingos have been slung for up to eight weeks continuously, with infrequent removal for routine weights or radiography. The Sea World sling now consists of fabric-covered foam pad with four webbing straps that end in D-rings. The bird is suspended using additional straps that thread through the D-rings on the sling pad and are attached with hook and loop tape. The hook and loop segment at the D-rings allow for easy height adjustments to the sling without necessitating the removal of the bird. This is vital to finding a comfortable posture and height for the bird. The other end of the suspension straps is then affixed to the interior of the caging ceiling. Once constructed, the sling can be machine-washed and dried, depending on the materials used. To place a bird in the sling, it is best to have two people, one to hold the body of the bird and a second person restraining and guiding the legs. Once in the sling, the flamingo must be wrapped into the sling around both the body and the sling. In general, the wrap should cover the body over the wings, as well as secure the bird from forward and backward movement within the sling.

Flamingos are susceptible to abrasions of the feet due to contact with hard or rough surfaces such as concrete, which may result in pododermatitis. The skin on the plantar surface of the foot becomes cracked and swollen. This may lead to bacterial invasion (*Staphylococcus* sp is most common) which eventually may lead to osteomyelitis, joint infections, renal disease (the kidney is the most common site for bacteria to lodge in a variety of avian species) and/or septicemia. Treatment may consist of one or all of the following depending on the severity:

- Local antiseptic cleaning and bandaging.
- Various lubricating or moisturizing ointments.
- Systemic antibiotics.
- Surgical debridement.

Prevention may include:

1) Providing a substrate of grass or similar short, dense vegetation and a shallow pool.
2) Good sanitation of the exhibit and water.
3) Winter housing should be well designed and allow birds access to outdoors when possible.
4) Increase the activity of the birds by stimulating them to forage, providing a larger flock size to stimulate normal social behaviors, and artificially increasing day length.

5) Salt water around the feeding areas may decrease the incidence of the condition.

6) When flamingos are confined for medical reasons, rubber or mesh mats should be used.

7) Regular examination of feet so treatment can be initiated early in the course of the disease.

Diet: A variety of commercially available, grain-based diets containing 20–40% protein, developed using the nutrient requirements of poultry and duck species, have proven successful for maintenance, growth, and reproduction of flamingos in captivity, and should be provided as the staple diet. Brands such as Mazuri, HMS and Zeigler are widely used. Their pink coloration depends on the ingestion of carotenoid pigments, which are chemically altered before deposition in the feathers.

The characteristic "bent" shape of the flamingo bill serves to keep the edges of the upper and lower mandible more or less parallel when the beak is opened, and especially when the beak is not fully opened. This facilitates the use of the lamellae on the bill surfaces as a filtration system or sieve by keeping the space between the mandibles constant. Withdrawal of the tongue produces a vacuum, drawing liquid into the mouth. The tongue then moves forward, forcing the liquid through the sieves formed by the. Size and shape of the tongue vary, but in all species the tongue carries an array of spines, presumed to move food particles from the bill back into the throat.

Leg problems and obesity in captive flocks may be associated with diets high in energy content that result in too rapid growth, but nutritional diseases/imbalances have rarely been reported in flamingos. Grit has been reported in digestive tract contents of all species. Unless it is available as part of the exhibit substrate, grit should be provided separately in the size range of 500–1000μm for the larger flamingo species, and 90–250μm for the smaller species.

Energetics calculations based on dry commercial feeds containing 3–3.5 kcal/g (12.5–14.6 kjoule/g) suggest that a 3 kg (6.6 lb) *Phoenicopterus ruber* requires 230–290 kcal per day (960–1200 kjoules per day) depending on activity level. This level would be supplied by consumption of 65–100 g (2.3–3.5 oz.) of dry food per individual (2–3% of body mass). If diets are presented as a slurry (up to 75% water), this may represent 260–400 g (9.2–14.2 oz.) of food offered daily, or up to 10% of body mass. It has been recommended that diets for captive flamingos contain not less than 13% protein. Diets, however, are invariably diluted with plant materials, mud, and other filtrates to provide a lower overall protein content, and commercially available diets or mixes used in zoos containing 20–45% protein have been shown to support reproduction. The majority of North American zoos use canthaxanthin as a pigment added to the diet: Roxanthin Red (10% dry canthaxanthin beadlets) or Carophyll Red (10% pure canthaxanthin) has been added in the range of 25 mg/kg food (Wildlife Conservation Society) to 1.0 g/kg food at Slimbridge. Approximately 10 g (0.3 oz.) of canthaxanthin has historically been fed on a daily basis to 275 birds at Slimbridge.

Breeding: It is recommended that all birds be sexed. The most commonly used sexing techniques are DNA analysis, surgical laparoscopy, body measurements and behavioral observations. DNA analysis and laparoscopy are the most reliable methods, with the former being the least invasive. Sexual and social behaviors are closely associated in flamingos, as the displays that precede breeding are performed in large groups. This helps with the synchronization of breeding, ensuring that all birds breed when environmental conditions are most favorable. Displays are performed by both sexes, but are usually initiated by males. Ritualized behavior is "contagious," with a particular display by one bird being followed by similar displays from others. Display may precede breeding by months (or may occur after breeding) and may occur at sites where breeding does not take place. Even though heterosexual pairings and fidelity are more common in flamingo social relationships, other relationships do exist. In captive populations, homosexual pairs may comprise 5% of the couples. Same-sex pairs will feed, call, sleep and deal with aggression by other birds together. Mountings and copulations may also take place.

In captivity, flock size seems to be an important factor in ensuring a group of flamingos will breed. Pickering et al (in press) showed that in Britain and Northern Ireland breeding flocks of Caribbean and Chilean flamingos are significantly larger than non-breeding flocks. Also, larger flocks breed more frequently than smaller flocks. All Chilean flocks containing more than 40 birds have bred successfully at least once, but one flock of only four Chilean flamingos has reared a chick. In contrast, all Caribbean flocks in Britain and Northern Ireland with more than 20 birds have successfully reared chicks yet the smallest Caribbean flock to rear a chick was one of 14 birds, although one flock with only four birds has laid eggs.

Flocks should ideally have equal sex ratios although female–female pairs appear to be able to rear chicks. Unpaired male greater flamingos may cause considerable disturbance to a breeding flock by persistently attempting to rape incubating birds. Unpaired male and female Chilean flamingos have been observed smashing and eating unguarded eggs at The Wildfowl & Wetlands Trust, Slimbridge. Unlike wild greater flamingos, captive flamingos tend to breed with the same partner each season, although the degree of fidelity appears to decrease as flock size increases. While paired birds are often seen associating in the non-breeding season, pairs can form very rapidly during the breeding season and birds losing their first egg may re-pair with a different bird in the same season.

The choice of substrate in a nesting area can be a very important decision. Some zoos have used clay, a clay/sand mixture, leaf mulch, and native soil from surrounding areas with success. The nesting area should be located in a secluded portion of the exhibit as far away from the general public as possible. Visitors should not be able to access the nest area. Several institutions have nesting areas located on islands, pool edges and on dry land in close proximity to the pool. The location of the nesting area in relation to daily exhibit servicing should be taken into account. Enough space should be provided for the keepers to do their daily work without disturbing nesting birds.

The egg-laying season extends for approximately half a year at any latitude. It has been suggested that flamingos are inhibited from breeding when there is less than 12 hours daylight. Clutch size of one is normal but birds will re-lay after losing an egg. A maximum of five eggs laid by one female Chilean flamingo has been recorded. The period between egg loss and re-laying is highly variable, 6–90 days, but is typically 10–15 days. Egg loss due to fighting can be very high. Vulnerable eggs can be removed to an incubator and replaced with dummy wooden or clay eggs. Although they have relatively thick shells and often become covered in mud, eggs can be easily candled at approximately 10 days. Infertile eggs can then be removed to encourage birds to re-lay. Pairs producing infertile first eggs will often produce fertile second eggs. It has been suggested that pinioned males may find difficulty balancing when mating. Captive pinioned male greater flamingos certainly fall off when mating more frequently than full-winged wild greater flamingos. There was an increase in hatching success in greater Chilean and Caribbean flamingos at Kobe Oji Zoo after the practice of feather cutting stopped when the birds were placed in an aviary. The share of incubation is almost equal between male and female parents. In trios one bird often takes a greater share of incubation than the other two. Incubation period is 28–30 days. Incubating flamingos stand and turn eggs 3–8 times during daylight hours.

It is essential that all routine maintenance work be carried out to a regular pattern in tune with the flock's breeding cycle. Greater flamingos seem to prefer open, sunny areas for nesting and resting. Caribbean and Chilean flamingos seem to prefer shaded areas. The nest site should also be separated from the resting and feeding areas. The same should apply to separation from the public, either by space or visual barriers. Once an established flock has begun breeding, indicating their choice of nest site, they will remain faithful to that site from season to season. A new flock, or one that hasn't nested yet, may choose sites that are inappropriate or difficult to manage. However, they may need to be encouraged at those sites to get them started.

The important criteria for flamingos to build nests are a suitable substrate and an appropriate moisture level. Where there are no suitable natural nesting sites, concrete-lined pits filled with mud and clay can be created as nest islands. The concrete lip of each nest site should be lower than the edges of ponds. This allows the nest sites to be regularly flooded by raising and lowering the water level in the pond; however, this may also bring risks in sites susceptible to botulism. An alternative method is to have a separate clean water supply to the nest so it can be flooded without the need to change the level of the water in the pond. Experiments have shown that rising water levels result in greater nest building activity than simple wetting of the nest site. At the San Antonio Zoo 50% sand/50% clay mix is used. Most managers will provide "starter" nest mounds to help stimulate the birds. Providing a nest start may be more important for young or newly formed flocks. It appears to act as a stimulus to entice first-time breeders into nest building. Flamingos tend to build their nest mounds as close together as possible.

Paired flamingos start nest building by standing at the site and begin pulling material to that site using their bills and reaching out with their long necks to extend the range of nest material gathering. The mated pairs work together. As the materials start to accumulate and pile up, the birds will alternately adopt a standing and a sitting position on top of the pile and continue pulling materials to them. With their breasts on the middle of the top, a shallow depression is formed with a lip. The mound will grow in diameter and height until the egg is laid. Nest building will continue after the egg is laid. The recommended space for each individual nest site is 1.15 m^2, or 1.07 x 1.07 m (12.25 sq. ft., or 3.5 x 3.5 feet). The size of the nesting area will be determined by the size of your flock.

Once chicks start hatching and leaving the nests, the hazards they may encounter if there is too much standing water around the nests become a concern. Careful water management is necessary to keep the substrate suitable for nest building without endangering chicks leaving the nests. Adding nest material if necessary, egg manipulation and management, chick management, and pinioning of chicks, all require considerable skill and experience to perform without undue disturbance of the colony. This disruption can cause egg breakage, injury to chicks, and potential abandonment of nests.

Pinioning flamingos definitely impacts fertility of eggs. The degree of reduction in the percentage of fertility in eggs can be quite variable, ranging from only 25% fertility in San Antonio Zoo's pinioned male Caribbean flamingos to higher percentages, like 100% fertility in the 14 eggs produced by the pinioned male Chilean flamingos at the San Antonio Zoo in 1997. Some individuals have "the knack" to achieve proper copulation and produce fertile eggs, while others never seem to get it. Careful egg management, such as egg pulling, dummies, recycling, egg substitution, candling and artificial incubation, can help but won't solve the ultimate problem of low fertility. Since 1988, the San Antonio Zoo's newer Caribbean flamingo exhibit with full-winged birds is enclosed but has an open front with curtains that can be lowered at night or whenever work has to be done in the exhibit. The birds, in general, have shown no inclination to fly out. Only five birds have flown out of the exhibit. All of these were one or two-year-olds and were recovered. The pay-off is nearly 100% fertility in the eggs.

Mean incubation periods

Species	Incubation Periods	Mean Egg Size (mm)
Caribbean	27–30 (28.9)	90.7 x 55.0
Greater	26–32 (29.7)	89.1 x 54.6
Chilean	28–31 (29.1)	92.0 x 53.1
Lesser	28	82.3 x 49.9
Andean	27–31	86.6 x 55.0
James'	27–31	82.8 x 51.0

The nest is similar in all flamingo species. It consists of a mud mound with a cupped top. The nest is built by both sexes, although the male often initiates building. Most building is performed by a bird sitting or standing on the nest, drawing up material with its bill, although its mate may add material while standing beside the nest. Building may start several weeks before laying. In suitable conditions the nest may be over 30 cm (12 in) high, though it is usually much lower at laying, with building continuing into and sometimes throughout incubation if suitable material is available. Breeding does not always follow nest building. In all species nest densities within colonies may be very high. Where mud is suitable, nests can be built in clusters with five nests per square meter. The normal clutch in all species is one. In studies of a number of species, frequencies of two-egg clutches have been found to be 0.2–2.0%, but in no case is it known whether these clutches result from one female laying two eggs, or from two females laying in the same nest.

In the lesser flamingo the incubation period is 28 days, 27–31 days for all other species. Incubation is shared by both sexes. Incubation shifts are variable in length, in some cases linked to the distance that the "off-duty" bird needs to fly to find food. Flamingo chicks may call loudly, often some time before the egg is pipped externally. Hatching usually takes 24–36 hours from external pip. Chicks of all species hatch with pink or red legs and bills, and white or pale gray down, and with eyes open. The chick stays on the nest for approximately 5–8 days and is fed on crop milk by both parents, and brooded by either parent for most of this time. Chicks often eat large quantities of eggshell in the first two days if this remains in the nest. Crop milk is a glandular secretion that is passed from the bill tip of either parent into that of the chick. In a captive greater flamingo crop milk contained 15% fat, 8–9% protein and 0.1–0.2% carbohydrate. It is initially dark red and later pink. On leaving the nest, chicks can walk and swim and are initially attended by one or both parents, and may return to the nest (if accessible) to be brooded. They later form large groups attended by small numbers of adults, although parent birds arrive regularly to feed their own chicks. Chicks may begin to exhibit feeding behavior at two weeks, and may be able to feed themselves effectively from about a month old, but are usually fed at least partly by their parents until fledging at 75 days. The bill is straight and shallow on hatching, gradually acquiring the typical deep, down-curved shape up to fledging. The first, pale gray down is replaced by a darker gray down at about two weeks. Legs and bill turn black within the first week. The juvenile plumage is gray marked with darker brown, and develops noticeably by 30 days, with the body feathering complete at 50 days. There is some pink suffusion in the juvenile plumage, particularly in those areas that are brightest in adults. The lesser flamingo chick is unable to stand for 24 hours. On the second day it can stand but remains on the nest. From the third day it will leave the nest if disturbed, but will otherwise not do so until the fifth or sixth day, as it cannot climb back onto the nest before this. Chicks leaving the nest early may come to no harm if the substrate at the base of the nest is reasonably solid, but they cannot cope with thick slushy mud. Large groups of chicks begin to form from about eight days.

Full adult coloration is usually attained at three years in all species, but this can be variable, and sexual maturity and the age of first breeding are not necessarily closely linked to plumage and leg color. In captivity, Caribbean flamingos have been recorded breeding at two years; there are records of lesser flamingos breeding in immature plumage in Namibia. Captive-bred Chilean flamingos at Slimbridge first breed between two and 10 years of age (mean = 4.8 years).

It seems to be a normal phenomenon for some egg breakage to occur in the gregarious, colonial nesting flamingo. Reports vary as follows: One Caribbean flamingo flock has suffered from 10–50% egg breakage, 13–37% for Slimbridge's greater, Caribbean and Chilean flamingos.

During the hatching process, the brooding parent will fidget and may rise (weather dependent) more often to check on the hatching chick, especially as the chick vocalizes. While eggs may be freely substituted, chicks cannot, nor can pipping eggs be switched. Once vocalizations have started, whether from the internally pipped egg, the externally pipped egg or the newly hatched chick, no exchanges can be made. Parents can help the chick emerge from the egg by gently pulling on shell fragments or membranes with their bills. An inexperienced parent may lack in some of these "solicitous" behaviors, actually appearing "alarmed" at the hatching events and easily displaced by curious or other aggressive flamingos. If an aggressive non-parent gets on the nest with the hatching egg or newly hatched chick, the response is usually immediate and fatal. Immediately after hatching, an exhausted chick will look wet, very "flattened" and immobile. Inexperienced observers can be alarmed and think the chick is dead or crushed. However, the chick should begin trying to lift its head and vocalize within an hour. The chick will fluff out and begin sticking its head up between the parent's body and wrist joint of the wing. This is usually how the chick will receive its first food, as the parent drools the red crop milk from its beak tip into the upraised bill of the chick.

Chicks will leave the nest as early as three days and as late as 11, with 5–7 days old the most common. When the chick leaves the nest, the parents will stand over it, or just to the side of it, and guard it from all other flamingos. They will brood and feed the chick off the nest just like they did on it. At this point, it is important to discontinue the water that has been supplied for nest building. A chick that leaves early is less able to walk to drier ground and may be vulnerable to drowning or being trampled. Aggression from other flamingos is another obstacle. First-time or novice parents may not be as adept or confident in caring for their chick and may suffer losses. Each season a map of the nest site should be drawn with the location of each nest marked prior to the first egg being laid. If the flock goes on to breed, it should be checked daily to record the location of each egg, lay date and identity of the parents. Care should be taken to check each nest every day as incoming pairs frequently displace birds from their nest.

It has been demonstrated that hand-reared flamingos can be weaned in as few as 23 days, but 30 days or longer is more typical. When being parentally reared in captivity, this process can take much longer, in fact, almost a year. The Basel Zoo reports that they regularly separate their greater flamingo parents and juveniles in the autumn. It is felt that when the juvenile stays too long with the parents (especially the male), it may adversely effect breeding in the following season. Adult Caribbean flamingos will lose their color while feeding their chick. One can detect the parents who are still feeding and the ones who are not, by the color in their plumage. At the San Antonio Zoo, some three-month-old parentally reared young were pulled from the flock to keep some hand-reared flamingos company and aid in their socialization and weaning. They were separated from their parents for 1.5 months. Within a day of their return to the flock, the juveniles were being fed by their respective parents. For Caribbean flamingos, fledging is 70–75 days, in the Camargue, greater flamingo chicks have been known to fledge as early as 62 days. If pressed, wild lesser flamingo chicks can fly at 65 days. Greater, Caribbean and Chilean flamingos are capable of breeding in their second year. Young, juvenile flamingos will participate in ritual displays and nest building. This practice probably hones these skills and prepares them for the future.

If one must remove eggs, for whatever reason, in order to minimize disturbance to the flock and other incubating birds, it helps to take the shortest route to the particular egg. Allow time, by moving slowly. If the birds stubbornly refuse to vacate the nest for the final grab, extend your hand, palm down over the egg as quickly as you can without totally spooking the bird. Keep your visits as brief as possible and do not make wholesale changes, only small limited changes. At the San Antonio Zoo our small auxiliary group of 20 Caribbean flamingos had three nests with eggs. It was determined all three eggs were infertile and were removed. Doing it all at once caused abandonment of the nest site.

If the egg is clean and free of dirt, determination of fertility may be made as early as four days following laying. At this stage, fertile eggs have a uniform "rosy glow" about them. After seven days of incubation, eggs free of dirt may be discarded if they are candled infertile, as by this time fertility, or lack thereof, is quite evident. As the time of hatching approaches, eggs should be carefully monitored. If the egg is to be returned to the parents (and not hand reared from day one), the optimal time is when the chick is just beginning to internally pip or just before. To increase the odds of survival for the chick, the egg should be placed under a pair that has demonstrated good parental skills in the past.

To increase the number of fertile eggs produced in one season, females can be encouraged to lay again. This is accomplished either by removing a dummy egg from the pair any time during the incubation period or by removing the nesting pair's egg for artificial incubation and not replacing it with a dummy egg. Although the production of a second egg will not always occur, if it is early enough in the nesting season, most pairs will lay a second egg, and even a third egg is possible.

If your management program calls for pinioning, this is something you can do during a catch-up to assess the chicks. One thing to consider when pinioning chicks is that it is better to do it when the chicks are younger (2–10 days old). During this catch-up you may also want to conduct other procedures such as weighing, taking blood for genetic sexing, tarsus measurement, banding, adding a transponder chip, egg candling of other eggs and nest management. Be aware of how long you are in the colony and try to keep all handling and disturbance to a minimum. If done properly, flamingos quickly adapt to the regular handling of young chicks and other nest management activities. A brooding parent may not even stand up when an egg or chick is removed from under him or her.

You can permanently band a chick at 60 days, but some people prefer to use temporary bands before that. Plastic poultry bands work well temporarily. Since there is some correlation between tarsus measurement and sex, you can also do a tarsus measurement (i.e., a larger tarsus = a male). The decision to pull a chick that is struggling should only be made by an experienced person who has spent a lot of time observing the situation.

Things to look for include:

- chick is isolated
- getting dirty
- not being fed by either parent
- shivering
- lethargic
- unable to stand
- lack of good vocalizations
- no contact with either parent

If you must pull a chick you should immediately provide supportive care, i.e., heat, food, and/or fluids. Do this until you can stabilize the chick, and then return it to its parents as soon as possible. Try to return it in the early morning so you have all day to watch. Chicks have been pulled for as long as two weeks and returned to the flock, and the parents have accepted them back. Your first choice should always be to return the chick to its parents. As the chick gets older and starts eating food on its own, you should add extra pans of food so the chicks don't have to compete with the adults. Also, if the adults are fed a dry pelleted food, providing pans of food with water in them can help encourage feeding.

Hand Rearing & Artificial Incubation: The Flamingo Husbandry Guidelines provide excellent advise on these fields but, as I was involved in the project described below, I have used this information for this section. I would, however, advise anybody embarking on an incubation and hand-rearing project to use the husbandry guidelines as a reference as well. This information has been provided to me by Michael Batty, who was the key person involved in this transfer from start to end.

Twenty greater flamingos originating from a captive flock held by WWT Slimbridge, UK were hatched in incubators and hand reared in pre-export isolation facilities prior to being successfully relocated by road and by air to Auckland Zoo, New Zealand when aged 33–71 days. In New Zealand the flamingos were held in a quarantine facility for a 30-day period prior to being introduced to an enclosure for public exhibition.

Inducing egg laying: On March 15, 2001, 50 nest sites were molded on the nest island of the "donor flock" by two keepers using buckets and spades. The nest sites were smooth sided, conical-shaped hillocks measuring approximately 0.3 m (1 ft.) high, 0.7 m (2.3 ft.) diameter at the base and 0.3 m (1 ft.) diameter at the top. Daily, from April 1, 2001, the nest-building and egg-laying activity of the flock was recorded at 10:00 a.m. and 2:00 p.m. Eggs were collected for artificial incubation daily at approximately 3:00 p.m. Eggs were replaced with resin dummy eggs painted white and fitted with a 10 cm (4 in.) metal spike to anchor them to nests. A unique alphanumeric code was marked on the side of each egg using a soft-leaded pencil. Eggs were carried 1.5 km (0.9 mi.) to the Slimbridge Centre Incubation Facility (SCIF) in a basket containing goose nest-down. Females were encouraged to re-lay by removing dummy eggs after 1–14 days of incubation.

Egg incubation: At the SCIF, eggshells were disinfected by exposure to UV light for 10 minutes. Egg mass was recorded to the nearest 0.1 g using an electronic balance. Eggs were incubated in Brinsea Polyhatcher incubators operating at 37.8ºC (100ºF) (top of egg), with a 65% relative humidity (RH) and automatically turned through 180 degrees every 30 minutes. To achieve a mass loss of 15.0–17.5% by the expected date of internal pipping (i.e. by day 27 of the 28–29 day incubation period), eggs were moved between three different incubators operating at different RH. Egg mass was recorded daily at 2:00 p.m. If eggs were below or above expected mass they were moved to a dryer (c. 55% RH) or wetter (c. 80% RH) incubator, respectively. Embryonic development was assessed using a Brinsea Superlume egg candler. Eggs were candled at 5, 10, 15, 20 and 25 days. At 26 days incubated eggs were transferred to an incubator operating at 36.5ºC (97.7ºF) (top of egg) with RH as before. From 26 days incubated eggs were candled up to seven times daily (at approximately three-hour intervals between 7:00 a.m. and 10:00 p.m.). If candling revealed that an embryo had entered the air space (i.e. internally pipped), the egg was transported immediately to the PEIF for hatching.

Prior to relocation, eggs were sprayed with 1% Virkon S solution at 40ºC (104ºF), dried with a paper towel and transferred to a sterile polystyrene box containing polystyrene beads to protect them during transit. All eggs were taken by car from the SCIF to the PEIF where they were transferred to a Brinsea Hatchmaker incubator operating at 37ºC (98.6ºF) (top of egg) and 70% RH. Eggs were placed on a clean cotton towel that was dampened with distilled water. Egg turning was discontinued.

Chick rearing procedure: Post-hatch and daily before the first feed, chick mass was recorded to the nearest 1.0 g using an electronic balance. The umbilicus of all hatchlings was swabbed with Novalssan antiseptic before they were placed in a "nest bowl" inside a Brinsea Hatchmaker incubator operating at 35ºC (95ºF). Nest bowls were 0.15 m (5.9 in.) diameter x 0.07 m (2.7 in.) deep plastic bowls lined with a cotton towel and paper towels. Chicks in their nest bowls were transferred to brooder boxes measuring 0.7 x 0.9 x 0.4 m (2.3 x 2.9 x 1.3 ft.) high, constructed from concrete and painted with water-resistant white paint, 4–12 hours post-hatch. Brooder boxes were maintained at 35ºC (95ºF) using red 250 watt lamps, suspended above each nest bowl. As flamingos eat eggshell post-hatch, their eggshells were moved with them to the brooder box. Brooder size was alterable using hard board dividers. The brooder box floor was covered with a woven nylon carpet to provide chicks with a non-slip surface on which to walk when they left the nest bowl. Carpets were replaced daily. Nest bowls and brooders were scrubbed with 10% Virkon S solution and rinsed with cold water daily. When aged 3–7 days, the chicks were pinioned using scissors and the amputated wing tip was used for chromosomal DNA gender determination. Brooder temperature was lowered to 30ºC (86ºF) and 20ºC (68ºF) when chicks were aged approximately seven and 14 days old, respectively. At age 11 days, 2–3 chicks were placed together in brooder boxes measuring 0.6 x 1.2 x 0.4 m (1.9 x 3.9 x 1.3 ft.) high, with lamps positioned at one end and a plastic water dish measuring 0.16 m (6.3 in.) diameter x 0.03 m (1.2 in.) deep, at the other. At this age, chick recognition was made possible using colored flat plastic leg-bands.

When 14–22 days old, chicks were moved to one of three rearing rooms measuring 1.8 x 2 x 2 m (5.9 x 6.5 x 6.5 ft) high. Rooms had smooth concrete floors and walls and contained a pool area measuring 0.7 x 0.7 x 0.15 m (2.3 x 2.3 x 0.5 ft) deep, overlaid with rubber pond liner. Floor areas, overlaid with woven nylon carpets, were cleaned and disinfected daily. In each room, single red 250 watt lamps were suspended above land areas so that chicks could loaf in ambient air temperature ranging 15–20ºC (59–68ºF). Rearing rooms were accessed via wire-fronted doors from a central corridor. When rooms were being cleaned, groups of chicks were temporarily confined in a 5 x 1.5 m (16.4 x 4.9 ft.) area of the central corridor. One week prior to export, 10 male and 10 female chicks had micro-chip transponders inserted intra-muscularly in the caudal superficial pectoral muscle, 10 mm lateral to the carina and 10 mm proximal to the caudal edge of the sternal keel.

Chick feeding: Chicks were syringe-fed 1–2 ml water at approximately 8 and 12 hours post-hatch. The hand-rearing diet, a modification of the diet developed by San Antonio Zoological Gardens and Aquariums (Kunnemann and Perry, 1990) (see Table 1), was prepared daily by mixing ingredients thoroughly in a food blender and freezing for use within 24 hours. Unused diet was discarded 12 hours post-thawing. The diet was heated to approximately 36.5ºC (36.5ºF) and first offered to chicks aged approximately 18 hours. Thereafter chicks received 2–6 feeds between 6:30 a.m. and 10:00 p.m. hours daily. Each chick's crop was externally palpated prior to each feed to ensure that food was moving through the digestive tract. The amount fed, and number of feeds given, was determined by each bird's appetite. Food consumption was recorded daily.

Chicks accepted the diet dribbled into their lower mandible from a 5 ml syringe at a rate determined by their ability to swallow. At seven days, chicks were tube-fed to save time and to minimize spillage and plumage soiling. Tube feeding involved inserting a soft 5 mm diameter plastic tube (which was attached to a syringe), 10–30 mm into the chicks' esophagus. Feeding equipment was washed in soapy water, disinfected with 10% Virkon S solution and rinsed with clean water after use. Chicks were offered Clark and Butcher Flamingo Breeder pellets when aged approximately 16 days. If a chick was observed eating pellets its next scheduled syringe feed was withheld to encourage further self-feeding. Thereafter, these birds were given tube-feeds only if their crop felt empty when palpated at a scheduled feed time. To encourage self-feeding, unweaned chicks were mixed with "pellet-eating chicks" whenever possible. Pre-export, each bird was given, by tube, 60 ml Poly-Aid and Guardian Angel solution 24 and 12 hours prior to, and at the time of, loading into crates.

The first egg to be hatched for hand rearing was laid on April 17, and the last 40 days later, on May 27. In total, 27 of 30 eggs (90%) were hatched successfully in the PEIF with an average incubation time of 27.8 days (range 26–29 days), and average mass loss of 17.0% fresh mass to the stage of internal pipping (range 13–22%). Two eggs from first-laid clutches failed to hatch. Post-mortem examination indicated embryo malpositioning: one embryo had its head between its legs while the other embryo had its head positioned under the left, rather than the right wing. One chick from a second-laid clutch was helped from the egg. This embryo had rotated along the longitudinal axis of the egg so that its head was away from the air cell and down the side of the egg.

Table 1

INGREDIENTS	CHICK AGE 1-30 DAYS	CHICK AGE >30 DAYS
Krill (or Shrimp)	75 g	75 g
Herring (heads, tails, fins removed)	75 g	75 g
Hard-boiled egg yolk	75 g	75 g
Oatmeal baby cereal	2 cups in 400 ml water	1 cup in 400 ml water
Nutrobal	5 g	5 g
Calcium carbonate	2.5 g	2.5 g
Vitamin B-1 (Thiamine)	125 mg	125 mg
Vitamin E (oil from capsule only)	100 I.U.	100 I.U.
Water	800 ml	500 ml
Clark & Butcher Flamingo Breeder Pellet (soaked)		1 cup

*Ingredients of blended diet for chicks aged 1–30 days and >30 days.

Table 3

Age (days)	No. Feeds	Amount Per Feed (ml)	Daily (ml)	Age (days)	No. Feeds	Amount Per (ml)	Feed Daily (ml)
1	0	0	0	**19**	5	49.4	247.0
2	5	4.0	20.0	**20**	5	52.2	261.0
3	6	6.1	36.6	**21**	5	54.3	271.5
4	6	8.2	49.2	**22**	5	57.4	287.0
5	6	10.8	64.8	**23**	5	60.2	301.0
6	6	12.1	72.6	**24**	5	63.0	315.0
7	6	15.0	90.0	**25**	5	65.3	326.5
8	6	18.3	109.8	**26**	4	66.8	267.2
9	6	21.2	127.2	**27**	4	68.4	273.6
10	6	23.5	141.0	**28**	4	70.0	280.0
11	6	26.3	157.8	**29**	4	86.0	344.0
12	6	29.3	175.8	**30**	4	90.5	362.0
13	6	32.2	193.2	**31**	4	92.2	368.8
14	6	35.5	213.0	**32**	4	95.0	380.0
15	5	38.7	193.5	**33**	4	97.0	388.0
16	5	41.6	208.0	**34**	4	103.0	412.0
17	5	44.3	221.5	**35**	4	105.0	420.0
18	5	46.3	231.5	**36**	4	105.0	420.0

*Average quantities of food given to 20 hand-reared greater flamingo chicks aged 0–35 days (i.e, day before average age of weaning).

Chick rearing procedure: The first chick, of the 20 relocated, hatched on May 17 and the last, 37 days later, on June 24. Mean hatch weight was 88.8 g (3.1 oz), [range: 83–105 g (2.9–3.7 oz)]. Chicks lost mass for approximately 60 hours post-hatch. Hatch mass was usually regained by six days. The daily rate of mass increase was 10% for chicks aged 5–15 days, declining and stabilizing at 6% at age 23 days until chicks were self-feeding at, on average, day 36.

To minimize aggressive interactions when chicks were first housed together in the same brooder box, they were "familiarized" with one another while in their nest-bowls prior to becoming mobile. Mesh screens were used to physically separate chicks: they were able to sit alongside each other, without physical contact. When put together at 11 days old, groups comprising three or more individuals showed fewer aggressive interactions than cohorts of two birds. Surviving chicks were DNA sexed as 11 males and 11 females. One male and one female were not transported to New Zealand and remained at Slimbridge.

The quantity of food given increased daily and was generally determined by each bird's appetite. Daily food consumption increased in a linear fashion from 0–300 g (0–10.6 oz) per day for 0–23 days of age. Thereafter food intake fluctuated from 300–350 g (10.6–12.3 oz) daily until birds were completely self-feeding. On average, the 19 chicks that went on to survive in New Zealand were weaned at age 36 days (range: 30–49 days). Average mass at weaning for eight chicks was 853 g (30.1 oz. [range: 461–941 g (16.3–33.2 oz.)].

The husbandry manual recommends:

> "Once the chicks are weaned and have fledged, weigh more than two kilograms, are more than 100 days old, and are self-feeding well (as determined by maintaining or gaining weight), they may be introduced to the flock. This is best accomplished on a day when the chicks can be put out in the morning (in good weather) so observations can be made all day and when staffing allows for intermittent observations. Prior to introduction, using the same or similar food pans in the back area as the colony feeds from will benefit the chicks, as they may be hesitant about eating from an unfamiliar food pan. There may be some initial aggression by the flock, and the chicks may remain on the outside fringe until the group accepts them."

Referencess:

Flamingo Husbandry Guidelines. A Joint Effort of the AZA and EAZA, in cooperation with WWT.

M. Batty, N.S. Jarrett, N. Forbes, M. Brown, T. Richardson, S. Oliver, B. Ireland, K.P. Chalmers & I. Fraser. Hand-rearing Greater Flamingos for translocation from WWT Slimbridge (UK) to Auckland Zoo (New Zealand). Unpublished as yet but sent to International Zoo Yearbook no. 40: 254–260.

20.12 Threskiornithidae Ibis and Spoonbills

Glen Holland

Ibis

Introduction: Ibis and spoonbills belong to the order Ciconiiformes and the family Threskiornithidae, which includes 27 species of ibis and six species of spoonbills. They are predominantly associated with water and wetlands. Most ibises forage in wet grasslands, swamps, marshlands, flooded agricultural fields and along the banks of streams and lakes, where they feed on small aquatic animals. The Madagascar crested, or white-winged, ibis, *Lophotibis cristata,* lives in the forest and feeds on insects from the forest floor. The hadada *Bostrychia hagedash*, bald *Geronticus calvus,* and wattled ibis *B. carunculata,* feed on insects in open, dry areas of short grass or cultivated fields.

Management: An article by Archibald, Lantis & Munetchika in the International Zoo Yearbook, number 20, gives a good historical overview of the breeding of endangered ibises. The most common species in aviculture, as well as in the wild state, is the sacred ibis *Threskiornis aethiopicus*. This species is bred in zoological collections throughout the world. They are hardy, adaptable birds whose dietary and breeding requirements are easy to meet. In fact, this species has been so prolific in some South African and Australian *(Threskiormis molucca)* collections that the free-flying birds have become a problem. A breeding colony produces rather pungent odors and the free-flying birds also visit all the enclosures, consuming any food available.

Sacred ibis require large aviaries of at least 3 m (10 ft.) in height, which should include a few large trees and plenty of additional thick dead branches on which the birds can perch. They are not particularly fussy about aviary landscape, but plenty of open space must be available for the birds to be able to walk and fly without hindrance. Their enclosure should be landscaped as described for Habitat F in the chapter on LANDSCAPE. A few large dead branches protruding from the water will display the birds naturally and make interesting photographic exhibits for park visitors. Enclosures should allow a minimum of about 8 m^3 of space per bird, as overcrowding will be detrimental to breeding results. Sacred ibis have also been kept and bred very successfully pinioned in open pens. This is, however, only an option with this common species and, due to the threat of an influx of wild birds, only advisable outside their natural distribution. A colony of 10 pairs is sufficient to represent the species without the problems of an excessive pungent smell. In general, however, it has been found that for the colonial nesters, colonies of 20–30 birds are ideal. In addition, rare species should not be mixed with other species, which may interfere with their breeding.

Diet: Their captive diet should include equal portions of diced fish, minced chicken, bird of prey diet, and soaked cat or dog kibble. Since much of their diet is stored frozen, a multivitamin and calcium supplement is added to the food twice a week. I witnessed interesting behavior by a sacred ibis kept on a natural pond. The bird was observed dropping a waterfowl laying pellet in the water to attract a shoal of small bream into the shallows. As the bream appeared, the ibis lunged into the water, retrieving a bream at about every third attempt. Food is given twice a day for maintenance and three times a day when breeding. For scarlet ibis, a pigment supplement is necessary to maintain the scarlet color. Most zoos supplement with canthaxanthin. Fungal skin disease (birds housed inside) is one of the reported health problems. There have been reports of obese individuals.

Breeding: Sacred ibis have been found to be sexually mature at 18–24 months of age. An interesting observation was made involving bald ibis that, when placed among experienced, older nesting birds, they reached sexual maturity at a much younger age than when housed with other younger birds of the same species. A mixed colony of sacred ibis, spoonbills, reed cormorants, and night herons makes an interesting public display, but care should be taken to ensure that the numbers of this ibis do not dominate the entire colony. When mixing species, there is a chance that hybridization will occur. At the World of Birds in Cape Town, sacred ibis have frequently hybridized with the African spoonbill *Platalea alba*. Other hybrids recorded include white *Eudocimus albus,* x scarlet *E. ruber*, white x straw-necked *Threskiornis spinicollis,* scarlet x sacred and black-headed *T. melanocephalus,* x Eurasian spoonbill *P. leucorodia*. Interaction within mixed flocks of ibises provides the best stimulus to breeding and nesting behavior, but there is a strong danger that under these circumstances the species will interbreed.

Keith L. Bildstein recommends the following for sacred ibis:

> "Nest are generally located about 2 m (6 ft.) above ground and within 2 m (6 ft.) of water. Preferred nest sites tend to be the highest ones. A perch near the nest allows birds to remain close to their nests, or to use the perch as a staging area for getting on, or off, the nest. Nests are built in the open; if built on top of vegetation, they may become somewhat obscured as plants grow during the nesting season.
>
> Nests in colonies are spaced 0.9–1.8 m (3–6 ft.) apart, as determined by available nest sites — not necessarily the birds' preference. Scarlet ibis will use artificial nests, built on a nest platform, or use a sturdy branch-crotch as a base. Nest materials include thin and pliable twigs, large leaves, and palm fronds.
>
> The average clutch has two eggs and replacement clutches are laid. Egg infertility and breakage are the most commonly cited reasons eggs fail to hatch. Incubation lasts approximately 23 days. Times for hatch to fledge/self-feed are about 50–90 days. Onset of breeding behavior occurs around the second year. Young birds tend to be unsuccessful parents, but some two-year-olds succeed in rearing offspring."

The hadada, the black, or red-naped ibis *Pseudibis papillosa,* and spot-breasted ibis *Bostrychia rara,* are solitary nesters, whereas the other species nest in colonies. The waldrapp *Gerontica eremita,* bald, wattled and sometimes buff-necked ibis *Theristicus caudatus,* nest on cliff ledges, which must be taken into consideration when they are kept in captivity. Wooden boxes are an acceptable alternative for these cliff-nesting species. The white-faced *Plegadis chihi,* Australian white *Threskiornis molucca,* and straw-necked ibis usually nest on the ground amid rushes. All other species nest in trees or low shrubs; the nests often overhang water and are usually inaccessible. To encourage breeding in tree-nesting species, nesting facilities consisting of concave wire platforms of about 70 cm (28 in.) in diameter or cane baskets of a similar size should be supplied, preferably positioned in a tree overhanging the water. These are installed at varying heights, about 1–2 m (3–6.5 ft.) apart, and a few sticks placed on the platforms or in the baskets. A plentiful supply of twigs should always be available for nesting and for the breeding stimulation associated with the manipulation of nesting material.

Males are first to occupy nest platforms, establishing small territories where females later join them. During courtship, the pairs of sacred ibis stand on the platforms with the plumes on their backs raised and their wings drooping slightly. They make wheezing, croaking, and squeaking calls while placing nest material and pecking at neighboring birds. The pair display by facing each other and then bowing, with the head and neck extended and pointed downward. After copulation, the male begins to collect nest material that is positioned by the female, who remains on the nest. The 2–4 eggs have a faint bluish or greenish tinge with rust-colored spots. Incubation is by both sexes and starts when the clutch is complete. The incubation period is 22–23 days, and the chicks leave the nest at about 18 days, forming loose flocks. Both parents feed the chicks by regurgitation, with the chick placing its beak into the adult's throat to receive food. They fledge from the colony at 6–7 weeks of age. Many pairs are double brooded.

Joseph Lindholm III gives the following account in the AFA Watchbird of a pair of hadada ibis that bred at Disney's Animal Kingdom:

> "In contrast to sacred, scarlet and American white ibises which re-pair each year and breed in large colonies, the non-colonial hadada mates for life, retaining a pair bond beyond the breeding season. In March the male was noted placing small twigs in the

Sacred Ibis *Threskiornis aethiopicus,* Hadedah Ibis *Bostrychia hagedash* , USA White *Eudocimus albus*

Days	Sacred Weight (g)	Hadedah Weight (g)	USA White Weight (g)	Days	Sacred Weight (g)	Hadedah Weight (g)	USA White Weight (g)
0	42	40	31	**8**	200	130	67
1	40	38	33	**9**	250	160	74
2	52	42	36.5	**10**	300	180	83
3	50	58	39.5	**11**	350	200	87.5
4	78	65	41.5	**12**	400	220	96
5	105	80	51	**13**	450	250	107
6	130	100	57	**14**	500	280	116
7	165	120	64				

Longevity Records of Ibis

Species	Age: yrs	mths	Species	Age: yrs	mths
American White	27	7	Wattled	12	
Scarlet	33	3	Spotted-breasted	8	4
Bare-faced	20		Madagascar Crested	1	
Glossy	26	9	Sacred	37	
White-faced	20	5	Oriental White	28	3
Puna	18		Australian White	33	3
Plumbeous	7		Straw-necked	39	5
Buff-necked	22		Waldrapp	32	10
Black-faced	34	6	Southern Bald	15	4
Buff-necked	26		Black	30	
Hadedah	16		Oriental Crested	25	9

Bald Ibis
Geronticus calvus

fork of an oak tree, spending almost an hour on the same twig. Five days later both birds were observed allo-preening and rattling bills together, the standard courtship display for the species. Thirteen days later a third and final nest site was chosen with the female now successful at placing branches, and working at nest building all afternoon. Nest construction progressed accompanied by bill clattering and evening vocalisations. On day 18 pre-copulatory behavior followed by copulation and a post-copulatory upward bill stretch. The female was observed to spend most of the day on the nest and the pair spent every night together on the nest. The birds begun to steal nesting material from the nest of a pair of hammerkops and continued, despite the hammerkops efforts to attempt to chase them off. Once incubating, staff recorded 'he was very aggressive towards keepers when nearby and he actually flew at keeper's head two times when approaching the nest.' When the birds relieved each other on the nest, 'Both the male and female arrived at the nest with dried grass that was presented to the other bird before duties were exchanged.' This appears to be a universal behavior among ibises in general, both during nest construction and incubation. A single chick fledged from the nest in June and was left with the adults during two subsequent nesting attempts in July and October. Due to the fact that the parents were siblings, the eggs of the second and third clutches were replaced by dummies that the birds incubated for up to two months."

Incubation & Hand Rearing: If required, ibis can be induced to lay a second clutch by removing the first clutch, which is then incubated artificially. At Walsrode, by manipulating two pairs of black-headed ibis, 20 young fledged in three years. Since breeding is influenced by visual contact with other incubating pairs, not necessarily of the same species, clutches should not be removed too soon after laying, as this will disrupt the breeding process within the rest of the flock. The best time for egg collection is when most, or all, pairs are already incubating. Incubators should be set at a temperature of 37.5°C (99.5°F), with a relative humidity of 60%. On average, 90% of ibis eggs are fertile. Of these, about 90% usually hatch successfully. When the chicks are hatching, the temperature should be lowered to 36.5°C (97.5°F) and the relative humidity increased to 70–80%. The chicks weigh between 40–50 g at hatching, and normally hatch unassisted within 24 hours of pipping. The altricial chicks are either covered with fine down or are naked, depending upon species. Two vestigial claws on each wing have been noted on scarlet, hadada, and sacred ibises. Immediately after hatching, the abdominal areas of the chicks can be swabbed with iodine solution to prevent the possibility of infection originating in the umbilical area. Chicks are placed in a small bowl lined with soft, absorbent paper towels and a few small twigs to simulate a nest. Without these twigs, the chicks will have difficulty in keeping their legs beneath themselves and may develop leg problems.

They are left in the incubator for the first 18–24 hours and then moved to a brooder unit set at a temperature of 34°C (93.2°F) and 60% relative humidity for a week. The eyes open at six days. Thereafter, a few chicks can be placed together in boxes with a brooder lamp supplying heat, which should be adjusted to about 30°C (86°F) and gradually decreased over time until the chicks are well feathered and comfortable in a room temperature of 25°C (77°F). Birds kept together should be of a similar size to avoid large chicks trampling smaller ones. At a month old they are moved to an outdoor aviary, where they can be mixed with others of a similar size. It is here that the chicks begin to feed themselves. This transition does not usually take long.

Initially, newly hatched chicks are fed on a diet of equal portions of minced ox heart, finely diced boiled fish, and soaked puppy pellets to which a small amount of water, a calcium and phosphorous supplement, and a multivitamin and trace element supplement are added. The calcium supplement is vital to meet the extraordinary rate of bone growth in young ibises.

At Chicago's Brookfield Zoo, the hand-rearing diet consists of equal proportions of the following:

dry dog food (Gaines Dog Meal)
ZuPreem Bird of Prey Diet
smelt (deboned, skinned, and eviscerated)

These ingredients are placed in a blender with sufficient water to produce a thick paste upon blending. This mixture is further supplemented with a powdered multivitamin, thiamin hydrochloride, and powdered bone ash. This diet can be prepared in a bulk concentrate and then frozen in portions as required. When feeding, hot water is added to produce a semi-liquid consistency of about body temperature. Initially, the chicks will reject food that has cooled. Additionally, one must be aware that the feeding of chilled food could result in a fatal lowering of the body temperature of young birds not yet capable of homeostasis. Young ibises will develop lethal impaction of the gut if fed the adult diet, as they cannot digest or regurgitate feathers and down.

For the first week, chicks are fed hourly between 7:00 a.m. and 7:00 p.m., and again at 10:00 p.m., and in the second week, every two hours between 7:00 a.m. and 7:00 p.m. Although it is possible to force-feed chicks, it is more desirable to stimulate normal begging and feeding behavior. A hand is held before the chick to elicit begging behavior. Chicks push their beak between the thumb and forefinger, and then open the beak while making swallowing movements. Food can be given with a bent spoon or a 2.5 ml syringe that has the entire tip section removed. Feeding is stopped once the begging behavior has significantly abated, as overfeeding can cause too rapid a weight gain, which can lead to bowing of the legs and constipation. Hand-reared ibises soon lose their affinity for humans, particularly when raised with other chicks. Once fledged, they can be introduced into existing colonies without any problems. Immature sacred ibis resemble the adults but have white feathers on the head and neck. At Brookfield Zoo, some of the sacred ibis are approaching 30 years of age.

Ring Size: bald, sacred, waldrapp, and straw-necked ibis and African spoonbill: 12.5 mm
hadada: 11 mm
glossy and scarlet ibis: 8–9 mm

Spoonbills

Introduction: Spoonbills belong to the order Ciconiiformes and the family Threskiornithidae that they share with the ibises. Spoonbills form the subfamily Plataleinae. There are six species, which are distributed through Africa, Australia, New Zealand, North and South America, Japan to China and Europe to South Asia. As with ibis, they are traditionally associated with aquatic habitats, usually breeding colonially, often in association with egrets, herons, and ibis. Nest sites vary from ground nests on islands to reedbeds, cliff faces, and trees overhanging water.

Management: The captive management and feeding requirements of spoonbills is considered to be very similar to ibises and is therefore not duplicated here. However, as hand rearing is an effective tool in increasing numbers of birds produced and also for saving clutches that are in trouble, I have included a paper resulting from work on the roseate spoonbill at Fort Worth Zoo. My personal experience with hand rearing African spoonbill *Platalea alba* was as a young boy with newly hatched chicks at the World of Birds in Cape Town. I vaguely remember using the same technique we used for ibis with a warm soup-textured food consisting of beef-flavored tinned dog food, flaked pets fish, multivitamins and a calcium supplement. As the birds matured, more solid items such as diced fish (with bones and scales) were added to the food. We managed to raise the birds very successfully, but fortunately the methods and diets are now better refined to ensure success.

Spoonbills breed in a very similar fashion to ibis, and flighted birds will breed on nest platforms such as dog baskets and wire platforms. I have seen both African and royal spoonbills breed on the ground.

Incubation & Hand rearing: The Fort Worth Zoo has held a breeding flock of roseate spoonbill *Ajaia ajaja* since the 1960s and has significant breeding success with the species. Roseate spoonbill eggs were incubated at 37.5°C (99.5°F) and a relative humidity of 84–86%. The incubation period is 27–28 days and the chicks are blind and have wispy bits of white down when they hatch. They are initially kept in artificial nests, usually a margarine tub or pint fruit basket with a towel lining at a temperature close to 37°C (98.6°F). for the first few days. Initially it was found that small brooders, 30 x 30 cm (12 x 12 in.) worked well for one or two chicks. Three or four chicks required a larger space. The temperature in the brooder is gradually decreased to ambient when the chicks are fully feathered. Behavior was the best guide for the temperature at which the chicks were comfortable. When overheated, chicks would pant, drop their wings, and drop their heads over the sides of the nest. Cold chicks would huddle together and their skin color would become paler. Humidity and airflow are important considerations since chicks dehydrate quickly if humidity is too low (below 82–86% RH) or if the nest is in the path of moving air from the brooder fan. It was determined that the combination of a baffle in the forced air machine and a tray of water kept the chicks well hydrated. Once additional heat is no longer required, the chicks can be housed in large aquariums or cardboard boxes, but the latter have to be discarded on a daily basis.

Nest sanitation is important, and for newly hatched chicks the containers described worked well. The towel was changed after each feed, and chicks soiled with feces were gently cleaned with a swab and warm water. It important not to rush this process as the top layer of skin can easily be removed. Routine swabbing of the inside of the bill prevented the build up of food and fecal matter, which can lead to bacterial growth and bill deformities. Each chick generally defecates at feeding time. Chicks tend to defecate over the edge of the nest, quickly developing quite a range. Initially, however, all the feces ends up in the nest. As the chicks grew, increasingly larger plastic or box containers made sturdy and easy-to-disinfect nests. Fresh, leafy twigs were added to the nests as the chicks became more active to help develop foot and leg muscles. Additionally, feces fell through the leaves leaving less to stick to the chicks. By this time chicks no longer needed towels under them for warmth, and cleaning time was further decreased by lining the box with newspaper so that excrement and plant material could be rolled up and tossed out.

Roseate Spoonbill
Platalea ajaja

Adult spoonbills feed their chicks by regurgitation, and to duplicate this a warm liquid diet was fed with a syringe and catheter tube. The size of the syringe was determined by the amount the chicks were consuming. For the first few feeds a 3 cc syringe was used. Syringe size was gradually increased to 6 cc, 12 cc and 20 cc syringe capacity. The catheter-tubing diameter was determined by the size of the syringe, not by the size of the chick. The tube was no more than 3.8 cm (1.5 in.) in length. Once a feeding routine was established, the chick would readily open its mouth to take the tube. Care was taken to avoid putting the formula in the trachea rather than the esophagus. The chicks moved away from the tube when they were full.

The hand-rearing diet used contained the following:

Ingredient	% of Diet
lake smelt	11.10
high protein dog chow	1.11
flamingo diet	11.10
Vionate (multi-vit)	0.30
bird of prey diet	11.10
calcium carbonate	0.15
shrimpmeal	2.22
Water	62.92

For the first 2–3 days, the formula was diluted 1:1 with Pedialyte and warmed to 39–40°C (102.2–104°F). During this period the chicks took no more than 1.5 cc and were fed five feeds, one every four hours between 8:00 a.m. and 12:00 p.m. The food was prepared in a large batch and frozen in smaller packages. Formula was thawed fresh for each feeding and all feeding utensils stored in a disinfectant between feeds.

By day four, the volume was up to 4 cc per feed and was diluted with distilled water at a ratio of 2:1. The chicks develop a good feeding response by day four, sitting up on their hocks, head and neck extended up, head bobbing and vocalizations of a high shrill trill.

Days 7–9, the formula was distilled 3:1 with water and by day 10 the full strength formula was fed. Care must be taken not to over feed the chick. Although they may continue to call and bob their heads, it will not be easy to insert the feeding tube. After each feed, the youngest chicks go straight to sleep, while older chicks stay up for a while and are at their nosiest.

Day 18, chicks were receiving four formula feedings of 60 cc daily. Additionally, small pieces of bird of prey diet mixed with Mazuri flamingo pellets were given. It is easier to feed using several 20 cc syringes than to struggle with a 60 cc syringe when the chicks are able to consume 60 ccs or more per feeding. It requires one hand to steady the chick. The plunger may be difficult to work on the 60 cc syringe depending on the size of the feeder's hands. If the formula goes in too quickly, it will come right back out. The chick will let the feeder know when it is full.

Day 30, the amount of formula fed was decreased to encourage consumption of solid food. Balls of the meat mixture were offered after each formula feeding.

By Day 35 only solid food was offered.

Hand-reared birds never reached weights comparable to parent raised birds at seven and 14 days of age. Hand raised chicks in 1987 and 1988 achieved 73–76% of adult weight by day 14. After a spell of no hand rearing, birds raised in 1993 only reached 30% of parent-raised chicks — this improved to 60% by 1997. By 1997, hand-reared birds were consuming as much as 100% of their body weight per day compared to 40% in 1993. In 1987 and 1988 solid food, such as the bird of prey diet, was incorporated into the diet prior to 18 days of age. It is possible that the solid food items are higher in energy content than the liquid diet.

Average weight gain of hand raised roseate spoonbill chicks:

Roseate Spoonbill *Ajaia ajaja*

Day	Weight (g)
1	55
7	150
14	390
21	800
28	1190

While a more concentrated formula may facilitate greater weight gains, increasing the concentration of the current formula to full strength in less than 10 days resulted in difficulties in keeping the birds hydrated. An African spoonbill hand reared at the North Carolina Zoological Park was introduced to solid food (Bird of Prey Diet & Flamingo Fancier) at 10 days of age and achieved weight gains similar to parent-reared birds at 28 days old. The African spoonbill was fed chopped pink rats at 70% of body weight up until the adult diet began to be incorporated at day 10. This bird was completely on the adult diet by day 15 consuming 50% of its body weight per day. The hand-reared roseate spoonbill did not begin consuming the adult diet until 18–21 days of age.

Until 1997 roseate chicks were independent by day 40, but in 1997 they still required to be partially fed at 80 days of age. In 1997 fish was not offered to the chicks during the weaning period. Fish was eliminated from the diet of the adult spoonbills earlier in the year. Rather than accustom chicks to fish and remove it from the diet, it was not offered. It is not clear how fish may play a role in the extended time taken for the 1997 birds to eat on their own compared to birds from previous years. Regardless of growth rate during the hand rearing, all birds hand reared at Fort Worth were able to attain adult weights although the process took longer than in parent-reared birds.

Longevity records for Spoonbills

Species	Age (yrs/months)	Species	Age (yrs/months)
Black-faced	15	Yellow-billed	28/7
White	30/1	Roseate	23/7
African	17/11	Royal	No data

As given in the International Zoo Yearbook, No. 33.

Contributions:

Ann Ward, Fort Worth Zoological Park.

W. Mangold, Managing Director, World of Birds, Cape Town, SA.

Species Requiring Similar Management:

Cormorants
Spoonbills
Herons
Egrets

References:

Archibald, G. W., S. D. Lantis, L. R. Lantis & I. Munetchika. 1980. "Endangered ibises — their future in captivity." *International Zoo Yearbook* 20.

AZAs Heron, Ibis and Hammerhead Advisory Groups' HUSBANDRY NOTEBOOK 2000.

Brouwer, K, H. Schifter & M.L. Jones, 1994. Longevity and Breeding records of Ibises and Spoonbills in Captivity. *International Zoo Yearbook* 33: 94–102.

Brown, C., C. Davis, C. Tucker, A. Ward & J. Wise. 1977. "Development of Hand-rearing Techniques for Roseate Spoonbills, Ajaia ajaja at the Fort Worth Zoo." Proceedings of the second conference of the Nutrition Advisory Group (NAG), American Zoo and Aquarium Association (AZA).

Brown, L. H., E. K. Urban & K. Newman.1982. The Birds of Africa. Vol I. London & NY: Academic Press.

Lindholm, J III. 2001. The Hadada Ibis. The AFA Watchbird No. 1.

Piland, R. M. & D. S. DeCourcey.1981. Hand-rearing techniques for ibis. AFA Watchbird 8 (3): 41–46.

20.13 Pelicanidae Shoebills & Pelicans

Glen Holland & Michael Mace

20.13a Shoebills

Balaenicipitinae

• Michael Mace

Introduction: The shoebill *Balaeniceps rex* has traditionally been aligned with the storks, but is now placed, on the basis of DNA evidence, in the subfamily Balaenicipitinae of the family Pelecanidae. Shoebills are arguably the most recognizable of the African birds. They occupy the vast areas of overgrown reeds, papyrus, and grasses of the Sudan. They are usually found in vegetation associated with shallow water, avoiding any high, dense cover that restricts their maneuverability. This is most likely due to their prey capturing technique, in which they either stand in waiting or slowly wade. When prey comes to the surface of the water, the birds collapse with their bill wide open, grabbing prey as well as surrounding vegetation in a manner similar to casting a fishing net. This material is manipulated until only the prey remains, which is then consumed. Shoebills predominately feed on lungfish, bichirs, *Tilapia,* water snakes, frogs, turtles, and monitor lizards.

Management: Because of their solitary nature, special care must be given to the design of an aviary or exhibit for shoebills. Pairs will spend a great deal of time at opposite ends of a territory. Considering this behavior, areas should be provided to give the birds an opportunity to have visual barriers. This has been effective in defusing combative incidences between members of a pair.

Shoebills are monomorphic, with males being slightly larger than the females. A pair can be kept in a mixed species exhibit. At the San Diego Wild Animal Park they have been exhibited with East African crowned cranes *Balearica regulorum,* Demoiselle cranes *Grus virgo,* black storks *Ciconia nigra,* African darters *Anhinga rufa,* and various African waterfowl. Because the species has been difficult to propagate, an effort should be made to duplicate as many features of their natural habitat as possible until the reproductive biology of the species is better understood. One feature included in exhibit design is to provide one or more pools that would allow the birds to hunt live prey items. The pool substrate should be constructed of soft material such as clay, so that when collapsing into the prey, the bird does not sustain damage to the bill. Their nests in the wild consist of platforms that are built in papyrus sudd. They are solitary nesters in the wild, and the nests are constructed on a true island or on floating vegetation. The clutch size is 1–3 eggs.

Diet: The maintenance diet is offered when the facility is serviced. The dietary items offered daily to each stork consist of three medium trout, five white bait, and five adult mice. The fish are supplemented with a vitamin supplement.

Breeding: To date there is no confirmed account of shoebills breeding in a zoological facility or in a private collection.

Ring size: 28 mm.

Species Requiring Similar Management:

Jabiru stork *Jabiru myteria*
Black-necked stork *Ephippiorhynchus asiaticus*

20.13b Pelicans

Pelecaninae

• Glen Holland

Introduction: Pelicans belong to the order Ciconiiformes and the family Pelecanidae, in which they occupy the subfamily Pelicaninae. There are only eight species of pelicans in the world, all belonging to the single genus *Pelecanus*. One species, the Dalmatian pelican *P. crispus,* is endangered. The survival of all pelican species can be ensured through captive breeding programs, adding to our understanding of this fascinating family and preventing the global reduction in numbers. Over the last few years, Jurong Bird Park in Singapore has successfully implemented such a program. So far, four of the eight species have been bred. These Four breeding species are the brown *P. occidentalis;* the Dalmatian; the great white, *P. onocrotalus;* and the pink-backed *P. rufescens*. Two of the remaining species, the American white *P. erythrorhynchos;* and the spot-billed *P. philippensis,* have yet to settle in, but are expected to breed in the near future. The park is still looking for a mate for its lone Australian pelican *P. conspicillatus*. Only the Peruvian pelican *P. thagus* is unrepresented in this remarkable collection. The Jurong Bird Park program has been exceptionally successful; over the last few years, 16 pairs of the breeding species have produced 37 baby pelicans.

Management: Pelicans are colonial birds, living and feeding together in large flocks. The males can be distinguished by their larger bodies and longer bills. The different species of pelicans can be easily identified by their size, color, and wing pattern. Wing pinioning of the founder stock of these breeding pelicans has curtailed their flying ability to short distances only. At Jurong, mature unpinioned offspring of the brown and Dalmatian pelicans can occasionally be seen flying high above the park and perching on trees bordering the lake island. Pelicans appear to be exceptionally sensitive to disturbance in both the wild and captivity, an important consideration in planning captive facilities. Ideally, their enclosures should not have more than two sides open to public approach.

Melbourne Zoo found that although this is an important requirement in the planning phase, birds will eventually become conditioned to a stable environment. Events in the surrounding areas have had little impact upon them so long as the imaginary flight-or-fright line is not crossed. Melbourne had their best breeding results during the demolition and construction of the Trail of the Elephants that was abutting the Pelican Lake.

It is important to understand the behavior and requirements of each species, and this can only be achieved by close observation. At Jurong Bird Park, the lake where the pelicans live is one of the main park attractions. It is shared with cormorants, ducks, geese, and swans. The enclosure is designed and constructed to afford an environment as close as possible to the natural habitat of water-loving birds. It covers an area approximately 140 m (460 ft.) in length by 65 m (215 ft.) in width. The average size of outdoor zoo enclosures where pelican breeding has occurred is 4,500 m^2 (48,438 sq. ft.). Enclosures need to be large enough to create a "safe" environment. The larger the enclosure, the more breeding and less incidence of stress -related disease. It is recommended that these birds be provided with open water at least 1 m (3 ft.) deep and at least 2–3 times larger than the land surface area. In addition to a main island, Jurong has installed 11 floating fiberglass islands to provide additional space for roosting and nesting. The land area should slope gently into the water, as young pelicans have drowned because of too steep a slope on the island margin. The feeding area should be situated close to the walkway to facilitate servicing and allow visitors a closer view. The Jurong lake is cleaned twice daily to remove scum and algae, and the water quality is closely monitored. Live freshwater fish, such as *Tilapia,* are released into the water to maintain the ecosystem and are an excellent indicator of water quality. The supply must be consistently maintained, as pelicans and cormorants, which are diving birds, eat a large number of fish.

Housing pelicans with other species, such as waterfowl, herons, spoonbills, ibis and cormorants, is common practice, but careful attention should be devoted to the selection of enclosure companions. Pelicans may eat small duck species, and other disturbances to waterfowl, herons, and egrets by pelicans have also been reported. Conversely, swans and large geese in breeding condition can pose serious threats to pelicans. All captive pelicans should be protected from freezing temperatures by having access to open water and dry land areas. Shelter, and possibly even heat, should be provided in extremely cold weather. Extra caution should be taken with tropical species kept in temperate-zone zoos and private collections. Pelicans are quite susceptible to chronic bumblefoot. Substrates such as cement and damp straw or hay, which are associated with this condition, should be avoided. Other health problems, such as aspergillosis and heavy infestations of pouch lice and roundworms, have been recorded. However, pelicans in captivity are normally long lived, and records of birds over 40 years of age are not unusual. The oldest bird at Melbourne Zoo is 43.

Diet: The brown pelican feeds on marine fish, while the other species feed predominantly on freshwater fish species. Food in captivity should always be of high quality, as fresh as possible, and fed in sufficient amounts. For details on the proper storage and thawing of fish, consult the diet section under the heading PENGUINS. It is advisable to feed different species of fish throughout the year, not only to reduce monotony and dependency on one food species, but also to take advantage of seasonal changes in availability and quality. Feeding of live fish on a regular basis is a valuable dietary addition, both nutritionally and behaviorally. The pelicans at Jurong Bird Park are fed as follows:

Species (adult)	Fish/bird/day
Brown	600 g
Dalmatian	1,000 g
Great white	1,300 g
Pink-backed	800 g

The birds are fed twice a day on a variety of salt-water fish and will stop eating when they have eaten a sufficient amount. Hand-raised chicks are given chopped anchovies for the first three weeks. Between three and six weeks, they are fed whole sprat fish. When six weeks old and beyond, they are fed a variety of saltwater fish. The diet is fortified with Mazuri Fish Eater Tablet–SDS vitamins to increase the general health of the birds. The vitamin tablets are crushed and mixed with the fish before they are fed to young birds up to three weeks of age. Adult birds are given whole tablets placed in the mouth of the fish they feed on. This diet has been analyzed to determine the nutrient composition.

The individual distribution of fish, tossed to the birds from a short distance or given to them directly by hand, is strongly recommended. It provides close contact between birds and keepers, thus guaranteeing a good animal–keeper relationship; it allows keepers an opportunity to inspect each individual closely and to monitor food intake; it provides an opportunity to give individual oral medication, vitamins or nutritional supplementation; it reduces both intra- and interspecific food competition; and it reduces the need for cleaning and, consequently, the opportunity for disease or infection, such as botulism, to occur. These considerations are particularly important if pelicans are housed in large, multispecies enclosures. An additional advantage is that hand feeding stimulates the interest of visitors and provides educational opportunities. At Melbourne, staff consider the hand feeding and opportunities that described above as paramount for the successful delivery of a "pelican" experience, for birds and visitors alike.

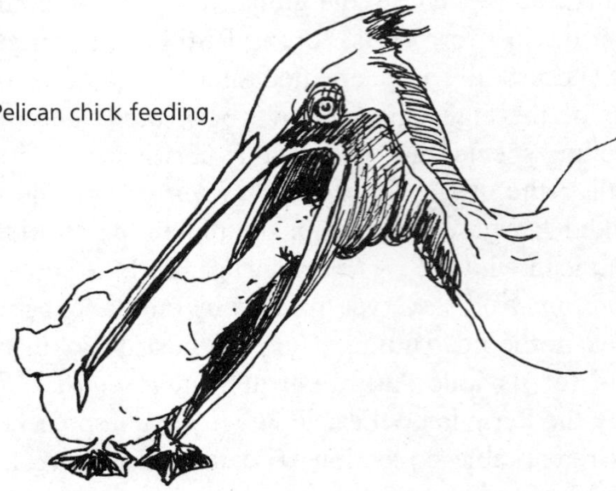
Pelican chick feeding.

Breeding: The eastern white pelican was the first species to be bred in captivity, at Rotterdam Zoo in 1872. A significant increase in the number of zoos breeding pelicans has occurred since 1982, although success within an institution is often sporadic. As pelicans are colonial breeders, at least a small group of birds is required for successful breeding. However, in small groups the breeding activity can be seriously compromised by the loss of even a single individual. Based on experience at Tierpark Berlin, Grummt (1983) suggested that a group of at least 20 birds is required for social stimulation and breeding synchronization. Unpaired birds, particularly males, can disturb the breeding activities of pairs by seizing nests and breaking eggs. In pelican groups, removal of these birds may be required in order for activities to proceed. Hybridization has occurred when several species of pelicans are kept together, and is particularly likely if unbalanced sex ratios or very low numbers of individuals per species limit mate choice. At Jurong, cross breeding is proving to be a problem with a mateless Australian pelican that has paired each year with Dalmatian or pink-backed pelicans. To prevent this problem, the staff have tried to encourage each species to nest on a different island by providing wooden frames for them to build their nests on. These frames are 60 x 60 cm x 35 cm (24 x 24 x 14 in.) high (Breeding data

Nutritional Analysis of Diet

Fish	Protein	Fat	Fiber	Ash	Calcium	Phosphorous
ANCHOVIES (up to 3 weeks)	63.5%	4.9%	0.3%	12.9%	2.86%	1.77%
SPRAT (3 to 6 weeks)	65.4%	6.7%	0.2%	15%	3.8%	2.76%
MARINE FISH (over 6 weeks)	67.3%	8.8%	0.3%	16.7%	4.26%	2.94%

Australian Birdkeeper No 6 (8) 1993

table below). The pink-backed have already established their exclusive territory. Interspecific disturbance may also be a problem in multispecies exhibits, particularly where species of varying sizes are mixed. Conversely, one species may encourage reproductive activity in others. For example, the breeding of brown pelicans at Jurong stimulated pairs of eastern white and Dalmatian pelicans to reproduce as well. As a rule, however, keeping several species together, especially when numbers of each species are very low or sex ratios unbalanced, is not recommended unless the formation of a reproductively active monospecific group is not possible. Provided there is sufficient space and suitable nest sites, breeding cormorants may assist in stimulating pelican reproduction in captivity. If competition for breeding space and/or food occurs, however, cormorants tend to dominate pelicans. They have also been recorded predating on young pelican chicks. Although the brown pelican may show signs of aggression if disturbed, it will not leave its nest. The other species are easily disturbed; on one occasion at Jurong, a Dalmatian pelican actually discharged its egg into the water when the island on which it nested was accidentally struck by a boat. The particular sensitivity of this species may partly explain its endangerment. If the parents do abandon the eggs or chicks, the eggs should be artificially incubated and hand raised.

Group size at Melbourne is 14. Best breeding results have occurred with numbers of 10–14. The arrival of a wild male pelican kicked off the first breeding success. This activity stimulated the resident collection of birds, which continued to breed after the wild male left. He still returns sporadically; his arrival, if it coincides with breeding preparation or the early stages of incubation, will usually upset all the birds, who leave nests or re align themselves within the colony as his influence dictates. If the birds are well into incubation, nest fidelity appears to be higher and his influence is lessened somewhat. The colony size and its ability to breed are directly calibrated to the size of the enclosure that has at least one solid side. Out of a total of seven eggs over the past four years at Melbourne, they have had six hatch. This is not much of a sample from which to achieve a representative figure, but one can deduce that, if the trend continues, the fertility rate will indeed be high.

Sexing pelicans by morphometric measurements may be suitable for some species but not for others. The only reliable means of sexing pelicans is by blood or feather pulp DNA work. Surgical sexing of these birds is difficult because of their extensive air-sac system. Captive-bred birds have started breeding at three years of age, and pelicans are known to breed until more than 30 years of age. During the breeding season, color changes have been observed in all four species being bred. The Dalmatian pelican's yellow pouch becomes orange, and the brown pelican's head changes to yellowish. The colors of both great white and pink-backed pelicans also change; the face of the white pelican becomes pinker and the end of its bill turns red, while the pouch and eyelids of the pink-backed change to yellow and its body turns pinker. These changes occur during the mating and incubation periods only. As soon as breeding signs are observed, nesting materials, such as elephant grass and twigs, are provided on the main island to be collected by the birds. The Dalmatian, pink-backed, and great white pelicans have shown a preference for nesting on one particular fiberglass island, which effectively isolates them from human disturbance. Since space is limited to only six nests, the birds will fight to gain this location and refuse to nest elsewhere. The brown pelican alone chooses to nest on the main island, which is nearer to the main road. Based on data for 1988–92, it appears that there is a recognizable breeding season for each species (Breeding Data Table). From this table, it can be seen that nest specifications vary considerably from species to species.

Paul Wexler describes the breeding management for pink-backed pelicans at Longleat Safari Park as follows:

> "Females indicate their readiness to breed by becoming increasingly red on the legs and inside the gular pouch. This normally reaches its reddest about 2–3 weeks before mating occurs. In the nuptial plumage the females also have a longer crest (on the nape), which the males lack. The male's role initially is to collect nesting material and build the bulk of the

Breeding Data of Four Species of Pelicans at Jurong Bird Park

Species	Breeding Season	Nest Measurement (cm)	No. Eggs/Clutch	Fresh Wt. (g)
Brown Pelican	Jan–April, Sept–Dec	L: 60, W: 30, H: 8	3	94.25
Dalmatian Pelican	Except July & August	L: 180, W: 165, H: 220	2	190.45
Great White Pelican	Apr–Aug	L: 135, W: 105, H: 18	2	211.72
Pink-backed Pelican	Mar–April, Oct–Dec	L: 130, W: 100, H: 160	2	133.54

Australian Birdkeeper No 6 (8) 1993

nest. Longleat staff help at this stage by collecting bins full of small twigs and unload them at the side of the pond. The male collects beaksful of these twigs and takes them to the nest. Any twigs that fall into the pond are not retrieved and as a result the males may gather twice as much nesting material as they need. The males perform their head bobbing and bill clapping display at the nest, before mating with the females, also on the nest platforms. Pairs which re-nest usually do so on the same nest platforms as they laid their previous clutches."

Fertility in wild populations appears to be much higher than in captive colonies. In captivity, infertility and egg breakage or disappearances are frequent problems. For example, of 81 eggs laid at Vogelpark Walsrode between 1986 and 1993, only six (7.4%) hatched. Egg breakage and rolling of eggs from nests is common. One of the problems observed was breakage of eggs by neighboring pairs when an incubating bird stood. It has been suggested that starting nests for pelicans in captivity may be a means of spacing nests to reduce egg loss. In the absence of serious reproductive activity, initiation of nest construction by zoo personal is recommended; it can have a stimulating effect and, in any case, is unlikely to depress breeding activity. Active sibling aggression and food monopolization by the oldest and, consequently, largest chick seem to be the major causes of brood reduction. Given the high rate of infertility of captive pelicans and low success of rearing two chicks in one nest, manipulation of eggs by giving one egg from a pair with two fertile eggs to a pair with infertile eggs or giving fertile eggs from an unreliable pair to more reliable pairs with infertile eggs may increase productivity. Pelicans often re-clutch following nesting failure.

Artificial Incubation & Hand Rearing: Records from the Alphen Vogelpark Avifauna reflect that from 1985 through 1992, four adult pairs of eastern white pelicans produced 30 young, 12 of which were hand reared to adulthood. Several losses of eggs were attributed to trampling by the adults or being fought over, and chick losses were caused by drowning when they were unable to climb out of the pond in the outdoor enclosure. Although parent rearing had been attempted, the high mortality of young resulted in the decision to remove all chicks for hand rearing. Initially, only 8% of eggs were fertile, and this was put down to insufficient water in which to mate. Once the water supply was increased, fertile eggs were produced at the end of the season.

The eggs were collected and placed in an incubator set at a temperature of 37.6°C (99.7°F) and a relative humidity of 64.5%. Brown pelican chicks from San Diego Zoo were hatched at 36.9°C (98.4°F) and a relative humidity of 69%. Newly hatched chicks were placed in a bowl lined with paper and put in a rearing box measuring 50 x 50 x 50 cm (20 x 20 x 20 in.), which was maintained at 36°C (96.8°F). The temperature was decreased by 2°C (3.6°F) each week. At birth the chicks were naked and pink, but after 2–3 days the skin turned dark blue-gray. The chicks were fed every two hours between 8:00 a.m. and 10:30 p.m. The diet was prepared in the morning, then refrigerated until needed. It was comprised of mashed smelt, vitamins, and a mineral supplement containing yeast, trace elements, calcium, and phosphorous. Prior to feeding, the food was warmed by placing it in a bowl inside a second bowl containing warm water. For the first week, the mixture was fed orally using a 2-ml syringe. In the second week, a 5-ml syringe was used, and the volume was increased to a 10-ml syringe in the third week. The quantity and texture of food offered is increased daily (see chart); at 14 days, the chicks were offered whole, small smelt with a vitamin and mineral supplement. As soon as the pelicans could eat whole fish, they were placed in a 2 x 2 x 1.8 m (6.5 x 6.5 x 6 ft.) high indoor enclosure with the temperature maintained by a heat lamp. In warm climates an outdoor, partly shaded facility can be used during the day. After a month the flight feathers begin to grow, and at four months feathers replace the gray down. Feathers on the back are the first to turn white. At three months the chicks are moved to a large outdoor enclosure with a grass base and a clean water supply. Single pelicans are highly susceptible to imprinting, and they should therefore be raised in groups of similar-aged birds where natural social interactions with others can occur.

Eastern White Pelican
Pelecanus onocrotalus

Day	Weight (g)	Food/feed (g)	Day	Weight (g)	Food/feed (g)
1	120	6	**9**	415	30
2	127	6	**10**	484	40
3	135	6	**11**	590	60
4	150	8	**12**	650	60
5	172	10	**13**	850	60
6	194	11	**14**	905	
7	319	20	**15**	1,010	
8	376	20	**16**	1,125	

As recorded in the International Zoo Yearbook No 36 (1998)

In an article by Y A. and A. Greichus, and D. J. Call in the *Avicultural Magazine* No 82, they describe the care and feeding of a group of white pelicans *P. erythorhynchos* that were hand raised to test for PCB's such as DDT.

> "The young birds collected from a wild colony were estimated to be 2–5 weeks of age. The young birds were transferred to a wildlife research area where they were held in cages. Heat lamps were used for the first few days for added warmth during the night. Twenty-seven pelicans were chosen for the experiment and placed in nine cages that contained three birds each. The cages were cleaned daily and fresh water was supplied to their swimming tanks as needed. The captive birds were fed black bullheads, carp, bigmouth buffalo and white suckers. Some fish flesh is low in thiamine content and contains a thiaminase enzyme. As a precautionary measure, therefore, a supplement of B vitamins was administered to the birds. This daily supplement contained 7.2 mg of thiamine mononitrate. Whole, not eviscerated, fish were fed to the pelicans. When the chicks were first received they had to be force-fed but within a few days they readily accepted the fish."

During this study it was noted that the skin of the young pelicans was very pale. Blood was drawn from six of the chicks and compared to blood samples from wild chicks. It was found that the captive birds had iron levels that were approximately half that of the wild birds. Approximately 15 mg of ferrous sulphate or ferrous gluconate was administered orally daily. After six weeks, the treatments were stopped and blood iron levels checked. At this stage the captive birds had considerably higher iron levels than the wild birds. Preventative measures against aspergillosis and secondary bacterial complications were taken. Gular pouch hematomas were caused by heavy infestations of the gular pouch louse, Piagetiella peralis. The hematomas gradually disappeared from all birds after the lice were removed with forceps and the infected areas disinfected with 80% ethanol. Two species of roundworms were found, and two birds died early in the trial and were found to be heavily infested with roundworms. Ketamine hydrochloride at a dosage of 7 mg/kg was found to be a very effective sedative when birds had to be handled. In one case, treatment for foot swelling was initiated. The swelling was found to be bumblefoot (caused by staphylococus). The infected bird was isolated in a carpeted cage (I have used artificial turf with flamingos) and treated with two weekly 0.05% copper sulfate footbaths. The swelling diminished and the bird was returned to its group.

Paul Wexler describes the hand rearing of a pink-backed pelican from an egg laid at Longleat Safari Park (note that incubation and hatching parameters used were very similar to those already described):

> "On hatching, the chick the chick weighed 95g (3.35 oz.). The first feed was attempted on the evening of the first day using liquidized sprat. The chick's head was supported with hand, while a spoon was used to dribble the food into the throat pouch. Unfortunately the chick seemed to have difficulty swallowing it and most if it was voided, leaving the chick in danger of suffocating as a result of the liquid food blocking its airways. After gently wiping the inside of the beak clean it was decided to leave it well alone until the following day.
>
> On the second day the chick was fed at two-hourly intervals between 8.00 am and 8.00 pm. After a re-think about the diet, we tried feeding it small, boned carp, about 3 cm (1.25 in.) in length. To prepare the meals the fish were first placed in hot water for a few minutes, which made the flesh soft and easy to extract by cutting off the tail and scraping out the body meat with a knife, leaving behind the skin and bones. With the addition of a little warm water the resulting moist mash could be fed to the chick, again using a spoon. This worked better than the fist attempt and the chick expelled some of the excess water form its mouth by squeezing the gular pouch against its neck (in a similar way that adult pelicans do when they have dredged a beakful of fish and water) before swallowing the food. The chicks weight increased by 10–20% per day. The diet was varied to include chopped pinkie mice and sprats, which were coarsely mashed and mixed with the carp. The young pelican was feeding itself by day eight and was encouraged to do so. A bowl containing a variety of chopped fish and chopped day-old chicks, dusted with Avimix for calcium and vitamins, was held towards the chick at the beginning of each feed and it confidently collected pouchfuls of food without needing too much assistance. Towards the end of the feed however, the food needed to be offered with tweezers, otherwise, over the coarse of the day, not enough food would be taken to produce a good weight gain.
>
> At 14 days old it weighed 80 g (2.8 oz.) and was eating about 600 g (21 oz.) of mixed fish that were given over the coarse of four feeds between 8:00 a.m. and 5:00 p.m. Within a week it weighed 1,950 g (4.3 lb.), although it was being fed only about 50 g more of food per day. At day 27 it weighed 3,000 g (6.6 lb.). At this stage the feeds averaged only 550 g

(1.21 lb.) of food per day and the weight gains had dropped to 2–7%, which were constant until it was 44 days old when the weight had reached 5,100 g (1.2 lb.). Sometimes the chick would eat only a single fish at a time and often feeds would end in the bird going into spasms. This was obviously a worrying time but with a little research it was found that throwing fits after feeding is really quite 'normal' in wild-reared chicks, even if there is no obvious reason. The chick's weight remained constant until day 54 at which time it declined by 300 g (10.6 oz.) over a period of five days."

Ring size: white pelican: 34 mm

Contribution: Paul Stuart-Higgs, Head of Bird Section, Melbourne Zoo.

References:

Brouwer, K., B. Hiddinga & C. E. King. 1994. Management and breeding of pelicans. *International Zoo Yearbook* 33: 24–39.

Greichus, Y.A., A. Greichus, A, and & D. Call. 1976. Care and Growth of Captive White Pelicans. *Avicultural Magazine* 82: 139–142.

Nyunt, K. M., H. Lim & K.M. Ng.1993. Pelican breeding. *Australian Birdkeeper* 6 (8). Ober, S.H. & R. Verkade. 1998. Hand-rearing the eastern or great white pelican at Vogelpark Avifauna, Alphen. *International Zoo Yearbook* 36: 171–173

Wexler, P. 1997. Breeding the Pink-backed Pelican. *Avicultural Magazine* Vol. 103 No.2: 80–84.

20.14 Ciconiidae Storks

20.14a Storks
Ciconiinae

• Michael Mace and Charles Siegel

Introduction: Storks belong to the order Ciconiiformes and the family Ciconiidae, in which they comprise the subfamily Ciconiinae. The shoebill *Balaeniceps rex,* which formerly occupied a family of its own but is now, on the basis of DNA evidence, relegated to the family Pelecanidae, where it is the sole member of the subfamily Balaenicipitinae, has often been associated with the storks, but is therefore discussed adjacent to pelicans. Another species covered in this chapter is the hammerkop *Scopus umbretta,* the sole member of the family Scopidae, which is sometimes considered with the storks. Among the true storks, the genus *Mycteria* contains four species: the wood stork *M. americana* of the Americas; the milky stork *M. cinerea* of Southeast Asia and Sulawesi; the yellow-billed stork *M. ibis* of Africa and Madagascar; and the painted stork *M. leucocephala* of southern and southeastern Asia. These species, along with the Asian openbill *Anastomus oscitans* and the African openbill *A. lamelligerus,* are medium-sized storks characterized by generalized courtship behavior and specialized feeding behavior. All storks within this group feed primarily by tactolocation.

The genus *Ciconia* contains perhaps the most typical storks, including the black stork *C. nigra* from Europe, Africa, and Asia; Abdim's stork *C. abdimii* from sub-Saharan Africa; the maguari stork *C. maguari* from South America; the white stork *C. ciconia* from Europe, Asia, and Africa; the wooly-necked stork *C. episcopus* from Africa, southern Asia, and the Philippines; the oriental stork *C. boyciana* from eastern and southern Asia; and Storm's stork *C. stormi* from Southeast Asia. All members of this genus have straight bills and primarily locate their food visually.

Among the larger storks, the genus *Ephippiorhynchus* contains two species, the black-necked stork *E. asiaticus,* with the nominate subspecies found in southern and southeastern Asia and the subspecies *E. a. australis* found in Australia and New Guinea, and the saddle-billed stork *E. senegalensis* from Africa. Other giant storks include the monotypic jabiru *Jabiru mycteria* and the three species that comprise the genus *Leptoptilos,* the marabou stork *L. crumeniferus* from Africa and two species of southern and southeastern Asian distribution, the lesser adjutant *L. javanicus* and the greater adjutant *L. dubius*. The storks in these genera are large in stature and have massive bills. Five species of storks are considered endangered or vulnerable.

With the exceptions of European white, marabou,

and painted storks, members of this subfamily have not been kept commonly in zoos. This trend has recently begun to change, with increasing numbers of individuals and species being added to various zoological collections. Although ten species have been bred successfully in captivity, only the European white, Abdim's, black, and painted storks are bred with any regularity at more than one institution. Breeding of the other species has been sporadic. Zoos have traditionally displayed storks either in mixed-species groupings or solitary, with little emphasis on their requirements for propagation in captivity. Most species of storks are monogamous, forming pairs that are territorial and breed seasonally. They have elaborate breeding displays such as up-down greeting and bill clattering. Males, particularly those of the larger species, can be very aggressive.

Yellow-billed Stork *Mycteria ibis*

• Michael Mace

Management: The yellow-billed stork *Mycteria ibis* is easily identified among the African storks. They are monomorphic, with males being slightly larger than the females. Their nests in the wild are found in trees such as acacias, baobabs, etc. The size of the nesting colony varies, and other species of birds, such as marabou storks, cormorants, and herons, also nest in the same trees. Multiple pairs have been kept in non-breeding groups in mixed species exhibits that have included mammals and other birds. Yellow-billed storks have been exhibited with saddle-billed storks *Ephippiorhynchus asiaticus,* vulturine guineafowl *Acryllium vulturinum,* various African waterfowl, and West African crowned cranes *Balearica pavonina.* At the San Diego Wild Animal Park, we have been unsuccessful in keeping more than a single pair of yellow-billed storks together in a breeding situation, although pairs have bred in exhibits with other species of birds. The yellow-billed storks have produced offspring in two situations. After trying unsuccessfully to breed this species for many years as a group and as individual pairs, a pinioned pair was placed with a breeding colony of painted storks *M. leucocephala,* in the hope that the social interactions would stimulate them to breed. This manipulation worked in the first season, resulting in the successful production of yellow-billed stork offspring. The following season, two additional pairs of yellow-billed storks were placed in an aviary next to a painted stork colony in the hope of expanding the breeding nucleus. Unfortunately, this strategy proved counterproductive, as the yellow-billed storks were very aggressive toward

Yellow-billed Stork
Mycteria ibis

each other during the breeding season. The management plan currently followed involves keeping a single breeding pair of yellow-billed storks in each aviary, mixed with other species such as Kori bustards *Ardeotis kori,* vulturine guineafowl, or painted storks.

The enclosed aviaries measure 23 x 16 x 5 m (75 x 52 x 16 ft.) high, and are planted with a variety of trees, shrubs, and groundcover. The open enclosures have been of similar size, or even larger. The fresh water source consists of a pool that measures 2 m (6 ft.) in diameter and 30 cm (12 in.) deep, which provides ample room for bathing. Fresh water should be constantly provided as storks have a propensity for taking their food to the pool before consuming it, a habit that contaminates the water supply quickly. Each pool is equipped with an overflow that provides a constant trickle of water.

Diet: The maintenance diet is offered each morning when the facility is serviced. The dietary items offered to each stork include 30 g (1.06 oz.) of Carnivore meat, one whole trout, one whole smelt, and one adult mouse. Additional food is provided when offspring are present, and the elements are proportional to the maintenance diet so that the nutritional requirements are maintained.

Breeding: Multiple nest platforms are provided for the each pair of storks. These platforms are constructed of metal and measure 60 cm (2 ft.) in diameter. The height of the platform varies depending on many factors. If the storks are free flight, the platform is placed as high as the aviary will allow. Careful consideration of the platform height must be taken to allow a pair to copulate on the platform. Enough room must be provided so that the male is unrestricted and does not come in contact with the aviary ceiling when mounting the female, which usually occurs on the nest. If the birds are pinioned, the platform is usually placed about 1 m (3 ft.) from the ground, with a small ramp installed to enable the storks to walk up to the nest site. The platform is made of small diameter bar stock that forms a grid. The grid has gapes that measure 10 cm (4 in.) between the bars. The purpose in the design is to simulate a natural branch pattern that the birds would encounter in the canopy of a tree. Some of these platforms are installed in the vegetation to provide as natural a setting as possible. At the onset of the breeding season, pliable branches are attached to the platform grid to create a foundation. Additional nesting material is placed in the aviary so that the birds can complete the nest to their standards. The typical clutch size is three eggs. Both parents share the incubation responsibilities. The incubation period is 24 days and begins with the first egg laid. The fledgling period ranges from 50–64 days.

With regard to managing painted storks in a colony, we have found that established breeding pairs have sometimes divorced for unknown reasons. In some instances, these proven breeding pairs have broken up and re-paired with other birds at the onset of the next breeding season. In a most interesting case, a female that had already produced offspring with one male re-paired with another male during the latter stages of chick development. She then produced eggs with the second male at another nest site. At the time of this writing, the eggs were being incubated. In both situations, the adult population within the colonies did not change. The causes for this behavior require further investigation.

Ring Size: 20–24 mm

Species Requiring Similar Management:

Painted stork *Mycteria leucocephala*
Milky stork *M. cinerea*
American wood stork *M. americana*
European white stork *Ciconia ciconia*

African Openbill *Anastomus lamelligerus*

• Michael Mace

The African openbill stork *Anastomus l. lamelligerus* is one of the most unique species of birds in the family Ciconiidae. Openbills inhabit aquatic areas such as swamps and floodplains in open country, and range from Senegal to Sudan and Zimbabwe. The most prominent feature of these birds is their extraordinary bill morphology. The mandibles meet at the tip, but there is a considerable gap at the center of the bill. This highly specialized bill is adapted to capturing their diet, which consists primarily of snails and freshwater mussels. Openbills will also consume frogs, crabs, worms, fish, and insects. The prey is manipulated with the tip of the bill. The bill tip is inserted the shell and the operculum is severed, releasing the mollusk from the shell. In studies conducted with the congeneric Asian openbill *A. oscitans,* the birds have been found to produce a narcotic secretion in their saliva that relaxes the snails muscle, making the task of extracting the snail from its shell easier.

Management: The openbills at the San Diego Wild Animal Park have produced in three different styles of aviaries. In each situation, the openbills were mixed with other species of birds. The largest aviary was a public walk-through exhibit that measured 43.3 x 13.7 x 9.1 m (145 x 45 x 32 ft.). This aviary housed 35 species of birds, ranging from ibis to whydahs, without any conflict observed between openbills and other species.

Diet: The captive diet of openbills at San Diego consists of a daily offering for each bird of 256 g (9.03 oz.) Bird of Prey Diet, one small trout, one smelt, and one adult mouse. In addition, fresh water mussels and snails have been offered as a maintenance item, but may not be necessary. However, it is important at this time to determine if the mussels and snails are an important component in the parent rearing of offspring. The exclusion of snails before the breeding season has had no impact on breeding.

Breeding: The first breeding record for this species at a zoological institution occurred at the San Diego Wild Animal Park on June 8, 1994. A second generation African openbill stork was produced in 1996. At the time of conception, the male was 16 months old and the female 14 months old. Both members of this pair were hand-raised individuals. Fertile openbill eggs were also produced in an aviary measuring 12.2 x 8.5 m x 3.7 m (42 x 28 x 16 ft.), which included a small pool and was sparsely planted with grasses as ground cover and a single Moreton Bay fig *Ficus macrophylla.* In this scenario, the small tree was unable to support a nest, so a platform was provided. The two larger aviaries were heavily planted with mature trees and an understory. In these enclosures the openbills nested in paper thorn *Acacia siebriana woodii,* Moreton Bay fig, and a fish-tailed palm *Caryota urens*. These nests varied in height from 3–7 m (10–23 ft.) above ground. Each enclosure contained a water feature or pool so that the birds could spend time probing and wading while searching for prey items.

Nest platforms measured 1 m (3.3 ft.) in diameter, and the bar stock is a grid not greater than 2 cm (0.8 in.) apart. The platform is designed to simulate the branches found in a tree canopy. In all cases, openbills constructed their own nests. As arboreal nesters, they built either in trees, as noted, or on the platforms provided. The nests are constructed of sticks or twigs and lined with finer material. Additional nesting material was scattered on the ground, and the birds would augment the nest when necessary. Openbills have readily bred in large and small aviaries in a season that extended from March to September. The normal clutch size is 2–5 eggs, which are laid every other day. Incubation is 25–30 days, and the chicks fledge at 50–55 days of age.

Historically adult openbills, for inconclusive reasons, have been largely ineffective in rearing their own young in captivity; to date, only one parent-reared chick has fledged. The openbills have displayed a consistent ability to build nests and incubate their eggs to full term, but have proved ineffective in rearing the chicks after they have hatched. This lack of success may be attributed to the specialized dietary requirements of the species when feeding young. Additional experimentation will be conducted in this area until a better understanding of the dietary requirements of parent-reared chicks results in more consistent success.

Artificial Incubation & Hand rearing: Due to the inability of parent birds to rear their own chicks in captivity under current management practices (97.7% failure rate), eggs and chicks were removed from nests at both facilities of the Zoological Society of San Diego for artificial incubation and hand rearing. A forced-air incubator was used, and the incubation parameters were calibrated at a temperature of 37.2°C (99°F) and a wet bulb reading of 60–64.5%. Eggs lost the desired 10–14% of their fresh weight during incubation, which was consistent with the target weight loss. They were mechanically rotated every hour; in addition, they were manually turned 180 degrees twice per day. The pip-to-hatch interval averaged 24 hours. The physical condition of each chick was assessed at hatching and the yolk sac seal treated with povidone iodine 3.5%.

After drying in the hatcher, chicks were housed in electronically controlled acrylic brooders at 35.6°C (96°F) with water pans added for humidity for one or two weeks. By two weeks of age they were transferred to wooden box brooders with the temperature set at 31.1°C (88°F). The rearing diet was based largely on prey items such as freshwater snails and mussels, neonate mice, fish, and meat, with electrolytes, vitamins, and minerals added as supplements.

Dosage: Viokase (pancreatic enzyme concentrate). Mix 1 g (0.035 oz.) per 100 g (3.5 oz.) food. Store in refrigerator all day; leave at room temperature 15–20 minutes before feeding. Apetate (high potency vitamins B1, B6, B12 for children). Use 1 cc per 50 g food fed.

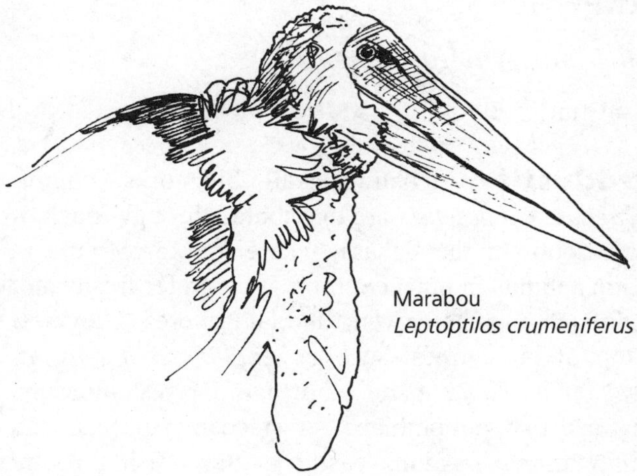

Marabou
Leptoptilos crumeniferus

Calcium carbonate and dicalcium phosphate. Use 1% of amount fed previous day. Calphos Vital E injection (to prevent metabolic bone disease.)

Between 1994 and 1999 the parent birds at the combined San Diego institutions hatched 55 chicks from a total of 59 eggs incubated, for a 93% hatch rate. Three intact fertile eggs were retrieved with late dead embryos. The parents reared only one chick out of 47 that were left with them (2%). Nine chicks were pulled for hand rearing, of which three survived (33%). Of 33 eggs that were artificially incubated, 28 chicks hatched (85%), of which 13 survived (46%). Four embryos died in the early stages of development and one was a late dead embryo.

Hand rearing was attempted with a total of 37 chicks, nine parent-hatched and 28 incubator hatched. Thirteen of the 17 surviving chicks (76%) were artificially hatched. Chicks experienced problems with regurgitation, dehydration, metabolic bone disease, and vitamin A imbalances. Regurgitation of predigested food items was viewed as an abnormality, which could result in dehydration problems. It was not always obvious in analyzing the data how much of the regurgitation was of undigested food items and how much was in the form of a pellet being cast in the normal process. It was common for seemingly healthy chicks to regurgitate undigested food. In some cases this food was re-consumed or even eaten by another chick housed in close proximity. In 1995, pathological evidence of excessive bone remodeling as seen in metabolic bone disease was noted, despite the presence of adequate calcium, phosphorous, dietary vitamin D, and UV-light exposure therapy. It is suspected that this problem was caused by excessive vitamin A in the diet due to over-supplementation. This supplement was subsequently removed from the diet. Research to perfect the rearing techniques for this species is ongoing.

Saddle-bill Stork

Ephippiorhynchus senegalensis

• Charles Siegel

Introduction: In nature, saddle-bill storks *Ephippiorhynchus senegalensis* are distributed through much of Africa south of the Sahara, where they are found in aquatic habitats in open country. They prefer freshwater floodplains, marshes, rivers, and lake shores. They nest on top of large trees, such as *Euphorbia ingens, E. dawei,* and *Acacia giraffe,* but avoid forest interiors. They eat fish, amphibians, crustaceans, insects, and small mammals. Adults easily swallow fish that are 15–30 centimeters in length and weigh 500 grams. Pairs are highly territorial. Adult males have brown eyes, and adult females have golden-yellow eyes. Adults do not vocalize, but do clatter their bills.

Management: Saddle-bill storks are large and strikingly marked and they make attractive birds for public exhibit. Most zoos usually pinion them and display them in pairs as added ornaments to hoofed mammal exhibits. This arrangement rarely provides saddle-bills with the opportunity to express their full behavioral repertoire. It excludes the possibility of flight, and interferes with breeding behavior. Zoos would best serve saddle-bill storks by keeping them fully flighted in their own enclosed exhibit. A suitable compromise would be to display them fully winged, but feather-clipped, in an open exhibit with small ducks and small herbivores such as Suni antelope *Neotragus moschatus*.

As of this writing, only five institutions, the Dallas Zoo, the San Diego Wild Animal Park, World of Birds in Cape Town, Audubon Park Zoo, and the Toledo Zoo have successfully reproduced the saddle-bill stork in captivity, with both hand-reared and parent-reared offspring. Zoo personnel attribute their success to an enclosed, off-exhibit breeding area and a fully flighted male. During copulation, the female stands on the nest. Then the long-legged male stands on the female and must flap his huge wings to stay balanced and achieve cloacal contact. Perhaps pinioned males are unable to balance well enough to mate successfully. Only time will tell.

Saddle-billed Stork
Ephippiorhynchus senegalensis

An exhibit for a pair of saddle-bills should have an outdoor area of at least 130 m² (1200 sq. ft.). Two-thirds of the area should have a soft substrate, such as short lawn. The outer perimeter should be planted to bushes or bamboo. The remainder should consist of a shallow pool, 20–40 cm (8–16 in.) deep, of fresh water exposed to direct sunlight. The birds will regularly bathe, drink, dip food, and display in the water. They also carry water to the nest and regurgitate it for both eggs and chicks. This behavior has also been recorded in the congeneric black-necked stork *E. asiaticus*. The outdoor exhibit should lead to two small outdoor holding pens about 3 x 3 m (10 x 10 ft.), one for each bird, which in turn lead to two small indoor holding pens of around 3 x 4 m (10 x 13 ft.). These extra pens are important when managing an injured or ill bird, and when sheltering the birds from bad weather. The birds should be habituated to walk inside the shelter on a daily basis, especially if nocturnal predators are a concern. More than one saddle-bill should never be locked together in a small shelter, as even members of a compatible pair may seriously injure each other in a confined space.

If a pool of cool water is present, saddle-bills can readily tolerate temperatures as high as 40.5°C (105°F). At the other extreme, their feet, legs and facial skin are very susceptible to frostbite, and it is therefore highly recommended that these birds be sheltered indoors when the wind-chill factor drops below –3.9°C (25°F). Natural sunlight is very important to the well-being of saddle-bills, especially during the growth and development of offspring. Parents stand in the nest several hours each day, and the chicks are exposed to a great deal of sunlight. This is crucial for the production of Vitamin D3, a molecule important in calcium metabolism. Deficiencies in this molecule result in metabolic bone disease, or rickets. Therefore, it is very important to expose chicks to a minimum of 10 minutes direct sunlight per day, and to give them adequate vitamin and mineral supplementation.

Diet: Captive diets for saddle-bill storks should minimally include a selection of fish, small rodents, large insects, and a vitamin and mineral supplement. The diet for adults with offspring should initially be modified to minimize indigestible material such as fur, feathers, chitin, and large bones. Offspring cannot cast until they are 30 days old, and proventricular impaction might otherwise result.

Vitamin/mineral supplementation. The amounts suggested below were established for nestlings in a covered area. In the Dallas Zoo's nest shelter, light could penetrate, but little ultraviolet light reached the nest. The Dallas Zoo selected RepCal* because it contains vitamin D and calcium, but no phosphorous. There are several companies with similar products. Di-Cal* tablets may also be supplemented if additional calcium is deemed necessary. The Dallas Zoo uses three tablets for chicks up to 1 kg, two tablets for chicks weighing 1–3 kg (2.2–6.6 lb.), and one tablet from three kg (6.6 lb.) to fledging.

Following is the daily ration used at the Dallas Zoo. The average weights of the various dietary items are as follows: whole mouse, 23 grams; skinned mouse, 19 grams; *Gallus gallus* whole chicks, 37 grams; skinned chicks, 34 grams. The birds are fed twice per day, with mice, chicks and Crown Carnivore Diet (ground meat) Bird of Prey Diet offered in the morning, and fish and insects in the afternoon.

Diet for one adult during non-breeding season (January 1–July 31).

- 3 mice
- 1 chick
- 4 oz. Bird of Prey diet
- 10 smelt three times each week
- 1 medium trout 226–340 g. (0.5 to 0.75 lb.)
- 2 SeaTabs inserted into the trout
- 15 giant mealworms or 30 live crickets
- ½ tablespoon Ca/Vionate in non-breeding ratios

Diet for one adult during breeding season (August 1–December 31): As above, but trout increased to two tablespoons Ca/Vionate in breeding ratios.

Diet for one adult during brooding of offspring: The adult brooding diet, along with offspring's portions, is divided into three equal portions fed out at 7:30 a.m., 11:30 a.m., and 3:30 p.m. The last bowl is left overnight. Mice, chicks and fish are most easily skinned or crushed while partly frozen. Parents eat their offspring's diet and regurgitate it to the chicks for the first 30–45 days, then begin carrying food items to the nest for the young. By the age of 60–80 days the young can feed from a bowl if they have left the nest. RepCal* is mixed with Vionate* in a 2:1 ratio.

Age of offspring 1–20 days:

- 3 skinned and crushed mice
- 1 skinned and crushed chick
- 4 oz. Crown Carnivore diet
- 10 crushed smelt three times each week
- 2 lb. crushed capelin
- 1 tsp. RepCal/Vionate mix
- 1½ crushed SeaTabs

Age of offspring 21–22 days:

- 2 skinned and crushed mice
- 1 cut mouse with pelt (3–4 pieces)
- 1 skinned and cut chick
- 4 oz. Crown Carnivore diet
- 10 whole smelt three times each week
- 2 lb. small capelin
- 2 SeaTabs inserted into the capelin
- 1 tsp. RepCal/Vionate mix

Age of offspring 23–24 days:

- 2 skinned and cut mice
- 1 cut mouse with pelt
- 1 cut chick with pelt
- 4 oz. Crown Carnivore diet
- 10 whole smelt three times each week
- 2 lb. small whole capelin
- 1½ SeaTabs inserted into capelin
- 1 tsp. RepCal/Vionate mix

Age of offspring 25–26 days:

- 1 skinned and cut mouse
- 2 cut mice with pelts
- 1 cut chick with pelt
- 4 oz. Crown Carnivore diet
- 10 whole smelt three times per week
- 2 lb. small capelin
- 1½ SeaTabs inserted into capelin
- 1 tsp. RepCal/Vionate mix

Age of offspring 27–29 days:

- 3 cut mice with pelts
- 1 cut chick with pelt
- 4 oz. Crown Carnivore diet
- 10 whole smelt three times each week

- 2 lb. small capelin
- 1½ SeaTabs inserted into capelin
- 1 tsp. RepCal/Vionate mix

Age of offspring 30–100 days:

- 3 whole mice
- 1 whole chick
- 4 oz. Crown Carnivore diet
- 10 whole smelt three times each week
- 2 lb. small capelin
- 1½ SeaTabs inserted into capelin
- 1 tsp. RepCal/Vionate mix

Diet for one (1) parent-reared chick: Metabolic bone disease is a serious concern for this species. It is very important, therefore, that each chick receives its specific vitamin amounts. The adults may dip the food in water before swallowing, rinsing off the vitamin powder. If this occurs, it may be necessary to hand feed the vitamins by placing them in food items given directly to the chicks. Dical-D* tablets may also be used if an additional calcium supplement is deemed necessary, at the same dosage used for hand-reared chicks.

The following are daily food amounts added to the adults' food bowl, which is then divided for multiple feedings.

Age of offspring 1–7 days:

- 4 mice with heads, feet, and tails removed, skinned and crushed
- 2 skinned and crushed chicks with heads and feet removed
- 2 oz. Crown Carnivore diet
- 12 oz crushed capelin with head and fins removed
- ½ crushed SeaTab mixed with the fish
- ¾ tsp. RepCal/Vionate mix

Age of offspring 8–14 days:

- 6 mice with heads, feet, and tails removed, skinned and crushed
- 2 skinned and crushed chicks with heads and feet removed
- 4 oz. Crown Carnivore diet
- 2 oz. crushed smelt
- 12 oz. crushed capelin with head and fins removed
- ½ crushed SeaTab mixed with fish
- ¾ tsp. RepCal/Vionate mix

Age of offspring 15–21 days:

- 10 mice with heads, feet, and tails removed, skinned and crushed
- 2 skinned and crushed chicks with heads and feet removed
- 8 oz. Crown Carnivore diet
- 4 oz. crushed smelt
- 16 oz. crushed capelin with heads and fins removed (heads left on from day 19)
- ½ crushed SeaTab mixed with fish
- ¾ tsp. RepCal/Vionate mix

  Note: Feed 2 mouse pelt directly to each chick on days 18–21

Age of offspring 22–28 days:

- 10 mice, skinned and crushed
- 2 skinned and crushed chicks
- 8 oz. Crown Carnivore diet
- 4 oz. whole smelt
- 16 oz. whole but crushed capelin
- ½ crushed SeaTab mixed with the fish
- 1½ tsp. RepCal/Vionate mix

  Note: Feed 1 mouse pelt directly to each chick on days 22–34; 12 per chick days 25–27; 2 per chick day 28.

Age of offspring 29–35 days:

- 5 skinned whole mice
- 5 whole mice with pelts
- 2 whole chicks with pelts
- 8 oz. Crown Carnivore diet
- 4 oz. whole smelt
- 16 oz. whole capelin
- ½ SeaTab inserted into capelin
- 1½ tsp. RepCal/Vionate mix
- 5 giant mealworms or 8 large crickets

Age of offspring 36–67 days:

- 10 whole mice
- 1 whole chick
- 6 oz. Crown Carnivore diet
- 4 oz. whole smelt
- 12 oz. whole capelin
- ½ SeaTab inserted into capelin
- 1½ tsp. RepCal/Vionate mix
- 5 giant mealworms or 8 large crickets

Age of offspring 68–100 days:

- 6 whole mice
- 1 whole chick
- 4 oz. Crown Carnivore diet
- 4 oz. whole smelt
- 12 oz. whole capelin
- ½ SeaTab inserted into capelin
- 1½ tsp. RepCal/Vionate mix
- 10 giant mealworms or 20 large crickets

Breeding: The nest is constructed by both male and female. A sturdy man-made nesting platform can aid the nesting effort. This platform should be elevated at least 1.5 m (5 ft.). The dimensions should be approximately 1.3 x 1.3 m (4 x 4 ft.). A sturdy staircase will help pinioned or feather-clipped birds reach the nest platform from ground-level, and will make it easier for the keeper to inspect eggs or chicks. Ceilings should be at least 3 m (10 ft.) above the nest platform. This will provide enough space so that the male will not strike it with his wings while copulating. The birds will continue to add nesting material even after chicks hatch, and an overzealous bird may accidentally cover a chick with grasses or hay. In North American zoos, nest building typically begins in September.

The parents demonstrate incubation behavior several days before the eggs are laid, sitting tightly on the empty nest. Adults will defend eggs and chicks, but can be pushed from the nest with a broom for inspections. The clutch size ranges from 1–5 eggs, laid at intervals of 36–48 hours. The eggs are oval and dull white, and each egg can be marked with a wax pencil for easy identification. Incubation, by both parents, lasts 31–32 days. Parents may regurgitate water on top of the eggs in warm weather. Chicks hatch asynchronously. Parents sometimes gently place one foot on a hatching egg, perhaps to feel vibrations from the hatching chick. A healthy chick will vocalize strongly while still encased in the shell.

If faced with limitations of space or personnel, the production of offspring can be regulated by adding two artificial eggs to the nest when the third real egg is laid. This results in the female realizing an apparent full clutch size of five eggs in the nest, and she will stop laying more eggs. On the other hand, to increase production of eggs, remove each egg from the nest as soon as it is laid. At the Dallas Zoo, one female produced nine eggs when each new egg was removed from the nest shortly after it was laid. In American zoos, egg laying typically begins in the last half of October.

Newly hatched chicks are covered with white down. A few days after hatching, the parents may spend several hours standing in the nest rather than brooding the chick. This behavior was observed at the Dallas Zoo, even at temperatures as low as 7.21°C (45.1°F), with no harmful effects to the chick; it is not known if this behavior would continue at even lower temperatures and become problematic. The parents take turns feeding the chick by regurgitating food close to it on the nest floor. The chick instinctively pecks at the food, and initially misses the food more often than making contact. The parent begins reingesting the food almost as soon as it is regurgitated. As the chick's coordination improves, it may eat enough to increase its body weight by 20–50% per day for the first 10 days.

Within 45 days after hatching, the parents will carry food items to the nest. By age 60–80 days, a captive-hatched chick may leave its nest and feed directly from the food bowl shared with its parents. In the wild, chicks may not fledge until they are 100 days old. At the Dallas Zoo, parents allowed their chick to feed before they themselves ate. This arrangement was still in place when the chick was one year old and the parents were incubating the next season's clutch of eggs.

Artificial Incubation: If eggs are removed for artificial incubation, they should be set at a dry bulb incubator temperature of 37.21°C (99.01°F). The wet bulb reading should be 31.11°C (88.01°F), providing a relative humidity of about 65%. The humidity level should be adjusted as necessary to achieve a 15% weight loss by the end of the 31–32-day incubation period. The eggs should be turned every hour. After the internal pip, when the chick punctures the air cell membrane, turning is no longer necessary and the humidity should be increased to saturation level in preparation for external pipping of the shell. Chicks emerge 18–48 hours after external pipping of the shell. At hatching, the humidity should be dropped, and the chick allowed to dry for 8–24 hours. Chicks should be completely dry and the yolk sac mostly absorbed before being placed in a brooder.

Hand Rearing: For the first week after hatching, gradually decrease the temperature of the brooder box from 35.1°C (95.1°F) to 32.1°C (88.1°F). As the chick gets older it will need less heat. Hand rearing is not recommended for these birds unless absolutely necessary. If used, every effort should be made to prevent imprinting on humans. Hand rearing was accomplished at the Dallas Zoo, where caretakers used a life-like puppet head to feed and preen the chick. Caretakers also wore a black-and-white fabric costume, complete with hood, to obscure the human form and loosely mimic an adult saddle-bill stork. No talking was allowed, although the puppet's bill should be clattered. If a puppet is not available, the chick can be fed and weighed from behind a blind. A mirror should be placed with a lone chick. It is important that the legs stay beneath the chick, and a small bowl padded with towels serves this purpose. Small sticks and hay can also be used to line the bowl.

Mean Hand-reared Weights

Day	Weight (g) Openbill	Weight (g) Marabou	Day	Weight (g) Openbill
1	40.6	90	32	754
2	40.18	93	34	824
4	44.56	115	36	838
6	57.2	132	38	905
10	86	260	42	1048
12	114	400	44	1105
14	150	620	46	1155
16	192	900	48	1182
18	240	1150	50	1205
20	296	1400	52	1212
22	368	1700	54	1217
24	455	2000	56	1310
26	542	2250	58	1360
28	614	2500	60	1356
30	682			

Openbill - *Anastomus oscitans*
Marabou - *Leptoptilos crumeniferus*

Some Longevity Records for Storks

Species	Age (yrs.)	Species	Age (yrs.)
Yellow-billed	19	White	35
Painted	28	Saddle-bill	30
Asian Open-billed	18	Marabou	44
Woolly-necked	30	Shoebill	35

It is essential that chicks be weighed each morning before the first feeding, as this weight provides the basis for the amount of food provided each day. All food items should be dipped in distilled water before being offered, to provide fluids for the chicks. Chicks can begin drinking from a shallow bowl when three weeks old. The first moistened food items should be rolled in and coated with the vitamin and mineral powder prior to being fed. Remaining food items are offered in a small bowl, and the chick can be stimulated to eat by tapping the food with tweezers. The crop should empty between feedings.

Diet for one (1) hand-reared saddlebill stork:

- Day 1: No food.
- Day 2: Offer 12.5% of total body weight for day, divided into five feedings, each consisting of 2.5% of the body weight. The hand-feeding diet at this stage should be three-quarters mouse, by weight, and one-quarter capelin. The mice are skinned and pulverized, with the bones crushed. The capelin should have the heads, fins, and tails removed. This diet is supplemented with 3 tsp. RepCal/Vionate mix added to the first, middle, and last feedings (: tsp. total for day), and three crushed SeaTab at the first and last feedings of the day.
- Day 3: Diet as outlined above, but increased to 17.5% of body weight total for day. Five feedings, 3.5% of body weight per feed.
- Day 4–7: Diet as above, but offer 25% of body weight total for day, given in five feedings of 5% of body weight per feeding.
- Day 8–13: As above, but change to one-eighth SeaTab per feeding, crushed.
- Day 14–20: As above, but mice chopped instead of pulverized. Insert two SeaTab into capelin at one feeding only.
- Day 21–29: Offer 20% of body weight total for day, reduced to four feedings of 5% of body weight per feeding. Proportions of mice and capelins as above, but increase RepCal/Vionate mix to 2 tsp. at first three feedings (total 12 tsp./day). Insert two SeaTab into capelin at one feeding only. Start adding mouse fur as casting material, and begin slowly increasing size of food pieces.

- Day 30–Day 39: Diet as above. Offer 20% of body weight total for day in three feedings of 7% of body weight per feeding. Increase amount of mouse fur added as casting material, and continue to slowly increase the size of food pieces.

- Day 40–Day 49: Offer 20% of body weight total for day divided into two feedings of 10% of body weight per feeding. The diet should now consist of the following:

    - ¼ (by weight) whole mice, offered in morning
    - ¼ (by weight) Carnivore diet, offered in morning
    - ¼ (by weight) whole capelin, offered in afternoon
    - ¼ (by weight) crushed chicks, offered in afternoon
    - ¾ tsp. RepCal/Vionate mix added to each feeding (total 12 tsp./day)
    - ½ SeaTab in capelin at one feeding only

- Day 50–Day 80: Offer 20% of body weight total for day. Two feedings, 10% of body weight per feeding. Diet as above, but begin offering 10 live mealworms and 10 live crickets with the rear legs removed, as forage.

- Day 81–Day 100: Provide adult diet and monitor intake. Adult saddle-bills weigh 4.5–6.35 kg (10–14 lb.) and consume 10–12% of their body weight per day.

Products mentioned in text:

RepCal, RepCal Research Labs, Los Gatos, CA 95031, www.repcal.com

Dical-D, Abbott Laboratories, North Chicago, IL 60064, Ca:P 1.3:1, Vitamin D 120 IU

Vionate (vitamin-mineral powder), Gimborn Pet Specialties, LLC Atlanta, GA

Crown Carnivore Diet, Dallas Crown Inc., Kaufman, TX

SeaTabs B.T.F.S., Pacific Research Laboratories, Inc., El Cajon, CA (619) 441-5840

Species Requiring Similar Management:

Black-necked stork *Ephippiorhynchus asiaticus*
Jabiru Jabiru *mycteria*
Woolly-necked stork *Ciconia episcopus*

References:

Brouwer, K., M.L. Jones, C.E. King & H. Schifter. 1992. Longevity and breeding records of storks (Ciconiidae) in captivity. *International Zoo Yearbook* 31: 131–139.

Mace, M. & P. Witman. 2000. Propagation of the African open-bill stork *Anastomas lamelligerus lamelligerus* at the Zoological Society of San Diego. *Der Zoologische Garten* N.F. 70 (5): 285–294.

20.15 Spheniscidae Penguins

Glen Holland

Introduction: Penguins belong to the order Ciconiiformes and the family Spheniscidae, with 17 species in six genera. They are primarily birds of the Southern Hemisphere, with species occurring from Antarctica and the sub-Antarctic islands to the slightly warmer waters of South Africa, southern South America, Tasmania and New Zealand. The exception is the Galapagos penguin *Spheniscus mendiculus*, which is endemic to the warmer waters of the Galapagos Islands. Penguins vary in size from the large emperor *Aptenodytes forsteri* and king penguins *A. patagonica* to the minute little blue penguin *Eudyptula minor*. All penguins species are flightless; their wings are modified for swimming, and do not fold. They have short necks and legs, spines on their tongues and palates for holding fish, and feet with three webbed front toes webbed and a rudimentary hind toe. The sexes are alike, and all species are monogamous, colonial nesters. Being flightless, they nest on the ground, laying one or two white eggs in shallow burrows lined with stones or grass. Penguin chicks are altricial, and receive parental care by both sexes. The 11 species that are featured in zoological/avicultural collections are the king, the emperor, the gentoo *Pygoscelis papua,* the Adelie *P. adeliae,* the chinstrap *P. antarctica,* the macaroni *Eudyptes chrysolophus,* the rockhopper *E. chrysocome,* the little blue, the Magellanic *Spheniscus magellanicus,* the Humboldt *S. humboldti,* and the African, or jackass *S. demersus*. Generally, the species held in captivity are not currently threatened in the wild, but as husbandry techniques improve and breeding successes increase, the knowledge can be applied to the endangered species of penguins.

Management: Much of the information in this chapter was obtained with kind permission from the AZA *Penguin Husbandry Manual*. Penguins are gregarious and best kept in groups; exhibits should therefore be designed to accommodate a minimum of six birds. Antarctic-dwelling penguins require a constant, cool temperature maintained with sophisticated air and water filtering systems. Failure of the cooling system compromises penguins survival, and a back-up system is essential.

Optimal air and water temperatures in indoor exhibits are outlined in the Penguin Husbandry Manual as follows:

Optimal Air & Water Temperatures

N Species	Air Temperature	Water Temperature
King, Gentoo, Chinstrap Macaroni & Rockhoppper	0–11°C (32–52°F)	4–13°C (39–55°F)
Adelie & Emperor	−7 to −1°C (19–30°F)	4–7°C (39–45°F)
Spheniscus spp. & Little Blue	3–29°F (37–72°F)	0–29°C (32–85°F)

Penguins tend to ingest odd items such as pieces of wire, brush bristles, etc., and after any maintenance work is performed, the enclosure should be carefully screened for such items. The enclosure pool should allow sufficient space for all the birds to be present in the pool at the same time. There should be sufficient land space to accommodate all the birds in the enclosure while allowing for territorial disputes. An isolation area that allows visual and vocal contact, such as a perspex screened area, is useful for forced pairings, the introduction of new birds, the hand rearing of chicks, or the separation of birds with non-contagious health problems.

Adelie Penguin
Pygoscelis adeliae

Species belonging to the genus *Spheniscus* can be housed either indoors or outdoors, provided the ambient temperature does not rise above 29.4°C (85°F). No penguins tolerate humid conditions, which cause a rise in the incidence of *Aspergillus* and malarial infections. Outdoor exhibits should allow for good air flow, but also provide shade for warm days and shelter from freezing winds in the winter months. Open water must always be available, and pools should not be allowed to freeze. Drainage systems are essential in the enclosure, but should be separate to avoid pool contamination from run-off of exhibit maintenance. All floor areas consisting of substrate such as concrete, astroturf, fiberglass, ice, rock and pea gravel, sand, etc. should be relatively smooth and slope toward the drain. Concrete has been associated with foot problems and should be covered with a smooth lacquer coat to reduce abrasiveness and fill in the small pores where bacterial colonies can become established. A broad-spectrum, non-toxic, disinfectant and fungicide should be used daily to clean penguin exhibits. Behavioral enrichment is important for these birds, and the addition of brightly-colored rubber balls, blocks of frozen fish in the water, varied feedings, etc. will improve their welfare.

The water in a penguin exhibit pool should be clear and of good color, with a low bacterial count. Power filtration systems are required for most species, and should be combined with adequate surface water skimming to remove any excess oil that builds up. Turbidity can be reduced by proper placement of water jets directed toward the pool bottom. Airflow is another important factor in indoor exhibits. An air turnover rate of 15 changes per hour has been recommended for laboratory animals, but this needs to be balanced with the following considerations: (a) filter efficiency and airflow or ventilation; and (b) fresh air exchange and temperature. Lighting that allows variation of the day length and intensity is reported to enhance reproductive success. The length of daylight should approximate that of the latitude from which the birds originate.

Cable ties placed around flippers, and metal or plastic leg bands can be used to identify individual birds, and may be color coded. Molt often necessitates replacing the flipper band with a larger size to accommodate swelling. The penguin TAG also recommends that transponders, placed in the loose skin of the back of the neck, also be used with these birds. Penguins are relatively hardy and can tolerate routine handling. The initial restraint is usually accomplished by grabbing the neck, followed by lifting the bird from behind. As a penguin's flippers are easily broken, they should never be grasped when restraining a bird. Once the bird has been properly grasped, it can easily be held between one handler's legs, with the body and flippers restrained. The head and neck are held by hand, allowing a second person to undertake work on the bird. Two people should always be employed when handling penguins, particularly *Aptenodytes* species, to prevent potential injury to the bird or the handler. Penguins should never be packed or crowded when moved or shipped, as they are likely to over-heat.

If the ambient temperature is above 4°C (40°F), all penguin species, except those in the genus *Spheniscus,* should be transported in a refrigerated truck. On a cool day, with a temperature of 0–21°C (32–70°F), *Spheniscus* species can be moved in plastic pet carrying cases, which should be divided to allow each bird standing space in its own compartment. Penguins should not be transported within the period commencing one month before and ending one month after the annual molt. Humboldt penguins should not be transported for two months after molting. Penguins usually molt after breeding, when the day length begins to shorten. Some species however, such as the African, molt and breed over most of the year. Conditions such as light cycles, temperature, and diet should be coordinated as closely as possible between the institutions shipping and receiving these birds. While being moved, the birds should have a constant source of light and a firm foot substrate, such as small rocks covered by ice, to prevent injuries caused by slipping.

Fertile eggs should be transported either when fresh or during the last one-third of the incubation period. Fresh eggs can be shipped cold, but partially incubated eggs must be kept warm. For shipment over short distances (taking up to five hours) a well-insulated cooler with a mounted temperature probe, furnished with a hot water bottle as a heat source, can be used. The temperature should be approximately 35.6°C (96°F).

Any introduction of adult or juvenile penguins into an existing group should be carefully monitored, with constant observations for at least a week after the birds are placed together. New birds can either be introduced gradually, beginning with a few hours at a time, then increasing the length of time with the existing group each day over a period of days, or by removing a few of the existing group and mixing them with the new birds in a separate enclosure prior to introduction into the main group. If such a small, separate enclosure allows visual and vocal contact between the old and new birds, it can also prove a successful means of gradual introduction.

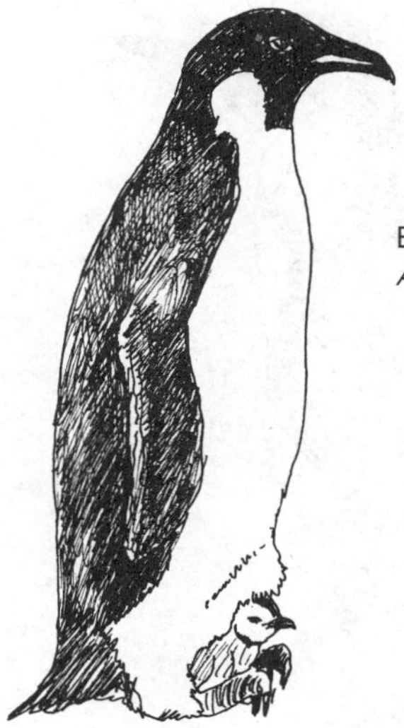

Emperor Penguin
Aptenodytes forsteri

Accurate records of all birds are best maintained on the SPARKS program. Weights are important for monitoring the health of individual animals and should be recorded on a regular basis whenever possible. Average weights (+10%) are as follows:

Average Penguin Weight

species	Avg. Wt. (Kg)	Species	Avg. Wt. (Kg)	Species	Avg. Wt. (Kg)
Emperor	26	**King**	15	**Gentoo**	5.4
Macaroni	4.5	**Adelie**	4.2	**Humboldt**	4.2
Chinstrap	3.7	**African**	3.2	**Little Blue**	1.1

The most commonly reported behavioral manifestations of illness are inappetence, lethargy, and isolation from the group. These symptoms, in combination with other indicators such as respiratory difficulty, weight loss, plumage condition, coloration of the mucous membranes, fecal appearance, etc., can lead to early diagnosis. Common disease problems are aspergillosis, malaria, and bumblefoot.

Diet: Prior to molt and breeding, there is a significant increase in appetite and weight gain in female penguins. African penguins have shown a pre-molt increase of 31% and a subsequent loss of 41% of their peak weight during molt. Breeding females increase their weight by as much as 20–25%. The penguins at Edinburgh are fed three times a day, with fish being hand fed to each bird. The gentoos are fed either fresh fish or thawed whiting, and the macaronis are fed thawed sprats.

It is important to consider a number of factors when developing a feeding regime for penguins. Given the perishable nature of fish and seafood, food handling pro-

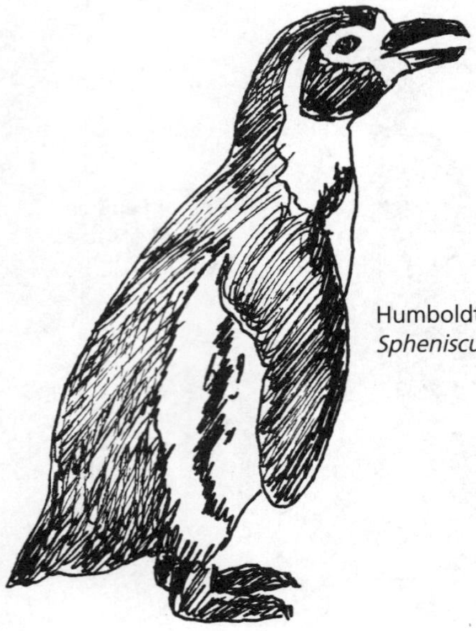

Humboldt Penguin
Spheniscus humboldti

cedures are critical. Fish should be kept frozen at a temperature of –2°C (–10°F) or lower, and rotated on a first in, last out basis. It should be thawed gradually in a refrigerator at 7°C (45°F) and fed soon after thawing. Penguins should be fed 2–3 times per day, and given only the amount that they will consume immediately. This will normally equal approximately 2–3% of their body weight for the two larger species, and 10–14% for the smaller ones. Thiamine and vitamin E are affected by storage, and require supplementation in the diet. Thiamine should be supplemented at 25–30 mg/kg and vitamin E at 0.1 IU/g of the diet on a wet basis. These can be hidden in the fish in pill form. If penguins are housed in fresh water, the supplementary salt at a rate of approximately 250 mg per bird per day is a consideration for the development and maintenance of the nasal glands.

Edinburgh Zoo Quantities Fed Per Bird Per Day

Species	Average (g)	Maximum (g)	Maintenance Supplement	Breeding Supplement
King	1000	1300	2 fish-eater	2 fish eater 100 IU vit. E 0.3 ml cod liver oil
Gentoo	790	1200	1 fish-eater	1 fish-eater
Macaroni	250	500	1 fish-eater 100 mg vit B1	1 fish-eater 100 mg vit. B1 0.3 ml cod liver oil

It is desirable to use a variety of whole fish, such as capelin, mackerel, herring, smelt, and trout. Anchovies should be avoided, as problems have resulted due to the high concentration of demoic acid they sometimes contain. A good variety of fish offers a greater nutritional range, and only thiamine and vitamin E will need to be added.

The diet should be supplemented as follows:

- Multi-vitamin/mineral supplement per animal/day:
  1) <15 kg: B3 tablet
  2) 15–35 kg B2 tablet
  3) >35 kg B1 tablet
- Sodium chloride:
  1) <5 kg: B3 tablet every other day
  2) 5–15 kg: B3 tablet every day
  3) 15–35 kg: B2 tablet every day
  4) >35 kg: B 1 tablet every day
- Vitamin E per kg fish fed for birds consuming:
  1) 500 g fish daily: B1 capsule every other day
  2) 1 kg fish daily: B1 capsule every day
  3) 2 kg fish daily: B2 capsules every day, etc.

Breeding: Hybridization among species, particularly in the genera *Spheniscus* and *Eudyptes,* has been recorded, and it is recommended that *Spheniscus* species be housed as single species populations. Penguins are monomorphic; the best method of sexing is through blood or feather DNA analysis. Numerous other means, such as surgical, cloacal and behavioral observations, have been used with less success. Courtship displays occur at the nest site. Aggressive behavior in penguins is most pronounced during courtship and pairing, and again when chicks are hatched. Excessive aggression or competition can deter breeding, and should be monitored by the keeper. Pairs generally stay together for many seasons, but mate fidelity can be affected by transfers, separation, or mortality. Extra-pair copulations after pairing are commonly observed in captive penguins and paternity should not be assumed based on pair bonds. To maintain cleanliness, each nest box should be set on a platform 10 cm (4 in.) above the ground, thus allowing air to circulate around the boxes.

Penguins of the genus *Aptenodytes* do not build nests, and therefore neither species requires nesting material. Emperors do not occupy a nest site. As soon as the egg is laid, the female passes it to the male, and he incubates it on top of his feet under a ventral flap of highly vascularized skin. Emperors tend to remain in one area of the exhibit, usually the darkest. They stand in pairs within groups until the egg is hatched. Creching does not occur in captive birds, where the chicks are brooded continuously for 4–6 weeks. Kings, however defend a small territory and share the incubation duties. As soon as a king chick fledges, the parents molt, court, and begin the breeding cycle again. Emperors have been recorded stealing eggs from other pairs. To prevent this, pairs with an egg can be separated from the colony with a plexiglass barrier. Once the female has passed the egg to the male, she can be released back into the main exhibit.

Those species in the genus *Pygoscelis,* the gentoo, adelie and chinstrap penguins, exhibit similar behavior. In nature, they all return to their natal rookeries every year. Prior to pairing, males construct a perfunctory nest within their territories, using small rocks. Captive birds tend to be fairly aggressive prior to incubation. Following egg laying, aggression falls to low levels. Captive pairs share incubation relatively evenly between the sexes. After hatching, chicks are brooded for approximately 15 days, after which both adults feed simultaneously, leaving the chicks alone at the nest site.

Members of the genus *Eudyptes,* the rockhopper and macaroni penguins, are naturally colonial, with some rookeries having enormous populations. Nests are built with rocks and sometimes a little grass. Rocks and stones supplied as nesting material should be too large for chicks to swallow. Two eggs are laid, the first smaller than the second egg. Both are viable, but the first egg is often lost. Rockhoppers often retain both eggs, but macaronis normally only retain second eggs. Both sexes incubate and feed the chicks. Aggressive tendencies during the early stages of the nesting process may be minimized through the provision of excess nesting sites.

The penguins of the genus *Spheniscus,* including the African, Humboldt, Magellanic and Galapagos, nest in holes dug in firm sand or clay. Generally two eggs are laid, with both parents sharing incubation and rearing. The monotypic genus *Eudyptula,* the little blue penguin and its subspecies, the white-flippered penguin, is similar in habits. These birds hide their nests in burrows among plants and under rocks, the incubating bird remaining in the burrow during the day. The mate returns at night until the chicks are half grown, after which both adults spend the day at sea. Burrows for *Spheniscus* species should measure approximately 64 x 43 x 24 cm (25 x 17 x 92 in.), with an opening of 20 x 23 cm (8 x 9 in.). Boxes for little blues should measure 20–25% less all round. Nests vary from natural excavations to wooden structures and small pet kennels. The roof should preferably be removable, or the burrow should have a rear entrance, to allow for keeper access without unnecessary disruption of the nest. Key components to consider include a proper burrow opening, neither too large nor too small; more nest sites than pairs of birds; good air circulation; accommodation for cleaning and disinfecting; and good drainage. The substrate below the nest should be absorbent; sand, pea gravel, and kitty litter have been used successfully. Vegetation as a nest base or for nest construction should be limited, as this material can result in a build up of fungal spores. It is further recommended that nest boxes be removed for annual disinfecting, painting or replacement. Rounded stones, which are too large for the birds to swallow, and a few dry broad-leaf leaves, should be provided as nesting material for birds maintained outdoors. Humboldt and African penguins regularly raise two broods per year.

Chicks should be carefully monitored while with the parents. If a problem is suspected, parent-reared chicks may be carefully removed from the nest, examined, and weighed. The weight gain within the first 5–7 days should be substantial. By three weeks of age, chicks of smaller species are able to swallow whole fish of moderate size. At this time, some institutions begin feeding their *Spheniscus* penguins on the nest to encourage acclimation to hand-feeding. This is, however, not recommended due to the disruption it may cause on the rookery, as well as the aggression of the parents, which is often extreme. At the time of fledging, chicks often

Recommended Minimum Exhibit Size Criteria

Kings/Emperors	Pool & Land Surface Area (per penguin)	Pool Depth	Isolation Pool & Land (per penguin)
Per/bird — for the first six birds	1.67 m^2 (18 sq. ft.)	1.22 m (4ft.)	0.84 m^2 (9 sq. ft.)
Each additional bird	0.84 m^2 (9 sq. ft.)	1.22 m (4 ft.)	
Other Species	**Pool & Land Surface Area (per penguin)**	**Pool Depth**	**Isolation Pool & Land (per penguin)**
Per bird — for the first six birds	0.74 m^2 (8 sq. ft.)	0.61 m (2 ft.)	0.37m^2 (4 sq. ft.)
Each additional bird	0.37 m^2 (4 sq. ft.)	0.61 m (2 ft.)	

As given in the Penguin Husbandry Manual

lose interest in feeding, and will show a corresponding loss in weight. Studies of captive Humboldt penguins revealed hatchability rates of 85% and 90% of fertile eggs, but survival rates to six weeks of age of only 31% and 54% of the chicks hatched. Similarly, king penguins at Whipsnade lost a high percentage of their eggs to breakage and other causes.

Artificial Incubation: The removal of eggs can be facilitated through the top or rear hatch as described in nest boxes. In gentoo and adelie, the first egg is larger than the second; in *Eudyptid* penguins, this is reversed. In both cases, the removal of the smaller egg for artificial incubation or fostering can greatly increase production. Two people are required to remove an *Aptenodytes* penguin egg, one to hold the bird and the other to lift the brood flap and retrieve the egg. For other species nesting in the open, an object can be dangled above the incubating bird, causing it to rise up and strike. As it does so, eggs can be inspected or removed. Penguin eggs are incubated at a temperature of around 35.8°C (96.5°F), with a variance between institutions ranging from 35.2–37.5°C (95.36–99.5°F), and a relative humidity of 55%. Magellanic eggs in particular were found to hatch better at 36.6°C (98°F). Eggs should be turned mechanically every 1–2 hours. In addition, some institutions also perform a 180 degree manual turn of the egg, which facilitates a more even development of vascularization in the egg. Eggs should be weighed on alternate days and their weight loss compared to a line representing a weight loss of 15% of the initial egg mass by the expected day of external pipping. Weight loss can be corrected by varying the humidity, as described for PHEASANTS. If excessive weight loss occurs, the functional porosity of a small portion of the pores can be waxed with melted candle wax and varnish, for a period as required. As soon as external pipping is observed, the egg is transferred to a hatcher maintained at the incubation temperature, but with the humidity increased to 60%. Turning is no longer necessary, and eggs remain in the hatcher until they have hatched. Internal pip generally occurs two days prior to hatching. Problem hatches should be dealt with as described for FALCONS.

Hand Rearing: The best policy is to hand rear penguin chicks only when absolutely necessary. Having said that, however, it is advisable to establish protocols and gain experience with hand rearing to anticipate any problem that may occur when hand rearing is required. King penguin chicks at Edinburgh are monitored, left with the adult birds and, while still feeding from the parent, are hand fed sprats as soon as possible. Gentoo chicks are weighed every other day; if losing weight, they are tube fed a liquified gruel for a few days, depending on weight gain. Small chicks are weighed on portable electronic scales but larger birds are placed in a bucket suspended from a spring balance. Adults allow keepers to feed the chicks and the chicks will accept whole fish from an early age, allowing keepers to hand feed them whitebait while they are still in the nest. This regime makes weaning the chicks onto whiting, when removed to the creche area, easier and less stressful for both birds and staff. Adult birds are fed on the nest site during the breeding season. Emperors generally eat from keeper's hands without difficulty during incubation. *Pygoscelis, Eudyptes* and *Spheniscus* penguins can be offered food on the nest provided it does not cause unnecessary stress for the birds.

Once chicks hatch in a brooder they should be allowed to dry for at least 12–24 hours. Immediately following hatching, the umbilical area should be swabbed with dilute iodine-based disinfectant such as Betadine. Once dry the chicks are moved to an open brooder with a heat source. Brooders can be made of painted wood, plexiglass, plastic or any other material that allows hygiene to be maintained. Adequate air circulation, ease of maintenance and disinfecting, size to allow for a temperature gradient and the number of chicks to be reared together are all factors that need to be considered when choosing a brooder. The heat source is placed over one end of the brooder to allow the chick to move away from

Incubation and Maturity Notes for Penguins

Species	Latin	Mean Incubation Period	Pip to Hatch Time	Fledging	Sexual Maturity female/male
Emperor	*Aptenodytes forsteri*	67 days	48–72 hours	4–6 months	5/6 years
King	*Aptenodytes patagonicus*	56 days	48–72 hours	4–8 months	5–7 years
Gentoo	*Pygoscelis papua*	38 days	36–48 hours	70–75 days	2–3 years
Adelie	*Pygoscelis adeliae*	37 days	24–48 hours	40–60 days	3–8 years
Chinstrap	*Pygoscelis antarctica*	37 days	36–48 hours	55–60 days	3 years
Macaroni	*Eudyptes chrysolophus*	36 days	24–48 hours	60–65 days	6 years
Rockhopper	*Eudyptes chrysocome*	36 days	24–48 hours	50–60 days	?

the heat source if desired. It is recommended that 2–5 chicks be raised together. Cloth toweling or brooder matting, which provides a firm footing for chicks and is easily laundered, are the most commonly used substrates. After the initial two weeks the chicks are moved to a larger brooder measuring about 2.4 x 1.2 x 0.6 m (7.9 x 3.9 x 2 ft.) with partitions for chicks.

The hand-rearing formula currently used at Sea World, USA consists of:

- 440 g herring fillet
- 473 g (ml) half and half
- 8 e, 7 grain brewer's yeast
- "6 lb." Mazuri Sea Bird Vitamins
- 2000 IU vitamin E.
- 4000 mg calcium carbonate
- 440 g krill
- 473 g water
- 500 mg Thiamine

Other institutions have used a squid, sprat, and prawn based diet successfully.

The temperature of the brooder is gradually lowered until, in the case of the African, at about 21 days they are removed to rubber-lined open containers at room temperature. Chicks are fed the warm (18°C) homogenized slurry via a clean 50 ml syringe chosen for its large nozzle, which does not become blocked, by the slurry. Many institutions begin with a 3 cc syringe with an attached feeding tube. The syringe is gradually increased to 6, 12, 20 then 50–60 cc. Placing a finger in a "V" formation on either side of the beak induces the innate begging response. Wiggling the fingers usually results in the chick responding by pushing into the crook of the fingers. The total amount given at each feed should be worked on 10% of the body weight. Overfeeding is a common problem.

Penguin's chicks require the following rearing protocol:

- Day 1: Following one or two 1-cc water feedings, supply formula that is diluted 50% with water. Feed slightly less than 10% of the first-day weight, five times per day, i.e. every three hours.
- Day 2: Feed formula diluted by 25% — slightly less than 10% of the first-day weight
- Day 3: Feed full strength formula — slightly less than 10% of the first day weight.
- Day 4: Formula fed at 10% of that morning's first weight.

When a weight of 200–300 g (7–10.5 oz.) is reached, slowly introduce fish cutlets of 2.5–3.8 cm (1–1.5 in.). As the chick matures, whole small fish are introduced. For day 1 on fish, 3–5 g (0.1–0.17 oz.) of fish is given in

proportion to formula and water following the 10% rule. On the second day increase fish to every other feed. At 300–400 g (10.5–14.1 oz.), increase fish to 10 g (0.35 oz.) every feeding following the 10% rule. At 400–500 g (14.1–17.6 oz.), increase fish to 10–15 g (0.35–). At 500–600 g (17.6–21.2 oz.) fish should be approximately 50% of the total intake, e.g. a 500 g (17.6 oz.) bird will receive 20 g (0.7 oz.) formula, 5 g (0.17 oz.) water and 25 g (0.7 oz.) fish. At this stage feeds are decreased to four per day, but staying with the 10% rule. At 700 g (24.7 oz.), fish is increased to two-thirds the total intake. At 1000 g (35.3 oz.) the transition to only fish begins.

Mean Hand-reared Weights - *Aptenodytes*

Day	Emperor	King	Day	Emperor	King
1	190	204	**50**	3210	2692
5	219	195	**55**	4065	3531
10	307	247	**60**	4921	4355
15	459	334	**65**	6171	5183
20	657	437	**70**	7188	5991
25	929	564	**75**	7728	7118
30	1260	740	**80**	8936	8553
35	1593	1000	**85**	10168	9770
40	1990	1368	**90**	14380	10907
45	2487	1948			

Aptenodytes differ as follows: The 10% rule and the described protocol for newly hatched species apply. The chicks tend, however, to lose weight in the first four days. Fish should be introduced as outlined for smaller species, beginning at 400 g (14.1 oz.) first morning weight for kings and 500 g (17.6 oz.) for Emperors. At this stage, add 10 g (0.35 oz.) of fish twice daily and then every second feed until weights of 550 and 650 g (19.4 & 22.9 oz.) respectively are reached, at which stage 10 g (0.35 oz.) of fish is fed at every feed. At 700 and 750 g (24.7 & 26.4 oz.) respectively, increase fish to

20 g (0.70 oz.), and increase again to 30 g (1.06 oz.) fish at 850 and 950 g (30 & 33.5 oz.), respectively. At 1000 and 1100 g respectively, decrease feeds to four per day and supplement with thiamine. At 1100 and 1500 g (38.3–52.9 oz.) respectively, 30 cc formula, 10 cc water, and 60 grams of fish four times daily would be an average. Begin supplementation at 1500 g (52.9 oz.).

Mean Hand-reared Weights

Day	Weight (g)			
	Macaroni	Humbolt	Magellanic	Rockhopper
1	107	68	81	82
5	165	96	121	86
10	401	203	313	144
15	650	392	539	229
20	959	564	799	313
25	1356	716	1076	445
30	1741	882	1422	504
35	2237	1088	1834	776
40	2669	1345	2252	974
45	2998	1490	2557	1197
50	3214	1561	2763	1487
55	3232	1565	2949	1847
60	3211	1517	3177	2145
65	3104	1453	3253	2362
70	3004	1255	3150	2625

African penguins at London Zoo showed a 60% hatch rate and a 69% hatchling to fledging survival rate for hand-raised birds. Only one of 48 birds showed any serious sign of imprinting. Imprinted *Pygoscelid* penguins do not breed, but highly imprinted *Spheniscus* penguins have bred well. Socialization with conspecifics should be ensured.

Lastly, I would strongly recommend that any institution investigating a penguin exhibit make contact with the USA Penguin TAG and obtain a copy of the excellent husbandry manual that forms the basis of this section, and which I believe is being upgraded at the present time.

References:

Beall, F., S. Branch, A. Cramm, S. Crissey, S. Ellis, L. Henry, P. McGill, T. Schneider, G. Sirpenski & M. Walsh. 1994. *Penguin Husbandry Manual*. F irst Edition. Publication of AZA

Clark, M., A. Ferguson, A. Maskell and F. Smith. 1994. Breeding the black-footed penguin at London Zoo. *International Zoo Yearbook* 33: 9–15.

Deeming, D.C., R.L. Harvey, L. Harvey, S. Carey & D. Leuchars. 1991. Artificial incubation and hand-rearing of Humbolt penguins at Birdworld, Farnham. *International Zoo Yearbook* 30: 165–173.

Stevenson, M.F., H. Dryden, A. Alabaster & C. Wren. 1994. The new penguin enclosure at Edinburgh Zoo: the Palace for the 1990's. *International Zoo Yearbook* 33: 9–15.

20.16 Procellariidae

20.16a Petrels

Procellariinae

Gray-faced Petrel *Pterodroma macroptera gouldie*

• Glen Holland

Background: The purpose of this research was to perfect a hand rearing diet, which could be used for the rescue process for chicks of the endangered taiko (magenta petrel) which may be in peril. In addition the diet could also be used in the future transfer of taiko and Chatham petrel chicks to founder new breeding populations. The analogue species chosen was the gray-faced petrel (GFP) of which the rearing of chicks to fledging had been rather unsuccessful with only one of twelve chicks fledged in 1996. It became apparent after preliminary trials, that one of the pre-requisites for the successful rearing of young GFP was the development of a nutritionally complete and balanced rearing diet suitable for each age class. Following research by the Monogastric Research Centre at Massey University of the nutrient composition of the crop contents of growing and adult GFP, they recommended Chef Salmon and Turkey diet (tinned cat food) with 5% oil which should have met all the nutritional needs of young GFP. This diet was compared against two more natural diets and the growth/survival of all chicks was compared against wild birds.

Twelve chicks of an estimated week old were received.

Principal Objectives:

- Transfer 12 recently hatched GFP of a similar age from the wild to captive for the trials.
- Divide the birds into three groups (2:4:6) and undertake hand-rearing trials using three different diets, one on each group.
- Compare the growth and survival of birds on each diet, between each group and with wild birds.
- Place fledglings in empty marked burrows and continue to feed until fledged.

Only boiled water was used in food preparation and chicks had access to crop stones/grit at all times. Food was fed warm – about 35 °C and once prepared and loaded into syringes, these were placed in a temperature controlled bath of warm water. Food was prepared fresh for each round of feeding. When initially received the chicks were fed twice per day and a judgment then made as to if this was to be reduced to once per day. Feeding once per day was planned (depending on weight gain), to be maintained through to fledging. Following on from a previous trial a recommendation if struggling to obtain sufficient weight gain was to rather feed less more often. If weight gains were insufficient on both or either of the natural diets but all the oil was being digested, the quantity of oil could be increased by 5% at a time until it was judged to be sufficient. Vitrate was used if birds were ill/dehydrated etc. Following on from an earlier trial where food quality was an issue, seafood products were sourced from a reliable supplier who could ensure that the products had not defrosted at any stage. Minced krill had been used in an earlier trial and on islands but in one case on the islands, during a transfer of storm petrels, the mixture that was held in flasks had become contaminated and resulted in high mortality.

Food Preparation: The following fish/squid handling protocols were adhered to:

- Fish was thawed in a fridge at 2-3°C or otherwise under cold running water. Under running water the fish should be removed once it becomes pliable. The primary disadvantage of the latter method is that nutrients, vitamins and minerals are leached from fish.
- Thaw only the quantity of fish required for one day.
- Fish should never be thawed or held for extended periods at room temperature.
- Do not attempt to use warm water to defrost fish as the fish often loose their skin and thus become difficult to handle and feed.
- In emergencies fish can be defrosted in a microwave set on defrost. Fish defrosted in this manner must be for immediate use only.
- The primary aim of ensuring the correct thawing and preparation is to avoid conditions, which lead to fat breakdown and nutrient loss.
- Thawed products should be refrigerated until preparation for actual feeding.
- Stored fish should always be rotated on a first in, first out basis. Fish that are purchased should be packaged in plastic lined boxes.
- Freezing tends to break down tissues, making the food much more susceptible to bacterial invasion after thawing. The preferred method is thus to plan ahead and defrost fish in a fridge. Once fish has been thawed it should not be refrozen. While thawing fish should be kept in a wrapping or container which will ensure that the fish thaws evenly.
- All food to be blended and if problems occur with chunks blocking the crop needles the food will be either strained or moulied.
- No food, which has been warmed for longer than one hour is to be used. Food is not to be re-heated for a later feed.
- Use a thermometer when checking the food temperature.

Hygiene: Start the day by washing hands with an anti-bacterial soap. Feeding of GFP is a messy business and it is advisable to dress for the occasion. A clean overall and apron are advised. Overalls, aprons and towels were washed on a daily basis. For those with a skin that reacts to continually using strong soap, gloves can be used while handling or feeding the birds. Staff wore a clean/dedicated pair of gumboots when entering the brooder room. No footbaths were used. Visitors used the rubber boot covers. When feeding seabirds always have spare towels available for cleaning up spilt food, wiping the bird and of course your hands. Crop needles were soaked in hibitane between feeds. A couple of needles were soaked and a clean one used for each bird. Hibitane in small quantities does not harm birds. All equipment, fridge, preparation surfaces and storage areas used were thoroughly cleaned between feeds. Only high quality plastic cutting board and either stainless steel or glass bowls were used and washed with hot soapy water between feeds. Birds were never handled with spilt food on hands. Any food, which spilled on the bird or around its gape, was cleaned immediately. If necessary a light saline solution (20g/litre water) was used to clean blocked nares etc. Brooders were cleaned every second day but due to the quantity of excrement from the birds on the tinned cat food, these were cleaned on a daily basis. Ideally half the brooders and the brooder room floor were cleaned each day. Brooders were cleaned with hot soapy water followed by a wipe with Virkon and a rinse with water at least ½ hour later. Soft rubber mats on the brooder floor were washed with a pressure hose and then placed in a washing machine with a little Napisan (bleach). A drainage system with a plug, to allow for easy disposal of waste and water was installed for each of the brooders.

Environmental factors: Six birds were housed in a brooder with visual barriers to avoid aggression, which would have otherwise occurred. A tape of GFP calls, from wild nests was played at feed times. As the chicks matured this was used for shorter periods. Good ventilation was maintained in the brooder room. Additional warmth was not necessary for the chicks but brooder temperatures ideally to be maintained between 13 and 18 °C. The room was maintained with very low light levels. Window covers were put in place and the feeding area was dimly lit (dull red bulb). Ideally no one person should feed for more than two consecutive days. The chicks can become too familiar with one person and then refuse to feed when the feeder changes. In addition most of the day was spent in low light levels which was unpleasant for staff. Two days, rather than alternative days allowed staff to get a feel for the chicks progress and ensured some continuity.

The birds were housed on brooder matting, which had been cut to size according to the bowl size. Birds progressed from small bowls to larger ones in the brooders and once wing exercises began the bowls were removed with each bird resting on brooder matting.

Methods: Chicks were weighed pre and post feeding. All three diets to be blended and fed warm (35 °C) by crop needle on a 60ml syringe. For those birds on the natural diet, once the bids were ¾ grown a few solid strips of food (including a few scales and bones) was fed after the crop feed using a forceps tweezer. The crop needle was inserted well beyond the trachea and preferably the entire quantity given without withdrawing the crop needle. If the crop needle was withdrawn and re-inserted it will first be dipped in hibitane and then wiped clean with tissue paper. Although initially the feeding was forced the chicks soon began to take the food willingly. He crop feeding technique was adopted for ease due to the large volumes given and the numbers of chicks being fed. The main risk associated with this type of feeding is ensuring the crop needle is inserted into the esophageal canal rather than the trachea leading to the lungs. The chicks were stimulated before feeding by gently massaging and touching the gape. If no response occurred the mandibles were gently separated.

The wing cord was measured every three days. This was rather difficult initially but became easier as the chick developed. The wing cord measurements were compared with field data. Birds were be transferred out when the wing cord measured approximately 250mm.

Nutrition: The Following Diets/groups Were Trialed

No. in group	Diet	Blending requirement
3	Tinned sardines in soy oil + 40% salt water + 5 % cod liver oil	Trial results: 1 tin sardines + 40ml salt water (18g/L) + 5 ml cod liver oil = 150 ml. Add Ca/P at 2.5% + Mazuri multivitamin.
3	40% fish, 30% squid, 10% cod liver oil and 20% salt water.	Remove the head of the fish. In trial: 2 pilchards + 2 pieces squid + 80ml (20%) salt water (18g/L) + 40ml (10%) cod liver oil = 375 ml. Add Ca/P at 2.5% + Mazuri multivitamin.
6	70% Chef salmon & turkey, 5% olive oil and 25% boiled water	½ tin cat food = 350g + 70ml water (15%) - not salted + 25ml olive oil (5%) = 450ml. No Ca or multivitamin to be added. This food is to be brought to the boil and allowed to cool before feeding.

All percent are by volume.

Fledging occurs at approximately 300 – 320 mm. Expected wing cord growth rates in the latter stages were: 2-3 mm/day @ 100 – 200 mm, 3-4 mm/day @ 200 – 250 mm, 5 mm/day @ > 250mm. The fledging wing cord length can sneak up quite quickly as happened with the only previously hand raised bird. A month prior to fledging some leaf litter was added to the nests and it was important to ensure that juveniles have sufficient space to exercise their wings – once this starts to happen. Weight peaks around 800-1000g at 80 days, and then drops down to 600g at fledge 115 days.

It was known that faeces may vary according to the diet being fed e.g. fish results in very white faeces and tinned cat food in rather pink faeces. Once the color was established for each diet the faeces were monitored and change taken as warning signals. Black faeces for example indicated a problem, which will be referred to the vet.

Results: Twelve chicks newly hatched chicks were subsequently brought into captivity. One chick died soon after arrival from dehydration. One chick died from a squid beak, which must have been fed to it by its parents in the nest that had pierced the bird's intestine. Another chick died from a large stone, which blocked the gut and once again it must have ingested while in its natural nest. One chick was considerably older than the others were when removed from its parents – this chick remained extremely aggressive towards staff throughout the trial.

Hydration and overfeeding: Complications after arrival during the rearing of young chicks, removed from wild nests, were considered to be as a result of dehydration. It is recommend that in future chicks are feed on fluids only for 2 days and the diluted diet gradually introduced thereafter.

It soon became apparent that there were some issues with the recommended tinned cat food. The birds did not gain weight as was required, had to be fed large quantities of food and in addition they excreted the food so fast that daily feeding was required. Naturally the adults may not feed their chicks for up to three days. Bacteria and invertebrates in the burrows decompose the chicks excretion but the quantity of food excreted by the group on the tinned cat food would have resulted in fouled burrows and ultimately either desertion and/or death of the chick. In addition another bird on this diet died due to gout, which was related to the diet. It is important not to overfeed which results in aspiration of food or the inhalation of regurgitated food. Three of the small chicks which died early in the trial showed symptoms of overfeeding – a case of too much too soon.

The tinned sardines and the liquidised and moulied diet of pilchard and squid both resulted in good weight gains with the birds only needing to be fed every 2-3 days. From approximately three weeks on it became apparent that the birds fed on tinned sardine achieved the best weight gain and maintained this trend throughout the trial. At three months old these birds had reached a similar weight to parent raised chicks. Quantities fed for the three diets varied considerably with approximately 50 – 70ml of the sardine diet fed every two to three days, 85 – 95ml of pilchard/squid fed two out of three days and 95 – 105ml of tinned cat food almost every day.

Following these trials, there is still need to investigate salt content of sardines and chick's salt requirements.

As the fledglings are naturally not fed once out at sea, it is vital that they have sufficient body reserves to survive to independence. Due to the practicalities involved in preparing the diets and considering that the trial was for a diet, which could be used to save threatened species on islands, shortly before moving the surviving three chicks to the island release site, all chicks were placed on the tinned sardine diet. Once on the island the birds were placed in empty gray-faced petrel burrows, which were securely closed with a piece of plastic mesh during the day. At night the mesh was removed allowing the chicks to exit and exercise their wings. Every second day the chicks were fed whole tinned sardines until they fledged.

Artificial Incubation, hatching and initial rearing: A number of generally unsuccessful attempts at petrel incubation have been completed. Some success has however resulted in a further trial and subsequent rearing in 2003 which is to be based on:

Incubation from 14 days natural incubation to hatch at 54 days:

- Use Curfew incubator. Aim: 37ºC; RH 55–60%
- Total weight loss of 16.5%
- Automatic roller turner set for hourly turning. Daily hand turn 180º once and back the same way the following day.
- Hand turning (for cracked egg) 90º (x 2) then back to top and the same the other way, turning every 1 – 1.5 hours, approx. 6 times daily.
- At internal pip turning stopped, move egg to Brinsea hatcher set at 36ºC; RH 75-80%.
- Internal pip to hatch approx. 4-5 days.
- When hatched, move to Brinsea dryer, set at 35ºC; RH 40-50%

When chicks dry:

- Day 1: move into brooder, 30 - 34ºC
- Day 2: 30ºC
- Day 3: 25-30ºC
- Day 4: 20ºC
- Day 5: Room temperature (up to 20ºC).

Feeding:

- Food warmed to 32ºC, warm to touch on wrist.
- Puree = 41.5% sardines + 50% salt water + 5% cod liver oil + 2.5% calcium + Mazuri vitamin tablet *
- Salt water = 18g per litre.
- Day 1: 5 feeds of 1-5ml water, electrolytes.
- Day 2: 5 feeds of 80:20 (1-5ml) water : puree (no salt/calcium/vit) + acidophilus
- Day 3: 4 feeds of 5-10ml 60:40 water: puree (no salt/calcium/vit) + acidophilus
- Day 4: 3 feeds of 15-20ml puree (no salt) /acidophilus
- Day 5: 2 feeds of 20-25ml puree/acidophilus
- Day 6: 1 feed 25-30ml puree (start to add some vitamin and calcium supplements).
- Day 7 – 25: 30-40ml puree
- Day 26 – 50: 60ml puree
- Day 51 – 70: 70-85ml puree
- Day 71 – 90: 70 – 90ml puree/sardine pieces
- At 80 days, start feeding sardine pieces + puree. Food now at room temperature.
- At 85 days, feed sardine pieces + water/vitamin/oil mix, no puree.

Graeme Taylor used these trials to provide protocols for the transfer of Pycroft's petrels and reports as follows:

> "The Pycroft's petrel project was carried out in March 2001, building on the results of the gray-faced petrel project. Sixty Pycroft's petrel chicks were located on Red Mercury Island. Thirty of these were selected as controls and left to be monitored by a university student. The other thirty chicks were transferred (in cardboard pet-carrying boxes) by Charter boat to Cuvier Island on 16 March. Each chick was fed 10mls of fresh water on the day of the transfer and the chicks kept in the boxes for two days until a suitable colony site was located. The chicks were then placed in artificial plastic nest chambers on a ridgetop under the forest . the birds were hand fed every second or third day for 8 days. Each feeding session the chicks were fed 5-10 mls of fresh water by crop needle and syringe, and one or two whole sardines (slightly warmed in the sun). These were fed directly into the throat and massaged down the neck to the proventriculus. The average meal per chick was 27g. Strict hygiene standards were maintained. Ten days later all thirty chicks were all alive and expected to fledge."

Reference:

Taylor, G. 2001. Comparison of survival rates of petrel chicks reared on artificial and natural diets as an analogue trial for endangered species. Interim Investigation Summary – unpublished. Biodiversity Unit, New Zealand Department of Conservation.

Thomas, D, Hendriks, W, Rutherford, S. 1998. Model rearing diets for the Grey-faced Petrel. Monogastric Research Centre. Massey University (New Zealand) publication.

Acknowledgements: Graeme Taylor, Jerry Pauli, Helen Gummer, Shaun O'Connor, Rose Collen and Raelene Berry of the Department of Conservation, New Zealand.

Pittas, Broadbills, Cotingas, Bowerbirds *et alia*

21 Passeriformes

21.01 Pittidae Pittas

Vikki G. Hall and Kevin Evans

Introduction: The pittas, family Pittidae, belong to the order Passeriformes. They are terrestrial birds with short wings and short tails, relatively strong legs and feet, and fairly stout bills. They are medium-sized, forest-dwelling birds superficially resembling the thrushes. Most species of pittas are extremely colorful, and the family has often been referred to in the past as jewel-thrushes. This is strictly an Old World family, with 31 species distributed through Africa south of the Sahara, tropical Asia, and Australia to the Solomon Islands. While all species are currently included in the single genus *Pitta*, some authorities believe there is sufficient diversity within the group to split them up into as many as six genera. All species are insectivorous, foraging on the ground in leaf litter and soil, and among rocks and rotting logs, for various invertebrates. Their prey may vary from ants to crickets and earthworms, and is usually battered against a rock or branch before being consumed (Lambert & Woodcock 1996). Pitta species are sometimes kept in zoological collections, but are not often available to private aviculture. They are, however, wonderful avicultural subjects and every effort should be made to establish those species that become available. This chaper was written by Kikki Hall with significant additions from the noisy pitta husbandry manual by Kevin Evans.

Noisy Pitta *Pitta versicolor*

The noisy pitta *Pitta versicolor* is a rainforest bird that was first described in Australia by Swanson in 1825 (Parish 1983). Three subspecies of the noisy pitta are currently recognized. The nominate subspecies, *P. v. versicolor,* is found throughout the east coast rainforests south to the Hunter Valley in New South Wales; the subspecies *simillima,* from the Cape York Peninsula to New Guinea. The subspecies *intermedia* is confined to the Cairns District in northern Queensland (Lambert & Woodcock 1996). The nominate form is the one most often kept in Australian zoos and is therefore the subject of this paper. Breeding programs for the noisy pitta are under way in numerous Australian zoos, as it is a studbook species. Currumbin Sanctuary has successfully bred this species on a number of occasions in recent years. The Currumbin breeding program began with a female donated about a decade ago by the public and a male that was donated by Taronga Zoo two years later. These two individuals were paired up in 1994, with breeding success achieved in following years. This success is attributed to paying close attention to natural history details, dietary requirements, and husbandry techniques such as careful introduction procedures.

Management: Noisy pittas are extremely territorial and have the potential to kill each other if correct introduction procedures are ignored. Suspended wire cage introductions are recommended. This involves setting up two portable cages within a new breeding enclosure, allowing for a familiarization period of three to four days. By using a new enclosure, neither bird has a territorial advantage, as both are unfamiliar with their new surroundings. The portable cages should be placed close enough for the birds to be in visual contact, but not close enough for them to inflict injury on each other. Females are aggressive toward males if placed in their established territories. The birds must be released at the same time, and closely monitored for the first few days. This method has been proven at Taronga Zoo, with all introductions over the past two years accomplished successfully.

Noisy pittas exhibit aggressive behavior at an early age. Taronga has experienced sibling perpetrated fatalities at less than three months of age. Once the chicks have been removed from their parents, it is advisable to house each individual separately until they reach breeding age. If two females are kept together in the same enclosure, one of them will undoubtedly become dominant over the other. This usually starts with a subtle posturing around the feed area, followed by full threat displays which will

Noisy Pitta
Pitta versicolor

eventually be acted upon, resulting in the death of the subordinate female if the birds are left together. When pittas engage in combat, they fly at each other, locking feet and stabbing at each other's heads with their sharp beaks. Death usually occurs extremely quickly. Even members of long-term compatible pairs have been known to suddenly become aggressive to their partners. In such cases, they should be separated for a period and then re-introduced using the technique described above.

In the past, captive pittas, particularly fledglings, have been susceptible to parasitic roundworm (ascarids and coccidiosis) and fungal disease (aspergillosis). Minimizing stress during removal from and introduction to aviaries is an important part of limiting disease. Adult pittas are also susceptible to ascarids due to their ground-dwelling habits. This problem can be detected through monthly fecal screening, and treated with any one of a number of anthelmintics.

The noisy pitta breeding aviary at Currumbin Sanctuary is rectangular in shape and faces northeast for maximum sunlight. The enclosure measures 4 x 4.55 x 1.9 m (13 x 15 x 6 ft.) high. At the back of the aviary, the wall and part of the roof is made of fiber material for weather protection, while the remainder of the aviary is enclosed by 5 x 2.5 cm (2 x 1 in.) weldmesh to prevent potential pests and predators from gaining access. Mulch and leaf litter are used as a substrate within the aviary, with large river rocks placed between plants such as Cunjevoi *Alocasia macrorrhiza,* palms, ferns, and vines. A pitta aviary should be landscaped as described for Habitat A in the chapter on LANDSCAPE. Overhead sprinklers are installed along the roof to help simulate a rainforest microclimate throughout the year. Noisy pittas at Currumbin Sanctuary have been housed with such other bird species as Wompoo pigeons *Ptilinopus magnificus,* superb fruit doves *P. superbus,* rose crowned fruit doves *P. regina,* silvereyes *Zosterops lateralis,* and crimson finches *Neochmia phaeton*. Noisy pittas have been known to predate fruit dove eggs, therefore care must be taken when positioning nesting trays in a multi-species enclosure. As a general rule, the noisy pitta should not be housed in a mixed species aviary with any valuable or high priority bird with a body weight of less than 100 g (3.5 oz.).

Diet: Noisy pittas forage entirely on the ground, locating invertebrates by shifting leaf litter with their bills in a sideways motion and stabbing into the soil. The mandibles become caked with soil in the process. Published wild foods include rainforest snails, beetles, chafers, weevils, fly larvae, sugar ants *Componotus* sp., wood lice, and spiders. The noisy pitta is an omnivore, with a significant preference for invertebrate food items, especially during the breeding season. They have also been recorded eating fallen fruit and berries. One of the fascinating features of this species is the unique manner in which it extricates the body parts from the shells of land snails by using stones or buttress roots as an anvil on which it strikes the snail shell. After the shell has been cracked, exposing the flesh, which is consumed, the shell fragments are discarded, forming a midden. Favored anvils invariably have a pile of snail shells at their base. Some pittas make such frequent use of their anvils that they become smooth from the constant pounding (Coutts 1997).

The diet for captive noisy pittas varies from a general maintenance diet to a specialized breeding diet used to help rear chicks successfully.

General maintenance diet:

- Fine meat mix: minced meat, insectivore mix, breadcrumbs, dogfood cubes, calcium powder, and vitamin E powder, with grated carrot and fly pupae on top.
- Soft fruit: watermelon, currants, sultanas, diced carrot, apple, rockmelon, pears, peas, and corn.
- Non-breeding greens: lettuce, silverbeet, grated carrot, peas and corn.

The daily diet includes one plate of meat mix, soft fruit, and non-breeding greens given in the morning, with a handful of mealworms given in the afternoon.

Breeding diet:

- Fine meat mix: as above
- Soft fruit mix: as above
- Breeding greens: as above, plus broccoli, parsley, millet seeds, mung bean seeds, and alfalfa seeds.

The specialized breeding diet includes two plates of fine meat mix, soft fruit, and breeding greens given morning and afternoon with the following supplements:

- Calcivet is placed in water (20 ml/1 liter).
- Mealworms given three times per day.
- Earthworms given three times per day.
- 8–10 snails given daily when available.
- 15–20 crickets given per day.
- Rep-cal powder sprinkled over fine meat mix.

At Taronga Zoo, dietary changes made in September, just prior to the breeding season, include additional leaf litter and an increase in the number of insects such as moths, locusts, crickets, cockroaches, snails, mealworms, wax moth caterpillars, stick insects, etc. While young are being reared, the amount of live insects should be increased according to the demand. The noisy pittas at Taronga are given a wide variety of food items during the breeding season, including pinkie mice.

Breeding: In the wild this species would have protection against intense sunlight, wind, and torrential rain from the dense sub-tropical and tropical rainforest canopy. This kind of dense planting is not easily replicated in a captive situation and must be provided in the structural design of the enclosure of holding/breeding facility. Based on the experiences of Taronga Zoo, there seems to be a correlation between breeding success and a heavily planted enclosure. Noisy pittas do not maintain their nests after incubation commences; the nests often fall into disrepair as the breeding season progresses. After the chicks have fledged, the parents will usually build a new nest in a different location. Provision should be made for an adequate supply of fresh nesting material, otherwise the pittas will use soiled material from the old nest. They are not particularly fussy in their choice of nesting material; grasses, moss, and fine roots are all suitable. Noisy pittas have been recorded as using dung of other animals to construct a doorstep in the entrance to their nests. The purpose of this behavior is not yet understood, however it may be a predator deterrant. Coutts (1997) reports that noisy pittas use decaying wood, possibly as a means of absorbing the feces of the young.

The advertising call, by males only, commences in late September. Between October and April three clutches may be reared by one pair, usual first and last clutches being October 10 and March 15, respectively. New nests may take anywhere from three days to a full month to complete. The call is a rich bi-syllabic sounding call often described as "walk to work" and is routinely uttered in often-repeated phrases, often at night during the breeding season.

At Taronga Zoo this species has nested in wooden nest boxes 1.5 m (5 ft.) above the ground, in the center of a bird's nest fern *Asplenium* sp., and on the aviary floor, favoring terraced corners or the base of trees. The more secluded the site, the better. The nest is a domed structure approximately 30 cm (12 in.) in diameter, with a small entrance and a ramp composed of sticks, often with the previously mentioned "dung doormat" or mud patio. The exterior nest chamber is composed of sticks and bark. The inner chamber is lined with fine grasses and tendrils. Only the male has been observed constructing the nest.

Noisy pittas at Currumbin Sanctuary usually produce a clutch of three eggs that are oval in shape and exhibit brown blotches or marks over the shell. Incubation is shared by the female and male, and takes 15–19 days. During this period and after the chicks hatch, the parents will permit nest inspection and/or manipulation. Both parents feed the chicks while they are in the nest and continue to do so even when the chicks fledge, which occurs about 19 days after hatching. Provision of live food at this stage is very important, as parents may attack weak or younger chicks to feed others if alternative live food is not readily available. Both parents remain protective of the fledglings throughout rearing, but after a period of 30 days the young noisy pittas are considered independent. In captivity, the noisy pitta has been recorded as breeding successfully at six months of age. Pairs have had at up to five clutches in a single breeding season under captive conditions at Taronga Zoo. It is not understood whether this is dictated by the constant availability of any type of food during this period, but this would certainly affect the success of each breeding event.

There are numerous problems that may arise when attempting to breed noisy pittas. The most commonly encountered problems, with suggested solutions, have been discussed under the appropriate headings above. The successes achieved by Currumbin Sanctuary with this species have resulted from a close attention to details and a willingness to experiment with dietary supplements to fill in gaps in our knowledge of their captive requirements. And while it seems that dietary supplements prevented developmental problems, more research is required for verification. Currumbin Sanctuary intends to extend its breeding program for the noisy pitta.

In the UK, Beeson (1999) wrote an account of the aviary breeding of the hooded pitta *P. sordida*.

> "The chosen nest site was at the base of an ivy-clad wall on the earth floor of the flight. Coconut fiber and pencil-thick sticks approximately 25 cm (10 in.) long were supplied as nesting material. They also used bamboo leaves and moss gathered from the surrounding undergrowth. The basic frame was formed of sticks that were then covered with dead leaves and moss. The inside of the nest was then lined with coconut fiber. The birds completed the nest, which was the size of a large ball, in seven days. Like the nest of the noisy pitta, the nest had a porch-like structure at the entrance. Four eggs were laid, but the chicks were ejected from the nest upon hatching.

The following day a new nest was started, situated approximately 20 cm (8 in.) below a plastic-covered section of the roof, in a basket formed of wire netting. The basket was filled with small pieces of wood and held together with mud to form the base upon which the nest was built. Material from the first nest was re-used to construct the new one. Again four eggs were laid, with a single chick hatching. This chick died at 14 days of age, two days after the birds were observed building their third nest in a rockery next to a pond and waterfall.

The adults were supplied with earthworms, white worms, waxworms, and crickets. In addition, ox heart was added to their soft food. When Beeson noticed both birds carrying waxworms in their beaks, he inspected the nest, discovered four chicks and one intact egg. The remaining egg hatched two days later, but the chick was found dead in the pond the following day. Two additional chicks were found dead outside the nest over the next three days. The two surviving chicks were fed by both parents, using earthworms, waxworms, and crickets. At seven days of age they resembled young hedgehogs, with growths of spiky black quills. Whenever the nest was inspected, they would retreat to the rear of the nest and present their backs to the observer. To prevent accidents, the pool was drained and a shallow dish of water substituted near the nest. The parents were given fresh food five times per day between 6:00 a.m. and 6:00 p.m. They showed no fear of the keeper and would feed the chicks while he stood and watched. The first chick left the nest at 13 days of age; the second, at 15. Their plumage at fledging was a drab version of the adult plumage."

Artificial Incubation & Rearing: Incubation temperature is 38°C (100.4°F) and the wet bulb should read 35°C (95°F). No information is available on egg weight loss, but a loss of approximately 16% would be consistent with the requirements of other species in a range of related genera. The internal pip to hatch is a critical period in the incubation process for pittas, and one should monitor the progress of the chick, and intervene if necessary. The normal pip-to-hatch period is from 24–30 hours. The chicks must be transferred from the incubator to the brooder at the same temperature in which they hatched, 38°C (100.4°F), after which the temperature can be gradually reduced until the chicks are comfortable. Generally, chicks will become restless and hold their wings away from their bodies when they are too warm. When cold, they become lethargic and feel cool to the touch, an observation skill that comes with experience. As the chicks develop, they begin to generate their own body warmth. Close monitoring is necessary to ensure that chicks do not overheat. The keeper must learn to gauge the proper time to intervene and make required alterations.

Feeding Frequency of Hand Reared Noisy Pitta *Pitta versicolor* Chicks

Age	Frequency	Duration
0–12 hours	Every ½ hours	6:00am–8:00pm
12–24 hours	Every ½ hours	6:00am–8:00pm
1–6 days	Every ½ hours	6:00am–8:00pm
7–13 days	Every ½ hours	6:00am–8:00pm
14–21 days	Every 2 hours	8:00am–4:00pm
22–28 days	Every 4 hours	8:00am–4:00pm
29 days	*1 feed per day	

*At this stage the chicks should be close to weaned, if not already weaned. Weights should be monitored, and hand-feeding resumed if the weight drops below 80 g (2.80 oz.).

Noisy pitta chicks have been hand-raised on a diet of chopped pinkie mice, paw-paw, carnivore diet, and insects — including mealworms, locusts, crickets, and moths.

Amount of Food Fed to Noisy Pitta Chicks During Hand-rearing Process.

Age	Amount Fed	Frequency
0–12 hours	0.5ml alectrolyte solution	Each feed (table 1)
12–24 hours	2 x 1.5 g	Each feed
1–6 days	6 x 1.5 g	Each feed
7–13 days	8 x 1.5 g	Each feed
14–21 days	10 x 1.5 g	Each feed
22–28 days	Until feeding response stops	Each feed
29 days on	Until feeding response stops	Once per day

Tweezers (forceps) are the standard utensil for feeding noisy pitta nestlings. Normally, the feeding response is strong and the chicks beg with the beak wide open, displaying their characteristic orange gape. Young birds may lose interest in feeding after just one swallow, and it is difficult to feed additional food when this occurs, for it is not a good practice to force-feed them. It is therefore essential that the first piece of food offered be of good size, but still small enough for the bird to swallow easily. The individual pieces of food should be about 1.5 grams in weight and approximately 3 cm (1.2 in.) square. Food items can increase in size as the chicks grow. As the chicks develop, they gain strength and beg progressively more vigorously, consuming more food each day. After being fed, each chick produces a fecal sac that should be removed from the nest container.

Weaning should start any time between 14–21 days of age, depending on the individual; 16 days is the average. Every effort should be made to encourage chicks to start eating independently. Weaning can be a stressful period for passerines and weight loss must be carefully monitored during this period. At Taronga Zoo we start introducing the chick to adult food at 18 days of age. In addition, insects are scattered in front of the chicks to further encourage self-feeding. It has been observed that the chicks respond quickly to moving insects and will attempt to eat them almost from the outset. However, it takes a few days before they can sufficiently coordinate their movements to kill and consume an insect. A weaning period to complete independence has been recorded as 30 days at Taronga Zoo.

Contributors:

Bruce Pascoe, Curator at Currumbin Sanctuary

Liz Romer, Senior Conservation Officer at Currumbin Sanctuary

Zannah Gubler, Wildlife Officer at David Fleay Wildlife Park.

References:

Beeson, B. 1999. Breeding the hooded pitta *Pitta sordida. Foreign Birds* 65 (4).

Evans, K. 2002. *Husbandry guidelines for the noisy pitta - Pitta versicolor*. Taronga Zoo, Sydney, Australia

Hobcroft, D. 1992. The husbandry of the noisy pitta *Pitta versicolor. Journal of the Avicultural Society of Australia* 46.4.

Lambert, F. & M. Woodcock. 1996. *Pittas, Broadbills and Asities*. Sussex: Pica Press.

Parish, R. 1983. Breeding the noisy pitta. *International Zoo News* 30 (5):

Woodall, P.F. 1994. Breeding season and clutch size of the noisy pitta *Pitta versicolor* in tropical and subtropical Australia. *Emu* 94: 273–277.

21.02 Eurylaimidae Broadbills

Robert Webster

The broadbills in the family Eurylaimidae are among the first birds listed on the taxonomic roll call of the great order Passeriformes. They are an ancient, Old World assemblage of amazing contrasts: lowland and highland residents, understory and canopy dwellers, frugivorous and insectivorous in diet, long-tailed and short-tailed, gaudy and plain. The division of the family's 15 extant species into eight genera is a reflection of this little-known group's amazing diversity.

The four African species exemplify much of the inherent conundrums facing those who study the broadbills. The three *Smithornis* species (insectivorous, predominantly brown, and lowland-dwelling), were not recognized as broadbills until 1914, having previously been classified as Old World flycatchers in the family Muscicapidae. The African green broadbill, *Pseudocalyptomena graueri* (omnivorous, predominantly green, and confined to highland forests in Uganda and the former Zaire), was only discovered in 1909 and not placed among the Eurylaimidae until 1931.

Southeast Asia is home to the remaining 11 species of broadbills, with the greatest concentration of species centered on peninsular Malaysia and Sumatra, each with seven species, and Borneo, with eight species, two of which are endemics. The ranges of the silver-breasted, *Serilophus lunatus,* and long-tailed, *Psarisomus dalhousiae,* broadbills extend from this epicenter northwestward among fragmented highland forests to the Himalayas, while two species are isolated in the Philippines.

Ranging in size from the black-and-yellow broadbill, *Eurylaimus ochromalus,* at 13.5 x 15 cm (5 x 6 in.), to the dusky broadbill, *Corydon sumatranus,* at 24 x 27.5 cm (9.5 x 11 in.), the Asian eurylaimids are predominantly forest-dwelling insectivores, though again, there are always exceptions among the broadbills.

The dusky broadbill, dark brown and somewhat frogmouth-like in appearance, is generally restricted to the deep forest canopy, while the dazzling black-and-red broadbill, *Cymbirhynchus macrorhynchus,* with its black, white, and crimson plumage and bright blue and yellow bill, seems able to adapt to some of the changes wrought by humans onto the landscape, unlike the majority of its kin. Naturally a bird of the forest edge, this broadbill also can be found in forest plantations and village gardens, as long as water is nearby. Commonly feeding in trees on caterpillars and other insects, it deviates from usual broadbill feeding strategies by regularly taking food such as crabs, fish, and mollusks from the ground. Though this species normally nests on low branches over waterways, it has adapted to building its nests on telephone wires over streams or forest clearings.

The black-and-red broadbill's nest was described by Madoc (1956) as "the most conspicuous of all Malayan birds," seems to have become the description of the broadbill nest that would later appear in many general bird texts: may be mistaken for a hanging mass of vegetation left by the last flood. It is a ragged egg-shaped mass of dead creepers, twigs, palm fiber and dead leaves, about two feet long and ten inches in diameter often with an untidy tail dangling another foot. In the center of this bundle is a small spherical egg chamber, with a nest entrance hole sheltered by a nave. Bamboo leaves are used extensively, particularly in the egg chamber. A lining of small green leaves is changed frequently during incubation (Madoc 1956).

The three species comprising the genus *Calyptomena* buck the trends seen in most of the other Asian broadbills. They are predominantly fruit eaters, strongly sexually dimorphic, and build more compact nests.

Sadly, with the loss of so much of Southeast Asia's primary forest, several broadbills are now species of concern for BirdLife International (2001). The African green broadbill has been considered ENDANGERED almost since its discovery, due to its very small zoogeographic range. The Philippine species, the Mindanao wattled broadbill, Eurylaimus steerii, and the Visayan wattled broadbill are listed as "Vulnerable" due to small fragmented populations and habitat loss. The lesser green broadbill, Calyptomena viridis, Hose's broadbill, C. hosii, and the black-and-yellow broadbill were listed as "Near-threatened" for the first time in 2001, also due to habitat loss. The lesser green and its Borneo-restricted congeners are considered tremendous seed-dispersers; their loss to these forests could have far-reaching and catastrophic effects (Lambert & Woodcock 1996).

Avicultural Management: A Tail of Two Broadbills

Historically, only two broadbill species, the lesser green and the long-tailed, have had any impact in aviculture. Luckily for those who crave diversity, these two species are physically and behaviorally very different. It is almost hard to believe that they are related at all.

The Lesser Green Broadbill

The lesser green broadbill is a master of camouflage. The females are a uniform grass green, while the males are a brilliant emerald with black wing and head markings. When discovered among the greenery of the forest or exhibit, however, a lasting impression is immediately made. (As a keeper at the San Antonio Zoo, I once heard a youngster misread the species identification graphic as "laser" green broadbill. Sure, that's technically wrong, but as a person with a passion for broadbills, I found the appellation quite accurate!)

Diet: Broadbills are not especially demanding diners. A proprietary softbill pellet, along with chopped fruit, forms a good basic diet. Small insects are readily accepted, especially at breeding season, and nectar is popular with some pairs. Avocado (sans pit and peel) is an excellent fruit to offer, as it is the most readily available high-fat fruit. Most of the natural fruits of Southeast Asia are rich in fats, and the lack of this item in the avicultural diet is one theory as to why so few green broadbills have successfully bred in captivity.

Though these birds have thus far proved reluctant to breed in aviary environments, they can experience a captive longevity of more than 15 years, offering hope that their riddles will continue to be revealed. I have tracked the North American population for 20 years, and no diseases or other health problems stand out as particularly pervasive.

Multiple pairs can be housed together for several months and this is an ideal situation for natural pairs to form. In the wild, many greens can be seen at a single fruiting tree, but pairs break away from the flock for nesting. In an aviary setting, I have never heard of a green broadbill antagonizing another bird, except for, occasionally, another green broadbill. They are, for the most part, extremely mild-mannered birds, so naturally they should not be mixed with aggressive birds like corvids or laughing thrushes. Green broadbills also do not seem to suffer busybodies such as leiothrix or bulbuls well. And if an earnest effort is being put forth to breed these demure

Lesser Green broadbill. *Calyptomena viridis*

Toledo Zoo Lesser Green Broadbill Hand-rearing Protocol

Day	# Feedings	Total daily intake	Incubator Temp./hum.	Comments	Weight	Weight gain
3	6	8 cricket abdomens				
		19 pcs papaya				
		25 pcs soaked dog food				
		15 waxworm guts	95 / 80%	Pulled from nest	5.7g	
4	7	26 pcs papaya		& brought to brooder		
		37 pcs soaked dog food				
		25 waxworm guts	95/ 80%		6.6g	+15.8%
5	7	30 pcs papaya				
		7 pcs banana				
		34 pcs soaked dog food				
		28 waxworm guts	95 / 80%	Pin feathers beginning to emerge	6.9g	+4.6%
8	6	31 papaya				
		30 soaked dog food				
		28 waxworm guts				
		2 cricket abdomen	92 / 85%		9.6g	+6.7%
11	6	19 papaya pcs				
		18 banana pcs				
		27 soaked dog food				
		33 waxworm guts				
		2 cricket abs				
		1 king mealworm gut				
		15 pinky pcs	89 / 83%		12.7g	+8.5%
14	6	20 papaya				
		36 banana				
		19 plantain				
		21 soaked dog food				
		7 cricket abs				
		38 waxworm guts				
		11 pinky pcs	90 / 83%	not digesting plantain well; discontinue for time-being	17.0g	+7.6%
17	5	23 papaya				
		19 banana				
		12 kiwi				
		3 blueberries				
		19 soaked dog food				
		4 cricket abs				
		32 waxworm guts				
		10 pinky pcs	89 / 82%	Tail feathers visible	21.7g	+10.2%
20	5	27 papaya				
		25 banana				
		21 kiwi				
		10 blueberries				
		5 avocado				
		22 soaked dog food				
		38 waxworm guts				
		9 pinky pcs	88 / 81%		25.8g	+7.0%
23	4	17 papaya				
		16 banana				
		2 kiwi				

		14 blueberries 1 avocado 17 soaked dog food 1 pc softbill pellet 34 waxworm guts 9 pinky pcs	88 / 70%		33.1g	+14.5%
26	5	13 papaya 21 banana 6 kiwi 7 blueberries 9 avocado 17 soaked dog food 4 softbill pellets 42 waxworms 4 pinky pcs	85 / 64%	Only interested in waxworms; midday set bird up in stack cage with a heat lamp	30.8g	+1.7%
29	5	20 papaya 21 banana / plantain 3 kiwi 15 blueberries 10 avocado 16 soaked dog food 12 softbill pellet 2 waxworms 46 waxworm guts 2 pinky pcs	84 degrees	Given own food bowl with some worms; practicing broadbill vocalizations	34.5g	-0.6%
32	3	17 papaya 23 banana / plantain 4 kiwi 8 blueberries 7 avocado 2 grape pcs 28 soaked dog food 8 softbill pellet 2 waxworms 6 waxworm guts 5 pinky pcs	80%	Trying to feed near food plate; in pm, ate 2 waxworms out of plate for first time; turn heat lamp off during day	36.6g	+3.1%
33	2	11 papaya 13 banana 5 kiwi 5 blueberries 1 avocado 13 soaked dog food 2 pinky pcs	84 degrees	Eating well on own from diet plate when coaxed	39.9g	+9.0%

				with paintbrush	
36	2	6 banana 3 kiwi 4 blueberries 2 grape pcs 6 soaked dog food 4 pinky pcs	83 degrees	Last day of hand-feeding; 47.8g discontinue soaking food in vitamin water and discontinue pinky pcs; provide am pan & pm pan & encourage him to eat a few times a day	+2.4%

Important notes:

Feeding schedule for first six days:

8:30 a.m. – use calcium carbonate, weigh chick
0:30 a.m.
1:00 p.m.
3:00 p.m.
4:45 p.m. – use calcium carb
7:00 p.m.
9:00 p.m.

- Soak/dip all food in vitamin water before feeding
- 1 cc Avitron : 1 qt water
- Put pinch of calcium carbonate in vitamin water at 8:30 a.m. feeding & 4:45 p.m. feeding
- mush food up a little before feeding
- 1 pinky should equal approximately 8 pieces for first offerings
- warm food in double waxworm container before feeding
- use quality paintbrush to feed chick, dabbing off excess water on a paper towel
- diet items should be introduced gradually
- check each fecal for consistency & digestion of food
- give 1– 2 paintbrushes of vitamin water after every feeding
- you may need to wipe face with a warm moist (not wet) cotton swab after feeding to keep nostrils clear
  do not use bleach as it is corrosive to paintbrushes and may leave a toxic residue

birds, hyperactive and nest-material thieves should not share their space. Another potential disturber of the broadbill's peace is a mid- to large-size pheasant species. All is well and good while the galliforms are walking about during the day, but a peacock pheasant roosting overnight on top of a carefully researched and human-constructed broadbill nest is not a welcome sight for broadbill or aviculturist. Successfully breeding green broadbills have been housed with honeycreepers, fruit doves, and roulroul partridge (Gossett and Besake 1997).

Lots of live plants and regular showers mimicking their rainforest home definitely seem to benefit lesser green broadbills. Though generally a quiet bird of the understory, content to sit for long periods of time observing the world, the male green broadbill is capable of some astounding courtship displays. Typically, the male hunches his posture into a neckless appearance and erects his crest forward over his beak, revealing the black and yellow feathers that otherwise remain hidden. He then ricochets around his space in a ritualized pattern, landing on low and exposed perches, generally 1 x 1.5 m (3 x 5 ft.) off the ground. The female will usually call softly to him, and he will alight near her, gaping in frenetic fashion. This performance, which accents the female's greater size, generally ends with the male regurgitating a berry for the hen (Webster 1991).

Another display, of unknown significance, has only been described from birds in aviculture. This involves a male in frenzied bumblebee-like flight alighting on a very low perch, about 0.5 m (20 in.) above the ground, and then rocketing to the ceiling of his exhibit where he spins around a few times, touching his beak to the ceiling with each turn. At San Antonio, females were observed watching this display, but they never seemed nearly as impressed as I was.

Successful breeding has occurred in indoor and outdoor enclosures measuring 3.7 x 6 m (12 x 20 ft.) x 6 m (20 ft.) x 3 x 6 m (10 x 20 ft.) high (Gossett & Besake 1997; Lewins 1996; Terri Heminger, pers. comm.). Only one female has built her own nests, while the other breeders have used hand-made artificial nests. Favored nests have been no more than 2 m (6.2 ft.) off the exhibit floor.

Spanish moss, coconut fibers, and excelsior are popular nesting materials, though few females seem adept at creating nests out of this material. Incubation is 17 days and is done by the female alone. The male pitches in with defending the nest and feeding the chicks. Fledging occurs at 22 days or so. Hand rearing was successful at the Toledo Zoo using a formula based on fruit and pinky mice.

Long-tailed broadbill.
Psarisomus dalhousiae

Toledo Zoo Lesser Green Broadbill Hand-rearing Protocol

In 1997, the Toledo Zoo bird department attempted to hand-rear five lesser green broadbill chicks, after previous attempts at parent rearing proved unsuccessful. Toledo managed to rear one of these chicks to fledging and beyond. The feeding chart on pages 399 — 401 is for the chick that thrived.

The Long-tailed Broadbill

The long-tailed broadbill is to the lesser green broadbill as a rocket is to a (very pretty) rock. Okay, this is a bit of a stretch, but the differences between these two species are quite striking. In contrast to the sedate green broadbill, the long-tailed is an extremely active, vocal, demonstrative bird that has a beautiful palette of greens, yellows, and blues comprising its plumage. The black pattern on the head almost makes the bird appear to be wearing a helmet. The sexes of at least some subspecies can be told apart in the hand by a seemingly minor difference in the nape feathers, the female having a normally concealed yellow spot in the blue region of its collar that is lacking in the male (Glenister 1951).

The long-tailed broadbill is a canopy bird of mountain forests, although in the aviary this species will investigate every nook of its home. They can be happy in the well-planted space described for lesser greens, but keep in mind that they seldom nest lower than 3 m (10 ft.) off the ground, and often much higher, so a taller enclosure may be optimal. They are very social, constantly calling and raising and lowering their tails in unison, and a number of birds can be housed in a very large aviary with little conflict. Nest building by more than two long-tails has also been observed, but subordinate individuals are usually driven away by the time the eggs are laid. They mix well with other species and can be housed with (and unintimidated by) birds as assertive as gaudy barbets.

Diets for long-tails should skew toward the insectivorous. They readily consume all commercially available invertebrates — the bigger the better. In the wild they eat insects as large as stick insects and mantids. They relish gleaning crickets of any size that have been thrown into the trees of their space by their keepers. A softbill pellet should also be provided, as should some fruit, avocado included.

Originating from higher altitudes may make this species more susceptible to aspergillosis than other species. Candida is also common, but this can be kept in check with regular prophylactic swabbing of the inside of the beak with an antifungal medication. The lifespan of this species has averaged 6–8 years in North American collections.

Yet another difference between almost all the long-tailed broadbills in the United States and most of the lesser green broadbills is that the long-tails seem to know what to do with the nest material provided for them. Beautiful nests have been built in nearly every collection that has held this species. A folded-over wicker-type welcome mat can encourage birds to nest within (Brown 1994). Three to seven eggs are laid, which are incubated for 17 days. At least seven zoos have hatched long-tails but, unfortunately, something seems to go wrong after hatching. To date, no North American collection has successfully parent reared this species. Two institutions have made great strides in broadbill husbandry, however, by developing a phenomenally results-proven hand-rearing protocol (R. Plasse and P. Witman, pers. comm.).

Other Broadbills

A handful of other broadbill species have been maintained in captivity for short periods of time. The African, or black-capped, broadbill *Smithornis capensis,* was maintained with seemingly little difficulty at one zoo and in the aviaries of Edward Marshall Boehm (Everitt 1973), and the black-and-red broadbill has appeared very irregularly in North American aviculture. As these two species are common in their respective ranges, well-organized breeding programs could be initiated to help aviculturists fill in our knowledge of this disparate family.

Species Requiring Similar Management:

Trogons
Cotingas

References

BirdLife International. 2001. *Threatened Birds of Asia: the BirdLife International Red Data Book*. Cambridge, UK: BirdLife International.

Brown, C. 1994. Another Quest: breeding long-tailed broadbills at the Fort Worth Zoo. *Caged Bird Hobbyist* 2 (2): 58–59.

Everitt, C. 1973. Birds of the Edward Marshall Boehm Aviaries. Lakeside Press, Chicago

Glenister, A.G. 1951. *The Birds of the Malay Peninsula, Singapore and Penang*. London: Oxford University Press.

Gossett, J. & C. Beseke. 1997. Breeding lesser green broadbills at Lincoln Park Zoo. *Avicultural Magazine* 103 (1): 15–19.

Lambert, F. & M. Woodcock. 1996. *Pittas, Broadbills and Asities*. Sussex, UK: Pica Press.

Lewins, E. 1996. Captive breeding of the lesser green broadbill (Calyptomena viridis continentis). *Avicultural Magazine* 102 (1): 1–5.

Madoc, G. C. 1956. Introduction to Malayan Birds. Malayan Nature Society.

Webster, R. 1991. The Broadbills: an overview of the Eurylaimidae with an emphasis on the lesser green broadbill (Calyptomena viridis) in the wild and in captivity.

21.03 Tyrannidae

21.03a Cotingas
Cotinginae

• Josef Lindholm III

Introduction: Profoundly separated by genetics and geography, the cotingas and birds of paradise none the less display remarkable parallels. Each group owes its amazing diversity to sexual selection. Fantastic extremes in form, color and voice have thus evolved, and both have developed complex lekking systems involving spectacular displays (Snow 1982). And cotingas and birds of paradise share a status in aviculture attained by few birds. Unusual publicity has greeted the introduction of novel species to collections, and often enormous effort has been invested in obtaining the first specimens in captivity. First captive breedings and efforts towards sustained propagation have likewise attracted far more attention than is usual in aviculture.

DNA hybridization conducted by Sibley, Monroe, and Ahlquist establishes cotingas as one of the five subfamilies composing the enormous, entirely New World family Tyrannidae. More traditional classification systems placed cotingas in their own family, together with becards and tityras (now placed in the subfamily Tityrinae). As recognized by Sibley et al, the Cotinginae are comprised of 69 species in 28 genera (including the plantcutters and sharpbills, both traditionally placed in separate families, and neither with any significant avicultural history). Thus defined, the Cotinginae range from tropical Mexico to Chile and Argentina, as well as Trinidad, although all species pertinent to aviculture are either tropical, or inhabitants of Andean forests.

Management: As with many soft-billed birds, documented aviculture of the cotingas commenced in the Victorian era. However, evidence suggests at least one species might have been maintained in pre-Columbian Mexico. The Codex Mendoza, an illustrated document of the tribute sent to the Aztec Capitol at Tenochtitlan before 1520, records (along with thousands of "handfuls of rich feathers" of specified colors, 40 jaguar skins, "800 bowls for drinking cacao," and two "lip plugs of clear amber, decorated with gold") an annual levy of "160 bird skins, of turquoise-blue feathers and purple breasts" from the province of Xoconochco, on the Pacific coast of Chiapas (Berdan & Anawalt 1992). The Codex includes two illustrations of these skins, quite clearly those of the lovely cotinga *Cotinga amiabilis,* well-known to the Aztecs (north of its range) as Xiuhtototl. Furthermore the annual tribute of "800 handfuls of rich blue feathers" is also identified as Xiuhtototl. The exquisite objects that the Aztecs crafted from cotinga feathers are well known, and several examples survive in museums.

While no direct evidence exists of living Xiuhtototl reaching Tenochtitlan (present day Mexico City), it is documented that a constant stream of ornamental living birds arrived as tribute. The sixteenth century scholar Fray Diego Duran documented: "Live birds, too, sent by these different nations, the most highly esteemed and those of the finest plumage…" (Duran 1994). There are definite records of the annual arrival of living Xochitenacaltototl (emerald toucanets) from Tuchpa, on the Gulf Coast (Berdan & Anawalt 1992). As the Totocalli (Montezuma's aviary) was totally destroyed on the orders of Cortes on August 13, 1521, less than two years after his arrival, there is only scanty documentation of its inhabitants. Cortes (1962), who recorded that it "was the whole task of three hundred men to look after these birds" and "others likewise were employed in ministering to those who were ill," mentions only aquatic birds and raptors. However, another member of the expedition, Bernal Diaz del Castillo (1996), wrote of "…tiny birds of many-colored plumage, also the birds from which they take the rich plumage which they use in their green feather work. The birds which have these feathers are about the size of the magpies in Spain, they are called in this country Quezales, and there are other birds which have feathers of five colors…". Diaz goes on to document that "from all these birds they plucked the feathers when the time was right to do so, and the feathers grew again". Thus we have circumstantial evidence that the Aztecs might have kept *Cotinga amiabilis* at their capitol.

Be that as it may, documented aviculture of cotingas certainly dates from 1866, when both naked-throated bellbirds *Procnias nudicollis* and eastern cocks-of-the-rock *Rupicola rupicola* arrived at the Gardens of the Zoological Society of London. Two *P. nudicollis* were received that year and, through 1882, a total of 17 specimens (most arriving as individuals) were obtained by that collection (Sclater 1883). However, from 1883 through 1895 only one further specimen arrived (Sclater 1896), while three more were received from 1996 through 1903. Examining the records of the Zoological Society of London, Peter Chalmers Mitchell (1911) found the average longevity of the 18 birds received after 1870 was 21 months, with a maximum longevity of four years and one month. (The first US captive cotinga record appears to be of two naked-throated bellbirds that arrived at the Central Park Menagerie in New York in 1879 (Marvin Jones, pers.com.).

From 1866 through 1881, nine eastern cocks-of-the rock arrived at the London Zoo (Sclater 1883), while from the period of 1883 through 1895, only two further birds were accessioned, in 1885 (Sclater 1896). Examining the records for the eight specimens received after 1870, P. Chalmers Mitchell (1911) determined an average longevity of eight-and-a-half months, with a maximum of only 20 months. Far better results were received with one of the two banded cotingas *Cotinga maculata* received at London in 1875 and 1877 (Sclater 1883). While Mitchell (1911) found one had died after five months, the other lived for four years and nine months, his best captive age record for any cotinga. The only other cotinga accessioned by London Zoo in the period 1866–1903, was a single white bellbird *Procnias alba,* acquired after 1895, which Mitchell (1911) records as living four years and five months. Through 1927, only two further species entered London Zoo's collection: The green-and-black fruiteater *Pipreola riefferii* and the orange-breasted fruiteater *P. jucunda*.

Little is recorded on how these specimens were exhibited and maintained, but I gather the general rule was to house them individually, in unplanted indoor cages. One of the earlier breaks with this practice was achieved in 1933 by the St. Louis Zoo, where Jean Delacour (1937) made the following observation at the famous bird house: "One very roomy and high compartment is the home of three magnificent male [Eastern] cocks of the rock, who have lived there over three years; it is a wonderful sight to watch them jump and fly among the limbs of a large banyan tree, and catch on the wing the grapes their keeper throws up in the air. To my amazement they never fight seriously."

Jean Delacour, himself a pioneer in the development of planted community aviaries, played an instrumental role in the progress of cotinga aviculture. Among his achievements as technical consultant to the New York Zoological Park (Bronx Zoo) from 1940 through 1947, was the engagement of Charles Cordier as collector for the New York Zoological Society. Cordier's first Bronx Zoo expedition returned from Ricaurte, Colombia on December 9, 1941 with one dozen scarlet cocks-of-the-rock *Rupicola peruviana sanguinolenta,* the first specimens of any subspecies of the Andean Cock-of-the-Rock held in captivity (Lindholm, 1988). From an expedition to Costa Rica, Cordier arrived at the Bronx Zoo on 9 October, 1942 with three bare-necked umbrellabirds *Cephalopterus glabricollis,* the first members of their genus collected alive (Cordier 1943). The two remaining members of the genus *C. ornatus* and *C. penduliger* were introduced to aviculture July 20, 1950, when two of the former and four of the latter arrived at the Bronx Zoo, along with 11 equatorial cocks-of-the-rock *Rupicola peruviana equatorialis,* three red-crested cotingas *Ampelion rubrocristata,* three purple-throated fruit crows *Querula purpurata,* and one barred fruiteater *Pipreola arcuata,* all collected in Ecuador (Crandall

1950) and likewise the first in captivity. Two other particularly spectacular cotingas Cordier introduced to aviculture via the Bronx Zoo were the three-wattled bellbird *Procnias tricarunculata,* collected in Costa Rica in 1947 (Cordier 194; Crandall 1948), and the calfbird *Perissocephalus tricolor,* from Surinam, in 1953 (Anon 1953). The Philadelphia Zoo subsequently received a Calfbird in 1953 (Marvin Jones, pers. comm.). Following Cordier's establishment of a Congolese base of operations, the Bronx Zoo achieved another avicultural first, in 1959, when its Curator of Birds, William G. Conway returned from a much-publicized expedition to Trinidad (Bridges 1974) with a male bearded or mossy-throated bellbird *Procnias averano,* which was exhibited with some fanfare alongside three-wattled and naked-throated bellbirds (Anon. 1959). This was, of course, a classic "stamp collection," but the term just as well applies to the entire Bronx Zoo series of Cotingas from the 1940s into the 60s.

All this activity in New York was not without influence elsewhere. Only three of the dozen scarlet cocks-of-the-rocks collected in 1941 were retained at the Bronx Zoo, two going to the private aviculturist Mrs. Milton Erlanger (who subsequently bred the pair), and the others to several US zoos. Jean Delacour (1948) commented: "Since Mr. Cordier's visit in 1941, the inhabitants of South-Western Colombia, who had learnt from him how to catch and to feed Scarlet Cocks-of-the-Rock, have regularly sent a few every year to the Louis Ruhe firm, in New York. The same thing happened with Quetzals in Costa Rica, so that these two marvelous species are well represented in American collections." This appears to have also been the case with the Ecuadorian cotingas. (The importation of scarlet cocks-of-the-rock by Louis Ruhe, Inc. is documented, with photographs, by Paul Zahl [1953]). Although none of the cotingas were ever commonplace or inexpensive, by end of the 1950s , and into the 1970s, a great many zoos and quite a few private collections around the world maintained a variety of commercially exported species and subspecies. Of course the advent of large scale air shipments accelerated this development.

This unprecedented availability of such species as Andean cocks-of-the-rocks and long-wattled umbrella birds, and the increased opportunity to acquire eastern cocks-of-the-rock and naked-throated bellbirds coincided with trend towards "walk-through" or open-fronted aviaries, which swept the avicultural world in the 1950s and 60s. Although the lushly planted or decorated, glass-fronted "jewel room" exhibits, pioneered by Jean Delacour in the 1940s and subsequently widely copied, frequently showcased individual specimens, the most common form of cotinga facilities were landscaped, mixed-species aviaries, often very large, and all too often overcrowded. (At one point in the 1960s the San Diego Zoo's "rainforest aviary" held a thousand specimens of two hundred species, among which both Andean and eastern cocks-of-the-rock were conspicuous). While those in Southern California were out door exhibits, in most places climate dictated indoor conservatory-type displays. In these situations stress from overcrowding and poorly chosen cagemates created less than optimal environments. At the same time, there were variously successful breeding attempts, and some satisfactory longevities were attained. For instance, in 1973 an eastern cock-of-the-rock was still living, after more than 20 years, at Delacour's Parc Zoologique de Cleres (Delacour 1974).

In the US, the exhibition of as many species of softbills as possible in a given exhibit was dealt a severe blow by the imposition of import restrictions and quarantine rerquirements following a 1971 outbreak of Exotic Newscastle's disease. Another factor affecting the availability of cotingas in particular was the increasing number of Central and South American countries prohibiting, or drastically limiting, bird exports commencing with Brazil in 1967. As of 2003, Surinam is the only country from which cotingas are regularly commercially exported. Until the imposition of the Wild Bird Conservation Act in 1993, CITES listings did not have a major effect on cotingas in US collections. Only two species are listed on Appendix I (prohibiting any commercial export), both endemic to Brazil's Atlantic Coastal Forest. I can find no captive records for the white-winged cotinga *Xipholena atropururea.* While, as noted above, the banded cotinga *Cotinga maculata* was maintained at the London Zoo as long ago as 1875, only occasional specimens have since been exported (one of the last living at St. Louis Zoo from 1964 until 1982). Although both species of cocks-of the-rock were placed on Appendix II in 1975, and the ornate and long-wattled umbrellabird on Appendix III in 1988, until 1993, their importation to the US was not impeded, as inclusion in Appendices II and III only requires documentation and does not prohibit commercial traffic. However, the Wild Bird Conservation Act prohibits commercial import of most birds listed on any CITES appendix, so that the only cock-of-the-rock importations to the US since 1993 have been under special exemptions for "zoological display." I do not believe any umbrellabirds, which were exhibited by a number of zoos in the 1970s, have arrived in the US since 1993.

As of 2003 cotingas from Surinam remain commercially available to both zoos and private aviculturists.

Birds reaching the US have been trans-shipped though Europe. The three most frequently exported species are the pompadour cotinga *Xipholena punicea* and spangled cotinga *Cotinga cayana,* both with brilliantly colored males, and the screaming piha *Lipaugus vociferans,* a gray, thrush-shaped bird with a remarkable call and lekking system. Other Surinamese species, such as the calfbird *Perrisocephalus tricolor* and the Guianan red cotinga *Phoenicircus carnifex* may be offered on dealer's lists from time to time. If the Surinamese extractive reserve proposed by the American zoo and Aquarium Association's Passerine Taxon Advisory Group is realized, a stable source of the above-noted birds as well as eastern cocks-of-the-Rrck, imported under special permit, may be anticipated for participating institutions. Paraguay has recently allowed limited export of birds, and small numbers of bare-throated bellbirds, mostly male, have appeared.

Lovely Cotinga
Cotinga amabilis

Andean Cock-of-the-Rock
Rupicola peruvianus

Management: While single species aviaries are ideal, quite satisfactory results may be attained in mixed species aviaries. Members of the genera *Pipreola, Cotinga, Xipholena,* and other cotingas of similar size have worked well with such birds as tanagers, honeycreepers, manakins, white-eyes, finches, and, when care is taken to allow for sufficient territory, hummingbirds or sunbirds. An exceptional case was a fatal scalping attack on a piping plover *Charadrius melodus* by a spangled cotinga female at the time a young cotinga fledged (Couch et al 2003). Bellbirds can be maintained with a wide variety of non-aggressive softbills. Cocks-of-the-rocks and umbrellabirds can prove dangerous to small birds and nestlings. Frank Todd documents that long-wattled umbrellabirds *Cephalopterus penduliger* "were seen to catch and consume small tanagers (genus Tangara)" (Todd et al 1972). A colleague told me of a small flock of white-eyes *Zosterops* sp. eaten in seconds after being liberated in a large walk-through aviary containing several Andean cocks-of-the-rock. Jan Van Oosten (1957) had no trouble maintaining a male scarlet cock-of-the-rock in the same aviary with white-eyes and yuhinas, which roosted near it at night. On the other hand he observed this bird hovering "like a humming bird" by the nest of a hybrid pair of bulbuls *Pycnonotus* sp. and removing a several-day-old chick. The cock-of-the-rock then "flew to a perch where he struck it a few time and then swallowed it whole." Van Osten (1957) witnessed this behavior repeatedly. While adult painted quail *Coturnix chinensis* in the same aviary were not molested, chicks disappeared at different times. There have been instances of cocks-of-the-rock displaying violently aggressive behavior towards humans. The dangerous eastern cock-of-the-rock mentioned by Bates & Busenbark (1963) was hand-tame, and thus possibly collected as a juvenile and imprinted. Cocks-of-the-rock are not immune to attacks from other birds. A particularly lamentable case was the killing of a male Andean cock-of-the-rock by a male golden pheasant *Chrysolophus pictus,* at a restaurant aviary in California. I imagine the bird's color may have made it a target.

A wide variety of cotingas have been maintained out of doors year round in southern California, where temperatures hardly ever drop below freezing. Long-wattled umbrellabirds at the Los Angeles Zoo were seen gasping when the temperature rose above 38°C (100.4°F). On the other hand, when ice formed on their drinking pools they did not appear to be uncomfortable (Todd et al 1972). In general, however, it is wise not to subject cotingas to extreme temperatures.

Cocks-of-the-rock are well known for enjoying waterfalls, misters, artificial rain storms and other water features, and it is likely most other cotingas also enjoy bathing. At least the green-and-black fruiteater *Pipreola riefferii* has been observed catching small fishes, while raising chicks, out of an aviary pool, where the birds were also noted to be "ardent bathers" (Everitt 1973). While spangled cotingas at the Milwaukee Zoo were never observed entering a pool, they regularly bathed on wet leaves following the daily hosing of their enclosure. Misting and light hosing of the Milwaukee birds also alleviated the tendency for food to cake on their beaks and feet (Couch et al 2003).

At the Milwaukee Zoo Spangled Cotingas appeared to be easily stressed by capture and such procedures as nail-clipping, and were found to tire quickly in capture attempts. This species, at least, appears accident prone. Both Milwaukee's breeding female and a recently fledged chick were found with their heads "stuck in palm fronds," with non-fatal results (Couch et al 2003).

Diet: For all their remarkable specializations, in general cotingas do well on a fairly standard diet for fruit-eating softbills and, once accustomed to eating in captivity, do not require diets consisting of items difficult to procure or requiring elaborate preparation. In the past, maintaining natural colors was a problem, especially with the various subspecies of Andean cocks-of-the-rock, which were inclined to fade out to a disappointing shade of pinkish-orange. Members of the genus *Pipreola* are also prone to severe fading. Don Bruning (1971) documents a Sclater's cotinga *Pipreola sclateri* (which I gather was a fiery-throated fruiteater *Pipreola chlorolepidota*) that, in the course of the year following its arrival at the Bronx Zoo, underwent dramatic color changes. The bright green faded to "a dull blue-gray, when it molted and grew new feathers", the bright red faded to yellowish-white, and the beak "dulled from a dark red-orange to a dull light yellow." The addition of canthaxanthin "beadlets" to its diet alleviated this fading to some extent, although results varied over a two-year period. The Bronx Zoo also attained satisfactory results with equatorial and scarlet cock-of-the-rocks though canthaxanthin, producing a shade "very near the natural color" (Bruning 1971).

Satisfactory color restoration in scarlet cocks-of-the-rock has also been achieved simply by feeding them tomatoes. The California private aviculturists George and Bessie Bray found that their birds initially refused to eat tomatoes but, by concealing avocado cubes beneath tomato cubes and withholding all other food, they eventually got them to eat tomatoes freely (Prestwich 1970). Of course since the 1980s tomatoes have been implicated as a food causing the iron storage diseases and hemosiderosis and hemochromatosis. I have been informed by Dr. Bernhard Neurohr of the Tiergarten Nurnberg that hemosiderosis is very common in cotingas, while some cases of hemochromatosis have been reported as well.

It is thus prudent to be careful in controlling the amounts of tomato, citrus, raisins, and other high-iron fruits in cotinga diets. At present, with the availability of commercially produced roxanthin, as well as proprietary diets containing color-enhancing compounds, feeding cotingas is fairly uncomplicated. Martin Vince (1996) recommends his fruit-based omnivore diet for all cotingas. This consists of 50% fruit, 33% proprietary softbill pellets, 10% chopped hardboiled egg, 5% vegetables and greens, 2% live food. Papaya, which is easily obtained wholesale in North America, comprises the major proportion of the diet at the Dallas World Aquarium, which currently maintains the most complete series of cotingas in the US.

At the Edward Marshall Boehm Aviaries, it was found that bare-throated bellbirds were very partial to newborn mice, so were each given one every other day; they also avidly consumed "as many mealworms and crickets as they could get hold of" (Everitt 1973). At the same collection, cocks-of-the-rock enthusiatically ate crickets. At the Los Angeles Zoo, umbrellabirds regularly received mice and day-old chicks.

Of course, as will be seen below, supplements of various kinds are generally required during nesting and chick rearing.

Breeding: To my knowledge, seven species of cotingas have been propagated in captivity. The bare-throated bellbird *Procnias nudicollis* was repeatedly bred at the Zoologischer Garten Wuppertal in Germany.

Bare-throated Bellbird
Procnias nudicollis

Volume 36 of the *International Zoo Yearbook* (Olney & Fisken 1998) indicates that the two unsexed birds hatched in 1995 and the male hatched in 1996 survived past fledging. I suspect the breeding took place in Wuppertal's Free Flight Hall, where a variety of hummingbirds and other softbills are exhibited. (Volume 36 of the *IZY* indicates Wuppertal also reared both taxa of Streamer-tailed hummingbirds *Trochilus polytmus* and *T. scitula* in 1996, while spangled cotingas, swallow-tailed manakins *Chiroxiphia caudata,* and vermillion flycatchers *Pyrocephalus rubinus* hatched in both 1995 and 1996.) One would imagine detailed accounts exist in German publications. The only extensive English language account of bare-throated bellbird aviculture is Lowe's (1979) observations of two males at the West Country Wildlife Park in the United Kingdom. To my knowledge, Wuppertal remains the only collection to have hatched any species of bellbird.

I have been informed by Marvin Jones that a private aviculturist in Chile recently hatched long-wattled umbrellabirds *Cephalopterus penduliger*, however, so far as I know, nothing has been published. On the other hand, detailed breeding accounts have been published in English for the other five species of captive-bred cotingas.

The first documented fully successful captive breeding of any cotinga is of the green-and-black fruiteater *Pipreola riefferii,* at the Edward Marshall Boehm Aviaries, at Trenton, New Jersey in 1962, discussed in great detail by the aviary's curator, Charles Everitt (1973). A male and two females were maintained in a planted aviary shared by six other species of birds. In April 1962, one female began nest construction in a basket in a hemlock, using dried grass, rootlets, and horsehair. Eggs were laid on April 17 and 18, incubation commencing with the first egg. Only the female incubated, but the male fed it on the nest, driving away the other female from the vicinity until she was transferred elsewhere. The male also stayed by the edge of the nest when the female left to bathe. Both eggs proved clear. The first of another clutch of two was laid June 12, the second three days later. Chicks hatched June 30 and July 2, the second chick dying after three days. Both parents fed the chick. For the first 10 days, only insects (mealworms, moths, and flies) were fed. After this, the parents began feeding fruit and raw ground beef, and the male caught small fish, which were fed both to the female and chick. Everitt (1973) noted the male swallowed the fish whole immediately after catching them. The chick was found dead at fourteen days. Following a division of the aviary to segregate the cotingas, the first of two eggs in a third clutch was laid July 24, the second on July 26. One egg did not hatch, and the second produced a chick that lived only two days. Shortly thereafter, while at work on a new nest, the female died entangled in horsehair. The female that had been previously removed was reintroduced to the male, that chased her for several days, but accepted her by the fifth day. This second female chose to build a nest in a box on a beam, just inside the aviary's entrance. Eggs were laid on of September 29 and 30, 1962. Everitt (1973) found this female would accept supplemental feedings by hand while on the nest, but not off it. A single chick hatched October 19, fledging November 9. Everitt (1973) provides much detail of this bird's development. While both parents had fed it on the nest, only the male did once it was fledged. At five months, when it had attained full adult male plumage, the male began harassing it, prompting its removal. No further breeding activity was documented.

Charles Everitt (1973) was perplexed that a pair of orange-breasted fruiteaters *Pipreola jucunda,* also at the Boehm aviaries, were never able to rear past 16 days, any of the chicks from several clutches. As it happened, the San Diego Zoo achieved full success with this species (exhaustively documented, with photographs, by K.C. Lint & James Dolan [1966]), in a very busy walk-through aviary containing a large number of birds. San Diego's famous hummingbird aviary measures 9.2 x 7.2 x 3.6 m high (64 x 24 x 12 ft.). At its opening on October 2, 1964, it held 164 birds of 56 species, including 20 species of hummingbirds, and 23 species of tanagers (Lint, 1966). Among the 13 remaining species of birds was a pair of orange-breasted fruiteaters, received June 29, 1964, and introduced to the hummingbird aviary October 28, 1964. Copulation was first observed June 26, 1965. The male was observed presenting mealworms to the female, a behavior seen only during the reproductive cycle. The male chose the nesting site, a hanging basket containing one of the many blooming fuchsias that have always been maintained in this aviary, eight feet above a visitor's bridge. While the male brought nesting materials to this site, the nest was actually constructed only by the female. Composed of grasses, rootlets, and sphangnum moss, the open-cup nest was completed in four days. The two eggs were laid on June 30 and July 1, and hatched July 16 and 17. Only mealworms were fed to the chicks for the first five days, after wich blueberries were also taken. While one chick disappeared at seven days, the other was fully parent raised. Its development is discussed in detail by Dolan and Lint (1966). *The International Zoo Yearbook* (Jarvis 1968) records that another chick was fully reared in 1967. To my knowledge, no further captive propagation of the genus *Piprola* has occurred. The last export of a member of this genus appears to have been in the early 1990s, when a small number of fiery-throated fruiteaters *P. chlorolepidota* became available.

It is interesting that while no captive breeding had been recorded for the genus *Cotinga* until the 1990s, the one species since bred has been comparatively prolific. Although the spangled cotinga *C. cayana* first appeared in aviculture in 1929 when the department store magnate J. Speden Lewis donated a pair to the London Zoo (Seth-Smith 1930), it remained an extreme rarity until the 1980s, when Dutch dealers began exporting it from Surinam. *International Zoo Yearbook* records (Zoological Society of London 1960–1998) commence in 1991 when one was hatched, but did not survive at Tierpark Berlin. The single birds hatched at Zoologishcher Garten Zurich in 1992 and 1993 likewise did not survive. Full success was achieved in 1995 when one was reared out of three hatched at Zoologischer Garten Wuppertal. One hatched but died at Wuppertal in 1996. The IZY also records two hatched and one reared at Burger's Zoo in Arnhem, the Netherlands in 1996. The first fully successful United Kingdom breeding for any cotinga was for this species at Amazon World on the Isle of Wight in 1999 (Coles 2003). The International Species Inventory System (ISIS) indicates one hatched at Zurich in 2003.

While American zoos began exhibiting spangled cotingas in the 1980s, the first three to achieve success did so within with in a few months from each other in 2001 (Azua, 2002). I do not know how the San Diego breeding birds were maintained, but at least one male chick was fully hand raised and then sent to the Milwaukee Zoo to be paired with a bird hatched there. San Diego's had-rearing techniques for this and other species are presented by Perretta (2003). Denver and Milwaukee parent raised their offspring. Denver's pair were housed by themselves in a one of a series of glass-fronted, planted aviaries in the entrance foyer of the Bird World building (Azua 2002). I am most grateful to Kim Smith, Curator of Birds at the Milwaukee County Zoological Gardens for making available a then-unpublished highly detailed account of spangled cotinga propagation there (Couch et al 2003). Milwaukee's breeding pair are maintained in an unusually spacious and private well-planted exhibit, 4.35 x 7.5 x 6 m high (14.5 x 25 x 20 ft). Visitor viewing extends only partway up the front of the exhibit, but this pair have proved to be excellent display animals, the male often perched only 1.5 m (5 ft.) off the ground. As of 2003, the only cagemates were a pair of double-striped thick-knees *Burhinus bistriatus,* although tanagers had been maintained in the past. Both cotingas were collected in Surinam in 1997, the female purchased in 1997, the male in 1998. They were introduced to each other in adjacent cages for 12 days in September 1998, then maintained together off exhibit for a further two months. In the spring of 1999, nests were placed in various sites in the aviary, and the female was observed placing material in one in March, but this did not progress. A "howdy cage" 6.7 m (22 ft.) above the floor was opened and outfitted with perching plants and a small cup nest. (Because this aviary was once part of an enormous free-flight hall, an upper level service area runs behind it, and the "howdy cage" is accessed from there.) An egg was laid at this site June 15, 1999 but, although fertile, did not hatch. No eggs were laid in 2000. In March of 2001, a shower curtain was installed behind the "howdy," concealing the service area behind it. An egg was laid April 7, 2001. (In contrast to *Pipreola,* a clutch consists of one egg.) Hatching occurred May 1, fledging on May 16. This chick was caught in the aviary's zoo mesh several times, prompting a nail trimming, and caught its head in a palm frond. "Some weakness in [the] wing" was noted, and attributed to an overabundance of fruit in the diet. However, this bird developed into a healthy adult and was shipped to the Toledo Zoo and introduced to a Denver Zoo-bred female. The second egg produced in 2001 was laid August 4. The chick hatched August 22 and fledged September 16. Although fledged with some wing problems and a crooked beak, it eventually developed normally as well. The third 2001 egg was laid October 22. This egg was removed by keepers November 20. Its failure to hatch was attributed both to the presence of the previously hatched chick, as well as increased nest checking and photography by staff. At the end of November, the female was removed from her enclosure (Couch et al 2003).

The female was reintroduced to the exhibit in early February 2002. Although the male began displaying at once (flying back and forth between high perches), the first egg was not laid until May 30, hatching June 14. This chick fledged July 12, and at seven months of age was paired with a San Diego hatched male, off-exhibit in Milwaukee. An egg laid August 8 was broken the same day. At this point the chick hatched in June was removed. Another egg was laid August 28, hatching 11 September 2002. That chick was still present when the fourth 2002 egg was laid October 24, and when it hatched November 15, producing a chick that lived only six days. In late November the "howdy cage" (where all eggs were laid) was closed until February 2003 and all other nest baskets removed, but the breeding pair was allowed to remain together, with more than one food station provided (Couch et al 2003). I have not heard if any reproduction occurred in 2003, nor am I aware of 2003 propagation in any collection in the US, where

ISIS records 14 males and 20 females were distributed among 11 institutions in November of 2003.

While, as previously noted, the calfbird or capuchinbird *Perissocephalus tricolor* debuted in aviculture in 1953 at the Bronx and Philadelphia Zoos, it has remained an extreme rarity in captivity and, as of 2003, none appear to be in collections, although Dutch dealers have recently offered specimens. This large and grotesque cotinga was hatched at the San Diego Zoo in 1992, and this breeding is thoroughly documented by Lewins et al (1994). San Diego procured a female in 1988 and a male and three females in 1990. Remarkably for a lekking species, the sexes cannot be readily visually determined, so surgical sexing was necessary. Subsequently, it was noted that the confirmed male weighed more than any of the females. In February 1991, a single pair was placed in an outdoor aviary measuring 3.7 x 6 x 3m (12.25 x 20 x 10 ft.). This exhibit contained a collection of tropical plants as well as three species of *Tangara* tanagers, orange-billed sparrows *Arremon aurantiirostris,* red-legged honeycreepers *Cyanerpes cyaneus,* and bare-faced ground doves *Metriopelia ceciliae.* No mention is made by Lewins et al (1994) of any of these birds being molested by the much larger cotingas. The male shortly commenced this species' famous vocalization displays, but the female did not respond in any observable way. However, on July 17, 1992, a nest was discovered in a Thunbergia vine roughly 2 m (6 ft.) off the ground at the back of the aviary. This nest was a platform of twigs, around 15 cm (6 in.) long, collected from the mulch used as a substrate. Further twigs were provided. The completed nest "remained a platform with only a slight indentation for an egg." The male continued to solicit the female. An egg was discovered August 7, but was broken the same morning following a skirmish, in the nest involving both birds. For the next week an absence of interest in the nest by the female and a decrease in courtship activities by the male were noted. When the female again displayed interest in the nest, the male resumed its earlier behavior. A second egg was discovered in the same nest August 24. When it was noticed the male was pursuing the female, which stayed clear of the nest area, the male was transferred to an adjoining enclosure. The female did not return to the nest until the male was moved out of sight and sound contact. Incubation lasted 26 days, during which the female invariably left the nest when staff entered the aviary, allowing for daily visual checks of the egg. Three days before the anticipated hatch date, a video camera and monitor were installed in an adjoining enclosure. Piping was noted on September 17. A weighing and close examination of the chick, afforded by the fact the female left the nest when anyone entered the aviary, was conducted the fourth day after hatching. Weights continued to be taken at five-day internals through day twenty, then two-or three-day intervals through day 29, when the chick left the nest. It was not observed to fly for a week, climbing through vines instead.

Throughout this time, all feeding was done by the female that was observed to feed "predominately protein," the main items being both common and giant mealworms, pinky mice, and anole lizards. The chick was not observed at a food pan until 61 days after hatching, when the female was observed feeding it fruit. The juvenile was first noted taking food itself at day 64, but it was not until January 27, when it was 131 days old, that the female refused to feed it. It was separated from its parent March 2, 2003, when it was also determined to be a female through surgical sexing. The male was returned to the breeding female in 2003, with another female housed in an adjoining aviary but, to my knowledge, no further reproduction occurred.

The first Andean cocks-of-the-rock *Rupicola peruviana* to breed in captivity were included among the first specimens to enter aviculture. Two of the twelve scarlet cocks-of-the-rock *R. p. sanguinolenta* brought to New York by Charles Cordier in 1941 went to the private aviculturist Mrs. Milton Erlanger, at Elberon New Jersey (Delacour 1942 & 1948). Eight years later, Jean Delacour found them "finger-tamed and quarrelling to a minimum" and noted, "The male, thanks to a careful diet of tomatoes and other appropriate food, has kept his brilliant deep red color, which in zoos too often turns dull orange. They have already played at nesting, and hopes are high for a brood some time!" (Delacour 1950). Delacour (1942) elsewhere mentions this pair was maintained in a converted kennel (Mrs. Erlanger bred Alsatians dogs). Delacour wrote nothing further of this pair. I had heard from K.C. Lint, long-time Curator of Birds at the San Diego Zoo, that a hatching had subsequently occurred at Elberon. However, the only published documentation I have read is a mention in passing by Jan Van Oosten (1957) that "Mrs. Erlanger some years ago had babies from cocks-of-the-rock and she fed them baby mice." Mr. Van Oosten informed me that he believed something was published in the avicultural society's *Journal of Aviculture,* which ceased publication in 1952. At any rate, this breeding, the first for any cotinga species in captivity, has generally been forgotten and is not mentioned in reviews of hatchings of this species (Berry 1980; Berry et al 1982).

From what I understand, the chicks hatched in Mrs. Erlanger's collection were not reared to maturity. This was to be the case for the next several collections hatching Andean cocks-of-the-rocks. At some date prior to 1979, the British softbill specialist Mrs. K.M. Scamell hatched, but did not raise, this species in her collection (Berry 1980; Rochelle Plasse, pers. com.). Prior to 1979, three US zoos bred it. The *International Zoo Yearbook* (Zoological Society of London, 1960–98) documents that the National Zoological Park in Washington DC hatched but did not raise one chick in 1972, hatched and lost one in 1974, and hatched and lost two in 1975. The same source indicates that one hatched but died at the Fort Worth Zoological Park in 1974, and one hatched but died at the Bronx Zoo in 1978. All of these breedings took place in indoor walk-through aviaries. Except for the Bronx bird, which hatched from an artificially incubated egg and was hand raised, all were parent reared until their deaths (Berry 1980). None of these chicks survived longer than 28 days (Berry et al 1982).

The first full success for this species was achieved at the Houston Zoological Gardens in 1979, and is exhaustively documented in two published accounts (Berry 1980; Berry et al 1982). The parent birds identified as the scarlet subspecies *R. peruviana sanguinolenta* arrived at a Florida dealer's compound in the same shipment. However the male was procured by Houston in 1975 (along with a female that died from a broken neck a month after arrival) and was then exhibited alone, while the female came to Houston in 1978. Since completing quarantine, the male had resided in Houston's indoor walk-through aviary, 40 ft wide, 70 ft long, and 25 ft high, shared with a wide variety of other birds. The new female was initially hosed in an exhibit aviary adjacent to the walkthrough "rainforest," with only a pire partition separating the two enclosures. After several days the female began plastering mud to the rockwork and the male courted her through the wire continuously with typical raucous calls. The female was released into the rainforest September 26, 1978, and began constructing a nest three days later, on top of a wooden nestbox 3.6 in. (12 ft.) above ground, in a recess behind an artificial rock wall. The nest could not be viewed by the public, but was visible from a service door. The nest was composed of mud, saliva, and plant material, and was lined with palm fibres, having "the appearance of the cone of a volcano" (Berry et al 1982). Despite the fact that the male's frenzied courtship intensified, a single egg laid November 23, 1978, and removed after thirty days, was found to be infertile. Another egg was discovered January 26, 1979, and a chick first observed February 14. Because the male was engaging in "incessant and continual courtship," it was removed to an adjacent exhibit.

In contrast to other cotingas that have bred in captivity, the female refused to feed mealworms or crickets to the chick, the only live food noted for the first several days being an occasional spider. It did accept Zu/Preem Bird of Prey diet and soaked dog food, but otherwise fed fish tail palm berries, grapes, soaked raisins, avocado, and papaya. However, the female was observed killing a live mouse, which it was not allowed to retain for fear it might have been poisoned. When staff then offered pinky mice, they were refused. It was discovered that small lizards (anoles and skinks) were avidly taken, the female feeding them whole to the chick after first crushing the lizard's bones. Five to six lizards were accepted daily. Nine days after hatching, the chick was found to have the toes of its right foot tangeled in the lining of the nest. Necrotic tissue from the foot was removed. From that day, the chick was removed from the nest daily for examinations and weighing. Through its 25th day, the chick steadily gained weight, but then began to experience fluctuations, so that by its 29th day, the chick had lost 30 g. Bob Berry, Curator of Birds at Houston, commenced hand feeding the chick at his house on the 30th day. Although the chick readily accepted food, it was obvious that the bird was ill when it passed an entire, only partially digested lizard in its feces. The bird died the following morning, and a necroopsy revealed gastro enteritis. In the meantime, the male had been returned to the rainforest, and two eggs were laid in April, but both were broken in the nest. It was suspected the male had done this.

Reproductive activity ceased when the female commenced molting in May. Shortly thereafter, it escaped out of the walk-through aviary and the bird house into the zoo grounds, but was enticed with food into "grabbing range" and then housed in a different exhibit until that autumn, when both birds were returned to the rainforest. Needless to say, the female's choice of a nesting site above the visitor's exit was discouraged. After experimenting with other sites, the female returned to the nest it had constructed in 1978, laying eggs on November 28 and 30. The male was at once transferred from the aviary. These eggs hatched on December 25 and 26. The female commenced feeding lizards immediately. Chick weighing commenced on the third day after this first chick hatched, and both showed excellent progress until the 23rd day, when the first chick showed no weight gain over the previous day. Fecal cultures were positive for salmonella for the older chick, although clear for the younger. It was decided to hand rear both chicks, Bob Berry again taking them to his house. As it was suspected the lizards might have been the source of salmonella, the chicks were fed cat food and Zu/Preem Bird of Prey diet rolled together into small balls. They also received diced grapes, applesauce and blueberries. After it was discovered grapes were not being digested, Berry reduced the diet to cat food, chopped pinky mice and blueberries. During this time, the older chick was being treated with gentamycin. The older chick died on the 27th day, and the following day, the younger chick, which had displayed no prior problems, exhibited diarrhea, a high temperature, and rapid respiration. Gentamycin treatment began at once. The chick accepted only minute amounts of food, which was fed at 20–30 minute intervals from 6 a.m. to 1 a.m. Injections of dextrose and electrolytes to counter dehydration were also administered. As the gentamycin appeared to be having no effect, oral tetracycline was prescribed instead, and within 12 hours, the bird exhibited marked improvement. Tetracycline was administered for five days, and buttermilk was added to the diet to restore microflora. By its 31st day, the chick was again at an hourly feeding schedule. At 46 days, it began picking up food, and began flight attempts at 49 days. This bird did not reach full independence until it was three months old. A female, it was sent to the San Diego Zoo sometime after 1981, when I saw it on public display at the Houston Zoo.

No further chicks were hatched by this bird's parents. Houston Zoo resumed Andean cock-of-the-rock propagation in 1987, this time with birds identified as the Bolivian subspecies *R. p. saturate*. The *International Zoo Yearbook* records a male hatched and fully reared in 1987, a male and an unsexed bird, both fully reared, in 1988, and a male, which did not survived in 1989. I have found nothing published on these hatchings. I am most grateful to Hannah Bailey, Zoo Supervisor, Birds at Houston, for providing me with information on these birds. All were parent raised in the rainforest. The male hatched on February 18, 1987 lived three years at Houston, as did the female hatched April 25, 1988. The bird hatched April 23, 1988 was sent to the National Zoological Park in October of that year.

1987 was also the year that breeding commenced at the only zoo outside the US that I know bred Andean cocks-of-the-rock. The Chiba Zoological Park is a municipal zoo on the outskirts of Tokyo. The *International Zoo Yearbook* records that two males and a female hatched there in 1987, the female failing to survive. Two females were hatched in 1988, one dying.

The final institution I am aware of having bred this species is the San Diego Zoo. This breeding is thoroughly documented (Kendall & Kuehler 1989). In the mid-1980s, three males and four females of the Bolivian subspecies *R. p. saturata* (including birds supplied by Charles Cordier in one of his last shipments) were established in the Scripps Aviary, a walkthrough exhibit 3.5 x 24 x 30 m (11.5 x 80 x 95 ft.) high (Lint 1960). A wide variety of neotropical birds were maintained there at that time, including curassows, oropendolas, herons, crested cariamas, and breeding toco toucans. So large a group of cocks-of-the-rock resulted in pronounced lekking behavior throughout the summer. In 1988 a nest composed of mud, saliva, dried grass, and maidenhair fern was constructed in an artificial rockwork cave (functioning as a staff service area). The single egg, discovered June 16, 1988, was at once removed for artificial incubation, due to the presence of so many other birds. It was placed in a Petersime, with a dry bulb temperature of 37.2°C (99°F), and wet bulb set at 30–31.1°C (86–88°F). Piping commenced July 4, but the chick did not complete hatching (which required assistance) until July 7. The initial diet was a slurry of Purina Cat Chow, Gerber's apple sauce and papaya, with water, fed hourly, from 7 a.m. to 7 p.m. On the third day, two feedings of soaked cat chow and papaya were added. The semi-liquid diet was discontinued by the seventh day, from which point an increasing variety of food items were provided, including fruits crickets, beef heart, chicken with meat and zebra finch meat. Through its 20th day, the chick was housed in a 10 gallon aquarium converted to a brooder. It was afterwards placed in a box brooder until it feathered out, then transferred to a small cage. The bird was considered fully weaned at 70 days. During this time, coccus infections of the skin were successfully treated. From the 49th day, toe curling was evident, but the removal of the infra red lamp and addition of further B vitamins prevented this condition from becoming serious.

This was to be the only cock-of-the-rock hatching at San Diego, as shortly thereafter the Scripps cage (built in 1922), and a large area surrounding it were renovated, and all birds were removed from this aviary. The bird hatched at Houston in 1989 was, to my knowledge, the last cock-of-the-rock hatched in captivity. Compiling ISIS statistics, Robert Webster (Pers. com.) found *Rupicola peruviana* to be the 20 most abundant passerine bird in North American Zoos as of December 31, 1984, with thirty birds distributed among 12 collections. By December 31, 1988, ISIS listed only 15 specimens distributed among five collections (with eight at San Diego). The enactment of the Wild Bird Conservation Act in 1973 ended commercial shipments to the US. As of January 1, 2003, ISIS listed only three specimens in North America: a pair at the Houston Zoo, and a female (recently imported under special permit) at the Dallas World Aquarium. However, in 2003 the Zoological Society of San Diego was permitted to import 15 *R. peruviana peruviana* through a Dutch collector. Groups have been set up at both the San Diego Zoo and Wild Animal Park, so captive propagation may resume shortly.

In marked contrast to the Andean cock-of-the-rock,

the eastern cock-of-the-rock *Rupicola rupicola* has never been bred in captivity. A highly detailed study of the behavior of two males and a female at the Kansas City Zoo has been published (Duce & Branian 1990) but all three birds died before mating was observed. Following a decade's absence of this species in US collections, the Dallas World Aquarium made a special importation of six males and six females in 2001. They are maintained in a specially designed off-exhibit rooftop aviary, generally entered by staff only twice daily. Their activities are continuously monitored with video cameras, so that any reproductive activity may be fully documented. As with other members their subfamily, any eventual success will be greeted with great interest by the avicultural community.

Postcript: Josef Lindholm, III, Senior Aviculturist, The Dallas World Aquarium

In the more than three years since the above document was completed, there have been significant developments in the aviculture of cotingas. At the same time, achievements of which I had been previously unaware have been brought to my attention.

Purple-breasted cotingas Cotinga cotinga, barenecked fruitcrows Gymnoderus foetidus, and purple-throated fruitcrows Querula purpurata have been exported in some numbers from Surinam, the later two being exhibited by several US collections.

The National Aquarium in Baltimore hatched a screaming piha Lipaugus vociferans each year from 2000 through 2003, as well as in 2005, though none of these chicks survived beyond one month. Further details, provided by the National Aquarium's senior aviculturist, Lori Smith, should appear the in 2007 volume of AFA Watchbird, in the second installation of an article I have written on contingas. Of particular interest, the male piha was discovered to be intolerant of male spangled cotingas Cotinga cayana (which was propagated at the aquarium), killing two in succession.

As of 2006, the hatching of three spangled cotingas at the National Aquarium in 2004 remains the most recent US breeding of which I am aware. A detailed account of the hatching and 2005 handrearing of this species by John Francis, a British private aviculturist, appears in the Avicultural Magazine, Volume 112 (2006), pages 19 to 23. Pompadour cotingas Xipholena punicea have produced eggs in the aviaries of a Florida private aviculturist, but none have hatched. Dr. Ulrich Schuerer, Director of the Wuppertal Zoo informs me that a presently unmated Guianan red cotinga Phoenicircus carnifex has repeatedly laid eggs. On the other hand Dr. Schuerer also informs me a male purple-throated fruitcrow was fully reared at Wuppertal in 2006. I believe this to be a world-first breeding. A pair in the huge indoor Mundo Maya exhibit at Dallas World Aquarium, have been exhibiting much courtship and nest material carrying as of spring 2007.

The first breeding of the long-wattled umbrella bird Cephalopterus penduliger took place in 2003 at Dr. Michael Durand's Centro de Aclimatacion Zoologica La Dehesa, in Santiago, Chile. This remains, to my knowledge, the only breeding record for the genus. This achievement has been documented by Luis J. Ortiz Merry, appearing in pages 108 to 116 of volume 110 (2004) of the Avicultural Magazine. This chick was hand-raised.

David Rimlinger, curator of birds at the San Diego Zoo, informed me of documentation of the laying of eggs in captivity of the Eastern (or Guianan) cock-of-the-rock, of which I had been previously unaware. This account, by the famed Italian aviculturist Paulo Bertagnolio appeared in the July 1970 number of the Avicultural Bulletin (Avicultural Society of America). A pair, imported from Brazil in 1963, did not appear to display reproductive behaviors until a second pair, of Bolivian origin, were placed in an adjoining aviary in the early summer of 1965. Shortly thereafter, the Brazilian hen began placing a variety of fibers in a "cement box 2x 2x 2ft. (0.6x 0.6x 0.6m) with a large opening" suspended from the roof of the aviary amidst shrubbery. Eggs were laid on 2 and 5 September 1965. Both were found broken on the ground after 10 days of incubation, following a violent nighttime storm. One of the eggs appeared fertile. In 1966 the breeding pair was transferred to another aviculturist, and the hen died that same year.

To my knowledge, no other eggs of Rupicola rupicola have been laid in captivity. At The Dallas World Aquarium, nest construction by two females has occurred each year commencing in December 2004, though no eggs have yet been laid. As of Spring 2007, five males and five females are maintained in an off-exhibit aviary. Until recently, all were kept together. In 2007, it has proved necessary to segregate some males, though they continue to be in visual and aural contact with the other birds. Favored display arenas are hotly contested among males, and females use the same nest site year after year. The existence of a ceiling or roof overhead appears necessary to conduce the commencement of nest construction. Nests are compact, solid low-sided bowls formed from mud and plant fibers. I have worked with this group since 2004, and have accumulated more than 70` pages of notes. At the time of their 2001 export from Venezuela, they were the only living specimens outside of South America, and remain so.

In 2004, The Dallas World Aquarium imported ten Andean Cocks of the Rock, through the agency of Parque de las Leyendas, in Lima, Peru. With the birds received by the San Diego Zoo and Wild Animal Park in 2003, and a female, previously held in Brazil, obtained by The Aquarium in 2000, they are, as of 2007, the only living specimens in North America. Eggs have been laid repeatedly at all three institutions, though, as the birds are mostly young, none have yet hatched. Since 2004, five Andean Cocks of the Rock, all male, have been fully reared at Wuppertal Zoo. Dr Schuerer informs me that all hatched in the free-flight tropical hall where Wuppertal's cotinga breedings have taken place. One was parent-raised, the others hand reared. The parents are on loan from the afore-mentioned Chilean collection of Dr. Durand. These are so far the only successes for a European zoo. The late Marvin Jones informed me that the Dutch aviculturist Jan van Osch has also hand-reared this species in recent years. Successes have also recently occurred at the Zoologico de Cali, Colombia.

References:

Anon. 1953. Fifty-eighth annual report for the year 1953. *New York Zoological Society*. 1960. Mossy-throated Bellbird at the Bronx Zoo, USA *International Zoo Yearbook* 1: 43.

Azua, J. 2002. Bird Division. Denver Zoo 2001 Annual Report. Denver Zoological Gardens.

Bates, H. & R. Busenbark. 1963. Finches and soft-billed birds. T.F.H. Publications.

Berdan, F. & P.R. Anawalt. 1992. *The Codex Mendoza*. University of California Press.

Berry, R.J. 1980 Unplanned parenthood. *Animal Kingdom* 83(No.4): 4–34.

Berry, R.J., W. Todd & R. Plasse 1982. Breeding the scarlet cock-or-the-rock *(Rupicola peruviana)* at the Houston Zoological Gardens. *International Zoo Yearbook*. 22: 171–175.

Bridges, W. 1974. *Gathering of animals* – An unconventional history of the New York Zoological Society. Harper & Row.

Bruning, D. 1971. Use of canthaxanthin to maintain the natural color of captive birds at Bronx Zoo. *International Zoo Yearbook* 11: 215–218.

Coles, D. 2003 First breeding records for birds reared to independence under controlled conditions in the United Kingdom. Dave Coles Books.

Cordier, C. 1943. The Umbrella Bird comes to the Zoo. Animal Kingdom 46: 2–10. 1947. A cargo of rarities from Costa Rica. *Animal Kingdom* 50: 171–178.

Cortes, H. 1962. Five letters – 1519–1526. W.W. Norton & Company.

Couch, H., M. O'Conner & E. Saksefski. 2003. Spangled Cotinga *(Cotinga cayana)*. AZA Regional Conference Proceedings 2003 (In litt.). AZA

Crandall, L.S. 1948. Notes on the display of the Three-wattled Bellbird *(Procnias Tricarunculata)*.Zoologica 33: 113–114.

1950. Charles Cordier's latest collection. *Avicultural Magazine*. 56: 205– 208.

Delacour, J.T. 1937. American Aviculture. *Avicultural Magazine*. Ser.5, Vol.2: 109–118.

1942. A collection of small birds in New Jersey. *Avicultural Magazine*. Ser. 5, Vol.7, 117–119.

1948. The Scarlet Cock-of-the-Rock. *Avicultural Magazine*. 50: 1–2.

1950. American Aviculture – 1949. *Avicultural Magazine*. 52: 62–66.

1974. The birds at Cleres in 1973. *Avicultural Magazine*. 80: 112–113.

Diaz del Castillo, B. 1996. The discovery and conquest of Mexico. Da Capo Press.

Duce, S. & J.D. Brannian. 1990. Social behavior of the Guianan Cock-of-the-Rock *(Rupicola rupicola)* in captivity. *Zoo Biology*. 9: 223–232.

Duran, D. 1994. The history of the Indies of New Spain. University of Oklahoma Press.

Everitt, C. 1973. Birds of the Edward Marshall Boehm aviaries. Edward Marshall Boehm, Inc.

Jarvis, C. 1968. Species of wild animals bred in captivity during 1966. *International Zoo Yearbook* 8.

Kendall, A. & C. Kuehler 1989. Artificial incubation and hand-rearing of an Andean Cock of the Rock *(Rupicola peruviana)*. *Avicultural Magazine*. 95: 153–158.

Lewins, E. L. Ordonez & W. Schulenburg 1994. Bredding the Capuchinbird. *Avicultural Magazine*. 100: 3–7.

Lindholm, J.H. 1988. Captain Delacour at the Bronx, 1940–1947. *Avicultural Magazine*. 94: 31–56.

Lint, K.C. 1960. Scripps flight cage at San Diego Zoo. *International Zoo Yearbook* 2: 72.

1966. Hummingbird diets and new walk-through aviary. *International Zoo Yearbook*. 6: 103–104.

Lint, K.C. & J.M. Dolan 1966. Successful breeding of the Orange-breasted Cotinga *(Pipreola jucunda)* in the San Diego Zoological Gardens. *Avicultural Magazine*. 72: 18–20.

Lowe, P.W. 1979. The introduction, management, care, and behavior of the Bare-throated Bellbird *(Procnias nudicollis)*. *International Zoo News* 160: 18–20

Mitchell, P.C. 1911. Longevity and relative viability in mammals and birds. Proceedings of the Zoological Society of London. 1911, Part 2: 425–555.

Perretta, N. 2003. Hand-rearing Passerines at the San Diego Zoo. American Federation of Aviculture Convention Proceedings 2003. (In litt.).

Prestwich, A.A. 1970. News and views. *Avicultural Magazine*. 76: 206–207.

Sclater, P.L, 1883. List of the vertebrated animals now or lately living in the Gardens of The Zoological Society of London. Eighth edition. Longmans, Green, Reader, and Dyer.

1896 List of the vertebrated animals now or lately living in the Gardens of the Zoological Society of London. Ninth edition. Longmans, Green, and Co.

Seth-Smith, D. 1930. The Purple-throated Cotinga. *Avicultural Magazine* Ser.4, Vol.8: 9–10.

Snow, D. 1982. The cotingas. British Museum (Natural History).

Todd, F.S., N.B. Gale, & J.R. Van Oosten 1972. El "Vaca del Monte". *Avicultural Magazine* 78:79–83.

Van Oosten, J.R. 1957. Adventures with softbills. *Avicultural Magazine* 63: 46–50.

Vince, M. (1996) *Softbills*. Hancock House.

Zahl, P.A. 1953 Exotic birds in Manhattan's Bowery. *National Geographic* 103: 77–98.

Zoological Society of London 1960–1998. Birds bred in captivity 1959–1996. *International Zoo Yearbook* 1–36.

21.04 Ptilonorhynchidae Catbirds & Bowerbirds

Neil Hamilton, Trena Carney & Simon Bruslund Jensen

Introduction: A total of 20 species of bowerbirds of the family Ptilonorhynchidae occur in Australia and New Guinea. All but the three monogamous species of catbirds in the genus *Ailuroedus* are polygynous birds, in which males are famed for constructing display-bowers and devoting a great deal of their time on maintaining and decorating these. Several species have been known to use or collect a wide variety of colorful objects and materials for their display bowers, preferences usually depending upon the individual species and where it is found. The females of these species care alone for the nest-building, incubation and rearing of the young.

Most species are plainly colored in brown tones; some are extremely beautiful birds with sharp contrasts of black and yellow or shiny gloss. They are all stocky build mid- to large-sized passerines that range from 25–35 cm (10–14 in.). Bowerbirds can be found in a wide range of habitats, from rainforest to eucalypt forest, mangroves, mallee, and even urban areas. The diet consists mainly of fruit, but also flowers, leaves, invertebrates and small vertebrates have been eaten. Currently none of the bowerbirds are considered endangered in the wild, although little is known about some New Guinean species. They may however be sensitive to habitat destruction in the future. (Frith & Frith)

A few species are now found in captivity, but very few are bred readily enough to no longer require collection from the wild. These are in captivity in Australia: the satin bowerbird, *Ptilonohynchus violaceus,* the regent bowerbird, *Sericulus chrysocephalus,* and the green catbird, *Ailuroedus crassirotris*. Outside Australia only the white-eared catbird seem to be really established in captivity. Many other species have been kept by a number of institutions, but primarily only for display purposes, with no serious breeding programs being undertaken.

Species like the great bowerbird *Chlamydera nuchalis,* spotted bowerbird *C. maculata,* western bowerbird *C. guttata,* fawn-breasted bowerbird *C. cerviniventris,* flame bowerbird *Sericulus ardens* and black-eared catbird, *A. melanotis,* have all been, or are currently being, kept by various institutions, and have occasionally been bred. The other 10 species have very rarely been held, let alone bred, in captivity. While many of these species have been housed at one time or another, there is very little recorded information on their management (Frith & Frith).

Maintaining Bowerbirds in Captivity in Australia

By Neil Hamilton & Trena Carney

Management: Captive requirements are similar for most species, except when bowerbird females have to be separated from the mature breeding males. However, this is not usually necessary when they are housed in a large, public, walk-through aviary. Satin bowerbirds are generally housed as pairs at Taronga Zoo, with the male being removed after egg laying has occurred. However, both Healesville Sanctuary and Currumbin Sanctuary have maintained groups of satin bowerbirds, each comprised of a mature male and female along with several uncolored males, in large walk-through aviaries, with successful breeding results. This arrangement is now being attempted at Taronga Zoo in a very large walk-through exhibit.

At Perth Zoo, the satin bowerbirds are kept in adult pairs or trios, one male to two females, using several side-by-side aviaries to facilitate introduction, breeding, and other management practices. This arrangement permits the quick separation of the males from the females, or vice versa, or the removal of offspring, etc., with minimal disturbance to the birds. When Perth Zoo first considered breeding bowerbirds, five aviaries were constructed for their breeding and long-term maintenance. These aviaries measure 2.7 x 4.5 x 2.7 m (9 x 16 x 9 ft.) in height. The aviaries are well planted as for Habitat A and, as bowerbirds are big plant eaters and destroyers, as many edible plant species as possible are used. It is important to provide plenty of decayed leaf mulch and fine sticks each season. Staff members replace all dead tea trees in the shelters, trim plants, and re-mulch these aviaries every autumn.

Introductions of bowerbirds can result in aggression, which is very stressful to the birds; this is why two well-planted, adjacent aviaries with interlocking doors are used. At least two feeding and drinking stations should be available. If an individual bird dominates one of these stations, the other bird will have an alternative food source. Close observation is very important during feeding. All species of bowerbirds are aggressive, not only to their own kind but also to other species found within their exhibit, especially at breeding time.

One of the main differences with the green catbirds at breeding time is that they are usually housed in pairs.

At this time captive birds should be monitored almost constantly. Once catbirds are paired they should not be moved, because they have been known to kill one another if unsettled or disturbed. This should be kept in mind whenever it is necessary to move them, especially when transferring, separating, or moving them to another exhibit or institution, as this is the time when most problems arise with these birds. A pair of catbirds requires a well-planted aviary and is best housed on their own. They can be kept in a mixed collection in walk-through aviaries for public display, but have been known to cause trouble with other birds.

Diet: The main diet of bowerbirds at the Perth Zoo consists of minced heart, ground dog kibble, cheese, egg mix (insectivore mix), SA 37 mineral and vitamin supplement, and diced fruits and vegetables such as bananas, papaya, rockmelon, grapes, pears, apples, oranges, tomatoes, watermelon, figs, currants, corn, carrots, lettuce, spinach, etc. A commercial finch or parrot soft food mixture is given with a few mealworms, fly pupae, and moths. This is fed to the birds twice a day. In addition, females with young receive live pinkie mice and moths three to five times per day. If the female does not have young, she will refuse the baby mice offered. With research indicating that cat food is more suitable for birds than dog food, one might consider replacing the dog kibble in this diet with cat kibble.

At Taronga Zoo, bowerbirds are offered equal amounts of insectivore, frugivore, and carnivore diets, as well as live insects daily. A more detailed breakdown of this diet follows.

Insectivore diet:

- 2.5 L Vetafarm Insectivore Diet
- 1.5 L fly pupae
- 34 mashed hardboiled eggs
- All ingredients are mixed to form a moist crumble.

Carnivore diet:

- 3 L Pal meaty bite minis (dog-kibble), soaked in 3 L water until soft
- 1 L Wombaroo Insectivore Rearing Powder
- After three hours soaking kibble, insectivore powder is mixed in.
- 1 kg kangaroo mince is mixed through the following morning while adding about 500 ml water.

Frugivore diet:

- 2.5 L finely diced apples
- 1 L currants
- 1 L sultanas
- 3 L finely chopped greens
- 2.5 L Paradise Pellets (manufactured by Zeigler Bros., USA)

The sultanas and currents are placed in the bottom of a bucket, and then apples and greens are added in separate layers, with the pellets placed on top. The following morning all ingredients are mixed together, with the following additions:

- 1 L finely chopped pear or grapes, in season
- 1 L frozen corn
- 1.5 L frozen berries
- 2.25 L finely chopped papaya

Live food used includes mealworms, crickets, native cockroaches, moths, and maggots. Regent bowerbirds are particularly fond of cockroaches when feeding nestlings. Pinkie mice are a popular food item when they have chicks in the nest.

Breeding: Bowerbirds need to have plenty of aviary space available, and aviaries should be located near each other to allow birds to interact and interchange to facilitate good breeding results. The bowerbirds also seem to do well in large walk-through aviaries at a ratio of one male to two females, but this is not ideal for maintaining the species for long-term breeding efforts. Immature males will build bowers prior to reaching breeding age. Two young male satin bowerbird siblings at Taronga Zoo constructed a bower cooperatively before they were 12 months old, while still in the same aviary as their mother. One side of the bower was constructed against a wall.

In addition, young, partially colored male regent and satin bowerbirds are capable of mating and siring young. At Taronga Zoo a young male regent bowerbird sired two separate clutches of chicks in one year. However, there were also several clutches of infertile eggs, suggesting that sperm count at this young age may be low, or some attempted matings unsuccessful. At the Healesville and Currumbin sanctuaries, partially colored male satin bowerbirds were also observed mating with females, with offspring being successfully produced.

Once full breeding plumage is attained, at 6–7 years

of age, the males will begin to chase females. At this stage, the male should have a beautiful bower already built and lined with blue, white, and yellow objects provided by the keeper. During this time, increased hassling and aggressive behavior between the male and the female can be observed, and it may be necessary to separate them for a while. The male continues to call and display through the wire. Due to aggression and possible infanticide, and also because the males take no part in rearing the chicks, males are generally removed from the aviary where the female is nesting as soon as incubation begins. However, at Taronga Zoo, it was found that it is not always necessary to remove male regent bowerbirds at this time, and multiple successful clutches of chicks of this species were produced with the male remaining in the same aviary throughout the breeding season.

The bower is constructed of very fine sticks closely knitted together. It is about 26 cm wide x 22 cm long (10 x 19 in.), with each wall about 20–30 cm (8–12 in.) in height, curving inward at the top, and about 8 cm (3 in.) thick at the base, facing north to south. We find the local Western Australian peppermint tree ideal for sticks for the male to build the bower and for the female to use as nest building material.

In captivity, male regent and satin bowerbirds will often maintain their bowers throughout the year. They do, however, devote more attention to it during the breeding season. Sometimes they ignore their bowers while they are molting, and in mixed-species aviaries this usually results in the bowers being damaged or destroyed by other aviary inhabitants. In Sydney, the molt normally occurs during late summer or early autumn. At this time, all of the head and neck feathers are replaced simultaneously, leaving the birds looking very tattered. It is possible that the high level of energy expended in completing the molt causes the temporary inability or lack of desire to maintain the bower or defend the site. At Taronga Zoo, the regent bowerbird males generally molted between late January and March, while the females generally molted between late January and April.

Regent Bowerbird
Sericulus chrysocephalus

When a male bowerbird is moved to a different exhibit, it will generally begin building a new bower almost immediately. In one instance at Perth Zoo, the male had built his bower and on August 20 the center dividing door was opened and the female was soon seen entering the male's aviary. The male immediately began following her around, calling and, at times, displaying to her. At one point the female was observed inspecting the bower, with the male following closely. The door was left open all day, but closed at night. On the following day, the partition was re-opened and the pair was observed together most of the day, with the male calling and displaying continually to the female. They were left together from then on. On September 6, the male was observed calling to the female, producing a number of soft rolling whistles, as she sat beside him on a branch above the bower. The female appeared to be mesmerized by the calls of the male and sat perfectly still. A short time later, the pair was observed tapping their bills together. The calling and displaying continued throughout the season with no breeding results.

The following year, the staff observed the female carrying sticks in her aviary. The resultant nest was 30 cm wide and 10 cm deep (12 x 14 in.), made of sticks only, and situated about 2.2 m (8 ft.) above the ground. It was only a few months later that the female eventually laid and incubated, at which stage the male was separated and locked into his adjacent aviary. The incubation period was 21 days. Eggshells on the ground in the far end of the aviary indicated hatching had occured; at the same time, the demand for live food increased. This particular female preferred to feed her chick moths, currants, soft fruits, and small pinkies. Other foods were offered but left untouched. Fresh food was offered up to six times per day. The chick fledged when 16 days old.

When a female has commenced incubation, the male can be used to mate with another female in an adjacent aviary. When the chick is independent, at about 45 days of age, it can be removed. The female then prepares a new nest, and the male is re-introduced when the nest is nearing completion. As soon as he senses that the female is ready to breed, his vocalizations and care of the bower increase again. It is important to re-introduce the birds prior to the completion of the nest, as infertile eggs might otherwise be produced. At Taronga Zoo, one satin bowerbird has bred in December on three occasions; another has bred in October, November, and December. In 1999, one female laid a single egg 41 days after the fledging of her first clutch, before the chick was fully independent. This egg was infertile, as the male had not been returned to the aviary following routine separation. The female satin bowerbird becomes extremely cryptic when incubating. This cryptic behavior continues until the eggs hatch, when the female can be observed foraging openly and carrying food in her bill back to the nest site.

During the 1999–2000 breeding season, a two-year-old, partially colored male regent bowerbird successfully sired two chicks. At only 12 months of age, this male was observed displaying to an adult female in an adjacent aviary. At 14 months, when several gold feathers were noted on the nape, he began building his first bower. At 18 months, he was decorating the bower and frequently displaying to the female in the adjoining aviary. At 19 months, the maxilla was changing in color from brown to yellow. The following month the female was introduced. By this time the veins in the primary flight feathers were yellow, and the maxilla was completely yellow. At 21 months of age the male was displaying to the adult female at the bower, and offering food. The maxilla and mandible were both fully colored at this time. At 30 months, the iris changed from brown to yellow.

The first bower constructed by this young male consisted of a platform of sticks approximately 10 cm high and 15 cm wide (4 x 6 in.). The sides of the bower were poorly constructed and emerged only a few centimeters above the base. The second and third bowers constructed by this bird were typical of those constructed by adult males of the species.

The female regent bowerbirds at Taronga Zoo have nested in August, October, November, December, January, February, and March. The March nest was a second clutch, following the post-fledging death of the first nestling. During the 1998–99 breeding season, one female nested in October, November, and March. All together, six eggs were laid but, as no male was present, none were fertile. During the next breeding season the same female nested in October, November, December, and February. A total of five eggs were laid, of which two eventually hatched. The time elapsing between nest construction and egg laying was 4–5 days. This female made use of the same nest on several occasions. The chosen nest site was generally about 1.8 m (6 ft.) above the ground, very near to the roof in the sheltered portion of the exhibit. The nest was constructed of small sticks and leafy brush from a *Melaleuca* sp. and various other plants, found in the leaf litter. Following a failed clutch, the hen would make a few repairs or alterations on the nest before laying again.

Chick Development and Fledgling Behavior: The staff at Taronga Zoo has kept records of the daily weight gain and development of several wild-hatched and captive-bred bowerbirds chicks (see table).

A newly fledged satin bowerbird at the Taronga Zoo was observed defending itself against two bush stone-curlews *Burhinus magnirostris* that shared the same enclosure, by throwing itself onto its back and kicking out with both feet.

Satin Bowerbird
Ptilonohynchus violaceus

Age (days)	Weight (g)	Stage of Development
1	-	Head and back covered in long black down at hatching.
3	26.93	
5	32.9	
6-7	33.5	
8	47.3	
9	60.0	
10	60.0	Pins emerging on wings and flanks.
11	79.1	
12	73.6	Eyes open.
13	80.0	Pins emerging on neck.
14	86.5	
15	97.7	Pins emerging on chest.
16	107.0	
17	107.2	
18	117.6	
19	122.3	
20	118.1	Tail feather emerging from pins.
21	117.5	Fledging.
22	119.1	
23	121.8	Pins emerging on scalp.
24	122.6	
25	127.4	
26	124.5	
27	133.0	
28	128.7	
29	131.4	
30	127.0	
31	125.0	
32	128.3	
33	127.2	Eating some food independently.
34	133.7	
35	140.0	
36	135.1	
37	135.7	
38	144.5	Head fully feathered.
39	145.0	
40	152.6	
41	141.5	
42	142.2	
44	138.0	Tail full length.
48	154.0	
51	170.0	
58	176.0	
76	193.0	

Daily weights of one captive bred, hand-reared satin bowerbird and the average development notes of three different chicks reared at the Toronga Zoo.

The development of one regent bowerbird chick has also been documented. However, as this bird was parent reared, the observations were made and recorded opportunistically, rather than regularly and systematically, and the data are consequently less precise.

At eight days of age the eyes were fully open. The head and back were covered in long white down. The primary and secondary flight feathers were emerging; these pinfeathers were approximately 1.5 cm (0.6 in.) in length. At this age the chick was very inactive, sitting quietly in the nest. At 18 days, the chest was fully feathered and the chick was perching on the edge of the nest. It fledged the following day, but perched quietly, still showing little activity. The contour feathers of the newly fledged chick completely covered its chest, belly, and wings, and the flight feathers were nearly full length. The head and the back, however, were still covered in down. Very short tail feathers were beginning to show, but had not yet emerged from the sheaths.

By 37 days of age, the fledgling was making very short flights, up to 1 m (3.3 ft.) in length. At this stage it actively followed the mother, constantly begging for food. The contour feathers were present on the back, but the head was still downy. The tail was still quite short, but the feathers were emerging from the sheaths. At 42 days, the chick was making flights of up to 10 m (33 ft.). The tail feathers were now 8–10 cm (3–4 in.) in length, and contour feathers now covered most of the head.

Four eggs of the regent bowerbird measured 39.2 mm x 26.3 mm (1.54 in. x 1.04 in.), 37.9 mm x 26.8 mm (1.49 in. x 1.05 in.), 36.5 mm x 26.8 mm (1.44 in. x 1.05 in.), 39.8 mm x 25.6 mm (1.57 in. x 1.00 in.), for an average size of 38.35 mm x 26.4 mm (1.51 in. x 1.04 in.).

Ring size: The proper ring size varies according to species. The following table gives the sizes recommended by the Australian Bird and Bat Banding Scheme 1997.

Band Sizes Recommended for Bowerbirds (ABBS 1997)

Species	ABBS Band	Type	Internal size diameter (mm)
Fawn Breasted	06 (07)	AY	4.5 (0.18 in)
Golden	05	AY	4.0 (0.16 in)
Great	08	SS	6.5 (0.26 in)
Regent	06	AY	4.5 (0.18 in)
Satin	08	SS	6.5 (0.26 in)
Green Catbird	08	SS	6.5 (0.26 in)
Spotted Catbird	07	SS	5.5 (0.22 in)

SS = stainless steel
AY = magnesium-aluminum alloy

Maintaining Bowerbirds, in Captivity, Outside Australia

By Simon Bruslund Jensen

There are old records of successful breeding satin and regent bowerbirds in the UK at the beginning of the 20th century. Satins have since been kept in zoos such as Rotterdam, Holland and San Diego, USA, but never breed well and have never established in aviculture outside Australia.

Today only a few species are kept, none are really common and only one can be considered established in captivity, and this only if care is taken to manage the captive populations better in the future.

Fawn-breasted Bowerbird *Chlamydera cerviniventris*

Chlamydera cerviniventris have been kept in recent years in several zoos and by a few private aviculturists. This species is comparatively drably colored even for a bowerbird and lacks the ornamental neck-feathers processed by its close relatives the greater, spotted and western bowerbirds. The male and female are identical in appearance. They favor an open habitat, preferably grasslands with some scattered trees or forest edges. They are distributed both in Northern Australia and New Guinea.

In captivity these fawn-breasted bowerbirds are somewhat overlooked, probably owing to their plain plumage and lack of knowledge about the family; in some instances they are wrongly identified, and have been exhibited as ovenbirds (pers. obs.). In captivity, as in the wild, males construct spectacular display bowers. They are typically elevated with a mat of sticks and debris up to 30 cm (12 in.). On top of that, they build an avenue bower in the same style as the satin and regent, however, with tighter walls and up to 90 cm (36 in.) long. The walls and the area in front of the bower are decorated with, in particular, green objects such as fresh leaves. In two facilities it has been observed that one male maintains two bowers at the same time, however, always with a strong preference for one of the two (pers. obs.). Females appear to be rather rare in captivity and the best way to identify one is through DNA sexing, although adult birds that do not produce a bower mostly also prove to be female. Most facilities keep the fawn-breasted bowerbird in a mixed aviary or a walk-through exhibit. They can be dominant towards other birds but physical aggression is rarely recorded; however, care should be taken with smaller birds.

Breeding: Little is known, but presumably it is very similar to that of other bowerbirds. At the San Diego Zoo two young were parent reared in two separate clutches in 2002. The incubation time seems to be unknown, but chicks fledge the nest after 22–25 days (Frith & Frith).

Diet: They do well on a fruit-based diet, preferably chopped into cubes with the addition of live food and commercial fruit-eater products such as mynah pellets or universal paste. No records of iron storage disease have been made yet, but it is advisable to maintain a low iron diet (pers. obs.).

White-eared Catbird *Ailuroedus buccoides*

Ailuroedus buccoides is regularly kept in Europe, North America and Asia, they breed fairly well, but there is no organized population management and frequently individual variation or even subspecies differences are considered as a sexual dimorphism leaving many "pairings" problematic. This species is distributed in New Guinea and nearby islands, and is found in lowland forests. Pairs seem to mate for several subsequent breeding seasons, possibly even for life. In captivity care should be taken when pairing up catbirds, on occasion some aggression can occur if the pair is not compatible. Preferably the birds are offered a multiple choice of partners, but they do not do well in a classic flocking situation. Ideally the birds are offered visual access through the mesh of several adjunct aviaries or are introduced to each other pair wise in turns. Males may be slightly larger on average but visual identification of sexes based on physical variation is frequently mistaking, there is evidence that all four geographical variations of this species have been available to aviculture in Europe (pers. obs.). Compatible pairs are best set up in large well planted flights, they can be kept with other birds and conflicts are rare although caution is advisable for smaller birds and nest predation have been suspected (pers. obs.).

Breeding: Pairs will construct the nest together, typically they will not accept nest platforms and will prefer to build their own nest in the thin branches or very often at the base of a thin stem towards the top of a low tree or bush. The nest is usually placed relatively close to the ground at around (1.6 – 2 meter up) even in a very high enclosure. It is well hidden but often leaves the bird exposed from above and with a good view of its surroundings. The nest itself is constructed by mainly sticks lined with grass, moss and some times even large green leaves, it can be rather fragile and often breeding attempts ends on the ground possibly due to the lack of a suitable nesting place. Speculations that catbirds like other bowerbirds make a display-bower or court of leaves on the ground have not been confirmed in captivity, catbirds can however be destructive to vegetation and seem to eat some parts of leaves and then drop the rest on the ground, some times leaving a pile under their favorite perches. In an aviary they are very sedentary to specific perches or trees and can often be seen using the same nesting spot again and again even if not successful (pers. obs., Frith & Frith).

The female incubates the eggs alone and is well camouflaged when sitting on the nest with the mottled green upper-side, in captivity one to three eggs have been laid and are incubated for 22 days. The young are covered in dark from day one and they develop fast and fledge after about three weeks. After fledging they are still cared for, for up to six weeks, however the pair may start a second clutch and care should be taken that the young are still feeding adequately and that they are isolated from the parents as soon as they are independent. The male remains close to the nest and often feeds the female during incubation and rearing. The young are mainly feed by the female until fledging after which the male feed the young routinely. During the time where young are being reared additional live food should be offered, most types are taken readily including mealworms, zophobas, crickets, rose-beetle larvae, fly larvae, wax-moths and grasshoppers also fruits and soft-food is used for feeding the young (pers. obs. Frith & Frith).

Diet: When not breeding the main diet is fruits, greens should also be offered regularly. They will readily take commercial soft-food or pellets and in particular live food. They are prone to become obese and care should be taken that vegetable matter form the vast majority of their diet (pers. obs., M. Edwards pers. comm.).

Breeding the flame bowerbird, *Sericulus ardens,* at the Al Wabra Wildlife Preservation, Qatar

By Simon Bruslund Jensen

Introduction: The spectacular looking flame bowerbird, *Sericulus ardens,* has long been considered a subspecies of the masked bowerbird, *Sericulus aureus,* but has recently been attributed species status based on its unique morphology (Frith & Frith). It is larger than the related regent bowerbird, *Sericulus chrysocephalus,* with a much shorter tail, and its soft silky plumage makes it look even bigger. The natural habits of the flame bowerbird are among the least known of any bird. It is rarely seen, let alone photographed, in the wild due to its shyness. Birds maintained at the Al Wabra Wildlife Preservation (AWWP) offered some of the first records of the species' nesting biology and behavior, as two different pairs reproduced successfully for the first time in 2003.

Management: At AWWP the flame bowerbirds are housed in breeding aviaries similar to those used for birds of paradise. Each aviary has access to an inside shelter that is climate controlled, maintaining around 22–25ºC (72–77 ºF) throughout the year. The outside facility is well-planted and covered with shade cloth from above; the front is not shaded and allows for some direct sunlight in the morning.

Birds are kept individually in adjunct aviaries, however, the birds do dominate each other so two birds at a time have visual access to each other. For periods of the year where breeding is likely to occur, the pairs are given access to each other. If too much aggression is observed or the birds appear disinterested, they are separated for some time.

The outside aviaries used for one bird are 18–35 m2 (59–115 sq ft.) and 3.5 m (11 ft.) high. Inside shelters are 10–14 m2 (33–46 sq. ft.) and 2.5 m (8 ft.) high and provided with daylight and artificial light 12–14 hours a day, depending on season. They are relatively inactive much of the time and usually stay in the upper regions of the aviary; but they will move freely around during the day and utilize all parts of the aviary, including the ground. When kept together with other bird species they are timid compared to other bowerbirds and are easily bullied even by smaller birds.

Diet: The birds are fed inside twice a day. The daily diet for one bird is:

Morning:

- 50 g papaya
- 30 g pear
- 5 g mango
- 2 g parsley or dandelion
- Dice and offer together with 15 g Ziegler bird of paradise pellets.
- For variety, interchange 10 g of orange, banana, chili, broccoli or seasonal fruits from day to day.

Afternoon:

- Small slice of papaya
- 15 g diced apple
- 10 g dry Wittemolen Mynah Granulate Care

Once a week, two mealworms are offered per bird. Vitamin and mineral supplements are offered four times a week alternating Nekton S, Nekton MSA, and Nekton Tonic F. The flame bowerbirds are particularly fond of fresh greens and consume large amounts of fresh leaves from a variety of trees and bushes. They have also been observed digging in the substrate of the outside aviary with their bills, and appear to occasionally eat some soil.

Breeding: The bowers produced by two different adult males at AWWP were small and mostly without additional decoration. Relatively little interest has been shown for colorful items placed in the aviary, apart from testing them for edibility. Occasionally, green leaves have been offered in or around the bower. The bowers only receive minimal attention from the females; in one

Two Hand Reared Flame Bowerbirds *Sericulus aureus* at AWWP

Day	Weight in grams	Development
1	11.9	Long silvery to brownish-grey duns covering whole body, skin bright orange, Duns dry 5–8 hours after hatching.
4	16.5	Pins on wings 2–4 mm dark grey, almost black.
9	32.5	Skin color becomes paler; pins on neck (dark) and on breast (pale white) emerge. Eyes halfway open.
11	36.3	Feathers start emerging from pins on wings.
16	53.8	Feathers start emerging from pins on breast, and pins visible under skin on head.
22	69,8	Fledging.
25	77.4	Pins on back emerging, feathers emerging from tail pins. Wings fully feathered.
35	107.9	Yellow feathers on breast emerge, head fully covered with pins.
45	126.4	Starts taking food from dish. Feathers start emerging from pins on head.
120	135	Start first molt into juvenile plumage.

case a pair mated and the female proceeded to rear a chick without the male having a bower.

Males have a very fast wing-flapping and head-bobbing display associated with varied vocalization that certainly seems to catch the attention of the females. This display is performed both on high perches and on the ground, and appears to be triggered by the presence of a female. If the female approaches a displaying male, the display is often interrupted by a short pursuit through the foliage.

The female constructs a large nest of twigs and vines lined with a thin layer of leaves and grasses. Square plastic baskets for aquatic plants, measuring 18 x 18 x 15 cm (7 x 7 x 6 in.) have been readily accepted for nesting when placed high in the aviary. The female will use the same nest site and even the same structure repeatedly without making significant additions to the nest. Egg laying has occurred in April, May, and July, the latter for second clutches. The egg is relatively large, averaging 41 x 28.5 mm (1.6 x 1 in.) and 15.8 g (0.6 oz.) fresh weight and, in appearance, very similar to the eggs of other bowerbirds, in particular those of the genus *Chlamydera*. In all cases at AWWP only a single egg have been laid per clutch. The incubation lasts 21–22 days.

A few days before a hatch is expected the female is offered fresh food several times daily and the diet is enriched with additional live food, soaked pellets, and meat from newborn rats. Initially six daily feedings are offered, with the frequency gradually reduced until the chick is almost independent. Parent-reared chicks fledge after 21–23 days. They are relatively active and move around the enclosure, however, they do not return to the nest. Newly fledged flame bowerbird chicks are only about a third of adult size and their plumage is poorly developed.

The female continues to care for the chick for up to eight weeks, but chicks have been left with their mother for as long as three months. Females will leave the nest due to the slightest disturbance both during incubation or chick brooding, however, they are not overly sensitive to nest inspections and soon return to the nest. The otherwise shy females become rather aggressive towards the keeper approaching the nest or fledged chick, and will call loudly and engage in agitated wing flapping to divert attention.

Hand Rearing: Newly hatched chicks have bright orange skin and remarkably long silvery duns covering the entire body. The head is comparatively large and the average weight is 11.9 grams (0.4 oz.). Initially, chicks are offered water and Lactated ringer 1:1; after the chick has absorbed the yolk (typically after 12–24 hours) food is offered. The chick has an audible begging pip and displays "bobbing" with the entire body. The gaping response is short, and the chick may not respond again until the next feeding if the opportunity to offer food is missed.

After 4–6 days the eyes "slit" and after a week the skin color becomes paler. After 22 days the chick will fledge, and shortly after starts pecking at food held with the forceps. At about 50 days the chick should be eating independently. If the brooder temperature is too high the chick will become restless in the nest-cup. Chicks are placed in small plastic bowls lined with paper tissue and carbon fiber filter material that ensures a firm grip; lining is exchanged daily. A fecal sack is not provided after every feeding.

Ring Size: 5.5 mm

Species Requiring Similar Management:

Cotingas Cotingidae
Satinbirds Cnemophilidae
Bowerbirds Ptilonorhynchidae
Birds of Paradise Paradiseaidae

References:

Forshaw, J.M. & W.T. Cooper. 1977. The Birds of Paradise and Bower Birds. Sydney: Collins.

Frith, C.B. & D.W. Frith. 2004. The Bowerbirds, Ptilonorhynchidae, Bird Families of the World, Oxford University Press.

Christidis, L. & W.E. Boles. 1994. The Taxonomy and Species of Birds of Australia and its Territories. Melbourne: Royal Australasian Ornithologists' Union. Monograph 2.

21.05 Maluridae Australian Wrens

Neil Hamilton, Shane Larsen & Richard Matkovics

The Australian wrens belong to the order Passeriformes and the family Maluridae, which contains two subfamilies: Malurinae and Amytornithinae. The first of these subfamilies is further divided into two tribes: Malurini, the fairywrens, and *Stipiturini,* the emuwrens, of which there are three species in the single genus *Stipiturus*. The second subfamily consists of the single genus *Amytornis,* the grasswrens, of which eight species are recognized. The emuwrens and grasswrens are confined to Australia, while nine species of fairywrens, all belonging to the genus *Malurus,* are found in Australia. An additional four species of Malurus fairywrens, as well as two species of fairywrens in the monotypic genera *Clytomyias* and *Sipodotus,* are found in New Guinea.

Each Australian species of *Malurus* is restricted to a particular habitat. An experienced birdwatcher can predict the presence of a particular species of by locating areas of vegetation that suit its requirements. Fairywrens occur in a variety of habitats, ranging from woodland margins to heathlands in the east of Australia, arid scrublands in central Australia, and rocky ridges in the north. They are often seen in family groups of up to seven birds, usually consisting of a pair with juveniles, which resemble the female. Sexes are dimorphic, with males having an eclipse plumage.

A significant reference for this chapter was the, as yet, unpublished *Draft Husbandry Manual for Maluridae* by Richard Matkovics of Taronga Zoo.

Management: The history of Maluridae in captivity has shown little long-term success achieved, with most species generally petering out over a relative short period. The key to success in housing and breeding of all Maluridae in captivity for long-term management is sufficient planning before acquiring any species. Many zoos and fauna parks have tried to keep many different species in small numbers at the same time in captivity; this has resulted in a high turnover of stock and a need to replace specimens from either the wild or other zoos/fauna parks regularly. If long-term sustainability is desired, it is advised to keep only one or two species of this group. This family has a high metabolic rate and losses will fluctuate between seasons, but will be high regardless of how good the husbandry.

The aviary must face away from direct weather such as prevailing winds, rain, drafts, etc., and still allow a little sun. The enclosure must be vermin proof and positioned so that disturbance is minimal. An aviary constructed of steel pipe frame covered with medium-gauge wire mesh, a metal roof, and a solid rear wall, with an adjoining safety alley, is recommended. Various substrates can be used, such as sand, soil, gravel, or mulch. Wrens do very well with only a few square meters of space, but will behave more naturally in a larger aviary. Some of our aviaries are hexagon shape with each side about 2.5 m (8 ft.) wide, providing a total floor area of approximately 16 m² (172 sq. ft.). The aviary is about 3 m (10 ft.) high. For Maluridae at Taronga Zoo, the smallest display exhibit is approximately 6 x 4 x 4 m high with a 2 x 3 x 3 m rear shelter. We furnish our enclosures to resemble the natural habitat while still creating an appealing display for public view. We provide old dead logs, rocks, leaf litter, etc. We also have specially made racks on which to hang *Melaleuca* branches, offering additional nesting sites. Generally stereotypic behaviors do not occur in captive wrens. If separate pairs are housed adjacent to one another, with only a weldmesh divider, males may develop stereotypic chasing behaviors. As they defend their territories and and females, they will chase one another along the divider. This behavior may then take up so much of their time that it can affect the success of their breeding. It is therefore suggested that a gap of at least 1 m (3 ft.) is left in between adjacent pairs. A pond is not essential, but always looks good in an aviary. A water receptacle should be provided and a feed station for live food and other feed. Generally, the plant life found in and around the natural habitat of the species is used in the aviary. Here we use bottlebrush, *Grevillea,* woolly bush, tea tree, Geraldton wax, *Melaleuca,* snake vine, and various species of grass tussocks. The plant density should provide an average of 0.5–0.6 m² (1.6–2 sq. ft.) per bird, arranged to allow for flight. It should be thickest toward the rear of the enclosure. When being transported, wrens are best confined to a carrying box with a fine gauze-covered window, containing some dense vegetation to provide security. Wire netting should not be used, as the birds will damage themselves on this. Preparation is vital for appropriate design of a malurid enclosure. The most commonly made mistake in exhibit design is lack of pest proofing. The number one rule to keeping small passerines alive is rodent prevention. Every exhibit must be rat-proof or savage loses will be incurred as these small birds are easy prey for large rodents. In other northern aareas of Australia, various other pest species, such as tree snakes, may have similar effect.

When introducing birds you must abide by some simple principles to alleviate fatalities. Introduce single-sex groups and pairs into a neutral territory whenever possible; although, in a heavily planted aviary breeding pairs can be introduced into each other's territory without much problem, careful observations will be necessary for the first few hours. Introducing and removing an individual bird or birds from an established aviary group is a tough assignment, as birds will not remember individual family members if removed and then re-introduced to the family group. If veterinarian procedures are required on individual birds, try to perform this in-situ or return it within 30 minutes or the bird could be killed.

Bird traps are the preferred method of capture of Maluridae for bird safety, but they can be time consuming. A small cage such as a suspended cage (Noegel cage) is an ideal way to capture Maluridae in an aviary, as it is less stressful to the animals. Capture is quite a simple procedure; just place a cage near the food station a few days before capture date, and put bird food in the cage. Attach a fishing line to the door of the cage and then through the back of cage to allow the door to be pulled closed from a distance. It is also advised to hold food for a few hours on capture day to make the birds hungry. When ready, find a comfy seat out of direct sight of the birds and wait patiently. Maluridae ideally should be transported singly to avoid aggression, but boxing established pairs or siblings is fine and may be preferred, as they will support each other. Birds from unrelated backgrounds should never be placed together for transport, as stress-related aggression can occur.

Maluridae do not have significant recurring health problems, but because of their high metabolic rate and their very small size (weights ranging from 7–14 g [0.24–0.49 oz.]) if a health problem does occur, the bird will generally die before disease is detected and/or treated. The most important factor in maintaining health (or treating ill-health) in high-metabolism insectivores is heat. Many of the wild-caught birds carry coccidia, which should be monitored and treated.

Wrens can be housed with other bird species such as small parrots, lorikeets, and finches, but for breeding purposes it is better to house them alone. For successful breeding results, wrens should be housed one pair per aviary. Keeping them with other wrens can lead to problems.

Diet: In the wild, fairy wrens collect a variety of small insects, including wasps, weevils, ants, termites, and other invertebrates. Termites might comprise up to 80% of the diet of some species. They will also eat a small quantity of seed. Maluridae are true insectivores (approximately 98% of diet is invertebrates) and have a high metabolic rate. They are active throughout the day gleaning off shrubs, rummaging amongst the leaf litter and hawking from perches. As mentioned, insect-attracting plants are an essential requirement for this genus, as this supplements their "artificial diet". Parents rearing chicks are often seen carrying very small insects to their nests, especially during the first week of rearing. They have never been observed carrying larger live foods such as mealworms. This demonstrates that vegetation is a very important feature when housing or breeding Maluridae in captivity. The most common mistake made when keeping and feeding maluridae in captivity is the assumption that wrens they will live on a diet that does not contain live foods. The addition of a provided diet is crucial to long-term survival of all species in this genus. A supplemented wild-insect diet is important but not enough to keep wrens alive year around. During cold snaps and harsh weather conditions, providing live food is essential. Without it, birds will tend to suffer and considerable numbers may be lost during such a period.

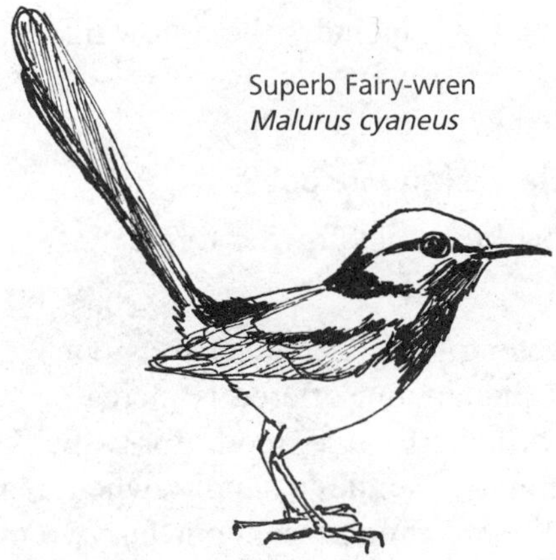

Superb Fairy-wren
Malurus cyaneus

At Perth Zoo, staff feed the fairywrens on fine mince, insectivore egg mix, mealworms, fly pupae, termites, moths, and a pinch of diced mixed fruit, which we consider optional. To attract insects to the aviary, a purple insect light may be installed, or old fruit skin like banana can be left in the aviary to decompose, attracting all sorts of insect life.

The fine mince mix we use consists of the following:

- Finely minced heart, with no fat
- Ground dog kibble, in powder form
- S.F.40 vitamin supplement
- Grated cheese, carrot, and sometimes apple

Our insectivore egg mix includes the following:

- Hard-boiled eggs, mashed
- Commercial insectivore powder
- Ground dog kibble, in powder form

The diet for breeding birds is the same as our regular captive diet, but the amount offered is increased when chicks hatch, particularly live food, for which they demand a seemingly endless supply when rearing young. Outside the breeding season, our fairywrens are on a twice-daily feeding regime, early morning and late afternoon. During the breeding season, we increase feeding up to six times per day, with the final feed being very heavy in order to last overnight.

Taronga Zoo Non-breeding Diet (maintenance diet): Food is provided for one pair of Maluridae in a flat dish. Only a small amount of dry food is presented with live food (mealworms). This ensures the live food will be easily sighted and eaten by the birds. Remember that movement triggers wrens' appetites, and if the mealworms are underneath the dry food then it is likely the birds will not eat the food.

Daily Requirements (Passerine Tray) A.M:

- 1/4 cup Taronga insectivore mix
- 10–20 mealworms
- For grasswrens also add a tablespoon of mixed finch seed.
- For emuwrens offer smaller sized (juvenile) meal worms

Daily Requirements P.M:

- 10–20 Mealworms, scattered throughout the open flight of the enclosure.
- Smaller mealworms for Emu-wrens

Weekly Requirements:

- 10–20 crickets (small to medium)
- Ad lib Maggots (house fly)
- Ad lib Moths

Mealworms must always be offered as a daily food requirement on morning food dishes as stated above. They are added to dishes in early a.m. food preparation to allow the worms to gut-load insectivore mix.

Breeding Diet: Note this applies to birds that are rearing chicks. If birds are housed in single-sex groups or not breeding, then non-breeding diet should be offered. The breeding diet live food amounts can vary depending on how many previous clutches and current chicks are with the parents. The key is to observe the birds, and if the parent wrens are extremely eager or waiting for food in the morning feeds, then increasing amounts of food offered should be considered. In the breeding season the diet should follow the maintenance diet as above, but increase the a.m. amount of mealworms to 20–30. Increase the crickets to 100–150 per week and offer daily.

Moth traps are very helpful to passerine management as they give zookeepers the opportunity to catch a live food diet, which is unavailable commercially and is a cost-efficient way to supplement the diet for Maluridae. Moth traps are used regularly throughout the year, especially during breeding season. They are of simple construction and use an exhaust fan placed in a PVC bend pipe and with a circular fluorescent UV tube placed at the front to attract moths to the entrance and a fan placed in reverse will suck them into a waiting cloth bag.

Breeding: Mating is usually preceded by a sequence of displays interspersed with chases; the male chases the female in a series of energetic flights. Displays involving plumage vary with different species; for those species where the males have elongated ear tuffs of contrasting colors, these are incorporated into displays. Well-planted flights are the key for success, with a dominance of grasses, sedges and low growing, insect-attracting native shrubs. Malurids build a dome shaped nest in low vegetation made of grasses and other vegetation. While the normal breeding season is August to January, fairywrens may breed throughout the year if conditions are right. Nesting materials used include grass, twigs, coconut fiber, palm fiber, strips of paperbark, spider webs, fine rootlets, cobwebs, emu feathers, and soft down feathers. Human hair, cotton, and clothing material should not be offered, as these materials are unsuitable for nests, being dangerous to young birds. Maluridae at Taronga Zoo have had the coincidental good fortune of having aviaries within close proximity of each other allowing many birds near each other calling in their own small territories (they can be single-sex groups). This stimulates the males that have females into courting their female more quickly (as males think the other males that are calling from adjacent territories can take their female). This has been a good accidental trigger for breeding at Taronga Zoo.

In the wild group, sizes vary between species. Generally, it is a family group comprised of one adult male, one adult female and their offspring, which may be comprised of males from the past few seasons and young of that year. In some species of wrens, like white-winged wrens *Malurus leucopterus,* they retain the young of many seasons and can form a group of up to and over 20 individuals. In the natural state, breeding groups may include young males, which are subservient to the dominant male, and these birds often retain their eclipse plumage longer than the adult breeding male. Unlike other malurids, white-winged wrens will suppress their breeding color for many seasons and will not color until they acquire a territory of their own. The molt of male fairywrens is directly dependent upon sexual activity. The breeding male of a family group retains his bright nuptial coloration longer than subservient males, and may pass an entire year without molting into the eclipse plumage. Adult breeding females are territorial and will not allow another female in their area. All young females raised are driven from the territory at the onset of the next breeding season, while young males may be tolerated.

All wrens are banded at Taronga Zoo generally in May/June when juveniles start to color. At the same time, all the family groups get processed into new aviaries awaiting new pairings for the next season. Family groups must be split after every season (just before new wild-breeding season starts. Times may vary depending on geographical location). This is extremely important, as adults will attack young very quickly and aggressively. In the wild, adults would chase young out of their territory. In captivity, as the young cannot get away, adults will continue to chase and attack often to the point of death. Young males have been left with their father for two seasons with the same female, but all other sibling juvenile females removed at the end of season each year. The young of the first few clutches of a season will help defend future clutches in the same season from attack and defend the territory. Single-sex grouping is one of the most important tools for long term management of Maluridae in captivity. This enables keeping large numbers of Maluridae in one institution without depending on other institutions for support, as malurids are small and susceptible to losses during the season.

Rufous-crowned Emuwren
Stipiturus ruficeps

There are some very important rules when establishing single-sex groups in captivity they are:

1) Keeping family groups, of, for example, father and sons, mother and daughters, together for the first six months until younger birds "find their feet" (although single birds can be taken from the groups if necessary for next season's breeding).

2) When separating single-sexed family groups (if required for pooling large numbers of single-sexed groups together) it is recommended to separate the family groups completely. They form sibling bonds and will gang-up and attack other wrens that are introduced. It is better to have single males from each breeding group, as there are no allegiances.

3) It is recommended, when establishing single-sex Maluridae groupings within an a single institution, to avoid keeping males and females in vocal and visual contact. At Taronga Zoo males are kept in several aviaries on display and the females are housed of off-display, as single-sex group birds have previously paired in separate aviaries then attacked siblings within their enclosures. Breeding pairs of birds can be housed in the same area as single-sex groups as long as the aviaries are separate and the wire does not meet, as the birds are already pair bonded and not calling for a mate.

4) When introducing birds, always do so in a neutral territory. Birds will remember their territory and defend it. If you take them away from their enclosure for more than a week they will lose their bond with their territory. Therefore, as long as bird movements are planned, single-sex groups can be housed in any location.

5) An all-male, single-sex group can not be placed in an adjacent aviary to an all-female, single-sex group, or in the near vicinity, as the birds will form a pair bond over quite some distances and then attack the other single-sexed bird in their

enclosures. It is advised, if possible, to keep all the males near each other and the females in another area. Males or females can be placed together in adjacent enclosures with their own sexes for long periods of time, as long as these points are followed.

6) Single-sexed grouping will only work in zoos and fauna parks that do not have similar species found within or around their grounds. The captive birds will form a pair bond with the wild birds and then kill the remaining birds housed within their enclosure. So the only other options are to keep species of Maluridae that are not similar to any wild malurids found in the area, or only keep pairs of the same species in which single sexing will not work.

Three eggs form the usual clutch, with incubation lasting 13–14 days, with only the female incubating. The chicks fledge at 13–14 days of age. Often 2–3 clutches are raised per season, with youngsters from early clutches assisting with the feeding of those from subsequent broods.

Fairywrens normally select a nesting site in thick live shrubbery, a dry bush or a dense clump of grass. The nest is usually positioned anywhere from 0.5 m (20 in.) to 4 m (13 ft.) above ground. The nest is a coarse, bulky, oval ball approximately 100–150 mm (4–6 in.) in diameter, with a side entrance. Nests are well hidden in foliage and can seldom be seen without disturbing the vegetation.

As each new breeding season approaches, all young from the previous season still remaining with the parents should be removed to another aviary. Preferably, they would already have been separated within two or three weeks of fledging, and banded with numbered metal or color-coded plastic rings for identification purposes. The enclosure should then be renovated for the new season by providing fresh tea trees or other bush cover as needed. Nesting material such as palm fiber, coconut fiber, grasses, and feathers should then be scattered over the aviary floor. Whenever possible, breeding aviaries for fairywrens should always have small water dishes with a stone in the center, as newly fledged young may drown in large water containers.

As they become older, the time in eclipse plumage gets less by the year until they don't lose color at all. At Taronga there is one individual of *M. l. lamberti* that is 12 years old and retains his breeding plumage throughout the year, only getting a dash of new brighter feathers and a slicker sheen in the breeding season.

Species Requiring Similar Management:

Splendid fairywren *Malurus splendens*
Red-winged fairywren *M. elegans*
White-winged fairywren *M. leucopterus*
Purple-crowned fairywren *M. coronatus*
Other fairywrens *Malurus* spp.
Cisticola species

Reference

Matkovics, R. 2003. *Draft Husbandry Manual for Maluridae* (unpublished).

21.06 Meliphagidae Honeyeaters

Glen Holland

Introduction: The honeyeaters comprise a large family of birds restricted to the Pacific region. Australia is particularly well endowed with honeyeaters, with every state in the country having at least one endemic species. They belong to the order Passeriformes and the family Meliphagidae, which encompasses 182 species in 42 genera. Among the typical honeyeater genera are *Lichmera, Meliphaga, Melithreptus, Phylidonyris, Xanthotis, Conopophila,* and numerous others. Many species in the family are called by common names other than honeyeater. The 30 species in the genus *Myzomela,* for example, are called myzomelas, and the 11 species in the genus *Melidectes* are referred to by that generic name. Species in the genera *Philemon* and *Melitograis* are known as friarbirds; those in the endemic Hawaiian genus *Moho* are oos. Other species and genera are known by such common names as straightbills, spinebills, miners, myzas, chats, and wattlebirds. Species in monotypic genera include the New Zealand bellbird *Anthornis melanura,* the tui *Prosthemadera novaeseelandiae,* the STITCHBIRD *Notiomystis cinctus* (see the section devoted to this species) and the Australian gibberbird *Ashbyia lovensis.*

Honeyeaters are found in habitats ranging from dry open forest, woodland and scrub to riverine woodland and tropical forests. The singing honeyeater *Lichenostomus virescens* is found over almost the entire Australian continent and, like many of its relatives, is common in parks and gardens. Generally, honeyeaters are territorial. Some species are found in groups, and their calls are loud. Much of the information obtained in this article was extracted from the *Husbandry Manual for the Yellow-tufted Honeyeater* prepared by Michael Miller of the Healesville Sanctuary, Victoria, with additional information obtained through personal communication with the staff of the Adelaide Zoo.

Management: Healesville Sanctuary has housed breeding birds in single pair aviaries of a number of different sizes. The smallest was 5 x 3 x 4 m (16 x 10 x 13 ft.) high and the largest was 10 x 3 x 4 m (33 x 10 x 13 ft.) high. A colony of four pairs was also established in an aviary 26 x 10 x 4 m (85 x 33 x 13 ft.) high. This aviary was heavily vegetated and had many corners and places into which less dominant birds could retreat. Heavily vegetated aviaries are preferred for a number of reasons. Live vegetation not only supplies plenty of cover for the birds, but also attracts insects into the aviary on which the birds can feed, supplies much nesting material for the birds, and provides many activities for the birds, to prevent boredom. The best types of plants will have twiggy growth to supply nesting sites, loose bark for nesting material, dense vegetation and flowers to supply nectar and attract insects. Plants of the genera *Eucalyptus, Banksia, Melaleuca,* etc., are ideal for these purposes. At Adelaide Zoo, three pairs of regent honeyeaters *Xanthomyza phrygia* were tried on the colony breeding system; all three females nested, although two of the females mated with one male. Large quantities of nesting material and bushes were supplied. Depending on the species, Habitats A or B, or a mixture of these, will suit honeyeaters.

The main diseases of honeyeaters in captivity are coccidiosis and aspergillosus. Although coccidia oocysts are generally shed in low numbers and *aspergillus* is present in most aviaries, these begin to affect the birds when stress is added to the equation. Most honeyeater species are noted for their aggressive temperament, and intra-specific trauma is a significant cause of mortality in captivity. Incompatible pairings, over-enthusiastic breeding males, or accidental pairings of same-sex birds all result in aggressive behavior and/or stress, which usually leads to dead birds. Aggression toward birds in adjoining enclosures has also contributed to mortality. Aspergillosus associated with the stress of constant harassment is often noted in post-mortem examinations. Coccidia are also prevalent when the birds are under stress.

Generally, it is possible to house yellow-tufted honeyeaters *Lichenostomus melanops* with other bird species that can either compete successfully with, or are larger than, the honeyeaters themselves. Due to the aggressive nature of the males, females are more easily managed in a mixed collection, but sufficient cover should always be available to allow birds under threat an opportunity to escape. Life threatening aggression is exhibited by the male relentlessly chasing the female and not allowing her to feed, whereas during normal courtship aggression he allows the female time to feed, rest, and collect nesting material. Aviaries with sufficient cover can house more than one pair, but it has been found that only the dominant breeding pair will nest successfully. In Australia these birds have been housed with lyrebirds, pigeons, doves, lapwings, fairy wrens, whistlers, and kingfishers.

Diet: Honeyeaters may gather in small flocks at suitable nectar flows. The main diet consists of plant exudates such as nectar or manna. The captive birds at Healesville are fed Leadbeater mix, insectivore mix, chopped soft fruit, mealworms, and maggots.

Leadbeater Mix:

- 1 liter honey
- 1 liter hot water
- 2 hard boiled eggs
- 0.5 teaspoon Sustagen powder
- 0.2 ml Pentavite
- 125 gm high protein baby cereal

Dissolve the honey in the hot water, place mixture in the food processor, add eggs and Sustagen. Blend together well. Add Pentavite and add high protein cereal to desired thickness. Fruit can be added to the mixture prior to blending.

The daily ration for each bird is:

- 15 ml Leadbeater mix, increased to 25 ml in summer
- 3 gm Insectivore Mix
- 3 gm chopped soft fruit
- 3–4 mealworms
- 5–6 maggots

At Adelaide Zoo the daily ration of live food is fed in the morning in a bucket with a perch fitted a few inches from the bottom. The birds can feed from the perch as necessary, but crickets and other live food cannot escape. This saves a lot of time by avoiding multiple feedings during the day.

During the breeding season, the insect supply must be greatly increased, as the chicks are fed almost exclusively on insects, and the parents will not rear the chicks if a sufficient number of insects is not provided. Live food in the form of moths, crickets, and flies should be provided at three-hour intervals during daylight hours.

Breeding: Courtship display consisting of increased vocalizations, the carrying of nesting material and a wing quivering display begins within a few days of the winter solstice. The male blue-faced honeyeater *Entomyzon cyanotis* has been observed flying to the female with a mealworm as she waited on a perch with her wings lowered and her quivering tail lifted. Following copulation, the male ate the mealworm. Successful copulation is usually indicated when the male dismounts from the female rather than the female disengaging from the male. During introduction to potential partners, multiple feed stations and refuge sites are required, as well as frequent keeper observation to detect incompatibility. Most effective pairing occurs by opening up a series of aviaries to allow birds to select their own mates. The yellow-tufted honeyeater and its subspecies, the helmeted honeyeater *L. m. cassidix,* are not sexually dimorphic in plumage, but males tend to be heavier, larger and more vocal. In the natural state there may be as many as four nest attempts per season. One pair of helmeted honeyeaters had nine unsuccessful attempts in the one season, while another reared eight young from four nests in one season. The birds nest in neighborhoods, sharing territory defense between a number of adjacent territories.

An increased supply of live food will act as a breeding trigger. The female constructs the nest and incubates

Blue-faced Honeyeater
Entomyzon cyanotis

the eggs. Both sexes care for the young and there may occasionally be assistance from other non-breeding and breeding birds from the same neighborhood. The cup-shaped nest is suspended by the rim in the outer branchlets or wedged in a fork in the inner branches of a shrub. It is constructed of bark and grasses, bound with cobwebs and decorated with spider egg sacs. It is lined with fur, feathers, fine grasses, leaves, or plant down. In captivity, nesting material such as cotton wool, coconut fiber, and feathers should be provided in addition to the materials already mentioned.

Usually two eggs, but occasionally three, are laid. The female alone incubates the eggs, which take 14 days to hatch. The chicks remain in the nest for 10–14 days, during which they are fed by both parents. The chicks usually fledge at about 12 days, and nest inspections should be avoided, as the chicks are likely to fledge too early if disturbed. At fledging, the chicks are incapable of controlled flight, and the maximum depth of any water in the aviary should be 10 mm. The chicks become independent at about 40 days. Live food should be provided a number of times daily. The re-nest period for honeyeaters is often within one week of a previous brood or clutch.

Helmeted honeyeaters may become sexually mature in their first year. First year females have laid eggs, but many have proven infertile. First year males have fathered chicks. Females breed better in their second year. Occasionally birds other than the parents may assist in feeding chicks. Common honeyeater species have been used as analogue species for the breeding of rare and endangered species, and cross fostering has also been used successfully.

Artificial Incubation & Rearing: At Healesville, a Marsh Rollex fan-forced air incubator is used. The eggs are placed in a small tray lined with cotton wadding and manually turned four times daily. Automatic turning is not used as it is likely to damage the thin shell of the small honeyeater eggs. Incubators are run at 37°C (98.6°F) with a relative humidity of 65%. The desired daily weight loss is 1.7% and the total weight loss is 22%. Hatching temperature is maintained at 37°C (98.6°F), but the relative humidity is increased to 80% two days prior to hatching to ensure the egg membrane does not dry and prevent hatching. Chicks at Healesville have all hatched within 24 hours of the first pip.

The keepers at Adelaide Zoo found that turning regent's eggs too frequently can be detrimental; turning the eggs manually 4–6 times per day is sufficient. They also found that the eggs would only lose 10–11% of their weight even when the incubator is run dry.

Their standard approach to incubating honeyeater eggs is now to run the unit dry as long as they can achieve a 10% weight loss, and the hatch is usually fine. Adelaide staff also noted that in clutches of three eggs incubated naturally, the embryo in the third egg usually dies from cooling during the hatch. For this reason they routinely pull the third egg once the first two begin pipping. It is hatched in an incubator and the chick is then returned to the nest.

The temperature in the hand-rearing room for yellow-tufted is initially set at 28°C (82.41°F) and gradually reduced so that when the chicks are 40 days old the ambient temperature prevails. Higher initial temperatures, i.e, 37°C (98.61°F), are required for newly hatched chicks. Blue-faced honeyeaters, however, require lower temperatures, and initial temperatures of 35°C (95°F) are reduced by one degree (1.81°F) per day until the ambient temperature of 24–27°C (75–80°F) is reached. Initially, the chicks are fed half-hourly through a 12-hour day, but by 20 days this is gradually reduced to 15 times per day and by 30 days of age to 10 feeds per day. The young are independent by 40 days. The hand-rearing diet consists of pawpaw, moth species captured in traps at night, freshly killed houseflies, and nectar. Nectar mix was usually introduced in small bowls when the chicks were 25 days old. Prior to this, moths dipped in nectar mix were occasionally fed to the chicks. Items that have been dipped should be wiped to avoid excess nectar, which can result in asphyxiation. Blue-faced are larger than the yellow-tufted, and have been raised on chopped pinkie mice, moths, and fly pupae.

I would recommend that the hand-rearing protocol developed by Adelaide Zoo, described below, be followed for all species of honeyeaters.

Four regent honeyeater chicks were hand reared from the egg by the bird department at Adelaide Zoo in 1999. Two chicks were from naturally incubated eggs, and two from eggs that had been artificially incubated full term. All chicks developed into healthy, quality birds with weight gains consistent with those of parent-reared birds. Initially, the parent-incubated chicks showed greater vitality and a better attitude than the incubator-hatched chicks. This was noticeable during the first 2–3 days in the form of stronger and more frequent gaping during the feeding procedure. For this reason, it is important to catch every gape, if possible, when feeding artificially-hatched chicks. Missing several feeding opportunities in a row could conceivably result in dehydration or in a failure to energize the chick sufficiently during the critical first few days. This problem diminishes as the chicks progress. However, to reduce the occurrence of this problem, it is recommended that, if possible, eggs should receive at least some natural incubation during development.

As with most passerine species, the regent honeyeater chicks responded to almost any noise or movement by gaping. It was found that during the first few days, unless activity within the nursery was performed quietly, the chicks were often "gaped out" by the time they were taken from the brooder to be fed. They were brooded in individual containers so that each chick could be taken out and fed without triggering a response from the others. Again, the problem of "gaping out" was reduced after the first few days as the chicks stabilized and became strong enough to gape when necessary for feeding. A high pitched whistle was used to elicit a gape response from chicks during feeding.

A still-air brooder, as opposed to a fan-forced unit, is strongly recommended for young passerines. When using a still-air brooder, it is important to take temperature readings at the site of the chick, due to the variable temperature zones found in such units. The regent chicks were brooded in a Brinsea Hatchmaker still-air brooder. Even a single degree of temperature over or under the optimum is significant during the first 4–6 days; indeed, many of the problems encountered with young passerine chicks are temperature related. Unless the nursery itself is heated, it is recommended that chicks be fed under a heat source while they are outside the brooder.

Brooding passerine chicks on tissues can lead to splayed legs, as their feet and legs continually push through to the side. A brooding material such as ChuxTM can be used, providing the feet a better surface to grip.

Honeyeater chicks are subject to dehydration, and it is felt that effective initial re-hydration, along with continued maintenance hydration for the first 5–7 days, is critical for success. The regent chicks were fed individual drops of fluid several times during day one, and each food piece was dipped into the fluid mix before being fed. Direct fluid feeding was discontinued after the first day, but each piece of food was dipped into the fluid before feeding until the chicks were almost fully feathered. Good initial hydration was further enhanced by not feeding the exoskeleton or outer shell of food items until day three.

The fluid mix, which should be kept warm during feeding, is as follows:

*10 ml Lactated ringers
0.5 TkS Wombaroo Lory Powder
0.5 TS Heinz applesauce
0.5 Hills A/D protein xxx feed

After Day 3 the ringers was replaced with water.

Yellow-tufted Honeyeater
Lichenostomus melanops

All food pieces were fed off the blunt end of a toothpick until the chicks were large enough to be fed with plastic, blunt-ended tweezers.

The diet consisted of 90% insects and 10% fruit. The insects used were mealworms, crickets, and large fishing maggots. All were prepared in some way until chicks were close to weaning. Mealworms had the heads removed, most of the internals squeezed out and small pieces of exoskeleton added. The crickets had the heads, legs, and wings removed, and the abdomen cut open. Maggots were prepared in the same way as the mealworms. Fruit included small pieces of grapes, bananas, and paw-paw.

The consumption at each feeding was affected by the volume taken at the previous feeding. This is acceptable and normal, and a chick must never be force-fed. It was also noted that feed response diminished anywhere between 7:00 p.m. and 11:00 p.m., as the chicks had consumed enough for that day and were ready for the night break.

Blue-faced *Entomyzon cyanotis* & Singing *Lichenostomus virescens* Honeyeaters:

Day	Weight (g) Blue-faced	Singing
1	7	2.32
5	20	6.18
10	51	14.6
15	74	17.2
20	82	21.2
25	92	22.8
30	100	23.4
36	103	

During hand rearing, strict hygiene is essential and equipment should be sterilized between each feed. Cages and brooders must be cleaned regularly. Feeding tweezers are used for one task only. Separate tweezers should be used for collecting fecal material. Aspergillosus is the most common problem encountered when hand raising honeyeaters, but can easily be eliminated with good hygiene. Chicks are best kept in plastic containers with a tissue paper base, which is comfortable, provides grip, and is easily replaced on a regular basis. Imprinting on humans has been a problem and puppet gloves have been used to avoid this, with some success. The best method to rectify or prevent this is by housing the birds in juvenile groups. Reduced human contact after subsequent placement in an aviary will also diminish imprinting. Hand-reared birds have subsequently nested and raised their chicks without any problems.

Ring Size: 3–4.5 mm

Stitchbird *Notiomystis cincta*

• Rose Collen, Glen Holland, Caroline Twentyman, Jerry Pauli & Betty Watt

Introduction: The stitchbird *Notiomystis cincta,* a New Zealand endemic, belongs to the order Passeriformes, the family Meliphagidae, and the genus *Notiomystis,* of which it is the sole member. The family Meliphagidae, which includes the honeyeaters, miners, friarbirds, and related species, is distributed primarily through Australia, New Guinea, Timor, and Fiji. Two additional endemics, the bellbird *Anthornis melanura* and the tui *Prosthemadura novaeseelandiae* are the only other members of this family occurring in New Zealand. The stitchbird, known locally by the Maori name *hihi,* is one of only two honeyeater species to nest in tree cavities. The name stitchbird originates from the contact call which resembles the word "stitch" but both sexes have a soft warbling song, an excited alarm call and males include a three note whistle and are known to mimic other species. In pre-European, times (Europeans introduced rodents and mustelids) they were distributed throughout the North Island and a number of the surrounding islands. Currently the only self-sustaining population resides on Little Barrier Island. The Department of Conservation has successfully re-introduced them to a further two islands where intensive management and monitoring is still underway.

The Mt. Bruce National Wildlife Centre, which is currently the only facility holding stitchbirds, is operated by the New Zealand Department of Conservation and is dedicated to the breeding and restoration of rare and endangered avifauna of New Zealand. The Department of Conservation stitchbird recovery plan established a captive breeding program in order to develop effective husbandry techniques to be used in the event of disaster in the wild population. Such trial techniques can assist in establishing new self-sustaining populations. The plan also acknowledges the environmental education role the species can play. Several challenges remain, such as further disease research, development of hand-rearing protocols, and testing designs for new aviaries that are easily cleaned and disease free.

In poor fruiting and flowering seasons, the natural reproductive rate of stitchbirds is low, with most second and third clutches lost or deserted. In the future, this phenomenon could be the key to providing a source of birds for hand rearing, by collecting these clutches that would otherwise be wasted. In addition, disease-free birds for further research could be hand raised from eggs collected from the captive population.

Management & Breeding: Stitchbirds at Mount Bruce are held in large bush aviaries, measuring at least 10 x 10 x 4 m (33 x 33 x 13 ft.) high. Each aviary has an interior divider, creating two flights. The aviary floor is natural, with fresh leaf-litter added regularly to boost insect numbers. Insects, particularly winged insects such as moths, are an important component of stitchbird diet, as attested by the amount of time they spend foraging in the vegetation for insects, especially when rearing chicks. The large size of the aviary allows the many plant species inside to grow to maturity, providing some of the fruits and flowers the stitchbirds would use in the wild. This large volume of vegetation also provides good habitat for an abundance of invertebrate life.

Adult birds have to be held in separate flights for most of the year, as their behavior is very aggressive and competitive, not only between individuals of the same sex, but also male toward female. This aggressive behavior is compounded in the captive situation, as subordinate birds in an aviary are unable to escape harassment by dominant birds. This social stress has led to health problems. Some individuals are compatible, however, usually older, experienced males with females. Any pairs held together require careful monitoring to ensure social problems don't develop. Juveniles are housed together until they are approximately two months old, when bickering among them will gradually escalate to the point where they must be separated. The requirement to house all birds individually has limited our ability to hold many birds, thus hampering our ability to research captive husbandry and disease.

The stitchbird breeding season begins in October, and final clutches are attempted by February. During this time, captive females can rear two to three clutches of up to four chicks per brood. They will make up to four nesting attempts if early clutches are unsuccessful. As stitchbirds are unusual cavity nesters in that they enter the cavity and build a platform upward, we provide nestboxes with the entrance hole at the base. The nest base consists of a tower of coarse twigs, grading into finer material such as fern rhizomes and leaf skeletons. The cup of the nest is lined thickly with tree fern scales. In the wild, some nest bases are up to 30 cm (12 in.) high. Each aviary is furnished with three nestboxes in the female's flight for her to choose from, and one in the male's flight, close to the dividing wall. In an effort to impress the female, the male will begin building a small nest in his box. This corresponds to their behavior in the wild, where males try to lead females to prospective nest sites. Accommodating this behavior sometimes helps stimulate nest building in the female next door. The nestboxes we provide measure 30 x 25 x 40 cm (12 x 10 x 16 in.) high.

The breeding season can be a very stressful time for stitchbirds. The males have high testosterone levels, and will pursue the female throughout the breeding months, mating with her many more times than is necessary. When the female solicits mating, with soft calls and tail raised, the male will usually copulate with her in the normal manner. However, copulation is often forced, with the male chasing the female down to the ground and mating with her in a face to face position, which is unique among birds, claws locked and wings spread over her to prevent escape. This constant pursuit and rape by the male is a cause of much stress for the female; if such attention persists for too many days, the female resorts to hiding from him. In the captive situation she may be harassed to the point where she is unable to feed sufficiently or even have time to build a

Stitchbird.
Notiomystis cincta

nest. This is why we now hold the male and female of a pair in adjacent flights during the season, thus stimulating the female to undertake nesting activities without harassment by the male. When she has nearly finished nest building, her weight is carefully monitored. As it begins to increase, indicating that she is preparing to lay, we then introduce the male to her flight. Depending on the degree of aggression shown by the individual male, we either separate the pair after copulation has been observed, at the end of each day, or if little aggression is observed, leave the male in the female's flight for a week, until egg laying is complete. One male, introduced to a female near the end of the season last year, was so courteous toward his mate that we allowed him to stay with her throughout incubation and assist with chick rearing. Females are capable of rearing chicks on their own, however, if the male is present, he will contribute approximately one third of the feeds.

Stitchbirds at Mt. Bruce usually lay 3–4 eggs in a clutch. On occasion, five eggs have been laid, but in these cases no more than four ever hatch. The small white eggs are incubated for 15 days. At around 10 days, it is possible to see the embryo developing inside the clear shell, without touching or candling the eggs, so they can be easily checked for fertility by opening the nestbox while the female is off the nest. When the eggs are due to hatch, the nest is checked twice daily to monitor the progress of the hatch, and any eggs that do not hatch within a day of the first are candled. Dead-in-shell or infertile eggs are removed immediately, as nests in the past have sometimes been abandoned if dead-in-shell eggs are left with young chicks. The nestling period is 28 days, during which time the chicks are checked daily. Fortunately, the parents will tolerate a fairly high level of disturbance at their nest. By day six, the chicks are passing relatively solid feces, which are collected daily for disease screening. The feces are examined for capillaria and coccidia loadings each day so that treatment can begin immediately if infection occurs. If left untreated, chicks infected with coccidia usually die near fledging time; this disease has been the number one killer of young chicks since the captive program began.

To guard against mites and fecal build-up in the nest, the nestbox is replaced with a clean one twice per week, which also provides an opportunity to weigh and dose the chicks as required. The chicks also become used to human contact and are quite tame and inquisitive as adults, making management practices, such as catching and moving between aviaries, as well as public viewing, easier. Inside the clean nestbox, the chicks are placed in an artificial nest once the natural nest becomes soiled. This consists of a sieve, with the handle removed, lined with polar fleece fabric to keep the chicks warm and snug, placed inside a raised wooden platform in the nestbox. The first time the natural nest is replaced, the female appears confused on re-entering the nest and hesitates at the door. Eventually the calls of the hungry chicks lure her in and she accepts the alteration from then on.

The chicks fledge at 28 days of age, and are fed by their parents for a further 1–2 weeks. They learn to use the artificial feed stations after about a week, and can be separated from their parents at around 40 days of age. Daily fecal sampling of each chick continues, whenever possible, until they are three months old. We assume that at this age they have built up a certain resistance to coccidia, which very rarely kills adult birds. At this time sampling is reduced to twice weekly until their feces are consistently clear of coccidia. Throughout the winter months, the adults and subadults are screened monthly. The past two breeding seasons have, in terms of long term juvenile survival, been the most successful to date, with four chicks fledged during 97/98 and 5 fledged in 98/99. Of these juveniles, the females have been retained for captive breeding and the males transferred and released onto Tiritiri Matangi Island in the Hauraki Gulf.

Causes of Deaths in captive Stitchbirds:

Aspergillosis: A fungal disease caused most frequently by *Aspergillus fumigatus*. It is seen most commonly in its chronic form at the NWC, and is characterized by rapid respiration, lethargy, anorexia, and weight loss. Aspergillus is a ubiquitous fungus that grows well in soil and organic matter. Transmission occurs by inhaling spores. Damp organic litter encourages production of large number of spores, for example damp nest box material, dusty environments and accumulation of moldy food scraps. Aspergillosis is acquired from environmental sources and is not an infectious disease. Factors that we decided were important in the epidemiology of the disease in the species in captivity include environmental conditions such as cage set up, access to leaf litter, local weather effects, and food hygiene. Bird factors include species susceptibility, diet, and stress. Management changes were implemented that we hoped would lessen the impact of the above factors and control this condition. When the disease is suspected, we treat affected birds with a combination of Ancotil and Diflucan.

Atoxoplasmosis: The first diagnosed case of coccidiosis in the stitchbirds at Mt. Bruce occurred in 1994. Analysis of mortalities show that it is now the most important disease entity in stitchbirds at Mt. Bruce, with all the deaths occurring in young birds. We are reasonably certain that the coccidial infections in stitchbirds at the NWC are due to Atoxoplasma organisms. These produce isospora-like oocysts that are passed in feces and are indistinguishable from other isosporan species. We have established the two times when deaths from coccidiosis are most likely to occur at Mt. Bruce. These are: 1) In very young birds where a whole brood has been lost. In one case the entire second clutch of the season for a pair was lost at 16 days of age. There was a concurrent severe red mite infection and capillaria was also present. 2) In fledged birds separated into a flocking aviary. The stress of movement can allow a low-level coccidial infection to flourish.

We have endeavored to establish a preventative program for coccidiosis to enable successful breeding to take place. This program revolves around hygiene, preventative treatment, and fecal monitoring. At present, treatment is based on Toltrazuril (Baycox), which unfortunately is only active against the gut forms of *Atoxoplasma*. We are also trialing a coccidiostat and a malarial treatment, Primequin, in combination with Baycox. We would be keen to make contact with and hear of any other treatments which aviculturists have used successfully in the treatment or control of Atoxoplasmosis.

Stress: Stress related deaths have been attributed to aggression from males toward females and to stress related to the capture and handling of wild-caught birds imported for new blood. The handling for transmitter fitting prior to transfer and release is also suspected to have caused the stress related death of one bird. In the another case, two six month males had a severe territorial dispute. Some of these birds had concurrent disease.

Asphyxia: Since 1990 there have been three deaths from asphyxia, all in birds less than a year old. Two involved arthropods, and in the third case, a whole kernel of corn was stuck in the pharynx.

Other: There have been a number of other causes of stitchbird mortality at the NWC, most of which have appeared as isolated cases and hopefully will not become bigger problems as a result of management changes put into place for the control of the diseases mentioned previously.

Coccidial Research: Because of the importance of coccidiosis as a cause of death in captive juvenile hihi, a group at Massey University, Palmerston North, has been carrying out research on this disease. It is believed that the organism responsible is a type of coccidia known as Atoxoplasma, however, the exact identification and the life cycle are unknown. This means there are many questions that need answering: How does infection occur and when? How does infection with this parasite cause disease? At what age does the bird become resistant? How common is this organism in free-living stitchbird and does it cause disease? Is this exact organism found in other species of birds or does it infect only the stitchbird? What treatments are effective?

The research group has tissues from all the Mt. Bruce stitchbirds that have died from this disease since 1994 and has been able to examine the pathology in detail. Regular examination of stitchbird feces from the captive population has been undertaken for evidence of the coccidial oocysts. In addition, examination of feces from other bird species, and from free-living stitchbirds, has been performed. So far, it has been found that free-living stitchbirds carry similar coccidia to those in captivity, but there has been no evidence of disease in wild birds. Transmission experiments using zebra finches have been conducted. In these experiments, which were unsuccessful, oocysts from stitchbirds were inoculated via the mouth into a number of parasite-free finches and feces from these finches were then tested twice daily for evidence of coccidia. At pre-determined times, the finches were euthanized and necropsied and the tissues examined microscopically in order to detect any sign of coccidia or coccidia-induced changes. It is necessary to sporulate the oocysts in the laboratory, i.e., encourage them to change their form into their infective stage, before identification can be attempted or transmission experiments undertaken. A major obstacle encountered is that the stitchbird oocysts from Mt. Bruce appear to be unresponsive to conventional methods used to stimulate sporulation, requiring stitchbird oocysts to be sourced from other populations. Further work is required in order to understand the mechanisms of infection and disease and minimize the effects of this parasite.

Diet: In the wild, stitchbirds feed naturally on a variety of nectar plants, fruit and invertebrates. The current artificial/captive diet is offered daily in three separate bowls (see photo), with a top up as required when rearing chicks. It consists of the following:

1) Jam water: 250 ml of apple & raspberry jam and a tablespoon each of bee pollen and Pronutro (22% grain based breakfast cereal) , blended into one liter of water. The pollen is a valuable nutritional supplement, which is not only consumed naturally, but also now a popular supplement with nectivores and species such as

lorikeets. Pronutro is a cereal-based South African product containing 22% protein, which I have used to raise everything from estrildids to psittacines. Although not currently used, I believe honey would be a valuable addition to this diet.

2) Wombaroo lorikeet nectar: A popular Australian lorikeet nectar product with <0.5% lactose.

3) A fruit/vegetable puree, finely blended, which includes sweetcorn, carrot, grapes, apple, pear, and orange. It is made in a large batch, then stored frozen in the amounts required daily.

In addition to the natural plantings in the aviary — and very importantly — the birds receive an almost daily supply of either nectar flowers or bunches of wild fruits such as Coprosma. These natural foods are placed in strategic places to enhance public viewing. Live food, particularly moths, are readily caught and consumed by the birds in the aviaries. The resulting overnight catch from two fan/light insect traps is distributed among the birds through the breeding season, with priority given to females with young. Small crickets, waxmoth larvae, and mealworms have been offered without success, but hand-raised birds will, in the future, be introduced to these at an early age.

Species Requiring Similar Management:
New Zealand bellbird *Anthornis melanura*
Tui *Prosthemadera novaeseelandiae*
Sugarbirds *Promerops* spp.
Fruitsuckers *Chloropsis* spp.

Contributor: Phil Digney of the Adelaide Zoo Bird Department

References:

Atchison, N. 1992. Breeding blue-faced honeyeaters at Taronga Zoo. *Australian Aviculture,* Feb. 1992: 29–35

Atchison, N. 1993. The spiny-cheeked honeyeater. *Australian Aviculture,* Sept. 1993: 234–237

Miller, M. 1994. *Husbandry manual for the yellow-tufted honeyeater.* Healesville Sanctuary.

21.07 Irenidae Leafbirds

Martin Vince

Introduction: Leafbirds and fairy bluebirds belong to the family Irenidae that falls within the order Passeriformes. This family includes eight species of leafbirds and two species of fairy bluebirds. The leafbirds form the single genus *Chloropsis,* and are also known as fruit-suckers, and sometimes as chloropsis, after their generic name. All species of leafbirds are basically green and have short, thick legs and medium-sized wings and tails. They are fairly common in the wild, ranging from northern India and China to the Philippines, Sumatra, Java and Borneo. Leafbirds are very adaptable, and can be found in various types of forest, as well as parks and orchards, which tend to provide an easy source of food. Soft fruits and flower nectar are the staple diets, with insects and their eggs being energetically sought after by peering under leaves and branches. Like their close relatives, the much larger fairy bluebirds, leafbirds have a loud and fairly attractive voice.

Management: All species of leafbirds are medium sized birds, 18–20 cm (7–8 in.) in length. Most species are dimorphic. However, six of the seven subspecies of the golden-fronted leafbird *Chloropsis aurifrons* are monomorphic, the exception being *C. a. media* from Sumatra, in which the female lacks the black throat (Delacour 1947). With the correct diet, healthy leafbirds are not difficult to look after, and often become quite tame. However, some leafbirds are prone to the disease atoxoplasmosis, a type of coccidia which also affects many other avian species. See the sections on STITCHBIRDS and TANAGERS for more information on this problem. The parasite is thought to be species-specific, with oocysts usually varying minutely in size between host species. Healthy birds can normally live with the infection, showing no symptoms whatsoever. It is the very young, old, sick or stressed birds that are most at risk. For example, major environmental changes, such as shipping and quarantine, can be so stressful that a bird's ability to suppress the parasite is impaired. Without treatment, death can occur, usually after several days of poor appetite and a depressed appearance.

"Atoxo" is usually identified on necropsy or on blood smear as it invades monocytes. Its ability to cause systemic infection rather than merely an intestinal infection is a special characteristic that separates it from other *Isospora* species. It is therefore often hard to identify, and in practice, Riverbanks Zoo treats any "atoxo-like" organisms as atoxoplasmosis. The European poultry medicine Baycox appears to be the best treatment. Although it is doubtful that the disease can currently be cured, it can be suppressed. Leafbirds are now so rare that for aviculturists with access to Baycox, prophylactic treatments in nectar should become standard procedure before and after shipment.

Newly acquired specimens should not be considered especially hardy, and are best housed indoors at temperatures of not less than 18.3°C (65°F). Birds that have been kept indoors through their first winter can be moved to an outdoor aviary in the summer. Thereafter, properly acclimated leafbirds are fairly hardy and can live outside year-round, provided they are protected from bad weather and shut into a frost-free shelter when necessary. In the northern United States and western Europe, where the winter days are as cold as the nights, small softbills such as leafbirds are safest indoors at a temperature of at least 11°C (55°F).

Leafbirds should be handled with care, as their soft feathers easily come out, especially in damp hands. Latex gloves can help keep plumage dry and undamaged, although fine control of the bird may be reduced.

The aviary should be planted in accordance with Habitat A in the chapter on LANDSCAPE. I have kept golden-fronted leafbirds, blue-winged leafbirds *C. cochinchinensis,* orange-bellied leafbirds *C. hardwickii,* and greater leafbirds *C. sonnerati* in mixed species aviaries. A single pair of these leafbirds will normally live amicably with other birds. However, additional leafbirds of any species are unlikely to live in the same aviary without fighting. Like any bird, leafbirds have the ability to injure smaller or feebler specimens; but if companions are chosen carefully, according to temperament and size, a mixed species community can usually be achieved.

Diet: Leafbirds require the same diet described in detail for TANAGERS. New arrivals can also be quickly encouraged to the food dish by providing a slice of orange or papaya. In people, it is thought that citrus fruits facilitate the uptake of iron. Since hemochromatosis, or iron storage disease, is a potential problem for many softbilled birds, the excessive use of orange is not recommended. However, leafbirds do appear to have a considerable liking, or perhaps even a need, for orange. My preference is to feed a quarter of one slice per pair of leafbirds per day, in addition to the normal diet.

There are bird species, such as the red-vented bulbul *Pycnonotus cafer,* that are thought to be unable to synthesize vitamin C (from glucose) themselves. Such birds need an almost daily intake of the vitamin in their food since it is not adequately stored in body tissues. It has been speculated that other birds, including perhaps leafbirds, also need a daily intake of vitamin C, more so than poultry or parrots, for which commonly available proprietary pellets tend to be designed.

Breeding: The clutch consists of 2–3 pinkish or white colored eggs, with black or red-brown blotches; incubation takes 13–14 days. Despite the fact that leafbirds thrive in aviaries, successful breeding is very rare and their long-term avicultural future is most uncertain. Nevertheless, it is quite easy to keep leafbirds in excellent condition with the diet previously described, and getting them into breeding condition is also relatively straightforward. For leafbirds in the wild, as well as all other Asian, North American, European and North African animals, the arrival of the breeding season is marked by a gradual increase in day length. This starts in the spring and reaches a maximum of 13–14 hours in the summer. At the same time, warmer weather brings blooming flowers, and a greater abundance of fruits and insects. For leafbirds kept in outside aviaries or in rooms lit naturally, the increasing day length is comparable to that found in Southeast Asia. This is an important breeding trigger that, combined with warmer, more humid weather, helps to bring the birds into breeding condition.

In colder climates, small softbills are often kept

Golden-fronted Leafbird
Chloropsis aurifrons

indoors where they can more easily be protected from bad weather and predators. For these birds, artificial lighting can be used successfully to regulate the photoperiod. A simple timer can be wired directly into a building's electrical system to automatically turn lights on and off at pre-set intervals. Timers are very worthwhile; they provide consistent day lengths that allow the maximum feeding time for a given season, promoting good health and breeding. During the winter, a photoperiod of about 12 hours is sufficient to keep the birds in good health. At the beginning of the breeding season, the photoperiod can be increased by two hour increments each week, until the lights are on for about 14 hours per day.

Leafbirds are so-called because of their green appearance and the acrobatic way in which they peer under leaves for insects and their larvae. Like other arboreal species, leafbirds appear to be calmest when given access to green foliage in which they can hide. Real or even artificial plants therefore help create a safe, secluded environment for breeding, and indeed a less stressful environment for quarantine. Natural plantings also offer nest sites and nest-building materials. Breeding can occur in mixed-species aviaries, although success is harder won, with other birds competing for food and creating a less stable environment. The aviculturist has more control when single pairs are the sole occupants of spacious, well-planted aviaries. A neat, cup-shaped nest is made of fine grasses, roots, animal hair and plant down. Nests are normally very difficult to see as they are fixed between twigs in dense foliage. In aviaries that do not provide adequate natural nest sites, half open-fronted boxes and cup-shaped wicker baskets should be provided.

Soft-bodied insects, such as waxworms and molted mealworms, are vital as a rearing food. Nectar and soft fruits are always an important part of the diet, and especially so when there are chicks in the nest. Once the young are independent, remove them from the aviary, because the adult male can become aggressive if he wishes to re-nest.

Artificial Incubation & Hand Rearing: I have never had to artificially incubate or hand rear a leafbird species. However, in the event that this becomes necessary, I would recommend following the incubation and hand-rearing procedures described for TANAGERS and HONEYEATERS.

Species Requiring Similar Management:

Fairy bluebirds *Irena* species
Ioras *Aegithina* species
Tanagers, all genera
Bananaquits *Coereba flaveola*

References:

Delacour, J. 1947. *Birds of Malaysia*. First edition NY. Macmillan.

21.08 Laniidae True Shrikes

Glen Holland

Introduction: The true shrikes, of which 30 species are recognized in three genera, belong to the family Laniidae and the order Passeriformes. The bushshrikes and related species, formerly included in the family Laniidae, are now placed in the family Corvidae, where they constitute the tribe Malaconotini in the subfamily Malaconotinae. This tribe includes 49 species in eight genera. Within this same subfamily, the helmetshrikes, genus *Prionops,* now occupy the tribe Vangini, which they share with vangas, batises, wattle-eyes, and similar birds. The bushshrikes are restricted to the African region. Of the true shrikes, two genera, *Eurocephalus* and *Corvinella,* each with two species, are strictly African, while the 26 species in the genus *Lanius* are distributed through Africa, Europe, the Middle East, Asia, Japan, the Philippines, Borneo, Sumatra, the Lesser Sundas, and North America. Habitats utilized by the shrikes and bushshrikes vary from semidesert, where species such as the crimson-breasted shrike or Burchell's gonolek *Laniarius atrococcineus* occur, to tropical forests, where a number of bushshrike species are found. Shrikes are extremely territorial and will not tolerate another of their species in their territories. Some species, such as the former mentioned, will spend much of their time hunting for insects on the ground while others, such as the bushshrikes of the genus *Telophorus,* are found in the forest canopy, seldom venturing to lower levels. Although most species occur in pairs, some, such as the helmetshrikes *Prionops* spp., and the long-tailed shrike *Corvinella melanoleuca,* are found in parties of 4–8 birds that breed cooperatively. Many species have rather unusual calls, and duets often occur between members of a pair. Calls of the long-tailed shrike and the bokmakierie *Telophorus zeylonus* are very pleasant, and the birds repeatedly whistle to each other. The common fiscal shrike *Lanius collaris* is often known as the "Jacky Hangman" after its habit of impaling its prey on thorns in trees and on barbed wire fences.

Management: The crimson-breasted shrike and the gorgeous bushshrike *Telophorus quadricolor* are two of the most striking of all African birds. Initially, wild shrikes are difficult to train onto artificial diets. To achieve this, live food such as termites, mealworms, and grasshoppers must be mixed with the substitute foods, with the percentage of live food gradually decreased over a period of 10 days. During this training period, the birds should be kept one, or two if a communal species, to a wooden box measuring 60 x 60 x 60 cm (24 x 24 x 24 in.), with the front covered with fine gauze to prevent injuries when the birds panic. An anti-stress formula should be added to the drinking water, and additional warmth supplied at night for the first five days. The birds must be left undisturbed, apart from feeding three times per day. Once feeding well, they are best introduced to a small flight with some dense vegetation for cover. A plentiful supply of insects should be maintained for a further five days and gradually reduced, but not excluded, to encourage birds to feed on artificial diets.

Shrikes become quite confiding and will learn in an aviary to take insects from the hand. They are best kept in a large planted aviary as described under Habitat B in the chapter on LANDSCAPE. Most shrikes show a preference for thorn trees and shrubs. Shrikes are fond of bathing and, particularly if a mist spray is placed over a shrub, they will hop about in the leaves of the shrub until the plumage is soaked. One of the most important factors to the survival of most species in captivity is to keep them dry, particularly in cold weather. In their natural habitat, thunderstorms of relatively short duration occur in the warm months. Shrikes are able to withstand night temperatures down to freezing, but they require warm days. In severe climates, particularly where cold and wet are combined with low daytime temperatures, they are best housed indoors in winter with an additional source of warmth.

Shrikes are predatory birds and the size of the prey varies from insects to bats and other birds. It is therefore essential, in order to ensure compatibility with other species in the aviary, to research the natural diet of the particular species which you decide to keep. In general, shrikes should be housed in pairs or with species of the same or larger size. I have kept species such as helmet-shrikes and crimson-breasted shrikes in a mixed collection of similar-sized and larger birds and have never observed aggression toward any other species or their chicks.

I have, however, had a report that in one aviary a crimson-breasted was particularly aggressive toward all newcomers and was also observed to steal and kill a clutch of chicks belonging to a pair of red-billed fire finches. I suspect that this was a particularly aggressive individual, but it would not be advisable to mix this species with small seedeaters. Fiscal, long-tailed, boubou *Laniarius* spp., bushshrikes, and most other species are killers when housed with birds smaller than themselves. Long-tailed shrikes will take larger prey, and I have had both a melba finch and a cordon blue waxbill killed by this species during capture in a mist net. Shrikes will rob nests and kill the fledglings of smaller species. In general, they are best housed with starlings, rollers, small hornbills, coucals and other large softbills. It is important to have the birds sexed to ensure a pair. Keeping two males together will ultimately result in the death of one.

With communal species such as long-tailed, white-crowned, *Eurocephalus anguitimens,* and helmet-shrikes, it is essential to be very careful when forming a new groups in an aviary. Birds coming into a group from different groups will initially be incompatible and will fight to the death. When introducing new birds to each other, they must all be released into an unfamiliar aviary at the same time and monitored until a harmonious relationship has been established as per FAIRY WRENS.

Diet: Captive shrikes require a diet of 60–70% substitute foods such as ox heart, minced bird of prey diet, minced chicken, and pet foods. The remainder of the

Hand feeding Shrike chick.

African Long-tailed Shrike
Lanius schach

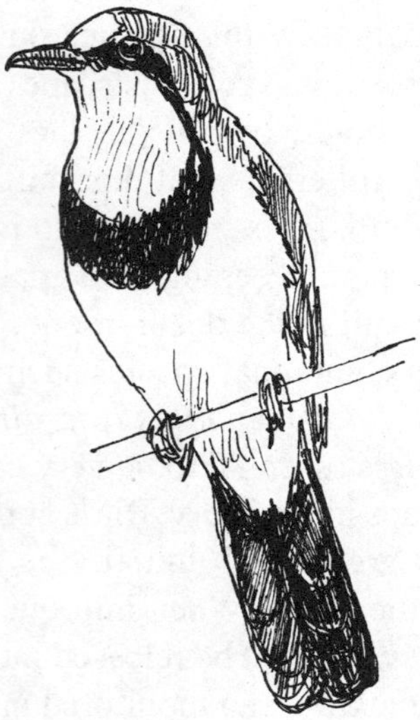

Gorgeous Bush Shrike
Telophorus viridis

diet should be natural food such as mealworms, crickets, gentles, grasshoppers, and flying ants. They soon learn to respond to a call at feeding time and will dash about their owner's feet seizing any insects which that may fall or be thrown to them. They will then retire to a dense shrub to devour their insects. Their natural diet consists of insects and arthropods, most of which are caught on the ground. If kept in a small aviary and fed largely on substitute foods, the color in those species thatwhich have red, orange, or yellow in their plumage will ultimately fade. Coloring agents added to their food will help to retain their original bright colors.

M. D. England recorded the following: "It may be of interest to note that Great Grey Shrikes *Lanius excubitor,* though eating mealworms with relish and disgorging pellets which contain only the skin, heads, and claws, do not eat maggots readily and seem quite unable to digest them, bringing up a mass of whole maggots shortly after swallowing them. On the other hand, they eat blowfly pupae and the resulting pellets contain cases only. My pair killed two broods of their nestlings by giving them nothing else. The young could not digest the cases and were not old enough to regurgitate pellets." Mr. England also had a pair of red-backed shrike, *Lanius scach,* that similarly fed their young on mealworms exclusively resulting in one chick dying and another being removed for hand rearing because of deformity relating to nutritional deficiency.

Breeding: Although they are hardy, robust birds, shrikes are not often bred in captivity. They require plenty of live food and no disturbance by other species before breeding will even be attempted. Nest inspections should be avoided, as this may cause nest desertion. Bokmakierie shrikes, are very susceptible to disturbance and will abandon the nest and even break their eggs if disturbed even slightly. In courtship, males will actively attract a mate with courtship feeding, vocalizations, aerial displays, and prominent body postureswhich that exhibit their colors well. In the fiscal shrike, for example, courtship consists of repeated courtship feeding, zigzag flight displays, and elaborate nest displays in involving head up postures, tail fanning, and extremely complex whistles and vocal mimicry. The southern boubou, in contrast, does not perform courtship feeding but undertakes a descending flight characterized by slow, exaggerated wing beats, rump fluffing, and metallic nasal calls. This is followed by hopping in tandem, the male chasing the female through the vegetation and frequently bowing his body and calling. In white helmetshrikes, the dominant pair display to each other by opening and slowly flapping their wings. Rump fluffing is common in the shrikes. Females often solicit copulation by crouching low, fluffing their body feathers, or fluttering wings, as would a juvenile begging to be fed.

Nests vary as much as the species themselves, but in general, shrikes build large nests with a central cup that is neatly lined with rootlets and some soft material. Nests of species such as bushshrikes, boubous, and fiscals are well concealed in a thicket, while others, such as the crimson-breasted and helmetshrikes, build their nests in exposed positions, although their nests are very well camouflaged with strips of bark, lichen, cobwebs, etc. For these species, a thorn tree with substantial branches should be included as a nesting site.

M. D. England gives the following account of breeding the red-backed shrike:

> "One day I found one of them sitting on the window ledge looking very sick. I was just turning away to

reach for an infrared lamp when I saw her do a little "shuffle": she was not sick but broody! A handful of grass proved it, for she immediately flew down and gathering an immense beakful, started looking for a nest site. I hastily fixed up a bundle of twigs and a rolled up piece of wire netting, and within a couple of hours the latter had a nest in it. Meanwhile the other bird next door had taken no notice at all. I thought that at least I might get some indication of sex if I put it in with the other, so, I prepared for murder, and I opened the communicating door. Within a very short while another nest had been built in the bundle of twigs. Two hens!

Before I could try another potential male, one of the pair was incubating three eggs in the nest in the wire and was being fed by the other bird. No signs of courtship or mating were observed and although by now we were calling the sitting bird she, the sex of the other bird was still in some doubt because, while it occasionally fed the sitting bird, it continued to add to the other nest in a desultory manner.

A fortnight later, doubts were dispelled by the sitting bird, on receipt of a mealworm from the other, rising and passing it on to something beneath it and when I cautiously put my head in the door to feed them I received a sharp blow on the nose from an enraged dive-bomber."

As mentioned under diet, this nesting failed after a week because of nutritional problems resulting from the high proportion of mealworms fed to the chicks.

The two to four eggs are laid on consecutive days and vary considerably in color, ranging from white to pale green or blue. Both sexes incubate for 16 days. The small, delicate chicks are rather sluggish feeders. The chicks make soft wheezing calls. Both adults tend to the nestlings, removing fecal sacs and provisioning the young. With newly hatched chicks that are continually brooded, the male will pass food to the female to then feed the chicks. They fledge at 15–19 days, and resemble the adults by three months, at which stage they are subject to aggression from the adults. With cooperative breeding species, such as long-tailed and helmetshrikes, only one nest is constructed by the breeding pair, and the entire group assists in feeding the female while she incubates and then in raising the chicks. Long-tailed chicks remain in the nest for 15 days and fledge with substantially shorter tails than the adults. Second broods are common with the chicks from the first brood also assisting in the rearing of the second. Fiscal shrikes are known to have up to three broods with chicks fledging at 19 days.

Hand-Rearing: Although somewhat delicate, shrike chicks are not difficult to hand rear. The diet should consist of equal portions of insects, small reptiles and rodents, diced pinkies and fuzzy mice, and ox heart fed with a forceps. A multivitamin and calcium/phosphorous supplement is added to the diet once a day. Feeding is done every 1½ hours from 6:00 a.m. to 10:00 p.m. Chicks are best kept on a pad of dry grass in a plastic bowl. Newly hatched chicks are maintained at 37°C (98.6°C), and the temperature reduced by 1°C (1.8°F) per day until a temperature of 25°C (77°F) is reached. By this time the chicks should be reasonably well covered in feathers. Within hours of hatching, chicks respond to vibration of the nest by begging silently. Their eyes open at about 5 days and they are completely feathered by 15 days. Additional warmth is necessary in cold weather until the chicks are fully feathered. The fecal sacs produced at feeding times are easily removed, ensuring that hygienic conditions are maintained. At fledging, the chicks are held in a cage measuring 1 m^3 (35 cu. ft.); once they begin to peck at food, they are introduced into a small outdoor flight. At this stage they

Fiscal Shrike
Lonius collaris

Crimson-breasted Gonolek
Laniarius atrococcineus

are fed every 2 hours from 8:00 a.m. to 6:00 p.m. They are brought indoors from this flight for the first few nights or during inclement weather. Once pecking at food, the hand-feeding frequency is gradually reduced until the chicks are eating well on their own; at this stage they will no longer beg energetically to be fed. The chicks become very tame and when fledged, they will fly onto their owner to fetch pieces of food. They make a high -pitched call when begging for food, accompanied by flitting the wings.

Growth chart for three species of shrikes

Day	Southern Boubou Weight (g)	Senegal Gonolek	Fiscal
1	3	2.7	2.4
2	3.24	2.8	6.6
4	4.85	3.63	7.28
6	7.75	5.15	9.39
8	11.64	8.4	11.93
10	16.58	12.39	15.47
12	22.73	16.7	18.8
14	30.14	23.58	21.71
16	38.7	26.67	23.03
18	41.23	30.45	22.8
20	39.9	33.36	24.18
22	40.98	32.98	25.35
24	40.73	32.85	25.29
26	44.75	35.03	27.91
28	46.25	37.72	29.3
30	45.63	41.2	31
32	45.63	42.56	32
34	46.53	41.4	32
36	45.5	36.3	32
38	48.1	38	33
40	48.95	41.52	32

Southern Boubou - *Laniarius ferrugineus*
Senegal Gonolek - *Laniarius barbarus*

*Infomation supplied by San Diego Zoo

An article by Kuehler, Lieberman, McIlraith, Everett, Scott, Morrison and Winchell titled: "Artificial Incubation and Hand Rearing of Loggerhead Shrikes provides" provides some excellent detail for success in had rearing with this group:

The San Clemente loggerhead shrike *(Lanus ludvicianus mearnsi)* was listed as a federally endangered species in 1977. This subspecies, endemic to the island of San Clemente, has the smallest range and population of any shrike subspecies in North America. The population has continued to decline throughout the 1980's, seemingly because of destruction of habitat by military training and feral goats *(Capra hircus)* and net losses caused by native and introduced predators. A survey conducted in March 1991 estimated the wild population at 14–20 adult birds. A non-endangered surrogate subspecies, mainland loggerhead shrikes *(L. l. gambeli)* was used as an experimental model to develop artificial incubation and hand rearing techniques for eggs and chicks removed from nesting wild San Clemente shrikes. Specific objectives were to (1) determine if artificial incubation parameters (dry and wet bulb temperatures) previously developed for related altricial bird species would hatch shrike eggs removed from the nest immediately after laying, (2) establish an optimum diet for shrike chicks immediately after they hatched, and (3) develop a feeding protocol that would promote adequate growth and health.

Food intakes (g as fed) for three experimental diets:

Diet	Food Items	Day 1–3	4–6	7–10
1	Iams caned cat food			1.1
	Crickets			0.3
	Pinkie mice			1.5
	Egg (hard-boiled)	1.4	2.2	1.6
2	Iams caned cat food	0.6	0.7	2.5
	Crickets	0.2	0.7	1.5
	Mealworms	0.1	0.2	1.1
	Pinkie mice	0.3	0.9	2.7
	Egg (hard-boiled)	0.3	0.7	2.3
3	Iams caned cat food		0.7	0.9
	Crickets		0.5	1.0
	Mealworms		0.1	0.2
	Pinkie mice		1.0	1.3
	Egg (hard-boiled)		0.8	1.7
	Papaya	0.1		
	Applesauce	0.1		
	Water	1.2		
	Purina dry cat food	0.2		

Sixteen mainland shrike eggs were collected from three wild nests. The eggs were candled on site; 15 were determined to be fertile and 1 contained a broken yolk and membranes. The eggs were positioned aircell up in a thermos containing small bird seed (millet) and a hot water bottle; temperature between 35–36.7°C (95–98.1°F). The eggs were transported by vehicle for 105 minutes to the Avian Propagation Center at San Diego Zoo. The eggs were set in a forced-air incubator (Humidaire model 20) and artificially incubated with protocols and parameters previously used to hatch similar birds: 38.1°C (100.6°F) (dry bulb) and 28.9°C (84°F) (wet bulb). The eggs were placed on their sides and automatically turned 24 times/day by the incubator. Additionally, eggs were turned 2 times/day by animal care staff to decrease the incidence of malpositioning of altricial embryos. Weighing the eggs daily monitored water loss. Cracked eggs were repaired with white glue. Because of the fragility and the incidence of cracks in mainland shrike eggs, in order to avoid such problems later with San Clemente eggs a spoon was used to manipulate eggs with no cracked eggs as a result.

When chicks pipped into the aircell they were transferred to a forced-air hatcher (Humidaire model 30). Incubator hatching parameters were set at 37.8 °C (100°F) (dry bulb) and 31.1–32.2 °C (88–90°F) (wet bulb). Chicks hatched on a substrate of crumpled tissue paper, and the pip-to-hatch interval was approximately 24–48 hours. Three diets were trialed:

Results:

Diet 1: Two of four chicks were successfully reared. This was an easy diet to feed but one chick died on day 4 and another on day 5 and the overall growth curve of chicks on this diet was the poorest of all experimental groups. Chicks were generally weak, showed signs of dehydration, and had difficulty passing feces. This diet was similar in nutrient composition; it had the lowest water content (10%) which may have accounted for the problems the chicks exhibited.

Diet 2: Six of 6 chicks were successfully reared on this diet. It was simple to feed and chicks had few problems with dehydration and passing feces. Additionally, chicks fed on diet 2 exhibited the greatest vitality and had the best growth curve of all the hand-reared chicks. Subsequent to the experiment 10 San Clemente shrikes were raised on this diet.

Diet 3: This diet had the lowest fat content and the highest water content of all the experimental diets (first 3 days) and the chicks gained a mass and achieved a satisfactory growth curve. The diet was difficult to feed; feeding incorrectly caused inadvertent choking and suffocation and the risk of asphyxiation was judged to great to continue feeding this diet. The 4 chicks assigned to this diet were reassigned to diet 2.

Discussion: The experimental artificial incubation techniques chosen for this project proved to be satisfactory for shrikes even when the eggs were removed from the parents early in incubation. The successful incubation and hand rearing trials of 10 mainland shrikes, the subsequent artificial hatching of 4 San Clemente shrike eggs and hand-rearing of 10 chicks, and the double clutching of one wild pair document the feasibility of incorporating the technique of multiple clutching and hand-rearing chicks as a viable part of the recovery plan for San Clemente shrikes. Preliminary observations indicate that hand-reared chicks, when raised with congeners of the same age, can be released into a natural environment with minimal behavioral disadvantages. Shrikes do not seem to require special consideration during hand-rearing process to produce "behaviorally acceptable releasable birds", unlike other bird species such as parrots.

M.D. England records a note from a red-backed shrike he hand raised:

> "Directly he was able to move about his cage he began trying to impale or wedge his food. Although various species of shrikes have been hand-reared before it seems unlikely that they learned impaling from seeing it done by their parents, it is probable that this is the first occasion when a youngster which was blind on removal from its parents has impaled immediately it was strong enough to do so, thus showing conclusively that the action is innate."

Ring Size: 3 mm for most species, 4.5 mm for larger species such as long-tailed and bokmakierie.

Species requiring similar management:

Rollers *Coracias* spp.
Australian Butcherbirds *Cracticus* spp.
Magpies and jays
Laughingthrushes *Garrulax* spp.

Reference

England, M. D. 1970. "The Rufous-backed Shrike." *Avicultural Magazine*. 76: 61–65.

Harris, T. and G. Arnott. 1988. *Shrikes of Southern Africa.*

Kuehler, C. M, A. Lieberman, B. McIlraith, W. Everett, T. A. Scott, M. L. Morrison and C. Winchell. 1993. "Artificial incubation and hand-rearing of Loggerhead Shrikes." *Wildlife Society Bulletin*. 21: 165–171.

21.09 Corvidae

21.09a Crows, Jays & Allies Corvinae

Patricia Cooper, Dave Rimlinger, Dick Schroeder, Tracey Goltz & Simon Bruslund Jensen

21.09ai Corvini

Jays, pies, magpies, crows, and ravens belong to the order Passeriformes and the family Corvidae. This large family is further broken down into seven subfamilies, several of which are divided into tribes. The typical crows, nutcrackers, magpies, and jays form the tribe Corvini in the subfamily Corvinae. They are large, hardy, long-lived birds with strong bills, nostrils covered by bristles, and strong legs with large feet. The sexes are similar in most species. In general, corvids are gregarious in nature and omnivorous in feeding habits. The nest is an open bowl, in a tree, on a ledge or in a hole; in some magpie species and in Stresemann's bush-crow *Zavattariornis stresemanni* the nest is roofed with sticks. Corvids may lay 2–9 eggs, either white or colored with markings, depending upon species. The chicks are naked or sparsely downy at hatching and require parental care by both parents. Well over 100 species are included in this tribe, with a worldwide distribution. Several species, such as the house crow *Corvus splendens* have been introduced and become established in areas to which they are not native. Due to their rather aggressive nature, birds of this family are not well represented in avicultural collections. For those who can afford the space, however, they are excellent avicultural subjects. Some species, such as the jays, are beautifully colored and others, such as ravens and crows, particularly if hand raised, learn to talk well and enjoy entertaining the public. Some of the larger species can become aggressive toward people. I once had a raven that would hold feathers out to children simply to lure them close enough to damage their unsuspecting fingers. Crows, magpies, and jays are intelligent and curious birds with amusing personalities. With the exception of the cissas, they adapt easily to captive diets.

White-Throated Magpie Jay *Calocitta formosa*

- Dick Schroeder

Introduction: The white-throated magpie-jay *Calocitta formosa* is a beautiful member of the Corvidae found throughout southern Mexico and through Central America to Costa Rica. This 45 cm (18 in.) long bird has a tail nearly as long as its blue and white body. This species has an interesting crest of loose feathers that curve forward on the head, giving a shaggy appearance. The crest feathers, as well as the forehead, may vary from blue-black to white. The face is mostly white, but each individual has various blue markings on the face. Also, the black necklace on the otherwise white throat and breast varies from bird to bird. In the closely related Collie's magpie-jay *C. colliei* from western Mexico, this area is solid black, as is the face, except from some blue or white on the cheeks. Juveniles of colliei have the dark crest feathers heavily tipped with white, whereas in formosa they are solid black. Upperparts are blue, of various shades, while underparts are white. Sexes are alike. Juveniles have a poorly defined neckband; the black on the head is reduced as well. In the wild, magpie-jays travel in small bands of up to a dozen birds, searching the ground for any kind of insect, frog, or lizard. They also eat many fruits, as well as cultivated corn. In captivity, I have found little that they will not eat.

Their vocalizations are wonderful! The do have a harsh call, similar to that of a crow, which is used as an alarm or to scold, but most of their sounds are soft and mellow. They have a variety of gurgles; at times they sound as if they are imitating human voices.

Management: These birds must be housed alone, one pair per flight. A planted aviary works well, as there is very minimal damage to the plants. They will certainly badger or even kill any other inhabitant, even birds larger than themselves. They are beautiful in flight, so the aviary should be of sufficient length to allow this, at least 4 m (12 ft.), but longer flights are even better. They have proven cold hardy here in southern California, where they see occasional frost in winter. They show no discomfort at temperatures below freezing, however, I have had no experience with them over a prolonged cold spell, as the daytime temperature here almost always exceeds 10°C (50°F).

Azure-winged magpie
Cyanopica cyana

These are interesting aviary subjects, as they will hide their favorite bits of food in various areas of the flight, usually stuffed between a perch and the wire. I always offer an old nest box to each pair, usually placed on the ground. They will gather up any number of items and stash them in this box. I have offered small plastic toys and they all end up in this toy box. Eventually they will destroy the box by pulling the layers of plywood apart. They should be provided with this type of enrichment, as they are intelligent birds and should not be forced to exist in an environment devoid of entertainment.

Diet: These birds are truly omnivores, eating nearly anything offered. Surprisingly, though, many will not eat crickets. This may be because I offer them crickets that have been frozen and then thawed. Live or freshly killed crickets might be taken with greater relish. Probably their favorite food is a mouse, and they are quite adept at catching them in the aviary. For this reason, I do not use poison but multiple catch traps instead. Trapped mice are drowned and fed to the birds. On occasion a snake will make its way into a flight where they are largely ignored; however, a small king snake was found suitably beaten and hidden near the end of a perch.

Their daily diet is made up a variety of chopped fruit and vegetables. I offer Kaytee Exact Softbill pellets both dry and soaked, mixed into the fruit/veggie mix. Mealworms that have had a vitamin/mineral supplement sprinkled on top are offered daily as well. I use Avico Vita Life for this. Live food must be increased when young are present, but on one occasion I had chicks that were nearly two weeks old when I discovered them, and they had done just fine with the normal, nonbreeding ration.

Breeding: Magpie-jays construct a nest of stout twigs, lined with finer grasses and rootlets. I provide a wire shelf about 30 cm (12 in.) square with 5 cm (2 in.) high sides, hung as high as possible in the sheltered part of the aviary. They will construct a rough nest fairly early in the season, mid-February usually. They refine it with the addition of finer materials provided, usually strips of palm frond, dried bamboo leaves, or even excelsior packing material. They will lay 3–5 eggs that are bluish gray, speckled with brown spots. Incubation is approximately 16 days and is carried out by the hen. In the wild they are cooperative nesters, with several additional birds helping feed the hen on the nest. This may work in captivity given a large enough enclosure, but male aggression toward newly weaned chicks has been a problem in my aviaries. It is always the young male birds that are killed or injured, and it happens quickly. One day they are fine, still being fed by the parents, the next day they can be found killed by the male. I always remove the chicks the day I see them feeding themselves. If the chicks are placed in an aviary with older siblings, care must be taken that the dominant bird in the group doesn't injure the newcomers.

When a pair has finished breeding for the season, after producing up to three clutches, the birds dismantle the nest, leaving only a few of the stout twigs in place. I have found that I can leave the last brood with the parents without the male aggression seen earlier in the season.

Rufous Tree-Pie *Dendrocitta vagabunda*

- Dick Schroeder

Introduction: The tree-pies are native to India and southeastern Asia. Rufous tree-pies *Dendrocitta vagabunda* range throughout India, as well as Sri Lanka, Assam, Burma, Thailand, Laos, Cambodia, and southern Vietnam. In India they are commonly found in parks and gardens, even entering porches and verandas in search of geckos or spiders. These are among the largest of the 10 species of tree-pies, being 45–50 cm (18–20 in.) long, including the tail. Slimmer than a magpie, they weigh approximately 120 g. The head, neck, and upper breast are soft, slate gray, darkest around the eyes. The scapulars and mantle are bright rusty brown. Upper wings are grayish white and lower wings, black. The long tail is bluish gray, except for a wide black band at the tip. The bill is dark gray with a large curve. Legs are blackish brown. Juveniles are much duller and have a sooty head and neck until the first molt.

Their voice is wonderful, sounding like a metallic

musical instrument, similar perhaps to a xylophone. They have a harsh alarm call, similar to a magpie. In the wild, they are primarily arboreal feeders taking fruits, berries, seeds, insects, including wasps, bees, spiders, and caterpillars, lizards, and eggs and young of small birds. However, this species has been observed nesting in the same tree as an oriole, both with young (Goodwin 1976). They are also known to feed on carrion, as well as human table scraps.

Management: I have always housed mine in planted flights. They are fairly shy birds, and appreciate the foliage to escape prying eyes. I usually plant with ficus trees. A spacious aviary is best, as these are good-sized birds. My aviaries are 1.2 x 3.6–4.8 x 2.4 m (4 x 12–16 x 8 ft.) high. Tree-pies are quite hardy, and here in southern California they can be housed outside year round with minimal shelter, even though I will have frost several times during winter. I have never housed them with other species, mainly because of their predatory nature, but they may get along with species as large as turacos, however breeding would be difficult under those conditions, definitely for the turacos.

Diet: Tree-pies in my collection feed primarily on various fruits. Those offered include papaya, apple, grapes, melon, figs, and other seasonal fruit. These are diced and combined with frozen mixed vegetables and soaked Kaytee Exact Softbill pellets. They will eat the pellets

Rufous treepie
Dendrocitta vagabunda

dry, but only when there is no other choice. They also receive about 50 mealworms and maybe a dozen crickets that have been frozen and then thawed, daily, along with the occasional mouse. When rearing young, the live food is increased greatly. Water is available at all times in a large crock. They seem to prefer bathing in the foliage rather than on the ground.

Breeding: Tree-pies are monomorphic, so some method must be used to determine sexes. I have been successful both with surgical sexing and with the DNA method. The nest is made of twigs of various lengths and diameters. It is built on a wire platform that is provided near the top of the aviary, but under the roof of the sheltered portion. The platform is about 30 cm (12 in.) square with wire sides 10 cm (4 in.) high. The birds built their own nest. When a jumble of sticks is in place, they line the nest with soft materials, including the dropped leaves from a weeping bamboo *Otatea acuminata aztecorum* and excelsior packing material. The eggs are greenish cream, with gray or brownish spots. The most eggs that I have had in a nest are four, but three seems to be the normal clutch. Nesting begins in early April here, and I have had chicks hatch as late as mid-July, but nothing after that. Usually two, sometimes three clutches, depending on the outcome of the prior nest, will be undertaken during the breeding season. If the chicks are reared successfully early in the season, the parents are certain to nest again. If the second nest fails, they may lay a third round, but if the second clutch is also raised to fledging, there most likely will not be a third clutch. It appears that both birds incubate, but they leave the nest as soon as anyone approaches, so it is difficult to verify this. The incubation period is around 18 days.

My tree-pies have sometimes raised three chicks without incident, but at other times the entire clutch will disappear shortly after hatching. The food supply is constant, so that is not the cause. I sometimes believed it was interference from the magpie jays *Calocitta formosa* in the adjoining flight, but a barrier was installed and there were still unexplained losses. I have hand reared them on occasion, mainly out of curiosity, to see what type of pet they would make. They were pulled at around 12–14 days and fed soaked Kaytee Softbill pellets and small pieces of papaya and other soft fruit. Mealworms were added at around 21 days. They did fine and made fair pets. While quite calm, they lacked the curiosity of some of the other corvids, such as crows and magpies. They also were unafraid and didn't seek shelter when people were around. The chicks were not closed banded, as I did not have the appropriate size available; rather a split, numbered aluminum band was used for identification purposes.

Hand Rearing: The following account is reproduced, with the kind permission of the author, Rosemary Low, from an article published in *Foreign Birds.*

> "On one occasion I was the reluctant receiver of a hunting Cissa *(Cissa chinensis)* egg for incubation. I say reluctant because my only experience of hand-raising softbills was with Toco Toucans. The egg hatched. After the initial feed, I gave the chick only Lactated Ringers because no faeces were produced. When they were I realized that I had been overcautious and had made my first mistake which, no doubt, any experienced rearer of softbills could have told me. I found that cissa's, unlike toucans, produce their faeces in little sacs. I then realized that a certain amount of food might have to be consumed before this little sac appeared. Thereafter I fed the chick about every ¾ hour, very small amounts, alternating tiny mealworms and tiny crickets. Faeces appeared, but after fairly long intervals, it seemed. From the following day I started to give a tiny piece of papaya with every other feed. I also noted down the time of the appearance of each faecal sac.
>
> This proved to be very useful in that it gave me a guide as to what was apparently normal. For example on the fourth day, faecal sacs were passed at 7.30 am, 12 noon, 1.30 pm, 4.30, 6.40, 7.35, 8.45, 11.20 and 11.35 pm. The chick usually seemed a little sluggish early in the morning, when I gave the first feed at 6.45 am, but thereafter always gaped as soon as I opened the brooder door. I offered the food with surgical scissors, which were blunt-ended. Several times daily I dipped the food in a small amount of Nekton MSA, a calcium and amino-acid supplement in powder form. Weight gains were much lower than in parent reared chicks and eventually, due to the diet suspected as being incorrect the chick was lost."

Rosemary goes on to mention a later visit to a breeder in the US who succeeded in rearing two young cissas on Roudybush Number 3. This was fed with a syringe into the mouth. At fledging, the chick was thriving on a commercial softbill mixture and beef-flavored tinned dog food.

The sections on hand rearing CROWS and SHRIKES provide additional advice on hand-rearing protocols that would be suitable for this species.

References

Davis, L.I. 1972. *Birds of Mexico and Central America.* Austin and London: University of Texas Press.

Goodwin, D. 1976. *Crows of the World.* Ithaca, NY: Cornell University Press.

Madge, S. & H. Burn. 1994. *Crows and Jays.* Boston and New York: Houghton Mifflin Co.

Stiles, F.G. & A.F. Skutch. 1989. *A Guide to the Birds of Costa Rica.* Ithaca, NY: Cornell University Press.

Hawaiian Crow *Corvus hawaiiensis*

• Tracey Goltz

Introduction: Tracey Goltz is employed by the Zoological Society of San Diego: Hawaiian Endangered Bird Conservation Program, Volcano, Hawaii.

The 'Alala is a forest dwelling, charismatic and largely frugivorous corvid. Five species of crows are thought to have evolved in the Hawaiian Islands, the most geographically isolated archipelago on earth. Fossils of three species of corvids, including the 'Alala have been found on the island of Hawaii, called the Big Island as it is the largest of the Hawaiian Islands. Possibly extinct in the wild, the 'Alala is endemic to the dry and mesic forests of the Big Island. Historically, the 'Alala may have lived on other Hawaiian Islands (James and Olsen 1991).

Since 1890, when 'Alala were described as abundant (Perkins 1898,1903), the number of birds in the wild decreased significantly. In 1987, two wild flocks existed on the western coast of the Big Island. One flock (North Kona) had only one bird remaining. In 1992, a survey of the South Kona flock on private land estimated a total of 11 birds. Captive breeding from 1976 -1995 had limited success fledging 13 chicks from 125 eggs laid in captivity during these19 years. (The Hawaiian Endangered Bird Conservation Program, unpublished Draft Report to the U.S. Fish and Wildlife Service, 2001)

In 1994 a public/private partnership (U.S. Fish and Wildlife Service, State of Hawaii, The Peregrine Fund, Zoological Society of San Diego) established the Hawaiian Endangered Bird Conservation Program (HEBCP) which operates a captive breeding facility in Volcano on the Big Island of Hawaii. In 1996, the program also assumed management of the previously state-run endangered species propagation facility on Maui.

The HEBCP's role in the recovery of the 'Alala is to manage the captive population of 'Alala in order to increase the production of chicks in captivity to provide birds for release. Equally important is the management of the remaining genetic diversity of the captive population to maximize the heterozygosity of the remaining population.

Release efforts between 1993-1999 reintroduced 27 captive-reared 'Alala to the South Kona area of the Big Island where the only remaining wild population existed. Short-term survival of 25 birds to 120 days post-release was good (92%). Long-term survival (of released birds as well as wild birds) was poor, with only six released birds surviving long-term (3 years survival post-release) (The Hawaiian Endangered Bird Conservation Program, unpublished Draft Report to the U.S. Fish and Wildlife Service, 2001). In 1999, the remaining released birds were returned to captivity and the wild flock had decreased to five individuals. These five individuals are now thought to be dead, resulting in the presumed extinction of the wild population.

Since the HEBCP's inception in 1994, the captive flock of 'Alala has grown from 12 birds to a current inventory of 56 birds. Fourteen breeding pairs are currently being managed in captivity. The remaining 28 birds in the flock are either too young or too old to breed or are behaviorally or physically non-reproductive.

Management: At the HEBCP's Keauhou Bird Conservation Center, where currently 11 breeding pairs of 'Alala are managed, each pair is housed in a 20'x40'x14'-18' (6.1m x 12.2 m x 4.3m - 5.5m) breeding aviary. The birds are given a nest platform at the highest point of the aviary overlooking the aviary's surrounding area and protected from the weather.

The birds are fed in a 6'x 6' (1.8m x 1.8m) "hack-box" that is equipped with sliding doors and one-way glass windows so that the birds can be caught easily and observed closely and anonymously.

All aviaries have wire "skirts" buried around the perimeter to prevent burrowing predators from entering the aviary. Each aviary has sheet metal flashing running from ground level to four feet (1.22m) up the outside walls to deter predators from climbing the aviary's mesh. All aviaries are covered in mosquito netting to prevent avian malaria and pox-carrying mosquitoes *(Culex quinquefasciatus)* from entering the aviary.

Aviary substrate consists of lava rock gravel which facilitates cleaning and drainage. When possible, live native plants are planted in the aviaries and fresh browse is provided.

Aviaries are divided lengthwise by a wire wall so that during the breeding season the pair can be separated to equal sides. This center wall is also very useful when introducing new pairs or groups as the birds can socialize through the wire and become acquainted.

Diet: The adult 'Alala diet consists of a mixture of fresh fruits, vegetables, egg, live insects, Zupreem ® primate chow (monkey biscuits), mice or rats, and vitamin and calcium supplements. A variety of native Hawaiian fruits that the birds would encounter in the wild are offered when in season. Diets are fed daily and quantities are adjusted according to daily consumption which can be seasonally dependent.

Breeding: In 1996, in order to better manage the breeding pairs of 'Alala, maximize egg production and confirm mate compatibility, a behavior monitoring program was established at the HEBCP in collaboration with behaviorists at the Center for Reproduction of

Hand raised azure-winged magpie *Cyanopica cyana* & Javan crested jay *Platylophus galericulatus*

Day	Weight/grams Azure-winged Magpie	Weight/grams Javan crested Jay	Day	Weight/grams Azure-winged	Weight/grams Javan crested Jay
1	5.01	5.42	26	57	65.19
2	5.66	5.58	28	59.5	68.75
4	8.13	8.8	30	60.5	69.92
6	12.5	14.35	32	56.5	70.9
8	18.8	22.35	34	56.7	74.5
10	23.22	29.19	36	62.5	75.5
12	33.29	34.6	38	63.5	76.75
14	40.11	44	40	66.5	73
16	42.75	48.5	42	67.5	
18	46.5	58.36	44	71.5	
20	51.2	60	46	72	
22	53.4	59.5	48	70	
24	57.5	60.08	50	70	

Based on the similarity of natural diet, Glen Holland recommends the same hand-feeding protocol for cissas and jays as described for STARLINGS.

Endangered Species (CRES) at the Zoological Society of San Diego (ZSSD). (Harvey et al 2002).

Breeding pairs of 'Alala are monitored remotely using video cameras that provide a view of each pair's nesting platform. Nesting areas are video-taped for several hours each day using time-lapse VCRs. Tapes are reviewed daily and behaviors are identified and documented using an ethogram that has been compiled by the behaviorists (See Alala Ethogram).

Frequency of cupforming and percent time spent sitting in the nest are the two behaviors most closely monitored in the daily management of 'Alala pairs. Data taken from the table below are graphed in order to illustrate a sharp rise and fall in number of cupforms per minute (intense nestbuilding) before an increase in percent time in nest (egg laying and incubation). From March to July, this female 'Alala produced four clutches of eggs and exhibited normal nesting behavior. Graphic illustrations of the season's nest building and laying events contribute to the individual bird profiles that are compiled from season to season so that breeding pairs can be managed effectively.

Pair bonding, courtship, nest building, nest attentiveness, copulation, egg laying and incubation are also monitored closely. Aberrant or disruptive behaviors around the nesting area are also monitored so that the appropriate measures can be taken to manage the pair.

Although birds were given the choice between several nesting sites throughout the aviary, every pair invariably chose a cantilever-type nesting platform that had been mounted on the transom wall of the highest point of the aviary with a view looking out over the aviary and surrounding area. Pairs are now offered only this preferred nesting opportunity. The video camera is positioned to give a clear view of the nest cup and surrounding perches.

By March 1 of each year, pairs are offered a variety of nesting materials in piles in their aviaries. Twigs (somewhat brittle so they are easily broken) are provided early in the breeding season. A variety of grasses, mosses and fern hairs are provided a month later after nest building has begun. A rubber nest tub is secured to the nest platform to entice the birds to start building. The rubber tub is 17" (43.2cm) in diameter and 4.5" (11.4cm) deep. Several 1" (2.54cm) holes are drilled into the bottom of the tub for adequate drainage and circulation. The bottom of the tub is lined with green matting (3M® nomad matting) to provide cushioning for any egg that might be laid in an unlined nest cup. A wreath made out of vines is secured with cable ties around the inside of the tub to provide the birds with a nest foundation. Some pairs require more assistance

Hawaiian crow
Corvus hawaiiensis

with nest building than others. Aviculturists will reinforce nests with grasses if the pair has not built an adequate nest by the date at which the pair has produced eggs in previous years.

In order to obtain as many fertile eggs from as many pairs as possible, pairs are allowed to nest-build together for as long as the pair displays normal behaviors. If any behavior by either bird appears to be hindering the nest-building process (social or solitary play, aggression, over-attentive behavior), the pair is separated. The female is left on the side with the nest platform to complete the nest-building process and to begin egg laying. The divided aviaries allow the male to carry on his daily territorial defense behaviors, stand sentinel in the nesting area, and often feed and preen the female through the wire. Once the male is separated from the nesting area, monitored "conjugal visits" are allowed daily or every other day to ensure fertility of the ensuing clutch. During this socialization period, aviculturists watch for the pair to copulate, then separate the male from the nesting area as soon as possible, using food as a lure. Once the female lays the first egg of the clutch, conjugal visits are discontinued and the female is observed for nest attentiveness.

Eggs remain under the females for a minimum of three days after the completion of the clutch only if the females have proven to be satisfactory incubators in previous seasons (either on dummy eggs or infertile eggs). Eggs are pulled at oviposition from females who have proven to be poor incubators or have historically laid thin shelled or abnormal eggs. These females are allowed to sit on wooden dummy eggs until their clutch is complete. Eggs are pulled from all pairs before the 21-22 day incubation period is complete so that females will lay again. Females will lay up to four clutches during a breeding season under this management routine.

Undisturbed, stable pairs will often exhibit similar behaviors from season to season. With historical data and pair "profiles," aviculturists can make reasonable predictions regarding the date of first clutch, interclutch recycle times, number of eggs in a clutch, and number of clutches per season.

The HEBCP manages three breeding pair of 'Alala at the smaller Maui Bird Conservation Center. Main aviaries are L-shaped (41'x 23' or 12.5m x 7.0m) and "cradle" a smaller nesting portion (15' x 24' or 4.6m x 7.3m). The same techniques are used in managing pairs for maximum production of fertile eggs.

As of breeding season 2007, HEBCP has not had the opportunity to parent rear 'Alala. If the situation presents itself, a genetically well represented female who exhibits excellent nesting behavior may be allowed to hatch and raise a chick. It is hoped that parent raised 'Alala would exhibit more natural behaviors earlier and make a better release candidates.

'Alala Ethogram

Behavior	Description
Offer Food	Bird will offer food with straight on, beak-to-beak contact (recipient may be begging).
Receive Food	Bird either accepts and swallows food inserted into its beak, or it stashes food and eats later. (This is usually done by sitting females who stash food in nest.)
Share Food	Bird allows mate to share a large food item (e.g., a mouse, rat or large insect) without fighting.
Feed Other	Includes inserting beak into another's open beak for feeding.
Food Beg	Head is tilted back, beak open, wings quivering and probably accompanied by vocalizations.
Preen Other	Bird preens, or nibbles, at any body part of another, includes allobilling.
Carry Nest Material	Bird carries nest materials in beak. Materials include sticks, twigs, moss, bark, fiber, grass strands, etc. *NOTE: Some females stash food in the nest, which they later eat; this is not nest material.*
Add Material to Nest	Bird places nest materials in nest. May jab or thrust stick, twig into the nest. May remove and re-insert if item is not secured properly. Usually see shuddering (side-to-side) movement on part of bird. Softer materials are bundled into nest.
Cup Formation	Bird crouches in nest, pressing against it with its breast and carpel joints of it's wings, and crouching and kicking with it's feet while pressing tail down on rim of nest. This movement is more pronounced when nest is first built, but is later reduced to a few sudden shoves. Count each cup formation movement in a bout of movements with use of ticks. This occurs during "Stand in Nest" bouts (see below).
Display	
Puff-Head-Bill-Down:	Bird approaches and/or circles mate with head down, body sleek and head feathers erect (in more intense displays — especially those leading to copulation — the proximate section of wings may also be lifted away from the body). Often accompanied by a soft repeating growl or coo.
Tail Quivering	Bird bends over forward but lifts beak toward the sky. Tail is lifted up, exposing the vent and flutters up and down. Seen before copulation as a pre-copulatory display by both male and female while perched side by side, often on the nest. Also has been seen as an "I give up" by a submissive bird towards a dominant bird following a chase.
Mount	Bird places both feet on the other's back. (Can occur after male display of head bobbing and calling as he hops from one side of the female to the other).
Copulation	Includes proper oriented mount for cloacal contact, accompanied by wing flapping.
Enter View	Bird comes into camera view on the nest platform.

Exit View	Bird leaves the nest platform (and the camera view).
Proximate Occupied Nest	Bird is on the nest platform within one bodylength of an occupied nest (i.e., bird's mate is standing or sitting in nest).
Stand in Nest	Bird has both feet inside bowl/cup of nest, but is NOT sitting in cup, or if sitting, is doing so only momentarily while manipulating nest material or forming cup. "Sitting in Cup" (next behavior) is reserved for real incubation, so "Stand in Nest" becomes anything else where the bird has both feet in cup. NOTE: Count all "Cup Formation" movements during a "Stand in Nest" Bout with checkmarks (see "Cup Formation" above).
Sit in Nest	Bird is sitting/lying down in nest without any type of nest manipulation or cup formation for at least 30 seconds. Until then score time in nest as "Stand in Nest". Once sitting has begun, if short bouts (less than 30 seconds) of manipulate cavity occur or bird stands momentarily to shift position and then resumes "Sit in Nest," DO NOT SCORE AS NEW BOUT. If bird resumes sitting in nest more than 30 seconds later, SCORE AS A NEW BOUT. NOTE: Sitting birds may get quite hot (bird will have beak held slightly open) and will crouch over eggs rather than sit, due to the heat. SCORE THIS AS "SIT IN NEST" not "Stand in Nest".
Manipulate Nest Cavity	Bird reaches into bottom of nest with beak while standing in nest, and may move head back and forth. This may represent egg turning, but eggs are not always visible from the camera's perspective.
Non Contact Aggression	Bird attempts to land on, strike with foot or beak, peck at, kick, chase in air or are on substrate, etc. WITHOUT making actual physical contact. This will also include the "forward threat" posture where bird stretches forward with feathers smooth, and bill sometimes opened.
Displace	Bird takes over space, spot or object in the camera view that leads to the other bird leaving the area within 30 seconds.
Contact Aggression	Bird actually MAKES PHYSICAL CONTACT with the other bird for any of the behaviors listed in "Non Contact Aggression" (above).
Stereotypy	Bird is engaged in repetitive, sometimes mutilating behaviors such as pulling out feathers of self or other, jabbing self with stick, biting toes of self.
Solo Play on Nest Platform	While mate is absent or more than one crow length away, bird crouches and flops or rolls onto its side or on its back on the nest platform while kicking its feet or manipulating objects such as sticks or other nest material. It may involve vigorous movements such as spinning around on its back.
Social Play on Nest Platform	Bird crouches and flops or rolls onto its side or on its back in the presence of its mate (mate is also playing or is less than one crow length away and oriented toward the focal bird). May include pulling gently on mate's toes or feathers or may involve a tug of war with mate over an object such as a stick.
Nudge Other in Nest Only	Male or female will use beak and head movement to nudge other off the nest. Do NOT score as "Contact Aggression", as behavior is neither intense nor is it accompanied by biting and pulling of feathers/skin. Include in this column all non-aggressive contact with the bird in the nest, such as tail pulling. Count each contact by the use of ticks. NOTE: If behavior becomes intense and is accompanied by any submissive posturing (e.g., head pulled in and fluffed with beak open) of bird in nest or rapid sudden departure of bird in nest, score as "Contact Aggression".

*Compiled in 1996–2003 by Nancy Harvey, Susan Farabaugh and Tracey Goltz

Artificial Incubation and Rearing: Incubation and brooder rooms are disinfected at the beginning of the breeding season by cleaning floors, counters and walls using Roccal® or Environ®. Rooms are then sterilized using a portable UV sterilizer (Hanovia Letharay 68 watts twin bulb UV sterilizer® - 30 minutes).

Throughout the breeding season, each room's floors and counters are disinfected daily using Roccal® or Environ®. Each room has a separate +/- airflow circulating fan system. High Efficiency Particulate Air (HEPA) filters are placed in critical hatchery and young chick brooder rooms.

Prior to the breeding season, all incubators (Humidaire Model 20 and 21®) and hatchers (AB Incubator®) are fumigated with potassium permanganate and formaldehyde, UV sterilized and disinfected with Environ®. Machines are turned on and run at 100 degrees Fahrenheit (F) (37.8degrees Celcius{C}) with the vents open for at least 24 hours before eggs are set.

Incubators and hatchers are disinfected between clutches with Environ® and Roccal®, respectively. Water pans and wet bulbs (and wicks) are changed every 48 hours.

Whenever possible, in addition to routine disinfection, incubators are run at high temperatures with opened vents to dry them out and decrease the possibility of water-borne fungus and bacterial infections.

Hands are washed with antibiotic soap and dried thoroughly before handling eggs.

Alala eggs are light blue/green with brown spots usually concentrated toward the aircell end of the egg. Between 1995 and 2003(n=87 clutches), captive female 'Alala laid an average of 2.2 eggs per clutch with the most being 4 and the least being 1. Eggs are laid every 30 to 48 hours until the clutch is complete.

Average length of an 'Alala egg is 42.7mm (max= 52.7mm, min=32.1mm) with an average width of 28.4 mm (max=34.5mm, min=17.8mm) from 117 measured eggs incubated by HEBCP. HEBCP has had the opportunity to incubate 29 freshly laid eggs from four different captive females. Average fresh weight of eggs from these females is 21.8grams with a high of 25.22g and a low of 15.71g.

'Alala eggs are incubated at 100 degrees F (37.8 degrees C) with the wet bulb being egg-dependant but ranging from dry to 90 degrees F (32.2 degrees C), before condensing occurs. Eggs are placed on a small towel in the incubator tray keeping the egg tilted at a 45 degree angle with the aircell angled up. Eggs are turned by hand every two hours from 6am-8pm and tilted by the incubator hourly, 24 hours a day.

Eggs are weighed at the same time every day (Scale: Ohaus Precision Standard TS400D®). Weight loss over the 21-22 day incubation period should equal approximately 14% of the egg's fresh weight. Weight loss is calculated daily and humidity is adjusted as necessary. When possible, the egg is weighed shortly after the chick has pipped through the shell. All eggs are measured (length and width) at the start of artificial incubation so that aviculturists can compare incubation parameters of similar sized eggs. Egg measurements also help in determining how much weight an egg should lose daily if no fresh weight is available.

'Alala eggs are candled daily after weights are taken (Candler: Lyon's Model HI-INT®). Detailed notes record the quality of the shell and membranes and development of circulation and embryo. Eggs are candled more frequently (three times a day) when the chick is expected to pip into the aircell.

Eggs are transferred from the incubator to the hatcher when the embryo is pushing or has pipped into the aircell. Time of transfer depends on the weight loss of that particular egg over the course of incubation or the size of the egg's aircell. Temperature of the hatcher is set at 99-99.5 degrees F (37.2-37.5 degrees C) with a wet bulb of 88-89 degrees F (31.1-31.7 degrees C). The egg is placed in a small cup on a substrate of crumpled tissue paper with the pip pointing up. Recorded adult 'Alala vocalizations are played during the pip to hatch interval. Pip to hatch interval for artificially incubated 'Alala eggs ranges from 13 to 48 hours

Eggs with shell or membrane abnormalities are monitored closely for fertility (often detected on day 6-7 of incubation). Recognizing the value of each 'Alala egg, those eggs with thin or chalky shells, abnormal shape or with wrinkled shells are all set and managed intensively.

Eggs with shell abnormalities require high wet bulb temperatures in order to decrease moisture loss through the thin, chalky, or wrinkled shell. If eggs continue to lose weight at an unacceptable rate, HEBCP aviculturists have used a mixture of Elmer's® glue and food coloring to "paint" weak spots on the shell. Food coloring allows aviculturists to track which areas have been painted. If the shell has no obvious weak spots, aviculturists begin by painting small areas of the egg, starting on the often-enlarged aircell end. Care is taken not to paint the expected pip and rotation area.

In extreme cases, clear fingernail polish has been painted over the glued areas to further decrease moisture loss. Waxed test tube paper may be wrapped around the egg, molding it to the shape of the egg, in order to create a removable substrate that prevents moisture loss during incubation.

Hand Rearing: After hatching, chicks remain in an AB Incubator® and are kept on tissue paper in plastic cups until day four when matting is placed on the bottom of the cup to encourage the chick to grasp the substrate. Small twigs are placed in the nest cup on day ten. The substrate of the nest cup is slanted down from the edges (funnel-like) so that the chick's legs sit under its body. Whenever possible, similar aged chicks are placed in the same nest cup for socialization, warmth and leg strengthening.

Weights are taken for each chick at hatch and each morning before the first feeding so that food intakes for that day can be calculated. Food intake begins at 25% of the chick's body weight in food on day one and increases to 70% by day eight. (Kuehler et al.1994)

Generally, the chick's food intake will peak at 70% and decrease once pinning begins and the chick gets finicky. At this stage, if the chick does not beg for the entire portion of food at a feeding, the feeder stops and returns at the next feeding. From day eight to weaning, the chick is allowed to eat as much as it will take at each feeding without feeding more than 70% for the day.

The actual weight of food per feeding varies from chick to chick. Once an AM weight is recorded, the feeder calculates how much food the chick will be offered over the course of the day by multiplying the weight of the chick by the percent intake for that age (See Table 1). The feeder divides that number by the number of feedings the chick is supposed to receive that day in order to determine how much to feed the chick at each feeding.

Chicks are fed every two hours from 6am to 8pm on day one. If the chick is weak, more frequent feedings are given in order to feed the entire amount of food over the course of the day. Once the chick's eyes open, the feeder will replace his/her blunt tipped feeding forceps with a puppet for feeding. On day 13, the chick is fed every three hours. By day 26, the chick is fed four times a day. By day 35, the chick is starting to show interest in a food pan. Chicks are slowly weaned off of puppet feedings until day 60-70 when they have weaned themselves.

Hawaiian Crow *Corvus hawaiiensis*

Day	Food Intake Per Day (% of A.M. Weight)
1	25%
2	32%
3	39%
4	46%
5	53%
6	60%
7	65%
8	70%

Brooder temperatures start at 97 degrees F on day one (36.1 degrees C) with a wet bulb of 87 degrees F (30.5 degrees C). Temperature is decreased by 1 degree F daily (which decreases the wet bulb) until the chick is comfortable at room temperature (approximately day 20). The chick is moved to a Lyons® Animal Intensive Care Unit (AICU) on day 4 where ventilation is better for the growing chick. By day 10 the chick has started to develop pin feathers and does well in a large brooder box, especially when several chicks occupy one nest cup.

The chicks are fed bee larvae, cricket guts or scrambled egg at every feeding for the first two days after hatch. Food items are floated in warmed pediatric electrolyte and Nekton I® (introduced gradually on day 2). Chopped pinkie and papaya are introduced on day three with supplements of calcium, bonemeal and brewer's yeast. Papaya is soaked in Nekton Lori®.

By day eight, chicks are fed a mix of the five items (bee larvae, cricket guts, scrambled egg, pinkie, papaya) throughout the day. Day nine, pinkies are replaced with mouse fuzzies (no head). Mouse meat, bone and fur are introduced gradually on day 12 and 13 and amounts of these items are increased over the next 4-5 days until the entire mouse is chopped and used in the mix.

Aviculturists at HEBCP take pains to prevent a chick from imprinting on its feeder. Chicks are raised together if possible and a feeding puppet is used once a chick's eyes have opened. Recordings of adult and juvenile vocalizations are played in the brooder rooms at all times and human contact and voices are kept to a minimum. Mirrors are placed in the brooder boxes of single chicks and young birds are introduced to other 'Alala as soon as safely possible.

Conservation and Recovery Strategy: The most urgent need for the continued existence of the `Alala as a species is to increase the number of birds as quickly as possible in order to retain maximum genetic diversity over time. No habitat of sufficient quality to support a release without high mortality is currently available. Therefore, captive propagation offers the safest means to safeguard the species' recovery potential. Within five years, according to mathematical models, the population needs to grow to approximately 77 birds in order to avoid further loss of irreplaceable genetic diversity.

Although this approach will be costly in terms of infrastructure development for the captive flock (aviaries, labor, operations), it is the only strategy at this time which does not jeopardize the long-term recovery potential of the species. Given the current reproductive parameters of the captive flock, surplus birds for release to the wild may be available in approximately 5 years or

RELATIONSHIP BETWEEN NEST BUILDING RATES, INCUBATION PERCENTAGE AND EGG PRODUCTION

Date	Cupform per minute	Percent time in nest	Eggs produced
4/18	0.37	0	
4/19	0.23	0	
4/20	0.34	0	
4/21	0.19	0	
4/22	0.51	0	
4/23	0.50	0	
4/24	0.72	0	
4/25	0.74	11.9	
4/26	0.46	22	
4/27	0.07	90	Clutch1 Egg1 (C1E1)
4/28	0.00	91	
4/29	0.00	89.8	C1E2
4/30	0.00	90.2	
5/1	0.00	88	C1E3
5/2	0.00	86	
5/3	0.00	59.6	C1E4eggs pulled
5/4	0.04	0	
5/5	0.09	0	
5/6	0.23	0	
5/7	0.12	0	
5/8	0.39	0	
5/9	0.30	0	
5/10	0.42	0	
5/11	0.37	0	
5/12	0.38	0	
5/13	0.77	0	
5/14	0.43	8.3	
5/15	0.38	15	
5/16	0.31	67.6	C2E1
5/17	0.00	89.2	
5/18	0.00	86.6	C2E2
5/19	0.00	87	
5/20	0.00	88	
5/21	0.00	93.8	Eggs pulled
5/22	0.06	0	
5/23	0.09	0	
5/24	0.44	0	
5/25	0.29	0	
5/26	0.34	0	
5/27	0.29	0	
5/28	0.42	0	
5/29	0.36	0	

possibly sooner. Concurrent to building the captive flock will be intensive habitat management in an area for the next release and management of all known mortality threats. To reduce threat from `io (Hawaiian hawk), when `Alala are released again on Hawai`i, birds will be parent-reared if possible and released in areas with adequate under-story cover. Limiting factors such as introduced predators (cats, rats, mongoose), diseases (malaria and pox) and habitat degradation (food plants, nesting sites) will be addressed as part of the overall habitat management strategy for species recovery.

Acknowledgements: I would like to acknowledge Cyndi Kuehler and Alan Lieberman for their invaluable contributions to this chapter. I would like to thank Nancy Harvey and Susan Farabaugh for their creation of our 'Alala behavior program and their guidance through each breeding season. I would also like to thank the staff and volunteers at the Maui and Keauhou Bird Conservation Centers whose commitment to the survival of the 'Alala is obvious in their daily work.

References:

The Hawaiian Endangered Bird Conservation Program, Zoological Society of San Diego, Alala Captive Breeding Project. Draft Report to the U.S. Fish and Wildlife Service Pacific Islands Ecoregion, unpublished. November 15, 2001.

Kuehler et al. 1994. Artificial incubation and Hand-Rearing of 'Alala *(Corvus hawaiiensis)* Eggs Removed From the Wild. Zoo Biology 13:257-266.

Harvey et al. 2002. Effects of Early Rearing Experience on Adult Behavior and Nesting in Captive Hawaiian Crows *(Corvus hawaiiensis)*. Zoo Biology 21:59-75.

James,H.F.,and S.L. Olson.1991. Descriptions of thirty-two new species of birds from the Hawaiian Islands: pt II. Passeriformes. Ornithol. Monogr.46.

Perkins, R.C.L. 1893. Notes on collecting in Kona, Hawaii. Ibis 5:101-114

Perkins, R.C.L. 1903. Fauna Hawaiiensis or the Zoology of the Sandwich Isles. Vol. 1, Part 4. The University Press, Cambridge.

21.09aii *Paradisaeini* Birds of Paradise

• Kurt Hundgen, with input from Patricia Cooper, Dave Rimlinger, Emily Ladwig, Pat Witman, James Mejeur and Simon Bruslund Jensen.

Introduction: The family Paradisaeidae has, in recent years, been the subject of major changes. In 2000, a paper based on DNA and morphological evidence placed three species — the yellow-breasted bird of paradise *Loboparadisea sericea,* crested bird of paradise *Cnemophilus Macgregori* and Loria's bird of paradise *Cnemophilus loriae,* into a new family, the so-called Satinbirds Cnemophilidea.

Another species traditionally considered a bird of paradise is Macgregor's bird of paradise *Macgregoria pulchra*. Convincing evidence placed it with the Honeyeaters Meliphagidae (Cracraft and Feinstein 2000).

A new species suggested is the growling riflebird *Ptiloris intercedens*, a previously subspecies of the magnificent riflebird *Ptiloris magnificus,* elevated to species status based on behavior and morphology (Beehler and Swaby 1991, Dickinson 2003, but rejected by Frith and Beehler 1998).

In light of the recent changes, 40 species are considered to belong to the birds of paradise, of which 34 are endemic to New Guinea and closely surrounding islands, two are endemic to the Northern Mollucans, and two exist only in eastern Australia. A further two species exist both in far northern Australia and in New Guinea. Each species occurs within a particular altitudinal range from sea level to 3000 m (10,000 ft). Generally, they have small home ranges within this distribution (Frith and Beehler 1998).

While males differ from species to species in respect to breast shields, tail feathers and flank plumes, females (and immature males) generally are similar in that they possess a relatively cryptic plumage. Highly metallic, iridescent feathers are common to adult males of most species. Adult males are slightly larger than adult females and juveniles. They range in body size from the 150 mm (5.9 in.) king bird of paradise *Cicinnurus regius,* to the 710 mm (27.9 in.) black sicklebill, *Epimachus fastosus*. Bill shape is variable, from short and stout in parotias *Parotia* sp., to long and decurved in sicklebills, *Epimachus* sp. and *Drepanornis* sp.

The 17 genera exhibit a range of mating systems, from monogamous and sexually monomorhpic manucodes *Manucodia* sp., and paradise crow *Lycocorax pyrrhopterus,* where the two sexes share parental duties, to the cryptically colored and mostly promiscuous females of other species that are alone responsible for the nest building, incubation, and rearing the young.

Courtship behaviors vary from ritualized chases in the pair-bonded manucodes, to the solitary court behavior of the magnificent bird of paradise *Cicinnurus magnificus* and the complex communal leks in most Paradisaea species. Clutch sizes are 1–2 eggs, and known incubation periods are 14–26 days.

The majority of species consume a wide variety of fruits, combined with arthropods. Two genera that are considered to be more insectivorous are *Ptiloris* and *Epimachus,* the riflebirds and sicklebills, respectively.

More inter-specific hybrids have been produced among the Paradisaeini than any other bird group. At least 21 distinct combinations have been identified in the wild from species that only exhibit polygynous mating systems.

Most species of birds of paradise are currently not threatened with extinction. However, several long-term threats have been identified, including large-scale forest disturbance from timber operations, regional development, human population growth, and deforestation from subsistence agriculture (Frith and Beehler 1998).

During the 1950s, E. Thomas Gilliard initiated the first significant field research for birds of paradise. Since then a variety of field research, especially since the early 1980s, has significantly increased our knowledge of these birds. Two books that are considered bibles by ornithologists and aviculturists working with these birds are, *Birds of Paradise and Bowerbirds* by E. T. Gilliard, published in 1969, and *The Birds of Paradise* by C. B. Frith and B. M. Beehler, published in 1998.

Since the turn of the century, birds of paradise have been highly prized in zoological and private collections. Their attractiveness stems from the elaborate courtship displays, vocalizations, and plumage possessed by adult males. Despite this popularity, and their longevity, their breeding success has been limited in captive collections. Only since the early 1990s have some species bred consistently in captivity.

Management: This section attempts to summarize current captive husbandry techniques used by the Wildlife Conservation Society's Bronx Zoo, the San Diego Zoo, and the Honolulu Zoo for the five species that breed with regularity in North America. The two smaller species with a solitary display are the magnificent bird of paradise *Cicinnurus magnificus,* and superb bird of paradise *Lophorina superba;* and the three larger species with a communal lek display are the lesser bird of paradise *Paradisaea minor,* the red bird of paradise, Paradisaea rubra, and the raggiana bird of paradise *Paradisaea raggiana.*

Housing: Birds of paradise in general are strong and fast flyers, with powerful feet and bills. They have inquisitive personalities. They manipulate objects, for example, food, plants, behavioral enrichment objects, and will use all areas and height of an enclosure. Enrichment items such as wire baskets stuffed with leaves have been developed and used to redirect their destructive behaviors and help minimize damage to plants (Maher and Sheppard, in press). Larger species aggressively compete with smaller birds in mixed species exhibits and may pose a threat to eggs and chicks.

Maintenance and exhibit housing for birds of paradise can be accomplished with various enclosures. The smallest flight cage to adequately maintain a single bird in good health for an extended period of time should measure 1.5 x 3 x 2.5 m (5 x 10 x 8 ft.). Smaller cages are used for short-term holding, and more commonly with juveniles. Communal lek species should be managed in larger aviaries in comparison to solitary species.

When possible, potted plants can be used to enhance comfort levels in stark enclosures. An ideal way to exhibit birds of paradise for public viewing is in heavily planted indoor or outdoor aviaries. Any number of plant species that are considered hardy with tough leaves can be used in bird of paradise enclosures. Typically, males are more destructive to exhibit vegetation for display purposes than females; in the wild, it is quite common for males to defoliate leaf vegetation to create display areas.

Breeding enclosures vary greatly in size and design among zoo programs. Several published articles (Hundgen et al 1991, Muller 1974, Searle 1980, Rimlinger et al 1997, Todd and Berry 1980) provide descriptions and measurements of successful enclosure designs.

The most important feature a breeding aviary should have is flexibility. A means to readily separate a female from a male at egg laying is particularly important. Left with a female, males can be disruptive, resulting in nest desertion or broken eggs. Since females are solitary nesters, the challenge is to provide access to a male for breeding, but isolation from the lek area so that she can nest, incubate, and rear young. In some species, separation can stimulate courtship upon reintroduction. If possible, an enclosure design should allow a female to be paired with different males over the course of a breeding season, or for the same male to be paired with multiple females.

Size of enclosure shift doors can vary from 30 x 30 cm (12 x 12 in.) to full-size keeper access doors. Enclosures should be designed with multiple ways of shifting birds for flexibility. All sectors of an enclosure should be accessible to each other. In general, birds of paradise will readily pick up shifting routines after a brief consistent training period.

In captivity, birds of paradise can tolerate temperatures down to 7°C (45°F) with access to some type of heat source. A higher incidence of aspergillosis has been observed in some collections during periods of constant high temperature (>34°C [95°F]) and high humidity, especially with birds in molt. Air conditioners have been effective in minimizing high temperatures during summer months for birds housed indoors.

For birds of paradise housed in more northern latitudes, a supplemental lighting system should be used to maintain a 12-hour photo period during the winter months. For birds of paradise that are housed in outdoor aviaries year round, supplemental lighting has not been used.

Proper perching is instrumental in maximizing male display behaviors for particular species. For communal lek species, perching should consist of an assortment of horizontal and vertical perches and vines situated in the upper strata of an enclosure. *Paradisaea* males housed in adjacent enclosures especially prefer mirror image perching. Branches and plants in close proximity to display areas have been observed to be popular perches for females prior to copulation. Males in the genus *Paradisaea* spend a considerable amount of time displaying while hanging from the wire, even when natural perching is plentiful.

Perching for the solitary displaying species, such as *L. superba* and *C. magnificus,* should be focused on the lower strata of an enclosure. In the wild a male *C. magnificus* establishes a terrestrial "display court" which he keeps clean of all litter. Several vertical saplings can be found within a cleared area and used by the male for courtship and copulation. In captivity, display courts can be constructed by placing several vertical poles within a clearing. Male *L. superba* also display close to and on the ground, usually within 2 m (6.5 ft.), on fallen logs or horizontal perches.

Soil, bark mulch, wood chips, and concrete are common substrates used in bird of paradise enclosures. Studies at the Bronx Zoo revealed high fungal growth in more moist substrates in indoor facilities, a contributing factor to aspergillosis (Raphael et al 1995).

Clean water should always be available. Birds of paradise will use streams, shallow pools, and stainless steel pans to bathe daily. Ideally, enclosures should be equipped with misters or a showering system. Food pans should be placed 1–2 m (3–6 ft.) off the ground. Birds will displace each other at pan sites, so in some cases at least one pan per bird may be necessary.

Diet: In spite of a considerable amount of effort made by zoo nutritionists to prevent iron storage disease (hemochromatosis/hemosiderosis), it still occurs in birds of paradise. However, birds of paradise are living longer with fewer instances of iron storage disease. Further studies are warranted to understand the dynamics involved with this problem.

Recent nutritional analysis suggests feeding a diet with iron concentrations of 50–100 mg/kg of dry matter. This low iron diet should also be low in vitamin C, which enhances iron absorption. Conversely, fruit pigments are known to reduce iron levels by acting as binding agents (Dierenfeld, pers. comm.). The recommended diet is comprised of approximately 50% Ziegler Bird of Paradise pellets (product # 65-2400-36-22), 50% fruits and vegetables, and a garnish of insects.

Some programs use a frugivore salad supplement to balance nutrients lacking in a produce-based mixed salad. Crickets, waxworms, and mealworms are the primary insects offered. Insects cultured for feeding purposes are known to require a calcium supplement. Care should be taken to ensure that such supplements do not contain high levels of iron and/or vitamin C.

Breeding: The management objectives for propagating communal lek species should be to minimize male access to female nesting area after mating has taken place and egg laying commences, to enhance male-to-male interactions, and to minimize interactions among mature females during the breeding season. For solitary displaying species, these objectives should be to minimize male access to female nesting areas upon egg laying and to minimize interactions among mature displaying males.

Captive-reared female birds of paradise become sexually mature at 3–4 years of age. At this point they generally become intolerant of being housed with or near other adult females. This is especially true in the genus *Paradisaea*.

Males typically take 4–7 years to develop full adult plumage. Immature males look identical to females in plumage. The male lesser bird of paradise has proven to be able to reproduce in immature plumage at three years of age (Laska et al 1992). Whether this holds for birds of paradise in general is yet to be proven. For a communal lek breeding program it is preferable to include at least two adult males with visual contact to maximize display behaviors. The use of mirrors and vocalization tapes has been effective in stimulating male display behaviors. However, breeding has occurred in single male situations. For solitary displaying species, adult males should be kept out of visual contact.

Superb bird-of-paradise
Lophorina superba

In North America, the breeding season for captive birds of paradise generally begins in October/November and ends in May/June. Most species molt during the late spring/summer months. There have been deviations from these generalizations. Mature males typically begin their molt before females. During this time, birds of paradise tend to remain quiet and sedentary in comparison to non-molting periods. Care should be taken to prevent stressful situations (for example, netting, introductions, and shipping) since these birds are particularly vulnerable to disease during this period.

Females use a variety of nest structures, sites, and heights for nest building. Those housed in lush outdoor exhibits will occasionally construct their own nests among bushes and trees, but usually prefer artificial nests such as baskets or open fronted nest boxes. Indoors, females readily accept artificial nests. In both cases, they usually select nests that are placed high in the enclosure.

A wide range of materials is used for nest building, such as thin flexible vines, dried grasses, various types of leaves, mosses, pine needles, and shredded cypress bark. Nests are completed within 2–4 days of being started, and egg laying occurs shortly thereafter. The clutch sizes for most birds of paradise is 1–2 eggs. In two-egg clutches, the second egg is laid 24–36 hours after the first. For the smaller, solitary species, egg weight range is 6–10 g (0.2–0.4 oz.). Egg weights for the larger, communal lek species are 9–13 g (0.3–0.5 oz.).

Management techniques for breeding vary between programs. Usually a female is given access to a male each day until an egg is laid, then daily introductions are discontinued. Other programs do not separate their birds on a daily basis, but keep them together continuously until the first egg is laid. Both methods have proven successful, and enclosure design usually dictates which method is used. All programs deny the male access to female nesting enclosures with the onset of incubation. Some pairs have been separated for several months after the breeding season to enhance courtship behavior upon reintroduction.

Parent Rearing: In general, birds of paradise have proven to be poor parents in captivity. Egg sabotage and chick abuse are common occurrences. Reasons for this are not fully understood, but it is likely in part due to a nesting female not feeling properly isolated and comfortable in certain situations. The comfort level may vary among individuals housed in the same type of enclosures. A female *P. raggiana* successfully raised her own chicks for three consecutive years in the early 1980s (Rimlinger 1984), whereas another female of the same species has consistently failed in recent years. Some instances of metabolic bone disease have occurred in several *P. rubra* chicks that have been parent reared. Females fed their chicks only live food and supplemental feedings in the nest were necessary. However, the *P. raggiana* chicks mentioned above were parent reared on a diet of crickets, mealworms, newborn mice, and fruit supplemented with Vionate, without any apparent bone problems. Further work is needed to refine parent-rearing techniques in captivity.

Incubation: Incubation periods for the five species commonly bred in captivity are 16–19 days. In cases where chicks are hand reared, eggs are generally incubated naturally for at least 10 days, and preferably up until 1–2 days before the expected hatch date. At this time, eggs are transferred to an artificial incubator for the remainder of the incubation period. Humidaire, Grumbach, and Alpha Genesis incubators have been used successfully. These incubators are maintained at a temperature of 37.5°C (99.5°F) and a relative humidity of 55% (wet bulb reading of 29°C/85°F). Once internally pipped, eggs are moved to a hatcher with a temperature of 37.5°C (99.5°F), and 85–95% relative humidity, (wet bulb reading of 34–37°C/94–99°F). Chicks have been lost after pipping, apparently due to the drying out of membranes in the hatcher. Increasing the humidity is recommended to prevent this condition.

Hand Rearing: After hatching, chicks should be examined for any physical problems, including unabsorbed yolk sac and abnormal umbilicus. After being weighed, chicks are placed in an isolette at a temperature of 33.3–33.9°C (92–93°F) and relative humidity of 51–73% (wet bulb: 25–30°C/78–86°F). Two chicks may be placed together in a small plastic cup lined with absorbent paper towels. Rubber mesh matting is used on top of the towels for a few days to help keep the chicks dry and clean. This matting also gives their feet something to grip until a small twig nest is added to the nest cup on day four. It is important that a chick's legs are properly positioned from hatching until it is perching independently. Improper positioning of the legs can generally be corrected by partially lifting a chick and allowing these appendages to be properly positioned.

Nest materials are changed frequently as they become soiled. These cup nests are approximately 6.35 cm (2.5 in.) diameter and are constructed using thin twigs woven through plastic mesh. In some instances, chicks have been hand reared in old nests originally constructed by egg-laying females. As feather growth progresses, chicks require less heat and humidity. Superb birds of paradise did have a more vigorous feeding response when the humidity was higher. Providing a chick is progressing well, brooder temperatures can be decreased by ~12°C (11°F) every two days until room temperature is reached. Close observation of a chick is the best guide to making changes in temperature and surroundings. A gaping chick can indicate heat stress, in which case the isolette temperature should be reduced by a few degrees.

At approximately 14–18 days, chicks become increasingly mobile and are moved to a medium brooder/holding cage. Temperature should be in the 29.4–30.6°C (85–87°F) range. Heat can be regulated using hood lights or an alternate heat source. Perches of appropriate diameter and varied heights are placed on either side of the brooder to encourage short flights. Corkbark or similar material can be placed on the floor to serve as an intermediate step for a chick that finds itself on the ground. Corkbark also provides an additional varied surface, which promotes development of the chick's feet. The food and water pan are placed close to a low perch to encourage chicks to start eating on their own. When chicks start perching, they have access to low perches only. As they become more stable from exercise and perching, higher perches are offered. As expected, chicks prefer to spend their time on higher perches. Placement of food and water near lower perches encourages movement and establishes a new habit for a chick to seek out a food source. In some cases, chicks that have been reared together up to this point may have to be separated due to the onset of aggression.

At approximately 28 days, fledglings can be placed in larger flight/holding cages that can be maintained at room temperature. The texture and thickness of perches should be varied to aid in coordination of legs and feet. Perch arrangement should leave adequate space for flight development. Food and water pans are set up close to perches. In some cases, fledglings of this age have been given access to outside facilities and provided with a heat source at night.

The hand-rearing diet for birds of paradise varies from program to program. Components of the diet for the first week can include pieces of pinkie (0.1–0.4 g and gradually increased), bee larvae, and two-week-old crickets. Total daily food intake begins at approximately

30% of hatch weight and is increased over the next 10 days to 65% of the body weight. Diet vitamin supplements have been used in quantities up to 1.0 ml per 50 g of food fed. Calcium carbonate or dicalcium phosphate has been used in quantities up to 1% of the daily food intake based on the previous day's intake.

Chicks should be kept well hydrated until their pinfeathers start opening. This can be accomplished by keeping the humidity high in the brooders (~50–80% RH) and also with oral fluids. Chicks suffering from dehydration are often lethargic with poor feeding responses and appear wrinkled. Dehydration can be treated with a small amount of water, electrolytes, or diluted dextrose. Once a chick has recovered, dipping food items in water will help prevent further dehydration.

The fruit and vegetable components of the adult diet can be introduced to chicks between day 7 and day 10. Soaked low-iron Ziegler Bird of Paradise pellets can be introduced on days 10–12. Pinkie mice are gradually eliminated from the diet by the time the chicks are weaned.

Chicks are usually not fed their first meal until they are at least 18 hours old. Even if a chick shows a feeding response right away, it still has a large yolk sac to absorb. This will easily sustain a chick overnight. Feeding responses can be encouraged with vocalizations and by touching the chick's beak with forceps.

Chicks are fed approximately eight times per day (every 1–2 hours) between 7:00 a.m. and 7:00 p.m. The frequency of daily feedings is reduced by one approximately every seven days until a chick is feeding on its own. A chick can be stimulated to eat on its own by using forceps to push food items around in its pan several times a day. Bird of paradise chicks begin eating on their own at approximately 18 days and generally are weaned by 35 days. Contact with keepers is cut back to minimize stress and imprinting. The level of imprinting for hand-reared birds of paradise seems to vary even among those that receive similar amounts of human contact. If other chicks are being hand-reared at the same time, placing them in adjacent cages encourages social interaction and aids in the weaning process.

Chick Development: Average weight gains are typically 14–27% (maximum 35%) during the first 10 days. After 10 days, pinfeathers begin to open and chicks' appetites begin to decrease, resulting in a slower growth rate. Chicks lose about 3–4% of their body weight between days 16–20, coinciding with the time they are fledging. Figure 1 contains average growth curves of 14 *P. minor* and 32 *P. rubra* hand-reared birds up to 35 days of age. *P. raggiana* chicks exhibit similar growth curves. Sex determination is possible when weight differences begin to become apparent at 14–16 days of age for all three *Paradisaea* species.

Juvenile Management: As breeding programs have become more successful in recent years, there has been a need to develop juvenile management techniques. For communal lek species such as *P. raggiana,* there have been several situations where groups of 3–5 male and female immatures have been kept together up to three years of age with minimal aggression. These groups have been housed in enclosures with approximate dimensions of 2.5 x 6.0 x 2.5 m (8 x 20 x 8 ft.) high. In other instances, immature *P. rubra* and *P. minor* have been housed with adult males in large enclosures. In these situations, immature males have been observed mimicking the displays and vocalizations of adult males at an early age. In one instance, development of display behaviors of juvenile *P. lesser* males was noticeably enhanced by visual access to a displaying adult *P. raggiana* male. For solitary species, juveniles have most commonly been housed in groups of two, preferably one-to-one. At 3–4 years of age, these immatures are separated and housed as singles until placed in potential breeding situations.

Conclusion: During the early 1990s, Honolulu Zoological Gardens, San Diego Zoo, and Wildlife Conservation Park/Bronx Zoo established propagation programs that began producing offspring consistently. While significant strides have been made in the captive husbandry and breeding of birds of paradise, challenges still exist. Bloodlines of several of these species have grown very thin in North American collections. The loss of birds to iron storage disease is still a problem to address, and parent-rearing techniques need to be developed. However, the captive management of birds of paradise in general has proven easier than originally thought. They are hardy and adapt well as long as certain needs are met. They can be housed and bred in small numbers. Wild-caught birds have transitioned calmly to captivity, and their hand-raised offspring have continued to breed and even raise young on their own. A Bird of Paradise Species Interest Committee was formed in 1997 with the goal of establishing representative groups of birds of paradise in captivity that will eventually be self-sustaining. As more institutions become involved in the captive management of birds of paradise, it is hoped this goal will be realized.

Breeding Birds of Paradise outside North America

By Simon Bruslund Jensen

Today birds of paradise are still rare in captivity. However, before World War II and again in the 1950s birds of paradise where regularly kept in zoos and by private collectors throughout Europe, North America, and Australia. Unfortunately there are only a few records of breeding attempts from this period. The rearing of two king birds of paradise *Cicinnurus regius* chicks by Sten Bergman in Sweden in 1956 (Bergman 1956) is often regarded to be one of the first successful captive breedings but, similar to other early successes, it was not repeated and failed to establish self-sustaining captive populations (Worth 1991).

The recent successful breeding programs have been a turning point in captive management of birds of paradise and, in particular, programs and strategies developed in leading North American Zoos have been instrumental for successes elsewhere.

Outside North America, captive breeding programs are also developing, aimed at self-sustaining programs. Some are more successful than others, but valuable experiences have been gained. Some of the more notable breeding successes and attempts are listed below.

Red Birds of Paradise *Paradisaea rubra*

This species has been bred at several institutions outside North American Zoos, including Vogelpark Walsrode, Germany 2001, Chester Zoo, UK 2003, Al Wabra Wildlife Preservation, Qatar 2004 and Caz, Chile 2005, thereby becoming the most widespread breeding species. The red bird of paradise typically lays a two-egg clutch, however females often have a nervous disposition that can cause problems with parent rearing. The male at Chester bred not fully plumed. (Wilkinson 2003, Michel Durant pers. comm., pers. obs.).

Greater Bird of Paradise *Paradisaea apoda*

One chick was parent reared in Walsrode in 2001 and repeated successes have occurred at Al Wabra since 2002 with both hand rearing and parent rearing. At Walsrode this species was found to cause severe problems in a mixed-species exhibit by destroying eggs and young of other species, in one case even the egg of a crowned pigeon *Goura scheepmakeri*. At Al Wabra a young male that had not yet attained male plumage fathered offspring that was parent reared. In all breeding successes the male was separated from breeding female before the nest building was finalized, but still a high incidence of nest destruction by males has been documented. Females may still produce a fertile egg up to seven days after being separated from the male.

At the Al Wabra Wildlife Preservation facility eight chicks were hand reared and two parent reared to independence during 2004, based on four founder females and six males. Two different subspecies have been identified in the population and are managed independently. Significant differences in biometrics, in particular bodyweight, between the two forms have been identified. Adult male Aru Island greater bird of paradise weigh up to 420 g (15 oz.) whereas males of the mainland form have not been recorded to exceed 275 g (10 oz.) (Jensen 2005 pers. obs.).

Raggiana Bird of Paradise *Paradisaea raggiana*

At Lae Rainforest Habitat in Papua New Guinea in 2001 a single chick was parent reared in a large well-planted flight, which also contained several other birds of paradise, including the male raggiana. The success was not repeated as the adult male was subsequently eaten by a fresh-water crocodile, also kept in the flight (David Whittaker pers. comm.). The Fukoaka Municipal Zoo, Japan hatched six parent-reared young in 1993–1995; unfortunately five died from trauma shortly after leaving the nest. One male was reared successfully in 1995 (Ryota Imada pers. comm.). The species has also been kept at the Tisch Family Zoo Jerusalem, Israel that received six males from a customs seizure (Beverly Burge pers. comm.).

Twelve-wire Bird of Paradise *Seleucidis melanoleuca*

The Jurong Birdpark, Singapore reared five chicks in 1995–1999. In the wild this species has a close affinity to pandanus palms both for food and likely also as a nesting place. Males typically display on vertical tree stumps and often advertise from the apex of such perches (Nyunt 2000). The species is also maintained at Walsrode and Al Wabra Wildlife Preservation, but without breeding success so far (pers. obs.).

Magnificent Riflebird
Ptiloris magnificus

Reportedly several were bred at CSCP, Rome, Italy in the early 1990s however nothing has been published to date. A single male offspring was kept at Vogelpark Walsrode for more than a decade (pers. obs.).

A hybrid offspring between a male trumpet manucode and a female crinkle-collared manucode was produced in Berlin Zoo in 2006, however, it only survived for three weeks. Parental duties were shared (Thomas Rempert pers. comm.).

King Birds of Paradise
Cicinnurus regius

They have been bred regularly at the Al Wabra Wildlife Preservation since 2002 and at Vogelpark Walsrode since 2004. Notable is that it is the only species in its group known to be a cavity nester. They consistently lay two eggs in a clutch and have reproduced on a consistent basis from several founders, both with parent rearing and hand rearing. Efforts for joint population management between the two institutions are underway (pers. obs.).

Notes on breeding the king bird of paradise *Cicinnurus regius regius* at Al Wabra Wildlife Preservation, Qatar

Simon Brusland Jensen

Introduction: The stunning adult male is scarlet red above, with a metallic green band across the breast and with fan like pectoral display plumes normally hidden under the wings. The male also sports elongated tail streamers with a curled tip resembling a coin. Lower breast and belly are white, the legs bright blue, and the partially hidden bill pale yellow. The female is well camouflaged in different shades of brown, finely barred below, the legs have the same bright blue, but the bill is darker. Body weights of the two sexes are similar.

Immature birds resemble the female. It appear that males maintain female-like plumage for at least four years, with only the paler bill giving them away as being males. However, older females may also frequently attain a yellowish bill. To positively identify females of unknown age, DNA sexing is the most functional option (Hammer 2003).

Males do an elaborate (with individualistic variation), display for the female, often reaching a state of ecstasy on her approach. They do part of the display inside vegetation and part on exposed branches, often seeking out rays of sunlight. This display is associated with loud, one-note contact calls and soft melodious singing, close to the peak of the display. On occasion females are observed performing some of the moves and calls of the males' display. Immature (female-plumed) males often repeat one fragment of the full display over and over again for hours. Young males have been observed practicing display behavior frequently, but not consistently, shortly after fledging on day 18 (pers. obs.). Both male and female utter a two-note contact call frequently throughout the day.

Management: It appears that nesting females are stressed by the presence of males in the same aviary, and on one occasion it was confirmed that a male removed and ate one freshly laid egg. Males have not been observed destroying nests made by the females, as seen in Paradisaea spp. As a result, breeding birds are kept singly in a well-planted outside aviary with an adjacent inside room where they are fed. The size of the enclosure is 8–15 m² (26–49 sq. ft.) and 3.5 m (11 ft.) high outside and 3–4 m² (10–13 sq. ft.) and 2.5 m (8 ft.) high inside. The birds easily pass through a 25 x 25 cm (10 x 10 in.) access hole to the inside room. The enclosures are placed in long rows next to each other.

Outside aviaries have mesh on front and sides and access hatches between cages. The roof is covered with a sun protective net, and built-in sprinklers shower the birds and plants from 1–5 times a day, depending on the season. The insides of the cages are covered with ceramic-fiber plates on three sides and mesh in the front. This allows high hygiene standards and prevents the birds from being disturbed by their neighbors. Substrate is gravel-sand for outside aviaries, and concrete inside. The females are offered several closed wooden nest-boxes in different shapes and sizes. Most preferred are small boxes with a large entrance hole.

The birds can get together through manually operated access doors in the mesh between the outside cages. These are only opened if the female seems to be interested in the male; the doors are left open as long as the female continues to show interest in the male by leaving her cage to visit his. The birds use most of the cage area but are rarely seen on the ground.

Small groups of non-breeding males, or immatures of both sexes, are kept together in larger enclosures. The birds will dominate each other and, even if no actual fighting is observed, it is possible that some birds stress each other, thus increasing the risk of infectious disease, or individuals being forced away from food or suitable resting places. Inter-specific aggression leading to death has been observed between two adult females in Vogelpark Walsrode (pers. obs.).

It has been speculated that king birds of paradise display in exploded leks in the wild (Frith and Beehler 1998). In captivity only one dominant male in a group will display, however, non-displaying males will display as soon as the dominant male is out of sight. King birds of paradise will inevitably try to dominate other birds, regardless of size, so care should be taken when keeping them in a mixed aviary.

Diet: The birds of paradise are believed to be mainly fruit-eaters. This is evident in their habits in the wild, as a displaying male would have little time to find the necessary number of insects every day. However, whenever possible, insects are consumed and animal protein should be offered in some form. Some birds have a high preference for live food, especially mealworms, and care should be taken not to overfeed them. The digestive tract is very short and intestines are only 20–30 cm (8–12 in.) long; experiments with colored food showed a passing time of as little as 20 minutes. The feces of the birds also indicate this, as a large proportion of food passes undigested.

At Al Wabra Wildlife Preservation the birds of paradise are fed twice a day. In the morning each bird receives one slice of papaya and one slice of pear, together with 18 Ziegler Bird of Paradise pellets. A different fruit is offered daily and if kept on a strict diet, most varieties will be readily taken. In the afternoon the food dish is exchanged and they receive a small slice of papaya, a few pieces of chopped apple and 20–22 Wittemolen Mynah Care Granulate pellets. Mealworms are feed very rarely, only two per week for non-breeding birds; and birds have commenced breeding without being offered any. A remarkably low incidence of iron storage disease in birds of paradise at Al Wabra Wildlife Preservation is attributed to the low percentage of live food and pellets, with less than 15% of the total diet consisting of these items. Vitamins and minerals are added to the food 2–3 times a week, rotating Nekton Tonic-F, Nekton MSA, and Nekton-S.

Breeding: Not much is known about the breeding times in the wild as only one nest has ever been found. Circumstantial evidence suggests that the most breeding activity occurs between April and October (Frith and Beehler 1998).

In captivity, king birds of paradise will attempt to breed at any given time of the year when the conditions are favorable. In Qatar, breeding attempts have occurred primarily in the spring and fall when the climate shows its friendliest side. Eggs have been laid in February, March, April, May, June, November, and December.

Nest building or interest in the nest boxes has been observed in February, March, April, September, and December without subsequent breeding attempts; the reminder of the year the females show no interest in the nest boxes whatsoever.

Several different cavity nest boxes have been offered, most frequently selected are rectangular boxes of 15 x 15 x 25 cm (6 x 6 x 10 in.) high with an entrance hole of 6 cm (2 in.) in diameter. The female uses nest material that she finds in the outside enclosure; additional grass and palm fibers elevated from the ground can be offered for her use, for example, in a branch fork. The nest itself is very similar to the cup-shaped nest of most other birds of paradise. When a female is expected to have laid the first egg, the male is confined to his own enclosure. Incubation period is 16 days and the two eggs typically hatch on the same day.

If chicks are expected or confirmed, the diet is changed and all fruit is diced into small squares, and mealworms that are in the process of shedding their exoskeleton or soft pupae, are feed up to six times a day. Additionally, cricket abdomens and rat organs are offered, all sprinkled with vitamin and mineral supplements.

The female tends to choose a new nest site for every clutch. Only very rarely has the same nest box placed in the same location been used two times consecutively. For this reason nest boxes are relocated and exchanged regularly until the female shows interest. After a completed breeding attempt, or as soon as the chicks fledge, the female destroys the old nest by pulling out all or most of the nesting material from the box and simply drops it on the ground.

The chicks fledge after 15–16 days and are about a third smaller than the adult. They will sit motionless in the vegetation for the first weeks and the female continues to feed for 4–6 weeks. The number of feedings and live food are gradually lowered.

Artificial Incubation: Eggs are dark creamy white to buff, with blotches and streaks of reddish- to grayish-brown. The average size is 27.38 mm (1.07 in.) long (25.8–29.5 mm) [1.01–1.16 in.] and 20.33 mm (0.8 in.) wide (20.0–20.7 mm) [0.78–0.81 in.]. The weight of freshly laid eggs average 6.23 g (0.21 oz.) (6.0–6.6 g) [0.21–0.23 oz.]. Profi and Grumbach incubators have been used with a temperature setting between 37.8–37.9°C (100–100.2°F) and relative humidity between 55–65%. The eggs were turned automatically eight times in a 24-hour cycles, and additionally turned by hand 4–5 times a day. Turning is suspended when an internal pip is observed or when the chick breaks the membrane to the air pocket inside the egg. By the external pip stage, the eggs are transferred to a hatching unit, with a temperature of 37.1°C (98.7°F). and a humidity of 80%. Average weight loss during incubation is 14.2%.

Hand-Rearing Techniques: The freshly hatched chicks are pale pink and naked. Within a few days the skin color becomes darker; feather tracts become visible at 4–5 days. Chicks are moved from the hatching unit to a brooder unit after 12–24 hours. We have used both the AB Newlife, and Profi and Grumbach. A day/night cycle with 14-hour daylight is maintained for the entire hand-rearing phase. At night all light is shut off, and care is taken not to disturb the chicks again before next morning. The feedings are at relatively short intervals throughout the day; all attempts at using longer intervals, for work/technical reasons, have led to problems such as weight loss or chick lethargy.

On the first two days, 14–16 feedings are done. Care should be taken to ensure that the yolk in the abdomen is almost completely absorbed. Initially, single drops of electrolyte solutions can be fed to avoid dehydration. If the chick is weak after a long hatching, glucose can be added. The number of feedings is gradually lowered throughout the period.

The chicks are placed individually in cup-shaped containers of fitting size. Initially, the lining is paper tissue, and after three days this is changed to a commercial filter material of carbon fiber that provides the chick with a firm grip and can be easily disinfected. On day 15 the chicks are moved to a fledging unit sized 100 x 100 x 100 cm (40 x 40 x 40 in.). Typically they will fledge the following day and will never return to the nest cup again. After fledging the chicks must be kept separate for at least four weeks as they will fight ferociously. Parent-reared chicks have not been observed fighting but do position themselves in opposite corners of the aviary. The fledglings are provided with a weak light during the night so that orientation is possible if they are disturbed.

The diet consists of 1–5 day-old mice or rats, of which only selected parts are fed. In the first week organs and pure muscle are used exclusively, and in the following weeks gradually more and more skin and bones are added with the pieces fed. Intestines and feet are never fed, as these are considered to be a potential infection risk. Freshly molted mealworm and fresh pupae are fed throughout the hand-rearing phase; occasionally the abdomens of crickets, up to third instars, are used. Initially all insects are chopped into small pieces.

Fruit is also fed every day and ripe papaya is given from day one. Initially fruit makes up only 5–10% of the hand-rearing diet but is gradually increased. Papaya is the primary choice but other fruits, such as mango, pear, grape, and banana, are used in turn as variation; only ripe fruits are used. Soaked Wittemolen Mynah Care Pellets are added after the first week. The food is presented in small pieces, with the size depending on the age of the chick. Problems have been experienced with pieces that are too small and become difficult to swallow. New food items are freshly cut for each feeding. The food is then slowly heated to approximately 40°C (104°F). Before feeding, vitamin and mineral powder is sprinkled over all food. Dry food items are soaked in water, with vitamin and mineral powder as well. Water is given to the chick with a 1.0 ml syringe at every feeding.

Excess water seems to be expelled with the feces. Healthy chicks will produce a fecal sack by almost every feeding after the second day. The fecal sack should normally be quite strong, but if it breaks on removal it could be a sign that the chick is dehydrated, the temperature is either too low or too high, or possibly that the chick suffers from an infection. Chicks are fed with forceps piece by piece.

Ring size: 4.5 mm

Acknowledgements: Thanks to Clifford Frith, Dean Tugade, Bernd Marcordes, Sven Hammer, and Sheikh Saoud Bin Mohammad Bin Ali Al-Thani.

References

Beehler, B.M. 1989. The birds of paradise. *Scientific American* 261: 116–123.

Beehler, B.M. & R.J. Swaby. 1991. Phylogeny and biogeography of the Ptiloris riflebirds (Aves: Paradisaeidae). *Condor* 93: 738–745.

Bergman, S. 1956. On the display and breeding of the King Bird of Paradise *Cicinurus regius rex* in Captivity. Nova Guinea, new ser. Vol. 7, part 2: 197–205.

Cooper, P. 1995. Observations of parent-rearing behavior in the lesser bird of paradise. *Avicultural Magazine* 101: 194–199.

Cracraft, J. & J. Feinstein. 2000. What is not a bird of paradise? Molecular and morphological evidence places Macgregoria in the Meliphagidae and the Cnemophilinae near the base of the corvoid tree. Proc. R. Soc. London B. 267: 233–241.

Dickinson, E.C. (Editor). 2003. *The Howard & Moore Complete Checklist of the Birds of the World*. 3rd Edition. London: Christopher Helm.

Frith, C.B. and B.M. Beehler. 1998. *The Birds of Paradise*. Oxford, UK: Oxford University Press.

Gilliard, E.T. 1969. *Birds of Paradise and Bowerbirds*. London: Weidenfeld and Nicholson.

Hammer S., S. Jensen, J. Balzer & D. Sandow. 2003. DNA sexing in Birds of Paradise and Bowerbirds. International Zoo News Vol. 50, No. 3: 156–159.

Hundgen, K., M. Hutchins, C. Sheppard & D. Bruning. 1991. Management and breeding of the red bird of paradise *(Paradisaea rubra)* at the New York Zoological Park. *International Zoo Yearbook* 30: 192–199.

Jensen S.B. 2005. Breeding Birds of Paradise at Al Wabra Wildlife Preservation, Qatar, AFA 31st Annual Conference Proceedings: 73–76.

Muller, K.A. 1974. Rearing the Count Raggis bird of paradise *(Paradisea raggiana)* at the Taronga Zoo, Sydney. *International Zoo Yearbook* 14: 102–105.

Nyunt, K. M., W. H. Mun & L.T. Meng. 2000. The world's first captive breeding of the Twelve-wired Bird of Paradise, Jurong Birdpar, Singapore. *International Zoo News* Vol. 47, No. 4: 215–221.

Raphael, B., C. Sheppard, D. Bruning, E. Mahre & P. Cooper. 1995. Management of substrate in aviaries to reduce exposure to fungus and mold. Proceedings Joint Conference AAZVA/WDA/AAWV: 220–222.

Rimlinger, D. 1984. The Empress of Germany's bird of paradise. *Zoonooz* 57 (2): 10–14.

Rimlinger, D., J. Azua, and E. Lewins. 1997. Observations on the breeding and hand rearing of the Empress of Germany's bird of paradise *Paradisaea raggiana augustaevictoriae*. *Zool. Garen* N.F. 67: 71–80.

Searle, K.C. 1980. Breeding Count Raggis bird of paradise *Paradisaea raggiana salvadori* at Hong Kong. *International Zoo Yearbook* 20: 210–214.

Todd, W. & R.J. Berry. 1980. Breeding the red bird of paradise Paradisaea rubra at the Houston Zoo. International Zoo Yearbook 20: 206–211.

Wilkinson R. 2003. Births and Hatchings, Chester Zoo. *EAZA News* no. 43 Sept 2003: 17–18.

Worth, W., M. Hutchins, C. Sheppard, J. Gonzalez & T. McNamara. 1991. Hand-rearing, growth and development of the red bird of paradise (Paradisaea rubra) at the New York Zoological Park. *Zoo Biology* 10: 17–33.

21.09aiii Drongos *Dicrurini*

• Glen Holland

Drongos constitute the tribe Dicrurini in the subfamily Dicrurinae in the family Corvidae of the order Passeriformes. While the pygmy drongo *Chaetorhynchus papuensis* of the mountains of New Guinea is given its own genus, the remaining 23 species of drongos all belong to the genus *Dicrurus*. Drongos are distributed through Africa and tropical Asia to Australia and the Solomon Islands. They occur in a variety of habitats including forest, woodland, thorn trees, and open grassland, where they reside in stands of trees planted by farmers. Their calls consist of a variety of twanging metallic sounds; they also imitate other birds. The sexes are alike. My avicultural experience is based on work with the African fork-tailed drongo *Dicrurus adsimilis*. This species is entirely black with a distinctly forked tail and bright red eyes. The toes are strong with sharp nails and the beak strong with a hooked tip. Having watched the spangled drongo *D. bracteatus* in the forests of Australia, I found their behavior very similar to that of the fork-tailed.

In Buddhist literature there are many references to the kalavinka bird (drongo) and its melodious voice. This wonderful singer appears, not for its own sake, but as a stereotyped image of Buddha and his voice, telling all beings the great truths about the nature of suffering and the impermanence of the imperial world. In India the voice of this beautiful and fearless bird, with its shinning plumage glossed in blue, ringing through the hill forests, became the symbol of the law-giving voice of The Enlightened One.

Management: Drongos can bite terribly when handled. Newly acquired birds take a long time to tame down and are extremely aggressive toward other species. They are not difficult to convert to artificial diets. Once they are taking insects in captivity, they learn to take substitute foods that have been mixed with the insects. Drongos are strongly territorial, and each bird has certain preferred perches within its territory from which it will hawk insects such as winged termites, butterflies, mantises, and bees.

Due to their aggressive nature, drongos are best kept with larger softbills such as hornbills and starlings. Black drongos *Dicrurus macrocercus* that have been introduced to the Micronesian islands are also suspected of being partially responsible for the decline of the rare Rota bridled white-eye *Zosterops rotensis*. They should never be mixed with small passerines in an aviary, as they will invariably corner, kill, and possibly eat them. I have read a report of a drongo that was seen to catch and kill a white-eye, which it then partially plucked before eating. Drongos will also rob the nests of smaller species. Only one pair of drongos can be kept per aviary, and it is necessary to have them surgically or DNA sexed. I have found them to be very inquisitive, and an established pair will chase any new birds introduced into their aviary to the point of exhaustion. They will mob new birds every time they attempt to land. They appear fearless, and in the wild they will often mob crows, hawks, and eagles. In a mixed collection, I eventually split a pair of these birds as they would gang up and attack an individual in the aviary. One drongo was left as a representative of the species and, although still aggressive, he did not pose as much of a threat.

Their aviary landscape should consist of Habitat B. Thorn trees such as *Acacia* species and exposed dead branches provide perches from which they will hawk insects both on the ground and in the air. As described under the section on BEE-EATERS, flowering plants that bloom at different times of the year will attract natural insect food into the aviary. Drongos are hardy birds that can withstand short periods of low temperatures. In areas where the temperature remains below 10°C (50°F) for extended periods, however, the birds will require protection from the elements. During freezing temperatures they will need to be brought indoors with a source of additional warmth supplied.

Diet: In the wild, drongos feed on insects that are caught on the wing and taken back to the perch to be eaten. They have been recorded killing small vertebrates including birds. The prey is held in the foot and broken up with the beak. When moving from perch to perch, they will often carry their prey in their talons like a raptor. They are swift flyers, able to pick insects out of the air with little effort. In captivity, 30% of their diet should consist of insects, and pinkie to young-age mice; the rest can be made up of minced bird of prey diet with a vitamin and mineral supplement added twice per week. They are able to catch food items that are tossed into the air, and are the only birds I have found that can compete with the speed of bee-eaters.

Breeding: Nesting drongos are a problem in that they become extremely aggressive toward other species. In the natural state, it is not uncommon to see a drongo fly up and harass a raptor that passes over the nesting territory. The nest is a small cup placed in the fork of a branch. It is built of animal hair, spider web, fine rootlets, and grasses. In an aviary, tame birds will often steal hair from the head of their owner. They do not like having the nest inspected and will mob the keeper during nest inspections. The 2–4 eggs are cream to white in ground color, with red-brown, pink or black spots. Incubation is by both sexes for 16 days, and the chicks fledge at 17–18 days. Drongos are often double brooded, particularly if clutches are removed for hand rearing. If the chicks are left for the adults to raise, they must be removed as soon as they become independent, which is about 3–4 weeks after fledging.

Hand Rearing: They are easy to hand rear on a diet consisting of 50% minced bird of prey diet, and beef-flavored tinned dog food. The food is slightly moistened with water and a calcium and multivitamin supplement added once a day. The remainder of the diet should be natural foods such as mealworm beetles and larvae, waxmoths, winged termites, crickets, diced pinkie and half-grown mice, grasshoppers, etc. These foods are fed with a forceps. The chicks will regurgitate pellets consisting of insect exoskeletons, bones, and fur. Hand-reared birds remain hand tame indefinitely, particularly if fed insects from the hand on a regular basis. Immatures are dark with a wavy gray tinge to the feathers. For further advice on hand rearing, read the section on hand rearing SHRIKES.

Ring Size: 4.5 mm

Species Requiring Similar Management:

Shrikes
Bee-eaters
Glossy Starlings
Flycatchers

21.10 Bombycillidae

21.10ai Waxwings *Bombycillini*

• Simon Brusland Jensen & Eddie Bach

Introduction: The true waxwings of the genus *Bombycilla* compromise three species. They are frequently linked to the four species of silky flycatchers *Ptilogonatinae,* the hypocolius *Hypocolius ampelinus,* and occasionally the palm chat *Dulus dominicus,* in one family. Only the *Bombycilla* waxwings of this group have regularly been maintained in captivity. They have proven to do well but still breed rarely.

All three species are distributed exclusively in the Northern Hemisphere. They are small to mid-sized soft-bills of a delicate and inviting beauty in soft colors of grey, buff and reddish brown. The plumage itself is soft and silky and all species possess a rounded crest.

In two species, the Bohemian waxwing, Bombycilla garrulous, and the cedar waxwing, Bombycilla cedroum, the tips of the secondary feathers are modified into a coral drop-like appendage that appears like sealing wax. The third species, the Japanese or Amur waxwing *Bombycilla japonica* are lacking modified feather tips but have a soft red color on the tip of the same feathers. The Japanese have a more erect crest and the tip of the tail is also red.

Bohemians are the largest species and are mainly different from the Japanese in the yellow tip of their tail feathers, which they also share with the cedar. Both the Japanese and the Bohemian have a yellow and white pattern in the flight feathers, which is missing in the cedar, the smallest of the three. Sexes are similar in appearance but it has been suggested that females have less, and perhaps smaller, wax-like feather appendages than males and that the colored tail tip in males is wider and more intensively colored. This, together with the visual size difference of the body or crest, cannot be definitively confirmed.

Natural Habits: All three species seem to have a close affinity with coniferous forests but they are found in various temperate forest categories, often in younger growth. Migrating birds are frequently seen in gardens and parks. All share highly gregarious behavior and are most frequently observed in small groups, family parties or larger foraging parties. During migration they can congregate in very large flocks and, in particular, the Bohemian is well known as an "invasion species" that occasionally (on average every decade or so) migrates in unusually large numbers. Flocks of several hundred birds can be observed throughout central Europe in winter. The reasons for the irregular migration habits are not well known, but may be related to hard winters and good breeding seasons.

These are confident birds and in some cases it appears that they are almost "naive" around humans; this is particularly true in the most northern breeding grounds. Berries constitute their main diet and they are well adapted to swallowing large items with their stout and wide bills. However, during breeding they switch to mainly an insectivore diet and use the wide bill to capture flying insects on the wing, much in the way of flycatchers.

The breeding grounds of Japanese and Bohemian waxwings are in the far north and their young are mainly reared on the mosquitoes and mayflies that are abundant during the long days of the northern summer. Cedar waxwings use a more varied array of insect prey for their young. The nest is concealed in coniferous, spruce, pine, larch or cedar trees, and may be placed either close to the trunk or further out in the thin branches. It has been speculated that the alignment of the nest may be important in cold areas where the nesting birds depend on the sun to keep the chicks warm. Their calls are melodic, moderate and appealing. Their short contact calls are uttered frequently, especially when the birds are in motion. The call of the cedar waxwing is more high-pitched than that of the other two species.

In the past cedar waxwings were frequently kept in Europe and North America, but after the introduction of legislation that prevents the keeping and exporting of native North American species, they have all but disappeared from private aviculture. The most likely way for aviculturists to be exposed to cedar waxwings is with rescued migratory birds that have either been injured or are too weak and are taken in for rehabilitation and given treatment or supplementary feeding.

Today the Japanese waxwing is the most frequently kept species, followed by the Bohemian, at least in Europe. Bohemian waxwings can only be kept legally in the European Union if they are registered as captive bred.

Management: The three species have very similar needs, all adapt well to captivity and, if well cared for, will be long-lived, curious, peaceful and attractive aviary birds. Bohemian waxwings have on several occasions been known to live for at least 14 years in aviaries. Waxwings quickly become accustomed to their keeper and need little training to start taking food from the hand, however they easily become obese in captivity. Still they do not breed readily in captivity, and may, as do many other migratory species, become restless in the migrating season.

Fairly large aviaries are necessary to accommodate their strong flying skills and to provide adequate exercise. Waxwings move mostly by flying rather than walking or climbing, and it is worthwhile to decorate the aviary in a fashion that facilitates more movement. They are competent, long-winged flyers with relatively small and short legs. They utilize the entire upper section of the aviary, rarely visiting the ground except when drinking or bathing.

The aviary should be well planted and the birds' preference for coniferous trees could be considered; however for perching, open trees and bushes seem to be preferred between which the birds often have a very fast and direct flight. The birds select thin perches for roosting, often high in the aviary.

Waxwings are very hardy and do better in outside aviaries than in heated surroundings during the winter months. It is advisable to provide them access to a shelter, and they will learn to enter a small flight hole if necessary. Although they eat frozen food, such as berries, and drink snow in the wild, in captivity it is better to keep their drinking water and food free from frost to ensure their well-being.

Waxwings are usually housed without problems in pairs or in small groups, although males have been known to fight on some occasions. When waxwings are in conflict, visual displays and displacement are the order of the day and only rarely does it escalate to physical aggression. The threat display of waxwings is unmistakable, with open bills, raised feathers on head and back, followed by bobbing head and lowered wings. Birds in dispute may keep each other from the food and care should be taken that all birds in a group get access to food.

Imbalance in sexes within a group may cause obvious problems when breeding. Ideally they are wintered in groups and pairs separated for breeding during spring; a transfer may also help by simulating the natural migration.

Waxwings are very peaceful towards other bird species and can even be kept in mixed aviaries with smaller birds. Mixed-species aviaries are particularly suitable as they may help to encourage exercise. Waxwings are easily pressed by other birds due to their passive character and should not be housed with ill-tempered species. They do well with species that mainly use the ground or the lower part of the aviary. Attention should be paid to ensure the waxwings receive a proper diet in a mixed aviary.

Diet: Waxwings will certainly often be among the first visitors to the food dish. They show strong preferences for certain food items and will search the dish and throw unwanted items on the ground in order to reach favorite foods. Due to their strong tendency to become overweight in captivity, it is advisable not to be fooled by the birds' appetite to increase the rations.

The diet should be based mainly on fruits, but waxwings will rapidly adapt to commercial substitutes such as mynah pellets or universal paste. They have a strong preference for all kinds of berries and it appears that rowan berries and pyracantha berries that have been frozen and defrosted are especially desirable. Sweet fruits and in particular bananas, as well as raisins and currants, are popular items. Fruits can be offered in cubes or in slices if soft. Live food is taken readily, and as a substitute they will also eat cubes of scraped meat such as beef heart.

A suggested diet plan: Fruit and berries should formulate 40–50% of the total diet in the summer and minimum 60–70% in the winter. A dry "universal food" (for example, Claus III) consisting of dried insects, dried fruit, and cereal can be mixed with finely chopped hard-boiled eggs and soaked dog kibble. Depending on the season, a quarter to a third soft food can be used. For the breeding season a small amount of scraped meat can be added to this mixture.

Mynah pellets such as Wittemolen Mynah Care Granulate can be used dry with fruit, or soaked in water or fruit juice. Live food should be offered restrictively in winter, but in the breeding season the birds' food preference markedly changes and adding live food and animal protein may be essential to encourage the birds to breed. It is possible to attract flying insects using flowering plants, UV lights and hoof-stock dung in or near the aviary; this may serve as stimulation as well as behavioral enrichment.

Waxwings have very large livers — almost double the size of the starling *Sturnus vulgaris* — which is possibly an adaptation to eating berries containing alcohol in winter. They do not seem to be very sensitive to iron storage disease. They have been reported to drink large amounts of water in captivity.

Breeding: Only a few waxwings are bred in captivity every year. Between 2001 and 2004 only 14 offspring of the Bohemian waxwing were registered in Germany. Increased efforts need to be made to maintain sustainable captive populations. Breeding has been recorded between May and September both in the wild and in captivity.

Wild Bohemian waxwings never nest far from water, and they may breed in small, dispersed colonies in clusters of trees. The cup-shaped nest is typically constructed of lichens and placed 2.5–4 m (8–13 ft.) above the ground, well hidden.

In captivity waxwings occasionally breed when kept in groups, but frequently only a single pair in the group reproduces. More often pairs kept alone during breeding season, and in groups during winter, become active. A pair spend much time together before breeding, and a mating display with posing males and repeated exchange of a piece of food between the two can be used to identify potential breeding pairs.

The male and female construct a nest together and readily accept a platform or basket placed in a high and protected location. Also very popular are "bouquets" of conifer branches that can be tied together and placed in different places attached to the mesh or perches in the aviary. The nest is often constructed of coconut fibers, rough grass and moss and is rather unorganized in appearance.

Bohemians, on average, lay five blue-grey eggs with black spots; 3–5 eggs are typical for cedars, and for the Japanese 4–6 eggs have been recorded. In all three species the female incubates alone for about two weeks (12–16 days have been documented) and the male will often feed the female on the nest throughout the period. Nesting time for chicks is normally 15–17 days, although some aviculturists have reported up to 25 days. Both parents care for the young, and it is not unusual for the female to start a second clutch shortly after the young have fledged.

When there are young in the nest it is likely that the parents will only eat live food, and large quantities of insects should be offered. Mealworms, crickets, wax moths, fly larvae, and grasshoppers are normally taken without problem; smaller wild-caught insects from mosquitoes to aphids are also eaten and present a welcome variation.

A pair of Japanese waxwings started posing and feeding each other a few weeks after being placed in their own aviary at the beginning of May and began nest building within days. Nesting material included twigs and moss and was placed in a large basket. The birds began eating increased numbers of mealworms, as many as 40–50 per day. The female was not observed to be incubating definitely until 25 days after nest building had started. Then she sat for days on end and the male was observed feeding her on the nest. After about 15 days of incubation eggshells were found on the ground and a single chick was confirmed.

After hatching the female stayed in the nest most of the time for the first seven days and the male fed both the young and the female. After the first week the female stopped brooding the young continuously, except at night. The parents took turns feeding the chick; the female seemed more selective in food choice, whereas the male supplied a wider range of items. Mainly white and soft mealworms were offered.

After nine days the female was observed collecting nesting material again and started construction of another nest right next to the 11-day-old, begging chick. The female started incubating the second clutch although the first chick had not yet fledged. At 20 days the chick had fledged and could fly well, and within 26 days started feeding itself and bathing in the water dish. Another two chicks hatched after 14 days of incubation, but both died within one week, likely due to recorded temperatures as high as 38°C (100°F) in the aviary. The first chick was not observed to either cause disturbance or to assist the parents during the second breeding attempt. Would suggest the hand rearing protocol for TANAGERS would be suitable for waxwings.

Hand rearing: Little is recorded with regard to hand rearing waxwings. It can be assumed that they rear well on an insectivores diet fed with forceps with the addition of fruit and water. It can be speculated that waxwing chicks could be adapted to long light hours and may be sensitive to overheating.

Ring size:

Cedar Waxwing 2.6 mm
Bohemian Eurasian Waxwing 3.3 mm

Species Requiring Similar Managment: The following species require similar care, but probably need more animal protein in their diet; and consideration should be made to the fact that their social structure is not widely understood.

Silky Flycatchers *Ptilogonatinae*
Hypocolius *Hypocolius ampelinus*
Palm xhat *Dulus dominicus*
New Guinea Berry-peckers, and to a lesser degree, Old World Thrushes and Mountain Tanagers.

Acknowledgement: Thanks to John Sidor.

References:

Altherton, D. 1987. *A Bird Keepers Guide to Softbills*. Salamander Books Ltd. (Can't find this author spelling.

Bach, E., T. Rafn & F. Nielsen. 2003. *Frugt og insektaedende fugle - Naturlig pasning*. Atelier.

Bates, H.J. & R.L. Busenbark. 1970. *Finches and Soft-Billed Birds*. New Jersey: T.F.H. Publications.

Eriksson, K. & H. Nummi. 1983. Alcohol accumulation from ingested berries and alcohol metabolism in passerine birds. *Ornis Fenn* 60: 2-0.

Parrins, C.M. 1990. *The Illustrated Encyclopedia of Birds*. Headline Book Publishing. TLC

Robiller, F. 2003. *Das grosse Lexicon der Vogelpflege*. Ulmer.

Woolham, F. 1974. *Aviary Birds in Color*. London: Blandford Press Ltd.

21.11 Muscicapidae

21.11a True Thrushes *Turdinae*

• David L. Bender

Introduction: Among the birds in the order Passeriformes, the true thrushes, along with the ROBINS, shamas, CHATS, and allied species, were formerly assigned to the family Turdidae. Now, under the S-A-M system of classification, they have all been transferred to the family Muscicapidae, in which the true thrushes, including the shortwings and alethes, occupy the subfamily Turdinae, while the robins, chats, redstarts, and related species have been relegated to the subfamily Muscicapinae, which they share with the Old World FLYCATCHERS, indicating a slightly closer relationship to the latter group of birds, although they are placed in a separate tribe, the Saxicolini.

The subfamily Muscicapinae in the family Muscicapidae, order Passeriformes, is split into two large tribes, Muscicapini, or Old World flycatchers, and Saxicolini, the chats and related birds. Included in this tribe are the African robins of the genera Pogonocichla, Stiphrornis, and Swynnertonia; the akalats, genus Sheppardia; the true robins, genus Erithacus; the nightingales, rubythroats, blue robins, and similar birds of the genus Luscinia; the Asian bush-robins, genus Tarsiger; the African robin-chats, genus Cossypha; the shamas and magpie-robins, genus Copsychus; the redstarts, genera Rhyacornis, Hodgsonius, Chaimarrornis, and Phoenicurus; the Asian robins, genus Cinclidium; the grandala Grandala coelicolor; the forktails, genus Enicurus; the cochoas, genus Cochoa; the bushchats and stonechats of the genus Saxicola; the wheatears, genus Oenanthe; and several additional genera of African chats. The African scrub-robins of the genus Cercotrichas are included in this family, while the Australian, Papuan, and New Guinea scrub-robins and other robins belong to the family Petroicidae and the Australian chats of the genus Epthianura are placed with the honeyeaters in the family Meliphagidae.

In his excellent monograph of the thrushes, Clement (2000) discusses the taxonomy of these birds in greater detail. Although he basically follows the S-A-M system, he has elected to treat the North American bluebirds of the genus *Sialia,* the monotypic black-breasted fruit-hunter *Chlamydochaera jefferyi* of northern Borneo, the shortwings of the genera *Brachypteryx* and *Heinrichia,* and the alethes of the genus *Alethe,* all of which Sibley and Monroe (1993) include in the Turdinae, in a future volume covering the chats and robins of the tribe Saxicolini. Clement also follows other recent authorities in merging the three species of the genus *Pseudocossyphus,* upheld by Monroe and Sibley, into the genus *Monticola* with the other rock-thrushes.

The true thrushes are divided at the generic level into several distinct groups, each with its own designation. Thus, the thrushes of the genera *Monticola* and *Pseudocossyphus* are known as ROCK-THRUSHES. This group includes many colorful and aviculturally desirable species from Europe, Asia, and Africa, several of which have been occasionally kept and bred in captivity. The four thrush species comprising the endemic African genus *Neocossyphus* are called flycatcher-thrushes and ant-thrushes, and are not aviculturally significant.

The WHISTLING-THRUSHES of the genus *Myophonus* (sometimes rendered *Myiophonus* or *Myiophoneus*), are large, aggressive thrushes, generally dark in color with an iridescent sheen when seen in good light. They are well worth keeping if they can be afforded an aviary to themselves. Captive breeding of whistling-thrushes is only rarely reported.

The solitaires are New World thrushes, several species of which are highly regarded for their singing ability. Included in this group are members of the genera *Entomodestes, Cichlopsis,* and *Myadestes*. The endemic Hawaiian thrush species are now assigned to the latter genus. In a similar manner, the former genus *Hylocichla* has been subsumed into *Catharus,* a genus that contains such well-known North American songbirds as the wood thrush *C. mustelinus,* the hermit thrush *C. guttatus,* and the veery *C. fuscescens,* as well as numerous Neotropical species known as nightingale-thrushes. This is basically a New World genus, with the range of only one species, the gray-cheeked thrush *C. minimus,* extending to Siberia, although vagrant specimens of Swainson's thrush *C. ustulatus,* have occasionally been collected in eastern Asia. See OTHER GENERA, below.

From an avicultural standpoint, the two most important groups of true thrushes are the TYPICAL THRUSHES of the genus *Turdus* and the GROUND THRUSHES of the genus *Zoothera,* which includes those species formerly placed in the obsolete genus

Geocichla. Encompassing at least 66 species, *Turdus* is by far the largest genus of thrushes. Geographically, it is also the most widespread, with representative species found throughout Europe, Asia, Africa, North and South America, the Caribbean islands, New Guinea, the Philippines, the islands of Melanesia, and various other islands throughout the world. The best-known thrush in North America is undoubtedly the American robin *T. migratorius* that was given the common name "robin" by early settlers out of nostalgia for the European robin *Erithacus rubecula,* which it superficially resembles in basic color pattern. Like many American aviculturists, my first experience in hand-feeding a baby bird involved a robin that had fallen (or perhaps was lifted; my memory is rather vague on this point) from the nest.

In Europe, the Eurasian blackbird *T. merula* and the European song thrush *T. philomelos* are perhaps the most popular members of this genus, followed closely by the mistle thrush *T. viscivorus,* the redwing *T. iliacus,* and the fieldfare *T. pilaris,* all of which have been bred in captivity. The blackbird and song thrush were introduced into Australia and New Zealand, where they have become established. Aviculturists throughout the world are probably familiar with these two species, but equally familiar with other *Turdus* species indigenous to their own local areas. The monotypic groundscraper thrush *Psophocichla litsipsirupa* from Africa is sometimes included in the genus Turdus; this species has been bred at the San Diego Zoo.

With about three dozen species, *Zoothera* is the second largest genus of thrushes. While this genus is centered in Southeast Asia and Wallacea, where many geographical races and isolated island endemics are found, the genus is also well represented in Asia and Africa, with outlying species occurring in North America. This genus includes many colorful and attractive species, as well as many excellent songsters. The orange-headed ground thrush *Z. citrina,* for example, is esteemed for both the beauty of its plumage and for its song. Like the typical thrushes, ground thrushes are popular and hardy avicultural subjects, and several species have proved to be fairly reliable breeders under captive conditions.

For several years I have kept and bred two species each from the genera *Turdus* and *Zoothera,* and have recently added pairs of the black-breasted thrush *Turdus dissimilis,* to my collection. I also have a single female eye-browed thrush *T. obscurus* shipped to me as a female Siberian thrush *Zoothera siberica,* which has thrived for several years under the same treatment as the other thrushes. The following remarks and observations are based on my experiences, supplemented by notes from the ornithological and avicultural literature, particularly in regard to the natural history and captive management of species that I have not been privileged to keep.

Typical Thrushes *Turdus* spp.

Introduction: Some *Turdus* species, including the blackbird and the black-breasted thrush, are dimorphic, while others, such as the song thrush and similar species, are monomorphic, and must be surgically or DNA sexed to accurately determine the gender of individual birds. The plumage of male blackbirds is entirely black, with a contrasting orange-yellow beak and eye ring; the legs are dark. The females are dark brown, slightly lighter on the underparts, with brown beaks. Song thrushes are basically brown, the underparts lighter in color but heavily spotted with black. Though not among the more colorful thrushes, they are, as the name implies, excellent singers.

Natural Habitat: With such an extensive geographical distribution, members of this genus occupy a wide variety of habitats. Many species are migratory, either latitudinally or altitudinally, and therefore frequent different habitats at different seasons of the year. An aviary environment similar to that provided for such temperate species as the blackbird and the song thrush should prove conducive to breeding for most species, even those that originate from tropical areas where they are found in habitats ranging from damp, broad-leaved evergreen forests to dry, open woodlands and savannas.

The blackbird, with 15 races scattered through Europe, northern Africa, the Middle East, central Asia to China and Southeast Asia, the Indian subcontinent and Sri Lanka, prefers deciduous woodlands and forests, but can also be found in more open areas such as parks, gardens, and orchards. The song thrush, with only three subspecies distributed from the British Isles through Europe and into western and central Siberia, inhabits both deciduous and coniferous woods, copses, and thickets. The black-breasted thrush *T. dissimilis,* which has been available to aviculturists for the past several years, ranges from northeastern India through northern Myanmar and southern China as far east as northern Vietnam, breeding in damp evergreen forests and wintering at lower elevations in scrub jungles.

Management: Pairs of both species were obtained on breeder loan when the retirement and relocation of a local aviculturist coincided with the reduction of the

Hancock Wildlife Research Center's extensive turaco and hornbill breeding program, placing several aviaries at my disposal. A pair of blackbirds was placed in an aviary measuring 3 x 5 x 3.6 m (10 x 16 x 12 ft.) high, including a 1.2 m (4 ft.) wide, 2.1 m (7 ft.) high safety alley extending the full width of one end, above which the upper portion of the flight projected. This aviary contained a clump of bamboo and several scattered bushes of red elderberry, with a dense tangle of English ivy filling one corner and a 1.2 m (4 ft.) square section of roof over several perches and a nest box to provide minimum shelter from the rain, a precaution that proved to be totally unnecessary due to the natural protection of thick vegetation. An old, proven pair of song thrushes was installed in private compartment measuring 3.6 x 3.6 x 3.6 m (12 x 12 x 12 ft.), with a 1.2 m (4 ft.) wide roof extending the length of one side. A cedar tree and a clump of red elderberry dominated this aviary.

Although open-fronted nest boxes were placed in several strategic locations in each aviary, and shallow baskets were attached to branches of the trees, both species opted to construct natural nests in the foliage, the song thrushes building in the cedar tree and the blackbirds selecting a site in the ivy vine. Since many thrush species use mud in nest building, a dish of mud is routinely maintained in each thrush aviary for the duration of the breeding season. Dried grass, as well as live green moss, is supplied for nesting material, with dead bamboo leaves and other detritus being available on the ground within the aviary. While half-open-fronted nest boxes had been recommended to me for thrushes, I have yet to have a single pair of thrushes of any species nest in such a box, although three species to date have nested on top of these boxes. In England, Raymond Sawyer noted that both of his female island thrushes *T. poliocephalus* nested in half-open-fronted boxes, but confirmed my own observation by remarking that this is not normal for thrushes (Sawyer 1997). Perhaps this species represents the exception that proves the rule, although many other species have yet to be tried in aviculture. In general, covered platforms, open on at least three sides, might be a better option. Baskets have been used at different times by all the thrush species in my possession except the blackbirds.

Breeding: The song thrush nest had already been completed, and the first egg deposited, before I realized the birds were nesting. The nest was placed about 7 ft. (2.1 m) above ground, well out in a branch of the cedar tree. Although situated within a few feet of access doors on two sides of the aviary, the nest was quite well concealed from view. Several days after discovering the nest, when trying to observe it from a different perspective, I found that it was actually the second nest constructed at the same site. An earlier nest, containing three partially incubated but no longer viable eggs, had become detached from its moorings, probably during a storm, and was hanging at a right angle beneath the occupied nest.

When completed, the second clutch consisted of four eggs, all of which hatched. One chick was found dead on the ground below the nest at two days of age, and the other three died at about a week of age. This loss was attributed to diet, or more precisely, to an insufficient amount of live food, particularly earthworms, in the diet. As anyone who has rehabilitated an American robin can attest, a generous supply of earthworms is indispensable in rearing Turdine thrushes. I had begun supplementing the diet of these thrushes as soon as chicks hatched with an unlimited supply of mealworms and what seemed at the time like an adequate supply of earthworms, their preferred food for feeding the young, but in retrospect I can only conclude that I had not provided sufficient earthworms to meet their needs. Ed Storozynski, who had reared young from this same pair of birds in previous years, had supplemented the diet provided by the parents with chopped-up slugs, which he hand-fed to the chicks in the nest. He recommended that I try this technique, but I opted to increase the supply of earthworms instead, allowing the parents to do all of the feeding, for the first few days, at least.

Fortunately, the birds recycled, building a new nest on top of a nest box that had been hung in the sheltered portion of the aviary at a height of about 10 ft. (3 m). This third clutch contained five eggs, all of which were fertile and hatched on schedule. I had been offering the birds about 80 earthworms per day while the previous brood was in the nest, but decided to provide as many as they would feed. Eventually, this pair and their nestlings were consuming more than 400 worms each day. The compost heaps from which the worms were being dug were taxed heavily at this time, as earthworms were also required for newly hatched blackbirds, as well as a brood of four orange-headed ground thrushes. It was therefore decided to pull three of the song thrush chicks and two of the ground thrush chicks for hand-rearing, leaving only two chicks with each set of parents.

The blackbirds, meanwhile, had hatched two chicks from a clutch of three eggs in their nest in a tangle of ivy. Unfortunately, the hen of this pair was a young, inexperienced bird, and apparently did not feed the chicks, as one was found dead on the ground at two days and an inspection of the nest revealed that the other chick had also died. The pair produced a second clutch, the first egg being laid within a week. Three chicks hatched, and careful observation indicated that the parents were feeding them regularly. One chick eventually disappeared, but the other two grew rapidly, only to be found dead on the ground at nine days of age. I then decided that if the birds nested again I would pull the chicks for hand-rearing. A third clutch was in fact laid, with all four eggs hatching. Two chicks were removed for hand-feeding at five days of age, and the other two were successfully reared by the parents.

Several breeding seasons later, I can state that all of the blackbird pairs that have nested for me have invariably used the same nest for subsequent broods. The song thrushes would occasionally use the same nest consecutively, but more often constructed a new nest for the second clutch, sometimes returning to the original nest for a third clutch, and alternating between the two nests if additional clutches were forthcoming. Since female song thrushes normally begin a new clutch before the current brood is weaned, the male continues to feed the fledglings while the hen tends to the new nest. Most song thrushes are very steady sitters, some females having to be literally lifted from the nest in order to inspect the eggs or chicks. Female blackbirds are reasonably steady, but will usually leave the nest upon being touched. Preliminary observations of the black-breasted thrush, based on three pairs, indicate that this species is far more nervous, leaving the nest whenever their aviary is entered or even approached.

All of the song thrushes I have raised, including both hand-reared and parent-reared specimens, have fledged at 14 days of age. The nestling period for blackbirds has been more variable, with chicks leaving the nest at 13–16 days after hatching. The incubation period for both species is 12–13 days, incubation apparently commencing with the laying of the penultimate egg.

Two male and one female black-breasted thrushes were purchased from an Ontario importer late in 2001. A trade was arranged to procure a surplus hen from another aviculturist, and an additional pair acquired from a more recent importation. These three birds arrived on April 30, 2002. The new pair, not expected to breed their first year, were placed in a large 11 x 15 x 3.6 m (36 x 50 x 12 ft.) high aviary that, despite its size, had been constructed to be finch-tight. They shared this aviary with a pair of fairy bluebirds *Irena puella* and a pair of song thrushes. The single female, contrary to normal introduction policy, was released into a 3.6 x 11.5 x 3.6 m (12 x 38 x 12 ft.) aviary in which the waiting male had already been established for several weeks. She flew directly from the shipping crate to a perch, and the male immediately flew up from the ground and briefly mounted her, then displayed to her, bowing toward her and seemingly shivering his feathers, but with no noticeable lowering of the wings or movement of the tail, from a distance of about 45 cm. (18 in.) He then approached her, mounted her again, and when she flew off, followed her around the aviary for several minutes before going his own way.

The original pair had earlier been set up in an aviary in which other thrushes had nested in previous years. Although this was the pair I had the greatest hope of breeding from, it was the newly-imported pair that made the first nesting attempt. The nest, which I thought was built by the song thrushes until I saw the female black-breasted carrying nesting material in her beak, was constructed over a period of two days in a basket placed at a height of 1.2 m (4 ft.) from the ground in a small cedar tree. The three eggs, laid on consecutive days, appeared to fit the description given by Clement (2000), buffy in ground color with darker spots. The first egg was laid just two weeks after the birds were shipped across the continent. Although the adults were never seen on the nest and only occasionally seen leaving the nest tree when the aviary was approached, the eggs were always warm to the touch, and apparently being incubated. However, they disappeared after about 10 days. The other birds were removed from the aviary and a second clutch was laid, but also came to grief.

Meanwhile, the second pair had constructed a nest in a cedar tree about 1.8 m (6 ft.) from the ground, and less than a foot from where a small basket had been wired in place. A clutch of five eggs was laid, from which I had hoped to ascertain the incubation period of the species, which is not given by Clement, and according to Ali & Ripley (1998), has never been recorded. However, this clutch eventually proved infertile. The eggs from this hen were light blue in ground color, fairly heavily spotted. The original pair, that had by this time occupied their breeding aviary for three months, finally built a nest in a basket which had been fastened about (1.8 m) 6 ft. above ground in a cedar. Four eggs were laid, which appeared to be an even deeper blue color but still spotted throughout. Unfortunately, exact egg dates and the beginning of incubation were not recorded, but on June 18 two chicks hatched, and by the following morning the remaining two eggs had also hatched. The estimated incubation period was 12–13 days, assuming that eggs were laid on the four days immediately after the nest was completed and incubation commenced with the laying of the last egg. The parents relied heavily on earthworms for feeding the chicks, which were removed for closed banding and hand feeding on June 24. Weights were recorded daily through fledging.

The hen that had laid the clutch of infertile eggs laid

again in the same nest, the first egg being noted on the morning of June 15. She began incubating on June 17, after the third of four eggs was laid. Three chicks hatched as anticipated on June 29, and the fourth had hatched by the morning of July 1, confirming an incubation period of 12–13 days. This figure was later corroborated by the second clutch of the original pair. The first egg was laid on July 2, and incubation again commenced with the laying of the penultimate egg, on July 4. One chick hatched on the afternoon of July 16; two more had hatched by the morning of July 17, and the final egg hatched during the ensuing night.

The second pair produced a third clutch, but the four chicks disappeared within a few days of hatching. This was followed by a fourth clutch, but none of the eggs from this clutch hatched. Then, to my surprise, a fifth clutch consisting of three eggs was discovered in the nest in early September. I assumed this clutch would prove infertile, as the male was in heavy molt by this time. However, an inspection of the nest on the evening of September 10 revealed two chicks had hatched. To generalize from observations made on a total of nine nesting attempts from three captive pairs during a single breeding season, the black-breasted thrush consistently uses the original nest for subsequent clutches, even after the eggs have disappeared or been removed when infertile, or after complete broods of young chicks have been taken for hand rearing or have been lost or thrown out of the nest at an early age.

In addition to this nest site tenacity, this species is also more aggressive in nest defense than the congeneric blackbirds and song thrushes I have kept, making a loud snapping or clicking sound with their beaks as a threat or warning signal, and then dive-bombing at the head of an intruder when their nest is approached. In the second pair, the male was the aggressive bird, flying into my head relentlessly until I departed from the vicinity of the nest, while the hen generally retreated to the rear of the aviary. In the original pair, the female was more aggressive in this regard than the male, but did not make actual contact during any of her attacks.

Diet: When I first started keeping insectivorous birds, I made every effort to provide as wide a variety of foods as possible, including various homemade and proprietary brands of softbill mixes, dog and cat food both dry and soaked, beef heart, and live food, as well as berries, grapes, and the fruit/vegetable mix we use for turacos and other frugivorous birds, on the theory that given a wide enough choice, the birds will select a balanced diet. Over the years I have reached the conclusion that for many species, such elaborate and expensive diets only represent time and money wasted, for invariably, most of the foods offered are neglected in favor of a few preferred items. In my experience, such birds as thrushes, babblers, and starlings soon make their preferences known, and refuse to eat other foods unless forced to do so by removing the favored items. Since we do not really have enough information to properly determine which items most species should be eating, it has been my policy to defer to the demonstrated preferences of the birds. With the exception of live food, which is always readily taken, my birds have consistently chosen, perhaps because of its conveniently edible size, Eukanuba small bites puppy chow (or more recently, cat food of a similar size and shape) over all other items. Even newly acquired birds that have been fed on other diets soon take this food almost exclusively, always preferring it in dry form rather than soaked.

For several years my basic maintenance diet for thrushes consisted of this brand of dry puppy chow supplemented with a few mealworms year round, and with additional live food and other items during the nesting season. For the hardy species that are wintered outdoors, this simple diet offers the benefit of remaining in palatable form even during subzero weather when soaked dog foods or commercial softbill mixes moistened with additives such as grated apples, carrots, or hard-boiled eggs soon freeze into a solid, inedible lump.

In the wild, blackbirds and many other thrushes are known to subsist largely on berries during the winter months, but none of my birds would touch mountain ash *Sorbus* sp., or firethorn *Pyracantha* sp. berries when offered.

Originally, I had tried to find a suitable dry cat food, as studies have indicated that cat food is better for birds than dog food. However, the various donut-, cross-, and star-shaped cat foods on the market did not seem small enough in size for the smaller softbills to easily consume, and the Eukanuba puppy chow was adopted more or less by default. Recently, however, Ed Storozynski, who uses the same basic diet for his thrushes, noted that the cat food his wife was buying for their cat was similar in size and shape to the Eukanuba puppy chow, and far less expensive. We then began mixing this cat chow (Kirkland Signature brand, purchased at Costco) with the Eukanuba. The small pellets were distinguishable only by their slightly larger size and lighter, more orangish color. At first the birds picked out the Eukanuba, leaving the cat food until last, but by the end of the first week, they were picking out the cat food and leaving the puppy chow. The proportion of cat food was gradually increased until the puppy chow was eliminated. We have now completed a full breeding season on this altered diet, and the cat food seems to work as well as the puppy chow, although most of the breeding successes described in this section were achieved using the Eukanuba and mealworms, supplemented with an unlimited supply of earthworms when chicks were being fed. Soaked Kaytee Exact softbill diet was routinely offered during the nestling and fledgling periods, but seldom touched. This year I started offering Pretty Bird brand softbill pellets at the weaning stage for variety, and the juvenile thrushes readily accept this food as well as the cat food and mealworms. Earthworms are discontinued shortly after the chicks leave the nest.

While I have been criticized for this uncomplicated diet, my thrushes have all remained healthy for several years, and the breeding results speak for themselves. I have not kept thrushes long enough to determine what, if any, effect this diet will have on their long-term health or longevity, but then, little data is available for comparison.

Hand Rearing: Thrush chicks for hand rearing are now removed from the nest at about 5–6 days of age, which seems to be the proper time for closed-ringing, as the feet of some chicks will be too large to slip the bands over as early as 7–8 days of age. Spending a great deal of time on the ground, thrushes have relatively large and strong legs and feet. Young thrushes were fed the same diet I have used successfully with various glossy starlings and mynahs, soaked Purina kitten chow supplemented with a lesser amount of soaked Kaytee Exact softbill pellets. This hand-feeding regime has worked well for blackbirds, song thrushes, black-breasted thrushes, orange-headed ground thrushes, and Siberian thrushes. Hand-fed chicks were weighed each morning before being given their first feed of the day, and progress recorded. Following are growth charts compiled from data collected

Siberian thrush
Zoothera sibirica

from six hand-reared song thrushes and eight hand-reared black-breasted thrushes.

Not only do the eggs of the black-breasted thrush vary greatly in color and shape, but in size as well. Although I did not weigh any of these eggs, I would estimate that the eggs of the original pair, from which chicks 1–4 hatched, were nearly twice as large as those of the second pair, that produced chicks 5–8. This discrepancy in egg size is reflected in the size and growth rate of the chicks from the two pairs. At six days of age, the chicks of the first clutch outweighed those of the second clutch by nearly 7 g (0.24 oz.). By 10 days of age the weight difference had increased to about 10 g (0.35 oz.), with chicks from the smaller eggs only reaching the six-day weight of those from the larger eggs. Yet just four days later, after the chicks had fledged, they were virtually equal in weight. Interestingly, the heaviest chick at 14 days of age was from one of the small eggs; the lightest chick, from one of the large eggs. Apparently the fledgling weight is of greater significance than the growth curve by which that weight is achieved.

A number of other species of *Turdus* thrushes have been kept and bred by various zoos and private aviculturists over the years. Charles Everitt reared several species at the Edward Marshall Boehm Aviaries in New Jersey, including the Kurrichane thrush *T. libonyanus,* from Africa, the pale-vented robin *T. fumigatus* from South America, and the gray-winged blackbird *T. boulboul* from Asia. The latter species nested in a community aviary with a pair of European blackbirds nesting in an adjacent tree; the two family groups lived in perfect harmony (Everitt 1973). Everitt also noted that the Cuban race of the red-legged thrush *T. plumbeus rubripes* had previously been bred in the US, and the Berlin Zoo also bred this species at least seven successive years in the 1960s. Herbert Murray gave an account of the breeding of the redwing *T. iliacus* in England. It is significant that while the parent redwings fed maggots, mealworms, and whatever insect life they could find in the aviary to their young, earthworms, although provided, "were not much used" (Murray 1960). The mistle thrush *T. viscivorus* was bred in England in 1970. Alex Isenberg reported breeding the Japanese thrush *T. cardis* in California, describing these birds as wonderful singers with great variety in their song (Isenberg 1960). They have the added advantage of an extended singing period. The gray-backed thrush *T. hortulorum,* once considered to be a subspecies of the black-breasted thrush, was bred at the London Zoo in 1968. The nest, made largely of mud with a small amount of grass, was built on a window ledge about 2.2 m (7 ft.) above ground (Yealland 1969). The nominate race of the olive thrush *T. olivaceus* a South African species known also as the Cape thrush, was bred at the West Country Wildlife Park in England. The nest was placed about 3 m (10 ft.) above ground in a canopy of Passiflora, the successful nesting taking place during December in a tropical house (Lowe 1981). This species has also been captive-bred at the zoo in Port Elizabeth, South Africa. The clay-colored thrush *T. grayi* a Central American species, has been bred on several occasions at the San Salvador, El Salvador, zoo. As mentioned earlier, two female island thrushes *T. poliocephalus* nested for Raymond Sawyer. Although he does not specify which of the 52 recognized subspecies of this highly variable species his birds represented, his brief description seems to fit the race inhabiting Taiwan *T. p. niveiceps*. An interesting aspect of this success is that the birds were kept as a trio, with both hens rearing young from the same male (Sawyer 1997).

Ground Thrushes *Zoothera* spp.

This large genus, not all of which are called ground thrushes, includes many of the most attractive species in the subfamily Turdinae, including the popular orange-headed ground thrush *Z. citrina*. For many years this species was regarded as an avicultural standard, being exported from India in large numbers. When that source was cut off, the species was seldom available until recent years, when birds from China or Southeast Asia have reached the markets. Another species that has been available in small numbers during the past few years is the Siberian thrush *Z. siberica*. Both of these species are dimorphic. Unfortunately, other species from this genus are seldom available. Many members of this genus are island endemics, occurring on only a single island or a small group of islands, from Indonesia through Borneo, the Philippines, the Ryukyus, Japan, New Guinea, and Melanesia.

Ed Storozynski and I were able to obtain a total of seven orange-headed thrushes, presumably wild-caught birds, from a Quebec importer over a period of two years. With the aid of racial descriptions from Clement's *Thrushes,* as well as several other references, along with tail and wing measurements, I have determined to my own satisfaction that these birds belong to the subspecies *Z. c. melli* from southeastern China. Eventually I hope to send specimens to the Field Museum in Chicago for confirmation. Two male and one female Siberian thrushes were also acquired from the same dealer, the second female of the two pairs ordered appearing to be a female eye-browed thrush *Turdus obscurus*.

The habitat of the orange-headed ground thrush is damp broad-leaved evergreen forest with an undergrowth of bushes and ferns, as well as bamboo thickets, woodlands, and the margins of cultivation. Some of the 12 recognized subspecies are migratory; others, sedentary. The Siberian thrush is migratory, breeding in eastern Russia, Siberia, northern Mongolia, and northeastern China, and wintering from Manipur through Myanmar and Thailand into peninsular Malaysia, Cambodia, southeastern China, and Vietnam, and as far south as Borneo and the Indonesian islands of Sumatra and Bali. The breeding habitat consists of lowland coniferous taiga with an undergrowth of alder and willow thickets. The preferred winter habitat is broad-leaved and coniferous montane forest at elevations of up to 2,565 m (8,460 ft.) (Clement 2000).

Management: See TYPICAL THRUSHES, above. I have not kept breeding pairs with other species, but during the nonbreeding season these birds have done well in mixed species flights. They have shown no aggression even toward smaller birds such as silver-eared mesias *Leiothrix argentauris,* or pekin robins *L. lutea.*

Breeding: A pair of the original birds purchased in 1999 nested for the first time in 2000. The nest was built in a fairly exposed location on top of a post, approximately 1.8 m (6 ft.) from the ground. When discovered,

European Song Thrush *Turdus philomelos*, 2001

Band	Date	Day 5	Day 6	Day 7	Day 8	Day 9	Day 10	Day 11	Day 12	Day 13	Day 14
21	30.06 g	32.1 g	39.5 g	48.2 g	48.2 g	48.2 g	51.4 g	52.2 g	55.1 g	56.6 g	
22	30.06 g	30.1 g	38.3 g	47.7 g	46.6 g	47.1 g	50.5 g	49.8 g	54.9 g	58.9 g	
23	01.07 g	22.9 g	31.3 g	39.0 g	41.9 g	42.1 g	44.3 g	46.2 g	51.2 g	54.4 g	55.5 g
24	03.07 g	25.6 g	30.7 g	38.2 g	45.3 g	44.3 g	46.9 g	49.1 g	54.0 g	54.4 g	57.3 g
25	03.07 g	29.3 g	35.5 g	40.7 g	45.6 g	44.1 g	45.2 g	46.7 g	47.9 g	50.2 g	51.1 g
26	03.07 g	29.0 g	32.7 g	38.2 g	43.2 g	42.9 g	44.4 g	44.9 g	49.7 g	51.5 g	52.5 g
	Average:	*26.7*	*32.1 g*	*39.0 g*	*45.3 g*	*44.7 g*	*46.0 g*	*48.1 g*	*50.8 g*	*53.4 g*	*55.3 g*

Compiled by David L. Bender, Hancock Wildlife Research Center, Surrey, BC, Canada.

the basic nest had already been constructed, and the hen was turning and twisting, shaping the interior of the bowl. The structure appeared quite shallow, and I assumed the sides would be further built up as work progressed. However, by noon, all work had ceased, and I expected to see the hen resume construction in the evening, in correspondence with the normal crepuscular activity pattern of the species. However, neither bird was seen near the nest until around 10:30 a.m. the next day, when the hen was observed sitting tightly on the nest. The nest was inspected that evening, and the cup proved deeper than a side view of the nest would suggest; apparently a depression had rotted into the top of the old post. The nest was constructed largely of grass and pieces of moss, with no lining or plastering of mud, which some subspecies are said to use. Four eggs were laid on consecutive days, but eventually proved infertile, the nest being abandoned on May 29, after eleven days of incubation.

Several days later the male was heard singing. A new nest was soon built on the opposite side of the aviary, on a 2 x 4 cross piece approximately 2.4 m (8 ft.) above ground, and in clear view from my picture window, facilitating observation. All incubation was carried out by the female, although the male would stand guard over the eggs when the female was off the nest. Three chicks hatched on June 26, and the fourth egg had hatched by the next morning. By July 1, the four chicks had grown sufficiently to completely fill the nest cup, and by July 5, they were literally overflowing the nest, and I decided to pull two chicks for hand feeding. All four chicks, both parent-reared and hand-reared, fledged at 13 days of age. At this stage the hand-fed chicks each weighed 44.5 g (1.56 oz.), and a parent-reared fledgling was caught and weighed for comparison, weighing in at 44.6 g (1.57 oz.). The young birds could be sexed within a few weeks of fledging, and within a few months had molted into adult plumage.

The following year, three pairs reared young, including a second wild-caught pair and a wild-caught hen mated to a young captive-bred male. By the end of the breeding season a total of 20 young had been reared, including the brood of four from the previous year. Unrelated pairs were sent to several zoos and private aviculturists, and at the time of writing, halfway through the 2002 breeding season, yearling pairs have produced young for me, as well as for the Assiniboine Park Zoo in Winnipeg.

Diet: See TYPICAL THRUSHES.

Hand Rearing: As for TYPICAL THRUSHES. A number of chicks of this species have now been hand-reared, and the following growth chart was compiled from the accumulated data. Several parent-reared chicks were weighed at fledging for comparative purposes. Fledging has occurred as early as 11 days of age for one chick, and as late as 14 days for several others, although most have fledged at 12–13 days.

A pair of Siberian thrushes also nested in 2001. They were transferred for the breeding season to a planted flight measuring1.2 x 4.8 x 3.3 m (4 x 16 x 11 ft.) high, extending from a building, with the top and sides of the first 1.2 m (4 ft.) covered. The partition between flights extends beneath the building, providing an additional ground area of the same width and a length of 3.6 m (12 ft.), the distance from the floor of the building to the natural earth substrate being about 1 m (3 ft.). Both the Siberians in this flight and a pair of orange-headed in an adjacent flight spent a great deal of time beneath the building. An open-fronted nest box and a large plastic canary nest were installed, and the plants growing in the aviary, including a small hemlock tree, a clump of red elderberry, and a tangle of ivy, provided natural nesting sites.

The nest was completed on June 12 in the plastic canary nest, constructed largely of mud with a small amount of dried grass and moss mixed in, and a thin lining of dried grass. The nest basket hung at a slight angle, and a definite lip of mud had been formed at the lower side of the nest. The first egg had already been laid, and the male thrush spent a great deal of time at the nest dur-

Black-breasted Thrush *Turdus dissimilis*

Band	Date	Day 4	Day 5	Day 6	Day 7	Day 8	Day 9	Day 10	Day 11	Day 12	Day 13	Day 14
DLB 1	18.06	33.1	34.2	40.1	42.5	44.0	44.6	44.2	46.5	46.0	47.5	
DLB 2	18.06	31.7	33.6	39.3	42.4	44.2	44.9	45.7	46.9	47.6	47.9	
DLB 3	19.06	27.7	28.5	33.8	38.0	41.2	41.9	42.8	43.9	44.4	44.2	45.1
DLB 4	19.06	23.0	24.0	30.6	33.5	36.7	38.5	40.1	40.7	41.4	40.7	41.5
DLB 5	29.06	27.0	28.7	28.4	28.2	31.2	35.4	35.8	39.7	43.6	45.9	
DLB 6	29.06	26.4	28.7	29.7	29.5	34.0	34.9	36.2	38.1	44.9	47.1	
DLB 7	29.06	23.4	27.4	27.6	27.8	28.2	29.8	31.1	32.7	38.7	42.0	
DLB 8	30.06	21.6	23.8	25.5	27.4	29.0	32.9	31.4	40.7	44.0	49.4	

Compiled by David L. Bender, Hancock Wildlife Research Center, Surrey, B.C., Canada.

Chestnut-capped Ground Thrush
Zoothera interpres

Age days	Weight (g)	Age days	Weight (g)	Age days	Weight (g)
1	3.5 g	17	26.7 g	25	36.9 g
2	4.7 g	18	29.8 g	26	40.1 g
3	5.6 g	19	31.2 g	27	37.9 g
7	17.4 g	20	32.7 g	28	39.6 g
8	26.3 g	21	34.5 g	29	41.7 g
12	25.9 g	22	34.9 g	31	42.1 g
15	26.8 g	23	34.6 g	32	43.3 g
16	27.3	24	38.0 g	33	45.1 g

Compiled by staff of Lowry Park Zoo, Tampa, Florida, and adapted from Salisbury (1991).

ing the day. However, I could not determine if actual incubation had commenced. Three additional eggs were laid on successive days.

On June 23, when attempting to point out the thrush nest to members of a visiting bird club, I was unable to see the nest from the safety alley. On entering the pen, I found that the plastic bracket holding the nest receptacle in place had broken, and the nest was hanging precariously in an almost vertical position. The male thrush flew from the nest as I approached, exposing two eggs that had fortuitously been saved by the prominent mud lip of the nest; the other two eggs were missing. As the nest was held in place by a mere thread of plastic, I immediately transferred the eggs, which resembled song thrush eggs, to the nest of an old, reliable pair of song thrushes that were fortunately nesting at the time. The Siberian eggs hatched on June 26 and 27. The chicks had much thicker down than any other thrush chicks I have seen, a possible adaptation to their northern nesting grounds in the wild. As Clement (2000) gives the incubation period of this species as 11–12 days, I assume that the third and fourth eggs laid were the survivors. The song thrushes fed their foster chicks dutifully, but when their own eggs began to hatch on July 2, I decided to pull the Siberians for hand feeding to prevent them from crushing their younger nest mates.

The chicks were closed-banded and weighed, and then weighed on each succeeding morning before being fed. They were heavier than orange-headed ground thrush chicks of the same age, and fledged at the remarkably early age of nine days. The adult Siberians renested, the first egg of the second clutch being laid only five days after disaster struck the first nest. Unfortunately, a second nest bracket gave way, and this clutch of three eggs was lost. The canary nest cup, still containing the original nest, was then nailed securely in place, and the first egg of a third clutch appeared on 11 July. Two additional eggs were laid on consecutive days, and all three eggs hatched on successive days, beginning on 22 July. This not only confirmed the 11-day incubation period, but also verified the fact that incubation begins with the first egg. All three chicks survived, fledging together when the oldest chick was 12 days of age. They were weighed once during the nestling period, when closed-banded, and again after fledging, and the resultant data incorporated into the following growth chart, along with weights from the hand-reared chicks (Bender 2002).

One of the smaller species of the genus, the chestnut-capped ground thrush, *Z. interpres,* from Asia, was reared at Lowry Park Zoo in Florida. After several unsuccessful nesting attempts by the parents, a clutch of eggs was artificially incubated and a single chick hand-raised. The adults shared a 1 x 3 x 2.4 m (5 x 10 x 8 ft.) high outdoor flight with a pair of roulroul partridge *Rollulus roulroul,* a pair of Gouldian finches *Chloebia gouldiae,* a pair of Leclancher's buntings *Passerina leclancheri,* a single male red-legged honeycreeper *Cyanerpes cyaneus,* a pair of blue-tongued skinks and four radiated tortoises. The flight was planted to bamboo, Ficus, Liriope, and various bromeliads. Nesting occurred from May through September, with two clutches of two egg and two clutches of three eggs being produced. The nest was built in a welded wire cup placed in the crotch of a Ficus tree about 1 m (3 ft.) from the ground, and constructed mainly from long-leaved pine needles. The hand-reared chick opened its eyes at seven days and was feeding independently at 28 days of age (Salisbury 1991).

Among the African species of the genus, the Abyssinian ground thrush *Z. piaggiae* is occasionally found in private or institutional collections. This species

Average Weights of Black-breasted thrush *Turdus dissimilis* Chicks at Selected Ages

	Day 6	Day 10	Day 14
Brood 1 (Pair 1, large eggs)	33.05 g	43.10 g	45.5 g
Brood 2 (Pair 2, small eggs)	25.90 g	33.25 g	46.1 g
Both broods	30.10 g	38.18 g	45.8 g

was reared at Winged World in England in 1971, following several years of frustrated nesting attempts. The incubation period was 15 days, as determined from several successful clutches. The male assisted in incubation but, after hatching, the chicks were brooded almost constantly by the female that was fed on the nest by the male. After 10 days, the male was observed feeding the chicks directly. A single chick survived, leaving the nest at 18 days of age (Ward 1971).

The varied thrush *Z. naevia* a handsome species from western North America is now placed in this genus, as is the Aztec thrush *Z. pinicola,* an attractive bird from the pine and pine-oak forests of Mexico.

Rock-Thrushes *Monticola* spp.

Natural Habitat: The 10 or 11 species of rock-thrushes of the genus *Monticola* are distributed in Europe, Asia, and Africa. Some authorities regard the Transvaal rock-thrush *M. pretoriae* as a full species; others, as a subspecies of the short-toed rock-thrush *M. brevipes*. Three additional species, often separated into the genus *Pseudocossyphus,* are endemic to the island of Madagascar, and can be used to illustrate the range of habitats occupied by rock-thrushes. The forest rock-thrush *M. sharpei* is confined to the rain forests of eastern Madagascar, while the littoral rock-thrush *M. imerinus* prefers savannas, grassy steppes, and sandy coastal areas, and Benson's rock-thrush *M. bensoni* is found in rocky terrain with rugged cliff-faces and sparse trees and scrub growth. Of the African species, some, such as the Cape rock-thrush *M. rupestris* inhabit mountainous areas, while others, like the sentinel rock-thrush *M. explorator* inhabit open hillsides and plateaus with large boulders and rocky outcrops. The miombo rock-thrush *M. angolensis* is usually found in Brachystegia woodlands, and the little rock-thrush *M. rufocinereus* in upland forests. The common rock-thrush *M. saxitalis,* as its name implies, has adapted to a wide variety of habitats.

Outside of Africa, the European and Asian species are more likely to be found in aviculture. The blue rock-thrush *M. solitarius,* with five subspecies ranging from southern Europe to eastern Russia and China southward through the Malay peninsula, is perhaps the best known. The chestnut-bellied rock-thrush *M. rufiventris* is a bird of the Himalayan region and China, wintering further south and at a lower altitude. The white-throated rock-thrush *M. gularis* migrates from its breeding grounds in eastern Siberia, Mongolia, China, Manchuria, and Korea to winter in southern China, Southeast Asia, and Myanmar. The blue-headed rock-thrush *M. cinclorhynchus* is another Himalayan migrant, wintering along the southwestern coast of India. All of these species are hardy, attractive, desirable avicultural subjects with the added advantage of being sexually dimorphic. The males of several species are fine singers. Everitt (1973) compared the song of the blue rock-thrush to that of the European blackbird, only shorter. Both the blue and blue-headed rock-thrushes have been bred in captivity.

Management and Breeding: Mrs. K.M. Scamell reported the successful breeding of the blue-

Orange-headed Ground Thrush *Zoothera citrina melli*

	Day 6	Day 7	Day 8	Day 9	Day 10	Day 11	Day 12	Day 13	Day 14
Band				(All weights in grams)					
Blue 15	38.4	43.5	44.5	44.5					
ES 3	36.3	40.3	42.0	44.5					
Blue 16 (Parent-reared)								44.6	
DLB 6	33.7	36.2	38.4	38.7	43.1	42.6	44.6	49.2	
DLB 7	31.0	32.9	36.2	37.9	42.9	44.6	45.0	47.0	
DLB 8	30.1	32.8	36.0	34.3	40.2	39.2	41.4	45.0	
DLB 9	26.5	30.4	33.0	34.2	39.8	40.0	41.4	43.9	46.8
DLB 14	36.7	42.1	45.1	46.0	48.2	44.2	48.8	50.0	
DLB 15	20.1	28.5	34.4	39.5	42.4	42.4	44.0	44.5	47.1
DLB 18	33.9	38.8	45.5	50.2	50.6	50.9	47.0	50.5	53.4
DLB 19	32.7	35.8	43.6	50.6	51.1	51.0	49.0	52.3	54.4
DLB 20	29.4	36.7	45.0	50.0	53.4	53.8	49.4	53.9	56.2
DLB 27 (Parent-reared)	46.3								
DLB 28 (Parent-reared)	38.5								
DLB 29 (Parent-reared)	42.3								

Compiled by David L. Bender, Hancock Wildlife Research Center, Surrey, B.C., Canada.

headed in England in 1960. Her pair was placed in a small, planted aviary situated in a birch woods in April, the hen being allowed to establish herself for a week before the male was introduced. The aviary measured 0.9 x 2.7 x 1.8 m (3 x 9 x 6 ft.) high, with a small, raised shelter at one end, and was furnished with a number of large rocks around which ferns grew. The thrushes shared this facility with two pairs of finches. After a few weeks the male began displaying and carrying leaves, and the hen selected a half-open-fronted nest box beneath the overhang of the shelter. The female alone incubated the eggs, sitting very tightly. She was seldom seen off the nest, and no examination of the eggs was undertaken. Both birds were seen flying excitedly about the aviary on July 2, and a quick check of the nest box revealed that three chicks had hatched in a cup-shaped nest at one end of the box. In addition to soft food and gentles, mealworms were supplied ad lib. Fresh ant eggs were given, and chopped boiled egg added to the soft food. The chicks, estimated at two days old when first observed, left the nest during a thunderstorm on July 14. One chick died after fledging, but two young hens were reared to independence. They were first seen feeding themselves two weeks after leaving the nest (Scamell 1960). Blue-headed rock-thrushes are relatively hardy birds. Everitt (1973) found that they can withstand temperatures well below freezing, but are subject to frost-bitten feet, which can be prevented by furnishing them with hay-filled boxes for roosting on winter nights.

Whistling Thrushes *Myophonus* spp.

The blue, or Himalayan whistling-thrush *Myophonus caeruleus,* is the largest species in the genus, reaching up to 330.5 in. (33.7 cm) in length. The endangered Sri Lanka whistling-thrush *M. blighi* at the other extreme, is similar in size to a rock-thrush, at 8–8 ½ in. (20–21.5 cm). Five other species fall between these two in size. All are bold, aggressive birds, usually from mountain habitats where they prefer forested areas with swift-flowing, rocky streams. The blue whistling-thrush, with seven subspecies, ranges from southern Kirgizstan, Tadzhikstan, and Afghanistan eastward through the Himalayas into central China, and southward through Myanmar, Thailand, Southeast Asia, and the Malay Peninsula to Java and Sumatra. The Malabar whistling-thrush *M. horsfieldi* is confined to western and central India, and the shiny whistling-thrush *M. melanurus* is endemic to Sumatra. The Sunda whistling-thrush *M. glaucinus* has subspecies on Borneo and Sumatra, while the nominate subspecies occurs on Java and Bali. The Malayan whistling-thrush *M. robinsoni* is restricted to a few areas in peninsular Malaysia, where its status is near-threatened. The Taiwan, or Formosan, whistling-thrush *M. insularis* is still fairly common in Taiwan. In most species, the sexes are alike or nearly so.

The blue whistling-thrush has appeared on dealers' lists in recent years. These large birds can become extremely belligerent toward smaller birds. If they cannot be given an aviary to themselves, they should be housed only with other strong and aggressive birds. At the Boehm aviaries a number of years ago they were kept with various laughingthrush species. Although nests were built and eggs laid, each predated on the eggs of the others, and no chicks were ever hatched. This species has a harsh, screeching alarm cry that is easily provoked, but also has a long, clear, melodious whistling song (Everitt 1973). To breed these birds, other than in a huge, walk-through aviary such as those found at zoos, would probably require a spacious, well planted aviary with running water and large boulders.

The first UK breeding took place in the Sawyer-Ezra collection in 1992, where a pair of these birds constructed "a typical thrush nest but larger than usual, due to their larger size" (Sawyer 1997). This pair bred again during the next few years, but the hen eventually killed her mate. They were aggressive toward other birds as large as scarlet ibis *Eudocimus ruber* and when nesting, the female would not hesitate to attack the keepers. The

Siberian Thrush *Zoothera siberica siberica*

Band	Day 5	Day 6	Day 7	Day 8	Day 9	Day 10	Day 11	Day 12	Day 13
DLB 16*	30.9	40.5	43.4	44.4	45.0	45.7	48.7	50.1	55.7
DLB 17*	24.1	34.1	38.0	43.4	44.5	44.7	45.5	46.7	49.3
DLB 30**	40.7	49.7							
DLB 31**	36.1	48.5							
DLB 32**	34.4	47.9							

***All weights in grams.**
*DLB 16 and 17 hatched under song thrush, pulled at 6 and 5 days of age, respectively, for hand feeding.
**DLB 30, 31, and 32 parent hatched and reared.
Compiled by David L. Bender, Hancock Wildlife Research Center, Surrey, BC, Canada.

fledglings resembled the adults, but were paler. This species had previously been bred at the San Diego Zoo (Lieberman 1991).

OTHER GENERA

The New World thrushes of the genus *Catharus* are comparatively drab in appearance, tending toward browns and grays, often with lighter colored underparts. The young share the speckling typical of most young thrushes. This genus includes such well-known American songbirds as the wood thrush, Swainson's thrush, and the hermit thrush, celebrated by the poet Walt Whitman. These birds are illegal to keep in North America, being protected by the migratory bird treaty, but are occasionally shipped to other parts of the world from their Central or South American wintering grounds. Although rarely kept, several species have been bred in captivity. Amsler (1927) reported on the breeding of the hermit thrush in his aviaries in England, and the olive-backed, or Swainson's thrush *C. ustulatus swainsoni* was bred by Charles Everitt in England in 1959. This breeding took place in a flight measuring 0.9 m (3 ft.) wide with an overall length of 4.2 m (14 ft.), of which 2.4 m (8 ft.) was inside a birdroom. The birds nested in the birdroom several times over a two-year period, were supplied with plenty of live food when the eggs hatched, but still failed to rear young. Success was finally achieved after the female was observed foraging for insects in the outside flight. Gentles were tossed on the ground, which she immediately collected and took to the young in their nest inside the birdroom. From that point on, live food was scattered regularly in the outside flight as well as provided in dishes in the indoor flight. The chicks were reared by the hen alone, the male having been removed prior to hatching when he became rather aggressive toward the hen whenever she left the nest (Everitt 1959).

Although seldom available to aviculturists, the solitaires of the genus *Myadestes* are noted for their wonderful singing ability, with at least two species, the Cuban solitaire *M. elisabeth* and the rufous-throated solitaire *M. genibarbis,* another Caribbean species, being rated exceptional in this regard (Clement 2000). Alex Isenberg reared three brown-backed solitaires *M. occidentalis* in California in 1955. In reporting this breeding, he referred to the species as the Mexican jilguero, giving it the scientific name *M. obscurus* then in use (Isenberg 1960). However, the specific name was subsequently changed to *occidentalis* when the endemic Hawaiian thrushes were transferred to the genus *Myadestes,* as the name *obscurus* was then preoccupied by the Omao, or Hawaiian thrush. The brown-backed solitaire was also bred in Oregon, with detailed accounts being published in both the *AFA Watchbird* and the *Avicultural Magazine* (Gibson 1993, 1994).

The Townsend's solitaire *M. townsendi* was kept in a mixed collection at the Boehm aviaries in New Jersey, where curator Charles Everitt described its song as loud, clear, and sweet. He was equally fascinated by its behavior, particularly the manner in which it maintained its solitary status. While never mixing with other birds, this solitaire was "always ready to snatch at any flying insects, almost flycatcherlike in its movements" (Everitt 1973). This species breeds in the northwestern US and western Canada, and winters from western Canada to northern Mexico. A South American species, the Andean solitaire *M. ralloides* had been bred by the Philadelphia Zoo.

Those familiar with the North American bluebirds of the genus *Sialia* do not always associate them with the true thrushes, for they are more like flycatchers in their habits, perching for long periods on a post and sallying forth to catch insects in flight. They will descend to the ground for insects but, unlike most thrushes, do not spend much time on the ground. They also differ from other true thrushes in being cavity nesters, a characteristic shared with certain members of the tribe Saxicolini, such as the shama *Copsychus malabaricus,* and the magpie robin or dhyal thrush *C. saularis*. Bluebirds are somewhat smaller than the *Catharus* thrushes, the largest of the three species, the mountain bluebird *Sialia currucoides,* measuring 15 cm (6 in.) overall, with both the eastern bluebird *S. sialis* and the western, or Mexican, bluebird *S. mexicana,* being about a centimeter shorter.

Ring Size: AACC size M for most species; size K or L for smaller species, and larger size for larger species.

References:

Amsler, M. 1927. Breeding the hermit thrush. *Avicultural Magazine*, 4th series, 5 (12): 334–38.

Bender, DL. 2002. Breeding the Siberian thrush. *Avicultural Journal* 25 (3): 9–14.

Clement, P. 2000. *Thrushes*. Princeton, New Jersey: Princeton University Press.

Everitt, C. 1959. The breeding of the olive-backed thrush. *Avicultural Magazine* 65 (4B5): 112–16.

Everitt, C. 1963. The Kurrichane thrush bred. *Foreign Birds* 29 (1): 14–16.

Everitt, C. 1964. Breeding the pale-vented robin. *Foreign Birds* 30 (2): 69–70.

Everitt, C. 1973. *Birds of the Edward Marshall Boehm Aviaries*. Trenton, New Jersey: Edward Marshall Boehm, Inc.

Gibson, L. 1993. Breeding the brown-backed solitaire. *Avicultural Magazine* 99 (3): 136–147.

Gibson, L. 1994. Breeding the brown-backed solitaire. *AFA Watchbird* 21 (5): 51–57.

Isenberg, A. 1960. Breeding results for the years 1955–60. *Avicultural Magazine* 66 (6): 219–20.

Lieberman, A. 1991. In News and Views. *Avicultural Magazine* 97 (4): 191–200.

Lowe, P. 1981. Breeding of the olive or cape thrush at the West Country Wildlife Park, Cricket St. Thomas. *Avicultural Magazine* 87 (2): 98.

Murray, H. 1960. Breeding of the redwing. *Avicultural Magazine* 66 (5): 166–67.

Salisbury, CL. 1991. Breeding the chestnut-capped ground thrush at the Lowry Park Zoological Garden. *AFA Watchbird* 18 (1): 7–9.

Sawyer, RJC. 1997. Twenty-five years at Cobham. *Avicultural Magazine* 103 (2): 49–66.

Scamell, KM. 1960. The breeding of the blue-headed rock thrush. *Avicultural Magazine* 75 (2): 51.

21.11b Old World Flycatchers, Chats & Robin-chats

Muscicapinae

• Glen Holland

21.11bi Old-World Flycatchers *Muscicapini*

Introduction: The flycatchers are a diverse group of birds, not all closely related, with hundreds of species scattered throughout the world. While all flycatchers belong to the order Passeriformes, the typical Old World species are classified with the thrushes in the family Muscicapidae, where they occupy the tribe Muscicapini in the subfamily Muscicapinae. This tribe encompasses 117 species in 18 genera. However, some species, such as the monarchs, crested flycatchers, paradise flycatchers, and various other flycatchers, are placed in the family Corvidae, where they comprise the tribe Monarchini in the subfamily Dicrurinae, which also includes the fantails and the drongos. The Monarchini includes 98 additional species, divided into 18 genera. The flycatcher species appearing most frequently in aviculture belong to these two tribes.

Most New World flycatchers, on the other hand, are placed with the manakins, cotingas, and similar birds in the family Tyrannidae, where they occupy two subfamilies, Pipromorphinae, the Mionectine flycatchers, with 53 species in eight genera, and Tyranninae, the tyrant flycatchers, with 341 species in 92 genera. To further confuse matters, however, the four species of silky-flycatchers, allocated to three genera, comprise the subfamily Ptilogonatini in the family Bombycillidae, and are therefore more closely related to the palmchats and waxwings.

Since my experience is limited to Old World species, this section concentrates on the management of these. Flycatchers can be found in a wide variety of habitats. Some species, such as the southern black flycatcher *Melaenornis pammelaina* and the spotted flycatcher *Muscicapa striata,* prefer open woodland or savannah where they frequent favorite hunting perches from which they sally forth to catch their prey. Others, such as the paradise flycatcher *Terpsiphone viridis,* prefer forests and forest margins. Black flycatchers have also adapted well to residential areas, where gardens and open lawns make it easy for them to locate prey. Flycatchers are gentle birds that either flutter down to the ground to capture insects or hawk them out of the air before returning to their perch.

Management: Newly acquired, wild-caught birds take some training, as described in the section on SHRIKES, before they accept an artificial diet, but hand-raised chicks make more suitable and steady avicultural subjects. The sexes are alike. They can be mixed with other softbills of a similar size or smaller, and also with seedeaters. Black flycatchers, however, must not be mixed with drongos, which attack them. Due to their territorial behavior, flycatchers should be kept one sexed pair per aviary, otherwise continual chasing will take place, eventually resulting in mortalities.

Habitat B, as described under the chapter on LANDSCAPE, will suit most flycatchers, with a few, such as the paradise, preferring an environment closer to Habitat A. The plants used in the aviary should include nectar and flowering plants, which will attract insects into the aviary for the birds to catch. Their aviary should have scattered clumps of dense shrubs, short grass, patches of bare soil, and branches and logs strategically placed to allow the birds to hunt insects on the aviary floor. Paradise flycatchers are the exception and will seldom drop to the ground to fetch insects. Their food is best placed 1.5 m (5 ft.) off the ground, with a perch placed conveniently nearby, permitting them to survey the food source.

As beautiful as paradise flycatchers are, they are not well suited to aviculture. Only the most determined aviculturist, with plenty of time and live food, will stand a chance of succeeding with these birds. The canopy feed-

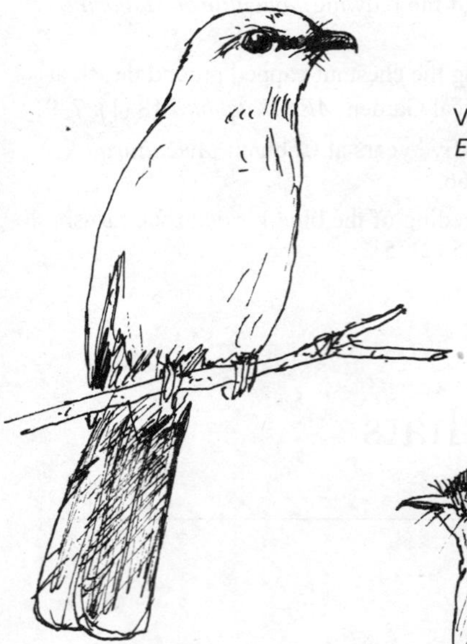

Verditer Flycatcher
Eumyias thalassina

Tickell's Blue-Flycatcher
Cyornis tickelliae

ing flycatchers such as these are far more difficult to establish than those species that take insects off the ground. Newly acquired birds are nervous and difficult, if not impossible, to convert to artificial diets, and even hand-raised paradise flycatchers are, unfortunately, not as easy to work with as typical hand-reared birds. To succeed with these birds, they should have their own very well planted aviary where no other birds can bully them or compete for live food.

Members of a pair are seldom found far apart and on a warm day will spend hours sitting in the sun with their eyes half closed. They bathe regularly on hot days. Most species bathe conventionally in a pool of water, but paradise flycatchers prefer to dip into a pool of water a few times while in flight. A perch placed strategically above a pool of water will be used by this species. All flycatchers enjoy a mist spray turned on over a dense shrub on a hot day.

Flycatchers are not particularly hardy; I would suggest that for most species, where temperatures remain below 10°C (50°F) for extended periods, sufficient shelter and even a source of warmth be supplied. Having said this, however, there are a few species, such as the mariqua flycatcher *Bradornis mariquensis* of South Africa, which are able to withstand lower temperatures. Paradise flycatchers are the most sensitive of this family; cold, wet conditions for extended periods will result in mortalities.

Diet: The captive diet should consist of commercial, fine grade softbill food, to which a little blended hard-boiled egg is added. Live food is essential, and should comprise about 30% of the captive diet. Some of this should be dispensed by scattering it on to the ground in open patches and allowing the birds to forage naturally. Alternatively, flycatchers can be taught to catch their live food in the air when it is thrown high. Species that feed primarily by hawking insects from the air, such as the paradise, should be offered the catch from insect light traps. An ultraviolet light can be switched on overnight in outdoor aviaries to attract insects. Additionally, insects such as small crickets, waxmoth and bee larvae, and gentles can be placed on feeding platforms situated among the vegetation. Some flycatcher species will take a small amount of diced fruit. Note in the account of breeding the verditer flycatcher *Eumyias thalassina* that Martin Vince used a container with a few holes to allow for the gradual release of the flies held within.

Breeding: The nests of most of the more common species are placed in the fork of a branch, in a cavity in a tree stump, among rocks, or in a large curled leaf of a plant. Nests are cup-shaped, constructed of fine twigs, rootlets, and vegetable fibers. The nest of paradise flycatchers differs. It is a tiny cup-shaped structure placed below a canopy of vegetation on an exposed branch. It is built of fine fibers and rootlets, bound together with cobwebs and lichen, with additional lichen placed around the outside of the nest. The 2–4 eggs are light green with brown spots. The incubation period is 13–14 days and the chicks fledge at 15–16 days of age. Large quantities of live food are essential if the adults are to rear the chicks. If the first brood is removed for hand rearing, a second clutch is usually produced in the same nest soon after the removal of the first.

An interesting account of the breeding of Tickell's blue flycatcher *Cyornis* (formerly *Muscicapa*) *tickelliae* is given by Rev. Canon Lowe (1966) of Chester, England, who noted that

> "when introduced into the aviary, both birds clicked a great deal and wagged their tails sideways. No aggression was shown. Early in June the male seemed to alter the whole contour of his body, creating the appearance of a round ball with a head and tail. At this time the hen began to carry grass roots. Several days later the nest site was discovered on a gorse branch in the back of the aviary, approximately 2 m (6.5 ft.) above ground. Shortly after the nest was completed the pair was observed mating for the first time. This consisted of a curious ritual in which the

female sang and called, then appeared to elongate herself with her tail raised and half fanned while she quivered her wings. The male flew straight on to her back. Following copulation, the male faced the female, threw his head up, and elongated his body. The chestnut throat patch was very visible. He held this posture for 10–15 seconds.

The female did all the nest building, while the male followed her about with his curious fluffed up, ball-like appearance. The first egg of a clutch of four was laid on June 9, and incubation was believed to commence on the 12th. The nest was not disturbed during the incubation period. On June 25, when the hen came off to feed, the interior of the nest was examined by touch and seemed to be full of chicks. The female did not brood the chicks after the fifth day. Both parents were very attentive, feeding the chicks assiduously on gentles and mealworms, the male feeding both items while the female only fed mealworms. Nest sanitation was performed by the female after about a week. The feces were carried to the far side of the aviary and placed on a large flat stone.

On the July 5, the nest was empty and the parents seemed nervous, clicking and singing incessantly with their beaks full of mealworms. At 5:30 p.m. the three fledglings were located and transferred to the aviary shelter, where the parents seemed grateful to have them all together. The young were largely independent by July 21, although the parents still fed them occasionally. By this time the adults were showing signs of starting a new nest."

Martin Vince (1966) wrote an interesting account of the breeding activities of a pair of Indian verditer flycatchers *Eumyias (Muscicapa) thalassina* while he was employed at the Sedgewick County Zoo in Kansas:

"This 15 cm (6 in.) long bird occupies a vast range stretching from the Himalayas through Thailand and Malaysia to Indochina, Sumatra, and Borneo. The entire plumage of the male is shades of bright green-blue. The female, although resembling the male, is duller colored throughout with small white spots peppering the throat and sides of the chin. Following acclimatization, verditer flycatchers prove reasonably hardy, but require heated winter accommodations if kept in an outside flight.

The breeding pair at Sedgewick occupied an indoor flight measuring 3 x 1 x 2 m (10 x 3 x 6 ft.) high in an off-exhibit aviary complex. The temperature in this facility was thermostatically controlled and the air humidified by a mister system programmed to come on for five minutes periods three times each week. The afternoon temperature would rise to about 28°C (82°F) during the summer, and the humidity level would reach about 65%. The winter temperature would not fall below 16°C (60°F).

The aviary was not heavily planted, containing only a small *Ficus* tree and two hanging baskets. Although the particular pair in question had been in captivity for some time, during which they had become quite steady in the presence of people, more nervous, newly imported birds would benefit from a heavily planted aviary. In the wild, veriter flycatchers retreat into dense forest to breed.

A substantial nest was constructed in a half open-fronted box measuring 15 x 15 x 23 cm (6 x 6 x 9 in.). The nest itself was 7 cm (3 in.) deep. It was comprised of sphagnum moss, pieces of fine string, leaf fragments, and fine grasses. The major component, however, was bison wool, which had been collected from the North American prairie exhibit. About 60% of the nest consisted of this material. The birds appeared to discriminate between nesting materials reflecting different levels of brightness, with the whiter colored wools and plant fibers being repeatedly ignored in favor of duller shades such as brown and green, suggesting that brighter materials are too conspicuous for nest safety. The female flycatcher proved very adept at concealing her work, and during the entire nest building process was not once observed actually taking material to the nest site. The nest building, which began on June 22, lasted for 13 days. The male seemed to take no active part in nest construction, although he stood guard and alerted the hen to the approach of danger by uttering a very rapid, loud rattle of a song, the only vocalization heard from these birds.

Clutches in this species range from 2–5 eggs. The female of the pair at Sedgewick County Zoo laid three eggs almost immediately after the nest was completed. Incubation began with the laying of the first egg, which hatched after an incubation period of 14 days. A second chick hatched the following day, and the third egg proved infertile. The younger chick died within a day of hatching, but the older chick thrived, fledging at only 17 days of age. The favorite rearing food was mini-maggots, but the parents only considered these after all the flies had been eaten. Flies were dispensed from a jar with a few holes perforated in the lid. To guarantee a sufficient supply, handfuls were periodically delivered to the aviary in addition to those emerging from the jar. Small mealworms were also provided, along with the usual insectivorous diet. All live food was sprinkled with calcium lactate and a multi-vitamin powder. The young flycatcher was slightly smaller than the parents, colored mostly dull gray-brown. It attained independence about three weeks after fledging."

Hand Rearing: Most flycatcher chicks are not difficult to hand rear. Rearing is as described for TANAGERS but without as much fruit in the diet. The rearing food is fed with forceps. Once the chicks have fledged, food can be thrown on to the ground below them and they soon learn to retrieve it. They will fly to the hand to fetch insects. The chicks are slightly delicate when they first fledge and must be watched carefully during bad weather. Other aviary inhabitants may bully them, and it is wise to watch for this if they are in a mixed collection. Ideally, they should be placed on their own in a flight for a month or two before being introduced into a mixed species aviary. At four months of age they resemble the adults.

The chicks of the paradise flycatcher are delicate creatures that need to be fed often, and in cool weather an additional source of warmth should be supplied. I have hand reared this species twice with a diet comprised half of insects such as bee larvae, waxmoths, and gentles and the other half of tiny pieces of pinkie mice dipped in moistened rearing food. The chicks are not able to swallow large pieces of food, and insects with a hard shell are first crushed with tweezers before being fed. Chicks will usually only take two small pieces of food at each feeding. The feeding schedule is every one and one-half hours. When the chicks fledge, they are best kept alone, as they are easily intimidated by other birds and will refuse to feed, eventually fading away. In some species, fledged chicks are only half the size of adults, but usually reach full adult size within 3–4 weeks.

Ring Size: 2.5–3 mm, depending on size of the species.

Species Requiring Similar Management:

Drongos *Dicrurus* species
Wattle-eyes *Platysteira* species
Niltavas *Niltava* species
Tits, family Paridae

References:

Lowe, C. 1966. The breeding of Tickell's blue flycatcher. *Avicultural Magazine* 72 (4): 115–117.

Vince, M. 1996. Breeding the Indian verditer flycatcher. *AFA Watchbird* 23 (1): 34–35.

21.11bii *Saxicolini* Chats & Robin-chats

Robin-chats

Introduction: Most of the species referred to as robins are classified in the tribe Saxicolini, subfamily Muscicapinae, of the family Muscicapidae. Notable exceptions include the robins of Australia, New Guinea, and New Zealand, which are placed in a separate family, the Petroicidae; the American robins, which are in fact typical thrushes that have been given the common name robin only because of their color patterns resembling the European robin *Erithacus rubecula;* and the so-called pekin robin *Leiothrix lutea,* which is actually a babbler, more properly named the red-billed leiothrix, belonging to the family Sylviidae. This species is briefly mentioned here but thoroughly covered under the section on BABBLERS.

Within the tribe Saxicolini, the members of the genus *Tarsiger* are called bush-robins; the members of the genus *Cossypha,* robin-chats; and those of the genus *Cercotrichas,* scrub-robins. The genus *Luscinia* includes species known variously as nightingales, robins, blue robins, rubythroats, bluethroats, and firethroats. In the genus *Copsychus,* some species are called magpie-robins and others shamas, or shama thrushes. Genera that are usually referred to as robins include *Irania, Erithacus, Pogonocichla, Swynnertonia, Stiphrornis, Saxicoloides,* and *Cinclidium.* Other species encompassed by this tribe are the CHATS, forktails, wheatears, akalats, redstarts, etc.

The African robin-chats are the group on which, together with the pekin robin and the New Zealand robin *Petroica australis,* my personal experience is based. For the purposes of this section I refer to these birds as robins only, which is the way they are usually named in their countries of origin. Pekin robins, Cape robin-chats *Cossypha caffra,* Natal robin-chats *C. natalensis,* and Heuglin's robin-chats *C. heuglini,* favor dense scrub, but not usually dense forest. In dry areas they are found in the lush vegetation along stream banks. Many species have become common garden birds, but seldom venture far from cover. Their habitats vary from coastal scrub to riverine vegetation, and for some species, high mountains. In contrast, however, white-throated robin-chats, *C. humeralis,* and red-backed scrub robins *Cercotrichus leucophrys* live in drier thorn bush areas, while New Zealand robin-chats *Petroica australis,* chorister robin-chats *Cossypha dichroa,* and starred robins *Pogonocichla stellata,* are strictly dense-forest birds. Kalahari scrub-robins *Cercotrichus paena* are rather unusual, occurring in dry semidesert habitats. Most robins are brightly colored birds, highly vocal, with some species able to mimic extremely well. The males are highly territorial during the breeding season.

Management: Due to their active nature and tendency to become tame in captivity, robins make interesting avicultural subjects. Sexes are generally alike, but females may appear a little duller than the male, particularly on the chest. They have a soft, pleasant warbling song, which in some species, such as the white-crowned robin-chat *Cossypha albicapilla* and Heuglin's robin-chat, may be almost deafening as the males burst into their morning or evening chorus. Their aviary should be well planted, in accordance with the following habitats, as described under the chapter on LANDSCAPE, for the various species:

Forest Habitat A

Natal *Cossypha natalenis*
Chorister *C. dichroa*
Starred *Pogonocichla stellata*
New Zealand *Petroica australis*

Woodland Habitat B

Pekin *Leiothrix lutea*
Cape *Cossypha caffra*
White-crowned *C. albicapilla*
Heuglin's *C. heuglini*
White-throated *C. humeralis*
Flame *Petroica phoenicea*
European *Erithacus rubecula*

Semidesert Habitat C

White-tailed *Cinclidium leucurum*
Kalahari *Cercotrichas paena*
Red-capped *Petroica goodenovii*

It is therefore recommended to research the natural habitat from which a given species originates before planning the aviary landscape for a pair of those birds. In general, robins are hardy birds, well able to withstand cold overnight temperatures, at least down to freezing. They do, however, require warmer daytime temperatures, and in climates with severe temperatures for extended periods, they are best housed indoors with some additional warmth.

Robins are fond of bathing, and a suitable bath dish or a shallow pond must be available. A mist spray or sprinkler placed over a shrub will also produce a flurry of activity from robins if made available for short periods on hot days. Robins enjoy scratching among fallen leaves on the aviary floor; if a layer of leaf litter is left, these fallen leaves will provide a source of live food for

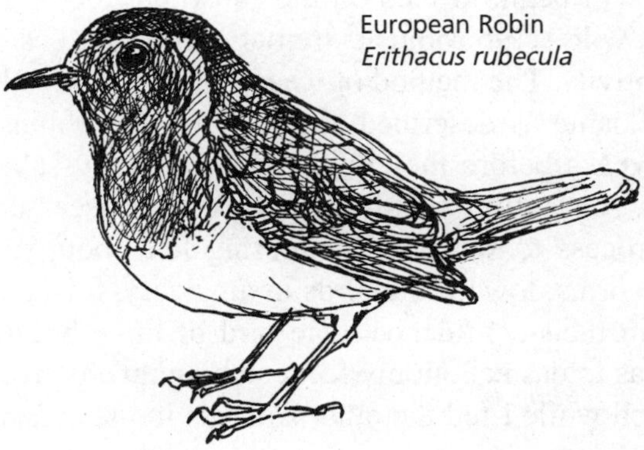

the birds. Due to their rather nervous nature, robins are not suited to life in a small cage. Smaller species, such as pekin, scrub, and Cape, are compatible with doves, large finches, and other small softbills, but the males should always be watched as they can be pugnacious toward other aviary inhabitants, and sometimes even to their own mates. Some of the larger species, such as white-crowned and Heuglin's, are particularly pugnacious, and should therefore be housed only with birds larger than themselves. It is generally possible to keep species of different genera together in large aviaries, but this practice needs to be assessed on a case-by-case basis. I have even kept scrub robins and Kalahari robins together in a very large aviary; although of the same genus, these species occupy different niches in their natural habitat.

I would recommend an aviary with a minimum surface area of 20 m^2 (215 sq. ft.) for a pair of robins. It must be sufficiently planted to allow the female to escape excessive chasing by the male. When introducing the members of a potential pair to each other, always allow the female to establish herself in the aviary, and then introduce the male, never vice versa! In aggressive encounters, they fly at each other and will even grasp an opponent's head or body by the claws while stabbing with their bills. Males will deliver the same treatment to independent young if they are not removed from the aviary in time. I have never actually known robins to steal eggs from the nests of another species, but would not put it past them, particularly from nests that are exposed.

Newly acquired robins are rather delicate and often reluctant to accept anything other than live food. Such birds need additional warmth and should be kept in a 60 x 60 x 60 cm (24 x 24 x 24 in.) box cage with a nylon gauze front to avoid damage to the beak and face. An anti-stress formula must be added to their drinking water, and a mixture of equal portions of live food and minced ox heart with moistened softbill food offered.

Initially, the robins will pick out only the live food, but as the quantity of live food is gradually reduced, the birds will begin to feed on the substitute foods. I have found wild scrub robins extremely difficult to establish in captivity. The method of weaning onto artificial diets is the same as described above, but scrub robins take 3–4 weeks before they begin to sample any substitute foods. A plentiful supply of live food is necessary for the process to succeed. By tossing live food, such as mealworms, toward the birds in an aviary, it is possible to gain their confidence. One bird of mine became so tame as to dash about my feet catching up any live food that fell while I fed the other softbills in the aviary.

Diet: Once established, their diet should include minced ox heart and/or bird of prey diet, tinned cat food (beef flavor), moistened commercial softbill food, softbill pellets, grated hard boiled egg with the shell, and diced fruits such as apples and grapes. Grated cheese is a favorite food item of robins, but should be limited, as too much may contribute toward obesity in the birds. A mixture containing all of these items can be prepared in bulk and stored frozen in the portions required daily. A multivitamin/trace element formula should be added to the soft food mixture. Live food is indispensable for keeping the birds in prime condition. Termites, mealworms, flying ants, and small insects exposed by turning the soil in the aviary or moving the water dish are relished. The number of insects offered should be increased during the breeding season, and decreased when breeding ends. This not only corresponds to the natural availability of insects, but will also stimulate the birds to turn more toward fruit during the non-breeding season, as many of these species do in the wild.

Although pekin robins have often been imported and fed primarily a seed diet by dealers, this is not natural for them; I hate to think of the losses which must occur when the birds are stressed from capture and then forced to feed on seed. If fed an incomplete diet, robins will gradually lose their bright colors. By using live food, the birds may become tame enough to take food from the hand. Planting the aviary with a variety of plants that attract insects will provide another source of natural live food. The addition of carophyll red, as used for canaries and flamingos, is necessary to maintain good color in a few species, but I would not recommend this as a standard additive for robin diets, as most species hold their color if fed the varied diet described above.

Breeding: Although breeding successes have been recorded, robins cannot be regarded as easy to breed in captivity. The first requirements are a large, undisturbed aviary and a plentiful supply of live food. Pair bonds become very close during the breeding season, with some species performing synchronized courtship display flights. While males play an active role in searching for and suggesting nest sites, the nest is constructed by the female. During this process, the males will usually courtship feed the females. They will also test the nest often, and despite vocal abuse from their mate, they will remove any odd piece of nesting material that they feel does not belong. Nests are sometimes placed in a dense shrub, but a half-open-fronted nest box placed in a shrub may also be used. I recall how, as a child, I once located and watched a wild Cape robin-chat that was rearing a brood of chicks in a rusty tin lodged in some dense vegetation. The nests of some species, such as scrub and white-throated robin-chats, are often placed on the ground among tufts of grass or brushwood, or even hidden in the curled dry leaves of an aloe.

The nest is cup-shaped, built mainly of coarse grass and lined with finer grass and rootlets. Some species build nests of twigs, grass, and leaves, lined with rootlets, spider webs, and animal hair. For species such as the red-capped robin *Petroica goodenovii,* which require spider webs, these can be collected on a dry twig and placed in the aviary. Nest inspections are tolerated, despite vocal objections from both members of a pair. The usual clutch is two or three eggs, which vary in color from light blue to pink and in some species are covered in rusty or pinkish speckles. The female alone incubates the eggs for 12–16 days. Both sexes feed the chicks, which fledge after 12–16 days. Newly hatched chicks are well covered in down. They are fed by both parents, with the female initially doing most of the feeding. The male takes on increased responsibility when the chicks are half grown, and it is he who cares for them post fledging, when the female is often starting a second brood, sometimes in the same nest, which she may first renovate slightly. Fledglings are usually independent within three weeks. It is important to monitor

Cape Robin-Chat
Cossypha caffra

them at this time to prevent the male from attacking them; if this happens before they become independent, he should be removed, but still kept in visual or vocal contact with the female. The chicks should be removed from the aviary as soon as they are independent. Some species, such as white-crowned robins, will produce up to three clutches per season if proper conditions have been provided. There is a record of a red-capped robin hatching ten chicks from six clutches.

Neil Hamilton (1980) wrote an interesting account of breeding the flame robin *Petroica phoenicea* in Australia:

> "The pair built a nest of bark and grasses they found in the aviary. They added material we supplied consisting of old bed mattress, coconut fiber and lengths of cotton and hair. Later the nest was found to be lined with spider webs and down feathers from the aviary birds. Nest building took approximately one week. The first clutch of three eggs hatched two chicks, both of which died within three days due to an insufficient supply of live food. The female carried out incubation with the male feeding her regularly. On the August 22, the female laid her first egg, and by the 24th had laid two more. This nest was located in the same area as the failed nest from the previous season. On the September 7 I observed two young in the nest. Consequently the bird staff began providing not only the regular food but also more live food, consisting of termites, mealworms and fly pupae. All the birds received the egg cake mixture as described for swallows. They had also been observed taking fruit and softbill nectar. One chick died of a suspected lack of calcium but the other fledged on the September 24. This fledgling was seen being harassed by other birds in the aviary and was therefore placed in a small holding cage at the back of the aviary where the parents continued to feed it through the wire (mainly by the male). After a period of two weeks the young bird was able to feed itself and was released back into the aviary. During this time the female had returned to the same nest laying another three eggs, starting on September 26. Prior to the eggs hatching the first fledged was removed to another aviary as it was observed to be interfering with the adults' behavior. Three eggs hatched and all fledged from this nest and at the time of writing the female is incubating a third clutch in the same nest ."

An excellent account of the successful breeding of British robins *Erithacus rubecula* was provided by Frank Meaden (1970):

> "The breeding pair had been hand-reared the previous year, and housed in separate but adjacent aviaries until they began to show signs of breeding behavior. In the female, this consisted of frequent visits to the nest log throughout the day; in the male, a readiness to mate was indicated by constant fighting with a resident wild male, hovering and fluttering up and down the wire mesh of the aviary. On the April 1, the male was placed in the aviary housing the female. Due to extreme aggression in previous attempts to breed this species, they were closely monitored, but no squabbling occurred. Their behavior consisted of the usual tail-flicking interrupted by an occasional chase of the female by the male. He always stopped a few inches short of the hen, concluding his pursuit with a little bow and tail flip. By April 5 the female was frequently carrying dead leaves and fine grass to the nest-hole, and five days later she was sitting tightly on five eggs, while the male sang almost constantly from a perch near the far corner of the aviary. On 26th April, broken eggshells were found on the opposite end of the aviary, and the male began taking food to the hen in the nest log."

Mr. Meaden describes his feeding regime for these birds as follows:

> "The normal daily food consisted of Avi-vite softbill food and finely grated cheddar cheese, and a ration of maggots powdered with Casilan.

European flour moth larvae in varying stages of growth were now included in the food, and a bucket of rotting fruit in their enclosure also attracted a certain amount of insect life, and since part of the enclosure covered a portion of the garden goldfish pond there may well have been other insect life taken but not recorded. With the knowledge that young now existed, we added Calcium Lactate as well to the maggots in an effort to obviate any deficiency and possible rickets, adding about five drops of Abidec to the maggots first, so giving them a slight film of the concentrated vitamin before adding this powder and the Casilan.

After the crucial first five days had elapsed and the chicks were still alive, a daily ration of about 20 mealworms was included in the diet. The robins fed most of these to the chicks, unlike many softbills, which quickly consume as many as possible when offered.

Four young robins, almost fully feathered, emerged from the nesting log on May 12. Within five days, the hen had refurbished the nest, but apparently did not immediately start laying a second clutch of eggs. The male seemed to ignore the chicks after they had become independent, but on May 29 he began attacking them, killing one within half an hour and then going after the next. The remaining young were immediately removed. It was later determined, when the remnants of egg shells again appeared on June 10, that the female had already been incubating a second clutch at the time of these attacks."

It is believed that among the factors contributing to the success of this nesting were the introduction of the male to the female's aviary, which probably served to reduce his initial dominance, and perhaps of even greater significance, the presence, during most of the nesting cycle, of the wild male, providing an alternative and more natural object for his aggression.

Several species in the genus *Luscinia* have been reared in captivity, with detailed accounts of the breeding of the Indian blue robin *L. brunnea* and the white-spotted bluethroat *L. svecica cyanecula* being published by K.M. Scamell and M. Barber, respectively. Mrs. Scamell's robins were given an aviary with an outside flight measuring 3.7 x 1 x 2 m (12 x 3 x 6 ft.) high with an attached, insulated shelter area about 1 x 1 x 2.3 m (3 x 3 x 7 ft.) high, a section of a larger building serving several flights. The shelter was heated, with a thermostatically controlled temperature that did not fall below 7.21°C (45.1°F)(Scamell 1969).

The white-spotted bluethroats were wintered in a birdroom maintained at a temperature of 7.2–10.1°C (45–50.1°F), then released in the spring into a flight measuring approximately 6 x 3 x 2.4 m (20 x 10 x 8 ft.) high, planted to elder, privet, and blackberry, and containing a small pool filled with about 10 cm (4 in.) of mud covered by 2.5 cm (1 in.) of water. Wild plants such as nettles also grew in the flight, and manure and animal carcasses were placed in the flight to attract insects. The pair did not breed the first year and, due to exceptionally mild weather conditions, were allowed to spend the following winter outside. They were fed on a diet of softbill mixture and mealworms (Barber 1975).

Breeding behavior was first observed during the following February and March. M. Barber (1975) describes the change in behavior as follows:

> "Fights broke out and on one occasion I picked up cock and hen locked in battle. A careful watch after this led me to believe that the cock was trying to become dominant and was ready to breed...To ease the tension I opened all the further half of my aviaries, which covered an area of approximately 70 ft. x 18 ft. x 8 ft. [21.3 m x 5.5 m x 2.4 m], and I think this did the trick, for there was less fighting and the cock began to sing."

Like the bluethroats, the Indian blue robins did not attempt to breed during their first spring in the aviary. Both birds molted in September, and during the following month the female became so aggressive toward the male that separation proved necessary, the hen remaining in the aviary while the male was transferred to a cage in the birdroom for the winter. Beginning in February, the hen carried dead leaves to various half-open nest boxes in the aviary, but her efforts did not extend beyond the construction of a pad of leaves. The male was reintroduced in early March, but was not tolerated by the hen; he was then moved to an empty, adjacent flight. Another male was tried and also had to be removed. Finally, during a spell of fine weather in April, the door between the two flights was opened. The birds now got along quite well, with no further fighting being observed. The female soon resumed her nest-building activities. Unlike wild birds of the species, which normally nest near the ground, she carried "masses of teased string and twine" to a nest box placed just under the roof of the flight, constructing a "very deep and large cup at the back of the nest-box." The male sang at intervals throughout the day.

The first egg was laid on May 20, and incubation commenced on the 24th. Extra mealworms were provided beginning on June 3, and an inspection of the nest the following day revealed the presence of chicks. Two days later the cock was very agitated, and was seen sparring with the hen. He was then confined to the original adjacent flight, where he continued to sing. The female was observed feeding the chicks a maggot and an earthworm. The consumption of mealworms also increased at this time. The chicks left the nest at about 12 days of age. A second clutch confirmed an incubation period of 14 days, and again the chicks left the nest when 12 days old (Scamell 1969).

The incubation period for the white-spotted bluethroats was 13–14 days, and the chicks left the nest at 13–15 days of age. At about 20 days of age they began picking up mealworms by themselves, and all were picking up mealworms within the next four days, although the male continued to feed them, the female having begun a new nest. They were self-supporting at seven weeks of age. The author, M. Barber (1975), offered the following hypothesis regarding the aggressive behavior of his bluethroats and other small softbills:

> "When pairs are put together, although some fighting will take place, provided that both birds are evenly matched and can give as good as they take, I believe this aggression to be natural and part of the breeding procedure. I have found with other species that if either the cock or the hen is dominant from the start, this results in the death of one of them. I believe that if the pair are to breed, the following should take place. The cock should make the first aggressive move and the hen retaliate, producing a state of "stalemate." The cock will then use other tactics such as song, food-offering, carrying of nesting material and displays. This occurred with my bluethroats and with other softbills that I have bred."

The author supplied additional details of the mating behavior and breeding performance of these birds. By the end of April the male's singing had intensified, being answered by the female, and the blue on his throat had darkened to a glossy vivid blue. In early May, the author witnessed the full display when he entered the aviary to find the two birds within a foot of each other. When he tossed a few mealworms to the ground, the male bluethroat picked one up and repeatedly dropped it in front of the female until she accepted it. He noted that during display:

> "The male draws himself up tall and thin with tail raised and spread to show all the colors that are present including the chestnut and brown tail. In this posture the cock moves around the hen, swaying from side to side. The movements seem to have a hypnotic effect on the hen who remains in one place. Both birds seem to be in a trance, and although the hen may peck at the ground she keeps watching her mate. The cock appears able to open and close the white breast spot at will and sings as he displays."

The nest was discovered in early May, built in a biscuit tin resting on its side on top of an oil drum that had previously been used as a feeding station. The nest was completed by May 9, and the first egg laid on the 11th. This egg disappeared, as did the second egg, which was not laid until six days later. However, from the 19th through the 23rd a clutch of five eggs was produced, incubation commencing when the final egg was laid. The male fed the female on the nest, and continued singing until the chicks hatched, the first two June 5 and three more a day later. The male continued to feed the brooding hen, that passed the food items on to the chicks. Waxmoth larvae, flour moths, moths, whiteworms, and small mealworms were supplied, all dusted with vitamin and calcium powder. Additional insect life was found in the aviary. After feeding, the parent birds would peck at the chicks' vents to stimulate the passage of fecal sacs that for the first few days after hatching were swallowed, but later removed to another part of the aviary and dropped. Three chicks left the nest on June 18, but the other two remained in the nest for three additional days.

The adult white-spotted bluethroats showed no aggression toward their offspring, even after a subsequent brood had hatched in mid-July. The two chicks from the second clutch left the nest at an earlier age, 11–12 days. Both brood of young bluethroats remained in the same aviary with their parents, and even after the five older chicks had molted into adult plumage no aggression was directed toward them by the parents. In this instance, the author attributed the remarkable lack of aggression to the free availability of live food.

Hand-Rearing: Robins are not difficult to hand rear and these hand-raised birds make good avicultural subjects. The technique for hand rearing is similar to that described for TANAGERS or the SHAMA. The hand-rearing diet should consist of equal portions of pinkie mice, commercial hand-rearing formula, and insects such as small crickets, mealworm and waxmoth larvae, beetles, moths and small grasshoppers. The diced pinkie mouse and hand-rearing formula are mixed together and water added until moist.

A few drops of multivitamin and a calcium/phosphorous supplement are then added to this mix. The food is fed with a forceps, hourly when chicks are small and then every two hours once they have are well feathered. If the food items are too large the chicks will not accept them. I have also replaced the hand-rearing formula with soaked cat kibble and have had equal success with this alternative diet. As soon as the tail reaches about 15 mm (0.6 in.) in length, the characteristic lifting of the tail occurs when the bird is excited. The chicks are rather delicate during the fledging stage and take some time before independence is reached. Initially, they should be placed in an aviary during the day, where they are still fed by hand, and brought in at night. Live food such as mealworms should be added to the substitute diet; the movement will attract the fledglings to the food and slowly they will learn to feed themselves. When they become self-feeding, they can be left outdoors at night, but should the weather turn foul during the first three days, they should be brought indoors.

Ring Size: 3 mm

White-capped Robin-thrush Hand Raised at San Diego Zoo

Day	Weight (g)	Day	Weight (g)
1	2.82	22	19.17
2	3.41	24	21.05
4	3.87	26	21.55
6	4.98	28	23.85
8	6.84	30	26.2
10	9.37	32	27.8
12	11.03	34	29.1
14	14.73	36	29
16	16.7	38	29.4
18	18.65	40	30.2
20	17.89		

White-Rumped Shama
Copsychus malabaricus

• Andrew Owen

The white-rumped shama is a widespread species with a large range and is categorized as common in many locations throughout South and South Eeast Asia. Eighteen subspecies are recognized, and the species is found in India, Sri Lanka, Nepal, Bhutan, Bangladesh, Myanmar (Burma), Thailand, Malaysia, Laos, Cambodia, Vietnam, China, Singapore, and Indonesia. It is found in primary and secondary evergreen forests in the lowlands and hills up to an altitude of 700 m (2,296 ft.); it favors dense cover such as bamboo thickets. Adult white-rumped shamas are easily distinguished; the male is primarily glossy black with a tan/orange belly and a bright white rump patch that may be fluffed up during bouts of display or excitement. Males are approx. 27 cm (11 in.) in length, of which 12–17 cm (4.5–6.5 in.) is tail. The eyes are dark brown, the bill is black, and the legs and feet are dark gray. The black plumage of the male is replaced with gray in the female. The tan underside is duller than the male, and females have a considerably shorter tail. There is a slight variability in plumage coloration and size between sub-species. There appears to be considerable variability in weight between birds from different sub-species. On average, males weigh 27–35 g (0.95–1.23 oz.), and females vary from 21–32 g (0.74–1.12 oz.) depending on race.

At Jersey Zoo, breeding white-rumped shamas are housed in pairs due to the aggressive territorial nature of the male — other males sharing the aviary would be attacked. They are housed in outdoor aviaries approximately 5 x 4 x 4 m (16 x 13 x 13 ft.) high. The aviaries are constructed using a treated timber frame and are covered with 0.7 mm (½ in.) 16-gauge galvanized wire mesh, painted with a matte black lead- free paint to aid viewing into the aviary. Part of the outside aviary roof is covered with Corrotherm sheeting, to provide protection against heavy rain and to provide a dry area where the birds' food dishes can be placed. Non-breeding birds can be housed in more modest aviaries of smaller dimensions. Minimum size 3 x 2 x 2 m high (10 x 6 x 6 ft.).

No environmental enrichment as such is provided, but each aviary is heavily planted to provide as natural an environment as possible for the birds, as well as making them esthetically pleasing for the public. A variety of plants, shrubs and vines are used to furnish the aviaries, particularly bamboo, as well as some thick evergreen shrubs, such as *Eliagnis,* which give cover

White-rumped Shama
Copsychus malabaricus

during the winter months. Natural perching is fixed at varying heights to allow the birds maximum use of the aviary. Each aviary has an indoor shelter attached to provide shelter and warmth during inclement weather. These shelters are of block or timber construction. Thermostatically controlled tubular aviary heaters provide a temperature of 25°C (77°F) in the indoor shelters. Florescent tubular lights attached to a timer switch set to come on at 6:00 a.m. and go off at 7:00 p.m. This provides the birds with at least 13 hours of daylight.

A deep layer of matured bark chips, mulch and mix and leaf litter of oak, magnolia and sweet chestnut cover the outside aviary floor. This gives the aviary a natural appearance and provides foraging opportunities for the birds. The floor of the indoor shelter has a layer of fine sand, which is sieved every week and replaced when necessary. All shamas obtained by Gibson had some intestinal parasites; Dwarf Tapeworm *Hymenolepis* spp. was the most common. Shamas may also become host to *Coccidia, Capillaria,* and *Strongiloid* worms.

Diet: In the wild, white-rumped shamas feed on a wide variety of insects and other invertebrates, which they pick from the forest floor or glean from foliage. Occasionally small fruits and berries may be consumed, although this does not form a major part of the birds' natural diet. Captive white-rumped shamas are offered an insectivorous mixture, in a small D-pot, which they receive between 10:00 and 10:30 a.m. This is made up of equal parts of the following: Witte Molen High insect diet, Witte Molen Low Iron diet, Witte Molen Mynah pellets (soaked), finely grated hard-boiled egg, finely grated carrot, wholemeal bread crumbs. This mixture is blended together to form a moist, pliable consistency, to which is added a small amount of finely diced fruit and a sprinkling of live insects (usually two or three wax moth larvae per bird). Very little fruit is consumed and if necessary may be omitted from the diet. Shamas receive two additional insect feeds, one at 8:00 a.m. and a last feed at 4:00–5:00 p.m. (winter/summer), consisting of a few mealworms, crickets, hatchling locusts or a combination of these scattered around the aviary to encourage the birds to forage. The birds also catch some natural prey, such as moths, spiders and earwigs. When chicks are being reared the adults are given a live food feed of soft-bodied insects every two hours throughout the day. The main food mixture and the insects are dusted with Nutrobal multi-vitamin powder. All insects offered are fed a diet of chick crumbs, progrub insect diet, cuttlefish bone and fruit and vegetables such as apple and cabbage. Fresh water is provided daily in ceramic bowls for drinking and bathing.

Breeding: At Jersey Zoo, this species has bred between May and August. Gibson et al found that the earliest clutch commenced April 19 and the last on August 22. A wooden nest box approximately. 15 cm (6 in.) square with an entrance hole 7 cm (3 in.) square is provided in which the birds construct their nest. This is completed within 1–3 days and is carried out by the female. A wide variety of fine nesting materials can be used in the construction of nests. At Jersey Zoo, we have found that the female initially lines the box with dried oak leaves, followed by a layer of bamboo leaves and a central cup of coconut fibers completes the nest. The normal clutch size is four. Eggs are laid at daily intervals and are pale green in color and are almost completely covered in mauve/brown speckles. Eggs weigh about 2.7 g (0.09 oz.) and range in size from 22–23 mm x 15–16.5 mm (0.866 x 0.59–0.649 in.). The incubation period is 11–13 days with incubation usually starting with the third egg. Incubation is carried out entirely by the female. Once chicks have fledged the female usually recommences nesting activities. A follow-up clutch is often laid within a week of the earlier brood fledging.

Copulation has only been observed once at Jersey Zoo, when the male flew onto the female from a high perch and copulation followed, lasting a few seconds. Inter-pair aggression can be a problem with this species. Males may become extremely aggressive towards the female, to the point where separation is required. At Jersey Zoo this has not been a problem, perhaps because the aviaries are large and heavily planted, females have not been too harassed. One male has proven to be extremely aggressive towards any female to which he is paired. In such an instance the male should be gradually introduced into the female's aviary, or both birds placed in a neutral aviary. New pairs should be formed when the birds are as young as possible, ideally when they are 3–4 months old. Males can also be aggressive towards fledged young. This behavior can be extremely sudden and result in fatalities. Keepers should be vigilant at this time, removing juveniles from the breeding aviary as soon as they are independent of their parents.

Chicks normally hatch within a day of each other; this is dependant upon when incubation started. On hatching chicks are altricial blind with black naked skin. Hatch weights for this species are not known, however one chick weighed 4.7 g (0.165 oz.) at two days of age. Chicks grow rapidly on a diet of soft-bodied insects, such as white mealworms, crickets and wax moth larvae fed to them by both adults, although the female carries out the majority of chick feeding duties. Pin feathers start to erupt and eyes begin to open when the chicks are 3–4 days old. Eyes are fully open at 4–5 days, at which time their bodies are totally covered in dark pinfeathers, apart from the central tail feathers, which are white. The chicks normally fledge at 11 days of age, when they are approximately 20 g (0.705 oz.) in weight and are covered in brown speckled feathers and have a very short tail. On fledging, the chicks are fully capable of flight but tend to sit motionless, often in thick vegetation. When an adult approaches, they will solicit for food by emitting a soft grating sound. Both adults feed the fledglings. If the adults re-nest, the male takes over the chick feeding, while the female rebuilds the nest and recommences incubation duties. The fledglings begin picking up food (insects) for themselves when they are about three weeks old, although they will continue to beg for food from their parents. When it can be ascertained that the fledglings are able to feed themselves, they should be removed from the breeding aviary, as there is a great risk that they will be attacked and killed by the adult male. Ideally the fledglings should be moved to an adjacent cage or enclosure to their parents, where any young males can learn the varied songs of their father. The sex of juveniles is discernible at their first molt, which is usually at three months of age. DNA or surgical sexing of this species is therefore not necessary.

Hand Rearing: One brood of four chicks was successfully hand reared in July 2001 at Jersey Zoo with chicks removed from the nest at two and three days of age. Initially, the chicks were placed in a ceramic pot, lined with tissue paper and a layer of coconut fiber to prevent foot problems or leg splaying. This was placed in a Curfew Brooder, set at a temperature of 35°C (95°F) with a humidity of approximately 50%. The chicks were fed every two hours throughout the day on a diet of pureed pinky mouse, dusted with Nutrobal multivitamin powder. This was fed to the chicks using tweezers. The brooder temperature was reduced by a degree each day until it reached 28°C (82°F). The pinky mouse puree was thickened after a few days, when the chicks were able to take chopped pieces. The chicks grew rapidly and fledged from the bowl at 11 days old. They were placed in a small budgie-type cage, and soft-bodied insects dusted in multi-vitamins gradually replaced the pinky mouse feeds. At this stage they were also offered a shallow bowl containing the insectivorous mix given to the adults and regular insect feeds. All four chicks were observed feeding themselves on insects when they were 16 days old, although additional feeds were still given for the next five days.

Birds are ringed using a numbered metal split ring size K on one leg and a colored plastic split ring 3.8 mm on the other leg.

Species Requiring Similar Management:

Silver-eared mesia *Leiothrix argentauris*
Flycatchers *Melaenornis, Muscicapa* and *Ficedula* spp.
Chats and similar species as described
Shamas and Magpie-robins *Copsychus* spp.
Niltavas *Niltava* spp.

Contributor: Neil Hamilton

References:

Barber, M. 1975. Breeding the white-spotted bluethroat. *Avicultural Magazine* 81(4):191–95.

Hamilton, N. 1980. Breeding the flame robin (*Petroica phoenicea*) in captivity. Printed in the proceedings of the first Captive management of wild animal workshop for zookeepers at Taronga Zoo.

Hutton, R. 1991. *Australian Softbill Management.*

Meaden, F. 1970. Breeding robins. *Avicultural Magazine* 76(4):158–60.

Scamell, KM. 1969. Breeding the Indian blue chat. *Avicultural Magazine* 75(7):265–70.

Chats

Natural Habits: While it is difficult to generalize from such a large tribe of birds, with over 150 species in 30 genera occupying a wide array of habitats, many of the species of chats found in aviculture prefer cliffs and old quarries where there are plenty of large boulders interspersed with clumps of bush. The mocking cliff-chat *Thamnolaea cinnamomeiventris* hops and glides from one boulder to another and, immediately after landing, raises its tail, giving the appearance that the bird has over balanced. The familiar chat *Cercomela familiaris* in turn flits its wings a few times after landing. They tend to sit on the top of the highest boulders or shrubs from where they have a clear view of their territory. Stonechats are birds of open scrub and grassland. The calls of chats vary greatly and some are excellent mimics with a loud song.

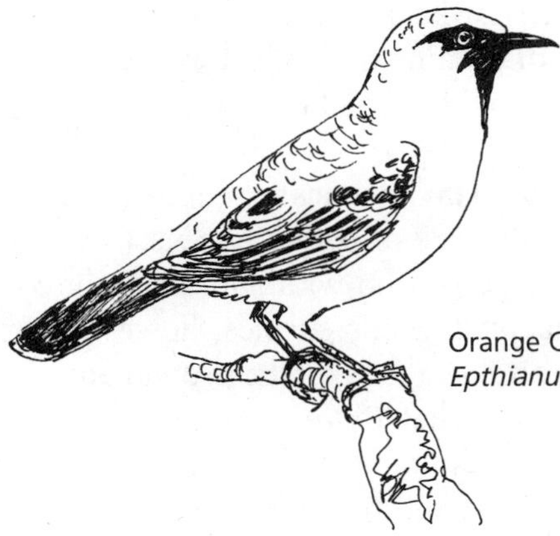

Orange Chat
Epthianura aurifrons

Management: Chats adapt well to captivity, but newly acquired adults can be a little delicate and stubborn when adapting to substitute foods. By mixing live food into their other foods, they will eventually learn to eat an artificial diet. As decided under the section on SHRIKES, wild-origin birds need to be carefully acclimated to aviary life. Wild birds may take many years to tame down, and the use of live food is recommended to gain their confidence. Even after years in captivity, chats will often fail to become as tame as other birds, such as robins. Hand-raised specimens, however, are easy to feed and will accept most softbill formulas. As chats are territorial, only one pair of a species can be housed in an aviary. It is possible however to mix species that are dissimilar in looks and habits. I have successfully mixed different species in the same genus on occasion, but in each case the two species involved occupied different niches in their environment. I have not noticed any serious aggression between chats and other occupants of an aviary, and the pairs tend to keep much to themselves. During courtship, the female sometimes shows limited aggression toward the male. They are extremely quick to snap up insects and this will sometimes result in larger species such as starlings chasing off the chats at feeding times.

Crimson Chat
Epthianura tricolor

They prefer a large planted aviary that includes rocky outcrops into which they will retreat if threatened, and where they will often enjoy basking in the sun. The landscape should be planted in accordance with Habitat B as described under the chapter on LANDSCAPE. In addition, a portion of Habitat D will provide exposed rocks and perches that chats prefer. From these vantage points they will hawk insects and also drop to the ground to pick up crawling insects. The males will spend much of their time displaying and calling from these favorite perches. Males also tend to sometimes chase females, but this does not result in serious fighting.

They are usually the first birds to give the alarm call if, for example, a raptor flies over, and the other birds in the aviary soon learn to take cover when the alarm call is heard. This call can be a little irritating, as the other birds tend to panic when hearing it. Chats are fond of bathing and a shallow pool of water should be available for this. They are willing nesters, but will not nest if the inhabitants of the aviary are incompatible, or if insufficient live food is supplied. Chats are relatively hardy birds, able to withstand overnight temperatures down to freezing. They do however require daytime temperatures in excess of 8°C (46.4°F). In severe climates that do not meet these requirements, they should be housed indoors with some additional warmth.

Diet: Chats are naturally almost entirely insectivorous, but in captivity they will feed on soaked cat kibble, commercial softbill food, minced ox heart, a variety of insects, and a little fruit. The same captive diet described for STARLINGS can be prepared in bulk and frozen in portions that can be thawed as required on a daily basis. Live food is always preferred by these birds, and can be fed by hand to assist in taming them. In time they will fly directly to the hand to receive these treats. Cheese is another popular food among chats, but should be fed sparingly to avoid obesity. I can recall an enjoyable visit by wild mocking chats to the breakfast table to steal butter. The Afrikaans name *spekvreeter* refers to the chats' habit of stealing the fat that was used to grease the wheels of the old ox wagons.

Breeding: For many species a wooden box with a half-open front, the size of budgerigar nest box, should be supplied. A large jam tin with half the front covered will also be accepted. Some species, however, such as the stonechats, prefer to nest on the ground among vegetation. The orange chat *Epthianura aurifrons,* one of the Australian species in the family Meliphagidae, nests in the fork of a dense shrub about 1.5 m (5 ft.) from the ground. Males locate a number of potential nest sites, from which the females make the final selection. The nest is cup shaped, built of grasses, coconut fibers, and rootlets, and lined with animal hair. The 3–4 eggs are pale greenish blue with some rust colored freckles. Both sexes incubate for 12–16 days, and both assist in feeding the chicks, which fledge at 14–20 days. During incubation, pairs are highly protective of the nest site, chasing away any intruder that ventures too close. They do not tolerate nest disturbance and are easily evicted from their nests if housed with larger softbills. To ensure success I would advise that the first clutch of chicks be removed for hand rearing at five days of age. The birds will soon lay a second clutch that can be left with them. If young birds are left for the adults to rear, a good supply of live food is essential. Without this the entire brood will usually be found dead by 4–5 days, scattered over the aviary floor. While rearing young, the parents will carry a number of insects to the nest neatly lined up in the bill. They also remove the fecal sacs from the nest site. At this stage of the reproductive cycle they can become quite pugnacious toward smaller species that compete for food. Rosemary Hutton describes how on one occasion a nest of three orange chat chicks at the age of 10 days, plus their parents, were consuming in one day approximately one cup full of moths, 40 large white mealworms, two dessert spoons of worker termites, one dessert spoon of termite alates, plus egg food, cheese and sundry caterpillars and other insects the parents caught for themselves. (Hutton 1991.)

For the first two or three days after fledging, the chicks will hide under the vegetation in the aviary. If the chicks are disturbed at this time, the parents become extremely perturbed and vocal; some species will use a broken wing distraction display when their chicks are threatened. Young chats should be removed to a separate aviary within six weeks of fledging to avoid persecution from the parents.

Hand Rearing: Chats are easy to hand raise, using a technique similar to that described for TANAGERS and SHRIKES. If chicks are removed from the nest when half feathered, they will begin to beg for food within 24 hours. A forceps and a thin syringe that has the narrow tip removed are used for feeding. Rearing-foods include commercial softbill formula, diced pinkie mouse, and a wide variety of insects, including moths, mealworms and winged termites. At least one third of the hand-rearing diet should be insects. The chicks are fed hourly when small, but this can be reduced to every two hours once they are well feathered. They will begin to peck at food about five days after fledging, but still require hand feeding for two weeks. A dish with a variety of foods, as mentioned above, should be supplied from five days after fledging onward. Half-feathered chicks should be kept at a temperature of 27°C (80.6°F), and this can slowly be reduced to 24°C (75.2°F) by the time they fledge. I have found that hand-raised chats do not remain hand tame when mature.

Ring Size: 3 mm

Species Requiring Similar Management:

Robin-chats *Cossypha* spp.
Akalats *Sheppardia* spp.
Eurasian robins *Erithacus, Cinclidium, Saxi-coloides* spp.
African robins *Pogonocichla*, *Swynnertonia*, *Stiphrornis* spp.
Nightingales, rubythroats, etc. *Luscinia* spp.
African scrub-robins *Cercotrichas* spp.
Redstarts *Phoenicurus, Hodgsonius* spp.
White-capped redstart *Chaimarrornis leucocephalus*
Water-redstarts *Rhyacornis* spp.
Bushchats and Stonechats *Saxicola* spp.
Wheatears *Oenanthe* spp.
Cliff-chats *Thamnolaea* spp.
Grandala *Grandala coelicolor*
Forktails *Enicurus* spp.

Reference:

Hutton, R. 1991. *Australian Softbill Management.*

21.12 Sturnidae Starlings & Mynahs

Glen Holland, Dick Schroeder & Andrew Owen

21.12ai *Sturnini* Starlings

• Glen Holland

Starlings

Introduction: Starlings belong to the family Sturnidae in the order Passeriformes. Under the Sibley-Ahlquist-Monroe classification system, the family Sturnidae includes two tribes, the Sturnini, which encompasses the starlings, mynas, and oxpeckers, and the Mimini, the catbirds, mockingbirds, thrashers, and tremblers. The tribe Sturnini contains 114 species in 27 genera, of which some are called starlings, some mynas, and some either starlings or mynas, as well as the two species of aberrant starlings called oxpeckers. Species of starlings are distributed through Africa, Eurasia, northern Australia and the Pacific Islands. The European starling *Sturnus vulgaris* has been introduced to many regions worldwide. Starlings are generally gregarious when not breeding, with some species breeding in solitary pairs and some species breeding in colonies. Starlings are common birds of open savanna, forests, and urban environments. Red-winged starlings *Onychognathus morio* are found on cliffs and have adapted well to urban environments, where they regularly visit bird tables and nest under house roofing. They often descend to the ground where they strut about like mynas while feeding. Some species, such as wattled starlings *Creatophora cinerea,* continually move in flocks from one food source to the next. Most starlings have rather pleasant, melodious calls, the most unusual being that of the wattled starling, which consists of a series of rasping and squeaky whistles. The songs uttered by breeding males are accompanied by bowing and wing flapping. To describe the captive management, I have concentrated on the superb starling *Lamprotornis superbus,* formerly included in the genus *Spreo* and often known aviculturally as the superb spreo, or simply, the spreo. I will also highlight particular management techniques required for other species. In addition, as the small amethyst starling *Cinnyricinclus leucogaster,* in particular, requires a different management technique, this species has been dealt with separately.

Superb Starling *Lamprotornis superbus*

Management: Superb starlings are popular avicultural subjects, and the most commonly kept of the African glossy starlings. Not only are they attractive, but they are hardy and well suited to aviculture. Newly imported birds are rather nervous and are best placed in a large planted aviary. Golden-breasted, or royal, starlings *Cosmopsarus regius* are rather delicate in nature compared with the other glossy starlings, and there is a reluctance among importers to import them due to the risk of high mortalities. Newly imported birds of this species require additional warmth and a plentiful supply of live food such as termites, crickets, and mealworms while in quarantine. Even after quarantine and initial acclimatization, they must be handled carefully during the introduction into an aviary. Several starling species, particularly golden-breasted, are susceptible to gape worm, with symptoms ranging from gaping to coughing and other respiratory problems.

Starlings adapt well to aviculture once they have learned to accept captive diets. This often requires a training period, which involves mixing live food into the substitute foods and gradually reducing the live food until the birds are feeding largely on the substitute foods. Outside of the breeding season, starlings can be kept in a colony in a large aviary, but in a small aviary or flight males may kill each other. As the breeding season approaches, each pair should be separated into a breeding flight with the same mate and nest box as the previous season. The aviary for starlings should be planted in accordance with Habitat B as described in the LANDSCAPE chapter. Superb starlings can be mixed with larger seedeaters and softbills without problems, but should not be mixed with similar cavity nesting species such as other glossy starlings. While building the nest and establishing the breeding territory they are particularly aggressive, but this aggression is reduced once breeding is under way. Other species of glossy starlings are similarly aggressive during the breeding season, when they are territorial, and are best kept as sexed pairs.

A pair of red-shouldered, or Cape glossy starlings *Lamprotornis nitens* will gain confidence by knowing they have a mate backing them up, and may fight viciously with other birds, usually related species. This

Superb Starling
Lamprotornis superbus

problem also occurs with other species such as the pied starling *Spreo bicolor,* and it may be necessary in a public display to keep only single representative specimens in order to reduce or eliminate aggression. In the breeding season, starlings will viciously attack any other species that may compete for food, especially insects, and nesting sites. Red-winged and pied starlings will swoop over people who approach the nest, and even give them an occasional peck on the head. This aggression can be serious; these birds may actually draw blood from the heads of unsuspecting visitors. When this occurs in public institutions, the chicks are best removed for hand rearing, which immediately alleviates the problem.

However, the hand-raised birds can also cause a problem in public walk-through aviaries, as some of them develop a tendency to peck at people's eyes. This is not a problem to the aviculturist who is aware of this possibility, but will catch park visitors by surprise. They also steal shiny objects such as earrings, and on one occasion we had a pearl earring stolen by one of these birds. It was never recovered! Starlings will also rob the nests of smaller birds for eggs, and are best kept with the larger softbills such as hornbills and rollers. They are fond of bathing, and also enjoy a mist spray placed over a shrub, where they will move about until thoroughly soaked.

Diet: Andrew Green of Durban, South Africa was a very successful breeder of superb starlings using the following mixture, which is prepared in bulk and then frozen in daily portions.

Mix together:

- 4 cups of commercial softbill mix/insectivore diet
- 2 tablespoons glucose
- 2 teaspoons calcium supplement (powder)
- 1 teaspoon multivitamins
- 2 teaspoon minerals and trace elements formulated for avian species
- 3 cups water
- 1 teaspoon cod liver oil

Add:

- One ox heart (or a similar amount of minced bird of prey diet), with all fat removed and then minced
- 4 hard-boiled eggs, grated with the shell finely crushed
- [GH Personally Iwould add 3–4 cups of soaked cat kibble to this diet.]

In addition, starlings will eat fruit such as soaked raisins, grapes, banana, papaya, apple, and orange, which can be cut in half and skewered on a nail. Insects are relished, and the birds will become reasonably tame if insects are fed from the hand. These are a must in the breeding season when a pair will easily take 200+ insects per day to their chicks. At the time of writing, commercial softbill pellets were difficult to source but diets should now include 30% of these.

Breeding: Courtship includes hopping about on the ground with wings drooping and neck outstretched. Superb starlings will use a nest box measuring 30 x 20 x 20 cm (12 x 8 x 8 in.) with an entrance hole of 65 mm (2.5 in.), but preferably, a hollow natural log should be supplied. Dried grass should be placed in the bottom of the nest, to which the birds will add a few leaves and feathers. In Canada, David Bender has found that his superb starlings will accept nest boxes ranging in size from a large finch box through a cockatiel size box to a parrot-type box roughly 25 cm (10 in.) square and 46 cm (18 in.) deep. The birds fill almost the entire box with dead grass, leaving only a small nesting chamber. When their box was removed to stop breeding, one pair built a large ball nest of dried grass in a fir branch hanging in the aviary. If sufficient nesting material is not made available, they will rob the nests of other species for material. The female alone incubates the 3–5 blue eggs for 14 days.

David Bender noted that most hens of this species, as well as purple glossy starling *Lamprotornis purpureus* hens, will leave the nest when it is approached by the keeper or other humans. His original pair hatched chicks before he even realized they were nesting. When the chicks hatch, a plentiful supply of live food is essential, which can include termites, small crickets, mealworms, and cleaned maggots. Without live food, the clutch may be deserted. In addition to live food, chicks are fed on softbill food and small quantities of fruit. Chicks resemble the adults after 6–8 months. Superb starlings usually raise two or three chicks per brood, and if live food is available they will breed through most of the year. Youngsters can be left with the adults for up to six months, and will act as helpers, assisting in the rearing of subsequent broods. To avoid confusion, the adult pair should be rung so that they are easily differentiated from the chicks. Two to three broods are raised per season.

David Bender has also pointed out that closed-ringing chicks in the nest can be disastrous, as the parents tend to remove all foreign objects from the nest during their routine nest hygiene activities. They may twist and mutilate the legs of their chicks in an attempt to remove the rings. In one instance, a closed-ringed superb starling chick was actually ejected from the nest by the parents. It has been suggested that stringing the rings on a wire and hanging them in the nestbox in such a way that they cannot be removed during the nesting process will familiarize the parents with the rings sufficiently to prevent this problem by the time the chicks have hatched and grown large enough to ring. It has also been suggested that smudging the rings with dirt to reduce the gloss will alleviate this problem, as the parents will be less likely to see them in the dark nest chamber. Of course, when closed-ringed offspring are required, the option of hand rearing remains. Most starlings are relatively easy to hand feed.

With colonial breeders, such as wattled and pied starlings, at least five pairs — to allow for their natural communal nature — are ideal in a large planted aviary. Wattled starlings require a large quantity of hard twigs, including thorn twigs, to build their huge nests. These nests are built in close proximity in a dense shrub or tree. Nest building is accompanied by the male singing, bowing, and flapping his wings, all of which helps to attract the female to the nest. They will nest readily in captivity, but if the colony is too small they will build without laying. A continual supply of nesting material is essential to ensure success. Nests often have two or more chambers. Pied starlings nest in burrows in the banks of streams, old excavation sites, etc. They have also been known to breed in cavities in the walls of buildings. The tunnels are about 1 m (3 ft.) in length with a padded nest at the end. Both sexes bring material to the nest, but only the female incubates.

Artificial Incubation and Hand rearing: As a child I regularly removed chicks from the nests of red-winged starlings for hand rearing as pets. While visiting a fellow bird breeder, I was shown a nest in his aviaries containing two newly hatched lesser blue-eared glossy starlings. The chicks appeared to be about two or three days old, and seemed very weak, with no indication that they were being fed by the parents. After some discussion and a period of observation to determine whether the adults were visiting the nest after consuming some live food, we concluded that the chicks had been deserted and decided to remove them for hand feeding. I have also used European starlings, at the Mt. Bruce National Wildlife Centre in New Zealand, as a staff training exercise in incubation and hand-rearing procedures for passerines. The following protocol is a combined record of these experiences. This section should be read in conjunction with the hand-rearing sections on SHRIKES and TANAGERS, which provide ideal diets for starling chicks.

Grosbeak Starling *Scissirostrum dubium* Hand Raised at San Diego Zoo

Day	Weight (g)	Day	Weight (g)
1	3.85	16	39.03
2	4.08	18	44.35
4	5.56	20	46.32
6	8.64	22	44.83
8	13.97	24	43.74
10	19.8	26	44.36
12	27.57	28	40.94
14	32.32	30	41.08

The incubator should be set at a temperature of 37.2°C (99°F) and a relative humidity of 57–62%. Eggs should be turned at least three times per day. In incubators that automatically turn eggs with rollers over a mat, such delicate eggs should be resting on paper towel. At the first external pip, the turner is switched off or the eggs moved to a separate hatcher. The same incubator settings are used for hatching as for incubation. Eggs usually hatch within eight hours of the first external pip, often within the first few hours.

About six hours after hatching, the chicks are removed from the incubator or hatcher. They are placed together in a bowl lined with paper towel, with some wood shavings added for them to grip. The bowl is placed in a brooder with a temperature of 36°C (96.8°F) and 65% humidity. Starling chicks are naked, and pink in color. Their first droppings are small and green or greenish blue in color, resulting from the yolk absorbed shortly before hatching.

- Day 1: The chicks are each given a small amount of commercial hand-rearing formula mixed with small pieces of pinkie mice, to which warm water has been added to give the mixture the consistency of a firm porridge, but still liquid enough to draw easily into the syringe. Feeding is accomplished using small forceps. Each feed totals 0.25 ml of the food mixture. The chicks are fed hourly from 6:00 a.m. until 10:00 p.m., and begin to voluntarily beg soon after hatching. Nest liners are changed three times per day.

- Day 2: The chicks should be begging eagerly for food. If the droppings are rather loose, begin adding waxmoths, mealworms (without the skin), and whole crushed pinhead crickets to the diet. These items are dipped in the moistened commercial rearing food and fed with forceps. One solid food item can be dipped in the rearing food given at each feeding. The brooder temperature remains the same as on Day 1.

- Day 3: On day three, the temperature is reduced to 35°C (95°F), and can be decreased from now on by 1°C (1.8°F) per day until an ambient temperature of 28°C (82.4°F) is reached, at which stage the chicks will be well feathered and able to regulate their own temperature. The chicks' droppings begin to show the typical 2/3 brown and 1/3 white color produced by healthy passerine chicks. The diet is gradually changed to include diced pinkie mouse in the moistened rearing food, and a portion of commercial softbill food. This food is fed with forceps and easily swallowed by the chicks. A soluble multivitamin and a calcium/phosphorous supplement are added to the rearing food. The chicks should now be taking two food items at each feed. A bowl of water is maintained in the brooder to ensure that the chicks do not dehydrate.

- Day 4: By day four, the droppings should be produced immediately after each feed, in a fecal sac which is approximately 2/3 white and 1/3 brown in color. These fecal sacs are strong enough to be removed with tweezers, and thus ensure good hygiene is maintained. Some fruit, in the form of grated apple, diced papaya, and banana, is added to the diet, and will be eagerly consumed by the chicks. At each feeding, 3–4 food items should be offered. Half of the diet should now be diced pinkie mouse dipped in rearing food; the other half should consist of fruit and insects such as diced mealworms, small crickets, and moths. The chicks are at an awkward stage where the belly and stomach have become large and they sway from side to side while begging for food. The feeding interval can now be reduced to every one and half-hours, and the chicks can be stimulated to beg by tapping or making a squeaking sound with the lips. While begging, they give a high pitched call typical of the starling family.

- Day 7: The brooder temperature should be 31°C (88°F). The eyes are beginning to open and the first quills are appearing on the back and wings. The nest should now be cleaned at least five times per day. Any soft food that accumulates on the chicks' gapes should be cleaned off twice a day with cotton dipped in warm water.

- Day 10: When the temperature has been reduced to 28°C (82°F), the chicks' eyes are wide open and quills have appeared on the flanks and crown as well. The size of the food items can be increased slightly. At this stage they are still fed every 12 hours, and consume up to four food items per feed. Their calls are quite loud at this stage.

- Day 13: With the temperature now maintained at around 28°C (82°F), the feathers start to break out of the quills over most of the body. The chicks become more active, and begin to stand up on their legs while begging. They are extremely vocal when approached.

- Day 16: Feathers are now covering most of the body and the green gloss has become apparent on the wings and back. The chicks begin to call when any movement or sound is made in the room. They are now fed every two hours from 6:00 a.m. to 8:00 p.m.

- Day 20: The chicks are giving indications that they are ready to fledge. They can now be placed in a small cage lined with teff grass and situated on top of the brooder. At this stage the chicks run forward to beg for food, but will only take 2–3 items per feed. They will also begin to sit on the perches provided, but will readily descend to the base of the cage while being fed. The wings are exercised vigorously while the birds grip the perch. Larger prey items, such as crickets, and a more varied diet, including pieces of fig, can now be included in the diet.

- Day 28: The plumage of blue-eared starling chicks is black with a slight iridescent green tinge above, and rufous brown below. The eye is gray. For the first time, the chicks begin to peck at food placed on the floor of the cage. In an effort to encourage the chicks to feed on their own, hand feedings are reduced to every three hours from 6:00 a.m. to 6:00 p.m.

- Day 34: The chicks are now feeding themselves well but still beg for food when approached. In the early morning and late evening, a few insects are given to help them on. The chicks no longer receive supplemental heat, and seem comfortable at a room temperature of around 24°C (75°F).

- Day 41: Three weeks after fledging, the chicks are entirely self sufficient. They can now be transferred to an outdoor flight. For the first two nights, they are brought indoors; thereafter, provided the weather is fine, they can remain outdoors.

Ring Size: Glassy Starlings 4.5 mm

Amethyst Starling *Cinnyricinclus leucogaster*

The amethyst starling, also known as the plum-colored, violet-backed, or Verreaux's starling, is one of the most beautiful members of the family. Only the males display the exquisite amethyst color on the head, neck and back. This color changes according to angle of light reflecting from it; in shade, it appears a blue-black. The chest and belly of the male are white. The female and chicks are a dull streaky brown color. The young males begin to color up after the first year.

Management: Amethyst starlings are delicate when newly acquired, but once accustomed to artificial foods, they acclimatize well. I class these birds as delicate because they must be kept with softbills their own size or smaller. If kept with large softbills, they show signs of stress and will spend most of the day hiding. Through the use of insects such as mealworms and crickets, they will become tame and confiding, and will even learn to fly to the hand to take live food. Their captive diet is as described for other starlings.

Breeding: This species uses old barbet and woodpecker nests, which they line with a few rootlets and green leaves. The males will often sit in the top of the tree in which they are nesting and whistle softly. In a large walk-through aviary, I housed two pairs of these birds together with no noticeable aggression between them. I have never seen amethyst starlings show any aggression toward another species. They make little attempt to defend their nest against intrusion by larger species. In an aviary, they should be supplied with a natural nest log in which to build their nest. Plastic or metal pipes with a 12 cm (4.7 in.) diameter, which are pushed into the ground and stand about 1.2 m (4 ft.) above the ground, are also accepted. A base should be placed about 30 cm (12 in.) below the top, and a wire ladder placed on the inside of the pipe. The nest is lined mainly with small green leaves. A plentiful supply of live food is essential if they are to raise their young. They are reliable parents if sufficient live food is available. Initially, the chicks are fed on insects, but after a few days, minced heart is also taken to the nest. After a week, fruit is added to the chicks' diet. It is best to ring the breeding female so that she can be easily distinguished from the fledged chicks. Amethyst starlings are usually double brooded, and will certainly produce a second clutch if the first brood is removed for hand rearing. Very often, however, only one or two chicks are raised per brood.

Hand Rearing: The chicks are not difficult to hand-raise as described for other starlings, SHRIKES and TANAGERS. They should not be removed from the nest before the flight feathers start to break through the quills, as I have found them to be a little delicate if removed too early. Once they have become well feathered, they begin to give a typical squeal of a starling chick being fed, but prior to this, they are quiet and rather sluggish feeders. Individually hand-raised birds, unlike those raised with others, remain hand tame.

Ring Size: 3 mm

Species Requiring Similar Management:

Bulbuls
Mynas
Barbets
Babblers

Contributors:

C. van Oostenrijk, Pretoria, S. Africa
A. Green, Durban, S. Africa
D. Bender, Surrey, BC, Canada

References:

Feare, C. & A. Craig. 1999. *Starlings and Mynas*. Princeton, New Jersey: Princeton University Press.

Mynahs

Dick Schroeder & Andrew Owen

Introduction: Mynahs belong to the tribe Sturnini in the family Sturnidae, along with the starlings and oxpeckers. Although the distinction between starlings and mynahs is somewhat arbitrary, the genera *Acridotheres, Ampeliceps, Mino, Basilornis, Streptocitta,* and *Gracula* are usually called mynahs in current terminology. The coleto *Sarcops calvus* is sometimes referred to as a mynah, sometimes as the bald starling. The highly endangered Bali mynah *Leucopsar rothschildi* is equally well-known as the Bali starling, Rothschild's mynah, or Rothschild's grackle. Similarly, the flame-browed starling *Enodes erythrophris* is also known as the flame-browed mynah, fiery-browed mynah, or red-browed starling, and the alternative name of the finch-billed mynah *Scissirostrum dubium* is the grosbeak starling. In the past, the names starling and mynah were virtually interchangeable, at least in regard to Asian species within the family. Apparently the African sturnids have always been referred to as starlings and glossy starlings.

Mynahs are distributed throughout Asia, Bali, New Guinea, the Celebes, and the Eastern islands. The common mynah *Acridotheres tristis* is now an introduced pest in many countries including South Africa, Malaysia, Australia and New Zealand. Mynahs are generally highly vocal and territorial, particularly when breeding. Those species that spend time walking on the ground do so with a bold swagger, with long strides and regular, deliberate head movements. The larger species of the genus *Gracula* do not walk, but tend to hop, and seldom venture to the ground from the forest canopy. Mynahs are cavity nesters, laying 3–5 green or blue eggs. Chicks are sparsely covered in down at hatching, and both sexes care for the young.

Hill Mynahs *Gracula religiosa*

- Dick Schroeder

Introduction: The mynahs in the genus *Gracula* are called hill mynahs. Most authorities recognize only two species, the common hill mynah *G. religiosa* with a wide geographical range and numerous subspecies, and the endemic Sri Lanka, or Ceylon, hill mynah *G. ptilogenys*. Other authorities, including Feare and Craig (1999) in their recent monograph on the starlings and mynahs, separate the genus into five species by restoring the southern hill mynah *G. indica* to its former specific status, and elevating two island forms, the Enggano hill mynah *G. enganensis* and the Nias hill mynah *G. robusta,* to full species. The talking mynahs of the pet trade belong largely to two subspecies of the common hill mynah, and the following remarks are based on my experiences with these two forms, the only ones commonly found in aviculture outside their native geographical ranges. The greater hill mynah *G. r. intermedia,* also called the greater Indian hill mynah, is distributed from northeastern India through the foothills of the Himalaya into Burma, Thailand, Indochina, and southern China. The Java hill mynah *G. r. religiosa* is found in Malaysia and on the Indonesian islands of Sumatra, Java, and Bali. The coloring of both races is the same. The main difference is in size, the Java hill being about an inch longer and more robust in stature. Also, the bright yellow wattle is divided behind the eye in the Java, while it is connected in this area in the greater. Both forms have black plumage with an iridescent sheen of purple and green, and a broad white stripe in the flight feathers. Their bills resemble candy corn, bright orange fading to yellow at the tip. Sexes are alike in appearance, and size is no indicator of gender, as the largest mynah in my collection is a female Java hill.

Management: Hill mynahs have been successfully housed and bred in a variety of sizes of cages or aviaries. Most are in either 1.2 x 1.2 x 1.8 m (4 x 4 x 6 ft.) or 0.9 x 1.8 x 1.8 m (3 x 6 x 6 ft.) small flight cages. Some are in 1.2 x 3.6 x 2.4 m (4 x 12 x 8 ft.) planted aviaries. The type of accommodation seems to have little bearing on the chances of a pair nesting successfully. One of my most prolific pairs is kept in one of the smaller cages. Mynahs are hardy birds and very cold tolerant; however, some sort of shelter from wind and rain must be provided. Mine tolerate frost without difficulty, even bathing in the water dishes on frosty winter mornings. In planted aviaries, Habitat A, as described in the chapter on LANDSCAPE, should be provided.

Diet: Proper diet is undoubtedly the major concern with these birds, as they are prone to hemochromatosis, or iron storage disease. The diet should be as low in iron as possible, which means that feeding dog or cat kibble is generally not advised, as such products are often meat based and therefore have a high iron content. It is much better to feed a pellet developed especially for softbilled birds. I use Kaytee Exact Softbill Pellets with good success (see the dietary section on TOUCANS and birds of PARADISE or additional information). I feed these pellets both dry, on the side, as well as soaked in warm water and added to my mix of fruits and vegetables. I use primarily papayas, some apple, melon, banana, and grapes, as well as occasional citrus. I also add frozen mixed vegetables, thawed, to the mix, most of which they will consume, with the exception of the lima beans. They receive live mealworms daily, perhaps 30–40 worms per pair when not breeding. Of course, much more live food is provided when they have chicks in the nest.

Breeding: I have found that mynahs will breed anytime of year between January and October, summer months being preferred. I use plywood nest boxes, usually 30 x 30 cm (12 x 12 in.) and 45–50 cm (18–20 in.) deep. The entrance hole is about 7.5 cm (3 in.) in diameter. I let the birds construct their own nest; both birds work at this task. For nesting material, they prefer something that is pliable, as they wrap it around the inside floor of the box, although some pairs will use dry twigs as a base and then place the more flexible material on top of this lower layer. I provide strips of palm fronds as nest material. This is used readily, as are any dropped feathers, or leaves from the overhanging acacia or eucalyptus trees.

Three, rarely four, eggs are laid. These are light blue with dark spots. Incubation is 14 days. The empty shells are immediately removed from the nest and dropped on the aviary floor. This makes it possible to determine when the chicks have hatched, as well as how many, so that live food may be increased accordingly. All of my mynahs feed their young on mealworms dusted with VitaLife calcium/vitamin supplement. Some mynahs will not eat crickets, but newborn mice are always taken. The chicks are fed primarily live food for the first week or so, after which the adult food is added. It is very important to keep live food before the parents at all times. If one of the chicks starts falling behind in development, it too becomes dinner.

Caution should be used when inspecting the nest, especially with a pair that is breeding for the first time. One can never be certain how a given pair will react to such nest disturbance. Some pairs will actually abandon the nest or throw out the eggs or chicks. For this reason I never allow visitors near nesting mynahs.

Chicks are often removed from the nest at around 10–14 days for hand rearing as pets. I have found that a parent-raised chick that is removed from the aviary as soon as it is seen eating on its own will make an equally good pet; it just takes a bit more handling. Early removal of the chicks will, however, promote renesting, and some pairs will respond by producing up to four clutches in a breeding season. Mynah chicks are easily hand fed on soaked pellets and bits of soft fruit. I do not give any live food when hand feeding as I have found it to be unnecessary. For detailed information on hand rearing, the section on hand rearing STARLINGS and TANAGERS provides additional information on diet and technique.

Common Hill Myna
Gracula religiosa

In conclusion, the speaking ability of the mynahs should be mentioned. Not all mynahs will learn to speak. All will mimic sounds, anything from a telephone ringing, car alarm, or the calls of other birds, but not all will mimic human speech. Hand rearing the birds does not guarantee that they will become talkers. However, mynahs that start speaking at an early age can acquire an incredible vocabulary, even speaking in several different voices. My aviary is filled with horns honking, dogs barking, and people speaking in various voices.

Bali Mynah *Leucopsar rothschildi*

- Dick Schroeder & Andrew Owen

Introduction: The Rothschild's, or Bali, mynah *Leucopsar rothschildi,* one of the most beautiful representatives of the family Sturnidae, is the only bird species endemic to the Indonesian island of Bali. The plumage is snow white with a mask of bare blue skin surrounding the eyes. There is a long, flimsy white crest, and the wings and tail are tipped in black. They are 25 cm (10 in.) long overall. Bali mynahs exhibit no obvious sexual dimorphism, but as a general rule mature males are slightly heavier and have a longer crest than mature females. In Indonesia, sexing is performed using a technique that involves studying the eye mask, behavior, and posture of a bird. Andrew Owen has provided additional information on the management of this species at the Durrell Wildlife Conservation Trust. We send breast feathers away for DNA testing to confirm a bird's sex. Alternatively, birds may be surgically sexed using a laparoscope.

In 1983, the wildlife division of the Indonesian government, the Directorate-General of Forest Protection and Nature Conservation (PHPA), asked Birdlife International to develop a conservation program for the Bali mynah. In 1987, the Durrell Wildlife Conservation Trust, together with the American Association of Zoological Parks and Aquaria and the organization mentioned above, created a five-year cooperative plan. The aim was to establish a viable and self-sustaining population of the Bali mynah in Bali Barat. This involved monitoring and protecting the wild population and coordinating a captive breeding program. To this end, 44 birds were shipped from various zoos to Surabaya Zoo in Java to create a captive breeding program in Indonesia. As a result, during the next 11 years groups of Bali starlings were released and monitored. Some of these birds integrated successfully with the remaining wild population, and several breeding successes were recorded between released and wild birds. Although they are completely protected by law, they are still being trapped by poachers because of the money they can bring, since they are regarded as a status symbol of the wealthy. Birds released from the captive-breeding program have also been lost to falcons. As with many other bird species around the world, there is a problem with loss of habitat as well. The savanna forests are still being converted to plantations and cropland. The future in the wild does not look good for this wonderful mynah. Outside of Bali, there are still over 1000 of these birds in captivity, with regional studbooks to help insure continued genetic diversity.

Maintenance: The Bali mynah is a very easy species to keep. While they do well in single-pair aviaries, they can also be mixed safely with other, smaller birds without a problem, since they are not aggressive. Dick Schroeder has a pair housed with a variety of other softbills in a 50 m^2 (500 sq. ft.) planted aviary, landscaped according to Habitat A, and they all get along well. The other inhabitants range from turacos to leafbirds and tanagers, with no signs of aggression on the part of the mynahs.

However, not all aviculturists have found the species to be quite that docile, particularly during the breeding season, when these birds become very territorial. Many breeders recommend that each pair be housed in a separate aviary free from other passerines and other Bali mynahs, as they can become aggressive toward cohabitants when they have chicks in the nest. Some pairs have also been prone to nest robbing. At Jersey, staff have even had Bali mynahs attack the feet of inquisitive birds in adjacent cages, with a yellow-throated laughingthrush *Garrulax galbanus simaoensis* losing a toe on one occasion. They have also been known to kill the chicks of Palawan peacock pheasants *Polyplectron emphanum* in the same aviary. We still house these species together, but substitute any Palawan eggs with dummy eggs. The young pheasants are then artificially hatched and hand reared.

Groups of up to five Bali starlings of the same sex can be kept together in an aviary similar in size to that used for a breeding pair, and even larger flocks could be kept in proportionately larger aviaries. For short periods of time, these starlings can be kept in smaller aviaries or even cages, although this practice in not encouraged, as obesity or boredom may prevail, the latter being a possible cause of feather plucking.

At Jersey, thermostatically controlled tubular aviary heaters provide a constant temperature of 25°C (77°F) in the indoor shelters. Flourescent tubular lights attached to a timer switch set to come on at 6:00 a.m. and go off at 7:00 p.m. provide the birds with a day length of at least 13 hours of daylight.

Some of the health issues encountered at Jersey include Atoxoplasma, a type of Isosporan coccidian that can disseminate to various body organs and white blood cells. In adult birds it seems to be fairly benign, but at the fledgling stage it is acute and, if present in a collection, can result in the loss of many fledgling birds. Transmission is fecal-oral, but adult birds will pass the organism to their chicks during feeding. Diagnosis involves collecting the feces (which would contain the oocysts) over a three-day period and placing them in potassium dichromate. To combat atoxoplasmosis, aviary substrates are changed regularly to rid the aviary of the oocysts, and nestboxes are thoroughly disinfected at the end of each breeding season. As a treatment, we use Baycox, although this drug has still to be clinically proven.

Other problems include additional coccidian parasites, also treated with Baycox. Our birds are also wormed twice a year with Flubenvet, which will control gapeworm and intestinal worms. A three-day course of 0.02 ml 10% Panacur injected into wax moth larvae is fed per bird to this species at Jersey Zoo to control parasites. These birds are also susceptible to the usual range of avian ailments, which can be diagnosed and treated by veterinarians.

In California, Dick's birds receive the same low iron diet that he offers to his other mynahs, as described under HILL MYNAHS, a variety of diced fruit consisting of papaya, apple, banana, and grapes, with the occasional addition of melons or other seasonal fruit. This is combined with frozen/thawed mixed vegetables and Kaytee Exact Softbill pellets soaked in warm water. The softbill pellets are also offered separately in dry form. As hemochromatosis can be a problem with these mynahs, as with the Gracula species, the importance of a low-iron diet cannot be overemphasized.

The diet used at Jersey Zoo has been equally successful. Bali starlings are offered an insectivorous mixture, in a small D-pot, which they receive between 10:00 a.m. and 10:30 a.m. This is made up of equal parts of the following:

- Witte Molen High insect diet
- Witte Molen Low iron diet
- Witte Molen Mynah pellets (soaked)
- Finely grated hard-boiled egg
- Finely grated carrot
- Wholemeal bread crumbs

This mixture is blended to form a slightly moist, pliable consistency, to which is added an equal amount of finely diced fruit comprised of apple, pear, grapes, banana, papaya, mango, and perhaps other fruits depending on the season. Finally, a mineral supplement called Avimix is finely dusted onto the fruit component.

In addition to their main meal, they also receive two additional insect feeds, one at 8:00 a.m. and a last feed at 4:00–5:00 p.m. (winter or summer), which consist of a small amount of either mealworms, crickets, adult locusts, or a combination of these, scattered around the aviary to encourage the birds to forage. Some natural prey, such as moths, spiders, and earwigs are also caught by the birds in their heavily vegetated aviaries. When chicks are being reared, the adults are given a live-food feed of soft-bodied insects every two hours throughout the day, starting as close to daybreak as practical and ending an hour before dusk.

Breeding: Surprisingly, Dick has found this endangered species much easier to breed than the hill mynahs. Several years ago he reared 15 youngsters in a single season from just two pairs of breeders. All of these birds were parent reared. He has even had them nest and successfully raise chicks in a small holding cage, 1.2 x 1.2 x 1.8 m (4 x 4 x 6 ft.) in size, which they temporarily shared with several other birds after his move from Los Angeles. There were a few cockatiel-sized nest boxes in place just for use as shelters. The other occupants included a bearded barbet *Lybius dubius* as well as a turaco and a pair of spur-winged plovers *Vanellus spinosus*. They went to nest while the new aviaries were under construction and fledged two chicks. They have nested successfully in large mixed-species aviaries on many occasions as well. Their success under these conditions is really dependent upon the other inhabitants. Being cavity nesters, the eggs and chicks are safe from the likes of the small hornbills *Tockus* spp. Dick has had trouble with birds breeding in this type of setup when

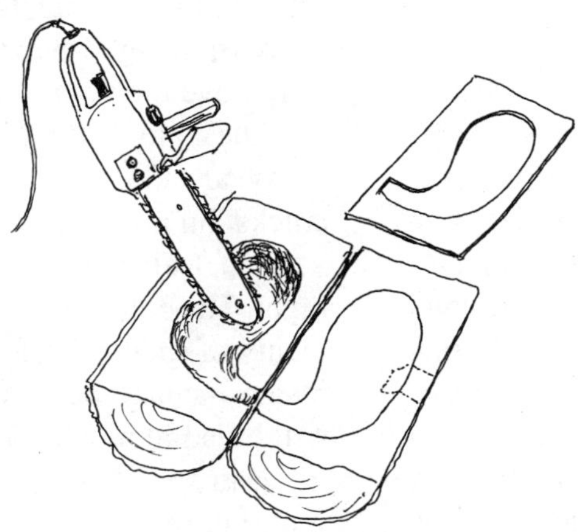

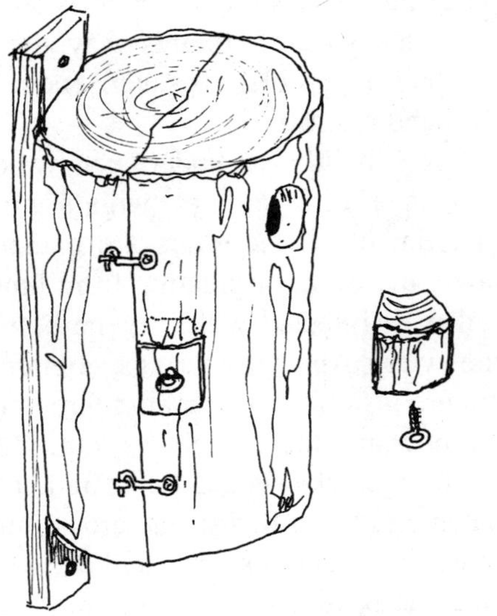

Hollowing out log for an artificial nest cavity.

the aviary also contains any of the small parrots, especially lories, which will often enter nest boxes and disturb the sitting hen. For this reason it is best not to include parrots in a mixed species aviary if breeding is part of the plan, although they pose no problem in a strictly ornamental flight.

At Jersey Zoo, nestboxes are provided for potential breeding pairs at the beginning of April, and removed again around the beginning of October, when there is an appreciable drop in temperature. This procedure is recommended for other Northern Hemisphere collections, as late chicks are subject to perishing in cold temperatures, especially if not brooded adequately by the parents. It is possible, however, to breed these birds at any time of the year if the nestbox is placed in the indoor portion of the enclosure. In their natural habitat Bali starlings nest in tree holes 4–10 m (13–33 ft.) above ground, and usually choose holes that they can just squeeze through. In a captive environment, therefore, we provide nestboxes that are constructed from treated wood, with an entrance hole diameter of 5.5 cm (2.16 in.) and a cavity measuring approximately 20 x 20 cm (8 x 8 in.) extending 30 cm (12 in.) deep from the entrance hole. A hinged door for cleaning and inspection purposes is also provided, as is a perch 4 cm (1.6 in.) below the entrance hole. The nestbox is positioned at a height of about 4 m (13 ft.) in a suitably sheltered part of the aviary flight. The nestbox should be hung near the top of smaller aviaries.

A further, and very useful, addition to some of our nestboxes is a compartment designed to house a camera, with one side of the nest cavity being constructed of glass. This feature has not deterred the starlings from using the box, and allows keepers to monitor incubation and feeding activity. Perhaps most importantly of all, it enables keepers to ascertain whether the parent birds are plucking the nestlings, which is sometimes a problem with this species. In this way the keeper can judge whether the problem is severe enough to remove chicks for hand rearing.

Nest building is undertaken by both birds, and takes 1–3 days. Fine springy twigs seem to be the nesting material of choice. Ours are gathered from a nearby lawn, the twigs originating from lime trees. This material is disinfected with Trigene before being placed in the aviary. Apart from twigs, conifer needles and feathers are also used. Generally, three bluish eggs are laid. On occasion there might be four, but it is unlikely that all four chicks would be reared. Dick has had three hatch and three fledge numerous times, but most often three will hatch and two will survive. The youngest or smallest becomes part of the food chain. Incubation is 14–16 days and, as with the hill mynahs, the shells are discarded a distance from the nest box. Dick has found that Bali mynahs will allow nest inspection without any problem. In fact, he has taken a box containing three young chicks about 10 days old from the aviary and placed it on the ground in order to take photographs, and the parents resumed caring for the chicks as soon as the box was hung back in place.

The chicks are fed nothing but live food for the first few days, but the parents soon begin offering their entire diet to them. All of Dick's Bali mynahs readily take crickets as well as mealworms. He feeds crickets that have been frozen and then thawed, as this keeps them from populating the neighborhood, and the birds do not seem to mind. The chicks grow rapidly on a diet of soft-bodied insects such as white mealworms, crickets, wax moth larvae, which are provided for them, and flies, moths, and other insects harvested from the aviary. Both adults carry out chick feeding duties, the male bird becoming the sole provider if the female is on eggs again.

The development and fledging of chicks has been recorded. Pin feathers start to erupt, and eyes begin to open, when the chicks are seven days old. The eyes are fully open at 10 days, at which time their bodies are totally covered in white pin feathers. Nestlings begin to exercise their wings, which are then fully feathered, at 14 days of age. At 20 days they are peering out of the nest hole, and normally fledge at 28 days of age. Both adults feed the fledglings. If the parents re-nest, the male takes over the chick feeding while the female rebuilds the nest, lays, and commences incubation duties. The chicks will continue to beg for food, and will be fed by both parents, although less and less, so that they are encouraged to pick up food for themselves. Fledglings will also return to the nestbox for periods during the day for the first few days after emerging. This is generally tolerated by the adults, even if the female has another clutch of eggs in the box. There will, however, come a point, anywhere from 7–17 or more days after fledging, when the adults start to chase the young away from any food provided. At this point it is prudent to remove the young birds from the aviary.

Dick has had eggs hatch as early as January 2nd one-year, and as late as mid-October. The very early chicks did not survive, as January is the coldest time of year in California, often dropping below freezing at night. The cold has no effect on the parents, as they have plenty of shelter. They are at their showy best in a large planted aviary but, as described earlier, they can and will breed in much starker surroundings.

Bali starlings are sexually mature at the age of one year, but may not necessarily breed successfully due to inexperience. A likely pair should be introduced to one another several months before the beginning of the breeding season. If compatible, courtship involves a head-bobbing display in which the bird erects its crest, points its bill upward and bobs up and down, emitting loud clucking vocalizations. Courtship may also include some allopreening.

A number of these starlings at Jersey Zoo are prone to feather plucking, either self-inflected, or by a mate. Most plucking seems to concentrate on the neck area, especially at the front. Caution should be exercised when pairing up birds that are prone to this vice. This is a phenomenon that occurs only with birds in captive environments. The exact cause is unknown, but it has been speculated that over-zealous courtship preening, poor diet, and boredom all contribute to its occurrence. It has also been suggested (but not substantiated) that birds displaying a penchant for plucking other adults will be more likely to pluck nestlings. However, there is a fair chance that nestlings will be plucked even by parents that are themselves in perfect feather.

If the adult birds are going to pluck the young, it will generally occur sometime around the 10-day mark, when the pin feathers are beginning to become established. The actual extent of such plucking can vary a great deal. Preening of the chicks is perfectly natural behavior and should not be mistaken for plucking. A small amount of plucking is acceptable, i.e., the chicks can still be raised by the parents without too much detriment to their health. However, if violent and persistent plucking is witnessed

Bali Mynah
Leucopsar rothschildi

or indicated, nestlings should be removed immediately for hand rearing, as such plucking can be so severe that a nestling's spinal cord is severed. If their eyes are already open, pulled chicks may have to be force fed until they start to gape, or even until they start to feed themselves. As long as the chick has been removed in good time, hand rearing seems to have no adverse effects, and a perfectly healthy juvenile should be the end result.

One other hazard worthy of mention is that at any stage during chick development, the adult birds may kill the young by pecking at them. Pecking may target the back, wing, or head, and may be mistaken for plucking. The most likely explanation of this behavior is that the chick has a medical complaint and the adult is either trying to get the chick to respond or realizes that the chick

Four Mynah Species Chicks from San Diego Zoo

Day	Gold-crested (g)	Sulawesi (King) (g)	Gold-breasted (g)	Southern white-collared (g)
1	5.75	5.57	5.85	8.35
2	5.73	5.55	5.95	8.5
4	7.23	7.6	7.05	9.7
6	9.48	11.03	8.5	11
8	13.7	18.12	12.8	15.75
10	22.03	29	19.65	19.8
12	32.8	38.47	31	26.75
14	42.08	52.77	44.5	32.2
16	48.3	63.2	64.15	40.4
18	53.75	70.7	87	45
20	57.38	73.46	96.5	57.2
22	59.45	74.7	99	63
24	62.2	83.7	102	61
26	62.63	80.3	112	67
28	62.58	80.8	122	68
30	63.6	80.5	120	80
32	66.5	79	121.5	87.5
34	75	78.5	125	102.6
36	75	79.3	117	110.2
38	73	82.1	112	120.4
40	74	85.7	117.5	128

will not survive, and kills it. During 2001, we found two 20-day-old chicks dead in the nest, apparently victimized by parents that had produced a flawless juvenile bird two months earlier. There was no evidence of feather plucking two days previously, and it seemed strange that plucking should begin so late. After post-mortem, it was learned that the chicks were suffering from a fatal infection which had most likely been detected by the adult birds.

One brood of three chicks was successfully hand reared at Jersey Zoo in October 2000, with the chicks being removed from the nest at five and six days of age. They were taken because they were hatched so late in the autumn that there was concern about them perishing due to exposure to cold temperatures. Initially, the chicks were placed in a ceramic pot lined with tissue paper and a layer of coconut fiber to prevent foot problems or leg splaying. The pot was placed in a Brinsea Brooder, set at a temperature of 35°C (95°F) with a relative humidity of approximately 50%. The chicks were fed every two hours throughout the day on a diet of pureed pinky mouse dusted with Nutrobal mult-vitamin powder. This was fed to the chicks using tweezers. The brooder temperature was reduced by 1° each day until it reached 28°C (82.4°F). The pinky mouse puree was thickened after a few days, when the chicks were able to take chopped pieces of pinky mice. They continued to be fed this diet until they refused to take pinky mice, preferring insects, soaked mynah pellets and papaya, at 22 days of age. They fledged at 27 days and, being too active for the brooder, were then placed in a budgie-type cage. At this stage they were also offered a container of adult diet, which they gradually started to take for themselves over the next few days. They were also being fed by the keeper, but not as frequently, four times per day between 8:00 a.m. and 5:30 p.m.

Bali mynahs have been known to breed up to 18 years of age, and probably survive to over 20.

Ring Size: Birds are ringed using a numbered metal split ring size F (6 mm) on one leg and a colored plastic ring 7 mm on the other leg.

Species Requiring Similar Management:

- Starlings
- Jays
- Tree pies
- Shrikes
- Babblers

References:

Collins, M. & T. Smith. 1995. Post-release behavior and movements of the Bali mynah *(Leucopsar rothschildi)* in Bali Barat National Park.

Feare, C. & A. Craig. 1999. *Starlings and Mynas*. Princeton University Press, Princeton, New Jersey.

MacKinnon, J. 1988. *Birds of Java and Bali*. Yogyakarta, Indonesia: Gadjah Mada University Press.

Owen, A. Management guidelines for *Leucopsar rothschildi,* the Bali mynah. Durrell Wildlife Conservation Trust, Jersey, Channel Islands.

Vince, M. 1996. *Softbills: Care, Breeding and Conservation.* Surrey, Canada and Blaine, USA: Hancock House Publishers.

Weil, M. 1981. *Mynahs*. Neptune, New Jersey: T.F.H. Publications.

Redbilled Oxpecker
Buphagus erythrorhynchus

• Glen Holland

Introduction: The two species of oxpecker form the sub-family Buphaginae of the family Sturnidae (starlings) and belong to the order Passeriformes. Both species occur in Africa, south of the Sahara and out of the breeding season they are gregarious found in groups of 2–12 birds, clinging (with sharp, curved claws) to game or domestic stock. They occur in savannah, thorn and open woodland. The flight is high and they are easily located in flight by their calls. The contact call in flight is a rattling *tsik, tsik* repeatedly. They also make a harsh *kssssk* call if alarmed. In the wild state this alarm call acts to warn game species of approaching danger.

Management: My experience with these species was with three hand reared red-billed oxpecker *Buphagus erythrorynchus* that had been rescued, by a friend, from a child who had removed them from a nest, and was about to feed them to his cat! Oxpeckers naturally feed on ectoparasites that they remove from their animal hosts. The beak is strong, and the toenails of these birds are extremely sharp, enabling them to cling onto their hosts when feeding. It has been found that captive birds that do not have access to a host are unable to perform the scissor-like movements with their beak which enables them to remove ectoparasites; the beak then becomes overgrown or deformed.

Once fledged, I released my hand-reared birds into a walk-through aviary where, after inflicting scratches on a number of our visitors, they were removed. (Their sharp nails are able to draw blood, particularly on bald heads!) These three birds were sent to the World of Birds in Cape Town, and a few days later I received a phone call to say that they now realized why I was so willing to give up such rare avicultural subjects.

This species was first bred at the Zurich Zoo where they were kept in a rhino exhibit, and at one stage with goats, which the birds accepted as hosts. It is vital that the yellow-billed *B. africanus* have a host as they roost on the host at night to keep warm. Dr. Kobus du Toit uses Cameroon dwarf goats as a host as they are inexpensive to maintain. A problem was experienced with wounds, which the oxpeckers do not allow to heal on their host, but this was remedied by spraying aerosol Necrospray containing chloramphenicol, which is bitter, and the birds then avoid the wound. Naturally wound tissue on hosts forms part the diet.

Oxpeckers should be housed in a large aviary with habitat B as described in the chapter on landscape.

Derek Goodwin describes his experience with a hand-raised captive red-billed oxpecker as follows:

> "The impression I had from the first, that the bird detached and swallowed tiny loose morsels of skin whilst nibbling the hand, was corroborated when I experimented. When I cut little bits of skin of my finger so that, while still partly attached, they stood up from the surface, the oxpecker nibbled, scissored, and tugged at them till (in a very few seconds) it had detached them, when they were greedily swallowed.
>
> I cut the back of my finger with a razor blade until a little blood showed. The oxpecker appeared even more excited by this than the fragments of loose skin. It consumed the blood with frantic haste and then nibbled avidly at the tiny wound. A few moments later I repeated the experiment (with a fresh incision) with the same result. It was noticeable that whereas only a very little blood oozed from the original wound (I waited for it to appear before offering it to the bird), after the oxpecker had nibbled for some moments the blood welled up more freely and in larger quantity. This was repeated an hour later. Although slightly painful and clearly effective in inducing freer flow of blood, the bird's nibbling at the cut place did not seem to result in any greater injury than the removal of cut bits of skin and, perhaps, some of the skin immediately around the cut."

Diet: Apart from food they will take from live hosts that are made available to them, the captive diet should consist of minced ox heart, insects, chopped pinkie mice, boiled egg, and a calcium and multivitamin supplement added once a week. Dr. du Toit is at present managing a breeding programme for yellow-billed oxpeckers and each day each of his birds receive frozen engorged ticks and 20 ml of blood with an anti-coagulant added.

Breeding: Courtship involves the pair lowering their bodies to the perch, opening their wings and then spreading the wings and tail horizontally. The pair then moves around each other with the wings vibrating. The female then moves under the male's wing and copulation follows. Limited courtship feeding occurs. Natural nest logs and a parakeet-sized box should be provided to allow the birds to chose a nest. The nest should have an internal diameter of 15 cm (6 in.), 35 cm (14 in.) deep and an entrance hole of 7.5 cm (3 in.). When breeding, they will fight with other cavity nesting species such as Starlings. A large proportion of the nest material is hair, with some grass and rootlets as well.

The 2–4 pinkish white eggs are speckled with rust and purple colored spots. The incubation period is 12–13 days, beginning with the first egg, and the chicks fledge 28 days later. During incubation only the breeding female incubates at night with both members of the breeding pair sharing incubation during the day. Both parents and up to three helpers from a previous clutch assist in rearing the chicks. A single pair will, however, breed successfully even if no helpers are available. Fecal sacs are removed from the nest.

They are usually double brooded. After a month the adult male becomes aggressive towards the immatures. This should be monitored and before it becomes serious, the immatures removed. Immatures are darker than the adult with a dark bill. Adult plumage is attained by six months.

Hand Rearing: Oxpeckers are raised with similar methods to those described for SHRIKES. The three

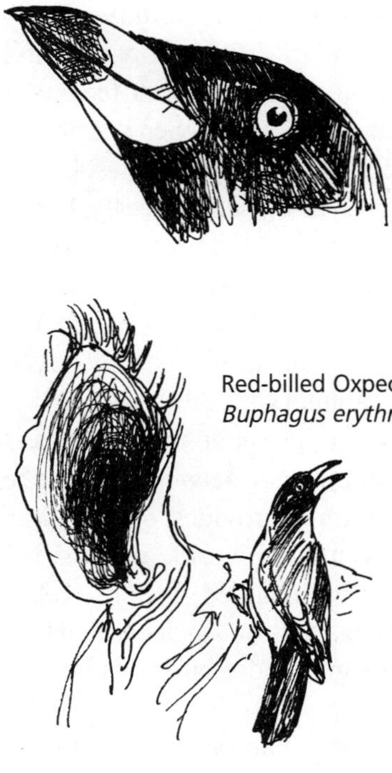

Red-billed Oxpecker
Buphagus erythrorhynchus

that I raised, were fed equal portions of ox heart strips to which a multivitamin and calcium supplement was added, and insects such as flying ants, mealworms and small crickets. The food was offered with forceps. Initially, the birds could not be identified but their plumage soon clarified it. The juveniles of red-billed nad yellow-billed are difficult to tell apart.

Ring Size: 4.5 mm

Contribution: Dr. Kobus du Toit, Du Toit Game Services, Onderstepoort, Pretoria.

Reference:

Goodwin, D. 1963. Some behavior of a captive red-billed oxpecker. *Avicultural Magazine* 69: 113–117.

Kaiser-Benz, M 1974. Breeding the Redbilled Oxpecker at Zurich Zoo. International Zoo Yearbook No

21.13 Hirundinidae Swallows

Glen Holland and Neil Hamilton

Swallows and martins form the family Hirundinidae within the order Passeriformes. Two species of river-martins, genus *Pseudochelidon,* are separated into the subfamily Pseudochelidoninae, while the remaining species of swallows, martins, rock-martins, house-swallows, sand martins, crag-martins, house-martins, and sawwings, a total of 87 species in 13 genera, are all included in the subfamily Hirundininae. This family is distributed through most regions of the world. Also included in this section are wood swallows of the family Artamidae, which includes 10 species, located in South-east Asia, Indonesia, and Australasia. Although wood swallows are not true swallows, they have been included here for avicultural purposes. A section on the masked wood swallow *Artamus personatus,* from Australia is included at the end of the swallow husbandry notes.

Swallows are primarily aerial feeders on insects and most are gregarious, at least when not breeding. They nest either colonially or solitarily. The nest consists of a pad of feathers and grass in a chamber at the end of a self-made burrow in a bank or in level ground, in a hole in a tree, or in a mud nest that may be either a half-cup or bowl-and-tunnel under an overhang. The 2–6 eggs are white, plain, or spotted, and the chicks are sparsely covered in down at hatching. The parental care is by both sexes. Many species are either local or international migrants, some moving thousands of kilometers between hemispheres, with the seasons. This family provides one of the biggest challenges in the avicultural world, and it is for this reason I felt they should be included. Wildlife rehabilitation centers often receive swallow chicks and, while not too difficult to raise, fledging these birds can be a true test.

Management: The welcome swallow *Hirundo neoxena* and the fairy martin *H. ariel*, have both been successfully kept in Australian aviculture. The welcome swallows inhabit eastern and southern Australia, New Zealand, and nearby islands. They have been particularly successful in aviculture and new colonies have been established in a number of Australian institutions. The project was initiated at Taronga Zoo, Sydney, and the original breeding stock was obtained as young birds mainly because, in the past, any adult birds caught or donated to the zoo did not survive long in captivity. I had the same experience in South African bird parks, primarily with lesser-striped swallows *H. abyssininca.* The reasons for this were the speed at which the birds fly, the fact that they feed on the wing, and our inability to supply sufficient flying insects. To establish the birds, instead of mist-netting adult birds, young birds were lifted from nests, taking only one or two from each clutch. These young birds are collected when half grown, at which stage they are relatively easy to raise.

As the hand-raised birds started to fledge, the lid was left off the brooder inside a temperature-controlled room. As the fledglings began to fly, they quickly recognized the different objects in the room, which they soon learned to avoid. At feeding time the older, stronger birds would eat and teach the others. At this stage the birds were fed every 2–3 hours and twice at night.

The first swallows we taught to feed from a dish at fledging age, once they were weaning. This was done by only offering live food first in the dish, e.g., mealworms and moths. As they started to take live food from the dish, the artificial diet was given and the live food eased off a little. Note that a portion of live food in the diet is essential at all times; do not attempt to completely take this species off live food.

Birds that you have feeding from a dish can then be used to teach any others you have coming in to your center. It is important that one identifies all birds in the group using color bands so observations of feeding can be done. A little later the birds became difficult to handle, so they were removed to a holding cage 1.2 x 0.9 x 0.6 m wide (4 x 3 x 2 ft.), giving them a little less flying room, and by doing this the birds were able to adopt different feeding and drinking methods. Once they were feeding well on substitute foods, out of a flat dish, they were introduced into a small flight and later into a larger aviary 7.5 x 1.8 x 3 m high (25 x 6 x 10 ft.).

Since swallows can be found in many different environments, one can landscape an exhibit to suit any particular needs. Requirements that must be provided are areas for nesting and roosting. Dead trees, rock or wooden ledges are important for these species. Staff must provide nesting materials such as wet mud, dry grasses and feathers for breeding. Mist sprinklers should be used, as well as a good water dish or shallow water pool as maintaining good plumage is extremely important for these species. Transporting requirements for swallows are similar to moving finches or softbills.

Initially the sexes appear alike but methods of sexing welcome swallows by the color and length of certain wing or tail feathers have been found. On the tail the two outermost feathers, numbers 5 and 6 vary between male and female as shown in the table in this section.

Diet: The initial diet used at Taronga consisted of an egg-cake mixture, so called because it contained sponge cake, mince, egg yolk, cheese and meat. Vitamin E, Pentavite (multivitamin) and calcium are also supplied. Their live food consisted of mealworms, fly pupae (which are added to the egg-cake mixture), termites, and moths. Perth Zoo uses a similar diet at present with the addition of ground dog kibble and commercial insectivore mix.

Breeding: Breeding of both the welcome swallow and fairy martins occurred in August and September. The welcome swallows built a cup-shaped nest of mud. To ensure sufficient nesting material is available, a shallow dish of wet mud should be supplied. The swallows and martins were observed rolling small clay balls in the shape of pellets and later carrying them to the nesting sites. The welcome swallows nested under a sandstone ledge, as they do in the wild. The fairy martin nested on a wall above the sandstone ledges, their nest being bottle-shaped. Fairy martins are colonial nesters and welcome swallows are solitary nesters.

Swallows are usually not aggressive to any smaller species they may be housed with. If one houses the welcome swallows in a small or medium exhibit, then two pairs will do. Martins breed in colonies so the aviary size is not as important. Any more than two pairs of welcome swallows will affect feeding and breeding, as they can become very territorial. In a larger exhibit, such as a walk-through, staff find about five pairs may allow successful breeding, but one has to supply plenty of live food, which is all ways hard to maintain due to the number of other species within that exhibit. Taronga have housed swallows with small parrots, wrens, dotterels, bee-eaters, terns, and stilts. Presently, they have swallows in the new wetland exhibit, which is 75 x 36 x 20 m (93.75 x 45 x 25 ft.) high. Species include ducks, stilts, glossy ibis, pied heron, brolga, black-necked stork, etc. If introducing swallows into larger exhibits then one must use an introduction-holding aviary, but only for a short period, say one-week maximum. Natural trees, dead trees, building entrances, wooden platforms all support the swallows in these exhibits. The species has yet to breed in the wetland exhibit.

All together in the first season there were three fairy martin and two welcome swallow nest sites. After most of the nesting sites were at the stage of completion, each species began an unusual chasing behavior, sometimes looking as if to kill one another. Later, upon observing this behavior, it was found that each bird was chasing his or her own mate. At this stage, the lining of the nest was carried out using feathers, cotton and other soft materials that were supplied.

The incubation period is 14 days and the chicks fledge at about 23 days. The chicks automatically follow the parents to the food dish. They are independent at around 14 days. We usually have two or three broods per pair a season depending on the success rate. Young can usually be left with the parents for a short period but this depends on the aviary size. Problems with aggression towards the fledglings will occur only if one does not supply enough food for the next group of young birds.

Hand Rearing: Young swallows are regulars among the numerous birds brought in for hand rearing at bird parks and rehabilitation centers in South Africa. I personally always dreaded receiving them as, although relatively easy to raise, I found many never had the strength to fledge, or those that did fly were unlikely to survive. The question in my mind was always the transition from hand feeding artificial diets to catching prey on the wing. Hopefully the work done by Australian aviculturists can teach us some valuable lessons.

The chicks I received, those collected to establish the Australian captive population, were always at least half grown. The younger birds were easier to feed and would

Welcome Swallow
Hirundo neoxena

beg for food within a day of being received. Older birds, near fledging age, proved rather stubborn and required days of force feeding before they would beg for food. I used a diet of equal proportions of diced heart, commercial rearing food, and insects, such as moths and waxmoth larvae. The insects and heart were held with forceps and dipped into the rearing food, which had a wet-to-loose porridge consistency. Feeding was done every one and a half hours from 6:00 a.m. to 8:00 p.m. The chicks were kept on a pad of tissue paper that was replaced three times per day. A similar diet, excluding the rearing food, was used at Taronga Zoo. Chicks are rather sluggish and after one or two beaksful they settle down again. If you need to force feed older birds, be gentle when forcing their rather delicate beak open, and place the food well back in their large gape. As mentioned earlier, the challenge is not in rearing the swallows, but rather fledging them and then training them onto artificial diets. See the notes on this under the management section.

E. Cowley of Doncaster in England wrote the following account of hand rearing sand martins *Riparia riparia*. It is, however, important to note that the modern diet should more closely follow the hand-rearing diets mentioned for passerines such as SHRIKES and TANAGERS.

> "Six sand martin young were hand reared to fledging. The birds were kept in a nest box constructed to resemble a normal nest chamber and for most of the period they were quite unrestricted. The rearing food was not measured but of 113 feeds, 98 (87%) were of maggots (gentles), the larvae of the bluebottle fly *Calliphore vomitoria,* nine of bread (8%) and six of earthworms (5%). Of these types of food, maggots were always split open before feeding and were eagerly accepted; bread (only given when other foods were not available) was accepted but was difficult to swallow and earthworms were not readily accepted. Ejected food left the gape at great speed and for some distance without any movement of the gaping bill.
>
> The birds were fed from 5–8 times per day and the maximum number of maggots given at one feed was 140, these being large ones up to 20 mm (.675 in.). The birds were normally fed one parcel of food each (consisting of six maggots) more or less in turn, until all had stopped soliciting. The nestlings were normally kept in artificial light in the evening to encourage them to feed later, and they ate very well at 11:45 p.m.; possibly because of this, the first feed in the morning, although given several hours after dawn, was not normally a good one. Once the birds had settled for the night, they did not solicit for food on being disturbed in artificial light. Little external

Dimorphic Differences in the Welcome Swallow *Hirundo neoxena*

Male	Female
The length of the outer tail feather no. 6 is usually 80mm+	The Length of the outer tail feather No. 6 is uasually 75mm+
The difference between the tips of the No. 6 and No. 5 is 17mm	The difference between the tips of No. 6 and No. 5 is <16mm+, usually 12–14mm.
Streamer No. 6 and the tip of No. 5 are longer and narrower than those of the female	Streamer No. 6 and the tip of No. 5 are broader than those of the female.
The white spot on the underside of No. 6 is usually larger and darker than in the female, is well marked, and generally reaches the shaft.	The white spot on the underside of No. 6 is usually small and does not reach the shaft.

warmth was given but the birds were kept well covered, and all would strive to gain a central position within the group.

All the chicks observed normal nest sanitation, some making strenuous efforts to get to the opening for this purpose. The smallest chick would always return to the back of the nest after this action, even though feeding was normally continuing at this time. The fecal sacs were quite normal and stayed intact when removed with a pair of tweezers (tweezers/forceps were also used to offer the food to the chicks). Although the chicks would generally crouch down in the nest when I approached, they had only to see my hand or the tweezers for soliciting to start. A hand placed over the entrance, darkening the nests as would the arrival of an adult bird, immediately caused soliciting, but as each chick neared fledging it became secretive, and one of them stopped soliciting, although it would feed after some enticement. The chicks flapped their wings vigorously as they prepared for flight and were even seen to push the wings onto the floor of the nest, holding the body off the nest. These chicks were eventually allowed to fledge at an existing sand martin colony."

Charles Everitt of Trenton, New Jersey, USA, wrote some interesting notes on the masked wood swallow in aviculture:

> "The masked wood swallow *Artamus pesonata* are gregarious in their habits, often associating with white-browed wood swallow. They generally collect in the southern parts of Australia during the breeding season, ranging from August to January. The sexes are dissimilar; the adult male being slate-gray above, with a jet-black facemask and throat patch, the latter edged with white. The underparts are pale silver-gray, the tail being forked with the feathers tipped with white. The female, of the same size, lacks the black face mask and throat patch, these being replaced by very similar markings, dark gray with the underparts a smoky-gray color.
>
> Although in their native habitats they normally select low bushes or trees for the location of their open, cup-shaped nests, in the planted aviary (shared with white-bearded honeyeaters, cedar waxwings and ruby-throats), a pair chose a most conspicuous position in one of the feed stands despite many admirable nest sites. It was a trough about 23 x 7 cm (9 x 3 in.), the sides being 5 cm (2 in.) high, and the swallows built in a piece of this trough measuring 7.5 x 7.5 cm (3 x 3 in.). In order to provide more space and to avoid interference from other birds, the feeding cups were removed from the trough and placed elsewhere in the aviary. A nest of fine twigs and rootlets was completed within three days and the first egg was laid on March 13 with a second the following day. Incubation, which was shared, lasted 12 days and began with the laying of the first egg. The eggs were light grayish-green, clouded and blotched with shades of brown and underlying spots of gray.
>
> The newly hatched chicks were very dark skinned, covered with gray down, streaked and sooty. Their gape was pale yellow. As the nest was so shallow and it was felt that a danger existed of the chicks being ejected or dragged out of it, a wire frame was fitted around it, raising the height to about 8.5 cm (3.5 in.). The parent birds resented our intrusions in a very physical manner, swooping down and striking us continually whilst it was being put in place, but went straight back to their young as soon as we were out of the aviary. As with the incubation, the rearing was shared, live food being the sole diet seen to be employed during the nesting period. A wide variety of this was provided in the form of mealworms, crickets, and whatever flying insects it had been possible to trap with night-lights. At six days old quill feathers were showing and their eyes were open.
>
> Vacating the nest at 14 days old, they were fairly strong on the wing for fledglings, and immediately portrayed the characteristic of the adults at feeding time by switching their tails from side to side as they begged for food. It was at this stage that the parent birds began to vary the diet by introducing them to the ground raw beef and the chopped fruits as a supplement to live food. The coloring of the young birds was basically gray, heavily flecked with sooty black, except for the flight and tail feathers, which were plain slate gray. One of them showed a much darker face and it was correctly assumed that there was one of

each sex. Actually this difference had been noted at 11 days when they were still in the nest. They were independent at 30 days old and adult plumage acquired by six and a half months old. In all they had four rounds that season, rearing seven young, four males and three females. In each instance the young had to be removed once the female went to nest again as the male began chasing them around the aviary. Although gregarious in the non-breeding season they did not tolerate other wood swallows in the aviary while breeding."

Ring size: 2.5mm

Contributions:
Neil Hamilton, Curator of Birds, Perth Zoo.

References:

Cowley, E. 1977. Hand-rearing Sand Martins *(Riparia riparia)*. *Avicultural Magazine* 83: 185 – 188.

Disney, HJ De S. 1991. Notes on Sexing and Plumages of the Welcome Swallows. *Corella* 15(1).

Everitt, C. 1964. The Masked Wood Swallow. *Avicultural Magazine* 70.

Hamilton, N. 1981. Breeding the Welcome swallow *(Hirundo neoxena)* and Fairy Martin *(Cecropis ariel)* in captivity. Thylacinus Vol 6 No 3 1981. Australian Society of Zoo Keeping.

21.14 Pycnonotidae Bulbuls

Glen Holland

Introduction: Bulbuls belong to the order Passeriformes and the family Pycnonotidae, which includes 138 species in 21 genera. This family is widespread in Asia, Africa, and Madagascar. In general, bulbuls are common birds, preferring areas of open bush or thorn savanna. Most species tend to avoid dense forest, but occur on forest margins and in dense vegetation. Some species, however, such as the terrestrial brownbul *Phyllastrephus terrestris,* can be found feeding on the leaf litter on the forest floor of tropical forests. Bulbuls occur from coastal areas to mountain scrub. The sexes are alike in most species, and both parents care for the young. When threatened by a predator, they group together and call excitedly. Their calls are loud, and are used to help maintain territories. The red-whiskered bulbul *Pycnonotus jocosus* has been introduced into Australia and Hawaii.

Many species of bulbuls are known by other names. The two species in the genus *Spizixos* are called finchbills, while those in the large genus *Andropadus,* as well as several smaller and lesser known genera, are called greenbuls. Another large genus, *Phyllastrephus,* includes species variously known as greenbuls, brownbuls, leaf-loves, and olive-greenbuls. Species in the genera *Bleda* and *Nicator* are called bristlebills and nicators, respectively. The best known genera of typical bulbuls include *Pycnonotus,* with 40 species; *Criniger,* with 5 species; and *Hypsipetes,* with nine species.

Several species of bulbuls are threatened with extinction in all or part of their ranges in the wild. The straw-headed bulbul *Pycnonotus zeylonicus,* which has been decreasing in numbers due to its popularity as a cage bird in Asia, has been singled out by the Bulbul Species Interest Committee of the PACT (Passeriforme, Apodiforme, Coliiforme, and Trogoniforme) TAG (Taxon Advisory Group) of the American Zoo and Aquarium Association for special attention in zoos (Lindholm 1996). In addition, captive breeding could help preserve Styan's bulbul *P. taivanus* from Taiwan. The integrity of this species is being compromised through hybridization with the Chinese, or light-vented, bulbul *P. sinensis,* which has been released in its habitat in large numbers in religious ceremonies (Severinghaus 1998).

Management: The sexes are alike, but in some species, such as the red-whiskered, the male has a flute-like whistle by which he can be vocally differentiated from the female. Bulbuls are easy to maintain in captivity and are hardy. This is one of the few families in which wild-caught birds can be introduced directly to an aviary without suffering losses from stress. The bulky, noisy, rather cheeky nature of greenbuls makes them interesting avicultural subjects. Newly acquired greenbuls need more attention initially than, for example, the black-eyed bulbul *Pycnonotus barbatus*. A longevity record for a hand-raised bird was 26 years and five months. Bulbuls are compatible with most other birds, but to avoid aggression in a flight, they should be kept as sexed pairs and not mixed with similar species. For example, black-eyed bulbuls should not share an aviary with the closely-related red-eyed bulbul *P. nigricans*. In a large planted aviary, however, a number of species that have differing natural habits can be kept together. For example red-whiskered, yellow-bellied *Chlorocichla flaviventris,* and terrestrial bulbuls will all live in harmony in a large aviary, each pair only defending the area in the immediate vicinity of their nest. During the breeding season, bulbuls can be rather pugnacious toward small birds.

Two orange-spotted bulbuls *Pycnonotus bimaculatus* owned by Ed Storozynski of British Columbia constructed a nest each year for several years, but when no eggs were produced they were DNA-sexed and proved to be both males. These birds were aggressive even outside the breeding season, and had killed a red-vented bulbul *P. cafer* and other small birds. They were even observed harassing a pair of much larger European blackbirds *Turdus merula* with which they shared an aviary, and had to be separated. They were later housed harmoniously with larger birds such as turacos and imperial pigeons.

The larger species of greenbuls, such as the yellow-bellied, should be mixed with other softbills of a similar size, or preferably, larger. They tend to be aggressive toward smaller softbills, and will often dominate feeding areas, particularly if live food is available. They are best kept in a well-planted aviary with plenty of dense shrubs, including plants that will attract insects into the aviary. Leaf litter should be left on the aviary floor as they enjoy scratching in the litter for small insects. Leaves on the ground are turned with the beak and then carefully inspected for insect life. A mixture of Habitats A and B, as described under the chapter on LANDSCAPE, will suit all bulbuls. They are fond of bathing in a pool of water or in a mist spray placed over a dense shrub. As most bulbul species naturally inhabit tropical and subtropical areas, in countries where temperatures remain below 10°C (50°F) for extended periods, these birds should be housed indoors.

Diet: The diet should consist of diced or segmented fruit such as apple, orange, papaya, soaked raisins, grapes, banana, guavas, etc. Fruit left in segments can be impaled on nails in a log. Berries and nectar-producing flowers such as cotoneaster, privet, coprosma, and honeysuckle should be offered as well. Commercial softbill foods with a cereal texture or pellets should be moistened with water. Grated boiled egg with the shell can also be added to moistened soft food. Minced ox heart mixed with a commercial softbill formula is another favorite. Soaked cat biscuits are eaten as well. A multivitamin/trace element powder should be added to the diet twice a week. Bulbuls are also fond of nectar such as described under the section on SUNBIRDS.

Bulbuls are also highly insectivorous, and will take insects off the ground and from the leaves of vegetation,

Red-whiskered Bulbul
Pycnonotus jocosus

or hawk these out of the air. Flying ants, mealworms, waxmoths, caterpillars, termites, etc. are relished by these birds. Live food is essential when they have chicks in the nest. By offering live food, the bird's confidence can be gained to the point where they will come to the hand to accept such tidbits.

Breeding: Bulbuls nest readily in captivity, often producing two or three clutches per season. During display, the male lifts his wings, spreads his tail, bows, and sings loudly. The nest is a cup placed in a fork, usually deep inside a dense shrub. It is often started with a single, basal piece of vegetation or material, and takes 2–3 days to complete. It is built of fine twigs and grasses, and lined with rootlets, animal hair, and wool. The 3–4 eggs are white and well covered in pinkish, rust, and gray spots. Incubation usually starts with the third egg. The incubation period is 12–13 days, and incubation is by the female alone. The chicks fledge at two weeks of age, and are independent two weeks later. The young remain near the nest for the first few days after fledging, and then begin to follow the adults, calling and begging for food continually. In a flight, the chicks should be removed once they are independent to avoid aggression and induce the parents to produce a second brood. Hand rearing is suggested when parents fail to raise their chicks, and the removal of a clutch will usually stimulate the pair to produce another clutch.

Hybridization is not uncommon among bulbuls, particularly within a genus, and care must be taken to avoid this problem. Even in the wild, hybrids are known to occur; for example, hybrids between the black-eyed bulbul and red-eyed bulbul have been recorded in Africa. Hybrids between either the white-eared bulbul *Pycnonotus leucotis* or the white-cheeked bulbul *P.*

leucogenys and the red-vented bulbul are found where these species overlap in Asia (Vaurie 1958). And, as mentioned earlier, the hybridization resulting from the release of Chinese bulbuls in areas of Taiwan where Styan's bulbul is prevalent poses threat to the continued existence of the latter species. The best ways to prevent unwanted hybridization in a mixed-species aviary are to associate only species of different genera and to only keep birds in sexed pairs, as single birds are more likely to attempt to pair up with a closely related species than are birds with a potential mate of their own kind.

Charles Everitt (1964) gave the following account of an aviary breeding of the red-eyed bulbul. This report refers to the African species mentioned above, rather than the Southeast Asian species *P. brunneus* which goes by the same common name.

> "In April they were seen to be constructing a nest of Spanish moss, grasses and fibers in one of the topmost branches of a crab-apple tree. It was only about 25 cm (9 in.) below the curved wire roof which, at that time, was still covered with visqueen, a clear plastic sheeting. As this most probably had deteriorated during the winter and was liable to let in the rain, a plywood cover was fitted over the nest after the first egg had been laid. The nest itself was an open cup, circular, with a central depression of about 5 cm (2 in.) and both birds had participated in the building of it.
>
> The first egg, pinkish-white, copiously marked with reddish-brown spots and flecks, and measuring about 21 x 15 mm, was laid on April 20, with a second the following day. The incubation, which was shared, lasted for 12–13 days and appeared to commence with the laying of the first egg. On the morning inspection of the nest on May 3, two nestlings were found. One obviously had just hatched as it was smaller than the other and there was still a piece of eggshell in the nest.
>
> The nestlings were devoid of all body covering and had bright pink gapes. Their eyes were open at five days old and quills were showing in their wings. By a further five days they were almost fully feathered, being dark greenish-brown above with darker heads. On May 15, at 12–13 days old, they vacated the nest. The upper plumage was dark greenish-brown, as with the adults, the head, face and neck being darker. The under-parts were a composite of clear yellowish-greyish-white with no fleckings. There was a complete absence of any colored eye-ring. Two weeks after vacating the nest they were feeding themselves, although not averse to still accepting anything offered to them by their parents. Four days later they were caught up and placed on their own. By July 1, yellowish-white eye-rings were discernible, these gradually changing to pink by the end of the month."

The young birds were in full adult plumage by eight months of age. I have read a number of other references on the breeding of other bulbuls in the genus *Pycnonotus,* such as the red-whiskered and red-vented, which are both fairly regularly available to aviculture, and successful husbandry techniques have differed little from those described above.

P.S. Clear (1963) of England reported the successful rearing of red-whiskered bulbuls in a small community aviary measuring 3.2 x 2 x 2.1 m (10.2 x 6.5 x 7 ft.). The other occupants of this aviary included one pair each of white-eyes *Zosterops* sp., Indian robins *Saxicoloides fulicata,* blue-naped mousebirds *Urocolius macrourus,* and white-breasted pygmy rails. The bulbuls were quite tolerant of the other birds, even allowing the mousebirds to preen one of their chicks.

At least two aviculturists have been successful in rearing the nominate subspecies of the Chinese, or light-vented, bulbul *P. sinensis* in England, beginning in 1975. R.E. Oxley of Essex placed a pair of these birds in a well-planted flight measuring approximately 2.75 x 1.5 x 1.8 m (9 x 5 x 6 ft.) high, along with two diuca-finches *Diuca* sp. and a pair of Chinese painted quail *Excalfactoria (Coturnix) chinensis*. After hatching chicks, the bulbuls became aggressive toward the diuca-finches, killing one and chasing the other until it was removed from the aviary. The bulbuls had constructed their nest in a half-open wooden nestbox 23 x 13 x 13 m (9 x 5 x 5 in.) high, fixed to the back of the aviary at a height of about 1.2 m (4 ft.). The nest was made entirely from dried grass. The two chicks were fed on maggots, mealworms, grasshoppers, and other insects, along with small quantities of bread and milk. The chicks left the nest at about 10 days of age. At this time the plumage was gray on the back and head, darker on the wings, with white underparts and a short gray tail. They molted into adult plumage at about eight weeks of age (Oxley 1976). The hen from this pair later died, and was subsequently replaced by another hen that nested in an old song thrush *Turdus philomelos* nest situated about 1 m (3 ft.) from the ground, laying three eggs that were off-white, heavily streaked and blotched with mauve. Two infertile eggs measured 23 x 17 and 24 x 17 mm. The single chick did not survive, but offspring were again reared the following year. The incubation period was 12 days, and the chicks fledged at 11 days of age. Three additional clutches were laid that year, with six young being reared from the four clutches (Oxley 1978).

William Lasky of California was successful in rearing the white-cheeked bulbul. When breeding in a mixed species aviary, this species proved extremely pugnacious toward other bulbuls, requiring their removal to another aviary. A pair of these birds nested in a canary nesting cup, laying three speckled eggs. Other clutches ranged from 1–4 eggs, and the incubation period was 13 days. A single chick hatched in 1961 grew and feathered fast, leaving the nest at 14 days of age. It was fed by the parents for some time after fledging, and was not completely independent until around a month of age (Lasky 1964).]

White-eared Bulbul
Pycnonotus leucotis

Age (days)	Weight (g)	Notes
3	6.4	Began feeding, peeled meal worms and soaked ZuPreem.
4	7.9	
5	10.6	
6	14.4	
7	16.0	
8	19.5	Chick accepted first mealworm.
9	21.0	
10	22.6	
11	22.8	Chick fledged. Perched on side of nest to sleep.
12	23.9	
13	23.1	Chick can fly across room, a distance of 4.6 m (15 ft.).
14	24.5	
15	24.7	
16	25.8	Chick began bathing behavior.
17	27.7	
18	28.8	
19	30.6	Chick observed nibbling on soaked ZuPreem.

Adapted from Huntington (2002). The chick was removed from the nest at 3 days of age:

More recently, the closely related white-eared bulbul was reared by another California aviculturist, Sally Huntington. Her pair nested in a covered flight measuring 1.8 x 1.5 x 2.1 m (6 x 5 x 7 ft.), which they shared with a pair of red-headed finches Amadina erythrocephala and four juvenile Chinese painted quail. The three chicks hatched 15 days after the first egg was laid. One chick was removed for hand-feeding at three days of age. When the remaining two chicks fledged, at about 15 days of age, they were only half the size of the parents. They had gray heads and tails and were in heavy pinfeather. The newly-fledged chicks were very inactive. At this time the adult bulbuls began to show aggression toward the red-headed finches. While never actually attacking or harming them, they would chase the finches away whenever they came near the young bulbuls. The ground-dwelling quail were totally ignored (Huntington 2002).

Mrs. K. Scamell of Surrey, England, provides an interesting account of breeding the black bulbul *Hypsipetes madagascariensis*. She noted that at breeding time, the pair became aggressive toward other bulbuls and small birds, but were not aggressive toward larger species such as superb starlings *Lamprotornis (Spreo) superbus* and long-tailed whydahs *Euplectes progne*. During some winters the bulbuls had heated quarters, but during mild winters they wintered without heat. After several years in a mixed species aviary in which the birds nested but never reared young, Mrs. Scamell decided to give the bulbuls an aviary to themselves to see if they could successfully breed when left undisturbed by other birds. Three bulbuls were placed in an aviary, and on May 18 one female was observed harassing the other, which was subsequently removed, leaving what was believed to be a true pair. Mrs. Scamell (1965) describes the aviary and the breeding results as follows:

> "This aviary is shaped like an "L" with a shelter at one end. At the other end (the base of the "L") is a well-pruned and old-established nut tree around which part of the flight had been built. A half-open nest box was fitted about 1.65 m (5 ft. 6 in.) up in the tree and a wicker nest-basket in the shelter. The hen took to the nest-box and after several false starts she finally built a deep nest inside the box. It was in late June when I realized that the hen was sitting and when she was off feeding I felt inside the nest to find the usual three eggs that are quite large in size having a pinky background covered with brown blotches and speckles. There was also a broken egg on the floor of the flight, which was grassed and contained a couple of shrubs. On July 7 both birds rushed into the shelter when I arrived with the daily feed, but it seemed to me that they ate all the live food and did not carry any to the nest. All the same I gave extra mealworms and gentles. The next morning the same thing happened again, but this time I opened the flight door and felt in the nest. There were three chicks probably two days old. I gave a pot of mealworms and gentles and a little later on I saw the hen take

some to the nest-box and feed. There was nothing reserved about this hen. After years of disappointment she now had a family and did not mind who knew it! One saw her either inside the box with her head and beak pointing upwards brooding the chicks or perched on the edge of the box cramming mealworms down their throats. For the next 10 days the main food taken was mealworms with a few gentles and practically no soft food. I occasionally saw the male bird take a mealworm out to the nest box, probably to feed the hen, but it seemed that all the feeding, at least when we were about, was done by the hen.

On Sunday, July 18, one young bulbul left the nest in the morning and by early afternoon all three youngsters were in the flight. In color they were slaty-grey, whitish underneath with dark beaks and eyes. The legs and feet were a dirty flesh color.

The young birds came on fast and I cut down on mealworms and increased the gentles. After a couple of days I reduced the gentles and forced the hen to feed the regular soft food which contained chopped dates, soaked figs cut into small pieces, soaked sultanas and shredded apple all added to the basic insectivorous food."

This female went on to hatch a second brood on August 18. The family of five had been occupying the nest and the newly hatched young were found dead on the floor with obvious wounds. Mrs. Scamell then removed the three young bulbuls. By early September, at about two months of age, they had grown full tails and were showing small crests. The beaks were still black and the legs flesh-colored, giving no indication of the brilliant coral red bills and legs of the adults.

Straw-headed Bulbul
Pycnonotus zeylanicus

Collared Finchbilled Bulbul
Spizixos semitorques

Among the other genera of bulbuls, Charles Everitt was successful in rearing the joyful greenbul *Chlorocichla latissima* at the Edward Marshall Boehm aviaries in New Jersey in 1963. He described this species as being about 25 cm (10 in.) in total length with the upper plumage golden green, the underparts golden yellow, and the crown feathers dark centered, giving a scaly effect. A pair of these birds constructed an open cup nest in a dense shrub in a 9.15 x 18.3 m (30 x 60 ft.) which also housed pairs of superb starlings, royal starlings *Cosmopsarus regius*, scarlet cocks-of-the-rock *Rupicola peruviana*, red-eyed bulbuls, and plumbeous redstarts *Rhyacornis fuliginosus*. The nest, which contained a single chick, was not discovered until the chick was perhaps a week old, fairly well-feathered on the back and head. The plumage was greenish-brown, the gape bright orange margined in pale yellow. The chick fledged a week later, fully feathered and able to fly. At this time the parents started feeding fruit along with the live food on which it had previously been fed. At six months of age the young greenbul resembled the adult male in color and in song (Everitt 1964).

Hand Rearing: Bulbul chicks are easy to hand rear on a diet comprised of 50% insects, small pieces of pinkie mice and commercial rearing formula that is moistened with water and a few drops of multivitamin. The pinkie mice segments are dipped in the moistened rearing formula before feeding. The remaining 50% of the hand-rearing diet consists of finely diced fruit such as apple, grapes, papaya, and banana. These foods are fed every two hours with forceps. The foods can be mixed at each feed or alternated from feeding to feeding, for example, one feed of insects followed by a feed of fruit. Newly hatched chicks should be kept in a brooder at 35°C (951F). The temperature can be lowered by 1°C (1.8°F) per day until a temperature of 25°C (77°F) is reached. For further information on hand rearing I would recommend that this be read in conjunction with the protocols for hand rearing HONEYEATERS and TANAGERS. Chicks become self-feeding within two weeks of fledging, and will only remain tame through hand feeding insects. Most young bulbuls resemble the adults by the time they are three months old.

When hand feeding a white-eared bulbul chick, Sally Huntington used the same basic diet fed to her adult birds, modified to accommodate the small size and high protein requirement of a nestling. For the first two days, she fed every 15–20 minutes, alternating one large mealworm which she had skinned, sometimes mixed with small chunks of hard-boiled egg, and a single Zupreem Cockatiel pellet soaked in Gatorade, fed whole. As the chick progressed, the feeding interval was reduced to hourly; feeding would continue until begging stopped. After 3–5 days, the chick was taking two soaked pellets and a whole mealworm at each feeding. Fruit was not offered until after the chick had fledged. Food was given only during the day; the chick was not fed for an 8–9 hour period each night (Huntington 2002). The following chart shows the daily weight gain of this chick.

Ring Size: 3 mm

Species requiring similar management:

Minivets *Pericrocotus* spp.
Pekin Robin
Starlings
White-eyes *Zosterops* spp.
Tanagers

Contributor: David Bender, Hancock Wildlife Research Center, British Columbia, Canada.

References:

Clear, P.S. 1963. Red-whiskered bulbuls bred. Foreign Birds 29 (2): 83.

Everitt, C. 1964. Breeding the joyful greenbul at the Edward Marshall Boehm Aviaries. *Foreign Birds* 30 (2): 56

Everitt, C. 1964. Breeding the red-eyed bulbul. *Avicultural Magazine* 70 (6): 214–216.

Huntington, S. 2002. Breeding the white-eared bulbul. *AFA Watchbird* 29 (1): 16–20.

Lasky, W.R. 1964. Notes on the breeding behaviour of a pair of white-cheeked bulbuls. *Foreign Birds* 30 (2): 57–60.

Lindholm, J.H., III. 1996. Reproduction of Asian passerines in American zoos. A.F.A. Annual Conference Proceedings 1996: 86–121.

Oxley, R.E. 1976. Breeding the Chinese (light-vented) bulbul. *Avicultural Magazine* 82 (3): 124–126.

Oxley, R.E. 1978. Further notes on the breeding of the Chinese (light-vented) bulbul. *Avicultural Magazine* 84 (2): 77–79.

Scamell, K.M. 1965. The breeding of the black bulbul. *Avicultural Magazine* 71 (6): 165–168.

Severinghaus, L.L. 1998. Religious bird releases and the extinction of Styan's bulbul *Pycnonotus taivanus*. Proceedings 22nd International Ornithological Congress, Durban. *Ostrich* 69 (3/4): 450.

Vaurie, C. 1958. Systematic notes on Palearctic Birds. No. 32 Oriolidae, Dicruridae, Bombycillidae, Pycnonotidae, Nectariniidae, and Zosteropidae. American Museum Novitates 1869.

21.15 Zosteropidae White-eyes

Glen Holland

Aviculturally, the term white-eyes is generally used to refer to any of the 72 species in the genus *Zosterops* of the family Zosteropidae in the order Passeriformes. Ornithologically, the Zosteropidae includes an additional 17 species of white-eyes in nine additional genera, mostly local island forms, as well as the mountain blackeye *Chlorocharis emiliae* from the mountains of northern Borneo, the cinnamon ibon *Hypocryptadius cinnamomeus* from the southern Philippines, and four species in the genus Speirops from Africa. The species *Zosterops lateralis,* with several subspecies in Australia, New Zealand, and southwestern Oceania, is known as the silver-eye. Other species within the genus *Zosterops* are distributed through Africa, Madagascar, the islands of the Indian Ocean, southern, eastern and southeastern Asia, Japan, the Philippines, Australia, New Guinea, Borneo, Java, Bali, Sulawesi, and many of the smaller islands of the Australasian and Pacific region as far as New Zealand. They utilize a wide variety of habitat types from mountain and coastal scrub to riverine bush, forest, and thorn scrub in dry areas. White-eyes are a common garden bird regularly visiting feed tables and water in the garden. Groups are found in winter but, as a rule, they occur in pairs during the summer breeding months. They are extremely agile, flitting among vegetation in search of the fruit, insects and nectar on which they feed, even hovering to reach the latter. The contact call of most white-eye species is a high-pitched, single-note call, and the males sing beautifully, particularly at dawn. This endearing song of the male birds is a key feature in the dawn chorus of forests throughout their geographical range; females do not sing. In some species, careful observation can reveal sexual dimorphism with males slightly brighter, particularly in those species with yellow in the plumage.

Management: My avicultural experience is primarily with the Cape white-eye *Zosterops pallidus* but management varies little between the species. These active little birds make very interesting avicultural subjects, adding life to any aviary. They are not difficult to maintain in good health. White-eyes can be mixed with small seed-eating birds without any problems. At Currumbin Sanctuary in Australia, gray-backed white-eyes *Zosterops lateralis* have been successfully mixed with species such as purple-crowned lorikeets, pigeons, dotterels, sunbirds, and pittas. This family has also been recorded to over-preen other species, resulting in bald patches. They can, however, become aggressive among themselves if they are kept as a group. In a large, planted aviary, however, about six of these birds provide an interesting display outside the breeding season, when they move about, feed, and bathe as a group. Plants that attract insects or provide nectar will assist to keep the birds in good condition and color. They are not particular about habitats and will utilize both forests and woodlands. Nectar-producing plants, such as those described for SUNBIRDS, will provide natural food in the form of nectar and insects that are attracted to the flowers. Either Habitat A or B as described in the LANDSCAPE chapter will be accepted. Ideally, however, a combination of the two will provide a better habitat for white-eyes than either type separately. In the natural state, over 60% of the insects in the diet are gleaned from the leaves of shrubs and trees. A shallow pool of water must be supplied for bathing, as these birds will bathe daily even in cold weather. A mist sprayer placed over a shrub is also enjoyed on a warm day, and the birds will bathe in the wet leaves. Although relatively hardy, in severe climates, where temperatures remain below 10°C (50°F) for extended periods, white-eyes should be housed indoors and extra warmth supplied during periods of low temperatures. In New Zealand, where gray-backed white-eyes, locally known as silver-eyes, are a self-introduced species, I have noticed groups of these little birds looking rather ill in very cold weather. I am sure that under such conditions there is a high natural mortality rate. I have also watched a pair of wild white-eyes go to roost. This involved the pair suddenly flitting onto a branch above where I was sitting, very quickly shuffling tightly up to each other and settling for the night. The pair bond appears very strong, and if a bird is lost or temporarily removed from an aviary, the remaining bird calls frantically for its mate.

Diet: White-eyes are inquisitive and will soon begin to sample any new foods offered to them. If the diet is not sufficiently varied, the bright colors on these birds, particularly the yellow, fade to a grayish color. The captive diet should include segments of apple, orange, pear, grapes, papaya, guavas, figs, mulberries, fruiting shrubs such as *Coprosma,* and a commercial softbill food moistened with water. Oranges in particular are a favorite food of white-eyes. Mealworms, termites, flying ants, fruit flies, bee larvae, and aphids are eagerly taken. Nectar such as described for SUNBIRDS is very popular, but to avoid having to mix nectar for these birds alone, a small amount of honey can be added to the moistened softbill food. White-eyes are capable of short bouts of hovering to gain access to nectar. It is not necessary to dice or grate fruit; the birds feed from segments that have been impaled on a nail on a branch. Flowers are pierced to gain access to nectar if it cannot be reached from the open end.

Breeding: A well-planted aviary is necessary if breeding is desired. Pairs are very affectionate, spending much time preening, sitting, and huddling tightly together. Little reproductive success results from a colony of birds, but by color banding the birds and removing bonded pairs to breeding aviaries, such pairs will often prove to be multiple clutched. Aggression amongst males can be severe and disruptive with continual chasing and bill snapping. This aggression, however, is limited to the breeding season, and the birds can safely be housed as a colony in the nonbreeding season. The nest is a small hammock-type structure slung in a fork in a dense shrub and fastened together with cobweb. The entire structure is rather flimsy and consists largely of cobweb and lichens, with other soft materials such as grass seed heads, wool, and hair. As described for pekin robins, they will usually accept and incorporate a sparse man-made support structure if provided early in the nest-building exercise. Both sexes build the nest, which takes about 4–7 days to complete. Incubation of the 2–3 light turquoise eggs is shared by both sexes for the 12-day incubation period. Both sexes feed the chicks, which fledge at 10–13 days. Newly fledged chicks cannot fly well and remain close to the nest for a few days. White-eyes are multiple brooded, with a new nest being built for each clutch. The second nest may be initiated by the male about a week after the chicks fledge. During this phase of nest construction, the female is primarily responsible for the feeding of the young until independent. Youngsters are best removed at three weeks to avoid excessive competition for live food, without which the parents will fail to rear their new broods. Birds of up to five years of age have been recorded breeding successfully.

R. A. Richardson gives a rather delightful report of a pair of Indian white-eye *Z. palpebrosa* that had been allowed liberty in Norfolk, England:

> "After a few unsuccessful breeding attempts in the aviary I decided to risk liberating this long-established and determined pair in the hope they would remain anchored about the garden by a spare single captive male's calls and their tubes of nectar. A small hole was beside one of the feeders and the pair soon flitted out and up an apple tree.

I had hardly finished re-closing the aperture before they returned to drink at the feeder that had been placed on the outside for their convenience. Thereafter their visits were at about half-hourly intervals and they roosted out of sight at night.

Two days later the hen was seen gathering spiders webs from the eaves of the bird house and to my delight I followed her to a half built cocoon-like nest 4 m (15 ft.) up in a neighbor's pear tree. Through binoculars I could see the bird at work through the gauzy, semi-transparent sides of the structure, which was a deep cup of billiard ball size slung oriole-like in the angle of a laterally forked twig. Round and round she shuffled, twining and threading, pressing with chin and breast and scrabbling with her claws, all the time shaping and consolidating the fairy-like cradle of cobwebs and grey synthetic material stolen from some female garment on a nearby washing line. The male took an equal share in building and when the inside walls of the "cocoon" were reinforced with short pieces of horizontally woven dry grass, it was further lined with downy feathers from the chicken run. The only display noticed was when the nest was complete the female suddenly left and her mate that had been singing his breezy little song in a nearby tree, dashed at her and chased her helter-skelter in and out of the garden for a few seconds before leaving her to preen and sunbathe. Incubation is shared, with the birds relieving each other at frequent intervals. The routine was identical on each occasion; the off-duty bird returning to the nest, which was vacated by the incubating bird as soon as its mate arrived. So rapid was the change over that the eggs were never left uncovered for longer than about two seconds at a time.

Shortly after the suspected hatch date the male disappeared and the female was clearly distressed. I decided to release the second, captive male in the hope that he might give the female some comfort by his near presence. Imagine my delight when this devoted little bird not only accompanied the female back to the nest but unhesitatingly started to alternately brood and feed his rival's offspring. Nest sanitation was meticulous and the new pair went on to fledge a single youngster."

I have read other reports that describe white-eyes as "not good homers with two escaped birds disappearing immediately." While this report gives an interesting insight into the breeding of white-eyes, the release of these birds would now be illegal in many countries.

Hand Rearing: White-eye chicks are extremely small, even when ready to fledge, and are rather delicate. When hand-rearing, the most important step is to ensure the chicks are kept warm, as they stop feeding and do not survive long if they become chilled. Secondly, the hand rearing diet must consist of 50% small insects such as diced mealworms, waxmoth larvae, gentles, flying ants, and tiny moths, which are fed with a forceps. The remainder of the diet should consist of tiny pieces of very young pinkie mouse, boiled egg, mashed fruit, such as apple, paw-paw, banana, and a 50/50 mix of commercial hand-rearing formula that should have a protein content of at least 24%, and pureed fruit baby food moistened with water and fed with a 1-ml syringe. A few drops of multi-vitamin and a calcium supplement are added to this soft food mix. Solid food such as and fruit and some of the insects can be dipped into the rearing formula/baby food mixture. Chicks should be fed every 1–1½ hours from 6:00 a.m. to 8:00 p.m., and they will typically take 3–4 food items per feeding. They beg with their wings and the head quivering continually. Hand-reared birds placed with others in an aviary do not remain tame.

This section should be read in conjunction with the section on hand rearing HONEYEATERS.

Ring Size: 2.5 mm

Species Requiring Similar Management:

- Sunbirds
- Spiderhunters
- Honeycreepers
- Chloropsis

Reference

Richardson, R.A. (1962) Liberty Breeding of the Indian White-eye (Zosterops palpebrosa). *Avicultural Magazine* 1962 Volume 68 page 51–54.

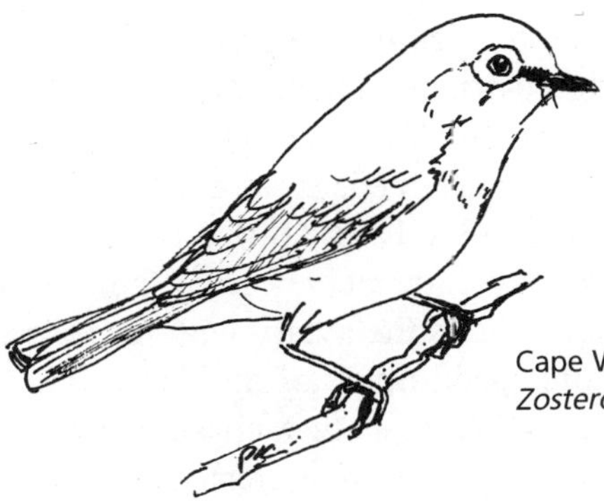

Cape White-eye
Zosterops pallidus

21.16 Sylviidae Laughingthrushes & Babblers

21.16a Garrulacinae Laughingthrushes

Michael Mace and Glen Holland

Introduction: There are more than 50 species of laughingthrushes in the genus *Garrulax* and another three species, often called liocichlas, in the genus *Liocichla*. These two genera comprise the subfamily Garrulacinae in the large family Sylviidae in the order Passeriformes. Their nearest relatives are the babblers and some of the Old World warblers. In general, laughingthrushes are large, rather boisterous birds, often loudly vocal. They are heavy-bodied birds with large heads and strong beaks, and will not hesitate to kill small mice or baby birds. They are seldom found far from dense cover and spend much of their time searching through the leaf litter on the forest floor. The various species are distributed through Borneo, Sumatra, Java, Thailand, Indochina, Vietnam, China, Myanmar, Assam and westward through the Himalayas and southward into India.

Management: Laughingthrushes tend to be rather domineering in the aviary, with some species, such as the white-crested *Garrulax leucolophus,* even known to catch and kill small birds. In one instance at Busch Bird Park, six new white-crested were introduced into a large walk-through aviary measuring 3000 m^2 (32,200 sq. ft.) and 23 m (75 ft.) high, where three established birds resided. "Within three days the six new birds were found dead, all apparently killed by the three senior residents which had begun nesting" (Young 1977). Tickell's laughingthrushes *G. strepitans* have been housed with Palawan peacock pheasants *Polyplectron emphanum* and gray-headed parrotbills *Paradoxornis gularis* without any problems. Andy Maycen of southern California has reared pekin robins *Leiothrix lutea* in flights shared with red-winged laughingthrushes *G. formosus*. Up to six different *Garrulax* species have been housed together, and many bred, in an extremely large aviary measuring 43 x 33 x 12 m (140 x 108 x 40 ft.). Laughingthrushes are extremely active and agile, bouncing and flying around the aviary with great speed when required. Pairs spend much of the day calling to each other and will also play together, chasing and mock fighting with each other. The white-crested has a cackling laugh and a variety of whistles. An aviary for a pair should measure at least 6 x 3 x 2.5 m (20 x 10 x 8 ft.) high, and should be well planted as described for Habitat A. The larger and better planted the aviary; the more likely the birds are to breed.

Diet: The maintenance diet should consist of mealworms, crickets, newborn mice, moistened dog chow or cat bisciuts, soaked softboll pellets, assorted diced fruits, including papaya, raisins, apple, figs, banana, and a multivitamin that is provided for the adults throughout the year. When the birds have young, additional live insect food is added to the diet.

Breeding: Laughingthrushes like to build high off the ground, very often choosing the highest possible nest site in the aviary. Generally nest sites are 1–3 m (3–10 ft.) above ground. Although nest baskets have been used, most pairs like to build their own nest. Tickell's have used woven wicker baskets with a circumference of 12.7 cm (5 in.) and a depth of 5 cm (2 in). Nesting material in the form of twigs, palm fibers, and pine needles were scattered on the ground. Nest building is always preceded by duet vocalizations. Both birds will readily carry twigs, which are used for the primary structure with finer material such as leaves; pine needles, and palm fibers used to line the interior. A nest is completed over a period of 7–10 days; a similar amount of time is taken whether a pair uses a nest basket or is building the complete structure. If a nest is destroyed, construction of a replacement can begin within 24 hours. Once the nest is complete, the thrushes will defend it vigorously from intruders. A pair of white-crested at the Hancock Wildlife Research Center hatched and reared two chicks in a nest built of grass and twigs in a basket hung 1.5 m (5 ft.) above ground in a service alley measuring approximately 90 x 3.6 x 3.6 m (300 x 12 x 12 ft.), which they shared with two dozen young turacos, and through which numerous visitors passed during the nesting period.

Observations made on parent-reared Tickell's

chicks revealed that high-protein foods such as mealworms, crickets, and newborn mice were fed to the young exclusively. The adults fed the chicks for about 40 days before they begin to forage for themselves. Although placing non-related pairs together can present problems with aggression, in one case in 1990 the parent-raised chicks were left in the same enclosure as the parents. In 1991, when the adults again hatched and reared chicks, the year-old juveniles were observed feeding insects to their newly fledged siblings, indicating that Tickell's laughingthrushes may employ helpers in rearing young. Another interesting note is that both white-throated laughingthrushes *G. albogularis* and yellow-throated laughingthrushes *G. galbanus* have also been observed feeding insects to the Tickell's fledglings, suggesting that these species too may use helpers at the nest.

Parent and Hand Raised Tickell's Laughingthrush *Turdus unicolor*

Day	Parent-raised Tickell's	Hand-raised Tickell's
1	6.4 g	6.2 g
2		
3	26.6 g	11 g
4		
6	34.5 g	26.4 g
7		
9	43.9 g	32.4 g
12	50 g	39.7 g
15	56.3 g (fledged)	46.3 g
18		53.2 g
21		58.3 g
24		61.4 g
27		71.9 g
30		78 g
33		87 g

Young (1977) also reported a spare adult white-crested assisting a pair to feed their chicks and added that, "We have since found in the four years of raising this species that its social structure is such that, regardless of how many adults in a given exhibit, only a single nest is produced and only one female lays eggs. While only the breeding pair will incubate the eggs, all adults in the group help protect the nest site and feed the young." Chicks have also been recorded assisting with nest construction. All four chicks in the first nesting were lost because they left the nest before they could fly and were either eaten by larger birds or drowned in the stream. A second nesting of that year produced one chick from three eggs, which again left the nest too young and perished. To avoid a recurrence of this circumstance, a cage was constructed of 25 x 25 mm (1 x 1 in.) weldmesh, measuring 76 x 76 x 76 cm (30 x 30 x 30 in.). Three fledglings were placed in the cage, within the aviary, and the adults immediately began feeding the fledglings through the mesh. After 12 days the chicks were released, but returned each evening, accompanied by the adults, to roost on top of the cage. An interesting observation was of a breeding male that escaped: while his female in the aviary constructed a nest and incubated three infertile clutches of three eggs each inside the aviary, he also constructed a nest and would sit on this nest and perform all the motions of parental care. Eventually the male was caught and reintroduced to the group, giving rise to noisy chorusing and aerial display, which continued sporadically throughout the day.

The table below summarizes the results from Busch Bird Park, in one large walk through aviary over five seasons:

	Nests	Eggs laid	Infertile/ lost	Hatched	Survived 7+ days	Raised
Total	27	81	35	46	37	32

Of the 32 birds fledged, three were parent raised and 29 hand raised.

Artificial Incubation and Hand Rearing:

The eggs of Tickell's are unpatterned and turquoise in color, those of the black-throated *G. chinensis* are porcelain white. One to four eggs, laid every other day, constitute a clutch. Incubation periods recorded, both parental and artificial, ranged from 14–16 days. Incubation begins with the first egg laid, both parents sharing in the incubation, and the chicks hatch asynchronously.

Artificial incubation of laughingthrush eggs has been successfully achieved in an incubator with the temperature maintained at 38°C (100.4°F) and the eggs mechanically turned every two hours around the clock. The period from pip to hatch was less than 24 hours. At pip the eggs were transferred to a hatcher set at 36.9°C (98.4°F). The chicks were left in the hatcher until their down had dried completely, then moved to a brooder set at 35°C (95°F).

The brooder temperature for laughingthrush chicks is maintained at 35°C (95°F) for the first few days, and then gradually decreased to ambient room temperature of about 25°C (77°F), by which time the chick is well feathered. Chicks are placed in a bowl lined with paper towel, with small twigs added to aid in the prevention of splayed legs. They are started on a diet of mealworms, minced crickets with head and limbs removed, and papaya supplemented with vitamins. On day 8, newborn mice are added to the diet. The chicks readily gape to be fed, vocalizing at the same time. As they develop, a rapid wing flutter is included in the begging response. Feeding intervals are nine times daily from 6:30 a.m. to 5:45 p.m.. The schedule is adjusted according to the average daily weight gain; after the first few days, and until fledging, the guideline is an expected rate of increase of about 10% of the chicks body weight per day. The frequency of feeding remains the same, but the amounts offered are increased on a ratio of seven mealworms to three cricket torsos. By day 7, a chick receives 70 mealworms, 30 crickets, and three newborn mice per day. Additional food items, such as papaya and apple, are offered and accepted from day 24. The chicks beg constantly and are rarely fed to satiation. They grow rapidly and are fledged by day 14, although flight is clumsy and they are still dependent.

Species Requiring Similar Management:

Laughingthrushes *Garrulax* spp.
Liocichlas *Liocichla* spp.
Silver-eared mesia *Leiothrix argentauris*
Shrike-babblers *Pteruthius* spp.
Sibias *Heterophasia* spp.
Fulvettas *Alcippe* spp.

References:

Mace, M.E. 1991. Breeding and hand-rearing Tickell's laughing thrush at the San Diego Wild Animal Park. *International Zoo Yearbook* 30: 187–191.

Mace, M.E. 1994. Notes on breeding the black-throated laughing thrush. *Avicultural Magazine* 100 (2): 98–100.

Young, R.A. 1977. Co-operative breeding in the white-crested laughing thrush with notes on its hand-rearing at Busch Bird Park, Houston.

21.16bi Babblers *Timaliini*

• Glen Holland

Introduction: The classification of the babblers has been changed a number of times over the years. While still falling within the order Passeriformes, they are now placed in the family Sylviidae, which includes the leaf-warblers and grass-warblers, as well as the laughing thrushes, with which they have long been associated. The true babblers now constitute the tribe Timaliini in the subfamily Sylviinae. This large tribe encompasses 236 species in 51 genera. Depending upon genus, these species are variously known as babblers, illadopsis, hill-babblers, scimitar-babblers, wren-babblers, striped-babblers, tit-babblers, chatterers, pied-babblers, babaxes, shrike-babblers, barwings, fulvettas, mountain-babblers, sibias, crocias, cutias, and others. This tribe also includes the aviculturally important yuhinas, minlas, and parrotbills, as well as the popular red-billed leiothrix, or pekin robin *Leiothrix lutea,* and the even more attractive silver-eared mesia *L. argentauris*. The babblers are birds of the Old World, with species widespread over Africa, Asia, the Indian subcontinent, and the eastern islands.

Natural Habits: The large genus *Turdoides* includes species native to southern Europe, the near east, Arabia, the Himalayan region, central and peninsular India, Burma, China, southern and southeastern Asia, and much of Africa. In this section I will be describing primarily the husbandry of the arrow-marked babbler *Turdoides jardineii,* an African species with a wide range. They are found in the drier thorn bush areas and woodlands, along riverbanks, and around homesteads on farms. Babblers are communal birds and usually occur in groups numbering 4–8 individuals. Turdoides babblers are called the "Seven Sisters" in India, and are also

Three Species of Laughingthrush

Day	Tickell's	Red-tailed Weight (g)	Yellow-bellied	Day	Tickell's	Red-tailed Weight (g)	Yellow-bellied
1	4	5.8	8.5	22	33	36	58
2	4.2	6.5	10	24	39	39.5	62
4	6.2	8.5	13	26	42.5	43	66
6	8.8	12	18.5	28	46	45	72
8	11.4	15	27	30	47	44.5	79
10	19	21	32	32	44	46	80
12	21.5	19	38	34	43.5	50.5	86
14	25.5	20.5	42	36	43	50	
16	28.5	28	47	38	42.5	54	
18	26.5	28.5	52	40	55		
20	27.5	31	55				

given the Hindi name of *Sat Bhai,* "Seven Brothers." They retreat to dense thickets when threatened, but will otherwise venture into open areas where they feed largely on the ground. This family is characterized by their raucous calls, started by one bird but soon echoed by the entire group. Most species have loud, grating *churr-churr-churr* calls. The sexes are similar.

Babblers are "contact" birds and will actively seek close contact with conspecifics. In babblers this tendency to clump is so strong that, in captivity, birds of the same sex freely associate in this way. In the absence of birds of the same species, individuals will attempt to clump with other babblers or even with unrelated species, passerine or non-passerine. Mutual preening is another social habit indulged in by babblers. The chin and throat receive special attention, the bird being preened remaining stock still, often with the bill elevated vertically. Frequently, one bird will preen the other hurriedly for a few seconds, as if to encourage it, and then hold its own head in position, inviting attention in return.

Anting, which involves the application of the defense fluids of ants (chiefly the formic acid of worker ants) to the bird's plumage, is also commonly practiced by these birds. Babblers of the genera *Leiothrix, Minla, Heterophasia,* and *Garrulax* often use the foot in feeding. The food may either be clamped against a perch with the front three toes while the hind toe grips the perch or, usually on the ground, the bird may extend the tarsus and grasp the food with all four toes while the other leg takes its weight. Vent pecking is another common behavior among babblers.

Management: Babblers are among my favorite softbills; with their interesting calls, behavior, and social structure, they make entertaining avicultural subjects. Most of my avicultural experience has been with arrow-marked babblers and pied babblers *Turdoides bicolor*. They are relatively hardy, easy to maintain, and not too difficult to feed. They do not like cold, wet climates for extended periods; although hardy to cold overnight temperatures, their days must be warm and dry. In areas where cold, wet, icy conditions occur, they should be housed indoors during the severe months. Les Gibson (1979), then living in Canada, describes wintering arrow-marked babblers in a small outdoor aviary with a bob-hole leading into a heated room into which they were locked each night, and then transferring them to a large planted aviary for the summer months. Raymond Franklin (1973) recommends housing black-headed sibias *Heterophasia capistrata* indoors during the winter in England.

Babblers are strongly territorial and a group of wild babblers regularly visited one of my large walk-through aviaries to fight with the occupants through the wire. The noise that accompanied these territorial disputes was deafening. While calling, the wings are held a little off the body, drooping slightly, and the tail is fanned. Clearly two groups of the same or similar species cannot be mixed in small or moderately sized aviaries. Arabian babblers *Turdoides squamiceps* have been noted in the wild, ganging up on creatures as diverse as jackals and tortoises, hounding them out of their territory. This species also exhibits a nightly pre-roosting dance in which a small group of birds takes turns dancing or parading to the admiring babbles of the audience.

Gibson describes testing his arrow-marked babblers with a mouse that was chased with great excitement around the aviary for fifteen minutes. Although the birds jumped at the mouse bravely and encouraged one another with a barrage of cackles, as far as could be seen, they did not actually touch the mouse. This species also performs a pre-roosting ceremony, which probably also functions as a proclamation of their territory. Gibson (1979) describes this display in an aviary setting:

> "Every evening at dusk the cock would fly to the far (non-roosting) end of the aviary, and sit on an exposed branch. Then he stretched up on his legs and extended his head to the full length of his neck. Clapping his wings hard to his sides he would give a full-throated babble. This was repeated a few times then the others joined him. After several choruses from all, they retired for the night. They sleep huddled together."

Gibson also mentions that his arrow-marked babblers, when housed with white-crested laughing thrushes *Garrulax leucolophus* were the only species he ever witnessed displace the laughing thrushes at the feed table. Jungle babblers *Turdoides striatus somervillei* in the natural state in India, have been observed periodically having free-for-all mock battles that start for no apparent reason.

If a new group is to be formed, all the birds should be placed at the same time in a temporary holding cage, rather than an aviary. In this way aggression can often be avoided, as the birds are under stress in an unfamiliar environment. Even under these conditions, however, they must be carefully monitored for signs of aggression, and any troublemakers immediately removed. They should be left together in the cage for a week before being transferred to a flight aviary, where again they should be carefully observed. I can recall one instance in which a well-established group of arrow-marked babblers turned on a member of their own group and killed it. Species that are markedly different, such as arrow-marked and pied babblers, can be mixed in a large aviary. Although I have never noticed aggression toward unrelated species, they will steal eggs and newly hatched chicks from the nests of smaller species.

Species in which the sexes appear similar should be DNA sexed or surgically sexed, but Gibson (1979) does note a visual method of sexing his arrow-marked babblers.

> "The cock has whiter and more extensive arrows. These are small downward pointing arrowheads on the tips of the breast feathers. On the hen, these do not go as far down the breast as on the cock, and she has scarcely any on the neck other than on the chin. The male has lighter markings going around the head and neck, giving him a more dapper silver-grey appearance than the browner hen. Although this difference is obvious, it can only be seen at close quarters. The remainder of the plumage of both birds is a greyish-brown."

Babblers require a large planted aviary, such as described under Habitat B in the LANDSCAPE chapter, with a patch of dense thicket or creeper such as jasmine, in which they can nest. They also need some patches of open ground, particularly with a layer of leaf litter, where they can scratch about, as a flock will spend much of its time searching for insects on the ground. While foraging in this manner, using soft calls, they are in constant contact with each other. If danger threatens, the alarm call is given and the group will retreat to a dense thicket.

Les Gibson also published the results of a four-year study of the red-billed leiothrix *Leiothrix lutea* or pekin robin, as it is commonly known in aviculture, which includes observations of this species made in the wild and in captivity. On one trip he heard 19 birds calling, but only caught a glimpse of two. The birds kept well hidden in thickets of bamboo and other dense undergrowth and their inconspicuous coloration made them invisible. "They are happiest in a well-planted aviary and will show this by nesting readily in the foliage." (Gibson 1978)

White-crested Laughingthrush *Garrulax leucolophus*

Black-throated Laughingthrush *Garrulax chinensis*

Like other babblers, pekin robins are very active and therefore not suited to life in a cage. They are not quarrelsome feeders and will not try to exclude or replace other species at the feeding station. Although they will not tolerate other Leiothrix in the same aviary, they ignore other species, even when breeding. A number of the same sex can be kept together, but introducing new birds to an established group results in persecution. Even attempts to switch mates will provoke attacks from resident birds of either sex. Established pairs regularly preen each other around the neck. The pair bond is close, and the birds stay together throughout the year. They bathe daily, and in warm weather will bathe several times each day.

The most reliable method of sexing pekin robins is to imitate the call of the female, or to play back a recorded call. This is a simple, two-syllable whistle repeated 3–5 times. Females will respond immediately with the same call, but males reply with a short burst of song, amounting to about half of the normal song. Birds can also be separated, and will start calling to each other in the characteristic manner.

Diet: Babblers are largely insectivorous and are particularly fond of insects and caterpillars, but will eat a small amount of fruit as well. In captivity they quickly accept substitute foods. Soaked dog cubes, fresh bonemeal, boiled egg, commercial softbill mixtures, diced fruit, and live food such as crickets and mealworms will suffice in captivity. Pekin robins are perhaps the easiest babblers to feed, and wild-caught birds are trained onto a seed diet pre-export. Two points of concern are the number of birds that must die during this process, and the number that initially survive but are expected to live the remainder of their lives on this inadequate diet. All too often these little babblers are kept with finch collections and not offered a sufficiently diversified diet.

Soaked cat kibble, egg food mixed with a commercial softbill diet, currants soaked overnight in water, diced grapes, slices of fruit pinned to perches, spiders, mealworms, and caterpillars are all taken. They do not like hard beetles, and only eat ants after they have vigorously rubbed them on their feathers four or five times before consuming them, presumably to remove some of the formic acid. Some species also regularly eat the feces from other species. Millet seed and grit should also be available. Sibias *Heterophasia* spp. favor grapes and nectar. They also hawk insects on the wing and will take wasps, often holding them down with one foot as the remove the sting, but sometimes swallowing them whole.

Breeding: Arrow-marked babblers build a cup-shaped nest in a dense thicket. Often it is placed close to the ground. The nest is constructed of fine twigs and grass; the interior is lined with a few rootlets and thin fibers. Clutches vary with 2–5 eggs, the average being three; the eggs are greenish-blue in color. Since it is normal for only one or two chicks from a clutch to survive, hand rearing should be considered for part, if not all, of the first clutch, particularly if only a single pair of birds is kept. These babblers are cooperative breeders, with only the dominant pair nesting, while the rest of the group assist in raising the chicks. By removing broods for hand rearing, up to three clutches will be produced in a season. The nest conditions are sanitary, with the nestlings' feces being eaten by the adults.

Gibson (1979) found that the time elapsing between nest construction and egg laying was longer than for any other species he had kept, and noted that this leisurely pace had also been observed among wild birds. The first nest was started on April 19 and completed by April 25, with a few additional strands added during the next few days. It was built in a cardboard box measuring 23 x 18 x 15 cm (9 x 7.5 x 6 in.) placed at a height of 1.5 m (5 ft.) in a laurel bush in the aviary shelter, which was constructed of clear acrylic sheeting. The nesting material was rather coarse, as was the construction, and long dry grasses, with a few twigs incorporated, formed the bulk of the structure. Later the box was filled with grass and few small leaves.

Laying did not commence until May 6, with three eggs being laid on consecutive days. As incubation progressed, the hen began to fill the nest box with grass. This continued throughout the incubation and nestling period, during which the nest cup was surrounded with grass. The box was completely filled by the time the chick fledged. Two additional nests were constructed. One was built on top of the old box, and the other in a wisteria vine. The male displayed and called loudly throughout the construction period. The male also performed at least 80% of the daytime incubation duty, with the female relieving him for 5–15 minute spells. It was later established that he was also in the nest at night. In contrast to their behavior during the building period, both birds became extremely quiet and relatively inconspicuous as soon as the eggs were laid.

The chicks were pink in color, and completely devoid of down. They showed no particular gape color. Like all softbills, they began begging almost immediately and were being fed within an hour or two of hatching. As with the incubation, the male did most of the brooding, and for the first day most of the feeding, after which both parents shared feeding duties. One chick died on the second day; it weighed 3.25 g (0.11 oz.). The remaining chicks were raised basically on mealworms and gentles, assisted by a hand-feeding regime using an egg-based soft food and minced meat (ox heart is recommended). The supplementary feeding was undertaken in the absence of the natural nest helpers that are so important to the successful rearing of such a communal species, and was done three times per day. The adults stopped eating their usual diet and fed solely on live food. Here again it would be advisable to add small crickets, grasshoppers, and waxmoths to the diet, all of which can be dusted with calcium powder.

Regardless of which parent was brooding, the female always made a noisy, close defense of the nest whenever the chicks were being examined or hand fed. Both parents cleaned the nest, usually consuming the fecal sacs. At seven days the remaining healthy chick weighed 18 g (0.63 oz.); it fledged at 14 days, weighing 46 g (1.62 oz.). The adults were very attentive, calling continually to the chick, which replied if undisturbed. When the chick fledged, the adults resumed their territorial cackling, which had halted with the onset of incubation. At 20 days of age the chick could fly fairly well, but remained hidden for almost another week, when it began sitting in the open. At 27 days it was seen eating from the feed table, but it was still be fed occasionally by the parents at 45 days of age. When the chick was 54 days old a fourth nest, containing two eggs, was discovered. Although these eggs disappeared a few days later, the number of nests does give an indication of the captive breeding potential of this species if properly manipulated.

A British account of the captive breeding of the rufous-bellied babbler *Dumetia h. hyperythra,* was provided by R.S. de Q. Quincey. (1961):

> "After being wintered in an indoor aviary, his birds, which he believed to be a pair although they seemed identical in appearance, were transferred to a very densely planted outdoor aviary which they shared with a pair of white-eyes *Zosterops* sp., a pair of red-flanked bush robins *Tarsiger cyanurus,* a pair of small tanagers, and a single hummingbird. Their first nesting attempt came early in the summer, when they constructed a dome-shaped nest and laid two eggs that subsequently hatched, although the chicks died at about five days of age and were thrown out of the nest. The adults appeared quiet for a while, but

during this time they had built three additional nests, all dome-shaped and wren-like. One was about 10 cm (4 in.) above ground in a grass tussock, and the other two were 0.5–1 m (2–3 ft.) above ground in a small planting of honeysuckle *Lonicera nitida*. The hen laid in one of these nests, but only a single chick was reared, leaving the nest early in July. Live ant eggs and wasp grubs were supplied, and apparently used as a rearing food.

Three weeks after leaving the nest the young bird was indistinguishable from its parents. The hen built another nest and laid a clutch of three eggs, which are described as being approximately the size of a wren's eggs, blotched and spotted more at the thick end with red-brown markings. The ground color is greyish. The incubation period was estimated to be 13 days. Two chicks hatched, but soon disappeared. The nestling period on the lone chick raised from the previous clutch was estimated to be 18 days."

In his previously cited study, Gibson (1978) provided details on the nesting of his red-billed leiothrix (pekin robins). Aviaries for breeding pairs should measure approximately 2 x 3 x 2.4 m high (6 x 10 x 8 ft.). This species should not be housed in adjacent aviaries, as they will spend much of their time trying to fight through the mesh. Same-sex pair bonds can be quite strong and all birds are best DNA sexed to ensure true pairs. If pairs are split, the birds are best well separated and left alone for 10 days before new mates are gradually introduced. The base diet for this species should be softbill pellets and moist cat biscuits. Although all of his young hens built nests the first year, not all of them laid eggs. One yearling hen laid but did not incubate.

The pattern of the nests was constant; all were loose, deep cups suspended from horizontal twigs about 12.5 cm (5 in.) long. Dry grass was invariably used as the main construction material, although an occasional long dried leaf, such as a bamboo leaf, was incorporated. The birds also used fine grasses, horsehair, and strands of sisal twine. Feathers, paper, etc. were consistently offered but never used. Although the nest is fairly thick, it is often so loosely woven that eggs work their way down to the bottom. To remedy this problem, a basket was pushed up around each nest and wired onto the twigs as soon as the nest was finished. Peter Karsten also found that once the birds begin to build their rather flimsy nest, a small hammock-type structure tied into the supporting vegetation as a brace will be accepted and incorporated into their nest. This hammock can be built of thin wire wrapped with raffia or similar strands of coarse string. (See Karsten notes below.)

Extra support branches were wired in as necessary. None of these alterations ever upset the birds. The top inside diameter of the nests measured 5 cm (2½ in.) and they were about 5 cm (2½ in.) deep. This is quite deep for the size of the nest, and only the beak and tail of the sitting bird stick out. The above data is based on 50 nests.

Nesting commences in March or early April in the Northern Hemisphere. Contrary to a long-standing belief that these birds prefer to nest near running water, Gibson found that his birds actually build their nests as far from running water as possible within the confines of the aviary. Nests were generally situated about 2 m (6.5 ft.) from the ground. There was some variation, but none were built at the maximum available height of 3 m (10 ft.) and none on or near the ground. No nests were built in boxes and none were built indoors. One young hen laid in a wicker basket in which no nesting material was added.

Further data from Peter Karsten in British Columbia confirms the conclusion that these birds prefer to nest in a thick evergreen shrub. He also advises that is essential that a breeding aviary for pekin robins has a thick understory of vegetation in which the chicks will hide for a week or so. Without this, chicks tend to exhaust themselves trying to gain elevation. This frantic activity, in the absence of dense undergrowth, may result in death. In cold weather newly fledged chicks should be provided with a heat lamp at their roost site. Chicks are only able to fly a week after they fledge

Although pekin robins have a reputation in aviculture as egg stealers, the evidence from Gibson's study does not support this claim. Cockatiels *Nymphicus hollandicus,* Bourke's grass parakeets *Neopsephotus bourkii,* diamond doves *Geopelia cuneata,* European blackbirds *Turdus merula,* cutthroat finches *Amadina fasciata,* red avadavats *Amandava amandava,* Gouldian finches *Chloebia gouldiae,* and domestic canaries all nested successfully in aviaries containing pekin robins, and in no instance was egg stealing recorded or even suspected.

Pekin robins lay a comparatively large egg for the size of the bird. The eggs are off-white with a tinge of green, lightly marked with reddish-brown blotches at the large end. Some eggs appear more bluish than green. The usual clutch is four eggs, but sometimes only three are laid. No clutch in the study exceeded four eggs. One pair laid seven clutches, totaling 26 eggs, in a single season. Incubation takes 12 days, and begins when the next-to-last egg is laid. Daytime incubation is shared by both sexes, which relieve each other at intervals ranging from one-half to two hours. The female incubates during the night, with the male roosting on a nearby twig or even on the edge of the nest itself. They have been observed feigning injury during nest inspections, but this behavior soon ceases when they realize that the nest is left unharmed. The brooding bird sits tightly, not moving from the nest until it is actually touched. While the nest is being inspected, both adults remain close by, chattering angrily. Nest inspection will not cause them to desert the nest.

When chicks hatch, the eggshells are deposited in the furthest corner of the aviary from the nest. The nestling period is 10 days; throughout this time, the parents keep the nest meticulously clean by removing all fecal sacs. Used nests generally look as good as new ones. It was speculated that this tidiness could cause a problem with close-banded chicks, and for this reason Gibson did not attempt to band nestlings. The female will often begin to refurbish the nest a week after the chicks fledge. If this happens before chicks are independent, which is about 10–14 days after fledging, it is advised to remove the nest lining and reduce the supply of nest material to slow down the female.

At the time of Gibson's study this species was freely imported in large numbers. Now that both species of Leiothrix have been designated for CITES protection, it becomes more imperative to close-band aviary-bred pekin robins and silver-eared mesias to allow for accurate record keeping and pedigree information, and to facilitate the exchange of birds between breeding programs. It is important when moving any softbills, and in particular pekin robins, that the transport cage does not have any open mesh but rather a fine gauze type mesh that will avoid damaging the birds' forehead and cere. They are also best housed in an aviary with a double-door system to avoid escapes.

Anticipating that close-banding of all CITES species may become mandatory in the future, Peter Karsten, in Canada, has been experimenting with close-banding procedures, and has achieved consistent success with a number of broods of pekins. He has found that nestling abandonment, or the throwing out of chicks by the parents, can be avoided if the parents are shifted out of sight, and preferably also out of sound, during the banding operation. He uses the AACC (Avicultural Advancement Council of Canada) size K band (3.94 mm inside diameter) that must be applied before the chicks are nine days old. The target day is eight days of age, at which time the foot is still small enough to receive the band, but large enough to prevent it from slipping off. In addition, the chicks at this age have a good covering of down and pinfeathers to help conceal the bands from the parent birds.

The slightly smaller size J bands, formerly recommended, can no longer be applied at the optimal age for banding this species. They are perhaps a better fit, but the risk of desertion is much greater when banding at an earlier age. If the chicks are handled gently during the procedure, they will not vocalize and alarm the parents. Karsten also feels that providing live food to the parents when they return to the nest area will divert their attention from the bands; they will be too busy delivering the food to the nest. This also tends to mobilize the chicks that are usually huddled motionless in the bottom of the nest after being handled.

Since incubation begins with the next-to-last egg, several chicks usually hatch simultaneously, with the final egg hatching a day later. In Gibson's experience, four chicks were never raised in a single clutch, the youngest usually dying first. During the first week only live food is fed by the parents, who carefully pick it into fine pieces before feeding for the first two days, after which it is fed whole. Peter Karsten reports that a single pair feeding three chicks require about 300 insects per day. He also injects extra calcium into a few wax moth larvae to ensure the chicks have sufficient calcium.

Inexperienced males have caused the death of chicks by feeding them whole mealworms at too young an age. To avoid this problem, mealworms can be chopped into small segments, or only offered after the chicks are 4–5 days of age. Spiders are a favorite rearing food, and success was enhanced when these were supplied. After the first week, egg food and currants were fed to the chicks. Vitamin D deficiencies, resulting in badly formed bones, can be prevented with a supplement, which can be as simple as a little cod liver oil added on to the soft food. Gibson also used mealworms that had been lightly coated with margarine. It should be noted that these birds require at least 12 hours of daylight to rear their chicks.

Growth is fast, with the eyes open at five days and the chicks fledging at 10, and sometimes only nine days. At this stage they almost all lack a tail and can only fly about a meter (3 ft.), but they can actively climb in the shrubbery. Within 2–3 weeks the female will start a second clutch, and lose interest in the fledglings. Their care is then taken over entirely by the male, who will begin to hold food that the chicks will take from his bill. He may also lose interest in feeding; if this occurs, he should be caged with the chicks within the aviary until they are at least 24 days old, which is the earliest age at which they should be separated from their parents. The young birds molt into adult plumage at about three months of age, but can be sexed by their calls only two weeks after fledging.

Raymond Franklin described the captive behavior and breeding of the black-headed sibia *Heterophasia capistrata,* in England. The hen of his pair had one or two light grey primary feathers in the wings; otherwise the sexes were difficult to tell apart. They jump from perch to perch like mesias and pekin robins, and their flight is short and jerky. The vocalized almost constantly, uttering small babbling notes. Vocalizations resembling a song were only heard before nesting. A peculiar noise, likened to the mewing of a cat, was heard two or three times in succession on several occasions, and was associated with what appeared to be a courtship display in which the male would chase the female at terrific speed in and out of a bamboo clump and around the aviary, attempting to peck at her cloaca.

After spending the winter in a bird room that was slightly heated during bad weather, the sibias were placed in an outside aviary containing a clump of bamboo and some ferns along with a trellis to which some wild hops vines were trained. The aviary measured 4 x 2 x 2 m high (13 x 6.5 x 6.5 ft.). Palm fibers were offered as nesting material, and at the end of May the male was observed carrying some fibers around the flight. On June 10 he displayed to the hen, raising his crest and vibrating his wings, while holding a long strand of palm fiber in his beak. This continued off and on until mid-July, when both birds were seen constructing a nest high up in the hops vine. The nest, completed on July 24, was described as being similar to that of a blackbird *Turdus merula* but with a smaller, deeper cup. The outer layer was comprised of palm fibers, partly woven around the hops vine. An inner layer consisted of fibers interspersed with bamboo leaves and other dried leaves, and the cup was lined with fiber.

The high-speed chase display described above was witnessed the following day, and the first egg, which resembled the egg of a greenfinch *Carduelis chloris* in size and color, was laid on July 28. The incubation period was 14 days, the first chick hatching on August 12 and a second appearing the following day. The chicks were fed largely on live food, and left the nest on August 28. The feather coloring of the fledglings was similar to that of the adults. (Franklin 1973)

Hand Rearing: The arrow-marked and pied babbler chicks are easy to hand rear and become very tame; but if mixed with others, some of their tameness will be lost as they mature. They can be hand raised on equal portions of commercial softbill formula, moistened beef-flavored cat food, and insects such as mealworm larvae and beetles, moths, grasshoppers, caterpillars, as well as strips of ox heart. A multivitamin/trace element and a calcium supplement are added to the moistened food once per day. Chicks should be fed with a forceps. For further details on hand-rearing procedures, read the section on STARLINGS and TANAGERS.

There is no aggression between chicks in the nest. Immatures are a duller version of the adults. Although I recommend hand rearing to ensure a good production of fledglings, a problem arises in that hand-reared young of cooperative flock species cannot be re-introduced into their parent group. If this is attempted, the youngster will be killed. Hand-raised chicks should be mixed with the young from other broods to form new groups. Sexes are alike, and the gender of individual birds must be determined through surgical procedures or DNA analysis.

Peter Karsten recorded breeding data of his pekin robins *Leiothrix lutea* on Denman Island on the West Coast of British Columbia, Canada between 1999 and 2005. The climate is conducive to housing the species outdoors year-round, with modest winter shelter. Minimum temperatures rarely fall below –8°C (18°F) and stay below 35°C (95°F) plus in the summer. Annual average precipitation is about 2,000 mm (79 in.).

Parent-raised Black-throated *Garrulax chinensis,* Red-tailed *Garrulax milnei* & Yellow-bellied *Garrulax galbanus* Laughingthrush

Day	Black-throated	Redtailed	yellow-bellied	Day	Red-tailed	yellow-bellied
	Weight				Weight	
1	5	2.8	8.5	22	36	58
2	7	6.5	10	24	39.5	62
4	17	8.5	13	26	43	66
6		12	18.5	28	45	72
8		15	27	30	44.5	79
10		21	32	32	46	80
12	28	19	38	34	50.5	86
14		20.5	42	36	50	
16		28	47	38	54	
18		28.5	52	40	55\	
20		31	55			

11.5 wild caught founders, 1.11 F1 and 1.0 F2 offspring formed 19 pair combination for the six breeding seasons.

Three pairs of silver-eared mesias *Leiothrix argentauris* performed similarly, although due to lower disturbance tolerance of this species only 65.8 % of 41 eggs laid hatched and only 14.6% of the total number of eggs laid produced successfully weaned chicks. Some chicks had been abandoned due to banding them in the nest, and subsequently required hand rearing.

Peter Karsten summarizes his experiences in breeding pekin robins as follows:

Pair Formation: Identification of sex is done by segregating birds, one at a time, from sight of others and testing vocalization. Hens do not have a contact song, but instead a plaintive call of 4–5 monotonous and repeated notes. Plumage differences are not 100% reliable for sex determination. DNA sexing is another, but unnecessary, alternative. Males can give the hen's plaintive call, but very rarely do, while hens never "sing" like males.

Introduction of mates is usually successful and low risk. Pair-forming behavior patterns are ritualized and rarely lead to serious confrontation. Switching of mates during breeding season can lead to failure, since nest building and egg laying may not be synchronized in the short term. Nesting delays and egg-binding can result. Pekins robins mate for life and may be reluctant to accept another mate for several days or weeks.

Pekins robins can be held in small flocks after the breeding season, when the fall molt sets in. After completion of the molt (November) conflict between pairs should be expected. At the beginning and during the breeding season one should not place two pairs in the same aviary or serious territorial fights can occur. This may result in the weaker pair, or at least the male, being killed. Several pairs or specimens can be held together in an unsuitable (non–planted) environment for nesting, since the drive to defend a nesting territory does not seem to develop.

Two pekins and even a pekin and a mesia will form roosting buddies, regardless of sex, and act like mated pairs except there will be no breeding. Always song-test a "mated" pair when setting it up for breeding. (I acquired more than one "guaranteed mated pair" only to discover by testing vocalization on their arrival that they were of the same sex.)

Breeding Aviary: Breeding aviaries should be no less than 2 x 3 x 2.25 m high (6 x 9 x 7 ft.). Birds that have been previously held in small birdcages may accept a smaller aviary for breeding, while birds bred in spacious aviaries generally do not. While nesting may commence, incubation and caring for chicks usually fails in small aviaries. Pairs that are not successful in a marginal aviary will often succeed when moved to a larger one.

Planted aviaries larger than 3 x 4 m (9 x 12 ft.) in floor space are more problematic for monitoring the birds. Dense planting with evergreens (bamboo, conifers, etc.) is absolutely essential to trigger nesting behavior. Adding additional plant material improves success rates, with the greater the abundance of plants, the greater the success (suggest 50% or more of aviary volume).

Wall-to-wall aviaries are not as effective as those with buffer space between them. Clustered aviary complexes are less effective than spread-out buffered aviaries. Partitions must be solid and opaque. Neighboring males will even try to harass each other through cracks between boards and open knotholes. This stalls nesting activity. Passageways between aviaries with screen fronts should be wider than 3.3 m (11 ft.) to reduce territorial anxiety.

Breeding Season and Nesting Behavior: Birds exposed to natural changes of seasons and related

Peter Karsten's Breeding Results for Pekin Robins *Leiothrix lutea* in 2005

Pekin robins	April	May	June	July	Aug.	Sept.	total
number of nests	4	10	14	23	12	6	69
total number of eggs laid	12	35	50	73	40	17	227
number eggs per nest	3	3.5	3.8	3.2	3.3	2.8	3.3
total number of chicks hatched	7	28	41	55	29	11	172
% of eggs hatched	58.3	80	82	75.3	76.3	64.7	75.8
total number of chicks fledged	1	13	22	47	19	9	111
total number of chicks weaned	1	8	18	37	16	5	85
% of success vs. numb. of eggs	8.3	22.8	36	50.7	42	29.4	37.4

Three pairs of silver-eared mesias *Leiothrix argentauris* performed similarly although, due to lower disturbance tolerance of this species, only 65.8 % of 41 eggs laid hatched and only 14.6% of the total number of eggs laid produced successfully weaned chicks. Some chicks had been abandoned due to banding them in the nest and required hand rearing.

Compiled by Peter Karsten, Silver Pine Aviaries

changes in day length breed better than those in environments with insufficient variation. Vernal equinox (spring) is the approximate start of the breeding activity.

First nests in the season are often unsuccessful. Low ambient temperatures, short day length and lack of supplementary wild insects in the enclosures are suspected as causes. The nesting season for the flock starts in May and last to mid September. Individual pairs may have 3–4 successful clutches about one month apart if they raise their chicks; pairs may be late or early nesters in a given season. Early nesters also start their fall molt earlier. When the breeding pair starts its molt, no further nesting behavior is noted.

More than 40% of the successful clutches were in August; 83.5% in June/July/August (based on a total of 69 nests.) Abandonment of eggs, expulsion of chicks (dead or alive) between the fourth and fifth day after hatching are more common in the shoulder season. The first nesting attempts of birds that hatched in the previous year are generally unsuccessful, but subsequent clutches may be raised in the same season. It appears that males are less successful in their first year than hens. Birds of unknown age imported before 1997 are still breeding. They would be more than nine years old.

Nest disturbance is tolerated in varying degrees. High-strung birds with persistent scolding behavior are generally less tolerant of disturbance. Males seem to govern the course of events. They select nest site and seemingly determine abandonment and expulsion of eggs or young. Hens can be observed returning to the empty nest to brood or feed, while the males show no further interest. I have never witnessed the interference with the nestlings or eggs by the parents and have now mounted closed circuit cameras for problem birds to monitor these events and facilitate possible rescue. Hens that were unsuccessful in raising chicks with irritable males did achieve success with less high-strung males for subsequent clutches.

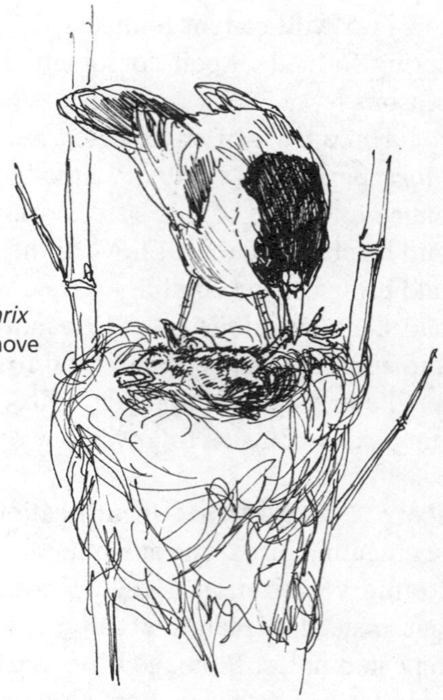

Silver-eared Mesia *Leiothrix argentauris* trying to remove a band of a chick's leg.

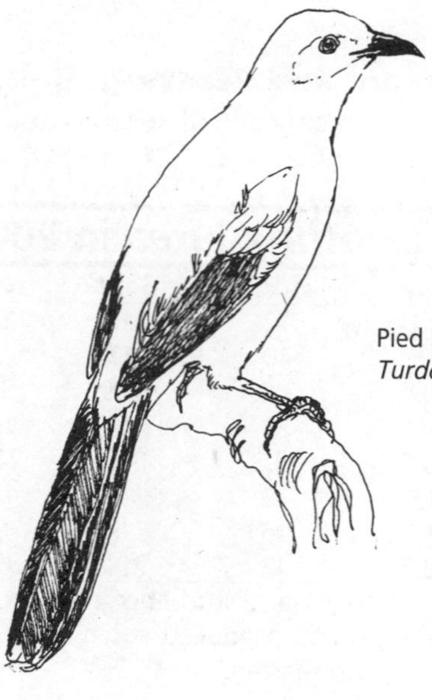

Pied Babbler *Turdoides bicolor*

Nests are built with strands of raffia cut to 10 cm (4 in.) lengths to construct the nest rim and frame, and are then filled with coconut fiber at times mixed with dry bamboo leaves. Nests are not lined with insulating material, as seen with northern bird species, hence my concern about low ambient temperatures in early spring. Some nests fail and begin to sag, causing the layering of the eggs and chicks. In some cases they may even drop through the bottom or side.

Nest supports are systematically added to the nests when the males begin to weave the nest rim. The birds are shifted out of the aviary, if at all possible, while the nest support is positioned. Nest supports are made of plastic covered wire (1.5–2 mm) by forming a ring with a 10 cm (4 in.) diameter, followed by crossing the ends into a bowl shape of about 4 cm (1.5 in.) depth. This is wrapped with raffia for better visual integration into the started nest. Rejection has only occurred with one male in 10, and such pairs are left to their own devices. Nest sites are protected from rain and direct sunlight, especially if the sunrays strike the nest through glass. Clear glass is painted with semi-transparent fence stain.

Egg Laying and Hatching: Eggs are laid one a day, usually in the morning. Three eggs per nest are typical; two and four are less common. Incubation period is 11 days from the day the parents sit solid and commences with the second egg laid.

Chicks of uneven size are common in clutches with more than two eggs. Chicks fledge on the same day as the rule. Smaller chicks remain on the ground after fledging longer than stronger siblings. Chicks are anxious to work their way up into the higher cover or branches for security and to join their siblings to roost in close body contact. A barren enclosure without cover in the lower region causes chicks to relentlessly try to jump onto higher perches, which results in muscle burn out (muscle miopathy) followed by shock and paralysis. Chicks must then be put into a hospital pen for slow recovery and hand feeding. Cover must be provided from the ground up, creating laddering structures, to prevent miopathy. Chicks that become lethargic and do not get off the ground for three or more days are prone to perish and should be taken out for hand rearing.

Hatching chicks normally survive the first three days. The reserves from the yolk are then exhausted and the chicks must survive entirely on the food brought by the parents. The hen will often hatch her next clutch before the chicks are weaned. This occurs when the hen uses the same nest after her previous clutch fledged. To mitigate this, the inner lining of the nest can be lifted out of the nest support to force the adults to work on rebuilding the nest or a new nest. It delays the laying of eggs and buys time for the previous clutch weaned. Removal of the inner nest lining has hygienic advantages as well.

Feeding Chicks: Parents only feed live insect food or insects killed on the spot just before feeding. Crickets, waxworms, and to a lesser degree, mealworms, have been offered as primary live food items. Crickets are the most balanced food of the three food items; still, mineral deficiency can be expected if Ca and vitamin D is not supplemented. Parent birds seek out variety in the insect food and additional wild-caught insects are fed to the chicks with enthusiasm. Light traps are employed to capture moths and other nocturnal insects.

The loss of chicks on the fourth to fifth day is widely known. Effective provisions of vitamin mineral supplements are the best defense. Vitamin/mineral supplements can be injected into waxworms or "fixed" to the live food by lightly coating the insects with margarine or vegetable oil. Dusting the dry insects with supplements is not effective, since the adults will strike the waxworm or mealworm repeatedly against the perch to prepare it for feeding and thus remove the additives. Liquid products are more suitable. The more wild-caught insects are fed, the better the results. Food insects must be "gut loaded," particularly if raised entirely on wheat bran, which interferes with calcium absorption.

Pekin Robin chicks after hatching.

A specially baked egg cake is the first non-insect food the chicks are fed by the parents. Once this takes place, about 7–10 days after fledging, a balanced nutrition is secured. Chicks begin to feed on their own when about 3 ½–4 weeks old. Parents wean their chicks when approximately 4–5 weeks old, however the chicks can be separated and weaned as soon as they feed on their own. Hand-raised chicks can become independent sooner (<4 weeks). Male chicks begin "practicing" their singing at about four weeks of age.

These are the most important observations with respect to breeding pekin robins in Peter Karsten's aviaries. Different environments may require different breeding regimes.

Ring Size: Arrow-marked and pied babblers, 4.5 mm; pekin robin 3.94 mm

Species Requiring Similar Management:

Liocichlas *Liocichla* species
Woodhoopoes *Phoeniculus* species
Sibias *Heterophasia* species
Shrike-babblers *Pteruthius* species
Jungle babbler *Turdoides striatus*
Laughing thrushes *Garrulax* species

Not all of the above-mentioned species are communal/cooperative breeders.

References

de Q. Quincey, R.S. 1961. The breeding of the rufus-bellied babbler *(Dumetia hyperythra)*.*Avicultural Magazine* 67 (2): 56–57.

Franklin, R. 1973. Breeding the black-headed sibia *(Heterophasia capistrata). Avicultural Magazine* 79 (4): 116–118

Gibson, L. 1978. The red-billed leiothrix: a four-year study. *Avicultural Magazine* 84 (1): 4–17.

Gibson, L. 1979. Breeding the arrow babbler Turdoides jardinei. *Avicultural Magazine* 85 (3): 112–120.

Karsten, P. 2006. *Pekin Robins and Other Softbills: management and breeding*. Surrey, Canada: Hancock House Publishing.

Karsten, P. 2000. Pekin Robins: Care and Breeding, Part I. *The Avicultural Journal* 23(1): 33–39. Avicultural Advancement Council of Canada.

Karsten, P. 2000. Pekin Robins: Care and Breeding, Part II. *The Avicultural Journal* 23(2): 21–27. Avicultural Advancement Council of Canada.

Karsten, P. 2002. Pekin Robins: Information on Their Care and Breeding. *The AFA Watchbird* 29(2): 56–68, Phoenix.

21.17 Alaudidae Sparrowlarks

Glen Holland

Natural Habits: Larks belong to the order Passeriformes and the family Alaudidae, which includes 82 species, of which the seven species known as sparrow-larks, are all placed in the genus *Eremopterix*. Sparrow-larks are widespread over the dry areas of Africa, from south of the Sahara to the Northern Cape Province. They are found in open savannas, particularly where there is some broken, stony ground or patches of bare soil. Sparrow-larks congregate in large flocks outside the breeding season. They have a rattling sparrow-like call, with a pleasant song made in flight. Flocks often occur at water holes during the dry months.

Management: Sparrow-larks closely resemble sparrows. The adult male of some species, such as the chestnut-backed *Eremopterix signata* is instantly recognizable by his white ear coverts, crown and flanks, which contrast sharply with the chestnut-colored mantle and back and the almost-black chest and belly. The female is a mottled brown color. Juveniles are similar to the female, with underparts whitish buff. Sparrow-larks are not easy to establish in captivity, and newly acquired stock should be carefully acclimatized. Initially, they are rather skittish and tend to fly straight up to the top of the aviary. For this reason, transport boxes for larks should always have foam lined tops to prevent head injuries. Initially it is advisable to clip one wing to avoid cranial damage. By the time the feathers have grown through, the birds will be settled and aware of their limits and the aviary roof. As they originate from dry areas, extended periods of cold, wet weather for extended periods of time will result in death.

Their aviary should be sparsely planted in accordance with Habitat E as described in the chapter on LANDSCAPE. They favor plenty of open, sandy areas between the vegetation, interspersed with a few flat rocks and stumps on which they like to perch and behind which they will find a shady spot in which to breed. Pairs spend much of the day perched on a flat rock. Small flights are not suitable for these rather restless birds, and I would recommend a minimum ground space of 20 m^2 (215 sq. ft.). and a height of 2.5–3 m (8–10 ft). They are fast, erratic flyers, and pairs held in small aviaries never seem to settle and will develop the habit of flitting their wings as though wanting, but unable, to take off. Breeding under such conditions is unlikely. One pair should be kept per aviary, as they become aggressive during the breeding season, particularly at the source of live food. Although they inhabit semi-desert areas that experience dry but extremely low night-time winter temperatures, so they are able to withstand overnight temperatures down to 0°C (32°F), provided daytime temperatures are warm and dry. Temperatures below 10°C (50°F), when combined with wet conditions, will stress these birds. In such climates where extended periods of cold are combined with wet, sparrow-larks should be housed indoors with additional warmth.

Diet: Sparrowlarks can be maintained on a basic finch seed mixture, but insects must be provided to ensure good health and promote breeding. They will take termites, mealworms, and any other small insects available, including small beetles. Insects are sometimes chased and caught at ground level. Although I have offered commercial soft foods, I have never observed them taking any of this.

Breeding: At the beginning of the breeding season the male becomes extremely vocal, with aerial display flights that are accompanied by song. As part of this display, the male often chases the female while in flight or runs in a crouched posture toward the female on the ground. The pair defends the immediate territory around the nest. The female selects the nest site; the nest itself is a shallow scrape in the ground, placed on the southern side of some vegetation or a rock. This instinct developed in response to the necessity of sheltering the nest from the extreme heat of the midday sun in the desert areas in which they breed. The female lines the scrape with a few pebbles, short pieces of grass, and rootlets. The usual clutch is two, sometimes three, light gray eggs that are covered in brown spots.

Incubation is shared by both sexes, but mainly performed by the female, for 11 days. The young beg with stretched necks and wings flapping, but do not call. During hot periods, the pair alternate in shading the chicks. At only eight days of age, the chicks leave the nest to hide at the base of a shrub from where they emerge to be fed. I have had chestnut-backed sparrow-larks breed on one occasion, and although the eggs hatched, the chicks were not raised, most probably due to an insufficient supply of live food. They have, however, been successfully bred in captivity. An interesting captive observation records a chestnut-backed male feeding a recently fledged quailfinch that was begging for food. I have never had the opportunity to hand-raise a sparrow-lark, but would recommend a high-protein diet as described for hand-rearing FINCHES, with a few small, crushed insects fed by tweezers.

Ring Size: 2.5 mm

Species Requiring Similar Management:

Larks *Mirafra* species
Bunting *Emberiza* species
Quailfinches *Ortygospiza* species

Contributors:

Neville Brickell, Avicultural Research Unit, Durban, South Africa.

21.18 Nectariniidae

Glen Holland

21.18ai *Nectariniini* Sunbirds

Introduction: The family Nectariniidae in the order Passeriformes is divided into two subfamilies. The two species of sugarbirds of South Africa, genus *Promerops*, are the only members of the subfamily Promeropinae. The subfamily Nectariniinae is further divided into two tribes, the tribe Dicaeini, the flowerpeckers, and the tribe Nectariniini, the sunbirds and spiderhunters. There is a total of 44 species of flowerpeckers, confined to two genera, *Dicaeum* and *Prionochilus.* Both the flowerpeckers and the sugarbirds were formerly separated into their own families, but their closer relationship has recently been established on the basis of DNA evidence.

In addition to the typical sunbirds, the tribe Nectariniini includes 10 species of spiderhunters in the genus *Arachnothera,* ranging from southern Asia and southwestern China through Southeast Asia, the Malayan archipelago, Borneo, Sumatra, the Philippines, and the Greater Sundas. The 114 species of sunbirds belong to only four genera. While the purple-naped sunbird *Hypogramma hypogrammicum* of China, Southeast Asia, Borneo, and Sumatra is the sole member of its genus, the other three genera have numerous representatives. The 14 species of the genus *Aethopyga* range from the Himalaya through India, Burma, China, Thailand, Southeast Asia, the Malay Archipelago, the Philippines, Borneo, Sumatra, Java, the Greater Sundas, and several of the smaller islands off northern Sulawesi. The genus *Anthreptes,* with 19 species, and the genus *Nectarinia,* with around 80 species, are widespread in some of the above areas, with representatives of the latter genus occurring in New Guinea and Australia. Both of these genera are heavily represented in Africa, with species occurring throughout the continent, primarily south of the Sahara, and on the island of Madagascar. Favored habitats include the protea scrub, or fynbos, of the Western Cape Province, forests and forest margins, thorn, and open woodland and coastal scrub.

The plumage of sunbirds is often brightly colored, with metallic green or bronze sheens, especially in males. Most species are sexually dimorphic, with the females lacking the bright colors of the males. The males of many species have red, orange, or yellow pectoral tufts used in display but otherwise concealed. The males spend much of their time chasing their mates and other sunbirds, which congregate at feeding sites. Males have favorite exposed perches where they sit for long periods, making their warbling song. At a good nectar source, they are often found in parties of up to 10 birds of various species together.

Management: Sunbirds adapt well to aviculture, but should be handled with great care when initially acquired. Due to their high energy requirements, these birds suffer badly after a period of stress with reduced feeding. Newly acquired birds are delicate and need special care during the first 10 days. An electrolyte anti-stress formula should be added to the nectar and additional warmth provided, particularly at night. Fruit flies and fruit must be offered in addition to the nectar. The birds should initially be kept in holding cages, covered in nylon gauze to avoid injuries. In these cages they adjust to captive conditions and diets. Newly captured birds will drink nectar from a tube if their beak is dipped into the nectar while they are being held in the hand. Once they have done this, they associate the tube with food and the same tube should be used continually. After about 10 days the birds will be ready to go into a flight aviary. When introducing a sunbird into an existing group, it is important to keep an eye on possible aggression from the established birds.

In a large planted aviary, some species, such as the collared *Anthreptes collaris,* greater double-collared *Nectarina afra,* scarlet-chested *N. senegalensis,* black *N. amethystina,* and white-bellied *N. talatala,* can be mixed. A few of the larger species, such as the malachite *N. famosa,* however, will kill other sunbirds and must therefore be housed separately. Most species will not tolerate any other birds, whether sunbirds or other species, feeding at their drinking tube. So if several species are to be kept together, each pair should be supplied with its own drinking tube. In a large aviary, more than one pair of black sunbirds can be kept together, but this is the exception to the general rule of one pair of a given species per aviary. Their habitat requirements will be met by Habitat B as described in the chapter on LANDSCAPE.

Bathing in a dish is accomplished either by the bird diving in and out of the water quickly or by dipping backward into the water while splashing the water with the wing tips. A favorite activity is dashing about and bathing in the wet leaves of a plant, and to accommodate this, a mist sprayer situated above a shrub can be switched on for 10–15 minutes on warm days. Sunbirds prefer to leaf bathe rather than use a pool of water whenever possible. Generally originating from warm climates, they cannot be described as hardy. In climates where temperatures fall below 10°C (50°F) with extended periods of cool, wet weather, sunbirds are best housed indoors, with additional warmth provided during the harsh winter months.

Diet: Sue Jackson and Susan Nicholson in their article titled "Sugar Junkies of the World" provide some interesting insights into sunbird nutrition:

> "There are three main types of sugar in the nectar of most flowers: sucrose, or ordinary table sugar; fructose, which is found in all fruit juices and may be bought in powder form in health shops; and glucose, the main sugar in most sports drinks. Sucrose consists of two molecules bound together, so one molecule of this sugar is twice the size of a molecule of glucose or fructose, and contains twice the energy. Sucrose must be broken in two before it can be absorbed into a bird's blood, and a special enzyme (sucrase) is required for this. Not all birds have this enzyme: starlings (both European and the indigenous southern African species), thrushes and flycatchers no longer produce sucrase and have therefore lost their ability to digest sucrose. Not only is table sugar distasteful to these birds, but it can make them ill. Starlings therefore avoid sucrose if it is offered to them in garden feeders. For many years, researchers in North America mistakenly thought that all passerines (perching birds) were unable to digest sucrose, because the researchers only compared the sugar preferences of hummingbirds (non-passerines which like sucrose), with those of starlings (passerines which can't tolerate this sugar). Recent South African studies have resulted in a revision of this generalization, however, as passerine sunbirds, sugarbirds and white-eyes of the region love sucrose, even preferring it to fructose and glucose. Australian honeyeaters are much the same. This is a wise preference, because the slightly bigger sucrose molecules are emptied more rapidly from the birds' stomachs into their intestines, whence they can be absorbed into the blood. How do we know which sugars birds prefer? Nectarivores are obliging birds and drink as willingly in a cage as they do when flying free in a garden. "Cafeteria-style taste tests" are therefore easy to conduct on captive birds. They clearly prefer stronger solutions with more energy. Also, when fed nectars of different strengths on different days, they are able to adjust the amount that they drink so that they get the same amount of energy each day: enough to last them through the night but not enough to make them too fat to fly. (Would that we all had the same restraint when it comes to sweet things)."

In the winter rainfall or Mediterranean climates of southern Australia and the south-western Cape province of South Africa, the matter of daily energy balance is critical for nectar-feeding birds' survival. This is because birds in these regions breed when the nectar of their chosen flowers is flowing most abundantly. This happens in late autumn, early spring or even midwinter, when it is cold, windy, and often raining. Parent birds have to keep themselves and their chicks warm, so their metabolism speeds up and they drink large quantities of nectar. It has been estimated that the female orange-breasted sunbird must drink up to three times her own body weight in nectar every day while she is incubating — the equivalent of 180 liters of double-concentrated Coke every day for an average-sized human. Sunbird and sugarbird kidneys must therefore be able to eliminate all the water that the birds drink with their food, because birds must stay as light as possible for flight. We are constantly being told that soft drinks and sugar aren't healthy, that they are bad for us. So how do nectar-feeding birds stay healthy and bright-eyed on this strange diet? They also eat insects for protein, and they feed more insects than nectar to their growing young. Interestingly, blood glucose levels that have been found to be normal for a hummingbird are high enough to cause the symptoms of severe diabetes in humans.

The colors of any sunbird that has yellow and/or red feathering will quickly fade in captivity unless a proper and varied diet is provided. The basis of the captive diet for sunbirds is nectar, which is offered in drinking tubes. The tubes should be evenly spaced through the aviary to minimize aggression between the birds. Other softbills are also fond of nectar and, if necessary, a wire sleeve should be placed over the tube, allowing only the long beaks of the sunbirds to reach the nectar, which should be replenished twice daily. The tubes should be thoroughly cleaned and sterilized each day. Poor hygiene results in yeast infections, which is the most common problem with these nectivores. Constant swallowing and yawning often indicate yeast infections such as Candida. If detected in the early stages, candidiasis can easily be treated with Nystatin. There are many different varieties of homemade nectar, as described below, and there are also nectars produced commercially such as Nekton and Biotropic products. A pair of sunbirds, depending on their size, will consume 60–100 ml (2–3 oz.) of nectar per day. Nectar consumption will increase by about 50% during the nesting period. As these birds become very tame, it is necessary to have a safety door system that will catch any birds that might otherwise escape in their hurry to reach the fresh nectar.

One example of a homemade diet that has resulted in regular breeding successes is Dirk Norval's recipe, which follows:

Add to 2 liters of warm water:

- 4 tablespoons brown sugar
- 4 tablespoons syrup
- 1 tablespoon Horlicks (energy/protein drink for humans)
- 1 tablespoon honey
- ¼ teaspoon Bovril (vegemite)
- yolk of one egg

The nectar should have a protein content of 65 g/kg of dry weight. Live food must be supplied to stimulate and ensure breeding success. Three commercial diets for sunbirds were tested and found to average only 24 g/kg dry matter. If maintained on a diet as low in protein as 22 g/kg or less, the birds will lose weight and become ill, but will soon recover when the percentage of protein is increased. Milk powder should not be used to boost the protein level in a sunbird diet, as these birds are unable to digest the lactose, and illness will result. Carbohydrates, in the form of sugars, give sunbirds their energy, and should be supplied at concentrations of about 150 grams per liter of nectar. These carbohydrates should be composed of 50% sucrose, 45% fructose, and 5% glucose. A kilogram of dry sunbird diet should consist of:

Gross energy	Mj	15.8
Carbohydrate	g	885
Crude oil	g	20
Crude protein	g	65
Vitamins, minerals, lineolic acid	g	5.6

The pH should be near neutral and the protein should be 100% water-soluble. Diets should not include yeasts or microbiological agents, which enhance fermentation.

Hand Rearing a Sunbird chick.

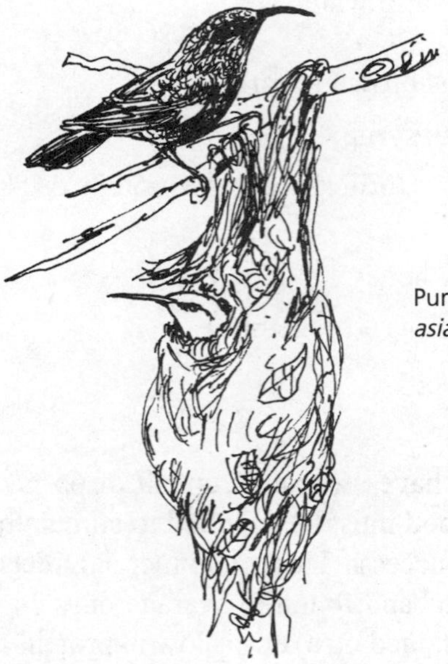

Purple sunbird *Cinnyris asiaticus* at nest.

One factor that is often ignored in nectivore diets is pollen, which the birds naturally and actively consume. Pollen is nature's wonder food, and apart from protein, it is packed with minerals and trace elements that are beneficial to the birds. I would recommend the addition of one tablespoon of dry pollen to a liter of any sunbird nectar. Pollen is a byproduct available from the honey industry. Sunbirds have also been recorded taking oyster shell grit as a calcium supplement. A soluble calcium supplement should be added to the nectar, particularly during the breeding season. Sunbirds are highly insectivorous, particularly when feeding young in the nest. The introduction of live food often stimulates courtship. When feeding naturally, they hunt for either insects or nectar, and while busy with one, they often ignore the other.

Fruit flies are commonly used for sunbirds. These are easily propagated and attracted by placing overripe fruit and fruit peels in a container with a wire gauze top to exclude the birds. As the flies escape from the container, they are caught by the sunbirds. It is advisable to have four such containers, which are put into the aviary on a rotational basis. A female feeding chicks will line up one fly after the other in her beak and then feed these to the chick. Spiders are a sunbird's favorite food, and their soft bodies are crushed in the tip of the bill and then tossed back and swallowed. Midges, pinhead crickets, beetles, flies, fly larvae, aphids, small waxworms, and flying ants are all eagerly taken. Their requirements for insects reach a peak when there are chicks in the nest. Food items are often dropped while the birds are feeding but they are remarkably quick to fly down and retrieve what has been dropped. The birds soon get to know when insects are to be offered and will eagerly await these, with the excited males sometimes performing their courtship dance in anticipation of the meal.

A successful breeder of sunbirds told me that he had not planted a single nectar-producing or insect-attractant type of plant in his aviary. I feel however that these natural food sources should not be excluded. Nectar-producing plants provide a simple and direct supply of natural food, including pollen, the latter assisting to meet the high protein requirements of sunbirds. If flowers are picked for the sunbirds, they are best placed with the stems in water overnight and fed to the birds in the morning. Doing this will ensure the birds have a fresh crop of pollen and nectar that has not been depleted by wild birds. Flowering plants also attract insects into the aviary, which are also eaten and thereby help to ensure that the sunbirds remain in good health. The following are some useful nectar-producing plants:

Aloes, *Aloe* spp.
Honeysuckle, *Lonicera* spp.
Hibiscus, *Hibiscus* spp.
Bottlebrush, Callistemon,
Tree Fuschia
Coral tree, *Erythrina* spp.
Schotia brachypetala
Silky oak, *Grevillea robusta*
Jade vine
Australian cherry
Watsonia spp.
Red hot poker, *Kniphophia*
Leonotis
Lantana, *Lantana camara*
Melaleuca spp.
Cotyledon spp.
Heath, *Erica* spp.
Protea spp.
Leucospermum
Bird-of-paradise, *Strelitzia* spp.
Marmalade tree

Sunbirds are also fond of fruit and will feed on segments of bananas, apple, oranges, grapes, and pear that have been spiked onto a branch. They will also feed on small quantities of moistened commercial softbill food. The birds enjoy a mixture of grated apple, pears, moistened softbill food and a multivitamin powder. Clearly sunbirds require much time and commitment from the aviculturist. Unlike some birds, such as finches for example, their food and water cannot be topped up and left for a weekend but must be provided fresh daily.

Breeding: An important part of the management of breeding sunbirds is to ensure that the female has enough cover to avoid the overzealous advances of the male. This is particularly a problem when the male is in breeding condition and the female is not. Fights can result in the pair lying on the floor locked in battle. The males show an elaborate display that begins with singing and swaying from side to side. If the female is ready to accept his advances, she makes a few short calls, which stimulate the male further, and he may display his pectoral tufts. Vent pecking, with the male pecking at the female's vent, is common, particularly while the female is building the nest. Nests are built in the upper portion of a tree, with the same nest site often used in successive years. Sunbirds have also been recorded nesting in a finch nest basket. When the female begins to collect nesting material, the male often becomes a hindrance to her with his continual chasing. The nest is placed at the end of an exposed branch at about 1.5 m (5 ft.) high. They also regularly use man-made objects for nesting sites, such as fence lines, hanging baskets, verandah structures, etc. Mating takes a few seconds and is repeated a number of times a day.

A few dense, dry heather-type branches attached under the roof of the shelter and hanging downward will often be accepted as a nest site. A few exposed dry perches are enjoyed, particularly by the males who perform on these to the females and other birds. Females will not hesitate to steal nesting material from other birds' nests. The nest is often constructed on the remains of a nest from the previous season or in the same tree. With some species, a feather is placed in the entrance to the nest and this bends out of the way as the birds enter or exit the nest.

Feathers are used to line the interior of the nest. Sheep wool, pet fur, shredded cotton balls, moss, hemp fibers, cob webs, leaves, thin strips of paper, bark, and feathers should be offered for nest building. Materials are often dipped in water before being added to the nest. Materials such as aquarium filter floss are readily added to the nest, but the danger of birds becoming entangled should be considered before such materials are supplied. Nest construction takes about two weeks. A protrusion over the nest entrance is built of grass inflorescences. The 2–3 white to off-white or even pale green eggs have spots and scrawls spread over the shell. The female alone incubates for 14–15 days. During this time the male spends much of his time singing near the nest. In some cases, due to aggression from the female, males have to be removed after mating and placed in an adjacent cage. All responsibility for raising the chicks is then left to the female.

Fecal sacs are removed from the nest. After 18–22 days the chicks fledge, and they move as a family group for at least a month, with the chicks calling *sweep, sweep* continually. It is possible to detect the plumage of a young male by eight months of age. An increase in insects will induce breeding, and these are essential when rearing young. The loss of sunbird fledglings due to an inadequate supply of live food has often been reported. Initially, chicks are fed by regurgitation; later, as the chicks increase in size, the insects are killed and fed from the beak whole.

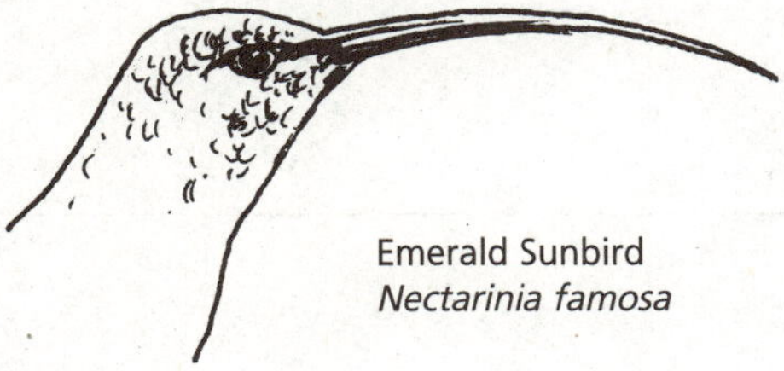

Emerald Sunbird
Nectarinia famosa

Golden-winged Sunbird
Nectarinia reichenowi

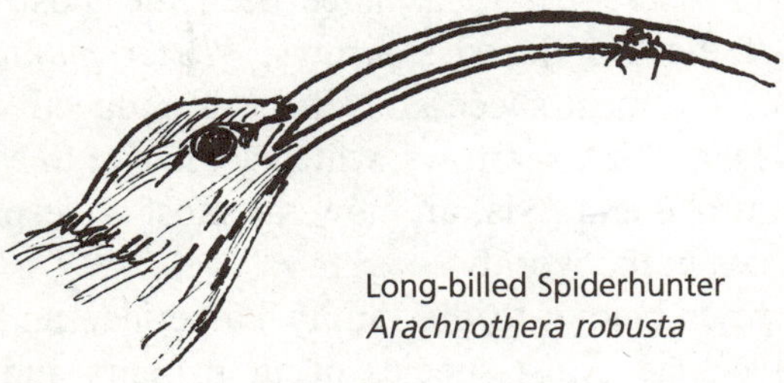

Long-billed Spiderhunter
Arachnothera robusta

Yellow-Bellied Sunbird
Neodrepanis hypoxantha

Age (days)	Mean weight (g)	Age (days)	Mean weight (g)
0	0.92	**7**	5.9
1	1.54	**8**	6.13
2	2.1	**9**	6.77
3	2.77	**10**	7.28
4	3.66	**11**	7.17
5	4.1	**12**	7.13
6	5.28	**13**	7.7

In the United States, sunbirds have been successfully maintained and bred in suspended cages. Such a sterile environment, however, results in the birds acquiring little of their own food, and insects such as spiders must be supplied in large quantities. These cages are surrounded with vegetation, and plants such as ferns and Spanish moss are allowed to hang into the cages. Plants are placed away from perches to avoid being soiled. Occasionally the plants are sprayed with water and the sunbirds enjoy bathing in these. Success was achieved at the Dallas Zoo with collared sunbirds, housed in fiberglass cubes measuring 1.75 m^3 (61.8 cu. ft.), which had glass fronts. The male parent was removed as the female began to lay her clutch and returned when the chicks had fledged and were old enough to be removed. Black sunbirds are often double clutched, using the same nest twice, and if it remains in good condition, they will use it again the following season. Males will aggressively chase off any intruders in the immediate area of their nest.

Hand Rearing: Sunbirds are not particularly difficult to hand rear if provided with a plentiful supply of small insects. The small chicks have a relatively short beak and large gape, which makes feeding easy. Initially, chicks are fed hourly from 6:00 a.m. to 10:00 p.m. Once quills and feathers begin to appear, this can be changed to every one and a half hours. It is essential to keep chicks warm; when allowed to become chilled, sunbird chicks will immediately refuse to eat. A thin forceps is used, and food items should include small moths, flying ants, diced mealworms, pinhead crickets and grasshoppers, flies, and tiny pieces of minced pinkie mouse. Insects should be slightly crushed with the forceps. These items can be dipped into nectar to which a few drops of a multivitamin solution have been added. As soon as the chicks fledge, nectar can be offered in a tube, but insects must still be fed on a regular basis. A malachite sunbird that I reared remained hand tame and would fly to greet me and steal some nectar as I entered the aviary. This particular bird was raised largely on flies that I caught in a flytrap.

Ring Size: 2.5 mm

Species Requiring Similar Management:

Spiderhunters *Arachnothera* spp.
Honeycreepers *Cyanerpes* spp.
Leafbirds *Chloropsis* spp.
White-eyes *Zosterops* spp.

References:

Dellinger, RM. and J. Eckart. Captive propagation of the olive sunbird. *AFA Watchbird* 18 (2).

Jackson, S. and N. Nicholson. Sugar Junkies of the Bird World. *Africa Birds & Birding*. Volume 6, Number 3. Page 33–35.

Maher, J. W. 1991. Growth and Development of the Yellow-bellied Sunbird. *EMU* Volume 91, part 1, page 60.

Peaker, M. 1990. Nutritional requirements and diets for hummingbirds and sunbirds. *International Zoo Yearbook* 29. Page 109-118.

Riphenburg, D. Observations of Sunbirds in Aviculture. *AFA Watchbird* 18 (5).

Skead, C. J. 1967. *Sunbirds of South Africa*.

VanDegrift, N. Care and Management of Sunbirds. *AFA Watchbird* 16 (4).

21.19 Passeridae

21.19a Passerinae Old World Sparrows

Glen Holland

Introduction: The family Emberizidae (sparrows, buntings and relatives) falls within the order Passeriformes (perching birds). Members of the Emberizidae family are commonly referred to as American or New World sparrows and buntings. The common name "sparrow" is actually a misnomer and is based on the superficial resemblance the members of the Emberizidae have to the family passeridae (Old World sparrows). Old World Sparrows have similar size, shape and colors as New World sparrows and were mistakenly called sparrows by people arriving in the New World. New World sparrows are not actually closely related to Old World sparrows despite their name. In this chapter I will be focusing on the Old World sparrows.

Old World sparrows were originally found in Europe, Asia and Africa. However, as a result of introductions by humans, today they have an almost worldwide distribution. Old World sparrows generally inhabit open areas. They are well adapted to urban landscapes and can be found alongside humans throughout the world. Humans introduced Old World sparrows to the Nearctic, Neotropical and Australian regions. House sparrows *Passer domesticus* have been the most successful of the introduced sparrows; *Passer montanus* have been the most successful of the introduced sparrows. In fact, house sparrows, which are native to North Africa, Europe and Asia, are now the most widespread bird species in the world

Sparrows belong to the family Passeridae and the order Ploceidae. Most species occur in pairs and are generally common through most of their range. Unlike house sparrows *Passer domesticus,* the wild species of sparrows are nervous birds that cannot be approached closely and are rather secretive by nature. Usually seen on the ground feeding on seeds, they inhabit a wide variety of habitats, avoiding dense forest. They are often found at garden bird food tables. Nests are rather large loose structure built either in vegetation or a cavity.

Management: My experience is based entirely on six African species that I have managed. They adapt well to captivity and soon tame down to make observation easier. In the breeding season they become aggressive towards others of their own kind and in flights they are best kept one pair per aviary. They are not suitable for a mixed aviary of small seedeaters and should be mixed with larger finches, weavers, and small softbills. In large aviaries, two or three pairs will live in harmony provided they are able to nest at opposite corners of the aviary. In our Pilanesberg aviary, as part of the education program, I wanted to exhibit all the indigenous sparrows together to enable people to view them at close quarters and be able to identify and compare them. Gray-headed *P. diffusus,* great *P. motitensis,* Cape *P. melanurus,* and yellow-throated *P. superciliaris,* were all kept together in a very large walk-through aviary, and no serious aggression was observed between the species. The pair is fond of bathing together on warm days after a bath; they will sit side-by-side sunning themselves. A planted aviary as described for Habitat B in the chapter on LANDSCAPE will satisfy the needs of these birds. In severe climates they will need to be brought indoors during winter.

Diet: The basic finch seed diet, together with some commercial soft food and insects, such as termites, waxmoths and mealworms, will cover their dietary needs. Insects are essential to raise the chicks, and soft food mixed with a grated boiled egg, including the crushed shell, is also a favorite. A high-protein finch cake, as described for seedcrackers, is also relished by sparrows. Eggshells, which have been sterilized by placing them in a microwave for a few minutes, are consumed in amazing quantities. Grit in the form of coarse river or beach sand is consumed in lesser quantities. In addition, sparrows take spinach, thistle, apple, pear, and are particularly fond of white grapes.

Breeding: Cape and great sparrows are great nest builders, and if sufficient nest material is not available they will rob the nests of other species for material. To avoid this, a large, loose ball of dry grasses mixed with a handful of feathers and some carpet under cover should be supplied. The nest is built by both sexes and is a large, untidy ball of dry grasses, with a side entrance. The nest chamber is lined with fine grasses, feathers, and any other soft material that is available. As gray-headed and yellow-throated sparrows are cavity-nesting species that use abandoned nests of others, more nests should be provided than the number of pairs of cavity nesters. These sparrows are usually the first birds to take up residence in a hollow log.

During the Pilanesberg project they caused us much frustration by nesting in half of the hollow logs we had erected hoping to harvest chicks from the more unusual softbill species. Naturally they use hollows made by birds such as barbets, but in captivity they will also use an L-shaped lovebird nest box in which to breed. The nest is lined with some grasses, feathers, and other soft material. The 3–5 eggs are spotted with blue-gray markings. The incubation period is 13–14 days by both sexes, and the chicks fledge at 19–22 days. They are usually double brooded, and once independent the immatures should be removed.

Jeffery Trollope of Middlesex, England, provides the following account of the breeding and behavior of the yellow-throated sparrow *P. xanthocollis:*

> "As noted with *P. dentata,* most live food is accepted, with the exception of white worms. In April a pair were released into an aviary measuring 3.25 x 1.5 x 2.25 m (10.6 x 5 x 7.5 feet). The aviary was planted with privet bushes and a wide variety of nesting receptacles from waxbill-type nesting baskets to budgerigar nest boxes. On May 14, the male bird was observed singing at the entrance hole of a budgerigar nest box hung on the aviary side some 2 m (6.5 ft.) from the ground level at the front of the aviary. During the next few days this activity continued and he was seen displaying to female. The display consisted of the male approaching the female with the head lifted, displaying the yellow throat patch to her. The wings are moved to display the chestnut patches.
>
> On May 19 the female was observed carrying dried grass to the nest box and both sexes spent some time inside the box. The pair were seen copulating on a perch on June 11, the nest contained one egg when I examined the box. The next two eggs were laid daily and as far as I could ascertain incubation commenced with the third egg. The nest consisted of a crude cup made of dried grass, paper, etc., that nearly filled the whole nest box; the cup was lined with feathers. The birds were very wary about enter-

ing the box if they thought that they were being observed and I used binoculars to watch them. The three chicks hatched on the June 28 and the parents displayed great excitement, flying up and down the aviary and continually flying back to the box and going inside. I supplied lots of small maggots (known as "pinkies" to fisherman) and tipped a collection of insects and spiders into the aviary twice a day; these were obtained by beating nettle heads with a stick, into a plastic bag. The parent birds soon learned that this bag contained live food, and would fly down to the aviary floor as soon as they saw the bag, wanting me to shake out the contents. Six days after the chicks were hatched I supplied the birds with mealworms but these were ignored.

The parents would never fly to the box directly if they thought they were being observed but would fly to the rear of the aviary and would work their way back to the box, using the gorse clumps on the side of the aviary as cover. Both parents were soon removing fecal sacs from the nest box, and the chicks hunger calls could be heard from the seventh day after hatching. Two of the chicks' left the nest on July 12 (at 15 days old), and the third chick the following day. The male appeared to then take over the care of the young and on the 13th the female laid the first of her next clutch of three eggs and was incubating on the 15th. The young birds of the first brood were flying on the 14th, and one was seen feeding itself on the 22nd, although they were still soliciting food from the male on the 24th. The chicks were completely independent and were removed from the aviary on the August 3. The brood did not return to the nest after fledging (as is common in many species)."

Derek Washington of Sussex, England, provides a detailed report on breeding the house sparrow *P. domesticus* and while I will not give the entire report there are a few important points from his article that add to the report above.

"The pair were housed in a mixed aviary that included bullfinch, goldfinch, greefinch, siskin and redpoll. The sparrows were seen to fed on chickweed, sow thistle and plantain.

Display and copulation with the female was noted. Seen a few feet from the nest box, she was quivering her wings. The male flew around the aviary several times, briefly pausing now and then on perches or on top of the nest box. He then quickly mounted the female from behind. During copulation she stopped quivering her wings, but the male flapped vigorously as if to keep balance. They remained like this for 2–3 seconds, then the male flew of for 5–10 seconds and returned and repeated the performance. This happened eight times in all. Incubation appeared to be by the female alone. The house sparrow is one of those species that exhibits "puppy fat," that is the legs of the chicks are thicker than those of the adults. Measurements of this showed that there was a reduction in tarsus thickness between the ages of nine and 20 days of 0.4 mm. The chicks were not covered in any down, they were simply bald and ugly.

Yellow-throated Sparrow
Petronia xanthocollis

Hand Rearing: As common and hardy as sparrows are, they are not the easiest of birds to hand raise. The chicks should be raised as described under the section on waxbills and finches WAXBILLS AND FINCHES. In addition, small insects such as moths, waxworms, and diced mealworms maggots should be fed with a tweezer. The chicks must be kept warm until fledged and fed every one and a half hours. Once fledged, they should be placed in a sunny spot during the day and kept warm at night. Some soaked seed mixed with moistened softbill food is placed on the floor of the cage. Once fledged, they should be feeding themselves by 10–15 days.

Ring Size: 3 mm

Golden Song Sparrow *Passer luteus*

Natural Habits: This species naturally occurs from western Sudan southeast to southwest Ethiopia, eastern Uganda and through Kenya to southwest Somalia and south to north-central Tanzania (Clement, Harris & Davis, 1993). Also known as the Sudan golden sparrow *Passer luteus* and the yellow sparrow, the golden song sparrow is very common in dry scrub and thornbush, sparse savannah, and cultivated fields. They are highly social and gregarious, breed colonially and generally occur in flocks of up to 100, but some roosts can number several hundred thousands. Song sparrows spend most of their time in feeding parties on the ground where they feed on seeds and insects. The breeding season is timed to coincide with the onset of rain.

Management: The sexes are diomorphic, with the male bird having the head and under parts a bright canary yellow, with the back and wings a chestnut color and the tail dark brown. The beak is horn colored, turning black in the breeding season. The female has a pale brown head and back, with pale grayish yellow underparts. The chicks are rather dull versions of the female and sexual dimorphism is apparent in the first season, but males are only fully colored in the second season. They acclimatize well but are rather skittish by nature. They are not at all suited to cages, where they develop the irritating habit of throwing the head backward before dashing to another perch. Mortalities are generally high in cages, due largely to stress and stress-induced illness. Unfortunately, there tends to be an abundance of male birds when imported. Possibly the trappers feel that there will be a better market for the brightly colored males. It is impressive that in Australia where imports have been banned for many years, captive-bred strains of song sparrows are well established. It is a pity, however, that because wild-caught birds are freely available, aviculturists in other countries do not follow suit. They are not an aggressive specie and a colony of these birds can be kept with other small finches and waxbills. Flowing water or a mist spray on a hot day stimulates these birds to bath and they also sand bath regularly. Song sparrows are not particularly hardy and will succumb to extreme cold or extended periods of cold and wet. It is advisable in areas that experience extreme cold in winter to ensure the birds have a fully covered shelter; the roof and sides of the aviaries are best covered in polythene or similar material to avoid draufts and wet.

Diet: As described for sparrows.

Breeding: During the breeding season the bill of the male turns jet black and both sexes become very active and vocal. Song sparrows will build their nests in boxes that have the front half open or in the fork of a dry thorny branch, the nests often being built very close together. They are quiet birds out of the breeding season, but during the breeding season they steal the show with liveliness and their constant chatter. The nest takes 3–7 days to build and is built of thin twigs, including thorn twigs and small pieces of dry bush, and it has a short tunnel entrance.

Cape Sparrow
Passer melanurus

Nest building is on the go continually and before using the twigs, they have a strange habit of dropping the twigs in the water dish, presumably to soften them. The entrance is at the top of the nest and slopes down. The interior of the nesting chamber is lined with feathers. Display consists of the male dropping his wings, and bouncing from branch to branch around the female, at the same time making his sparrow-like chatter. The hen does not invite copulation but sits still while copulation takes place a number of times in quick succession. The three or four light blue eggs are white blotched with grayish brown, and hatch in 12 days. The chicks fledge at 16 days. Incubation is shared, but the female does the larger share.

It is not uncommon for chicks to fledge prematurely and such chicks are best returned to their nest until they are clearly old enough to fledge. Song sparrows are tolerant of nest inspections. Once the chicks are about a week old, the male will sometimes take another mate, build a nest, and hatch another brood. This can happen continually and a single male will cover up to three females. They nest repeatedly, often in the same nest that is refurbished between clutches and the chicks are usually raised well if sufficient live food is available. The male assists with some feeding at the nest initially but soon moves off to nest with the next female. Chicks are fed constantly, the parent flying between the nest and the food continually, each time returning to the nest with a full beak. Even after fledging, the chicks will die if a good supply of live food and high-protein rearing food is not supplied. Chicks can be left with the colony and even though not fully colored, they will breed in their first season. Due to the thorny nature of their nest it is not recommended that the birds be rung, as the rings will undoubtedly result in birds getting stuck in their

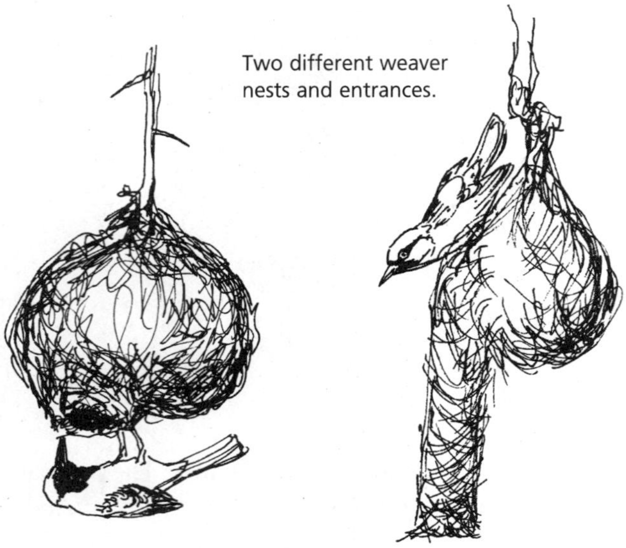

Two different weaver nests and entrances.

nest material. It would also be extremely difficult to gain access to the chicks in their thorny nest. Immatures are very similar to the adult female but grayer on the head, nape, and face and underparts also much paler.

Marcus Pollard records his experience with this species in Australia as follows:

> "At the onset of the breeding season, usually September in Australia, the males emit a constant machine-gun like 'chipping'. If you are unsure of this description, listen to a house sparrow and multiply the noise by 10. They leave you in no doubt that they are a member of the sparrow family. We supply our birds with pliable willow and gum twigs that are about 18 cm (7 in.) long. Not all the pairs are avian engineers and some just use a humble budgie-type nest box. With those that construct their own nests the actual nesting chamber sits on the very top of the structure and not, as you might expect, in the middle. There is very little material that they won't use to construct their nests so it is vital that you supply copious amounts of feathers, cotton wool, shredded tissue, coconut fiber in order to prevent them from robbing neighbors to complete their own nests (this applies to all sparrows). Regardless of the size of the nest, the inner chamber is filled with soft feathers and cotton wool that provides a very comfortable and insulated bed for the developing chicks. Parents are devoted carers of their chicks if they have plenty of live food and you can expect 3–4 nests per season per pair. The same nest is often reused for subsequent clutches but some pairs will construct a new nest for each clutch. A fellow breeder calculated that two breeding pairs with young would consume around 250 mealworms per day plus an unknown (but very large) number of maggots. Crickets are also readily taken when available. The birds do not search for live food and for example when mealworms are covered by bran they will be left. I have lost chicks when parents were fooled into believing I had not left them in live food in this manner.
>
> Certain males have been known to show overt aggression around their nest site, which is often directed towards other, usually more placid species. I have kept them in mixed collections for many years without any trouble, but little apart from song sparrows ever bred in the aviary. Their constant agitation often proves too much for smaller waxbills and grassfinches and little breeding is achieved. My birds are now housed only with diamond sparrows and breed happily but even diamonds are ill at ease around song sparrows. This season I saw the end result of an attack by a male song sparrow that left a red avadavat and pied red-faced parrot finch on the floor of the aviary. These birds had shown very little aggression in the past but one day was all it took to revise our opinion of that. Another two rather annoying traits are their habits of dust bathing in powdered vitamin supplements and carrying large shell grit all over the aviary. Despite this, their magnificent color, gregarious nature and constant chatter make them high on my list of favorites."

Ring Size: 3 mm

Species Requiring Similar Management:

Weavers
Arabian golden sparrow
Red-headed & cut-throat finch
Bishops

Contributions:

Peter Bender, Cape Town, South Africa
Marcus Pallard, Tasmania

References:

Clement, H., A. Harris, J. Davis. 1994. *Finches & Sparrows*: 442–473.

Maclean, GL. 1993. *Robert's Birds of Southern Africa*: 709–713.

Pollard, M. 2003. The Golden Song Sparrow. *New Zealand BirdZ* Issue 13 Nov–Dec 2003.

Trollope, J. 1979. The Breeding Behavior of the Yellow-throated Sparrow. *Avicultural Magazine* No 85, pp. 135–138.

Washington, D. 1973. Breeding the House Sparrow. *Avicultural Magazine* No 79, pp. 109–115.

21.19b Motacillinae Wagtails

Glen Holland

Natural Habits: Wagtails belong to the order Passeriformes, forming, along with the pipits, the subfamily Motacillinae in the family Passeridae, which includes the sparrows, accentors, weavers, estrildine finches, and whydahs. Wagtails occur in gardens, and along the edges of ponds, dams, and streams. They are unmistakable, walking along with the tail wagging continually, the bird moving quickly among pebbles on the water's edge, dashing after insects and even jumping into the air to catch them. They are usually found in pairs, but sometimes singularly. The subfamily Motacillinae contains only five genera, with *Dendronanthus* and *Tmetothylacus* being monotypic. The 11 species of wagtails in the genus *Motacilla* are widespread throughout the northern hemisphere. Seven species of longclaws are all included in the genus *Macronyx,* which is confined to the continent of Africa. The pipits, with 45 species in the genus *Anthus,* include representatives on every continent except Antarctica. The Willie wagtail *Rhipidura leucophrys* of Australia is not a true wagtail, but a fantail. Its common name was bestowed by early European settlers in recognition of its wagtail-like behavior.

Management: Wagtails adapt readily to captivity. Newly acquired birds need to be taught to feed on substitute foods, as described in greater detail under SHRIKES & ROBINS, by mixing these with live food. They soon learn to take the substitute foods as well. Hand-raised chicks are ideal avicultural subjects and take a wide variety of foods. The aviary for these birds should have a shallow stream or pond where the birds can walk in the shallow water and catch small insects. Habitat F, as described under the LANDSCAPE chapter, with a few dense shrubs for nesting, will supply the ideal habitat. Wagtails are not, however, particular about habitat, provided it includes flowing water, which they favor. The aviary floor should be natural. A few logs and boulders in the water are greatly appreciated by these birds, as they hop from one to the other while hunting. Although hardy and able to withstand periods of cold, wet weather, during freezing temperatures they should be housed indoors with some additional warmth.

They are best kept in sexed pairs, as during the breeding season they are strictly territorial. If more than one pair is kept together they will chase each other continually, leaving little time for successful breeding. Birds should always be introduced into a new aviary simultaneously to prevent aggression from an established resident toward a new introduction. Being extremely active birds, one pair in a large aviary will be visible most of the day as the birds dash about hunting prey. I have never noticed any sign of aggression toward any other species, and they can safely be housed with other softbills, or even finches and waxbills. I once placed a hand-reared green fruit pigeon in the same fledging cage as a hand-reared wagtail. Months later, long after being transferred to a large walk-through aviary, these two birds would often still perch side by side.

Diet: They naturally feed largely on insects, but in captivity they will also take minced ox heart, soaked cat kibble and commercial softbill food moistened with water. Insects such as waxmoths, mealworms, small moths, termites and pinhead crickets are firm favorites, and by using these to gain their confidence, the birds will learn to take food from the hand. In England, R. McCluskey maintained several species of wagtails on a diet consisting a homemade soft food mixture, maggots, mealworms, and stick insects. A red color food was added to the soft food and fed to the mealworms and maggots to help to prevent various softbills from losing color. When small minnows and other minnow-sized fish taken from a local river were added to this diet, the wagtails relied on this food item heavily, particularly when rearing young (McCluskey 1972).

Breeding: A true pair will usually nest very successfully in captivity. It is not uncommon for them to produce two clutches in a season. The nest is a thick structure of grass and sedges, with a deep cup lined with fine rootlets and hair. Both sexes build the nest and form a bulky base up to 200 mm (8 in.) in diameter, cup 55-65 mm (2–2.5 in.) across, 45-50 mm. Nests may be built in as little as 2 days or may take up to 3 weeks. Nest sites, not nests, are often re-used for successive clutches. The 3–4 cream-colored eggs are covered in fine brown spots. Incubation and feeding is shared by both sexes, the incubation period being 14 days, with the chicks fledging 14–17 days later. Newly hatched young have pink skin, orange mouth and are covered with tufts of gray down and are brooded for the first few days by both parents and at night for at least a week.

Nests are often constructed in the debris along the

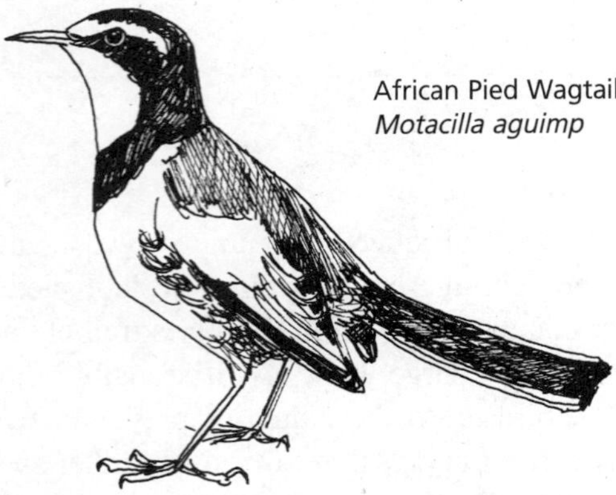

African Pied Wagtail
Motacilla aguimp

water side, but these birds are not fussy in their choice of nest sites, and I have seen them use a half open-fronted nest box in captivity, as well as a plant pot in a garden. I once had a pair of African pied wagtails *Motacilla aguimp* nest in the nose of a boat on the shore of a lake. Even after their chicks had fledged, this pair would fly out to escort the boat back to shore. The 3–4 cream-colored eggs are covered in fine brown spots. The chicks are fed roughly every 15 minutes, and initially small insects are fed by the beakful. As the chicks mature, larger food items are brought to the nest. Chicks fledge at three weeks and are fed by the parents for up to a month after fledging, and follow the adults calling continually and dashing forward to receive any prey they see the adults catch. The chicks are easily recognized by their short tails and continual cries for food. Once independent, they should be removed, particularly if the adults show signs of nesting a second time. The fledglings soon resemble the adults, which makes it difficult to differentiate them from the breeding pair when removing them after two months. It is therefore advisable to close ring the chicks in the nest at eight days of age.

A pair of gray wagtails *M. cinerea* bred in the aviaries of R. McCluskey in England. The nest was situated in a stone wall hidden by nettles and tall grass. It was constructed of grass, rootlets, moss, and horsehair, and was only discovered when the female wagtail was witnessed carrying a bill full of fish to the chicks. Four chicks fledged, but as the parents were largely feeding them on minnows, they were transferred to a large stock cage and gradually weaned on to a diet of soft food and insects (McCluskey 1972).

Hand Rearing: Wagtail chicks are easy to hand rear using equal portions of insects, minced ox heart and/or pinkie mice, and commercial rearing food. The ox heart/pinkie mice and rearing food are mixed together and a little water added. The food is fed using a forceps, with a mixture of natural and substitute food given at each feeding. If hand-rearing from the egg, the protocol described for SHRIKES should be followed. Chicks are fed hourly until half feathered, when the feeding interval can be extended to every 12 hours. Newly hatched chicks should be kept at a temperature of 36°C (96.8°F), which can be decreased by 1°C (1.8°F) per day until an ambient temperature of 28°C (84.2°F) is reached. Feeding starts at 6:00 a.m. and stops at 8:00 p.m. Once the chicks fledge, they should be offered a plate of very wet food in which they can walk; they will soon begin to peck at this. The addition of small live food will stimulate them to start eating. Hand-reared wagtail chicks do not remain hand tame when mature. They resemble the adults by three months of age.

Ring Size: 3 mm, placed on the chick at eight days.

Species Requiring Similar Management:

- Longclaws *Macronyx* species
- Pipits *Anthus* species
- Golden pipit *Tmetothylacus tenellus*
- Forest wagtail *Dendronanthus indicus*

Reference:

McCluskey, R. 1972. Breeding grey wagtails, *Motacilla cinerea*, using fish as food. *Avicultural Magazine* 78 (1): 20–22.

21.19c Ploceinae Weavers

Glen Holland

Introduction: Weavers belong to the family Passeridae in the order Passeriformes, and are further separated into the subfamily Ploceinae, which includes 118 species in 17 genera. More than half of the weavers are placed in the genus *Ploceus,* which contains 63 species. While the weaver subfamily is concentrated in continental Africa, two species in the genus *Ploceus* are endemic to Madagascar and another five species occur in India and southern and southeastern Asia. The genus *Foudia,* which covers the six species known as fodies, is confined to Madagascar and the islands of the Indian Ocean. The second largest genus of weavers, *Euplectes,* with 17 species, includes both the short-tailed BISHOPS and the long-tailed WIDOWBIRDS. The three species of BUFFALO-WEAVERS are found in two genera, *Bubalornis,* with two species, and the monotypic genus *Dinemellia*. The three species in the genus *Plocepasser* are known as SPARROW-WEAVERS, and the two members of the genus *Pseudonigrita* are known as social-weavers, not to be confused with the much-studied SOCIABLE WEAVER, *Philetairus socius,* which occupies a genus of its own. The SCALY WEAVER *Sporopipes squamifrons* is discussed in the chapter on WAXBILLS & FINCHES, which it resembles in size, habits, and captive requirements. The congeneric speckled-fronted weaver *S. frontalis* is somewhat larger, and usually kept with other weavers. The 11 species of malimbes comprising the genus *Malimbus* are birds of the rainforest. Another well-known genus, *Quelea,* contains three species of QUELEAS. The remaining six species of weavers are confined to monotypic genera. Several species of weavers were introduced and have subsequently become established in other parts of the world.

Most species of weavers are sexually dimorphic. In many species, the males have an eclipse plumage, resembling the rather drab females during the nonbreeding season, but assume bright colors when breeding. In general, weavers are highly gregarious, and breeding colonies of thousands of birds have been recorded in some species. Nest structures are comparatively elaborate and are always either roofed or constructed in a cavity. The 2–6 eggs are incubated by the females alone, but in most species the young are fed by both sexes. One species, the cuckoo finch or parasitic weaver *Anomalospiza imberbis* is brood-parasitic.

Buffalo Weavers *Bubalornis albirostris*

Introduction: Red-billed buffalo-weavers *Bubalornis niger* are most often seen on the ground in small groups of about a dozen birds. These parties of birds stay close together as they walk about looking for seeds and insects. When disturbed, the flock moves as one to a high perch, where the birds can be extremely vocal. Their calls consist of a variety of loud, rather metallic calls; they are particularly noisy around the nest sites.

Management: This rather large weaver is not well represented in avicultural collections. The adult male is

Buffalo Weaver
Bubalornis albirostris

black with some white on the flanks and on a wing patch, which is most obvious when the bird is in flight. The beak is red, and legs are orange. The female is brown with a mottled appearance on the under parts. They are quite aggressive, tending to dominate feeding stations in the aviary; for this reason they are best mixed only with larger seedeaters such as other weavers, and larger softbills such as glossy starlings. They are particularly aggressive when breeding, chasing any birds that venture too close to their nests.

I personally know of two instances in captivity where their aggressive nature became a problem. In the first, the weavers killed a sparrow; in the second, they evicted two newly hatched chicks from the nest of a green-crested turaco. Their aviary should be rather arid, with plenty of open, sandy areas on the floor, and should include thorn trees, both alive and dry dead thickets, in which the birds can build their nests. Either Habitat B or Habitat E, as described in the chapter on LANDSCAPE, will suit these birds. Buffalo-weavers should be kept in flocks of at least four pairs. Much of their day is spent collecting nest material, which should always be freely available to them. They will not hesitate to steal material from the nests of other species to line their own. Their ability to withstand severe climates is similar to that described for SOCIABLE WEAVERS.

Diet: As described for SOCIABLE WEAVERS.

Breeding: The nests are built high among thorny branches. Materials used include dry twigs, including thorn twigs, and the chamber is lined with grass and leaves. A good supply of thorny twigs about 20 cm (8 in.) in length will greatly assist the birds in completing their nests. There are usually two entrances to the nest chamber. The male builds the exterior stick structure, and the female adds the lining. The 3–4 eggs are greenish white in ground color, speckled with olive and gray. The female alone incubates the eggs and feeds the chicks. The incubation period is 12 days and the chicks fledge three weeks later. Initially, the chicks are fed largely on insects. Without such live food, the clutch will be deserted.

Hand Rearing: As described for the MASKED WEAVER and MADAGASCAR FODY.

Ring Size: 3 mm

Species Requiring Similar Management: White-headed buffalo-weaver *Dinemellia dinemelli*

Sociable Weaver *Philetairus socius*

Introduction: The sociable weaver *Philetairus socius* is a common species from dry areas of southwestern Africa, where they occur in small flocks in the neighborhood of their large communal nests. These nests have been known to cause electrical shorts on telephone lines. To prevent this problem, the telephone companies now provide platforms on the telephone poles from which the birds can suspend their nests, thus avoiding entanglement in the lines. Sociable weavers spend much of their time on the ground, where they feed in small groups on seeds and insects. They seldom venture far from the nest site and after feeding often return carrying material to add to the nest. This species prefers naturally dry and semi-desert regions, and seems to exist independent of water. In addition to the short, sharp *chik, chik* contact call, a rather metallic chatter is often carried on between birds.

Management: Newly acquired birds are slightly delicate, requiring extra attention during the period of acclimatization. Additional warmth should be supplied. As these birds naturally roost in their nests, a number of budgie-type nest boxes, each containing some dry grass, should be supplied. At least one nest box should be provided for each pair of birds in the aviary. Nesting material in the form of grass stalks should always be available on the aviary floor. Although sociable weavers are not generally as aggressive as sparrow-weavers, for example, they will fight off intruders at their nests, even falling to the ground in combat. They are best kept with softbills and larger finches, such as the red-headed finch *Amadina erythrocephala*. One problem that does occur is that sociable weavers are compulsive nest builders; if a ready supply of nest material is not maintained, they will steal all the material that other aviary occupants have used to build nests. Their aviary should be landscaped as described for Habitat E in the chapter on LANDSCAPE. They are relatively cold hardy and tolerant of cold nights as long as the weather remains dry and the daytime temperature exceeds 15°C (59°F). In more severe climates, where temperatures remain lower than recommended or where cold temperatures are combined with wet conditions, sociable weavers should be housed indoors with some additional warmth.

Diet: A mixture of finch and canary seeds, soft foods with grated boiled egg, which can be blended into commercial softbill food, blended soaked cat biscuit, insects, and sterilized eggshells should be supplied. Sprouted millet and sunflower seed are also favorite foods of most weavers. Sociable weavers will also take fruit, and a half apple or orange should be offered on a spike. They are highly insectivorous at all times, and a constant supply of live food such as termites, gentles, locusts, crickets, and mealworms is essential for the successful rearing of chicks. Grit and cuttlebone should always be available and a multivitamin/trace element powder added to the soft food twice per week.

Breeding: A constant supply of thorny twigs, dry grass stems 10–20 cm (4–8 in.) in length, and a few feathers for lining the nest should be placed on the aviary floor. The nest consists of two layers, a roof built of thorn twigs and a lower portion containing the nest chambers. Each chamber has its own entrance tunnel up to 25 cm (10 in.) in length and 6–7 cm (2.5–2.75 in.) in diameter leading vertically into a nest chamber measuring about 15 cm (6 in.) in diameter set off to one side of the tunnel. The nest entrance is lined with softer green plant material. The chamber is lined with soft dry materials such as blades of grass, grass inflorescences, and some feathers. Straws are carried to the nest individually while soft lining material is carried in large quantities in the beak. It is not essential to have a large colony to promote breeding and a few pairs put together will nest successfully. The sexes are alike.

The colonial nests are built from the top down and a wire platform on a strong frame should be supplied as an anchor for the nest. At Frankfurt Zoo, the weavers have built a colonial nest on an artificial tree designed with some support structures for the nest. Over a period of 14 years, the zoo bred 44 birds from a colony that averaged about 35 birds. Sharp grass stalks are placed into the entrance tunnel to the nests and these make the removal of chicks difficult. When entering their nests, the flock arrives together and the birds fly straight into the nest tunnels. They will also make use of nest boxes, often also building on top of the box. I once had a single female that laid and incubated clutches continually in a budgie box that she lined.

At the top of the tunnel is a round nest chamber where the 3–4 eggs are laid. The eggs are a dull white color covered in slate gray speckles. The incubation period is 13–14 days. The chicks fledge after 22 days, but are unable to fly well for the first few days. Without insects the birds will nest and hatch chicks, only to desert them after a few days. Dead chicks found on the floor will indicate that the birds are not satisfied with the diet offered. The young have less defined patterns on the back and on the flanks. Immatures from one brood will assist in feeding chicks and maintaining the nest during the next brood. By three months of age, they already begin to show the urge to build nests.

Hand Rearing: As described for MASKED WEAVERS and MADAGASCAR FODY.

Ring Size: 3 mm

Species Requiring Similar Management:

Speckle-fronted weaver, *Sporopipes frontalis*
Scaly-feathered finch, *S. squamifrons*

White-Browed Sparrow Weaver
Plocepasser mahali

Introduction: The white-browed sparrow-weaver *Plocepasser mahali* is a common species of the dry acacia woodland. These weavers live in colonies of 2–20 birds and their large nests are a common sight in the region they inhabit. They spend much of their time strutting about on bare patches of soil, picking up insects and seeds. The contact call is *chick, chick* and they also have a melodious warbling song. This species is often found in close association to human habitation.

Management: Their loud calls and interesting nests make them an ideal species for public display. Their one drawback is that they are extremely aggressive, particularly during the breeding season. This aggression is not limited to their own kind, and they will not hesitate to attack larger species such as crested barbets. I have had them kill a pied kingfisher and a black-headed oriole. Extreme caution must be taken when introducing new birds to an established colony of this species; the entire group will normally turn on and kill the new birds. In the

Sociable Weaver
Philetairus socius

Pilanesberg Aviary, I found that even birds collected from the same colony will suddenly turn on, and kill, a member of their own group. They also tend to dominate feeding areas, chasing off other species. Sparrow-weavers are best kept with species such as rollers, hornbills, and gray go-away birds. If a tree is not available in the aviary, a large thicket of dry sticks will provide a nesting site. The aviary should be landscaped as described for Habitat B in the chapter on LANDSCAPE. They are equally as hardy as sociable weavers.

Diet: As described for SOCIABLE WEAVERS.

Breeding: During courtship, both birds lower their wings and raise their head and tail, exposing the rump. Body feathers are puffed out. The male then performs a dance that involves swaying from side to side as he moves closer to, then away from, the female. During this dance he quivers his wings. The female remains motionless except for her rapidly quivering tail. After mating, the male chases the female for a short distance. The nest is an untidy round ball of grass with stiff stalks protruding outward at angles. It is constructed at the tips of a thorn tree branch. There are usually far more nests than pairs of birds, and I have had a single captive bird maintain seven nests. This attribute makes it possible to have a single specimen represent the species for public display, thus reducing the aggression in the aviary that can be expected with a larger number of these birds. A plentiful supply of stiff, dry grass stems is essential, as the birds spend much of the day on nest maintenance. Collection of nest material is accompanied by loud calling. A handful of feathers should be supplied periodically for nest lining. Roosting nests have an entrance at each end, while breeding nests have a single entrance and are lined with soft grasses and feathers. The 2–3 whitish pink eggs hatch after 14 days and the chicks fledge at 22 days. Immatures resemble the adults but have pale bills. In common with most weavers, the newly hatched chicks will be deserted if sufficient live food is not available to the parents.

Hand Rearing: As described for the MASKED WEAVER and MADAGASCAR FODY.

Ring Size: 4.3 mm

African Masked Weavers *Ploceus* spp.

Introduction: The true weavers share the genus Ploceus, in which all species except the forest weaver Ploceus bicolor, the olive-headed weaver *P. olivaceiceps,* and the spectacled weaver *P. ocularis,* are gregarious. A number of species are called masked weavers in various parts of Africa, including the lesser masked weaver *P. intermedius,* the Victoria masked weaver *P. victoriae,* Heuglin's masked weaver *P. heuglini,* the vitelline masked weaver *P. vitellinus,* the northern masked weaver *P. taeniopterus,* the Katanga masked weaver *P. katangae,* Ruwet's masked weaver *P. ruweti,* the Tanzania masked weaver *P. reichardi,* and the southern masked weaver *P. velatus,* to which these notes refer. They occur in small groups at all times of the year. The favored habitat is open woodland and scrub. They are seldom found far from a source of water and are common in wetlands. Flocks descend to the ground as a group to feed on seeds and insects, and then take off as a group. The contact call is a short, sharp chik, chik, but when displaying, a zwizz, zwizz call is made continually.

Management: Masked weavers are dimorphic during the summer breeding season with males assuming brighter yellow hues and the dark facemask. Outside the breeding season they are monomorphic, but males are larger and tend to show more white on the belly. Southern masked weavers are hardy birds that adjust well to captivity. They are confident, even cheeky birds, and are best mixed with larger softbills and doves. If mixed with small seedeaters, they dominate feeding areas aggressively. They will not hesitate to destroy a clutch of eggs or steal nesting material, such as feathers, from existing nests to add to their own nests. Species such as waxbills are clearly nervous when housed with these birds. Larger seedeaters, such as red-headed finches and sparrows, are better able to hold their own when housed with masked weavers. Their aviary should be landscaped as Habitat B, with a section of Habitat F to provide the required wetland nesting sites. Masked weavers are as hardy as sociable weavers. They are fond of bathing, and this is also a communal activity. In breeding condition, the male has red eyes. Immatures are dull olive green versions of the female.

Diet: As described for the SOCIABLE WEAVER. Masked weavers are also fond of green grass seed heads, which can be grown in the aviary. Insects such as termites, locusts, and mealworms are cherished. Flying insects such as moths are also actively pursued and sometimes caught in the air. They also enjoy nectar, such as described under the section on SUNBIRDS, but this is not necessary. In the wild state, they often appear to have an orange facial mask, due to the accumulation of pollen from the aloe flowers from which they obtain nectar and pollen.

Breeding: The nests are constructed by the males, and usually built between two reeds or on the tip of a branch overhanging water. They are oval structures built of coarse strips of grass and reeds. Once the nest is complete, males spend much of their day hanging at the nest entrance singing their "sizzling" calls and displaying with wings flapping slowly while they sway from side to side. A visit from a female will send all the males to their nests with much frenzied display and calling. Only after a female has accepted a nest will it be lined for breeding. Nests that have been rejected are usually bitten off at the supporting twigs and dropped to the ground, although it is not uncommon to have a new nest attached to an earlier one that was considered inadequate. Masked weavers usually nest in small loose colonies of about a dozen birds. A single male will mate with more than one female, unlike the lesser masked and spotted-backed weavers *P. cucullatus spilonotus,* which nest in large tight colonies.

The female alone incubates the 2–3 eggs that vary greatly in color, from white to pink, pale green, or blue, well marked with rust and gray spots. The incubation period is 13 days and the chicks fledge at 17–19 days. Chicks are fed largely on soft food, with some seeds and insects. Fledglings follow the adults begging for food, calling, and vibrating their wings continually. I had a group breed successfully in a mixed aviary, and as soon as the nests had been abandoned they were taken over by cut-throat finches *Amadina fasciata,* which successfully reared their chicks in the same nests.

Hand Rearing: Weavers are not difficult to hand-raise. Their hand rearing is similar to the detailed section on HAND REARING ESTRILDIDS in the chapter on WAXBILLS AND FINCHES. When the chicks are small, a syringe is used to feed moistened commercial hand-rearing formula with a 24% protein content. This should be just moist enough to be drawn up into a syringe. A few drops of vegetable oil and a multivitamin are added to the food. Small pieces of ox heart, pinkie mice and insects, such as small moths, flying ants, or diced mealworms, are fed daily with tweezers. Once the chicks are well feathered, the diet can be changed to half beef-flavored tinned cat food or minced ox heart, and half psittacine hand-rearing formula with the addition of a few insects at each feed. The mixture is moistened with water, and a few drops of multivitamin and a calcium supplement are added once a day. The chicks are fed every two hours from 6:00 a.m. to 10:00 p.m. This should be read in conjunction with the section on MADAGASCAR FODY.

When the chicks are well feathered, the mixture should not be too wet, as it is coarse and must therefore be fed with tweezers. Chicks respond energetically at feeding time. Tiny chicks shake the head vigorously, while larger chicks and fledglings vibrate their head and wings continually while begging for food. This begging is accompanied by continual loud cries, which can become extremely irritating, as they begin to cry with every movement made by the feeder. Care must be taken not to overfeed these chicks, as they will beg vigorously even when full. The bedding material on which they are kept is quickly soiled and should be changed at least twice a day. Alternatively, it is possible to remove the fecal sacs after each feeding. Once fledged, the chicks should be offered a dish of soft food, which they will approach, begging, and eventually start to peck at. If individually raised and handled regularly, the chicks will remain tame. I had an adult Cape weaver *P. capensis* (aptly named "The Vulture") that would fly to my shoulder and sing his "sizzling" song while pecking vigorously at my neck or hand. In this excited state his pupils were dilated, making his appearance even nastier than his behavior.

Ring Size: 3 mm

Species Requiring Similar Management:

- Typical weavers, *Ploceus* spp.
- Fodies, *Foudia* spp.
- Malimbes, *Malimbus* spp.
- Bishops and widowbirds, *Euplectes* spp.

Spectacled Weaver *Ploceus ocularis*

Introduction: Spectacled weavers *Ploceus ocularis* occur along forest margins, riverine bush, and well-planted gardens. They are territorial and, in common with the other insectivorous weavers, are found in pairs. Territories are easily located by the call of the birds. The call starts with a loud note and then fades away *tsee, tsee, tsee, tsee, tsee*.

Management: These weavers are not easy to establish in captivity. They are rather nervous and largely insectivorous. Newly acquired birds should be treated in the same manner as any other insectivorous bird. Spectacled weavers will also learn to respond to a call, coming closer to receive live food. As with other territorial species, the rule of one pair per aviary should be strictly followed. They can, however, be mixed with other birds and are best kept with softbills and large seedeaters. They are active birds that spend much of their time clambering about in dense vegetation in search of insects. I have also watched them drop to the floor to fetch an insect and then fly back up to a perch to eat it. Their aviary should be well planted with shrubs and a few creepers, where the birds will find some of their own live food. Landscape should consist of a mixture of Habitats A and B. These birds originate from subtropical climates and will suffer at temperatures below 16°C (61°F) for extended periods. In severe climates, they should be housed indoors during the winter months, with an additional source of warmth.

Diet: As described for the SOCIABLE WEAVER, but with greater emphasis on soft food, live food, and a daily ration of nectar. Initially, half of their diet should consist of live food such as mealworms, and the other

Grenadier Weaver
Euplectes Orix nest

half, softbill food. A little millet can be sprinkled on the food. They are fond of nectar, and the nectar described in the chapter on SUNBIRDS should be offered daily. As they become more established, the quantity of live food can be gradually decreased but never entirely excluded from their diet.

Breeding: Courtship includes wing shivering by both sexes. The nest is built of thin strips of grass and leaves such as flax. The male fetches most of the material but the female also collects some. The female spends most of her time organizing the interior of the nest, while the male weaves on the outside, particularly strengthening the area where the nest is bound to the tip of the branch.

The tunnel entrance is about 30 cm (12 in.) long. The 2–4 eggs are light green with gray and brown spots. Both sexes incubate and feed the chicks. The incubation period is 13–14 days and the chicks fledge after 18–19 days. Chicks are initially fed entirely on insects. I witnessed one instance in which a nest containing chicks fell apart. A nest from another species was put in place of the broken nest, and the chicks inserted. In no time the adults were back feeding their chicks, and within a few days the male had extended the entrance tunnel to his satisfaction.

Hand Rearing: The technique described for hand rearing MADAGASCAR FODY should be used, but this species is rather difficult to hand raise. To achieve success, the chicks should be fed insects such as winged termites, wax moths, and mealworms, which are dipped in a commercial rearing food and fed with tweezers. Chicks must be given additional warmth until fully feathered. They are fed every two hours from 6:00 a.m. to 10:00 p.m., and will take up to three insects per feed.

Ring Size: 3 mm

Species Requiring Similar Management:

Red-headed weaver *Anaplectes rubriceps*
Forest weaver *Ploceus bicolor*
Olive-headed weaver *P. olivaceiceps*

Bishops and Widowbirds *Euplectes* spp.

Introduction: Although the bishops and widowbirds share the single genus *Euplectes,* the differences between the two groups are almost as great as their similarities. Bishops have short tails, while widowbirds have long tails similar to whydahs, with which they are often associated. However, in contrast to the true whydahs of the tribe Viduini, the widowbirds are not brood parasitic. In general, the *Euplectes* weavers prefer aquatic habitats such as marshes, wetlands, and reedbeds. Bishops may be found in pairs, but occur more often in loose colonies, with the breeding groups well spaced. Outside the breeding season these birds form tight flocks that are seldom found far from water. In grain-producing areas, large flocks of bishops can cause considerable crop damage. Widowbirds are polygamous, with each male guarding a territory that may contain 2–4 females. Even small patches of rank grass, marsh, and reeds will often accommodate a few pairs.

Red Bishop *Euplectes orix*

Natural Habits: The red bishop *Euplectes orix,* or grenadier weaver (as they are known in Australia), is one of the most striking of the weavers and, consequently, one of the most popular among aviculturists. There are few birds that match the beauty of a male red bishop or golden bishop *E. afer* in breeding plumage. The bright red of the former or bright yellow of the latter contrasts sharply against a background of pitch black, and the birds are conspicuous at some distance. Male bishops have an eclipse plumage; outside the breeding season, they lose their gorgeous nuptial colors and resemble the drab, streaky brown females. Immature birds also resemble females, young males only assuming adult plumage in their second year. The contact call is a short *chip* and, during display, the male makes sizzling and rattling calls. In the natural state, they are found in reed patches during the breeding season, sometimes occurring in pairs, but more often in loose colonies in which breeding groups are well spaced. Even small reedbeds will often house a few pairs. In grain-producing areas, large flocks sometimes assemble and can cause considerable damage to crops. Outside the breeding season, these birds form tight flocks, which are seldom found far from water.

Management: While bishops are monomorphic outside the summer breeding season, males are generally larger than females, particularly across the crown of the head. Out-of-color males can be identified by color-ringing all of the birds in a group, listening for their sizzling calls, and noting which birds are doing the calling. First-year males may show a few bursts of color in their plumage, make the typical breeding calls, and even build the beginnings of a nest. Females, however, have been known to breed in their first year. Bishops are relatively hardy and can withstand temperatures as low as 10°C (50°F) with no problems. When temperatures drop below this level, particularly during wet weather, they should be housed indoors with a supplemental source of heat.

These delightful little birds are amusing to watch and ideal for a public display. They are best kept in a large, planted aviary that includes a spongy area with a reedbed. Habitat C, with a section of the wetland Habitat F, will suit these birds. Although generally not an aggressive species, red bishops will bully small seedeaters at feeding stations in flight aviaries. If they are kept in a mixed collection with such birds, the aviary should be large and well planted and the number of bishops limited to a few pairs. Alternatively, they can be kept in a large colony and mixed with larger birds. Other species of weavers, cutthroat finches, green singing finches, and buntings cope well when housed with bishops. The sight of a breeding colony of red bishops is particularly impressive in a public display. Weavers in general — and the red bishop is no exception — are not suitable for cages. A breeding aviary for these birds should measure at least 6 m (20 ft.) square, with a minimum height of 2 m (6.5 ft.)

Diet: Bishops will feed on a mixed seed diet supplemented, particularly during the breeding season, with insects such as termites, locusts, and mealworms. Bishops will learn to take a finch-type soft food and are also very fond of seeding grass heads. Seeds of wild grasses, such as *Panicum* and *Setaria* species, should be planted to supply this valuable food source. Sprouted seed, consisting of a mix of sunflower and white and Japanese millets, is popular and should be supplied at the onset of the breeding season. Some birds will take a small amount of fruit and sweet corn. If maintained on an inadequate diet of dry seed alone, these birds will fail to breed and males will not come into full color.

Breeding: Bishops are polygamous, each male constructing several nests to accommodate 3–5 females. Depending on the size of the aviary, a number of these breeding groups can be kept together. When males are housed in the same aviary, they will chase each other in short bursts, but this aggression only becomes serious when the aviary is too small to allow the birds sufficient hiding places or escape routes. During aggression, males will puff themselves up in between short chases of neighboring birds. There appears to be a hierarchy among the males, but this changes during the season with dominant males becoming submissive and viceversa as the season progresses. Courtship includes a swaying movement made on the reeds next to the nest, and a rattling call. The display flight of males involves flying airily over the reedbeds in a slow, bumblebee-like flight, with their plumes puffed out, while making sizzling sounds. The females play no active part during courtship. As noted above, each male will construct more than one nest, and when females approach, he will provide his finest performance by puffing out his plumage, swaying to and fro, and calling. Copulation follows after a female has shown acceptance of the nest.

Males are not particular about where they construct their nests, and in captivity will even build on the aviary wire. If, however, a pool of water is surrounded by vegetation, they will usually instinctively build their nests there. The oval nests have the opening on one side of the top and are made of thin strips of reed blades. Using his bill, a male will usually snip the base of a leaf and then fly away with it, tearing a strip from the leaf as he flies. Most nests are so thinly constructed initially that the eggs can be seen inside the nest. Incubation is by the female only and, as incubation progresses, she adds a lining of soft grass seed heads to the base of the nest. The eggs are pale greenish blue and the clutch normally consists of three eggs. The incubation period is 13–14 days and the chicks fledge at 14–16 days. As with most polygamous species, the chicks are fed by the female only. If a proper diet is available, two broods per season are common.

The secret to encouraging these birds to breed in an aviary is to provide the correct habitat. In a large, walk-through aviary, I constructed a shallow pool of water surrounded by *Cyperus* reeds, bordered in turn by dense clumps of seeding grasses, and the red bishops bred very well in this environment. Although a hardy species that has to naturally survive some cold winter temperatures, the red bishop should not be subjected to cold and wet conditions for extended periods. Night temperatures should not drop below 5°C (41°F); minimum daytime temperature should be 12°C (53.6°F). In addition, it is

Red Bishop
Euplectes orix

important to have sufficient shelter and cover for the birds, with approximately one-third of the aviary covered. I have known these birds to survive suboptimal conditions, but it should be kept in mind that after surviving the winter, the birds are expected to have a sufficient energy reserve to undertake breeding. In harsh climates, such as winter in the UK and much of the USA, I would recommend that these birds be confined to indoor enclosures. During periods of confinement, they will still need a constant supply of greens, both to eat and to shred. In nature, deserted nests of this species are often used by orange-breasted waxbills *Amandava subflava* for nesting. An exceptional longevity record in Britain for the red bishop is 28 years, but they will often live 15–18 years in captivity. With the mass importation of wild-caught stock hopefully reaching the end of its reign, it becomes even more important that those keeping the red bishop in captivity develop pure breeding strains of the species.

Ring Size: 3 mm

Species Requiring Similar Management:

Golden bishop *Euplectes afer*
Red-crowned bishop *E. hordeaceus*
Yellow-rumped bishop *E. aureus*
Orange bishop *E. franciscanus*

Widowbirds *Euplectes* spp.

Introduction: Widowbirds occur in tall, rank grass and areas bordering old cultivated lands, sedges, reedbeds, and wetlands. A number of males are often found in one area, each having his own favorite display perches. Females are seldom seen during the breeding season, spending most of their time below the grass canopy. The female is a drab bird, buff below, and slightly darker and streaked above. They strut about on open ground where they feed communally in the nonbreeding season. Males call *tizzk, tizzk* in flight, and a sizzling call is given during the display flight.

Management: Widowbirds, particularly the long-tailed *Euplectes progne* are large birds that are best mixed with birds of similar size, such as weavers or softbills. Although not aggressive, breeding males will intimidate smaller birds with their bulky plumage. They are best kept in a large, planted aviary, where a number of males can be kept together with no serious aggression. The aviary landscape for widowbirds should be a combination of Habitats C and F as described in the chapter on LANDSCAPE. This will result in a mixture of aquatic and grassland habitats, which should be ideal for these birds. They are not suited to small flights where their courtship display will be impaired. Widowbirds are excellent for public exhibits in large aviaries, where the displaying males are an impressive sight. Like the bishops, widowbirds should not be kept in cages or small flights. Each male should be housed with three females. Patches of rank grasses and reeds, particularly tall grass and *Cyperus* reeds, should be planted adjacent to a pool of water, and it is here that the females will build their nests. Two or three grass thickets should be provided in different areas of the aviary. Widowbirds are not difficult to breed, but success depends upon the correct habitat and a good supply of green seeding grasses. They are able to withstand short periods of cold and wet weather, and even nighttime temperatures as low as freezing. In the natural state, however, the days are warm, and if kept in climates that are more severe, these birds should be housed indoors during the winter months.

Diet: As for RED BISHOP.

Breeding: Courtship involves flights by the male over the females and potential nest sites. These displays include slow, deliberate wing-beats, during which the shoulder patches flash brightly. Courtship also includes the male sitting on a high perch with his tail fanned. During flight, the tail is thrust forward, arching toward the ground. During flight displays the male gives his sizzling call continually. In a large aviary, three or more breeding groups, each consisting of a male accompanied by two or three females, can be kept together. During courtship, the males will chase each other almost constantly, but these chases do not include serious aggression and serve to enhance the breeding instinct.

In the aviary in which I bred the white-winged widowbird *E. albonotatus,* I planted dense clumps of *Panicum* grass among a few dry branches; when thickets developed, the birds built their nests and produced young. The female builds most of the nest out of grass seed heads. It is placed just above the ground. The clutch numbers 2–4 light green eggs, heavily spotted with gray and brown. Incubation is by the female alone and lasts for 14 days. The chicks are fed by the female and fledge after 17 days.

Hand Rearing: As described for WAXBILLS AND FINCHES.

Ring Size: 3 mm

Species Requiring Similar Management:
Jackson's widowbird *Euplectes jacksoni*
Fan-tailed widowbird *E. axillaris*
Red-collared widowbird *E. ardens*
Yellow-mantled widowbird *E. macrourus*

Queleas *Quelea quelea*

Introduction: The red-billed quelea *Quelea quelea* is a nomadic species inhabiting savanna areas. Queleas occur in large flocks, which, after a good season, may number hundreds of thousands of birds. They roost communally in reedbeds. Due to the damage that these birds cause to crops such as wheat, they are often referred to as "feathered locusts." In one year in Zimbabwe, an estimated 120 million birds were destroyed. Unfortunately, control methods are not very selective, and many other species, including raptors, which feed on poisoned birds, are killed.

Management: Queleas are hardy birds, and one of the few species that will survive being introduced directly into an aviary immediately after capture. The breeding male has a black mask over the forehead, face, and throat. Nonbreeding males resemble the female, and both sexes have red bills. Queleas are only really vocal in the breeding colony, and sometimes in flight. They are rather nervous birds that take a long time to settle in flights, but they soon settle down in larger planted aviaries. They should be kept as a small colony of at least 10 birds. The aviary should be planted as described for Habitat B or C, with clumps of reeds or bamboo in which they will nest. They are not particularly fussy about their nest site. Queleas should not be mixed with small seedeaters, but rather with more robust species, such as cut-throat finches *Amadina fasciata,* Java sparrows *Padda oryzivora,* and golden song sparrows *Passer luteus*. Although hardy birds, able to withstand cold temperatures down to 0°C (32°F), in severe climates with extended periods of cold, wet, and low temperatures, they should be housed indoors with additional warmth.

Diet: Queleas will thrive on a finch seed mixture, but will also take live food and soft food that is mixed with sprouted seed, particularly when there are chicks in the nest. They have a habit of jumping forward with both feet to move the earth and seeds on which they are feeding. They are fond of half-ripe seeds, and rows of millet seed should be planted in their aviary. Cuttlefish bone, sterilized egg shells, and grit should always be available.

Breeding: The nest is rounded, with a large entrance, and is constructed by the male alone. It is woven from strips of grass and is a rather feeble structure. Each male maintains a small territory around his nest where he displays. Once a female has accepted the male and his nest she crouches and calls like a chick, and copulation soon follows. The 2–5 eggs are pale greenish blue, and are incubated by both sexes for 10–11 days. Egg laying is usually synchronized in the colony, females often laying the first egg before the nest is complete. The chicks are fed by both sexes and fledge at 12–13 days, but remain close to the nest for several days, until they are able to fly well. Insects are essential in the diet for the first few days after the chicks hatch. They are usually multiple brooded, and their productivity can cause a problem, particularly in large planted aviaries. Their numbers have to be controlled to avoid their complete domination of the aviary. A large colony of these birds will consume vast quantities of seed, which will prove very expensive.

Hand Rearing: As described for WAXBILLS AND FINCHES.

Ring Size: 2.5 mm

Mauritius Fody *Foudia rubra*

- Andrew Owen

Introduction: The Mauritius fody *Foudia rubra* is one of only six endemic passerine species remaining on the island of Mauritius. Some fody species are generalist feeders, eating a variety of food items such as seeds, fruit, insects, and nectar. The Mauritius fody has a more specialized diet, feeding mainly on insects and nectar. It has evolved a brush-tipped tongue for sipping nectar. Classified as critically endangered, the declining population of Mauritius fodies is estimated to be between 100–120 pairs. Loss of suitable native habitat throughout the island and predation from introduced exotic mammals, primarily black rats *Rattus rattus* and crab-eating macaques *Macaca fascicularis,* has confined it to small areas of upland forest within the Black River Gorges National Park.

Within the park, a six-hectare grove of exotic Japanese cedar *Cryptomeria japonica* surrounded by and inter-planted with native forest, known as Pigeon Wood, is of particular importance to the species. The prickly, sap-laden conifers provide some security for nesting Mauritius fodies and, as they produce no edible fruit, are less frequently used by arboreal mammals. Pigeon

Wood holds one of the largest concentrations of Mauritius fodies (9–10 pairs), and studies have shown that nesting attempts are considerably higher in *C. japonica* — 45% success to fledging, compared to only 6% in other (native and exotic) tree species. As the Mauritius fody remains extremely vulnerable, the Mauritian Wildlife Foundation (MWF) passerine team has concentrated its attention on the species in recent years and is investigating ways to increase the dwindling population.

The planting of additional areas with the non-invasive *C. japonica* and the control or eradication of introduced mammalian predators would play an important role in the long-term survival of the Mauritius fody. Other intensive management techniques may have to be implemented in the short-to-medium term to ensure the survival of the species. One option is to harvest or rescue chicks from wild nests and hand rear them in captivity. The hand-reared young birds could then be moved to other locations, such as safe predator-free offshore islands. Some of the hand-reared birds could be established in captivity to act as founder captive breeding stock to produce young birds for future release. From October to February of the 2002/03 fody breeding season, an experimental trial was implemented to harvest the nests of wild Mauritius fodies, hand rear the young, and establish them in captivity at the aviaries of the Gerald Durrell Endemic Wildlife Sanctuary, Black River.

Chick Collection: The nests were collected and driven to the Black River aviaries, 45 minutes away. As the chicks were at least three days old and the nests were heavily lined with pink pigeon feathers, it was thought unnecessary to transport them in a portable incubator. On arrival at the aviaries, the chicks were removed from the nest, weighed, and placed in their sibling groups in small plastic tubs lined with tissue paper and part of the nest substrate (fine twigs, grasses, and plant rootlets). This material provided a firm, non-slip surface which the chicks could grip. As Mauritius fody nests are domed structures, a sheet of paper towel was placed over the nest bowl to resemble the darkened nest and cover the chicks.

Brooding: Depending on their estimated age, the chicks were placed in one of three forced-air Roll-X incubators set at different temperatures. The incubators were maintained in an air-conditioned room at a temperature of 25°C (77°F). Young chicks, estimated to be three or four days old were brooded at a temperature of 34°C (93°F). Older chicks, 6–7 days old were placed in an incubator a degree cooler. As the chicks developed, the temperature was gradually reduced. When they were 8–10 days old, the temperature was 32°C (90°F). 31°C (87°F) from 11–14 days and maintained at 29–30°C (84–86°F) from day 14 until the chicks fledged at 16–18 days. Humidity was maintained at 50–65% RH throughout the rearing period using shallow trays containing deionized water.

Feeding: Chicks were given their first feeds two hours after collection. By this time they were hungry and eager to feed from the forceps. Vibrations from opening the incubator lid were usually enough to trigger a begging response. Chicks reluctant to beg during their first feeds were stimulated to feed by tapping on the side of the nest bowl. Food offered comprised of bee larvae, scrambled egg (plain egg, beaten and microwaved for two minutes) and small pieces of papaya. This was lightly dusted with Nutrobal multi-vitamin powder, every second feed. Chicks reared later in the season were also fed pureed or finely chopped pieces of pinkie mouse as part of their diet, replacing the bee larvae with pinkies during alternate feeds. (This food source was not available during the rearing of earlier chicks and is not included in the first table). Typical feeds on the rearing sheet are as follows:

3 sml bee larvae. Chick begging well.
4 sml bee larvae, 1 piece scrambled egg.
3 bee larvae, 1 egg, 1 cricket abdomen, Nutrobal.
5 bee larvae, 1 egg, 1 papaya. Fecal sac.
5 bee, 1 cricket abdomen, 2 egg, Nutrobal.

Twelve feeds per day were given to young chicks (three days old) at one-and-a-half-hourly intervals. The first daily feed was given at 6:00 a.m. and the last feed at 10:30 p.m. The interval between feeds was increased to every two hours and the number of feeds was gradually decreased as the chicks developed. Between days four and five, chicks received 11 feeds per day. Between days seven and nine, chicks were receiving nine feeds a day, with the last feed at 10:00 p.m. By 18 days of age or until they fledged, if sooner, chicks received eight feeds, with the last feed at 9:00 p.m. Mauritius fody chicks have an elongated crop that is clearly visible when full, on the right side of the neck. It was important that the crop was not over-filled and had emptied prior to the next feed. Large fecal sacs, up to 2 g (0.07 oz.) in weight from older chicks, were usually produced by each chick after the first morsel of food was swallowed. Fecal sacs produced by young chicks were removed from the nest after feeding, older chicks were encouraged to defecate over the edge of the nest onto a piece of tissue paper, maintaining a clean environment for the young birds. After feeding, the forceps were washed and placed in a solution of Virkon S. virucidal disinfectant at a dilution rate of 1:100. Forceps were rinsed prior to the next feed.

Approximate Amount of Food Offered at Each Feeding.

Chick Age	Bee larvae	Egg	Papaya	Weight of food (g)
3	3–4	1	1	.648–1.068
4–5	4–5	1–2	1–2	1.068–1.226
6	5–6	2–3	2–3	1.22–1.73
7–8	6–7	3	3	1.73–1.80
9–10	7–8	3	3	1.80–1.87
11	8	3–4	3	1.87–1.91
12	7–8	3–4	3	1.73–1.91
13–14	6–7	3	3	1.73–1.80
15	6	2–3	2–3	1.29–1.73
16–18	6	2	2	1.73

Note: 1 bee larvae = approx. 0.07g, 1 piece egg (approx. 5mm³) = 0.038g, 1 piece papaya (approx. 5mm³) = 0.4g.

Weighing: Chicks were weighed before every second feed using Acculab digital scales, accurate to 0.1 g (0.004 oz.). Three of the 14 Mauritius fody chicks reared were harvested from the wild when they were 13–14 days old. These birds were initially very nervous and were therefore not put through the additional stress of weighing.

Chick Fledging and Weaning: Chicks fledged at 16–18 days old. At this time they became very active and were able to fly and perch well. All chicks left the nest at the same time, even though there may have been an age difference between them of up to two days. Once they had left the nest bowl, they were removed from the incubator and placed in a wire mesh cage 60 x 40 x 40 cm high (24 x 16 x 16 in.) furnished with natural branches for perching. The fledging cage was kept in a room maintained at a temperature of 28–29°C (82–86°F). Once chicks had been placed in their fledging cage, they were offered a shallow dish of insectivorous mixture, to which was added finely chopped egg, bee larvae and finely chopped fruit such as grape, banana, papaya, and apple. In addition, pieces of papaya and mango were spiked onto perches within the cage. A shallow dish of water was provided for drinking and bathing. Fledglings were observed pecking at the spiked fruit within hours of being placed in the cage and bathing was observed two days after being placed in the cage (18–20 days old). Forceps feeds of bee larvae, egg, and papaya were offered every two hours throughout the day, with the last feed given at 6:00 p.m. After two days, the number of forceps feeds was reduced to four per day to encourage the chicks to feed themselves. After four or five days in the cage (20–23 days old) they became reluctant to feed from the forceps and began picking at the food offered in the dish, this was further encouraged by dropping bee larvae into the dish from above on a regular basis throughout the day.

Daily Growth Rates for Three Hand Reared Mauritius Fodies *Foudia rubra*

Temp.	RH	days	Chick #1	Chick #2	Chick #3
34		1			
34		2			
34		3			4.70
34		4	6.40	5.60	6.32
34		5	8.04	7.02	8.85
33	51	6	9.88	9.55	11.25
33	55	7	11.48	11.14	12.82
32	54	8	13.19	12.62	14.36
32	57	9	14.86	14.39	15.09
32	55	10	17.03	16.11	17.04
31	51	11	17.22	17.20	16.70
31	55	12	16.96	17.01	17.45
30	56	13	17.71	17.34	17.53
30	55	14	18.03	18.20	18.30
29	54	15	18.70	19.60	18.90
29	54	16	18.50	19.70	18.70
29	54	17	18.10	20.30	17.40
29		18			

Mauritius fody chicks are usually independent and self-feeding after 9–10 days in the cage (25–28 days old), although they may still solicit for food. At this time they also received a tube of artificial nectar (Avesnectar) and flowers of *Hybiscus* sp. and Bbottle-brush, *Callistemon citrinus,* when available. Once they were independent, the young birds were moved to a larger cage 90 x 50 x 50 cm (35 x 20 x 20 in.) where they had more room to exercise. They were placed in an outside aviary, within their weaning cage during the day, to acclimatize to the hotter conditions. The cage was covered with a cloth and placed in a shady position to prevent the birds from overheating and was returned to the rearing room at night, where a low wattage light was left on overnight.

Outside Aviaries: After 5–7 days acclimatizing to the outside temperatures, the young birds were moved to one of the aviary shelters 2 x 1.5 x 2 m (6.5 x 5 x 6.5) deep, where they were confined for a period of 24–48 hours. Once accustomed to their surroundings and feeding well, the young fodies were given access to an outside flight. These enclosures were 8 x 2.5 x 3 m high (26 x 8 x 9.75 ft.) and were covered in 1.4 cm^2 (1/2-in.) wire mesh. Each aviary was connected to a shelter and had the back quarter of the roof covered in shade cloth. Each aviary had a substrate of soil, grass, and leaf litter and was planted with a number of shrubs and small trees. A concrete food platform surrounded by a shallow moat of water prevented ants from gaining access to the birds' food. The fodies received a dish of food (insectivorous mix, egg, grated carrot, finely diced fruit, and bee larvae) and pieces of fruit spiked on branches throughout the aviary at 7:30 a.m. Additional egg, bee larvae, or insect feeds of small- to medium-sized cockroaches, crickets, and leaf-hoppers were given every two hours during the day. A tube of nectar was provided at 2:00 p.m. Insects form a major part of the fodies' diet and the birds were frequently seen searching for them among the foliage and leaf litter. Flying insects such as moths and cockroaches were chased and caught on the wing. Night-lights were left on in the aviary shelters to attract night-flying insects into the aviaries.

Durrell Wildlife Conservation Trust supported this project, allowing me leave to carry out this important initial husbandry work on the Mauritius fody.

Contributor: Dr. S.G. Stadler, Curator of Birds, Frankfurt Zoological Gardens.

References:

Maclean, GL. 1972. The sociable weaver — nest architecture and social organization. *Ostrich* 44 (3 & 4).

Maclean, GL. 1993. *Roberts Birds of Southern Africa*. Cape Town, SA: John Voelcker Bird Book Fund.

Mundy, PJ. & MJF. Jarvis. 1989. *Africa's Feathered Locust*.

Owen., A. 2003. "Hand-rearing the Mauritius Fody, *Foudia rubra*." Unpublished report by the Bird Department Durrell Wildlife Conservation Trust.

21.19di Waxbills & Finches *Estrildini*

• Glen Holland

Introduction: Finches and waxbills belong to the order Passeriformes, the family Passeridae, the subfamily Estrildinae, and the tribe Estrildini, which encompasses all of the species referred to as estrildids. This includes the typical waxbills *Estrilda* spp., the CORDON-BLEUS and GRENADIERS, the FIREFINCHES, the TWINSPOTS, the PYTILIAS, the crimsonwings, *Cryptospiza* spp., the SEEDCRACKERS, the bluebills, *Spermophaga* spp., the OLIVEBACKS, the negrofinches *Negrita* spp., the antpeckers *Parmoptila* spp., the AVADAVATS and the GOLDBREASTED WAXBILL, *Amandava spp.,* the QUAILFINCHES and locustfinch, *Ortygospiza* spp., the MANNIKINS and munias, *Lemuresthes, Lonchura, Heteromunia,* and *Padda* spp., the RED-HEADED FINCH and the CUT-THROAT FINCH, *Amadina* spp.; the AUSTRALIAN GRASSFINCHES, the GOULDIAN FINCH, the PARROTFINCHES, and the mountain firetail *Oreostruthus fuliginosus*. The subfamily Estrildinae also includes the tribe Viduini, the indigobirds and whydahs, with the single genus *Vidua* containing 16 species, as compared to 140 species in 29 genera in the tribe Estrildini. The indigobirds and whydahs are parasitic on other estrildids. These birds range in size from the tiniest waxbills at 9 cm (3.5 in.) in length to the Java sparrow *Padda oryzivora* at 17 cm (6 in.), excluding the whydahs that are small birds whose males develop nuptial plumage and extremely long tails during the breeding season.

Outside the breeding season, estrildids form flocks of 12–200 birds. They are active birds, often spotted feeding on the tips of seeding grasses. They feed primarily on seeds and insects on the ground, seldom venturing far from cover. Many species flick the tail from side to side. Species such as grenadiers, grassfinches, painted firetails and cordon bleus occur in short, dense and thorny scrub in dry areas. Waxbills and pytilias are particularly fond of acacia thorn thickets, which have dense clumps of seeding grasses growing between them. Tall grass and sedges in moist areas are the preferred habitat of some, such as gold-breasted *Amandava subflava* and common waxbills *Estrilda astrild*. Many species, such as parrotfinches, twinspots, mannikins, and firefinches, live in seeding grass patches on forest margins. During the dry season the estrildids are found in flocks in the dry areas; these flocks regularly visit water holes to drink and bathe. Wild populations of many of the waxbills are said to be declining as a result of habitat degradation and large scale capture. Habitat C, grass and shrub thickets, with a small reed/wetland patch, as described under the chapter on LANDSCAPE, will suit most species. They have very involved courtship displays, as well as ritualized nest-building behavior and responses to nestlings. The nestling of each species has characteristic markings on the palate and sides of the tongue, with often brightly colored gape tubercles. Although not estrildids, I am including the Cuban finch *Tiaris canora,* a grassquit, and the scaly weaver *Sporopipes squamifrons,* in this chapter, due to their size and avicultural requirements, and the fact that they are often kept in mixed collections of estrildids and other small seedeaters.

Diet: A standard finch seed mix, along with a portion of wild grass seeds, provides the basic diet. To maintain good condition and promote breeding sprouted seeds, green grass seed heads, and insects such as fruit flies, termites, gentles, and mealworms are essential for some species. The bunches of freshly picked grass seeds will remain fresh in the aviary if the stems are placed in a bottle of water. The neck of the bottle should be narrow to prevent access by the birds. Green seeding grasses with the seeds in the milky stage are a favorite natural food that will also stimulate many of the finches to breed. Millet can also be grown either inside or outside the aviary for feeding to the birds while the seeds are still moist. Spray millet is enjoyed by most finches but has no more dietary value than the dry seed purchased at a shop. Dry seeds are high in calories and carbohydrates, up to 73%, but have a very low fat value, around 2.5%. However, insects have a high fat content. Termite alates, for example, have a fat content of 25%; the workers, 12%. The birds soon get to know the feeding routine and eagerly await the presentation of their live food. Some species, such as red-eared waxbills *Estrilda troglodytes* and orange- cheeked waxbills *E. melpoda,* show little or no interest in live food. I have had grenadiers take moistened brown bread and recently saw them feeding in an aviary on boiled yellow corn on the cob. Commercial canary soft food, particularly when mixed with grated hard-boiled egg, sprouted seed, and grated apple, is another favorite among many of the waxbills. Many aviculturists add a few drops of cod liver to the softfood mix. Finch cake, as described for SEEDCRACKERS and PARROT FINCH, will also be taken by most of the waxbills. Eggshells that have been sterilized by cooking for two minutes in the microwave, cuttlebone, and grit should always be available to waxbills.

Estrildids are generally not comfortable at daytime temperatures below 16°C (61°F). Overnight temperatures can drop to single figures provided the birds face a warm, dry morning. In severe climates, where cold temperatures persist for extended periods, particularly if combined with wet conditions, these small birds should be housed in a heated indoor aviary or birdroom. Even in more favorable climates, sufficient shelter should always be available to allow these birds to avoid short spells of cold, wet weather. Another important factor is the length of the birds' daylight hours. Many estrilids begin to suffer when the daylight drops below 12 hours, resulting in insufficient food and body reserves to get them through long nights. I personally ensure that the birds have a minimum of 12 hours of light.

Olivebacks *Nesocharis* spp.

Introduction: Three species of olivebacks form the genus *Nesocharis*. One species, the gray-headed oliveback *Nesocharis capistrata,* also known as the white-cheeked olive weaver, has occasionally appeared in aviculture as an uncommon addition to mixed shipments of waxbills and finches. The other two species, the white-collared oliveback *N. ansorgei* and the Fernando Po oliveback *N. shelleyi,* are virtually unknown in aviculture. Olivebacks occur in small parties of 2–6 birds. They forage among the leaves of plants in a rather

Long-tailed Widowbird
Euplectes progne

warbler-like way, feeding on seeds, as well as wild figs. They are often found in the canopy of trees, where they search for insects, but also descend to the ground to feed on seeding grasses. Their calls include a soft *chwee, chwee* and an excited chattering alarm call.

Management: Olivebacks are best kept as sexed pairs and can be mixed with other waxbills and finches. However, they will kill others of their kind once breeding commences. In the aviaries of Chris Koen in Durban, South Africa, a male oliveback attacked and killed two spare males as soon as incubation began. The sexes are alike, and surgical DNA sexing is recommended. Their aviary should be planted in accordance with Habitat C as described in the chapter on LANDSCAPE. A few dense shrubs and clumps of wild grasses provide a habitat where they will spend much of their time in search of insects. The aviary should not, however, be too lush or damp. As this species originates from a tropical climate, they will not be comfortable in temperatures below 12°C (53.6°F); in severe climates they are best housed indoors during the cold winter months. A combination of extended periods of cold and wet will prove fatal to this species.

Breeding: Courtship involves the male taking a grass stem in his beak and bobbing up and down alongside the female. They are secretive nesters and, in captivity, like to build in a dense, dry bush suspended in the aviary shelter. The nest is a large round ball with an outer lining of broad-leafed material and an inner lining of fine teff grass. The nest is usually constructed about 1 m (3.3 ft.) from the ground, and has a side entrance. The 3–4 eggs are white and hatch after an incubation period of 15 days. A plentiful supply of live food is essential if the birds are to rear the chicks successfully. Without this live food, the nest is deserted after a few days. When nesting, olivebacks become aggressive and will chase any species that ventures close to the nest site. They become extremely agitated during nest inspections, dashing about excitedly while giving their alarm call. It is advisable, particularly with the first clutch, to avoid nest inspections. Chicks fledge after 23 days and resemble the adults after four months. They should be removed as soon as they are independent, to prevent interference with subsequent nests and to keep them from being subjected to aggression by the adults. Immatures can be distinguished by their dark gray head pattern, lack of yellow on the flanks, generally darker plumage, and whitish bill with a gray tip, which darkens with age.

Ring Size: 2.5 mm

Species Requiring Similar Management:

Crimson-wings *Cryptospiza* species
Parrotfinches *Erythrura* species
Seedcrackers *Pyrenestes* species
Bluebills *Spermophaga* species

Contributor: Neville Brickell, Avicultural Research Unit, Durban, South Africa

Pytilias *Pytilia* spp.

Introduction: The genus *Pytilia* includes five species, of which the green-winged pytilia, *Pytilia melba,* more commonly known as the Melba finch, most frequently appears in avicultural collections. The female is easily differentiated from the male, as the red on the head and throat and the yellow on the chest of the male are replaced by a wavy gray pattern in the female. Immatures are olive on the wings, gray below, with the rump and tail pale red. The Melba finch is a common species of the dense, dry thorn scrub. They prefer areas of short thorn scrub to large woodland trees.

Management: The aviary for these birds should be planted with clumps of grasses with open patches of soil left in between. Dense clumps of dry twigs are a favorite hideaway and nest site for Melba finches. Their call is a short *wick* and a soft warbling song, usually made from dense cover. Melba finches are extremely fast flying, and I have found them the most difficult finch to capture with a net in an aviary.

As pytilias are delicate when first acquired, they should be treated for stress with an electrolyte solution in the drinking water. Live food and additional heat should also be supplied at this time. In common with the other pytilias and as indicated by their long, thin bill, Melbas are highly insectivorous; however, they will also learn to take soft food that has been moistened and mixed with grated boiled egg. The pair bond is very strong in this species. They are aggressive birds and should be kept one pair to a flight. Since their aggression is normally only directed at other pytilias, they can safely be mixed with other finches and waxbills. During the breeding season, however, the Melbas may even chase these birds away from a source of insects or a nest site. They are very willing to nest in captivity, but success will depend on a constant supply of live food. A study in nature has shown that they feed almost entirely on the ground; 50% of the crop contents of specimens examined consisted of insects.

Breeding: The nest is built in a dense shrub or a wicker basket. In the natural state, 81% of nests were found 51–200 cm (20–80 in.) from the ground. The nest is constructed of grass stems that protrude around the side like bristles, and can be distinguished from waxbill nests by the rather coarse blades of grass used on the exterior. The male collects material and builds the nest, with the female regularly checking on his progress. It is well lined with feathers that are placed over the entrance, making nest inspection difficult. Both sexes often bring a feather to the nest when relieving the mate during incubation, which happens roughly every one-and-a-half hours. The eggs are pure white. See the table for clutch sizes as given by Skead (1975):

EGGS PER CLUTCH	2	3	4	5	6
PERCENTAGE OF NESTS	5.9	21.6	33.3	35.3	3.9

Both sexes incubate, starting with the first egg; the incubation period is 12–13 days. Newly hatched chicks have dark skin and are covered with sandy-white down. The gape markings are brightly colored. Older chicks are covered with gray feathers and fledge at 21 days, spending the first few days on the aviary floor. The adults become extremely agitated if the chicks are approached or interfered with. Initially, the fledglings are gray with light orange tails and a touch of olive on the back. By the age of 25 days, the young males begin to show their first red feathers; at a month, the barring on the lower parts begins to appear. The adults continue to feed the young for 2–3 weeks after fledging. By two months of age, however, the chicks must be removed. If left with their parents, the young males in particular will be persecuted by the adult male.

Two to three broods are usually raised each season, pairs often nesting again within two weeks of the chicks fledging, sometimes using the same nest. I have found that if the parents lay too soon after a clutch has fledged, the newly hatched chicks of the second brood are deserted and the subsequent brood then raised. Breeding seasons for this species are separated by a six-month period of nonbreeding activity. The hatching date can be determined by observing the behavior of the adults. When the chicks hatch, the adults immediately lose their shy nature and come down to feed as soon as live food is offered. Sprouted seeds are also a valuable food source when chicks are in the nest. The Melba finch is parasitized by the paradise whydah, and up to one-third of all nests in the wild have been found to contain from one to three whydah eggs.

Among the other pytilia species, the red-winged *P. phoenicoptera,* also known as the aurora finch or aurora waxbill, is often available, and is an equally free breeder in captivity. The orange-winged *P. afra* and red-faced *P. hypogrammica* pytilias have also been bred in captivity, although less frequently.

Ring Size: 2.5 mm

Species Requiring Similar Management:
- Orange-winged pytilia *Pytilia afra*
- Red-winged pytilia *P. phoenicoptera*
- Lineated pytilia *P. lineata*
- Red-faced pytilia *P. hypogrammica*

Twinspots

Introduction: Twinspots are closely related to PYTILIAS, FIREFINCHES, bluebills, SEEDCRACKERS, and crimson-wings. The six recognized species of twinspots are assigned to four genera, *Mandingoa, Clytospiza, Hypargos,* and *Euschistospiza*.

Natural Habits: In the natural state, twinspots are found in small parties outside of the breeding season. They spend much of their time feeding on the ground and are associated with forest margins, woodland thickets and watercourses in dry areas. They are nervous birds, and when disturbed they retreat into thickets of vegetation. Their constant calls consist of trills and soft tinkling notes.

Management: When newly acquired, twinspots are delicate and should be given additional warmth with an anti-stress treatment added to the drinking water. Initially, food must be loosely scattered to ensure that the birds find it. Twinspots are best kept in a sunny, planted aviary where they will find some live food. They enjoy sunbathing. The Peters' twinspot *Hypargos niveoguttatusi* is a large bird in comparison to the other twinspots and has a reputation of being aggressive, so it is best to keep one pair per aviary. Provided enough nesting space is available, this aggression should not be a problem in a small mixed collection. Other twinspots, pytilias, firefinches, and other birds with a red coloration should not be mixed with either Peters' twinspots or Dybowski's twinspots *Euschistospiza dybowskii*. Although breeding successes with the green twinspot *Mandingoa nitidula* have been few and far between, this species is much sought after by aviculturists. They are delicate when newly acquired, and should be kept in a warm secluded spot for the first 10 days or so.

Twinspots will thrive in a planted aviary with plenty of dense shrubs interspersed with grass clumps in which they can clamber about. A mixture of Habitats B and C will favor most species except the green twinspot, which, originating from forests, requires a mixture of Habitat A and C. Although captive-bred twinspots are less skittish than wild-caught birds, they are still very nervous in cages, and should preferably be given a planted aviary. Having said that, however, I have seen Peter's bred very successfully in cages. Green twinspots can be kept in a mixed collection of similar-size birds, and in a large aviary more than one pair of this species can be kept together. Pairs form a tight bond and if one bird is lost, the mate may be lost shortly thereafter. To encourage breeding, the number of other aviary inhabitants should be kept to a minimum, and limited to species selected for compatibility. A mist spray placed over a dense shrub is enjoyed by the birds on a hot day. The aviary floor should be well planted with grasses and dense shrubs where the birds will spend much of their time searching for insects. They are also fond of sunbathing. In severe climates where cold and wet conditions prevail, or where temperatures remain below 16°C (60°F) for extended periods, twinspots should be housed indoors, with an additional source of warmth during the winter months.

I have also seen pairs of Peter's housed indoors with a matt of long fine grass on the aviary floor. The birds breed in this, on the ground.

Diet: Twinspots can be maintained on a mixed seed diet, along with some insects, such as mealworms and termites, and sprouted Japanese millet. They are also fond of feeding on the seed heads of wild grasses, and these can be grown in the aviary to cater for their needs. Cuttlefish bone should always be available. They also take commercial soft food for finches, to which a grated boiled egg with the shell can periodically be added. Although I have not had a chance to trial this, finch cake as described under the section on SEEDCRACKERS will be beneficial if twinspots can be trained to accept it.

Pink-throated Twinspot
Hypargos margaritatus

Breeding: Peters' and Dybowski's twinspots have undoubtedly been bred more often in captivity than any other twinspots, but I would not consider either species well established in captivity. Captive-bred strains of Dybowski's were established in Australia and New Zealand, but even there problems are being experienced with inbreeding. One of the problems with the Peters' is the loss of females due to egg binding, as a result of hens being either too young, allowed to breed too often and/or breeding into the cold months. Females need to be watched carefully and should not be used for breeding until a year old.

The display is performed on the ground; the male will carry a feather or a grass seed head in his beak, and then dance around the female with wings held away from his body and the neck outstretched. The female responds by moving her tail from side to side. The male also bobs up and down alongside the female with a piece of grass in his beak. Occasionally, the female will take the grass from the male. The nest is a dome-shaped structure of grass, fibers, rootlets, and feathers, and is placed low down in a bush or on the ground. Most pairs prefer to construct their own nest, but some will make use of a nest box. Both sexes incubate the 3–6 white eggs that hatch after 12–13 days. A sudden lack of fear, combined with a keen interest in live food, is a sure indication that their chicks have hatched. As mentioned, this species will often choose to nest on the ground.

Initially, the chicks are pink, but they turn dark after a few days. They fledge at 20–24 days, but return to the nest for the first few nights after fledging. The adults continue to feed them for another 12–14 days. The chicks are fully independent at 18 days. The parents will refuse to raise the chicks, evicting them from the nest, if sufficient live food is not available on a regular basis. As the chicks grow, the demand for live food increases sharply, peaking when the chicks are 7–10 days of age. Peters' are sensitive to nest inspections, but sometimes a pair that has bred successfully previously will tolerate limited disturbance. Once weaned, the chicks are best removed to avoid interference with the adults that may be starting a second clutch. There is always a risk of the male bird attacking the immature birds. Young birds will begin to assume adult plumage in their second month.

Hand Rearing: The only hand-raised twinspot I have come across is a Dybowski's that was artificially incubated from an egg and hand-raised using the same technique I describe in the section on HAND REARING ESTRILIDS at the end of this chapter. Having been raised alone, the bird, a male, is rather imprinted and whenever its owner enters the aviary the bird immediately fetches a piece of nesting material and then proceeds to busily look for a nest site!

Ring Size: 2.5 mm

Species requiring similar management:

Rosy twinspot *Hypargos margaritatus*
Dusky twinspot *Euschistospiza cinereovinacea*
Brown twinspot *Clytospiza monteiri*
Crimson-wings *Cryptospiza species*
Firefinches *Lagonosticta* spp.
Pytilias *Pytilia* spp.
Grenadiers *Uraeginthus granatina, U. ianthinogaster*

Seedcrackers *Pyrenestes* spp.

Introduction: The three species of seedcrackers in the genus *Pyrenestes* originate from humid sub-tropical regions of Africa, where they are found largely near forest edges and in marshes. Seedcrackers are unique in having variable bill sizes. Although always large for the size of the birds, the beaks of some individuals appear massive; in others, they are comparatively small. This phenomenon has been noted in all three species, and it has been speculated that the variations have evolved to allow maximum feeding efficiency on two types of sedges that grow within the range of these species, one a large, hard-seeded variety, and the other a smaller, soft-seeded variety. Their calls consist of a variety of short, high-pitched contact calls and a similar song, which reminds one of the call of a blue waxbill. The alarm call is a grating *chirr*.

Management: In 1985, the Riverbanks Zoological Park & Botanical Garden in Columbia, South Carolina, collaborated with Dr. Thomas Bates-Smith on a research project to study bill size polymorphism in the black-bellied seedcracker *Pyrenestes ostrinus*. Thirty pairs of birds were involved in the project. Twenty-four flight cages measuring 1.2 x 0.9 x 2 m (4 x 3 x 6.2 ft) high were set up indoors. Nest boxes measuring 13 cm^3 (5 cu. in.) were partly hidden in vegetation. A dark covering was placed over the back half of the top of the flight cages to simulate the darkness of the forest. A mister hose was run along the top of the flight cages to create an artificial rainy season. The birds are also fond of bathing. The birds were rather delicate when newly acquired and were placed in indoor cages to become established before being released into breeding aviaries.

Even when established they cannot be described as hardy; they are unable to withstand cold weather for any period of time. Robert Seibels and Tracy Frampton of Riverbanks Zoo note that temperatures below 26.7°C (80.1°F) are cold to these birds, with temperatures below 21.1°C (70.1°F) representing "a comparatively hard freeze." It is advisable for each pair to have a well-planted flight to themselves, as they are territorial and aggressive toward others of their own species, as well as toward other small seedeaters, particularly other species having red in the plumage. They can be mixed with small softbills, doves, and quail.

Breeding: The male carried out his nuptial dance with a blade of grass in the beak, singing simultaneously. Several types of nesting material were offered, including dried grasses and paper strips. The aviaries were separated with opaque green plastic sheeting, to give the birds a sense of their own space and privacy. Cleaning the aviaries was reduced to once a week during the breeding season.

Clutches varied from 1–4 eggs. Most nestings resulted in two or three chicks, but occasionally four were produced. The incubation period is 14 days, and fledging occurs at approximately 20 days.

Immatures are chocolate brown and molt into adult plumage at four months. White wing patches showed in many of the chicks, but these were the result of a vitamin deficiency in the diet. The young were removed two weeks after fledging to avoid potential aggression from the male parent. The young males in particular are prone to attack.

Young birds first showed an interest in breeding at eight months; however, none were successful when younger than a year. There were usually 2–3 successful clutches per experienced pair, along with one or two unsuccessful clutches.

Light was provided in the building by skylights, broad spectrum and standard fluorescent tubes. During the first few months, a tape recording of sounds in the Cameroon jungle was played to break the silence in the building. Later this was replaced by a radio playing softly, which broke the silence and put the birds more at ease. The temperature in the building ranged from 22–34°C (71.6–93.2°F). The humidity levels ranged from 55–70% relative humidity. A dehumidifier was installed to control humidity. After a year, the first egg was laid, and within a few years more than fifty chicks had been hatched and raised.

Diet: The diet included parakeet and finch seed, oyster shell, and Vionate, a multivitamin. When rearing chicks, pairs were offered a finch protein mix consisting of insectivore mix; Petamine, a dietary supplement containing essential vitamins, minerals and amino acids; high-protein baby cereal; boiled egg; and Vionate, along with soaked seed. Insects caught overnight in a live insect trap were readily hawked by the seedcrackers and fed to their young. Mealworms were offered daily, but were taken only reluctantly by wild-caught birds; experienced breeders and captive-bred specimens would utilize them when there were chicks to be fed. Occasionally, lettuce or spinach was given. When feeding chicks, adults are offered a portion of protein finch cake along with the protein mix.

Finch Cake:

- 1 cup each of corn meal, whole-wheat flour, and soya flour
- 2 cups of wheat germ oil, corn oil, and Vionate
- 2 teaspoon baking powder
- 10 eggs, with the shells
- 12 cups hulled millet
- 1 cup of chopped sunflower seed
- Mix all ingredients well and spread into an oblong pan
- Baked at 150°C (300°F) for 20 minutes.
- After cooling, cut the cake into 3-cm (13-in.) squares.
- Half a square is crumbled into the adults' food cup daily during chick rearing.

Substitute finch cake:

- 1 cup self-raising flour
- 1 cup whole meal flour
- 1/3 cup wheatgerm
- 1/3 cup insectivore mix
- 1/3 cup ground cat kibble
- 1/4 cup sugar and
- 2 tablespoons honey
- 6 eggs with shells
- Wash shells in warm water.
- Microwave for two minutes then finely crush with hand and add to mix.
- 2 tablespoons each of margarine, maw seed and rolled oats
- 2 teaspoons baking powder
- 1 teaspoon each: multivitamin and cod liver oil
- Beat all ingredients well.
- Pour into two round hole (donut) cake pans and cook in the microwave for 9 minutes.
- Allow to cool then cut into squares. – I allow 1 x ½ inch square for a pair.
- Freeze squares. Remove daily portion and defrost by moistening with a little warm water.

Ring size: 3 mm

Species Requiring Similar Management:
Bluebills *Spermophaga* spp.
Crimsonwings *Cryptospiza* spp.
Twinspots, all four genera
Parrot finches *Erythrura* spp.

Contributor: Reese Bradshaw, Riverbanks Zoological Park & Botanical Gardens, Columbia, South Carolina, USA.

Firefinches *Lagonosticta* spp.

Introduction: The 11 species of firefinches belong to the genus *Lagonosticta*. Firefinches occur naturally in pairs during the breeding season, or as small family groups when not breeding. Red-billed firefinches *Lagonosticta senegala* sometimes form large flocks. Firefinches are seldom found far from dense cover; some species, such as the red-billed, are found in dry areas of thorn scrub, whereas others, such as the blue-billed *L. rubricata* occur in forests and on forest margins.

Management: The red-billed firefinch is also known in aviculture as the little ruddy firefinch, or ruddy waxbill. This is a common species, relatively hardy under captive conditions. It mixes well with other small seed-eating birds, and in a large aviary a number of pairs of this species can be kept together without any serious aggression. In flights they must be kept as single pairs to avoid aggression. Other species of firefinch, however, such as the blue-billed and black-bellied *L. rara* are strictly territorial and should be housed one pair per aviary. Firefinches are particularly fond of live food, such as termites. In Africa, where termites are frequently used in feeding small birds, the firefinches are the first to appear when a termite mound is opened. Their aviary should be sparsely planted, with open sandy patches where they will spend much of their time. They are fond of bathing in shallow water.

Breeding: The red-billed firefinch is one of the easiest of all the small African seedeaters to breed. The nest is built in a dense dry bush, box, or wicker basket, and is a small ball of grass slightly pointed toward the entrance. It is often built less than a meter (3 ft.) above ground and is well lined with feathers. The exact location of a firefinch nest can be determined by simply placing a handful of feathers in the aviary, then standing back to watch the male carry these off to the nest site. The 3–5 tiny white eggs hatch after an incubation period of 11–12 days and the chicks fledge after 16–18 days. The chicks are dark to light brown with a maroon patch on the upper tail coverts and often still show some white down on the top of the head and back.

Initially the chicks are brooded continually by one of the parents. From my observations, this is usually the female. When feeding chicks, a good supply of live food is essential. Having said that, however, in New Zealand I have raised red-billed on soft food alone but in South Africa I found live food essential for the blue-billed firefinch. In additional to the other offerings, the parents will also enjoy searching through soil that has been turned. Droppings are not removed from the nest, and the entrance is soon lined with the chicks' feces. Nest inspections will be tolerated, but should be kept to a minimum and strictly avoided once the chicks begin to feather, as they will fledge prematurely if disturbed. When rearing a brood, adults visit the nest every two hours to feed the chicks. They land near the nest entrance and, after checking that all is in order, quickly enter the nest, feed the chicks, and then disappear again. The chicks do not return to the nest after fledging. For the first four days, fledglings are unable to fly well; if the chicks are approached, the parents become extremely agitated and scold continually.

Soon after their chicks have fledged, the parents build a new nest and a second brood is started. The chicks are fed for two weeks after fledging and young males have their full plumage by four months. The red-billed firefinch is the host species for the village combassou *Vidua chalybeata;* one-third of all nests in the wild have been found to contain eggs of this parasitice species.

Ring Size: 2 mm

Cordon-Bleus *Uraeginthus* spp.

Introduction: The three species of cordon-bleus and two species of grenadiers form the genus *Uraeginthus*. They originate from dry areas of thorn scrub and thickets in various parts of Africa. Found in pairs in the breeding season, they form flocks of 20 or more out of the breeding season. Blue waxbills are often observed on the ground, where they feed on seeds and insects. They are usually the most common species around African water holes. The contact call is a high-pitched *tsit, tseet,* with the males also having a soft, melodious song. The calls of the three cordon-bleu species vary slightly in pitch, with the blue-capped *Uraeginthus cyanocephala* having the highest pitched call. The alarm call is a harsh rattling sound.

Management: The female blue-breasted cordon-bleu *U. angolensis* is a pale version of the male, but particularly dull-colored males or especially brightly colored females make accurate sexing difficult. Immatures resemble females, but are paler, with blackish bills. Female blue-capped cordon-bleus lack the bright blue head of the male and have whitish buff feathers on the breast. In the red-cheeked cordon-bleu *U. bengalus* the females are duller than the males, have a buff-colored breast and lack the red cheek patches of the males. A lack of white on the breast distinguishes them from female blue-capped cordon-bleus, which are otherwise quite similar.

Cordon-bleus are delicate when newly acquired and should be provided with additional warmth and a stress/vitamin supplement in the drinking water. It is advisable to keep them in a cage environment for the first week before release into the aviary. While in this cage, both live food and green grass seed heads should be included in their diet. Habitat C, as described under LANDSCAPE, is ideal for all three species. Cordon-bleus mix well with other waxbills and finches of a similar size, but in the breeding season will become very aggressive toward opposing males of the same species, although in a large, planted aviary two or three pairs can be kept together, provided they are introduced simultaneously. As with most finches, it is often a problem when two pairs are housed together as aggression is aimed at a single individual, but the aggression subsides with three pairs. The different cordon-bleu species should never be mixed, as they will hybridize, contaminating the captive gene pool. It should not be presumed that imports of such species will happen in the long-term. Pairs spend much time mutually preening, which serves to strengthen the pair bond.

Breeding: Blue waxbills can be very willing nesters, but to ensure success it is essential that plenty of live food is supplied, particularly for the first 10 days after the chicks hatch. In Africa, I have found the blue-capped to be the most reliable nester of all the cordon-bleus. In New Zealand and Australia, however, captive strains of the red-cheeked are very well established and breed freely. Their breeding seasons are separated by a period of seven months with no breeding activity. During display, the male will bob up and down on the perch next to the female. He calls softly and holds a dry grass seed head or feather in his beak while displaying. The nest is a dome-shaped structure built in a dry bush or occasionally in a basket or nest box. I have found that a dense dry bush, hung in a sheltered spot, provides the most acceptable aviary nesting site to the birds. In nature, 67% of the nests were found 1–2 m (3–6.2 ft.) from the ground (Skead 1975). The nest is constructed of dry grass stems and seed heads, and lined with a few feathers. It is often constructed close to a wasp nest for protection. Both sexes assist in nest building, the male providing the nesting material while the female positions the material that he delivers to the nest every few minutes.

The 3–6 white eggs are incubated by both sexes for 11–12 days. Incubation starts with the first egg. During the day, the incubating birds relieve each other hourly, but only the female incubates at night.

Clutch sizes for blue-breasted, as given by Skead (1975) are indicated in the table below:

EGGS PER CLUTCH	2	3	4	5	6
PERCENTAGE OF NESTS	5.5	15.6	45.3	29.7	3.9

Newly hatched chicks are blind, with flesh-colored skin that is well covered with long, almost golden down. Their gape markings are not brightly colored. Both sexes continue to line the nest with feathers throughout the period that chicks are in the nest. The chicks fledge at 16–18 days and, in my experience, are fed by the adults for a further 18–22 days, not two weeks as is often stated, and as is the case with some waxbill species. They should be left with the parents for a month. I have had a pair of blue-breasted cordon-bleus produce two clutches, resulting in nine fledged chicks. Immatures closely resemble their parents, but can be distinguished from adults by their dark beaks.

For a day or two after fledging, adults give their chattering alarm call when the aviary is approached. Newly fledged chicks often spend 2–3 days on the aviary floor before flying with the parents. Fledglings can be fairly reliably sexed; the males show blue on the flanks and head, whereas the females show grayish brown with a light bluish tinge. While being fed, chicks emit a mixed chattering call similar to that of the adults. As mentioned, the adults will often desert chicks if insufficient live food is available. Grit is included in the diet when feeding chicks. Under ideal situations it is not unusual for a pair to nest three or four times per season. They are capable of breeding at six months of age. Hybrids have been recorded between all three cordon-bleu species and also with common waxbills.

Hand Rearing: On one occasion I had a nest collapse, with two half-grown chicks falling out. With little hope of success, I began hand rearing the chicks. They responded very well and proved easy to feed, using a 1-ml syringe and a diet of commercial canary food and the tip of a teaspoon of boiled egg yolk, which was finely blended and moistened with water. A few drops of multivitamin were added to this solution. Commercially produced hand-rearing formulas for birds are now available, and a formula with 24% protein should be used. The chicks begged by leaning forward and then looking back over their shoulders, waving their heads and tongues, and crying at the same time. Every two hours they were fed until their crops were visibly full. Additional warmth was supplied in cold weather and at night. These hand-reared birds did not remain tame. For further information, see the section on HAND REARING ESTRILIDS at the end of this chapter.

Ring Size: 2 mm

Species Requiring Similar Management:

Lavender finch *Estrilda caerulescens*
Black-cheeked waxbill *E. erythronotos*
Grassfinches *Peophila* spp.

Grenadiers *Uraeginthus* spp.

Introduction: The two species of grenadiers, formerly separated into the genus *Granatina,* are now included with the blue waxbills in the genus *Uraeginthus*. The common grenadier *Uraeginthus granatina* better known as the violet-eared waxbill, and the purple grenadier *U. ianthinogaster* are among the most prized avicultural subjects. In both species, the sexes are dimorphic. The male violet-eared waxbill is colored in rich shades of chestnut and reddish brown, with bright blue upper tail coverts, blue on the forehead, and a brilliant purplish blue facial patch. In the female, the rufous brown colors of the male are replaced by beige. The male purple grenadier also has rich brown upper parts, but the underparts, from the chest area through the under tail coverts, are purplish blue with irregular chestnut markings. The rump is cobalt blue. The under parts of the female are light brown with off-white barring and spots. Immatures are dull versions of the female. The contact call in the violet-ear is a soft *tit-tit* made between birds in a party, and the male has a soft warbling song, with a rattling alarm call. The purple grenadier is more vocal than the violet-eared waxbill, with a metallic tinkling contact call and a rattling alarm call. Both sexes sing a rather musical song, which ends in a trill.

Management: These beautiful species have always been favorites in aviculture. They are, however, rather delicate, particularly when newly acquired. Losses can be minimized if birds are kept warm initially and gradually acclimatized before being introduced into an aviary. A wild-caught grenadier that is introduced directly into an aviary will probably be found dead the following day. While working in Botswana I was requested to capture a number of native species for a large walk-through aviary for the conservation departments education staff. I included species such as violet-ears and black-cheeked waxbills and losses were negligible. This was achieved by placing newly captured birds in tea chests that measured approximately 60 x 60 x 90 cm long (24 x 24 x 36 ft. long). The front was covered with shade netting, lights were left on around the clock for the first two days, an electrolyte/stress solution included in the drinking water and apart from seed, dishes of termites were offered twice per day. At all other times the birds were left undisturbed. Additional warmth was supplied overnight. After a week the birds were introduced into a small, densely planted flight and after a month they were introduced into the large walk-through. These birds cannot withstand cold, wet conditions for long periods and their aviary must be well sheltered. They are able to withstand fairly cold nights provided the days are warm and dry. If the temperature drops below 16°C (61°F), the birds will show signs of stress and require warm sheltered areas. In severe climates, grenadiers should definitely be housed indoors with additional warmth supplied. Even mildly cold and wet conditions for extended periods may result in death.

Thickets of dense, dry bush should be suspended in the shelter and the aviary planted as described for Habitat C. They are fond of bathing and have the strange habit of often choosing the worst possible weather for this activity. This habit is most likely a natural instinct to bathe after warm summer rainfall in their natural habitat. Grenadiers can be kept in a mixed collection, but their aviary should not be crowded. The two species of grenadier should never be mixed, as this will result in mortalities, usually of the violet-eared waxbill. Pairs placed in adjacent flights, however, will stimulate each other to breed.

Breeding: In courtship, the male will approach the female with his beak and tail pointed toward her and the feathers on the back of his crown slightly raised. This is often accompanied by bill wiping on the perch. If the female responds, the male will fly to the ground and fetch a feather or dry grass inflorescence in his beak, then fly back to his mate to complete a bobbing dance, singing softly during his performance. Copulation takes place a number of times in quick succession. I had a male that could not resist white feathers on the aviary floor, and within minutes of these being made available, he would be dancing next to his mate with one of these in his beak. The introduction of a pile of feathers mixed with dry seeding grass will stimulate most waxbills to display and build nests.

They breed throughout most of the year. The nest is constructed in a dense, dry, tumbleweed-type bush or wicker basket, both of which should be hung under the shelter. It is a spherical structure made of fine grasses and lined with feathers and inflorescences of soft grasses such as pampas grass. The birds will continue to add this lining even after incubation has commenced. In the natural state, 60% of violet-eared nests were found within 50–100 cm (20–40 in.) of the ground (Skead 1975). Both sexes incubate the 3–6 white eggs for 13 days, and the chicks fledge at 18 days of age. Toward hatching, both birds are often found together in the nest. They are not particularly jittery and do not get off their nest for every disturbance. Clutch sizes for violet-ears as given by Skead (1975) are indicated in the table below:

EGGS PER CLUTCH	2	3	4	5	6
PERCENTAGE OF CLUTCHES	0	21.4	42.9	28.6	7.1

Incubation begins with the first egg, and is usually indicated by the tips of the birds' tails being clearly bent from sitting in the nest. The incubating birds change approximately every one and a half hours and are silent during the change. For the first four days, the chick's droppings are removed from the nest, and thereafter, the chicks foul the nest. By day 11, the chicks begin to defecate on the lower lip of the nest entrance. The birds often bring a feather to the nest when relieving a mate during incubation and while there are chicks in the nest. These layers of feathers, in combination with the loose, aerated structure of the nest, help to prevent damp, unhygienic conditions from developing in the nest. A plentiful supply of live food is essential for the chicks to be raised. Ideally, live food should be supplied three times per day. Insufficient live food at this time will invariably result in nest desertion.

Newly hatched chicks are blind, with dark, almost black, skin covered with long, sparse grayish white down. The eyes begin to open by day 4, and the quills begin to show on the wings and tail. White quills appear on the abdomen by day 6. Chicks fledge at 16–18 days. At this stage they are light brown on the wings and tail and buff below, with some blue on the rump. The beak is dark, and small light blue tubercles are still visible on the gape. The parent birds become extremely agitated if the chicks are approached. In the evenings they call and fly continually back and forth between the nest and the chicks, encouraging them to return to the nest. On the first night, if the weather looks threatening, the chicks can be caught by hand and returned to the nest.

The coloring on the head is partially visible on the juvenile after 25 days, and at this stage the tail is equal in length to that of the adults. By 48 days, the young males can clearly be identified, as their adult plumage begins to show. They are fully colored by five months. A second brood is often produced after the immatures are removed from the aviary.

The violet-eared waxbill is parasitized by the shaft-tailed whydah *Vidua regia*. In a study by Skead (1975) in nature, one-third of the nests violet-eared nests examined contained one or more eggs from this parasitic species. The purple grenadier is parasitized by Fischer's whydah *V. fischeri*.

Ring Size: 2.3 mm

Species Requiring Similar Management:

Pytilias *Pytilia* spp.
Twinspots, all genera
Firetails, all genera

Waxbills *Estrilda* spp.

Introduction: The typical waxbills belong to the genus *Estrilda*. Waxbills normally occur in pairs or family groups when breeding, and in flocks of 20 or more birds during the remainder of the year. They roost communally in reed beds and grass thickets. During the day, they are seldom found far from rank vegetation along watercourses and around cultivated fields, where they can often be seen flying down to feed on seeding grasses. The management of the common, or St. Helena, waxbill *Estrilda astrild* and the red-eared, or black-rumped, waxbill *E. troglodytes* is so similar that I will discuss the two species together. Their calls consist of a loud *ping, ping* and the alarm and display call a *dee-dee, dur, dit*. In the case of the red-eared waxbill, the call is higher pitched, almost squeaky, in comparison to that of the common waxbill.

Management: The sexes are alike in the red-eared waxbill, but with the common waxbill, the vent area is black in the male and ash gray in the female. This is even evident in newly fledged chicks. Male common waxbills also show far more red coloration on the breast than females. Red-ears are often sexed on the belief that the male has a larger and brighter rosy patch around the vent. This is untrue, as both sexes exhibit a great deal of variation in color. The red-eared waxbill represents something of a black mark against aviculture. Tens of thousands of these birds have been exported from Africa to bird markets worldwide, and yet for most species viable captive-bred strains have yet to be established. The attitude that these waxbills are not worth breeding because they have always been so cheap to replace needs to be reconsidered. With legislation being called for worldwide to curb the trade in birds, it is time that aviculturists embark on serious breeding programs for species such as the red-ear. Personally, I believe that a ban on the mass capture of waxbills is just what is needed to make aviculturists put something back into the hobby. As Ian Hinze states in his article on this species, "Those in the hobby who have kept birds merely for a rosette-embellished ego-trip are gradually coming to terms with the fact that they owe a duty to aviculture by way of embarking on serious breeding programs."

In Australia, where imports have been banned for many years, captive-bred common waxbills are now well established. A fawn mutation has also been produced there. Birds with a few white feathers, particularly in the wings, are relatively common, even in the natural state. The red-eared requires similar management to the common waxbill, but is gregarious even when breeding.

Violet-eared Waxbill
Uraeginthus granatina

Common waxbills are best housed in pairs in flights, or a few pairs in a larger planted aviary; red-ears can be kept in larger groups. Habitat C is ideal for both these species. Both species can be kept in a mixed collection with birds of a similar size, which are unlikely to interfere with their nesting. When housed in a group, excited chases and bursts of song are common between the males. They are fond of bathing, and a shallow bowl of clean water should always be available. They are not suited to cold, damp climates, and in such climates plenty of dry sheltered area should be available. The common waxbill is a little hardier than the red-ear, but the rule of thumb for most waxbills is that daytime temperatures below 16°C (61°F) for extended periods will stress the birds. In severe climates these birds should be housed indoors with additional warmth during the cold months.

Breeding: Both species are extremely fond of wild grasses such as *Eragrostis* and *Panicum* species, and various weeds that have the seed in the milky stage. These green seeds are the catalysts for successful breeding, and once the birds get to know the routine, they eagerly await the arrival of these bunches of seed heads. In addition, common waxbills take small insects when raising their chicks. The courtship is a typical waxbill display in which the male bobs up and down alongside the female on a branch while holding a piece of nesting material, usually a dry grass inflorescence, in his beak and singing at the same time. This display does not always precede copulation, which in the red-ear is thought to take place in the nest. The nest is a pear-shaped structure started with a platform base and then shaped upward. It features a tubular entrance and a cock's nest, which is positioned on top of the actual nest. This cock's nest is thought to be a device that has evolved to deceive predators searching for eggs or chicks. The cock's nest is not used by either the male or the chicks to roost in, as was commonly believed. The nest is usually situated at ground level and built between grass and a dense dry thicket placed on the ground. A soft, dense heather type bush is ideal for this.

The nest is built in 3–8 days. It consists largely of dry grass and grass inflorescences that are pushed into position around the tunnel entrance. It is lined with soft plant material and pale-colored feathers. The cock's nest is also usually lined. I have found these waxbills to be particularly fond of both white and guineafowl feathers. Common waxbills will also build off the ground in a suspended dry bush. Aggression between pairs of red-ears nesting adjacent to each other is minimal, and only minute territories immediately around the nest are defended. Ian Hinze found his red-ears to be easily bullied; even a pair of orange-cheeked waxbills *E. melpoda* was able to evict them from their nest.

Both sexes incubate the 3–5 white eggs and assist in rearing the chicks. During the incubation period, and for the first five days after hatching, one of the adults is always found in the nest. On approaching the nest to relieve its mate or feed the chicks, the arriving bird gives a soft call to indicate its presence. The mate will then either leave the nest immediately or depart shortly after the relieving bird has entered the nest. Should one approach the nest, the outside bird gives an alarm call, which immediately brings its mate off the nest. Although occasional nest inspections are tolerated, these should not be attempted with older chicks in the nest, as any disturbance will cause them to fledge prematurely. They normally fledge at 17–21 days, at which stage they are about two-thirds the size of the adults, drab brown above and pale buff below. The eye stripe is present, but so pale that it is hardly visible. The bill is black and has bluish white tubercles on the gape. The tail is shorter than that of the adults; in the red-ear, the tail already shows the white outer tail feathers that are diagnostic of this species. For the first few nights, the adults entice the chicks back to the nest with much excited calling.

Hand Rearing: Many years ago, as a child, I removed three small common waxbills from a nest in the wild for hand rearing. Rearing techniques are exactly as described under the general heading HAND-REARING ESTRILDIDS, which follows the species accounts in this chapter. Of interest, however, in the instance mentioned, is the fact that all three of these chicks turned out to be pin-tailed whydahs *Vidua macroura* a specie that parasitizes the common waxbill. The fact that they were whydahs only became apparent a few weeks after they had fledged.

Ring Size: 1.8 mm

Species Requiring Similar Management:

Sydney waxbill *Neochmia temporalis*
Black-crowned waxbill *Estrilda nonnula*
Crimson-rumped waxbill *E. rhodopyga*
Black-capped waxbill *E. atricapilla*

Orange-cheeked Waxbill *Estrilda melpoda*

Introduction: The orange-cheeked waxbill *Estrilda melpoda* occurs in a variety of habitats, including grass thickets on the margins of forest and cultivated fields to swamps and dense patches of weeds. Usually they are found in pairs or small family groups, but form flocks in excess of 30 birds out of the breeding season. In nature, this species tends to avoid areas where common waxbills occur in numbers. They feed on tall grass seed heads and small seeds on the ground. The call is a short, high-pitched squeak, and the male has a soft song similar to that of the common waxbill.

Management: Sexes are similar but males may show a larger and brighter orange patch on the belly. They must be carefully acclimatized when initially obtained but soon settle down. To keep the birds in good condition and promote breeding, seeding grasses must be available. Grasses such as *Panicum* and *Eragrostis* can be planted in the aviary, and bunches of fresh seeding grasses can be collected in the wild and then hung in the aviary where the birds will feed on them. Soft food and large insects such as mealworms are ignored. Small insects such as termites and fruit flies will occasionally be taken.

The orange-cheeked waxbill is so common in aviculture that they are usually considered beginner's birds. This is unfortunate, especially when one stops to wonder where all those thousands of orange-cheeked, and of course all their aviary-bred offspring, are today? Perhaps as aviculturists it is easier to turn a blind eye to such things, to rationalize that they are such common birds anyway! Like the red-eared waxbill, the orange-cheeked represents little more than an embarrassment to modern aviculture. There is no justifiable excuse for the failure of aviculturists to establish self-sustaining breeding populations of this species in captivity. A self-sustained population is now established in Australian aviculture where they are priced at $1500 per pair. They should be kept as a colony and can be mixed with other small seedeaters of a similar size. Habitat and temperature requirements are exactly as described for the red-eared and common waxbills.

Breeding: During courtship, the male does a bobbing dance next to the female and sings loudly. During this display he holds a blade of dry grass in his beak and puffs out the feathers on his rump and belly. The song is a series of short trills. The nest is placed on or close to the ground in a clump of grass. It is a bottle-shaped structure made of fine grass seed heads and lined with feathers. A cock's nest is placed on top of the breeding nest. Orange-cheeks will occasionally build in a basket or box, but it is advisable to provide the breeding habitat as described, on the aviary floor. Large bunches of teff grass and feathers scattered on the ground work well to entice most waxbills to begin building nests.

Having often wondered why we do not have an oversupply of these birds, I decided to obtain 10 orange-cheeked and attempt to breed them. The chosen aviary measured 6 x 1.8 x 2 m (20 x 6 x 6.2 ft.) high and contained 12 other small seedeaters. A few of the orange-cheeked were clearly lacking in condition and had been slightly plucked. Following their introduction, there was much excitement as they explored the grass thickets in their new home. After a while, the group descended to enjoy a bath. Over the next two weeks, three of the birds that had looked in bad condition when obtained, died. I attribute these losses to stress, combined with incorrect feeding and resulting loss of condition from the time of their capture until I acquired them for my avicultural experiment. It should not come as any surprise that dry seed would not keep any waxbill in good condition, particularly through a period of stress. A bunch of *Panicum* grass seed, with the seeds still in the milky stage, is clearly a favorite with these waxbills; when all the seeds are eaten and the stalks begin to dry, the seed heads are used for nest construction.

In the first summer season, I noticed that two pairs had begun building nests. One pair chose a site in a dense dry thicket hung about a meter (3 ft.) from the ground, and the second pair nested about 20 cm (8 in.) off the ground in a *Setaria* grass thicket. As with a number of other waxbill species common and red-eared waxbills, the nest is a double story affair with the lower breeding nest and the upper, rather incomplete, cock's nest. Initially, all effort is put into completing the breeding nest, which is a pear-shaped affair with a narrow tunnel entrance pointing slightly downward. The tunnel in these nests was not as long as in the case of common and red-ear waxbills. Only after the breeding nest is completed, which takes 8–12 days, is the cock's nest slowly added. The latter is not really a complete nest, as it lacks a complete roof and has a large, open-fronted entrance.

As soon as I noticed the nest building in progress, a hide was placed close to the nest in the grass thicket. The following remarks summarize my observations of these magnificent little birds. The group of orange-cheeks had clearly split into pairs. Each pair moved about the aviary together, their high pitched "squeaks" accompanied by excited flicking of the tail from side to side. Males were identified by slightly brighter facial and vent color. When approaching the nest, the birds would land on top of a grass stem and then clamber and slide down to the nest. I could not help but compare them to little firemen being called for duty. The male was often seen singing excitedly in the grass thicket, his call sounding very similar to the *tsit, tsit, tseeet,* of the common waxbill, but higher pitched and softer. While singing, the feathers of his crown and lower parts were puffed out. The pair then met at the nest, with the female, carrying a grass seed head, entering the lower nest.

The male then entered the cock's nest and spent a few minutes sorting out the existing nest material. Although a third orange-cheeked that approached to within a foot of the nest was driven off for a short distance by the male, the chase seemed rather halfhearted. Both birds then left the nest, only to appear again a few minutes later, the male with a dry grass seed head in his beak. He then took this material into the breeding nest, only to be replaced after a minute by the female. The female then spent some time in the nest while the male brought additional material for her to arrange. Both birds then spent some time together in the breeding nest. They left the nest together, and soon the female returned for a vigorous bout of knocking the nest into shape. During this procedure, the nest could was visibly shaking about as the female perfected her home.

A short while later, she was again joined by the male with more nesting material. The male then left to spend a while checking the condition of the cock's nest. Again, a third orange-cheek arrived to inspect the progress, but was chased away by the male. After an hour on a hard bench I had to move. The male saw my movement and gave an excited high-pitched call, which resulted in the female leaving the nest. Both birds then flew off together.

This pair went on to fledge two chicks that unfortunately died as the result of a tremendous storm the day after they fledged. The other pair abandoned their nest before the eggs hatched. The three eggs were removed and measured, averaging 12.8 x 9.7 mm. By February of this season I was down to five adult orange-cheekeds, but so far one pair has fledged two chicks that were colored as follows: crown, back, wings, and tail brown, forehead and small cheek patches creamy yellow, slightly brighter on the forehead, and the bill black. Underparts light gray and rump yellow-orange. After three months, they resembled the adults. All four chicks I have fledged to date have had white windows in their wings and obvious white tubercles on the gape. Initially, the fledglings were unable to fly well. They call and flit their tails in a manner similar to the adults. While being fed, fledged chicks lean forward with the beak pointing back over the shoulder to solicit feeding. I have never seen the adult birds show any interest in live food, and none was offered during this nesting period.

The 4–6 white eggs hatch in 11 days, and the chicks fledge at 16–19 days of age. Orange-cheeks are now relatively well established in Australian aviculture with all birds captive-bred.

Ring Size: 1.8 mm

Avadavats *Amandava* spp.

Introduction: The genus *Amandava* consists of three species, including the two species of avadavats from Asia and the gold-breasted waxbill from Africa. The name "avadavat" is a corruption of Ahmadabad, the Indian city from which the red avadavat *Amandava amandava* was first sent to Europe. Other names for the species are tiger finch, strawberry finch, and red waxbill. The green avadavat *A. formosa* also known as the green munia or green waxbill, is seen in aviculture much less frequently than the red avadavat. This species originates from central India. The single African representative of the genus, the GOLD-BREASTED WAXBILL *A. subflava,* a common avicultural subject, will be discussed separately from the avadavats.

Red avadavats are strikingly colored. The male is coppery to bright red with black underparts, dark brown wings and tail, and a reddish-brown crown. The back, rump, wings, and belly are dashed with white spots. The female is more somber, being dark brown with paler under parts, yellow on the belly, having some red on the tail and white spots on the wings. The subspecies from Indochina, Java, and Bali *A. a. puniceus* is slightly smaller than the nominate form from India and Pakistan, and the males in breeding plumage are brighter red with proportionately smaller white spots; they lack, or have only a suggestion, of the black stripe from gape to eye. Some bird dealers use the name strawberry finch to designate this race only, referring to the nominate form as the Bombay avadavat or tiger finch. The subspecies from southwestern China, Burma, and the islands of Lombok, Flores, Sumba and Timor *A. a. flavidiventris* is similar to *puniceus* in size, but the red on the breast of the breeding male is less brilliant, shading through pinkish orange to golden or pale buff on the belly. This subspecies is often called the golden-bellied avadavat.

Male red avadavats are unique among the waxbills in having a nonbreeding eclipse plumage; at the end of the breeding season, they lose their brilliant color, molting into a plumage very much like that of the females. Males can, however, always be distinguished from females by the brighter and more extensive red on their rumps. The natural song of red avadavats is short, but sweet and musical, given on a descending scale. The song of the female is even shorter, and, generally speaking, less often uttered.

Management: JR. Waldon (1997) provides some excellent information on the red avadavat in aviculture:

> "Avadavats live in warm, damp areas where reeds and tall grasses flourish. They have a long avicul-

tural history, and thrive in captivity. Their management is similar to that of the gold-breasted waxbill, to which they are related. They are not ideally suited to cage breeding, but require ample space, which is best provided by housing them in an aviary. They are reasonably hardy and can be left outside year-round in warm areas with mild winters, but it is wise not to let the temperature drop below 7°C (45°F). Green avadavats are less hardy, and are best housed indoors when temperatures fall below 12°C (54°F). When they are not feeding or busy with nesting, avadavats spend most of their time sitting together in small groups, usually of three or four, and sometimes up to nine, birds. Avadavats are very tolerant of each other's company, and they do more than merely sit together; they actively push or lean against each other, so much so that a bird on the outside may have to lower its wing as a prop in order to prevent itself from being pushed over. This behavior is known as clumping. Clumps usually start to form around a bird that is sitting quietly. Other birds approach it carefully in a submissive attitude. Normally, when one bird approaches another bird that is not already its mate or its regular companion, the bird being approached will display signs of aggression. A show of submission by the intruder quickly calms its fears so that it is not provoked into attacking or fleeing.

In the wild, avadavats feed on the ground, picking up seeds and insects. In aviaries, the selection of food for red avadavats is very important, for if they are not well fed, they soon lose the brilliant red plumage, becoming coppery brown or even black. They need a lot of sunshine, as this also seems to be essential for color retention. Traditionally, these birds were trapped in great numbers throughout their native India, and generally sent to Europe from Calcutta. If the birds were out of color when caught, the dealers would often spray or dip them with yellow or green vegetable dye to make up for their lack of color."

JH. Sparks (1964) provides some very interesting comments on the behavior of the red avadavat:

"Avadavats and other birds that live together in close groups can often be seen preening each other. This behavior is called allopreening, to distinguish it from autopreening, the term used when a bird preens its own feathers. In allopreening, the preening bird grasps a feather of the other bird at the base and draws its beak along the shaft, gently nibbling as it goes, precisely as it would do if preening its own feathers. Allopreening is, however, usually confined to the partner's head, and the bird being preened assists the preener by rolling and twisting its head to present different parts to the preener's bill. Allopreening takes place when the avadavats are perching in their tight clumps. Sometimes a bird will lean over its neighbor to preen the bird beyond it. A more careful watch on the clump will reveal that a bird invites allopreening by adopting a special invitation posture, in which it ruffles the plumage around its head and rapidly opens and shuts its beak, sometimes uttering a very high-pitched, rapidly repeated call. If very keen to be preened, it will even push against the bird it is inviting, repeatedly bringing the top of its head in front of the latter's bill. At first sight, it would seem that allopreening evolved to enable a bird to get its head preened, an action it cannot perform for itself. This idea was once tested by putting flour on the head feathers of some avadavats, but there was no difference in the time spent being allopreened before and after soiling, and it was concluded that this behavior must serve some other function. Further observations suggested that allopreening is used to reduce aggression between two birds, for whenever two animals come together, they will normally often react with either aggression or fear. In a clump, the head of an avadavat is exposed, and is the place most likely to be attacked by another bird. The ruffling of the head feathers in the invitation posture could therefore be construed as a gesture of appeasement, used to alter a confrontation from one of aggression to one of friendly preening. Further allopreening serves to cement the friendly relationship. It is very likely that allopreening did in fact originate as a means of getting inaccessible parts of the body cleaned, but gradually assumed an added, more important, social function. Birds will spend much more time allopreening that is actually necessary for cleaning the feathers, which tends to confirm its role in ensuring group harmony.

Clumping and related allopreening behavior seem far less developed in green avadavats, which are more inclined to pluck each other badly when closely confined in cages or aviaries."

Diet: A recommended diet would include various millets and canary seed, offered both dry and soaked. Avadavats will take both the green and ripe seeds of grasses, soaked spray millet, and a certain amount of green food. Most of these foodstuffs contain carotenoids, members of the group of chemicals that include carotene, which gives the orange color to carrots. The birds convert these chemicals into brighter forms that become concentrated in their feathers. Once a week, and more often in winter, the seed should be treated with a little halibut oil. Bird sand, mineralized grit, cuttlefish bone, and eggshells should also be given.

While feeding chicks, the pair will actively search for live food. I regularly bred this species in South Africa using termites for live food.

Breeding: In the wild, breeding takes place after the monsoons. Before mating, avadavats go through a series of displays. One is the straw display, in which either the male or the female picks up a feather or piece of grass by the shaft and, holding it out in front of itself, fluffs out its feathers and bows slowly. The displaying bird also sings during the display, a high-pitched warble that slowly descends the scale. Each male defends an area extending several yards in each direction from his nest. Somewhere in this area will be a special perch from which he sings. He shows aggression toward any male that comes into his territory by either displaying in the intruder's direction or actually attacking him. The signal that releases this aggressive behavior is the red plumage on the other male. Females, and cocks in eclipse plumage, are ignored. Females, however, will attack other females.

In the wild, red avadavats usually build their nests at the base of trees, within a few feet of the ground or even almost on it, in a clump of grass or similar vegetation. In captivity, the nest is often placed high in a tree. The nest itself is untidy and not neatly woven. The avadavats take bundles of material to the site. First, a flat platform is made among twigs, then grass is added, and finally, a lining of feathers. Four to six pure white eggs are laid and incubated for 11–12 days. Both parents feed the young. When a young avadavat is approached by one of its parents, it crouches and opens its bill, at the same time waving its head. Often it will also open or flutter both wings. This food begging, and the form it takes, depends on how hungry the chick is. If it is well fed, the chick merely opens its bill without begging. Live food in the form of termites, mini-mealworms, and ant cocoons are essential for the successful rearing of the young.

Ring Size: 2.5 mm

Gold-Breasted Waxbill
Amandava subflava

Introduction: The gold-breasted waxbill *Amandava subflava* also known as the orange-breasted waxbill or the zebra waxbill, is widespread in Africa south of the Sahara. This species is usually gregarious, often occurring in flocks of up to 20 or more birds. Members of a flock call continually, both in flight and while perched. These tiny waxbills are commonly found near water, inhabiting tall rank grass, sedges, and reed beds in wetlands and along stream banks. They feed largely on grass seeds, and are very active, constantly flicking the wings and tail and restlessly hopping about. When disturbed, they move off a short distance in small straggling groups. They are easily identified by their "tinkling" calls. The contact call is a rather metallic tinkling note, and males sing beautifully at dawn. A total of 13 different calls were identified in studies on this species.

Gold-breasted waxbills are sexually dimorphic. The males are basically olive brown, somewhat darker on the back, with deep scarlet red on the rump and upper tail coverts. They also have a red superciliary stripe, a small buff stripe beneath the eye, and a blackish loral stripe. The breast and underparts vary from rich yellow and orange to bright orange-red. Females lack the red eye stripe and rump, with the underparts very light yellow with less obvious stripes on the flanks and belly. Immatures are olive brown with a black beak. They resemble the adults by four months of age, but males can be distinguished by the red eye stripe at two months. The northern races have brighter and more extensive red markings on the breast.

Management: Gold-breasted waxbills adapt and breed well in captivity. In comparison to many of the waxbills, they are a relatively hardy species. They should be kept in pairs in flights, but a number of pairs can be kept together in a large, planted aviary. They are active birds that like to bathe regularly. On a warm day they are fond of bathing in a light water spray. I believe this bathing in a light spray of water plays an important role in successful breeding of this species. When moving along a perch, the tail is flicked sideways. They feed largely on the ground.

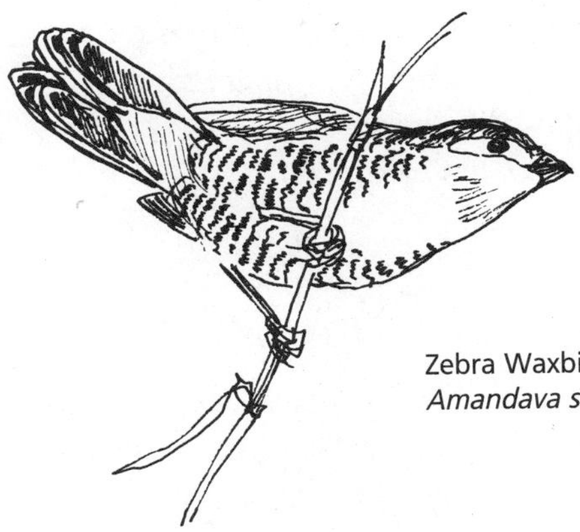

Zebra Waxbill
Amandava subflava

Breeding: During courtship, the male hops slowly back and forth on the ground next to the female. His body is held in an upright posture, bill pointed slightly downward and tail pointed toward the female. The feathers on the flanks and belly are fluffed out, and the entire performance is accompanied by a soft song. Males sometimes also bow a number of times next to the female, usually on the ground. When ready to mate, the female crouches with her tail quivering slightly. Gold-breasted waxbills are strictly territorial in the immediate vicinity of the nest and will not hesitate to leave the nest to chase off intruders. Nests are built in a thick clump of grass or a low bush and are often flimsy and exposed. Although I have always found my birds to prefer building their own nests, in the wild they favor other birds' nests, particularly those of the red bishop *Euplectes orix*. The male gold-breasted searches for nests while the female follows. If a nest is unoccupied, the birds will enter to investigate and inspect it. Males do most of the building and lining of nests, often adding material when relieving the incubating female. Self-built nests are made of coarse grass externally and finer grasses internally. Guineafowl feathers are a favorite for the lining of the nests.

An average clutch consists of 4–5 white eggs laid on consecutive days, with incubation starting after the second or third egg has been laid. Incubation, which takes 13 days, is shared during the day, but only the female incubates at night. Newly hatched chicks are pink, but turn blackish by the fifth day. They are fed on dry, sprouted, and green seeds, with a few insects such as termites. If disturbed, nestlings will fledge early, then scatter and hide in vegetation. Normally, they fledge at 18 days of age. Upon fledging, the chicks remain near the nest initially and are noisy when being fed. Fledglings still show a few down feathers on the crown. They return to the nest at night for 1–2 weeks after fledging and are independent within three weeks.

Ring Size: 1.8 mm

Species Requiring Similar Management:

Strawberry finch *Amandava amandava*
Green avadavat *A. formosa*
Quailfinches *Ortygospiza* spp.

African Quailfinch *Ortygospiza spp.*

Introduction: The genus *Ortygospiza* includes two species of quail finch and the locust finch. The African quailfinch *Ortygospiza atricollis* enjoys a wide range in western, central, and southern Africa, with several subspecies recognized. The red-billed quailfinch *O. gabonensis* is confined to the grasslands of central Africa. Also known as the black-chinned quailfinch, this species is slightly smaller and more delicate than the common quailfinch and lacks the bold white eye patches and chin spot of that species. It also lacks some of the barring of the East African bird. A narrow white circle surrounds the eye. The locust finch *O. locustella* is found in south-central and southeastern Africa, with a subspecies in Zaire.

Quailfinches occur as pairs in the breeding season and as flocks of up to 50 birds when not breeding. They are naturally partial to short grasslands, especially short sedge-type vegetation on the edges of pans of water. When disturbed, they rise steeply with an erratic jerky flight and utter a distinctive bell-like *tirrilink*. They fly high between feeding areas, and in display the male also flies high, then plummets down with a clicking song.

Management: These rather unusual little birds make interesting subjects for the aviculturist. They enjoy an exposed flat perch such as a rock or tree stump on which to sit. The aviary floor should be planted with clumps of grasses, interspersed with patches of dry, sandy soil. They do not like perching on high branches, as do other finches. They like to bathe in shallow water and sand bathe as well. They are definitely not suited to a cage, as their habit of flying straight upward when disturbed can lead to damaged heads. They are delicate when newly acquired and need to be treated for stress by providing additional warmth, as well as an anti-stress solution in the drinking water. I have always clipped one wing of newly acquired quailfinches. By the time the feathers have grown, the birds have tamed down sufficiently that they are unlikely to panic and suffer damaged crowns from collision with the aviary roof. When moved, the transport boxes should have a thin layer of sponge on the inside of the roof to prevent possible head damage from panicking. They are not aggressive, and a few pairs can be kept together in an aviary. The nests are often found close together. These birds do not like cold, wet conditions and in severe climates they should be housed indoors with additional warmth through the winter. Aviculturists in Zimbabwe have reportedly failed to establish the locust finch in captivity for any length of time.

Breeding: The nest is a small ball of grass placed on the ground under an overhanging tuft of grass. It is constructed by the male and is lined with fine grasses and feathers. Both birds incubate, sometimes together, but usually individually. When kept as a colony, it is best not to remove birds until the end of the breeding season, as the activity involved in catching the birds will disturb the colony, possibly resulting in a cessation of breeding. The 4–6 white eggs are incubated for 14 days, and the young fledge at 18 days. While feeding their chicks, quailfinches are largely insectivorous. Initially, the chicks make soft wheezing sounds, but by two weeks of age their calls resemble those of the adults. The chicks return to the nest at night for about a week after fledging. Parents continue to feed the chicks for some time after fledging. I had one pair that raised three broods in a season. Immatures lack the stripes and have a black bill that gradually assumes the red coloration of the adults.

Ring Size: 2 mm

African Quail Finch
Ortygospiza atricollis

Australian Grassfinches *Poephilia* spp.

Introduction: The five species of Australian grassfinches form the genus *Poephila,* with two of the species often separated into the genus *Taeniopygia*. The three undisputed *Poephila* species include the long-tailed grassfinch *P. acuticauda* and its well-known subspecies, Heck's grassfinch *P. a. hecki,* the masked finch *P. personata,* and the black-throated, or parson finch *P. cincta*. The two species usually separated into their own genus are the bicheno or double-barred finch *Taeniopygia bichenovii,* better known outside Australia as the owl finch, and the spotted-sided finch *T. guttata,* known aviculturally as the zebra finch.

The long-tailed grassfinch is a bird of streamlined proportions, a feature unique to only a few finch species. The nominate race has a yellow bill, and Heck's has a red bill. The total length of the bird is around 15 cm (6 in.), including the tail with its long-shafted central rectrices. The minor sexual differences are displayed most obviously by the shape of the black bib. The male's bib is triangular in shape, whereas the female's is more teardrop shaped. The crown of the female is sootier blue-gray than that of the male, and the width of the crown is greater on the male than the female. The males generally have bolder heads. Immature birds attain adult coloration at around three months of age. It is commonly believed that the orange or coral-billed grassfinches found in aviary are produced by hybridizing the yellow-billed and red-billed subspecies. Indeed this may be the case, however, this interbreeding does occur in the wild in the overlapping areas of distribution. The natural distribution of the long-tailed grassfinch is virtually right across northern Australia from around Broome in the west through to the western and central areas of Cape York Peninsula in the east. The majority of this area is inhabited by yellow-billed birds, with the red-billed (Heck's) longtails being confined to the eastern Queensland areas.

Management: In captivity, long-tailed grassfinches can become quite time. When approaching my aviaries, I used to whistle to my birds, to which they would all respond by rushing forward to greet me, calling and bobbing up and down. Long-tails require draft-free, dry conditions. The aviary should be designed to maximize the benefits of natural sunlight. Long-tailed grassfinches may be housed as a colony of their own kind in a large aviary, or as single pairs in small aviaries or breeding cages. It should be noted that in particular the black-throated (Parson's) are more aggressive and, as a rule, should be housed as pairs. I have also found it interesting that they appear to have very tight pair bonds and a requirement to be housed with one of their own species, even if of the same sex. Birds of this species kept singularly soon look miserable. Russell Kingston provides some very good advice on the longtailed finch:

> "As roundworms and coccidia are common problems in captive-bred long-tails, the aviary floor must be kept as dry as possible. A concrete floor or a layer of 7 mm diameter washed gravel will provide dry conditions. However, I have found that these grassfinches like to forage on a natural earth floor, so for this reason I personally prefer to provide a substrate of soil, which I regularly dig over and lightly lime. A suspended, all-wire breeding cage is also acceptable for long-tails, provided it is kept in a dry and draft-free position. They can be mixed with other finches, but need to be carefully observed for aggressive behavior toward individuals. A

planted aviary for grassfinches should include tall grasses such as green *Panicum* as well as dense bushes. Lemon-scented ti-tree and *Melaleuca* are recommended species. Long-tails like nothing better than to bathe regularly, and for this reason clean, fresh water must be available at all times. Greens such as half-ripe seed heads must be fed consistently, as infrequent or irregular feedings can very easily induce scours in grassfinches. Pied, fawn, and white mutations have been produced. These birds should be carefully selected; by using the birds with the darker pigments, stronger mutations are achieved."

Breeding: In the wild, the breeding season is precipitated by the monsoonal or wet season, which ensures a proliferation of seeding grasses; it is this, more than any other factor, that stimulates the birds into breeding. Their natural breeding season is January through June. In captivity, however, they have two breeding seasons, with the second, shorter season occurring in October and November. During the breeding season, wild long-tailed grassfinches consume copious quantities of live food, the bulk of which consists of termites. In captivity, they are certainly successful in rearing their young on termites and medium-sized mealworms, supplemented by daily quantities of half-ripe grass seed. It is not uncommon, however, for aviary specimens to successfully rear their young on half-ripe seed heads and a protein source such as described under SEEDCRACKERS, without the availability of live food. I currently breed about 30 parson finches per annum without the use of any live food. I have a also succesfully fostered this species under Bengalese finch.

In the wild, they breed in loose colonies, in which one nest is usually located per tree, and nesting trees are some distance apart. Pair bonding in long-tailed grassfinches is very strong, usually for life, although widowed individuals do re-mate. Single pairs should be housed in flights or cages, but in large aviaries the colony system can be used. To ensure best breeding results by this method, it is important that birds be kept in an even sex ratio. Long-tails are opportunistic nesters; in the wild they will nest in tree hollows, palms, and even fence posts. In captivity, they will breed in open or closed nest boxes, gourds, woven baskets, tins, and dry bushes. Nesting materials provided should consist of lengths of dried grass and grass seed heads. Softer grasses and feathers are used to line the nest, which is bottle shaped, with a substantial entrance tunnel. Five to seven white eggs are laid, with the fertility rate being usually around 85%. The incubation period is 14 days and the chicks fledge after three weeks. Weaning takes a further 21–25 days. Bengalese finches can be used to foster grassfinches, but due to the size difference, only three or fewer chicks should be given to each pair of Bengalese. Three or four successive nests are not uncommon with this species. They are known to live up to seven years, with an active breeding life of five years. I have bred numerous black-throated using only seed and the finch cake described under PARROT FINCHES.

Ring size: 2.5 mm

Species Requiring Similar Management:

Masked grassfinch *Poephila personata*
Black-throated grassfinch *P. cincta*
Bicheno finch *Taeniopygia bichenovii*
Star finch *Neochmia ruficauda*
Diamond sparrow *Stagonopleura guttata*

Parrotfinches *Erythrura* spp.

Management: The parrotfinches are among the most attractive of all small birds. The 12 species of parrotfinches all belong to the single genus *Erythrura*, which is distributed in the Malay Archipelago, Southeast Asia, the Philippines, the Greater and Lesser Sundas, New Guinea, Australia, and various islands in Wallacea and western Oceania. The red-headed parrot finch *Erythrura cyaneovirens* has held my fascination since childhood. This species is not only beautiful, with rich colors from the bright red head and tail contrasting sharply with the dark green body, but they are also lively, hardy subjects in the aviary. I have kept this species both in planted aviaries and suspended cages, and to date I have had more successes in the planted aviaries. It is also a pleasure to watch the natural habits of this species and to hear their trilling calls as they clamber about in vegetation.

There has been some confusion in aviculture over the common names of this species and the red-throated parrotfinch *E. psittacea*, which much of the avicultural literature has in the past also referred to (and some aviculturists therefore continue to refer to) as the red-headed parrotfinch. This was perhaps less confusing when the only true red-headed parrotfinches available were various subspecies known by their own common names, e.g., *E. c. peali* Peale's parrotfinch; *E. c. regia* the royal parrotfinch; or the nominate race *E. c. cyaneovirens* the Samoa red-headed parrotfinch, etc.

Diet: My birds are fed an equal mix of dry white, red and Japanese millet. In addition, they receive a soft food mixture that consists of:

- cup whole-meal flour
- 50% soaked, slightly sprouted Japanese millet
- 20% sprouted oats & sunflower
- 20% commercial soft food, with some blended boiled egg, including the shell
- 10% finely diced apple

More recently I substitute the soft food with a finch cake that contains the following ingredients:

- 1 cup self -raising flour
- 1/3 cup wheatgerm
- 1/3 cup insectivore mix
- cup ground cat kibble
- 1/4 cup sugar and 2 tablespoons honey
- 6 eggs with shells (wash shells in warm water, finely crush with hand and add to mix.)
- 2 tablespoons of each of margarine, maw seed and rolled oats
- 2 teaspoons baking powder.
- 1 teaspoon each multivitamin and cod liver oil
- Beat it all well; pour into two round-hole (donut) cake pans and cook in the microwave for 5 minutes. Allow to cool then cut into squares.
- I allow 1 x ½ inch square for a pair.
- The bulk quantity is frozen; squares as required daily are then defrosted for 25 seconds in a microwave and moistened with a little water.

This mixture is given to all my finches twice per day. The sprouted sunflower and oats were originally added for the Madagascar lovebirds Agapornis canus, which share the aviary, but I have observed the parrotfinches feeding on these as well. Fruit is an essential part of any parrotfinch diet, and in addition to the apple contained in the softbill mix; I periodically give mine quartered oranges as well. Live food is relished by this species, and when incubating, diced mealworms are offered in the soft food twice per day. To date, I have only kept my parrotfinches one pair per planted flight, but other aviculturists have bred them colonially, even in suspended cages. I have not experienced any serious problems with aggression against their mates or other finches in the flights.

Breeding: I have experienced no problems in getting the birds to breed, and mine have nested through most of the year. In the suspended cages, my birds had no option but to nest in finch boxes, which they did, but in the flights I have found they prefer to build their own nests. The nest is a large ball constructed of rather coarse blades of grass and then lined with some teff and a few feathers. The entrance is rather large and oval and lacks the tunnel shape that is a feature of many estrildid nests. The normal clutch consists of 3–4 white eggs. Incubating birds become obvious by the bent tips to their tails.

In common with many other estrildids, parrotfinches sometimes desert a brood at about 6–10 days. I believe this is primarily because the adults are unable to find sufficient live food to meet the demands of their growing brood. As a result of this, many successful parrotfinch breeders rely on society finches, or Bengalese, as foster parents. In one recent case, I lost a parrotfinch female that had a brood of young in the nest. Not having any potential foster parents available, I decided to hand raise the chicks. This was successfully achieved using methods as described under HAND REARING ESTRILDIDS at the end of this chapter. In New Zealand, I and others have bred a number of red-headed and blue-faced without any live food.

Gouldian Finch *Poephile* spp.

Introduction: Most readers will be familiar with gouldian finches, and because of their dazzling beauty just about every finch breeder would probably like to own one or several pairs of these birds. The following article appeared in *Foreign Birds,* the magazine of The Foreign Bird League and its title was a "Success Secrets with Gouldians — Their Care and Management" by Rae V. Anderson of California. I will attempt to give a summary of this fairly long article and mention the important points, which will hopefully be useful to gouldian breeders. Gouldians appear in three primary color variations; black-headed *Poephile gouldiae,* red-headed *Poephile mirabilis,* and yellow-headed *Poephile armitiana.* At the time of writing there are numerous other color variations, albino, lutinistic, blue-backed and dilute-backed; albino and lutino being mutations. Female gouldians have all the colors of the male but they are not brilliant and sharply defined.

Management: From Rae Anderson: I live in Southern California with fairly substantial extremes in temperature. Winter temperatures drop down below freezing on successive nights but by noon the temperatures are well above that range. In a flock of more than 100 birds, not one very few gouldians were was lost during these cold spells. In the morning frozen water had to be replaced each day until the weather warmed. In summer most days are 33–40°C (90–105°F) and even reach 46°C (115°F) and yet I have not lost a gouldian during or shortly after these extremes. It is necessary to buy birds of good stock, i.e., one must know something about the history of the birds one buys to ensure that they are not inbred and weakened, badly nourished, etc.

The aviaries are largely open, with 1.2–1.8 m (3.5–6 ft.) of the top being covered at each end. The back is covered from roof to the ground. About 1.2 m (3.5 ft.) of the side adjacent to the back of the cages at the extreme ends of the battery of aviaries is also covered; the shelter is actually part of the flight with no doors, partitions, or curtains of any kind to close it off or separate it from the flight. Otherwise, the sides and top are aviary wire with no other protection. Birds are offered only enough protection to keep the rain and dew off them and to prevent the wind from blowing through the shelter area. Although these birds are hardy they would probably not survive temperatures near freezing without some form of additional protection. Personally, I have found gouldians to really enjoy a rain shower in warm weather, but if we experience extended periods of cold and damp, the gouldians adopt a rather hunched posture, but return to a more sleek posture in warmer weather.

Newly imported birds must certainly not be put into unheated outdoor aviaries in the late autumn or winter otherwise they will probably die. Locally raised birds that have acclimatized for a couple of months in outdoor aviaries with temperature conditions similar to where they will be living should have little or no trouble. Ideally, one should obtain birds and put them into aviaries in late spring or early summer. Newly acquired birds should not be offered any green food for 10 days, and it is important to check the source of the water that they have been on. A change in water and green food, when combined with stress can cause severe gut upsets and the loss of birds. In the Northern Hemisphere it is found that even after 10 generations, gouldians are not adapted to the change in seasons after having been shifted from the South to the North. Unless forced to do so, they will continue to attempt to breed through the months of their natural spring and summer, even though it may be late autumn and winter where they are living now. The way to force them to change would be by removing all nest boxes and nesting materials in autumn and separating the sexes.

Gouldians should not be crowded. A minimum size outdoor aviary should be 1.8 x 1.5 x 1.8 m (6 x 4 x 6 ft.). In a cage of this size there should not be more than one pair of birds. In a larger cage, e.g., 3.6 x 1.2 x 1.6 m (12 x 3.5 x 5 ft.), a maximum of three or four pairs of compatible birds will breed successfully. However, the best result will be obtained if each of the pairs of birds in the one cage is of a different species. Birds that are stronger and more aggressive than gouldians frequently interfere with the gouldian breeding. This includes birds such as shaft tails, diamond sparrows, Parsons, etc. You can only be certain that the small, weak species will not interfere.

Adequate nest boxes are necessary; preferably two boxes per pair of birds and a few spares so that there is no fighting over a particular box. It definitely seems to be better to allow birds to be able to choose between various nests. Both male and females incubate eggs and feed the young. Natural branches and twigs of various diameters, but generally smaller than a pencil, are preferred to commercially made perches. It is good to provide birds with changes in foot posture.

Reducing Losses: I have had postmortem examinations or autopsies made of every bird of mine that died. In a larger percentage of cases these examinations showed no conclusive cause of death. This, of course, is the gouldian story with which all too many of us are familiar. Some of the reports showed cranial hemorrhage, probably because the bird was hit with a net when caught or frightened at night by cats or an owl on the aviary and consequently hit its head on perches in the aviary. Some showed extreme constipation and some cases of egg binding. In a few instances there were a few possible contributing factors, such as an enlarged spleen. One can, of course, try to avoid accidents and keep away cats and owls from all one's birds. Sometimes gouldians will look dumpy and show some signs of distress. These birds can usually be saved with antibiotics and heat or with conventional methods in the case of egg binding.

Breeding: Normally only about 50% of the first-year birds breed. If the same is experienced with older birds it can be almost completely overcome by two means. One option is to house them in single-pair breeding cages with no other gouldians around. The hens do not always breed with your choice of mate for them. Also, one can normally get 100% breeding with birds two years old or more by letting the female choose her own mate. It is easy to do but requires a few extra males. After birds have completed their molt simply put one female and three or four males, each with a plastic leg band of a different color, in the desired breeding flight gage. Within a week or 10 days, sometimes even within a day, you should see the female accepting the display of one male. He will be dancing and singing close to her while she will pick at or fly away from other males as they display. Prior to the breeding season all other males must be removed except the one that the female has accepted, and these other males can then be presented to another female. One compatible pair of these birds in a cage will usually raise as many young as will 4–6 pairs that are all sharing the same cage. They simply breed better when they do not have company of their own kind. Gouldian females are in breeding condition when their beaks become dark blackish gray instead of the usual shell color. Following the molt, males also show a bright tip to the bill.

These birds seem to prefer a closed type nest. A suitable size has inside dimensions of 10 x 10 cm or 12.5 x 12.5 cm (4 x 4 in. or 5 x 5 in.) with a top on. The nest opening should be 3.75–5 cm (1.5–2 in.) in diameter and at the center of one of the vertical sides. There should be a small perch below the entrance. Another alternative to a perch would be a projection of the board, which forms the bottom. Some males do not build nests so the hens end up laying eggs directly on the wooden bottom of the nest box. Others build a complete dome-type nest from long pieces of grass and line the bottom with short 1.25–2.5 cm (1/2–1 in.) pieces. The reason why some birds do not build nests could be the result of many generations in captivity or simply lack of natural type nesting facilities. Yet, even when I had gouldians in an aviary, I found that they preferred to use a nest box rather than build their own bottle-shaped nest of dried grass in bushes, trees, or hollow limbs of trees as they do in the wild. Nesting materials should never consist of strong fine fibers such as string, thread, nylon, etc., as these will most likely tangle around the feet of the chicks, cut off the blood circulation and cause deformed feet.

Gouldians tend to be quite tolerant of nest inspection with their first nest each season. After that it is more risky to inspect unless there may be trouble. Some pairs will abandon their chicks or carry them out and drop them on the aviary floor. It is best to minimize interference. Once they are out of the nest, young gouldians seldom return even to sleep. Young gouldians reach independence quickly so that it is perfectly safe to remove them from their parents 14–21 days after they leave the nest. At this stage they can eat by themselves and should survive easily. Parents would usually start their next clutch of eggs within two or three weeks after the chicks leave the nests. If young birds are not removed from the breeding cages they will disturb the hen because of their curiosity to see what is going on inside the nest box, and the hen may be inclined to keep hopping out of the nest to see what her subadults are up to. Nest boxes should be removed preferably in autumn in the Northern Hemisphere. Gouldians appear to complete their molt quicker and go to nest earlier after the molt if the sexes have been separated during winter. This separation serves a number of purposes. It prevents young from being hatched when the days are short and the nights cold. Winter breeding means less food to the babies because the days are shorter. It is believed that gouldians feed their young by regurgitation through the night during the first week or so after hatching but the writer is not sure that this is in fact so. However, the adult birds often stop brooding the young at night when they are one to two weeks old and thus cold winter nights result in dead chicks. Winter-raised babies often take longer to leave the nest than those hatched in summer. Separation forces the hen to take a rest from raising families and permits both sexes to build themselves up for preparation for the complete molt which they undergo in spring. This will reduce loss in birds.

Breeders from previous years go back to nest very soon after being reunited. You should have identified your pairs with rings so that you put the same birds back together that bred for you the previous year. They appear to recognize each other and seem anxious to set up house keeping again.

Separation prevents egg binding that would otherwise be experienced by some birds in cold weather. As

Gouldian Finch
Chloebia gouldiae

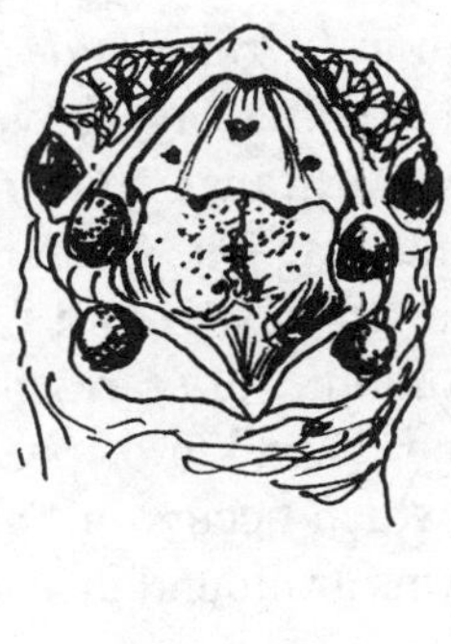

Gouldian Finch chick gape.

noted, Gouldians, if permitted, would frequently nest from October through to March in the Northern Hemisphere. This retards the molt, puts the birds into molt in a weakened condition and reduces the chances of summer breeding for that year. I do not permit any pair of birds to raise more than three clutches of young per season and I consider 3–5 young in the nest as ideal. The birds should also have a reduced range in diet outside the breeding season and be fed primarily dry seed. This will match the natural availability of food, and at the onset of the breeding season the addition of finch cake and sprouted seed provide a breeding stimulus. I personally do not stick to this absolutely and still offer the birds some sprouted seed and soluble vitamins in their water once a week.

Captive gouldians are as susceptible to internal parasites as are other finches, and fecal checks should be conducted on the flock before and after the breeding season. In addition, gouldians are particularly susceptible to air sac mite, scaly mite, and coccidia.

The young can take up to 5–6 months to molt and show their head color and sex. I usually steal a few feathers from the head and chest of each bird. These feathers are soon replaced indicating head color and sex.

African Mannikins *Lonchura* spp.

Introduction: While the distribution of the genus *Lonchura,* which includes most of the species of mannikins, or munias, is centered in the Australasian region, Africa is home to six species, several of which have distinct geographical races or subspecies. The best known of this entire group is the bronze manikin *Lonchura cucullatus,* commonly known in South Africa by as the Afrikaans name *Fret* and in aviculture as the bronze-winged mannikin. This small mannikin is widespread through most of Africa south of the Sahara, including urban environments. The other African species include the black-and-white manikin *L. bicolor,* of which the nominate subspecies is known as the red-backed mannikin, or the brown-backed manikin *L. nigriceps* in aviculture, and the pied mannikin *L. fringilloides,* which is also called the magpie mannikin. The above species, like most of their Asian and Australian congenerics, are alternatively known as munias. The remaining two African mannikins are the African silverbill *L. cantans,* which is widespread across the dry areas of North Africa, and the gray-headed mannikin or pearl-headed silverbill *L. griseicapilla,* which occurs in East Africa. In the natural state, mannikins are found in small flocks of 4–20 birds. These flocks often include both adults and immatures. Mannikins can be found in pairs when breeding.

Habitat preferences vary greatly, with the bronze mannikin being a common garden bird, preferring more open habitats, including thorn scrub. Habitats B and C will suit these species. The pied and brown-backed species, however, prefer forest margins and open patches within forests, interspersed with the seeding grasses on which they feed. When disturbed, they retreat deep into the thickets. The brown-backed will visit garden feed trays if placed close to their forest habitat, offering an easy escape route. The bronze mannikin and the silverbills have short sharp *zit-zit* calls, while the pied, black-and-white, and brown-backed have whistling *peeo-peeo* calls, which are a little louder in the case of the pied.

Management: Mannikins can safely be mixed with other estrildids and similar species without serious problems. In large aviaries, they are best kept in colonies of about four pairs, with the independent chicks being periodically removed to start new colonies. Brown-backed and black-and-white mannikins are rather nervous species; for breeding purposes, they are best housed in a well-planted aviary with dense thickets of shrub and dry bush, in which they will build their nests. The aviary should be planted in accordance with Habitat C as described under the LANDSCAPE section. Small groups can be added to existing flocks, but single additions are often not tolerated. Brown-backed can be pugnacious to others of their won species and I even heard of a case where birds were lost after fighting occurred while in a transport cage. I have also observed brown-backed to chase the larger pied mannikins away from a food source. Pied mannikins will nest in breeding cages, and I believe that this species is an ideal candidate for domestication. Bathing is a regular activity of all mannikins, and they will often bathe several times a day. The topping up of a water container stimulates them to bathe, and within minutes the entire group descends for a communal bath. A flock of mannikins is a pleasure to watch in the aviary; the birds resemble a swarm of bees as they move about together. This activity is followed by the flock ascending to a branch for a bout of preening and sunning.

Diet: As described for common waxbills and parrot finches.

Breeding: Courtship consists largely of the male closely approaching the perched female with the feathers of his head and body puffed out. In addition, he sings a faintly audible wheezing song while bobbing up and down on the perch. In the case of silverbills, the male will hold a dry grass stalk in his beak while the head is flicked upward. The silverbill's song is a soft chattering sound, which is more audible than that of most other mannikins. During courtship, the male also points his tail toward the female.

Mannikins are compulsive nest builders, with two different nests being constructed for different purposes. The first is a roosting nest, a rather loose, bulky structure with a large entrance. Such nests are built and used by a number of birds. Secondly, a breeding nest is built by a pair, sometimes with one or two helpers. This breeding nest is more compact, contains far more material than the roosting nest and has a tighter, slightly tunnel-shaped entrance. While one bird remains in the nest pushing it into shape, the other bird provides material. The nest is built of grass stems, dry grass seed heads, and flowering reed inflorescences. It is quite amusing in the wild to see a large inflorescence traveling through the air with a tiny mannikin towing in front. They prefer green nest material, particularly the inflorescences of grasses. A breeding nest is lined with soft grasses and feathers. I encountered a problem with the captive breeding of these birds using nest boxes. As the chicks' feces are not removed by the parents, the nest soon becomes fouled. Unlike a grass nest, through which dry air can freely flow, the nest boxes become rather wet and unhygienic. This encourages flies, and maggots (gentles) soon appear in the nest, which will then be deserted. If nest boxes are used it would, therefore, be advisable to first dust them with a chemical powder such as Karbadust, which should eliminate the problem. In addition, a number of pencil-thick holes should be drilled on two sides of the nest box. This will allow some free flow of air, ensuring that the box remains relatively dry. Most mannikins, particularly the African silverbill, will nest and raise young without insects.

Often pairs will breed in close proximity to each other. A territory of about a meter (3 ft.) around the nest is maintained. Some inter- and intra-specific aggression does occur. Brown-backed are particularly aggressive and, while watching them from a hide, I have seen them chase off blue-billed firefinches, green twinspots, and the larger pied mannikin when these birds were feeding from the same food source. When bickering, the birds make a high-pitched *swizz-swizz* call, which is sometimes accompanied by lifting one wing to add to their size and aggressive posture. I have found that they do not tolerate interference with the first clutch, but thereafter, nest inspections are tolerated. However, nest inspections should be limited to the first two weeks following hatching; from day 14 on, the chicks will fledge prematurely if disturbed. If this occurs, the parents will then spend some time trying to lure the chicks back to the nest. A pair of bronze mannikins I owned bred throughout the year, producing 37 chicks in a period of 24 months.

Hand rearing a mannikin chick.

The clutch consists of *four to seven* white eggs, and incubation varies from 13–14 days. Both sexes incubate, often together. All mannikins relish insects such as termites when they have chicks in the nest. Newly hatched chicks are bare and a grayish pink color. They are fed by regurgitation, and their crops bulge tremendously with seeds and green food, which is clearly visible through the thin skin covering the crop. Chicks fledge at 18–22 days of age and are independent three weeks later.

Initially, fledglings return to the nest to roost. They are brown in color, slightly lighter on the underparts. At a month of age the first dark feathers appear on the face, throat, and head. Adult plumage is attained 3–4 months of age. Brown-backed chicks are more beige in color on the underparts than the bronze, which are darker. Bronze mannikins have hybridized with black-and-white and pied mannikins, both African and Indian silverbills *L. malabarica* and the bib finch or Madagascar munia *Lonchura* or *Lemuresthes nana*. However, hybrids are unlikely to occur if birds are able to choose mates of their own species. I have often been handed bronze mannikin nests that have blown out of trees, still containing chicks. In addition, I have a healthy population of captive pied mannikins that are proving quite prolific. I recently had to hand raise seven of these and found them identical in behavior to the bronze mannikin chicks I have regularly raised.

I have hand raised a number of mannikin species and have found them a pleasure to work with. When stationed in Zululand, South Africa, I was known as the parent to all orphaned bronze-winged mannikins, often entire clutches in the nest that were picked up in town gardens. I did, however, also use hand rearing as a technique to bolster numbers of the pied mannikin that were very scarce in aviculture. These hand raised young bred very successfully, but I have seen a bronze-winged mannikin that was raised alone and even as an adult was completely imprinted on its owner and would settle quite happily as if nest building in a cupped hand. Mannikins are hand reared as described in the section on HAND REARING ESTRILIDS at the end of this chapter.

Ring Size: 2 mm, except pied, which require 2.5 mm. Rings are fitted at 10 days of age.

Species Requiring Similar Management:

Zebra finch *Taeniopygia guttata*
Munias *Lonchura* and *Heteromunia* spp.
Star finch *Neochmia ruficauda*
Cut-throat finch *Amadina fasciata*

Red-headed Finch *Amadina erythrocephala*

Introduction: The red-headed finch *Amadina erythrocephala* also known as the paradise sparrow or aberdeen, and the cut-throat finch *A. fasciata* are the only two members of their genus. The red-headed finch originates from the dry thorn scrub areas of central South Africa. Found in pairs during the breeding season, they form flocks of up to 50 birds during the remainder of the year. They are common birds around farm homesteads, where they nest under eaves or in the abandoned nests of sparrows, which also favor human habitation. They feed on dry seeds on the ground.

Management: The red-headed finch is sexually dimorphic, with the adult male having a rusty red colored head, while the female has a grayish brown head, although some have rust spots on the crown. Immatures resemble females, but young males show more red, even in the nest. They are fully colored within four months. Red-headed finches are bulky finches and should not be mixed with small waxbills and finches that they are likely to bully. They are best housed with larger finches and doves. In flight-type aviaries, they are best kept one pair per aviary. They have also been bred in indoor cages, which should measure at least 120 x 40 x 40 cm (48 x 16 x 16 in.) and are an ideal species for domestication.

In the natural state, they breed in the cold winter season at which time they take over disused sparrow nests. This habit of nesting in discarded nests makes it necessary to provide a teff fine grass base in their nest box, which they will then re-arrange and supplement to suit their needs. Some pairs refuse to raise chicks if live food is not available. Should this occur, it is advisable to foster clutches of eggs from wild-caught birds under society finches. Due to the size of the chicks, the society finches should not be allowed more than two chicks per brood. These captive-bred chicks will later willingly use soft food to feed their chicks, and thereafter, the strain will breed without insects, provided a high-protein source such as the finch cake described under SEEDCRACKERS is available. Red-headed finches are hardy, but in severe climates where temperatures stay below 10°C (50°F) for extended periods, they are best housed indoors with additional warmth supplied. In the natural state, the cold months are also the dry months and, as with most finches, this species will struggle in extended periods of cold and wet.

Breeding: They nest readily in captivity and will either build their own nest or accept a budgie-type nest box. The nest is built of grass stems and lined with feathers. The 3–6 white eggs are incubated by both sexes for 13 days, and the chicks fledge at 20 days of age. The chicks are very vocal. In David Dennison's experiment with this species he found that by removing clutches of eggs for fostering under societies, four clutches can be produced per season, the final two clutches being left for the parents to rear. An average of 2–3, and occasionally four, chicks are raised per clutch. Chicks beg with a chattering call, and the tongue is waved continually. They have five black spots in the mouth, surrounded by a white border. They are not difficult to hand raise, following the procedures outlined at the end of this section. The chicks breed in their first season and, depending on the size of the aviary, parent-reared chicks can be left with the parents with no aggression. They breed in their first season. Cut-throat finches have been hybridized with the red-headed finch, with the hybrids proving fertile. The two species should therefore not be kept together. They have also hybridized with the African silverbill.

Ring Size: 2.5 mm placed on the chicks at five days.

Species Requiring Similar Management:

Cut-throat finch *Amadina fasciata*
Sparrows *Passer* spp.
Speckle-fronted weaver *Sporopipes frontalis*
Diamond firetail *Stagonopleura guttata*

Cuban Finch *Tiaris canora*

Introduction: The Cuban finch, *Tiaris canora,* also known as the melodious grassquit or Cuban melodious finch, originates from Cuba and the Isle of Pines. It is in fact a member of the South American grassquits and not an estrildid at all. The grassquits are included in the tribe Thraupini, subfamily Emberizinae, of the family Fringillidae, and are therefore more closely related to the TANAGERS, honeycreepers, flower-piercers, and seedeaters. The Cuban finch inhabits mainly scrub and woodland. It is a very small bird of only 8–10 cm (3.2–4 in.) in overall length. The male is mainly olive green above; the forehead, chin, face, and throat are black, bordered by a bright yellow crescent marking starting above the eye and continuing around the face, broadening across the breast to almost meet at the front. The breast is black, fading to gray on the lower belly. The female is duller overall, being brownish on the face, grayish on the crown, and with the yellow much paler.

Management: Cubans are among the most attractive finches; with their small size and beautiful markings, they are often mistaken for waxbills, which they resemble not only in size but also in avicultural requirements. They are hardy birds and even in cool wet weather in New Zealand I have found that of all the finches they remain the most active and vocal. Cubans can be quite prolific, but pair compatibility plays an important role; some birds simply will not tolerate random pairings and refuse to breed. On occasion, they have even been known to harass or kill their intended partners, but they are generally harmless to other species. Many have been bred in large cages with various waxbills, their nests only inches apart, with no apparent aggression. Such tolerance does not apply to their own species, however, and pairs should be kept well separated. They should not even occupy adjoining flights where they can see each other, as they will spend most of their time sparring through the netting instead of settling down to breed. In such cases, males show damaged feathering on the head.

Breeding: While Cubans are a delight to keep and breed in an aviary, some of the best breeding results with this species have been achieved with the birds housed in large cages. They will use a variety of substances to line the nest chamber, but almost all pairs will choose coconut fiber if it is available. Cubans are extremely active birds and may construct several nests before finally breeding. It has been reported that when hens are in breeding condition, they have a strong musky odor. This smell has been confirmed by several breeders. On occasion, males have been observed feeding the hens females egg food during courtship.

A normal clutch usually contains four eggs that are pale grayish green in color with brown specks. The hen alone incubates the eggs for an average of 12–13 days. When they hatch, the chicks have little or no down and have a rather large yellow gape. They beg with the head and neck stretched upwards, not sideways or over their shoulder as in many estrildid finches. As much egg food, or finch cake as described under SEEDCRACKERS, as the birds will consume should be fed at this time. Although live food will also be taken, Cuban finches are not dependent on live food for rearing their young. Too much live food can, in fact, overstimulate the male, resulting in problems of aggression, as he will attempt to drive the hen to nest again before the youngsters have fledged. While feeding chicks, they are extremely fond of sprouted seed and fresh green food, particularly seed that has grown to about an inch-long stem with a green leaf tip. Some pairs will throw out the first couple of rounds of chicks before they settle down to rear young successfully; apparently this is something that also happens in the wild. Pairs that continue to eject their chicks on a regular basis should be split apart and, after a period of rest, paired with different partners. They will then probably go on to rear young without further problems.

The youngsters tend to leave the nest at about 16–18 days of age and are then left for the male to rear, as the hen will often have started the next round of eggs. The young should be closely watched at this time to prevent bullying or feather plucking on the part of the parents. They should be removed from their parents' aviary around four weeks after leaving the nest. At this stage, the young males usually have a few black feathers showing around their face and chest, allowing the sexes to be easily separated. Keeping just two young males together should be avoided, as fighting may occur when they molt out and come into breeding condition. If three or more are

Cuban Finch
Tiaris canora

kept together, the danger is greatly reduced provided no females are present or visible. Another reason for moving the young birds soon after fledging is that adult males have been known to kill young males as they start to sing. I have had youngsters of five months old breed.

Ring Size: 2.5 mm

Species Requiring Similar Management:

Grassquits *Tiaris* spp.
South American seedeaters *Sporophila* spp.
Jacarini finch *Volatinia jacarini*

Scaly feathered finch

Sporopipes squamifrons

Introduction: The scaly-feathered finch *Sporopipes squamifrons* is also known as the scaly weaver or scaly-crowned weaver. It is not an estrildid, but belongs to the subfamily Ploceinae of the family Passeridae. These weavers inhabit the dry central areas of thorn scrub in southern Africa, where they occur in pairs when breeding, and flocks of 10–20 birds when not breeding. They are common visitors to waterholes. The rather apt Afrikaans name *Baardmannetjie* means Bearded Man.

Management: The sexes are alike. In an aviary they are best mixed with larger finches. They are aggressive at all times and will fight with others of their own species, or any smaller finches or waxbills that land near their nest. They will also take over the nests of other small birds, evicting the occupants. I have seen them fall to the ground while locked in a battle with each other. The best way to select a pair is by color ringing a group of birds and observing their behavior. It is sometimes possible, however, to hold two birds in the hand and compare their facial markings. The males usually have bolder markings than the females.

Breeding: Insects are essential if breeding is to be successful. The nest is an untidy round structure built of grass stems. It has a pointed entrance, and is lined with feathers and other soft material, such as carpet under cover (felt). The nest will be built in a dense dry bush or a nest box. The clutch numbers 3–6 eggs that are pale green with grayish brown blotches. The incubation period is 12 days, with incubation duties being shared by both sexes. The chicks fledge at 16–18 days. Although they will readily go to nest, scaly-feathered weavers are not an easy species to breed successfully. This is mainly due to the quantities of insects that they require. I had a pair that built in a nest box, and only after two unsuccessful attempts did they finally manage to raise a single chick. They fed the chick largely on termites and mealworms. A second pair, however, raised two chicks; then, without any indication that they were breeding again, three additional chicks appeared two months after the first brood had fledged. So far they have raised 12 chicks from four broods and, at the time of writing, are incubating again. Fledglings spend their first few days on the aviary floor, but return to the nest at night. They are independent by 16 days and at this stage they begin to show an interest in insects. The bills of immatures are black on the top and yellow on the lower mandible. They have little scaling on the forehead, and the colors on the wings more buff than those of the adults. As they soon resemble the adults, they should be removed by six weeks of age to avoid confusion and possible splitting of the breeding pair. The adult pair is best color rung to avoid this possibility. Scaly-feathered weavers are multiple brooded and will nest up to four times a season.

Ring Size: 2 mm

Hand-Rearing Estrildids

- Glen Holland

If society finches, also known as Bengalese, are available, they can be used to foster the chicks of waxbills and other estrildids. When fostering species that are larger than the society finches, the foster parents should be given only two chicks to rear; with species of equal or smaller size, they can manage three chicks with no difficulty. Successful rearing by this method will depend on the willingness of the society finches to take and feed a high protein diet to the chicks. They usually feed their chicks on a basic seed diet, and this will not suffice for most estrildids. If the society finches are introduced to the rearing food before the fostering procedure begins, they are much more likely to use it when the foster chicks hatch. I have found societies very willing to take and feed to their chicks the moistened finch cake as described in the section on SEEDCRACKERS.

If society finches are unavailable or impractical, then hand feeding using a syringe can be employed. For hand rearing very young finches, the opening of the syringe is decreased in diameter by placing a pin in the opening and then melting the tip of the syringe slightly so as to leave an opening the size of the pin when it is removed. Crop needles are not recommended for hand-rearing finches. Finch eggs can be hatched in an incubator set at a temperature of 37.2°C (99°F), with a humidity of between 55–60%. Eggs are turned five times per day, ideally more often over 24 hours, and chicks seldom need any help when hatching. Newly hatched birds remain in an incubator at the same temperature, but are placed in a small bowl lined with paper towel.

Those species that sway their heads while feeding are best restrained with the spare hand while being fed. The formula is fed warm; if the food is allowed to cool, the chicks will immediately cease feeding. During feeding, the chicks must be allowed to swallow completely and then resume begging, before an attempt is made to feed the next portion. These newly hatched chicks are fed hourly from 6:00 a.m. to 10:00 p.m. Chicks that are initially too small to receive the tip of a syringe can be fed with the flat tip of a toothpick. The entire tip of the toothpick, with a little rearing food, is placed in the mouth. When the chick begins to swallow, the toothpick is gently removed, allowing the chick to swallow the food.

At this stage, a full crop will appear as a fat neck. Some chicks will continue to beg even when the crop is full, and care should be taken not to over feed. As newly hatched chicks are extremely susceptible to bacterial and other infections, hygiene is extremely important. Always refrigerate any unused portions of formula in a covered container. The food should only be warmed twice and a new batch then made. Unless it is kept frozen, do not keep food overnight. After the final feed at night, the chicks should be left in darkness because in the light the crop will empty more quickly. The first week is the most difficult when hand rearing estrildids; beyond this point, one can deservedly begin to feel a little more confident. The temperature should be decreased by 1°C (1.8°F) per day until a temperature of 25°C (77°F) is reached.

At a week of age, the chicks should be separated into groups of 2–3 birds of similar size, and each group placed in a plastic container lined with soft, dry grass. The grass is replaced every three days. A bowl of water can be placed in the brooder to provide humidity and prevent dehydration. Once the chicks reach a week of age, the interval between feeds can gradually be increased from one to every two hours, still feeding between 6:00 a.m. and 10:00 p.m. Within an hour of first being placed in the brooder, the chicks begin to beg for food when any vibration occurs at the nest. Cries from the first chick soon stimulate the others to begin soliciting food. This food is fed warm, not hot, and the mixture must have a loose porridge consistency.

The rearing food I prepared consists of the following:

- One tablespoon of Kaytee parrot hand-rearing formula.
- 3 drops of sunflower oil.
- Sufficient hard-boiled egg yolk to just cover the tip of a teaspoon, and an equal amount of peanut butter, are added.
- For the first five days for chicks hatched from the egg, add half a milliliter (if a measure is not available, cover the tip of a teaspoon) of acidophilus natural yogurt to about a tablespoon of the hand-rearing formula.
- To prepare, hot water is added to the peanut butter and boiled egg yolk. When these ingredients are dissolved, a heaped teaspoon of dry rearing food and the other the ingredients are added. A multivitamin is added once a day.

In addition to meeting the dietary fat requirement, a further advantage of a little vegetable oil is that it lubricates the syringe, alleviating the problems of a jammed plunger. The food is prepared fresh twice per day and re-warmed by soaking a syringe full in a bowl of warm water before each feed. After taking the syringe out of the warm water for a minute, the side of the syringe can be pressed to one's cheek to check if the temperature is not too hot or too cold. Alternatively, place a few drops on the back of a hand and touch this to your lip. The entire tip of the syringe is placed well into the chick's gape when feeding. Each bird is given one short gentle squirt into the back of the gape, and feeding rotated among the group, thus giving each bird a chance to swallow the food properly before additional food is given. When a week old, the small chicks take about 1 ml of food, administered 0.35 ml at a time, as they are not capable of swallowing the full 1 ml at one time. Birds of an estimated 10 days take 1.5 ml per feed and by 14 days, the amount is increased to 2 ml per feed, given 0.5 ml at a time.

The chicks of some estrildids, such as parrotfinches, beg with the head and neck pointing directly upward, not twisted down and backward as with many of the estrildids, including the mannikins. Small mannikin chicks beg with the head pointing upward until they are well feathered, when their begging posture changes to a head down position with the neck twisted so that the beak faces upward. In the African mannikins, once this head down posture is adopted, their habit of lifting one wing vertically into the air also begins. The wing lifted is usually the one on the opposite side from which food is being offered. This is done to exclude siblings and ensures that the individual blocking the others is fed first. This habit makes it essential to separate the chicks while feeding. Feeding a group of more than four birds is difficult as they all lift one wing and clamber over one another to be the first chick fed. Most estrildid chicks make a similar *chit-chit-chit* call when soliciting food.

At the age of 7–9 days, the chicks are fitted with closed rings. After ringing, it is important to inspect the rings at feeding time for the first few days to make sure they have not slipped down, locking all four toes in the forward position. By 12 days of age, they should have grown enough that this can no longer happen. By two weeks, the temperature in the brooder should have been gradually reduced to 25°C (77°F). At this stage, feathers begin to show well, and the chicks are fed every two hours. I once had to leave chicks for four hours without a feed, with no detrimental effects, but this would soon prove fatal if done on a regular basis.

The bright tubercles of some estrildids are still obvious at 18 days of age, at which time the chicks are ready to fledge and begin to jump out of the containers. For about two days prior to fledging, the chicks become extremely alert and also begin to mandibulate on the coarse pieces of food in their mixture. Immediately after fledging, their demands for food decrease for a while, but they soon begin to make up for this. They can then be placed together. After about five days, they will begin to peck at the food, and they will also attempt their first bath. At this stage, the fledglings should be offered dry and sprouted seed, along with a little moistened soft food and a bowl of water. By 12 days, they are eating well enough to slowly reduce feeds to three per day and then to two. They are then placed in a cage that has a source of warmth. A nest receptacle should also be available, and the chicks will usually sleep in this at night. Often one or two birds sleep on top of the rest of the group in the nest. Green grass seed heads and sprouted seed should also be scattered on the cage floor; the young birds enjoy not only the seeds, but they also chew on the stalks and thus feed on the juice from these as well. The floor of the cage is covered in teff fine grass, the seeds of which they also eat.

Once the chicks are feeding well, they lose interest in the syringe and can then be weaned over a period of a few days. Although many popular books suggest that estrildid chicks can be weaned in two weeks, it usually takes longer than this, and unnecessary losses can occur if attempts are made to force weaning before the chicks are ready to start eating on their own. It is also advisable not to be in too much of a hurry to either remove the extra warmth or to keep them too warm, both of which will result in soft chicks that will never adapt successfully to aviary life. While rearing the chicks, I regularly had to travel and even stay away from home overnight. On these trips, the chicks and their nest, covered by a cloth, were placed in a small cooler box containing a hot water bottle wrapped in a towel.

Three weeks after fledging the chicks can be placed, together with their nest, in an outdoor suspended cage of 1.2 m^2 (13 sq. ft.). Seed is initially scattered on the aviary floor to ensure the birds do not struggle to find food. Soft food is supplied in the bowl with which they are familiar. As with adult birds, the soft food mixture and seeding grasses are favored over the dry seed mix. Most species are fully colored by four months of age.

References:

Anderson, RV. 1993. Gouldian finches. *Avizundum*, February 1993: 12–15.

Brickell, N. 1976. Observations on wild and captive magpie manikins.

Clement, P. 1993. *Finches & Sparrows*. London: Christopher Helm Ltd.

Colahan B. D. 1982. Biology of the orange-breasted waxbill. *Ostrich.*

Gabel, R. 1990. Breeding rufous-backed mannikins. *AFA Watchbird* 17(2).

Goodwin, D. 1960. Observations on avadavats and golden-breasted waxbills. *Avicultural Magazine* 66(5): 174–199.

Hinze, I. 1993. Common red-ears heading for rarity status. *AFA Watchbird* 20(3): 34–37.

Kingston, R. 1993. The streamlined longtailed finch. *Australian Birdkeeper* 6(9): 442–444.

Kroner, K. 1990. Breeding the magpie mannikin. *AFA Watchbird* 17(2).

Kruger and Du Toit.1993. The "twinspot" family. *Avizundum*: July: 4–8 & August: 24–25.

Maclean, GL. 1993. *Roberts' Birds of Southern Africa*. Cape Town: 733–773. John Voelcker Bird Book Fund.

Newman, M. 1992. Purple grenadier. *AFA Watchbird* 19(1).

Seibels, RE. & T. Frampton. 1989. Maintaining and breeding seedcrackers. *AFA Watchbird* 16(2): 34–35.

Skead, DM. 1975. Ecological studies of four estrildines in the Central Transvaal. *Ostrich Supplement* 11.

Sparks, J.H. 1964. Flock structure of the red avadavat with particular reference to clumping and allopreening. *Animal Behaviour* 12(1): 125–136.

Waldon, JR. 1997. Red avadavats. *The Estrildian, Breeding Waxbills and Finches* 5(2).

21.19dii Viduini Whydah and Indigobirds

Glen Holland

Introduction: The subfamily Viduidae includes whydahs and combassous (also known as widow finches or indigobirds) and belongs to the family Ploceidae and the order Passeriformes. This includes nine species of whydahs (five paradise whydahs) and 10 combassou species. However, new species are still being found. Males have an eclipse plumage that, out of the breeding season, resembles the rather drab females. The males' breeding plumage is primarily black in the combassous and black white and fawn in the whydahs. The male whydahs have long tails of up to 30 cm (12 in.) in length during the breeding season. They have elaborate aerial displays where males call and fly in jerky movements above perched females. Combassous often have a calling perch from where they sing to attract females. They are polygamous and brood-parasitic on waxbills and finches (Estrildids). Amazingly the chicks gape markings are identical to those of their host species. DNA work has shown that the Viduidae are related to the Estrildids, which are their hosts.

Introduction: The pin-tailed whydah *Vidua macroura* is a common specie found in open scrub and grassland, often seen near the wetland habitat favored by its host, the common waxbill. Males have favorite display perches in their territory. Pin-tails are a pest at garden birdfood tables where they spend the entire breeding season chasing other birds away. In the nonbreeding season they form flocks of up to 40 birds. Nonbreeding males resemble the female. Males in breeding call continually *tseek, tsip, tseek, tsip, chee, chee, chee,* particularly while displaying or chasing other birds in their territory.

Management: Pin-tailed whydahs are not a popular bird among aviculturists in their home range. The aggressive nature of the males in breeding condition and the reputed egg-eating habits of most whydahs makes them incompatible in a mixed collection of small seedeaters. Unless the aviary is large, only one male should be kept, as frequent fights with subordinate males will occur. Each male should be housed with 2–4 females, and in a large aviary it is possible to keep two such groups together. Pintails are best kept with species that are able to hold their own, such as red-headed and scaly-feathered finches. If breeding is desired, they should be housed with a few pairs of their host specie, the common waxbill.

The pin-tailed is an exception amongst whydahs, as unlike the other species it does not learn the song and other vocalizations of the host waxbill. In each of the other whydah species they imitate the host species. The song of the host acts as a stimulus for the whydahs, and when Bob Payne (1973) played taped melba finch songs to a female paradise whydah she eventually laid an egg. Other whydah species are not as aggressive as the pin-tailed and can be kept in a mixed collection with no problem. The following are the other whydah species and their respective hosts:

Species	Scientific	Host
Fischer's	*Vidua fischeri*	Purple grenadier
Shaft-tailed	*V. regia*	Grenadier
Paradise	*V. paradisea*	Melba finch
Sahel paradise	*V. orientalis*	West African Melba
Exclamatory	*V. paradisea*	Aurora finch paradise
Togo paradise	*V. togoensis*	Yellow-winged pytilia
Broad-tailed paradise	*V. orientalis*	Orange-winged pytilia
Steel-blue	*V. hypocherina*	Black-faced waxbill
Pin-tailed	*V. macroura*	Common, orange-cheek & black-rumped waxbill
Cuckoo finch	*Anomalospiza imberbis*	Cisticola & prinia species

Like their hosts, with the exception of the pin-tailed, most whydahs originate from subtropical climates and are uncomfortable at temperatures below 12°C. In severe cold, wet climates with low temperatures for extended periods, they should be housed indoors with a source of warmth during the severe weather.

Diet: Their diet should consist of a finch seed mix, and to avoid their domineering the food table, a number of feeding areas should be provided. They are also fond of green seeding grasses such as *Panicum* and *Eragrostis* but prefer to feed on the ground where these hang down. Whydahs are also fond of small insects and the paradise in particular will readily take white ants.

Breeding: Successful breeding is most likely to succeed in a large planted aviary, with plenty of grass thickets as described for Habitat C in the chapter on LANDSCAPE. Whydahs are promiscuous; there are no pair bonds between male and female. Males often gather in an area to sing and display. The females then come to mate and then leave to search out hosts' nests to lay their eggs. During the breeding season, males spend much of their time courting the females. Courtship includes the male flying and hovering over the females singing and deliberately waving his long tail over the females with a jerky flight. Males also spend long periods singing on favorite exposed perches in their territory. During their courtship flights, the females take very little notice.

Courtship often takes place while the females are feeding on the ground. Each female lays 1–4 white eggs; the eggs are a little larger than those of the host and their eggs tend to hatch in 10–11 days, a day or so earlier than the eggs of their hosts, which take 12 days. Whydah chicks have almost identical spots in the mouth to those of the host chicks, and unlike cuckoos they are reared together with the host's chicks. One difference in the nest is that viduine chicks sit straight up and open their beaks when begging for food but estrilid chicks twist and lower their heads below the horizontal so as to expose the mouth markings. The fledgling viduine chick twists its head like the host nestling when begging for food. There are records of pin-tailed whydahs using other species such as orange-breasted waxbills as hosts, but they show a preference for common waxbills. Very seldom do whydahs lay their eggs in the nest of the wrong host but this does happen occasionally and, thus, they learn the incorrect host's song.

Two studies in the wild have shown that in the first instance one-third and in the second, three-quarters of all grenadiers and melba finch nests contained one or more eggs from their prospective parasitic whydahs as listed above. A nest usually contains 2–3, but may contain as many as five whydah eggs. Payne (1977) found that a single female paradise whydah laid up to 26 eggs per season. Interestingly, a female shaft-tailed whydah was seen to enter a grenadier nest while the pair was in the nest, and without any sign of aggression the female grenadier was seen to leave the nest. The whydah spent 45 seconds in the nest with the male host and, as she left, the female host re-entered the nest.

While studying wild birds, both Nicolai (1966) and Bob Payne (1977) used Bengalese (society) finches to raise whydah chicks. Viduines do not imprint sexually on Bengalese finches. Bob Payne used Bengalese in the field to see if larger, rounder eggs found in estrilid nests were those of viduines. Eight large eggs collected in the wild produced viduines and six small eggs produced host species.

Louis Baptista has suggested that since whydahs learn songs of hosts, and are sexually stimulated by their songs and displays, it is theoretically possible to imprint whydahs on Bengalese or zebra finches as hosts that raise them in captivity and, thus, establish domestic strains. These will be domestically raised whydahs independent of the need for their sometimes harder-to-acquire natural hosts, the fire finches and waxbills. This would result in more freely breeding viduines in aviculture, i.e., strains of whydahs stimulated to breed by hearing songs of Bengalese or zebra finches.

David Holmes of New South Wales in Australia explains how he has succeeded in breeding the pin-tailed whydah:

> "After selecting an aviary of at least 3.6 m^2 (11.3 sq. ft.) you then start the steps needed for success. The layout has to suit the waxbills because they have to rear their own chicks along with a possible young pintail. The whydahs don't worry as all the female does is lay in a working nest and leave to look for another nest. The waxbills prefer to nest in dry brush hanging on the shelter wall or thick grass and plants on the ground. I normally keep all nesting material away and keep their diet very basic until the whydahs come into breeding condition. When the whydah hens are ready, fine long grass is introduced and a better diet variation is put into place. It staggers me how much nesting grass will be used in a matter of three or four days. It is quite easy to see when the pintail whydahs are in breeding condition, the male has changed feather pattern from a brown pattern like the hen to a attractive black and white garb with four long black tail feathers and a bright red beak. He will display to the hen virtually all day. Hovering over her using the long tail feathers as counter weights. The female will move him all around the aviary until she is ready to mate. She will not change feather color but remains the same; her beak is the key, as it will change color. When the female is coming into breeding condition her beak changes from red to black. Eventually when she is about to lay you will see her checking the nests out and if one is to her liking she will lay in it and leave the rest to the pair of waxbills. In stocking the minimum 3.6 x 3.6 meter square aviary I would put 8–10 pair of waxbills and a pair of whydahs. I would only put one hen whydah in, as two will put too much pressure on the waxbills, the male whydah will become quite territorial and will seem to dive bomb the birds at the seed tray and scatter them. But I can't remember ever seeing the

male damage any birds. Unlike the young cuckoo that ejects its hosts, the young whydah is reared along side its hosts. When they fledge they have a totally different plumage pattern and color and are easily identified. In conclusion, if you can breed Saint Helena waxbills and your aviary is big enough you can be successful in breeding pin-tailed whydahs."

Reports, by Mr. and Mrs. JG. Weeks, on the first breeding of both paradise and shaft-tailed whydahs are given as follows:

"On entering the aviary I saw what I thought was a young melba finch fly out from under the shelter. The following evening I saw the bird again and noticed no red on the tail feathers or green on the back like a normal young Melba finch. The aviary houses a mixed collection of 45–50 birds including violet-ears, Cubans, fire finch, cordon-bleus, shaft-tailed whydah, melba finches and a pair of paradise whydah. On February 18 I saw the melba male feeding two young paradise whydahs. In the meantime the melba female had gone to nest again in a cestrum bush. The paradise whydah female was very interested in the new nest. On the 19th the female whydah was observed flying from a perch fastened to the top of the aviary and into the red cestrum in front of the nest and back to the perch again. The female melba refused to move from the nest. The birds in this aviary were supplied with live termites a number of times each day. On March 4 I inspected the nest and found nine eggs – five (the melba's) smaller than the other four. This nest was unsuccessful but another pair, in a separate aviary, fledged two more young paradise whydahs. The following evening both chicks were seen being fed by the male melba. At no time did I notice the female melba near, or feeding the youngsters. It takes 2–3 months for the young to molt from their juvenile feathers to the adult plumage of the female whydah. The four captive-bred birds were in full breeding plumage in January the following year."

Breeding the Shaft-tailed Whydah: This breeding took place in the same aviary as the successful breeding of the paradise whydah described above.

"On entering the aviary on March 2, I noticed in a small cotoneaster shrub situated near the door three young birds that appeared to be young violet-eared waxbills, and on having a closer look I noticed that one young bird had no blue on the rump, as do young violet-ears. On the following evening three members of our cage bird club came to inspect the birds and verified them to be young a shaft-tailed whydah and two violet-ears. On Sunday March 5, while watching the birds, I saw two young birds in the cotoneaster; a pair of violet-ears flew into the shrub followed by three young birds and the parents then commenced to feed all five young birds. We then realized that the pair had fledged five chicks, three shaft-tailed whydah and two violet-ears from one clutch. This was thought to be the first breeding of a shaft-tailed whydah in captivity. The birds were given live termites three times per day and, as with the paradise whydah, the three chicks were in full breeding plumage in January of the following year."

Hand Rearing: Many years ago, as a school child, I collected three young St. Helena waxbills from a wild nest. I then hand raised three young supposed common

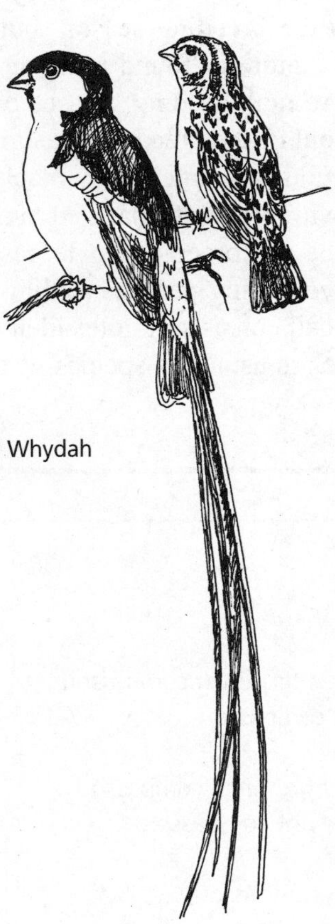

Male & female Pin-tailed Whydah
Vidua macroura

St. Helena waxbill chicks that sounded and looked like the young of common St. Helena waxbills, but soon after fledging it became apparent that they were pin-tailed whydahs. They were raised using a syringe and moistened Pronutro (cereal with 24% protein), which had a multivitamin added to it. For further information on hand-rearing, see the methods as described under the hand rearing section on waxbills and finches WAXBILLS/FINCHES.

Ring Size: 2.5 mm

Species Requiring Similar Management:

Combassous
Widows

Combassou *Vidua* spp.

Introduction: When courting or looking for a female, combassous sing the estrilid finch song of their host. Many new species of combassous have been discovered when song mimicry of host species other than fire finches was recognized. Here the village combassou is an exception as, unlike the other combassous, it does not learn the song of its host. They are resident and territorial in the breeding season, but nomadic when not breeding. Naturally found in open areas with scattered trees where rank seeding grasses occur, they are found in flocks out of the breeding season. The male in breeding condition is a dark glossy black and the female is a light brown color with dark streaks over the head and wings. The nonbreeding male resembles the female. Males have calling perches within their breeding territory. The call consists of notes identical to the call of the host, which is usually a species of fire finch.

Management: The management of combassous is similar to that as described for the pin-tailed whydah, which is discussed in detail and should be read in conjunction with this section. Combassous are not very popular avicultural subjects, being kept largely for showing. They are pugnacious in the breeding season, will eat any eggs found in the aviary and will often dominate a food table if housed with smaller species. Bob Payne also experienced the egg eating problem and despite trying mustard-filled eggs and using calcium powder instead of eggshells in the diet, the problem continued. Due to their aggressive behavior in the breeding season it is better to keep one male per aviary, but each male can have 2–3 females. They are best kept with species that are able to hold their own such as song sparrows and cutthroat finches and, if breeding is desired, their natural host specie. Habitat C as described under the chapter on LANDSCAPE is ideal for combassous.

Diet: Combassous are hardy birds that will thrive on a finch seed mixture and some small insects. Most of their feeding is done on the ground.

Breeding: Males spend much of the breeding season singing and courting the females. As noted, when courting or trying to advertise for a female, combassous sing the estrildid finch song of their host. Males tend to congregate in an area and display. The females visiting these areas mate with the dominant male and then move off to find a host nest.

In common with the whydahs, combassous are parasitic. The village combassou uses the red-billed fire finch as its host, and if breeding is desired these two species must be housed together. After a study in the wild, a third of all red-billed fire finches were found to have one or more eggs from this parasite. They have been bred successfully on a number of occasions and,

Combassou	Scientific name	Host
Village	*V. chalybeata*	Red-billed fire finch
Dusky	*V. funereal*	Blue-billed fire finch
Purple	*V. purpurascens*	Jameson's fire finch
Bar-breasted fire finch combassou	*V. wilsoni*	Bar-breasted fire finch
Cameroon combassou	*V. camerunensis*	Blue-billed, black-billed fire finch, Dybowski's brown twinspot
Black-faced fire finch combassou	*V. larvaticola*	Black-faced fire-x finch
Peters twinspot combassou	*V. codringtoni*	Peter's twinspot
Jos Plateau	*V. maryae*	Rock firefinch
Quail finch combassou	*V. nigeriae*	Quail finch
Goldbreast combassou	*V. raricola*	Golden-breasted waxbill

due to the ease with which the red-billed fire finch breeds, the village is the most likely of all the combassous to breed. The song of the host species acts as a stimulus for the female to ovulate, and Payne (1983) found that only by playing a recording of the correct host, the female combassou stimulated to develop ovarian follicles. He also now has a population of *V. chalybeata,* which are Bengalese-imprinted, and despite the presence of their natural fire finch hosts, they choose to lay in the Bengalese nests. The female lays one or more eggs in each nest of the host, but more than one female may lay in the same nest. Bob Payne describes his breeding aviaries as "Big – 8 x 4 x 3 m (26 x 13 x 9.75 ft.), divided into three with screens between and a hole for the birds to move between breeding chambers. We give them lots of pairs of host species and the female indigobirds always like the kind of bird that reared them, though sometimes they lay in empty nests which may have had the host go in and out."

Each nest may contain up to three combassou eggs and these are easily recognized, as they are larger and rounder than the eggs of the host. Bob Payne (1977) housed this specie with their host and found that a single combassou could lay up to 22 eggs in a season. The incubation period is 11 days, and chicks resemble those of the host identically. The following are the various species of combassous and their respective hosts:

Bob Payne mentioned that no combassous are yet known for the Mali fire finch.

Ring size: 2.5 mm

Contribution: Robert B. Payne, Curator of Birds and Professor of Zoology, University of Michigan, USA.

References:

Baptista, LF. The Biology and Husbandry of Whydahs and Combassous. *AFA Watchbird* 19 (1).

Payne, RB. and Payne, K. (1977). Social organization and mating success in local song populations of Village Indigobirds, *Vidua chalybeata. Z.Tierpsychol.* 45: 113–73.

Payne, RB. 1998. A new species of firefinch Lagonostica from northern Nigeria and its association with the Jos Plateau Indigobird (*Vidua marayae*). Ibis 140: 368–81.

Weeks, Mr. & Mrs. JG. 1968. Breeding of the Paradise Whydah. *Avicultural Magazine* 74: 5–6.

Weeks, Mr. & Mrs. JG. 1968. Breeding of the Shaft-tailed Whydah. *Avicultural Magazine* 74: 6–7.

21.20 Fringillidae Canaries, Buntings, Tanagers & Allies

Glen Holland and Charles E. Siegel

21.20a Fringillinae

21.20ai Canaries *Carduelini*

Introduction: Canaries belong to the order Passeriformes, the family Fringillidae, the subfamily Fringillinae, and the tribe Carduelini, which includes 137 species in 20 genera. The genus *Serinus* alone includes 46 species variously known as serins, canaries, citril finches, seedeaters, and siskins, depending upon individual species and locality. This genus is largely African, with a few representatives occurring in Europe, Asia, the Middle East, Indonesia, and the Philippines. The wild canary *Serinus canaria,* from which the domesticated stock originated, occurs on the Canary, Azores and Madeira Islands. The 31 species in the genus *Carduelis* include the greenfinches of Europe and Asia; the siskins and goldfinches of Eurasia, North America, and South America; the twite, from Europe; the redpolls of Eurasia and North America; and the linnets of Europe, Asia, and Africa. This tribe also includes the rosy-finches and mountain-finches of the genus *Leucosticte,* with six species in the Himalaya, Europe, and western North America; the rose-finches, with one species in the genus *Uragus* and 21 in the genus *Carpodacus,* centered in Asia, but with species radiating as far as Europe and North America; and the bullfinches of the genus *Pyrrhula.* Other well-known birds encompassed by this tribe include several genera of finches, grosbeaks, and crossbills.

Wild canaries occur in a wide variety of habitats, from coastal scrub and dry woodland to alpine scrub. I will discuss my experiences working with the African serins, and Charles Siegel of the Dallas Zoo will recount his work with the endangered red siskin *Carduelis cucullata* of South America.

The black-headed canary *Serinus alario,* also known in aviculture as the alario finch, and the brimstone canary *S. sulphuratus* are found in barren, dry areas. The yellow-eye, or yellow-fronted canary *S. mozambicus,* which is the popular green singing finch of aviculture, and the black-throated canary *S. atrogularis,* the gray singing finch of aviculture, occur in thorn scrub and woodlands, while the Cape canary *S. canicollis* and forest canary *S. scotops* favor forest margins. The goldfinch *Carduelis carduelis,* greenfinch *C. chloris,* redpoll *C. (Acanthis) flammea,* and various siskin species tend to inhabit the margins of scrub and forests from where they venture to feed on seeding and flowering weeds and grasses. Out of the breeding season, canaries form flocks of 10–300 birds. While feeding, they descend to grasslands and weeded areas such as overgrown cultivated lands. Both sexes have very pleasant calls, with the males singing beautifully, particularly during the breeding season.

Management: The pleasant songs and attractive colors of these birds have made them favorites in aviculture. It is sad, however, that even though in the past they were trapped in large numbers for the cage bird trade, few viable captive populations have been established in aviculture. In New Zealand and Australia, where the importation of wild-caught or foreign-bred birds has long been prohibited, the yellow-eyed canary and red siskin are well established as captive-bred populations. In England, captive-bred and close-rung strains, including many mutations, of siskins, linnets, greenfinches, and goldfinches are well established. Canaries can be mixed with small finches, waxbills, and doves, but breeding is unlikely in a crowded aviary. Most species are, however, aggressive toward others of their own kind, and even in a large aviary are best restricted to pairs. If a number are kept together, the males spend much of their time fighting, chasing, and gaping at each other. Yellow canaries *Serinus flaviventris,* for example, should be housed one pair per aviary as the males fight to the death. If possible, pairs should not be housed in adjacent flights, as males will damage their foreheads while attempting to fight through the wire. A breeding pair will chase off any birds that venture too close to their nest. I know of one breeder who now breeds all his canaries as pairs in cages to prevent them from spending most of their time chasing other birds from their nest sites. If kept in aviaries where the birds can hear opponents, the competition in song between the males will stimulate nesting activity.

European Bullfinch
Pyrrhula pyrrhula

The black-headed canary, or alario finch, is a much sought-after species in aviculture, but unfortunately, despite its willingness to breed under aviary conditions, self-sustaining captive strains have not yet been established. There has been a wasted emphasis on hybridizing males of this species with domestic canaries to produce mules for show purposes. This is a great pity, and contributes nothing toward the establishment of the species in aviculture. This situation is even more unfortunate because the species has a limited distribution and is uncommon within its range, and therefore seldom available to aviculture. Most canaries require a habitat similar to that described for FINCHES and WAXBILLS, with a few dense shrubs to provide nesting sites. Habitat C should therefore be combined with a portion of Habitat B, as described in the chapter on LANDSCAPE. Canaries are generally hardy birds, but for those species originating from warm climates, temperatures below 10°C (50°F) combined with wet weather for extended periods will result in sick birds. In such climates they are best housed indoors during the winter months, with some additional warmth supplied.

Diet: Canaries thrive on a finch seed mixture, niger and canary seed, sprouted seeds such as Japanese millet and sunflower, and seeding grasses and weeds such as *Panicum* and *Eragrostis* and thistle with the seeds still in the milky stage. They have also been recorded feeding on flower petals and insects. The latter are important if the birds are to maintain good color and attempt breeding. Sprouted sunflower is too large for the smaller species to crack, and should therefore be slightly crushed with a rolling pin before being fed.

A commercial canary soft food, to which some grated boiled egg with the ground shell and apple are added, will be taken by the birds, and also used to feed chicks in the nest. A high protein cake such as described for SEEDCRACKERS will also be accepted by canaries. Most canaries are not reliant on live food to raise their chicks, but will desert the nest if sprouted seed, green grass and weed seed heads are not available. Most species can, through captive breeding, be taught to move away from live food even when raising chicks. However, a substitute food with a protein percentage of at least 24% is necessary to replace the live food. Segments of fruit such as oranges, sliced apples, or grapes can be spiked on a perch. Green food such as chickweed, thistle, spinach, and silverbeet is a must in the daily diet of canaries. Grit, cuttlefish bone, and sterilized eggshells should be supplied, and are taken regularly.

Breeding: When a pair decides to breed, the male will be seen feeding the female and carrying nesting material. A suitable nesting site will be selected in a dense, dry bush or shrub. The members of the pair will twitter softly to each other continually. During courtship, the male will chase the female and sing loudly. When the female is ready to accept his advances, she solicits for food. Courtship feeding is sometimes followed by copulation. The female alone constructs the cup-shaped nest, using soft grasses and fine twigs. Soft materials such as feathers, coconut fiber, teased hessian, pampas grass heads, dry flower heads, carpet underfelt, and cotton wool are used to line the nest. Most canaries are willing nesters in captivity, often producing a second, and occasionally a third, brood. A piece of dense dry bush, an open box, and a wicker or wire basket should be made available, allowing the birds a selection of potential nest sites. A pad of felt is sewn into the base of the nesting basket.

The 3–5 eggs are generally light cream to greenish blue with some fine speckles at the broad end. The male feeds the female on the nest, and she calls loudly when he approaches the nest. When he flies toward her she shivers her wings, cocks her tail, and stands up or hops onto the side of the nest, soliciting for food. She is fed a number of times in quick succession. When the chicks hatch, the male initially feeds only the brooding female and she in turn feeds the chicks, but a week after hatching the male feeds both the female and the chicks. If excessively disturbed, they may abandon the nest, but they will tolerate quick inspections and the ringing of the chicks, during which the female will perch close by calling loudly. Chicks should either be rung by the time they are half feathered or otherwise post fledging; if disturbed when close to fledging they are likely to fledge prematurely. The incubation period is 13–14 days. Yellow canary and white-throated seedeater *S. albogularis* females sing loudly while incubating on the nest. As the chicks grow, the female begins to leave the nest, and both parents then feed the young and carry away the chicks' feces. The chicks fledge at 16–19 days, and are best removed at six weeks of age to encourage the adults to produce another clutch. While males should be watched for aggression toward the young, with most species this problem is over-emphasized. Fledglings of most species are dull versions of the female and often show a few wispy down feathers on the head for up to a week after fledging. With many species, such as the green singer, the male will continue to feed newly fledged young while the female relines the nest to prepare for the next brood. Two to three broods, per season, are not uncommon with many species.

Wild canaries deserve more effort from aviculturists to produce viable captive-bred strains. In recent times, there have been ever increasing successes recorded in breeding canaries in domestic canary double breeder cages. I have seen this achieved with yellow-eye (green singer), lemon-breasted *S. citrinipectus,* and black-headed canaries. In England, most of the *Carduelis* and *Acanthis* species are bred in breeding cages. More information on cage breeding is contained in the RED SISKIN section by Charles Siegel, but in most species it is not necessary to move the male out of the breeding cage. Some redpoll breeders have set up breeding cages with only a female in each. Selected males are then moved from one cage to the next, being left with the female only until she is incubating. While this may prove to be one method of producing captive populations of these birds, planted aviaries are still the best method of maintaining wild species in captivity. I have seen hybrids involving the yellow-eyed, black-throated, Cape, black-headed, yellow, and domestic canaries. In England mules are a popular feature of many of the cage-bird shows. With species such as black-headed canaries that are not well established in aviculture, the efforts involved in producing these mules would be better spent on developing pure, captive-bred, viable populations of these birds.

Hand rearing: A teaspoon of commercial, 22–24% protein hand-rearing formula, moistened with water and with 2–3 drops of sunflower oil added, should be used, and can easily be fed with a 1 ml syringe. In addition, I add a small amount of the mashed white and a little yolk from a hard-boiled egg (about 10%) by dissolving it in the warm water used to moisten the dry rearing food. The mixture should be as thick as possible, but liquid enough to draw into the syringe. If chicks are artificially hatched from the egg, a small amount of natural acidophilus yogurt (about 10%) should be added to the rearing mix for the first three days. Chicks that have been partially parent reared soon begin to beg for food if the edge of the nest or the tip of the bill is lightly tapped. Placing a rather dilute drop of the rearing mix on the tip of the bill often stimulates swallowing which turns into a full blown begging response.

Newly hatched chicks should be fed hourly from 6:00 a.m. to 9:00 a.m., but from the age of four days on they should gradually be fed less often, until at about a week of age they are being fed every two hours, and given enough at each feeding that the crop can be seen bulging slightly. However, overfeeding should be avoided. The chicks will call constantly, even when well fed. Newly hatched chicks are maintained at a temperature of 36°C (96.8°F). This temperature can be reduced by 1°C (1.8°F) per day until an ambient temperature of 25°C (77°F) is reached. At this stage the chicks should be well feathered. Soon after fledging they will begin to pick at food, although they will still require hand feeding. To wean the chicks, a small amount of sprouted and dry seed can be mixed with the rearing food; gradually, the rearing food is decreased until the chicks are on seed alone, which should be accomplished within three weeks of fledging. Sprouted seed works well when training chicks to feed themselves. Hand-raised individuals remain tame for some time.

This section should be read in conjunction with the section on HAND REARING ESTRILDID finches.

Ring Size: 2.5 mm placed on the chicks at 7–8 days of age, when the flight feathers are starting to show through the quills.

Species Requiring Similar Management:

Domestic canaries, all varieties
Goldfinch *Carduelis carduelis*
Siskins *Carduelis* spp.
Redpolls *C. hornemanni, C. flammea*
Eurasian linnet *C. cannabina*
Greenfinches *Carduelis* spp.

Red Siskin *Carduelis cucullata*

Introduction: The rare red siskin *Carduelis cucullata* belongs to the Fringillidae, a family of finches that reaches its greatest diversity in Eurasia. Due, in part, to its popularity in aviculture, the red siskin has several common names in a variety of languages. Spanish names include Cardenalito, Lugano cardenalito, and Lugano encapuchado; French, Chardonneret rouge, Tatin rouge du Venezuela; and English, Venezuelan black-hooded red siskin. These lively little birds are 10 cm (4 in.) in length, with an average weight of 9–10 grams. The sexes are different. The unique male is mostly bright orange-vermilion, somewhat darker on the back; the hood, wings, and tail are black, with a pale orange-vermilion band across the base of the primaries. The female is brownish-gray above, redder on the rump; grayish-white below, with a salmon red wash on the sides of the breast; the wings and tail are dusky, the wings marked as in the male, but paler. The thin bill is pointed, slightly larger and deeper than that of typical siskins, and is paler on the lower mandible than the upper.

This species is facing a very high risk of extinction in the wild in the near future (1996 IUCN Red List Category EN C2a). The wild population is severely fragmented, with no subpopulation estimated to contain more than 250 individuals. The total wild population is estimated to number less than 2500 mature individuals. This sad state is the result of excessive, long-term persecution by bird-trappers to satisfy the demands of aviculture and the cage bird trade. Although trapping of the species has been illegal since the 1940s, poaching continues. The red siskin hybridizes with domesticated canaries to form "red-factor" variants that are sometimes fertile. There is concern that many red siskins in aviculture are actually hybrids. The American Federation of Aviculture's Red Siskin Project (RSP) uses SPARKS software to manage its studbook. The RSP is officially recognized by the United States Fish and Wildlife Service as a Cooperative Breeding Group devoted to protecting the red siskin by means of aviculture. Many aviculturists in Germany, Spain, Australia, and other countries are also devoted to breeding pure red siskins, and most are members of the International Cardueline Society.

The red siskin was once distributed throughout the foothills of northern Venezuela extending west into eastern Colombia and east into Trinidad. Its range is now very reduced and restricted to fragments of the upper tropical zone of the northern Cordillera in extreme northeastern Colombia and northern Venezuela between 300 m (1000 ft.) and 1200 m (4000 ft.). A small population may remain in Norte de Santander, Colombia. It was last seen in the wild on Trinidad in 1960. Descendants of introduced birds may still exist in the dry hills of southern Puerto Rico, but trapping and hurricane damage may have eliminated the viability of this population. In northern Venezuela, the species inhabits open country or edges of moist evergreen forest, dry scrub or grassy areas with occasional low trees, bushes or shrubs, and moves semi-nomadically and altitudinally on a seasonal and even daily basis between approximately 300 m (1000 ft.) and 1300 m (4265 ft.). The red siskin prefers semi-humid woodland and forest borders when breeding. After the breeding season, family groups used to combine to form large flocks in drier, open habitat, and were easily captured with bait by trappers. The red siskin feeds on a wide variety of seeds, mostly of trees and shrubs, but also on seeds of grasses and flower-heads. The birds in Trinidad sometimes ate the fruit of cacti.

Management: Pairs are typically housed in canary breeding cages with dimensions 90 x 45 x 40 cm (36 x 18 x 16 in.) high, placed no lower than 90 cm (3 ft.) above the floor. The door, food and water bowls are set toward one end of the cage to allow the birds to fly to the opposite end during servicing. The plastic food and water bowls should be placed at perch height. Perches should be varied in size, typically one 9.5 mm (3/8 in.) and one 12.7 mm (2 in.) perch per cage. Red siskins may be aggressive when incompatible or out of breeding condition, so provisions should be made to physically separate the male from the female by dividing the cage in half with a see-through mesh divider. The birds should still be able to see and feed each other through the wire mesh divider. The divider may again be useful once the female has young in the nest, as males have been known to kill offspring. Some breeders do house and breed them in planted flights.

Temperature is best kept at 21–24°C (70–75°F) during the non-breeding season, and 24–27°C (75–80°F) during the breeding season, which lasts from January to June. The warmer breeding season temperatures are necessitated by the behavior of the hens, which may stop brooding the young four days after hatching. The young are still blind and naked at this point, and cannot thermoregulate well. Humidity during non-breeding season should be near 50%, and increased to 60–70% during breeding season. Lighting during the non-breeding season can be as low as 10 hours per day in October, gradually increased by February to 15 hours per day during breeding season. In May, begin gradually dropping the photoperiod to the non-breeding length of 10 hours by October. Experienced hens may start laying at 11.5 hours of daylight, while first year females may need 14–15.5 hours before being stimulated to lay. Red siskins need intense lighting conditions artificially supplied by several 120 cm (48 in.) full spectrum fluorescent lights placed 90–120 cm (3–4 ft.) distance from the cages. Night lights are important to prevent birds from killing themselves if frightened during darkness.

European Goldfinch
Carduelis carduelis

Breeding and Nesting: When males are in breeding condition, they will actively try to get into the female's cage. The males will also sing while dropping their wings to expose the wing bars, or while dancing back and forth on the perches. The female may solicit the male to mount her by squatting on her perch while flapping her wings and twittering. At this point, the cage divider should be removed and a plastic canary nest cup measuring approximately 9.4 cm (3 in.) in diameter and 5.6 cm (23 in.) deep should be installed in the cage as high as possible while still allowing the male sufficient clearance to mount the female while she is perched on the nest. A distance of about 12.5 cm (4 in.) from the top of the cage is ideal. An all-natural felt pad should be sewn to the inside bottom of the nest. Short strands of grass or washed burlap material, 5 cm (2 in.) in length, can be provided for nesting material. Hair or artificial fibers that may get wound around toes or legs should be avoided. The female's abdomen will swell slightly around two days before she begins laying. The usual clutch consists of 4–5 eggs. The incubation period is 12–12.5 days.

Diet: Red siskins do not have powerful bills and therefore require relatively small seeds. They flourish on a diet with both dry and soaked seed mixes, greens, egg food, and some insects. The basic dry seed mix should contain equal parts of canary seed, niger (thistle) seed, and hulled oat groats. They may not eat other seeds, although they are very fond of spray millet. The soaked seeds and egg food should be kept separate from the dry seed using a small dish attached to the cage door for easy removal. Egg food usually consists of hard-boiled egg mashed together with high protein baby cereals, with one teaspoon being fed three times per day during the breeding and molting seasons, and once per day during resting periods. Dark, leafy greens contain carotenoids that will be metabolized into rich plumage coloration; pieces of kale, endive, romaine lettuce, or cabbage can be hung in the cage. Thin slices of orange are also relished. The birds also enjoy small, soft-bodied insects such as aphids and fruitfly larvae. Cuttlebone and crushed, oven-sterilized eggshells should be available at all times for calcium content. This diet should be supplemented with a commercial avian vitamin-mineral mixture as directed.

The American Federation of Aviculture's Red Siskin Project uses both dry and soaked seed mixes custom prepared and donated by Kellogg's Inc. (P.O. Box 44220, Madison, Wisconsin, 53744-4220 USA):

Dry seed mix:

- (Crude protein, not less than 17%; crude fat, not less than 20%; crude fiber, not more than 10%; moisture, not more than 10%):
- Niger seed, canary grass seed, hulled oats, turf type ryegrass, sunflower seed meat fines, small german yellow millet, sesame seed, flax seed, canola rape seed, hemp seed, and teasel seed.
- They will usually eat the niger first, so some aviculturists blow the hulls off the uneaten dry seeds daily, adding fresh mix only every two or three days. This ensures the birds eat a variety of seeds.

Soaked seed mix: (Crude protein, not less than 21.4%; crude fat, not less than 31%; crude fiber, not more than 13%; moisture, not more than 11%): Niger seed, rape seed, sunflower meats, Japanese millet, lettuce seed, finch millet, sesame seed, anise seed. In a jar with a seed-sprouting lid, rinse three days worth of seed with cold water until the drained water runs clear. Soak the seeds in water overnight (8–12 hours) at room temperature. Rinse the seeds again. Drain off all water and refrigerate the seeds. Rinse twice daily until the soaked seeds are used up. Feed soaked seeds three times daily to breeding and molting birds. Some females feed only soaked seeds to their young. Feed soaked seed once per day during other times of the year.

Ring size: 2.5 mm

Species Requiring Similar Management:

Hooded siskin *Carduelis magellanica*
Black-headed siskin *C. notata*
Yellow-bellied siskin *C. xanthogastra*
Pine siskin *C. pinus*
American goldfinch *C. tristis*
European goldfinch *C. carduelis*
European siskin *C. spinus*
European serin *Serinus serinus*

21.20b Emberizinae

21.20bi Buntings *Emberizini*

• Glen Holland

Introduction: Buntings belong to the order Passeriformes and the family Fringillidae and the subfamily Emberizidae. They are divided into five genera with 47 species. Those that feature predominantly in aviculture are of the genus *Emberiza* with 38 species and *Passerina* from the Americas with six species. Most of the species are sexually dimorphic with the females lacking the bright colors of the males. Some species form large, loose congregations out of the breeding season with local migrations. Males are vocal and territorial in the breeding season. Cup-shaped nests built of fine twigs, rootlets, and leaves are generally positioned either at, or within, two meters (6.5 ft.) of the ground.

Management: Buntings are generally rather nervous and take a while to adapt to captive life. They should be carefully acclimatized and trained into captivity. Due to their nervous nature they are not suited to flights, and planted aviaries measuring a minimum of 6 x 3 x 2 m (20 x 10 x 6.5 ft.) should be built for them. The ideal aviary for buntings should be landscaped according to Habitat B as described in the LANDSCAPE chapter. One of the exceptions would be the African rock bunting, which requires a more open habitat with rocky mounds and stacks where they will perch and nest. Buntings are relatively compatible in a mixed aviary provided only one pair of a species is housed together. Different birds can be housed together, but it is important to monitor, for aggression, those with similar coloration and the same genus. They are also compatible with a mixed finch collection, which includes a few

Painted Bunting
Passerina ciris

small doves. For the purpose of this section I have chosen to relate the writing to two species of different genera, with which I have had the most experience. These are the golden-breasted bunting *Emberiza flaviventris,* which originates from east and central Africa, and the painted bunting *Passerina ciris* also known as the nonpareil, which originates from southern USA, Mexico and Central America. With both of these species, their bright color has made them favorites in aviculture and although well known, their breeding history is disappointing. The painted bunting tends to lose some of its color in captivity, particularly the red, which tends to turn yellow, and for this reason a natural planted aviary with plenty of live food is essential.

Buntings enjoy bathing in shallow water, often joining in when a group of finches or waxbills begin to bathe. They also like to sunbathe and can often be seen "lying" on the ground with the tail fanned and a wing outstretched. Both these species originate from warm climates and in the case of the golden-breasted, a dry warm climate. They are therefore prone to suffer in extreme climates and either cold below 10°C (50°F) for extended periods, or cold combined with wet will require that the birds have a dry, well-sheltered section in the aviary. Unfortunately, even in foul weather, they tend to spend their time outdoors and may need to be closed into a sheltered section. Pairs spend much of their day searching for insects amongst leaf litter and plants. As the breeding season approaches the male will often find a high perch and vocally proclaim his territory.

Diet: Buntings should be fed a mixed finch seed diet and insects such as white ants and mealworms are essential. Naturally they feed on seeds, termites, flying ants, mealworms, grasshoppers, moths, butterflies, and small beetles. I have watched them catch and devour insects that appeared far too large for a bird of their size. Live insects are squashed in the beak and are then rubbed on the ground to kill them, then finally swallowed whole. An ultraviolet light switched on in the aviary overnight or the catch from a light trap will provide a variety of insects, which the buntings will eat. Buntings are quite adept at catching flying insects, such as moths, and will often be seen to skip into the air to snap up an insect. Sprouted seed mixed into a moistened soft food as described for SEEDCRACKERS will also be taken by some species. Fruit segments such as apple and orange that are spiked onto a branch are enjoyed particularly by the American bunting species. The dry, sandy floor suggested for their aviary meets buntings' grit requirements. They will also take eggshell, which should be sterilized for two minutes in a microwave oven before being placed in the aviary.

Breeding: Few bunting species can be described as free breeders. In Britain, captive-bred strains of yellowhammer are well established and breeding relatively freely. The African and American species are, however, bred less often and breeding successes have lacked continuity. The first step to achieve success is to provide the birds with the correct landscape, a good-sized aviary, compatible species that will not compete with the buntings for live food, and ideal nest sites. The golden-breasted bunting is relatively willing to build a nest and lay eggs; the real challenge is in supplying sufficient live food to ensure the parents feel that their requirements are met. The nest is a cup of grasses and rootlets and positioned in the fork of a shrub. They will also use platforms such as bark peeling off a tree stump in which to nest. The 2–3 eggs vary in color from pale cream to green or blue with dark scrolls and spots over the thick end of the shell. The eggs are incubated by the female alone and hatch after 12–13 days. Both parents feed the chicks almost entirely on insects. The insects are fed directly from the adult's bill, not regurgitated. Chicks fledge after 17 days and it is not unusual for a female to lay a second clutch shortly thereafter. Juveniles are dull versions of the adult female. Immatures are best removed once independent, as they soon resemble the adults, and they will be persecuted if the adults start a second brood.

None of the Passerina buntings can be described as free breeders, and for the best chance of success they should be housed with very few other species. The female painted bunting is solely responsible for nest construction, which, like the golden-breasted, is placed low in a dense shrub. Some pairs will build a nest in a half-open nest box. The clutch consists of 4–5 pale blue eggs with red-brown spots. The female alone incubates for 13 days; she alone feeds the chicks, which fledge after 13 days. The male spends most of his day guarding the nest site and can be rather pugnacious towards birds that approach the nest site. Once a pair starts to nest, up to three broods may be raised in a season.

William H. Timmis described a successful breeding of the red-headed bunting *E. bruniceps* at Chester Zoo:

> "Two pairs of red-headed buntings were placed in an outside aviary measuring 15.5 x 3.25 x 1.8 m high (49 x 10.5 x 5 ft.). The outside flight was heavily planted with shrubs such as golden privet, holly, laurel, gorse and honeysuckle. Other species in the aviary included a variety of weavers, Java sparrows, cut-throats, green singers, diamond doves and various finches, waxbills and whydahs. In May of their second year in the aviary, one of the females was observed carrying nesting material into a dense clump of honeysuckle. This was abandoned two days later due to the weavers taking a great deal of the nest material. The female started building a second nest a few days later, this being well concealed in honeysuckle that had grown over a short gorse bush. The nest was completed in a little over four days and was a very untidy cup of grass mixed with dead leaves and long stems of Lucerne (alfalfa). It was lined with hair and fine grassy fibers.
>
> The work of building appeared to be the work of the female alone. I noticed on several occasions that if nesting material was collected near to the site, the male would perch high on a branch nearby, calling softly unless feeding or engaged in territorial disputes. As with the female yellow-breasted buntings, when the female red-headed wondered further from the nest she was generally accompanied by the male, but he was never seen to take an active part in nest building. The female carried out incubation alone. In a total of over 60 visits to the nest I always found the female incubating. Three eggs were laid at daily intervals and the hatching was complete in just under 22 hours after an incubation period of 14 days. I never observed the male feed the female on the nest during incubation. The chicks were brooded almost continually for three or four days and during this period the male was seen to fed the female on the nest. After the period of close brooding the female was occasionally seen to brood the chicks but this brooding did not seem to correlate to air temperature.
>
> The food given to the parents to feed the young consisted of small insects and soaked seed fed by regurgitation. After about three days the young were fed on caterpillars, mealworms, maggots and soaked seed. The nestlings demanded the constant attention of both parents, calling loudly for more food the moment their bills were emptied. Nest sanitation was carried out by both sexes although it appeared that most was carried out by the female, and feces lying around the rim of the nest was removed by the first parent to arrive. On some visits to the nest both sexes would pick up a fecal sac and swallow it but sometimes they would carry it some distance down the aviary before dropping it.
>
> We found that the male took very little part in feeding the young until they were about five days old, after which the number of visits in which food was brought to the nest was approximately equal by the two sexes. If the female was on the nest when the male arrived, or if the two birds arrived together, the female generally took the food from the male to give to the young. The fledging period was 15 days and at the stage the young looked very much like dull versions of the female. Two chicks fledged and for several days perched quite near to the nest site, occasionally fluttering down to the ground where they would peck at items. The adults continued to feed them for almost two weeks after fledging, after which time they had gained strength and ability to fly around the aviary and fend for themselves. Both red-breasted *E. bruniceps* and yellow-breasted *E. aureola* buntings were bred in a similar fashion at Chester Zoo."

Hand Rearing: I have only once ever raised a single golden-breasted bunting that was removed from a deserted nest in one of my aviaries. The chick was approximately five days old when taken for hand rearing and was fed largely on insects including pieces of flying ants, moths, and small mealworms, the latter of which had both ends cut off. Prior to feeding the chick, the insects were dipped in a rearing food as described for rearing ESTRILDID FINCHES. The food was not warmed prior to feeding and the chick took 2–3 insects at each feed. It did not beg continually as do many passerines but would take a little food and then settle back into the tissue paper on which it rested. The chick was prompted to solicit for food by tapping gently on the side of the nest. Initially the chick was placed under a brooder lamp that supplied heat at 30°C (86°F). This proved too warm, and the chick was more comfortable at 28°C (83°F). This temperature was gradually reduced to 25°C (77°F) and held at that until the chick fledged. At no stage was the chick tame and once fledged it was extremely nervous at feeding times. It would rush forward to receive food and then retreat once fed. Within a week of fledging the chick was seen pecking at seed on the floor of the cage and was independent 15 days after fledging. At this stage, in addition to millet, it took small live food that was dropped into the cage.

I would recommend the hand rearing section on the MAURITIUS FODY for further advice on hand rearing diets and techniques for rearing buntings.

Ring size: 2.5 mm

Species Requiring Similar Management:

Bunting species
Cardinals
Larks
Red-crested and black-crested finch
Saffron finch

Reference:

Timmis, WH. 1973. Breeding the Red-headed Bunting at Chester Zoo. *Avicultural Magazine* 79: 3–7.

Cirl Bunting *Emberiza cirlus*

• Jo Gregson

Introduction: The Cirl bunting *(Emberiza cirlus)* is a sedentary bird and rarely strays more than a mile or so from its winter and summer territories. Changes in farming practices have caused small flocks of these birds to become isolated along the south Devon coastline. There are other suitable habitats along Britain's coast line but no habitat corridors for the Cirls to move along. While Cirls have maintained and in many cases increased in numbers there has been little range expansion.

During 2001 Paignton Zoo, the RSPB, The National Trust and English Nature combined forces to set up the "Cirl Bunting Re-introduction Project", with the aim of re-establishing Cirl buntings in other parts of southern England. The first phase of the project was to ascertain whether or not Cirls buntings could actually be hand reared from a young age without becoming imprinted, and of course to research the behaviour of the chicks in captivity. During 2001, 2003 and 2004 chicks were pulled from the wild population. Over three years the partners finely tuned nest finding and hand rearing skills. Once independent the hand reared Cirls behaved like wild birds, and were found to be very alert and quick to react to any perceived threat.

Five donor sites were selected for the project. RSPB field workers closely monitored the wild population ensuring that birds were not under any local pressures. Just one brood was taken from any one pair. Cirls lay up to three clutches and the first brood of the year is invariably lost due to heavy rain fall and lack of insect life, this is the brood that was targeted. In total 22 broods were harvested from a possible 67 pairs so it is unlikely that the relocation will have any effect on the south Devon population. The project was put on hold during 2002 due to a foot and mouth outbreak.

At the end of 2004 twelve young Cirl buntings were released into an area of South Devon from where they had previously been taken. The birds were kept in an aviary and soft-released being fed each day at the site until eventually dispersing. These birds were closely monitored over the winter and on into 2005. The birds mixed well with wild wintering Cirl flocks. During 2005 four of these birds paired with wild Cirl buntings and three of those produced young. 2006 saw the project take off, 72 Cirls were reared of which 70 were released into an area in Cornwall. Two aviculturists were employed for the task of hand rearing and release, several RSPB field workers undertook the monitoring of the birds in the wild and will continue to do so well into 2007. The project will be repeated for the next three years. With at least 60 birds being released annually it is more than likely that the Cirl bunting will once again become self-sustaining along the Cornish coast and surrounding farmlands.

Rearing: We found that the ideal age to start hand rearing a Cirl bunting is at six to seven days. When taken at eight days or older the chicks are less confiding and more difficult to feed. Any younger than 6 days the birds are very small and difficult to manage in a practical sense though birds both older and younger have been reared successfully. The chicks are kept in their birth broods and placed in a brooder set at 28°C. They are nestled in a round dish and bowled together using soft tissue paper which prevents legs from splaying. Chicks are not fed for the first hour allowing them time to settle. After this time they are fed every one and a half to two hours from 6am to 12pm.

The birds are weighed each day and kept to a weight gain of approx 10%. Some older chicks did prove to be stubborn at feeding times, for these birds we used a soft bristle paint brush to tease the bill open and place food in the bird's mouth. After a few feeds all slipped into the routine very well. Another trick we have used on stubborn feeders is to feed another brood at the same time. The calls of other Cirls feeding is too much for the chick to bear causing it to automatically open its bill, all be it somewhat reluctantly.

The Diet consist of (60%) Mazuri diet A pellets soaked until soft then mixed with (35%) grated boiled egg. A small amount of banana (5%) is added to bind the mix so it can be easily formed into balls for feeding. A light sprinkling of SA 37 a vitamin mineral supplement is added. Mealworms are offered throughout the day.

Baycox (Bayer) 2.5% Toltrazuril solution is added to the diet as a prophylactic. The oral dose rate for passerines is 12.5 mg/kg body weight, once daily for two consecutive days each week. Throughout the rearing regular visual and faecal checks are undertaken with all chicks. From our experience during the 2004 release we learned that Cirl bunting will not cope with handling after fledging, and have shown many stress responses. Therefore unless signs of disease are obvious all checks are hands off.

Equipment: Brinsea TLC 4 Brooders were used for the rearing period. Fledging birds were weaned in standard sized double wooden breeding cages. Once weaned the birds were place in aviaries. These release Aviaries are designed with a release door in the roof as well as at the front. It was noticed during the first trial that some birds preferred to enter and leave the aviary via the roof. To prevent birds of prey entering the aviaries two inch mesh is placed over both release doors. Fixed at the back of each aviary under a shelter is a wooden breeding cage. The weaned chicks are kept in this small cage for a couple of days before being let into the aviary. This allows the birds a chance to take in their new surroundings before being given flight space. Hopefully has also reduced the incidence of traumas. Each aviary is heavily perched and placed facing a thick hedgerow with an open stubble field beyond.

Fledging: The chicks are only in the brooders for a short while. At 12 days they start to show signs of fledging. At this time they will not gain as much weight and their appetite will drop. Becoming more observant they will sit on the edge of the nest bowl and even jump out if disturbed unexpectedly. Between days twelve and fourteen they are moved into standard size wooden breeding cages. As would be expected on the first day of being moved the chicks become quiet and subdued. They are moved early in the morning and closely watched during the day. If by the end of the day a bird is not relaxed it is placed back in the brooder for another night.

A lot of changes are made at this time. The cages are furnished with leafy branches to allow them to perch and hide. Wild seeds and millet are added to the diet, and a dish of water is provided. Fledging is a very stressful time for all birds and great care is taken to ensure that they are kept in a comfortable routine. With stress can come disease so this is a time when all the birds could be at risk. They must be kept pristine clean, but it is also the time when they make the most mess when they learn to eat and jump in the food dish! Gradually mixed seeds are added to make up half of the diet.

Weaning: From around day 16 the chicks begin to pick up food. At first they will only pick up from the hand, but after a few days of repeatedly dropping the food in front of them they are persuaded to pick it from the floor or dish. Live food is not picked up until a few days later though dead live food will be taken. As they grow more independent they become hand shy and weighing is stopped. The birds stay in cages until all are feeding, bathing, and preening efficiently. By day 25 they are moved to the release aviaries.

Release: Once the birds have been released from the breeding cages they became noticeable to other predatory animals within the area. Several methods were used to deter birds of prey and domestic cats. An electric fence was placed around all ten aviaries, more usually used to keep domestic stock in it worked very well at keeping cats out. Also a sonic sound device was used, but the effectiveness of this could not be measured since the electric fence was already in place. Several large balloons adorned with two large target eyes were tied to trees and shrubs around the aviaries to put off any lurking Sparrow hawks.

Wild Cirls spend a lot of time foraging on the ground so some seed and mealworms are placed on the floor of the aviary. Most food however is placed on the inside ledge of the escape hatch. During this time the rearing mix is slowly withdrawn and replaced with a mixture of millets and wild seeds. The Cirls spend about twelve days confined to the aviaries after when, providing they are fit and healthy, the release hatches are opened. All food is then placed on the outside ledge of the escape hatch and after one week the birds are shut out of the aviary, but still fed on the ledge. As late as October some birds were still returning for food and for this reason one aviary has been left standing as a feeding station. Seed is also scattered in the adjacent stubble field.

RSPB conservation officer Chris Townend reported on the first breeding from the released birds in June 2007: 'It is very symbolic that the birds have chosen this particular back garden to nest in. What better tribute to the efforts of farmers in supporting the species than to be rewarded not just by a pair of cirl buntings in your garden – but the first confirmed breeding pair in Cornwall for such a long time.'

Acknowledgements: The Paignton Zoo staff involved in this project comprises the Curator of Birds, Colin Bath, the Senior Head Keeper of Birds, Jo Gregson and aviculturists, Carl Laven and Julie Wallace.

Reference: http://www.rspb.org.uk:80/news/details.asp?id=tcm:9-162110 New bunting family is cause for celebration. June 2007.

21.20bii Parulini Wood Warblers

Jeff Claffy

Introduction: The North American warblers, often called wood warblers, are intriguing avicultural subjects. The New World warblers occupy the tribe Parulini in the subfamily Emberizinae in the family Fringillidae, and are therefore more closely related to the buntings and tanagers than to the Old World warblers of the families Cisticolidae and Sylviidae. A total of 115 species of wood warblers are recognized, assigned to 25 genera. Approximately half of these species reach North America, with the remaining species distributed throughout Central and South America and the Caribbean. Wood warblers are small, insectivorous, migratory songbirds. As the native species are illegal to keep in North American collections except by special permit, very few individuals or institutions have worked extensively with these birds in captivity. A few species are currently being housed in several zoos, and data on their maintenance and husbandry are gradually accumulating. The same issues that discourage most aviculturists from keeping these specialized birds encourage a few bird keepers to try, but the rare person who attempts to breed these birds should expect a challenge.

The term wood warbler is something of a misnomer, as the Old World warblers also occur in wooded habitat. The term North American warblers is equally inappropriate, as many of the species migrate to and from Central or South America and cannot be considered exclusively North American. Perhaps the best designation for this group of birds would be neotropical migratory warblers, however, for our purposes here we will refer to them simply as warblers.

The scarcity of warblers in aviculture is no mere coincidence. Protective legislation at local, state, and federal levels is the biggest obstacle in keeping these birds. Acquiring government permission to keep native birds can be an arduous and dissuasive task. Fortunately, these restrictions ensure that only qualified people are allowed to collect these protected species for public display or avicultural purposes.

Once permission has been granted to keep native warblers, the problem of obtaining them still exists. It is generally acknowledged that for most species, the direct collection of a limited number of specimens from a wild population for research and zoological purposes will have a negligible effect on the overall survival of the species. The prevalent and optimal method of collecting warblers is by mist-netting them, but this is not an easy task. A second collection method involves finding, collecting, and hand rearing nestlings of the species desired. The challenge here is to locate an active nest from which to remove chicks. This can be a time-consuming process, further compounded if several nests are required to provide unrelated birds for an eventual breeding project. If this option is selected, it is beneficial to have previous experience hand rearing related species or species with similar captive requirements. Another possibility is to acquire specimens from a wildlife rehabilitation center. This is the least likely alternative. Even when the proper paperwork has been provided, the chances of receiving healthy, rehabilitated birds from a wildlife rescuer are slim. Normally, only disabled individuals, not suitable for release, are placed in captivity.

Aside from the legal and logistical aspects, there are several other reasons why captive warblers are difficult to find. Warblers are often perceived, incorrectly, as being common and boring. Most bird keepers would rather devote their resources to exotic, rare, and dramatic species. Another cause for the scarcity of warblers in captivity is their specialized husbandry requirements. However, if more institutions were to exhibit and breed warblers it would allow for an increase in knowledge of their specific husbandry needs.

Currently, the majority of work being performed with captive warblers is of a short-term nature, with the shortest-term work being conducted by Certified Wildlife Rehabilitators. The work of rehabilitators should not be discounted. Rehabilitators, like private aviculturists, are a source for a wealth of practical bird-keeping information. Academic pursuits account for the second largest group of people keeping captive warblers. Many colleges and universities use warblers for research on a multitude of topics. These short-term studies require that birds be held captive from a few hours to a few weeks, rarely longer.

Zoos, aquaria, and public aviaries comprise the smallest group of aviculturists focusing on warblers. Their work strives for the long-term maintenance and propagation of these valuable birds. Even among this very large and diverse community, few avian collections include warblers. Within Canada and the United States, for instance, there are only a dozen institutions that possess these birds. Of these 12, just two or three zoos are concentrating on keeping warblers in significant numbers to achieve consistent, appreciable breeding and longevity success.

Birds of the genus *Dendroica,* such as the yellow-rumped warbler *D. coronata* and the magnolia warbler *D. magnolia* are the most common warblers in American aviculture. Species from the genus *Seiurus* are the next most common. These species include the ovenbird *S. aurocapillus* and both the northern and Louisiana waterthrush, *S. noveboracensis* and *S. motacilla,* respectively. Rounding out the commonly kept warblers are the American redstart *Setophaga ruticilla,* the yellow-breasted chat *Icteria virens,* and several species of yellowthroats of the genus *Geothlypis*.

Natural Habits: Warblers range through a wide spectrum of wooded and semi-wooded habitats. Although they are generally birds of woodlands, forests, or dense brush, specific habitats vary widely among species, as well as geographically and seasonally within species (Dunn & Garrett 1977). These habitat preferences are summarized in the Peterson Field Guides. For instance, the edition on eastern birds notes that the prothonotary warbler *Protonotaria citrea* lives in wooded swamps, the black-throated blue warbler *Dendroica caerulescens* inhabits undergrowth of deciduous and mixed woodlands, Kirtland's warbler *D. kirtlandii* can be found in groves of young jack pines 5–18 feet high with a ground cover of blueberries, bearberry, or sweet-fern, and the prairie warbler *D. discolor* occurs in brushy slashings, bushy pastures, and low pines. Clearly, the diversity of habitat types used by warblers is impressive. Within these diverse habitats, foraging for insects occurs in every possible fashion, as these birds are primarily insectivorous. In addition to insects, fruit and nectar are occasionally eaten.

Management: The difficulty of acquiring warblers is surpassed only by the difficulty of converting them to a traditional softbill diet. Most warblers will never adjust to eating commercial fare and persist almost entirely on insects. Broadly speaking, the period of acclimation from a wild state to captive conditions should be regarded as an especially delicate one. Warblers are particularly difficult to acclimate and need to be monitored thoroughly and consistently. One must avoid being loud and disruptive while tending these birds. Kelly Vineyard, an experienced warbler keeper from the Columbus Zoo, emphasizes that the second 24-hour period after capture is the most critical phase. After this period, Kelly will release non-eating birds back to the wild if they have shown significant weight loss for two consecutive days. In fact, it is a common practice to force-feed Nekton Nektar, a sucrose solution, or a watery mix of Emeraid to the warblers during the first day of captivity. These ultra-high calorie mixtures help to sustain the stressed birds. The ordeal of acclimating adult birds can be avoided entirely by hand rearing chicks, even though this method has serious drawbacks. Hand rearing is not always successful, and if it is, may result in imprinted birds.

There are various techniques used for acclimating and quarantining these fragile songbirds, but the different methods have two themes in common. The first and foremost consideration is to prevent the bird from harming itself within the cage. As wild birds forage at a feverish pace, being placed in confinement is a tremendous shock to their nature. A small cage 0.61 x 0.61 x 0.61 m (2 x 2 x 2 ft.) will significantly limit their mobility, reducing their frantic attempts at escape and thereby preventing opportunities for the birds to hurt themselves. Alternatively, larger cages constructed of softer, more pliable materials can be used. For example, warblers have been acclimated in appliance-sized cardboard boxes retrofitted with wire fronts and feeding doors. Secondly, newly caged birds must learn to eat commercially available insects fed in unnatural dishes. By supplying multiple bowls filled with insects, water, and also a pan of frugivore diet in the very small enclosure, the birds are literally forced to focus on and interact with this foreign diet. Insects should also be broadcast on the floor of the cage in case the birds refuse to eat from the dishes. One theory pertaining to collecting warblers is to capture them immediately preceding or during the fall migration. At this time the birds have the instinctual behavior to consume phenomenal quantities of food in preparation for their journey. It is thought that this drive to eat will make acclimation and diet conversion less traumatic. However, the reaction of individual birds to captivity runs the gamut. Some warblers will obviously need to be released; others are very complacent and peaceful. I even know of a prothonotary warbler that took a bath in the water vessel of its transport container just seconds after capture. Also, I have worked with a common yellowthroat *Geothlypis trichas* that accepted a dish of mealworms immediately after being removed from the net and placed in a box.

Due to the extreme variation of the natural habitat types occupied by warblers, the captive requirements of different species may also vary, making it difficult to generalize the essential requirements of the group as a whole. The natural history and unique habitat preferences should be taken into consideration when designing accommodations for a particular species. Exhibits that replicate natural habitat will result in birds that are more comfortable under captive conditions. Strategic planning of plant species, plant location, and water features are of paramount importance. For example, ground-dwelling ovenbirds should not be placed in an exhibit with no ground cover. Jean-Philippe Gagnon of the Biodôme de Montréal wisely advises the aviculturist to never put anything in the aviary with the hope that some of the birds may survive and at the end this will be your collection. He adds that success can best be achieved by making a collection plan and sticking to it as closely as possible.

There are, however, some general requirements that all warbler accommodations should include. The importance of natural lighting cannot be over emphasized. Providing an outdoor enclosure, or an indoor enclosure with outside access, is the recommended method to supply these sun-loving birds with satisfactory amounts of natural light. One of the many benefits for providing unfiltered natural sunlight is the pleasure of watching your birds enjoying and sunbathing in it. Additionally, well-planted outdoor aviaries will attract a larger variety of insects, making this important food source available to your collection. This is not to say that warblers cannot be kept indoors, as long as the indoor habitat provides a sufficient source of ultraviolet light in the proper spectrum for avian species. A prothonotary warbler faded to a very dull yellow color after being kept indoors without proper lighting. Attempts to brighten the yellow by adding coloring agents to the diet yielded an orange-yellow bird. After this bird was placed in an outdoor aviary, and allowing sufficient time to molt, the colors were restored to their proper brilliant yellow shade. Color loss can be a dietary issue as well.

Any outdoor facility should be partially covered to protect the birds from the elements. In addition, aviary plants should be selected and planted in a manner that will allow the birds to seek refuge within dense foliage. Flowering vines such as honeysuckle *Lonicera* spp. or clematis *Clematis* spp. provide good cover. A broad-leafed, densely foliated shrub such as autumn olive *Elaeagnus umbellata* is an excellent windbreak. One should try to choose non-toxic fruiting or flowering plants that will attract more insects into the enclosure. This will create more opportunities for the birds to behave naturally, foraging for bugs and possibly even drinking nectar. An indoor flight with outside access will provide the best of both worlds, natural light and plenty of insects coupled with the protection of a building. However, not all birds will use indoor housing even when it is provided. I have often found warblers outside even in the worst weather, when I expected to find them perching indoors. On some occasions, the opposite also held true. On many gorgeous summer days I found the common yellowthroats hiding indoors in potted dwarf umbrella trees *Schefflera arboricola*. During very severe weather, it may be necessary to force these birds into indoor quarters.

Inside or out, any warbler housing should include a water feature. Warblers love to bathe. Shallow water dishes are best, but frequently need refilling. A shallow running stream or pool of water suits these birds very well, as does an overhead misting system. These very active creatures will exploit every type of water feature. They are sometimes seen bathing by darting through wet foliage.

The temperature requirements of warblers in aviculture are not yet well established. Each institute tends to be treating them differently. An ideal temperature would fall between 12.8–21.1°C (55–70°F). Since warblers migrate to warmer southern climates for the winter, there is a tendency to treat these birds similarly to any other tropical softbill. Despite this notion, these amazing birds can tolerate colder temperatures than one might expect, providing they have access to supplemental heat. Of course, an increase in diet is in order when subjecting the birds to near freezing temperatures. The zoos that have been most proficient in warbler husbandry do indeed subject the birds to cold weather. At the Columbus Zoo, the native bird aviary, which includes warblers, is open all winter. However, a series of 10 heated wooden boxes is placed throughout the exhibit. These boxes are open fronted or slotted and are heated with either a spotlight or infrared lamp. The birds also have free access to an indoor holding building, although most of the birds prefer not to use it. Additionally, the amount of insects offered is increased, and precautions are taken to keep the insects from freezing. Since 1998 the Columbus Zoo has displayed warblers outdoors all winter long, with temperatures regularly dipping to 0°C (32°F), with no temperature-related problems. The Biodôme de Montréal maintains their warblers in a completely indoor Laurentian Forest Ecosystem exhibit that artificially simulates a winter season with nighttime temperatures dropping to 4°C (39.2°F).

Warblers may be able to endure cold temperatures but they are still fragile, easily stressed birds. Handling should be kept to an absolute minimum and blood samples should be taken only under the rarest of circumstances. Unnecessary medications must be avoided because warblers can be intolerant of the medications commonly used on other types of birds. Lastly, warblers exhibit fewer symptoms of illness than most other species of birds, often giving no noticeable indications when they are not well. Birds have been found dead despite appearing completely normal the previous day.

Enclosures for warblers tend to be constructed as native North American exhibits, which is of course their normal summer habitat. In this situation they can be exhibited with other North American avifauna. They do mix well with other types of birds, although they do not necessarily mix well together. Aggression between warblers is a serious concern during breeding season. Finches, buntings, sparrows, quail, flickers, orioles, bluebirds, catbirds, and assorted waterfowl species have all been successfully housed with warblers in very large free-flight, walk-through exhibits. In smaller flight cages I have seen warblers held with a variety of tropical birds including *Ptinilopus* fruit doves, *Tangara* and *Thraupis* tanagers, as well as with snowy-headed robin chats *Cossypha niveicapilla*. Neotropical migratory warblers utilize tropical habitat for portions of the year, so they are equally suitable for display in tropical exhibits. Regardless of the enclosure type, attention must be given to the number of warblers kept together. The Laurentian Forest Ecosystem exhibit at the Biodôme de Montréal covers roughly 2.5 acres (1.0 hectare) and contains many types of birds, but just eight species of warbler. Even in this large area, a total of only three breeding pairs have been safely managed during the spring breeding period. Jean-Philippe Gagnon, a birdkeeper at the Biodôme, feels that it is better if all of the mate-less warblers in the exhibit are females. He offers the following advice based on his experience:

> "Every bird needs an aerial space to feel comfortable and fly freely in the habitat. When we introduce a large bird like a thrush or an oriole, those birds take a place that is no longer available for the small warblers. If you want them to sing, they have to fit in the environment and fit in the avian community you try to reproduce. I will add to this that the landscape of the exhibit needs to fit with the avian community."

Kelly Vineyard from the Columbus Zoo works in an aviary that is 25,603.2 m^3 (84,000 cu. ft.) and contains 11 species of warblers, but no more than two males of any given species.

Maintaining warblers in captivity is certainly a challenge. Few private individuals are likely to obtain legal permission to keep these delightful little birds. A person needs the fiscal and human resources of an entire organization to be successful. Acclimating the birds to captivity is the most difficult part. Novice aviculturists would not be well advised to attempt keeping warblers. Supplying the requisite insect diet is expensive and time consuming. Coaxing the warblers to eat essential fruits and vegetables is nearly impossible. Building an adequate enclosure takes thorough planning and plenty of practice. Warblers are not the kind of birds one can simply add to an existing aviary for embellishment. To avoid losses, one must be very dedicated to the warblers and design a living space exclusively for them. Assorted types of birds can be added to a warbler aviary, but not vice versa. The aviculture of warblers is still in its infancy and may never progress past this point. Pioneering work is currently taking place but there are no steadfast and guaranteed recipes for success.

Despite this situation, anyone who is fortunate enough to work with these intriguing birds will be rewarded. Warblers are amazing to observe and glorious to hear singing. Their diminutive stature is awe-inspiring. Their size does not hinder them; they are packed full of energy. Sunbathing is especially enjoyed and bathing in water is always undertaken enthusiastically. Trips to and from the food dish are relentless. Observing these small birds resting among dense foliage is always difficult, but once they are located, they are relatively approachable. Except during the breeding season, these small insectivores are usually very calm and peaceful. Warblers simply have a vital, vibrant, and pleasing nature. To top it off, very few bird songs can rival the sound of a singing male warbler.

Diet: Live insects form the bulk of the captive diet. It is therefore important to provide as wide a variety as possible. Mealworms *Tenebrio molitor* and crickets *Acheta domestica* of assorted sizes, cleaned maggots *Musca domestica,* and fruitflies *Drosophila* spp. of different sizes comprise the largest percentage of the captive insect diet. All small-size commercially available insects will be readily eaten. Warblers cherish waxworms *Galleria mellonella*. The problem is the high fat content of waxworms. Some keepers refuse to feed them entirely. If waxworms are fed, they should comprise only a very small part of the diet. In addition, nontraditional methods of supplying insects are recommended. Setting up ultraviolet insect traps or sweep netting for bugs will increase the diversity of insects that can be offered. Live food can be presented in several ways. Filling small dishes with insects is the easiest method. All food should be offered on elevated food platforms. Except for the ovenbird and waterthrushes, warblers do not often feed on the ground. Broadcasting bugs throughout the habitat is fine as long as one remembers that the bugs that reach the ground may be wasted. Puzzle and slow-release feeders are another excellent method to keep these birds psychologically stimulated. Placing a rotted, insect-infested log within a flight cage was one of the most satisfying things I was able to do for the warblers under my care. Rotting tree trunks can be replicated by mounting a log several feet off the ground, drilling holes in it, and filling them with bugs. The hardest thing about feeding live food to the warblers is finding ways to present the food so that the birds can eat it before it escapes. Although living food is preferable, some warblers will accept insects that were previously frozen and thawed. Chilling live insects briefly in a refrigerator is a simple way to slow them down for a few minutes. This is a great trick for very hot days when the birds are lethargic and the insects are warm and lively. The quantity of insects provided will need to be increased during the breeding season, especially when chicks are in the nest. Insect colonies should be fed high quality food in order to "gut load" them. The insects can only provide proper nutrients to your birds if they themselves have been given a nutritional diet.

Aside from insects, softbill diet should be offered. Whether a commercial brand or a homemade preparation is used really does not matter. What does matter is that the birds have something in addition to insects available for consumption. Some warblers will eat minuscule amounts of finely diced fruit and vegetables. Among the more commonly offered items are apple, pear, carrot, grapes, peas, corn, raisins, and berries. Grated hard-boiled egg can be fed as well. Pelleted softbill diets are not readily consumed by these birds, although there are mixed reports of success with them. Personally, I have never seen a warbler willingly eat softbill pellets, although I have watched them cleverly eat all the fruit bits from around the pellets and inadvertently consume a few pellets stuck to the moist fruit.

A high quality nectar mixture such as Nekton Nektar-Plus should be given regularly as well. Wild-caught birds may take a very long time to become accustomed to the nectar feeders, and should be fed accordingly. If the birds only drink the nectar once every other week, it need not be prepared every day. Conversely, if birds drink nectar daily, it should be offered daily. All dietary items should be fed in moderation.

The use of vitamin supplements is recommended for warblers. A variety of high quality powdered vitamins are available. Vitamins should be added to the food rather than the water, as the birds will bathe in it. The manufacturer's guidelines should be strictly followed to avoid overdosing with vitamins. Vitamins containing coloring agents are perfectly acceptable, however, it is inadvisable to purchase pure color enhancing powders and carelessly dump them on the birds' diet.

Breeding: Reproduction in captivity is not rare, but it is not common, either. The most commonly kept warbler species all build cup nests. Some of these nests are tidy and symmetrical, whereas others are somewhat bulky. What they all have in common is the range of materials used in their construction. Twigs, fine and coarse grasses, roots, and rootlets are routinely used. Moss, spiderwebs, and bark bits have also been recorded. The nests are usually lined with very fine grass, feathers, or hair. The birds should therefore be provided any of these materials that are not already growing in their enclosure.

Nest placement varies among species. The common yellowthroat nests among low, dense, vegetation at or nearly at ground level. Most other captive warblers nest in trees, at varying heights. The types of trees selected for nesting sites will also depend upon the species. Clutches range from 2–5 eggs, with 3–4 egg clutches occurring most frequently. Incubation can last as long as 14 days, or be as short as 11 days, again depending upon the particular species. Fledging age also varies among species. For instance, American redstart chicks fledge in 8–9 days, compared to 12–14 days for those of the yellow-rumped warbler.

Hybridization between species in the closely related genera *Dendroica* and *Mniotilta* have been documented in captivity. A male black-and-white warbler *M. varia* at the Biodôme de Montréal has bred with a female black-throated green warbler *D. virens,* as well as with a female yellow-rumped warbler, producing live chicks. So far in the literature, there are no recorded breeding accounts of captive ovenbirds or

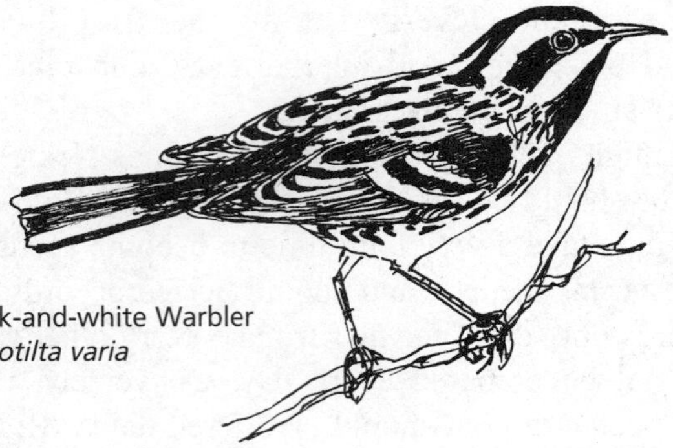

Black-and-white Warbler
Mniotilta varia

waterthrushes. The breeding cycle is very straightforward, providing that the birds feel comfortable and secure within their enclosure. The key is providing the proper habitat, the necessary nest materials, and plenty of insects for the parents to feed hungry chicks. It is not yet clear if double clutching can be achieved regularly in captivity by removing eggs for artificial incubation, or by pulling chicks for hand rearing.

Hand Rearing: If at all possible, hand rearing should be avoided. If hand feeding becomes necessary, it is still best to leave the chicks with the parents for a few days. Unfeathered chicks are extremely delicate and difficult to work with. Feeding the naked birds can be frustrating if they do not exhibit a strong feeding response. They should be placed in a small bowl lined with a soft paper towel and kept in a warm brooder. The paper towel should be changed at each feeding. The initial feedings should be scheduled at one-hour intervals during daylight hours. Defecation usually occurs during or right after each feeding, and should be performed without difficulty.

As the chicks grow, the interval between feedings can be increased accordingly. Several liquid formulas for hand feeding softbills are commercially available. In addition, almost any aviculturist or rehabber can supply a homemade recipe. The formula is administered with an eyedropper or even a small paintbrush. The urge to overfeed should be curbed; overfeeding may aspirate the chick or cause an impaction. In addition to the liquid formula, a plentiful supply of live food is indispensable. Insects can be fed to the chicks with tweezers that should first be used to crush the insects. Newly shed mealworms, with a soft, white skin, are ideal. Waxworms should have the hard mouthparts and tough skin squished. Clean maggots and small, gut-loaded crickets can be fed sparingly. Insects with hard exoskeletons should be avoided as much as possible. Lastly, pieces of well-diced fruit and vegetables and moistened pellets should be offered. If observation of the feces reveals that whole insects are being passed, then they have not be crushed sufficiently before being fed. As the birds grow, the liquid formula should be decreased and the whole foods correspondingly increased. Warbler chicks grow rapidly and fledge at an early age. Their growth should be monitored by weighing them daily. After they leave the nest, they must be housed alone until they learn to recognize the diet pans as a food source. Hand-reared juvenile warblers need time alone to develop sufficient flying skills to evade aggression by future cagemates.

It is recommended that this be read in conjunction with the HONEYEATER hand rearing protocol.

Band Size: 2.0-2.8 mm, depending on species

Suppliers:

Emeraid
Lafeber Company
24981 North 1400 East Road
Cornell, Illinois 61319
Tel. 1-800-842-6495
Fax. (815)-358-2352
Internet: http://www.lafeber.com

Nektar-Plus
Nekton-Produkte
Guenter Enderle Enterprises, Inc.
27 West Tarpon Avenue
Tarpon Springs, Florida 34689
Tel. (727) 938-1544
Fax. (727) 938-1545
Internet: http://www.nekton.de

Contributors:

Kelly Vineyard, Columbus Zoo; Columbus, Ohio, USA.

Jean-Philippe Gagnon, Biodôme de Montréal; Montréal, Québec, Canada.

References:

Beadle, D., J. Curson & D. Quinn. 1994. *Warblers of the Americas: an identification guide*. New York: Houghton Mifflin.

Dunn, J. & K. Garrett. 1997. *Peterson Field Guides: Warblers*. New York: Houghton Mifflin.

Peterson, R.T. 1980. Peterson Field Guide to Eastern Birds. Houghton Mifflin, New York

Plante, A.-M. & J.-P. Gagnon. 2001. Breeding warblers successfully at the Biodôme de Montréal. American Association of Zookeepers National Conference, Toronto, Ontario, September 30BOctober 4, 2001.

21.20biii Thraupini Tanagers

Martin Vince with contributions on handrearing from the staff of Brookfield Zoo

Introduction: Under the Sibley-Ahlquist-Monroe classification, tanagers belong to the order Passeriformes, the family Fringillidae, the subfamily Emberizinae, and the tribe Thraupini, which encompasses 413 species in 105 genera, of which approximately 242 species are regarded as tanagers. All are confined to Central or South America except the four *Piranga* species, which breed in the United States and migrate to the tropics in the autumn. The tanager group also includes the aviculturally important euphonias, chlorophonias, honeycreepers, and dacnis, as well as less familiar species such as the flower-piercers, conebills, and orangequit.

Honeycreepers: The six species of honeycreepers average 10–13 cm (4–5 in.) in length. All species are dimorphic. The eclipse plumage of the male red-legged honeycreeper *Cyanerpes cyaneus* assumed during the nonbreeding season, has the appearance of the female, although the male retains his black wing and tail feathers as well as his bright red leg coloration.

Euphonias: The euphonias are even smaller than the honeycreepers, at 9–11 cm (3.5–4.5 in.). All 27 species are dimorphic. In all species that appear in aviculture, males are yellow and glossy purplish black; females are shades of dull olive.

Chlorophonias: The five species in this group are slightly larger, 10–14 cm (4–5.5 in.), and are also dimorphic. Chlorophonias are considerably greener than euphonias, with yellow or chestnut underparts. Females are similar to the males but not as vivid, generally duller green with less yellow.

Dacnis: The nine species of dacnis are 10–13 cm (4–5 in.) in length, and all are dimorphic. The blue dacnis *Dacnis cayana* and the black-faced dacnis *D. lineata* are a vibrant sky blue with black markings. Females are greenish olive in both species.

Flower-piercers: Most flower-piercers are monomorphic. The 18 species, which comprise the genera *Diglossa* and *Diglossopis,* range in size from 10–16 cm (4–6.5 in.).

Orangequit: The 13 cm (5 in.) orangequit *Euneornis campestris* is dimorphic.

Other fairly distinct groupings among the tanagers include the bush-tanagers, with nine species in the genus *Chlorospingus* and one in the genus *Cnemoscopus;* 12 species of hemispingus in the genus *Hemispingus;* four species of shrike-tanagers, genus *Lanio;* five species of ant-tanagers, genus *Habia;* seven species of mountain-tanagers, five in the genus *Anisognathus* and two in monotypic genera; and the monotypic swallow-tanager *Tersina viridis*. Of the remaining tanagers, many species occur in monotypic genera. Most of the typical tanagers that reach avicultural collections belong to the genera *Thraupis, Tachyphonus, Ramphocelus,* or *Tangara*.

The smallest tanager species, in the genera *Euphonia* and *Cyanerpes,* are about 9 cm (3.5 in.) long and weigh about 9 g (0.31 oz.). The longest species, the magpie tanager *Cissopis leveriana* is around 26 cm (10.25 in.) in total length. The heaviest species, the white-capped tanager *Sericossypha albocristata* weighs around 114 g (4.02 oz.) (Isler & Isler 1999).

Among the more common species appearing in aviculture, the blue-gray tanager *Thraupis episcopus* a monomorphic species with many geographic races, is 16 cm (6.3 in.) in length, with an average weight of around 35 g (1.23 oz.). The blue-and-yellow tanager *T. bonariensis* is slightly larger, weighing just a gram or two more, on average. This species is dimorphic, as is the silver-beaked tanager *Ramphocelus carbo,* which is similar in length to the preceding species, 16–17 cm (6.5 in.), but weighs considerably less, with several subspecies ranging from 24–28 g (0.84–0.98 oz.), on average. Another commonly kept dimorphic species, the white-lined tanager *Tachyphonus rufus* reaches 17 cm (7 in.) in length, with an average weight of around 33 g (1.16 oz.).

Nearly all tanagers feed in mixed species flocks, usually at mid-heights in the forest canopy, on fruits, insects and nectar. In the wild, fruit is the staple for the vast majority of species, although a small number are purely insectivorous while others are specially adapted to eat nectar. Habitat A, with small sections of Habitat B, will provide the ideal habitat for these birds.

Management: Newly acquired, healthy tanager species are usually easy to acclimate using the diets shown below. A clearly visible piece of orange or papaya often brings new arrivals to the food dish more quickly, and helps them to settle down. However, citrus fruits have been shown, at least in people, to enhance the absorption of dietary iron; and since iron-storage disease, or hemochromatosis, is a potential problem in most softbilled birds, the prolonged use of large amounts of orange is not recommended.

Another interesting, and emerging, problem is atoxoplasmosis. This protozoan disease is a type of coccidia and is suspected in certain red-legged honeycreepers at Riverbanks Zoo. Atoxoplasmosis does not usually threaten the life of a healthy bird, although the sick, stressed, old or young are most at risk from severe complications. Indeed, high levels of neonatal mortality are typical of the disease. The European poultry medicine called Baycox™ is the best treatment. It can suppress the symptoms, allowing the bird to live a normal life, although it is doubtful that atoxoplasmosis can currently be cured. See also the section on LEAFBIRDS and STITCHBIRDS for more information on atoxoplasmosis.

When acclimated, species of the genera *Thraupis, Ramphocelus,* and *Anisognathus* are capable of living outdoors year-round with minimal heat. Examples of hardy tanagers are the blue-gray, the sayaca *Thraupis sayaca*, the palm *T. palmarum,* the silver-beaked, the Brazilian *Ramphocelus bresilius,* and the blue-winged mountain-tanager *Anisognathus flavinuchus*. In the northern United States and Western Europe, where winter temperatures fall below freezing for prolonged periods, a frost-free shelter and a heat lamp should be provided. In climates where heat lamps are not necessary, it is helpful to install a 10-watt light bulb on the ceiling of the shelter to simulate moonlight. This will not only allow the birds to feed after dark, but also will enable them to become reoriented if panicked by storms or predators.

Euphonias are also quite hardy and tend to do well in aviaries. However, since their bodies are smaller than the species mentioned above, I prefer to offer additional heat during the coldest months, making sure that the temperature in their quarters does not fall below 13°C (55°F). Chlorophonias are also fairly hardy once acclimated, but they can be difficult to establish and are not ideal for the beginner. All other tanager species should be considered somewhat delicate, especially the *Tangara* species, honeycreepers, dacnis, flower-piercers, and the orangequit. I prefer to maintain such birds at temperatures above 16°C (60°F). No tanager species should be kept under conditions where they could be subjected to sudden, drastic temperature changes.

Ramphocelus, Thraupis, and *Anisognathus* species sometimes harass smaller birds or birds of any size that do not stand up for themselves. Most small, omnivorous tanagers, forest euphonies, and dacnis are often found in flocks that seem to form and disband irregularly at certain times of the year. In aviaries of 140 m² (1500 sq. ft.) to 3/4 acre, I have seen blue-and-yellow, turquoise *Tangara mexicana,* blue-gray, silver-beaked, and blue-winged mountain-tanagers persistently chase species such as fairy bluebirds *Irena puella,* leafbirds *Chloropsis* spp., and broadbills. Brookfield Zoo staff members suggest that to prevent interspecific competition, these flocks should be limited to house single pairs of each species. *Tangara* species work particularly well together due to their less aggressive dispositions. On the other hand, some *Ramphocelus* species are reputed to be somewhat aggressive toward other birds. Injuries do not normally occur, and certainly not all specimens of these species make a nuisance of themselves. But it seems that many of the now commonly kept tanagers have the ability to be troublesome in mixed species aviaries. Multiple pairs of tanagers do not usually live together successfully, unless in very large enclosures. Honeycreepers and dacnis are also best kept in pairs; when kept in groups, the dominant male tends to confront other males, chasing them away to hide in a bush and cutting off access to food and water. Brookfield Zoo staff found that during nesting season, disturbances from other birds may limit breeding efforts more than tangible objects such as food, nest sites, and materials. Managers and keepers should therefore be aware of the carrying capacity of enclosures. Highly inquisitive species or unpaired tanagers, even if they are nonaggressive, can hinder breeding. They steal nest material, intrude in territories, and harass breeding and nonbreeding birds.

At Brookfield Zoo, tanagers are also bred as single pairs in indoor aviaries that vary from 90–180 cm (3–6 ft.) wide and all 180 cm (6 ft.) tall. The cement floor is easily disinfected. An overhead misting system, operated manually, can increase humidity and simulate rain. The entire roof has skylights and variable timers regulate the fluorescent lights and can be adjusted to vary the day length. Artificial lights on timers supplement the day length for 10–13.5 hours of light, simulating day length at 20° latitude. Disturbance from people is kept to a minimum and the environment is strictly controlled. As observation of pairs and behavior is difficult in the situation, video cameras are now used to record behavior. A disadvantage of this system is that the birds' lack of privacy from neighbors may hinder breeding in some species.

Diet: Tanagers and similar softbills should be fed a fresh diet in the morning and late afternoon. It is important to thoroughly clean and disinfect food dishes, especially nectar containers; and it is very important to stack them inverted so that they can dry thoroughly before reuse. Two separate containers of food should be provided in the morning and again in the afternoon. The first, a shallow dish or hummingbird tube of proprietary lory or hummingbird nectar, should be available throughout the daylight hours. Nectar should not be left in the aviary overnight because it can turn bad. Nectar is most important for honeycreepers, dacnis, flower-piercers, and the orangequit. For other tanagers, nectar is beneficial, although not as essential. The second dish should contain 45% small proprietary pellets, such as those marketed for finches and canaries, and 55% finely chopped fruit. The fruit and pellets should be thoroughly mixed together. A light sprinkling of grated hard-boiled egg is beneficial, as well as about six mealworms or waxworms per bird per day, depending on the size of the bird. Live food can be increased for the breeding season to help stimulate breeding activity. Commercially available live food, such as mealworms and waxworms, are high in fat and will cause life-threatening obesity if overfed year-round. Many tanagers eat the leaves of plants, often favoring *Ficus* trees. *Thraupis* tanagers, with their slightly hooked upper mandibles, appear to do the most damage. Providing a chunk of lettuce alongside the usual diet may help to reduce damage to plants but is not always effective.

Breeding: Honeycreepers normally lay two eggs, white or pale blue with brown spots; incubation takes 12–13 days, solely by the female. Euphonias and chlorophonias lay 3–5 eggs, usually pale with dark blotches, with incubation lasting approximately 15 days, also solely by the female. Most of the tropical tanager species commonly found in aviculture lay 2–3 eggs, ranging in ground color from white to pale blue, usually with heavy dark blotches. Incubation usually takes 12–14 days, solely by the female. Most tanagers are monogamous. Unless placed in an especially large aviary, more than one pair of the same species will usually find it difficult to live and breed together amicably. Pairs of different tanager species may coexist and breed in large aviaries, but this is also risky and must be undertaken with care.

Establishing a compatible pair is the first step toward any successful breeding program. In the past, birds were simply housed together with the hope that they would form a bond. Although often successful, it sometimes resulted in aggression (from threat gaping to actual fighting) or complete indifference to one another. Different combinations of individuals had to be tried before a suitable pair was eventually established. Brookfield Zoo employed a method of pairing whereby the birds were allowed to choose a mate before being introduced to each other in the same enclosure. By housing the female next to more than one potential mate, they observe her behavior to determine to which bird she is more attracted. Generally, if a male and female constantly perch near each other 15 cm (6 in.) approximately or less, the male sings frequently, he feeds her through the wire and/or the female solicits the male, it is a good indication that the two would make a compatible breeding pair. All the above general courtship activities will continue once the pair is housed together.

The majority of tanager species build a cup-shaped nest in a bush or tree using dead leaves, rootlets, grasses, and plant fibers; honeycreepers may also use cobwebs to anchor their nest to thin branches. By contrast, euphonias and chlorophonias build a domed nest, usually with a side entrance that has a tiny overhang. The nest is normally well hidden behind foliage in the crotch of a tree, or in cavities in rock or woodwork. A pair of tanagers should be given an aviary with plenty of bushes and vegetation. A variety of nesting receptacles should be provided for all tanager species until the preference of a given pair has been determined. Half or fully open-fronted boxes, planted hanging baskets, clumps of epiphytic plants, and cup-shaped wicker canary baskets can be located both in secluded locations and in the open. As many different nesting materials as possible should be provided. Palm fibers taken from the leaves, and especially from the trunk, of the tree; partly decomposed leaf fragments, which can be found on the ground of undisturbed wooded areas; moss, burlap, animal hair, which must be used sparingly to help prevent entanglements; fine grasses of varying lengths; spiders' webs; and plant down should offer enough variety for the most discriminating pair of birds.

Chicks of most tanager species are fed live food for approximately the first week of their lives. Waxworms tend to be the favored food and are probably safer than mealworms, since they are much softer-skinned. Live food, such as waxworms, should be constantly available to adults that are rearing young. As the chicks grow, they will start to receive pieces of the adult diet. The chicks of most species fledge at 2–3 weeks of age and are fed by their parents for a similar period after fledging. Euphonias and chlorophonias are fed by regurgitation and take several days longer to fledge. They may accept the usual range of live food, but most wild-caught specimens kept in aviculture will only feed live spiders to their young. This continues for about the first two weeks of the chicks' lives, after which time they become independent.

Brookfield Zoo staff use an alternative to hand rearing fledglings that are thought to be at risk. To keep fledglings in the aviary where they can learn from and be fed by the parents, a special fledgling cage is used. Constructed of 25 x 13 mm (1 x 0.5 in.) weldmesh, the cage measuring 58 x 45 x 48 cm (23 x 18 x 19 in.) has perches at varying heights to allow the chick to move from the top to the bottom of the cage with ease. The ceiling of the cage is slightly concave, which allows the chick to receive food from the parent when perched directly beneath it. This cage has been used quite successfully with turquoise tanagers and purple honeycreepers. It was tried with golden tanagers but failed; their behavior suggested that this species may need a larger mesh size through which to feed their young. Cross fostering of eggs has also been successful but should only be attempted as a last measure and only between similar species.

Artificial Incubation and Hand Rearing: As the hand rearing of passerines from the egg has been perfected by relatively few to date, I have also added the hand-rearing protocol from Brookfield Zoo. For further passerine hand-rearing information, an informative description is given of the rearing HONEYEATERS by Adelaide Zoo. The artificial incubation of such tiny and difficult eggs has the greatest chances of success with the use of a good quality incubator such as a Grumbach. Tanager eggs should be incubated at 38.1°C (100.6°F) with a relative humidity of approximately 55%. Set the incubator to turn the eggs frequently, such as every 1–2 hours. Although it takes more time, I recommend carefully monitoring the eggs' progress during incubation by using a weight-loss chart. Carefully weigh the egg throughout incubation to ensure that it hatches at approximately 12–15% less than the freshly laid weight.

Violet Euphonia
Tanagra violacea

Blue-grey Tanager
Thraupis episcopus

I once incubated a blue-gray tanager egg incorrectly at 37.5°C (99.5°F) while achieving the correct weight loss of 12–15%. The chick developed properly, but upon hatching, the yolk sac was not fully retracted. I left the chick in the incubator overnight, still partly in the shell, to allow its body to absorb the yolk. By the following morning the yolk had been absorbed and the chick was fine.

Day 1 to Day 7: Keep the chick on a towel surface in a small dish at 37.5°C (99.5°F). Feedings must be very small and given every 1.5–2 hours from about 7:00 a.m. to 11:00 p.m. As the chick grows, the time between feeds can be increased along with the amount per feed; the chick does not need to be fed through the night.

I recommend using a 1-cc syringe to feed one of the proprietary hand-rearing formulas originally developed for parrots. My personal preference is Kaytee Exact Original Formula, although any of the other major brands is likely to be equally good. The very first feed should be mostly liquid, preferably distilled water or an oral electrolyte such as Pedialyte, with only a very light sprinkle of Kaytee. For all of the other feeds, the Kaytee must be prepared, and progressively thickened, according to the manufacturer's instructions; and it should be mixed with a small amount, by volume, of pureed fruits. The mix I use consists of 80% Kaytee, 10% pureed apple, and 10% pureed papaya.

The chicks should be weighed at the same time each day, and the average daily weight gain should be 10–15%. If the weight gain is less than about 6% or more than 18% for two consecutive days, the food intake should be adjusted accordingly. This can be accomplished by changing the amount per feed and/or the time interval between feeds.

After each feed, it is important to clean food residue from inside and outside the chick's mouth with a small piece of moistened cloth or a Q-Tip. If this is not done, food residue can easily spoil and help give rise to the fungal infection *Candida*. Some aviculturists use the drug Nystatin as a prophylaxis in the hand-rearing formula of many avian species. The use of Nystatin for the first 2–3 weeks seems to do no harm and can reduce the risk of *Candida,* especially when combined with proper oral hygiene.

Day 8 onwards: As the chick enters its second week, forceps can be used to offer small pieces of the adult diet. Waxworms and thin slices of frozen pinkie mice are also excellent foods at this stage. To ensure balanced nutrition, all foods should be dipped, just prior to being given to the chick, in warmed Kaytee hand-rearing formula that has already been mixed with water. I also feed a calcium/mineral supplement, such as Nekton MSA, once a day. Growth is rapid from this time onward. As the chick reaches fledging age, it is important to provide thin, low perches next to the dish in which the chick is being reared. When the bird is ready, it will fledge and start to perch for itself. Food, and a very shallow dish of water, should be provided; within a few weeks the bird will start to feed independently and can be moved to a small aviary or large cage.

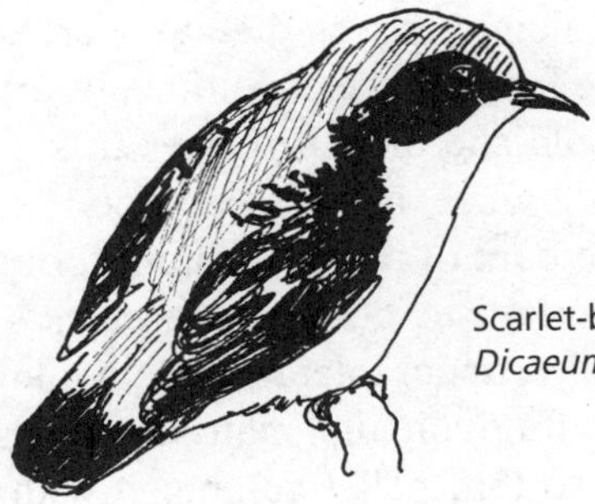
Scarlet-backed Flowerpecker
Dicaeum cruentatum

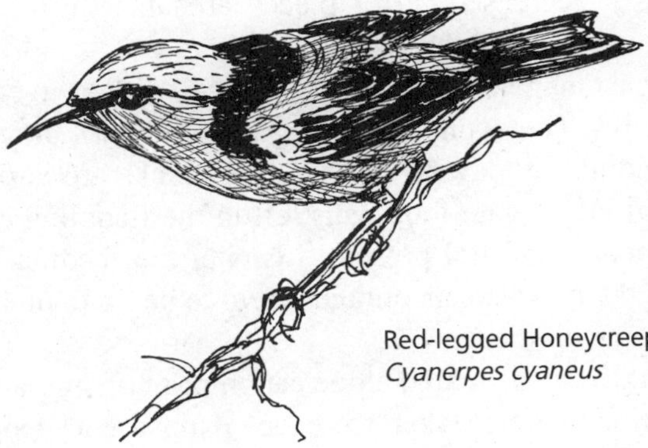
Red-legged Honeycreeper
Cyanerpes cyaneus

Paradise Tanager
Tangara chilensis

Brookfield Zoo Bird Department Tanager Hand-Rearing Guidelines

by Anne Oiler and Patty McGill

At Brookfield Zoo, the Bird Department staff has established general guidelines for hand rearing small passerines including tanagers, and this protocol is modified based on the chick's development. Neonates are placed in a small glazed crockery bowl with soft tissue paper and housed in a brooder until they are completely feathered. As the chick grows, small pieces of rubber matting or small sticks can be placed in the crock to provide surfaces for the chicks to grip with their feet. Humidity should be maintained at 30–50%. For newly hatched chicks the temperature is held at 36.7°C (98°F) and gradually decreased to 35.5°C (96°F) by day 7. After that point the temperature is decreased one degree on alternate days until the brooder is at room temperature. Then the chick is moved to a fledgling cage.

Hand Raised Silver-throated Tanager
Tangara icterocephala

Day	Weight (g)	Day	Weight (g)
1	2.1	22	13.49
2	2.36	24	14.38
4	3.21	26	15.2
6	4.25	28	15.96
8	6.04	30	16.37
10	8.89	32	15.6
12	10.68	34	16.9
14	12.41	36	16.4
16	11.61	38	17.8
18	11	40	18.1
20	12.89		

Keepers feed the newly hatched chick every 45 minutes for the first 1–5 days post-hatching, for a total of 10 feedings per day. We do not feed the chicks overnight. This interval is increased to every hour between days 6–12, and then adjusted to every 1.25–1.5 hours until the chick is self-feeding. The chick will usually defecate a fecal sac after every feeding without manual stimulation.

The diet consists of pinkie mice viscera or larger pieces depending on the size of the chick, and waxmoth larvae for the first 11 days. The pinkie pieces are dipped into a warmed electrolyte solution during the first 48 hours and then water is used after that period. By removing the heads of the waxmoth larvae, the chicks are better able to digest the insects. These larvae are also dipped in a high-calcium insect diet to increase the nutrient levels of the food source. Additionally, after the chick is three days old and until it fledges it is given a vitamin supplement. Once a day, one food item is dipped in a small amount of a multivitamin formulated for birds. It is important to monitor the chick's fecal matter, as some chicks cannot tolerate certain multivitamins. At day 12, the frugivore diet matrix, composed of a crushed low-iron softbill pellet combined with additional minerals, is added to the diet and the pinkie is gradually decreased. Fruit pieces are also introduced slowly when the chick is 14 days old.

Most tanagers fledge at 10–16 days, with an average of 14 days. They continue to beg for food until they are self-feeding, by day 24–27. Tanager chicks are curious and will investigate food pans left in the fledgling cage when people are not present. Leaving the feeding forceps in the pan also encourages them to begin to pick up food items on their own.

Single chicks reared alone can imprint on humans if precautions are not taken to reduce exposure to people. Provide mirrors, use a puppet for feeding, and limit human contact as much as possible. Older chicks that have had some parental care are less at risk of imprinting than chicks reared from a young age. For additional socialization, the fledgling can be housed in a howdy cage within the enclosure of adults and continue to be hand fed through the wire mesh.

Nectar-dependent species, such as purple honeycreepers and *Violaceous euphonias,* will require more frequent feedings (hourly feedings and more per day). Omit the pinkie from the diet and feed nectar specifically formulated for chick rearing, that is, one that has a 9–15% protein level. Fruit flies and soft-bodied insects or mealworm viscera should also be fed. *Violaceous euphonias* require fruit mixed into the nectar and this can be done in a food processor. Alternate between feeding plain nectar and nectar mixed with papaya to decrease the possibility of the fruit causing constipation. Due to the high fruit content of the euphonia diet, the chicks may appear lethargic at the initial feeding of the day. They usually respond to feedings and increase activity by the third feeding. Initially these chicks may only defecate every third feeding and require manual stimulation to induce fecal sac production.

Species Requiring Similar Management:

White-eyes *Zosterops* spp.
Leafbirds *Chloropsis* spp.
Fairy bluebirds *Irena* spp.
Ioras *Aegithina* spp.
Bananaquit *Coereba flaveola*

Reference

Oiler, A., P. McGill, and S. Schiller. 2000. The biology and husbandry of tanagers in captivity. *The AFA Watchbird.* 27(1): 8–16.

21.20biv Cardinalini Cardinals

Glen Holland

Introduction: Cardinals belong to the order Passeriformes, the family Fringillidae, and the subfamily Emberizinae, which encompasses a wide array of birds including, but not limited to, finches and sparrows, tanagers, buntings, grosbeaks, wood warblers, and icterids, such as the New World blackbirds, orioles, caciques, and oropendolas. The species of cardinals in the South American genera *Paroaria* and *Gubernatrix* are placed in the tribe Emberizini, along with most of the buntings and the brush-finches, while the species of the genus *Cardinalis* are placed in the tribe Cardinalini, which is sequenced several tribes later, following the wood warblers, tanagers, warbling finches and seedeaters, and along with the New World grosbeaks and saltators.

In North America, the common cardinal *Cardinalis cardinalis* is known variously as the Virginian cardinal, the northern cardinal, or by any of several other names, depending upon subspecies and locality. Neither this species nor the *C. sinuatus,* a resident of the southwestern US and Mexico, can legally be kept in North American collections due to restrictions on keeping native birds in captivity; however, these species sometimes appear on dealers' lists in other parts of the world. The third species in the this genus, the vermilion cardinal *C. phoenicea,* originating from Colombia and Venezuela, is occasionally available to aviculture and highly prized.

The genus Paroaria consists of five species, of which one, the red-crested cardinal *P. coronata,* is still fairly common in aviculture, but concentrated efforts should be made to breed this species in captivity, due to its continued decline in the wild. Although it has been established in Hawaii and on some of the Caribbean islands, this native of southeastern Brazil, Paraguay, Bolivia, Uruguay, and northern Argentina is becoming scarce in some of these areas due to overtrapping and habitat destruction. Other members of this genus have been kept in aviculture less frequently, although several have been bred in zoos and private collections, including the red-cowled, or pope, cardinal *P. dominicana,* the yellow-billed cardinal *P. capitata,* and the red-capped cardinal *P. gularis*. The sexes of Paroaria cardinals are similar in appearance, while species in the genus Cardinalis are sexually dimorphic. The monotypic green cardinal *Gubernatrix cristata,* sometimes referred to as the yellow cardinal, is also dimorphic, but to a lesser degree than the red cardinals. Green cardinals, endemic to Uruguay and parts of Argentina, are featured in many collections; they are, like all cardinals, excellent avicultural subjects. Green cardinals can also be relatively reliably sexed on plumages with the female visually duller than the male.

Although not a true cardinal, the black-crested finch *Lophospingus pusillus* also falls into the subfamily Emberizinae; this species is often referred to aviculturally as the pygmy cardinal or cardinelli.

Red-Crested Cardinal *Paroaria coronata*

Management: Red-crested cardinals, often known aviculture as Brazilian cardinals, will live in harmony with other species such as finches, canaries and small parakeets, but pairs are strictly territorial and intraspecifically aggressive, not tolerating others of their own species in their territories, fighting to the death with odd birds. Preferably they should not be housed in adjacent aviaries, as males will spend more time fighting with their neighbors than settling down to breed. The birds are difficult to sex and the easiest way is by observation. The male sings, whistles, and dances enthusiastically to the female as part of the courtship display. The female responds with interest, but displays far less enthusiasm than the male. Alternatively, the sex of these birds can be determined through DNA-sexing. Ideally, pairs should be allowed to choose their own mates; by placing one male and two color-ringed females in an aviary; the odd bird will soon become apparent. The birds should be closely monitored during this procedure to ensure the safety of the odd bird.

Once they have settled in, red-crested cardinals are hardy birds, able to withstand temperatures down to 5°C (41°F). In such severe climates, however, they should have sufficient shelter to avoid combined cold and damp conditions. Although they can be housed in flights, success is more likely in larger aviaries, measuring 4 x 4 m (13 x 13 ft.), containing some well-planted thickets, as described for Habitat B in the chapter on LANDSCAPE. Cardinals will enjoy a mist spray placed over a dense shrub and will also bathe regularly. They tend to be rather nervous birds and should not be housed where regular disturbance might take place. I have read reports of this species living in excess of 22 years in captivity.

Diet: A mixture of sunflower seed, pinhead oats, paddy rice, hemp seed, mixed millets, and canary seed form the basis of the diet. A high-protein soft food, such as that described for SEEDCRACKERS, is also taken by these birds. Sprouted millet and sunflower; diced fruits such as apple, grapes, and orange; and sweetcorn, can all be mixed into the soft food. A multivitamin and trace element supplement is sprinkled over the soft food three times per week. About two-thirds of their diet is made up of seed and the remainder of fresh foods. Green millet sprays with the seeds still in the milky stage, chickweed, and thistle are all consumed by the birds. Cuttlebone and eggshell, which has been sterilized in a microwave, are also consumed. Roulston describes using universal food and crushed mynah bird pellets for red-capped cardinals *P. gularis*. Egg food was also supplied, the amount being increased during the breeding season. Outside the breeding season, these birds are offered some live food once per day. This can be in the form of a mixture of termites, house crickets, cleaned

Northern Cardinal
Cardinalis cardinalis

maggots (house fly larvae), mealworms or waxmoth larvae (about 12 mealworms and six waxmoth larvae). Some pairs will take larger prey, such as crickets or grasshoppers. Once courtship or nest building commences, live food is offered twice per day, and three times per day after the eggs have hatched. When I bred this species in 1982, I placed the live food in a container near the nest and the protective male ensured that the live food was not taken by other species.

Breeding: Red-crested cardinals are relatively willing nesters provided a true pair has been established in an aviary providing the proper habitat. They are multiple clutched, and I have read of one pair producing 10 youngsters, two clutches of three, and two clutches of two, in a single season. If chicks are abandoned or rejected a few days after hatching, it is usually because the parents have lost interest as a result of an inadequate or improper food supply. The nest is a rather untidy structure. Pairs should be provided with a wicker basket that they will line with coconut fiber and soft materials such as grasses, fiber from the trunk of palms, moss and feathers. I have also had pairs construct their own nests in dense shrubs. Generally they build about 1–1.5 m (3–5 ft.) from the ground. Males tend to be rather hard on the females and drive them to the nest. Strict hygiene is maintained in the nest with the young producing a gelatinacious fecal sack that is dropped well away from the nest by the parents. Roulston and Harris describe breeding red-capped cardinals:

> "If all the eggs hatch, the parents often have a difficult time raising the babies. The size difference places the second and third offspring at a disadvantage. A two-day-old cardinal will be double the size of a hatchling. If there isn't enough food to go around, the first hatched will get most of it. By substituting the first two eggs with dummies and then placing all them back in the nest when the third egg is laid, more young are able to survive."

The female incubates the 3–5 eggs for about 13–14 days and the chicks fledge after about 14 days. The young are fed large quantities of live food by their parents. It is fun to watch both parents load up with insects after a new batch is provided for them. The bugs are loaded in a puffin like manner into the bill before returning to the nest to feed the young. Once the young have eaten enough, the parents swallow the remaining insects. The young are not able to fly well at fledging and spend up to 10 days on the aviary floor. It is important at this time to ensure that no water deeper than 15 mm (0.625 in.), in which the chicks could drown, remains in the aviary. Parents are very attentive to the newly fledged chicks, becoming very excited if the chicks are approached or handled. After the young leave the nest, the parents may want to start with a new clutch, which can be prevented by destroying any new nest construction that occurs. This ensures the parents do not stop feeding the fledglings prematurely. Once the chicks are established, they often become aggressive, even toward their parents, and are best moved to another aviary about a month after fledging. Chicks from a previous clutch have, however, been observed to feed chicks from later clutches.

The chicks differ from the parents in lacking the bright scarlet feathering on the head and crest. For the first year their heads are bronze-brown and the chicks have a fluffy appearance. After the first year, having acquired their adult plumage, they are mature enough to breed.

Artificial Incubation and Rearing: Roulston and Harris provide the following information for the artificial incubation and rearing of red-capped cardinals:

> "We have found that the best is to remove the eggs before hatching and hatch them in an incubator, if they are to be hand fed. If the parents have fed them at all they will have developed a feeding response that is different than the one we need for hand feeding. After hatching, our young are placed in a small cup that is prepared ahead of time and left in the incubator. Soft bedding is molded into shape to provide good footing for the young chicks. The young are left in the incubator for 12–24 hours to absorb the yolk sac. They are then placed cup and all, in a brooder set at 37°C (97°F).
>
> For the first few feedings they are fed a liquid mixture of half Ziegler hand feeding formula and half Heinz No. 2 turkey and turkey gravy (baby food) to which we add a little water. Ziegler hand feeding formula is wheat based and more easily digested by hatchlings than some of the corn based formulas. The Heinz baby food is high in protein and contains gelatin that allows the formulation of a fecal sack, which is important for good digestive health. The feeding utensil of choice is a small, quality fine tip artist paintbrush. This is dipped in the formula and placed in the open mouth. The young bird is allowed to swallow before the next dollop is offered. At three days old we

begin to add cricket abdomens, in conjunction with the formula, every other feeding. Best results have been obtained by feeding every hour between the hours of 7:00 a.m. to 11:00 p.m. with another feeding at about 3:00 a.m. for the first 10 days. To reduce the likelihood of human imprinting, interactions with them are kept to the minimum, and they are placed in aviaries adjoining other cardinals as soon as they are weaned."

I would suggest the hand rearing protocol described for MADAGASCAR FODY would suite cardinals.

Green Cardinal *Gubernatrix cristata*

Management: The green cardinal is the least colorful of the cardinals and not as vocal as common cardinals, but make up for this by being relatively free breeders with interesting patterns of behavior. They may become delightfully tame, particularly when their favored live food is offered regularly from the hand. Sexes can be determined by plumage comparisons. The female is paler than the male. The streak over the eye and the throat are yellow in the male and white in the female. The breast feathers are greenish yellow in the male and brownish gray in the female. They are relatively hardy and can be accommodated in a draft-free and damp-proof aviary throughout the winter, even in relatively harsh climates. Where temperatures remain below freezing for any length of time, however, I would recommend housing them indoors with slight heat supplied in extreme temperatures. They are relatively compatible in a mixed aviary and can be housed with small softbills, finches, doves, and grass parakeets. They may become pugnacious when breeding, and should be carefully observed at this time to ensure they do not harm other birds in the aviary. Breeding pairs should be housed separately, preferably with no visual contact. As noted for the red-crested cardinal, males spend much of their time attempting to fight when housed in adjacent aviaries. Their aviary should be landscaped as described for the red-crested cardinal. However, if on public exhibit, shrubs with brown or grayish leaves will improve the chances of seeing this largely green species, which will otherwise blend in with the vegetation.

Breeding: The breeding habits of green cardinals are similar to those of red-crested, but they are usually more free-breeding than other cardinal species. While most pairs prefer to build their own nest in a dense shrub, some pairs will also nest in half-open-fronted nest boxes or mesh baskets. The nest is built of rather coarse material, including twigs, and then lined with fine plant material. The incubation and fledging periods are both 14 days. Once chicks have hatched, the demand for live food is quite obvious; the parent birds will greet the keeper at the aviary door and excitedly follow him around until live food is supplied. Mealworms, termites, waxmoth larvae, and anything caught in an overnight insect trap will all be consumed. I remember one pair I had that was particularly fond of small moths; when rearing chicks, they would wipe the moths on the ground, presumably to remove excess powder, and then fly straight to the nest to feed these to the nestlings. Insectivorous and softbill foods will also be taken in small quantities. As described for the red-crested cardinal, live food should be offered three times per day when the eggs have hatched. Freshly grown millet sprays are also taken, and are a valuable addition to the diet when chicks are being reared. Upon fledging, the chicks should be carefully monitored to prevent persecution by the male. They can be removed and placed in another aviary at 25–30 days of age. The parents will usually build a nest and lay a new clutch within two weeks after the chicks are removed. It is not unusual for this species to produce three clutches in a season.

Diet and Artificial Rearing: As described for the red-crested cardinal.

Ring Size: 3 mm

(Cardinelli) Black-crested Finch *Lophospingus pusillis*

As a teenager I owned a pair of these delightful birds, but was never successful in getting them to nest. I do remember, however, that in the cold and wet winters of Cape Town they often looked miserable. I would recommend that in harsh climates these birds be wintered indoors, or at least in a situation where they are only allowed outdoors through a bob-hole in fine weather.

The black-crested finch is a native of Argentina and Paraguay, patterned in black, white, and gray. The colors of the male are more sharply defined than those of the female, and he also has a deep black bib, which the female lacks. Both sexes have a black, cockatiel-like crest. They are extremely attractive birds with clean lines.

Management: Despite my own lack of success in breeding this species, a number of accounts have been published over the years. C. af Enehjelm bred them in Finland in the 1950s, first in a mixed species aviary with a 6 x 2.1 x 2.1 m (20 x 7 x 7 ft.) outside flight and a 2.1 x.2.1 x 2.1 m (7 x 7 x 7 ft.) heated shelter, then in a smaller mixed species aviary measuring 2.4 x 1.8 x 2.3 m (8 x 6 x 7.5 ft.), which allowed for closer observation, and finally, after they showed aggression toward smaller finches, in a flight measuring 1.8 x 0.9 x 2.3 m (6 x 3 x 7.5 ft.), shared only with a pair of painted quail *Excalfactoria (Coturnix) chinensis,* which they ignored. In one year, five consecutive clutches were produced. The birds were supplied with 12.7 x 12.7 x 12.7 cm (5 x 5 x 5 in.) half-open-fronted plywood boxes and Harzer wicker cages for nesting, and generally preferred the latter, in which they constructed a deep cup-shaped nest of coconut fibers and dog hair (Enehjelm 1954). Several years later the species was bred in England by C. Everitt, again in a mixed-species aviary whose pair nested first in the topmost fork of an old Christmas tree, and then in the roots of a privet bush that had been used to replace the tree the following year. The nest was constructed of dried grasses, feathers, and pieces of wool; it was about 10 cm (4 in.) across at the top, with a 5 cm (2 in.) depression lined with wool and feathers (Everitt 1957). A pair of black-crested finches owned by RM. Low (1963) nested in an old blackbird *Turdus merula* nest placed among twigs in the corner of the flight, which they relined with feathers.

Another successful breeder was Leila Leitch, whose pair was housed in a small planted aviary with a heated shelter, along with other small seed-eaters. Her pair chose a well-screened nest site in the shelter, building the nest of hemp and wool in "a of finch-basket with two entrances" (Leitch 1970).

Breeding: The normal clutch consists of 2–3 eggs, gray-green with brownish spots. The incubation period is about 12 days, and incubation is by the female alone, while the male defends the nesting area and sings from a nearby perch. The chicks usually leave the nest at about two weeks of age, but sometimes even earlier. Apparently the hen alone feeds the chicks for the first day or several days after hatching, although the male will feed the hen on the nest; after a few days he assists in feeding the nestlings directly. Live food is essential in rearing the chicks. Enehjelm's pairs fed only fresh ant eggs to their chicks, although the English successes were achieved with a combination of mealworms and maggots, the latter being fed only sparingly by the parents, although always available (Leitch 1970). Everitt supplemented the mealworm and maggots with live ant cocoons. John and Kristin Rounds, in New Jersey, noted that black-crested finch chicks are extremely quiet in the nest, the first hint of nestlings being the female's eagerness for mealworms that she would take to the nest three or four at a time. Their black-crests showed no interest, however, in crickets (Rounds & Rounds 1982).

The consensus seems to be that a more generous supply of live food results in stronger, healthier, more vital chicks.

The rearing diet used by Leitch, in addition to the live foods mentioned, included a commercial insectivorous food that was fed to the chicks but otherwise not eaten by the adults. Sponge cake mixed with honey, beef extract, rose-hip syrup, and a liquid multivitamin was a favorite rearing and maintenance food. She also offered a soft cheese salad consisting of cottage cheese, finely chopped hardboiled egg, grated apple, and finely chopped greens.

Males showed a preference for feeding certain individuals among the fledged chicks. Because of this tendency, fledglings should be carefully monitored; those that seem to be struggling can be hand fed egg and commercial rearing food. Occasionally nestlings will be plucked by the parents, their downy breast feathers used to line a new nest. This problem can be avoided by supplying soft feathers and wool for this purpose. Although the parents are relatively tolerant of their independent offspring in the company of successive broods, eventually problems may develop, and for this reason fledged chicks should be removed as soon as they are independent, allowing the adults to renest in peace. Enehjelm noted that a young male molted into adult plumage over a six-week period beginning at about five months of age, however, according to Smith, young birds can be easily distinguished from adult females by their blotchy gray-black, as opposed to solid gray, chests. Young females seem to be more delicate and difficult to rear than young males.

Diet: Black-crested finches were described by Everitt as being basically seed-eaters, although very insectivorous, particularly during the breeding season. There can be little doubt that the breeding successes cited above can be attributed, at least in part, to the varied diet available to the birds as part of mixed species collections including both seed-eaters and softbills. More recently, Ken Smith of Florida reported a breeding success in a small 1.8 x 0.9 x 2.1 m (6 x 3 x 7 ft.) high aviary that also contained breeding pairs of African firefinches *Lagonosticta* sp., green singing finches *Serinus mozambicus,* Bourke's parakeets *Neopsephotus bourkii,* diamond doves *Geopelia cuneata,* and painted quail. The diet offered year round to this collection included finch and parakeet seed mixes, small black oily sunflower seed, safflower seed, and greens, daily. On alternate days sprouted seed is given, and canned corn is offered three times per week. An egg food consisting of hard-boiled eggs mixed with Vionate and other supplements is fed daily, along with a few mealworms. A commercial insectile mixture, grit, and cuttlefish bone are always available. The Bourke's parakeets crack the sunflower and safflower seeds, after which the finches help themselves. However, breeding success was credited mostly to the "Supplement for the Birds" egg food mix, which the black-crested finches would devour to feed their young (Smith 1982).

A finch cake as described for SEEDCRACKERS is also suitable for this species.

Species Requiring Similar Management:

Buntings *Emberiza* and *Passerina* species
Grosbeaks *Caryothraustes* and *Pheucticus* species
Red-crested finch *Coryphospingus cucullatus*
Saffron finch *Sicalis flaveola*

References

Enehjelm, C. af. 1954. The breeding of the black-crested finch. *Avicultural Magazine* 60(1): 11–15.

Everitt, C. 1957. Breeding the black-crested finch. *Foreign Birds* 23(4): 160–162.

Leitch, L. 1970. Breeding black-crested finches or pigmy cardinals. *Avicultural Magazine* 76(5): 174–176.

Low, RM. 1963. The black-crested finch–a near miss. *Foreign Birds* 29(1):17.

Roulston, G and Harris, R. 2001. Breeding the Red-capped Cardinal. *The AFA Watchbird Magazine* 27(4): 8–14.

Rounds, J. & K. Rounds. 1982. Three successes with exotic finches. AFA Watchbird 8(6):14–17.

Smith, K. 1982. Black-crested finch. *AFA Watchbird* 9(2): 30–33.

21.20v Icterini Old and New World Orioles

Glen Holland and Andrew Owen (Durrell Wildlife Conservation Trust)

Introduction: There are two distinct groups of birds known as orioles. Although both groups belong to the order Passeriformes, they are not closely related. The New World orioles, from the Americas, formerly placed in their own family, the Icteridae, are now included in the family Fringillidae where, along with the oropendolas, caciques, New World grackles and blackbirds, marshbirds, meadowlarks, cowbirds and bobolink, they comprise the tribe Icterini in the subfamily Emberizinae. There is a total of 24 or 25 species of New World orioles, depending upon the authority cited, all belonging to the genus *Icterus*. The Old World orioles, with 27 species in the single genus *Oriolus,* were previously assigned to their own family, the Oriolidae, which also included the two species of figbirds. These orioles and figbirds are currently lumped, along with the cuckooshrikes, trillers, minivets, and flycatcher-shrikes, in the tribe Oriolini of the subfamily Corvinae in the family Corvidae. Species of Old World orioles occur in Africa, Australia, Asia, Europe and many of the eastern islands including Borneo, New Guinea, Sumatra, the islands of Malaysia, and Timor.

A representative species of Old World oriole, the African black-headed oriole *Oriolus larvatus,* is found in a variety of habitats from open savanna to dense evergreen forest, and even exotic plantations. These orioles feed on fruits, particularly wild figs, and are also often seen feeding on nectar from aloe plants, which flower during the winter. They occur individually or in pairs. The contact call is a loud *cheeow,* and the male has a short melodious call in the breeding season. The name oriole comes from a Latin word meaning yellow. My experience with orioles is based primarily on work with this species, and to a lesser degree with the Asian black-naped oriole *O. chinensis,* a widespread species that is frequently available in aviculture. Some species such as the European oriole *O. oriolus* are naturally migratory.

A representative species of New World oriole, the Montserrat oriole *Icterus oberi* is a critically endangered species endemic to the Caribbean island of Montserrat, where it is confined to the Centre Hills mountain range. Montserrat orioles live in moist tropical forest, favoring damp gullies known locally as "Ghauts" where they frequent stands of *Heliconia caribea* plants, weaving their nests under the leaves. A special thanks goes to Andrew Owen of the Durrell Wildlife Conservation Trust, who provided husbandry notes on this species.

The Jersey Zoo, operated by the Durrell trust, became involved with the Montserrat oriole in 1997 at the request of the government of Montserrat for assistance in saving the island's national bird and the only endemic avian species following severe volcanic activity from the Soufriere Hills volcano. Population survey work and nest monitoring have been carried out by the RSPB since 1997 to determine population size and chick survivability. In 1998 Owen carried out a study on the feasibility of collecting a small number of Montserrat orioles for husbandry trials at Jersey Zoo. In July 1999, eight Montserrat orioles were captured from the wild and transferred to the zoo for this purpose.

Management: Orioles, being strong fliers, are not suited to small aviaries. Wild caught, imported specimens are nervous birds that may take more than a year to begin to settle down, and even then never lose their fear of humans, and cannot be approached too closely. A large, well-planted aviary is ideal, and breeding will only be attempted under such conditions. Orioles are not habitat specific and their aviary should be planted in accordance with either Habitat A or B, or a mixture of the two, as described in the chapter on LANDSCAPE. Nectar plants, such as those mentioned for SUNBIRDS, will provide a valuable natural food source for orioles. Orioles are strongly territorial and only one sexed pair should be kept per aviary. I have never noticed aggressive behavior between orioles and another species in the aviary. In one extremely large, walk-through, Asian aviary I kept a number of black-naped orioles that lived together in harmony, however, this would probably have changed if any of the birds had nested. They are shy birds, unlikely to breed if they do not have a secluded spot where they will not be disturbed by other species. Orioles are very fond of bathing, particularly if a mist spray is provided over some vegetation, where they will bathe and hang upside down until soaked. They also enjoy sunbathing, during which they will stretch out one wing at a time and pant with the beak open.

The melodious calls are often made between members of a pair, or otherwise by individual birds. The beauty of these birds matches their beautiful calls, and their quiet nature makes them pleasing aviary subjects. Orioles are birds of the sun and naturally occur in warm climates. They are able to withstand night temperatures down to zero, but in countries with severe, cold climates for extended periods, they are best housed indoors with a little additional warmth during cold spells. Extended periods of cold wet weather, even at temperatures of 14°C (57°F) or lower, will prove detrimental to their health.

At the Jersey Zoo, the Montserrat orioles were individually ringed using a numbered metal size K ring on one leg and a colored plastic 5.5 mm split ring on the other leg for easy identification. They were placed in pairs in outside aviaries measuring 5 x 4 x 4 m (16 x 13 x 13 ft.), with no other occupants. The aviaries are constructed from a treated timber frame and are covered with 5 cm (2 in.) 16-gauge galvanized wire mesh, painted with a matte black lead-free paint to facilitate observation of the birds. Part of the outside aviary roof is covered with a material such as Corrotherm sheeting to provide protection against heavy rain and ensure a dry area for the food dishes. Aviaries for this species should be as large as possible, as these birds are prone to obesity when housed in small areas.

Each aviary is heavily planted to create as natural an environment as possible for the birds and to make the aviaries aesthetically pleasing for zoo visitors. A variety of plants, including vines, shrubs, and small trees, are selected to resemble the oriole's rainforest home, Habitat A in the chapter on LANDSCAPE. Plants such as banana, ginger, bamboo, clematis, and various ferns and bromeliads provide the birds with an abundance of nest building, foraging, and feeding opportunities. In addition, the aviaries are furnished with natural perches at varying heights to allow the birds maximum use of the available space. Each aviary has an attached indoor area to provide shelter and warmth during inclement weather. These shelters are of block or timber construction.

Thermostatically controlled tubular aviary heaters maintain a temperature of 25°C (77°F.) in the indoor shelters. Flourescent tubular lights are attached to a timer switch set to come on at 6:00 a.m. and go off at 7:00 p.m., providing the birds with a minimum of 13 hours of daylight. The birds may be confined to the warm indoor quarters during periods of cold winter weather.

A deep layer of matured bark chips, mulch, and leaf litter of oak, magnolia, and sweet chestnut covers the ground in the outside aviary. This not only contributes to the natural appearance of the aviary but also affords additional foraging opportunities for the occupants. Some aviaries have pebbles and rocks to resemble dry river beds, although these are more for aesthetic reasons than for the benefit of the birds. Fresh drinking and bathing water is offered to the birds daily in ceramic bowls. In addition, some pairs have small streams or waterfalls in their aviaries in which they can bathe.

Few health problems have been encountered during the short time this species has been kept in captivity. Montserrat orioles seem prone to obesity, particularly when housed in small cages or aviaries that do not provide sufficient opportunities for exercise and foraging. Rich, fatty foods such as insects should be strictly rationed, particularly when birds are held in small cages. Orioles housed in large planted aviaries do not seem susceptible to large weight increases. A small number of chicks have suffered from bone deformities or fractures during the nestling period. The exact cause of this problem has not been determined, and it has been speculated that inadequate quantities of calcium and vitamins are being fed to the chicks by their parents. It is therefore important that all live insects cultured for these birds receive a balanced diet that is high in calcium.

Diet: Oriolus species are fed on a diet of mixed fruits such as apple, oranges, papaya, soaked raisins, and diced grapes, along with commercial softbill mixtures and the same nectar, including pollen, as described for SUNBIRDS. They also take live food such as mealworms, waxmoth larvae, crickets, and winged termites. In some areas, breeding occurs during the peak of the cicada beetle season. Chicks are fed largely on insects.

In the wild, Montserrat orioles are thought to be specialist insectivores, feeding almost exclusively on insects, insect larvae, and other invertebrates. The species is known to take a wide variety of small and large invertebrates, such as crickets, beetles, spiders, and moths, which are gleaned from vegetation at various levels of the forest strata from the canopy to the forest floor. Fruit is not thought to form a major part of the normal diet; however, one group of these orioles was observed feeding on mangoes on the forest floor (Owen, pers. obs. 1999). Whether this is natural behavior or resulted from a lack of invertebrates following the volcanic activity is not known.

Montserrat orioles are offered an insectivorous mixture in a small D-pot between 10:00 a.m. and 10:30 a.m. The mixture consists of equal parts of the following:

- Witte Molen Universal food or High insect diet
- Witte Molen Low Iron diet
- Witte Molen Mynah pellets (soaked)
- Finely grated hard-boiled egg
- Finely grated carrot
- Wholemeal bread crumbs

This mixture is blended to a moist, pliable consistency, and a small amount of finely diced fruit and a sprinkling of live insects (usually 2–3 waxmoth larvae per bird) is added. One or two pieces of fruit are spiked onto perches in the aviary. Various fruits are eaten in this way, including papaya, mango, kiwi, melon, pear, and prickly pear. In addition to their main meal, the birds also receive two insect feeds, at 8:00 a.m. and again at 4:00 or 5:00 p.m. (winter or summer). These feeds consist of a small amount of insects, usually crickets or locusts, which the birds will accept at any stage of development, from hatchlings to adults. The insects are scattered around the aviary to encourage foraging. The birds also catch some natural prey such as moths, spiders, and earwigs, as they actively forage in their heavily vegetated aviaries.

When chicks are being reared, the adults are given a live-food feed of soft-bodied insects, dusted with a multivitamin powder, every two hours throughout the day. Unlike the Old World orioles, or even other *Icterus* species, wild Montserrat orioles are not thought to eat nectar and pollen. In captivity, however, they have been observed feeding on the nectar or pollen of several species of flowering plants, including New Zealand flax, honeysuckle, bromeliads, and canna lilies. Food is dusted with Nutrobal multi-vitamin powder once daily. All live insects are dusted with Nutrobal before being offered to the birds.

Breeding: Due to their rather specialized nesting requirements, orioles are not often bred in captivity. They are not aggressive enough to ward off other inquisitive softbills such as starlings, which are likely to interfere with their nesting attempts. Ideally, therefore, a pair should be housed with small softbills and seedeaters. The sexes of most species are alike and, if breeding is desired, the birds must be either surgically sexed or DNA sexed. Adult Montserrat orioles are sexually dimorphic, so there is no need for DNA or surgical sexing techniques to be employed for this species.

A breeding pair is strictly territorial and will not tolerate another oriole in the aviary. The nest is a hammock type structure slung between a forked twig in a broadleafed tree. It is constructed of fine grasses, cobwebs, and "Old Man's Beard" lichen, usually 2–10 m (62–33 ft.) from the ground. The eggs number 2–4 and are a beautiful coral color with a few brown wavy marks concentrated in a ring at the thick end of the egg. Incubation

Montezuma's oropendola
Gymnostinops montezuma
and nest.

Golden Oriole on the nest.

takes 14–15 days, and the chicks, initially sparsely covered in down at hatching, fledge at 16–18 days of age. I have hand raised black-headed orioles on three occasions, and have found that the chicks remain tame enough to take an insect from the fingers, but they will not perch on people as many hand-fed birds will do. The young are rather sluggish feeders, but they have a large gape, which makes feeding easy.

During 2000, Montserrat orioles bred at Jersey Zoo between May and September, the first nest being started May 15 and the last chick fledging September 23. In the wild, these birds are specialized nesters; using plant fibers, they weave a cup-shaped nest that is normally attached to the underside of a large leaf of a *Heliconia caribaea* plant. In captivity, they are not so conservative in their choice of nesting sites. Nests are woven almost entirely by the female at a height of at least 2.5 m (8 ft.) above ground, and may be built in either the outside aviary or the inside shelter. These nests are constructed almost entirely of fine strands of sisal rope cut to approximately 30 cm (12 in.) lengths. Nests may be attached to branches, aviary mesh, vines, or other living plants. Several "typical" Montserrat oriole nests have been woven to the undersides of large banana plant leaves.

Clutch size is 1–3 eggs, however, two-egg clutches have been the average from captive nests at Jersey Zoo. Eggs are pale whitish-cream in ground color with faint reddish brown spots and speckles, mainly at the thick end. Dimensions for two eggs that failed to hatch were 23.9 mm x 18.3 mm and 25.1 mm x 17.7 mm. Eggs are usually laid on consecutive days, although in one nest a second egg was laid two days after the first. In this instance, the resultant chick did not survive. Incubation is carried out entirely by the female and takes 13–14 days. There is considerable variation in time between nesting attempts, but this may depend in part upon whether or not the previous nest was successful. After its first nest failed, one pair built a replacement nest and laid the first egg of the second clutch within six days. Pairs that have successfully raised young have started new nests within 5–25 days of the earlier chicks fledging.

Both sexes sing, although the male's song is louder and more varied. Display has been observed occasionally in male Montserrat orioles. The wings are lowered and the yellow flank and rump feathers are raised while the male utters a quiet twittering sound. Copulation is brief, with the female fluttering her wings and uttering a high-pitched squeak call. The male mounts the female for about 2–3 seconds.

Montserrat orioles are kept as pairs and, although they do not form strong pair bonds and are not sociable, they tolerate each other. During the nesting period, females that leave the nest to feed are often ushered back to the nest by the male, who utters a high-pitched, scolding chatter. Some aggression through the aviary mesh toward conspecifics or white-rumped shamas *Copsychus malabaricus* in adjacent aviaries has been observed.

Montserrat oriole chicks are blind and helpless when they hatch. Their pink skin is covered with fine, wispy, pale gray down. They grow fast on a diet of insects, which is fed by both parents. The female alone broods the chicks, the male taking no part in this activity. As the chicks grow, their fine down is replaced with slightly thicker gray down. Their eyes open at seven days of age, and at eight days, gray pinfeathers begin to erupt along their backs and undersides.

When the chicks fledge, at 14–16 days of age, they are fully feathered with dull gray-green plumage. At this time, the tail is very short. Initially, they sit quietly among the thick vegetation, only uttering a high-pitched twittering call when an adult approaches with an item of food. Within a few days, the fledglings

African Golden Oriole
Oriolus auratus

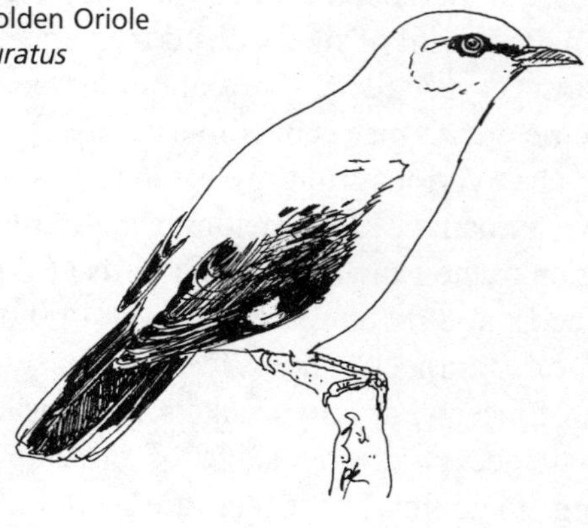

become more active and can be seen following their parents around the aviary, begging for food by uttering the same call and shaking their wings vigorously. At approximately six weeks of age, fledglings have been observed feeding themselves on insects and the cut pieces of fruit provided, although they continue to solicit food from their parents for several weeks afterward. Juveniles molt from the grayish-green color to a brighter olive yellow plumage similar to that of the adult female at approximately 32 months of age. By this time, sex can usually be determined, as young males begin to show a few black throat feathers. They complete a second molt when they are approximately one year of age. Following this molt, the females resemble adult females, while males attain a sub-adult plumage with an increased feathering of black and yellow, giving them a spotted appearance. Virtual adult male plumage is completed at two years of age, a tawny olive brown nape being the only distinguishing feature differentiating them from older birds.

Hand Rearing: The techniques and diets used for orioles are similar to those described for TANAGERS. They are hand raised on a diet consisting of equal portions of insects (including bee larvae, waxmoths, and moths caught around a night light), psittacine hand-rearing formula that has been moistened with water, and diced fruits such as figs, papaya, apples, etc. The insects are fed with forceps, and the rearing food with a thin syringe from which the tip has been removed. Both the insects and fruit can be dipped into the moistened rearing formula before being fed. Nestling feces are produced in fecal sacs that are easily removed, thus maintaining good hygiene standards.

Ring Size: 4.5 mm

Species Requiring Similar Management:

Turacos
Bulbuls
Starlings
Tanagers
Leafbirds

Reference:

Owen, A. 2001. Species management guideline for the Montserrat oriole. Durrell Wildlife Conservation Trust, Jersey, Channel Islands.

INDEX

B

C

D

E

F

G

H

M

Q

R

V

W

X

Y

Z